Analysis, Geometry, Nonlinear Optimization and Applications

Series on Computers and Operations Research

Series Editor: Panos M Pardalos *(University of Florida, USA)*

Published

Vol. 9 *Analysis, Geometry, Nonlinear Optimization and Applications*
edited by P. M. Pardalos and T. M. Rassias

Vol. 8 *Network Design and Optimization for Smart Cities*
edited by K. Gakis and P. M. Pardalos

Vol. 7 *Computer Aided Methods in Optimal Design and Operations*
edited by I. D. L. Bogle and J. Zilinskas

Vol. 6 *Recent Advances in Data Mining of Enterprise Data: Algorithms and Applications*
edited by T. Warren Liao and Evangelos Triantaphyllou

Vol. 5 *Application of Quantitative Techniques for the Prediction of Bank Acquisition Targets*
by F. Pasiouras, S. K. Tanna and C. Zopounidis

Vol. 4 *Theory and Algorithms for Cooperative Systems*
edited by D. Grundel, R. Murphey and P. M. Pardalos

Vol. 3 *Marketing Trends for Organic Food in the 21st Century*
edited by G. Baourakis

Vol. 2 *Supply Chain and Finance*
edited by P. M. Pardalos, A. Migdalas and G. Baourakis

Vol. 1 *Optimization and Optimal Control*
edited by P. M. Pardalos, I. Tseveendorj and R. Enkhbat

Series on Computers and Operations Research Vol. 9

Analysis, Geometry, Nonlinear Optimization and Applications

Editors

Panos M Pardalos

University of Florida, USA

Themistocles M Rassias

National Technical University of Athens, Greece

World Scientific

NEW JERSEY • LONDON • SINGAPORE • BEIJING • SHANGHAI • HONG KONG • TAIPEI • CHENNAI

Published by

World Scientific Publishing Co. Pte. Ltd.
5 Toh Tuck Link, Singapore 596224
USA office: 27 Warren Street, Suite 401-402, Hackensack, NJ 07601
UK office: 57 Shelton Street, Covent Garden, London WC2H 9HE

Library of Congress Cataloging-in-Publication Data
Names: Pardalos, P. M. (Panos M.), 1954– editor. | Rassias, Themistocles M., 1951– editor.
Title: Analysis, geometry, nonlinear optimization and applications / editors,
Panos M Pardalos, University of Florida, USA,
Themistocles M Rassias, National Technical University of Athens, Greece.
Description: New Jersey : World Scientific, [2023] | Series: Series on computers and operations research, 1793-7973 ; vol. 9 | Includes bibliographical references and index.
Identifiers: LCCN 2022047089 | ISBN 9789811261565 (hardcover) |
ISBN 9789811261572 (ebook for institutions) | ISBN 9789811261589 (ebook for individuals)
Subjects: LCSH: Mathematical optimization. | Differential equations, Nonlinear. |
Geometry. | Control theory.
Classification: LCC QA402.5 .A464 2023 | DDC 510--dc23/eng20230123
LC record available at https://lccn.loc.gov/2022047089

British Library Cataloguing-in-Publication Data
A catalogue record for this book is available from the British Library.

For any available supplementary material, please visit
https://www.worldscientific.com/worldscibooks/10.1142/13002#t=suppl

Desk Editors: Soundararajan Raghuraman/Steven Patt

Typeset by Stallion Press
Email: enquiries@stallionpress.com

https://doi.org/10.1142/9789811261572_fmatter

Preface

Analysis, Geometry, Nonlinear Optimization and Applications publishes papers which are devoted to both classical as well as very current domains of research. Specifically topics treated within this book include optimization of control points, graph pursuit games, game theory models, Nash equilibrium, motion around equilibrium points in the photogravitational R3BP, quantitative methods, linearization methods, dynamic geometry, superquadratic functions, semicontinuous relations in relator spaces, generalized Cartan matrices, multivalued mappings, integral inequalities, norm inequalities, variational inequalities, variational inclusions, generalized Ostrowski and trapezoid type rules, stability of functional equations, hyperstability of ternary Jordan homomorphisms, nonlinear evolution PDEs in Biomechanics, Volterra integro–differential equations, interpolation functions and hypergeometric series, Caristi–Kirk theorems, metrical coercivity, low rank affine groups, wavelet transforms, wavelet fluctuation analysis and related subjects.

Effort has been made for the problems studied within this volume to have an interdisciplinary flavor as well as extend to a broad spectrum of domains, presenting the state of the art on the problems treated.

We would like to express our warmest appreciation to all the authors who contributed their valuable works to be published in this volume. We would also like to extend our sincere thanks to the staff of World Scientific Publ. Co. for their help throughout the preparation of this book.

Panos M. Pardalos
Themistocles M. Rassias
Florida, USA
Athens, Greece

https://doi.org/10.1142/9789811261572_fmatter

About the Editors

Panos M. Pardalos serves as Professor Emeritus of industrial and systems engineering at the University of Florida. Additionally, he is the Paul and Heidi Brown Preeminent Professor of industrial and systems engineering. He is also an affiliated faculty member of the Computer and Information Science Department, the Hellenic Studies Center, and the biomedical engineering program, as well as the Director of the Center for Applied Optimization. Pardalos is a world-leading expert in global and combinatorial optimization. His recent research interests include network design problems, optimization in telecommunications, e-commerce, data mining, biomedical applications and massive computing.

Themistocles M. Rassias is Professor at the National Technical University of Athens, Greece. He has published more than 300 papers, 10 research books and 45 edited volumes in research Mathematics as well as four textbooks in Mathematics (in Greek) for university students. He serves as a member of the Editorial Board of several international mathematical journals. His work extends over several fields of mathematical analysis. It includes nonlinear functional analysis, functional equations, approximation theory, analysis on manifolds, calculus of variations, inequalities, metric geometry and their applications. He has contributed a number of results in the stability of minimal submanifolds, in the solution of Ulam's Problem for approximate homomorphisms in Banach spaces, in the theory of isometric mappings in metric spaces and in complex analysis (Poincaré's inequality and harmonic mappings).

https://doi.org/10.1142/9789811261572_fmatter

Contents

Preface v

About the Editors vii

1. Optimization of Control Points for Controlling the Heating Temperature of a Heating Medium in a Furnace 1
   Vagif Abdullayev

2. Monotonicity of Averages of Superquadratic and Related Functions 19
   Shoshana Abramovich

3. Upper and Lower Semicontinuous Relations in Relator Spaces 41
   Santanu Acharjee, Michael Th. Rassias, and Árpád Száz

4. Geometric Realization of Generalized Cartan Matrices of Rank 3 113
   Abdullah Alazemi, Milica Anđelić, and Kyriakos Papadopoulos

5. Dynamic Geometry Generated by the Circumcircle Midarc Triangle 129
   Dorin Andrica and Dan Ştefan Marinescu

6. A Review of Linearization Methods for Polynomial Matrices and their Applications 157

E. Antoniou, I. Kafetzis, and S. Vologiannidis

7. A Game Theory Model for the Award of a Public Tender Procedure 189

G. Colajanni, P. Daniele, and D. Sciacca

8. Perov-Type Results for Multivalued Mappings 215

Marija Cvetković, Erdal Karapinar, Vladimir Rakočević, and Seher Sultan Yeşilkaya

9. Some Triple Integral Inequalities for Bounded Functions Defined on Three-Dimensional Bodies 255

Silvestru Sever Dragomir

10. Generalized Ostrowski and Trapezoid Type-Rules for Approximating the Integral of Analytic Complex Functions on Paths from General Domains 279

Silvestru Sever Dragomir

11. Hyperstability of Ternary Jordan Homomorphisms on Unital Ternary C^*-Algebras 307

Madjid Eshaghi Gordji and Vahid Keshavarz

12. Analytic and Numerical Solutions to Nonlinear Partial Differential Equations in Biomechanics 331

Anastasios C. Felias, Konstantina C. Kyriakoudi, Kyriaki N. Mpiraki, and Michail A. Xenos

13. Localized Cerami Condition and a Deformation Theorem 405

Lucas Fresse and Viorica V. Motreanu

14. A Two-phase Problem Related to Phase-change Material 417

D. Goeleven and R. Oujja

15. Motive of the Representation Varieties of Torus Knots for Low Rank Affine Groups 435

Ángel González-Prieto, Marina Logares, and Vicente Muñoz

16. Quaternionic Fractional Wavelet Transform 453

Bivek Gupta, Amit K. Verma, and Carlo Cattani

17. From Variational Inequalities to Singular Integral Operator Theory — A Note on the Lions–Stampacchia Theorem 481

Joachim Gwinner

18. Graph Pursuit Games and New Algorithms for Nash Equilibrium Computation 491

Athanasios Kehagias and Michael Koutsmanis

19. Geometry in Quantitative Methods and Applications 535

Christos Kitsos and Stavros Fatouros

20. Norm Inequalities for Fractional Laplace-Type Integral Operators in Banach Spaces 565

Jichang Kuang

21. Hyers–Ulam–Rassias Stability of Functional Equations in G-Normed Spaces 589

Jung Rye Lee, Choonkil Park, and Themistocles M. Rassias

22. Wavelet Detrended Fluctuation Analysis: Review and Extension to Mixed Cases 599

Anouar Ben Mabrouk, Mohamed Essaied Hamrita, and Carlo Cattani

23. Stability of Some Functional Equations on Restricted Domains 625

Abbas Najati, Mohammad B. Moghimi, Batool Noori, and Themistocles M. Rassias

24. System of General Variational Inclusions 641

Muhammad Aslam Noor, Khalida Inayat Noor, and Michael Th. Rassias

25. Analytical Solution of nth-Order Volterra Integro-Differential Equations of Convolution Type with Non-local Conditions 659

E. Providas and I.N. Parasidis

26. Some Families of Finite Sums Associated with Interpolation Functions for many Classes of Special Numbers and Polynomials and Hypergeometric Series 677

Yilmaz Simsek

27. Transitive Pseudometric Principles and Caristi–Kirk Theorems 751

Mihai Turinici

28. Metrical Coercivity for Monotone Functionals 787

Mihai Turinici

29. Motion Around the Equilibrium Points in the Photogravitational R3BP under the Effects of Poynting–Robertson Drag, Circumbinary Belt and Triaxial Primaries with an Oblate Infinitesimal Body: Application on Achird Binary System 839

Aguda Ekele Vincent and Vassilis S. Kalantonis

Index 871

https://doi.org/10.1142/9789811261572_0001

Chapter 1

Optimization of Control Points for Controlling the Heating Temperature of a Heating Medium in a Furnace

Vagif Abdullayev

Azerbaijan State Oil and Industry University,
Azadlig ave. 20, Baku, AZ1010, Azerbaijan
Institute of Control Systems of Ministry of Science and Education Republic of Azerbaijan,
Vahabzade str., 9, Baku, AZ1141, Azerbaijan
vaqif_ab@rambler.ru

We consider a problem of controlling a heating appliance used for heating a heat-transfer agent, which delivers heat into a closed system. To control the process, we use feedback, under which information on the process state is continuously or discretely received from individual points of the appliance with installed temperature sensors. The mathematical model of the controlled process is described in both cases by a pointwise-loaded first-order hyperbolic equation. We have obtained formulas for the gradient of the objective functional of the problem. These formulas allow us to use numerical first-order optimization methods for solving the problems. Numerical experiments have been carried out by the example of solving several test problems.

1. Introduction

The chapter proposes an approach to construction of a feedback control system for a distributed-parameter object. The considered object selected is a heat supply system fed with fluid, which is heated in a heat exchanger constituting a steam jacket [1]. There are temperature sensors installed at some points of the heat exchanger, based on the readings of which the heat is supplied to the heat exchanger. The heat exchange process in heat exchanger follows a first-order hyperbolic (transfer) equation [1].

In boundary conditions, a time lag argument is present due to the time it takes the heated fluid to pass through the heat supply system.

It should be noted that there has been increased interest in the problems of optimal control of distributed-parameter objects following different types of partial differential equations with different kinds of initial-boundary conditions over recent years [1–8]. The feedback control (regulation) problems are particularly challenging. While these problems are rather well-studied for lumped-parameter objects [9,10], the feedback control problems for objects following partial differential equations, by contrast, are substantially under-explored [4–8,11–13]. Firstly, this is associated with the complexity of the practical implementation of space- and time-distributed object control systems [14,15]. This complexity is due to the unfeasibility of continuous or even discretely-timed prompt acquisition of information on the state of the entire object (at all its points). Secondly, there are mathematical and computational issues, since solving initial boundary-value problems for partial differential equations requires, to an extent, a long time, which often makes it impossible to construct distributed-parameter object control systems in real time scale.

The approach to the optimal control synthesis of heat supply process proposed in the chapter is based on using process state information at a finite number of sensor points. In addition, the problem provides for optimization of both locations of sensor points themselves and their number. Formulas have been obtained for the gradient of the objective functional with respect to the optimizable feedback control parameters, which have been used in the numerical solution of the problem using first-order iterative optimization methods. These formulas make it possible to derive necessary optimality conditions similar to Pontryagin's maximum principle.

2. Problem Statement

The process of heating a heat-carrying agent in the furnace of the heated apparatus (heat exchanger) of the heating system can be described by the following transport equation (see Refs. [1,16]):

$$\frac{\partial u(x,t)}{\partial t} + a\frac{\partial u(x,t)}{\partial x} = \alpha\left[\vartheta(t) - u(x,t)\right],\ (x,t) \in \Omega = (0,l) \times (0,T], \quad (1)$$

where $u = u(x,t)$ is the temperature of the heat-carrying agent at the point x of the heat exchanger at the point of time t; l, the length of the heating tube, in which the heat-carrying agent is heated; a, the velocity of the heat-carrying agent in the heat supply system, the value of which is constant for all points of the heat supply system, i.e. the motion is assumed to be steady (stationary); α, the given value of the heat transfer coefficient between the furnace and the heat-carrying agent in the heating apparatus; $\vartheta(t)$, the temperature inside the furnace, by means of which the heating process is controlled, subject to the technological limit

$$\underline{\vartheta} \leq \vartheta(t) \leq \bar{\vartheta}. \tag{2}$$

Let L be the linear length of the whole heat supply system, and L far exceeds l, i.e. $L >> l$. Then the heat-carrying agent heated in the furnace needs the time $T^d = L/a$ in order to return to the beginning of the furnace, i.e.

$$u(0,t) = (1-\gamma)\, u(l, t-T^d), \quad t > 0, \tag{3}$$

where γ is the constant value that determines the heat lost during the motion in the heating system, which, in essence, depends considerably on the temperature of the external environment. On the basis of practical considerations, we have the obvious condition:

$$0 \leq \gamma \leq 1. \tag{4}$$

Denote by Γ the set of all possible values of γ, determining the amount of lost heat, satisfying (3) and (4). It is assumed that a density function $\rho_\Gamma(\gamma)$ is given on this set satisfying the following condition:

$$\rho_\Gamma(\gamma) \geq 0, \;\; \gamma \in \Gamma, \int_\Gamma \rho_\Gamma(\gamma) d\gamma = 1,$$
$$\rho_\Gamma(\gamma) \in [0,1], \quad \gamma \in \Gamma.$$

Let the initial history be given by

$$u(x,t) = \text{const}, \;\; x \in [0,l], \;\; -T^d < t < 0. \tag{5}$$

The problem of controlling the heating of the heat-carrying agent consists in maintaining the furnace temperature at such a level that provides a certain temperature $u(l,t) = V = \text{const}$, $t \in (0,T]$, of the heat-carrying agent at the exit of the furnace under all possible admissible values of the heat lost by the heat-carrying agent when it moves in the heat supply system, determined by the values $\gamma \in \Gamma$.

Let sensors be installed at M arbitrary points $\xi_i \in [0,l]$, $i = 1, 2, \ldots, M$, of the heating apparatus (Fig. 1), at which temperature measurements are taken continuously:

$$u_i(t) = u(\xi_i, t), \quad t \in [0,T], \tag{6}$$

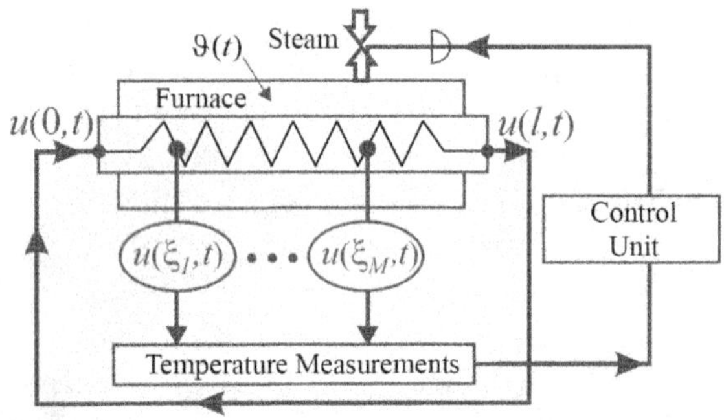

Fig. 1: Steam-heated shell-and-tube heat exchanger control diagram.

or at discrete points of time

$$u_{ij} = u(\xi_i, t_j), \quad t_j \in [0, T], \; j = 1, 2, \ldots, m. \tag{7}$$

To construct a heating control system with a continuous feedback, consider the following variant of the temperature control system:

$$\vartheta(t) = \frac{1}{l} \sum_{i=1}^{M} \lambda_i \tilde{k}_i [u(\xi_i, t) - z_i], \tag{8}$$

where $\tilde{k}_i$ is the amplification coefficient; z_i, the effective temperature at the point ξ_i, at which we need to control the amount of deviation from this value; $\lambda_i = \text{const}$, the weighting coefficient, determining the importance of taking a measurement at the point ξ_i, $i = 1, 2, \ldots, M$,

$$\lambda \in \Lambda = \left\{ \lambda \in E^M : \; 0 \le \lambda_i \le 1, \;\; i = 1, 2, \ldots, M, \; \sum_{i=1}^{M} \lambda_i = 1 \right\}.$$

We introduce complex parameters

$$k_i = \frac{\lambda_i \tilde{k}_i}{l}, \quad i = 1, 2, \ldots, M.$$

In this case, the formula for the temperature in the furnace (8) takes the form

$$\vartheta(t) = \vartheta(t; y) = \sum_{i=1}^{M} k_i [u(\xi_i, t) - z_i]. \tag{9}$$

Here $y = (\xi, k, z)^* \in R^{3M}$ is the vector of parameters of the feedback that determines the current control value (furnace temperature) depending on the measured temperature values at the heat exchanger measurement points; "*" is the transposition sign.

Substituting (8) into (1), we obtain

$$\frac{\partial u(x,t)}{\partial t} + a\frac{\partial u(x,t)}{\partial x} = \alpha\left[\sum_{i=1}^{M} k_i[u(\xi_i,t) - z_i] - u(x,t)\right],$$

$$(x,t) \in \Omega = (0,l) \times (0,T]. \tag{10}$$

The minimized criterion of the control quality is given by the following form:

$$J(y;\gamma) = \int_{\Gamma} I(y;\gamma)\rho_{\Gamma}(\gamma)\,d\gamma, \tag{11}$$

$$I(y;\gamma) = \beta_1 \int_0^T [u(l,t;y,\gamma) - V]^2 dt + \sigma\,\|y - \hat{y}\|^2_{R^{3M}}, \tag{12}$$

where $u(x,t;y,\gamma)$ is the solution of initial boundary-value problem (1), (3), (5) under specified feasible values of feedback parameters y and heat loss parameters $\gamma \in \Gamma$; $\hat{y} = (\hat{k}, \hat{z}, \hat{\xi}) \in R^{3M}$ and σ, a small positive quantity, are the regularization parameters. Thus, initial feedback control problem reduces to a parametric optimal control problem. The peculiarity of the problem is that it follows a point wise-loaded differential equation (10) (due to the presence of the spatial variable of a value of the unknown function $u(x,t)$ at given points $\xi_i, i = 1,2,\ldots,M$ in the equation, under boundary conditions with the lagging argument (3) [17–25].

Taking into account transformations (8), the optimizable feedback control parameters y can be constrained based on technical and process considerations:

$$0 \le \xi_i \le l, \quad \underline{k}_i \le k_i \le \bar{k}_i, \quad \underline{z_i} \le z_i \le \bar{z}_i, \quad i = 1,2,\ldots,M. \tag{13}$$

Here, $\underline{k}_i, \bar{k}_i, \underline{z}_i, \bar{z}_i, i = 1,2,\ldots,M$, are given values. The values $\underline{k}_i, \bar{k}_i$, $i = 1,2,\ldots,M$, are derived from formula (9) taking into account constraints (2) the *a priori* information on possible and permissible values of steam and coolant temperature. The values $\underline{z}_i, \bar{z}_i, i = 1,2,\ldots,M$, are mainly determined by the desired value of coolant temperature V at the heater outlet.

3. Derivation of the Formulas for the Gradient of the Functional

For numerical solution to the obtained parametric optimal control problem for a loaded system with distributed parameters, we propose to use first-order methods, for example, the gradient projection method (see Ref. [26]).

To construct a minimizing sequence $y^\nu, \nu = 0, 1, \ldots,$ an iterative process is constructed as follows:

$$y^{\nu+1} = P_{(13)}[y^\nu - \mu_\nu \operatorname{grad} J(y^\nu)], \quad \nu = 0, 1, \ldots, \tag{14}$$

Here, $P_{(13)}(y)$ is the projection operator of a $3M$ — dimensional point $y = (\xi, k, z)^*$ on the set defined by the constraints (13); $\mu_\nu > 0$ the step in the direction of the projected anti-gradient. The initial approximation y^0 can be arbitrary, satisfying, in particular, the conditions (13). Considering the simplicity of the structure of the admissible set of optimizable parameters defined by the constraints (13), the projection operator has a constructive character and is easy to implement.

The criterion for stopping the iterative process (14) can be the fulfillment of one of the following inequalities:

$$J(y^\nu) - J(y^{\nu+1}) \leq \epsilon_1 \quad \text{and} \quad \left\|y^\nu - y^{\nu+1}\right\| \leq \epsilon_2, \tag{15}$$

where ϵ_1 and ϵ_2 are given positive numbers.

To build the procedure (14), we obtain formulas for the components of the gradient of the functional (11), (12) with respect to the optimizable parameters:

$$\operatorname{grad} J(y) = \left(\frac{\partial J(\xi,k,z)}{\partial \xi}, \frac{\partial J(\xi,k,z)}{\partial k}, \frac{\partial J(\xi,k,z)}{\partial z}\right)^*.$$

To this end, we use the well-known technology of obtaining formulas for an increment of the functional obtained at the expense of the increment of the optimizable arguments of the functional (see Ref. [26]). In this case, the linear part of the increment of the functional with respect to each of the arguments will be the desired component of the gradient of the functional with respect to the corresponding argument.

Before obtaining formulas for the gradient components of the functional, we note the following. Taking into account that the parameter $\gamma \in \Gamma$, determining the amount of lost heat, does not depend on the process of heating the heat-carrying agent in the heat exchanger, from (11), (12) it follows that:

$$\operatorname{grad} J(y) = \operatorname{grad} \int_\Gamma I(y;\gamma)\rho_\Gamma(\gamma)d\gamma = \int_\Gamma \operatorname{grad} I(y;\gamma)\rho_\Gamma(\gamma)d\gamma. \tag{16}$$

Therefore, we obtain the formula $\operatorname{grad} I(y;\gamma)$ for any one arbitrarily given parameter $\gamma \in \Gamma$.

Let $u(x,t;y,\gamma)$ be the solution to the loaded initial- and boundary-value problem (10), (3), (5) for an arbitrary chosen vector of the optimizable parameters $y = (\xi, k, z)^*$ and for a given value of the parameter $\gamma \in \Gamma$. For brevity, where this does not cause ambiguity, the parameters y, γ will be omitted from the solution $u(x, t; y, \gamma)$.

Let the parameters $y = (\xi, k, z)^*$ have obtained some admissible increments $\Delta y = (\Delta\xi, \Delta k, \Delta z)^*$, and $\tilde{u}(x,t) = \tilde{u}(x,t;\tilde{y}) = u(x,t) + \Delta u(x,t)$ be the solution to the problem (10), (3), (5) that corresponds to the incremented vector of arguments $\tilde{y} = y + \Delta y$.

Substituting the function $\tilde{u}(x,t)$ into the conditions (10), (3), (5), we obtain the following initial- and boundary-value problem accurate within the terms of the first order of smallness with respect to the increment $\Delta u(x,t)$ of the phase variable:

$$\begin{aligned}\Delta u_t(x,t) + a\Delta u_x(x,t) = \alpha \sum_{i=1}^{M} [&k_i \Delta u(\xi_i, t) + k_i u_x(\xi_i, t)\Delta\xi_i \\ &+ (u(\xi_i, t) - z_i)\,\Delta k_i - k_i \Delta z_i] \\ &-\alpha\Delta u(x,t), \quad (x,t) \in \Omega, \end{aligned} \tag{17}$$

$$\Delta u(x,0) = 0, \ x \in [0,l], \tag{18}$$

$$\Delta u(0,t) = \begin{cases} 0, & t < T^d, \\ (1-\gamma)\Delta u(l, t - T^d), & t \geq T^d. \end{cases} \tag{19}$$

In obtaining formula (17), we used the relation

$$u(\xi_i + \Delta\xi_i, t) = u(\xi_i, t) + u_x(\xi_i, t)\Delta\xi_i + o(|\Delta\xi_i|).$$

For the increment of the functional (12), it is not difficult to obtain directly the representation

$$\begin{aligned}\Delta I(y;\gamma) &= I(\tilde{y};\gamma) - I(y;\gamma) = I(y + \Delta y;\gamma) - I(y;\gamma) \\ &= 2\beta_1 \int_0^T [u(l,t;y,\gamma) - V]\,\Delta u(l,t)dt \\ &\quad + 2\sigma \sum_{i=1}^{3M} (y_i - y_i^0)\Delta y_i.\end{aligned}$$

$$\sum_{i=1}^{3M} (y_i - y_i^0)\Delta y_i = \sum_{i=1}^{3M} \left[(\xi_i - \xi_i^0)\Delta\xi_i + (k_i - k_i^0)\Delta k_i + (z_i - z_i^0)\Delta z_i\right].$$

Let $\psi(x,t) = \psi(x,t;y,\gamma)$ be yet an arbitrary function continuous everywhere on Ω, except at the points $x = \xi_i$, $i = 1, 2, \ldots, M$, differentiable with respect to x for $x \in (\xi_i, \xi_{i+1}), i = 0, 1, |\ldots, M, \xi_0 = 0, \ \xi_{M+1} = l,$

differentiable with respect to t for $t \in (0,T)$. The presence of the arguments y and γ in the function $\psi(x,t;y,\gamma)$ indicate that it can vary when the feedback parameter vector y and the parameter γ change. Where it is possible, we will omit the parameters y and γ in the function $\psi(x,t;y,\gamma)$. We multiply equation (17) by $\psi(x,t)$ and integrate it over the rectangle Ω. Taking into account the assumed assumptions and conditions (18), (19), we have

$$\begin{aligned}
&\int_0^T\int_0^l \psi(x,t)\Delta u_t(x,t)dxdt + a\sum_{i=0}^{M}\int_{\xi_i}^{\xi_{i+1}}\int_0^T \psi(x,t)\Delta u_x(x,t)dtdx \\
&\quad -\alpha\int_0^T\int_0^l \psi(x,t)\sum_{i=1}^{M}[k_i\Delta u(\xi_i,t) + k_i u_x(\xi_i,t)\Delta\xi_i \\
&\qquad\qquad + (u(\xi_i,t)-z_i)\,\Delta k_i - k_i\Delta z_i] \\
&\quad +\alpha\int_0^T\int_0^l \psi(x,t)\Delta u(x,t)dx\,dt = 0. \qquad (20)
\end{aligned}$$

Using integration by parts for the first and second terms of (20) separately, and taking (18) and (19) into account, we obtain

$$\begin{aligned}
&\int_0^T\int_0^l \psi(x,t)\Delta u_t(x,t)dxdt \\
&\quad = \int_0^l \psi(x,T)\Delta u(x,T)dx - \int_0^T\int_0^l \psi_t(x,t)\Delta u(x,t)dxdt, \qquad (21)
\end{aligned}$$

$$\begin{aligned}
&a\sum_{i=0}^{M}\int_{\xi_i}^{\xi_{i+1}}\int_0^T \psi(x,t)\Delta u_x(x,t)dtdx \\
&\quad = a\int_0^T [\psi(l,t)\Delta u(l,t) - \psi(0,t)\Delta u(0,t)]dt \\
&\qquad + a\sum_{i=1}^{M}\int_0^T \left[\psi(\xi_i^-,t) - \psi(\xi_i^+,t)\right]\,\Delta u(\xi_i,t)dt \\
&\qquad - a\int_0^T\int_0^l \psi_x(x,t)\Delta u(x,t)dxdt \\
&\quad = a\int_0^T \psi(l,t)\Delta u(l,t)dt - a(1-\gamma)\int_{T^d}^T \psi(0,t)\Delta u(l,t-T^d)dt
\end{aligned}$$

$$
\begin{aligned}
&+a\sum_{i=1}^{M}\int_0^T\left[\psi(\xi_i^-,t)-\psi(\xi_i^+,t)\right]\ \Delta u(\xi_i,t)\,dt\\
&-a\int_0^T\int_0^l \psi_x(x,t)\Delta u(x,t)dxdt\\
&=a\int_0^T\psi(l,t)\Delta u(l,t)dt-a(1-\gamma)\int_0^{T-T^d}\psi(0,t+T^d)\Delta u(l,t)dt\\
&+a\sum_{i=1}^{M}\int_0^T\left[\psi(\xi_i^-,t)-\psi(\xi_i^+,t)\right]\ \Delta u(\xi_i,t)\,dt\\
&-a\int_0^T\int_0^l \psi_x(x,t)\Delta u(x,t)dxdt.
\end{aligned}\tag{22}
$$

Here, we have used the notation

$$\psi(\xi_i^-,t)=\psi(\xi_i-0,t),\ \psi(\xi_i^+,t)=\psi(\xi_i+0,t).$$

Taking (20)–(22) into account, we obtain for the increment of the functional:

$$
\begin{aligned}
\Delta I=\beta_1&\int_{T-T^d}^{T}\left[a\psi(l,t)+2(u(l,t)-V)\right]\Delta u(l,t)dt+\int_0^l\psi(x,T)\Delta u(x,T)dx\\
&+\int_0^{T-T^d}\left[a\psi(l,t)+a(1-\gamma)\psi(0,t+T^d)+2(u(l,t)-V)\right]\Delta u(l,t)dt\\
&+\int_0^T\int_0^l[-\psi_t(x,t)-a\psi_x(x,t)+\alpha\psi(x,t)]\Delta u(x,t)dxdt\\
&+a\sum_{i=1}^{M}\int_0^T\left[\psi(\xi_i^-,t)-\psi(\xi_i^+,t)-\frac{\alpha}{a}k_i\int_0^l\psi(x,t)dx\right]\ \Delta u(\xi_i,t)dt\\
&-\alpha\int_0^T\int_0^l\psi(x,t)\sum_{i=1}^{M}\left[k_iu_x(\xi_i,t)\Delta\xi_i+\left(u(\xi_i,t)-z_i\right)\Delta k_i\right.\\
&\qquad\qquad\left.-k_i\Delta z_i\right]dx\,dt\\
&+2\sigma\sum_{i=1}^{M}\left[(\xi_i-\xi_i^0)\Delta\xi_i+(k_i-k_i^0)\Delta k_i+(z_i-z_i^0)\Delta z_i\right].
\end{aligned}\tag{23}
$$

Since the function $\psi(x,t)$ is arbitrary, we require that it be almost everywhere a solution of the following adjoint initial- and boundary-value problem:

$$\psi_t(x,t)+a\psi_x(x,t)=\alpha\psi(x,t),\quad (x,t)\in\Omega,\tag{24}$$

$$\psi(x,T)=0, \quad x\in[0,l], \tag{25}$$

$$\psi(l,t)=-\frac{2}{a}(u(l,t)-V), \quad t\in(T-T^d,T], \tag{26}$$

$$\begin{aligned}\psi(l,t)=&-\frac{\alpha}{a}(1-\gamma)\psi(0,t+T^d)\\&-\frac{2}{a}(u(l,t)-V), \quad t\in(0,T-T^d],\end{aligned} \tag{27}$$

and at the points $\xi_i, i=1,2,\dots,M$ for $t\in[0,T]$, it satisfy the condition

$$\psi(\xi_i^-,t)=\psi(\xi_i^+,t)+\frac{\alpha}{a}k_i\int_0^l\psi(x,t)dx, \quad i=1,2,\dots,M. \tag{28}$$

Taking into account that the components of the gradient of the functional are determined by the linear part of the increment of the functional under the increments of the corresponding arguments, we obtain

$$\mathrm{grad}_{\xi_i}I=-\alpha k_i\int_0^T\left(\int_0^l\psi(x,t)dx\right)u_x(\xi_i,t)dt+2\sigma(\xi_i-\xi_i^0), \tag{29}$$

$$\mathrm{grad}_{k_i}I=-\alpha\int_0^T(u(\xi_i,t)-z_i)\left(\int_0^l\psi(x,t)dx\right)dt+2\sigma(k_i-k_i^0), \tag{30}$$

$$\mathrm{grad}_{z_i}I=\alpha k_i\int_0^l\psi(x,t)dx+2\sigma(z_i-z_i^0), \tag{31}$$

where $i=1,2,\dots,M$. The adjoint problem (24)–(28) can also be represented in another equivalent form, without the jump conditions (28). To do this, using the property of the δ function, we reduce the third term in (20) to the form

$$\begin{aligned}&a\int_0^T\int_0^l\psi(x,t)\sum_{i=1}^M(k_i\Delta u(\xi_i,t)+k_iu_x(\xi_i,t)\Delta\xi_i\\&\qquad+(u(\xi_i,t)-z_i)\Delta k_i-k_i\Delta z_i)dxdt\\&=a\sum_{i=1}^L k_i\int_0^T\int_0^l\int_0^l\psi(\zeta,t)\,\delta(\zeta-\xi_i)\Delta u(\zeta,t)d\zeta dxdt\\&+a\int_0^T\int_0^l\psi(x,t)\sum_{i=1}^M(k_iu_x(\xi_i,t)\Delta\xi_i\\&+(u(\xi_i,t)-z_i)\Delta k_i-k_i\Delta z_i)dxdt.\end{aligned}$$

Changing the order of integration with respect to ζ and x in the first triple integral and renaming again the integration variables with respect to ζ and x between each other, we obtain

$$\int_0^T \int_0^l \int_0^l \psi(\zeta,t)\,\delta(\zeta-\xi_i)\Delta u(\zeta,t)d\zeta dx dt$$
$$= \int_0^T \int_0^l \left(\int_0^l \psi(\zeta,t)d\zeta\right)\delta(x-\xi_i)\Delta u(x,t)dxdt. \qquad (32)$$

Taking (32) into account in (20), after making a rearrangement of the terms, instead of (24), (27), we obtain the following form of the adjoint problem:

$$\psi_t(x,t)+a\psi_x(x,t)=\alpha\psi(x,t)-\alpha\int_0^l \psi(\zeta,t)d\zeta\sum_{i=1}^{M}k_i\,\delta(x-\xi_i), \quad (x,t)\in\Omega, \qquad (33)$$

while preserving the initial and boundary conditions (25)–(27), but without the jump conditions (28).

Thus, we can consider the following theorem to be proved.

Theorem 1. *For the optimality of the vector of parameters $y^* \in R^{3M}$ in the problem* (10), (3)–(5), (11) *and* (12), *it is necessary and sufficient that*

$$(\mathrm{grad}J(y^*),\, y^*-y)\le 0$$

for all admissible control parameters $y\in R^{3M}$ satisfying conditions (13). *The components of the gradient vector grad$J(y)$ are defined by the following formulas:*

$$\mathrm{grad}_{\xi_i}J(y)=\int_\Gamma\left\{-\alpha k_i\int_0^T\left(\int_0^l\psi(x,t;y,\gamma)dx\right)u_x(\xi_i,t;y,\gamma)dt\right.$$
$$\left.+2\sigma(\xi_i-\xi_i^0)\right\}\rho_\Gamma(\gamma)d\gamma, \qquad (34)$$

$$\mathrm{grad}_{k_i}J(y)=\int_\Gamma\left\{-\alpha\int_0^T(u(\xi_i,t;y,\gamma)-z_i)\left(\int_0^l\psi(x,t;y,\gamma)dx\right)dt\right.$$
$$\left.+2\sigma(k_i-k_i^0)\right\}\rho_\Gamma(\gamma)d\gamma, \qquad (35)$$

$$\mathrm{grad}_{z_i}J(y)=\int_\Gamma\left\{\alpha k_i\int_0^l\psi(x,t;y,\gamma)dx+2\sigma(z_i-z_i^0)\right\}\rho_\Gamma(\gamma)d\gamma, \qquad (36)$$

where $i=1,2,\ldots,M$, $u(x,t;y,\gamma)$; $\psi(x,t;y,\gamma)$ are the solutions to the direct and adjoint boundary-value problems (10), (3)–(5) *and* (24)–(28), *resp.*

In numerical solution of the initial problem of optimizing the parameters to be synthesized, each iteration of procedure (14) involves solving direct (8)–(13) and adjoint (19)–(23) boundary-value problems with the specifics described above. Numerical solution of loaded boundary-value problems can be obtained using methods of meshes or lines. Their application to solutions of similar problems was studied, for example, in Refs. [27,28]. Lagging under boundary conditions can be taken into account using the "step method" [29].

4. Optimization of the Number of Observation Points

It is possible that the number of observation points is not specified, but has to be optimized, together with their places. In this context, the following approach is proposed for choosing an optimal number of observation points. Obviously, the optimal number of observation points has to be minimized to some extent.

Let $J_M^* = J^*(y^M; M)$ be the minimum value of the functional of problem (3), (5), (10)–(13) for a given number M of observation points, and let y^M denote the values of the optimal parameters of the synthesized feedback control (see Fig. 2). Obviously, $J_M^* = J^*(y^M; M)$ regarded as a composite function of M is a non-increasing function, i.e. in the general case, we have the inequality

$$J^*(y^*; \cdot) \leq J^*(y^{M_1}; M_1) \leq J^*(y^{M_2}; M_2), \tag{37}$$

for $M_2 > M_1$. Here, $J_M^* = J^*(y^M; M)$ denotes the optimal value of the functional in the original problem (1)–(5) (11) and (12) with M observation points and $J^* = J^*(y^*; \cdot)$ is the optimal value of the functional in the

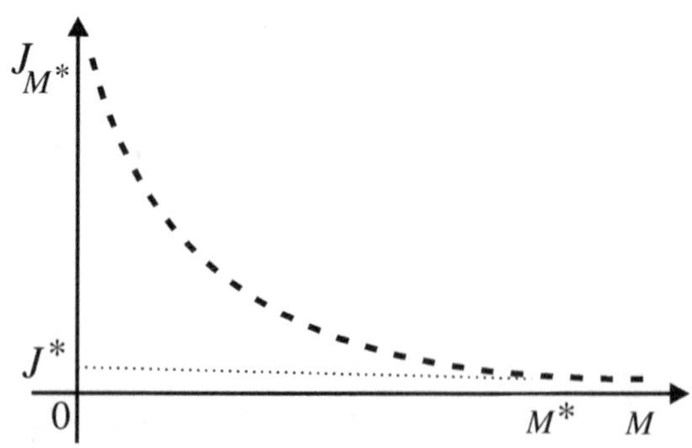

Fig. 2: Optimal value of the cost functional against the number M of observation points.

heating problem with feedback distributed over the entire systems, which corresponds to observations of the current state at all points of the heating system.

$$\lim_{M \to \infty} J^*(y^M; M) = J^*.$$

It follows from (37) that, as the number of observation points increases, the optimal values of the objective functional decrease and approach arbitrarily close to J^*.

It may happen that there exists a finite value M^* such that

$$J^*(y^M; M) = J^*_M, \ \ M > M^*.$$

As an optimal number of observation points, we use the minimum M^* for which one of the following inequalities is satisfied for the first time:

$$\Delta J^*(y^{M^*}; M^*) = |J^*(y^{M^*+1}; M^* + 1) - J^*(y^{M^*}; M^*)| \leq \delta, \tag{38}$$

$$\Delta J^*(y^{M^*}; M^*)/J^*(y^{M^*}; M^*) \leq \delta, \tag{39}$$

where δ is a given positive number determined by the required accuracy of finding the optimal number of observation points.

After finding control parameters for heating with M observation points, the number M can be reduced if the resulting optimal vector ξ^M is such that, for two neighboring observation points,

$$|\xi^M_{j+1} - \xi^M_j| \leq \delta_1, \quad j = 1, 2, \ldots, M - 1, \tag{40}$$

here $\delta_1 > 0$ is sufficiently small. Under condition (40), we can retain one of the two neighboring observation points, thus reducing their total number by one.

5. Results of the Numerical Experiments

In this section, we present the results of the solution of the following model problem. The process is described by the boundary-value problem (1)–(5). It is required to design an optimal control (regulation) system for the coolant heating process, first, with two feedback points, i.e. $M = 2$. Thus, it is required to determine $\xi = (\xi_1, \ \xi_2)$, that is, locations of two temperature sensors, as well as feedback parameters $k, z \in R^2$. Hence, the total number of parameters to be synthesized is six.

The problem is considered solved at the following values of parameters comprising its statement: $l = 1$; $a = 1$; $\alpha = 0, 1$; $T^d = 0.2$, $T = 5$, $V = 70$, $\Gamma = [0; 0.2]$, $\underline{\vartheta} = 55, \bar{\vartheta} = 75$, $\bar{k}_1 = \bar{k}_2 = 8$, $\underline{k}_1 = \underline{k}_2 = 1$, $\bar{z}_1 = \bar{z}_2 = 75$,

$\underline{z}_1 = \underline{z}_2 = 57$. In calculations, the density function $\rho_\Gamma(\gamma)$ has been taken uniformly distributed on [0; 0.2], while approximation of the integral over Γ was performed using the method of rectangles with the step 0.05. It should be noted that values $\bar{k}_1$ and $\bar{k}_2$ were chosen using the results of trial calculations performed, which required the process constraint (2) to be true for given $\underline{\vartheta}$ and $\bar{\vartheta}$.

The numerical experiments have been carried out for different initial values of parameters $(y^0)^j = (k_1^0, k_2^0, z_1^0, z_2^0, \xi_1^0, \xi_2^0)^j, j = 1, 2, \ldots, 5$, used in iterative optimization procedure (14). Table 1 gives these values, as well as the corresponding values of the functional at these points.

Table 2 gives the values of parameters $(y^{(*)})^j = (k_1^{(*)}, k_2^{(*)}, z_1^{(*)}, z_2^{(*)}, \xi_1^{(*)}, \xi_2^{(*)})^j$ and the functional $J(y^*)^j$ obtained using the gradient projection method, (14), (15) at $\delta_1 = 0.005$ and $\delta_2 = 0.001$ starting from the initial points $(y^0)^j, j = 1, 2, \ldots, 5$, specified in Table 1.

As can be seen from Table 3, which gives the results of experiments for optimization of M, that is, the number of sensor points, at $M = 6$ and

Table 1: Initial values of the parameters $(y^0)^j, j = 1, 2, \ldots, 5$, to be optimized and the corresponding values of the functional.

	Values of the parameters to be optimized						Value of the functional
j	$(k_1^0)^j$	$(k_2^0)^j$	$(z_1^0)^j$	$(z_2^0)^j$	$(\xi_1^0)^j$	$(\xi_2^0)^j$	$J(y^0)^j$
1	4	6	61	63	0,1	0,8	363.210004
2	3	5	65	60	0,2	0,9	357.150011
3	1	8	62	63	0,4	0,8	257.310003
4	5	2	63	66	0,5	0,7	165.150016
5	6	4	66	62	0,2	0,7	205.190007

Table 2: Values of parameters and the functional obtained at the sixth iterations of process (14) for different initial values $(y^0)^j, j = 1, 2, \ldots, 5$.

	Values of the parameters to be optimized						Value of the functional
j	$(k_1^*)^j$	$(k_2^*)^j$	$(z_1^*)^j$	$(z_2^*)^j$	$(\xi_1^*)^j$	$(\xi_2^*)^j$	$J(y^*)^j$
1	5.9956	3.9952	66.9945	68.9949	0.2994	0.5994	0.3422
2	5.9977	3.9983	66.9978	68.9954	0.3000	0.6000	0.3259
3	5.9962	3.9988	66.9951	68.9948	0.2971	0.5971	0.3538
4	5.9978	3.9971	66.9991	68.9975	0.3000	0.6000	0.3145
5	5.9991	3.9961	66.9964	68.9973	0.3000	0.6000	0.3062

Table 3: Results of the problem solutions at different numbers of observation points.

M	$(\xi^{(0)})\,;(k^{(0)})\,;(z^{(0)})$	$J(y^0)$	$(\xi^{(*)})\,;(k^{(*)})\,;(z^{(*)})$	$J(y^*)$
3	(0.1, 0.4, 0.7);	333.46	(0.300, 0.600, 0.899);	0.3456
	(3, 4, 8);		(5.002, 4.201, 4.002);	
	(61, 65, 67)		(66.998, 67.998, 68.998)	0.3456
4	(0.1, 0.5, 0.7, 0.8);	323.64	(0.150, 0.300, 0.600, 0.849);	0.3449
	(1, 4, 8, 2);		(5.001, 4.102, 4.006, 3.999);	
	(60, 63, 66, 67)		(66.996, 67.999, 68.001, 68.999)	
5	(0.1, 0.2, 0.5, 0.7, 0.8);	368.54	(0.250, 0.300, 0.610, 0.800, 0.896);	0.3436
	(3, 5, 7,8, 3);		(5.101, 4.126, 4.106, 4.012, 3.9982);	
	(61, 63, 64, 66, 67)		(66.987, 67.979, 68.201, 68.571, 68.989)	
6	(0.1, 0.2, 0.5, 0.6, 0.7, 0.8);	408.37	(0.208, 0.305, 0.481, 0.605, 0.805, 0.900);	0.3234
	(3, 5, 6, 7, 8, 3);		(5.003, 4.086, 4.015, 4.013 , 3.906, 3.999);	
	(58, 61, 64, 65, 66, 68)		(66.997, 67.999, 68.121, 68.571, 68.989, 68.999)	
7	(0.1, 0.2, 0.3, 0.5, 0.6, 0.7, 0.8);	217.23	(0.198, 0.303, 0.307, 0.491, 0.62, 0.791, 0.901);	0.3023
	(3,4, 5, 6, 7, 8,3);		(5.003, 4.086, 4.015, 4.013 , 3.906, 3.912, 3.998);	
	(58, 60, 63, 64, 66, 67, 70)		(66.998, 68.003, 68.323, 68.772, 68.979, 69.002, 69.012)	

Table 4: Values of the functional and relative deviations between the obtained and the desired temperature at the unit outlet for different noise levels in measurement.

Noise level χ	Relative deviation $\max_{t\in[0,5]}\lvert u(l,t)-V\rvert/\lvert V\rvert$	Value of the functional $J^*(y)$
0.00	0.021941	0.3023
0.01	0.033052	0.3543
0.03	0.038311	0.3762
0.05	0.064574	0.3916

$M = 7$, minimum values of the functionals satisfy the condition (38), while at $M = 7$ optimal values of the second and the third components of vector ξ satisfy condition (40) at $\delta = \delta_1 = 0.005$. Then, $M^* = 6$ can be taken as the optimal number of sensor points.

Numerical experiments have been performed, in which exact values of the process states observed at sensor points $u(\xi_1, t)$ and $u(\xi_2, t)$ have been perturbed with random noise as follows:

$$u(\xi_i, t) = u(\xi_i, t)\,(1 + \chi(2\theta_i - 1))\,, \quad i = 1,\, 2\,,$$

where θ_i is a random value uniformly distributed on the interval [0, 1], and χ is the noise level.

Table 4 gives the obtained values of the functional and relative deviations between the obtained and the desired temperature at the unit outlet for noise levels equal to 0% (no noise), 1%, 3%, and 5%, which correspond to values $\chi = 0$ (no noise), 0.01, 0.03, and 0.05.

As can be seen from Table 4, feedback control of the coolant heating process in the furnace of the heated apparatus is quite resistant to measurement errors.

6. Conclusion

Automatic feedback control systems for technical objects and technological processes with distributed parameters are widely used owing to the significantly increased capabilities of measuring and computing facilities. In the chapter, we studied the problem of controlling a heating apparatus for heating a heat-carrying agent, which ensures the supply of heat to a closed heat supply system. The specificity of the investigated problem, described by the first-order hyperbolic equation, lies in the fact that in its boundary conditions a delayed in time argument is present. The mathematical model of the controlled process is reduced to a pointwise-loaded hyperbolic equation, and the problem under consideration is reduced to the parametric optimal control problem. In order to use first-order optimization methods for numerical solution to the problem of optimizing the locations of sensors and the parameters of feedback control actions, we obtained formulas for the gradient of the objective functional. The formulation of the problem and the approach used in this chapter to obtain computational formulas for its numerical solution can be extended to cases of feedback control by many other processes described by other types of partial differential equations.

References

[1] W.H. Ray, *Advanced Process Control* (McGraw-Hill, New York, 1981).

[2] A.G. Butkovskii, *Methods of Control for Systems with Distributed Parameters* (Nauka, Moscow, 1984).

[3] B.T. Polyak and P.S. Shcherbakov, *Robust Stability and Control* (Nauka, Moscow, 2002).

[4] K.R. Aida-zade and V.A. Hashimov, Optimizing the arrangement of lumped sources and measurement points of plate heating, *Cybern. Syst. Analysis.* **55**(4), 605–615, (2019).

[5] S.Z. Guliyev, Synthesis of zonal controls for a problem of heating with delay under nonseparated boundary conditions, *Cybern. Syst. Analysis.* **54**(1), 110–121, (2018).

[6] H. Shang, J.F. Forbes, and M. Guay, Feedback control of hyperbolic PDE systems, *IFAC Proceedings.* **33**(10), 533–538, (2000).

[7] J.M. Coron and Zh. Wang, Output feedback stabilization for a scalar conservation law with a nonlocal velocity, *SIAM J. Math. Anal.* **45**(5), 2646–2665, (2013).

[8] L. Afifi, K. Lasri, M. Joundi, and N. Amimi, Feedback controls for exact remediability in disturbed dynamical systems, *IMA J. Math. Control Inf.* **35**(2), 411–425, (2018).

[9] W. Mitkowski, W. Bauer, and M. Zagórowsk, Discrete-time feedback stabilization, *Arch. Control Sci.* **27**(2), 309–322, (2017).

[10] A.S. Antipin and E.V. Khoroshilova, Feedback synthesis for a terminal control problem, *Comput. Math. Math. Phys.* **58**(12), 1903–1918, (2018).

[11] K.R. Aida-zade and V.M. Abdullaev, On an approach to designing control of the distributed-parameter processes, *Autom. Remote Control.* **73**(9), 1443–1455, (2012).

[12] K.R. Aida-zade and V.M. Abdullayev, Optimizing placement of the control points at synthesis of the heating process control, *Autom. Remote Control.* **78**(9), 1585–1599, (2017).

[13] V.M. Abdullayev and K.R. Aida-zade, Numerical solution of the problem of determining the number and locations of state observation points in feedback

control of a heating process, *Comput. Math. and Math. Phys.* **58**(1), 78-89, (2018).

[14] J.L. Lions, *Controle optimal des syst'emes gouvern'es par des 'equations aux deriv'ees partielles* (Dunod Gauthier-Villars, Paris, 1968).

[15] A. Hamidoglu and E.N. Mahmudov, On construction of sampling patterns for preserving observability/controllability of linear sampled-data systems, *Int. J. Control.* (2020). Doi: 10.1080/00207179.2020.1787523.

[16] A.N. Tikhonov and A.A. Samarskii, *Equations of Mathematical Physics* (Nauka, Moscow, 1966).

[17] M.T. Dzhenaliev, Optimal control of linear loaded parabolic equations, *Differ. Equations.* **25**(4), 641–651, (1989).

[18] A.M. Nakhushev, *Loaded Equations and Their Application* (Nauka, Moscow, 2012).

[19] I.N. Parasidis and E. Providas, Closed-Form Solutions for Some Classes of Loaded Difference Equations with Initial and Nonlocal Multipoint Conditions, In: N. Daras, T. Rassias, (Eds.), *Modern Discrete Mathematics and Analysis*, Vol. 131, Springer Optimization and Its Applications, pp. 363–387, (2018).

[20] I.N. Parasidis and E. Providas, An exact solution method for a class of nonlinear loaded difference equations with multipoint boundary conditions, *J. Differ. Equ. Appl.* **24**(10), 1649–1663, (2018).

[21] V.M. Abdullayev and K.R. Aida-zade, Approach to the numerical solution of optimal control problems for loaded differential equations with nonlocal conditions, *Comput. Math. and Math. Phys.* **59**(5), 696–707, (2019).

[22] A.A. Alikhanov, A.M. Berezgov, and M.Kh. Shkhanukov-Lafshiev, Boundary value problems for certain classes of loaded differential equations and solving them by finite difference methods, *Comput. Math. and Math. Phys.* **48**(9), 641–651, (2008).

[23] K.R. Aida-zade, An approach for solving nonlinearly loaded problems for linear ordinary differential equations, *Proc. of the Institute of Mathematics and Mechanics.* **44**(2), 338–350, (2018).

[24] V.M. Abdullayev, Numerical solution to optimal control problems with multipoint and integral conditions, *Proc. of the Institute of Mathematics and Mechanics.* **44**(2), 171–186, (2018).

[25] V.M. Abdullayev, Identification of the functions of response to loading for stationary systems, *Cybern. Syst. Analysis.* **53**(3), 417–425, (2017).

[26] F.P. Vasil'ev, *Optimization Methods.* (Factorial Press, Moscow, 2002).

[27] V.M. Abdullayev and K.R. Aida-zade, Optimization of loading places and load response functions for stationary systems, *Comput. Math. Math. Phys.* **57**(4), 634–644, (2017).

[28] V.M. Abdullayev and K.R. Aida-zade, Finite-difference methods for solving loaded parabolic equations, *Comput. Math. Math. Phys.* **56**(1), 93–105, (2016).

[29] L.E. Elsgolts and S.B. Norkin, *Introduction to the Theory and Application of Differential Equations with Deviating Arguments* (Nauka, Moscow, 1971).

https://doi.org/10.1142/9789811261572_0002

Chapter 2

Monotonicity of Averages of Superquadratic and Related Functions

Shoshana Abramovich
Department of Mathematics University of Haifa, Haifa, Israel
abramos@math.haifa.ac.il

The purpose of this work is to focus on various **successive differences of averages** and their bounds when f is a superquadratic, a subquadratic or an 1-quasiconvex function. These inequalities refine inequalities obtained when f is a non-negative convex function and lead to Alzer-type inequalities.

1. Introduction

The definition of superquadracity as brought here appeared first in 2004 in the papers [1] and [2]. Since then superquadracity and its applications were dealt with in numerous papers. Overviews of this subject were presented in Refs. [3] and [4], in particular results related to Jensen's inequality in one and several variables, to Jensen–Steffensen's inequality, to Hardy inequality and to Hermite–Hadamard and Fejer's inequalities.

Our purpose is to focus here on various **successive differences of averages** and their bounds when f is a superquadratic, a subquadratic or a 1-quasiconvex function.

It is clear that many more results of these types may be obtained by using superquadracity and its closely related operators. We encourage the interested readers to continue researching in these directions.

One application of difference of averages inequality is Alzer-type inequality. For instance, from the difference of averages inequality

$$\frac{1}{n}\sum_{i=1}^{n} f\left(\frac{k}{n}\right) - \frac{1}{n+1}\sum_{i=1}^{n+1} f\left(\frac{k}{n+1}\right) \geq 0,$$

where f is a convex function, Alzer's inequality:

$$\frac{n}{n+1} \leq \left(\frac{(n+1)\sum_{i=1}^{n} i^r}{n\sum_{i=1}^{n+1} i^r}\right)^{\frac{1}{r}}, \quad r > 0, \quad n \geq 1,$$

is obtained when $f(x) = x^r$.

The high interest in the monotonicity of averages and Alzer-type inequalities can be seen in Refs. [2–35]. The results in those papers are obtained mainly through convex functions.

All the inequalities derived by using superquadratic functions and in particular for successive differences are analogs of inequalities satisfied by convex functions. In those cases that the superquadratic functions are also non-negative, refinements of inequalities of successive differences satisfied by convex functions are obtained and as a consequence, the resulting Alzer's-type inequalities are stronger than those for convex functions (see Example 5).

We start with some definitions, notations and lemmas used in the sequel.

Definition 1 ([1], Definition 2.1). A function $f : [0,\infty) \rightarrow \mathbb{R}$ is superquadratic provided that for all $x \in [0,\infty)$ there exists a constant $C_f(x) \in \mathbb{R}$ such that the inequality

$$f(y) - f(x) - C_f(x)(y-x) - f(|y-x|) \geq 0 \tag{1}$$

holds for all $y \in [0,\infty)$.

f is called subquadratic if $-f$ is superquadratic.

Corollary 1 ([1], Lemma 2.1). *When the function f is non-negative superquadratic, and therefore it is also convex, $C_f(x) = f'(x)$ and $f'(0) = f(0) = 0$. In particular, the functions $f(x) = x^p$, $x \geq 0$ are superquadratic for $p \geq 2$, subquadratic for $0 < p \leq 2$, for which $C_f(x) = px^{p-1} = f'(x)$. When $p = 2$ (1) is an equality.*

Corollary 2 ([1,2], Lemma 2.4). *Suppose that f is superquadratic. Let $\xi_i \geq 0$, $i = 1,\ldots,m$, and let $\overline{\xi} = \sum_{i=1}^{m} p_i\xi_i$ where $p_i \geq 0, i = 1,\ldots,m$, and $\sum_{i=1}^{m} p_i = 1$. Then*

$$\sum_{i=1}^{m} p_i f(\xi_i) - f(\overline{\xi}) \geq \sum_{i=1}^{m} p_i f(|\xi_i - \overline{\xi}|)$$

and in the special case that $m = 2$, $0 \leq \lambda \leq 1$ *and* $0 \leq x, y < \infty$

$$\begin{aligned}(1-\lambda) f(y) + \lambda f(x) \\ \geq f((1-\lambda) y + \lambda x) + (1-\lambda) f(\lambda |x-y|) + \lambda f((1-\lambda)|x-y|)\end{aligned}$$

hold.

Definition 2 ([7], Definition 2). A function $f : [0, b) \to \mathbb{R}$ that satisfies $f = x\varphi$, where φ is a convex function is called 1-quasiconvex function.

Lemma 1 ([7], Lemma 2). *Let* x_i, $0 \leq x_i < b$, $0 < b \leq \infty$, $0 \leq \alpha_i \leq 1, 1 = 1, \ldots, m$, $\sum_{i=1}^{m} \alpha_i = 1$, $\overline{x} = \sum_{i=1}^{m} \alpha_i x_i$. *Let* $\varphi : [0, b) \to \mathbb{R}$, $0 < b \leq \infty$ *be a differentiable convex function, and* f *be* 1*-quasiconvex, where* $f = x\varphi$. *Then*

$$\sum_{i=1}^{m} \alpha_i f(x_i) - f(\overline{x}) \geq \sum_{i=1}^{m} \varphi'(\overline{x}) \alpha_i (x_i - \overline{x})^2, \tag{2}$$

holds. Moreover, when φ *increases* (2) *is a refinement of Jensen's inequality.*

Lemma 2 ([7], Lemma 4). *Let* $\varphi : [0, b) \to \mathbb{R}_+$, $0 < b \leq \infty$, *be a differentiable convex increasing function satisfying*

$$\varphi(0) = \lim_{x \to 0+} x\varphi'(x) = 0.$$

Then the 1*-quasiconvex function* f, *where* $f = x\varphi$, *is also superquadratic and convex.*

In Section 2, basic results are quoted from Ref. [2], which is the first publication dealing with successive differences of averages of superquadratic functions.

In Section 3, the results obtained in Section 2 are refined and besides upper (lower) bounds which appear in Section 2, lower (upper) bounds of the successive differences of averages are added. The results are quoted from Ref. [5].

In Section 4, the bounds of difference of averages between the closely related superquadratic functions and 1-quasiconvex functions are obtained and compared. The results are quoted from Ref. [7].

2. Basic Inequalities for Averages for Superquadratic Functions

In this section, we quote results from Ref. [2], which is the first publication dealing with successive differences of averages of superquadratic functions.

For a function f, let

$$A_n(f) = \frac{1}{n-1}\sum_{r=1}^{n-1} f\left(\frac{r}{n}\right) \quad (n \geq 2)$$

and

$$B_n(f) = \frac{1}{n+1}\sum_{r=0}^{n} f\left(\frac{r}{n}\right) \quad (n \geq 1).$$

In Ref. [10], the results related to $A_n(f)$ and $B_n(f)$ are obtained when the functions involved are convex. Theorems 1 and 2 extend and refine these results when the functions involved are superquadratic:

Theorem 1. *If f is superquadratic on $[0,1]$, then for $n \geq 2$*

$$A_{n+1}(f) - A_n(f) \geq \sum_{r=1}^{n-1} \lambda_r f(x_r),$$

where

$$\lambda_r = \frac{2r}{n(n-1)}, \quad x_r = \frac{n-r}{n(n+1)}.$$

Further,

$$A_{n+1}(f) - A_n(f) \geq f\left(\frac{1}{3n}\right) + \sum_{r=1}^{n-1} \lambda_r f(y_r),$$

where

$$y_r = \frac{|2n-1-3r|}{3n(n+1)}.$$

If f is also non-negative, then for $n \geq 3$,

$$A_{n+1}(f) - A_n(f) \geq f\left(\frac{1}{3n}\right) + f\left(\frac{16}{81(n+3)}\right),$$

and as f is also convex, we get a refinement of the result $A_{n+1}(f) \geq A_n(f)$ in Ref. [10] when f is a convex function.

This theorem is proved by dealing with the identity

$$\begin{aligned}\Delta_n &= \frac{n-1}{n}\sum_{r=1}^{n} f\left(\frac{r}{n+1}\right) - \sum_{r=1}^{n-1} f\left(\frac{r}{n}\right) \\ &= \sum_{r=1}^{n-1} \frac{r}{n}\left[f\left(\frac{r+1}{n+1}\right) - f\left(\frac{r}{n}\right)\right] + \sum_{r=1}^{n-1} \frac{n-r}{n}\left[f\left(\frac{r}{n+1}\right) - f\left(\frac{r}{n}\right)\right]\end{aligned}$$

and by using the basic Jensen inequality for superquadratic functions, taking into account that

$$\frac{r+1}{n+1} - \frac{r}{n} = \frac{n-r}{n\left(n+1\right)}.$$

Theorem 2. *If f is superquadratic on $[0,1]$, then for $n \geq 2$,*

$$B_{n-1}\left(f\right) - B_n\left(f\right) \geq \sum_{r=1}^{n} \lambda_r f\left(x_r\right),$$

where

$$\lambda_r = \frac{2r}{n\left(n-1\right)}, \quad x_r = \frac{n-r}{n\left(n+1\right)}.$$

Further,

$$B_{n-1}\left(f\right) - B_n\left(f\right) \geq f\left(\frac{1}{3n}\right) + \sum_{r=1}^{n} \lambda_r f\left(y_r\right),$$

where

$$y_r = \frac{\left|2n+1-3r\right|}{3n\left(n-1\right)}.$$

The opposite inequalities hold if f is subquadratic.

If f is also non-negative, then for $n \geq 2$,

$$B_{n-1}\left(f\right) - B_n\left(f\right) \geq f\left(\frac{1}{3n}\right) + f\left(\frac{16}{81n}\right),$$

and as f is also convex, we get a refinement of the result $B_{n-1}(f) \geq B_n(f)$ in Ref. [10] when f is a convex function.

The proof of this theorem uses the identity

$$\begin{aligned}\Delta_n &= \left(n+1\right)\left[B_{n-1}\left(f\right) - B_n\left(f\right)\right] \\ &= \sum_{r=1}^{n} \frac{r}{n}\left[f\left(\frac{r-1}{n-1}\right) - f\left(\frac{r}{n}\right)\right] + \sum_{r=0}^{n-1} \frac{n-r}{n}\left[f\left(\frac{r}{n-1}\right) - f\left(\frac{r}{n}\right)\right].\end{aligned}$$

Theorems 3 and 4 are obtained by replacing in Theorem 1 and in Theorem 2 $f(r/n)$ by $f(a_r/a_n)$ and $1/(n\pm1)$ by $1/c_{n\pm1}$, and under suitable conditions on the sequences (a_n) and (c_n):

Theorem 3. *Let $(a_n)_{n\geq1}$ and $(c_n)_{n\geq0}$ be sequences such that $a_n > 0$ and $c_n > 0$ for $n \geq 1$ and:*

(A1) *$c_0 = 0$ and c_n is increasing,*
(A2) *$c_{n+1} - c_n$ is decreasing for $n \geq 0$,*
(A3) *$c_n(a_{n+1}/a_n - 1)$ is decreasing for $n \geq 1$.*

Given a function f, let

$$A_n\left(f,(a_n),(c_{n-1})\right) = \frac{1}{c_{n-1}}\sum_{r=1}^{n-1} f\left(\frac{a_r}{a_n}\right).$$

Suppose that f is superquadratic and non-negative. Then

$$\begin{aligned}&A_{n+1}\left(f,\left(a_{n+1}\right),\left(c_n\right)\right) - A_n\left(f,\left(a_n\right),\left(c_{n-1}\right)\right)\\ &\geq \frac{1}{c_nc_{n-1}}\sum_{r=1}^{n-1} c_r\left(\left|\frac{a_{r+1}}{a_{n+1}} - \frac{a_r}{a_n}\right|\right)\\ &\quad+\frac{1}{c_nc_{n-1}}\sum_{r=1}^{n-1}(c_n - c_r)\, f\left(\left|\frac{a_r}{a_n} - \frac{a_r}{a_{n+1}}\right|\right).\end{aligned}$$

Theorem 4. *Let $(a_n)_{n\geq0}$ and $(c_n)_{n\geq0}$ be sequences such that $a_n > 0$ and $c_n > 0$ for $n \geq 1$ and*

(B1) *$c_0 = 0$ and c_n is increasing,*
(B2) *$c_n - c_{n-1}$ is increasing for $n \geq 1$,*
(B3) *$c_n\left(1 - \frac{a_{n-1}}{a_n}\right)$ is increasing for $n \geq 1$,*
(B4) *either $a_0 = 0$ or (a_n) is increasing.*

Given a function f, let

$$B_n\left(f,\left(a_n\right),\left(c_{n+1}\right)\right) = \frac{1}{c_{n+1}}\sum_{r=0}^{n} f\left(\frac{a_r}{a_n}\right), \qquad n \geq 1.$$

Suppose that f is superquadratic and non-negative. Then

$$\begin{aligned}&B_{n-1}\left(f,\left(a_{n-1}\right),\left(c_n\right)\right) - B_n\left(f,\left(a_n\right),\left(c_{n+1}\right)\right)\\ &\geq \frac{1}{c_nc_{n+1}}\sum_{r=1}^{n-1} c_r f\left(\left|\frac{a_r}{a_n} - \frac{a_{r-1}}{a_{n-1}}\right|\right)\\ &\quad+\frac{1}{c_nc_{n+1}}\sum_{r=1}^{n-1}(c_n - c_r)\, f\left(\left|\frac{a_r}{a_n - 1} - \frac{a_r}{a_n}\right|\right).\end{aligned}$$

[2] ends with specific examples related to superquadratic functions which satisfy the conditions on $(\mathbf{a}_n)$ and $(\mathbf{c}_n)$ dealt with in Theorems 3 and 4.

3. More on Superquadracity, Subquadracity and Bounds of Differences of averages

In this section, we present the results, quoted from Ref. [5], about the general sequence $S_n(f) = 1/c_n \sum_{r=1}^{n} f(a_r/b_n)$ in which the authors obtain lower bounds for the successive differences for various cases of $(a_n)_{n\geq 0}$, $(b_n)_{n\geq 0}$, $(c_n)_{n\geq 0}$ when f is either a superquadratic or subquadratic function. These results, besides refining known results for convex functions, also extend the basic results in Section 2.

Subsection 3.1 deals with superquadracity and upper bounds of successive differences of averages.

In Subsection 3.2, additional refined results from Section 2 are presented for lower bounds of various successive differences when the functions f are superquadratic.

In Subsection 3.3, bounds of various successive differences when f is an increasing subquadratic function are presented.

3.1. *Superquadracity and upper bounds of averages*

An upper bound for the difference $A_{n+1}(f) - A_n(f)$ for superquadratic functions is shown in Theorem 5:

Theorem 5 ([5],Theorem 1). *Let f be a superquadratic function on $[0,1]$. Then for $1 \leq r \leq n$, $n \geq 3$*

$$\begin{aligned}
&A_{n+1}(f) - A_n(f) \\
&= \frac{1}{n}\sum_{r=1}^{n} f\left(\frac{r}{n+1}\right) - \frac{1}{n-1}\sum_{r=1}^{n-1} f\left(\frac{r}{n}\right) \\
&\leq \frac{1}{2}\left[f\left(\frac{1}{n+1}\right) + f\left(\frac{n}{n+1}\right)\right] \\
&\quad - \sum_{r=1}^{n-1}\left[\frac{2r}{n(n-1)} f\left(\frac{n-r-1}{n+1}\right) + \frac{1}{n-1} f\left(\frac{r}{n}\right)\right] \qquad (3)
\end{aligned}$$

is obtained.

Moreover, if f is also positive, then

$$A_{n+1}(f) - A_n(f) \leq \frac{1}{2}\left[f\left(\frac{1}{n+1}\right) + f\left(\frac{n}{n+1}\right)\right] - \left[f\left(\frac{n-2}{3(n+1)}\right) + f\left(\frac{1}{2}\right)\right]. \quad (4)$$

An upper bound for the difference $B_{n-1}(f) - B_n(f)$ is quoted in Theorem 6 from Ref. [5].

Theorem 6 ([5], Theorem 2). *Let f be a positive superquadratic function on $[0,1]$. Then when $n \geq 3$*

$$\begin{aligned} &B_{n-1}(f) - B_n(f) \\ &= \frac{1}{n}\sum_{r=0}^{n-1} f\left(\frac{r}{n-1}\right) - \frac{1}{n+1}\sum_{r=0}^{n} f\left(\frac{r}{n}\right) \\ &\leq \frac{n-1}{2n}\left[f\left(\frac{1}{n-1}\right) + f(1)\right] - \frac{n-3}{n} f\left(\frac{1}{3}\right) - f\left(\frac{1}{2}\right), \end{aligned}$$

holds.

Remark 1. The arguments in Theorems 5 and 6 are related to an upper bound of $\sum_{i=1}^{m} f\left(\frac{a_i}{an}\right)$ where $(a_i)_{i\geq 1}$ is positive increasing and f is superquadratic. To get an upper bound, insert in Corollary 2 $\lambda = \frac{a_n - a_i}{a_n - a_1}$, $x = \frac{a_i}{a_n}$, $y = \frac{a_n}{a_n} = 1$, and obtain the inequality

$$f\left(\frac{a_i}{a_n}\right) \leq \frac{a_n - a_i}{a_n - a_1} f\left(\frac{a_1}{a_n}\right) + \frac{a_i - a_1}{a_n - a_1} f(1) - \frac{a_n - a_i}{a_n - a_1} f\left(\frac{a_i - a_1}{a_n}\right) - \frac{a_i - a_1}{a_n - a_1} f\left(\frac{a_n - a_i}{a_n}\right).$$

Therefore,

$$\sum_{i=1}^{m} f\left(\frac{a_i}{a_n}\right) \leq f\left(\frac{a_1}{a_n}\right)\sum_{i=1}^{m}\frac{a_n - a_i}{a_n - a_1} + f(1)\sum_{i=1}^{m}\frac{a_i - a_1}{a_n - a_1} - \sum_{i=1}^{m}\left[\frac{a_n - a_i}{a_n - a_1} f\left(\frac{a_i - a_1}{a_n}\right) + \frac{a_i - a_1}{a_n - a_1} f\left(\frac{a_n - a_i}{a_n}\right)\right].$$

If f is also positive and therefore convex, then from the last inequality the following inequality is obtained:

$$\sum_{i=1}^{m} f\left(\frac{a_i}{a_n}\right) \leq f\left(\frac{a_1}{a_n}\right)\sum_{i=1}^{m}\frac{a_n-a_i}{a_n-a_1}+f(1)\sum_{i=1}^{m}\frac{a_i-a_1}{a_n-a_1}$$
$$-mf\left(\sum_{i=1}^{m}\frac{2(a_n-a_i)(a_i-a_1)}{(a_n-a_1)a_n m}\right). \tag{5}$$

If $a_i, i = 1, \ldots, n$ is a general positive increasing sequence, then by using $2(a_n-a_i)(a_i-a_1) > 2(a_n-a_{n-1})(a_2-a_1)$, and by choosing $m = n$ inserted in (5), the inequality

$$\sum_{i=1}^{n} f\left(\frac{a_i}{a_n}\right) \leq f\left(\frac{a_1}{a_n}\right)\left(\frac{na_n-\sum_{i=1}^{n}a_i}{a_n-a_1}\right)+f(1)\left(\frac{\sum_{i=1}^{n}a_i-na_1}{(a_n-a_1)}\right)$$
$$-nf\left(\frac{2(a_n-a_{n-1})(a_2-a_1)}{(a_n-a_1)a_n}\right) \tag{6}$$

holds.

A result from (6) is as follows:

Theorem 7 ([5], Theorem 3). *Let f be a positive superquadratic function on $[0.1]$, (and therefore also convex) and let $a_i > 0$, $i = 1, \ldots, n+1$ be an increasing sequence. Then*

$$\frac{1}{n}\sum_{i=1}^{n} f\left(\frac{a_i}{a_n}\right)-\frac{1}{n+1}\sum_{i=1}^{n+1} f\left(\frac{a_i}{a_{n+1}}\right)$$
$$\leq \frac{na_n-\sum_{i=1}^{n}a_i}{n(a_n-a_1)} f\left(\frac{a_1}{a_n}\right)+\frac{\left(\sum_{i=1}^{n}a_i\right)-na_1}{n(a_n-a_1)} f(1)$$
$$-f\left(\frac{2(a_n-a_{n-1})(a_2-a_1)}{(a_n-a_1)a_n}\right)-f\left(\frac{\sum_{i=1}^{n+1}a_i}{(n+1)a_{n+1}}\right).$$

Theorem 8. *Let f be a positive superquadratic function on $[0.1]$. Let $a_i > 0$ be increasing and $c_i > 0$, $i = 1, \ldots, n+1$. Then*

$$\frac{1}{c_n}\sum_{i=1}^{n} f\left(\frac{a_i}{a_n}\right)-\frac{1}{c_{n+1}}\sum_{i=1}^{n+1} f\left(\frac{a_i}{a_{n+1}}\right)$$
$$\leq \frac{na_n-\sum_{i=1}^{n}a_i}{(a_n-a_1)}\cdot\frac{f\left(\frac{a_1}{a_n}\right)}{c_n}+\frac{\sum_{i=1}^{n}a_i-na_1}{n(a_n-a_1)}\cdot\frac{f(1)}{c_n}$$
$$-\frac{n}{c_n} f\left(\frac{2(a_n-a_{n-1})(a_2-a_1)}{(a_n-a_1)a_n}\right)-\frac{n+1}{c_{n+1}} f\left(\frac{\sum_{i=1}^{n+1}a_i}{(n+1)a_{n+1}}\right).$$

3.2. *More on superquadraticity and lower bounds of averages*

The first theorem refines the results of [28] for convex functions which are also superquadratic (like $f(x) = x^m,\ m \geq 2$).

Theorem 9. *Let f be a positive superquadratic function on $[0,1]$ and let $a_i > 0$ and $i\left(1 - \frac{a_i}{a_{i+1}}\right)$, $i = 1, \ldots, n+1$ be increasing sequences. Then for $n \geq 2$*

$$\begin{aligned}\Delta &:= \frac{1}{n}\sum_{i=1}^{n} f\left(\frac{a_i}{a_n}\right) - \frac{1}{n+1}\sum_{i=1}^{n+1} f\left(\frac{a_i}{a_{n+1}}\right) \\ &\geq \frac{n-1}{n+1}\left[f\left(\frac{a_2 - a_1}{na_n}\right) + f\left(\frac{(n-2)(a_2 - a_1)}{2(n-1)na_n}\right)\right. \\ &\qquad \left. + f\left(\frac{(n-2)(a_2 - a_1)}{3n^2 a_n}\right)\right] \end{aligned} \tag{7}$$

holds.

The proof of Theorem 9 follows from the relation

$$\begin{aligned}\Delta &= \frac{1}{n}\sum_{i=1}^{n} f\left(\frac{a_i}{a_n}\right) - \frac{1}{n+1}\sum_{i=1}^{n+1} f\left(\frac{a_i}{a_{n+1}}\right) \\ &= \frac{1}{n+1}\sum_{i=1}^{n}\left[\frac{(i-1)}{n} f\left(\frac{a_{i-1}}{a_n}\right) + \frac{(n-i+1)}{n} f\left(\frac{a_i}{a_n}\right) - f\left(\frac{a_i}{a_{n+1}}\right)\right], \end{aligned} \tag{8}$$

and by using superquadraticity of f.

Theorem 10 deals with a lower bound for

$$\frac{1}{c_n}\sum_{i=1}^{n} f\left(\frac{a_i}{a_n}\right) - \frac{1}{c_{n+1}}\sum_{i=1}^{n+1} f\left(\frac{a_i}{a_{n+1}}\right),$$

under the same conditions as in Ref. [2, Theorems 5 and 6], where a different lower bound is obtained for positive superquadratic f. Also, in Ref. [2, Theorem 5.4(i)] it is proved that $\frac{1}{c_n}\sum_{i=1}^{n} f(\frac{a_i}{a_n})$ is decreasing with n when f is convex non-negative and increasing, under the same conditions on the sequences c_n and a_n as in Theorem 10.

Theorem 10. *Let $(a_n)_{n\geq 0}$ and $(c_n)_{n\geq 0}$ be sequences such that $a_n > 0$, $c_n > 0$ for $n \geq 1$ and*

(I) c_n *is increasing,* $c_0 = 0$
(II) $c_n - c_{n-1}$ *is increasing for* $n \geq 1$
(III) $c_1\left(1 - \frac{a_1}{a_2}\right) \leq c_{i-1}\left(1 - \frac{a_{i-1}}{a_i}\right) \leq c_n\left(1 - \frac{a_n}{a_{n+1}}\right)$, $n \geq 1$
(IV) $a_0 = 0$ *and* (a_n) *is increasing.*

Given a function f, *let*

$$S_n(f, (a_n), (c_n)) = S_n(f) := \frac{1}{c_n}\sum_{r=1}^{n} f\left(\frac{a_r}{a_n}\right), \quad n \geq 1.$$

If f *is superquadratic and non-negative on* $[0,1]$, *then*

$$D := S_n(f) - S_{n+1}(f) \geq \frac{n-1}{c_{n+1}} f\left(\frac{2c_1(a_2 - a_1)(c_n - c_{n-1})}{c_n^2 a_n}\right).$$

Remark 2. From Theorem 10 a refinement of Theorem 2 in Ref. [28] when $c_n = a_n$ is obtained for convex functions that are also superquadratic.

Theorem 11 extends the investigation to three sequences (see for instance Theorems 3 and 4 in Section 2 here).

Theorem 11. *Let* f *be a positive superquadratic function on* $[0, L]$. *Let* $(a_n)_{n\geq 0}$, $(b_n)_{n\geq 0}$, $(c_n)_{n\geq 0}$ *be sequences such that* $a_n > 0$, $b_n > 0$, $c_n > 0$ *for* $n \geq 1$ *and*

(a) a_n, b_n, c_n *are increasing,* $a_0 = c_0 = 0$,
(b) $c_n - c_{n-1}$ *is increasing for* $n \geq 1$,
(c) $c_n\left(1 - \frac{b_n}{b_{n+1}}\right) \geq c_r\left(1 - \frac{a_r}{a_{r+1}}\right)$, *for* $r \leq n$.

Then

$$H := S_n(f) - S_{n+1}(f) = \frac{1}{c_n}\sum_{r=1}^{n} f\left(\frac{a_r}{b_n}\right) - \frac{1}{c_{n+1}}\sum_{r=1}^{n+1} f\left(\frac{a_r}{b_{n+1}}\right)$$
$$\geq \frac{n-1}{c_{n+1}} f\left(\frac{2c_1(c_n - c_{n-1})A}{c_n^2 b_n}\right), \tag{9}$$

where $A := \min\{a_{i+1} - a_i : i = 1, \ldots, n\}$.

3.3. *Subquadraticity and new bounds of differences of averages*

In this subsection, the authors of [5] deal with functions that are increasing and subquadratic on $[0,1]$, like $f(x) = x^m$, $0 \leq m \leq 2$.

Theorem 12. *Let f be an increasing subquadratic function on $[0,1]$. Let $(a_i)_{i\geq 0}$ satisfy*

(A) $a_i > 0$, a_i *increases*, $i = 1, \ldots, n+1$,
(B) $i\left(\frac{a_{i+1}}{a_i} - 1\right) \leq n\left(\frac{a_{n+1}}{an} - 1\right)$, $i = 1, \ldots, n$.

Then

$$\begin{aligned} E &:= \frac{1}{n+1}\sum_{i=1}^{n+1} f\left(\frac{a_i}{a_{n+1}}\right) - \frac{1}{n}\sum_{i=1}^{n} f\left(\frac{a_i}{a_n}\right) \\ &\leq \sum_{i=1}^{n}\left[\frac{i}{n(n+1)} f\left(\frac{n-i+1}{n+1} \cdot \frac{a_{i+1}-a_i}{a_{n+1}}\right)\right. \\ &\qquad\left. + \frac{n-i+1}{n(n+1)} f\left(\frac{i}{n+1} \cdot \frac{a_{i+1}-a_i}{a_{n+1}}\right)\right]. \end{aligned} \tag{10}$$

Moreover, if in addition

(C) $\frac{a_{i+1}}{a_i} \leq 2$, $i = 1, \ldots, n$

holds too, then

$$\frac{1}{n+1}\sum_{i=1}^{n+1} f\left(\frac{a_i}{a_{n+1}}\right) \leq \frac{2}{n}\sum_{i=1}^{n} f\left(\frac{a_i}{a_n}\right). \tag{11}$$

Theorem 13. *Let f be an increasing subquadratic function on $[0,1]$. Let $(a_i)_{i\geq 0}$ satisfy*

(i) $a_i > 0$, $i = 1, \ldots, n+1$, $a_0 = 0$,
(ii) a_i, $a_i - a_{i-1}$ *increase*, $i = 1, \ldots, n+1$.

Then

$$\begin{aligned} R &:= \frac{1}{a_{n+1}}\sum_{i=1}^{n+1} f\left(\frac{a_i}{a_{n+1}}\right) - \frac{1}{a_n}\sum_{i=1}^{n} f\left(\frac{a_i}{a_n}\right) \\ &\leq \frac{1}{a_n}\sum_{i=1}^{n}\left[\frac{a_{n+1}-a_i}{a_{n+1}} f\left(\frac{a_i}{a_n} \cdot \frac{a_{i+1}-a_i}{a_{n+1}}\right)\right. \\ &\qquad\left. + \frac{a_i}{a_{n+1}} f\left(\frac{a_n-a_i}{a_{n+1}} \cdot \frac{a_{i+1}-a_i}{a_{n+1}}\right)\right]. \end{aligned} \tag{12}$$

Moreover, if in addition

(iii) $\frac{a_{i+1}}{a_i} \leq 2$, $i = 1, \ldots, n$,

then

$$\frac{1}{a_{n+1}}\sum_{i=1}^{n+1} f\left(\frac{a_i}{a_{n+1}}\right) \le \frac{2}{a_n}\sum_{i=1}^{n} f\left(\frac{a_i}{a_n}\right). \tag{13}$$

4. Comparisons between Differences of Averages for Superquadratic and 1-quasiconvex Functions

In this section, the bounds of differences of averages for superquadratic functions are compared with the bounds of differences of averages for 1-quasiconvex functions. The results are quoted from Ref. [7].

One basic idea for the investigations in this section is to show that simultaneously superquadratic and 1-quasiconvex functions can be used for comparing bounds of the difference

$$\begin{aligned} &B_{n-1}\left(f,(a_{n-1}),(a_n)\right) - B_n\left(f,(a_n),(a_{n+1})\right) \\ &:= \frac{1}{a_n}\sum_{i=0}^{n-1} f\left(\frac{a_i}{a_{n-1}}\right) - \frac{1}{a_{n+1}}\sum_{i=0}^{n} f\left(\frac{a_i}{a_n}\right), \end{aligned} \tag{14}$$

and also the bounds of the difference

$$\begin{aligned} &A_{n+1}\left(f,(a_{n+1}),(a_n)\right) - A_n\left(f,(a_n),(a_{n-1})\right) \\ &:= \frac{1}{a_n}\sum_{i=1}^{n} f\left(\frac{a_i}{a_{n+1}}\right) - \frac{1}{a_{n-1}}\sum_{i=1}^{n-1} f\left(\frac{a_i}{a_n}\right). \end{aligned} \tag{15}$$

The following example was one of the main motivations to introduce in Ref. [7] this research on bounds of differences of averages:

Example 1. In Ref. [10], it was noted that if f is convex, then

$$\frac{1}{n}\sum_{i=0}^{n-1} f\left(\frac{i}{n-1}\right) \ge \frac{1}{n+1}\sum_{i=0}^{n} f\left(\frac{i}{n}\right). \tag{16}$$

In particular, if $f(x) = x^p$, $x \ge 0$, $p \ge 1$, then this inequality can be rewritten as

$$\left(\frac{(n+1)\sum_{i=1}^{n-1} i^p}{n\sum_{i=1}^{n} i^p}\right)^{\frac{1}{p}} \ge \frac{n-1}{n}, \quad n \ge 2. \tag{17}$$

As further applications, a strict improvement of (16) and similar inequalities are obtained when $f(x) = x^p$ for $p \ge 2$, $x \ge 0$, that is, when the function f is superquadratic.

The first main result in this section reads:

Theorem 14. *Let $\varphi : [0, b) \to \mathbb{R}_+$, $0 < b \le \infty$ be differentiable convex increasing function, and let $f = x\varphi$. Let the sequence $\{a_i\}$ be such that $a_0 = 0$, a_i, $a_{i+1} - a_i$, $i = 1, \ldots$ are increasing. Then, for $n \ge 2$ by* (14), *using* (2) *for the* 1*-quasiconvex function f, the inequalities*

$$\begin{aligned} & B_{n-1}\left(f, (a_{n-1}), (a_n)\right) - B_n\left(f, (a_n), (a_{n+1})\right) \\ & \ge \sum_{i=1}^{n-1} \frac{(a_i - a_{i-1})^2 a_i (a_n - a_i)}{a_{n+1} a_n^2 a_{n-1}^2} \varphi'\left(\frac{a_i (a_n + a_{i-1} - a_i)}{a_n a_{n-1}}\right) \\ & \ge \sum_{i=1}^{n-1} \frac{(a_i - a_{i-1})^2 a_i (a_n - a_i)}{a_{n+1} a_n^2 a_{n-1}^2} \varphi'\left(\frac{a_i}{a_n}\right) \ge 0 \end{aligned} \tag{18}$$

are obtained.

If in addition φ' is convex, then

$$\begin{aligned} & B_{n-1}\left(f, (a_{n-1}), (a_n)\right) - B_n\left(f, a_n), (a_{n+1})\right) \\ & \ge \sum_{i=1}^{n-1} \frac{(a_i - a_{i-1})^2 a_i (a_n - a_i)}{a_{n+1} a_n^2 a_{n-1}^2} \varphi' \\ & \quad \times \left(\frac{\sum_{i=1}^{n-1} a_i^2 (a_n - a_i)(a_i - a_{i-1})^2}{a_n \sum_{i=1}^{n-1} a_i (a_n - a_i)(a_i - a_{i-1})^2}\right) \ge 0. \end{aligned} \tag{19}$$

Next, a similar result, but with an additional condition guaranteeing that $f(x) = x\varphi(x)$ is superquadratic, is stated (see Lemma 2).

Theorem 15. *Let $\varphi : [0, b) \to \mathbb{R}_+$, $0 < b \le \infty$ be a differentiable convex increasing function, $\varphi(0) = 0 = \lim_{x \to 0^+} x\varphi(x)$, and let $f = x\varphi$. Let the sequence $\{a_i\}$ be such that $a_0 = 0$, $a_i > 0$, $a_{i+1} - a_i$, $i = 1, \ldots,$ are increasing and let $B_n(f, (a_n), (a_{n+1}))$ be as defined in* (14). *Then, for $n \ge 2$*

$$\begin{aligned} & B_{n-1}\left(f, (a_{n-1}), (a_n)\right) - B_n\left(f, (a_n), (a_{n+1})\right) \\ & \ge \frac{1}{a_{n+1}} \sum_{i=1}^{n-1} \left(\frac{a_i}{a_n} f\left(\frac{a_n - a_i}{a_n}\left|\frac{a_{i-1}}{a_{n-1}} - \frac{a_i}{a_{n-1}}\right|\right)\right. \\ & \qquad \left. + \frac{a_n - a_i}{a_n} f\left(\frac{a_i}{a_n}\left|\frac{a_{i-1}}{a_{n-1}} - \frac{a_i}{a_{n-1}}\right|\right)\right) \end{aligned}$$

$$
\begin{aligned}
&= \sum_{i=1}^{n-1} \frac{(a_i - a_{i-1}) a_i (a_n - a_i)}{a_{n+1} a_n^2 a_{n-1}} \\
&\quad \times \left(\varphi\left(\frac{a_i (a_i - a_{i-1})}{a_{n-1} a_n} \right) + \varphi\left(\frac{(a_n - a_i)(a_i - a_{i-1})}{a_{n-1} a_n} \right) \right) \\
&\geq \sum_{i=1}^{n-1} \frac{2a_i (a_n - a_i)(a_i - a_{i-1})}{a_{n+1} a_n^2 a_{n-1}} \varphi\left(\frac{a_i - a_{i-1}}{2a_{n-1}} \right) \\
&\geq \sum_{i=1}^{n-1} \frac{2a_i (a_n - a_i)(a_i - a_{i-1})}{a_{n+1} a_n^2 a_{n-1}} \varphi \\
&\quad \times \left(\frac{\sum_{i=1}^{n-1} a_i (a_n - a_i)(a_i - a_{i-1})^2}{\sum_{i=1}^{n-1} 2a_{n-1} a_i (a_n - a_i)(a_i - a_{i-1})} \right) \geq 0 \qquad (20)
\end{aligned}
$$

holds.

In Theorem 16, it is shown by comparing (20) with (18) that the lower bound of $B_{n-1}(f) - B_n(f)$ obtained by the 1-quasiconvexity of f is better than by using its superquadracity:

Theorem 16. *Let $\varphi : [0, b) \to \mathbb{R}_+$ $0 < b \leq \infty$ be a differentiable convex increasing function satisfying $\varphi(0) = 0 = \lim_{x \to 0^+} x\varphi'(x)$ and let $f = x\varphi$. Let the positive sequences $\{a_i\}$, and $\{a_i - a_{i-1}\}$, $i = 1, 2, \ldots$, be increasing and let $a_0 = 0$. Then*

$$
\begin{aligned}
&B_{n-1}(f, (a_{n-1}), (a_n)) - B_n(f, (a_n), (a_{n+1})) \\
&\geq \sum_{i=1}^{n-1} \frac{(a_i - a_{i-1})(a_n - a_i) a_i}{a_{n-1}^2 a_n^2 a_{n+1}} \varphi'\left(\frac{a_i (a_n + a_{i-1} - a_i)}{a_n a_{n-1}} \right) \\
&\geq \sum_{i=1}^{n-1} \frac{(a_i - a_{i-1})(a_n - a_i) a_i}{a_{n-1} a_n^2 a_{n+1}} \\
&\quad \times \left(\varphi\left(\frac{a_i (a_i - a_{i-1})}{a_n a_{n-1}} \right) + \varphi\left(\frac{(a_n - a_i)(a_i - a_{i-1})}{a_n a_{n-1}} \right) \right) \\
&\geq \sum_{i=1}^{n-1} \frac{2a_i (a_n - a_i)(a_i - a_{i-1})}{a_{n+1} a_n^2 a_{n-1}} \varphi \\
&\quad \times \left(\frac{\sum_{i=1}^{n-1} a_i (a_n - a_i)(a_i - a_{i-1})^2}{\sum_{i=1}^{n-1} 2a_{n-1} a_i (a_n - a_i)(a_i - a_{i-1})} \right) \geq 0
\end{aligned}
$$

hold, which means that the bound obtained by the 1-quasiconvexity of f is better than the bound obtained by its superquadracity.

In the three theorems above, the authors dealt with and compared the lower bounds derived for the differences defined by (14). Using similar arguments, analogous results for the differences defined by (15) can be derived, too.

Theorem 17. *Let $\varphi : [0, b) \to \mathbb{R}_+$, $0 < b \leq \infty$ be a differentiable convex increasing function and let $f = x\varphi$. Let the sequence $\{a_i\}$, $i = 1, \ldots,$ be increasing and such that $\{a_{i+1} - a_i\}$ is decreasing and let $a_0 = 0$. Let $A_n(f, (a_n), (a_{n-1}))$ be as in (15). Then, for $n \geq 2$*

$$
\begin{aligned}
&A_{n+1}(f, (a_{n+1}), (a_n)) - A_n(f, (a_n), (a_{n-1})) \\
&\quad \geq \sum_{i=1}^{n-1} \frac{(a_{i+1} - a_i)^2 a_i (a_n - a_i)}{a_{n-1} a_n^2 a_{n+1}^2} \varphi' \left(\frac{a_i (a_n + a_{i+1} - a_i)}{a_n a_{n+1}} \right) \\
&\quad \geq \sum_{i=1}^{n-1} \frac{(a_{i+1} - a_i)^2 a_i (a_n - a_i)}{a_{n-1} a_n^2 a_{n+1}^2} \varphi' \left(\frac{a_i}{a_n} \right) \geq 0. \qquad (21)
\end{aligned}
$$

If φ satisfies also that $\varphi(0) = 0 = \lim_{x \to 0^+} x\varphi'(x)$, then f is also superquadratic and

$$
\begin{aligned}
&A_{n+1}(f, (a_{n+1}), (a_n)) - A_n(f, (a_n), (a_{n-1})) \\
&\quad \geq \sum_{i=1}^{n-1} \frac{(a_{i+1} - a_i) a_i (a_n - a_i)}{a_{n-1} a_n^2 a_{n+1}} \\
&\qquad \times \left(\varphi \left(\frac{(a_n - a_i)(a_{i+1} - a_i)}{a_n a_{n+1}} \right) + \varphi \left(\frac{a_i (a_{i+1} - a_i)}{a_n (a_{n+1})} \right) \right) \\
&\quad \geq \sum_{i=1}^{n-1} \frac{2 (a_{i+1} - a_i) a_i (a_n - a_i)}{a_{n-1} a_n^2 a_{n+1}} \varphi \left(\frac{a_{i+1} - a_i}{2 a_{n+1}} \right) \\
&\quad \geq \sum_{i=1}^{n-1} \frac{2 (a_{i+1} - a_i) a_i (a_n - a_i)}{a_{n-1} a_n^2 a_{n+1}} \varphi \\
&\qquad \times \left(\frac{\sum_{i=1}^{n-1} (a_{i+1} - a_i)^2 a_i (a_n - a_i)}{\sum_{i=1}^{n-1} 2 a_{n+1} (a_{i+1} - a_i) a_i (a_n - a_i)} \right) \geq 0 \qquad (22)
\end{aligned}
$$

is obtained. Moreover, the two inequalities above can be compared. In fact,

$$\begin{aligned} &A_{n+1}\left(f,\left(a_{n+1}\right),\left(a_{n}\right)\right)-A_{n}\left(f,\left(a_{n}\right),\left(a_{n-1}\right)\right) \\ &\geq \sum_{i=1}^{n-1} \frac{\left(a_{i+1}-a_{i}\right)^{2} a_{i}\left(a_{n}-a_{i}\right)}{a_{n-1} a_{n}^{2} a_{n+1}^{2}} \varphi^{\prime}\left(\frac{a_{i}\left(a_{n}+a_{i+1}-a_{i}\right)}{a_{n} a_{n+1}}\right) \\ &\geq \sum_{i=1}^{n-1} \frac{\left(a_{i+1}-a_{i}\right) a_{i}\left(a_{n}-a_{i}\right)}{a_{n-1} a_{n}^{2} a_{n+1}} \\ &\quad \times\left(\varphi\left(\frac{\left(a_{n}-a_{i}\right)\left(a_{i+1}-a_{i}\right)}{a_{n} a_{n+1}}\right)+\varphi\left(\frac{a_{i}\left(a_{i+1}-a_{i}\right)}{a_{n}\left(a_{n+1}\right)}\right)\right) \geq 0 \quad (23) \end{aligned}$$

hold.

Further, if φ' is also convex, then

$$\begin{aligned} &A_{n+1}\left(f,\left(a_{n+1}\right),\left(a_{n}\right)\right)-A_{n}\left(f,\left(a_{n}\right),\left(a_{n-1}\right)\right) \\ &\geq \sum_{i=1}^{n-1} \frac{\left(a_{i+1}-a_{i}\right)^{2} a_{i}\left(a_{n}-a_{i}\right)}{a_{n-1} a_{n}^{2} a_{n+1}^{2}} \varphi^{\prime} \\ &\quad \times\left(\frac{\sum_{i=1}^{n-1} a_{i}^{2}\left(a_{n}-a_{i}\right)\left(a_{i+1}-a_{i}\right)^{2}}{a_{n} \sum_{i=1}^{n-1} a_{i}\left(a_{n}-a_{i}\right)\left(a_{i+1}-a_{i}\right)^{2}}\right) \geq 0 \quad (24) \end{aligned}$$

is derived.

The following example of Theorem 16 demonstrates a refinement of (16):

Example 2. By the 1-quasiconvexity and superquadracity of f, where φ is increasing, convex and 3-convex satisfying $\varphi(0)=\lim_{x\rightarrow 0^{+}} x\varphi'(x)=0$, it is deduced in Ref. [7] that the bound obtained by the 1-quasiconvexity of $f=x\varphi$ is better than by its superquadracity and that for $a_i=i$, $i=0,1,\ldots$

$$\begin{aligned} &\frac{1}{n} \sum_{i=0}^{n-1} f\left(\frac{i}{n-1}\right)-\frac{1}{n+1} \sum_{i=0}^{n} f\left(\frac{i}{n}\right) \\ &\geq \frac{1}{6 n(n-1)} \varphi^{\prime}\left(\frac{1}{2}\right) \geq \frac{1}{3 n} \varphi\left(\frac{1}{2(n-1)}\right) \geq 0 \end{aligned}$$

holds.

The following example presents a natural choice of the basic a_n sequence in (14). In this case the requirement $a_n \geq 2\left(a_{n-1}-a_{n-2}\right)$ holds for $n \geq 3$, $a_0=0$ and the lower bound that is obtained by the quasiconvexity of $f=x\varphi$ when $\varphi(0)=0=\lim_{x\rightarrow 0^{+}} x\varphi'(x)$ is better than the lower bound obtained by its superquadracity which means that:

Example 3. From Theorem 16 we get the inequalities

$$\begin{aligned}
&\frac{1}{2n-1}\sum_{i=1}^{n-1} f\left(\frac{2i-1}{2n-3}\right)-\frac{1}{2n+1}\sum_{i=1}^{n} f\left(\frac{2i-1}{2n-1}\right)\\
&\geq \frac{8}{(2n+1)(2n-1)^2(2n-3)^2}\sum_{i=1}^{n-1}(2i-1)(n-i)\varphi'\left(\frac{2i-1}{2n-1}\right)\\
&\geq \sum_{i=1}^{n-1}\frac{4(2i-1)(n-i)}{(2n+1)(2n-1)^2(2n-3)}\varphi\left(\frac{4(n-i)}{(2n-3)(2n-1)}\right)\\
&\quad+\sum_{i=1}^{n-1}\frac{4(2i-1)(n-i)}{(2n+1)(2n-1)^2(2n-3)}\varphi\left(\frac{2(2i-1)}{(2n-3)(2n-1)}\right)\\
&\geq \frac{4n(n-1)}{(2n+1)(2n-1)(2n-3)}\varphi\left(\frac{1}{2n-1}\right)\geq 0.
\end{aligned}$$

In the next example, a similar application of Theorem 17 is obtained:

Example 4. In this example we use Theorem 17 to see that in such a case all the lower bounds obtained by the 1-quasiconvexity of f, when $\varphi(0) = 0 = \lim_{x\to 0^+} x\varphi'(x)$, are better that those obtained by its superquadracity, for the sequence $a_i = i$, $i = 0, 1, \ldots, n$ where $A_n(f)$ is

$$A_n(f) = \frac{1}{n-1}\sum_{i=1}^{n-1} f\left(\frac{i}{n}\right).$$

That is, we get that the inequalities

$$A_{n+1}(f) - A_n(f) \geq \frac{1}{6n(n+1)}\varphi'\left(\frac{1}{2}\right) \geq \frac{1}{3n}\varphi\left(\frac{1}{2(n+1)}\right) \geq 0$$

hold.

Finally, a refinement of (17) is presented:

Example 5. The functions $f(x) = x^p$, $p \geq 2$, $x \geq 0$, are the basic cases of 1-quasiconvex functions as well as superquadratic functions. Therefore from Example 2 it is obtained that the ratio $\left(\frac{(n+1)\sum_{i=1}^{n-1} i^p}{n\sum_{i=1}^{n} i^p}\right)^{\frac{1}{p}}$ is not only bounded below by $\frac{n-1}{n}$ but by strictly better lower bounds when $p \geq 2$ instead of $p \geq 1$. Indeed, from Example 2 and from the 1-quasiconvexity of $f(x) = x^p$, $x \geq 0$, $p \geq 2$ when $n \geq 2$ the first inequality in (25) holds

$$\left(\frac{(n+1)\sum_{i=1}^{n-1} i^p}{n\sum_{i=1}^{n} i^p}\right)^{\frac{1}{p}} \geq \frac{n-1}{n}(1+\Delta_1)^{\frac{1}{p}} \geq \frac{n-1}{n}(1+\Delta_2)^{\frac{1}{p}} \geq 0, \quad (25)$$

where Δ_1 and Δ_2 are defined by (26).

The second inequality in (25) holds since $0 < \frac{i}{n} \le 1$, $i = 1, \ldots, n$, $\sum_{i=1}^{n} \left(\frac{i}{n}\right)^p \le \sum_{i=1}^{n} \left(\frac{i}{n}\right)^2$, when $p \ge 2$ and, therefore,

$$\Delta_1 = \frac{(p-1)(n+1)}{2^{p-2} 6n (n-1) \sum_{i=1}^{n} \left(\frac{i}{n}\right)^p} \ge \Delta_2 = \frac{p-1}{2^{p-2} (n-1)(2n+1)} \ge 0$$

is satisfied.

Similarly from Example 4 the following is obtained:

Example 6. For $p \ge 2$ when $n \ge 2$

$$\left(\frac{(n-1) \sum_{i=1}^{n} i^p}{n \sum_{i=1}^{n-1} i^p}\right)^{\frac{1}{p}} \ge \frac{n+1}{n} (1+\Delta_1)^{\frac{1}{p}} \ge \frac{n+1}{n} (1+\Delta_2)^{\frac{1}{p}} \ge 0 \qquad (26)$$

holds where

$$\Delta_1 = \frac{(p-1)(n-1)}{2^{p-2} 6n (n+1) \sum_{i=1}^{n-1} \left(\frac{i}{n}\right)^p} \ge \Delta_2 = \frac{p-1}{2^{p-2} (n+1)(2n-1)} \ge 0.$$

References

[1] S. Abramovich, G. Jameson, and G. Sinnamon, Refining Jensen's inequality, *Bull. Sci. Math. Roum.* **47**, 3–14, (2004).

[2] S. Abramovich, G. Jameson, and G. Sinnamon, Inequalities for averages of convex and superquadratic functions, *JIPAM*, **5**(4), Article 91, (2004).

[3] S. Abramovich, *New Applications of Superquadracity, Analytic Number Theory, Approximation Theory and Special Functions*, 365–395 (Springer, New York, 2014).

[4] S. Abramovich, Refined inequalities via superquadracity, overview, *Nonlinear Stud.* **26**(4), 723–740, (2019).

[5] S. Abramovich, S. Banić, M. Matić, and J. Pečarić, Superquadratic functions and refinements of inequalities between averaged, arXiv:1110.5217v1 [math.NA] 24 Oct 2011.

[6] S. Abramovich, J. Barić, M. Matić, and J. Pečarić, On the Van De Lune-Alzer's inequality, *J. Math. Inequal.* **1**(4), 563–587, (2007).

[7] S. Abramovich and L.-E. Persson, Inequalities for averages of quasiconvex and superquadratic functions, *Math. Inequal. Applic*, **19**(2), 535–550, (2016).

[8] H. Alzer, On an inequality of H. Minc and L. Sathre, *J. Math. Anal. Appl.*, **179**, 396–402, (1993).

[9] G. Bennett, Meaningful sequences, *Houston J. Math.* **33**(2), 555–580, (2007).

[10] G. Bennett and G. Jameson, Monotonic averages of convex functions, *J. Math. Anal. Appl.* **252**, 410–430, (2000).

[11] I. Brnetić and J. Pečarić, Comments on some analytic inequalities, *J. Inequal. Pure Appl. Math.* **4**(1), Article 20, (2003).

[12] C.-P. Chen and F. Qi, Notes on proofs of Alzer's inequality, *Octogon Math. Mag.* **11**(1), 29–33, (2003).

[13] C.-P. Chen and F. Qi, The inequality of Alzer for negative powers, *Octogon Math. Mag.* **11**(2), 442–445, (2003).

[14] C.-P. Chen and F. Qi, On integral version of Alzer's inequality and Martins' inequality, *RGMIA Research Report Collection* **8**(1), Article 13, (2005). http://rgmia.vu.edu.au/v8n1.html.

[15] C.-P. Chen and F. Qi, Monotonicity properties for generalized logarithmic means, Austral. *J. Math. Anal. Appl.* **1**(2), Article 2, (2004).

[16] C.-P. Chen and F. Qi, Extension of an inequality of H. Alzer for negative powers, *Tamkang J. Math.* **36**(1), 69–72, (2005).

[17] C.-P. Chen and F. Qi, Generalization of an inequality of Alzer for negative powers, *Tamkang J. Math.* **36**(3), 219–222, (2005).

[18] C.-P. Chen and F. Qi, Note on Alzer,s inequality, *Tamkang J. Math.* **37**(1), 11–14, (2006).

[19] C.-P. Chen, F. Qi, P. Cerone, and S. S. Dragomir, Monotonicity of sequences involving convex and concave functions, *Math. Inequal. Appl.* **6**(2), 229–239, (2003).

[20] N. Elezović and J. Pečarić, On Alzer's inequality, *J. Math. Anal. Appl.* **223**, 366–369, (1998).

[21] B. Gavrea and I. Gavrea, An inequality for linear positive functionals, *J. Inequal. Pure Appl. Math.* **1**(1), Article 5, (2000).

[22] B.-N. Guo and F. Qi, Inequalities and monotonicity of the ratio for the geometric means of a positive arithmetic sequence with arbitrary difference, *Tamkang J. Math.* **34**(3), 261–270, (2003).

[23] G. O. Jameson, Monotonicity of weighted averages of convex functions, *MIA*, **23**(2), 425–432, (2020).

[24] A.-J. Li, X.-M. Wang, and C.-P. Chen, Generalizations of the Ky Fan inequality, *J. Inequal. Pure Appl. Math.* **7**(4), Article 130, (2006).

[25] F. Qi, Generalizations of Alzer's and Kuang's inequality, *Tamkang J. Math*, **31**(3), 223–227, (2000).

[26] F. Qi, An algebraic inequality, *J. Inequal. Pure Appl. Math.* **2**(1), Article 13, (2001).

[27] F. Qi and L. Debnath, On a new generalization of Alzer's inequality, *Int. J. Math. Math. Sci.* **23**(12), 815–818, (2000).

[28] F. Qi and B.-N. Guo, Monotonicity of sequences involving convex function and sequence, *Math. Inequal. Appl.* **9**(2), 247–254, (2006).

[29] F. Qi, B.-N. Guo, and L. Debnath, A lower bound for ratio of power means, *Int. J. Math. Math. Sci.* **2004**(1–4), 49–53, (2004).

[30] F. Qi and Q.-M. Luo, Generalization of H. Minc and Sathre's inequality, *Tamkang J. Math.* **31**(2), 145–148, (2000).

[31] J. Sándor, On an inequality of Alzer, II, *O. Math. Mag.* **11**(2), 554–555, (2003).

[32] J. Sándor, On an inequality of Alzer for negative powers, *RGMIA Research Report Collection* **9**(4), (2006).

[33] J. S. Ume, An inequality for a positive real function, *Math. Inequal. Appl.* **5**(4), 693–696, (2002).

[34] Z. Xu and D. Xu, A general form of Alzer's inequality, *Comput. Math. with Appl.* **44**, 365–373, (2002).

[35] S.-L. Zhang, C.-P. Chen, and F. Qi, Continuous analogue of Alzer's inequality, *Tamkang J. Math.* **37**(2), 105–108, (2006).

https://doi.org/10.1142/9789811261572_0003

Chapter 3

Upper and Lower Semicontinuous Relations in Relator Spaces

Santanu Acharjee[*,§], Michael Th. Rassias[†,¶], and Árpád Száz[‡,||]

[*]*Department of Mathematics, Gauhati University*
Guwahati 781014, Assam, India
[†]*Department of Mathematics and Engineering Sciences*
Hellenic Military Academy
16673 Vari Attikis, Greece
[‡]*Institute of Mathematics, University of Debrecen*
H–4002 Debrecen, Pf. 400, Hungary
[§]*sacharjee326@gmail.com*
[¶]*mrassias@sse.gr, mthrassias@yahoo.com*
[||]*szaz@science.unideb.hu*

In 2018, P. Thangavelu, S. Premakumari and P. Xavier called a multifunction F of one topological space $X(\tau)$ to another $X(\sigma)$ to be

(1) upper mixed continuous if for every $x \in X$ and $V \in \sigma$, with $F(x) \subseteq V$, there exists $U \in \tau$, with $x \in U$, such that for every $u \in U$ we have $F(u) \cap V \neq \emptyset$;
(2) lower mixed continuous if for every $x \in X$ and $V \in \sigma$, with $F(x) \cap V \neq \emptyset$, there exists $U \in \tau$, with $x \in U$, such that for every $u \in U$ we have $F(u) \subseteq V$.

In this chapter, we shall show that these strange continuity properties can also be nicely generalized to relations between relator spaces.

By a relator space, in a narrower sense, we mean an ordered pair $X(\mathcal{R}) = (X, \mathcal{R})$ consisting of a set X and a family $\mathcal{R}$ of relations on X.

Thus, relator spaces are generalizations of not only ordered sets and uniform spaces, but also proximity, closure, topological and convergence spaces.

1. Introduction

Here, following the notations of Thangavelu *et al.* [1], we shall assume that $X(\tau)$ and $X(\sigma)$ are topological spaces, and moreover F is a function of X to $\mathcal{P}(Y)$ such that $F(x) \neq \emptyset$ for all $x \in X$.

Thus, in accordance with the standard definitions of upper and lower semicontinuous set-valued functions [2,3], we may also naturally consider the following global version of [1, Definition 4].

Definition 1. The multifunction F is called

(1) *upper continuous* if for every $x \in X$ and $V \in \sigma$, with $F(x) \subseteq V$, there exists $U \in \tau$, with $x \in U$, such that for every $u \in U$ we have $F(u) \subseteq V$;
(2) *lower continuous* if for every $x \in X$ and $V \in \sigma$, with $F(x) \cap V \neq \emptyset$, there exists $U \in \tau$, with $x \in U$, such that for every $u \in U$ we have $F(u) \cap V \neq \emptyset$.

Moreover, much less naturally, we may also consider the following global version of [1, Definition 6]. For another strange possibility, see also [72].

Definition 2. The multifunction F is called

(1) *upper mixed continuous* if for every $x \in X$ and $V \in \sigma$, with $F(x) \subseteq V$, there exists $U \in \tau$, with $x \in U$, such that for every $u \in U$ we have $F(u) \cap V \neq \emptyset$;
(2) *lower mixed continuous* if for every $x \in X$ and $V \in \sigma$, with $F(x) \cap V \neq \emptyset$, there exists $U \in \tau$, with $x \in U$, such that for every $u \in U$ we have $F(u) \subseteq V$.

Now, in addition to the above two definitions, we may also naturally consider the following:

Definition 3. The multifunction F is called

(1) *openness preserving* if $A \in \tau$ implies $F[A] \in \sigma$;
(2) *closedness preserving* if $A^c \in \tau$ implies $F[A]^c \in \sigma$.

Moreover, if in particular f is a function of X to Y, then by using some plausible notations, we may also naturally consider the following:

Definition 4. The function f is called

(1) *closure preserving* if $x \in \mathrm{cl}_\tau(A)$ implies $f(x) \in \mathrm{cl}_\sigma\big(f[A]\big)$;
(2) *interior preserving* if $x \in \mathrm{int}_\tau(A)$ implies $f(x) \in \mathrm{int}_\sigma\big(f[A]\big)$.

Some basic properties of upper and lower mixed continuities have been established in Ref. [1]. Moreover, in two subsequent papers [4,5] several stronger and weaker forms of these continuities have also been investigated.

However, in the present chapter, we shall only be interested in some straightforward generalizations of the above four fundamental definitions to relations between relator (generalized uniform) spaces [6,7].

Relational characterizations of the above continuity properties naturally lead us to the definitions of four general continuity properties of ordered pairs of relators [29] whose particular cases can be studied by hundreds of mathematicians.

For this, instead of relator spaces, one can even use birelator spaces [87] which are natural generalizations of not only bitopological spaces, but also ideal topological spaces.

To keep the chapter completely self-contained, the necessary prerequisites on relations and relators, which may be useful for the reader, will be briefly laid out in the subsequent preparatory sections with only a few clarifying proofs.

2. A Few Basic Facts on Relations

A subset R of a product set $X \times Y$ is called a *relation* on X to Y. In particular, a relation R on X to itself is simply called a relation on X. And, $\Delta_X = \{(x,x) : \ x \in X\}$ is called the *identity relation* of X.

If R is a relation on X to Y, then for any $x \in X$ and $A \subseteq X$ the sets $R(x) = \{y \in Y : (x,y) \in R\}$ and $R[A] = \bigcup_{a \in A} R(a)$ are called the *images* or *neighborhoods* of x and A under R, resp.

If $(x,y) \in R$, then instead of $y \in R(x)$, we may also write $x R y$. However, instead of $R[A]$, we cannot write $R(A)$. Namely, it may occur that, in addition to $A \subseteq X$, we also have $A \in X$.

Now, the sets $D_R = \{x \in X : R(x) \neq \emptyset\}$ and $R[X]$ are called the *domain* and *range* of R, resp. If in particular $D_R = X$, then we say that R is a *relation of X to Y*, or that R is a *total (or non-partial) relation on X to Y*.

In particular, a relation f on X to Y is called a *function* if for each $x \in D_f$ there exists $y \in Y$ such that $f(x) = \{y\}$. In this case, by identifying singletons with their elements, we may simply write $f(x) = y$ in place of $f(x) = \{y\}$.

Moreover, a function $\star$ of X to itself is called a *unary operation* on X. While, a function $*$ of X^2 to X is called a *binary operation* on X. And, for any $x, y \in X$, we usually write $x^\star$ and $x * y$ instead of $\star(x)$ and $*(x,y)$.

If R is a relation on X to Y, then a function f of D_R to Y is called a *selection function* of R if $f(x) \in R(x)$ for all $x \in D_R$. Thus, by the Axiom of Choice, we can see that every relation is the union of its selection functions.

For a relation R on X to Y, we may naturally define two *set-valued functions* φ_R of X to $\mathcal{P}(Y)$ and Φ_R of $\mathcal{P}(X)$ to $\mathcal{P}(Y)$ such that $\varphi_R(x) = R(x)$ for all $x \in X$ and $\Phi_R(A) = R[A]$ for all $A \subseteq X$.

Functions of X to $\mathcal{P}(Y)$ can be naturally identified with relations on X to Y. While, functions of $\mathcal{P}(X)$ to $\mathcal{P}(Y)$ are more general objects than relations on X to Y. In Refs. [8–10], they were briefly called *correlations* on X to Y.

However, a relation on $\mathcal{P}(X)$ to Y should be rather called a *super relation* on X to Y, and a relation on $\mathcal{P}(X)$ to $\mathcal{P}(Y)$ should be rather called a *hyper relation* on X to Y [11,12]. Thus, closures (proximities) [13] are super (hyper) relations.

If R is a relation on X to Y, then we have $R = \bigcup_{x \in X} \{x\} \times R(x)$. Therefore, the values $R(x)$, where $x \in X$, uniquely determine R. Thus, a relation R on X to Y can be naturally defined by specifying $R(x)$ for all $x \in X$.

For instance, the *complement relation* R^c can be defined such that $R^c(x) = R(x)^c = Y \setminus R(x)$ for all $x \in X$. Thus, we also have $R^c = (X \times Y) \setminus R$. Moreover, we can note that $R^c[A]^c = \bigcap_{a \in A} R(a)$ for all $A \subseteq X$ [14].

While, the *inverse relation* R^{-1} can be defined such that $R^{-1}(y) = \{x \in X : \ y \in R(x)\}$ for all $y \in Y$. Thus, we also have $R^{-1} = \{(y, x) : (x, y) \in R\}$. And, we can note that $R^{-1}[B] = \{x \in X : R(x) \cap B \neq \emptyset\}$ for all $B \subseteq Y$.

Moreover, if in addition S is a relation on Y to Z, then the *composition relation* $S \circ R$ can be defined such that $(S \circ R)(x) = S[R(x)]$ for all $x \in X$. Thus, it can be easily seen that $(S \circ R)[A] = S\big[R[A]\big]$ for all $A \subseteq X$.

While, if S is a relation on Z to W, then the *box product* $R \boxtimes S$ can be defined such that $(R \boxtimes S)(x, z) = R(x) \times S(z)$ for all $x \in X$ and $z \in Z$. Thus, it can be shown that $(R \boxtimes S)[A] = S \circ A \circ R^{-1}$ for all $A \subseteq X \times Z$ [14].

Hence, by taking $A = \{(x, z)\}$, and $A = \Delta_Y$ if $Y = Z$, one can at once see that the box and composition products are actually equivalent tools. However, the box product can be immediately defined for any family of relations.

Now, a relation R on X may be briefly defined to be *reflexive* if $\Delta_X \subseteq R$, and *transitive* if $R \circ R \subseteq R$. Moreover, R may be briefly defined to be *symmetric* if $R^{-1} \subseteq R$, and *antisymmetric* if $R \cap R^{-1} \subseteq \Delta_X$.

Thus, a reflexive and transitive (symmetric) relation may be called a *preorder* (*tolerance*) *relation.* And, a symmetric (antisymmetric) preorder relation may be called an *equivalence* (*partial order*) *relation.*

For any relation R on X, we may also define $R^0 = \Delta_X$, and $R^n = R \circ R^{n-1}$ if $n \in \mathbb{N}$. Moreover, we may also define $R^\infty = \bigcup_{n=0}^{\infty} R^n$. Thus, it can be shown that R^∞ is the smallest preorder relation on X containing R [15].

Now, in contrast to $(R^c)^c = R$ and $(R^{-1})^{-1} = R$, we have $(R^\infty)^\infty = R^\infty$. Moreover, analogously to $(R^c)^{-1} = (R^{-1})^c$, we also have $(R^\infty)^{-1} = (R^{-1})^\infty$. Thus, in particular R^{-1} is also a preorder on X if R is a preorder on X.

For $A \subseteq X$, the *Pervin relation* $R_A = A^2 \cup (A^c \times X)$ is an important preorder on X [16]. While, for a *pseudometric* d on X, the *Weil surrounding* $B_r = \{(x, y) \in X^2 : d(x, y) < r\}$, with $r > 0$, is an important tolerance on X [17].

Note that $S_A = R_A \cap R_A^{-1} = R_A \cap R_{A^c} = A^2 \cap (A^c)^2$ is already an equivalence relation on X. And, more generally if $\mathcal{A}$ is a *cover* (*partition*) of X, then $S_{\mathcal{A}} = \bigcup_{A \in \mathcal{A}} A^2$ is a tolerance (equivalence) relation on X.

As an important generalization of the Pervin relation R_A, for any $A \subseteq X$ and $B \subseteq Y$, we may also naturally consider the *Hunsaker–Lindgren relation* $R_{(A,B)} = (A \times B) \cup (A^c \times Y)$ [18]. Namely, thus we evidently have $R_A = R_{(A,A)}$.

The Pervin relations R_A and the Hunsaker–Lindgren relations $R_{(A,B)}$ were actually first used by Davis [19] and Császár [20, p. 42 and 351] in some less explicit and convenient forms, resp.

3. A Few Basic Facts on Relators

A family $\mathcal{R}$ of relations on one set X to another Y is called a *relator on X to Y*, and the ordered pair $(X, Y)(\mathcal{R}) = \big((X, Y), \mathcal{R}\big)$ is called a *relator space.* For the origins of this notion, see Refs. [23,26].

If in particular $\mathcal{R}$ is a relator on X to itself, then $\mathcal{R}$ is simply called a *relator on X*. Thus, by identifying singletons with their elements, we may naturally write $X(\mathcal{R})$ instead of $(X, X)(\mathcal{R})$. Namely, $(X, X) = \{\{X\}, \{X, X\}\} = \{\{X\}\}$.

Relator spaces of this simpler type are already substantial generalizations of the various *ordered sets* [22,23] and *uniform spaces* [24,25]. However, they are insufficient for some important purposes. (See Refs. [21,26–30].)

A relator $\mathcal{R}$ on X to Y, or the relator space $(X, Y)(\mathcal{R})$, is called *simple* if $\mathcal{R} = \{R\}$ for some relation R on X to Y. Simple relator spaces $(X, Y)(R)$ and $X(R)$ were called *formal contexts* and *gosets* in Refs. [23,26], resp.

Moreover, a relator $\mathcal{R}$ on X, or the relator space $X(\mathcal{R})$, may, for instance, be naturally called *reflexive* if each member of $\mathcal{R}$ is reflexive on X. Thus, we may also naturally speak of *preorder, tolerance and equivalence relators.*

For instance, for a family $\mathcal{A}$ of subsets of X, the family $\mathcal{R}_{\mathcal{A}} = \{R_A : A \in \mathcal{A}\}$, where $R_A = A^2 \cup (A^c \times X)$, is an important preorder relator on X. Such relators were first used by Pervin [16] and Levine [31].

While, for a family $\mathcal{D}$ of *pseudometrics* on X, the family $\mathcal{R}_{\mathcal{D}} = \{B_r^d : r > 0, d \in \mathcal{D}\}$, where $B_r^d = \{(x, y) : d(x, y) < r\}$, is an important tolerance relator on X. Such relators were first considered by Weil [17].

Moreover, if $\mathfrak{S}$ is a family of *covers* (*partitions*) of X, then the family $\mathcal{R}_{\mathfrak{S}} = \{S_{\mathcal{A}} : \mathcal{A} \in \mathfrak{S}\}$, where $S_{\mathcal{A}} = \bigcup_{A \in \mathcal{A}} A^2$, is an important tolerance (equivalence) relator on X. Equivalence relators were first studied by Levine [32].

If $\star$ is a unary operation for relations on X to Y, then for any relator $\mathcal{R}$ on X to Y we may naturally define $\mathcal{R}^\star = \{R^\star : R \in \mathcal{R}\}$. However, this plausible notation may cause confusions if $\star$ is a set-theoretic operation.

For instance, for any relator $\mathcal{R}$ on X to Y, we may naturally define the *elementwise complement* $\mathcal{R}^c = \{R^c : R \in \mathcal{R}\}$, which may easily be confused with the *global complement* $\mathcal{R}^c = \mathcal{P}(X \times Y) \setminus \mathcal{R}$ of $\mathcal{R}$.

However, for instance, the practical notations $\mathcal{R}^{-1} = \{R^{-1} : R \in \mathcal{R}\}$, and $\mathcal{R}^\infty = \{R^\infty : R \in \mathcal{R}\}$ whenever $\mathcal{R}$ is only a relator on X, will certainly not cause confusions in the sequel.

In particular, for a relator $\mathcal{R}$ on X, we may also naturally define $\mathcal{R}^\partial = \{S \subseteq X^2 : S^\infty \in \mathcal{R}\}$. Namely, for any two relators $\mathcal{R}$ and $\mathcal{S}$ on X, we evidently have $\mathcal{R}^\infty \subseteq \mathcal{S} \iff \mathcal{R} \subseteq \mathcal{S}^\partial$. That is, ∞ and ∂ form a Galois connection [22, p. 155].

The operations ∞ and ∂ were first introduced by Mala [33,34] and Pataki [35,36], resp. These two former PhD students of the third author, together with János Kurdics [37,38], made substantial developments in the theory of relators.

Moreover, if $*$ is a binary operation for relations, then for any two relators $\mathcal{R}$ and $\mathcal{S}$ we may naturally define $\mathcal{R} * \mathcal{S} = \{R * S : R \in \mathcal{R},\ S \in \mathcal{S}\}$.

However, this notation may again cause confusions if $*$ is a set-theoretic operation.

Therefore, in the former papers, we rather wrote $\mathcal{R} \wedge \mathcal{S} = \{R \cap S : R \in \mathcal{R},\ S \in \mathcal{S}\}$. Moreover, for instance, we also wrote $\mathcal{R} \triangle \mathcal{R}^{-1} = \{R \cap R^{-1} : R \in \mathcal{R}\}$. Thus, $\mathcal{R} \triangle \mathcal{R}^{-1}$ is a symmetric relator such that $\mathcal{R} \triangle \mathcal{R}^{-1} \subseteq \mathcal{R} \wedge \mathcal{R}^{-1}$.

A function $\Box$ of the family of all relators on X to Y is called a *direct* (*indirect*) *unary operation for relators* if, for every relator $\mathcal{R}$ on X to Y, the value $\mathcal{R}^{\Box} = \Box(\mathcal{R})$ is a relator on X to Y (on Y to X).

More generally, a function $\mathfrak{F}$ of the family of all relators on X to Y is called a *structure for relators* if, for every relator $\mathcal{R}$ on X to Y, the value $\mathfrak{F}_{\mathcal{R}} = \mathfrak{F}(\mathcal{R})$ is in a power set depending only on X and Y.

Concerning structures and operations for relators, we can freely use some basic terminology on set-to-set functions. However, for closures and projections, we can now also use the terms *refinements and modifications*, resp.

For instance, c and -1 are *involution operations* for relators. While, ∞ and ∂ are *projection operations* for relators. Moreover, the operation $\Box = c$, ∞ or ∂ is *inversion compatible* in the sense that $\left(\mathcal{R}^{\Box}\right)^{-1} = \left(\mathcal{R}^{-1}\right)^{\Box}$.

While, if for instance $\operatorname{int}_{\mathcal{R}}(B) = \{x \in X : \exists\ R \in \mathcal{R} :\ R(x) \subseteq B\}$ for every relator $\mathcal{R}$ on X to Y and $B \subseteq Y$, then the function $\mathfrak{F}$, defined by $\mathfrak{F}(\mathcal{R}) = \operatorname{int}_{\mathcal{R}}$, is a union-preserving structure for relators.

The first basic problem in the theory of relators is that, for any increasing structure $\mathfrak{F}$, we have to find an operation $\Box$ for relators such that, for any two relators $\mathcal{R}$ and $\mathcal{S}$ on X to Y we could have $\mathfrak{F}_{\mathcal{S}} \subseteq \mathfrak{F}_{\mathcal{R}} \iff \mathcal{S} \subseteq \mathcal{R}^{\Box}$.

By using *Pataki connections* [35,39], several closure operations can be derived from union-preserving structures. However, more generally, one can find first the *Galois adjoint* $\mathfrak{G}$ of such a structure $\mathfrak{F}$, and then take $\Box_{\mathfrak{F}} = \mathfrak{G} \circ \mathfrak{F}$ [40].

By finding the Galois adjoint of the structure $\mathfrak{F}$, the second basic problem for relators, that which structures can be derived from relators, can also be solved. However, for this, some direct methods can also be well used [41,42,70].

Now, for an operation $\Box$ for relators, a relator $\mathcal{R}$ on X to Y may be naturally called $\Box$-*fine* if $\mathcal{R}^{\Box} = \mathcal{R}$. And, for some structure $\mathfrak{F}$ for relators, two relators $\mathcal{R}$ and $\mathcal{S}$ on X to Y may be naturally called $\mathfrak{F}$-*equivalent* if $\mathfrak{F}_{\mathcal{R}} = \mathfrak{F}_{\mathcal{S}}$.

Moreover, for a structure $\mathfrak{F}$ for relators, a relator $\mathcal{R}$ on X to Y may, for instance, be naturally called *$\mathfrak{F}$-simple* if $\mathfrak{F}_{\mathcal{R}} = \mathfrak{F}_R$ for some relation R on X to Y. Thus, singleton relators have to be actually called *properly simple.*

4. Structures Derived from Relators

Notation 1. In this section, we shall assume that $\mathcal{R}$ is a relator on X to Y.

Definition 5. For any $x \in X$, $A \subseteq X$ and $B \subseteq Y$, we define:

(1) $A \in \operatorname{Int}_{\mathcal{R}}(B)$ if $R[A] \subseteq B$ for some $R \in \mathcal{R}$;
(2) $A \in \operatorname{Cl}_{\mathcal{R}}(B)$ if $R[A] \cap B \neq \emptyset$ for all $R \in \mathcal{R}$;
(3) $x \in \operatorname{int}_{\mathcal{R}}(B)$ if $\{x\} \in \operatorname{Int}_{\mathcal{R}}(B)$; (4) $x \in \operatorname{cl}_{\mathcal{R}}(B)$ if $\{x\} \in \operatorname{Cl}_{\mathcal{R}}(B)$;
(5) $B \in \mathcal{E}_{\mathcal{R}}$ if $\operatorname{int}_{\mathcal{R}}(B) \neq \emptyset$; (6) $B \in \mathcal{D}_{\mathcal{R}}$ if $\operatorname{cl}_{\mathcal{R}}(B) = X$.

Remark 1. The relations $\operatorname{Int}_{\mathcal{R}}$ and $\operatorname{int}_{\mathcal{R}}$ are called *the proximal and topological interiors* generated by $\mathcal{R}$, resp. While, the members of the families, $\mathcal{E}_{\mathcal{R}}$ and $\mathcal{D}_{\mathcal{R}}$ are called the *fat and dense subsets* of the relator space $(X,Y)(\mathcal{R})$, resp.

The origins of the relations $\operatorname{Cl}_{\mathcal{R}}$ and $\operatorname{Int}_{\mathcal{R}}$ go back to Efremović's proximity δ [43] and Smirnov's strong inclusion $\Subset$ [44], resp. While, the convenient notations $\operatorname{Cl}_{\mathcal{R}}$ and $\operatorname{Int}_{\mathcal{R}}$, and the family $\mathcal{E}_{\mathcal{R}}$, together with its dual $\mathcal{D}_{\mathcal{R}}$, were first explicitly used by the third author in Refs. [6,41,45,46,80].

The following theorem shows that the corresponding closure and interior relations are equivalent tools. Moreover, in a relator space the topological closure of a set can be more nicely described than in a topological one.

Theorem 1. *For any $B \subseteq Y$, we have*

(1) $\operatorname{Cl}_{\mathcal{R}}(B) = \mathcal{P}(X) \setminus \operatorname{Int}_{\mathcal{R}}(B^c)$;
(2) $\operatorname{cl}_{\mathcal{R}}(B) = X \setminus \operatorname{int}_{\mathcal{R}}(B^c)$; (3) $\operatorname{cl}_{\mathcal{R}}(B) = \bigcap_{R \in \mathcal{R}} R^{-1}[B]$.

Remark 2. By using appropriate complementations, assertion (1) can be expressed in the more concise form that $\operatorname{Cl}_{\mathcal{R}} = \left(\operatorname{Int}_{\mathcal{R}} \circ \mathcal{C}_Y\right)^c = \left(\operatorname{Int}_{\mathcal{R}}\right)^c \circ \mathcal{C}_Y$.

Moreover, by defining the *infinitesimal closure* $\rho_{\mathcal{R}}$ such that $\rho_{\mathcal{R}}(y) = \operatorname{cl}_{\mathcal{R}}(\{y\})$ for all $y \in Y$, from assertion (3) we can easily see that $\rho_{\mathcal{R}} = \bigcap \mathcal{R}^{-1} = \left(\bigcap \mathcal{R}\right)^{-1}$.

In addition to Theorem 1, it is also worth noticing that the small closure and interior relations are usually much weaker tools than the big closure and interior ones. Namely, in general, we can only state the following:

Theorem 2. *For any $A \subseteq X$ and $B \subseteq Y$,*

(1) $A \in \operatorname{Int}_{\mathcal{R}}(B)$ *implies* $A \subseteq \operatorname{int}_{\mathcal{R}}(B)$;
(2) $A \cap \operatorname{cl}_{\mathcal{R}}(B) \neq \emptyset$ *implies* $A \in \operatorname{Cl}_{\mathcal{R}}(B)$.

Remark 3. Later, we shall see that if in particular $\mathcal{R}$ is topologically fine, then the converse implications are also true.

The following theorem shows that, in contrast to their equivalence, the big closure relation is usually a more convenient tool than the big interior one.

Theorem 3. *We have*

(1) $\operatorname{Cl}_{\mathcal{R}^{-1}} = \operatorname{Cl}_{\mathcal{R}}^{-1}$; (2) $\operatorname{Int}_{\mathcal{R}^{-1}} = \mathcal{C}_Y \circ \operatorname{Int}_{\mathcal{R}}^{-1} \circ \mathcal{C}_X$.

Concerning the small interior and closure relations, we can also easily prove:

Theorem 4. *If $R \in \mathcal{R}$, then for any $A \subseteq X$ and $B \subseteq Y$ we have*

$$A \subseteq \operatorname{int}_R(B) \iff \operatorname{cl}_{R^{-1}}(A) \subseteq B.$$

Proof. For instance, if $A \subseteq \operatorname{int}_R(B)$, i.e. $A \subseteq \operatorname{int}_{\{R\}}(B)$, then for each $x \in A$ we have $x \in \operatorname{int}_R(B)$, and thus $R(x) \subseteq B$. Moreover, by Theorem 1, we can see that $\operatorname{cl}_{R^{-1}}(A) = R[A] = \bigcup_{x \in A} R(x)$. Therefore, $\operatorname{cl}_{R^{-1}}(A) \subseteq B$ also holds.

Remark 4. This theorem shows that the mappings $A \mapsto \operatorname{cl}_{R^{-1}}(A)$ and $B \mapsto \operatorname{int}_R(B)$ form a Galois connection between the posets $\mathcal{P}(X)$ and $\mathcal{P}(Y)$.

Later, we shall see that the above *closure-interior Galois connection*, used first in Ref. [47,86], is not independent from the well-known upper and lower bound one [48,81].

By using Theorem 1 and Definition 5, we can easily establish the following:

Theorem 5. *We have*

(1) $\mathcal{D}_{\mathcal{R}} = \{B \subseteq Y : \forall\, R \in \mathcal{R} : X = R^{-1}[B]\}$;
(2) $\mathcal{E}_{\mathcal{R}} = \bigcup_{x \in X} \mathcal{U}_{\mathcal{R}}(x)$, *where* $\mathcal{U}_{\mathcal{R}}(x) = \operatorname{int}_{\mathcal{R}}^{-1}(x)$.

Remark 5. Note that thus $\mathcal{U}_{\mathcal{R}}(x) = \operatorname{int}_{\mathcal{R}}^{-1}(x) = \{B \subseteq Y : x \in \operatorname{int}_{\mathcal{R}}(B)\}$ is just the family of all *neighborhoods* of the point x of X in Y.

Having in mind the ideas of Doičinov [71], we can quite similarly define the neighborhoods of a subset A of X in Y too.

Note that, according to [78], neighborhood properties of relations and relators can also be investigated.

The following theorem shows that the families of fat and dense sets are also equivalent tools.

Theorem 6. *We have*

(1) $\mathcal{D}_{\mathcal{R}} = \{D \subseteq Y : D^c \notin \mathcal{E}_{\mathcal{R}}\}$;
(2) $\mathcal{D}_{\mathcal{R}} = \{D \subseteq Y : \forall E \in \mathcal{E}_{\mathcal{R}} : E \cap D \neq \emptyset\}$.

Remark 6. If $\leq$ is a relation on X, then for any $A \subseteq X$ we have

(1) $A \in \mathcal{E}_{\leq}$ if and only if there exists $x \in X$ such that $y \in A$ for all $y \geq x$;
(2) $A \in \mathcal{D}_{\leq}$ if and only if for each $x \in X$ there exists $y \in A$ such that $y \geq x$.

Therefore, $\mathcal{E}_{\leq}$ and $\mathcal{D}_{\leq}$ are just the families of all *residual* and *cofinal* subsets of the goset (generalized ordered set) $X(\leq)$, resp.

Finally, we note that, by Definition 5 and [8, Theorem 3], the following theorem is also true.

Theorem 7. *The structures* Int, int *and* $\mathcal{E}$ *are union-preserving.*

Remark 7. In the sequel, instead of the union-preservingness of Int, we shall only need that $\operatorname{Int}_{\mathcal{R}} = \bigcup_{R \in \mathcal{R}} \operatorname{Int}_R$. Therefore, [8, Theorem 3] is only of some terminological importance for us.

5. Some Further Families Derived from Relators

Notation 2. In this section, we shall assume that $\mathcal{R}$ is a relator on X.

Definition 6. For any $A \subseteq X$, we define:

(1) $A \in \tau_{\mathcal{R}}$ if $A \in \operatorname{Int}_{\mathcal{R}}(A)$; (2) $A \in \tau\!\!\!-_{\mathcal{R}}$ if $A^c \notin \operatorname{Cl}_{\mathcal{R}}(A)$;
(3) $A \in \mathcal{T}_{\mathcal{R}}$ if $A \subseteq \operatorname{int}_{\mathcal{R}}(A)$; (4) $A \in \mathcal{F}_{\mathcal{R}}$ if $\operatorname{cl}_{\mathcal{R}}(A) \subseteq A$;
(5) $A \in \mathcal{N}_{\mathcal{R}}$ if $\operatorname{cl}_{\mathcal{R}}(A) \notin \mathcal{E}_{\mathcal{R}}$; (6) $A \in \mathcal{M}_{\mathcal{R}}$ if $\operatorname{int}_{\mathcal{R}}(A) \in \mathcal{D}_{\mathcal{R}}$.

Remark 8. The members of the families, $\tau_{\mathcal{R}}$, $\mathcal{T}_{\mathcal{R}}$ and $\mathcal{N}_{\mathcal{R}}$ are called the *proximally open, topologically open and rare (or nowhere dense) subset* relator spaces $X(\mathcal{R})$, resp.

The families $\tau_{\mathcal{R}}$ and $\tau\!\!\!-_{\mathcal{R}}$ were first explicitly used by the third author in Refs. [41,45]. While, the practical notation $\tau\!\!\!-_{\mathcal{R}}$ has been suggested by J. Kurdics who first noted that connectedness is a particular case of well-chainedness [36–38].

By using Definition 6 and the corresponding results of Section 4, we can easily establish the following two theorems:

Theorem 8. *We have*

(1) $\tau\!\!\!-_{\mathcal{R}} = \tau_{\mathcal{R}^{-1}}$; (2) $\tau\!\!\!-_{\mathcal{R}} = \{A \subseteq X : A^c \in \tau_{\mathcal{R}}\}$;
(3) $\mathcal{F}_{\mathcal{R}} = \{A \subseteq X : A^c \in \mathcal{T}_{\mathcal{R}}\}$; (4) $\mathcal{M}_{\mathcal{R}} = \{A \subseteq X : A^c \in \mathcal{N}_{\mathcal{R}}\}$.

Theorem 9. *We have*

(1) $\tau_{\mathcal{R}} \subseteq \mathcal{T}_{\mathcal{R}}$; (2) $\mathcal{T}_{\mathcal{R}} \setminus \{\emptyset\} \subseteq \mathcal{E}_{\mathcal{R}}$; (3) $\mathcal{D}_{\mathcal{R}} \cap \mathcal{F}_{\mathcal{R}} \subseteq \{X\}$.

Remark 9. In addition to assertion (1), it also worth noting that $\tau_R = \mathcal{T}_R$ for any $R \in \mathcal{R}$.

Moreover, from assertion (3), by using global complementations, we can easily infer that $\mathcal{F}_{\mathcal{R}} \subseteq (\mathcal{D}_{\mathcal{R}})^c \cup \{X\}$ and $\mathcal{D}_{\mathcal{R}} \subseteq (\mathcal{F}_{\mathcal{R}})^c \cup \{X\}$.

However, it is now more important to note that we also have the following:

Theorem 10. *For any* $A \subseteq X$, *we have*

(1) $\mathcal{P}(A) \cap (\mathcal{T}_{\mathcal{R}} \setminus \{\emptyset\}) \neq \emptyset$ *implies* $A \in \mathcal{E}_{\mathcal{R}}$;
(2) $\bigcup \mathcal{T}_{\mathcal{R}} \cap \mathcal{P}(A) \subseteq \operatorname{int}_{\mathcal{R}}(A)$; (3) $\mathcal{P}[\tau_{\mathcal{R}} \cap \mathcal{P}(A)] \subseteq \operatorname{Int}_{\mathcal{R}}(A)$.

Remark 10. The fat sets are frequently more convenient tools than the open ones. For instance, if $\leq$ is a relation on X, then $\mathcal{T}_{\leq}$ and $\mathcal{E}_{\leq}$ are the families of all *ascending and residual subsets* of the goset $X(\leq)$, resp.

Fat and dense sets in normed groups and vector relator spaces [84] can also be applied to nicely characterize continuity properties of additive and linear functions and relations [27,83].

Some further remarkable applications of relator spaces to the extensive theory of additive functions and relations were also given in [83,85].

The importance of fat sets, stressed first by the third author at the Seventh Prague Topological Symposium in 1991, can also be well seen from the following:

Example 1. If in particular $X = \mathbb{R}$ and $R(x) = \{x-1\} \cup [x, +\infty[$ for all $x \in X$, then R is a reflexive relation on X such that $\mathcal{T}_R = \{\emptyset, X\}$, but $\mathcal{E}_R$ is quite a large family.

Remark 11. However, if the relator is *topological or proximal* in the sense that:

(1) for each $x \in X$ and $R \in \mathcal{R}$ there exists $V \in \mathcal{T}_{\mathcal{R}}$ such that $x \in V \subseteq R(x)$;
(2) for each $A \subseteq X$ and $R \in \mathcal{R}$ there exists $V \in \tau_{\mathcal{R}}$ such that $A \subseteq V \subseteq R[A]$;

resp., then the converses of the assertions (1)–(3) of Theorem 10 can also be proved. Therefore, in these cases, the families $\mathcal{T}_{\mathcal{R}}$ and $\tau_{\mathcal{R}}$ are also quite powerful tools.

Finally, we note that, by Definition 6 and [8, Theorem 3], the following theorem is also true:

Theorem 11. *The structure τ is also union-preserving.*

The following simple example shows that the increasing structure $\mathcal{T}$ need not be union-preserving. This is a serious disadvantage of the topologically open sets.

Example 2. If $\operatorname{card}(X) > 2$ and $x_1, x_2 \in X$ such that $x_1 \neq x_2$, and

$$R_i = \{x_i\}^2 \cup \left(\{x_i\}^c\right)^2,$$

for all $i = 1, 2$, then $\mathcal{R} = \{R_1, R_2\}$ is an equivalence relator on X such that $\{x_1, x_2\} \in \mathcal{T}_{\mathcal{R}} \setminus \left(\mathcal{T}_{R_1} \cup \mathcal{T}_{R_2}\right)$, and thus $\mathcal{T}_{\mathcal{R}} \not\subseteq \mathcal{T}_{R_1} \cup \mathcal{T}_{R_2}$.

Remark 12. Later, by using the topological closure (refinement) $\mathcal{R}^{\wedge}$ of $\mathcal{R}$, we can see that $\mathcal{T}_{\mathcal{R}} = \bigcup_{R \in \mathcal{R}^{\wedge}} \mathcal{T}_R$.

6. Some Further Relations Derived from Relators

By using Definition 5, we may easily introduce some further important notions. For instance, we may naturally have the following:

Definition 7. If $\mathcal{R}$ is a relator on X to Y, then for any $B \subseteq Y$, we define

(1) $\operatorname{bnd}_{\mathcal{R}}(B) = \operatorname{cl}_{\mathcal{R}}(B) \setminus \operatorname{int}_{\mathcal{R}}(B)$.
Moreover, if in particular $\mathcal{R}$ is a relator on X, then for any $A \subseteq X$ we also define
(2) $\operatorname{res}_{\mathcal{R}}(A) = \operatorname{cl}_{\mathcal{R}}(A) \setminus A$; (3) $\operatorname{bor}_{\mathcal{R}}(A) = A \setminus \operatorname{int}_{\mathcal{R}}(A)$.

Remark 13. Somewhat differently, the *border, boundary and residue of a set* in neighborhood and closure spaces were already introduced by Hausdorff and Kuratowski [49, pp. 4–5]. (See also Elez and Papaz [50] for a recent treatment.)

Concerning the above definition, we shall only mention here the following:

Theorem 12. *If $\mathcal{R}$ is a reflexive relator on X, then for any $A \subseteq X$ we have*

$$\operatorname{bnd}_{\mathcal{R}}(A) = \operatorname{res}_{\mathcal{R}}(A) \cup \operatorname{bor}_{\mathcal{R}}(A) = \operatorname{res}_{\mathcal{R}}(A) \cup \operatorname{res}_{\mathcal{R}}(A^c).$$

Proof. Now, under the convenient notations $A^\circ = \operatorname{int}_{\mathcal{R}}(A)$ and $A^- = \operatorname{cl}_{\mathcal{R}}(A)$, we have $A^\circ \subseteq A \subseteq A^-$. Therefore,

$$\operatorname{res}_{\mathcal{R}}(A^c) = A^{c-} \setminus A^c = A^{c-} \cap A^{cc} = A^{\circ c} \cap A = A \setminus A^\circ = \operatorname{bor}_{\mathcal{R}}(A).$$

Remark 14. Note that if in particular $A \in \mathcal{T}_{\mathcal{R}}$, then $\operatorname{bor}_{\mathcal{R}}(A) = \emptyset$. Therefore, in this particular case, we can simply state that $\operatorname{bnd}_{\mathcal{R}}(A) = \operatorname{res}_{\mathcal{R}}(A)$.

Now, having in mind the convergence and adherence of ordinary and generalized sequences and an observation of Efremović and Švarc [51], we may also naturally introduce the following:

Definition 8. If $\mathcal{R}$ is a relator on X to Y, and moreover φ and ψ are functions of a relator space $\Gamma(\mathcal{U})$ to X and Y, resp., and $(\varphi, \psi)(\gamma) = \big(\varphi(\gamma), \psi(\gamma)\big)$ for all $\gamma \in \Gamma$, then we define:

(1) $\varphi \in \operatorname{Lim}_{\mathcal{R}}(\psi)$ if $(\varphi, \psi)^{-1}[R] \in \mathcal{E}_{\mathcal{U}}$ for all $R \in \mathcal{R}$;
(2) $\varphi \in \operatorname{Adh}_{\mathcal{R}}(\psi)$ if $(\varphi, \psi)^{-1}[R] \in \mathcal{D}_{\mathcal{U}}$ for all $R \in \mathcal{R}$.
Moreover, if $x \in X$ and $x_\Gamma(\gamma) = x$ for all $\gamma \in \Gamma$, then we also define
(3) $x \in \lim_{\mathcal{R}}(\psi)$ if $x_\Gamma \in \operatorname{Lim}_{\mathcal{R}}(\psi)$; (4) $x \in \operatorname{adh}_{\mathcal{R}}(\psi)$ if $x_\Gamma \in \operatorname{Adh}_{\mathcal{R}}(\psi)$.

Remark 15. This definition can be immediately generalized to the case when Φ and Ψ are relations on Γ to X and Y, resp. and $(\Phi \otimes \Psi)(\gamma) = \Phi(\gamma) \times \Psi(\gamma)$ for all $\gamma \in \Gamma$. Moreover, $A \subseteq X$ and $A_\Gamma(\gamma) = A$ for all $\gamma \in \Gamma$.

However, to make the above big limit and adherence relations be stronger tools than the big closure and interior ones, it is sufficient to consider only a *proset* (*preordered set*) $\Gamma(\leq)$ instead of the relator space $\Gamma(\mathcal{U})$.

Theorem 13. *If $\mathcal{R}$ is a relator on X to Y, then for any $A \subseteq X$ and $B \subseteq Y$, the following assertions are equivalent:*

(1) $A \in \mathrm{Cl}_{\mathcal{R}}(B)$;

(2) *there exists a poset $\Gamma(\leq)$ and functions φ and ψ of Γ to A and B, resp. such that $\varphi \in \mathrm{Lim}_{\mathcal{R}}(\psi)$;*

(3) *there exists a non-partial relator space $\Gamma(\mathcal{U})$ and functions φ and ψ of Γ to A and B, resp. such that $\varphi \in \mathrm{Lim}_{\mathcal{R}}(\psi)$.*

Proof. For instance, if (1) holds, then for each $R \in \mathcal{R}$ we have $R[A] \cap B \neq \emptyset$. Therefore, there exist $\varphi(R) \in A$ and $\psi(R) \in B$ such that $\psi(R) \in R(\varphi(R))$. Hence, we can already infer that $(\varphi, \psi)(R) = \big(\varphi(R), \psi(R)\big) \in R$, and thus

$$R \in (\varphi, \psi)^{-1}[R].$$

Now, by taking $\Gamma = \mathcal{R}$ and $\leq = \supseteq$, we can see that $\Gamma(\leq)$ is poset (partially ordered set) such that $\Gamma \neq \emptyset$ if $\mathcal{R} \neq \emptyset$. Moreover, if $R \in \mathcal{R}$, then $R \in \Gamma$ such that for any $S \in \Gamma$, with $S \geq R$, we have $S \subseteq R$, and thus

$$S \in (\varphi, \psi)^{-1}[S] \subseteq (\varphi, \psi)^{-1}[R].$$

This shows that $(\varphi, \psi)^{-1}[R]$ is a residual, and thus a fat subset of the poset $\Gamma(\leq)$. Therefore, $\varphi \in \mathrm{Lim}_{\mathcal{R}}(\psi)$, and thus (2) also holds.

Remark 16. To prove an analogous theorem for the relation $\mathrm{Adh}_{\mathcal{R}}$, on the set $\Gamma = \mathcal{R}$ we have to consider the preorder $\leq = \Gamma^2$. Therefore, posets are usually not sufficient.

However, if $\mathcal{R}$ is *uniformly filtered* in the sense that for every $R, S \in \mathcal{R}$ there exists $T \in \mathcal{R}$ such that $T \subseteq R \cap S$, then in the proof of the corresponding theorem we can also take $\leq = \supseteq$.

By using the convergence and adherence of nets, completeness, compactness and the Lebesque property can also be nicely treated in relator space [75,76,79].

7. Some Algebraic Structures Derived from Relators

Notation 3. First, we shall assume that $\mathcal{R}$ is a relator on X to Y,

Now, according to [7], we may also naturally introduce the following:

Definition 9. For any $A \subseteq X$, $B \subseteq Y$, $x \in X$ and $y \in Y$, we define

(1) $A \in \mathrm{Lb}_{\mathcal{R}}(B)$ and $B \in \mathrm{Ub}_{\mathcal{R}}(A)$ if $A \times B \subseteq R$ for some $R \in \mathcal{R}$;

(2) $x \in \text{lb}_{\mathcal{R}}(B)$ if $\{x\} \in \text{Lb}_{\mathcal{R}}(B)$; (3) $y \in \text{ub}_{\mathcal{R}}(A)$ if $\{y\} \in \text{Ub}_{\mathcal{R}}(A)$;
(4) $B \in \mathfrak{L}_{\mathcal{R}}$ if $\text{lb}_{\mathcal{R}}(B) \neq \emptyset$; (5) $A \in \mathfrak{U}_{\mathcal{R}}$ if $\text{ub}_{\mathcal{R}}(A) \neq \emptyset$.

Thus, for instance, we can easily prove the following three theorems:

Theorem 14. *We have*

(1) $\text{Ub}_{\mathcal{R}} = \text{Lb}_{\mathcal{R}^{-1}} = \text{Lb}_{\mathcal{R}}^{-1}$; (2) $\text{ub}_{\mathcal{R}} = \text{lb}_{\mathcal{R}^{-1}}$.

Theorem 15. *We have*

(1) $\text{Lb}_{\mathcal{R}} = \text{Cl}_{\mathcal{R}^c}^{c} = \text{Int}_{\mathcal{R}^c} \circ \mathcal{C}_Y$; (2) $\text{lb}_{\mathcal{R}} = \text{cl}_{\mathcal{R}^c}^{c} = \text{int}_{\mathcal{R}^c} \circ \mathcal{C}_Y$.

Proof. By Definitions 5 and 9, for any $A \subseteq X$ and $B \subseteq Y$ we have

$$
\begin{aligned}
A \in \text{Lb}_{\mathcal{R}}(B) &\iff \exists\, R \in \mathcal{R} : A \times B \subseteq R \\
&\iff \exists\, R \in \mathcal{R} : \forall\, (a,b) \in A \times B : \ (a,b) \notin R^c \\
&\iff \exists\, R \in \mathcal{R} : \forall\, a \in A,\ b \in B : b \notin R^c(a) \\
&\iff \exists\, R \in \mathcal{R} : R^c[A] \cap B = \emptyset \iff A \notin \text{Cl}_{\mathcal{R}^c}(B) \\
&\iff A \in \text{Cl}_{\mathcal{R}^c}(B)^c \iff A \in \text{Cl}_{\mathcal{R}^c}^{c}(B).
\end{aligned}
$$

Therefore, $\text{Lb}_{\mathcal{R}}(B) = \text{Cl}_{\mathcal{R}^c}^{c}(B)$ for all $B \subseteq Y$, and thus the first part of (1) is true. The second part of (1) follows from Theorem 1.

Remark 17. The above two theorems show that, for instance, the relations $\text{Lb}_{\mathcal{R}}$, $\text{Ub}_{\mathcal{R}}$, $\text{Cl}_{\mathcal{R}}$ and $\text{Int}_{\mathcal{R}}$ are also equivalent tools in the relator space $(X,Y)(\mathcal{R})$.

Moreover, from Theorem 15, we can see that some algebraic and topological structures are just as closely related to each other by equalities (1) and (2) as the exponential and trigonometric functions are by the famous Euler formulas.

Theorem 16. *For any $R \in \mathcal{R}$, $A \subseteq X$ and $B \subseteq Y$, we have*

$$A \subseteq \text{lb}_R(B) \iff B \subseteq \text{ub}_R(A).$$

Remark 18. This shows that the mappings $A \mapsto \text{ub}_R(A)$ and $B \mapsto \text{lb}_R(B)$ establish a Galois connection between the poset $\mathcal{P}(X)$ and the dual of $\mathcal{P}(Y)$.

Notation 4. Next, we shall assume that $\mathcal{R}$ is a relator on X.

Now, in addition to Definition 9, we may also naturally introduce the following:

Definition 10. For any $A \subseteq X$, we define

(1) $\min_{\mathcal{R}}(A) = A \cap \operatorname{lb}_{\mathcal{R}}(A)$; (2) $\max_{\mathcal{R}}(A) = A \cap \operatorname{ub}_{\mathcal{R}}(A)$;
(3) $\operatorname{Min}_{\mathcal{R}}(A) = \mathcal{P}(A) \cap \operatorname{Lb}_{\mathcal{R}}(A)$; (4) $\operatorname{Max}_{\mathcal{R}}(A) = \mathcal{P}(A) \cap \operatorname{Ub}_{\mathcal{R}}(A)$;
(5) $\inf_{\mathcal{R}}(A) = \max_{\mathcal{R}}\big(\operatorname{lb}_{\mathcal{R}}(A)\big)$; (6) $\sup_{\mathcal{R}}(A) = \min_{\mathcal{R}}\big(\operatorname{ub}_{\mathcal{R}}(A)\big)$;
(7) $\operatorname{Inf}_{\mathcal{R}}(A) = \operatorname{Max}_{\mathcal{R}}\big[\operatorname{Lb}_{\mathcal{R}}(A)\big]$; (8) $\operatorname{Sup}_{\mathcal{R}}(A) = \operatorname{Min}_{\mathcal{R}}\big[\operatorname{Ub}_{\mathcal{R}}(A)\big]$;
(9) $A \in \ell_{\mathcal{R}}$ if $A \in \operatorname{Lb}_{\mathcal{R}}(A)$; (10) $A \in \mathcal{L}_{\mathcal{R}}$ if $A \subseteq \operatorname{lb}_{\mathcal{R}}(A)$.

Remark 19. Thus, for instance, for any $x \in X$ and $A \subseteq X$ we also have $x \in \min_{\mathcal{R}}(A) \iff x \in A,\ x \in \operatorname{lb}_{\mathcal{R}}(A) \iff \{x\} \in \mathcal{P}(A), \{x\} \in \operatorname{Lb}_{\mathcal{R}}(A) \iff \{x\} \in \operatorname{Min}_{\mathcal{R}}(A)$.

However, a similar statement for the relations $\sup_{\mathcal{R}}$ and $\operatorname{Sup}_{\mathcal{R}}$ seems not to be true. Therefore, the definition of $\operatorname{Sup}_{\mathcal{R}}$, and thus also that of $\operatorname{Inf}_{\mathcal{R}}$, is perhaps not the most convenient one.

In Ref. [7], the third author, for instance, proved the following theorems:

Theorem 17. *We have*

(1) $\max_{\mathcal{R}} = \min_{\mathcal{R}^{-1}}$; (2) $\operatorname{Max}_{\mathcal{R}} = \operatorname{Min}_{\mathcal{R}^{-1}}$.

Theorem 18. *For any* $A \subseteq X$, *we have*

(1) $\operatorname{Max}_{\mathcal{R}}(A) \subseteq \mathcal{P}\big(\max_{\mathcal{R}}(A)\big)$;
(2) $\operatorname{Max}_{\mathcal{R}}(A) = \big\{B \subseteq X :\ P(A) \subseteq \operatorname{Lb}_{\mathcal{R}}(B)\big\}$.

Remark 20. Concerning the relation $\operatorname{Max}_{\mathcal{R}}$, it also worth mentioning that $\operatorname{Max}_{\mathcal{R}} = \mathcal{P} \circ \operatorname{Max}_{\mathcal{R}}$.

Theorem 19. *For any* $A \subseteq X$, *we have*

(1) $\min_{\mathcal{R}}(A) = A \setminus \operatorname{cl}_{\mathcal{R}^c}(A) = A \cap \operatorname{int}_{\mathcal{R}^c}(A^c)$;
(2) $\operatorname{Min}_{\mathcal{R}}(A) = \mathcal{P}(A) \setminus \operatorname{Cl}_{\mathcal{R}^c}(A) = \mathcal{P}(A) \cap \operatorname{Int}_{\mathcal{R}^c}(A^c)$.

Remark 21. The latter assertion can be expressed in the more concise form that $\operatorname{Min}_{\mathcal{R}} = \mathcal{P} \setminus \operatorname{Cl}_{\mathcal{R}^c} = \mathcal{P} \cap \big(\operatorname{Int}_{\mathcal{R}^c} \circ \mathcal{C}_X\big)$.

Theorem 20. *For any* $\emptyset \neq A \subseteq X$, *we have*

(1) $\max_{\mathcal{R}}(A) = \bigcup_{R \in \mathcal{R}} \big(A \setminus R^c[A]\big) = \bigcup_{R \in \mathcal{R}} \bigcap_{a \in A} A \cap R(a)$;
(2) $\operatorname{Max}_{\mathcal{R}}(A) = \bigcup_{R \in \mathcal{R}} \mathcal{P}\big(A \setminus R^c[A]\big) = \bigcup_{R \in \mathcal{R}} \bigcap_{a \in A} \mathcal{P}\big(A \cap R(a)\big)$.

Theorem 21. *For any $A \subseteq X$, the following assertions are equivalent:*

(1) $A \in \ell_{\mathcal{R}}$; (2) $A \in \mathrm{Ub}_{\mathcal{R}}(A)$; (3) $A \in \mathrm{Min}_{\mathcal{R}}(A)$; (4)/ $A \in \mathrm{Max}_{\mathcal{R}}(A)$.

Remark 22. By using Theorem 15, assertion (1) can also be reformulated in the form that $A \notin \mathrm{Cl}_{\mathcal{R}^c}(A)$, or equivalently $A \in \mathrm{Int}_{\mathcal{R}^c}(A^c)$.

Moreover, by using the corresponding definitions, we can also easily prove that $\ell_{\mathcal{R}} = \mathrm{Min}_{\mathcal{R}}[\mathcal{P}(X)] = \mathrm{Max}_{\mathcal{R}}[\mathcal{P}(X)]$.

Theorem 22. *For any $A \subseteq X$, the following assertions are equivalent:*

(1) $A \in \mathcal{L}_{\mathcal{R}}$; (2) $A = \mathrm{min}_{\mathcal{R}}(A)$; (3) $A \in \bigcap_{a \in A} \mathrm{lb}_{\mathcal{R}}^{-1}(\{a\})$.

Remark 23. By using Theorem 15, assertion (1) can also be reformulated in the form that $\mathrm{cl}_{\mathcal{R}^c}(A) \subseteq A^c$, or equivalently $A \subseteq \mathrm{int}_{\mathcal{R}^c}(A^c)$.

Theorem 23. *We have*

(1) $\ell_{\mathcal{R}} = \ell_{\mathcal{R}^{-1}}$; (2) $\ell_{\mathcal{R}} \subseteq \mathcal{L}_{\mathcal{R}} \cap \mathcal{L}_{\mathcal{R}^{-1}}$; (3) $\mathcal{L}_{\mathcal{R}} = \{\mathrm{min}_{\mathcal{R}}(A) : A \subseteq X\}$.

8. Closure Operations for Relators

Notation 5. In this section, we shall assume that $\mathcal{R}$ is a relator on X to Y.

Some of the following operations were already considered by Kenyon [52] and Nakano and Nakano [53]. Some further developments can be found in [6,35,45,77].

Definition 11. The relators

$$\mathcal{R}^* = \{S \subseteq X \times Y : \exists\ R \in \mathcal{R} : R \subseteq S\},$$
$$\mathcal{R}^\# = \{S \subseteq X \times Y : \forall\ A \subseteq X : \exists\ R \in \mathcal{R} :\ R[A] \subseteq S[A]\},$$
$$\mathcal{R}^\wedge = \{S \subseteq X \times Y : \forall x \in X : \exists\ R \in \mathcal{R} :\ R(x) \subseteq S(x)\},$$
$$\mathcal{R}^\triangle = \{S \subseteq X \times Y : \forall\ x \in X : \exists\ u \in X : \exists\ R \in \mathcal{R} :\ R(u) \subseteq S(x)\}$$

are called the *uniform, proximal, topological and paratopological closures* (*or refinements*) of the relator $\mathcal{R}$, resp.

Remark 24. Thus, we evidently have $\mathcal{R} \subseteq \mathcal{R}^* \subseteq \mathcal{R}^\# \subseteq \mathcal{R}^\wedge \subseteq \mathcal{R}^\triangle$. Moreover, if $\mathcal{R}$ is a relator on X, then we can easily prove that $\mathcal{R}^\infty \subseteq \mathcal{R}^{*\infty} \subseteq \mathcal{R}^{\infty *} \subseteq \mathcal{R}^*$.

However, it is now more important to note that, by definitions, we also have

$$\mathcal{R}^{\#} = \{ S \subseteq X \times Y : \forall\, A \subseteq X : A \in \operatorname{Int}_{\mathcal{R}}(S[A]) \};$$
$$\mathcal{R}^{\wedge} = \{ S \subseteq X \times Y : \forall x \in X : x \in \operatorname{int}_{\mathcal{R}}(S(x)) \};$$
$$\mathcal{R}^{\vartriangle} = \{ S \subseteq X \times Y : \forall\, x \in X : \; S(x) \in \mathcal{E}_{\mathcal{R}} \}.$$

Moreover, by using *Pataki connections* [35,39,54], the assertions of the following theorem and its corollary can be proved in a unified way.

Theorem 24. *$\#$, $\wedge$ and $\vartriangle$ are closure operations for relators on X to Y such that, for any relator $\mathcal{S}$ on X to Y,*

(1) $\mathcal{S} \subseteq \mathcal{R}^{\#} \iff \operatorname{Int}_{\mathcal{S}} \subseteq \operatorname{Int}_{\mathcal{R}} \iff \operatorname{Cl}_{\mathcal{R}} \subseteq \operatorname{Cl}_{\mathcal{S}}$;
(2) $\mathcal{S} \subseteq \mathcal{R}^{\wedge} \iff \operatorname{int}_{\mathcal{S}} \subseteq \operatorname{int}_{\mathcal{R}} \iff \operatorname{cl}_{\mathcal{R}} \subseteq \operatorname{cl}_{\mathcal{S}}$;
(3) $\mathcal{S} \subseteq \mathcal{R}^{\vartriangle} \iff \mathcal{E}_{\mathcal{S}} \subseteq \mathcal{E}_{\mathcal{R}} \iff \mathcal{D}_{\mathcal{R}} \subseteq \mathcal{D}_{\mathcal{S}}$.

Remark 25. The statement that $\#$, $\wedge$ and $\vartriangle$ are closure operations can be derived from assertions (1)–(3).

While, assertions (1)–(3) can be derived from the fact that the structures Int, int and $\mathcal{E}$ are union-preserving.

Corollary 1. *The following assertions are true:*

(1) $\mathcal{S} = \mathcal{R}^{\#}$ *is the largest relator on X to Y such that* $\operatorname{Int}_{\mathcal{S}} = \operatorname{Int}_{\mathcal{R}}$ $(\operatorname{Cl}_{\mathcal{S}} = \operatorname{Cl}_{\mathcal{R}})$;
(2) $\mathcal{S} = \mathcal{R}^{\wedge}$ *is the largest relator on X to Y such that* $\operatorname{int}_{\mathcal{S}} = \operatorname{int}_{\mathcal{R}}$ $(\operatorname{cl}_{\mathcal{S}} = \operatorname{cl}_{\mathcal{R}})$;
(3) $\mathcal{S} = \mathcal{R}^{\vartriangle}$ *is the largest relator on X to Y such that* $\mathcal{E}_{\mathcal{S}} = \mathcal{E}_{\mathcal{R}}$ $(\mathcal{D}_{\mathcal{S}} = \mathcal{D}_{\mathcal{R}})$.

Remark 26. To prove some similar statements for the operation $*$, the intersection-preserving structures Lim and Adh have to be used.

Concerning the above basic closure operations, we can also prove the following:

Theorem 25. *We have*

(1) $\mathcal{R}^{\#} = \mathcal{R}^{*\#} = \mathcal{R}^{\#*}$;
(2) $\mathcal{R}^{\wedge} = \mathcal{R}^{\Diamond\wedge} = \mathcal{R}^{\wedge\Diamond}$ *with* $\Diamond = *$ *and* $\#$;
(3) $\mathcal{R}^{\vartriangle} = \mathcal{R}^{\Diamond\vartriangle} = \mathcal{R}^{\vartriangle\Diamond}$ *with* $\Diamond = *, \#$ *and* $\wedge$.

Proof. To prove (1), note that, by Remark 24 and the closure properties, we have $\mathcal{R}^{\#} \subseteq \mathcal{R}^{\#*} \subseteq \mathcal{R}^{\#\#} = \mathcal{R}^{\#}$ and $\mathcal{R}^{\#} \subseteq \mathcal{R}^{*\#} \subseteq \mathcal{R}^{\#\#} = \mathcal{R}^{\#}$.

Remark 27. By using Remark 24, we can also easily prove that $\mathcal{R}^{*\infty} = \mathcal{R}^{\infty*\infty}$ and $\mathcal{R}^{\infty*} = \mathcal{R}^{*\infty*}$ for any relator $\mathcal{R}$ on X.

However, it is now more important to note that now we also have the following:

Theorem 26. *We have*

(1) $\mathcal{R}^{*-1} = \mathcal{R}^{-1*}$; (2) $\mathcal{R}^{\#-1} = \mathcal{R}^{-1\#}$.

Proof. To prove (2), note that by Theorem 3 and Corollary 1 we have $\mathrm{Cl}_{\mathcal{R}^{\#-1}} = \mathrm{Cl}_{\mathcal{R}^{\#}}^{-1} = \mathrm{Cl}_{\mathcal{R}}^{-1} = \mathrm{Cl}_{\mathcal{R}^{-1}}$, and thus in particular $\mathrm{Cl}_{\mathcal{R}^{-1}} \subseteq \mathrm{Cl}_{\mathcal{R}^{\#-1}}$. Hence, by using Theorem 24, we can infer that $\mathcal{R}^{\#-1} \subseteq \mathcal{R}^{-1\#}$. Now, by writing $\mathcal{R}^{-1}$ in place of $\mathcal{R}$, we can see that assertion (2) is also true.

Remark 28. We can note that the elementwise operations c and ∞ are also inversion compatible. Moreover, the operation ∂ is also inversion compatible.

Namely, for any relator $\mathcal{R}$ and relation S on X, we have $S \in \mathcal{R}^{-1\partial} \iff S^{\infty} \in \mathcal{R}^{-1} \iff S^{\infty-1} \in \mathcal{R} \iff S^{-1\infty} \in \mathcal{R} \iff S^{-1} \in \mathcal{R}^{\partial} \iff S \in \mathcal{R}^{\partial-1}$.

However, for instance, the operations $\wedge$ and $\vartriangle$ are not inversion compatible. Therefore, in addition to Definition 11, we must also have the following:

Definition 12. We define $\mathcal{R}^{\vee} = \mathcal{R}^{\wedge-1}$ and $\mathcal{R}^{\triangledown} = \mathcal{R}^{\vartriangle-1}$.

Remark 29. The latter operations have very curious properties. For instance, if $\mathcal{R} \neq \emptyset$, then $\mathcal{R}^{\vee\wedge} = \{\rho_{\mathcal{R}}\}^{\wedge}$, and thus in particular the relator $\mathcal{R}^{\vee}$ is topologically simple. (For some generalizations, see Ref. [55].)

Moreover, the operations $\vee\vee$ and $\triangledown\triangledown$ already coincide with the extremal closure operations $\bullet$ and $\blacklozenge$, defined for any relator $\mathcal{R}$ on X to Y such that $\mathcal{R}^{\bullet} = \left\{\bigcap \mathcal{R}\right\}^{*}$ and

$$\mathcal{R}^{\blacklozenge} = \mathcal{R} \quad \text{if} \quad \mathcal{R} = \{X{\times}Y\} \qquad \text{and} \qquad \mathcal{R}^{\blacklozenge} = \mathcal{P}(X{\times}Y) \quad \text{if} \quad \mathcal{R} \neq \{X{\times}Y\}.$$

The importance of the operation $\blacklozenge$ lies in the fact that it is the ultimate *stable* unary operation for relators on X to Y in the sense that $\{X \times Y\}^{\blacklozenge} = \{X \times Y\}$.

9. Some Further Theorems on the Operations $\wedge$ and Δ

A preliminary form of the following basic theorem was already proved in Ref. [6].

Theorem 27. *If $\mathcal{R}$ is a non-void relator on X to Y, then for any $B \subseteq Y$ we have*

(1) $\operatorname{Int}_{\mathcal{R}^\wedge}(B) = \mathcal{P}\big(\operatorname{int}_{\mathcal{R}}(B)\big)$; (2) $\operatorname{Cl}_{\mathcal{R}^\wedge}(B) = \mathcal{P}\big(\operatorname{cl}_{\mathcal{R}}(B)^c\big)^c$.

Proof. To prove (1), note that if $A \in \operatorname{Int}_{\mathcal{R}^\wedge}(B)$, then by Theorem 2 and Corollary 1 we have $A \subseteq \operatorname{int}_{\mathcal{R}^\wedge}(B) = \operatorname{int}_{\mathcal{R}}(B)$, and thus $A \in \mathcal{P}\big(\operatorname{int}_{\mathcal{R}}(B)\big)$. Therefore, $\operatorname{Int}_{\mathcal{R}^\wedge}(B) \subseteq \mathcal{P}\big(\operatorname{int}_{\mathcal{R}}(B)\big)$.

While, if $A \in \mathcal{P}\big(\operatorname{int}_{\mathcal{R}}(B)\big)$, then $A \subseteq \operatorname{int}_{\mathcal{R}}(B)$. Therefore, for each $x \in A$, there exists $R_x \in \mathcal{R}$ such that $R_x(x) \subseteq B$. Now, by defining

$$S(x) = R_x(x) \quad \text{for all} \quad x \in A \qquad \text{and} \qquad S(x) = Y \quad \text{for all} \quad x \in A^c,$$

we can at once state that $S[A] \subseteq B$. Moreover, by using that $\mathcal{R} \neq \emptyset$, we can also easily note that $S \in \mathcal{R}^\wedge$. Therefore, we also have $A \in \operatorname{Int}_{\mathcal{R}^\wedge}(B)$. Consequently, $\mathcal{P}\big(\operatorname{int}_{\mathcal{R}}(B)\big) \subseteq \operatorname{Int}_{\mathcal{R}^\wedge}(B)$, and thus (1) also holds.

Remark 30. By assertion (2), for any $A \subseteq X$, we have $A \in \operatorname{Cl}_{\mathcal{R}^\wedge}(B)$ if and only if $A \cap \operatorname{cl}_{\mathcal{R}}(B) \neq \emptyset$.

From the above theorem, by using Definition 6, we can immediately derive

Corollary 2. *If $\mathcal{R}$ is a non-void relator on X, then*

(1) ${\scriptstyle\mathcal{T}}_{\mathcal{R}^\wedge} = \mathcal{T}_{\mathcal{R}}$; (2) ${\scriptstyle\mathcal{F}}_{\mathcal{R}^\wedge} = \mathcal{F}_{\mathcal{R}}$.

Remark 31. Recall that, by Theorem 11 and Remark 9, we have ${\scriptstyle\mathcal{T}}_{\mathcal{R}} = \bigcup_{R\in\mathcal{R}} {\scriptstyle\mathcal{T}}_R = \bigcup_{R\in\mathcal{R}} \mathcal{T}_R$. Hence, by writing $\mathcal{R}^\wedge$ in place of $\mathcal{R}$ and using Corollary 2, we can immediately infer that $\mathcal{T}_{\mathcal{R}} = \bigcup_{R\in\mathcal{R}^\wedge} \mathcal{T}_R$.

From Corollary 2, by using Theorem 25, we can also immediately derive

Corollary 3. *If $\mathcal{R}$ is a non-void relator on X, then*
(1) ${\scriptstyle\mathcal{T}}_{\mathcal{R}^\Delta} = \mathcal{T}_{\mathcal{R}^\Delta}$; (2) ${\scriptstyle\mathcal{F}}_{\mathcal{R}^\Delta} = \mathcal{F}_{\mathcal{R}^\Delta}$.

Concerning the operation Δ, we can also prove the following:

Theorem 28. *If $\mathcal{R}$ is a non-void relator on X to Y, then for any $B \subseteq Y$ we have*

(1) $\operatorname{Int}_{\mathcal{R}^\triangle}(B) = \{\emptyset\}$ *if* $B \notin \mathcal{E}_{\mathcal{R}}$ *and* $\operatorname{Int}_{\mathcal{R}^\triangle}(B) = \mathcal{P}(X)$ *if* $B \in \mathcal{E}_{\mathcal{R}}$;
(2) $\operatorname{Cl}_{\mathcal{R}^\triangle}(B) = \emptyset$ *if* $B \notin \mathcal{D}_{\mathcal{R}}$ *and* $\operatorname{Cl}_{\mathcal{R}^\triangle}(B) = \mathcal{P}(X) \setminus \{\emptyset\}$ *if* $B \in \mathcal{D}_{\mathcal{R}}$.

Proof. If $A \in \operatorname{Int}_{\mathcal{R}^\triangle}(B)$, then there exists $S \in \mathcal{R}^\triangle$ such that $S[A] \subseteq B$. Therefore, if $A \neq \emptyset$, then there exists $x \in X$ such that $S(x) \subseteq B$. Hence, by using that $S(x) \in \mathcal{E}_{\mathcal{R}}$ and $\mathcal{E}_{\mathcal{R}}$ is ascending, we can infer that $B \in \mathcal{E}_{\mathcal{R}}$. Therefore, if $B \notin \mathcal{E}_{\mathcal{R}}$, then we necessarily have $\operatorname{Int}_{\mathcal{R}^\triangle}(B) \subseteq \{\emptyset\}$. Moreover, since $\mathcal{R} \neq \emptyset$, we can also note that $\mathcal{R}^\triangle \neq \emptyset$, and thus $\emptyset \in \operatorname{Int}_{\mathcal{R}^\triangle}(B)$. Therefore, the first part of assertion (1) is true.

On the other hand, if $B \in \mathcal{E}_{\mathcal{R}}$, then by defining $R = X \times B$ and using Remark 24, we can see that $R \in \mathcal{R}^\triangle$. Moreover, we can also note that $R[A] \subseteq B$, and thus $A \in \operatorname{Int}_{\mathcal{R}^\triangle}(B)$ for all $A \subseteq X$. Therefore, the second part of assertion (1) is also true.

From this theorem, by Definition 5, it is clear that in particular we also have

Corollary 4. *If $\mathcal{R}$ is a non-void relator on X to Y, then for any $B \subseteq Y$:*

(1) $\operatorname{cl}_{\mathcal{R}^\triangle}(B) = \emptyset$ *if* $B \notin \mathcal{D}_{\mathcal{R}}$ *and* $\operatorname{cl}_{\mathcal{R}^\triangle}(B) = X$ *if* $B \in \mathcal{D}_{\mathcal{R}}$;
(2) $\operatorname{int}_{\mathcal{R}^\triangle}(B) = \emptyset$ *if* $B \notin \mathcal{E}_{\mathcal{R}}$ *and* $\operatorname{int}_{\mathcal{R}^\triangle}(B) = X$ *if* $B \in \mathcal{E}_{\mathcal{R}}$.

Hence, by using Definitions 5 and 6, we can immediately derive

Corollary 5. *If $\mathcal{R}$ is a relator on X, then*

(1) $\mathcal{T}_{\mathcal{R}^\triangle} = \mathcal{E}_{\mathcal{R}} \cup \{\emptyset\}$; (2) $\mathcal{F}_{\mathcal{R}^\triangle} = \big(\mathcal{P}(X) \setminus \mathcal{D}_{\mathcal{R}}\big) \cup \{X\}$.

Remark 32. Note that if in particular $\mathcal{R} = \emptyset$, then $\mathcal{E}_{\mathcal{R}} = \emptyset$. Moreover, $\mathcal{R}^\triangle = \emptyset$ if $X \neq \emptyset$, and $\mathcal{R}^\triangle = \{\emptyset\}$ if $X = \emptyset$. Therefore, $\mathcal{T}_{\mathcal{R}^\triangle} = \{\emptyset\}$, and thus assertion (1) is still true.

Now, since $\emptyset \notin \mathcal{E}_{\mathcal{R}}$ if $\mathcal{R}$ is non-partial, we can also state the following:

Corollary 6. *If $\mathcal{R}$ is a non-partial relator on X, then*

(1) $\mathcal{E}_{\mathcal{R}} = \mathcal{T}_{\mathcal{R}^\triangle} \setminus \{\emptyset\}$; (2) $\mathcal{D}_{\mathcal{R}} = \big(\mathcal{P}(X) \setminus \mathcal{F}_{\mathcal{R}^\triangle}\big) \cup \{X\}$.

10. Projection Operations for Relators

Notation 6. In this section, we shall assume that $\mathcal{R}$ is a relator on X.

The importance of the operation ∞ lies mainly in the following:

Theorem 29. *∞ is a closure operation for relations on X such that, for any $R, S \in \mathcal{R}$, we have*

$$S \subseteq R^{\infty} \iff \tau_R \subseteq \tau_S \iff \tau\!\!\!\!-_R \subseteq \tau\!\!\!\!-_S .$$

Proof. If $x \in X$, then because of the inclusion $R \subseteq R^{\infty}$ and the transitivity of R^{∞}, we have

$$R[R^{\infty}(x)] \subseteq R^{\infty}[R^{\infty}(x)] = \big(R^{\infty} \circ R^{\infty}\big)(x) \subseteq R^{\infty}(x).$$

Thus, by the definition of τ_R, we have $R^{\infty}(x) \in \tau_{\mathcal{R}}$. Now, if $\tau_R \subseteq \tau_S$ holds, then we can see that $R^{\infty}(x) \in \tau_S$, and thus $S[R^{\infty}(x)] \subseteq R^{\infty}(x)$. Hence, by using the reflexivity of R^{∞}, we can already infer that $S(x) \subseteq R^{\infty}(x)$. Therefore, $S \subseteq R^{\infty}$ also holds.

While, if $A \in \tau_R$, then by the definition of $\tau_{\mathcal{R}}$ we have $R[A] \subseteq A$. Hence, by induction, we can see that $R^n[A] \subseteq A$ for all $n \in \mathbb{N}$. Now, since $R^0[A] = \Delta_X[A] = A$ also holds, we can already state that

$$R^{\infty}[A] = \left(\bigcup_{n=0}^{\infty} R^n\right)[A] = \bigcup_{n=0}^{\infty} R^n[A] \subseteq \bigcup_{n=0}^{\infty} A = A.$$

Therefore, if $S \subseteq R^{\infty}$ holds, then we have $S[A] \subseteq R^{\infty}[A] \subseteq A$, and thus $A \in \tau_S$. Consequently, $\tau_R \subseteq \tau_S$ also holds.

Now, analogously to our former similar results, we can also state

Corollary 7. *For any $R \in \mathcal{R}$, just $S = R^{\infty}$ is the largest relation on X such that $\tau_R = \tau_S$ $\big(\tau\!\!\!\!-_R = \tau\!\!\!\!-_S\big)$.*

Remark 33. Preliminary forms of the above theorem and its corollary were first proved by Mala [33].

Moreover, he also proved that $R^{\infty}(x) = \bigcap\{A \in \tau_R : x \in A\}$ for all $x \in X$, and thus $R^{\infty} = \bigcap\{R_A : A \in \tau_R\}$.

By using Theorem 29, as an analogue of Theorem 24, we can also prove

Theorem 30. *$\#\partial$ is a closure operation for relators on X such that, for any relator $\mathcal{S}$ on X, we have*

$$\mathcal{S} \subseteq \mathcal{R}^{\#\partial} \iff \tau_{\mathcal{S}} \subseteq \tau_{\mathcal{R}} \iff \tau\!\!\!\!-_{\mathcal{S}} \subseteq \tau\!\!\!\!-_{\mathcal{R}} .$$

Thus, analogously to Corollary 1, we can also state the following:

Corollary 8. *$\mathcal{S} = \mathcal{R}^{\#\partial}$ is the largest relator on X such that $\tau_{\mathcal{S}} = \tau_{\mathcal{R}}$ $\big(\tau\!\!\!\!-_{\mathcal{S}} = \tau\!\!\!\!-_{\mathcal{R}}\big)$.*

By using the Galois property of the operation ∂, Theorem 30 can be reformulated in a more convenient form.

Theorem 31. *$\#\infty$ is a projection operation for relators on X such that, for any relator $\mathcal{S}$ on X, we have*

$$\mathcal{S}^{\infty} \subseteq \mathcal{R}^{\#} \iff \mathscr{T}_{\mathcal{S}} \subseteq \mathscr{T}_{\mathcal{R}} \iff \mathscr{F}_{\mathcal{S}} \subseteq \mathscr{F}_{\mathcal{R}}.$$

Remark 34. It can be shown that the inclusions $\mathcal{S}^{\infty} \subseteq \mathcal{R}^{\#}$, $\mathcal{S}^{\#\infty} \subseteq \mathcal{R}^{\#\infty}$, $\mathcal{S}^{\#\infty} \subseteq \mathcal{R}^{\#}$ and $\mathcal{S}^{\infty\#} \subseteq \mathcal{R}^{\infty\#}$ are also equivalent.

Now, analogously to our former corollaries, we can also state the following:

Corollary 9. *$\mathcal{S} = \mathcal{R}^{\#\infty}$ is the largest preorder relator on X such that $\mathscr{T}_{\mathcal{S}} = \mathscr{T}_{\mathcal{R}}$ $\big(\mathscr{F}_{\mathcal{S}} = \mathscr{F}_{\mathcal{R}}\big)$.*

Remark 35. The advantage of the projection operation $\#\infty$ over the closure operation $\#\partial$ lies mainly in the fact that, in contrast to $\#\partial$, the operation $\#\infty$ is stable.

Since the structure $\mathcal{T}$ is not union-preserving, by using some parts of the theory of Pataki connections [35,39,54], we can only prove the following:

Theorem 32. *$\wedge\partial$ is a preclosure operation for relators such that, for any relator $\mathcal{S}$ on X, we have*

$$\mathcal{T}_{\mathcal{S}} \subseteq \mathcal{T}_{\mathcal{R}} \iff \mathcal{F}_{\mathcal{S}} \subseteq \mathcal{F}_{\mathcal{R}} \implies \mathcal{S}^{\wedge} \subseteq \mathcal{R}^{\wedge\partial}.$$

Remark 36. If $\operatorname{card}(X) > 2$, then by using the equivalence relator $\mathcal{R} = \{X^2\}$ Mala [33, Example 5.3] proved that there does not exist a largest relator $\mathcal{S}$ on X such that $\mathcal{T}_{\mathcal{R}} = \mathcal{T}_{\mathcal{S}}$.

Moreover, Pataki [35, Example 7.2] proved that $\mathcal{T}_{\mathcal{R}^{\wedge\partial}} \not\subseteq \mathcal{T}_{\mathcal{R}}$ and $\wedge\partial$ is not idempotent. (Actually, it can be proved that $\mathcal{R}^{\wedge\partial\wedge} \not\subseteq \mathcal{R}^{\wedge\partial}$ also holds [40, Example 10.11].)

Fortunately, as an analogue of Theorem 31, we can also prove

Theorem 33. *$\wedge\infty$ is a projection operation for relators on X such that if $\mathcal{R} \neq \emptyset$, then for any non-void relator $\mathcal{S}$ on X, we have*

$$\mathcal{S}^{\wedge\infty} \subseteq \mathcal{R}^{\wedge} \iff \mathcal{T}_{\mathcal{S}} \subseteq \mathcal{T}_{\mathcal{R}} \iff \mathcal{F}_{\mathcal{S}} \subseteq \mathcal{F}_{\mathcal{R}}.$$

Thus, in particular, we can also state

Corollary 10. *If $\mathcal{R} \neq \emptyset$, then $\mathcal{S} = \mathcal{R}^{\wedge\infty}$ is the largest preorder relator on X such that $\mathcal{T}_{\mathcal{S}} = \mathcal{T}_{\mathcal{R}}$ $\big(\mathcal{F}_{\mathcal{S}} = \mathcal{F}_{\mathcal{R}}\big)$.*

Remark 37. In the light of the several disadvantages of the structure $\mathcal{T}$, it is rather curious that most of the works in general topology and abstract analysis have been based on open sets suggested by Tietze [56] and Alexandroff [57], and standardized by Bourbaki [58] and Kelley [24]. (See Thron [13, p. 18].)

Moreover, it is a striking fact that, despite the results of Davis [19], Pervin [16], Hunsaker, and Lindgren [18] and the third author [41,42], generalized proximities and closures, minimal structures, generalized topologies and stacks (ascending systems) are still intensively investigated by a great number of mathematicians without using generalized uniformities.

11. Some General Theorems on Unary Operations for Relators

Notation 7. In this and the next section, we shall assume that $\square$ and $\Diamond$ are unary operations for relators.

However, the forthcoming definitions and theorems can be easily extended to some more general settings.

Definition 13. The operation $\square$ will be called *$\Diamond$–dominating, $\Diamond$–invariant, $\Diamond$–absorbing, and $\Diamond$–compatible*, resp., if for any relator $\mathcal{R}$ we have

$$\mathcal{R}^{\Diamond} \subseteq \mathcal{R}^{\square}, \quad \mathcal{R}^{\square} = \mathcal{R}^{\square\Diamond}, \quad \mathcal{R}^{\square} = \mathcal{R}^{\Diamond\square} \quad \text{and} \quad \mathcal{R}^{\square\Diamond} = \mathcal{R}^{\Diamond\square}.$$

Remark 38. From Theorem 25, we can see that if $\Diamond, \square \in \{*, \#, \wedge, \vartriangle\}$ such that $\Diamond$ precedes $\square$ in the above list, then $\square$ is both $\Diamond$–invariant and $\Diamond$–absorbing. Thus, in particular it is also $\Diamond$–compatible.

Moreover, from Theorem 26 and Remark 28, we know that the operations $*$, $\#$, c, ∞, and ∂ are inversion-compatible. However, the important closure operations $\wedge$ and $\vartriangle$ are not inversion-compatible.

By using Definition 13, we can also easily prove the following two theorems:

Theorem 34. *If $\Diamond$ is extensive and $\square$ is $\Diamond$–dominating and idempotent, then $\square$ is $\Diamond$–invariant. Moreover, if in addition $\square$ is increasing, then $\square$ is $\Diamond$–absorbing and $\Diamond$–compatible.*

Remark 39. In this respect, it is also worth mentioning that if $\Diamond$ is extensive and $\square$ is $\Diamond$-dominating, then $\square$ is also extensive.

Moreover, if $\Diamond$ is increasing and $\square$ is extensive such that $\mathcal{R}^{\square\Diamond} \subseteq \mathcal{R}^{\square}$ for every relator $\mathcal{R}$, then $\square$ is $\Diamond$–dominating.

Theorem 35. *If* $\Box$ *and* $\Diamond$ *are inversion-compatible, then their compositions* $\Box\Diamond$ *and* $\Diamond\Box$ *are also inversion-compatible.*

Remark 40. Note that if $\Box$ is an inversion-compatible operation for relations, then the elementwise operation defined by it for relators is also inversion-compatible.

Or, somewhat differently, if $\Box$ is a union-preserving operation for relators, then $\Box$ is inversion-compatible if and only if $\{R\}^{\Box-1} = \{R^{-1}\}^{\Box}$ for every relation R.

Now, concerning closure, projection and involution operations for relators, we can also prove the following theorems:

Theorem 36. *The following assertions are equivalent:*

(1) $\Box$ *is an involution operation;*
(2) *for any two relators* $\mathcal{R}$ *and* $\mathcal{S}$ *on* X *to* Y, *we have*

$$\mathcal{R}^{\Box} \subseteq \mathcal{S} \iff R \subseteq \mathcal{S}^{\Box}.$$

Proof. For instance, if assertion (2) holds (i.e. $\Box$ and $\Box$ form a Galois connection), then for any relator $\mathcal{R}$ on X to Y

$$\mathcal{R}^{\Box} \subseteq \mathcal{R}^{\Box} \implies R \subseteq \mathcal{R}^{\Box\Box}, \mathcal{R}^{\Box\Box} \subseteq \mathcal{R} \implies \mathcal{R} = \mathcal{R}^{\Box\Box}.$$

Therefore, $\Box$ is involutive in the sense that $\Box \circ \Box$ is the identity operation Δ for relators.

Moreover, for any two relators $\mathcal{R}$ and $\mathcal{S}$ on X to Y,

$$\mathcal{R} \subseteq \mathcal{S} \implies R^{\Box\Box} \subseteq \mathcal{S}^{\Box\Box} \implies R^{\Box} \subseteq \mathcal{S}^{\Box\Box\Box} \implies R^{\Box} \subseteq \mathcal{S}^{\Box}.$$

Therefore, $\Box$ is increasing, too, and thus assertion (1) also holds

Theorem 37. *The following assertions are equivalent:*

(1) $\Box$ *is a closure operation;*
(2) *for any two relators* $\mathcal{R}$ *and* $\mathcal{S}$ *on* X *to* Y, *we have*

$$\mathcal{R} \subseteq \mathcal{S}^{\Box} \iff R^{\Box} \subseteq \mathcal{S}^{\Box}.$$

Remark 41. Now, instead of the equivalence of (1) and (2), it is more convenient to prove that assertion (1) is equivalent to the statement that there exists a structure $\mathfrak{F}$ for relators such that $\mathfrak{F}$ and $\Box$ form a Pataki connection in the sense that, for any two relators $\mathcal{R}$ and $\mathcal{S}$ on X to Y, we have $\mathfrak{F}_{\mathcal{R}} \subseteq \mathfrak{F}_{\mathcal{S}} \iff R \subseteq \mathcal{S}^{\Box}$.

Theorem 38. *If $\Box$ and $\Diamond$ are compatible closure (projection) operations, then $\Box\Diamond$ is also a closure (projection) operation.*

Proof. To prove that $\Box\Diamond$ is also idempotent, note that

$$(\Box\Diamond)(\Box\Diamond) = (\Diamond\Box)(\Box\Diamond) = \Diamond(\Box\Box)\Diamond$$
$$= \Diamond\Box\Diamond = (\Diamond\Box)\Diamond = (\Box\Diamond)\Diamond = \Box(\Diamond\Diamond) = \Box\Diamond.$$

Remark 42. In this respect, it is also worth noticing that the composition of two union-preserving operations for relators is also union-preserving.

It can be easily seen that the operations c, -1, ∞, ∂, and $*$ are union-preserving. However, the important closure operations $\#$, $\wedge$, and $\triangle$ are not union-preserving. Concerning them, we can only make use of the following:

Theorem 39. *If $\Box$ is a closure operation, then for any family $(\mathcal{R}_i)_{i\in I}$ of relators we have*

(1) $\bigcap_{i\in I}\mathcal{R}_i^{\Box} = \left(\bigcap_{i\in I}\mathcal{R}_i^{\Box}\right)^{\Box}$; (2) $\left(\bigcup_{i\in I}\mathcal{R}_i\right)^{\Box} = \left(\bigcup_{i\in I}\mathcal{R}_i^{\Box}\right)^{\Box}$.

Proof. To prove (1), note that if $\mathcal{R} = \bigcap_{i\in I}\mathcal{R}_i$, then for each $i \in I$ we have $\mathcal{R} \subseteq \mathcal{R}_i$, and hence also $\mathcal{R}^{\Box} \subseteq \mathcal{R}_i^{\Box}$. Therefore,

$$\left(\bigcap_{i\in I}\mathcal{R}_i\right)^{\Box} = \mathcal{R}^{\Box} \subseteq \bigcap_{i\in I}\mathcal{R}_i^{\Box}.$$

Hence, by taking $\mathcal{R}_i^{\Box}$ in place of $\mathcal{R}_i$, the converse inclusion can also be proved.

Remark 43. From assertion (2), we can see that $\left(\bigcup_{i\in I}\mathcal{R}_i\right)^{\Box} = \bigcup_{i\in I}\mathcal{R}_i^{\Box}$ if and only if the relator $\bigcup_{i\in I}\mathcal{R}_i^{\Box}$ is $\Box$–invariant.

While, from assertion (1), we can see that the relator $\bigcap_{i\in I}\mathcal{R}_i^{\Box}$ is always $\Box$–invariant. Moreover, if each $\mathcal{R}_i$ is $\Box$-invariant, then the relator $\bigcap_{i\in I}\mathcal{R}_i$ is also $\Box$–invariant.

Note that the proofs of the above three theorems can also be used to establish some useful statements on preclosure, semiclosure and modification operations.

12. Some Further Theorems on Unary Operations for Relators

In addition to Theorem 38, we can also easily prove the following:

Theorem 40. *If $\Box$ is a closure (projection) and $\diamond$ is an involution operation for relators, then $\diamond\Box\diamond$ is also a closure (projection) operation for relators.*

Proof. To prove that $\diamond\square\diamond$ is also idempotent, note that

$$\begin{aligned}(\diamond\square\diamond)(\diamond\square\diamond) &= (\diamond\square)\big((\diamond\diamond)(\square\diamond)\big)\\ &= (\diamond\square)\big(\Delta(\square\diamond)\big) = (\diamond\square)\big(\square\diamond\big) = \diamond\big((\square\square)\diamond\big) = \diamond(\square\diamond),\end{aligned}$$

where Δ is the identity operation for relators on X to Y.

Because of this theorem, we may also naturally introduce the following:

Definition 14. For the operation $\square$, we define

$$\textcircled{$\square$} = c\square c \qquad\qquad \text{and} \qquad\qquad \boxdot = -1\square-1.$$

Remark 44. Thus, by Theorem 40, for instance $\circledast$ is also a closure operation for relators.

This is also quite obvious from the fact that, for any relator $\mathcal{R}$ on X to Y, we have

$$\mathcal{R}^{\circledast} = \bigcup_{R\in\mathcal{R}} \mathcal{P}(R) = \big\{S\subseteq X\times Y: \quad \exists\, R\in\mathcal{R}: \quad S\subseteq R\big\}.$$

Namely, if for instance $S\in\mathcal{R}^{\circledast}$, then $S\in\mathcal{R}^{c*c}$, and thus $S^c\in\mathcal{R}^{c*}$. Therefore, there exists $R\in\mathcal{R}$ such that $R^c\subseteq S^c$. Hence, it follows that $S\subseteq R$, and thus $S\in\mathcal{P}(R)$. Therefore, $S\in\bigcup_{R\in\mathcal{R}}\mathcal{P}(R)$ also holds.

However, the importance of Definition 14 lies mainly in the following counterpart of Theorem 24:

Theorem 41. $\textcircled{\#}$ *and* $\textcircled{$\wedge$}$ *are closure operations for relators such that, for any two relators* $\mathcal{R}$ *and* $\mathcal{S}$ *on* X *to* Y,

(1) $\mathcal{S}\subseteq\mathcal{R}^{\textcircled{\#}} \iff \mathrm{Lb}_{\mathcal{S}}\subseteq\mathrm{Lb}_{\mathcal{R}} \iff Ub_{\mathcal{S}}\subseteq\mathrm{Ub}_{\mathcal{R}}$;
(2) $\mathcal{S}\subseteq\mathcal{R}^{\textcircled{$\wedge$}} \iff \mathrm{lb}_{\mathcal{S}}\subseteq\mathrm{lb}_{\mathcal{R}}$.

Proof. By the corresponding definitions and Theorems 24, 15 and 14, we have

$$\begin{aligned}\mathcal{S}\subseteq\mathcal{R}^{\textcircled{\#}} &\iff \mathcal{S}\subseteq\mathcal{R}^{c\#c} \iff \mathcal{S}^c\subseteq\mathcal{R}^{c\#} \iff \mathrm{Int}_{\mathcal{S}^c}\subseteq\mathrm{Int}_{\mathcal{R}^c}\\ &\iff \mathrm{Int}_{\mathcal{S}^c}\circ\mathcal{C}_Y\subseteq\mathrm{Int}_{\mathcal{R}^c}\circ\mathcal{C}_Y \iff \mathrm{Lb}_{\mathcal{S}}\subseteq\mathrm{Lb}_{\mathcal{R}}\\ &\iff \mathrm{Lb}_{\mathcal{S}}^{-1}\subseteq\mathrm{Lb}_{\mathcal{R}}^{-1} \iff Ub_{\mathcal{S}}\subseteq\mathrm{Ub}_{\mathcal{R}}.\end{aligned}$$

Therefore, assertion (1) is true. The proof of assertion (2) is quite similar.

Now, analogously to Corollary 1, we can also state the following:

Corollary 11. *For a relator $\mathcal{R}$ on X to Y, the following assertions are true:*

(1) $\mathcal{S} = \mathcal{R}^{\textcircled{\#}}$ *is the largest relator on X to Y such that* $\mathrm{Lb}_{\mathcal{S}} = \mathrm{Lb}_{\mathcal{R}}$ $(\mathrm{Ub}_{\mathcal{S}} = \mathrm{Ub}_{\mathcal{R}})$;
(2) $\mathcal{S} = \mathcal{R}^{\textcircled{\wedge}}$ *is the largest relator on X to Y such that* $\mathrm{lb}_{\mathcal{S}} = \mathrm{lb}_{\mathcal{R}}$.

Remark 45. Concerning the structure ub, by using Theorems 14 and 41, we can only note that

$$\mathrm{ub}_{\mathcal{S}} \subseteq \mathrm{ub}_{\mathcal{R}} \iff \mathrm{lb}_{\mathcal{S}^{-1}} \subseteq \mathrm{lb}_{\mathcal{R}^{-1}} \iff \mathrm{S}^{-1} \subseteq \mathcal{R}^{-1\textcircled{\wedge}}$$
$$\iff \mathrm{S} \subseteq \mathcal{R}^{-1\textcircled{\wedge}-1} \iff \mathrm{S} \subseteq \mathcal{R}^{\boxed{\textcircled{\wedge}}}$$

In this respect, it is also worth mentioning that, by using the associativity of composition and the inversion compatibility of c, we can also easily see that

$$\boxed{\textcircled{\wedge}} = -1\textcircled{\wedge} - 1 = -1c \wedge c - 1 = c - 1 \wedge -1c = c\boxed{\wedge}c = \textcircled{\boxed{\wedge}}.$$

13. Reflexive, Non-Partial and Non-Degenerated Relators

Definition 15. A relator $\mathcal{R}$ on X is called *reflexive* if each member R of $\mathcal{R}$ is a reflexive relation on X.

Remark 46. Thus, the following assertions are equivalent:

(1) $\mathcal{R}$ is reflexive;
(2) $x \in R(x)$ for all $x \in X$ and $R \in \mathcal{R}$;
(3) $A \subseteq R[A]$ for all $A \subseteq X$ and $R \in \mathcal{R}$.

The importance of reflexive relators is also apparent from the following two obvious theorems:

Theorem 42. *For a relator $\mathcal{R}$ on X, the following assertions are equivalent:*

(1) $\rho_{\mathcal{R}}$ *is reflexive;* (2) $\mathcal{R}$ *is reflexive;*
(3) $A \subseteq \mathrm{cl}_{\mathcal{R}}(A)\big(\mathrm{int}_{\mathcal{R}}(A) \subseteq A\big)$ *for all* $A \subseteq X$.

Proof. To see the equivalence of (1) and (2), recall that $\rho_{\mathcal{R}} = \left(\bigcap \mathcal{R}\right)^{-1}$.

Remark 47. Therefore, if $\mathcal{R}$ is a reflexive relator on X, then for any $A \subseteq X$ we have $A \in \mathcal{T}_{\mathcal{R}}$ $(A \in \mathcal{F}_{\mathcal{R}})$ if and only if $A = \mathrm{int}_{\mathcal{R}}(A)$ $(A = \mathrm{cl}_{\mathcal{R}}(A))$.

Theorem 43. *For a relator $\mathcal{R}$ on X, the following assertions are equivalent:*

(1) *$\mathcal{R}$ is reflexive;*
(2) *$A \in \operatorname{Int}_{\mathcal{R}}(B)$ implies $A \subseteq B$ for all $A, B \subseteq X$;*
(3) *$A \cap B \neq \emptyset$ implies $A \in \operatorname{Cl}_{\mathcal{R}}(B)$ for all $A, B \subseteq X$.*

Remark 48. In addition to the above two theorems, it is also worth mentioning that if $\mathcal{R}$ is a reflexive relator on X, then

(1) $\operatorname{Int}_{\mathcal{R}}$ is transitive;
(2) $B \in \operatorname{Cl}_{\mathcal{R}}(A)$ implies $\mathcal{P}(X) = \operatorname{Cl}_{\mathcal{R}}(A)^c \cup \operatorname{Cl}_{\mathcal{R}}^{-1}(B)$;
(3) $\operatorname{int}_{\mathcal{R}}(\operatorname{bor}_{\mathcal{R}}(A)) = \emptyset$ and $\operatorname{int}_{\mathcal{R}}(\operatorname{res}_{\mathcal{R}}(A)) = \emptyset$ for all $A \subseteq X$.

Thus, for instance, for any $A \subseteq X$ we have $\operatorname{res}_{\mathcal{R}}(A) \in \mathcal{T}_{\mathcal{R}}$ if and only if $A \in \mathcal{F}_{\mathcal{R}}$.

In contrast to the reflexivity property of a relator $\mathcal{R}$ on X, we may naturally introduce a great abundance of important symmetry and transitivity properties of $\mathcal{R}$ [45,59,60].

However, it is now more important to note that, analogously to Definition 15, we may also naturally introduce the following:

Definition 16. A relator $\mathcal{R}$ on X to Y is called *non-partial* if each member R of $\mathcal{R}$ is a non-partial relation on X to Y.

Remark 49. Thus, the following assertions are equivalent:

(1) $\mathcal{R}$ is non-partial;
(2) $R^{-1}[Y] = X$ for all $R \in \mathcal{R}$;
(3) $R(x) \neq \emptyset$ for all $x \in X$ and $R \in \mathcal{R}$.

The importance of non-partial relators is apparent from the following:

Theorem 44. *For a relator $\mathcal{R}$ on X to Y, the following assertions are equivalent:*

(1) *$\mathcal{R}$ is non-partial;*
(2) $\emptyset \notin \mathcal{E}_{\mathcal{R}}$; (3)/ $\mathcal{D}_{\mathcal{R}} \neq \emptyset$; (4) $Y \in \mathcal{D}_{\mathcal{R}}$; (5) $\mathcal{E}_{\mathcal{R}} \neq \mathcal{P}(Y)$.

Sometimes, we also need the following localized form of Definition 16.

Definition 17. A relator $\mathcal{R}$ on X is called *locally non-partial* if for each $x \in X$ there exists $R \in \mathcal{R}$ such that for any $y \in R(x)$ and $S \in \mathcal{R}$ we have $S(y) \neq \emptyset$.

Remark 50. Thus, if either $X = \emptyset$ or $\mathcal{R}$ is non-void and non-partial, then $\mathcal{R}$ is locally non-partial.

Moreover, by using the corresponding definitions, we can also easily prove

Theorem 45. *For a relator $\mathcal{R}$ on X, the following assertions are equivalent:*

(1) *$\mathcal{R}$ is locally non-partial;* (2) $X = \operatorname{int}_{\mathcal{R}}\big(\operatorname{cl}_{\mathcal{R}}(X)\big)$.

Proof. To prove the implication (1) $\Longrightarrow$ (2), note that if (1) holds, then for each $x \in X$ there exists $R \in \mathcal{R}$ such that for any $y \in R(x)$ and for any $S \in \mathcal{R}$ we have $S(y) \cap X = S(y) \neq \emptyset$, and thus $y \in \operatorname{cl}_{\mathcal{R}}(X)$.

Therefore, for each $x \in X$ there exists $R \in \mathcal{R}$ such that $R(x) \subseteq \operatorname{cl}_{\mathcal{R}}(X)$, and thus $x \in \operatorname{int}_{\mathcal{R}}\big(\operatorname{cl}_{\mathcal{R}}(X)\big)$. Hence, we can already see that $X \subseteq \operatorname{int}_{\mathcal{R}}\big(\operatorname{cl}_{\mathcal{R}}(X)\big)$, and thus (2) also holds.

Remark 51. Thus, the relator $\mathcal{R}$ is locally non-partial if and only if X is a *topologically regular open subset* of the relator space $X(\mathcal{R})$.

In addition to Definition 16, it is also worth introducing the following:

Definition 18. A relator $\mathcal{R}$ on X to Y is called *non-degenerated* if both $X \neq \emptyset$ and $\mathcal{R} \neq \emptyset$.

Thus, analogously to Theorem 44, we can also easily establish the following:

Theorem 46. *For a relator $\mathcal{R}$ on X to Y, the following assertions are equivalent:*

(1) *$\mathcal{R}$ is non-degenerated;*
(2) $\emptyset \notin \mathcal{D}_{\mathcal{R}}$; (3) $\mathcal{E}_{\mathcal{R}} \neq \emptyset$; (4)] $Y \in \mathcal{E}_{\mathcal{R}}$; (5) $\mathcal{D}_{\mathcal{R}} \neq \mathcal{P}(Y)$.

Remark 52. In addition to Theorems 44 and 46, it is also worth mentioning that if the relator $\mathcal{R}$ is *paratopologically simple* in the sense that $\mathcal{E}_{\mathcal{R}} = \mathcal{E}_R$ for some relation R on X to Y, then the stack $\mathcal{E}_{\mathcal{R}}$ has a base $\mathcal{B}$ with $\operatorname{card}(\mathcal{B}) \leq \operatorname{card}(X)$. (See Ref. [61, Theorem 5.9] of Pataki.)

The existence of a non-paratopologically simple (actually finite equivalence) relator, proved first by Pataki [61, Example 5.11], shows that in our definitions of the relations $\operatorname{Lim}_{\mathcal{R}}$ and $\operatorname{Adh}_{\mathcal{R}}$ we cannot restrict ourselves to functions of gosets (generalized ordered sets) without some loss of generality.

14. Topological and Quasi-Topological Relators

Notation 8. In this and the next section, we shall assume that $\mathcal{R}$ is a relator on X.

The following improvement of [59, Definition 2.1] was first considered in Ref. [45]. (See Ref. [62] for a subsequent treatment.)

Definition 19. The relator $\mathcal{R}$ is called:

(1) *quasi-topological* if $x \in \operatorname{int}_{\mathcal{R}}\big(\operatorname{int}_{\mathcal{R}}(R(x))\big)$ for all $x \in X$ and $R \in \mathcal{R}$;
(2) *topological* if for any $x \in X$ and $R \in \mathcal{R}$ there exists $V \in \mathcal{T}_{\mathcal{R}}$ such that $x \in V \subseteq R(x)$.

The appropriateness of this definition is already quite obvious from the following four theorems:

Theorem 47. *The following assertions are equivalent:*

(1) *$\mathcal{R}$ is quasi-topological;*
(2) $\operatorname{int}_{\mathcal{R}}\big(R(x)\big) \in \mathcal{T}_{\mathcal{R}}$ *for all* $x \in X$ *and* $R \in \mathcal{R}$;
(3) $\operatorname{cl}_{\mathcal{R}}(A) \in \mathcal{F}_{\mathcal{R}}\big(\operatorname{int}_{\mathcal{R}}(A) \in \mathcal{T}_{\mathcal{R}}\big)$ *for all* $A \subseteq X$.

Remark 53. Hence, we can see that the relator $\mathcal{R}$ is quasi-topological if and only if the super relation $\operatorname{cl}_{\mathcal{R}}$ is upper semi-idempotent ($\operatorname{int}_{\mathcal{R}}$ is lower semi-idempotent).

Theorem 48. *The following assertions are equivalent:*

(1) *$\mathcal{R}$ is topological;* (2) *$\mathcal{R}$ is reflexive and quasi-topological.*

Remark 54. By Theorem 47, the relator $\mathcal{R}$ may be called *weakly* (*strongly*) *quasi-topological* if $\rho_{\mathcal{R}}(x) \in \mathcal{F}_{\mathcal{R}}\big(R(x) \in \mathcal{T}_{\mathcal{R}}\big)$ for all $x \in X$ and $R \in \mathcal{R}$.

Moreover, by Theorem 48, the relator $\mathcal{R}$ may be called *weakly* (*strongly*) *topological* if it is reflexive and weakly (strongly) quasi-topological.

The following theorem shows that in a topological relator space $X(\mathcal{R})$, the relation $\operatorname{int}_{\mathcal{R}}$ and the family $\mathcal{T}_{\mathcal{R}}$ are equivalent tools.

Theorem 49. *The following assertions are equivalent:*

(1) *$\mathcal{R}$ is topological;*
(2) $\operatorname{int}_{\mathcal{R}}(A) = \bigcup \mathcal{T}_{\mathcal{R}} \cap \mathcal{P}(A)$ *for all* $A \subseteq X$;
(3) $\operatorname{cl}_{\mathcal{R}}(A) = \bigcap \mathcal{F}_{\mathcal{R}} \cap \mathcal{P}^{-1}(A)$ *for all* $A \subseteq X$.

Now, as an immediate consequence of Theorems 49 and 6, we can also state

Corollary 12. *If $\mathcal{R}$ is topological, then for any $A \subset X$, we have*

(1) $A \in \mathcal{E}_{\mathcal{R}}$ *if and only if there exists $V \in \mathcal{T}_{\mathcal{R}} \setminus \{\emptyset\}$ such that $V \subseteq A$;*
(2) $A \in \mathcal{D}_{\mathcal{R}}$ *if and only if for all $W \in \mathcal{F}_{\mathcal{R}} \setminus \{X\}$ we have $A \setminus W \neq \emptyset$.*

However, it is now more important to note that we can also prove the following:

Theorem 50. *The following assertions are equivalent:*

(1) *$\mathcal{R}$ is topological;*
(2) *$\mathcal{R}$ is topologically equivalent to $\mathcal{R}^{\wedge\infty}$;*
(3) *$\mathcal{R}$ is topologically equivalent to a preorder relator.*

Proof. To prove the implication (1) $\Longrightarrow$ (3), note that if (1) holds, then by Definition 19, for any $x \in X$ and $R \in \mathcal{R}$, there exists $V \in \mathcal{T}_{\mathcal{R}}$ such that $x \in V \subseteq R(x)$. Thus, by using the Pervin preorder relator

$$\mathcal{S} = \mathcal{R}_{\mathcal{T}_{\mathcal{R}}} = \big\{ R_V : V \in \mathcal{T}_{\mathcal{R}} \big\}, \qquad \text{where} \qquad R_V = V^2 \cup V^c \times X,$$

we can show that $\operatorname{int}_{\mathcal{R}}(A) = \operatorname{int}_{\mathcal{S}}(A)$ for all $A \subseteq X$, and thus $\operatorname{int}_{\mathcal{R}} = \operatorname{int}_{\mathcal{S}}$.

In addition to Theorem 50, it is also worth proving the following:

Theorem 51. *The following assertions are equivalent:*

(1) *$\mathcal{R}$ is quasi-topological;* (2) $\mathcal{R} \subseteq \big(\mathcal{R}^{\wedge} \circ \mathcal{R}\big)^{\wedge}$; (3) $\mathcal{R}^{\wedge} \subseteq \big(\mathcal{R}^{\wedge} \circ \mathcal{R}^{\wedge}\big)^{\wedge}$.

Remark 55. By [59], a relator $\mathcal{R}$ on X may be naturally called *topologically transitive* if, for each $x \in X$ and $R \in \mathcal{R}$ there exist $S, T \in \mathcal{R}$ such that $T[S(x)] \subseteq R(x)$.

This property can be reformulated in the concise form that $\mathcal{R} \subseteq \big(\mathcal{R} \circ \mathcal{R}\big)^{\wedge}$. Thus, the equivalence (1) and (3) can be expressed by saying that $\mathcal{R}$ is quasi-topological if and only if $\mathcal{R}^{\wedge}$ is topologically transitive.

In particular, we can easily prove the following:

Theorem 52. *For any $R \in \mathcal{R}$, the following assertions are equivalent:*

(1) *R is quasi-topological;* (2) *R is transitive.*

Hence, it is clear that, even more specially, we can also state

Corollary 13. *An $R \in \mathcal{R}$ is topological if and only if it is a preorder relation.*

Remark 56. Analogously to Definition 19, the relator $\mathcal{R}$ may be called *proximal* if for any $A \subseteq X$ and $R \in \mathcal{R}$ there exists $V \in \tau_{\mathcal{R}}$ such that $A \subseteq V \subseteq R[A]$.

Thus, in addition to the counterparts of Theorems 49 and 50, we can prove that $\mathcal{R}$ is topological if and only if its topological closure (refinement) $\mathcal{R}^{\wedge}$ is proximal.

15. The Pointwise Interior and Closure of Relations

Definition 20. If S is a relation on X, then for any $x \in X$ we define

$$S^{\circ}(x) = \operatorname{int}_{\mathcal{R}}\big(S(x)\big) \qquad \text{and} \qquad S^{-}(x) = \operatorname{cl}_{\mathcal{R}}\big(S(x)\big).$$

Remark 57. Thus, by Theorem 1, we have

$$S^{-}(x) = \operatorname{cl}_{\mathcal{R}}\big(S(x)\big) = \operatorname{int}_{\mathcal{R}}\big(S(x)^{c}\big)^{c} = \operatorname{int}_{\mathcal{R}}\big(S^{c}(x)\big)^{c} = S^{c\circ}(x)^{c} = S^{c\circ c}(x)$$

for all $x \in X$, and hence $S^{-} = S^{c\circ c}$.

Therefore, the properties of the relation S^{-} can, in principle, be immediately derived from those of S°.

The importance of the relation S° is apparent from the following:

Theorem 53. *For any relation S on X, the following assertions are true:*

(1) $S^{\circ}(x) \in \mathcal{T}_{\mathcal{R}^{\circ}}$ *for all* $x \in X$;
(2) *if $\mathcal{R}$ is reflexive, then* $S^{\circ} \subseteq S$;
(3) *if $S \in \mathcal{R}^{\wedge}$, then* S° *is reflexive*;
(4) *if $\mathcal{R}$ is quasi-topological, then* $S^{\circ}(x) \in \mathcal{T}_{\mathcal{R}}$ *for all* $x \in X$.

Proof. If $x \in X$ and $y \in S^{\circ}(x)$, then by the corresponding definitions there exists $R \in \mathcal{R}$ such that $R(y) \subseteq S(x)$. Hence, we can infer that

$$R^{\circ}(y) = \operatorname{int}_{\mathcal{R}}\big(R(y)\big) \subseteq \operatorname{int}_{\mathcal{R}}\big(S(x)\big) = S^{\circ}(x).$$

Hence, we can see that $y \in \operatorname{int}_{\mathcal{R}^{\circ}}\big(S^{\circ}(x)\big)$, and thus $S^{\circ}(x) \subseteq \operatorname{int}_{\mathcal{R}^{\circ}}\big(S^{\circ}(x)\big)$. Therefore, $S^{\circ}(x) \in \mathcal{T}_{\mathcal{R}^{\circ}}$, and thus assertion (1) is true.

The remaining assertions (2), (3) and (4) are immediate from Remark 24 and Theorems 42 and 47.

Remark 58. Now, in accordance with our former notations, we may also define

$$\mathcal{R}^\circ = \{R^\circ : R \in \mathcal{R}\} \qquad \text{and} \qquad \mathcal{R}^- = \{R^- : R \in \mathcal{R}\}.$$

Thus, by Remark 57, we can at once state that $\mathcal{R}^- = \mathcal{R}^{coc}$.

Therefore, the properties of the relator $\mathcal{R}^-$ can, in principle, be immediately derived from those of $\mathcal{R}^\circ$.

By using Theorem 53, in addition to Theorem 50, we can easily prove

Theorem 54. *$\mathcal{R}^\circ$ is a strongly topological relator on X, such that the following assertions are equivalent:*

(1) *$\mathcal{R}$ is topological;*
(2) *$\mathcal{R}$ and $\mathcal{R}^\circ$ are topologically equivalent;*
(3) *$\mathcal{R}$ is topologically equivalent to a strongly topological relator.*

Proof. If $R \in \mathcal{R}$, then $R \in \mathcal{R}^\wedge$, and thus by Theorem 53 the relation R° is reflexive. Therefore, the relator $\mathcal{R}^\circ$ is also reflexive. Moreover, if $x \in X$, then by Theorem 53 we have $R^\circ(x) \in \mathcal{T}_{\mathcal{R}^\circ}$. Thus, by Remark 54, the relator $\mathcal{R}^\circ$ is strongly topological.

Therefore, if (2) holds, then (3) also holds. Moreover, if (3) holds, then $\mathcal{R}$ is, in particular, topologically equivalent to a topological relator. Hence, by Definition 5 and Corollary 1, it is clear that (1) also holds.

Now, to complete the proof, it remains only to show that (1) also implies (2). For this, note that if (1) holds, then by Theorem 48 the relator $\mathcal{R}$ is reflexive and quasi-topological. Thus, in particular, by Theorem 53 we have $R^\circ \subseteq R$ for all $R \in \mathcal{R}$, and thus $\mathcal{R} \subseteq (\mathcal{R}^\circ)^* \subseteq (\mathcal{R}^\circ)^\wedge$.

Moreover, by Theorem 53 and the corresponding definitions, we have

$$x \in R^\circ(x) \subseteq \operatorname{int}_{\mathcal{R}}(R^\circ(x))$$

for all $x \in X$ and $R \in \mathcal{R}$. Hence, by Remark 24, we can see that $R^\circ \in \mathcal{R}^\wedge$ for all $R \in \mathcal{R}$, and thus $\mathcal{R}^\circ \subseteq \mathcal{R}^\wedge$. Now, since $\wedge$ is a closure operation, it is clear that $(\mathcal{R}^\circ)^\wedge = \mathcal{R}^\wedge$, and thus (2) also holds.

16. Composition-Compatible Unary Operations for Relators

Notation 9. In this section, we shall assume that $\square$ is a unary operation for relators.

Composition-compatibility properties of $\square$ have been first considered in Ref. [63] in somewhat different forms.

Definition 21. The operation $\square$ will be called

(1) *left composition-compatible*, if $\big(\mathcal{S} \circ \mathcal{R}\big)^{\square} = \big(\mathcal{S} \circ \mathcal{R}^{\square}\big)^{\square}$ for any two relators $\mathcal{R}$ on X to Y and $\mathcal{S}$ on Y to Z;
(2) *right composition-compatible*, if $\big(\mathcal{S} \circ \mathcal{R}\big)^{\square} = \big(\mathcal{S}^{\square} \circ \mathcal{R}\big)^{\square}$ for any two relators $\mathcal{R}$ on X to Y and $\mathcal{S}$ on Y to Z.

Remark 59. Now, the operation $\square$ may be naturally called *composition-compatible* if it is both left and right composition-compatible.

Note that, actually, this is also a very weak composition-compatibility property. However, by the next theorems, it will be sufficient for our subsequent purposes.

Theorem 55. *If $\square$ is left (right) composition-compatible, then $\square$ is, in a certain sense, idempotent.*

Proof. If $\square$ is left composition-compatible, then for any relator $\mathcal{R}$ on X to Y

$$\mathcal{R}^{\square\square} = \big(\mathcal{R}^{\square}\big)^{\square} = \Big(\{\Delta_Y\} \circ \mathcal{R}^{\square}\Big)^{\square} = \big(\{\Delta_Y\} \circ \mathcal{R}\big)^{\square} = \mathcal{R}^{\square}.$$

Theorem 56. *If $\square$ is composition-compatible, then for any two relators $\mathcal{R}$ on X to Y and $\mathcal{S}$ on Y to Z we have*

$$\big(\mathcal{S} \circ \mathcal{R}\big)^{\square} = \big(\mathcal{S}^{\square} \circ \mathcal{R}\big)^{\square} = \big(\mathcal{S} \circ \mathcal{R}^{\square}\big)^{\square} = \big(\mathcal{S}^{\square} \circ \mathcal{R}^{\square}\big)^{\square}.$$

Proof. Namely, for instance, we have $\big(\mathcal{S} \circ \mathcal{R}\big)^{\square} = \big(\mathcal{S} \circ \mathcal{R}^{\square}\big)^{\square} = \big(\mathcal{S}^{\square} \circ \mathcal{R}^{\square}\big)^{\square}$.

Theorem 57. *If $\square$ is composition-compatible, then for any three relators $\mathcal{R}$ on X to Y, $\mathcal{S}$ on Y to Z, and $\mathcal{T}$ on Z to W, we have*

$$\begin{aligned}\big(\mathcal{T} \circ \mathcal{S} \circ \mathcal{R}\big)^{\square} &= \big(\mathcal{T}^{\square} \circ \mathcal{S} \circ \mathcal{R}\big)^{\square}\\ &= \big(\mathcal{T} \circ \mathcal{S}^{\square} \circ \mathcal{R}\big)^{\square} = \big(\mathcal{T} \circ \mathcal{S} \circ \mathcal{R}^{\square}\big)^{\square} = \big(\mathcal{T}^{\square} \circ \mathcal{S}^{\square} \circ \mathcal{R}^{\square}\big)^{\square}.\end{aligned}$$

Proof. By using Theorem 56, for instance, we can see that

$$\begin{aligned}\big(\mathcal{T} \circ (\mathcal{S} \circ \mathcal{R})\big)^{\square} &= \big(\mathcal{T}^{\square} \circ (\mathcal{S} \circ \mathcal{R})^{\square}\big)^{\square}\\ &= \big(\mathcal{T}^{\square} \circ (\mathcal{S}^{\square} \circ \mathcal{R}^{\square})^{\square}\big)^{\square} = \big(\mathcal{T}^{\square} \circ (\mathcal{S}^{\square} \circ \mathcal{R}^{\square})\big)^{\square}.\end{aligned}$$

Theorem 58. *If $\square$ is a preclosure operation, then*

(1) *$\square$ is left composition-compatible if and only if $\left(\mathcal{S}\circ\mathcal{R}^{\square}\right)^{\square}\subseteq\left(\mathcal{S}\circ\mathcal{R}\right)^{\square}$ for any two relators $\mathcal{R}$ on X to Y and $\mathcal{S}$ on Y to Z;*

(2) *$\square$ is right composition-compatible if and only if $\left(\mathcal{S}^{\square}\circ\mathcal{R}\right)^{\square}\subseteq\left(\mathcal{S}\circ\mathcal{R}\right)^{\square}$ for any two relators $\mathcal{R}$ on X to Y and $\mathcal{S}$ on Y to Z.*

Proof. If $\mathcal{R}$ and S are as above, then we have $\mathcal{R}\subseteq\mathcal{R}^{\square}$, and thus $\mathcal{S}\circ\mathcal{R}\subseteq\mathcal{S}\circ\mathcal{R}^{\square}$, and thus $\left(\mathcal{S}\circ\mathcal{R}\right)^{\square}\subseteq\left(\mathcal{S}\circ\mathcal{R}^{\square}\right)^{\square}$.

Corollary 14. *If $\square$ is a closure operation, then*

(1) *$\square$ is left composition-compatible if and only if $\mathcal{S}\circ\mathcal{R}^{\square}\subseteq\left(\mathcal{S}\circ\mathcal{R}\right)^{\square}$ for any two relators $\mathcal{R}$ on X to Y and $\mathcal{S}$ on Y to Z;*

(2) *$\square$ is right composition-compatible if and only if $\mathcal{S}^{\square}\circ\mathcal{R}\subseteq\left(\mathcal{S}\circ\mathcal{R}\right)^{\square}$ for any two relators $\mathcal{R}$ on X to Y and $\mathcal{S}$ on Y to Z.*

Remark 60. In addition to the above results, it is also worth noticing that if $\square$ is an involution operation, then $\square$ is left composition-compatible if and only if $\mathcal{S}\circ\mathcal{R}=\mathcal{S}\circ\mathcal{R}^{\square}$ for any two relators $\mathcal{R}$ on X to Y and $\mathcal{S}$ on Y to Z.

Moreover, since $\mathcal{S}\circ\mathcal{R}=\bigcup_{S\in\mathcal{S}}S\circ\mathcal{R}$ holds, we can also at once state that if $\square$ is an involution operation, then $\square$ is left composition-compatible if and only if $S\circ\mathcal{R}=S\circ\mathcal{R}^{\square}$ for any relator $\mathcal{R}$ on X to Y and relation S on Y to Z.

17. Some Further Theorems on Composition-Compatible Operations

Now, by using Corollary 14 and Theorem 39, we can also prove the following:

Theorem 59. *If $\square$ is a closure operation, then*

(1) *$\square$ is left composition-compatible if and only if $S\circ\mathcal{R}^{\square}\subseteq\left(S\circ\mathcal{R}\right)^{\square}$ for any relator $\mathcal{R}$ on X to Y and relation S on Y to Z;*

(2) *$\square$ is right composition-compatible if and only if $\mathcal{S}^{\square}\circ R\subseteq\left(\mathcal{S}\circ R\right)^{\square}$ for any relation R on X to Y and relator $\mathcal{S}$ on Y to Z.*

Proof. If $\square$ is left composition-compatible, then by Corollary 14, for any relator $\mathcal{R}$ and relation S on Y to Z, we have $\{S\}\circ\mathcal{R}^{\square}\subseteq\left(\{S\}\circ\mathcal{R}\right)^{\square}$, and thus $S\circ\mathcal{R}^{\square}\subseteq\left(S\circ\mathcal{R}\right)^{\square}$. Therefore, the "only if part" of (1) is true.

Conversely, if $\mathcal{R}$ is a relator on X to Y and $\mathcal{S}$ is a relator on Y to Z, and the inclusion $S \circ \mathcal{R}^{\square} \subseteq (S \circ \mathcal{R})^{\square}$ holds for any relation S on Y to Z, then by using the corresponding definitions and Theorem 39 we can see that

$$\mathcal{S} \circ \mathcal{R}^{\square} = \bigcup_{S \in \mathcal{S}} S \circ \mathcal{R}^{\square} \subset \bigcup_{S \in \mathcal{S}} (S \circ \mathcal{R})^{\square}$$

$$\subseteq \left(\bigcup_{S \in \mathcal{S}} (S \circ \mathcal{R})^{\square} \right)^{\square} = \left(\bigcup_{S \in \mathcal{S}} S \circ \mathcal{R} \right)^{\square} = (\mathcal{S} \circ \mathcal{R})^{\square}.$$

Therefore, by Corollary 14, the "if part" of (1) is also true.

By using this theorem, we can somewhat more easily establish the composition compatibility properties of the basic closure operations considered in Section 8.

Theorem 60. *The operations $*$ and $\#$ are composition-compatible.*

Proof. To prove right composition compatibility of $\#$, by Theorem 59, it is enough to prove only that, for any relation R on X to Y and relator $\mathcal{S}$ on Y to Z, we have $\mathcal{S}^{\#} \circ R \subseteq (\mathcal{S} \circ R)^{\#}$.

For this, suppose that $W \in \mathcal{S}^{\#} \circ R$ and $A \subset X$. Then, there exists $V \in \mathcal{S}^{\#}$ such that $W = V \circ R$. Moreover, there exists $S \in \mathcal{S}$ such that $S\big[R[A]\big] \subseteq V\big[R[A]\big]$, and thus $(S \circ R)[A] \subseteq (V \circ R)[A] = W[A]$. Hence, by taking $U = S \circ R$, we can see that $U \in \mathcal{S} \circ R$ such that $U[A] \subseteq W[A]$. Therefore, $W \in (\mathcal{S} \circ R)^{\#}$ also holds.

Theorem 61. *The operations $\wedge$ and Δ are left composition-compatible.*

Proof. To prove left composition compatibility of Δ, by Theorem 59, it is enough to prove only that, for any relator $\mathcal{R}$ on X to Y and relation S on Y to Z, we have $S \circ \mathcal{R}^{\Delta} \subseteq (S \circ \mathcal{R})^{\Delta}$.

For this, suppose that $W \in S \circ \mathcal{R}^{\Delta}$ and $x \in X$. Then, there exists $V \in \mathcal{R}^{\Delta}$ such that $W = S \circ V$. Moreover, there exist $u \in X$ and $R \in \mathcal{R}$ such that $R(u) \subseteq V(x)$. Hence, we can infer that

$$(S \circ R)(u) = S\big[R(u)\big] \subseteq S\big[V(x)\big] = (S \circ V)(x) = W(x).$$

Now, by taking $U = S \circ R$, we can see that $U \in S \circ \mathcal{R}$ such that $U(u) \subset W(x)$. Therefore, $W \in (S \circ R)^{\Delta}$ also holds.

Instead of the right composition-compatibility of the operations $\wedge$ and Δ, we can only prove the following:

Theorem 62. *For any two relators $\mathcal{R}$ on X to Y and $\mathcal{S}$ on Y to Z, we have*

(1) $(\mathcal{S} \circ \mathcal{R})^{\wedge} = (\mathcal{S}^{*} \circ \mathcal{R})^{\wedge} = (\mathcal{S}^{\#} \circ \mathcal{R})^{\wedge}$;
(2) $(\mathcal{S} \circ \mathcal{R})^{\Delta} = (\mathcal{S}^{*} \circ \mathcal{R})^{\Delta} = (\mathcal{S}^{\#} \circ \mathcal{R})^{\Delta}$.

Proof. By using Theorems 25 and 60, for instance, we can at once see that

$$(\mathcal{S} \circ \mathcal{R})^{\wedge} = ((\mathcal{S} \circ \mathcal{R})^{\#})^{\wedge} = ((\mathcal{S}^{\#} \circ \mathcal{R})^{\#})^{\wedge} = (\mathcal{S}^{\#} \circ \mathcal{R})^{\wedge}.$$

Hence, it is clear that assertion (1) is true. Assertion (2) can be immediately derived from (1) by using that $\mathcal{U}^{\wedge\Delta} = \mathcal{U}^{\Delta}$ for any relator $\mathcal{U}$ on X to Z.

From this theorem, by using Theorem 61, we can immediately derive

Corollary 15. *For any two relators $\mathcal{R}$ on X to Y and $\mathcal{S}$ on Y to Z, we have*

(1) $(\mathcal{S} \circ \mathcal{R})^{\wedge} = (\mathcal{S}^{*} \circ \mathcal{R}^{\wedge})^{\wedge} = (\mathcal{S}^{\#} \circ \mathcal{R}^{\wedge})^{\wedge}$;
(2) $(\mathcal{S} \circ \mathcal{R})^{\Delta} = (\mathcal{S}^{*} \circ \mathcal{R}^{\Delta})^{\Delta} = (\mathcal{S}^{\#} \circ \mathcal{R}^{\Delta})^{\Delta}$.

Remark 61. By using Theorem 59, we can also somewhat more easily prove that the operation, considered in Remark 44, is also composition-compatible.

18. Quasi-Topologically Upper and Lower Continuous Relations

Notation 10. In this and the next nine sections, we shall assume that $\boldsymbol{F}$ is a relation on one relator space $\boldsymbol{X}(\mathcal{R})$ to another $\boldsymbol{Y}(\mathcal{S})$.

Thus, analogously to Definition 1, we may naturally introduce the following:

Definition 22. We shall say that relation

(1) F is *quasi-topologically upper continuous* if for each $x \in X$ and $V \in \mathcal{T}_{\mathcal{S}}$, with $F(x) \subseteq V$, there exists $U \in \mathcal{T}_{\mathcal{R}}$, with $x \in U$, such that for each $u \in U$ we have $F(u) \subseteq V$;
(2) F is *quasi-topologically lower continuous* if for each $x \in X$ and $V \in \mathcal{T}_{\mathcal{S}}$, with $F(x) \cap V \neq \emptyset$, there exists $U \in \mathcal{T}_{\mathcal{R}}$, with $x \in U$, such that for each $u \in U$ we have $F(u) \cap V \neq \emptyset$.

The above definition can be reformulated in the following more concise form.

Theorem 63. *The following assertions are true:*

(1) *F is quasi-topologically lower continuous if and only if for each $V \in \mathcal{T}_{\mathcal{S}}$ and $x \in F^{-1}[V]$ there exists $U \in \mathcal{T}_{\mathcal{R}}$ such that $x \in U \subseteq F^{-1}[V]$;*
(2) *F is quasi-topologically upper continuous if and only if for each $V \in \mathcal{T}_{\mathcal{S}}$ and $x \in F^{-1}[V^c]^c$ there exists $U \in \mathcal{T}_{\mathcal{R}}$ such that $x \in U \subseteq F^{-1}[V^c]^c$.*

Hint. For any $x \in X$ and $V \subseteq Y$, we have

$$F(x) \cap V \neq \emptyset \iff x \in F^{-1}[V].$$

Moreover, quite similarly, we also have

$$F(x) \subseteq V \iff F(x) \cap V^c = \emptyset \iff x \notin F^{-1}[V^c] \iff x \in F^{-1}[V^c]^c.$$

Thus, since $V \in \mathcal{T}_{\mathcal{S}}$ if and only if $V^c \in \mathcal{F}_{\mathcal{S}}$, we may also naturally introduce the following generalization of Definition 3:

Definition 23. We shall say that the relation

(1) F is *topological openness preserving* if $A \in \mathcal{T}_{\mathcal{R}}$ implies $F[A] \in \mathcal{T}_{\mathcal{S}}$;
(2) F is *topological closedness preserving* if $A \in \mathcal{F}_{\mathcal{R}}$ implies $F[A] \in \mathcal{F}_{\mathcal{S}}$.

Remark 62. Quite similarly, the relation F may, for instance, be naturally called *topological openness reversing* if $A \in \mathcal{T}_{\mathcal{R}}$ implies $F[A] \in \mathcal{F}_{\mathcal{S}}$.

Such relations were first investigated in Ref. [29], to put the notion of contra continuous functions of Dontchev [64] into a more general setting.

Now, by using Theorem 63, we can prove the following two theorems:

Theorem 64. *If $\mathcal{R}$ is topological, then the following assertions are equivalent:*

(1) *F is quasi-topologically lower continuous;*
(2) *F^{-1} is topological openness preserving.*

Proof. If assertion (1) holds, and moreover $V \in \mathcal{T}_{\mathcal{S}}$ and $x \in F^{-1}[V]$, then by Theorem 63 there exists $U \in \mathcal{T}_{\mathcal{R}}$ such that $x \in U \subseteq F^{-1}[V]$. Moreover, there exists $R \in \mathcal{R}$ such that $R(x) \subseteq U$. Therefore, $R(x) \subseteq F^{-1}[V]$. This shows that $F^{-1}[V] \in \mathcal{T}_{\mathcal{R}}$, and thus assertion (2) also holds.

Conversely, suppose now that assertion (2) holds, and moreover $V \in \mathcal{T}_{\mathcal{S}}$ and $x \in F^{-1}[V]$. Then, by assertion (2), we have $F^{-1}[V] \in \mathcal{T}_{\mathcal{R}}$. Thus, there exists $R \in \mathcal{R}$ such that $R(x) \subseteq F^{-1}[V]$. Moreover, since $\mathcal{R}$ is topological, there exists $U \in \mathcal{T}_{\mathcal{R}}$ such that $x \in U \subseteq R(x)$. Therefore, we also have $U \subseteq F^{-1}[V]$. Thus, by Theorem 63, assertion (1) also holds.

Theorem 65. *If $\mathcal{R}$ is topological, then the following assertions are equivalent:*

(1) *F is quasi-topologically upper continuous;*
(2) *F^{-1} is topological closedness preserving.*

Proof. Suppose that assertion (1) holds, and moreover $V \in \mathcal{F}_{\mathcal{S}}$ and $x \in F^{-1}[V]^c$. Then $V^c \in \mathcal{T}_{\mathcal{S}}$ and $x \in F^{-1}[(V^c)^c]^c$. Thus, by Theorem 63, there exists $U \in \mathcal{T}_{\mathcal{R}}$ such that $x \in U \subseteq F^{-1}[(V^c)^c]^c = F^{-1}[V]^c$. Moreover, there exists $R \in \mathcal{R}$ such that $R(x) \subseteq U$. Therefore, we also have $R(x) \subseteq F^{-1}[V]^c$. This, shows that $F^{-1}[V]^c \in \mathcal{T}_{\mathcal{R}}$, and thus $F^{-1}[V] \in \mathcal{F}_{\mathcal{R}}$. Therefore, assertion (2) also holds.

Conversely, suppose now that assertion (2) holds, and moreover $V \in \mathcal{T}_{\mathcal{S}}$ and $x \in F^{-1}[V^c]^c$. Then, we have $V^c \in \mathcal{F}_{\mathcal{S}}$. Hence, by assertion (2), we can see that $F^{-1}[V^c] \in \mathcal{F}_{\mathcal{R}}$, and thus $F^{-1}[V^c]^c \in \mathcal{T}_{\mathcal{R}}$. Thus, there exists $R \in \mathcal{R}$ such that $R(x) \subseteq F^{-1}[V^c]^c$. Moreover, since $\mathcal{R}$ is topological, there exists $U \in \mathcal{T}_{\mathcal{R}}$ such that $x \in U \subseteq R(x)$. Therefore, we also have $U \subseteq F^{-1}[V^c]^c$. Thus, by Theorem 63, assertion (1) also holds.

Remark 63. Note that the implications (1) $\Longrightarrow$ (2) in the above two theorems do not require the relator $\mathcal{R}$ to be topological.

19. Some Useful Reformulations of the Results of Section 18

Remark 64. By Theorem 1, we have

$$F^{-1}(V) = \mathrm{cl}_F(V) \qquad \text{and} \qquad F^{-1}[V^c]^c = \mathrm{cl}_F(V^c)^c = \mathrm{int}_F(V)$$

for all $V \subseteq Y$.

Therefore, as a useful reformulation of Theorem 63, we can at once state

Theorem 66. *The following assertions are true:*

(1) *F is quasi-topologically lower continuous if and only if for each $V \in \mathcal{T}_{\mathcal{S}}$ and $x \in \mathrm{cl}_F(V)$ there exists $U \in \mathcal{T}_{\mathcal{R}}$ such that $x \in U \subseteq \mathrm{cl}_F(V)$;*
(2) *F is quasi-topologically upper continuous if and only if for each $V \in \mathcal{T}_{\mathcal{S}}$ and $x \in \mathrm{int}_F(V)$ there exists $U \in \mathcal{T}_{\mathcal{R}}$ such that $x \in U \subseteq \mathrm{int}_F(V)$.*

Moreover, as some similar reformulations of the corresponding parts of Definition 23, we can also easily establish the following two theorems:

Theorem 67. *The following assertions are equivalent:*

(1) F^{-1} *is topologically openness preserving;*
(2) $\operatorname{cl}_F(V) \subseteq \operatorname{int}_{\mathcal{R}}\big(\operatorname{cl}_F(V)\big)$ *for all* $V \in \mathcal{T}_{\mathcal{S}}$;
(3) $\operatorname{cl}_{\mathcal{R}}\big(\operatorname{int}_F(V)\big) \subseteq \operatorname{int}_F(V)$ *for all* $V \in \mathcal{F}_{\mathcal{S}}$.

Theorem 68. *The following assertions are equivalent:*

(1) F^{-1} *is topologically closedness preserving;*
(2) $\operatorname{cl}_{\mathcal{R}}\big(\operatorname{cl}_F(V)\big) \subseteq \operatorname{cl}_F(V)$ *for all* $V \in \mathcal{F}_{\mathcal{S}}$;
(3) $\operatorname{int}_F(V) \subseteq \operatorname{int}_{\mathcal{R}}\big(\operatorname{int}_F(V)\big)$ *for all* $V \in \mathcal{T}_{\mathcal{S}}$.

Proof. By Definition 6 and Theorem 1, for any $V \subseteq Y$ we have

$$F^{-1}[V] \in \mathcal{F}_{\mathcal{R}} \iff \operatorname{cl}_{\mathcal{R}}\big(F^{-1}[V]\big) \subseteq F^{-1}[V] \iff \operatorname{cl}_{\mathcal{R}}\big(\operatorname{cl}_F(V)\big) \subseteq \operatorname{cl}_F(V).$$

Hence, by Definition 23, it is clear that assertions (1) and (2) are equivalent.

Moreover, by using Theorem 1, we can also see that

$$\begin{aligned} \operatorname{cl}_{\mathcal{R}}\big(\operatorname{cl}_F(V)\big) \subseteq \operatorname{cl}_F(V) &\iff \operatorname{cl}_F(V)^c \subseteq \operatorname{cl}_{\mathcal{R}}\big(\operatorname{cl}_F(V)\big)^c \\ &\iff \operatorname{cl}_F(V)^c \subseteq \operatorname{int}_{\mathcal{R}}\big(\operatorname{cl}_F(V)^c\big) \\ &\iff \operatorname{int}_F(V^c) \subseteq \operatorname{int}_{\mathcal{R}}\big(\operatorname{int}_F(V^c)\big). \end{aligned}$$

Hence, by Theorem 8, it is clear that assertions (2) and (3) are also equivalent.

Now, as some useful reformulations of Theorems 64 and 65, we can also at once state the following two theorems:

Theorem 69. *If* $\mathcal{R}$ *is topological, then the following assertions are equivalent:*

(1) F *is quasi-topologically lower continuous;*
(2) $\operatorname{cl}_F(V) \subseteq \operatorname{int}_{\mathcal{R}}\big(\operatorname{cl}_F(V)\big)$ *for all* $V \in \mathcal{T}_{\mathcal{S}}$;
(3) $\operatorname{cl}_{\mathcal{R}}\big(\operatorname{int}_F(V)\big) \subseteq \operatorname{int}_F(V)$ *for all* $V \in \mathcal{F}_{\mathcal{S}}$.

Theorem 70. *If* $\mathcal{R}$ *is topological, then the following assertions are equivalent:*

(1) F *is quasi-topologically lower continuous;*
(2) $\operatorname{cl}_{\mathcal{R}}\big(\operatorname{cl}_F(V)\big) \subseteq \operatorname{cl}_F(V)$ *for all* $V \in \mathcal{F}_{\mathcal{S}}$;
(3) $\operatorname{int}_F(V) \subseteq \operatorname{int}_{\mathcal{R}}\big(\operatorname{int}_F(V)\big)$ *for all* $V \in \mathcal{T}_{\mathcal{S}}$.

Remark 65. Note that the implications (1) $\Longrightarrow$ (2) $\Longleftrightarrow$ (3) in the above two theorems do not require the relator $\mathcal{R}$ to be topological.

20. Relational Reformulations of Quasi-Topological Upper Continuity

Theorem 71. *If both $\mathcal{R}$ and $\mathcal{S}$ are topological and $\mathcal{S} \neq \emptyset$, then the following assertions are equivalent:*

(1) *F is quasi-topologically upper continuous;*

(2) *for each $x \in X$ and $S \in \mathcal{S}^\wedge$ there exists $R \in \mathcal{R}$ such that*

$$F[R(x)] \subseteq S[F(x)];$$

(3) *for each $x \in X$ and $S \in \mathcal{S}^\wedge$ there exists $R \in \mathcal{R}$ such that*

$$(F \circ R)(x) \subseteq (\mathcal{S} \circ F)(x).$$

Proof. Suppose that assertion (1) holds, and moreover $x \in X$ and $S \in \mathcal{S}^\wedge$. Define

$$S^\circ(y) = \operatorname{int}_{\mathcal{S}}\big(S(y)\big)$$

for all $y \in Y$, and moreover $V = S^\circ[F(x)]$.

Then, by Theorem 53, the relation S° is reflexive, and thus

$$F(x) \subseteq S^\circ[F(x)] = V.$$

Moreover, by Theorem 53, we have $S^\circ(y) \in \mathcal{T}_{\mathcal{S}}$ for all $y \in Y$. Therefore,

$$V = S^\circ[F(x)] = \bigcup_{y \in F(x)} S^\circ(y) \in \mathcal{T}_{\mathcal{S}}.$$

Now, by using Definition 22, we can see that there exists $U \in \mathcal{T}_{\mathcal{R}}$, with $x \in U$, such that $F[U] \subseteq V$. Moreover, we can also note that there exists $R \in \mathcal{R}$ such that $R(x) \subseteq U$. Thus, in particular, we also have

$$F[R(x)] \subseteq V.$$

Moreover, by Theorem 53, we also have $\mathcal{S}^\circ \subseteq S$. Therefore,

$$V = S^\circ[F(x)] \subseteq S[F(x)].$$

Thus, in particular, we also have $F[R(x)] \subseteq S[F(x)]$. Consequently, assertion (2) also holds.

Conversely, suppose now that assertion (2) holds, and moreover $x \in X$ and $V \in \mathcal{T}_{\mathcal{S}}$ such that $F(x) \subseteq V$. Then, $V \subseteq \operatorname{int}_{\mathcal{S}}(V)$, and thus $F(x) \subseteq \operatorname{int}_{\mathcal{S}}(V)$. Hence, by using Theorem 27, we can infer that $F(x) \in \operatorname{Int}_{\mathcal{S}^\wedge}(V)$.

Therefore, there exists $S \in \mathcal{S}^\wedge$ such that $S[F(x)] \subseteq V$. Now, by assertion (2), we can state that there exists $R \in \mathcal{R}$ such that $F[R(x)] \subseteq S[F(x)]$, and thus also $F[R(x)] \subseteq V$. Moreover, since $\mathcal{R}$ is topological, there exists $U \in \mathcal{T}_{\mathcal{R}}$ such that $x \in U \subseteq R(x)$. Therefore, we also have $F[U] \subseteq V$. Thus, by Definition 22, assertion (1) also holds.

Theorem 72. *If both $\mathcal{R}$ and $\mathcal{S}$ are topological and $\mathcal{S} \neq \emptyset$, then the following assertions are equivalent:*

(1) *F is quasi-topologically upper continuous;*
(2) $\mathcal{S}^\wedge \circ F \subseteq (F \circ \mathcal{R})^\wedge$; (3) $(\mathcal{S}^\wedge \circ F)^\wedge \subseteq (F \circ \mathcal{R}^\wedge)^\wedge$.

Proof. By Theorem 71, Definition 11 and Theorem 37, it is clear that

$$(1) \iff (2) \iff (\mathcal{S}^\wedge \circ F)^\wedge \subseteq (F \circ \mathcal{R})^\wedge.$$

Moreover, from Theorem 61, we can see that

$$(F \circ \mathcal{R})^\wedge = (\{F\} \circ \mathcal{R})^\wedge = (\{F\} \circ \mathcal{R}^\wedge)^\wedge = (F \circ \mathcal{R}^\wedge)^\wedge.$$

Therefore, assertions (2) and (3) are also equivalent.

From the above two theorems, by using Theorem 65, we can immediately derive

Corollary 16. *If both $\mathcal{R}$ and $\mathcal{S}$ are topological and $\mathcal{R} \neq \emptyset$, then the following assertions are equivalent:*

(1) *F is topological closedness preserving;*
(2) *for each $y \in Y$ and $R \in \mathcal{R}^\wedge$ there exists $S \in \mathcal{S}$ such that*

$$F^{-1}[S(y)] \subseteq R[F^{-1}(y)];$$

(3) $\mathcal{R}^\wedge \circ F^{-1} \subseteq (F^{-1} \circ \mathcal{S})^\wedge$; (4) $(\mathcal{R}^\wedge \circ F^{-1})^\wedge \subseteq (F^{-1} \circ \mathcal{S}^\wedge)^\wedge$.

21. Relational Reformulations of Quasi-Topological Lower Continuity

Theorem 73. *If both $\mathcal{R}$ and $\mathcal{S}$ are topological and $\mathcal{R} \neq \emptyset$, then the following assertions are equivalent:*

(1) *F is quasi-topologically lower continuous;*
(2) *for each $y \in Y$ and $S \in \mathcal{S}$ there exists $R \in \mathcal{R}^\wedge$ such that*

$$R\left[F^{-1}(y)\right] \subseteq F^{-1}[S(y)];$$

(3) *for each* $y \in Y$ *and* $S \in \mathcal{S}$ *there exists* $R \in \mathcal{R}^\wedge$ *such that*

$$\big(R \circ F^{-1}\big)(y) \subseteq \big(F^{-1} \circ S\big)(y).$$

Proof. Suppose that assertion (1) holds, and moreover $y \in Y$ and $S \in \mathcal{S}$. Define

$$S^\circ(v) = \operatorname{int}_{\mathcal{S}}\big(S(v)\big),$$

for all $v \in Y$. Then, from Theorem 53, we can see that S° is a reflexive relation on Y such that $S^\circ \subseteq S$ and $S^\circ(y) \in \mathcal{T}_{\mathcal{S}}$. Hence, by using Theorem 64 and Corollary 2, we can infer that

$$F^{-1}[S^\circ(y)] \in \mathcal{T}_{\mathcal{R}} = \tau_{\mathcal{R}^\wedge}.$$

Thus, there exists $R \in \mathcal{R}^\wedge$ such that $R\left[F^{-1}[S^\circ(y)]\right] \subseteq F^{-1}[S^\circ(y)]$. Now, by using our former observations, we can also see that

$$R\left[F^{-1}(y)\right] \subseteq R\left[F^{-1}[S^\circ(y)]\right] \subseteq F^{-1}[S^\circ(y)] \subseteq F^{-1}[S(y)].$$

Therefore, assertion (2) also holds.

To prove the converse implication (2) $\Longrightarrow$ (1), by Theorem 64, it is enough to show only that if (2) holds, then F^{-1} is topological openness preserving. That is, if $V \in \mathcal{T}_{\mathcal{S}}$, then $F^{-1}[V] \in \mathcal{T}_{\mathcal{R}}$ also holds. For this, note that if $x \in F^{-1}[V]$, then there exists $y \in V$ such that $x \in F^{-1}(y)$. Thus, in particular, there exists $S \in \mathcal{S}$ such that $S(y) \subseteq V$. Moreover, if (2) holds, then there exists $R \in \mathcal{R}$ such that $R\left[F^{-1}(y)\right] \subseteq F^{-1}[S(y)]$. Hence, we can already see that

$$R(x) \subseteq R\left[F^{-1}(y)\right] \subseteq F^{-1}[S(y)] \subseteq F^{-1}[V].$$

Therefore, $F^{-1}[V] \in \mathcal{T}_{\mathcal{R}}$ also holds.

Now, analogously to the corresponding results of Section 20, we can also easily establish the following theorem and its corollary:

Theorem 74. *If both* $\mathcal{R}$ *and* $\mathcal{S}$ *are topological and* $\mathcal{R} \neq \emptyset$, *then the following assertions are equivalent:*

(1) F *is quasi-topologically lower continuous;*
(2) $F^{-1} \circ \mathcal{S} \subseteq \big(\mathcal{R}^\wedge \circ F^{-1}\big)^\wedge$; (3) $\big(F^{-1} \circ \mathcal{S}^\wedge\big)^\wedge \subseteq \big(\mathcal{R}^\wedge \circ F^{-1}\big)^\wedge$.

Corollary 17. *If both* $\mathcal{R}$ *and* $\mathcal{S}$ *are topological and* $\mathcal{S} \neq \emptyset$, *then the following assertions are equivalent:*

(1) F *is topological openness preserving;*
(2) *for each* $x \in X$ *and* $R \in \mathcal{R}$ *there exists* $S \in \mathcal{S}^\wedge$ *such that*

$$S\left[F(x)\right] \subseteq F[R(x)].$$

(3) $F \circ \mathcal{R} \subseteq \big(\mathcal{S}^\wedge \circ F\big)^\wedge$; (4) $\big(F \circ \mathcal{R}^\wedge\big)^\wedge \subseteq \big(\mathcal{S}^\wedge \circ F\big)^\wedge$.

22. Some Further Theorems on Quasi-Topological Upper and Lower Continuities

Theorem 75. *If both $\mathcal{R}$ and $\mathcal{S}$ are non-void, then the following assertions are equivalent:*

(1) *F is quasi-topologically upper (lower) continuous with respect to the relators $\mathcal{R}$ and $\mathcal{S}$;*
(2) *F is quasi-topologically upper (lower) continuous with respect to the relators $\mathcal{R}^{\wedge\infty}$ and $\mathcal{S}^{\wedge\infty}$.*

Proof. By Corollary 10, we have

$$\mathcal{T}_{\mathcal{R}} = \mathcal{T}_{\mathcal{R}^{\wedge\infty}} \qquad \text{and} \qquad \mathcal{T}_{\mathcal{S}} = \mathcal{T}_{\mathcal{S}^{\wedge\infty}}.$$

Hence, by Definition 22, it is clear that assertions (1) and (2) are equivalent.

Note that the preorder relators $\mathcal{R}^{\wedge\infty}$ and $\mathcal{S}^{\wedge\infty}$ are topological. Therefore, from the above theorem, by using Theorems 72 and 74, we can immediately derive the following two theorems.

Theorem 76. *If both $\mathcal{R}$ and $\mathcal{S}$ are non-void, then the following assertions are equivalent:*

(1) *F is quasi-topologically upper continuous;*
(2) $\mathcal{S}^{\wedge\infty\wedge} \circ F \subseteq \left(F \circ \mathcal{R}^{\wedge\infty}\right)^{\wedge}$; (3) $\left(\mathcal{S}^{\wedge\infty\wedge} \circ F\right)^{\wedge} \subseteq \left(F \circ \mathcal{R}^{\wedge\infty\wedge}\right)^{\wedge}$.

Remark 66. Note that, by Theorem 61, in assertion (2) we may also write $\mathcal{R}^{\wedge\infty\wedge}$ instead of $\mathcal{R}^{\wedge\infty}$. Moreover, if (1) holds, then we can also state that

$$\left(\mathcal{S}^{\wedge\infty\wedge} \circ F\right)^{\wedge\infty\wedge} \subseteq \left(F \circ \mathcal{R}^{\wedge\infty\wedge}\right)^{\wedge\infty\wedge}.$$

However, it is now more important to note that, from assertion (2) of Theorem 76, by using the inclusion $\mathcal{S}^{\wedge\infty} \subseteq \mathcal{S}^{\wedge\infty\wedge}$ and the corresponding properties of the operations $\wedge$ and ∞, we can also immediately derive the following:

Corollary 18. *If F is quasi-topologically upper continuous and both $\mathcal{R}$ and $\mathcal{S}$ are non-void, then*

(1) $\mathcal{S}^{\wedge\infty} \circ F \subseteq \left(F \circ \mathcal{R}^{\wedge\infty}\right)^{\wedge}$; (2) $\left(\mathcal{S}^{\wedge\infty} \circ F\right)^{\wedge\infty} \subseteq \left(F \circ \mathcal{R}^{\wedge\infty}\right)^{\wedge\infty}$.

Remark 67. Unfortunately, the fundamental inclusions (1) and (2) seem not to imply the quasi-topological upper continuity of F.

Theorem 77. *If both $\mathcal{R}$ and $\mathcal{S}$ are non-void, then the following assertions are equivalent:*

(1) *F is quasi-topologically lower continuous;*
(2) $F^{-1} \circ \mathcal{S}^{\wedge\infty} \subseteq \left(\mathcal{R}^{\wedge\infty\wedge} \circ F^{-1}\right)^{\wedge}$;
(3) $\left(F^{-1} \circ \mathcal{S}^{\wedge\infty\wedge}\right)^{\wedge} \subseteq \left(\mathcal{R}^{\wedge\infty\wedge} \circ F^{-1}\right)^{\wedge}$.

Remark 68. Thus, if (1) holds, then

$$\left(F^{-1} \circ \mathcal{S}^{\wedge\infty\wedge}\right)^{\wedge\infty\wedge} \subseteq \left(\mathcal{R}^{\wedge\infty\wedge} \circ F^{-1}\right)^{\wedge\infty\wedge}$$

also holds.

However, assertion (1) seems not to be equivalent to the fundamental inclusions

$$F^{-1} \circ \mathcal{S}^{\wedge\infty} \subseteq \left(\mathcal{R}^{\wedge\infty} \circ F^{-1}\right)^{\wedge} \quad \text{and} \quad \left(F^{-1} \circ \mathcal{S}^{\wedge\infty}\right)^{\wedge\infty} \subseteq \left(\mathcal{R}^{\wedge\infty} \circ F^{-1}\right)^{\wedge\infty}.$$

23. Quasi-Topologically Mixed Upper and Lower Continuous Relations

Now, as a straightforward generalization of Definition 2, we may also naturally introduce the following:

Definition 24. We shall say that the relation

(1) F is *quasi-topologically mixed upper continuous* if for each $x \in X$ and $V \in \mathcal{T}_{\mathcal{S}}$, with $F(x) \subseteq V$, there exists $U \in \mathcal{T}_{\mathcal{R}}$, with $x \in U$, such that for each $u \in U$ we have $F(u) \cap V \neq \emptyset$;
(2) F is *quasi-topologically mixed lower continuous* if for each $x \in X$ and $V \in \mathcal{T}_{\mathcal{S}}$, with $F(x) \cap V \neq \emptyset$, there exists $U \in \mathcal{T}_{\mathcal{R}}$, with $x \in U$, such that for each $u \in U$ we have $F(u) \subseteq V$.

Thus, analogously to the results of Sections 18 and 19, we can prove the following theorems:

Theorem 78. *The following assertions are true:*

(1) *F is quasi-topologically mixed lower continuous if and only if for each $V \in \mathcal{T}_{\mathcal{S}}$ and $x \in \operatorname{cl}_F(V)$ there exists $U \in \mathcal{T}_{\mathcal{R}}$ such that $x \in U \subseteq \operatorname{int}_F(V)$;*
(2) *F is quasi-topologically mixed upper continuous if and only if for each $V \in \mathcal{T}_{\mathcal{S}}$ and $x \in \operatorname{int}_F(V)$ there exists $U \in \mathcal{T}_{\mathcal{R}}$ such that $x \in U \subseteq \operatorname{cl}_F(V)$.*

Theorem 79. *If $\mathcal{R}$ is topological, then the following assertions are equivalent:*

(1) *F is quasi-topologically mixed lower continuous;*
(2) $\operatorname{cl}_F(V) \subseteq \operatorname{int}_{\mathcal{R}}\big(\operatorname{int}_F(V)\big)$ *for all* $V \in \mathcal{T}_{\mathcal{S}}$;
(3) $\operatorname{cl}_{\mathcal{R}}\big(\operatorname{cl}_F(V)\big) \subseteq \operatorname{int}_F(V)$ *for all* $V \in \mathcal{F}_{\mathcal{S}}$.

Proof. Suppose that assertion (1) holds and $V \in \mathcal{T}_{\mathcal{S}}$. Then, by Theorem 78, for each $x \in \operatorname{cl}_F(V)$, there exists $U \in \mathcal{T}_{\mathcal{R}}$ such that $x \in U \subseteq \operatorname{int}_F(V)$. Moreover, there exists $R \in \mathcal{R}$ such that $R(x) \subseteq U$. Therefore, $R(x) \subseteq \operatorname{int}_F(V)$, and thus $x \in \operatorname{int}_{\mathcal{R}}\big(\operatorname{int}_F(V)\big)$. This shows that

$$\operatorname{cl}_F(V) \subseteq \operatorname{int}_{\mathcal{R}}\big(\operatorname{int}_F(V)\big),$$

and thus assertion (2) also holds.

Conversely, suppose now that assertion (2) holds, and moreover $V \in \mathcal{T}_{\mathcal{S}}$ and $x \in \operatorname{cl}_F(V)$. Then, by using assertion (2), we can see that

$$x \in \operatorname{cl}_F(V) \subseteq \operatorname{int}_{\mathcal{R}}\big(\operatorname{int}_F(V)\big).$$

Therefore, there exists $R \in \mathcal{R}$ such that $R(x) \subseteq \operatorname{int}_F(V)$. Moreover, since $\mathcal{R}$ is topological, there exists $U \in \mathcal{T}_{\mathcal{R}}$ such that $x \in U \subseteq R(x)$. Therefore, we also have $U \subseteq \operatorname{int}_F(V)$. Thus, by Theorem 78, assertion (1) also holds.

Thus, we have proved that assertions (1) and (2) are equivalent. Moreover, by using Theorems 1 and 8, we can easily see that assertions (2) and (3) are also equivalent.

Theorem 80. *If $\mathcal{R}$ is topological, then the following assertions are equivalent:*

(1) *F is quasi-topologically mixed upper continuous;*
(2) $\operatorname{int}_F(V) \subseteq \operatorname{int}_{\mathcal{R}}\big(\operatorname{cl}_F(V)\big)$ *for all* $V \in \mathcal{T}_{\mathcal{S}}$;
(3) $\operatorname{cl}_{\mathcal{R}}\big(\operatorname{int}_F(V)\big) \subseteq \operatorname{cl}_F(V)$ *for all* $V \in \mathcal{F}_{\mathcal{S}}$.

Proof. Suppose that assertion (1) holds and $V \in \mathcal{T}_{\mathcal{S}}$. Then, by Theorem 78, for each $x \in \operatorname{int}_F(V)$ there exists $U \in \mathcal{T}_{\mathcal{R}}$ such that $x \in U \subseteq \operatorname{cl}_F(V)$. Moreover, there exists $R \in \mathcal{R}$ such that $R(x) \subseteq U$. Therefore, we also have $R(x) \subseteq \operatorname{cl}_F(V)$, and thus $x \in \operatorname{int}_{\mathcal{R}}\big(\operatorname{cl}_F(V)\big)$. This shows that

$$\operatorname{int}_F(V) \subseteq \operatorname{int}_{\mathcal{R}}\big(\operatorname{cl}_F(V)\big),$$

and thus assertion (2) also holds.

Conversely, suppose now that assertion (2) holds, and moreover $V \in \mathcal{T}_{\mathcal{S}}$ and $x \in \operatorname{int}_F(V)$. Then, by using assertion (2), we can see that

$$x \in \operatorname{int}_F(V) \subseteq \operatorname{int}_{\mathcal{R}}\big(\operatorname{cl}_F(V)\big).$$

Therefore, there exists $R \in \mathcal{R}$ such that $R(x) \subseteq \mathrm{cl}_F(V)$. Moreover, since $\mathcal{R}$ is topological, there exists $U \in \mathcal{T}_\mathcal{R}$ such that $x \in U \subseteq R(x)$. Therefore, we also have $U \subseteq \mathrm{cl}_F(V)$. Thus, by Theorem 78, assertion (1) also holds.

Thus, we have proved that assertions (1) and (2) are equivalent. Moreover, by using Theorems 1 and 8, we can easily see that assertions (2) and (3) are also equivalent.

Remark 69. Note that the implications (1) $\Longrightarrow$ (2) $\Longleftrightarrow$ (3) in the above two theorems do not require the relator $\mathcal{R}$ to be topological.

Moreover, Theorem 80 is a straightforward generalization of [1, Propositions 2 and 3] of Thangavelu *et al.*

24. Relationships Among the Former Four Continuity Properties

Remark 70. By using Definitions 5 and 16, we can easily see that the following assertions are equivalent:

(1) F is non-partial;
(2) $\mathrm{int}_F(V) \subseteq \mathrm{cl}_F(V)$ for all $V \subseteq Y$.

To prove the implication (2) $\Longrightarrow$ (1), note that if assertion (1) does not hold, then there exists $x \in X$ such that $F(x) = \emptyset$. Therefore, $F(x) = \emptyset \subseteq V$, and thus $x \in \mathrm{int}_F(V)$ for all $V \subseteq Y$. However, $F(x) \cap V = \emptyset \cap V = \emptyset$, and thus $x \notin \mathrm{cl}_F(V)$ for all $V \subseteq Y$. Thus, in particular, assertion (2) does not also hold.

Now, by using the above remark and Theorems 66 and 78, we can easily prove the following generalizations of the implications summarized in Ref. [1, Diagram 1] of Thangavelu *et al.*

Theorem 81. *If F is non-partial and quasi-topologically lower continuous, then F is also quasi-topologically mixed upper continuous.*

Proof. Suppose that $V \in \mathcal{T}_\mathcal{S}$ and $x \in \mathrm{int}_F(V)$. Then, by Remark 70, we also have $x \in \mathrm{cl}_F(V)$. Thus, by Theorem 66, there exists $U \in \mathcal{T}_\mathcal{R}$ such that $x \in U \subseteq \mathrm{cl}_F(V)$. Therefore, by Theorem 78, F is quasi-topologically mixed upper continuous.

Theorem 82. *If F is non-partial and quasi-topologically upper continuous, then F is also quasi-topologically mixed upper continuous.*

Proof. Supposed that $V \in \mathcal{T}_{\mathcal{S}}$ and $x \in \operatorname{int}_F(V)$. Then, by Theorem 66, there exists $U \in \mathcal{T}_{\mathcal{R}}$ such that $x \in U \subseteq \operatorname{int}_F(V)$. Hence, by using Remark 70, we can infer that $U \subseteq \operatorname{cl}_F(V)$. Therefore, by Theorem 78, F is quasi-topologically mixed upper continuous.

Theorem 83. *If F is non-partial and quasi-topologically mixed lower continuous, then F is also quasi-topologically upper continuous.*

Proof. Supposed that $V \in \mathcal{T}_{\mathcal{S}}$ and $x \in \operatorname{int}_F(V)$. Then, by Remark 70, we also have $x \in \operatorname{cl}_F(V)$. Thus, by Theorem 78, there exists $U \in \mathcal{T}_{\mathcal{R}}$ such that $x \in U \subseteq \operatorname{int}_F(V)$. Therefore, by Theorem 66, F is quasi-topologically upper continuous.

Now, as an immediate consequence of the latter two theorems, we can also state

Corollary 19. *If F is non-partial and quasi-topologically mixed lower continuous, then F is also quasi-topologically mixed upper continuous.*

Remark 71. Note that if in particular $\mathcal{R}$ is topological, then the above statements can also be derived from Theorems 69, 70, 79 and 80 by using Remark 70.

Now, for an easy illustration of our former Definitions 22 and 24, we can use the following:

Example 3. Define $X = \{1, 2\}$,

$$R(1) = \{1\}, \qquad R(2) = X \qquad \text{and} \qquad F(1) = X, \qquad F(2) = \{1\}.$$

Then, R is a preorder relation on X and F is a symmetric, non-partial relation on $X(R)$ such that:

(1) F is quasi-topologically lower continuous;
(2) F is not quasi-topologically upper continuous;
(3) F is quasi-topologically mixed upper continuous;
(4) F is not quasi-topologically mixed lower continuous.

To check the above assertions, it is convenient to note that

$$R = \{1\}^2 \cup \big((X \setminus \{1\}) \times X\big)$$

is just the Pervin relation associated with the subset $\{1\}$ of X. Therefore, R is a preorder relation on X, and thus $\{R\}$ is a topological relator on X. Moreover, we have

$$\mathcal{T}_R = \{\emptyset, \{1\}, X\}, \qquad \text{and thus} \qquad \mathcal{F}_R = \{\emptyset, \{2\}, X\}.$$

On the other hand, we can also easily note that F is non-partial, $F^{-1} = F$,

$$F^{-1}[\{1\}] = F(1) = X \in \mathcal{T}_R, \qquad \text{but} \qquad F^{-1}[\{2\}] = F(2) = \{1\} \notin \mathcal{F}_R.$$

Thus, the relation F^{-1} is openness-preserving, but not closedness-preserving. Therefore, by Theorems 64 and 65, assertions (1) and (2) are true. Hence, by Theorems 81 and 83, we can see that assertions (3) and (4) are also true.

25. Relational Reformulations of Quasi-Topological Mixed Upper Continuity

Theorem 84. *If both $\mathcal{R}$ and $\mathcal{S}$ are topological and $\mathcal{S} \neq \emptyset$, then the following assertions are equivalent:*

(1) *F is quasi-topologically mixed upper continuous;*
(2) *For each $x \in X$ and $S \in \mathcal{S}^\wedge$ there exists $R \in \mathcal{R}$ such that*

$$R(x) \subseteq F^{-1}\big[S[F(x)]\big];$$

(3) *For each $x \in X$ and $S \in \mathcal{S}^\wedge$ there exists $R \in \mathcal{R}$ such that*

$$R(x) \subseteq \big(F^{-1} \circ S \circ F\big)(x).$$

Proof. Suppose that assertion (1) holds, and moreover $x \in X$ and $S \in \mathcal{S}^\wedge$. Define

$$S^\circ(y) = \operatorname{int}_{\mathcal{S}}\big(S(y)\big)$$

for all $y \in Y$, and moreover $V = S^\circ[F(x)]$.

Then, by the proof of Theorem 71, we can state that

$$F(x) \subseteq S^\circ[F(x)] = V \in \mathcal{T}_{\mathcal{S}} \qquad \text{and} \qquad V = S^\circ[F(x)] \subseteq S[F(x)].$$

Hence, by using Definition 24, we can see that there exists $U \in \mathcal{T}_{\mathcal{R}}$, with $x \in U$, such that for all $u \in U$ we have $F(u) \cap V \neq \emptyset$, and thus $u \in F^{-1}[V]$. Therefore, $U \subseteq F^{-1}[V]$. Moreover, we can also note that there exists $R \in \mathcal{R}$ such that $R(x) \subseteq U$. Thus, in particular, we also have

$$R(x) \subseteq U \subseteq F^{-1}[V] \subseteq F^{-1}\big[S[F(x)]\big].$$

Therefore, assertion (2) also holds.

Conversely, suppose now that assertion (2) holds, and moreover $x \in X$ and $V \in \mathcal{T}_{\mathcal{S}}$ such that $F(x) \subseteq V$. Then, $V \subseteq \operatorname{int}_{\mathcal{S}}(V)$, and thus $F(x) \subseteq \operatorname{int}_{\mathcal{S}}(V)$. Hence, by using Theorem 27, we can infer that $F(x) \in \operatorname{Int}_{\mathcal{S}^\wedge}(V)$. Therefore, there exists $S \in \mathcal{S}^\wedge$ such that $S[F(x)] \subseteq V$. Now, by assertion

(2), we can state that there exists $R \in \mathcal{R}$ such that $R(x) \subseteq F^{-1}\big[S[F(x)]\big]$. Moreover, since $\mathcal{R}$ is topological, there exists $U \in \mathcal{T}_{\mathcal{R}}$ such that $x \in U \subseteq R(x)$. Therefore, we also have

$$U \subseteq R(x) \subseteq F^{-1}\big[S[F(x)]\big] \subseteq F^{-1}[V],$$

and thus $F(u) \cap V \neq \emptyset$ for all $u \in U$. Hence, by Definition 24, we can see that assertion (1) also holds.

Remark 72. Assertion (3) can also be reformulated in the form that for each $x \in X$ and $S \in \mathcal{S}^{\wedge}$ there exists $R \in \mathcal{R}$ such that

$$R(x) \subseteq \big(F \boxtimes F\big)^{-1}[S](x).$$

That is,

$$\mathrm{cl}_{F \boxtimes F}(S) = \big(F \boxtimes F\big)^{-1}[S] \in \mathcal{R}^{\wedge}$$

for all $S \in \mathcal{S}^{\wedge}$.

Now, analogously to Theorem 72, we can also easily establish the following:

Theorem 85. *If both $\mathcal{R}$ and $\mathcal{S}$ are topological and $\mathcal{S} \neq \emptyset$, then the following assertions are equivalent:*

(1) *F is quasi-topologically mixed upper continuous;*
(2) $F^{-1} \circ \mathcal{S}^{\wedge} \circ F \subseteq \mathcal{R}^{\wedge}$; (3) $\big(F^{-1} \circ \mathcal{S}^{\wedge} \circ F\big)^{\wedge} \subseteq \mathcal{R}^{\wedge}$.

26. Relational Reformulations of Quasi-Topological Mixed Lower Continuity

Theorem 86. *If both $\mathcal{R}$ and $\mathcal{S}$ are topological and $\mathcal{S} \neq \emptyset$, then the following assertions are equivalent:*

(1) *F is quasi-topologically mixed lower continuous;*
(2) *for each $y \in Y$ and $S \in \mathcal{S}$ there exists $R \in \mathcal{R}^{\wedge}$ such that*

$$R\big[F^{-1}(y)\big] \subseteq F^{-1}\big[S(y)^c\big]^c;$$

(3) *for each $y \in Y$ and $S \in \mathcal{S}$ there exists $R \in \mathcal{R}^{\wedge}$ such that*

$$\big(R \circ F^{-1}\big)(y) \subseteq \big(F^{-1} \circ S^c\big)^c(y).$$

Proof. Suppose that assertion (1) holds, and moreover $y \in Y$ and $S \in \mathcal{S}$. Define

$$S^\circ(v) = \operatorname{int}_{\mathcal{S}}\big(S(v)\big) \tag{1}$$

for all $v \in Y$. Then, from Theorem 53, we can see that S° is a reflexive relation on Y such that $S^\circ \subseteq S$ and $S^\circ(y) \in \mathcal{T}_{\mathcal{S}}$. Hence, by using Theorem 79 and Remark 64, we can infer that

$$F^{-1}[S^\circ(y)] \subseteq \operatorname{int}_{\mathcal{R}}\big(F^{-1}[S^\circ(y)^c]^c\big).$$

Thus, by Theorem 27, we can also state that $F^{-1}[S^\circ(y)] \in \operatorname{Int}_{\mathcal{R}^\wedge}\big(F^{-1}[S^\circ(y)^c]^c\big)$. Therefore, there exists $R \in \mathcal{R}^\wedge$ such that

$$R\big[F^{-1}[S^\circ(y)]\big] \subseteq F^{-1}[S^\circ(y)^c]^c.$$

Hence, by using the inclusions $y \in S^\circ(y)$ and $S^\circ(y) \subseteq S(y)$, and the implication

$$\begin{aligned} S^\circ(y) \subseteq S(y) &\implies S(y)^c \subseteq S^\circ(y)^c \\ &\implies F^{-1}[S(y)^c] \subseteq F^{-1}[S^\circ(y)^c] \\ &\implies F^{-1}[S^\circ(y)^c]^c \subseteq F^{-1}[S(y)^c]^c, \end{aligned}$$

we can already see that

$$R\big[F^{-1}(y)\big] \subseteq R\big[F^{-1}[S^\circ(y)]\big] \subseteq F^{-1}[S^\circ(y)^c]^c \subseteq F^{-1}[S(y)^c]^c.$$

Therefore, assertion (2) also holds.

To prove the converse implication (2) $\implies$ (1), by Theorem 79 and Remark 64, it is enough to prove that if (2) holds, then for any $V \in \mathcal{T}_{\mathcal{S}}$ we have $F^{-1}[V] \subseteq \operatorname{int}_{\mathcal{R}}\big(F^{-1}[V^c]^c\big)$. For this, note that if $x \in F^{-1}[V]$, then there exists $y \in V$ such that $x \in F^{-1}(y)$. Moreover, there exists $S \in \mathcal{S}$ such that $S(y) \subseteq V$. Hence, by using that

$$S(y) \subseteq V \implies V^c \subseteq S(y)^c \implies F^{-1}[V^c] \subseteq F^{-1}[S(y)^c],$$

we can infer that

$$F^{-1}[S(y)^c]^c \subseteq F^{-1}[V^c]^c.$$

Moreover, by (2), we can state that there exists $R \in \mathcal{R}^\wedge$ such that

$$R\big[F^{-1}(y)\big] \subseteq F^{-1}[S(y)^c]^c.$$

Therefore, we also have

$$R(x) \subseteq R\big[F^{-1}(y)\big] \subseteq F^{-1}[S(y)^c]^c \subseteq F^{-1}[V^c]^c.$$

Hence, by using Corollary 1, we can already see that

$$x \in \operatorname{int}_{\mathcal{R}^\wedge}\big(F^{-1}[V^c]^c\big) = \operatorname{int}_{\mathcal{R}}\big(F^{-1}[V^c]^c\big).$$

Therefore, $F^{-1}[V] \subseteq \operatorname{int}_{\mathcal{R}}\big(F^{-1}[V^c]^c\big)$ also holds.

Now, analogously to Theorem 72, we can also easily establish

Theorem 87. *If both $\mathcal{R}$ and $\mathcal{S}$ are topological and $\mathcal{S} \neq \emptyset$, then the following assertions are equivalent:*

(1) *F is quasi-topologically mixed lower continuous;*
(2) $\big(F^{-1} \circ \mathcal{S}^c\big)^c \subseteq \big(\mathcal{R}^\wedge \circ F^{-1}\big)^\wedge$; (3) $\big(F^{-1} \circ \mathcal{S}^{\wedge c}\big)^{c\wedge} \subseteq \big(\mathcal{R}^\wedge \circ F^{-1}\big)^\wedge$.

Proof. From Theorem 86, by the definition of the operation $\wedge$, it is clear that assertions (1) and (2) are equivalent. Moreover, by Theorem 37, it is clear that inclusion (2) is equivalent to the inclusion

(a) $\big(F^{-1} \circ \mathcal{S}^c\big)^{c\wedge} \subseteq \big(\mathcal{R}^\wedge \circ F^{-1}\big)^\wedge$.

Therefore, assertion (1) is equivalent to inclusion (a), too.

Furthermore, from Corollary 8.6, we can see that $\mathcal{T}_{\mathcal{S}^\wedge} = \mathcal{T}_{\mathcal{S}}$. Thus, by Definition 24, the following assertions are equivalent:

(b) F is quasi-topologically mixed lower continuous with respect to the relators $\mathcal{R}$ and $\mathcal{S}$;
(c) F is quasi-topologically mixed lower continuous with respect to the relators $\mathcal{R}$ and $\mathcal{S}^\wedge$.

Therefore, in inclusions (a) and (2) we may write $\mathcal{S}^\wedge$ in place of $\mathcal{S}$. Thus, inclusions (2) and (3) are also equivalent.

Remark 73. Unfortunately, if G is a relation on Y to Z, then we can only prove that

(1) $(G \circ F)^c \subseteq G^c \circ F$ if $X = F^{-1}[Y]$;
(2) $(G \circ F)^c \subseteq G \circ F^c$ if $Z = G[Y]$.

Therefore, the inclusions (2) and (3) in Theorem 81 cannot, in general, be simplified.

27. Some Further Theorems on Quasi-Topological Mixed Upper and Lower Continuities

Analogously to Theorem 75, we can also easily establish the following:

Theorem 88. *If both $\mathcal{R}$ and $\mathcal{S}$ are non-void, then the following assertions are equivalent:*

(1) *F is quasi-topologically mixed upper (lower) continuous with respect to the relators $\mathcal{R}$ and $\mathcal{S}$;*
(2) *F is quasi-topologically mixed upper (lower) continuous with respect to the relators $\mathcal{R}^{\wedge\infty}$ and $\mathcal{S}^{\wedge\infty}$.*

Hence, by using Theorems 85 and 87, we can immediately derive the following two theorems:

Theorem 89. *If both $\mathcal{R}$ and $\mathcal{S}$ are non-void, then the following assertions are equivalent:*

(1) *F is quasi-topologically mixed upper continuous;*
(2) $F^{-1} \circ \mathcal{S}^{\wedge\infty\wedge} \circ F \subseteq \mathcal{R}^{\wedge\infty\wedge}$; (3) $\left(F^{-1} \circ \mathcal{S}^{\wedge\infty\wedge} \circ F\right)^{\wedge} \subseteq \mathcal{R}^{\wedge\infty\wedge}$.

Remark 74. Thus, if (1) holds, then by the increasingness of $\infty\wedge$ and the idempotency of $\wedge\infty$ we also have $\left(F^{-1} \circ \mathcal{S}^{\wedge\infty\wedge} \circ F\right)^{\wedge\infty\wedge} \subseteq \mathcal{R}^{\wedge\infty\wedge}$.

Moreover, from the above theorem, we can also easily derive the following:

Corollary 20. *If F is quasi-topologically mixed upper continuous and both $\mathcal{R}$ and $\mathcal{S}$ are non-void, then*

(1) $F^{-1} \circ \mathcal{S}^{\wedge\infty} \circ \mathcal{F} \subseteq \mathcal{R}^{\wedge}$; (2) $\left(F^{-1} \circ \mathcal{S}^{\wedge\infty} \circ F\right)^{\wedge\infty} \subseteq \mathcal{R}^{\wedge\infty}$.

Proof. By using Theorem 89 and the corresponding properties of the operations $\wedge$ and ∞, we can see that

$$F^{-1} \circ \mathcal{S}^{\wedge\infty} \circ F \subseteq F^{-1} \circ \mathcal{S}^{\wedge\infty\wedge} \circ F \subseteq \mathcal{R}^{\wedge\infty\wedge} \subseteq \mathcal{R}^{\wedge*\wedge} = \mathcal{R}^{\wedge\wedge} = \mathcal{R}^{\wedge}.$$

Therefore, assertion (1) is true.

Moreover, from assertion (1), we can see that $\left(F^{-1} \circ \mathcal{S}^{\wedge\infty} \circ F\right)^{\wedge} \subseteq \mathcal{R}^{\wedge\wedge} = \mathcal{R}^{\wedge}$. Therefore, assertion (2) is also true.

Remark 75. However, the fundamental inclusions (1) and (2) seem not to imply the quasi-topological mixed upper continuity of F.

Theorem 90. *If both $\mathcal{R}$ and $\mathcal{S}$ are non-void, then the following assertions are equivalent:*

(1) *F is quasi-topologically mixed lower continuous;*
(2) $\left(F^{-1} \circ \mathcal{S}^{\wedge\infty c}\right)^{c} \subseteq \left(\mathcal{R}^{\wedge\infty\wedge} \circ F^{-1}\right)^{\wedge}$;
(3) $\left(F^{-1} \circ \mathcal{S}^{\wedge\infty\wedge c}\right)^{c\wedge} \subseteq \left(\mathcal{R}^{\wedge\infty\wedge} \circ F^{-1}\right)^{\wedge}$.

Remark 76. This theorem also shows that, quasi-topological mixed lower continuity is a less convenient property than its upper counterpart.

28. Specializations to Functions

Notation 11. In this and the next two sections, we shall assume that f is a function of one relator space $X(\mathcal{R})$ to another $Y(\mathcal{S})$.

Thus, in particular, we can easily establish the following two theorems:

Theorem 91. *The following assertions are equivalent:*

(1) *f is quasi-topologically lower continuous;*
(2) *f is quasi-topologically upper continuous;*
(3) *f is quasi-topologically mixed lower continuous;*
(4) *f is quasi-topologically mixed upper continuous.*

Proof. For any $y \in X$ and $V \subseteq Y$, we have

$$\{y\} \cap V \neq \emptyset \iff y \in V \iff \{y\} \subseteq V.$$

Hence, by Definitions 22 and 24, and the usual identifications of singletons with their elements, it is clear that the above assertions are equivalent.

Theorem 92. *The following assertions are equivalent:*

(1) *f^{-1} is topological openness preserving;*
(2) *f^{-1} is topological closedness preserving.*

Proof. To prove this, in addition to Definition 23, we have to use Theorem 8 and the basic fact that now, for any $V \subseteq Y$, we have $f^{-1}[V^c] = f^{-1}[V]^c$.

Namely, if for instance assertion (1) holds, then we can see

$$\begin{aligned} V \in \mathcal{F}_{\mathcal{S}} &\implies V^c \in \mathcal{T}_{\mathcal{R}} \implies f^{-1}[V^c] \in \mathcal{T}_{\mathcal{R}} \\ &\implies f^{-1}[V]^c \in \mathcal{T}_{\mathcal{R}} \implies f^{-1}[V] \in \mathcal{F}_{\mathcal{R}}. \end{aligned}$$

Therefore, assertion (2) also holds.

Remark 77. In this respect, it is worth noting that, for a relation F on X to Y, we have

(1) $X = F^{-1}[Y]$ if and only if $F^{-1}[B]^c \subseteq F^{-1}[B^c]$ for all $B \subseteq Y$;
(2) F is a function if and only if $F^{-1}[B^c] \subseteq F^{-1}[B]^c$ for all $B \subseteq Y$.

To prove the "if part" of (2), note that if F is not a function, then there exist $x \in X$ and $y_1, y_2 \in F(x)$ such that $y_1 \neq y_2$. Hence, we see that

$$x \in F^{-1}(y_2) \subseteq F^{-1}[\{y_1\}^c], \text{ but } x \in F^{-1}[\{y_1\}], \text{and thus } x \notin F^{-1}[\{y_1\}]^c.$$

Now, as an immediate consequence of Theorems 91, 84 and 85, we can also state the following:

Theorem 93. *If both $\mathcal{R}$ and $\mathcal{S}$ are topological, and $\mathcal{S} \neq \emptyset$, then the following assertions are equivalent:*

(1) *f is quasi-topologically upper continuous;*
(2) *for each $x \in X$ and $S \in \mathcal{S}^\wedge$ there exists $R \in \mathcal{R}$ such that*

$$R(x) \subseteq f^{-1}\big[S\big(f(x)\big)\big];$$

(3) $f^{-1} \circ \mathcal{S}^\wedge \circ f \subseteq \mathcal{R}^\wedge$; (4) $\big(f^{-1} \circ \mathcal{S}^\wedge \circ f\big)^\wedge \subseteq \mathcal{R}^\wedge$.

Remark 78. This theorem can be more naturally derived from Theorems 71 and 72 by using that $\Delta_X \subseteq f^{-1} \circ f$ and $f \circ f^{-1} \subseteq \Delta_Y$.

From the above theorem, by using Theorem 65, we can immediately derive

Corollary 21. *If F is a relation on X onto Y such that F^{-1} is a function, both $\mathcal{R}$ and $\mathcal{S}$ are topological, and $\mathcal{R} \neq \emptyset$, then the following assertions are equivalent:*

(1) *F is topological closedness preserving;*
(2) *for each $y \in Y$ and $R \in \mathcal{R}^\wedge$ there exists $S \in \mathcal{S}$ such that*

$$S(y) \subseteq F\big[R[F^{-1}(y)]\big];$$

(3) $F \circ \mathcal{R}^\wedge \circ F^{-1} \subseteq \mathcal{S}^\wedge$; (4) $\big(F^\wedge \circ \mathcal{R}^\wedge \circ F^{-1}\big)^\wedge \subseteq \mathcal{S}^\wedge$.

Remark 79. Assertion (2) can also be reformulated in the form that for each $y \in Y$ and $R \in \mathcal{R}^\wedge$ there exists $S \in \mathcal{S}$ such that

$$S(y) \subseteq \big(F \boxtimes F\big)[R](y).$$

That is, $\big(F \boxtimes F\big)[R] \in \mathcal{S}^\wedge$ for all $R \in \mathcal{R}^\wedge$.

29. Topological Closure and Interior Preserving Functions

Now, as a straightforward generalization of Definition 4, we may also naturally introduce the following:

Definition 25. We say that the function

(1) f is *topological closure preserving* if for all $x \in X$ and $A \subseteq X$

$$x \in \mathrm{cl}_{\mathcal{R}}(A) \implies f(x) \in \mathrm{cl}_{\mathcal{S}}\big(f[A]\big);$$

(2) f is *topological interior preserving* if for all $x \in X$ and $A \subseteq X$

$$x \in \mathrm{int}_{\mathcal{R}}(A) \implies f(x) \in \mathrm{int}_{\mathcal{S}}\big(f[A]\big).$$

Remark 80. Quite similarly, the function f may, for instance, be naturally called *topological interior reversing* if $x \in \mathrm{int}_{\mathcal{R}}(A)$ implies $f(x) \in \mathrm{cl}_{\mathcal{S}}\big(f[A]\big)$ for all $x \in X$ and $A \subseteq X$. Such functions were also first investigated in Ref. [29].

The above definition can be reformulated in the following more concise form.

Theorem 94. *The following assertions are true:*

(1) *f is topological closure preserving if and only if for all $A \subseteq X$ we have*

$$f\,[\mathrm{cl}_{\mathcal{R}}(A)] \subseteq \mathrm{cl}_{\mathcal{S}}\big(f[A]\big);$$

(2) *f is topological interior preserving if and only if for all $A \subseteq X$ we have*

$$f\,[\mathrm{int}_{\mathcal{R}}(A)] \subseteq \mathrm{int}_{\mathcal{S}}\big(f[A]\big).$$

Now, by using this theorem, we can easily prove the following:

Theorem 95. *If $\mathcal{R}$ is topological, then the following assertions are equivalent:*

(1) *f is topological interior preserving;*
(2) *f is topological openness preserving.*

Proof. Suppose that assertion (1) holds and $A \in \mathcal{T}_{\mathcal{R}}$. Then, by Definition 6, we have $A \subseteq \mathrm{int}_{\mathcal{R}}(A)$. Hence, by using Theorem 94, we can infer that

$$f[A] \subseteq f\,[\mathrm{int}_{\mathcal{R}}(A)] \subseteq \mathrm{int}_{\mathcal{S}}\big(f[A]\big).$$

Therefore, by Definition 6, we also have $f[A] \in \mathcal{T}_{\mathcal{S}}$. Thus, by Definition 23, assertion (2) also holds.

Conversely, suppose now that assertion (2) holds and $A \subseteq X$. Then, by Theorem 47, we have $\operatorname{int}_{\mathcal{R}}(A) \in \mathcal{T}_{\mathcal{R}}$. Hence, by using Definition 23, we can infer that $f[\operatorname{int}_{\mathcal{R}}(A)] \in \mathcal{T}_{\mathcal{S}}$. Now, by Definition 6 and Theorem 42, we can also see that

$$f[\operatorname{int}_{\mathcal{R}}(A)] \subseteq \operatorname{int}_{\mathcal{S}}\big(f[\operatorname{int}_{\mathcal{R}}(A)]\big) \subseteq \operatorname{int}_{\mathcal{S}}\big(f[A]\big).$$

Therefore, by Theorem 94, assertion (1) also holds.

Remark 81. Note that the implication (1) $\Longrightarrow$ (2) does not require the relator $\mathcal{R}$ to be topological.

Concerning topological closure preserving functions, instead of an analogue of the above theorem, we can only prove the following:

Theorem 96. *The following assertions are equivalent:*

(1) *f is topological closure preserving;*
(2) *for all $B \subseteq Y$ we have*

$$f^{-1}[\operatorname{int}_{\mathcal{S}}(B)] \subseteq \operatorname{int}_{\mathcal{R}}\big(f^{-1}[B]\big);$$

(3) *for all $x \in X$ and $B \subseteq Y$*

$$f(x) \in \operatorname{int}_{\mathcal{S}}(B) \implies x \in \operatorname{int}_{\mathcal{R}}\big(f^{-1}[B]\big).$$

Proof. By using Theorem 1, we can see that, for any $x \in X$ and $A \subseteq X$, the following implications are equivalent:

$$\begin{aligned} x \in \operatorname{cl}_{\mathcal{R}}(A) &\implies f(x) \in \operatorname{cl}_{\mathcal{S}}\big(f[A]\big), \\ f(x) \notin \operatorname{cl}_{\mathcal{S}}\big(f[A]\big) &\implies x \notin \operatorname{cl}_{\mathcal{R}}(A), \\ f(x) \in \operatorname{cl}_{\mathcal{S}}\big(f[A]\big)^{c} &\implies x \in \operatorname{cl}_{\mathcal{R}}(A)^{c}, \\ f(x) \in \operatorname{int}_{\mathcal{S}}\big(f[A]^{c}\big) &\implies x \in \operatorname{int}_{\mathcal{R}}\big(A^{c}\big). \end{aligned}$$

Therefore, assertion (1) is equivalent to the assertion:

(a) for any $x \in X$ and $A \subseteq X$

$$f(x) \in \operatorname{int}_{\mathcal{S}}\big(f[A]^{c}\big) \implies x \in \operatorname{int}_{\mathcal{R}}(A^{c}).$$

Now, if assertion (a) holds, moreover $x \in X$ and $B \subseteq Y$, then by using the inclusion

$$f\big[f^{-1}[B]^{c}\big] \subseteq B^{c},$$

derivable from the proof of Theorem 63, we can see that

$$f(x) \in \text{int}_{\mathcal{S}}(B) \implies f(x) \in \text{int}_{\mathcal{S}}\left(f\left[f^{-1}[B]^c\right]^c\right) \implies x \in \text{int}_{\mathcal{R}}\left(f^{-1}[B]\right).$$

Therefore, assertion (3) also holds.

Conversely, if assertion (3) holds, moreover $x \in X$ and $A \subseteq X$, then by using the inclusion

$$f^{-1}\left[f[A]^c\right] \subseteq A^c,$$

derivable from the proof of Theorem 63, we can see that

$$f(x) \in \text{int}_{\mathcal{S}}(f[A]^c) \implies x \in \text{int}_{\mathcal{R}}\left(f^{-1}[f[A]^c]\right) \implies x \in \text{int}_{\mathcal{R}}(A^c).$$

Therefore, assertion (a) also holds.

Thus, we have proved that (1) $\iff$ (a) $\iff$ (3). Therefore, to complete the proof, it remains only to note that assertion (2) is only a simple reformulation of assertion of (3).

Remark 82. Now, analogously to this theorem, we can also prove that f is topological interior preserving if and only if $f^{-1}[\text{cl}_{\mathcal{S}}(B)] \subseteq cl_{\mathcal{R}}\left(f^{-1}[B]\right)$ for all $B \subseteq Y$.

30. Relational Reformulations of Topological Interior and Closure Preservingness Properties

Theorem 97. *The following assertions are equivalent:*

(1) *f is topological interior preserving;*

(2) *for any $x \in X$ and $R \in \mathcal{R}$ there exists $S \in \mathcal{S}$ such that*

$$S(f(x)) \subseteq f[R(x)];$$

(3) *for any $x \in X$ and $R \in \mathcal{R}$ there exists $S \in \mathcal{S}$ such that*

$$(S \circ f)(x) \subseteq (f \circ R)(x).$$

Proof. Suppose that assertion (1) holds, moreover $x \in X$ and $R \in \mathcal{R}$. Then, by Definition 5, we have $x \in \text{int}_{\mathcal{R}}(R(x))$. Hence, by using Definition 25, we can infer that $f(x) \in \text{int}_{\mathcal{S}}(f[R(x)])$. Therefore, by Definition 5, there exists $S \in \mathcal{S}$ such that $S(f(x)) \subseteq G[R(x)]$. Thus, assertion (2) also holds.

Conversely, suppose now that assertion (2) holds, moreover $x \in X$ and $A \subseteq X$ such that $x \in \text{int}_{\mathcal{R}}(A)$. Then, by Definition 5, there exists $R \in \mathcal{R}$ such that $R(x) \subseteq A$. Hence, we can infer that $f[R(x)] \subseteq f[A]$. Moreover, by assertion (2), we can state that there exists $S \in \mathcal{S}$ such that $S(f(x)) \subseteq f[R(x)]$. Therefore, we also have $S(f(x)) \subseteq f[A]$. Hence, by Definition 5, we can already infer that $f(x) \in \operatorname{int}_{\mathcal{S}}(f[A])$. Therefore, by Definition 25, assertion (1) also holds.

Thus, we have proved that assertions (1) and (2) are equivalent. Now, to complete the proof, it remains to note only that assertion (3) is only a reformulation of assertion (2).

Theorem 98. *The following assertions are equivalent:*

(1) *f is topological interior preserving;*
(2) $f \circ \mathcal{R} \subseteq (\mathcal{S} \circ f)^{\wedge}$; (3) $(f \circ \mathcal{R}^{\wedge})^{\wedge} \subseteq (\mathcal{S}^{\wedge} \circ f)^{\wedge}$.

Proof. By Theorem 97 and Definition 11, it is clear that assertions (1) and (2) are equivalent. Moreover, from Theorems 24 and 37, we can see that assertion (2) is equivalent to the inclusion $(f \circ \mathcal{R})^{\wedge} \subseteq (\mathcal{S} \circ f)^{\wedge}$.

On the other hand, from Corollary 1, we can see that $\operatorname{int}_{\mathcal{R}} = \operatorname{int}_{\mathcal{R}^{\wedge}}$ and $\operatorname{int}_{\mathcal{S}} = \operatorname{int}_{\mathcal{S}^{\wedge}}$. Therefore, by Definition 25, we can state that f is topological interior preserving with respect to the relators $\mathcal{R}$ and $\mathcal{S}$ if and only if f is topological interior preserving with respect to the relators $\mathcal{R}^{\wedge}$ and $\mathcal{S}^{\wedge}$. Therefore, by the equivalence of assertion (1) to the inclusion $(f \circ \mathcal{R})^{\wedge} \subseteq (\mathcal{S} \circ f)^{\wedge}$, the equivalence of assertions (1) and (3) can also be stated.

Now, analogously to Theorem 97, we can also prove the following:

Theorem 99. *The following assertions are equivalent:*

(1) *f is topological closure preserving;*
(2) *for all $x \in X$ and $S \in \mathcal{S}$ there exists $R \in \mathcal{R}$ such that*

$$R(x) \subseteq f^{-1}\left[S(f(x))\right];$$

(3) *for all $x \in X$ and $S \in \mathcal{S}$ there exists $R \in \mathcal{R}$ such that*

$$R(x) \subseteq \left(f^{-1} \circ S \circ f\right)(x).$$

Proof. Suppose on the contrary that assertion (1) holds, but (2) does not hold. Then, there exist $x \in X$ and $S \in \mathcal{S}$ such that for any $R \in \mathcal{R}$ we have

$$R(x) \not\subseteq f^{-1}\left[S(f(x))\right], \qquad \text{and thus} \qquad R(x) \cap f^{-1}\left[S(f(x))\right]^{c} \neq \emptyset.$$

Hence, by using Definition 5, we can infer that

$$x \in \operatorname{cl}_{\mathcal{R}}\left(f^{-1}\left[S(f(x))\right]^{c}\right).$$

Now, by using assertion (1) and the inclusion $f \circ f^{-1} \subseteq \Delta_Y$, we can see that

$$f(x) \in \mathrm{cl}_{\mathcal{S}}\left(f\left[f^{-1}\left[S(f(x))\right]^c\right]\right) = \mathrm{cl}_{\mathcal{S}}\left(\left(f \circ f^{-1}\right)\left[S(f(x))\right]^c\right)$$
$$\subseteq \mathrm{cl}_{\mathcal{S}}\left(\Delta_Y\left[S(f(x))\right]^c\right) = \mathrm{cl}_{\mathcal{S}}\left(S(f(x))^c\right).$$

Hence, by using Definition 5, we can infer that $S(f(x)) \cap S(f(x))^c \neq \emptyset$. This contradiction proves that (1) implies (2).

To prove the converse implication (2) $\Longrightarrow$ (1), suppose now that assertion (2) holds, moreover $x \in X$ and $A \subseteq X$ such that $x \in \mathrm{cl}_{\mathcal{R}}(A)$, and $S \in \mathcal{S}$. Then, by assertion (2), there exists $R \in \mathcal{R}$ such that

$$R(x) \subseteq f^{-1}\left[S(f(x))\right].$$

Moreover, by Definition 5, we have $R(x) \cap A \neq \emptyset$. Thus, there exists $u \in R(x)$ such that $u \in A$. Now, we can see that

$$u \in f^{-1}\left[S(f(x))\right], \qquad \text{and thus} \qquad f(u) \in S(f(x)).$$

Hence, by using that $f(u) \in f[A]$, we can infer that $S(f(x)) \cap f[A] \neq \emptyset$. Therefore, by Definition 5, we also have $f(x) \in \mathrm{cl}_{\mathcal{S}}(f[A])$. Thus, by Definition 25, assertion (1) also holds.

Thus, we have proved that assertions (1) and (2) are equivalent. Now, to complete the proof, it remains to note only that assertion (3) is only a reformulation of assertion (2).

From the above theorem, by the proof of Theorem 98, it is clear that the following theorem is also true:

Theorem 100. *The following assertions are equivalent:*

(1) *f is topological closure preserving;*
(2) $f^{-1} \circ \mathcal{S} \circ f \subseteq \mathcal{R}^\wedge$; (3) $\left(f^{-1} \circ \mathcal{S}^\wedge \circ f\right)^\wedge \subseteq \mathcal{R}^\wedge$.

Now, as an immediate consequence of Theorems 93 and 100, we can also state

Corollary 22. *If both $\mathcal{R}$ and $\mathcal{S}$ are topological, and $\mathcal{S} \neq \emptyset$, then the following assertions are equivalent:*

(1) *f is topological closure preserving;*
(2) *f is quasi-topologically upper continuous.*

31. Four General Continuity Properties of Relators

Notation 12. In this and the next section, we shall assume that

(a) $(X, Y)(\mathcal{R})$ **and** $(Z, W)(\mathcal{S})$ **are relator spaces;**
(b) $\mathcal{F}$ **is a relator on** X **to** Z **and** $\mathcal{G}$ **is a relator on** Y **to** W;
(c) $\square = \left(\square_i\right)_{i=1}^{4}$ **is a family of direct unary operations for relators.**

Remark 83. To keep in mind assumptions (a) and (b), for any $R \in \mathcal{R}$, $S \in \mathcal{S}$, $F \in \mathcal{F}$ and $G \in \mathcal{G}$, we may use the illustrating diagram:

$$\begin{array}{ccc} X & \xrightarrow{\;F\;} & Z \\ {\scriptstyle R}\big\downarrow & & \big\downarrow{\scriptstyle S} \\ Y & \xrightarrow{\;G\;} & W \end{array}$$

Now, by pexiderizing [69] several former compositional inclusions on relations and relators, and using relators instead of relations, we may naturally introduce the following general unifying definition, which has been mainly developed by the third author, in a Debrecen program of continuity [6,21,27,30,65–68,73].

Definition 26. We shall say that the ordered pair

(1) $(\mathcal{F}, G)$ is *upper right* $\square$*-continuous* if $\left(\mathcal{S}^{\square_1} \circ \mathcal{F}\right)^{\square_2} \subseteq \left(\mathcal{G} \circ \mathcal{R}^{\square_3}\right)^{\square_4}$;
(2) $(\mathcal{F}, G)$ is *mildly right* $\square$*-continuous* if $\left(\mathcal{G}^{-1} \circ \mathcal{S}^{\square_1} \circ \mathcal{F}\right)^{\square_2} \subseteq \mathcal{R}^{\square_3 \square_4}$;
(3) $(\mathcal{F}, G)$ is *vaguely right* $\square$*-continuous* if $\mathcal{S}^{\square_1 \square_2} \subseteq \left(\mathcal{G} \circ \mathcal{R}^{\square_3} \circ \mathcal{F}^{-1}\right)^{\square_4}$;
(4) $(\mathcal{F}, G)$ is *lower right* $\square$*-continuous* if $\left(\mathcal{G}^{-1} \circ \mathcal{S}^{\square_1}\right)^{\square_2} \subseteq \left(\mathcal{R}^{\square_3} \circ \mathcal{F}^{-1}\right)^{\square_4}$.

Remark 84. Here, according to the corresponding definitions in topological spaces [2,3], we should again use the terms "upper semicontinuous" and "lower semicontinuous" instead of "upper continuous" and "lower continuous", resp.

Moreover, according to some similar definitions [30,68], the term "right" should be deleted. However, having in mind Galois connections, it seems convenient to define the corresponding left continuity properties by reversing the above inclusions.

Remark 85. Now, for any $F \in \mathcal{F}$ and $G \in \mathcal{G}$, the pair (F, G) may, for instance, be naturally called *upper right* $\square$*-continuous*, if the pair $(\{F\},\{G\})$ is upper right $\Box$–continuous. That is, $\left(\mathcal{S}^{\Box_1} \circ F\right)^{\Box_2} \subseteq \left(G \circ \mathcal{R}^{\Box_3}\right)^{\Box_4}$.

Remark 86. Moreover, the pair $(\mathcal{F},\mathcal{G})$ may, for instance, be naturally called *right $\Box$–continuous* if it is both upper and lower right $\Box$–continuous. For functions, upper and lower right continuities will usually coincide.

Thus, for any $F \in \mathcal{F}$ and $G \in \mathcal{G}$, the pair (F,G) may, for instance, be naturally called *selectionally right $\Box$–continuous* if for any selection functions f of F and g of G the pair (f,g) is right $\Box$–continuous.

Moreover, the pair $(\mathcal{F},G)$ itself may, for instance, be naturally called *elementwise right $\Box$–continuous* if for any $F \in \mathcal{F}$ and $G \in \mathcal{G}$, the pair (F,G) is right $\Box$–continuous.

Remark 87. If in particular $\Box$ is a single direct unary operation for relators, then the pair $(\mathcal{F},G)$ may, for instance, be naturally called upper right $\Box$–continuous if it is upper right $(\Box)_{i=1}^{4}$–continuous. That is,

$$\left(\mathcal{S}^{\Box} \circ \mathcal{F}\right)^{\Box} \subseteq \left(\mathcal{G} \circ \mathcal{R}^{\Box}\right)^{\Box}.$$

Remark 88. Thus, the pair $(\mathcal{F},G)$ may, for instance, be naturally called *properly upper right continuous* if it is upper right $\Box$–continuous with $\Box = \Delta$ being the identity operation for relators. That is, $\mathcal{S} \circ \mathcal{F} \subseteq \mathcal{G} \circ \mathcal{R}$.

Moreover, the pair $(\mathcal{F},G)$ may, for instance, be also naturally called *uniformly, proximally, topologically and paratopologically upper right continuous* if it is upper right $\Box$–continuous with $\Box = *, \#, \wedge$ and $\vartriangle$, resp.

Thus, by using the operations $\Box\infty$ and $\Box\partial$ instead of $\Box$, we can quite similarly speak of the corresponding *quasi-continuity and pseudo-continuity properties* of $(\mathcal{F},G)$. However, this terminology may certainly cause confusions.

Remark 89. Furthermore, we note that if $X = Y$ and $Z = W$, then the relator $\mathcal{F}$ and a relation $F \in \mathcal{F}$ may, for instance, be naturally called *upper right $\Box$–continuous* if the pairs $(\mathcal{F},\mathcal{F})$ and (F,F) are upper right $\Box$–continuous, resp.

That is,

$$\left(\mathcal{S}^{\Box_1} \circ \mathcal{F}\right)^{\Box_2} \subseteq \left(\mathcal{F} \circ \mathcal{R}^{\Box_3}\right)^{\Box_4} \qquad \text{and} \qquad \left(\mathcal{S}^{\Box_1} \circ F\right)^{\Box_2} \subseteq \left(F \circ \mathcal{R}^{\Box_3}\right)^{\Box_4},$$

resp.

Remark 90. In this respect, it is also worth mentioning that, because of the observations of [68], in Definition 26 we may sometimes also naturally write "increasing" instead of "continuous".

However, having in mind set-valued functions, a relation F on a goset $X(\leq)$ to a set Y may be naturally called increasing if $u \leq v$ implies $F(u) \subseteq F(v)$ for all $u, v \in X$.

Thus, it can be easily shown that the relation F is increasing if and only if its inverse F^{-1} is *ascending-valued* in the sense that $F^{-1}(y)$ is an ascending subset of $X(\leq)$ for all $y \in Y$.

By using the better notation $R = \leq$, the latter statement can be reformulated in the form that $R[F^{-1}(y)] \subseteq F^{-1}(y)$ for all $y \in Y$. That is, $R \circ F^{-1} \subseteq F^{-1}$, and thus $F^{-1} \circ \Delta_Y \subseteq \left(R \circ F^{-1}\right)^*$.

Remark 91. Finally, we note that if $\mathcal{V}$ is a relator on W to Y such that

$$\left(\mathcal{S}^{\square_1} \circ \mathcal{F}\right)^{\square_2} \subseteq \left(\mathcal{V}^{-1} \circ \mathcal{R}^{\square_3}\right)^{\square_4},$$

then according to [28, Definition 15.1] we may also say that the relator $\mathcal{F}$ is *upper right* $\square$–$\mathcal{V}$*–normal*, or that the relators $\mathcal{F}$ and $\mathcal{V}$ form an *upper right* $\square$*–Galois connection.*

Moreover, if $Z = W$ and Φ is a relator on X to Y such that

$$\left(\mathcal{F}^{-1} \circ \mathcal{S}^{\square_1} \circ \mathcal{F}\right)^{\square_2} \subseteq \left(\Phi^{-1} \circ \mathcal{R}^{\square_3}\right)^{\square_4},$$

then according to [28, Definition 16.1] we may also say that the relator $\mathcal{F}$ is *upper right* $\square$–Φ*–regular*, or that the relators $\mathcal{F}$ and Φ form an *upper right* $\square$*–Pataki connection.*

32. Relationships Among the Above Four Continuity Properties

Theorem 101. *If the operations* $\square_i$ *are inversion compatible, then the following assertions are equivalent:*

(1) $(\mathcal{F}, \mathcal{G})$ *is lower right* $\square$*–continuous with respect to* $\mathcal{R}$ *and* $\mathcal{S}$;
(2) $(\mathcal{G}, \mathcal{F})$ *is upper right* $\square$*–continuous with respect to* $\mathcal{R}^{-1}$ *and* $\mathcal{S}^{-1}$.

Proof. By the corresponding definitions, we have

$$\begin{aligned}
(1) &\iff \left(\mathcal{G}^{-1} \circ \mathcal{S}^{\square_1}\right)^{\square_2} \subseteq \left(\mathcal{R}^{\square_3} \circ \mathcal{F}^{-1}\right)^{\square_4} \\
&\iff \left(\mathcal{G}^{-1} \circ \mathcal{S}^{\square_1}\right)^{\square_2 -1} \subseteq \left(\mathcal{R}^{\square_3} \circ \mathcal{F}^{-1}\right)^{\square_4 -1} \\
&\iff \left(\mathcal{G}^{-1} \circ \mathcal{S}^{\square_1}\right)^{-1\square_2} \subseteq \left(\mathcal{R}^{\square_3} \circ \mathcal{F}^{-1}\right)^{-1\square_4} \\
&\iff \left(\mathcal{S}^{\square_1 -1} \circ \mathcal{G}\right)^{\square_2} \subseteq \left(\mathcal{F} \circ \mathcal{R}^{\square_3 -1}\right)^{\square_4} \\
&\iff \left(\mathcal{S}^{-1\square_1} \circ \mathcal{G}\right)^{\square_2} \subseteq \left(\mathcal{F} \circ \mathcal{R}^{-1\square_3}\right)^{\square_4} \iff (2).
\end{aligned}$$

Remark 92. This simple theorem shows again a remarkable advantage of relator spaces over the topological ones.

In particular, it will help us to easily keep in mind the definition of the lower right $\square$–continuity of the pair $(\mathcal{F}, \mathcal{G})$.

Analogously to Theorem 101, we can also prove the following two theorems.

Theorem 102. *If the operations $\square_i$ are inversion-compatible, then the following assertions are equivalent:*

(1) $(\mathcal{F}, \mathcal{G})$ *is mildly right $\square$-continuous with respect to $\mathcal{R}$ and $\mathcal{S}$;*
(2) $(\mathcal{G}, \mathcal{F})$ *is mildly right $\square$-continuous with respect to $\mathcal{R}^{-1}$ and $\mathcal{S}^{-1}$.*

Theorem 103. *If the operations $\square_i$ are inversion-compatible, then the following assertions are equivalent:*

(1) $(\mathcal{F}, \mathcal{G})$ *is vaguely right $\square$-continuous with respect to $\mathcal{R}$ and $\mathcal{S}$;*
(2) $(\mathcal{G}, \mathcal{F})$ *is mildly vaguely right $\square$-continuous with respect to $\mathcal{R}^{-1}$ and $\mathcal{S}^{-1}$.*

Remark 93. Note that the above three theorems cannot be applied in the particular case whenever some or all of the operations $\square_i$ are either $\wedge$ or $\triangle$.

However, for instance, in addition to Theorem 101, we can also prove

Theorem 104. *If the operations $\square_2$ and $\square_4$ are inversion compatible, moreover $X = Y$, $Z = W$, and the relators $\mathcal{R}$ and $\mathcal{S}$ are $\square_3$-symmetric and $\square_1$-symmetric, resp., then the following assertions are equivalent:*

(1) $(\mathcal{F}, \mathcal{G})$ *is lower right $\square$-continuous with respect to $\mathcal{R}$ and $\mathcal{S}$;*
(2) $(\mathcal{G}, \mathcal{F})$ *is upper right $\square$-continuous with respect to $\mathcal{R}$ and $\mathcal{S}$.*

Remark 94. In the $X = Y$ particular case, the relator $\mathcal{R}$ may be called $\square_3$–symmetric if $\mathcal{R}^{\square_3\,-1} = \mathcal{R}^{\square_3}$.

Thus, if the operation $\square_3$ is inversion compatible, then the relator $\mathcal{R}$ is $\square_3$–symmetric if and only if the relators $\mathcal{R}$ and $\mathcal{R}^{-1}$ are $\square_3$–equivalent.

Now, in addition to Theorem 101, we can also easily prove the following:

Theorem 105. *Under the notation $\Diamond = (\square_3, \square_4, \square_1, \square_2)$, the following assertions are equivalent:*

(1) $(\mathcal{F}, \mathcal{G})$ *is vaguely right $\square$-continuous with respect to $\mathcal{R}$ and $\mathcal{S}$;*
(2) $(\mathcal{F}^{-1}, \mathcal{G}^{-1})$ *is mildly left $\Diamond$-continuous with respect to $\mathcal{S}$ and $\mathcal{R}$.*

Proof. By the corresponding definitions, we have

$$(1) \iff \mathcal{S}^{\square_1 \square_2} \subseteq \left(\mathcal{G} \circ \mathcal{R}^{\square_3} \circ \mathcal{F}^{-1}\right)^{\square_4}$$

$$\iff \mathcal{S}^{\square_1 \square_2} \subseteq \left(\left(\mathcal{G}^{-1}\right)^{-1} \circ \mathcal{R}^{\square_3} \circ \mathcal{F}^{-1}\right)^{\square_4} \iff (2)$$

33. Continuity Properties with respect to a Single Operation

Notation 13. In this section, in addition to the assumptions (a) and (b) of Notation 12, we shall assume that $\square$ is a single direct unary operation for relators.

Now, according to Remark 87, we can use the following particular case of Definition 26.

Definition 27. We shall say that the ordered pair

(1) $(\mathcal{F}, G)$ is *upper right $\square$–continuous* if $\left(\mathcal{S}^{\square} \circ \mathcal{F}\right)^{\square} \subseteq \left(\mathcal{G} \circ \mathcal{R}^{\square}\right)^{\square}$;
(2) $(\mathcal{F}, G)$ is *mildly right $\square$–continuous* if $\left(\mathcal{G}^{-1} \circ \mathcal{S}^{\square} \circ \mathcal{F}\right)^{\square} \subseteq \mathcal{R}^{\square\square}$;
(3) $(\mathcal{F}, G)$ is *vaguely right $\square$–continuous* if $\mathcal{S}^{\square\square} \subseteq \left(\mathcal{G} \circ \mathcal{R}^{\square} \circ \mathcal{F}^{-1}\right)^{\square}$;
(4) $(\mathcal{F}, G)$ is *lower right $\square$–continuous* if $\left(\mathcal{G}^{-1} \circ \mathcal{S}^{\square}\right)^{\square} \subseteq \left(\mathcal{R}^{\square} \circ \mathcal{F}^{-1}\right)^{\square}$.

This definition can be further simplified when the operation $\square$ has some useful additional properties. For instance, by using the corresponding definitions and Theorems 37 and 36, we can easily establish the following three theorems:

Theorem 106. *If $\square$ is a projection operation, then*

(1) *$(\mathcal{F}, \mathcal{G})$ is mildly right $\square$–continuous if and only if* $\left(\mathcal{G}^{-1} \circ \mathcal{S}^{\square} \circ \mathcal{F}\right)^{\square} \subseteq \mathcal{R}^{\square}$;
(2) *$(\mathcal{F}, \mathcal{G})$ is vaguely right $\square$–continuous if and only if* $\mathcal{S}^{\square} \subseteq \left(\mathcal{G} \circ \mathcal{R}^{\square} \circ \mathcal{F}^{-1}\right)^{\square}$.

Proof. Because of the idempotency of $\square$, we now have $\square\square = \square$.

Theorem 107. *If $\square$ is a closure operation, then*

(1) *$(\mathcal{F}, \mathcal{G})$ is upper right $\square$–continuous if and only if,* $\mathcal{S}^{\square} \circ \mathcal{F} \subseteq \left(\mathcal{G} \circ \mathcal{R}^{\square}\right)^{\square}$;
(2) *$(\mathcal{F}, \mathcal{G})$ is mildly right $\square$–continuous if and only if* $\mathcal{G}^{-1} \circ \mathcal{S}^{\square} \circ \mathcal{F} \subseteq \mathcal{R}^{\square}$;

(3) $(\mathcal{F}, \mathcal{G})$ *is vaguely right* $\square$*–continuous if and only if* $\mathcal{S} \subseteq \left(\mathcal{G} \circ \mathcal{R}^{\square} \circ \mathcal{F}^{-1}\right)^{\square}$;
(4) $(\mathcal{F}, \mathcal{G})$ *is lower right* $\square$*–continuous if and only if* $\mathcal{G}^{-1} \circ \mathcal{S}^{\square} \subseteq \left(\mathcal{R}^{\square} \circ \mathcal{F}^{-1}\right)^{\square}$.

Proof. By Theorem 37, for any two relators $\mathcal{U}$ and $\mathcal{V}$ on M to N, we now have

$$\mathcal{U}^{\square} \subseteq \mathcal{V}^{\square} \iff U \subseteq \mathcal{V}^{\square}.$$

Theorem 108. *If* $\square$ *is an involution operation, then*

(1) $(\mathcal{F}, \mathcal{G})$ *is upper right* $\square$*–continuous if and only if* $\mathcal{S}^{\square} \circ \mathcal{F} \subseteq \mathcal{G} \circ \mathcal{R}^{\square}$;
(2) $(\mathcal{F}, \mathcal{G})$ *is mildly right* $\square$*–continuous if and only if* $\mathcal{G}^{-1} \circ \mathcal{S}^{\square} \circ \mathcal{F} \subseteq \mathcal{R}^{\square}$;
(3) $(\mathcal{F}, \mathcal{G})$ *is vaguely right* $\square$*–continuous if and only if* $\mathcal{S} \subseteq \mathcal{G} \circ \mathcal{R}^{\square} \circ \mathcal{F}^{-1}$;
(4) $(\mathcal{F}, \mathcal{G})$ *is lower right* $\square$*–continuous if and only if* $\mathcal{G}^{-1} \circ \mathcal{S}^{\square} \subseteq \mathcal{R}^{\square} \circ \mathcal{F}^{-1}$.

Proof. By Theorem 36, for any two relators $\mathcal{U}$ and $\mathcal{V}$ on M to N, we now have

$$\mathcal{U}^{\square} \subseteq \mathcal{V}^{\square} \iff \mathcal{U} \subseteq \mathcal{V}^{\square\square} \iff \mathcal{U} \subseteq \mathcal{V}.$$

If the operation $\square$ is in addition composition-compatible, the above inclusion can be further simplified. For instance, we can easily establish the following:

Theorem 109. *If* $\square$ *is a composition-compatible closure operation, then*

(1) $(\mathcal{F}, \mathcal{G})$ *is upper right* $\square$*–continuous if and only if* $\mathcal{S} \circ \mathcal{F} \subseteq (\mathcal{G} \circ \mathcal{R})^{\square}$;
(2) $(\mathcal{F}, \mathcal{G})$ *is mildly right* $\square$*–continuous if and only if* $\mathcal{G}^{-1} \circ \mathcal{S} \circ \mathcal{F} \subseteq \mathcal{R}^{\square}$;
(3) $(\mathcal{F}, \mathcal{G})$ *is vaguely right* $\square$*–continuous if and only if* $\mathcal{S} \subseteq \left(\mathcal{G} \circ \mathcal{R} \circ \mathcal{F}^{-1}\right)^{\square}$;
(4) $(\mathcal{F}, \mathcal{G})$ *is lower right* $\square$*–continuous if and only if* $\mathcal{G}^{-1} \circ \mathcal{S} \subseteq \left(\mathcal{R} \circ \mathcal{F}^{-1}\right)^{\square}$.

Proof. To prove (1), note that by Definition 27 the pair $(\mathcal{F}, \mathcal{G})$ is upper right $\square$–continuous if and only if $\left(\mathcal{S}^{\square} \circ \mathcal{F}\right)^{\square} \subseteq \left(\mathcal{G} \circ \mathcal{R}^{\square}\right)^{\square}$.

Moreover, by Definition 21 and Theorem 37, we now have

$$\left(\mathcal{S}^{\square} \circ \mathcal{F}\right)^{\square} \subseteq \left(\mathcal{G} \circ \mathcal{R}^{\square}\right)^{\square} \iff (\mathcal{S} \circ \mathcal{F})^{\square} \subseteq (\mathcal{G} \circ \mathcal{R})^{\square} \iff \mathcal{S} \circ \mathcal{F} \subseteq (\mathcal{G} \circ \mathcal{R})^{\square}.$$

References

[1] P. Thangavelu, S. Premakumari, and P. Xavier, Mixed continuous multifunctions, *Int. J. Mech. Eng. Tech.* **9**, 776–781, (2018).

[2] R.E. Smithson, Multifunctions, *Nieuw Arch. Wiskunde* **20**, 31–53, (1972).

[3] E. Klein and A.C. Thompson, *Theory of Correspondences* (Wiley and Sons, New York, 1984).

[4] P. Thangavelu and S. Premakumari, Strong forms of mixed continuity, *Int. J. Math. Trends Tech.* **61**, 43–50, (2018).

[5] P. Thangavelu and S. Premakumari, Weak forms of mixed continuity, *Int. J. Adv. Res. Sci. Eng. Tech.* **5**, 6563–6570, (2018).

[6] Á. Száz, Basic tools and mild continuities in relator spaces, *Acta Math. Hungar.* **50**, 177–201, (1987).

[7] Á. Száz, Upper and lower bounds in relator spaces, *Serdica Math. J.* **29**, 239–270, (2003).

[8] Á. Száz, A particular Galois connection between relations and set functions, *Acta Univ. Sapientiae, Math.* **6**, 73–91, (2014).

[9] Á. Száz, Correlations are more powerful tools than relations. In: Th.M. Rassias (Ed.), *Applications of Nonlinear Analysis*, Vol. 134, Springer Optimization and Its Applications, pp. 711–779, (2018).

[10] Á. Száz, Relationships between inclusions for relations and inequalities for correlations, *Math. Pannon.* **26**, 15–31, (2018).

[11] Á. Száz, Super and hyper products of super relations, *Tatra Mt. Math. Publ.*, **78**, 1–34, (2021).

[12] Th.M. Rassias and Á. Száz, Ordinary, super and hyper relators can be used to treat the various generalized open sets in a unified way. In: N.J. Daras and Th.M. Rassias (Eds.), *Approximation and Computation in Science and Engineering*, Vol. 180, Springer Optimization and Its Applications, pp. 709–782, (2022).

[13] W.J. Thron, *Topological Structures* (Holt, Rinehart and Winston, New York, 1966).

[14] Á. Száz, Inclusions for compositions and box products of relations, *J. Int. Math. Virt. Inst.* **3**, 97–125, (2013).

[15] T. Glavosits, Generated preorders and equivalences, *Acta Acad. Paed. Agriensis, Sect. Math.* **29**, 95–103, (2002).

[16] W.J. Pervin, Quasi-uniformization of topological spaces, *Math. Ann.* **147**, 316–317 (1962).

[17] A. Weil, Sur les espaces á structure uniforme et sur la topologie générale, *Actual. Sci. Ind.* Vol. 551 (Herman and Cie, Paris, 1937).

[18] W. Hunsaker and W. Lindgren, Construction of quasi-uniformities, *Math. Ann.* **188**, 39–42, (1970).

[19] A.S. Davis, Indexed systems of neighborhoods for general topological spaces, *Amer. Math. Monthly* **68**, 886–893, (1961).

[20] Á. Császár, *Foundations of General Topology* (Pergamon Press, London, 1963).

[21] Á. Száz, Somewhat continuity in a unified framework for continuities of relations, *Tatra Mt. Math. Publ.* **24**, 41–56, (2002).

[22] B.A. Davey and H.A. Priestley, *Introduction to Lattices and Order* (Cambridge University Press, Cambridge, 2002).

[23] Á. Száz, Basic Tools, increasing functions, and closure operations in generalized ordered sets. In: P.M. Pardalos and Th.M. Rassias (Eds.), *Contributions in Mathematics and Engineering: In Honor of Constantin Caratheodory* (Springer, 2016), pp. 551–616.

[24] J.L. Kelley, *General Topology*, Van Nostrand Reinhold Company, New York, 1955.

[25] P. Fletcher and W.F. Lindgren, *Quasi-Uniform Spaces* (Marcel Dekker, New York, 1982).

[26] B. Ganter and R. Wille, *Formal Concept Analysis* (Springer-Verlag, Berlin, 1999).

[27] Á. Száz, Lower semicontinuity properties of relations in relator spaces, *Adv. Stud. Contemp. Math. (Kyungshang)* **23**, 107–158, (2013).

[28] Á. Száz, Generalizations of Galois and Pataki connections to relator spaces, *J. Int. Math. Virt. Inst.* **4**, 43–75, (2014).

[29] Á. Száz, Contra continuity properties of relations in relator spaces, *Tech. Rep., Inst. Math., Univ. Debrecen* **2017/5**, 48 pp.

[30] Á. Száz and A. Zakaria, Mild continuity properties of relations and relators in relator spaces. In: P.M. Pardalos and Th.M. Rassias (Eds.), *Essays in Mathematics and its Applications: In Honor of Vladimir Arnold*, (Springer, 2016), pp. 439–511.

[31] N. Levine, On Pervin's quasi uniformity, *Math. J. Okayama Univ.* **14**, 97–102, (1970).

[32] N. Levine, On uniformities generated by equivalence relations, *Rend. Circ. Mat. Palermo* **18**, 62–70, (1969).

[33] J. Mala, Relators generating the same generalized topology, *Acta Math. Hungar.* **60**, 291–297, (1992).

[34] J. Mala and Á. Száz, Modifications of relators, *Acta Math. Hungar.* **77**, 69–81, (1997).

[35] G. Pataki, On the extensions, refinements and modifications of relators, *Math. Balk.* **15**, 155–186, (2001).

[36] G. Pataki and Á. Száz, A unified treatment of well-chainedness and connectedness properties, *Acta Math. Acad. Paedagog. Nyházi. (N.S.)* **19**, 101–165, (2003).

[37] J. Kurdics, A note on connection properties, *Acta Math. Acad. Paedagog. Nyházi.* **12**, 57–59, (1990).

[38] J. Kurdics and Á. Száz, Well-chainedness characterizations of connected relators, *Math. Pannon.* **4**, 37–45, (1993).

[39] Á. Száz, Galois and Pataki connections on generalized ordered sets, *Earthline J. Math. Sci.* **2**, 283–323, (2019).

[40] Á. Száz, Galois-type connections on power sets and their applications to relators, *Tech. Rep., Inst. Math., Univ. Debrecen* **2005/2**, 38 p.

[41] Á. Száz, Structures derivable from relators, *Singularité* **3**, 14–30, (1992).

[42] Á. Száz, Minimal structures, generalized topologies, and ascending systems should not be studied without generalized uniformities, *Filomat (Nis)* **21**, 87–97, (2007).

[43] V.A. Efremovič, The geometry of proximity, *Mat. Sb.* **31**, 189–200, (1952) (Russian).

[44] Yu. M. Smirnov, On proximity spaces, *Math. Sb.* **31**, 543–574, (1952) (Russian).

[45] Á. Száz, *Relators, Nets and Integrals,* Unfinished doctoral thesis, Debrecen, 1991, 126 pp.

[46] Á. Száz, Rare and meager sets in relator spaces, *Tatra Mt. Math. Publ.* **28**, 75–95, (2004).

[47] Á. Száz, Remarks and problems at the Conference on Inequalities and Applications, Hajdúszoboszló, Hungary, 2014, *Tech. Rep., Inst. Math., Univ. Debrecen* **2014/5**, 12 pp.

[48] Á. Száz, Galois type connections and closure operations on preordered sets, *Acta Math. Univ. Comen.* **78**, 1–21, (2009).

[49] K. Kuratowski, Sur l'opération $\overline{A}$ de l'analysis situs, *Fund. Math.* **3**, 182–199, (1922).

[50] N. Elez and O. Papaz, The new operators in topological spaces, *Math. Moravica* **17**, 63–68, (2013).

[51] V.A. Efremović and A.S. Švarc, A new definition of uniform spaces, Metrization of proximity spaces, *Dokl. Acad. Nauk. SSSR* **89**, 393–396, (1953) (Russian).

[52] H. Kenyon, Two theorems about relations, *Trans. Amer. Math. Soc.* **107**, 1–9, (1963).

[53] H. Nakano and K. Nakano, Connector theory, *Pacific J. Math.* **56**, 195–213, (1975).

[54] M. Salih and Á. Száz, Generalizations of some ordinary and extreme connectedness properties of topological spaces to relator spaces, *Elec. Res. Arch.* **28**, 471–548, (2020).

[55] J. Mala and Á. Száz, Properly topologically conjugated relators, *Pure Math. Appl. Ser. B* **3**, 119–136, (1992).

[56] H. Tietze, Beiträge zur allgemeinen Topologie I. Axiome für verschiedene Fassungen des Umgebungsbegriffs, *Math. Ann.* **88**, 290–312, (1923).

[57] P. Alexandroff, Zur Begründung der n-dimensionalen mengentheorischen Topologie, *Math. Ann.* **94**, 296–308, (1925).

[58] N. Bourbaki, *General Topology, Chapters 1–4* (Springer-Verlag, Berlin, 1989).

[59] Á. Száz, Directed, topological and transitive relators, *Publ. Math. Debrecen* **35**, 179–196, (1988).

[60] Á. Száz, Inverse and symmetric relators, *Acta Math. Hungar.* **60**, 157–176, (1992).

[61] G. Pataki, Supplementary notes to the theory of simple relators, *Radovi Mat.* **9**, 101–118, (1999).

[62] Á. Száz, Topological characterizations of relational properties, *Grazer Math. Ber.* **327**, 37–52, (1996).
[63] Á. Száz, Galois-type connections and continuities of pairs of relations, *J. Int. Math. Virt. Inst.* **2**, 39–66, (2012).
[64] J. Dontchev, Contra-continuous functions and strongly S-closed spaces, *Int. J. Math. Math. Sci.* **19**, 303–310, (1996).
[65] Á. Száz, An extension of Kelley's closed relation theorem to relator spaces, *Filomat* **14**, 49–71, (2000).
[66] Cs. Rakaczki and Á. Száz, Semicontinuity and closedness properties of relations in relator spaces, *Mathematica (Cluj)* **45**, 73–92, (2003).
[67] Á. Száz, Four general continuity properties, for pairs of functions, relations and relators, whose particular cases could be investigated by hundreds of mathematicians, *Tech. Rep., Inst. Math., Univ. Debrecen* **2017/1**, 17 pp.
[68] Á. Száz, A unifying framework for studying continuity, increasingness, and Galois connections, *MathLab J.* **1**, 154–173, (2018).
[69] J. Aczél, On a generalization of the functional equations of Pexider, *Publ. Inst. Math. N.S.* **4**, 77–80, (1964).
[70] H. Arianpoor, Preorder relators and generalized topologies, *J. Lin. Top. Algebra* **5**, 271–277, (2016).
[71] D. Doičinov, A unified theory of topological spaces, proximity spaces and uniform spaces, *Dokl. Akad. Nauk SSSR* **156**, 21–24, (Russian) (1964).
[72] J. Dontchev and T. Noiri, Contra-semicontinuous functions, *Math. Pannon.* **190**, 159–168, (1999).
[73] M.Th. Rassias and Á. Száz, Basic tools and continuity-like properties in relator spaces, *Contribut. Math.* **3**, 77–106, (2021).
[74] Á. Száz, Projective and inductive generations of relator spaces, *Acta Math. Hungar.* **53**, 407–430, (1989).
[75] Á. Száz, Lebesgue relators, *Monatsh. Math.* **110**, 315–319, (1990).
[76] Á. Száz, Cauchy nets and completeness in relator spaces, *Colloq. Math. Soc. János Bolyai* **55**, 479–489, (1993).
[77] Á. Száz, Refinements of relators, *Tech. Rep., Inst. Math., Univ. Debrecen* **76**, 19 pp, (1993).
[78] Á. Száz, Neighbourhood relators, *Bolyai Soc. Math. Stud.* **4**, 449–465, (1995).
[79] Á. Száz, Uniformly, proximally and topologically compact relators, *Math. Pannon.* **8**, 103–116, (1997).
[80] Á. Száz, An extension of Baire's category theorem to relator spaces, *Math. Morav.* **7**, 73–89, (2003).
[81] Á. Száz, Supremum properties of Galois–type connections, *Comment. Math. Univ. Carolin.* **47**, 569–583, (2006).
[82] Á. Száz, Applications of fat and dense sets in the theory of additive functions, *Tech. Rep., Inst. Math., Univ. Debrecen* **2007/3**, 29 pp.
[83] Á. Száz, Applications of relations and relators in the extensions of stability theorems for homogeneous and additive functions, *Aust. J. Math. Anal. Appl.* **6**, 1–66, (2009).

[84] Á. Száz, Foundations of the theory of vector relators, *Adv. Stud. Contemp. Math.* **20**, 139–195, (2010).

[85] Á. Száz, An extension of an additive selection theorem of Z. Gajda and R. Ger to vector relator spaces, *Sci. Ser. A Math. Sci. (N.S.)* **24**, 33–54, (2013).

[86] Á. Száz, The closure-interior Galois connection and its applications to relational inclusions and equations, *J. Int. Math. Virt. Inst.* **8**, 181–224, (2018).

[87] Á. Száz, Birelator spaces are natural generalizations of not only bitopological spaces, but also Ideal Topological Spaces, In: Th.M. Rassias and P.M. Pardalos (Eds.), *Mathematical Analysis and Applications*, Springer Optimization and Its Applications Vol. 154, pp. 543–586, (Springer Nature Switzerland AG, 2019).

https://doi.org/10.1142/9789811261572_0004

Chapter 4

Geometric Realization of Generalized Cartan Matrices of Rank 3

Abdullah Alazemi*, Milica Anđelić†, and Kyriakos Papadopoulos‡

Department of Mathematics, Kuwait University, Safat 13060, Kuwait

* *abdullah.alazemi@ku.edu.kw*

† *milica.andelic@ku.edu.kw*

‡ *kyriakos.papadopoulos@ku.edu.kw*

The primary aim of this chapter is to review the classification of generalized Cartan matrices of rank 3, as it was introduced by V.A. Gritsenko and V.V. Nikulin, in their seminal paper of 1998. In this classification, the authors introduced a conjecture, which we study in detail. In particular, we propose an algorithm in four steps, that could be used (given an appropriate computer program) to solve this conjecture, giving justification on why it should work.

1. Introduction

Infinite dimensional Lie algebras have found many applications in Mathematical Physics like, for instance, in conformal field theory.[a] Kac–Moody algebras are Lie algebras that are usually infinite dimensional (see Refs. [1,2]) and have a rich structure, which gives them a central role, both in modern mathematics and physics. Generalized Cartan matrices are used to define Kac–Moody Lie algebras. In particular, properties of such matrices are used to develop root systems of Kac–Moody algebras (see Ref. [1]). Gritsenko and Nikulin took full advantage of the close link between algebra and geometry by pointing out that a reflexive hyperbolic lattice with the group of reflections and with a lattice Weyl vector defines a corresponding

[a]For more details, the reader is referred to Ref. [7] and in particular chapter [8].

hyperbolic Kac–Moody Lie algebra which is graded by the lattice, and they developed a general theory of Lorentzian Kac–Moody algebras. They also constructed and classified a great number of them for rank greater or equal than 3 (see Ref. [3]). The significance of Lorentzian Kac–Moody algebras is based on the fact that they are automorphic corrections of hyperbolic Kac–Moody algebras. An important example is the Fake Monster Lie algebra (see Refs. [3–5]), which was used for the solution of the Moonshine conjecture.

This chapter serves as a review to the classification of generalized Cartan matrices of rank 3, with the lattice Weyl vector, which was presented by Gritsenko and Nikulin in 1998 (see Ref. [6]). We focus on a conjecture that the authors presented in this classification, and we propose a four-step algorithm for its solution. We solve Step 1 explicitly and give a code that could be used in Matlab and solves Step 2. A suitable computer program for Steps 3 and 4, which employs the results from 1 and 2, will hopefully give a complete answer.

2. Preliminaries

In this section, we state some basic definitions on generalized Cartan matrices, and show their relation to reflection groups in integral hyperbolic lattices (for a more detailed treatment, see Refs. [6,9,10]).

Definition 1. For a countable set of indices I, a finite-rank matrix $A = (a_{ij})$ is called *generalized Cartan matrix*, if and only if:

(1) $a_{ii} = 2\,;$
(2) $a_{ij} \in \mathbb{Z}^-$, $i \neq j$ and
(3) $a_{ij} = 0 \Rightarrow a_{ji} = 0\,.$

We consider such a matrix to be *indecomposable*, i.e. there does not exist a decomposition $I = I_1 \cup I_2$, such that $I_1 \neq \emptyset$, $I_2 \neq \emptyset$ and $a_{ij} = 0$, for $i \in I_1$, $j \in I_2$.

Definition 2. A generalized Cartan matrix A is *symmetrizable*, if there exists an invertible diagonal matrix $D = \text{diag}(\ldots \epsilon_i \ldots)$ and a symmetric matrix $B = (b_{ij})$, such that:

$$A = DB \quad \text{or} \quad (a_{ij}) = (\epsilon_i\, b_{ij}),$$

where $\epsilon_i \in \mathbb{Q}$, $\epsilon_i > 0$, $b_{ij} \in \mathbb{Z}^-$, $b_{ii} \in 2\mathbb{Z}$ and $b_{ii} > 0$.

The matrices D and B are defined uniquely, up to a multiplicative constant, and B is called *symmetrized generalized Cartan matrix*.

Remark 1. A symmetrizable generalized Cartan matrix $A = (a_{ij})$ and its symmetrized generalized Cartan matrix $B = (b_{ij})$ are related as follows:

$$(a_{ij}) = \left(\frac{2\,b_{ij}}{b_{ii}}\right),$$

where $b_{ii}|2\,b_{ij}$.

Definition 3. A symmetrizable generalized Cartan matrix A is called *hyperbolic*, if its symmetrized generalized Cartan matrix B has exactly one negative square or equivalently A has exactly one negative eigenvalue.

We recall the definition of *hyperbolic integral quadratic form S*.

Definition 4. A *hyperbolic integral symmetric bilinear form* (or hyperbolic integral quadratic form) on a finite rank free $\mathbb{Z}$-module M of dimension n, over the integers, is a map:

$$S : M \times M \to \mathbb{Z},$$

satisfying the following conditions:

(1) $S(\alpha\, m_1 + \beta\, m_2, m_3) = \alpha\, S(m_1, m_3) + \beta\, S(m_2, m_3)$;
(2) $S(m_3, \alpha\, m_1 + \beta\, m_2) = \alpha\, S(m_3, m_1) + \beta\, S(m_3, m_2)$;
(3) $S(m_1, m_2) = S(m_2, m_1)$ (symmetry) and
(4) signature $= (n, 1)$, i.e. in a suitable basis, the corresponding matrix of S is a diagonal matrix with n positive squares and one negative square in the diagonal;

where α, $\beta \in \mathbb{Z}$ and $m_1, m_2, m_3 \in M$.

We now consider an *integral hyperbolic lattice* (M, S), i.e. a pair of a free $\mathbb{Z}$-module M and a hyperbolic integral symmetric bilinear form S. By considering the corresponding cone:

$$V(M) = \{x \in M \otimes \mathbb{R} : (x, x) < 0\},$$

and by choosing its half-cone $V^+(M)$, we can define the corresponding *hyperbolic* (*or Lobachevskii*[b]) *space*:

$$\Lambda^+(M) = V^+(M)/\mathbb{R}^{++},$$

as a section (slice) of the cone, by a hyperplane.

[b]For a historic exposition and more technical details on hyperbolic geometry, see Ref. [11].

After defining the hyperbolic space, we can work in hyperbolic geometry, by defining the *distance* ρ between two points $\mathbb{R}^{++}x$ and $\mathbb{R}^{++}y$ in $\Lambda^+(M)$, as follows:

$$\cos\mathrm{h}\,\rho\,(\mathbb{R}^{++}x, \mathbb{R}^{++}y) = \frac{-S(x,y)}{\sqrt{S(x,x)S(y,y)}}.$$

Obviously, these two points in hyperbolic space are rays in the half-cone $V^+(M)$.

Remark 2. By definition, when we use signature $(n, 1)$, the square of a vector which is outside the cone $V(M)$ is strictly greater than zero and the square of a vector which is inside the cone is strictly negative. Furthermore, if a vector lies on the surface of the cone, then its square is zero. Obviously, we put a minus sign in the numerator of the definition of distance in hyperbolic space, because the hyperbolic cosine should be always positive.

From now on, instead of writing $S(x, y)$, we will simply write (x, y).

We observe that each element $\alpha_i \in M \otimes \mathbb{R}$, with $(\alpha_i, \alpha_i) > 0$, defines the half spaces:

$$H^+_{\alpha_i} = \{\mathbb{R}^{++}x \in \Lambda^+(M) : (x, \alpha_i) \leq 0\},$$
$$H^+_{\alpha_i^-} = \{\mathbb{R}^{++}x \in \Lambda^+(M) : (x, \alpha_i) > 0\},$$

which are bounded by the hyperplane:

$$H_{\alpha_i} = \{\mathbb{R}^{++}x \in \Lambda^+(M) : (x, \alpha_i) = 0\},$$

where $\alpha_i \in M \otimes \mathbb{R}$ is defined up to multiplication on elements of $\mathbb{R}^{++}$. The hyperplane H_{α_i} is also called *mirror of symmetry.*

Let us denote by $O(M)$ the *group of automorphisms*, which preserves the cone $V(M)$. Its subgroup $O^+(M) \subseteq O(M)$ is of index 2, and fixes the half-cone $V^+(M)$. Furthermore, $O^+(M)$ is discrete in $\Lambda^+(M)$, and has fundamental domain of finite volume.

Definition 5. For $(\alpha_i, \alpha_i) > 0$, by $s_{\alpha_i} \in O^+(M)$ we define *reflection* in a hyperplane H_{α_i}, of $\Lambda^+(M)$, as follows:

$$s_{\alpha_i}(x) = x - \frac{2(x, \alpha_i)}{(\alpha_i, \alpha_i)}\alpha_i,$$

where $x \in M$ and $\alpha_i \in M$.

Remark 3. One may ask the following question: Why does the equation we gave for reflection in our hyperbolic lattice, work? The answer relies on two facts:

(1) For $x = \alpha_i$, we get that $s_{\alpha_i}(\alpha_i) = -\alpha_i$ and
(2) For x perpendicular to α_i, we have that $s_{\alpha_i}(x) = x$.
Both of them show that our formula cannot work if we omit number 2 from the numerator.

An obvious remark is that a reflection s_{a_i} changes place between the half-spaces $H^+_{a_i}$ and $H^+_{a_i^-}$.

In addition, if an orthogonal vector $\alpha_i \in M$, $(\alpha_i, \alpha_i) > 0$, in a hyperplane H_{a_i} of $\Lambda^+(M)$, is a *primitive root*, i.e. its coordinates are coprime numbers, then:

$$\frac{2(M, \alpha_i)}{(\alpha_i, \alpha_i)}\alpha_i \subseteq M \Leftrightarrow (\alpha_i, \alpha_i) | \, 2(M, \alpha_i).$$

Definition 6. Any subgroup of $O(M)$ (the corresponding *discrete* group of motions of $\Lambda(M)$), generated by reflections, is called *reflection group.*

We denote by $W(M)$ the subgroup of $O^+(M)$ generated by *all* reflections of M, of elements with positive squares (and for signature $(n, 1)$).

We will also denote by W the subgroup of $W(M)$ generated by reflections in a set of elements of M. Clearly, $W \subseteq W(M) \subseteq O^+(M)$ is a subgroup of finite index.

Definition 7. A lattice M is called *reflexive* if the index $[O(M) : W(M)]$ is finite. In other words, $W(M)$ has a fundamental polyhedron of finite volume, in $\Lambda(M)$.

Talking a bit more about lattices, we consider again our integral hyperbolic lattice $S : M \times M \to \mathbb{Z}$. For $m \in \mathbb{Q}$, we denote by $S(m)$ the lattice which one gets if multiplying S by m. If S is reflective, then $S(m)$ is reflective as well.

Furthermore, S is called an *even lattice* if $S(x, x)$ is even for any $x \in M$. Otherwise, S is called *odd*.

Last, but not least, S is called a *primitive lattice* (or an *even primitive*, resp.), if $S(\frac{1}{m})$ is not a lattice (or even lattice, resp.), $m \in \mathbb{N}$, $m \geq 2$.

Definition 8. A convex polyhedron $\mathcal{M}$, in $\Lambda(M)$, is an intersection:

$$\mathcal{M} = \bigcap_{\alpha_i} H^+_{\alpha_i}$$

of several half-spaces orthogonal to elements $\alpha_i \in M$, $(\alpha_i, \alpha_i) > 0$.

This convex polyhedron is the fundamental chamber for a reflection group $W(M)$, with reflections generated by primitive roots, in M, each of them being orthogonal to exactly one side of this polyhedron. Moreover, we obtain this chamber if we remove all the mirrors of reflection, and take the connected components of the complement (with the boundary). The fundamental chamber acts *simply transitively*, because if we consider an element $w \in W(M)$, then $w(\mathcal{M}_1) = \mathcal{M}_2$. In other words, it fills in our hyperbolic space with congruent polyhedra.

The polyhedron $\mathcal{M}$ belongs to the cone

$$\mathbb{R}^+\mathcal{M} = \{x \in V^+(M) : (x, \alpha_i) \leq 0\},$$

where $\alpha_i \in P(\mathcal{M}) = \{\alpha_i : i \in I\}$; the set of orthogonal vectors to $\mathcal{M}$, where exactly one element α_i is orthogonal to each face of $\mathcal{M}$.

We say that $P(\mathcal{M})$ is *acceptable* if each of its elements is a (primitive) root, which is perpendicular to exactly one side of a convex polyhedron, in $\Lambda^+(M)$.

Also, $\mathcal{M}$ is *non-degenerate* if it contains a non-empty open subset of $\Lambda^+(M)$ and *elliptic* if it is a convex envelope of a finite set of points in $\Lambda^+(M)$ or at infinity of $\Lambda^+(M)$.

We now return back to the theory for generalized Cartan matrices, and relate it with the material that we have introduced for reflection groups of integral hyperbolic lattices.

Definition 9. An indecomposable hyperbolic generalized Cartan matrix A is equivalent to a triplet, as follows:

$$A \sim (M, W, P(\mathcal{M})),$$

where $S : M \times M \to \mathbb{Z}$ is a hyperbolic integral symmetric bilinear form, $W \subseteq W(M) \subseteq O^+(M)$, W is a subgroup of reflections in a set of elements of M, with positive squares, $W(M)$ is the subgroup of reflections in all elements of M (with positive squares), $O^+(M)$ is the group of automorphisms, which fixes the half-cone $V^+(M)$, and $P(\mathcal{M}) = \{\alpha_i : i \in I\}$, $(\alpha_i, \alpha_i) > 0$, $A = \left(2\frac{(\alpha', \alpha)}{(\alpha, \alpha)}\right)$, where $\alpha, \alpha' \in P(\mathcal{M})$, $\mathcal{M}$ is a locally finite polyhedron in $\Lambda^+(M)$, $\mathcal{M} = \bigcap_{\alpha_i} H^+_{\alpha_i}$ and $H^+_{\alpha_i} = \{\mathbb{R}^{++}x \in \Lambda^+(M) : (x, \alpha_i) \leq 0\}$.

The triplet $(M, W, P(\mathcal{M}))$ is called the *geometric realization of* A.

Let us now consider $\lambda(\alpha) \in \mathbb{N}$, $\alpha \in P(\mathcal{M})$ and the greatest common divisor $\gcd(\{\lambda(\alpha) : \alpha \in P(\mathcal{M})\}) = 1$.

Setting $\tilde{\alpha} = \lambda(\alpha)\,\alpha$, $(\tilde{\alpha},\tilde{\alpha})|\,2(\tilde{\alpha}',\tilde{\alpha}) \Leftrightarrow \lambda(\alpha)(\alpha,\alpha)|\,2\lambda(\alpha')(\alpha',\alpha)$, then

$$\tilde{A} = (2(\tilde{\alpha}',\tilde{\alpha})/(\tilde{\alpha},\tilde{\alpha})) = (2\lambda(\alpha')(\alpha',\alpha)/\lambda(\alpha)(\alpha,\alpha)),\ \alpha,\alpha' \in P(\mathcal{M}),$$

where $\tilde{A}$ is *twisted to* A, hyperbolic generalized Cartan matrix, and $\lambda(\alpha)$ are called *twisted coefficients of* α.

Also, $\tilde{A} = (\tilde{M},\tilde{W},\tilde{P}(\tilde{\mathcal{M}})) = (M \supseteq [\{\lambda(\alpha)\,\alpha : \alpha \in P(\mathcal{M})\}\,], W, \{\lambda(\alpha)\,\alpha : \alpha \in P(\mathcal{M})\})$. In other words, W and $\mathcal{M}$ are the same for $\tilde{A}$ and A.

Obviously, A is *untwisted* if it cannot be twisted to any generalized Cartan matrix, different from itself.

Our aim is to work on hyperbolic generalized Cartan matrices of *elliptic type*. Therefore, we need to introduce some additional material.

Definition 10. Let A be a hyperbolic generalized Cartan matrix, $A \sim (M, W, P(\mathcal{M}))$. We define the *group of symmetries of* A (or $P(\mathcal{M})$) as follows:

$$\text{Sym }(A) = \text{Sym }(P(\mathcal{M})) = \{g \in O^+(M) : g(P(\mathcal{M})) = P(\mathcal{M})\}.$$

Definition 11. A hyperbolic generalized Cartan matrix A has *restricted arithmetic type* if it is not empty and the semi-direct product of W with $\text{Sym}(A)$, which is equivalent to the semi-direct product of W with $\text{Sym}(P(\mathcal{M}))$, has finite index in $O^+(M)$.

Remark 4. For $W = (w_1, s_1)$, $S = (w_2, s_2)$ $\mathbb{Z}-$modules, the semi-direct product of W with S is equal to $((s_2 w_1)w_2, s_1 s_2)$.

Definition 12. A hyperbolic generalized Cartan matrix A has a *lattice Weyl vector* if there exists $\rho \in M \otimes \mathbb{Q}$ such that:

$$(\rho,\alpha) = -(\alpha,\alpha)/2, \quad \alpha \in P(\mathcal{M}).$$

Additionally, A has *generalized lattice Weyl vector*, if there exists $\mathbf{0} \neq \rho \in M \otimes \mathbb{Q}$, such that for a constant $N > 0$:

$$0 \leq -(\rho,\alpha) \leq N.$$

We can think of ρ in $\Lambda^+(M)$ as being the center of the inscribed circle to $\mathcal{M}$, where $\mathcal{M}$ is the fundamental chamber of a reflection group W.

Definition 13. A hyperbolic generalized Cartan matrix A has *elliptic type* if it has restricted arithmetic type and generalized lattice Weyl vector ρ, such that $(\rho,\rho) < 0$. In other words, $[O(M) : W] < \infty$ or $\text{vol}(\mathcal{M}) < \infty$ or $P(\mathcal{M}) < \infty$.

3. The Classification of Generalized Cartan Matrices of Rank 3, of Elliptic Type, with the Lattice Weyl Vector, Which are Twisted to Symmetric Generalized Cartan Matrices

In this section, we describe how Gritsenko and Nikulin classified in Ref. [6] the generalized Cartan matrices of rank 3, of elliptic type (in the sense that they have the generalized lattice Weyl vector), which are twisted to symmetric generalized Cartan matrices.

Let A be a generalized Cartan matrix of elliptic type, twisted to a symmetric generalized Cartan matrix $\tilde{A}$. This automatically implies that $\tilde{A}$ is of elliptic type, too. Let also $G(A) = (M, W, P(\mathcal{M}))$ be the geometric realization of A, where the rank of A is equal to 3.

Furthermore, for $\alpha \in P(\mathcal{M})$, we set:

$$\alpha = \lambda(\alpha)\,\delta(\alpha),$$

where $\lambda(\alpha) \in \mathbb{N}$ are the twisted coefficients of α, and $(\delta(\alpha), \delta(\alpha)) = 2$. So, $\tilde{P}(\mathcal{M}) = \{\delta(a) = \alpha/\lambda(\alpha) : \alpha \in P(\mathcal{M})\}$.

Throughout the remaining content of the chapter, we use relaxed notation:

$$\delta(\alpha_i) = \delta_i \quad \text{and} \quad \lambda(\alpha_i) = \lambda_i.$$

Now, A and its geometric realization are equivalent to a $(1 + [n/2]) \times n$ matrix $G(A)$:

$$\begin{aligned} \text{1st raw: } & \lambda_1, \ldots, \lambda_n, \\ (i+1)\text{th raw: } & -(\delta_1, \delta_{1+i}), \ldots, -(\delta_n, \delta_{n+1});\ 1 \leq i \leq [n/2], \\ j\text{th column: } & (\lambda_j, (\delta_j, \delta_{j+1}), \ldots, (\delta_j, \delta_{j+[n/2]}))^t;\ 1 \leq j \leq n (\operatorname{mod} n). \end{aligned}$$

We illustrate this by providing a specific example.

Example 1. Let δ_1, δ_2, δ_3, δ_4, δ_5 be elements with positive squares, each of them orthogonal to exactly one side of a convex polytope of five sides in hyperbolic space. So, $1 \leq i \leq [5/2]$ and

$$G(A) = \begin{pmatrix} \lambda_1 & \lambda_2 & \lambda_3 & \lambda_4 & \lambda_5 \\ \delta_1\,\delta_2 & \delta_2\,\delta_3 & \delta_3\,\delta_4 & \delta_4\,\delta_5 & \delta_5\,\delta_1 \\ \delta_1\,\delta_3 & \delta_2\,\delta_4 & \delta_3\,\delta_5 & \delta_4\,\delta_1 & \delta_5\,\delta_2 \end{pmatrix}.$$

Our problem is to find all matrices $G(A)$, having the lattice Weyl vector ρ, $\rho \in M \otimes \mathbb{Q}$, such that $(\rho, \alpha) = -(\alpha, \alpha)/2 \Leftrightarrow (\rho, \delta_i) = -\lambda_i$, $i = 1, \ldots, n$. The answer has been presented by Gritsenko and Nikulin in Ref. [6, Theorem 1.2.1].

Theorem 1. *All geometric realizations $G(A)$, of hyperbolic generalized Cartan matrices A of rank 3, of elliptic type, with the lattice Weyl vector, which are twisted to symmetric generalized Cartan matrices, all twisting coefficients λ_i satisfying*

$$\lambda_i \leq 12$$

are given in Ref. [6, *Table* 1].

Remark 5. Table 1 in Ref. [6] lists 60 matrices; seven of them are of the compact case (they represent a convex polytope of finite volume in hyperbolic space), with four being untwisted. The remaining 53 matrices are of the non-compact case.[c]

The main aim of the next section will be to discuss the conjecture which follows Theorem 1.3.1.

Conjecture 1 (Gritsenko–Nikulin). *Table* 1 *in Ref.* [6] *gives the complete list of hyperbolic generalized Cartan matrices A of rank* 3, *of elliptic type, with the lattice Weyl vector, which are twisted to symmetric generalized Cartan matrices.*

In other words, one can drop the inequality $\lambda_i \leq 12$ from Ref. [6, Theorem 1.2.1].

In the same paper, Gritsenko and Nikulin presented the following arguments for supporting the conjecture:

(1) The number of all hyperbolic generalized Cartan matrices of elliptic type, with the lattice Weyl vector, is finite, for rank greater than or equal to 3.
So, there exists an absolute constant m, such that $\lambda_i \leq m$.
(2) Calculations were done for all $\lambda_i \leq 12$, but the result has only matrices with all $\lambda_i \leq 6$.
So, there definitely do not exist new solutions in between 6 and 12.

[c]All of these matrices can be found in Ref. [6], 166–168.

4. Conjecture of Gritsenko–Nikulin

In this section, we propose a four-step algorithm for approaching the mentioned conjecture, by giving the complete solution for the first proposed step and by suggesting algorithmic procedure for the remaining steps. (A complete solution will be hopefully presented in a future work.)

STEP 1

We consider a triangle in hyperbolic space, with sides a, b and c. The angle between a and b is $\pi/2$, between a and c is $\pi/3$ and between c and b is 0 radians, that is, the vertex which is created from the intersection of c and b is at infinity of our space. This triangle will be the fundamental chamber for reflection in $\Lambda^+(M)$, and these reflections will cover the entire space, tending to infinity.

We bare in mind the following relations, presented in the proof of [6, Theorem 1.2.1]

$$0 \leq (\delta_1, \delta_2) \leq 2,\ 0 \leq (\delta_1, \delta_3) < 14,\ 0 \leq (\delta_2, \delta_3) \leq 2, \tag{1}$$

for δ_1, δ_2, $\delta_3 \in \tilde{P}(\mathcal{M})$ being orthogonal vectors to three consecutive sides, of a polygon $A_1 \ldots A_n$ in $\Lambda^+(M)$, namely A_1A_2, A_2A_3, A_3A_4.

One should find all these δ_1, δ_2, δ_3 satisfying (1) for the group generated by reflections in $\Lambda^+(M)$, with a fundamental chamber in the shape of a triangle with sides a, b, c.

We fix $\delta_2 = a$. Then, we have the following possibilities for δ_1:

$$\delta_1 = c,$$
$$s_c(b) = b - 2((b,c)/c^2)c = b + 2c,$$
$$s_{b+2c}(a) = a - 2((a, b+2c)/(b+2c)^2)(b+2c) = a + 2b + 4c,$$

etc.

In other words:

$$\delta_1 = na + (n+1)b + 2(n+1)c.$$

Now, the possibilities for δ_3 are

$$s_b(c) = c - 2((c,b)/b^2)b \Rightarrow \delta_3 = 2b + c,$$
$$s_{2b+c}(-b) = -b - 2((b, 2b+c)/(2b+c)^2)(2b+c) \Rightarrow \delta_3 = 3b + 2c.$$

Also,

$$s_{3b+2c}(a) = a - 2((a, 3b+2c)/(3b+2c)^2)(3b+2c) = a+6b+4c,$$

$$s_{a+6b+4c}(-3b-2c) = (-3b-2c) - 2((-3b-2c, a+6b+4c)/ (a+6b+4c)^2)(a+6b+4c) = 2a+9b+6c,$$

etc.

So,

$$\delta_3 = (n+1)a + (3n+6)b + (2n+4)c.$$

Remark 6. We fixed $\delta_2 = a$, and we looked for possible δ_1 and δ_3, such that the angle between δ_2 and δ_1 is acute. The same applies for the angle between δ_2 and δ_3.

For $\delta_2 \in \{a, b, c\}$, Table 1 presents all possible values for δ_1 and δ_3 so that $(\delta_1, \delta_3) < 14$. For the sake of simplicity, an entry (k_1, k_2, k_3) in Table 1 corresponds to the expression $k_1\, a + k_2\, b + k_3\, c$.

STEP 2

For each of the 115 triples of elements δ_1, δ_2, δ_3, given above, we need to find the corresponding twisting coefficients, λ_1, λ_2, λ_3. Therefore, we introduce three new elements, $\tilde{\delta_1} = \lambda_1\, \delta_1$, $\tilde{\delta_2} = \lambda_2\, \delta_2$, $\tilde{\delta_3} = \lambda_3, \delta_3$.

As we mentioned in the previous section, any $\tilde{\delta_i}$ and $\tilde{\delta_j}$ should satisfy the following relations:

$$\begin{aligned} \tilde{\delta_i}^2 \,|\, 2(\tilde{\delta_i}, \tilde{\delta_j}) \quad &\Rightarrow \quad 2\lambda_i |\, 2\lambda_j(\delta_i, \delta_j), \\ \tilde{\delta_j}^2 \,|\, 2(\tilde{\delta_i}, \tilde{\delta_j}) \quad &\Rightarrow \quad 2\lambda_j |\, 2\lambda_i(\delta_i, \delta_j). \end{aligned} \tag{2}$$

By $\nu_p(\lambda)$, we denote the power of the prime number p, in the prime factorization of the natural number λ. For example, $\nu_3(21) = 1$ and $\nu_2(21) = 0$.

Applying this notation to our case, we obtain that for $\tilde{\delta_1}$, $\tilde{\delta_2}$, $\tilde{\delta_3}$, the following conditions should be satisfied:

$$|\nu_p(\lambda_1) - \nu_p(\lambda_2)| \leq \nu_p(g_{12}),$$

$$|\nu_p(\lambda_1) - \nu_p(\lambda_3)| \leq \nu_p(g_{13}),$$

$$|\nu_p(\lambda_2) - \nu_p(\lambda_3)| \leq \nu_p(g_{23}),$$

where g_{12}, g_{13}, g_{23} are the elements of the Gram matrix of δ_1, δ_2, δ_3, with diagonal equal to 2 and (by definition) $g_{ij} = g_{ji}$.

Table 1: Feasible triples $(\delta_1, \delta_2, \delta_3)$.

$\delta_2 = (1,0,0)$		$\delta_2 = (0,1,0)$		$\delta_2 = (0,0,1)$	
δ_1	δ_3	δ_1	δ_3	δ_1	δ_3
(0, 0, 1)	(0, 1, 0)	(0, 0, 1)	(1, 0, 0)	(0, 1, 0)	(1, 0, 0)
(0, 1, 2)	(0, 1, 0)	(0, 1, 2)	(1, 0, 0)	(0, 2, 1)	(1, 0, 0)
(1, 2, 4)	(0, 1, 0)	(0, 2, 3)	(1, 0, 0)	(0, 3, 2)	(1, 0, 0)
(2, 3, 6)	(0, 1, 0)	(0, 3, 4)	(1, 0, 0)	(0, 4, 3)	(1, 0, 0)
(3, 4, 8)	(0, 1, 0)	(0, 4, 5)	(1, 0, 0)	(0, 5, 4)	(1, 0, 0)
(4, 5, 10)	(0, 1, 0)	(0, 5, 6)	(1, 0, 0)	(0, 6, 5)	(1, 0, 0)
(5, 6, 12)	(0, 1, 0)	(0, 6, 7)	(1, 0, 0)	(0, 7, 6)	(1, 0, 0)
(0, 0, 1)	(0, 2, 1)	(0, 7, 8)	(1, 0, 0)	(0, 8, 7)	(1, 0, 0)
(0, 0, 1)	(0, 3, 2)	(0, 8, 9)	(1, 0, 0)	(0, 9, 8)	(1, 0, 0)
(0, 0, 1)	(1, 6, 4)	(0, 9, 10)	(1, 0, 0)	(0, 10, 9)	(1, 0, 0)
(0, 0, 1)	(2, 9, 6)	(0, 10, 11)	(1, 0, 0)	(0, 11, 10)	(1, 0, 0)
(0, 0, 1)	(3, 12, 8)	(0, 11, 12)	(1, 0, 0)	(0, 12, 11)	(1, 0, 0)
(0, 1, 2)	(0, 2, 1)	(0, 12, 13)	(1, 0, 0)	(0, 13, 12)	(1, 0, 0)
(1, 2, 4)	(0, 2, 1)	(0, 0, 1)	(1, 0, 1)	(0, 14, 13)	(1, 0, 0)
(2, 3, 6)	(0, 2, 1)	(0, 0, 1)	(2, 1, 2)	(0, 1, 0)	(2, 1, 2)
(3, 4, 8)	(0, 2, 1)	(0, 0, 1)	(3, 2, 3)	(0, 1, 0)	(8, 6, 9)
(0, 1, 2)	(0, 3, 2)	(0, 0, 1)	(4, 3, 4)	(0, 1, 0)	(12, 9, 14)
(1, 2, 4)	(0, 3, 2)	(0, 0, 1)	(5, 4, 5)	(0, 1, 0)	(16, 12, 19)
(2, 3, 6)	(0, 3, 2)	(0, 0, 1)	(6, 5, 6)	(0, 2, 1)	(2, 1, 2)
(0, 1, 2)	(1, 6, 4)	(0, 0, 1)	(7, 6, 7)	(0, 2, 1)	(3, 2, 3)
(0, 1, 2)	(2, 9, 6)	(0, 0, 1)	(8, 7, 8)	(0, 2, 1)	(4, 3, 4)
		(0, 0, 1)	(9, 8, 9)	(0, 2, 1)	(5, 4, 5)
		(0, 0, 1)	(10, 9, 10)	(0, 2, 1)	(6, 5, 6)
		(0, 0, 1)	(11, 10, 11)	(0, 2, 1)	(7, 6, 7)
		(0, 0, 1)	(12, 11, 12)	(0, 2, 1)	(8, 7, 8)
		(0, 0, 1)	(13, 12, 13)	(0, 2, 1)	(9, 8, 9)
		(0, 0, 1)	(14, 13, 14)	(0, 2, 1)	(10, 9, 10)
		(0, 0, 1)	(15, 14, 15)	(0, 2, 1)	(11, 10, 11)
		(0, 1, 2)	(1, 0, 1)	(0, 3, 2)	(2, 1, 2)
		(0, 1, 2)	(2, 1, 2)	(0, 3, 2)	(3, 2, 3)
		(0, 1, 2)	(3, 2, 3)	(0, 3, 2)	(4, 3, 4)
		(0, 1, 2)	(4, 3, 4)	(0, 3, 2)	(5, 4, 5)
		(0, 1, 2)	(5, 4, 5)	(0, 4, 3)	(2, 1, 2)
		(0, 1, 2)	(6, 5, 6)	(0, 4, 3)	(3, 2, 3)
		(0, 1, 2)	(7, 6, 7)	(0, 5, 4)	(2, 1, 2)
		(0, 2, 3)	(1, 0, 1)	(0, 5, 4)	(3, 2, 3)
		(0, 2, 3)	(2, 1, 2)	(0, 6, 5)	(2, 1, 2)
		(0, 2, 3)	(3, 2, 3)		
		(0, 2, 3)	(4, 3, 4)		
		(0, 2, 3)	(5, 4, 5)		
		(0, 3, 4)	(1, 0, 1)		
		(0, 3, 4)	(2, 1, 2)		
		(0, 3, 4)	(3, 2, 3)		
		(0, 4, 5)	(1, 0, 1)		
		(0, 4, 5)	(2, 1, 2)		
		(0, 4, 5)	(3, 2, 3)		

(Continued)

Table 1: (*Continued*)

$\delta_2 = (1,0,0)$		$\delta_2 = (0,1,0)$		$\delta_2 = (0,0,1)$	
δ_1	δ_3	δ_1	δ_3	δ_1	δ_3
		$(0,5,6)$	$(1,0,1)$		
		$(0,5,6)$	$(2,1,2)$		
		$(0,6,7)$	$(1,0,1)$		
		$(0,6,7)$	$(2,1,2)$		
		$(0,7,8)$	$(1,0,1)$		
		$(0,7,8)$	$(2,1,2)$		
		$(0,8,9)$	$(1,0,1)$		
		$(0,9,10)$	$(1,0,1)$		
		$(0,10,11)$	$(1,0,1)$		
		$(0,11,12)$	$(1,0,1)$		
		$(0,12,13)$	$(1,0,1)$		
		$(0,13,14)$	$(1,0,1)$		
		$(0,14,15)$	$(1,0,1)$		

The information we have collected is sufficient for calculating λ_1, λ_2, λ_3, for each of the 115 triples that we determined in STEP 1, simply by working on the Gram matrix for each case.

An alternative way to do these calculations is via programming. Here we present Algorithm 1, which can be implemented for instance in Matlab.

Running Algorithm 1 with $(2,4;2,2;4,2)$ we get the following nine solutions.

$$
\begin{array}{lll}
1,1,1; & 1,1,2; & 1,2,1; \\
1,2,2; & 1,2,4; & 2,1,1; \\
2,1,2; & 2,2,1; & 4,2,1;
\end{array}
$$

We now explain how this algorithm works. For each δ_1, δ_2, δ_3 we apply (2) and obtain six relations of the type:

$$
\begin{array}{lll}
x \big| x_y * y; & y \big| y_x * x; & z \big| z_x * x; \\
x \big| x_z * z; & y \big| y_z * z; & z \big| z_y * y;
\end{array}
$$

where $\gcd(x,y,z) = 1$. It is sufficient to examine the values of x, which belong to the set S_x, of the divisors of the least common multiple $\operatorname{lcm}(x_y, x_z)$. In other words, $\operatorname{lcm}(x,y,z) = x$. Similarly for y and z, we have the sets S_y and S_z, resp., and check for which triples (x,y,z) (with x, y, z belonging to S_x, S_y, S_z, resp.) the conditions hold.

Algorithm 1 STEP-2

INPUT: $(x_y, x_z; y_x, y_z; z_x, z_y)$
OUTPUT: Finds x, y, and z, where $\gcd(x, y, z) = 1$.

```
procedure FIND FEASIBLE SOLUTIONS (x, y, z)
    count ← 0.
    for x ∈ divisor(lcm(x_y, x_z)) do
        for y ∈ divisor(lcm(y_x, y_z)) do
            for z ∈ divisor(lcm(z_x, z_y)) do
                # Consider all distinct pairs a, b ∈ {x, y, z}.
                if mod(a_b b, a) = 0 and gcd(x, y, z) = 1 then
                    # Producing a new solution.
                    count ← count + 1
                    Print solution (x, y, z).
                end if
            end for
        end for
    end for
end procedure
```

STEP 3

The third step serves to find for each triple δ_1, δ_2, δ_3 the lattice Weyl vector. As we already mentioned, in the previous section the following relation should be satisfied:

$$(\rho, \delta_i) = -\lambda_i, \quad 1 \leq i \leq 3.$$

Furthermore, $(\rho, \rho) < 0$ should always hold.

This calculation should be implemented into the computer program as STEP 3, after STEP 1 and STEP 2. This is a crucial step, since it is expected that the final number of matrices will be reduced significantly.

STEP 4

At this stage, we try to find more elements δ_i, which should be orthogonal to sides of polytopes in $\Lambda^+(M)$; polytopes to which δ_1, δ_2, δ_3 are orthogonal (each one perpendicular to a side of the polytope, resp.).

Consequently, for each case (from STEP 1 to STEP 3) we should examine if one more side exists, where a new element δ_4 should be orthogonal to this side. Again it will be of vital interest to prove the existence of the lattice Weyl vector.

What we already know is that the determinant of the Gram matrix of δ_1, δ_2, δ_3, δ_4 should be equal to zero, since δ_1, δ_2, δ_3 should come from a 3D lattice. Therefore, the four elements δ_1, δ_2, δ_3, δ_4 should be linearly independent.

Also, (2) should take the following form:

$$(\lambda_4\,\delta_4)^2 \,|\, 2(\lambda_4\,\delta_4, \lambda_i\,\delta_i),$$

where $\delta_4 = x_1\delta_1 + x_2\delta_2 + x_3\delta_3$, $x_i \in \mathbb{Q}$.

In case δ_4 does not exist, we stop. Otherwise, we repeat the same procedure for a new element, δ_5, and so on.

In this way, we will have proved the conjecture to [9, Theorem 1.1.2], when we have found all convex polytopes in the hyperbolic space, which correspond to the matrices from Ref. [6, Table 1].

Acknowledgment

The third author would like to express his gratitude to V.V. Nikulin, for the introduction to this field.

References

[1] V.G. Kac, *Infinite Dimensional Lie Algebras* (Cambridge University Press, 1990, 3rd edn.).

[2] V.V. Nikulin, A lecture on Kac–Moody lie algebras of arithmetic type, Preprint alg-geom/9412003.

[3] V.A. Gritsenko and V.V. Nikulin, Lorentzian Kac–Moody algebras with Weyl groups of 2-reflections, *Proc. Lon. Math. Soc.* **116**, 485–533, (2018).

[4] R. Borcherds, The monster Lie algebra, *Adv. Math.* **83**, 30–47, (1990).

[5] R. Borcherds, The monstrous moonshine and monstrous Lie superalgebras, *Invent. Math.* **109**, 405–444, (1992).

[6] V.A. Gritsenko and V.V. Nikulin, Automorphic forms and Lorentzian Kac–Moody algebras, part I, *Int. J. Math.* **9**(2), 201–275, (1998).

[7] J.-P. Françoise, G.L. Naber, and T.S. Tsun (Eds.), *Encyclopedia of Mathematical Physics* (Elsevier, 2006).

[8] E. Date, Solitons and Kac–Moody Lie algebras, *Encyclopedia of Mathematical Physics* (Elsevier, 2006).

[9] V.V. Nikulin, Reflection groups in hyperbolic spaces and the denominator formula for Lorentzian Kac–Moody algebras, *Izv. Math.* **60**(2), 305–334, (1996).

[10] E.B. Vinberg, Hyperbolic reflection groups, *Russ. Math. Surv.* **40**(1), 31–75, (1985).
[11] V.V. Nikulin and I.R. Shafarevich, *Geometries and Groups* (Springer-Verlag, 1994).

https://doi.org/10.1142/9789811261572_0005

Chapter 5

Dynamic Geometry Generated by the Circumcircle Midarc Triangle

Dorin Andrica[*,‡] and Dan Ştefan Marinescu[†,§]

[*] *Babeş-Bolyai University, Faculty of Mathematics and Computer Science, Cluj-Napoca 400084, Romania*

[†] *"Iancu de Hunedoara" National College, Hunedoara, Romania,*

[‡] *dandrica@math.ubbcluj.ro*

[§] *marinescuds@gmail.com*

We present some new aspects involving some geometrical iterative processes mainly focused on the iteration generated by the circumcircle midarc triangle. Applications to the interpolation of the classical Euler's inequality are provided and some extensions of the results to arbitrary polygons are also given.

1. Introduction

Given a fixed plane configuration $\mathcal{F}_0$ and a sequence of plane transformations $(T_n)_{n\geq 0}$, we consider the iterative process described by

$$\mathcal{F}_0 \xrightarrow{T_0} \mathcal{F}_1 \xrightarrow{T_1} \mathcal{F}_2 \xrightarrow{T_2} \cdots \xrightarrow{T_{n-1}} \mathcal{F}_n \xrightarrow{T_n} \mathcal{F}_{n+1} \xrightarrow{T_{n+1}} \cdots$$

This means that $\mathcal{F}_0$ is transformed by T_0 in $\mathcal{F}_1$ which is transformed by T_1 in $\mathcal{F}_2$, etc. Clearly, after n steps the initial configuration $\mathcal{F}_0$ is transformed in $\mathcal{F}_n$ by the composition $T_{n-1} \circ T_{n-2} \circ \cdots \circ T_0$. We call such an iterative process the *dynamic geometry* generated by $\mathcal{F}_0$ and the sequence $(T_n)_{n\geq 0}$. The initial configuration $\mathcal{F}_0$ could be any standard configuration defined using polygons (triangles, quadrilaterals, etc.), circles, and associated geometric elements.

There are some general problems arising in the study of a dynamic geometry:

(1) Describe the n-step configuration $\mathcal{F}_n$ and its geometric elements;
(2) Study the convergence of the sequence $(\mathcal{F}_n)_{n\geq 0}$;
(3) Study the convergence in shape of the sequence $(\mathcal{F}_n)_{n\geq 0}$;
(4) If the above sequence is convergent, find the convergence order of some of its elements;
(5) Obtain properties of the initial configuration $\mathcal{F}_0$ from the study of the geometry of $\mathcal{F}_n$ for some $n \geq 1$.

In case of convergence, it seems that finding the limit of a certain iterative geometric process is a much more difficult problem than the one concerning the limiting shape. One reason is that in some configurations the angle computations are easier than computations involving distances, ratios, etc.

The present chapter is organized into six sections. In Section 2, we review some examples of iterative processes inspired by simple geometrical configurations: the Kasner triangles in various situations, the dynamic geometry generated by the incircle and the circumcircle of a triangle, the pedal triangle, the orthic triangle and the incentral triangle. The possible extensions to the arbitrary polygons are discussed in the last subsection. Let us mention that such recursive systems describing some dynamic geometries are considered by Abbot [1], Chang and Davis [2], Clarke [3], Ding *et al.*, Hitt and Zhang [5], and Ismailescu and Jacobs [6]. Section 3 contains some new results involving the iteration process generated by the circumcircle midarc triangle and the main result is contained in Theorem 5. Some new results on the interpolation of the classical Euler inequality $R \geq 2r$ are given in Theorem 6 and Corollary 1 of Section 4. The complete asymptotic expansion of the sequence $(r_n)_{n\geq 0}$ of inradii generated by this process is obtained in Theorem 7 of Section 5. Section 6 is devoted to the extension of some of the above results to polygons.

In order to study an iterative geometric process, a useful method is to associate complex coordinates to the points and to transfer the geometrical problem in one in the complex plane. We illustrate this idea in few of the examples in Section 2. Also, this is the principal technique used in Sections 3–6.

2. Examples of Geometrical Iterative Processes

2.1. *Classical Kasner triangles*

Perhaps the simplest example of this type of construction uses the so-called Kasner triangles (named after E. Kasner (1878–1955)). In this construction, we consider the initial configuration $\mathcal{F}_0$ to be the triangle $A_0B_0C_0$ and one forms $\mathcal{F}_1$ which is the triangle $A_1B_1C_1$ whose vertices are the mid-points of the edges of $\mathcal{F}_0$. Then the third configuration $\mathcal{F}_2$ is the triangle $A_2B_2C_2$ formed by the midpoints of the edges of $\mathcal{F}_1$ (see Fig. 1). Continuing this process one obtains a sequence of triangles connected by the corresponding transformations, which in this case are affine transformations.

The geometry of $\mathcal{F}_n$ is very simple since the triangle $A_nB_nC_n$ is similar to $A_0B_0C_0$ with the similarity ratio $1/2^n$.

Considering the complex coordinates of the vertices of the triangle $\mathcal{F}_n$, we obtain that the process is described by the points $A_n(a_n), B_n(b_n), C_n(c_n)$, where for $n = 0, 1, \ldots$, we have

$$\begin{cases} a_{n+1} = \dfrac{1}{2}(b_n + c_n) \\[2mm] b_{n+1} = \dfrac{1}{2}(c_n + a_n), \\[2mm] c_{n+1} = \dfrac{1}{2}(a_n + b_n) \end{cases} \tag{1}$$

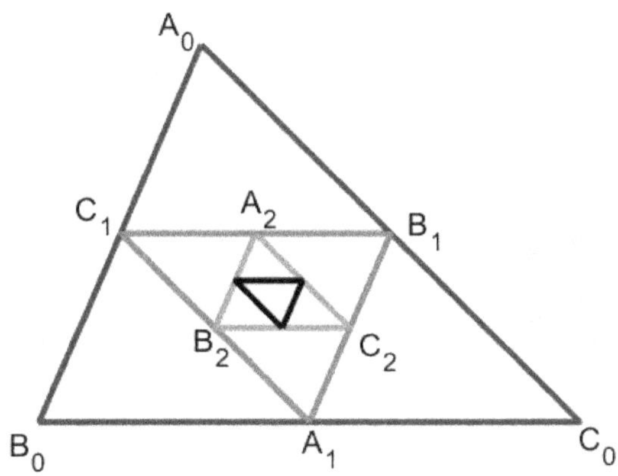

Fig. 1: Sequence of median triangles.

and a_0, b_0, c_0 are fixed complex numbers. An elementary approach of system (1) is to add the equations and obtain

$$a_{n+1} + b_{n+1} + c_{n+1} = a_n + b_n + c_n = \cdots = a_0 + b_0 + c_0 = 3g_0,$$

where g_0 is the complex coordinates of the centroid G of the initial triangle $A_0B_0C_0$ (see the book of T. Andreescu and D. Andrica [7]).

By replacing in the first equation of (1), it follows $b_n + c_n = 3g_0 - a_n$, hence

$$a_{n+1} = \frac{1}{2}(3g_0 - a_n), \quad n = 0, 1, \ldots,$$

That is

$$a_{n+1} - g_0 = -\frac{1}{2}(a_n - g_0), \quad n = 0, 1, \ldots.$$

From the last relation we obtain

$$a_n - g_0 = -\frac{(-1)^n}{2^n}(a_0 - g_0), \quad n = 0, 1, \ldots, \tag{2}$$

and similar formulas for the sequences $(b_n)_{n\geq 0}$ and $(c_n)_{n\geq 0}$. These formulas show the convergence

$$a_n \to g_0, \; b_n \to g_0, \; c_n \to g_0,$$

that is the limit of sequence $(\triangle A_nB_nC_n)_{n\geq 0}$ is the degenerated triangle at G. The order of convergence of $(a_n)_{n\geq 0}, (b_n)_{n\geq 0}$ and $(c_n)_{n\geq 0}$ is obtained also from (2)

$$\begin{aligned}\lim_{n\to+\infty} 2^n|a_n - g_0| &= \lim_{n\to+\infty} 2^n|b_n - g_0| \\ &= \lim_{n\to+\infty} 2^n|c_n - g_0| = |a_0 - g_0|.\end{aligned}$$

2.2. *Kasner triangles with a fixed weight*

An extension of the previous construction is the following. Consider a real number $\alpha \in \mathbb{R}^*, \alpha \neq \frac{1}{2}$ and the initial triangle $A_0B_0C_0$. Now construct the triangle $A_1B_1C_1$ such that A_1, B_1, C_1 divides the segments $[B_0C_0], [C_0A_0]$, resp., $[A_0B_0]$ in the ratio $1-\alpha : \alpha$ (see Fig. 2). Continuing this process one obtains a sequence of triangles $A_nB_nC_n$ connected by the corresponding transformations which in this case are also affine transformations.

The complex coordinates a_n, b_n, c_n of the vertices of the triangle $A_nB_nC_n$ satisfy the recurrent system

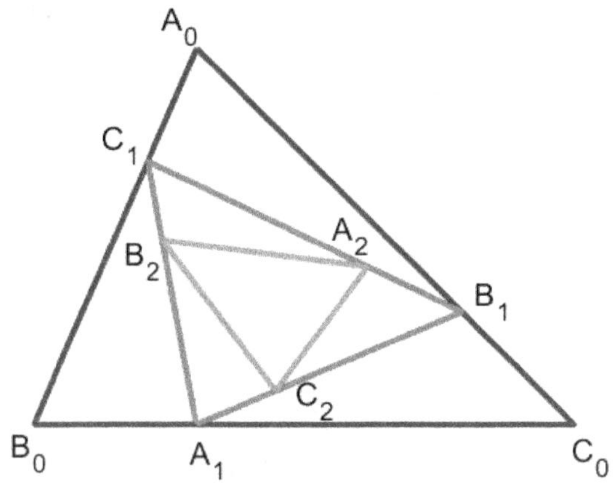

Fig. 2: Sequence of triangles for $\alpha = 0.3$.

$$\begin{cases} a_{n+1} = \alpha b_n + (1-\alpha)c_n \\ b_{n+1} = \alpha c_n + (1-\alpha)a_n, \\ c_{n+1} = \alpha a_n + (1-\alpha)b_n \end{cases} \tag{3}$$

where a_0, b_0, c_0 are fixed complex numbers. If we add the equations of system (3), one obtains again that for every integer $n \geq 0$ we have $a_n + b_n + c_n = a_0 + b_0 + c_0 = 3g_0$, where g_0 is the complex coordinate of the centroid G of the initial triangle $A_0B_0C_0$, but the argument discussed in Example 1 does not easily work. That is why we need a higher level approach.

It is easy to see that system (3) is equivalent to the following matrix relation:

$$\begin{pmatrix} a_{n+1} \\ b_{n+1} \\ c_{n+1} \end{pmatrix} = \begin{pmatrix} 0 & \alpha & 1-\alpha \\ 1-\alpha & 0 & \alpha \\ \alpha & 1-\alpha & 0 \end{pmatrix} \begin{pmatrix} a_n \\ b_n \\ c_n \end{pmatrix}, \tag{4}$$

therefore,

$$\begin{pmatrix} a_n \\ b_n \\ c_n \end{pmatrix} = U^n \begin{pmatrix} a_0 \\ b_0 \\ c_0 \end{pmatrix}, \tag{5}$$

where U is the circulant double stochastic matrix (see the book of Davis [8]) given by

$$U = \begin{pmatrix} 0 & \alpha & 1-\alpha \\ 1-\alpha & 0 & \alpha \\ \alpha & 1-\alpha & 0 \end{pmatrix}. \tag{6}$$

The matrix U has the characteristic polynomial $p_U(t) = (t-1)(t^2+t+3\alpha^2-3\alpha+1)$, hence its eigenvalues are $t_1 = 1$ and $t_{2,3} = \frac{1}{2}(-1 \pm i|2\alpha-1|\sqrt{3})$. It follows that we have

$$U = P \begin{pmatrix} 1 & 0 & 0 \\ 0 & t_2 & 0 \\ 0 & 0 & t_3 \end{pmatrix} P^{-1}, \tag{7}$$

for some non-singular matrix P, hence for every positive integer $n \geq 0$ we have

$$U^n = P \begin{pmatrix} 1 & 0 & 0 \\ 0 & t_2^n & 0 \\ 0 & 0 & t_3^n \end{pmatrix} P^{-1}. \tag{8}$$

The relation (5) shows that the sequences $(a_n)_{n\geq 0}, (b_n)_{n\geq 0}, (c_n)_{n\geq 0}$ are convergent if and only if the matrix sequence $(U^n)_{n\geq 0}$ is convergent. But the convergence of $(U^n)_{n\geq 0}$ is reduced by relation (8) to $t_2^n \to 0$ and $t_3^n \to 0$, which is equivalent to $|t_{2,3}| < 1$. On the other hand, we have $|t_2|^2 = |t_3|^2 = t_2 t_3 = 3\alpha^2 - 3\alpha + 1$, and we have obtained the following result:

Theorem 1. *The sequence $(\triangle A_n B_n C_n)_{n\geq 0}$ is convergent if and only if $\alpha \in (0,1)$. When the sequence is convergent, its limit is the degenerated triangle at G, the centroid of $\triangle A_0 B_0 C_0$.*

In order to find the limit of this sequence of triangles, let us consider $\alpha \in (0,1)$ and $a_n \to a, b_n \to b, c_n \to c$. Passing to the limit in system (3), we obtain

$$\begin{cases} a = \alpha b + (1-\alpha)c \\ b = \alpha c + (1-\alpha)a \\ c = \alpha a + (1-\alpha)b, \end{cases} \tag{9}$$

hence $a - c = \alpha(b-c), b - a = \alpha(c-a), c - b = \alpha(a-b)$. If $a \neq b, b \neq c, c \neq a$, then by multiplying the above relation it follows $\alpha^3 = -1$, a contradiction.

If we have one equality, for instance $a = b$, then we get $b = c$, and finally $a = b = c$. Now, passing to the limit in $a_n + b_n + c_n = 3g_0$, one obtains $a = b = c = g_0$. The conclusion is that the limit of sequence $(\triangle A_nB_nC_n)_{n\geq 0}$ is the degenerated triangle at point G.

Concerning this process, the following interesting result is stated in the paper of Ismailescu and Jacobs [6]. Let

$$\theta = \cos^{-1}\left(-1 + \frac{1}{2(1-3\alpha+3\alpha^2)}\right),$$

If $\theta = (p/q)\pi$, where p and q are positive integers, then the shape sequence is periodical. More precisely, for every $k \geq 0$, triangles $A_kB_kC_k$ and $A_{k+2q}B_{k+2q}C_{k+2q}$ are similar.

2.3. *Kasner triangles with three fixed weights*

A natural extension of the previous construction is the following. Consider real numbers $\alpha, \beta, \gamma \in \mathbb{R}$ and the initial triangle $A_0B_0C_0$. Now construct the triangle $A_1B_1C_1$ such that A_1, B_1, C_1 divides the segments $[B_0C_0]$, $[C_0A_0]$ and, resp., $[A_0B_0]$ in the ratios $1-\alpha : \alpha, 1-\beta : \beta$ and $1-\gamma : \gamma$ (see Fig. 3). Continuing this process one obtains a sequence of triangles $A_nB_nC_n$ connected by the corresponding transformations, which in this case are also affine transformations.

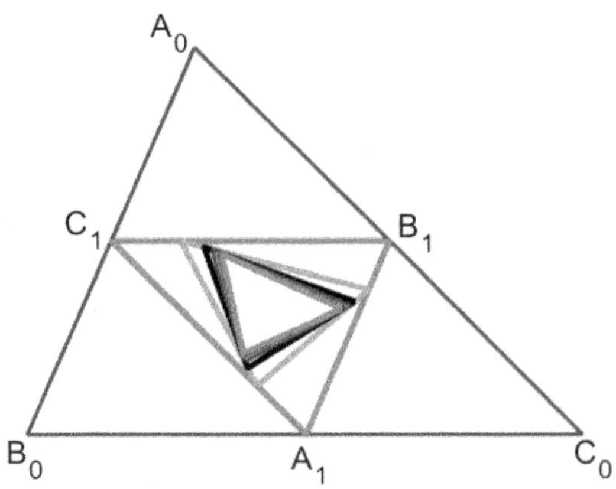

Fig. 3: Sequence of triangles with $t_n = \frac{1}{2^n}, n = 1, 2, \ldots$.

The complex coordinates a_n, b_n, c_n of the vertices of the triangle $A_nB_nC_n$ satisfy the recurrent system

$$\begin{cases} a_{n+1} = \alpha b_n + (1-\alpha)c_n \\ b_{n+1} = \beta c_n + (1-\beta)a_n, \\ c_{n+1} = \gamma a_n + (1-\gamma)b_n \end{cases} \tag{10}$$

where a_0, b_0, c_0 are fixed complex numbers. As in the previous example, the behavior of this dynamic process is controlled by the row stochastic matrix

$$\begin{pmatrix} 0 & \alpha & 1-\alpha \\ 1-\beta & 0 & \beta \\ \gamma & 1-\gamma & 0 \end{pmatrix}. \tag{11}$$

In the forthcoming paper [9], some graphical simulations are given and the following result is proved.

Theorem 2. *The sequence $(\triangle A_nB_nC_n)_{n\geq 0}$ is convergent if and only if the following inequality holds*

$$|1 \pm \sqrt{4(\alpha+\beta+\gamma-\alpha\beta-\beta\gamma-\gamma\alpha)-3}| < 2. \tag{12}$$

In case of convergence, the limit is the degenerated triangle at G, the centroid of $\triangle A_0B_0C_0$.

2.4. *Kasner triangles with a fixed sequence of weights*

A very general extension of the construction in Subsection 1.2, is the following. Consider a sequence $(t_n)_{n\geq 0}$ of real numbers and the initial triangle $A_0B_0C_0$. Now construct the triangle $A_1B_1C_1$ such that A_1, B_1, C_1 divides the segments $[B_0C_0], [C_0A_0]$ and $[A_0B_0]$ in the ratio $1-t_0 : t_0$, resp. Continuing this process one obtains a sequence of triangles $A_nB_nC_n$ connected by the corresponding transformations, which in this case are also affine transformations.

The complex coordinates a_n, b_n, c_n of the vertices of the triangles $A_nB_nC_n$ obtained in this way satisfy the recurrent system

$$\begin{cases} a_{n+1} = t_n b_n + (1-t_n)c_n \\ b_{n+1} = t_n c_n + (1-t_n)a_n, \\ c_{n+1} = t_n a_n + (1-t_n)b_n \end{cases} \tag{13}$$

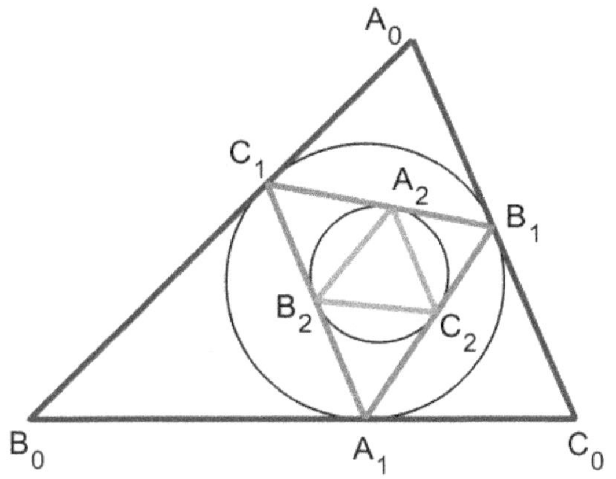

Fig. 4: Triangles generated by incircle–circumcircle iteration.

where a_0, b_0, c_0 are fixed complex numbers. The system (13) is equivalent to the matrix relation

$$\begin{pmatrix} a_{n+1} \\ b_{n+1} \\ c_{n+1} \end{pmatrix} = \begin{pmatrix} 0 & t_n & 1-t_n \\ 1-t_n & 0 & t_n \\ t_n & 1-t_n & 0 \end{pmatrix} \begin{pmatrix} a_n \\ b_n \\ c_n \end{pmatrix}, \tag{14}$$

therefore, for every positive integer n we have

$$\begin{pmatrix} a_n \\ b_n \\ c_n \end{pmatrix} = T_{n-1}T_{n-2}\dots T_0 \begin{pmatrix} a_0 \\ b_0 \\ c_0 \end{pmatrix}, \tag{15}$$

where T_n is the 3×3 circulant double-stochastic matrix in (14). In the forthcoming paper [9], so-called discrete Fourier transform is used to obtain some results concerning the system (13).

2.5. *The incircle–circumcircle dynamic geometry*

Given a triangle $A_0B_0C_0$ label by A_1, B_1, C_1 the points where the incircle touches the sides $[B_0C_0]$, $[C_0A_0]$ and $[A_0B_0]$, resp. Consider the new triangle $A_1B_1C_1$ (see Fig. 4). Similarly, we can form the triangle $A_2B_2C_2$ using triangle $A_1B_1C_1$. Continuing in this process we construct the sequence $(\triangle A_nB_nC_n)_{n\geq 0}$.

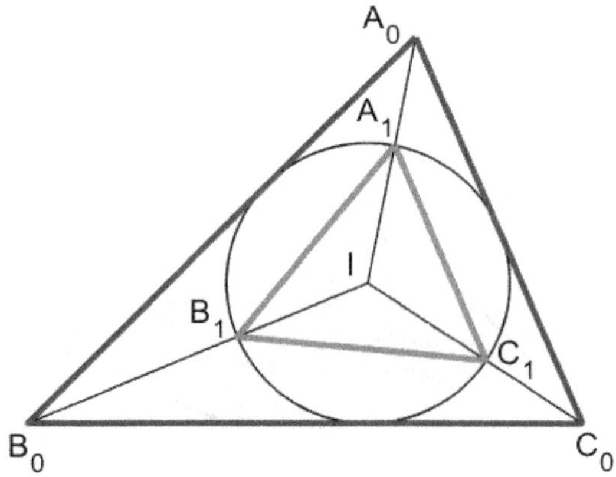

Fig. 5: Triangles obtained by intersections of bisectors with the incircle.

Hitt, Zhang [5] and Stewart [10] have shown that this sequence is convergent in shape and its limit is an equilateral triangle. A simple proof of this property is given by Ismailescu and Jacobs [6].

2.6. *The incircle dynamic geometry*

Given the initial triangle $A_0B_0C_0$ with incenter I, denote by A_1, B_1, C_1 the points where the line segments A_0I, B_0I and C_0I, resp., intersect the incircle of triangle $A_0B_0C_0$, and consider the triangle $A_1B_1C_1$ (see Fig. 5). In the same way, we can construct the triangles $A_2B_2C_2, \ldots, A_nB_nC_n, \ldots$.

Also, in the paper [6] it is proved that this sequence is convergent in shape and its limit is an equilateral triangle.

2.7. *The pedal triangle dynamic geometry*

Let $A_0B_0C_0$ be an arbitrary triangle and let P be a point inside the triangle. Drop the perpendiculars from the point P onto the lines, A_0B_0, A_0C_0 and B_0C_0. We label the points of intersection C_1, B_1 and A_1, resp. We can now form a new triangle, $A_1B_1C_1$. Similarly, we drop the perpendiculars from P onto A_1B_1, A_1C_1 and B_1C_1. We label the points of intersection C_2, B_2 and A_2, resp. Thus, we can form the triangle $A_2B_2C_2$ (see Fig. 6). In the same way, we can construct the triangles $A_3B_3C_3, \ldots, A_nB_nC_n, \ldots$

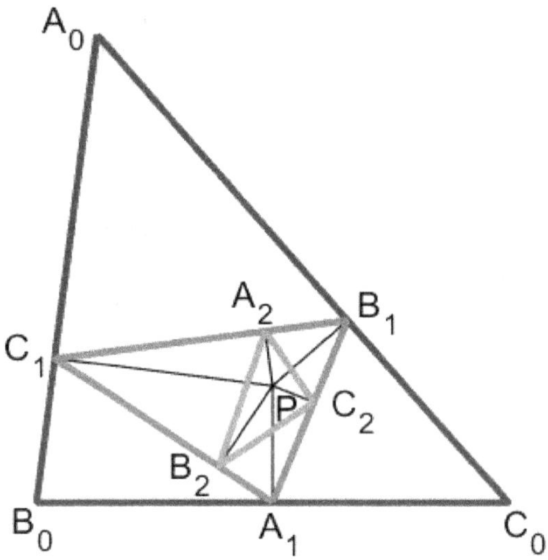

Fig. 6: Sequence of pedal triangles of the point P.

The following result, mentioned in Refs. [10] and [11], is important to understand the convergence in shape of this process:

Theorem 3 (Neuberg). *The triangles $A_0B_0C_0$ and $A_3B_3C_3$ are similar.*

The above result implies that the shape sequence of $(\triangle A_nB_nC_n)_{n\geq 0}$ is periodical of period 3.

2.8. *The orthic triangle dynamic geometry*

It is well known that the *orthic triangle* of a given triangle is the triangle whose vertices are the feet of altitudes from the vertices. Given the initial triangle $A_0B_0C_0$ denote by $A_1B_1C_1$ its orthic triangle. In the same way, we consider $\triangle A_2B_2C_2$ to be the orthic triangle of $\triangle A_1B_1C_1$. Continuing this process we obtain the sequence of triangles $(\triangle A_nB_nC_n)_{n\geq 0}$.

Kingston and Synge found in Ref. [12] necessary and sufficient conditions for the shape-sequence of $(\triangle A_nB_nC_n)_{n\geq 0}$ to be periodical for any given period p. Moreover, they showed that there are triangles $A_0B_0C_0$ for which the periodicity phenomenon appears only after an arbitrarily large number of iterations (they call this periodicity with delay). In other words, they show that given any positive integers p and d, there is a choice for $\triangle A_0B_0C_0$

such that no two triangles in the list $\triangle A_0B_0C_0, \triangle A_1B_1C_1, \ldots, \triangle A_dB_dC_d$ are similar to each other but $\triangle A_kB_kC_k$ is similar to $\triangle A_{k+p}B_{k+p}C_{k+p}$ for every $k \geq d$.

2.9. *The incentral triangle dynamic geometry*

Let $A_0B_0C_0$ be an arbitrary triangle, and let $\triangle A_1B_1C_1$ be the incentral triangle of the initial triangle. That is the triangle formed by the intersection points of the internal angle bisectors on its three sides (see Fig. 7). Construct $\triangle A_2B_2C_2, \triangle A_3B_3C_3, \ldots$ in the same manner.

The following result is proved by Ismailescu and Jacobs [6]:

Theorem 4. *The sequence* $(\triangle A_nB_nC_n)_{n\geq 0}$ *converges in shape to an equilateral triangle.*

2.10. *Extensions to polygon dynamic geometry*

Given an m-sided polygon $\mathcal{P}_0$, we build another $\mathcal{P}_1$ and iterate the construction to get sequences $(\mathcal{P}_n)_{n\geq 0}$. Some iteration processes described in the Subsections 2.1–2.4 can be extended for an arbitrary m-polygon to obtain classical Kasner polygons, Kasner polygons with a fixed weight, Kasner polygons with m fixed weights, as well as Kasner polygons with a fixed sequence of weights. For results in this direction, we mention here the papers of Chang and Davis [2], Donisi, Martini, Vincenzi, Vitale [13], Clarke [3], Ding, Hitt, and Zhang [4], Hitt and Zhang [5], Pech [14], Roeschel [15], Schoenberg [16], Stewart [10], de Villiers [11], and Ziv [17].

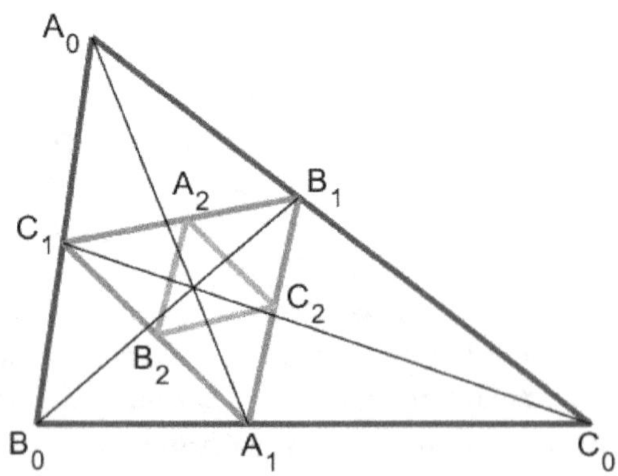

Fig. 7: Sequence of incentral triangles.

3. Iterating the Circumcircle Midarc Triangle

The midpoint-stretching polygon of a cyclic polygon $\mathcal{P}$ is another cyclic polygon inscribed in the same circle, the polygon whose vertices are the midpoints of the circular arcs between the vertices of $\mathcal{P}$. The midpoint-stretching polygon is also called the shadow of $\mathcal{P}$. When the circle is used to describe a repetitive time sequence and the polygon vertices on it represent the onsets of a drum beat, the shadow represents the set of times when the drummer's hands are highest, and has greater rhythmic evenness than the original rhythm. Given an m-sided cyclic polygon $\mathcal{P}_0$, we build its midpoint-stretching polygon $\mathcal{P}_1$ and then iterate the construction to get the sequences $(\mathcal{P}_n)_{n\geq 0}$.

In the case $m = 3$, we will illustrate the above general iterative process by considering the following special geometric situation also studied by Abbot [1] and Marinescu, Monea, Opincariu and Stroe [18]. Recall that, if P is a point in the plane of the triangle ABC, the circumcevian triangle of P with respect to $\triangle ABC$ is the triangle defined by the intersections of the Cevians AP, BP, CP with the circumcircle of $\triangle ABC$. We consider $\triangle A_1B_1C_1$ to be the circumcevian triangle of the incenter I of $\triangle ABC$, i.e. the circumcircle midarc triangle of $\triangle ABC$.

In this case, the points A_1, B_1, C_1 are the midpoints of the arcs $\overset{\frown}{BC}, \overset{\frown}{CA}, \overset{\frown}{AB}$ not containing the vertices A, B, C, resp., and the angles of triangle $A_1B_1C_1$ are given by the formulas

$$A_1 = \frac{1}{2}(\pi - A), B_1 = \frac{1}{2}(\pi - B), C_1 = \frac{1}{2}(\pi - C).$$

Define recursively the sequence of triangles $\mathcal{T}_n$ as follows: $\mathcal{T}_{n+1}$ is the circumcevian triangle with respect to the incenter of $\mathcal{T}_n$ and $\mathcal{T}_0$ is the triangle ABC (see Fig. 8). The angles of triangles $\mathcal{T}_n$ are given by the recurrence relations $A_{n+1} = \frac{1}{2}(\pi - A_n), B_{n+1} = \frac{1}{2}(\pi - B_n), C_{n+1} = \frac{1}{2}(\pi - C_n)$, where $A_0 = A, B_0 = B, C_0 = C$. Solving these recurrences, after easy computations, we get

$$A_n = \left(-\frac{1}{2}\right)^n A + \frac{\pi}{3}\left(1 - \left(-\frac{1}{2}\right)^n\right),$$

$$B_n = \left(-\frac{1}{2}\right)^n B + \frac{\pi}{3}\left(1 - \left(-\frac{1}{2}\right)^n\right),$$

$$C_n = \left(-\frac{1}{2}\right)^n C + \frac{\pi}{3}\left(1 - \left(-\frac{1}{2}\right)^n\right).$$

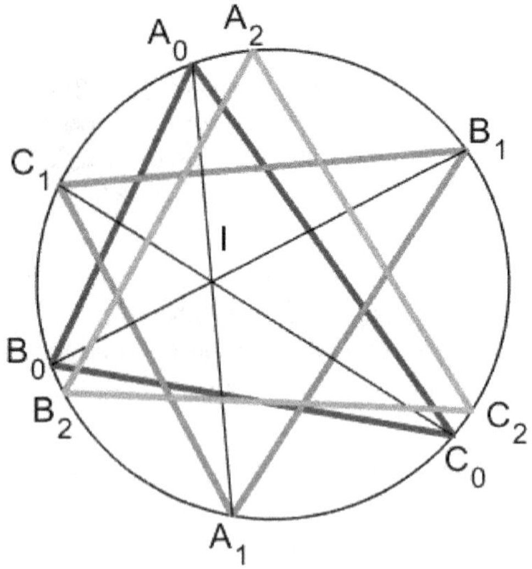

Fig. 8: Sequence of triangles obtained by midarc iteration.

The above formulas show that the sequences $(A_n)_{n\geq 0}, (B_n)_{n\geq 0}, (C_n)_{n\geq 0}$ of angles of $(\mathcal{T}_n)_{n\geq 0}$ are convergent and they have the same limit $\frac{\pi}{3}$. This remark implies that every convergent subsequence of the sequence of triangles $(\mathcal{T}_n)_{n\geq 0}$ has as its limit an equilateral triangle.

The following result clarifies what are the convergent subsequences of $(\mathcal{T}_n)_{n\geq 0}$ and it gives their limits:

Theorem 5. *The sequence $(\mathcal{T}_n)_{n\geq 0}$ has two convergent subsequences, namely $(\mathcal{T}_{2n})_{n\geq 0}$ and $(\mathcal{T}_{2n+1})_{n\geq 0}$. If a, b, c are the complex coordinates of the A, B, C, then the limit of $(\mathcal{T}_{2n})_{n\geq 0}$ is the equilateral triangle having the vertices coordinates given by the cubic roots of abc. The limit of $(\mathcal{T}_{2n+1})_{n\geq 0}$ is the equilateral triangle with the vertices coordinates given by the cubic roots of $-abc$.*

Proof. Considering a_n, b_n, c_n the complex coordinates of A_n, B_n, C_n, a simple geometric argument shows that $|A_nA_{n+1}| = |a_{n+1} - a_n| = 2R\cos\frac{B_n-C_n}{2}$, implying that the sequence $(|a_{n+1} - a_n|)_{n\geq 0}$ is not convergent.

Now,

$$|A_nA_{n+2}| = |a_{n+2} - a_n| = 2R\sin\frac{B_n - C_n}{4} = 2R\sin\frac{B-C}{2^{n+2}}.$$

Taking into account that the series $\sum_{n\geq 0}\sin\frac{B-C}{2^{n+2}}$ is convergent, it follows that the sequences $(a_{2n})_{n\geq 0}$ and $(a_{2n+1})_{n\geq 0}$ are fundamental, hence they are convergent. Similarly, we obtain the convergence of the sequences $(b_{2n})_{n\geq 0}$, $(b_{2n+1})_{n\geq 0}$ and $(c_{2n})_{n\geq 0}$, $(c_{2n+1})_{n\geq 0}$.

Let u, v, w be the limits of $(a_{2n})_{n\geq 0}, (b_{2n})_{n\geq 0}, (c_{2n})_{n\geq 0}$ and let u', v', w' be the limits of $(a_{2n+1})_{n\geq 0}, (b_{2n+1})_{n\geq 0}, (c_{2n+1})_{n\geq 0}$. Combining the above two remarks, we have $|u-u'| = 2R, |v-v'| = 2R, |w-w'| = 2R$, that is, the pairs $(u,u'), (v,v'), (w,w')$ are antipodal in the circumcircle of triangle ABC.

On the other hand, the sequences $(a_n)_{n\geq 0}, (b_n)_{n\geq 0}, (c_n)_{n\geq 0}$ satisfy the recursive relations

$$a_{n+1}^2 = b_n c_n, b_{n+1}^2 = c_n a_n, c_{n+1}^2 = a_n b_n, n = 0, 1, \ldots,$$

where $a_0 = a, b_0 = b, c_0 = c$. From these recursive relations by multiplication the equality $(a_{n+1}b_{n+1}c_{n+1})^2 = (a_n b_n c_n)^2, n = 0, 1, 2, \ldots$ follows. Therefore, for $n = 0, 1, 2, \ldots$, we have

$$(a_n b_n c_n)^2 = (a_{n-1}b_{n-1}c_{n-1})^2 = \cdots = (a_1 b_1 c_1)^2 = (a_0 b_0 c_0)^2 = (abc)^2.$$

Considering $n \to \infty$ in $(a_n b_n c_n)^2 = (abc)^2$ and $(a_{2n+1}b_{2n+1}c_{2n+1})^2 = (abc)^2$, it follows that $(uvw)^2 = (abc)^2$ and $(u'v'w')^2 = (abc)^2$.

But we know that the triangle UVW with the vertices of complex coordinates u, v, w is equilateral, hence u, v, w are the roots of a cubic equation $t^3 - z = 0$. We get $z = uvw$, that is $z^2 = (uvw)^2 = (abc)^2$. We obtain that u, v, w are the roots to the equation $t^6 - (abc)^2 = 0$.

With a similar argument, the triangle $U'V'W'$ with the vertices of complex coordinates u', v', w' is equilateral, hence u', v', w' are the roots of a cubic equation $t^3 - z' = 0$. We get $z' = u'v'w'$, that is $(z')^2 = (u'v'w')^2 = (abc)^2$. We obtain that u', v', w' are the other roots to the equation $t^6 - (abc)^2 = 0$. From the factorization $t^6 - (abc)^2 = (t^3 - abc)(t^3 + abc)$, the conclusion follows. □

4. The Sequence of Inradii and Euler's Inequality

Let ABC be a triangle with the angles A, B, C measured in radians, the circumradius R, the inradius r and the semi-perimeter s. These numbers are called the symmetric invariants of the triangle.

Recall the following important double inequality known as Gerretsen inequalities of the triangle (see Ref. [27, page 45]):

$$16Rr - 5r^2 \leq s^2 \leq 4R^2 + 4Rr + 3r^2, \tag{16}$$

which simply follow from the computation of the distances GI and HI in terms of the symmetric invariants, and getting $GI^2 = \frac{1}{9}(s^2 + 5r^2 - 16Rr)$ and $HI^2 = 4R^2 + 4Rr + 3r^2 - s^2$.

In what follows we need an equivalent trigonometric version of the right inequality of (16).

Lemma 1. *In every triangle we have*

$$(1 - \cos A)(1 - \cos B)(1 - \cos C) \geq \cos A \cos B \cos C. \tag{17}$$

Proof. It is well known that the roots of the cubic polynomial

$$P(t) = 4R^2t^3 - 4R(R + r)t^2 + (s^2 + r^2 - 4R^2)t + (2R + r)^2 - s^2$$

are $\cos A$, $\cos B$, $\cos C$ (see Ref. [19, page 7, Property 6]), hence, we have the factorization

$$P(t) = 4R^2(t - \cos A)(t - \cos B)(t - \cos C).$$

We obtain

$$\begin{aligned} P(1) &= 4R^2(1 - \cos A)(1 - \cos B)(1 - \cos C) \\ &= 4R^2 - 4R(R + r) + s^2 + r^2 - 4R^2 + (2R + r)^2 - s^2 = 2r^2. \end{aligned}$$

Now, by the relation

$$\cos A \cos B \cos C = \frac{s^2 - (2R + r)^2}{4R^2},$$

it follows

$$\begin{aligned} &(1 - \cos A)(1 - \cos B)(1 - \cos C) - \cos A \cos B \cos C \\ &= \frac{P(1)}{4R^2} - \frac{s^2 - (2R + r)^2}{4R^2} \\ &= \frac{1}{4R^2}(4R^2 + 4Rr + 3r^2 - s^2) \geq 0, \end{aligned}$$

where we have used the right inequality in Lemma 1. □

Applying the inequality (17) for the triangle with the angles $\frac{\pi}{2} - \frac{A}{2}$, $\frac{\pi}{2} - \frac{B}{2}$, $\frac{\pi}{2} - \frac{C}{2}$, we get

Lemma 2. *In every triangle we have*

$$\left(1-\sin\frac{A}{2}\right)\left(1-\sin\frac{B}{2}\right)\left(1-\sin\frac{C}{2}\right)\geq\sin\frac{A}{2}\sin\frac{B}{2}\sin\frac{C}{2}. \tag{18}$$

Consider the sequence of triangles $(\mathcal{T}_n)_{n\geq 0}$ defined by the process in the previous section and let r_n be the inradius of $\mathcal{T}_n, n=0,1,\ldots$ The following result is proved using convexity arguments by Andrica and Marinescu [20].

Theorem 6. *With the above notations, the sequence* $(r_n)_{n\geq 0}$ *is increasing and we have*

$$\lim_{n\to\infty} r_n=\frac{R}{2}. \tag{19}$$

Proof. We present a proof, using the trigonometric inequality in Lemma 2, for the property that the sequence $(r_n)_{n\geq 0}$ is increasing. In this respect, we note that

$$\begin{aligned}\left(\frac{r_{n+1}}{r_n}\right)^2&=\left(\frac{4R\sin\frac{A_{n+1}}{2}\sin\frac{B_{n+1}}{2}\sin\frac{C_{n+1}}{2}}{4R\sin\frac{A_n}{2}\sin\frac{B_n}{2}\sin\frac{C_n}{2}}\right)^2\\&=\frac{\sin^2(\frac{\pi}{4}-\frac{A_n}{4})\sin^2(\frac{\pi}{4}-\frac{B_n}{4})\sin^2(\frac{\pi}{4}-\frac{C_n}{4})}{\sin^2\frac{A_n}{2}\sin^2\frac{B_n}{2}\sin^2\frac{C_n}{2}}\\&=\frac{\frac{1}{8}(\cos\frac{A_n}{4}-\sin\frac{A_n}{4})^2(\cos\frac{B_n}{4}-\sin\frac{B_n}{4})^2(\cos\frac{C_n}{4}-\sin\frac{C_n}{4})^2}{\sin^2\frac{A_n}{2}\sin^2\frac{B_n}{2}\sin^2\frac{C_n}{2}}\\&=\frac{\frac{1}{8}(1-\sin\frac{A_n}{2})(1-\sin\frac{B_n}{2})(1-\sin\frac{C_n}{2})}{\sin^2\frac{A_n}{2}\sin^2\frac{B_n}{2}\sin^2\frac{C_n}{2}}\geq 1,\end{aligned}$$

where we have applied the well-known inequality $\sin\frac{A_n}{2}\sin\frac{B_n}{2}\sin\frac{C_n}{2}\leq\frac{1}{8}$ and the inequality (19).

The limit of the sequence $(r_n)_{n\geq 0}$ follows from the relation

$$r_n=4R\sin\frac{A_n}{2}\sin\frac{B_n}{2}\sin\frac{C_n}{2}$$

and from the property $\lim_{n\to\infty}A_n=\lim_{n\to\infty}B_n=\lim_{n\to\infty}C_n=\frac{\pi}{3}$. □

This property was used to construct interpolating sequences for Mitrinović's inequality $s\leq\frac{3\sqrt{3}}{2}R$, Weitzenbőck's inequality [3,21–25], Gordon's inequality [26], Curry's inequality [27], Finsler–Hadwiger inequality [3,28,37,38], the Pólya and Szegő inequality [29], and the Chen [30] mentioned in the recent paper of Wu, Lokesha and Srivastava [31].

Recall that in any triangle ABC, Euler's inequality $2r \leq R$ holds. Euler's inequality is a central result in triangle geometry (see Andreescu, Mushkarov and Stoyanov [32], Andrica [33], Mitrinovic, Pecaric and Volonec [19], and Popescu, Maftei, Diaz-Barrero and Dincă [34]). It is a direct consequence of Blundon's inequality and it has numerous and various refinements (see for instance Andrica [33], Andrica and Marinescu [35], and Mitrinovic, Pecaric and Volonec [19]).

Writing this inequality in the form $r \leq \frac{R}{2}$, the following result is a direct consequence of the results in Theorem 6.

Corollary 1. *With the above notations, the sequence $(r_n)_{n\geq 0}$ is interpolating Euler's inequality, that is we have*

$$r = r_0 \leq r_1 \leq r_2 \leq \ldots \leq r_n \leq \ldots \leq \frac{R}{2}. \tag{20}$$

5. The Complete Asymptotic Expansion of the Sequence $(r_n)_{n\geq 0}$

From the well-known formula $1 + \frac{r_n}{R} = \cos A_n + \cos B_n + \cos C_n$, we obtain

$$r_n = -R + R(\cos A_n + \cos B_n + \cos C_n).$$

Observe that we have

$$\cos A_n = \cos\left(\left(A_n - \frac{\pi}{3}\right) + \frac{\pi}{3}\right) = \frac{1}{2}\cos\left(A_n - \frac{\pi}{3}\right) - \frac{\sqrt{3}}{2}\sin\left(A_n - \frac{\pi}{3}\right),$$

and similarly for the angles B_n and C_n. Because $A_n, B_n, C_n \to \frac{\pi}{3}$, for n sufficiently large we have $|A_n - \frac{\pi}{3}| = \frac{1}{2^n}|A - \frac{\pi}{3}| < 1$, $|B_n - \frac{\pi}{3}| = \frac{1}{2^n}|B - \frac{\pi}{3}| < 1$, $|C_n - \frac{\pi}{3}| = \frac{1}{2^n}|C - \frac{\pi}{3}| < 1$, hence, we can expand, using the cos and sin series, the expression $\cos A_n + \cos B_n + \cos C_n$. It follows

$$r_n = -R + \frac{R}{2}\sum_{k=0}^{\infty}\frac{(-1)^k}{2^{2kn}(2k)!}\sigma_{2k} - \frac{R\sqrt{3}}{2}\sum_{k=0}^{\infty}\frac{(-1)^{n+k}}{2^{(2k+1)n}(2k+1)!}\sigma_{2k+1},$$

where

$$\sigma_j = \left(A - \frac{\pi}{3}\right)^j + \left(B - \frac{\pi}{3}\right)^j + \left(C - \frac{\pi}{3}\right)^j, \quad j = 0, 1, 2, \ldots.$$

We have $\sigma_0 = 3$ and $\sigma_1 = 0$, hence the above formula gives the complete asymptotic expansion of the sequence $(r_n)_{n\geq 0}$ in terms of the angles of triangle ABC.

Theorem 7. *The following expansion holds:*

$$r_n = \frac{R}{2} + \frac{R}{2}\sum_{k=1}^{\infty}\frac{(-1)^k}{2^{2kn}(2k)!}\sigma_{2k}$$
$$-(-1)^n\frac{R\sqrt{3}}{2}\sum_{k=1}^{\infty}\frac{(-1)^k}{2^{(2k+1)n}(2k+1)!}\sigma_{2k+1}. \tag{21}$$

For instance, from this formula we obtain the first order of convergence of the sequence $(r_n)_{n\geq 0}$ is 4^n and

$$\lim_{n\to\infty} 4^n\left(r_n - \frac{R}{2}\right) = -\frac{R}{4}\sigma_2. \tag{22}$$

Equation (21) shows that the second order of convergence of $(r_n)_{n\geq 0}$ must be 8^n, but surprisingly the sequence $(8^n(4^n(r_n - \frac{R}{2}) + \frac{R}{4}\sigma_2)))_{n\geq 0}$ is not convergent. It has two convergent subsequences according to the parity of n. We obtain

$$\lim_{m\to\infty} 8^{2m}\left(4^{2m}\left(r_{2m} - \frac{R}{2}\right) + \frac{R}{4}\sigma_2\right) = \frac{R\sqrt{3}}{12}\sigma_3 \tag{23}$$

and

$$\lim_{m\to\infty} 8^{2m+1}\left(4^{2m+1}\left(r_{2m+1} - \frac{R}{2}\right) + \frac{R}{4}\sigma_2\right) = -\frac{R\sqrt{3}}{12}\sigma_3. \tag{24}$$

Corollary 2. *For every sufficiently large positive integer n, the following inequality holds*

$$r \leq \frac{R}{2} - \frac{R}{16^n}\sigma_2. \tag{25}$$

Proof. From (24) it follows that for every sufficiently large positive integer m we have

$$\left(4^{2m+1}\left(r_{2m+1} - \frac{R}{2}\right) + \frac{R}{4}\sigma_2\right) \leq 0,$$

hence,

$$r_{2m+1} \leq \frac{R}{2} - \frac{R}{16^{m+1}}\sigma_2.$$

Combining with the inequality $r \leq r_{2m+1}$ contained in Corollary 1 and replacing $m+1$ by n, the inequality (25) follows. □

Remark 1. The inequality (25) completes from the right side the inequality (2.15) in Andrica, Rădulescu and Rădulescu [36], therefore for every sufficiently large positive integer n we have

$$\frac{1}{2}-\frac{1}{6}\sigma_2 \leq \frac{r}{R} \leq \frac{1}{2}-\frac{1}{16^n}\sigma_2.$$

In case of a non-equilateral triangle, the smallest positive integer n_0 with the above property is

$$n_0=\left\lfloor \log_{16}\frac{2R\sigma_2}{R-2r}\right\rfloor+1.$$

6. Extension to Polygons

6.1. *The extension of Theorem* 5

Let $\mathcal{P}=P_1,\ldots,P_k$ be a polygon inscribed in the circle $\mathcal{C}$ of center O and radius R and consider the sequence of polygons $(\mathcal{P}_n)_{n\geq 0}$, where $\mathcal{P}_n = P_1^{(n)},\ldots,P_k^{(n)}$ is defined as follows: For each $j=1,\ldots,k$, we define the vertex $P_j^{(n)}$ recursively as $P_j^{(0)}=P_j$ for all j, and for $n\geq 0$, $P_j^{(n+1)}$ is the middle point of the arc not containing any other vertices of $\mathcal{P}_n$, $P_j^{(n)}P_{j+1}^{(n)}$ (with the usual convention $P_{k+1}^{(n)}=P_1^{(n)}$).

For each $n\geq 0$ and $j=1,\ldots,k$, let $x_j^{(n)}$ denote the measure of the angle $\angle P_j^{(n)}OP_{j+1}^{(n)}$. Then it is easy to see that we have the recursive formula

$$x_j^{(n+1)}=\frac{1}{2}\left(x_j^{(n)}+x_{j+1}^{(n)}\right),\quad j=1,\ldots,k,$$

Note also that

$$\left|P_j^{(n)}P_{j+1}^{(n)}\right|=2R\sin\left(\frac{x_j^{(n)}}{2}\right),\quad j=1,\ldots,k,$$

so to study the nature of the polygons $\mathcal{P}_n=P_1^{(n)},\ldots,P_k^{(n)}$ it suffices to study the sequence

$$\left(x_1^{(n)},x_2^{(n)},\ldots,x_k^{(n)}\right)\in\mathbb{R}^k.$$

Let us write $x_j^{(0)}=x_j$ and

$$x_i^{(n)}=a_{i1}^{(n)}x_1+a_{i2}^{(n)}x_2+\cdots+a_{ik}^{(n)}x_k,$$

i.e. express each term of the nth sequence in terms of the original sequence, where $\left(a_{ij}^{(n)}\right)_{\substack{1\le i\le k\\ 1\le j\le k}}$ are rational numbers. Then for each $i = 1, 2, \ldots, k$ we have the recurrence

$$a_{ij}^{(n+1)} = \frac{1}{2}\left(a_{ij}^{(n)} + a_{i+1,j}^{(n)}\right),$$

with the usual convention that $a_{k+1,j}^{(n)} = a_{1j}^{(n)}$. Let

$$A_n = \begin{pmatrix} a_{11}^{(n)} & a_{12}^{(n)} & \ldots & a_{1k}^{(n)} \\ a_{21}^{(n)} & a_{22}^{(n)} & \ldots & a_{2k}^{(n)} \\ \vdots & \vdots & \vdots & \vdots \\ a_{k1}^{(n)} & a_{k2}^{(n)} & \ldots & a_{kk}^{(n)} \end{pmatrix}.$$

Then

$$A_{n+1} = \frac{1}{2}\begin{pmatrix} 1 & 1 & 0 & 0 & \ldots & 0 \\ 0 & 1 & 1 & 0 & \ldots & 0 \\ \vdots & \vdots & \vdots & \vdots & \ldots & \vdots \\ 1 & 0 & 0 & 0 & \ldots & 1 \end{pmatrix} A_n,$$

so

$$A_n = \frac{1}{2^n}\begin{pmatrix} 1 & 1 & 0 & 0 & \ldots & 0 \\ 0 & 1 & 1 & 0 & \ldots & 0 \\ \vdots & \vdots & \vdots & \vdots & \ldots & \vdots \\ 1 & 0 & 0 & 0 & \ldots & 1 \end{pmatrix}^n.$$

Now write

$$\begin{pmatrix} 1 & 1 & 0 & 0 & \ldots & 0 \\ 0 & 1 & 1 & 0 & \ldots & 0 \\ \vdots & \vdots & \vdots & \vdots & \ldots & \vdots \\ 1 & 0 & 0 & 0 & \ldots & 1 \end{pmatrix} = I_k + P,$$

where

$$P = \begin{pmatrix} 0 & 1 & 0 & 0 & \dots & 0 \\ 0 & 0 & 1 & 0 & \dots & 0 \\ \vdots & \vdots & \vdots & \vdots & \dots & \vdots \\ 1 & 0 & 0 & 0 & \dots & 0 \end{pmatrix}.$$

Note that $P^k = I_k$, so

$$\begin{pmatrix} 1 & 1 & 0 & 0 & \dots & 0 \\ 0 & 1 & 1 & 0 & \dots & 0 \\ \vdots & \vdots & \vdots & \vdots & \dots & \vdots \\ 1 & 0 & 0 & 0 & \dots & 1 \end{pmatrix}^n = (I_k + P)^n = \sum_{l=0}^{n} \binom{n}{l} P^{n \bmod k}$$

$$= \sum_{i=0}^{k-1} P^i \left(\binom{n}{i} + \binom{n}{k+i} + \dots + \binom{n}{\lfloor \frac{n-i}{k} \rfloor k + i} \right).$$

Let

$$S_r^{(n)} := \binom{n}{r} + \binom{n}{k+r} + \dots + \binom{n}{\lfloor \frac{n-r}{k} \rfloor k + r}.$$

(Note that $S_i^{(n)}$ is also the coefficient of ζ_k^r in the expansion $(1+\zeta_k)^n$, where ζ_k is a kth root of unity).

We find that

$$a_{ij}^{(n)} = \frac{1}{2^n} S_{(j-i)}^{(n)},$$

where for $m \leq 0$, we let $S_m^{(n)} := S_{k-m}^{(n)}$.

To calculate $\frac{1}{2^n} S_{(j-i)}^{(n)}$, we use our previous observation, namely that

$$\binom{n}{r} + \binom{n}{k+r} + \dots + \binom{n}{\lfloor \frac{n-r}{k} \rfloor k + r}$$

is the coefficient of ζ_k^i in the expansion $(1+\zeta_k)^n$, for ζ_k^r a kth root of unity. We use the following well-known result: if $\omega_j := e(\frac{2j\pi}{k})$, $j = 0, \dots, k-1$ are all the kth roots of unity, then

$$\sum_{j=0}^{k-1} \omega_j^m = \begin{cases} 0 & \text{if } k \nmid m; \\ k & \text{if } k \mid m. \end{cases}$$

Thus,

$$\binom{n}{r} + \binom{n}{k+r} + \cdots + \binom{n}{\lfloor \frac{n-r}{k} \rfloor k + r} = \frac{1}{k}\sum_{j=0}^{k-1} \omega_j^{-r}(1+\omega_j)^n$$

$$= \frac{1}{k}\sum_{j=0}^{k-1} \omega_j^{-r}\left(2\cos\frac{\pi j}{k} e\left(\frac{\pi j}{k}\right)\right)^n,$$

thus,

$$\frac{1}{2^n} S_r^{(n)} = \frac{1}{k}\sum_{j=0}^{k-1} \left(\cos\frac{\pi j}{k}\right)^n e\left(\frac{(n-2r)j\pi}{k}\right).$$

Now note that $\cos\frac{\pi j}{k} = 1$ precisely when $j = 0$. Therefore, for $j \neq 0$, since $|e(\frac{(n-2r)j\pi}{k})| = 1$, we have that

$$\left(\cos\frac{\pi j}{k}\right)^n e\left(\frac{(n-2r)j\pi}{k}\right) \to 0 \quad \text{as} \quad n \to \infty.$$

Hence, as we are summing a finite number of terms (more precisely $k-1$), we have

$$\lim_{n\to\infty} \frac{1}{2^n} S_r^{(n)} = \frac{1}{k}.$$

This shows that

$$\lim_{n\to\infty} a_{ij}^{(n)} = \frac{1}{k}, \quad \text{for all} \quad i, j,$$

hence,

$$\lim_{n\to\infty} x_j^{(n)} = \frac{x_1 + x_2 + \cdots + x_k}{k}.$$

In particular, the limit is a regular k-gon which has all the angles at the center equal to the arithmetic mean of angles at the center of the original polygon.

Remark 2. While the above argument shows that the limit is a regular k-gon, it does not say that this regular k-gon has a "fixed position". But it is easy to see that the position of the vertices is determined by the parity of n: indeed, once we have a regular k-gon, at the next iteration we obtain another regular k-gon which rotated the initial one by $\frac{\alpha}{2}$, where α is the angle at the center. Hence, in two iterations we overlap with the previous k-gon (but the order of the vertices has been shifted by 1).

Let $z_1, \ldots, z_k$ be the complex coordinates of the vertices $P_1, \ldots, P_k$ of the k-gon $\mathcal{P}$. Considering $z_1^{(n)}, \ldots, z_k^{(n)}$ the complex coordinates of the vertices $P_1^{(n)}, \ldots, P_k^{(n)}$ of the k-gon $\mathcal{P}_n$, a simple geometric argument shows that the sequences $(z_1^{(n)})_{n\geq 0}, \ldots, (z_k^{(n)})_{n\geq 0}$ are defined recursively by

$$z_j^{(n+1)} = \pm\sqrt{z_j^{(n)} z_{j+1}^{(n)}}, \quad j = 1, \ldots, k, n = 0, 1, \ldots,$$

where $z_j^{(0)} = z_j, j = 1, \ldots, k$, for some choice of signs $+$ and $-$, where $\sqrt{z}$ is the first complex square root of the complex number z. From these recursive relations it follows by multiplication the equality $z_1^{(n+1)}, \ldots, z_k^{(n+1)} = \pm z_1^{(n)}, \ldots, z_k^{(n)}, n = 0, 1, 2, \ldots$. Therefore, for $n = 0, 1, 2, \ldots$, we have

$$z_1^{(n)} \ldots z_k^{(n)} = \epsilon_n z_1, \ldots, z_k,$$

where $\epsilon_n = \pm 1$. Assume that $u_1, \ldots, u_k$ are the limits of the sequences $(z_1^{(\alpha_n)})_{n\geq 0}, \ldots, (z_k^{(\alpha_n)})_{n\geq 0}$, where $(\alpha_n)_{n\geq 0}$ is the sequence of positive integers with the property $\epsilon_{\alpha_n} = 1$. Passing to the limit, it follows $u_1 \cdots u_k = z_1, \ldots, z_k$. But we know that the limit k-gon $U_1, \ldots, U_k$ is regular, hence $u_1, \ldots, u_k$ are the roots of a binomial equation $t^k - z = 0$. We get, $z = z_1, \cdots, z_k$, therefore in this case the complex coordinates of the vertices of the limit k-gon are the kth roots of the complex number $z_1, \ldots, z_k$.

Similarly, considering $(\beta_n)_{n\geq 0}$, the sequence of the positive integers with the property $\epsilon_{\beta_n} = -1$, we obtain that the in this case the complex coordinates of the vertices of the limit k-gon are the kth roots of the complex number $-z_1 \ldots z_k$.

Putting together the above results, one obtains the following:

Theorem 8. *The sequence $(\mathcal{P}_n)_{n\geq 0}$ has two convergent subsequences and the limits are regular polygons. If $z_1, \ldots, z_k$ are the complex coordinates of the vertices $P_1, \ldots, P_k$ of $\mathcal{P}$, then the limits are the regular polygons having the vertices coordinates given by the kth roots of $z_1, \ldots, z_k$ and of $-z_1, \ldots, z_k$, resp.*

6.2. *The sequence (p_n) of perimeters*

Denote by (p_n) the perimeter of the polygon $\mathcal{P}_n, n = 0, 1, \ldots$. Because the limits are regular polygons, it follows that the center O of the circumcircle belongs to the interior of $\mathcal{P}_n$, for every sufficiently large n. Without loss of generality, we may assume that this property holds for all polygons in the sequence $(\mathcal{P}_n)_{n\geq 0}$.

Theorem 9. *The sequence $(p_n)_{n\geq 0}$ is increasing.*

Proof. Clearly, we have

$$p_n = \sum_{j=1}^{k} \left| P_j^{(n)} P_{j+1}^{(n)} \right| = 2R \sum_{j=1}^{k} \sin \frac{x_j^{(n)}}{2}.$$

Using the recursive formula for the sequences $(x_j^{(n)})_{n\geq 0}$, we can write

$$\begin{aligned} p_{n+1} &= 2R \sum_{j=1}^{k} \sin \frac{x_j^{(n+1)}}{2} = 2R \sum_{j=1}^{k} \sin \frac{1}{2} \left(\frac{x_j^{(n)}}{2} + \frac{x_{j+1}^{(n)}}{2} \right) \\ &\geq 2R \sum_{j=1}^{k} \frac{1}{2} \left(\sin \frac{x_j^{(n)}}{2} + \sin \frac{x_{j+1}^{(n)}}{2} \right) = 2R \sum_{j=1}^{k} \sin \frac{x_j^{(n)}}{2} = p_n, \end{aligned}$$

where we have used the concavity of the function sinus on the interval $[0, \pi]$. □

6.3. *The complete asymptotic expansion*

In this subsection, we will obtain the complete asymptotic formula for the sequence $(p_n)_{n\geq 0}$. For this purpose we observe that

$$\begin{aligned} p_n &= 2R \sum_{j=1}^{k} \sin \frac{x_j^{(n)}}{2} = 2R \sum_{j=1}^{k} \sin \left(\frac{x_j^{(n)}}{2} - \frac{\pi}{k} + \frac{\pi}{k} \right) \\ &= 2R \sin \frac{\pi}{k} \sum_{j=1}^{k} \cos \left(\frac{x_j^{(n)}}{2} - \frac{\pi}{k} \right) + 2R \cos \frac{\pi}{k} \sum_{j=1}^{k} \sin \left(\frac{x_j^{(n)}}{2} - \frac{\pi}{k} \right). \end{aligned}$$

Because the limit polygons of the sequence $(\mathcal{P}_n)_{n\geq 0}$ are regular polygons, it follows that $x_j^{(n)} \to \frac{2\pi}{k}$ for $n \to \infty$ and $j = 1, \ldots, k$, hence for n big enough we have

$$\left| \frac{x_j^{(n)}}{2} - \frac{\pi}{k} \right| < 1, \quad j = 1, \ldots, k.$$

We obtain

$$\begin{aligned} p_n &= 2R \sin \frac{\pi}{k} \sum_{j=1}^{k} \cos \left(\frac{x_j^{(n)}}{2} - \frac{\pi}{k} \right) + 2R \cos \frac{\pi}{k} \sum_{j=1}^{k} \sin \left(\frac{x_j^{(n)}}{2} - \frac{\pi}{k} \right) \\ &= 2R \sin \frac{\pi}{k} \sum_{j=1}^{k} \sum_{s=0}^{\infty} \frac{(-1)^s}{(2s)!} \left(\frac{x_j^{(n)}}{2} - \frac{\pi}{k} \right)^{2s} \end{aligned}$$

$$+2R\cos\frac{\pi}{k}\sum_{j=1}^{k}\sum_{s=0}^{\infty}\frac{(-1)^s}{(2s+1)!}\left(\frac{x_j^{(n)}}{2}-\frac{\pi}{k}\right)^{2s+1}$$

$$=2R\sin\frac{\pi}{k}\sum_{s=0}^{\infty}\frac{(-1)^s}{(2s)!}\sum_{j=1}^{k}\left(\frac{x_j^{(n)}}{2}-\frac{\pi}{k}\right)^{2s}$$

$$+\,2R\cos\frac{\pi}{k}\sum_{s=0}^{\infty}\frac{(-1)^s}{(2s+1)!}\sum_{j=1}^{k}\left(\frac{x_j^{(n)}}{2}-\frac{\pi}{k}\right)^{2s+1}$$

$$=2R\sin\frac{\pi}{k}\sum_{s=0}^{\infty}\frac{(-1)^s\Omega_{2s}^{(n)}}{(2s)!}+2R\cos\frac{\pi}{k}\sum_{s=0}^{\infty}\frac{(-1)^s\Omega_{2s+1}^{(n)}}{(2s+1)!},$$

where $\Omega_r^{(n)}=\sum_{j=1}^{k}\left(\frac{x_j^{(n)}}{2}-\frac{\pi}{k}\right)^r$. Clearly, for all n we have $\Omega_0^{(n)}=k$ and $\Omega_1^{(n)}=0$.

References

[1] S. Abbot, Average sequences and triangles, *Math. Gaz.* **80**, 222–224, (1996).

[2] G.Z. Chang and P.J. Davis, Iterative processes in elementary geometry, *Amer. Math. Monthly* **90**(7), 421–431, (1983).

[3] R.J. Clarke, Sequences of polygons, *Math. Mag.* **90**(2), 102–105, (1979).

[4] J. Ding, L.R. Hitt, and X-M. Zhang, Markov chains and dynamic geometry of polygons, *Linear Algebra Its Appl.* **367**, 255–270, (2003).

[5] L.R. Hitt and X-M. Zhang, Dynamic geometry of polygons, *Elem. Math.* **56**(1), 21–37, (2001).

[6] D. Ismailescu and J. Jacobs, On sequences of nested triangles, *Period. Math. Hung.* **53**(1-2), 169–184, (2006).

[7] T. Andreescu and D. Andrica, *Complex Numbers from A to...Z* (Birkhauser, 2nd edn. 2014).

[8] P.J. Davis, *Circulant Matrices* (AMS Chelsea Publishing, 1994).

[9] D. Andrica, D. Şt. Marinescu, and O. Bagdasar, *Dynamic Geometry of Kasner Triangles with a Fixed Sequence of Weights*, *Int. J. Geom.* **11**(2), 101–110, (2022).

[10] B.M. Stewart, Cyclic properties of Miguel polygons, *Amer. Math. Monthly* **47**(7), 462–466, (Aug. Sep., 1940).

[11] M. de Villiers, From nested Miguel triangles to Miguel distances, *Math. Gaz.* **86**(507), 390–395, (2002).

[12] J.G. Kingston and J.L. Synge, The sequence of pedal triangles, *Amer. Math. Monthly* **95**(7), 609–620, (1988).

[13] S. Donisi, H. Martini, G. Vincenzi, and G. Vitale, Polygons derived from polygons via iterated constructions, *Electron. J. Differ. Geom. Dyn. Syst.* **18**, 14–31, (2016).

[14] P. Pech, The harmonic analysis of polygons and Napoleons theorem, *J. Geom. Gr.* **5**(1), 13–22, (2001).
[15] O. Roeschel, Polygons and iteratively regularizing affine transformations, *Beitr Algebra Geom.* **58**, 69–79, (2017).
[16] I.J. Schoenberg, The finite Fourier series and elementary geometry, *Am. Math. Monthly* **57**, 390–404, (1950).
[17] B. Ziv, Napoleon-like configurations and sequences of triangles, *Forum Geom.* **2**, 115–128, (2002).
[18] D. Şt. Marinescu, M. Monea, M. Opincariu, and M. Stroe, A sequence of triangles and geometric inequalities, *Forum Geom.* **9**, 291–295, (2009).
[19] D.S. Mitrinovic, J. Pecaric, and V. Volonec, *Recent Advances in Geometric Inequalities (Mathematics and its Applications)*, Springer; Softcover reprint of the original 1st ed. 1989 edition (Sep. 17, 2011).
[20] D. Andrica and D. Şt. Marinescu, Sequences interpolating some geometric inequalities, *Creat. Math. Inform.* **28**, 9–18, (2019).
[21] R. Weitzenböck, Uber eine ungleichung in der dreiecksgeometrie, *Math. Zeitschrift* **5**(12), 137–146, (1919).
[22] C. Alsina and R. Nelsen, Geometric proofs of the Weitzenbock and Finsler–Hadwiger inequality, *Math. Mag.* **81**, 216–219, (2008).
[23] C. Lupu, C. Mateescu, V. Matei, and M. Opincariu, A refinement of the Finsler-Hadwiger reverse inequality, *Gaz. Mat. Ser. A* **28**, 130–133, (2010).
[24] C. Lupu and C. Pohoaţă, Sharpening the Finsler-Hadwiger inequality, *Crux Mathematicorum* **34**(2), 97–101, (2008).
[25] E. Stoica, N. Minculete, and C. Barbu, New aspects of Weitzenböck's inequality, *Balkan J. Geom. Appl.* **21**(2), 95–101, (2016).
[26] V.O. Gordon, *Matematika v Skole*, 1, p. 89, (1966).
[27] T.R. Curry, Problem E 1861, *Am. Math. Monthly* **73**(2), (1966).
[28] P. von Finsler and H. Hadwiger, Einige relationen im Dreieck, *Commentarii Mathematici Helvetici* **10**(1), 316–326, (1937).
[29] G. Pólya and G. Szegő, Problems and Theorems in Analysis, Vol. II. In *Grundlehren der Mathematischen Wissenschaften, Band 20* (Springer-Verlag, New York, Heidelberg and Berlin, 1976), (Translated from the revised and enlarged Fourth German Edition by C. E. Billigheimer).
[30] S.-L. Chen, An inequality chain relating to several famous inequalities, *Hunan Math. Comm.*, **1**, 41, (1995) (in Chinese).
[31] Y-D. Wu, V. Lokesha, and H.M. Srivastava, Another refinement of the Pólya-Szegő inequality, *Comput. Math. Appl.* **60**, 761–770, (2010).
[32] T. Andreescu, O. Mushkarov, and L. Stoyanov, *Geometric Problems on Maxima and Minima*, Birkauser (Boston-Basel-Berlin, 2006).
[33] D. Andrica, *Geometry* (Romanian) (Casa Cărţii de Ştiinţă, Cluj-Napoca, 2017).
[34] P.G. Popescu, I.V. Maftei, J.L. Diaz-Barrero, and M. Dincă, *Inegalităţi Matematice: Modele Inovatoare* (Romanian) (Editura Didactică şi Pedagogică, Bucureşti, 2007).
[35] D. Andrica and D. Şt. Marinescu, *New Interpolation Inequalities to Eulers* $R \geq 2r$, *Forum Geometricorum* **17**, 149–156, (2017).

[36] D. Andrica, S. Rădulescu, and M.S. Rădulescu, Convexity revisited: Methods, results, and applications. In: D. Andrica, Th.M. Rassias, (Eds.), *Differential and Integral Inequalities*, pp. 49–134, Springer Optimization and its Applications (Springer, 2019).
[37] A. Cipu, Optimal reverse Finsler-Hadwiger inequalities, *Gaz. Mat. Ser. A* **30**(3–4), 61–68, (2012).
[38] D. Şt. Marinescu, M. Monea, M. Opincariu, and M. Stroe, Note on Hadwiger-Finsler's inequalities, *J. Math. Inequal.* **6**(1), 57–64, (2012).

https://doi.org/10.1142/9789811261572_0006

Chapter 6

A Review of Linearization Methods for Polynomial Matrices and their Applications

E. Antoniou[*,§], I. Kafetzis[†,‖], and S. Vologiannidis[‡,¶]

[*]*Department of Information and Electronic Engineering, International Hellenic University, 57400 — Thessaloniki, Greece*
[†]*Department of Physics, Aristotle University of Thessaloniki, 54006 — Thessaloniki, Greece*
[‡]*Department of Computer, Informatics and Telecommunications Engineering, International Hellenic University, Terma Magnisias, 62124 — Serres, Greece*
[§]*antoniou@ihu.gr*
[‖]*kafetzii@math.auth.gr*
[¶]*svol@ihu.gr*

Matrix polynomials are a prominent tool in the representation and study of systems of differential or difference equations. Systems of such equations can be described via polynomial matrices, whose algebraic properties are inextricably linked with the dynamics of the system. In this context, the degree of the polynomial entries represents the orders of differentiation of the forward shift operator. This is the foundation of the theory of linearizations of matrix polynomials where the goal is, given a matrix polynomial, to construct a new matrix polynomial in such a way that the maximum degree occurring among its entries equals one while maintaining structural aspects of the original matrix. In this work we present the most influential methods for the linearization of matrix polynomials that have been developed over the past two decades, based on the algebraic properties preserved. Thus, we thoroughly discuss a variety of methods, present the ideas that led to their definition and highlight the relations among them.

1. Introduction

Studying any physical system usually starts by applying various physical laws to derive the mathematical equations describing it. Those equations can take multiple forms, such as integral, differential or difference equations, linear or nonlinear. System analysis and design originally focused on using single input single output transfer functions or on the state space approach for both SISO and MIMO cases. Although polynomial techniques use simpler concepts, the development of mathematical tools and the examination of the underlying numerical aspects matured only in the past few decades. Second, third or even fourth degree polynomial matrices arise in the study and control of several systems such as large flexible space structures, mechanical multi-body systems, damped gyroscopic systems, robotics, structural dynamics, aero-acoustics, fluid mechanics, social, biological or economic systems. The study of systems using the polynomial approach goes hand in hand with the solution of several fundamental polynomial matrix problems, one of the most important ones being the polynomial eigenvalue problem, with several practical applications that can be found in Refs. [1,2].

A common tool for the study of polynomial eigenvalue problems is linearizations. Traditionally, the first and second Frobenius companion forms are the most common examples of linearizations of a given polynomial matrix. However, issues related to the conditioning of the associated numerical problem, that is the lack of robustness of the linearized system to tolerate even small perturbations and their inability to preserve the structure present in the original polynomial matrix, may be a serious drawback for their use in real-world situations (see Ref. [3]). Yet, many linearizations exist, and other than the convenience of their construction "by inspection" of the coefficients of the polynomial matrix, there is no apparent reason for preferring the Frobenius companion forms over other choices available. Notably, block symmetric companion matrices appear in the literature even in the early 1960s (see Refs. [4,5]), but their advantages over traditional companion matrices did not receive much attention until the mid 2000s when the subject of structured linearizations came into focus (see Refs. [3,6–11]). The goal of the present chapter is to present the key developments in the area of linearizations and provide an overview of their applications in the study of high-order linear dynamical systems.

In Ref. [12], a new family of companion matrices for a given scalar polynomial, parametrized by products of elementary constant matrices, was provided. This seminal paper, included three main ideas. The first one is the observation that the companion form can be written as a product of elementary matrices, the second one is the fact that the order of the elementary matrices can be changed giving rise to very interesting companion matrices. This family was shown to have companion matrices that could not be obtained by permutational similarity from the usual companion matrix. In Ref. [6], the results of [12] were generalized for regular polynomial matrices, showing that the new family of companion forms, named Fiedler pencils, preserves both the finite and infinite elementary divisors structure of the original polynomial matrix and thus are indeed linearizations. In Ref. [6], it was first noted that members of this family of linearizations, can preserve symmetries of the original polynomial matrix such as being self-adjoint. In Ref. [13], De Terán, Dopico and Mackey revisited Fiedler pencils, examining the case where the original polynomial matrix is singular. An overview of the work in Fiedler linearizations was given in Ref. [14], which was an overview of the talk given at the "Minisymposium in Honor of Miroslav Fiedler" at the 17th ILAS Conference, held at TU Braunschweig, Germany, on Thursday 25 August 2011.

An alternative approach for the construction of linearizations was achieved by generalizing the first and second companion form. In Refs. [3, 11,15], a method which produces large classes of linearizations of non-singular polynomial matrices is introduced. To this end, for any non-singular polynomial matrix P, two vector spaces of matrix pencils, namely $\mathbb{L}_1(P)$ and $\mathbb{L}_2(P)$, which generalize the first and second companion form, resp., are defined. It is proved that almost all of the pencils contained in those vector spaces constitute linearizations of P. Furthermore, the intersection of these two vector spaces, that is, $\mathbb{DL}(P) = \mathbb{L}_1(P) \cap \mathbb{L}_2(P)$, is shown to contain pencils of particular significance. This work was the stepping stone for several generalizations. Most noticeably, in Ref. [16] similar results with the use of bivariate polynomials are derived. Another important extension of this work came from Ref. [17], where the definition of the above vector spaces is extended so that the original polynomial matrix is expressed in terms of any orthogonal polynomial basis, rather than the classical one.

When it comes to applications, linearizations of polynomial matrices play a key role in the study of the behavior of linear dynamical systems. In many scenarios, modeling of such systems leads naturally to nonlinear eigenvalue problems associated to a structured polynomial matrix. For example, the coefficient matrices of the underlying polynomial matrix may be all symmetric, alternate between symmetric and skew-symmetric, or even have palindromic structure (see Ref. [18] for a comprehensive collection of examples). Usually, the presence of such structure on the coefficients of the polynomial matrix results in structural symmetries on the respective spectra that have physical significance. It is therefore crucial to be able to apply numerical techniques that take into account the special structure of the polynomial matrix when such spectra are under investigation. Since linearization of the polynomial matrix is usually the first step toward this direction, it is important to have access to linearizations reflecting the structure of the original matrix, and then apply numerical methods for the corresponding linear eigenvalue problem that properly take into their structure as well. This has been the subject of several research works during the last two decades (see Refs. [3,7–9,11,15,19–24] and references therein).

The manuscript is organized as follows: In Section 2, we review the necessary mathematical framework for the study of polynomial matrices. Section 3 is devoted to the presentation of the three mainstream approaches for the linearization of polynomial matrices developed (mostly) during the decade 2000–2010. In Section 4, a number of selected examples of dynamical systems taken from Ref. [18] are revisited, and the respective structured linearizations employed for the solutions of the associated eigenvalue problems are discussed. Finally, in Section 5, we draw our conclusions and discuss potential extensions of the theory presented.

2. Polynomial Matrix Equivalence and Linearizations

In what follows, we denote the fields of real and complex numbers by $\mathbb{R}$ and $\mathbb{C}$ resp., while $\mathbb{F}$ will be used to denote either of them. The set of $p \times m$ matrices with entries in the field $\mathbb{F}$ are denoted by $\mathbb{F}^{p\times m}$. Given a square matrix A, $\det A$ or $|A|$ stands for the *determinant* of A. The *rank* of a matrix $A \in \mathbb{F}^{p\times m}$, denoted by $\operatorname{rank}_{\mathbb{F}} A$, is the maximal number of linearly independent columns of A, when the latter are considered as vectors of $\mathbb{F}^p$. Finally, I_p stands for the $p \times p$ *identity matrix.*

The ring of polynomials in the indeterminate λ with coefficients from the field $\mathbb{F}$ will be denoted by $\mathbb{F}[\lambda]$. The quotient field of $\mathbb{F}[\lambda]$ is the set of rational functions $\mathbb{F}(s)$. The set of *polynomial matrices* whose entries are polynomials in $\mathbb{F}[\lambda]$ with dimensions $p \times m$ is denoted by $\mathbb{F}[\lambda]^{p\times m}$. Let

$$P(\lambda) = P_n\lambda^n + P_{n-1}\lambda^{n-1} + \cdots + P_0 \in \mathbb{F}[\lambda]^{p\times m}, \tag{1}$$

where $P_i \in \mathbb{F}^{p\times m}$, $i = 0, 1, \ldots n$ and $P_n \neq 0$ is a polynomial matrix. Following the terminology used in Ref. [25], the *degree* or *order* of the polynomial matrix $P(\lambda) \in \mathbb{F}[\lambda]^{p\times m}$, denoted by $\deg P(\lambda)$, is the highest among the degrees of the polynomial entries of $P(\lambda)$, that is $\deg P(\lambda) = n$. A square polynomial matrix $P(\lambda) \in \mathbb{F}[\lambda]^{p\times p}$ is called *regular* if its determinant is not identically equal to zero, or equivalently if there exists $\lambda_0 \in \mathbb{C}$ such that $\det P(\lambda_0) \neq 0$. It can be easily seen that if $P(\lambda)$ is a regular polynomial matrix, then it is invertible for almost all $\lambda \in \mathbb{C}$. The *finite eigenvalues* or *zeros* of a regular $P(\lambda)$ are the points $\lambda_i \in \mathbb{C}$, for which $\det P(\lambda_i) = 0$. In the general, non-regular case, the finite zeros of $P(\lambda)$ are the points $\lambda_i \in \mathbb{C}$, for which $\text{rank}_{\mathbb{F}} P(\lambda_i) < \text{rank}_{\mathbb{F}(\lambda)} P(\lambda)$, that is the points $\lambda_i \in \mathbb{C}$ at which the rank of the constant matrix $P(\lambda_i)$ is less than the normal rank of $P(\lambda)$. The *normal rank* of $P(\lambda)$, denoted by $\text{rank}_{\mathbb{F}(\lambda)} P(\lambda)$, is the maximal number of linearly independent columns of $P(\lambda)$, when the latter are considered as vectors of the rational vector space of $\mathbb{F}(\lambda)^p$.

A square polynomial matrix $U(\lambda)$ is called *unimodular* if $\det U(\lambda) \neq 0$, for all $\lambda \in \mathbb{C}$. Equivalently, $U(\lambda)$ is unimodular if and only if it has no finite zeros. Moreover, it can be easily seen that the inverse of a unimodular matrix exists and it is unimodular, and thus polynomial, as well. Pre and post multiplication of a polynomial matrix by a unimodular one gives rise to a new polynomial matrix sharing the same zero structure with the original one. In view of this fact, the following definition has been given.

Definition 1. (Unimodular equivalence [26, Vol. 1, Definition 2, p. 133]). Two polynomial matrices $P_i(\lambda) \in \mathbb{F}[\lambda]^{p\times m}$, $i = 1, 2$ are *unimodular equivalent*, if there exist unimodular matrices $U(\lambda) \in \mathbb{F}[\lambda]^{p\times p}$, $V(\lambda) \in \mathbb{F}[\lambda]^{m\times m}$, such that

$$P_1(\lambda) = U(\lambda)P_2(\lambda)V(\lambda). \tag{2}$$

Unimodular equivalence defines an equivalence relation on $\mathbb{F}[\lambda]^{p\times m}$. The members of each equivalence class of unimodular equivalence share the *Smith canonical form*. Particularly, if $\lambda_i, i = 1, 2, \ldots, \mu$ are the distinct finite zeros of $P(\lambda)$, then there exist unimodular matrices $U(\lambda), V(\lambda)$ of appropriate dimensions such that

$$S^{\mathbb{C}}_{P(\lambda)}(\lambda) = U(\lambda)P(\lambda)V(\lambda), \tag{3}$$

where $S^{\mathbb{C}}_{P(\lambda)}(\lambda) = diag\{f_1(\lambda), f_2(\lambda), \ldots, f_r(\lambda), 0_{m-r,n-r}\}$ is the *Smith canonical form* of $A(s)$ in $\mathbb{C}$, $f_j(\lambda) = \prod_{i=1}^{\mu}(\lambda - \lambda_i)^{\sigma_{ij}}$ are the *invariant polynomials* of $A(s)$ and the *partial multiplicities* of λ_i satisfy $\sigma_{ij} \leq \sigma_{i,j+1}$, hence $f_i(s) \mid f_{i+1}(s)$ [25, pp. 9–14], while $r = \text{rank}_{\mathbb{R}(s)} P(\lambda)$ is the normal rank of $P(\lambda)$. Additionally, the factors $(\lambda - \lambda_i)^{\sigma_{ij}}$ are the *finite elementary divisors* (*f.e.d.'s*) of $P(\lambda)$.

The *reverse* or *dual* of a polynomial matrix $P(\lambda) = \sum_{i=0}^{n} P_i\lambda^i \in \mathbb{F}[\lambda]^{p\times m}$, with $P_n \neq 0$, is given by

$$\mathrm{rev}P(\lambda) = \lambda^n P(\lambda^{-1}) = \sum_{i=0}^{n} P_{n-i}\lambda^i. \tag{4}$$

It can be easily verified that if $\lambda_0 \neq 0$ is a finite eigenvalue of $\mathrm{rev}P(\lambda)$, then λ_0^{-1} is a finite eigenvalue of the original $P(\lambda)$. In case $\mathrm{rev}P(\lambda)$ has the eigenvalue $\lambda_0 = 0$, the polynomial matrix $P(\lambda)$ is said to have an *infinite eigenvalue*. The algebraic, geometric and partial multiplicities of the infinite eigenvalue of $P(\lambda)$ are defined to be those of the zero eigenvalue of $\mathrm{rev}P(\lambda)$. In this respect, the *infinite elementary divisors* (*i.e.d.s*) of $P(\lambda)$ are the finite elementary divisors of $\mathrm{rev}P(\lambda)$ at $\lambda = 0$.

We now recall some facts related to the concept of linearization of a regular polynomial matrix. A linearization is essentially a *matrix pencil*, that is a first-order polynomial matrix, capturing the finite eigenstructure of the polynomial matrix being linearized. Its definition is given as follows:

Definition 2 (Linearization [9,27,50]). A matrix pencil $L(\lambda) = \lambda L_1 + L_0$, where $L_i \in \mathbb{F}^{np\times np}$, $i = 0, 1$, is a *linearization* of $P(\lambda) \in \mathbb{F}[\lambda]^{p\times p}$, with $\deg P(\lambda) = n$, if there exist unimodular matrices $U(\lambda) \in \mathbb{F}[\lambda]^{np\times np}$, $V(\lambda) \in \mathbb{F}[\lambda]^{np\times np}$, such that

$$U(\lambda)(\lambda L_1 + L_0)V(\lambda) = \begin{bmatrix} P(\lambda) & 0 \\ 0 & I_{(n-1)p} \end{bmatrix}. \tag{5}$$

It is worth noting that under certain assumptions, namely when a non-trivial infinite eigenstructure is present in $P(\lambda)$, it is possible to obtain matrix pencils $L(\lambda)$ of dimensions smaller than $np \times np$, satisfying (5).

We now focus our attention on the block versions of the well-known *first and second Frobenius companion forms* of $P(\lambda)$, given by

$$F_1(\lambda) = \lambda \begin{bmatrix} I_p & 0 & \cdots & \cdots & 0 \\ 0 & I_p & \ddots & & \vdots \\ \vdots & \ddots & I_p & \ddots & \vdots \\ \vdots & & \ddots & \ddots & 0 \\ 0 & \cdots & \cdots & 0 & P_n \end{bmatrix} + \begin{bmatrix} 0 & -I_p & 0 & \cdots & 0 \\ 0 & 0 & \ddots & \ddots & \vdots \\ \vdots & & \ddots & -I_p & 0 \\ 0 & \cdots & \cdots & 0 & -I_p \\ P_0 & P_1 & \cdots & P_{n-2} & P_{n-1} \end{bmatrix}, \tag{6}$$

and

$$F_2(\lambda) = \lambda \begin{bmatrix} I_p & 0 & \cdots & \cdots & 0 \\ 0 & I_p & \ddots & & \vdots \\ \vdots & \ddots & I_p & \ddots & \vdots \\ \vdots & & \ddots & \ddots & 0 \\ 0 & \cdots & \cdots & 0 & P_n \end{bmatrix} + \begin{bmatrix} 0 & 0 & 0 & \cdots & P_0 \\ -I_p & 0 & \ddots & \ddots & P_1 \\ 0 & \ddots & \ddots & 0 & \vdots \\ \vdots & \ddots & -I_p & 0 & P_{n-2} \\ 0 & \cdots & 0 & -I_p & P_{n-1} \end{bmatrix} \tag{7}$$

resp. These matrix pencils are known [27] to be linearizations of $T(\lambda)$.

A linearization $L(\lambda)$ and the original polynomial matrix $P(\lambda)$ have identical (up to trivial expansion) finite eigenstructures. However, in many applications it is desired to obtain linearizations of a given polynomial matrix, preserving both the finite and infinite eigenstructures of the original matrix. This is the key feature of *strong linearizations* introduced as follows:

Definition 3 (Strong Linearization [10]). A linearization $L(\lambda)$ of $P(\lambda)$ is *strong,* if the matrix pencil $\text{rev}L(\lambda) = L_1 + \lambda L_0$ is a linearization of the polynomial matrix $\text{rev}P(\lambda) = \sum_{i=0}^{n} P_{n-i}\lambda^i$.

Strong linearizations of the same polynomial matrices are related via *strict equivalence,* which is defined as follows:

Definition 4 (Strict equivalence [26, Vol. 2, Definition 1, p. 24]). Two matrix pencils $\lambda L_1 + L_0$, $\lambda M_1 + M_0$ with $L_i, M_i \in \mathbb{F}^{p\times m}$, $i = 0, 1$, are *strictly equivalent* if there exist non-singular matrices $U \in \mathbb{F}^{p\times p}$, $V \in \mathbb{F}^{m\times m}$, such that

$$\lambda L_1 + L_0 = U(\lambda M_1 + M_0)V. \tag{8}$$

Strictly equivalent matrix pencils share identical finite and infinite elementary divisor structures and also left and right null space structures [26, Vol. 2, p. 39]. Their overall algebraic structures are rendered through their common *Kronecker canonical form.*

Both the first and second Frobenius companion forms are known to be strong linearizations of the polynomial matrix $P(\lambda)$ (see for instance [28]). Furthermore, as a direct consequence of the results in Ref. [28], every strong linearization is strictly equivalent to the first Frobenius companion form, that is, there exist constant invertible matrices U, V such that

$$L(\lambda) = UF_1(\lambda)V. \tag{9}$$

It is clear that since a strong linearization $L(\lambda)$ is an ordinary linearization as well, it preserves the finite eigenstructure of the original polynomial matrix $P(\lambda)$. The preservation of the infinite eigenstructure is evident from the fact that $\text{rev}L(\lambda)$ is a linearization of $\text{rev}P(\lambda)$ and the zero eigenvalue of $\text{rev}P(\lambda)$ reflects the infinite eigenvalue of $P(\lambda)$.

A serious drawback of Definition 3 is that in order to check whether a matrix pencil is a strong linearization of a given polynomial matrix, one has to verify that two distinct ordinary linearization definitions are satisfied. A more compact characterization of pairs of polynomial matrices sharing isomorphic finite and infinite elementary divisors structures can be found in Refs. [29] and [30], where the notion of *divisor equivalence* is introduced.

Definition 5 (Divisor equivalence [30]). Two regular matrices $P_i(\lambda) \in \mathbb{F}[s]^{p_i \times p_i}$, $i = 1, 2$ with $p_1 \deg P_1(\lambda) = p_2 \deg P_2(\lambda)$ are said to be *divisor equivalent* if there exist polynomial matrices $U(\lambda), V(\lambda)$ of appropriate dimensions, such that

$$U(\lambda)P_1(\lambda) = P_2(\lambda)V(\lambda) \tag{10}$$

is satisfied and the composite matrices

$$\begin{bmatrix} U(\lambda) & P_2(\lambda) \end{bmatrix}, \begin{bmatrix} P_1(\lambda) \\ -V(\lambda) \end{bmatrix}$$

have no finite, nor infinite elementary divisors.

It can be shown that the above defined relation between polynomial matrices of appropriate degrees and dimensions is indeed an equivalence relation. The key feature of divisor equivalence is that polynomial matrices related through it share common finite and infinite elementary divisor structures. It is worth noting that if $P_i(\lambda), i = 1, 2$ are first-order matrix pencils, then $P_i(\lambda)$ are strictly equivalent (see Definition 4) if and only if they are divisor equivalent. The notion of divisor equivalence has been generalized to allow comparison of matrices of arbitrary dimensions in Ref. [31]. The interested reader is referred to Ref. [32] for a comprehensive review of polynomial matrix and system theoretic notions equivalences.

3. Linearization Methods

3.1. *Lancaster's companion matrices*

Given a $p \times p$ square polynomial matrix $P(\lambda) = \sum_{i=0}^{n} \lambda^i P_i$, an interesting family of linearizations of $P(\lambda)$ is proposed in Refs. [4,5,9]. The construction of the family is based on the introduction of block matrices

$$S_i = \left[\begin{array}{cccc|cccc} 0 & \cdots & 0 & -P_0 & 0 & \cdots & \cdots & 0 \\ \vdots & \cdot^{\cdot^{\cdot}} & \cdot^{\cdot^{\cdot}} & \vdots & \vdots & & & \vdots \\ 0 & \cdot^{\cdot^{\cdot}} & & \vdots & \vdots & & & \vdots \\ -P_0 & \cdots & \cdots & -P_{i-1} & 0 & \cdots & \cdots & 0 \\ \hline 0 & \cdots & \cdots & 0 & P_{i+1} & \cdots & \cdots & P_n \\ \vdots & & & \vdots & \vdots & & \cdot^{\cdot^{\cdot}} & 0 \\ \vdots & & & \vdots & \vdots & \cdot^{\cdot^{\cdot}} & \cdot^{\cdot^{\cdot}} & \vdots \\ 0 & \cdots & \cdots & 0 & P_n & 0 & \cdots & 0 \end{array}\right], \tag{11}$$

and pencils of the form

$$L_i(\lambda) = \lambda S_{i-1} - S_i, \tag{12}$$

where $i = 0, 1, \ldots, n$. With this setup, assuming that $\det P_n \neq 0$, it is shown in Ref. [9] that the pencils $L_i(\lambda)$, $i = 0, 1, \ldots, n$ generate an n-dimensional subspace $\mathcal{V}_0$, of the vector space $\mathcal{V}$ of all linearizations of $P(\lambda)$. Thus, the following results:

Theorem 6 ([9, Theorem 11]). *Assume that* $\det P_n \neq 0$. *A pencil* $L_c(\lambda) = \sum_{i=0}^{n} \lambda^i L_i(\lambda)$ *is a linearization of* $P(\lambda)$ *if and only if the polynomial* $p_c(\lambda) = \sum_{i=0}^{n} \lambda^i L_i(\lambda)$ *is non-zero at all the eigenvalues of* $P(\lambda)$.

Some noteworthy consequences of Theorem 6 are the following:

- The pencil $L_1(\lambda)$ is a linearization of $P(\lambda)$, if $\det P_n \neq 0$.
- The pencils $L_i(\lambda)$, for $i = 2, 3, \ldots, n$ are linearizations of $P(\lambda)$, if and only if both $\det P_n \neq 0$ and $\det P_0 \neq 0$.

Notably, the block symmetric companion pencils (12) are known in the literature since the early 1960s (see Refs. [4,5]), but their advantage over traditional companion matrices for the computation of the spectra of symmetric polynomial matrices did not receive much attention until recently. Moreover, the family of linearizations (12) may serve as a basis of the vector space $\mathbb{DL}(P)$ presented in Section 3.3.

3.2. *Fiedler Linearizations*

The key idea behind Fiedler linearizations first appears in the seminal paper of Fiedler [12] where a new family of companion matrices for a given scalar polynomial, parametrized by products of elementary constant matrices was introduced. This family of linearizations, originally intended to serve as companion matrices for scalar polynomials, was generalized to fit the polynomial matrix case in Refs. [6,24]. The key aspects of this generalization are outlined in the present section. In the following, we will adopt the terminology of [24], where the most general form of the family of Fiedler linearizations was obtained.

Definition 7 ([6,24]). Let $P(\lambda) = P_n\lambda^n + P_{n-1}\lambda^{n-1} + \cdots + P_0 \in \mathbb{C}[\lambda]^{p\times p}$: be a polynomial matrix. Define the following elementary matrices:

$$A_k = \begin{bmatrix} I_{p(k-1)} & 0 & \cdots \\ 0 & C_k & \ddots \\ \vdots & \ddots & I_{p(n-k-1)} \end{bmatrix}, \quad k = 1, 2, \ldots, n-1, \tag{13}$$

$$C_k = \begin{bmatrix} 0 & I_p \\ I_p & -P_k \end{bmatrix}, \tag{14}$$

and

$$A_0 = diag\{-P_0, I_{p(n-1)}\}. \tag{15}$$

Note that $A_i, i = 1, ..., n-1$ are non-singular and A_0 is non-singular if and only if P_0 is, resp., non-singular. Before we proceed to the main outcome of this section it is necessary to introduce a series of definitions and intermediate results.

Definition 8. Let $\mathcal{I} = (i_1, i_2, \ldots, i_m)$ be an ordered tuple containing indices from $\{0, 1, 2, \ldots, n-1\}$. Then $A_{\mathcal{I}} := A_{i_1}A_{i_2}\cdots A_{i_m}$.

Lemma 9 ([6,12]). *Let $i, j \in \{0, 1, 2, \ldots, n-1\}$. Then $A_iA_j = A_jA_i$ if and only if $|i-j| \neq 1$.*

Definition 10 ([24]). Let $\mathcal{I}_1$ and $\mathcal{I}_2$ be two tuples. $\mathcal{I}_1$ will be termed equivalent to $\mathcal{I}_2$ ($\mathcal{I}_1 \sim \mathcal{I}_2$) if and only if $A_{\mathcal{I}_1} = A_{\mathcal{I}_2}$.

It can be seen that $\mathcal{I}_1 \sim \mathcal{I}_2$ if and only if $\mathcal{I}_1$ can be obtained from $\mathcal{I}_2$ using a finite number of allowable (in the sense of Lemma 9) transpositions. As it will be evident in the next pages, each equivalent class of index tuples defines uniquely one product of elementary matrices and vice versa.

Definition 11 ([24]). Let $k, l \in \mathbb{Z}$ with $k \leq l$. Then we define

$$(k:l) := \begin{cases} (k, k+1, ..., l), k \leq l \\ \varnothing, k > l \end{cases}. \tag{16}$$

Next we define the juxtaposition of two tuples.

Definition 12 ([24]). Let $\mathcal{I}_1$ and $\mathcal{I}_2$ be two tuples. By $(\mathcal{I}_1, \mathcal{I}_2)$ we denote the juxtaposition of $\mathcal{I}_1$ and $\mathcal{I}_2$.

It is easy to see that $A_{(\mathcal{I}_1,\mathcal{I}_2)} = A_{\mathcal{I}_1} A_{\mathcal{I}_2}$.

Definition 13 ([24]). Given an index tuple $\mathcal{I} = (i_1, i_2, \ldots, i_m)$. We define the reverse tuple $(i_m, i_{m-1}, \ldots, i_1)$, which will be denoted as $\bar{\mathcal{I}}$.

One of the main concepts introduced in Refs. [6,24] is the notion of operation-free products of elementary matrices. Operation-free products are essentially products resulting in block matrices containing only trivial blocks such as 0 or I_p and P_i. By avoiding operations between the coefficients P_i, the companion-like structure of the linearizations is guaranteed and the numerical data of the original problem are not perturbed. The important notion of operation-free products of elementary matrices is given as follows:

Definition 14 ([24]). A product of two elementary matrices A_i, A_j with indices $i, j \in \{0, 1, 2, \ldots, n-1\}$ will be called *operation-free* iff the block elements of the product are either $0, I_p$ or $-P_i$ (for generic matrices P_i).

In the next Lemma, some products of two or three elementary matrices are checked to see if they are operation-free or not.

Lemma 15 ([24]). *The product $A_i A_i$ is not operation-free for $i = 0, \ldots, n-1$. The product $A_i A_{i+1} A_i$ is operation-free, while $A_{i+1} A_i A_{i+1}$ is not for $i = 0, \ldots, n-2$.*

An important property of non-operation-free products is that they cannot be extended to operation-free ones.

Lemma 16. *Let $\mathcal{M}$ be an index tuple such that $A_\mathcal{M}$ is not operation-free. Then for any two other index tuples $\mathcal{L}$ and $\mathcal{R}$, $A_\mathcal{L} A_\mathcal{M} A_\mathcal{R}$ is not operation-free.*

If $A_\mathcal{I}$ is an operation-free product of elementary matrices and $\mathcal{M}$ is any index tuple such that $\mathcal{I} = (\mathcal{L}, \mathcal{M}, \mathcal{R})$ for some tuples $\mathcal{L}, \mathcal{R}$, then $A_\mathcal{M}$ is operation-free. Next, we introduce the notion of block transposition of a block matrix. If $A = [A_{ij}]_{n\times m}$ is a block matrix consisting of block elements $A_{ij} \in \mathbb{C}^{p\times p}$, then its *block transpose* is defined by

$$A^{\mathcal{B}} = [A_{ji}]_{m\times n}.$$

The next lemma describes the form of $A_{(k:l)}$.

Lemma 17. *The product* $A_{(k:l)}$ *is of the form*

$$A_{(k:l)} = \begin{cases} \left[\begin{array}{c|c|c|c} I_{k-1} & & & \\ \hline \cdots & 0_{1\times(l-k+1)} & I & \\ & & -P_k & \\ & I_{l-k+1} & \vdots & \\ & & -P_l & \\ \hline & & & I_{n-l-1} \end{array}\right], k > 0 \\ \left[\begin{array}{c|c|c} 0_{1\times l} & -P_0 & \\ \vdots & -P_1 & \\ I_l & \vdots & \\ & -P_l & \\ \hline & & I_{n-l-1} \end{array}\right], \quad k = 0 \end{cases}, k \leq l \leq n-1, \tag{17}$$

where the dimensions appearing in the zero and identity matrices are block dimensions.

The following Theorem characterizes some operation-free products of elementary matrices.

Theorem 18 ([24]). *Every product of the form*

$$\prod_{i=n-1}^{0} A_{(c_i:i)}, \text{ for } c_i \in (0:i) \cup \{\infty\} \tag{18}$$

is an operation-free product. Form (18) *of a product of elementary matrices will be called column standard form.*

The products of the form (18) are completely characterized by the ordered set of indices $\mathcal{C} = (c_{n-1}, c_{n-2}, \ldots, c_0)$.

Definition 19 ([24, Successor Infix Property (SIP)]). Let $\mathcal{I} = (i_1, i_2, \ldots, i_k)$ be an index tuple. $\mathcal{I}$ will be called successor infixed if for every pair of indices $i_a, i_b \in \mathcal{I}$, with $1 \leq a < b \leq k$, satisfying $i_a = i_b$, there exists at least one index $i_c = i_a + 1$, such that $a < c < b$.

Theorem 20 ([24]). *Let $\mathcal{I}$ be an index tuple. The following statements are equivalent:*

(1) $A_{\mathcal{I}}$ is operation-free.
(2) $\mathcal{I}$ satisfies the successor infixed property (SIP).
(3) $A_{\mathcal{I}}$ can be written in the column standard form (18) as $\prod_{i=n-1}^{0} A_{(c_i:i)}$, for $c_i \in (0:i) \cup \{\infty\}$.

Similar results are obtained using products of A_k^{-1}

$$A_{-k} := A_k^{-1} = \begin{bmatrix} I_{p(k-1)} & 0 & \cdots \\ 0 & C_k^{-1} & \ddots \\ \vdots & \ddots & I_{p(n-k-1)} \end{bmatrix}, \quad k = 1, \ldots, n-1, \tag{19}$$

with

$$C_{-k} := C_k^{-1} = \begin{bmatrix} P_k & I_p \\ I_p & 0 \end{bmatrix},$$

while

$$A_{-n} = diag\{I_{p(n-1)}, P_n\}.$$

Note that products of $A_{(-k:-l)}$ where $1 \le l \le k \le n$ have the form

$$A_{(-k:-l)} = \begin{cases} \left[\begin{array}{c|cc|c} I_{l-1} & & & \\ \hline & P_l & & \\ & \vdots & I_{(k-l+1)} & \\ & P_k & & \\ \cline{2-3} & I & 0_{1\times(k-l+1)} & \\ \hline & & & I_{n-k-1} \end{array}\right], & l \le k < n \\[2ex] \left[\begin{array}{c|c|c} I_{l-1} & & \\ \hline & P_l & \\ & \vdots & I_{(n-l)} \\ & P_{n-1} & \\ \hline & P_n & 0_{1\times(n-l)} \end{array}\right], & k = n. \end{cases}$$

Similarly, let $\mathcal{I} = (i_1, i_2, \ldots, i_k)$ be an index tuple with elements from $\{-n, ..., -1\}$. $\mathcal{I}$ will be called successor infixed if and only if for every pair of indices $i_a, i_b \in \mathcal{I}$, with $1 \le a < b \le k$, satisfying $i_a = i_b$, there exists at least one index $i_c = i_a + 1$, such that $a < c < b$.

Theorem 21 ([24]). *Let $\mathcal{I} = (i_1, i_2, \ldots, i_m)$ be an index tuple from the set $\{-n, ..., -1\}$. The following statements are equivalent:*

(1) $A_{\mathcal{I}}$ is operation-free.
(2) $\mathcal{I}$ satisfies the SIP.
(3) $A_{\mathcal{I}}$ can be written in the column standard form $\prod_{i=-1}^{-n} A_{(c_i:i)}$, for $c_i \in (-n : i) \cup \infty$.
(4) $A_{\mathcal{I}}$ can be written in the row standard form $\prod_{j=-n}^{-1} A_{\overline{(r_j:j)}}$, for $r_j \in (-n : j) \cup \infty$.

The main result of [24], generalizing the family of linearizations described in Refs. [6,7], is shown in what follows.

Theorem 22 ([24]). *Let $P(\lambda)$ be a regular $p \times p$ polynomial matrix with degree n with P_0, P_n non-singular, $k \in \{1, 2, \ldots, n\}$. Let $\mathcal{P}$ be a permutation of the index tuple $(0 : k-1)$ where $k \in \{1, 2, \ldots, n\}$. Let also $\mathcal{L_P}, \mathcal{R_P}$ be index tuples with elements from the set $\{0, 1, \ldots, k-2\}$ such that $(\mathcal{L_P}, \mathcal{P}, \mathcal{R_P})$ satisfies the SIP. Let $\mathcal{N}$ be a permutation of the index tuple $(-n : -k)$ where $k \in \{1, 2, \ldots, n\}$. Let also $\mathcal{L_N}, \mathcal{R_N}$ be index tuples with elements from the set $\{-n, -n+1, \ldots, -k-1\}$ such that $(\mathcal{L_N}, \mathcal{N}, \mathcal{R_N})$ satisfies the SIP. Then the matrix pencil*

$$sA_{(\mathcal{L_N}, \mathcal{L_P}, \mathcal{N}, \mathcal{R_P}, \mathcal{R_N})} - A_{(\mathcal{L_N}, \mathcal{L_P}, \mathcal{P}, \mathcal{R_P}, \mathcal{R_N})} \tag{20}$$

is a linearization of the polynomial matrix $P(\lambda)$ using operation-free products as coefficients.

The linearizations in (20) include all the symmetric linearizations that form a basis of the symmetric linearizations vector space $\mathbb{L}_1 \cap \mathbb{L}_2$ in (43). The extended family of linearizations introduced in Ref. [24], later termed "Fiedler pencils with repetitions", allow more complex multiplications of elementary matrices than those in Ref. [6], thus giving rise to a richer family of companion-like linearizations. In Ref. [33], the notion of Fiedler pencils was extended from square to rectangular matrix polynomials, and it is shown that minimal indices and bases of polynomials can be recovered via the same simple procedures developed previously for square polynomial matrices. In Refs. [34,35], the concept of linearization has been generalized to rational matrices and Fiedler-like matrix pencils for rational matrix functions have been constructed. Fiedler linearizations were used to develop structure-preserving companion forms for several classes of structured matrix polynomials such as alternating, palindromic, Hermitian, skew-symmetric, etc. Additionally, recovery formulas for eigenvectors and minimal bases were identified, see for example [13,19,20]. Finally, Fiedler pencils were adapted to matrix polynomials expressed in a Bernstein and Newton polynomial bases in Refs. [36,37].

3.3. *Vector spaces of linearizations*

In this section, the key components for the investigation of vector spaces of linearizations introduced in Ref. [3] are investigated. Following the exposition path in Ref. [3], we first focus on the derivation of the proposed family of linearizations from the first and second Frobenius companion forms in conjunction with the concept of *ansatz* vectors. The resulting vector spaces of linearizations are in turn investigated.

Consider the matrix $P(\lambda) \in \mathbb{R}^{n\times n}[\lambda]$ and write it as matrix polynomial of the form $P(\lambda) = \sum_{i=0}^{k} \lambda^i A_i$ with $A_i \neq 0_{n\times n}$. The eigenvalue problem of interest is $P(\lambda) \cdot x = \left(\sum_{i=0}^{k} \lambda^i A_i\right) \cdot x = 0$. Setting $x_1 = \lambda^{k-1}x$, $x_2 = \lambda^{k-2}x$, $\ldots, x_{k-1} = \lambda x$, $x_k = x$ leads to the First Companion Form as in (6), which is

$$\underbrace{\left(\begin{bmatrix} A_k & 0 & \cdots & 0 \\ 0 & I_n & \cdots & 0 \\ \vdots & & \ddots & \vdots \\ 0 & \cdots & 0 & I_n \end{bmatrix} \lambda + \begin{bmatrix} A_{k-1} & A_{k-2} & \cdots & A_0 \\ -I_n & 0 & \cdots & 0 \\ \vdots & & \ddots & \vdots \\ 0 & \cdots & -I_n & 0 \end{bmatrix}\right)}_{C_1(\lambda)} \begin{pmatrix} x_1 \\ x_2 \\ \vdots \\ x_k \end{pmatrix} = 0. \tag{21}$$

The definition of the vectors $x_1, \ldots, x_k$ combined with the Kronecker product of two matrices allows writing

$$\begin{bmatrix} x_1 \\ x_2 \\ \vdots \\ x_k \end{bmatrix} = \begin{bmatrix} \lambda^{k-1}x \\ \lambda^{k-2}x \\ \vdots \\ x \end{bmatrix} = \Lambda \otimes x \tag{22}$$

for some $x \in \mathbb{C}$, where $\Lambda(\lambda) = \begin{bmatrix} \lambda^{k-1} & \lambda^{k-2} & \cdots & 1 \end{bmatrix}$. Observe that $\Lambda(\lambda)$ is the standard basis for polynomials of degree $k-1$. Let now $x \in \mathbb{C}^n$ and consider the matrix $\Lambda \otimes x$ as in (22). Then (21) can be written as

$$C_1(\lambda) \begin{bmatrix} x_1 \\ x_2 \\ \vdots \\ x_k \end{bmatrix} = C_1(\lambda)\,(\Lambda \otimes x) = \left[(P(\lambda)x)^T \; 0 \cdots 0\right]^T, \tag{23}$$

which in turn leads to

$$C_1(\lambda)\,(\Lambda \otimes I_n) = C_1(\lambda) \begin{bmatrix} \lambda^{k-1} I_n \\ \lambda^{k-2} I_n \\ \vdots \\ \lambda I_n \\ I_n \end{bmatrix} = \begin{bmatrix} P(\lambda) \\ 0 \\ \vdots \\ 0 \\ 0 \end{bmatrix} = e_1 \otimes P(\lambda), \tag{24}$$

where e_1 denotes the first column of the standard basis of $\mathbb{R}^n$. The general form of the above for $L(\lambda) = \lambda X + Y \in \mathbb{R}^{kn\times kn}$ a matrix pencil and vector $v = \left[\begin{smallmatrix} v_1 \\ v_2 \\ \vdots \\ v_k \end{smallmatrix}\right]$ is

$$L(\lambda)\,(\Lambda \otimes I_n) = L(\lambda) \begin{bmatrix} \lambda^{k-1} I_n \\ \lambda^{k-2} I_n \\ \vdots \\ \lambda I_n \\ I_n \end{bmatrix} = \begin{bmatrix} v_1 P(\lambda) \\ v_2 P(\lambda) \\ \vdots \\ v_{k-1} P(\lambda) \\ v_k P(\lambda) \end{bmatrix} = v \otimes P(\lambda). \tag{25}$$

The connection between (25) and the first companion form is clear, since this relation holds for $L(\lambda) = C_1(\lambda)$ and $v = e_1$. Using the above, the vector space $\mathbb{L}_1$ can be defined in the following.

Definition 23 ([3]). For any polynomial matrix $P(\lambda) = \sum_{i=0}^{k} \lambda^i A_i \in \mathbb{F}^{n\times n}$ with $A_i \neq 0_{n\times n}$, the vector space $\mathbb{L}_1(P)$ is defined as

$$\begin{aligned} \mathbb{L}_1(P) = \{ & L(\lambda) = \lambda X + Y | X, Y \in \mathbb{F}^{kn\times kn}\ \exists v \in \mathbb{F}^k : L(\lambda)(\Lambda \otimes I_n) \\ & = v \otimes P(\lambda) \}. \end{aligned} \tag{26}$$

Any matrix pencil $L(\lambda) \in \mathbb{L}_1(P)$ is said to *satisfy the right ansatz with vector* v or equivalently v *is the right ansatz vector for* $L(\lambda)$", The fact that $C_1(\lambda) \in \mathbb{L}_1(P)$, for $C_1(\lambda)$ as in (23), guarantees that $\mathbb{L}_1$ is non-trivial. Another characterization of the vector space $\mathbb{L}_1(P)$ can be given by utilizing the operation of *Column Shifted Sum*, which is defined as follows:

Definition 24 ([3]). Consider the block matrices

$$X = \begin{bmatrix} X_{11} & \cdots & X_{1k} \\ \vdots & & \vdots \\ X_{k1} & \cdots & X_{kk} \end{bmatrix} \text{ and } Y = \begin{bmatrix} Y_{11} & \cdots & Y_{1k} \\ \vdots & & \vdots \\ Y_{k1} & \cdots & Y_{kk} \end{bmatrix}, \tag{27}$$

where $X_{ij}, Y_{ij} \in \mathbb{F}^{n\times n}$, $i,j = 1,\ldots,k$. Then, the Column Shifted Sum of X and Y is defined as

$$X \boxplus\!\!\rightarrow Y = \begin{bmatrix} X_{11} & \cdots & X_{1k} & 0 \\ \vdots & & \vdots & \vdots \\ X_{k1} & \cdots & X_{kk} & 0 \end{bmatrix} + \begin{bmatrix} 0 & Y_{11} & \cdots & Y_{1k} \\ \vdots & \vdots & & \vdots \\ 0 & Y_{k1} & \cdots & Y_{kk} \end{bmatrix}. \tag{28}$$

Consider the matrix $C_1(\lambda)$ as in (21) and write it as $C_1(\lambda) = \lambda X_1 + Y_1$. Then the column shifted sum of the coefficient matrices of $C_1(\lambda)$ is

$$X_1 \boxplus\!\!\rightarrow Y_1 = \begin{bmatrix} A_k & 0 & \cdots & 0 \\ 0 & I_n & \cdots & 0 \\ \vdots & & \ddots & \vdots \\ 0 & \cdots & 0 & I_n \end{bmatrix} \boxplus\!\!\rightarrow \begin{bmatrix} A_{k-1} & A_{k-2} & \cdots & A_0 \\ -I_n & 0 & \cdots & 0 \\ \vdots & \ddots & & \vdots \\ 0 & \cdots & -I_n & 0 \end{bmatrix}, \tag{29}$$

which allows writing $X_1 \boxplus\!\!\rightarrow Y_1$ as

$$X_1 \boxplus\!\!\rightarrow Y_1 = \begin{bmatrix} A_k & A_{k-1} & \cdots & A_0 \\ 0 & 0 & \cdots & 0 \\ \vdots & \ddots & & \vdots \\ 0 & \cdots & 0 & 0 \end{bmatrix} = e_1 \oplus \begin{bmatrix} A_k & A_{k-1} & \cdots & A_0 \end{bmatrix}. \tag{30}$$

The definition of the Shifted Column Sum operation allows the following equivalence:

$$(\lambda X + Y)(\Lambda \otimes I_n) = v \otimes P(\lambda) \Leftrightarrow X \boxplus\!\!\rightarrow Y = v \otimes [A_k \quad A_{k-1} \quad \cdots \quad A_0], \tag{31}$$

which in turn leads to an equivalent definition of the vector space $\mathbb{L}_1(P)$, which is

$$\mathbb{L}_1(P) = \left\{\lambda X + Y | X \boxplus\!\!\rightarrow Y = v \otimes [A_k\ A_{k-1} \cdots A_0], v \in \mathbb{F}^k\right\}. \tag{32}$$

The importance of (32) lies in the fact that it allows one to characterize the pencils contained in $\mathbb{L}_1(P)$. This characterization is given in the following theorem:

Theorem 25 ([3]). *Let $P(\lambda) = \sum_{i=0}^{k} \lambda^i A_i \in \mathbb{F}^{n\times n}[\lambda]$ and $v \in \mathbb{F}^k$. Then $L(\lambda) = \lambda X + Y$ lies inside $\mathbb{L}_1(P)$ with ansatz vector v if*

$$X = \begin{bmatrix} v \otimes A_k & W \end{bmatrix}, \quad Y = \begin{bmatrix} W + (v \otimes [A_{k-1} \cdots A_1]) & v \otimes A_0 \end{bmatrix}, \tag{33}$$

with $W \in \mathbb{F}^{kn\times(k-1)n}$ and otherwise arbitrary.

An almost immediate result that arises from Theorem 25 is that the dimension of dim $(\mathbb{L}(P)) = k(k-1)n^2 + k$. In what follows, it is shown that pencils of $\mathbb{L}_1(P)$ with ansatz vector $v = e_1$ have a distinguishable form

Corollary 26 ([3]). *Let $P(\lambda) = \sum_{i=0}^{k} \lambda^i A_i \in \mathbb{F}^{n\times n}[\lambda]$ with $A_k \neq 0$ and $L(\lambda) = \lambda X + Y \in \mathbb{L}_1(P)$ has a non-zero ansatz vector $v = ae_1$. Then*

$$X = \begin{bmatrix} aA_k & X_{12} \\ 0 & -Z \end{bmatrix} \text{ and } Y = \begin{bmatrix} Y_{11} & aA_0 \\ Z & 0 \end{bmatrix}, \tag{34}$$

for some $Z \in \mathbb{F}^{(k-1)n\times(k-1)n}$.

Having determined the general form of a matrix pencil inside $\mathbb{L}_1(P)$, the focus is now turned to the original problem of connecting the eigenstructure of the original matrix P, with that of its linearization. It can be seen that there is a clear connection between the aforementioned structures, described next.

Theorem 27 ([3]). *Let $P(\lambda) = \sum_{i=0}^{k} \lambda^i A_i$ be an $n \times n$ polynomial matrix and $L(\lambda) \in \mathbb{L}_1(P)$ with non-zero right ansatz vector v. Then $x \in \mathbb{C}^n$ is an eigenvector for $P(\lambda)$ with finite eigenvalue $\lambda \in \mathbb{C}$ if and only if $\Lambda \otimes x$ is an eigenvector for $L(\lambda)$ with eigenvalue λ. If in addition P is regular and $L \in \mathbb{L}_1(P)$ is a linearization for P, then every eigenvector of L with finite eigenvalue λ is of the form $\Lambda \otimes x$ for some eigenvector x of P.*

Given the analytical expressions above, a question that still remains is, "Which of the pencils contained in $\mathbb{L}_1(P)$ are actually linearizations for P?" This question is answered by taking into consideration the special structure that a pencil has when e_1 is its ansatz vector.

Theorem 28 ([3]). *Let $P(\lambda) = \sum_{i=0}^{k} \lambda^i A_i$ with $A_k \neq 0$ being an $n \times n$ matrix, and $L(\lambda) = \lambda X + Y \in \mathbb{L}_1(P)$ having a non-zero ansatz vector $v = ae_1$. Then, according to Corollary 26, the matrices X and Y are of the form*

$$X = \begin{bmatrix} aA_k & X_{12} \\ 0 & -Z \end{bmatrix} \text{ and } Y = \begin{bmatrix} Y_{11} & aA_0 \\ Z & 0 \end{bmatrix}, \tag{35}$$

where $Z \in \mathbb{F}^{(k-1)n\times(k-1)n}$. Then, Z being non-singular implies that $L(\lambda)$ is a strong linearization of $P(\lambda)$.

Furthermore, it is proved that if the original matrix $P(\lambda)$ is regular, then all of its linearizations inside $\mathbb{L}_1(P)$ are regular matrices and actually are strong linearizations.

Theorem 29 ([3]). *Let $P(\lambda)$ be a regular matrix polynomial and let $L(\lambda) \in \mathbb{L}_1(P)$. Then the following statements are equivalent.*

1. *$L(\lambda)$ is a linearization for $P(\lambda)$.*
2. *$L(\lambda)$ is a regular pencil.*
3. *$L(\lambda)$ is a strong linearization for $P(\lambda)$.*

A question that naturally arises from the above is the determination of a statement similar to Theorem 27 regarding the recovery of infinite eigenvalues and eigenvectors.

Theorem 30 ([3]). *Let $P(\lambda)$ be an $n \times n$ matrix polynomial of degree $k \geq 2$, and $L(\lambda) \in \mathbb{L}_1(P)$ with non-zero right ansatz vector v. Then $x \in \mathbb{C}^n$ is a right eigenvector for $P(\lambda)$ with eigenvalue ∞ if and only if $e_1 \otimes x$ is a right eigenvector for $L(\lambda)$ with eigenvalue ∞. If in addition P is regular and $L(\lambda) \in \mathbb{L}_1(P)$ is a linearization for P, then every right eigenvector of L with eigenvalue ∞ is of the form $e_1 \otimes x$, for some right eigenvector x of P, with eigenvalue x.*

Observe that the conditions of Theorem 30 can be expressed via the structure of the zero eigenvalue of the reverse matrix polynomials.

Theorem 28 constitutes a linearization condition but only for those pencils in $\mathbb{L}_1(P)$ that have non-zero right ansatz vector of the form $v = ae_1$. This can easily be extended to the general case where v is an arbitrary, non-zero vector. To do so, consider that for any non-zero vector v there exists a non-unique, non-singular constant matrix M, such that $Mv = ae_1$. Since M is non-singular, then so is the block transformation matrix $M \otimes I_n$ and thus, if one defines the pencil $\widetilde{L}(\lambda) = (M \otimes I_n) L(\lambda)$, then, according to Corollary 26, $\widetilde{L}(\lambda)$ is of the form

$$\widetilde{L}(\lambda) = \lambda \begin{bmatrix} \widetilde{X}_{11} & \widetilde{X}_{12} \\ 0 & -Z \end{bmatrix} + \begin{bmatrix} \widetilde{Y}_{11} & \widetilde{Y}_{12} \\ Z & 0 \end{bmatrix}. \tag{36}$$

Now (36) allows the extraction of the matrix Z, which constitutes the linearization condition. Indeed, if Z is non-singular, then $L(\lambda)$ is a linearization of P. Since interest lies only on the matrix Z, then the computation of the matrix $\widetilde{Y}(\lambda) = (M \otimes I_n)Y$, where Y, is the constant coefficient matrix of $L(\lambda)$ suffices.

It has already been discussed that not all pencils $L(\lambda) \in \mathbb{L}_1(P)$ are linearizations of an $n \times n$ polynomial matrix $P(\lambda)$, and an algorithmic method for determining whether $L(\lambda)$ is a linearization has been described directly above. The importance of the vector space $\mathbb{L}_1(P)$ is demonstrated from the fact that almost every matrix pencil contained in it is a linearization for $P(\lambda)$, as stated in the next theorem.

Theorem 31 ([3]). *For any regular $n \times n$ matrix polynomial $P(\lambda)$ of degree k, almost every pencil in $\mathbb{L}_1(P)$ is a linearization of $P(\lambda)$, where "almost every" implies that the set of matrix pencils in $\mathbb{L}_1(P)$ that are not linearizations of $P(\lambda)$ form a closed, nowhere dense set of measure zero in $\mathbb{L}_1(P)$.*

A second vector space, namely $\mathbb{L}_2(P)$, which is actually the dual space of $\mathbb{L}_1(P)$ is also of importance. This vector space arises if the Second Companion Form of a polynomial matrix P is considered to be the starting point. Indeed, let $C_2(s)$ denote the Second companion form of $P(\lambda)$ as in (7). The correspondence with the previous results can be seen in the equality

$$\left[\lambda^{k-1} I_n \ \cdots \ \lambda I_n \ I_n\right] C_2(\lambda) = \left[P(\lambda) \ 0 \cdots 0\right], \tag{37}$$

which can equivalently be written as $\left(\Lambda^T \otimes I_n\right) C_2(\lambda) = e_1^T \otimes P(\lambda)$. Thus, it feels natural to consider matrix pencils of the form $L(\lambda) = \lambda X + Y$ that satisfy the *"left ansatz"*

$$\left(\Lambda^T \otimes I_n\right) L(\lambda) = w \otimes P(\lambda), \tag{38}$$

where w shall be referred to as "the left ansatz vector for $P(\lambda)$". Once again, the dependency on the standard basis for polynomials, namely $\Lambda(\lambda)$, can be clearly seen in (37) and (38). Furthermore, the definition of the vector space $\mathbb{L}_2(P)$ arises naturally.

Definition 32 ([3]). For any polynomial matrix $P(\lambda) = \sum_{i=0}^{k} \lambda^i A_i \in \mathbb{F}^{n\times n}$ with $A_i \neq 0_{n\times n}$, the vector space $\mathbb{L}_2(P)$ is defined as

$$\begin{aligned}\mathbb{L}_2(P) = \{L(\lambda) = \lambda X + Y | X, Y \in \mathbb{F}^{kn\times kn} \ \exists w \in \mathbb{F}^k : (\Lambda^T \otimes I_n)L(\lambda)\\ = w^T \otimes P(\lambda)\}.\end{aligned} \tag{39}$$

The duality of the vector spaces $\mathbb{L}_1(P)$ and $\mathbb{L}_2(P)$ becomes clear when taking into consideration Definitions 23 and 32, resp. Furthermore, the vector space $\mathbb{L}_2(P)$ can be defined using the *Row Shifted Sum*, described next, which is a fact further highlighting that the two structures are dual.

Definition 33 ([3]). Let X and Y be block matrices defined as

$$X = \begin{bmatrix} X_{11} & \cdots & X_{1k} \\ \vdots & & \vdots \\ X_{k1} & \cdots & Y_{kk} \end{bmatrix} \text{ and } Y = \begin{bmatrix} Y_{11} & \cdots & Y_{1k} \\ \vdots & & \vdots \\ Y_{k1} & \cdots & Y_{kk} \end{bmatrix}, \tag{40}$$

where $X_{ij}, Y_{ij} \in \mathbb{F}^{n\times n}$, for all $i, j = 1, \dots, k$. Then the row shifted sum of X and Y is defined as

$$X \boxplus_{\downarrow} Y = \begin{bmatrix} X_{11} & \cdots & X_{1k} \\ \vdots & & \vdots \\ X_{k1} & \cdots & X_{kk} \\ 0 & \cdots & 0 \end{bmatrix} + \begin{bmatrix} 0 & \cdots & 0 \\ Y_{11} & \cdots & Y_{1k} \\ \vdots & & \vdots \\ Y_{k1} & \cdots & Y_{kk} \end{bmatrix} \in \mathbb{F}^{(k+1)n\times kn}, \tag{41}$$

where the dimensions of the zero matrices are $n \times n$.

It is proved that, considering the definition of the row shifted sum, the following relation holds:

$$\left(\Lambda^T \otimes I_n\right)(\lambda X + Y) = w^T \otimes P(\lambda) \Leftrightarrow X \boxplus_{\downarrow} Y = w^T \otimes \begin{bmatrix} A_k \\ \vdots \\ A_0 \end{bmatrix}. \tag{42}$$

This relation allows for an equivalent definition of the vector space $\mathbb{L}_2(P)$.

An immediate result of (42) is that it allows the determination of an explicit connection between the vector spaces $\mathbb{L}_1(P)$ and $\mathbb{L}_2(P)$ for a polynomial matrix $P(\lambda)$, which highlights the duality between these two spaces.

Proposition 34 ([3]). *The connection between the vector spaces $\mathbb{L}_1(P)$ and $\mathbb{L}_2(P)$ for a matrix polynomial P is described as*

$$\mathbb{L}_2(P) = \left[\mathbb{L}_1\left(P^T\right)\right]^T.$$

The duality of the two vector spaces leads to the definition of the left eigenvector associated with a finite eigenvalue or to the eigenvalue at infinity.

Definition 35 ([15]). A left eigenvector of an $n \times n$ matrix polynomial P associated with a finite eigenvalue λ is a non-zero vector $y \in \mathbb{C}^n$ such that $y \cdot P(\lambda) = 0$. A left eigenvector for P corresponding to the eigenvalue ∞ is a left eigenvector of rev$P$ associated with eigenvalue 0.

Having defined the left eigenvectors for both finite and infinite eigenvalues, their recovery using matrix pencils inside of $\mathbb{L}_2(P)$ can be described. The result follows immediately from the recovery of the right eigenstructure via $\mathbb{L}_1(P)$ and the duality of the two vector spaces.

Theorem 36 ([3]). *Let $P(\lambda)$ be an $n \times n$ matrix polynomial of degree k, and $L(\lambda)$, any pencil in $\mathbb{L}_2(P)$ with non-zero ansatz vector w. Then $y \in \mathbb{C}^n$*

is a left eigenvector for $P(\lambda)$ *with finite eigenvalue* $\lambda \in \mathbb{C}$ *if and only if* $\bar{\Lambda} \otimes y$ *is a left eigenvector for* $L(\lambda)$*, with eigenvalue* λ*. If in addition* P *is non-singular, and* $L \in \mathbb{L}_2(P)$ *is a linearization for* P*, then every left eigenvector of* L *with finite eigenvalue* λ *is of the form* $\bar{\Lambda} \otimes y$ *for some left eigenvector* y *of* P*.*

In a similar fashion, the recovery of eigenvectors at ∞ can be achieved through pencils in $\mathbb{L}_2(P)$.

Theorem 37 ([3]). *Let* $P(\lambda)$ *be an* $n \times n$ *matrix polynomial of degree* k*, and* $L(\lambda)$ *any pencil in* $\mathbb{L}_2(P)$ *with non-zero left ansatz vector* w*. Then* $y \in \mathbb{C}^n$ *is a left eigenvector for* $P(\lambda)$ *with eigenvalue* ∞ *if and only if* $e_1 \otimes y$ *is a left eigenvector for* $L(\lambda)$*, with eigenvalue* ∞*. If in addition* P *is non-singular, and* $L \in \mathbb{L}_2(P)$ *is a linearization for* P*, then every left eigenvector of* L *with eigenvalue* ∞ *is of the form* $e_1 \otimes y$ *for some left eigenvector* y *of* P *with eigenvalue* ∞*.*

Considering the definition of the vector spaces $\mathbb{L}_1(P)$ and $\mathbb{L}_2(P)$, studying their intersection, which will by default be a vector space itself, feels natural. This idea leads to the definition of the *double ansatz space* of a matrix polynomial $P(\lambda)$.

Definition 38 ([3]). Let P be any $n \times n$ polynomial matrix of degree k. Then the double ansatz space of P, denoted by $\mathbb{DL}(P)$, is defined as

$$\mathbb{DL}(P) = \mathbb{L}_1(P) \cap \mathbb{L}_2(P), \tag{43}$$

which is actually the set of pencils $L(\lambda)$ that satisfy simultaneously the right ansatz for some vector $v \in \mathbb{F}^n$ and the left ansatz for some vector $w \in \mathbb{F}^n$.

The first question that has to be answered is whether these vector spaces are trivial or not. It is actually proved that in order for a matrix pencil to satisfy both the left and right ansatz, the vectors v and w of Definition 38 have to coincide. This is described in the following Theorem.

Theorem 39 ([3]). *Let* $P(\lambda) = \sum_{i=0}^{k} \lambda^i A_i$ *be a matrix polynomial with coefficients in* $\mathbb{F}^{n\times n}$ *and* $A_k \neq 0$*. Then for vectors* $v, w \in \mathbb{F}^n$*, there exists a* $kn \times kn$ *matrix pencil* $L(\lambda) = \lambda X + Y$ *that simultaneously satisfies*

$$L(\lambda)\,(\Lambda \otimes I_n) = v \otimes P(\lambda) \text{ and } \left(\Lambda^T \otimes I_n\right) L(\lambda) = w^T \otimes P(\lambda), \tag{44}$$

if and only if $v = w$*.*

Thus, in the case of double ansatz vector spaces, it suffices to talk about the "ansatz vector" without having to specify if it is right or left ansatz. An important property of the double ansatz vector space is that it preserves symmetry. Indeed, if $P(\lambda)$ is a symmetric matrix polynomial, then every pencil in $\mathbb{DL}(P)$ is also symmetric.

Next up, we can establish conditions under which a matrix pencil $L(\lambda) \in \mathbb{DL}(P)$ with ansatz vector v is a linearization of P. To do so, we begin by associating each vector $v \in \mathbb{F}^k$ with a polynomial of degree $k-1$.

Definition 40 ([3]). Any vector $v = \begin{bmatrix} v_1 \, v_2 \, \cdots \, v_k \end{bmatrix} \in \mathbb{F}^k$ has an associated scalar polynomial, called the *v-polynomial of* v, defined as

$$p(x;\, v) = v_1 x^{k-1} + v_2 x^{k-2} + \cdots + v_{k-1} x + v_k.$$

Furthermore, we use the convention that the polynomial $p(x;\, v)$ has a root at ∞ whenever $v_1 = 0$.

Definition of v-polynomials allows for a very elegant method for checking whether a matrix polynomial in $\mathbb{DL}(P)$ is a linearization for P.

Theorem 41 ([15]). *Let $P(\lambda)$ be a non-singular polynomial matrix and $L(\lambda) \in \mathbb{DL}(P)$ with ansatz vector v. Then $L(\lambda)$ is a linearization for $P(\lambda)$ if and only if no root of the v-polynomial $p(x;\, v)$ is an eigenvalue of $P(\lambda)$. This statement includes ∞ as one of the possible roots for $p(x;\, v)$ is an eigenvalue of $P(\lambda)$.*

Closing this section, it is important to note that $\mathbb{DL}(P)$ maintains the nice property of $\mathbb{L}_1(P)$ and $\mathbb{L}_2(P)$, that almost all of the pencils in it constitute linearizations of P.

4. Structured Linearizations and Polynomial Eigenvalue Problems

Polynomial eigenvalue problems (PEP) have received attention in many research works in the recent years due to their applications in areas such as vibration analysis of mechanical systems, in acoustics, linear stability of flows in fluid mechanics and many others. The standard form of a polynomial eigenvalue problem is

$$P(\lambda)x = 0, \tag{45}$$

where $P(\lambda)$ is a polynomial matrix of the form (1) and one seeks to find a scalar $\lambda \in \mathbb{C}$ and a non-zero vector $x \in \mathbb{C}^m$ satisfying (45). In this setup, λ is an eigenvalue of $P(\lambda)$ and x is the associated eigenvector. Polynomial eigenvalue problems can be seen as a generalization of the *standard eigenvalue problem* (SEP)

$$Ax = \lambda x, \tag{46}$$

and the *generalized eigenvalue problem* (GEP)

$$Ax = \lambda Ex, \tag{47}$$

where both can be treated as a PEP associated to the first-order polynomial matrices $\lambda I - A$ and $\lambda E - A$, resp. Well-established numerical approaches for the solution of these two problems involve the *Schur canonical form* of A for the SEP or the *generalized Schur form* of the matrix pencil $\lambda E - A$ in the case of GEP (see for instance [38,39]).

Another class of PEP that has received special attention in the literature is that of quadratic eigenvalue problems (QEP) (see Ref. [2] for a comprehensive survey). Quadratic polynomial matrices provide the natural framework for the description of second-order linear dynamical systems. An area of particular importance where second-order differential equations arise is the field of study of mechanical vibrations. In this framework, mechanical systems are often modeled by differential equations of the form

$$M\ddot{q}(t) + C\dot{q}(t) + Kq(t) = f(t), \tag{48}$$

where M, C, K are $n \times n$ matrices representing the generalized mass, damping and stiffness coefficients, $q(t)$ is the displacement vector and $f(t)$ is the external force acting on the system. The solution of the differential equation (48) is closely related to the solution of the quadratic eigenvalue problem

$$(\lambda^2 M + \lambda C + K)x = 0. \tag{49}$$

A major difficulty in the study and solution of QEPs is that there is no simple analogue to the Schur canonical form for the SEP or the generalized Schur form for the GEP. This complication is also present in polynomial eigenvalue problems of order greater than two. A common approach to overcome this difficulty is the use of linearization techniques to effectively reduce any high — order PEP to an equivalent GEP or SEP.

In general, polynomial eigenvalue problems occur in a wide variety of situations where linear dynamics of order greater than one are under consideration. A large collection of models associated to polynomial eigenvalue problems can be found in Ref. [18]. The problems presented therein are classified according to certain properties of the associated polynomial matrix such as its shape, domain of its entries (real or complex), its order (degree) and possible symmetries. Some indicative examples in this collection are reproduced in the following:

acoustic_wave_1d: This is a quadratic eigenvalue problem which arises from the finite element discretization of the time-harmonic wave equation $-\Delta p-(2\pi f/c^2)p=0$, where p is the acoustic pressure in a bounded domain subject partly to Dirichlet conditions ($p=0$) and partly to $\frac{\partial p}{\partial n}+\frac{2\pi i f}{\zeta}p=0$. In this setting, f is the frequency, c is the speed of sound in the medium, and ζ is the (possibly complex) impedance. For $c=1$ in the 1-D domain $[0,1]$, the polynomial matrix associated with the discretized wave equation is

$$Q(\lambda)=\lambda^2 M+\lambda C+K, \tag{50}$$

where

$$M=\frac{-4\pi^2}{n}\left(I_n-\frac{1}{2}e_n e_n^T\right),\quad C=\frac{2\pi i}{\zeta}e_n e_n^T,\quad K=n\begin{bmatrix} 2 & -1 & \cdots & 0 \\ -1 & \ddots & \ddots & \vdots \\ \vdots & \ddots & 2 & -1 \\ 0 & \cdots & -1 & 1 \end{bmatrix}. \tag{51}$$

The eigenvalues of the polynomial matrix Q are the resonant frequencies of the system, which for the given formulation lie in the upper half of the complex plane. Notably, the coefficient matrices M,C,K are symmetric. A similar quadratic eigenvalue problem arises in the 2-D case on the unit square $[0,1]\times[0,1]$ (see Ref. [18]).

railtrack: The railtrack model stems from a model of the vibration of rail tracks under the excitation of high speed trains, discretized by classical mechanical finite elements. The 1005×1005 polynomial matrix associated to the resulting quadratic eigenvalue problem has the form

$$Q(\lambda)=\lambda^2 A^T+\lambda B+A, \tag{52}$$

where $B^T=B$,

$$A=\begin{bmatrix} 0 & 0 \\ A_{21} & 0 \end{bmatrix}\in\mathbb{C}^{1005\times 1005},$$

and $A_{21}\in\mathbb{C}^{201\times 67}$ is a full column rank matrix. Clearly, due to the special structure of the coefficients A,B, the polynomial matrix $Q(\lambda)$ is T–palindromic, i.e. $\text{rev}Q^T(\lambda)=Q(\lambda)$. Moreover, in view of the rank deficiency of A^T, non-trivial eigenstructure at infinity is expected.

plasma_drift: This is a cubic polynomial eigenvalue problem of dimension 128 or 512 resulting from the modeling of drift instabilities in the plasma edge inside a Tokamak reactor. The associated polynomial matrix has the form

$$P(\lambda) = \lambda^3 A_3 + \lambda^2 A_2 + \lambda A_1 + A_0, \tag{53}$$

where A_0 and A_1 are complex, A_2 is complex symmetric, and A_3 is real symmetric. The desired eigenpair is the eigenpair corresponding to the eigenvalue with the largest imaginary part being the one of particular interest.

orr_sommerfeld: This quartic polynomial eigenvalue problem arises from the spatial stability analysis of the Orr–Sommerfeld equation. The Orr–Sommerfeld equation is the result of the linearization of the incompressible Navier–Stokes equations in which the perturbations in velocity and pressure are assumed to have the form $\Phi(x, y, t) = \phi(y)e^{i(\lambda x - \omega t)}$, where λ is a wavenumber and ω is a radian frequency. Given the Reynolds number R, the Orr–Sommerfeld equation takes the form

$$\left[\left(\frac{d^2}{dy^2} - \lambda^2\right)^2 - iR\left\{(\lambda U - \omega)\left(\frac{d^2}{dy^2} - \lambda^2\right) - \lambda U''\right\}\right]\phi = 0. \tag{54}$$

The spatial stability analysis parameter λ appears to the fourth power in the above equation, hence the quartic polynomial eigenvalue problem. The eigenvalues λ of the polynomial matrix associated to (54) must satisfy $\mathrm{Im}(\lambda) > 0$ to guaranteed stability, with those closest to the real axis being the most interesting ones.

planar_waveguide: The 129×129 fourth-order polynomial matrix of the form

$$P(\lambda) = \lambda^4 A_4 + \lambda^3 A_3 + \lambda^2 A_2 + \lambda A_1 + A_0, \tag{55}$$

where

$$A_1 = \frac{\delta^2}{4}\mathrm{diag}(-1, 0, 0, \ldots, 0, 1), \qquad A_3 = \mathrm{diag}(1, 0, 0, \ldots, 0, 1),$$

$$A_0(i, j) = \frac{\delta^4}{16}(\phi_i, \phi_j), \quad A_2(i, j) = (\phi_i', \phi_j') - (q\phi_i, \phi_j),$$

$$A_2(i, j) = (\phi_i, \phi_j),$$

arises from a finite element solution of the equation for the modes of a planar waveguide using piecewise linear basis functions $\phi_i, i =$

$0, 1, 2, \ldots, 128$. Clearly, A_1, A_3 are diagonal and A_0, A_2, A_4 tridiagonal, hence, the polynomial matrix $P(\lambda)$ is symmetric. The parameter δ describes the difference in refractive index between the cover and the substrate of the waveguide, while q is a function used in the derivation of the variational formulation and is constant in each layer.

As mentioned earlier, a very common approach to solving polynomial eigenvalue problems is through linearizations. The PEP $P(\lambda)x = 0$ is transformed into a larger size GEP of the form $(\lambda E + A)z = 0$ with the same eigenvalues, so that standard linear eigenvalue techniques can be employed. In this respect, some straightforward candidates for the substitution of the polynomial matrix $P(\lambda)$ could be the first and second Frobenius companion forms given in (6) and (7). However, as shown in the above indicative examples, modeling of physical processes often results in eigenvalue problems involving polynomial matrices with special structure (see Refs. [11,18]). Such polynomial matrices may posses symmetric, skew-symmetric, alternating symmetric/skew-symmetric, palindromic or other types of internal structures, which reflect certain types of symmetries or constraints on their spectra. Both from a theoretical and computational point of view, it is essential that the symmetries of the spectra of the original matrix be also present in the spectra of the respective linearization. It is therefore important to construct linearizations preserving the possible structure of a given matrix polynomial and apply numerical methods for the corresponding linear eigenvalue problems that properly take into account these structures as well.

Polynomial matrices with symmetric coefficients are probably the most common cases occurring in many modeling scenarios. Families of block symmetric linearizations have been proposed by many authors (see [3,6,7,9,19,20,24,40]). Given the QEP of the form (49) or (50) with $M = M^T$, $C = C^T$, $K = K^T$ and the additional assumption $\det M \neq 0$, the symmetric Fiedler linearization proposed in Ref. [6,7] has the form

$$L(\lambda) = \lambda \begin{bmatrix} 0 & I \\ I & C \end{bmatrix} - \begin{bmatrix} M^{-1} & 0 \\ 0 & -K \end{bmatrix}. \tag{56}$$

Using the same family of linearizations the cubic PEP (53) can be reduced to a linear block symmetric GEP of the form

$$L(\lambda) = \lambda \begin{bmatrix} A_3 & 0 & 0 \\ 0 & 0 & I \\ 0 & I & A_1 \end{bmatrix} - \begin{bmatrix} -A_2 & I & 0 \\ I & 0 & 0 \\ 0 & 0 & -A_0 \end{bmatrix}. \tag{57}$$

More block symmetric linearizations may be derived for the cubic case using any of the approaches proposed [9,24,40]. For instance, it can be easily verified that the matrix pencil

$$L(\lambda) = \lambda \begin{bmatrix} A_1 & A_2 & A_3 \\ A_2 & A_3 & 0 \\ A_3 & 0 & 0 \end{bmatrix} - \begin{bmatrix} -A_0 & 0 & 0 \\ 0 & A_2 & A_3 \\ 0 & A_3 & 0 \end{bmatrix} \tag{58}$$

is a block symmetric linearization of (53) which can be derived using any of the techniques presented in Sections 3.1, 3.2 and 3.3.

The QEP associated to the T–palindromic matrix $Q(\lambda)$ in (52) can be linearized using any of the techniques proposed in Refs. [11,21,23]. Applying the "ansatz vectors" method presented in Ref. [11], one may obtain the T-palindromic linearization

$$\lambda Z + Z^T = \lambda \begin{bmatrix} A^T & B - A \\ A^T & A^T \end{bmatrix} + \begin{bmatrix} A & A \\ B - A^T & A \end{bmatrix}, \tag{59}$$

or the T–anti-palindromic linearization

$$\lambda \begin{bmatrix} A^T & B + A \\ -A^T & A^T \end{bmatrix} + \begin{bmatrix} -A & A \\ -B - A^T & -A \end{bmatrix}. \tag{60}$$

It is clear that the possibilities for the construction of structured linearizations depending on the structure and constraints of the underlying PEP are by no means limited to the cases presented above. The interested reader is referred to the cited references for detailed descriptions of the methods available in the literature.

5. Conclusion

Linearizations of polynomial matrices provide a valuable tool for the study of polynomial eigenvalue problems. Having the traditional Frobenius companion forms as prototypes for the study of the roots of scalar polynomials, several authors have addressed the more general problem of reducing the polynomial eigenvalue problem, which arises naturally in the modeling of high-order linear dynamical systems, to a generalized first-order eigenvalue problem, through the introduction of new families of "block companion" matrices. The main motivation behind this workaround is dictated by the fact that ordinary or generalized eigenvalue problems have been extensively studied in the past and well-established numerical techniques for their solutions are available in the literature. Beyond the advantage of using proven numerical tools for the solutions of polynomial eigenvalue problems, linearizations benefit from the fact that in most cases their construction can be accomplished simply "by inspection" of the coefficients of the original polynomial matrix, avoiding this way the introduction of unnecessary round-off errors during the reduction process. However, the key advantage of linearization techniques lies in their flexibility of choice from a variety of companion forms, among which are those with the desired structure, which reflects the symmetries of the spectrum of the underlying polynomial matrix.

Current and future research directions on the topic include linearizations expressed in polynomial bases other than the monomial one [17,36,37,41, 42]) linearizations of non-regular or rectangular polynomial matrices and the recovery of their structural invariants [43–45] or even linearizations of rational matrices [35,46]. Another interesting and noteworthy direction of research is related to the notion of ℓ-ification of the polynomial matrix that has been proposed as a generalization to linearizations in Refs. [47–49]. In view of the above, it is clear that linearizations of polynomial matrices remain an active research field with many prospects, both from a theoretical and an applications point of view.

References

[1] V. Mehrmann and D. Watkins, Polynomial eigenvalue problems with Hamiltonian structure, *Electr. Trans. Num. Anal.* **13**, 106–113, 2002.

[2] F. Tisseur and K. Meerbergen, The quadratic eigenvalue problem, *SIAM Rev.* **43**(2), 235–286, (January 2001).

[3] D.S. Mackey, *Structured Linearizations for Matrix Polynomials.* PhD, University of Manchester, 2006.

[4] P. Lancaster, Symmetric transformations of the companion matrix, *NABLA: Bull. Malayan Math. Soc.* (8), 146–148, (1961).

[5] P. Lancaster, *Lambda-matrices and Vibrating Systems*, 1st edn. (Pergamon Press Inc, Oxford, 1966).

[6] E.N. Antoniou and S. Vologiannidis, A new family of companion forms of polynomial matrices, *Electron. J. Linear Algebra* **11**, 78–87, (2004).

[7] E.N. Antoniou and S. Vologiannidis, Linearizations of polynomial matrices with symmetries and their applications, *Electron. J. Linear Algebra* **15**, 107–114, (2006).

[8] N.J. Higham, D.S. Mackey, N. Mackey, and F. Tisseur, Symmetric linearizations for matrix polynomials, *SIAM J. Matrix Anal. Appl.* **29**(1), 143–159, (2006).

[9] P. Lancaster and U. Prells, Isospectral families of high-order systems. *ZAMM Zeitschrift fur Angewandte Mathematik und Mechanik* **87**(3), 219–234, (2007).

[10] P. Lancaster and P. Psarrakos, A Note on Weak and Strong Linearizations of Regular Matrix Polynomials. MIMS EPrint 2006.72, Manchester Institute for Mathematical Sciences, University of Manchester, Manchester, UK, May 2006.
[11] D.S. Mackey, N. Mackey, C. Mehl, and V. Mehrmann, Structured polynomial eigenvalue problems: Good vibrations from good linearizations, *SIAM J. Matrix Anal. Appl.* **28**(4), 1029–1051, (2006).
[12] M. Fiedler, A note on companion matrices, *Linear Algebra Appl.* **372**, 325–331, (October 2003).
[13] F. De Terán, F.M. Dopico, and D.S. Mackey, Fiedler companion linearizations and the recovery of minimal indices, *SIAM J. Matrix Anal. Appl.* **31**(4), 2181–2204, (January 2010).
[14] D.S. Mackey, The continuing influence of Fiedler's work on companion matrices, *Linear Algebra Appl.* **439**(4), 810–817, (August 2013).
[15] D.S. Mackey, N. Mackey, C. Mehl, and V. Mehrmann, Vector Spaces of Linearizations for Matrix Polynomials. MIMS EPrint 2005.26, Manchester Institute for Mathematical Sciences, University of Manchester, Manchester, UK, December 2005.
[16] Y. Nakatsukasa, V. Noferini, and A. Townsend, Vector spaces of linearizations for matrix polynomials: A bivariate polynomial approach, *SIAM J. Matrix Anal. Appl.* **38**(1), 1–29, (January 2017).
[17] H. Faßbender and P. Saltenberger, On vector spaces of linearizations for matrix polynomials in orthogonal bases, *Linear Algebra Appl.* **525**, 59–83, (July 2017).
[18] T. Betcke, N.J. Higham, V. Mehrmann, C. Schröder, and F. Tisseur, NLEVP: A Collection of Nonlinear Eigenvalue Problems. MIMS EPrint 2011.116, Manchester Institute for Mathematical Sciences, University of Manchester, Manchester, UK, December 2011.
[19] M.I. Bueno, K. Curlett and S. Furtado, Structured strong linearizations from Fiedler pencils with repetition I, *Linear Algebra Appl.* **460**, 51–80, (November 2014).
[20] M.I. Bueno and S. Furtado, Structured strong linearizations from Fiedler pencils with repetition II. *Linear Algebra Appl.* **463**, 282–321, (December 2014).
[21] F. De Terán, F.M. Dopico, and D. Steven MacKey, Palindromic companion forms for matrix polynomials of odd degree, *J. Comput. Appl. Math.* **236**(6), 1464–1480, (2011).
[22] F. De Terán, A. Dmytryshyn, and F. Dopico, Generic symmetric matrix polynomials with bounded rank and fixed Odd grade, *SIAM J. Matrix Anal. Appl.* **41**, 1033–1058, (January 2020).
[23] F. De Teran, F.M. Dopico, and D.S. Mackey, Structured linearizations for palindromic matrix polynomials of odd degree, MIMS Preprint, Manchester Institute for Mathematical Sciences, The University of Manchester, April 2010.
[24] S. Vologiannidis and E.N. Antoniou, A permuted factors approach for the linearization of polynomial matrices, *Math. Control, Signals, Syst.* **22**, 317–342, (April 2011).

[25] A.I.G. Vardulakis, *Linear Multivariable Control: Algebraic Analysis and Synthesis Methods* (J. Wiley, 1991).

[26] F.R. Gantmacher, *The Theory of Matrices* (Chelsea Publishing Company, New York, 1959).

[27] I. Gohberg, P. Lancaster, and L. Rodman. *Matrix Polynomials* (Academic Press Inc., New York, 1982).

[28] A.I.G. Vardulakis and E. Antoniou. Fundamental equivalence of discrete-time AR representations, *Int. J. Control* **76**(11), 1078–1088, (2003).

[29] N.P. Karampetakis and S. Vologiannidis, Infinite elementary divisor structure-preserving transformations for polynomial matrices, *Int. J. Appl. Math. Comput. Sci.* **13**(4) 493–504, (2003).

[30] N.P. Karampetakis, S. Vologiannidis, and A.I.G. Vardulakis, A new notion of equivalence for discrete time AR representations, *Int. J. Control* **77**(6), 584–597, (2004).

[31] A. Amparan, S. Marcaida, and I. Zaballa, On matrix polynomials with the same finite and infinite elementary divisors, *Linear Algebra Appl.* **513**, 1–32, (January 2017).

[32] S. Vologiannidis, E.N. Antoniou, N.P. Karampetakis, and A.I.G. Vardulakis, Polynomial matrix equivalences: System transformations and structural invariants, *IMA J. Math. Control Inf.* **38**(1), 54–73, (March 2021).

[33] F. De Terán, F.M. Dopico, and D.S. MacKey, Fiedler companion linearizations for rectangular matrix polynomials, *Linear Algebra Appl.* **437**(3), 957–991, (2012).

[34] R. Alam and N. Behera, Linearizations for rational matrix functions and Rosenbrock system polynomials, *SIAM J. Matrix Anal. Appl.* **37**(1), 354–380. (January 2016).

[35] N. Behera, *Fiedler Line Arizations for LTI State-space Systems and for Rational Eigenvalue Problems.* Thesis, 2014. Accepted: 2015-09-22T10:16:17Z.

[36] D.S. Mackey and V. Perović, Linearizations of matrix polynomials in Bernstein bases, *Linear Algebra Appl.* **501**, 162–197, (2016).

[37] V. Perović and D.S. Mackey, Linearizations of matrix polynomials in Newton bases, *Linear Algebra Appl.* **556**, 1–45, (2018).

[38] J. Demmel, J. Dongarra, A. Ruhe, and H. van der Vorst, *Templates for the Solution of Algebraic Eigenvalue Problems: A Practical Guide.* Society for Industrial and Applied Mathematics, Philadelphia, PA, USA, 2000.

[39] G.H. Golub and C.F. Van Loan, *Matrix Computations* (Johns Hopkins University Press, Baltimore, 1996).

[40] N.J. Higham, D.S. Mackey, N. Mackey, and F. Tisseur, Symmetric linearizations for matrix polynomials, *SIAM J. Matrix Anal. Appl.* **29**(1), 143–159, (January 2007).

[41] A. Amiraslani, R.M. Corless, and P. Lancaster, Linearization of matrix polynomials expressed in polynomial bases, *IMA J. Numer. Anal.* **29**(1), 141–157, (February 2008).

[42] A.S. Karetsou and N.P. Karampetakis, Linearization of bivariate polynomial matrices expressed in non monomial basis, *Multidimens. Syst. Signal Process* **26**, 503–517, (2015).

[43] M.I. Bueno and F.D. Terán, Eigenvectors and minimal bases for some families of Fiedler-like linearizations, *Linear Multilinear Algebra* **62**(1), 39–62, (2014).
[44] F. De Teran, F. Dopico, and D.S. Mackey, Linearizations of singular matrix polynomials and the recovery of minimal indices, *Electron. J. Linear Algebra* **18**, 371–402, (July 2009).
[45] F.D. Terán, F.M. Dopico, and D.S. Mackey, Fiedler companion linearizations and the recovery of minimal indices, *SIAM J. Matrix Anal. Appl.* **31**(4), 2181–2204, (2009).
[46] R. Alam and N. Behera, Generalized fiedler pencils for rational matrix functions, *SIAM J. Matrix Anal. Appl.* **39**(2), 587–610, (January 2018). Publisher: Society for Industrial and Applied Mathematics.
[47] D.A. Bini and L. Robol, On a class of matrix pencils and ℓ-ifications equivalent to a given matrix polynomial, *Linear Algebra Appl.* **502**, 275–298, (August 2016).
[48] F. De Terán, C. Hernando, and J. Pérez, Structured strong ℓ-ifications for structured matrix polynomials in the monomial basis, *Electron. J. Linear Algebra* **37**, 35–71, (2021).
[49] F.M. Dopico, J. Pérez and P. Van Dooren, Block minimal bases ℓ-ifications of matrix polynomials, *Linear Algebra Appl.* **562**, 163–204, (2019).
[50] P. Lancaster, Linearization of regular matrix polynomials, *Electron. J. Linear Algebra* **17**, 21–27, (2008).

https://doi.org/10.1142/9789811261572_0007

Chapter 7

A Game Theory Model for the Award of a Public Tender Procedure

G. Colajanni*, P. Daniele†, and D. Sciacca‡

Department of Mathematics and Computer Science, University of Catania, Viale Andrea Doria, 6, 95125 Catania, Italy

**colajanni@dmi.unict.it*

†patrizia.daniele@unict.it

‡daniele.sciacca@unipa.it

We propose a game theory model consisting of N suppliers competing in a non-cooperative manner for the award of a public tender based on the "Most Economically Advantageous Tender" (MEAT) criterion. According to the European legislation, the total score of each supplier is given by a weighted sum of the criteria set by the tender, which can be of a quantitative or qualitative nature. Suppliers are assumed to be smart and rational and, furthermore, scoring formulas are common knowledge. These formulas are assumed to be interdependent, so that each supplier cannot establish his own score *a priori* because his score depends on other suppliers' offers. We show that this problem can be formulated as a Nash Equilibrium model and that the governing Nash equilibrium conditions can be formulated as a variational inequality problem, for which we provide a result for the existence and uniqueness of the solutions.

1. Introduction

Government procurement or public procurement is the procurement of goods, services and works on behalf of a public authority, such as a government agency and it is necessary because governments cannot produce all the inputs for the goods they provide themselves.

In general, public tenders concern the supply of goods (office supplies and equipment, furniture, IT equipment, vehicles, medical supplies), the execution of works (new construction of structures, renovations, extensions and repairs) and the provision of services (feasibility studies, project management, engineering services).

An award criterion is established for each call for tenders. The criteria provided by the current European legislation are as follows:

- "Lower price" criterion.
- "Most Economically Advantageous Tender" (MEAT) criterion.

For the first criterion, the contract is won by the supplier who proposes the lowest realization price. Whereas, the idea underlying the MEAT criterion is that, when the public administration purchases works, services or supplies to directly satisfy its own needs or to offer certain services to users, it must not only look at cost savings but must also consider the quality of what is purchased. Basically, a trade-off is usually created between cost and quality and the tender is considered the most suitable way to ensure the best balance between these two needs. In the tender design phase, the (qualified) contracting authority must concretely identify its objectives (usually multiple), attribute a relative weight to each of them, define the methods by which the degree of adequacy of each offer with respect to the individual objective is assessed, as well as summarize the information relating to each offer into a single final numerical value.

In this model, the MEAT criterion will be chosen and only an economic offer and quantitative evaluation criteria will be considered. In this context, for each potential supplier a total score is calculated and it is given by a weighted sum of the score attributed to the supplier's economic offer (the price proposed by the supplier to complete the objective of the tender) and the score attributed to the technical/social offer. Usually, the technical/social offer is made up of elements of a quantitative nature such as, for instance, the execution time of the works, the yield, the duration of the concession and the level of tariffs, of elements referring to the absence or presence of a specific characteristic, such as, for instance, possession of quality certification or the legality rating and, finally, of qualitative elements, on which the tender commission must express its opinion, according to the criteria or subcriteria established in the tender notice.

This chapter is organized as follows. In Section 2, we present the mathematical game theory model, providing a Nash Equilibrium framework and deriving the associated variational inequality formulation, for which we discuss some qualitative properties in terms of existence and uniqueness of the solutions. In Section 3, we provide an example of normalization functions, and in Section 4, we derive an alternative variational inequality formulation of the problem and we describe the computational procedure. In Section 5, we propose some illustrative numerical examples for whose solution we use an iterative algorithm based on the Euler Method and we provide a comparison between the results obtained by our formulas and the interdependent ones commonly used in practice. Finally, in Section 6, we summarize our results, present our conclusions and also give suggestions for future research.

2. The Mathematical Model

The game theory model of public requirement tenders (see also [14]) consists in N suppliers who compete in a non-cooperative manner to win the established offer by the selection procedure. As said before, we suppose that tenders are awarded through the MEAT criterion.

We assume that the public requirement tender is launched for the construction of public works or for the acquisition of goods or services and that the suppliers competing for the award are able to satisfy this request. We denote by Π the price imposed by the public tender and by π_i the price proposed by each supplier, $i = 1, \ldots, N$, and we group these quantities into the vector $\pi \in \mathbb{R}^N_+$. In order for a supplier to be admitted to the selection, it must be

$$\Gamma \leq \pi_i \leq \Pi, \quad \forall i = 1, \ldots, N. \tag{1}$$

Thereby, we are assuming that the procedure is not with rising prices, that is, the supplier cannot bargain on the price specified by the announcement. Moreover, we also suppose that the price is no less than a quantity $\Gamma > 0$ to exclude the unrealistic case where a supplier makes a null price offer.

We are supposed to consider M quantitative evaluation criteria. Among the quantitative criteria, we can distinguish those for which the best offer is the one with the highest value and those for which the best offer is the one with the lowest value. Therefore, we suppose that $M_1 \leq M$ quantitative criteria are of the first type, with a typical one denoted by j, and $M_2 \leq M$ quantitative criteria are of the second type, with a typical one denoted by k, where $M_1 + M_2 = M$, $j = 1, \ldots, M_1$, $k = 1, \ldots, M_2$.

Let $q_{ij} \in \mathbb{R}_+$ be the quantity that supplier i decides to propose for his project to satisfy the quantitative criterion j for which the best offer is the one with the highest value such as, for instance, the post-delivery free maintenance time, and we group these quantities, for all j, into the vector $q_i \in \mathbb{R}_+^{M_1}$. In turn, we group these quantities into the vector $q \in \mathbb{R}_+^{NM_1}$. We suppose that

$$\underline{Q}_j \leq q_{ij} \leq 2\underline{Q}_j, \quad \forall i = 1, \ldots, N, \ \forall j = 1, \ldots, M_1, \tag{2}$$

where $\underline{Q}_j$ is the value placed at the basis of the tender for quantitative criterion j.

Let $\tilde{q}_{ik} \in \mathbb{R}_+$ be the quantity that supplier i decides to propose for his project to satisfy the quantitative criterion k for which the best offer is the one with the lowest value such as, for example, the execution time of works, and we group these quantities, for all k, into the vector $\tilde{q}_i \in \mathbb{R}_+^{M_2}$. In turn, we group these quantities into the vector $\tilde{q} \in \mathbb{R}_+^{NM_2}$. We suppose that

$$\frac{1}{2}\tilde{Q}_k \leq \tilde{q}_{ik} \leq \tilde{Q}_k, \quad \forall i = 1, \ldots, N, \ \forall k = 1, \ldots, M_2, \tag{3}$$

where $\tilde{Q}_k$ is the value placed at the basis of the tender for quantitative criterion k. Finally, we group the vectors q and $\tilde{q}$ into the NM-dimensional vector $\varrho = (q, \tilde{q}) \in \mathbb{R}_+^{NM}$.

Constraints (2) ensure that, for each quantitative evaluation criterion j, $j = 1, \ldots, M_1$, the value proposed by each supplier is no less than the value placed at the basis of the tender. Moreover, suppliers cannot propose a rise greater than 100%, because it represents an unrealistic scenario.

Constraints (3) ensure that, for each quantitative evaluation criterion k, $k = 1, \ldots, M_2$, the value proposed by each supplier is not greater than the value placed at the basis of the tender. Moreover, we suppose that suppliers cannot propose a discount greater than 100%, because it represents an unrealistic scenario.

With the quantities q_{ij} and $\tilde{q}_{ik}$, we associate the investment cost functions γ_{ij} and $\tilde{\gamma}_{ik}$, $i = 1, \ldots, N$, $j = 1, \ldots, M_1$ and $k = 1, \ldots, M_2$, which are assumed to be continuously differentiable and convex. We assume that

$$\gamma_{ij} := \gamma_{ij}(q_{ij}), \quad \forall i = 1, \ldots, N, \ j = 1, \ldots, M_1, \tag{4}$$

$$\tilde{\gamma}_{ik} := \tilde{\gamma}_{ik}(\tilde{q}_{ik}), \quad \forall i = 1, \ldots, N, \ k = 1, \ldots, M_2. \tag{5}$$

In our model, each supplier is faced with a limited budget for all quantitative criteria. Moreover, she/he seeks to obtain a positive profit. Hence, the following constraint must be satisfied:

$$\sum_{j=1}^{M_1} \gamma_{ij}(q_{ij}) + \sum_{k=1}^{M_2} \tilde{\gamma}_{ik}(\tilde{q}_{ik}) \leq \pi_i, \quad i = 1, \ldots, N, \tag{6}$$

that is, each supplier can't invest more than its own economic offer.

As said before, a weight is associated with each evaluation criterion. Let w be the weight assigned to the price. Moreover, let $w_j \in \mathbb{R}_+$ and $\tilde{w}_k \in \mathbb{R}_+$, be the weights assigned to the quantitative criterion j of the first type, $j = 1, \ldots, M_1$ and to the quantitative criterion k of the second type, $k = 1, \ldots, M_2$, resp.

We denote by $\Lambda \in \mathbb{R}_+$ the maximum achievable score set by the public tender for quantitative criteria and the price criterion. According to [1], the sum of the weights attributed to the price criterion and to each quantitative criterion must be equal to the maximum score set by the procedure for the economic and quantitative offer, i.e.

$$w + \sum_{j=1}^{M_1} w_j + \sum_{k=1}^{M_2} \tilde{w}_k = \Lambda. \tag{7}$$

Let p_i be the score that the awarding commission assigns to the supplier i for his price and let p_{ij} and $\tilde{p}_{ik}$ be the scores that the awarding commission assigns to the supplier i for the quantitative evaluation criterion j, $j = 1, \ldots, M_1$, and quantitative evaluation criterion k, $k = 1, \ldots, M_2$, resp.

The score attributed to each criterion and to the price is given by the product of a coefficient and the weight assigned to the criterion and to the price, resp. Current legislation requires that the coefficients must be such as to ensure that the maximum score established for each criterion, represented by the weight, is reached. Therefore, for each criterion the associated coefficient must be a value between 0 and 1. Furthermore, in order to reach the maximum score for each criterion, the formulas chosen for the calculation of such coefficients must guarantee that for at least one supplier the value 1 is assumed.

In the case of quantitative criteria, current legislation provides different formulas for calculating the scores attributable to each criterion (see Ref. [1]). These formulas can be independent or interdependent. In this model, we will assume that such formulas are interdependent. In this way, the score to be attributed to each supplier for their offers depends on the offers of all the other suppliers. In the next section, we will provide new formulas (also named *normalization functions*) to calculate the suppliers' scores for both economic and quantitative criteria and, in the section devoted to the numerical illustrations, we will compare these new formulas with some independent formulas commonly used in practice (see Ref. [1]).

We suppose that

$$p_i := p_i(\pi) = C_i(\pi) \times w, \quad \forall i = 1, \ldots, N.$$

Moreover, we assume that

$$p_{ij} := p_{ij}(q_j) = C_{ij}(q_j) \times w_j, \quad \forall i = 1, \ldots, N, \ \forall j = 1, \ldots M_1,$$

and

$$\tilde{p}_{ik} := \tilde{p}_{ik}(\tilde{q}_k) = \tilde{C}_{ik}(\tilde{q}_k) \times \tilde{w}_k, \quad \forall i = 1, \ldots, N, \ \forall k = 1, \ldots M_2,$$

where $q_j = (q_{ij})_{i=1,\ldots,N}$, for all $j = 1, \ldots, M_1$ and $\tilde{q}_k = (\tilde{q}_{ik})_{i=1,\ldots,N}$, for all $k = 1, \ldots, M_2$.

We require that $C_i : \mathbb{R}^N_+ \to [0,1]$, $C_{ij} : \mathbb{R}^N_+ \to [0,1]$ and $\tilde{C}_{ik} : \mathbb{R}^N_+ \to [0,1]$ are continuously differentiable and concave functions with respect to each variable which they depend on, for all $i = 1, \ldots, N$, $j = 1, \ldots, M_1$ and $k = 1, \ldots, M_2$. This assumption ensures that, for each quantitative criterion, no supplier may exceed the maximum score.

Following the MEAT criterion with compensatory aggregative method, the utility, $E(U_i)$, of supplier i, $i = 1, \ldots, N$, which corresponds to his total economic and quantitative score, is

$$E(U_i) = C_i(\pi) \times w + \sum_{j=1}^{M_1} C_{ij}(q_j) \times w_j + \sum_{k=1}^{M_2} \tilde{C}_{ik}(\tilde{q}_k) \times \tilde{w}_k. \tag{8}$$

Specifically, the first term of (8) represents the score associated with the economic offer, the second term of the sum of the scores is associated with the quantitative criteria of the first type and the third one, the sum of the scores associated with the quantitative criteria of the second type.

We group the expected utilities of all suppliers into the N-dimensional vector $E(U)$ with components: $(E(U_1), \ldots, E(U_N))$.

Let K^i denote the feasible set corresponding to retailer i, where

$$K^i = \Big\{ (\pi_i, q_i, \tilde{q}_i) | \Gamma \le \pi_i \le \Pi, \ \underline{Q}_j \le q_{ij} \le 2\underline{Q}_j, \ \frac{1}{2}\tilde{Q}_k \le \tilde{q}_{ik} \le \tilde{Q}_k, \ \forall j, k \\ \text{and (6) holds for } i \Big\}, \tag{9}$$

and define $K = \prod_{i=1}^N K^i$.

We observe that the feasible set K is a compact and convex set; indeed, all variables are bounded on both sides and the functions involved in constraint (6) are convex by assumption.

Each supplier competes non-cooperatively to win the contract, offering his price and his own quantitative parameters and tries to maximize his own expected score. We seek to determine a non-negative vector $(\pi^*, q^*, \tilde{q}^*) \in K$

for which the N suppliers will be in a state of equilibrium as defined in what follows. Nash generalized Cournot's concept (see Refs. [2–4]) of an equilibrium for a model of several players in a non-cooperative game is presented. Decision-makers act in their own self-interest and each player is assumed to know the equilibrium strategies of the other players, and no player has anything to gain by changing only his own strategy.

We now want to give the following definition of a public procurement tender Nash equilibrium in price and quantitative criteria.

Definition 1 (a public procurement tender Nash equilibrium in price and quantitative criteria). A vector $(\pi^*, q^*, \tilde{q}^*) \in K$ is said to constitute a public procurement tender Nash Equilibrium if for each retailer i, $i = 1, \ldots, N$,

$$E(U_i(\pi_i^*, q_i^*, \tilde{q}_i^*, \pi_{-i}^*, q_{-i}^*, \tilde{q}_{-i}^*)) \geq E(U_i(\pi_i, q_i, \tilde{q}_i, \pi_{-i}^*, q_{-i}^*, \tilde{q}_{-i}^*)),$$
$$\forall (\pi_i, q_i, \tilde{q}_i) \in K^i, \tag{10}$$

where

$$\pi_{-i}^* = (\pi_1^*, \ldots, \pi_{i-1}^*, \pi_{i+1}^*, \ldots, \pi_N^*),$$
$$q_{-i}^* = (q_1^*, \ldots, q_{i-1}^*, q_{i+1}^*, \ldots, q_N^*), \tag{11}$$
$$\tilde{q}_{-i}^* = (\tilde{q}_1^*, \ldots, \tilde{q}_{i-1}^*, \tilde{q}_{i+1}^*, \ldots, \tilde{q}_N^*). \tag{12}$$

Following [5,6], we can now deduce the variational inequality formulation of Nash equilibrium conditions in Definition 1 (see, for instance, [6,7]). The following theorem holds true.

Theorem 1 (variational inequality formulation). *Assume that, for each retailer i, $i = 1, \ldots, N$, the expected utility function $E(U_i)$ is concave with respect to the variables π_i, $\{q_{i1}, \ldots, q_{iM_1}\}$ and $\{\tilde{q}_{i1}, \ldots, \tilde{q}_{iM_2}\}$, and is continuously differentiable. Then $(\pi^*, q^*, \tilde{q}^*) \in K$ is a Public Procurement Tender Nash Equilibrium according to Definition 1 if and only if it is a solution to the variational inequality*

$$-\sum_{i=1}^{N} \frac{\partial E(U_i(\pi^*, q^*, \tilde{q}^*))}{\partial \pi_i} \times (\pi_i - \pi_i^*)$$
$$-\sum_{i=1}^{N}\sum_{j=1}^{M_1} \frac{\partial E(U_i(\pi^*, q^*, \tilde{q}^*))}{\partial q_{ij}} \times (q_{ij} - q_{ij}^*)$$
$$-\sum_{i=1}^{N}\sum_{k=1}^{M_2} \frac{\partial E(U_i(\pi^*, q^*, \tilde{q}^*))}{\partial \tilde{q}_{ik}} \times (\tilde{q}_{ik} - \tilde{q}_{ik}^*) \geq 0, \quad \forall (\pi, q, \tilde{q}) \in K, \tag{13}$$

or, equivalently, $(\pi^*, q^*, \tilde{q}^*) \in K$ *is a Public Procurement Tender Nash Equilibrium if and only if it satisfies the variational inequality:*

$$-\sum_{i=1}^{N}\left[\frac{\partial C_i(\pi^*)}{\partial \pi_i} \times w\right] \times (\pi_i - \pi_i^*) - \sum_{i=1}^{N}\sum_{j=1}^{M_1}\left[\frac{\partial C_{ij}(q_j^*)}{\partial q_{ij}} \times w_j\right]$$

$$\times (q_{ij} - q_{ij}^*) - \sum_{i=1}^{N}\sum_{k=1}^{M_2}\left[\frac{\partial \tilde{C}_{ik}(\tilde{q}_k^*)}{\partial \tilde{q}_{ik}} \times \tilde{w}_k\right]$$

$$\times (\tilde{q}_{ik} - \tilde{q}_{ik}^*) \geq 0, \quad \forall (\pi, q, \tilde{q}) \in K. \tag{14}$$

We now put variational inequality (13) into standard form, that is: determine $X^* \in \mathcal{K} \subset \mathbb{R}^{\mathcal{N}}$, such that

$$\langle F(X^*), X - X^* \rangle \geq 0, \quad \forall X \in \mathcal{K}, \tag{15}$$

where F is a given continuous function from $\mathcal{K}$ to $\mathbb{R}^{\mathcal{N}}$ and $\mathcal{K}$ is a closed and convex set.

We set $\mathcal{N} = (N + NM_1 + NM_2)$ and we define the $\mathcal{N}$-dimensional column vector $X \equiv (\pi, q, \tilde{q})$ and $\mathcal{N}$-dimensional column vector $F(X) \equiv (F^1(X), F^2(X), F^3(X))$ with the i-th component, F_i^1, of $F^1(X)$ given by

$$F_i^1(X) \equiv -\frac{\partial E(U_i(\pi, q, \tilde{q}))}{\partial \pi_i} = -\frac{\partial C_i(\pi)}{\partial \pi_i} \times w, \quad \forall i \tag{16}$$

the (i,j)-th component, F_{ij}^2, of $F^2(X)$ given by

$$F_{ij}^2(X) \equiv -\frac{\partial E(U_i(\pi, q, \tilde{q}))}{\partial q_{ij}} = -\frac{\partial C_{ij}(q_j)}{\partial q_{ij}} \times w_j, \quad \forall i, \ \forall j \tag{17}$$

and, finally, the (i,k)-th component, F_{ik}^3, of $F^3(X)$ given by

$$F_{ik}^3(X) \equiv -\frac{\partial E(U_i(\pi, q, \tilde{q}))}{\partial \tilde{q}_{ik}} = -\frac{\partial \tilde{C}_{ik}(\tilde{q}_k)}{\partial \tilde{q}_{ik}} \times \tilde{w}_k, \quad \forall i, \ \forall k \tag{18}$$

and with the feasible set $\mathcal{K} \equiv K$. Then, clearly, variational inequality (13) can be put into standard form (15).

We now provide a qualitative property in terms of existence and uniqueness of a solution to variational inequality (13).

The following result holds true in virtue of [8].

Theorem 2 (existence and uniqueness). *A solution* $(\pi^*, q^*, \tilde{q}^*)$ *to variational inequality* (13) *is guaranteed to exist since the feasible set* $\mathcal{K}$ *is compact and convex and the function* F *is continuous. Moreover, if the function that enters variational inequality* (15) *is strictly monotone, that is*

$$\langle F(X^1) - F(X^2), X^1 - X^2 \rangle > 0, \quad \forall X^1, X^2 \in \mathcal{K}, \ X^1 \neq X^2,$$

then the solution $(\pi^*, q^*, \tilde{q}^*)$ *to variational inequality* (13) *is unique.*

3. Normalization Functions

As stated in Section 2, for each criterion involved in the tender procedure, a function must be defined to calculate the scores attributable to each criterion, namely $C_i(\pi)$, $C_{ij}(q_j)$ and $\tilde{C}_{ik}(\tilde{q}_k)$, for all $i = 1, \ldots, N$, $j = 1, \ldots, M_1$ and $k = 1, \ldots, M_2$. As said before, such functions, called also *normalization functions*, must guarantee that for at least one supplier the value 1 is assumed, that is, must assign score 1 to the best offer. These properties lead us to consider functions that admit intervals with null derivative, since improving a best offer does not improve the score. Consequently, to ensure that the utility function of each supplier, $E(U_i)$, $i = 1, \ldots, N$, is continuously differentiable, it is necessary to choose carefully the normalization functions. For instance, none of the function expressions suggested in Ref. [1] has this property. Moreover, we recall that the function F defined in (16)–(18) must be a continuous function to guarantee the existence of a solution to variational inequality (13).

In this context, we consider the following normalization functions:

$$C_i(\pi) = \begin{cases} \sqrt{1 - \dfrac{(\pi_i - \pi_{AVE})^2}{(\Pi - \pi_{AVE})^2}} & \text{if } \pi_i \geq \pi_{AVE} \\ 1 & otherwise \end{cases}, \quad \forall i, \tag{19}$$

$$C_{ij}(q_j) = \begin{cases} \sqrt{1 - \dfrac{(q_{ij} - q_{j,AVE})^2}{(q_{j,AVE} - \underline{Q}_j)^2}} & \text{if } q_{ij} \leq q_{j,AVE} \\ 1 & otherwise \end{cases}, \forall i, \ \forall j, \tag{20}$$

$$\tilde{C}_{ik}(\tilde{q}_k) = \begin{cases} \sqrt{1 - \dfrac{(\tilde{q}_{ik} - \tilde{q}_{k,AVE})^2}{(\tilde{Q}_k - \tilde{q}_{k,AVE})^2}} & \text{if } \tilde{q}_{ik} \geq \tilde{q}_{k,AVE} \\ 1 & otherwise \end{cases}, \quad \forall i, \ \forall k, \tag{21}$$

where $\pi_{AVE}, q_{j,AVE}$ and $\tilde{q}_{k,AVE}$ represent the average values of prices proposed by all the suppliers, of the quantities that all the suppliers propose to satisfy a criterion for which the best offer corresponds with the highest or lowest value, resp., namely:

$$\pi_{AVE} = \frac{1}{N}\sum_{l=1}^{N} \pi_l, \ q_{j,AVE} = \frac{1}{N}\sum_{l=1}^{N} q_{lj}, \ \tilde{q}_{k,AVE} = \frac{1}{N}\sum_{l=1}^{N} \tilde{q}_{lk}.$$

The normalization function (19) is constructed by a plateau which guarantees that for at least one supplier the value 1 is assumed, indeed the function $C_i(\pi)$ assumes value 1 for each element belonging to the interval

$[0, \pi_{AVE}]$ in which the derivative is null. Moreover, considering a standard ellipse with semi-axes of lengths 1 and $\Pi - \pi_{AVE}$, resp., for each π_i greater than or equal to the average π_{AVE}, the function is represented by a branch of such an ellipse translated by a vector $\overrightarrow{v} = (\pi_{AVE}, 0)$, and, therefore, it is decreasing and concave.

To ensure the continuity of the previous functions, we assume that

$$\Pi \neq \pi_{AVE}, \; q_{j,AVE} \neq \underline{Q}_j, \; \tilde{Q}_k \neq \tilde{q}_{k,AVE}, \quad \forall j, \; \forall k.$$

These assumptions are not restrictive, indeed the suppliers compete in a non-cooperative way to win the tender object and to do this they adapt their economic offer and their quantitative criteria to maximize their expected score. Therefore, it is legitimate to assume that there is at least one supplier who proposes offers that differ from the quantities set at the auction basis. Thereby, the average offer can be supposed to be not equal to the worst offers.

These functions, whose expressions are inspired by Average Scoring formula (see Ref. [9]), represent interdependent formulas since they depend on the average price or on the average quantity proposed for evaluation criteria. Fig. 1 illustrates not only the generic trend of a score normalization function associated with the economic offer $C_i(\pi)$, as previously described (in which we fixed $\Pi = 9$), but also the behavior of such a function varying the average values of prices proposed by all the suppliers π_{AVE}.

With these functions, all prices (equivalently all offers for evaluation criteria) below the average price (equivalently average offer) obtain the maximum price score (equivalently the maximum evaluation criteria's score). Moreover, the economic offer Π receives a null score. Particularly, low prices are not rewarded in terms of a price score since such bids get the same price score as a bid, which is just below the average price. Moreover, if one of the submitted prices is much larger than all other submitted prices, it can have very serious effect on the final ranking. Opposite considerations can be made for the functions defined in (20) by a branch of a standard ellipse with semi-axes of lengths $q_{j,AVE} - \underline{Q}_j$ and 1, resp., centered at $(q_{j,AVE}, 0)$ for each $q_{ij} \leq q_{j,AVE}$ and by a plateau with value 1 for quantities proposed by a supplier for the criterion associated with the highest value, q_{ij}, greater than the mean $q_{j,AVE}$ (see, Fig. 2 in which we fixed $\underline{Q}_j = 2$).

Analogous considerations to functions (19) can be stated for the score normalization functions associated with quantitative criteria of the second type (21) since they have the same analytic structure.

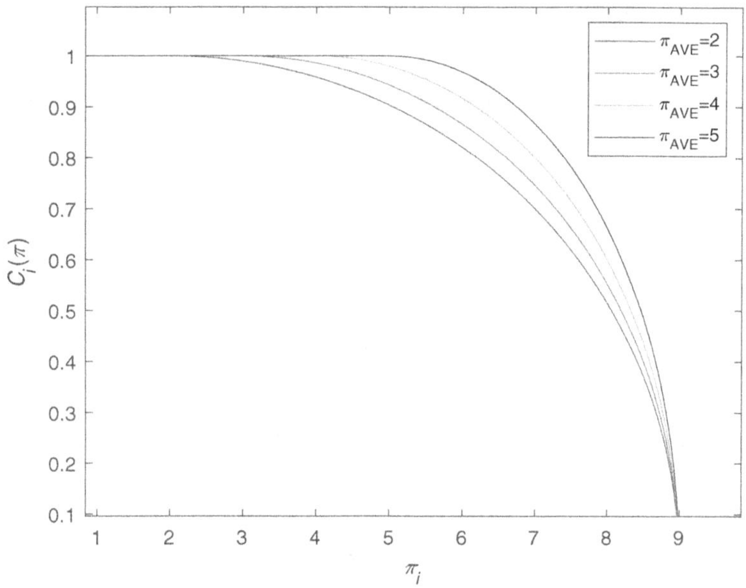

Fig. 1: Score normalization functions associated with the economic offer.

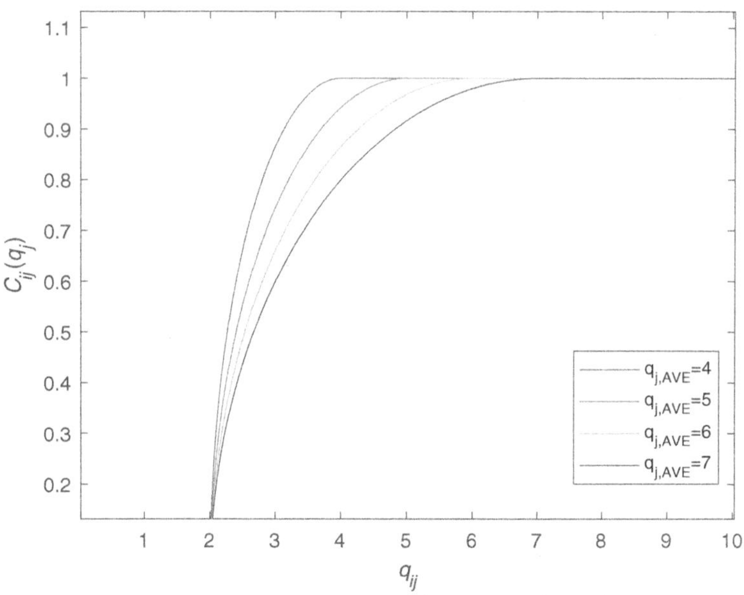

Fig. 2: Score normalization functions associated with quantitative criteria of the first type.

Moreover, we observe that such functions are continuously differentiable. Indeed, we have

$$\frac{\partial C_i(\pi)}{\partial \pi_i} = \begin{cases} \dfrac{-\dfrac{2}{N}\dfrac{(\pi_i-\pi_{AVE})^2}{(\Pi-\pi_{AVE})^3} - \dfrac{2(N-1)}{N}\dfrac{\pi_i-\pi_{AVE}}{(\Pi-\pi_{AVE})^2}}{2\sqrt{1-\dfrac{(\pi_i-\pi_{AVE})^2}{(\Pi-\pi_{AVE})^2}}} & \text{if } \pi_i \geq \pi_{AVE} \\ 0 & otherwise \end{cases}, \quad \forall i,$$

$$\frac{\partial C_{ij}(q_j)}{\partial q_{ij}} = \begin{cases} \dfrac{\dfrac{2}{N}\dfrac{(q_{ij}-q_{j,AVE})^2}{(q_{j,AVE}-\underline{Q}_j)^3} - \dfrac{2(N-1)}{N}\dfrac{q_{ij}-q_{j,AVE}}{(q_{j,AVE}-\underline{Q}_j)^2}}{2\sqrt{1-\dfrac{(q_{ij}-q_{j,AVE})^2}{(q_{j,AVE}-\underline{Q}_j)^2}}} & \text{if } q_{ij} \leq q_{j,AVE} \\ 0 & otherwise \end{cases}, \quad \forall i,\ j,$$

$$\frac{\partial \tilde{C}_{ik}(\tilde{q}_k)}{\partial \tilde{q}_{ik}} = \begin{cases} \dfrac{-\dfrac{2}{N}\dfrac{(\tilde{q}_{ik}-\tilde{q}_{k,AVE})^2}{(\tilde{Q}_k-\tilde{q}_{k,AVE})^3} - \dfrac{2(N-1)}{N}\dfrac{\tilde{q}_{ik}-\tilde{q}_{k,AVE}}{(\tilde{Q}_k-\tilde{q}_{k,AVE})^2}}{2\sqrt{1-\dfrac{(\tilde{q}_{ik}-\tilde{q}_{k,AVE})^2}{(\tilde{Q}_k-\tilde{q}_{k,AVE})^2}}} & \text{if } \tilde{q}_{ik} \geq \tilde{q}_{k,AVE} \\ 0 & otherwise \end{cases}, \quad \forall i,\ k.$$

Figures 3 and 4 show the derivative of score normalization functions associated with the economic offer and the quantitative criteria of the first type, resp. Obviously, the derivative of score normalization functions associated with the quantitative criteria of the second type has a similar trend as the derivative score normalization functions associated with the economic offer.

The differentiable continuity of the proposed functions allows us to affirm that the utility of each supplier i, $i = 1, \dots, N$, is also a continuously differentiable function. Furthermore, it is straightforward to verify that, with this choice, F is a concave function with respect to its strategic variables, i.e.

$$\frac{\partial^2 E(U_i)}{\partial \pi_i^2} = \frac{\partial^2 C_i(\pi)}{\partial \pi_i^2} \times w \leq 0, \quad \forall i. \tag{22}$$

$$\frac{\partial^2 E(U_i)}{\partial q_{ij}^2} = \frac{\partial^2 C_{ij}(q_j)}{\partial q_{ij}^2} \times w_j \leq 0, \quad \forall i, \forall j. \tag{23}$$

$$\frac{\partial^2 E(U_i)}{\partial \tilde{q}_{ik}^2} = \frac{\partial^2 \tilde{C}_{ik}(\tilde{q}_k)}{\partial \tilde{q}_{ik}^2} \times \tilde{w}_k \leq 0, \quad \forall i, \forall k. \tag{24}$$

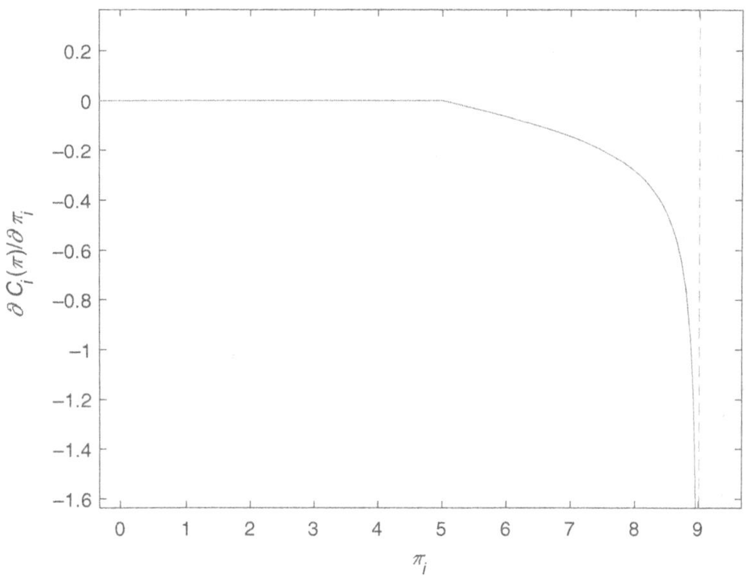

Fig. 3: Derivative score normalization functions associated with the economic offer.

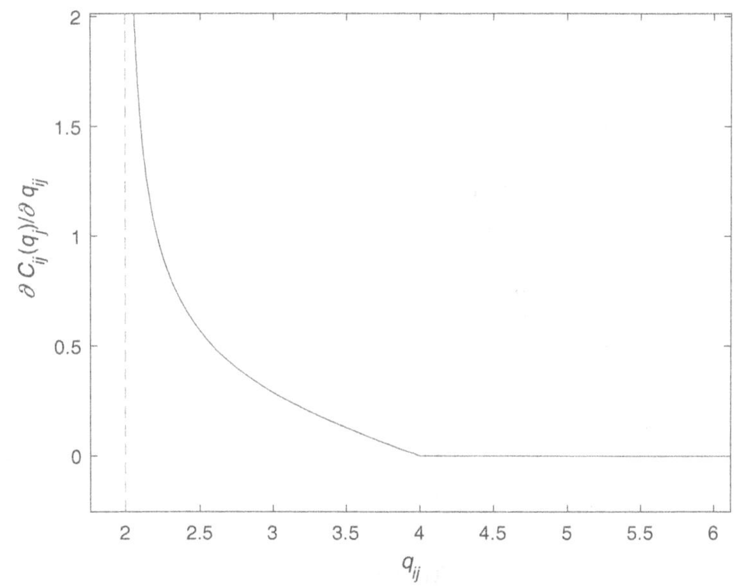

Fig. 4: Derivative score normalization functions associated with quantitative criteria of the first type.

These properties, by virtue of Theorem 2, guarantee the existence of at least one solution to the variational inequality (15) and, therefore, the existence of at least one Nash Equilibrium of the model.

4. Alternative Variational Inequality and Computational Procedure

In this section, we propose a computational procedure, based on Euler Method (see, for instance, [10]), to obtain Nash Equilibria of the game described in Section 2.

To derive explicit formulas for the computation of Nash Equilibria, we derive an alternative formulation of variational inequality (14), that includes Lagrange multipliers associated with constraints (6) (see, for instance, [6,11]). This alternative variational inequality allows us to derive an iterative scheme to Euler Method (see Ref. [10] for a detailed description).

Theorem 3. *A vector* $(\pi^*, q^*, \tilde{q}^*) \in \mathcal{K}$ *is a solution to variational inequality* (14) *if and only if it is a solution to the following variational inequality:*

$$\sum_{i=1}^{N}\left[-\frac{\partial C_i(\pi)}{\partial \pi_i} \times w - \lambda_i^*\right] \times (\pi_i - \pi_i^*)$$
$$+\sum_{i=1}^{N}\sum_{j=1}^{M_1}\left[-\frac{\partial C_{ij}(q_j^*)}{\partial q_{ij}} \times w_j + \lambda_i^* \frac{\partial \gamma_{ij}(q_{ij}^*)}{\partial q_{ij}}\right] \times (q_{ij} - q_{ij}^*)$$
$$+\sum_{i=1}^{N}\sum_{k=1}^{M_2}\left[-\frac{\partial \tilde{C}_{ik}(\tilde{q}_k^*)}{\partial \tilde{q}_{ik}} \times \tilde{w}_k + \lambda_i^* \frac{\partial \tilde{\gamma}_{ik}(\tilde{q}_{ik}^*)}{\partial \tilde{q}_{ik}}\right] \times (\tilde{q}_{ik} - \tilde{q}_{ik}^*)$$
$$+\sum_{i=1}^{N}\left[\pi_i - \sum_{j=1}^{M_1}\gamma_{ij}(q_{ij}^*) - \sum_{k=1}^{M_2}\tilde{\gamma}_{ik}(\tilde{q}_{ik}^*)\right]$$
$$\times (\lambda_i - \lambda_i^*) \geq 0, \quad \forall (\pi, q, \tilde{q}, \lambda) \in \mathcal{K} \times \mathbb{R}_+^N. \tag{25}$$

Proof. Each retailer i, $i = 1, \ldots, N$, according to Definition 1, seeks to maximize his utility function, that is seeks to solve the maximization problem:

$$\text{Maximize}_{(\pi_i, q_i, \tilde{q}_i)} E(U_i) = C_i(\pi) \times w + \sum_{j=1}^{M_1} C_{ij}(q_j)$$
$$\times w_j + \sum_{k=1}^{M_2} \tilde{C}_{ik}(\tilde{q}_k) \times \tilde{w}_k, \tag{26}$$

subject to

$$\sum_{j=1}^{M_1} \gamma_{ij}(q_{ij}) + \sum_{k=1}^{M_2} \tilde{\gamma}_{ik}(\tilde{q}_{ik}) \leq \pi_i, \quad i = 1, \dots, N,$$

$$\Gamma \leq \pi_i \leq \Pi, \quad \forall i = 1, \dots, N,$$

$$\underline{Q}_j \leq q_{ij} \leq 2\underline{Q}_j, \quad \forall i = 1, \dots, N, \ \forall j = 1, \dots, M_1,$$

$$\frac{\tilde{Q}_k}{2} \leq \tilde{q}_{ik} \leq \tilde{Q}_k, \quad \forall i = 1, \dots, N, \ \forall k = 1, \dots, M_2.$$

Converting the maximization problem into a minimization one, the above optimization problem becomes

$$-\text{Min} \hat{f}_i(X_i, \hat{X}_i^*), \tag{27}$$

subject to

$$g_i(X_i) \leq 0, \tag{28}$$

$$X_i \in \mathcal{K}_i^1, \tag{29}$$

where

- $X_i \equiv (\pi_i, q_i, \tilde{q}_i)$ and $\hat{X}_i^* \equiv (X_1^*, \dots, X_{i-1}^*, X_{i+1}^*, \dots, X_N^*)$, for all $i = 1, \dots, N$;
- $\hat{f}_i(X_i, \hat{X}_i) = -E(U_i)$, for all $i = 1, \dots, N$;
- $g_i(X_i) = \sum_{j=1}^{M_1} \gamma_{ij}(q_{ij}) + \sum_{k=1}^{M_2} \tilde{\gamma}_{ik}(\tilde{q}_{ik}) - \pi_i$, for all $i = 1, \dots, N$;
- $\mathcal{K}_i^1 \equiv \{(\pi_i, q_i, \tilde{q}_i) : \Gamma \leq \pi_i \leq \Pi, \ \underline{Q}_j \leq q_{ij} \leq 2\underline{Q}_j, \ \frac{\tilde{Q}_k}{2} \leq \tilde{q}_{ik} \leq \tilde{Q}_k, \ \forall i = 1, \dots, N, \ \forall j = 1, \dots, M_1, \forall k = 1, \dots, M_2\}$, for all $i = 1, \dots, N$.

The Lagrangian function is

$$\mathcal{L}(X_i, \hat{X}_i^*, \lambda_i) = \hat{f}_i(X_i, \hat{X}_i^*) + \lambda_i g_i(X_i).$$

The following assumption is easy to verify:

Assumption 1 (Slater condition). There exists a Slater vector $\tilde{X}_i \in \mathcal{K}_i^1$, $i = 1, \dots, N$, such that $g_i(\tilde{X}_i) < 0$.

Following [12], since $\hat{f}_i$ is convex in X_i and continuously differentiable and g_i is also convex and continuously differentiable, and the feasible set is non-empty, closed and convex, vector $(X_i^*, \lambda_i^*) \in \mathcal{K}_i^1 \times \mathbb{R}_+$ is a solution to the above minimization problem (27), subject to (28)–(29), if and only if it is a solution to the variational inequality:

$$\nabla_{X_i} \mathcal{L}(X_i^*, \hat{X}_i^*, \lambda_i^*) \times (X_i - X_i^*) + (-g_i(X_i^*)) \times (\lambda_i - \lambda_i^*) \geq 0,$$
$$\forall (X_i, \lambda_i) \in \mathcal{K}_i^1 \times \mathbb{R}_+.$$

Equation (25) is obtained by making explicit the calculations of partial derivatives. □

We observe that variational inequality (25) can be rewritten in standard form: determine $X^* \in \mathcal{K}^1$ such that $\langle F(X^*), X - X^* \rangle \geq 0$, $\forall X \in \mathcal{K}^1$, by putting:

- $X \equiv (\pi, q, \tilde{q}, \lambda) \in \mathbb{R}^{NM_1+NM_2+2N}$;
- $F(X) = (\hat{F}^1(X), \hat{F}^2(X), \hat{F}^3(X), \hat{F}^4(X))$ where:

$$\hat{F}_i^1(X) \equiv \left[-\frac{\partial C_i(\pi)}{\partial \pi_i} \times w - \lambda_i \right], \quad \forall i,$$

$$\hat{F}_{ij}^2(X) \equiv \left[-\frac{\partial C_{ij}(q_j)}{\partial q_{ij}} \times w_j + \lambda_i \frac{\partial \gamma_{ij}(q_{ij})}{\partial q_{ij}} \right], \quad \forall i, \forall j,$$

$$\hat{F}_{ik}^3(X) \equiv \left[-\frac{\partial \tilde{C}_{ik}(\tilde{q}_k)}{\partial \tilde{q}_{ik}} \times w_k + \lambda_i \frac{\partial \tilde{\gamma}_{ik}(\tilde{q}_{ik})}{\partial \tilde{q}_{ik}} \right], \quad \forall i, \forall j_2,$$

$$\hat{F}_i^4(X) \equiv \left[\pi_i - \sum_{j=1}^{M_1} \gamma_{ij}(q_{ij}) - \sum_{k=1}^{M_2} \tilde{\gamma}_{ik}(\tilde{q}_{ik}) \right], \quad \forall i,$$

- $\mathcal{K}^1 \equiv \prod_{i=1}^N \mathcal{K}_i^1$.

4.1. *Explicit formulas for Euler method*

The explicit formulas for the Euler method applied to variational inequality (25) are as follows:

$$\pi_i^{\tau+1} = \max\left\{ \Gamma, \min\left\{ \Pi, \pi_i^\tau + a_\tau \left(\frac{\partial C_i(\pi^\tau)}{\partial \pi_i} \times w + \lambda_i^\tau \right) \right\} \right\}, \quad \forall i,$$

for the price of supplier i and

$$q_{ij}^{\tau+1} = \max\left\{ \underline{Q}_j, \min\left\{ 2\underline{Q}_j, q_{ij}^\tau + a_\tau \times \left(\sum_{l=1}^{M_1} \frac{\partial C_{il}(q^\tau)}{\partial q_{il}} \times w_l - \lambda_i^\tau \frac{\partial \gamma_{ij}(q_{ij}^\tau)}{\partial q_{ij}} \right) \right\} \right\}, \quad \forall i, \ \forall j,$$

$$\tilde{q}_{ik}^{\tau+1} = \max\left\{ \frac{\tilde{Q}_k}{2}, \min\left\{ \tilde{Q}_k, \tilde{q}_{ik}^\tau + a_\tau \times \left(\sum_{m=1}^{M_2} \frac{\partial \tilde{C}_{im}(\tilde{q}^\tau)}{\partial \tilde{q}_{im}} \times \tilde{w}_m - \lambda_i^\tau \frac{\partial \tilde{\gamma}_{ik}(\tilde{q}_{ik}^\tau)}{\partial \tilde{q}_{ik}} \right) \right\} \right\}, \ \forall i, \ \forall k,$$

for the values proposed by supplier i to satisfy quantitative evaluation criteria of the first type and the second type, resp., and

$$\lambda_i^{\tau+1} = \max\left\{0, \lambda_i^\tau + a_\tau \left(-\pi_i^\tau + \sum_{j=1}^{M_1} \gamma_{ij}(q_{ij}^\tau) + \sum_{k=1}^{M_2} \tilde{\gamma}_{ik}(\tilde{q}_{ik}^\tau)\right)\right\}, \quad \forall i$$

for the Lagrange multipliers.

5. Illustrative Numerical Examples

In this section, we propose two illustrative numerical examples to validate the effectiveness of our model. To solve the variational inequalities governing the Nash equilibria of the model, we use the computational procedure described in Section 4. The calculations are performed using the MATLAB program. The algorithm has been implemented on a HP laptop with an AMD compute cores 2C+3G processor, 8 GB RAM. For the convergence of the method, a tolerance of $\epsilon = 10^{-4}$ is fixed. The method has been implemented with a constant step $\alpha = 0.1$.

5.1. *Example* 1

First, we consider $N = 2$ suppliers, two quantitative criteria of the first type and one quantitative criterion of the second type. The parameters of the numerical example are reported in Table 1.

Economic offers are expressed in tens of thousands of euros. Moreover, we put $\Gamma = 5$ and $\Pi = 10$, $w = 30$, $w_1 = 15$, $w_2 = 10$ and $\tilde{w}_3 = 15$, therefore we set $\Lambda = 70$.

The normalization functions proposed in Section 3 are shown in Figs. 5(a)–5(d).

Table 1: Quantitative criteria for the numerical example.

Criterion	Description	Lower bound	Upper bound
$j = 1$:	post-delivery free maintenance	$\underline{Q}_1 = 1$ yr.	$2\underline{Q}_j = 2$ yr.
$j = 2$:	worker hourly rate	$\underline{Q}_2 = 7.5$ €/h	$2\underline{Q}_2 = 15$ €/h
$k = 1$:	execution time of work	$\frac{\tilde{Q}_1}{2} = 0.5$ yr.	$\tilde{Q}_1 = 1$ yr.

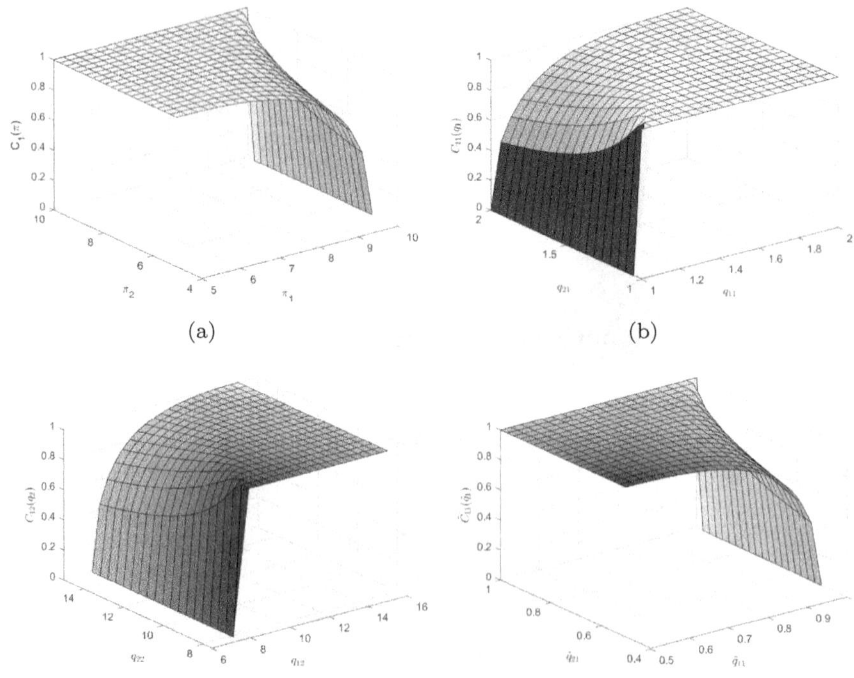

Fig. 5: Score normalization functions for Example 1. Score normalization function associated with (a) the economic offer, $C_1(\pi)$, (b) the first ($j = 1$) quantitative criteria of the first type (the best offer is the one with the highest value), $C_{11}(q_1)$, (c) the second ($j = 2$) quantitative criteria of the first type (the best offer is the one with the highest value), $C_{12}(q_2)$, (d) the quantitative criteria of the second type (for which the best offer is the one with the lowest value), $\tilde{C}_{11}(\tilde{q}_1)$.

The investment cost functions, expressed in tens of thousands, are as follows:

$$\textbf{Supplier 1:} \quad \begin{aligned} \gamma_{11}(q_{11}) &= (q_{11})^2 + 0.5q_{11} + 0.2 \\ \gamma_{12}(q_{12}) &= 0.01(q_{12})^2 + 0.04q_{12} + 0.2 \\ \tilde{\gamma}_{11}(\tilde{q}_{11}) &= (\tilde{q}_{11})^2 - 2.1\tilde{q}_{11} + 0.6; \end{aligned}$$

$$\textbf{Supplier 2:} \quad \begin{aligned} \gamma_{21}(q_{21}) &= 3(q_{21})^2 + q_{21} + 0.2 \\ \gamma_{22}(q_{22}) &= 0.01(q_{22})^2 + 0.04q_{22} + 0.3 \\ \tilde{\gamma}_{21}(\tilde{q}_{21}) &= 0.1(\tilde{q}_{21})^2 - 0.3\tilde{q}_{21} + 0.5. \end{aligned}$$

Solving variational inequality (25), we obtain the equilibrium solution reported in Table 2. We observe that the total score of supplier 1 is positively affected by his own offers proposed for the price, the post-delivery free

Table 2: Example 1: Equilibrium solution.

Suppliers	Price π_i^*	Post-delivery free maintenance q_{i1}^*	Worker hourly rate q_{i2}^*	Execution time of work $\tilde{q}_{i1}^*$	Score $E(U_i^*)$	Ranking
1	7.16	1.67	10.95	0.63	70.00	1
2	7.72	1.56	10.40	1	56.92	2

maintenance, the worker hourly rate and the execution time of work. Indeed, for each of these criteria, the associated score for supplier 1 is the greatest, because she/he offers the best offer, as required by the normalization functions chosen in Section 3 and she/he obtains the maximum score $E(U_1^*) = \Lambda = 70$.

5.2. *Example* 2

We now show another example whose data are the same as in the previous example but we consider $N = 4$ suppliers. The parameters of the numerical example are the same as reported in Table 1.

The investment cost functions, expressed in tens of thousands, are as follows:

$$\textbf{Supplier 1:} \quad \begin{aligned} \gamma_{11}(q_{11}) &= (q_{11})^2 + 0.5q_{11} + 0.2 \\ \gamma_{12}(q_{12}) &= 0.01(q_{12})^2 + 0.04q_{12} + 0.2 \\ \tilde{\gamma}_{11}(\tilde{q}_{11}) &= (\tilde{q}_{11})^2 - 2.1\tilde{q}_{11} + 0.6; \end{aligned}$$

$$\textbf{Supplier 2:} \quad \begin{aligned} \gamma_{21}(q_{21}) &= 3(q_{21})^2 + q_{21} + 0.2 \\ \gamma_{22}(q_{22}) &= 0.01(q_{22})^2 + 0.04q_{22} + 0.3 \\ \tilde{\gamma}_{21}(\tilde{q}_{21}) &= 0.1(\tilde{q}_{21})^2 - 0.3\tilde{q}_{21} + 0.5; \end{aligned}$$

$$\textbf{Supplier 3:} \quad \begin{aligned} \gamma_{31}(q_{31}) &= 3(q_{31})^2 + 1.5q_{31} + 0.1 \\ \gamma_{32}(q_{32}) &= 0.01(q_{32})^2 + 0.08q_{32} + 0.1 \\ \tilde{\gamma}_{31}(\tilde{q}_{31}) &= 0.01(\tilde{q}_{31})^2 - 0.2\tilde{q}_{31} + 0.4; \end{aligned}$$

$$\textbf{Supplier 4:} \quad \begin{aligned} \gamma_{41}(q_{41}) &= (q_{41})^2 + 0.4q_{41} + 0.2 \\ \gamma_{42}(q_{42}) &= 0.01(q_{42})^2 + 0.05q_{42} + 0.1 \\ \tilde{\gamma}_{41}(\tilde{q}_{41}) &= 0.1(\tilde{q}_{41})^2 - 0.4\tilde{q}_{41} + 0.3. \end{aligned}$$

Note that the investment cost functions $\gamma_{ij}(q_{ij})$ are increasing functions, while $\tilde{\gamma}_{ik}(\tilde{q}_{ik})$ are decreasing functions, in accordance with the fact that a higher score is obtained for higher values of q_{ij} and for lower values of

Table 3: Equilibrium solution.

Suppliers	π_i^*	Post-delivery free maintenance q_{i1}^*	Worker hourly rate q_{i2}^*	Execution time of work $\tilde{q}_{i1}^*$	Score $E(U_i^*)$	Ranking
1	8.11	2	10.65	1	55.98	3
2	7.18	1	10.53	0.87	56.33	2
3	7.00	1	10.47	0.96	52.88	4
4	7.66	2	10.82	0.69	69.93	1

$\tilde{q}_{ik}$, resp. Moreover, it is straightforward to verify that such functions are convex.

Solving variational inequality (25), we obtain the equilibrium solution shown in Table 3.

We observe that the total score of supplier 4 is positively affected by his own offers proposed for the post-delivery free maintenance, the worker hourly rate and the execution time of work. Particularly, supplier 4 obtains the greatest score associated to each of these criteria, because she/he offers the best offer (that is the highest value for post-delivery free maintenance and the worker hourly rate and the lowest value for the execution time of work), as required by the normalization functions chosen in Section 3. Furthermore, despite the supplier 4 having not proposed the best economic offer, he is the winner of the tender and this is in accordance with the MEAT criterion. Indeed, the total score of supplier 3, who proposes the best economic offer, is penalized by the scores assigned to the post-delivery free maintenance, the worker hourly rate and the execution time of work, for which she/he proposes the worst offers, since the first two are below the average and the last, above the average.

5.2.1. *Example 2: Comparison with the most commonly used interdependent formulas*

As mentioned in Section 2, we now compare the interdependent formulas provided in this work with some interdependent formulas commonly used in practice. Particularly, we take into account four different formulations denoted by

- *linear to the best offer*;
- *nonlinear with* $\alpha = \frac{1}{2}$;

- *minimum–maximum offer*;
- *inverse proportionality*.

The *linear to the best offer formulas* attribute scores by means of the linear interpolation method between the presented best offer (to which the maximum score is attributed) and the worst admissible offer (to which a score of zero is attributed). In other words, the formula assigns proportional scores to the discounts offered with respect to the auction base, with a proportionality coefficient. Such a proportionality coefficient is higher the smaller the maximum discount offered in the tender. It is a formula capable of guaranteeing high price competition, as it tends to generate high differences between the scores attributed to the offered prices offered, especially in cases where the best price offered is just below the auction base. The general mathematical expressions for these formulas are as follows:

$$C_i(\pi) = \left(\frac{\Pi - \pi_i}{\Pi - \min_{j=1,\ldots,N} \pi_j}\right)^{\alpha}, \quad \forall i = 1, \ldots, N, \tag{30}$$

$$C_{ij}(q_j) = \left(\frac{q_{ij} - \underline{Q}_j}{\max_{l=1,\ldots,N} q_{lj} - \underline{Q}_j}\right)^{\alpha}, \quad j = 1, \ldots, M_1, \; i = 1, \ldots, N, \tag{31}$$

$$\tilde{C}_{ik}(\tilde{q}_k) = \left(\frac{\tilde{Q}_k - \tilde{q}_{ik}}{\tilde{Q}_k - \min_{l=1,\ldots,N} q_{lk}}\right)^{\alpha}, \quad k = 1, \ldots, M_2, \; i = 1, \ldots, N, \tag{32}$$

where $\alpha = 1$. If $\alpha \neq 1$, in this form the formula becomes in fact a nonlinear one. The choice of the coefficient α is essential, in relation to the objective pursued:

- For values of α between 0 and 1 (excluded), the formula provides concave downward curves, discouraging higher declines.
- For values of $\alpha > 1$, concave curves upwards (or convex) are, rewarding the highest discounts and creating greater competition on the price.

Specifically, when $\alpha = \frac{1}{2}$, the linear to the best offer formulas become the *nonlinear with* $\alpha = \frac{1}{2}$ *formulas*.

The *inverse proportionality formulas*, in the case of criteria whose best offer is the lowest, assign the maximum score to the lowest offer and a score inversely proportional to other offers, with proportionality coefficient given by the lowest offer in the tender. The slope of the curve (and, therefore, the differences in the scores attributed) is less than that of the linear formula for the best offer, except for special cases in which the discounts offered are very high (higher than 50% compared to the basis of auction). As a consequence, such a formula induces smaller differences between the scores attributed and, therefore, less competition. Opposite considerations can be drawn for the criteria whose best offer is the highest. For these formulas, the general mathematical expressions are as follows:

$$C_i(\pi) = \frac{\min_{l=1,\dots,N} \pi_l}{\pi_i}, \quad \forall i = 1,\dots,N, \tag{33}$$

$$C_{ij}(q_j) = \frac{q_{ij}}{\max_{l=1,\dots,N} q_{lj}}, \quad \forall i = 1,\dots,N,\ \forall j = 1,\dots,M_1, \tag{34}$$

$$\tilde{C}_{ik} = \frac{\min_{l=1,\dots,N} \tilde{q}_{lk}}{\tilde{q}_{ik}}, \quad \forall i = 1,\dots,N,\ \forall k = 1,\dots,M_2. \tag{35}$$

Finally, the *Minimum–Maximum offer formula* assigns scores by linear interpolation between the best offer presented (to which the maximum score is attributed) and the worst offer presented (to which a score of zero is attributed). In other words, the formula assigns proportional scores to the discounts offered with respect to the auction base, with the proportionality coefficient greater the smaller the difference between the best and worst offer presented in the tender. It is a formula capable of guaranteeing particularly high competition, as it tends to generate high differences between the scores attributed even to quite similar offers, especially in cases where the best and worst offers are close enough. The general mathematical expressions for this formula are as follows:

$$C_i(\pi) = \frac{\max_{l=1,\dots,N} \pi_l - \pi_i}{\max_{l=1,\dots,N} \pi_l - \min_{l=1,\dots,N} \pi_l}, \quad \forall i = 1,\dots,N, \tag{36}$$

$$C_{ij}(q_j) = \frac{\min_{l=1,\dots,N} q_{lj} - q_{ij}}{\min_{l=1,\dots,N} q_{lj} - \max_{l=1,\dots,N} q_{lj}}, \tag{37}$$

$$\forall i = 1,\dots,N,\ \forall j = 1,\dots,M_1, \tag{38}$$

$$\tilde{C}_{ik}(\tilde{q}_k) = \frac{\max_{l=1,\dots,N} \tilde{q}_{lk} - \tilde{q}_{ik}}{\max_{l=1,\dots,N} \tilde{q}_{lk} - \min_{l=1,\dots,N} \tilde{q}_{lk}}, \tag{39}$$

$$\forall i = 1,\dots,N,\ \forall k = 1,\dots,M_2. \tag{40}$$

We want to stress that none of the formulas presented here satisfies the regularity hypotheses (to guarantee the existence of a solution to the variational inequality (15) or (25)), which instead are satisfied by our formulas. In fact, the presence of the max or min functions makes the normalization functions presented in this section non-continuous and, therefore, non-differentiable.

In order to provide a comparison between our formulas and the ones described in this section, in Table 4 we reported the scores attributed to

Table 4: Scores attributed to each supplier and ranking obtained using each of the interdependent commonly used formulas (*linear to the best offer*, *nonlinear with* $\alpha = \frac{1}{2}$, *inverse proportionality*, *minimum–maximum offer formulas*) and our formulas.

Formulas	Linear to the best offer	Nonlinear with $\alpha = \frac{1}{2}$	Inverse propor-tionality	Minimum–maximum	Our formulas
Supp. 1	43.4	48.5	61.4	20.1	55.9
Supp. 2	43.6	53.6	58.4	32.9	56.3
Supp. 3	40.7	44.8	57.9	31.9	52.8
Supp. 4	63.5	66.5	67.4	52.2	69.9
Ranking	4,2,1,3	4,2,1,3	4,2,1,3	4,2,3,1	4,2,1,3

each supplier obtained using each of the previously mentioned formulas and, in the last row, we reported the final ranking which is expressed by the supplier's identification number sorted from the first (the winner, who obtained the highest score) to the last.

As we can observe from the data, for each of the considered formulas, supplier 4 results to be the winner of the tender. Moreover, the ranking of our formulas and the other ones are the same, except for the *Minimum–Maximum formula*, for which suppliers 3 and 1 are inverted. This difference can be justified by the fact that supplier 1 presents the worst offer for the price criterion and, consequently, the *minimum–maximum formula* assigns a null score for this criterion. In addition, the "execution time of work" criterion is assigned a null score, as supplier 1 offers the worst offer.

The results obtained by this comparison allow us to state that our formulas are appropriate and consistent with the MEAT criterion.

6. Conclusion

In this model, we presented a game theory model whose aim is to describe the behavior of N potential suppliers who compete in a non-cooperative manner to win a public tender according to the MEAT criterion. We demonstrated that the N suppliers seek to maximize their expected utility, which corresponds to their expected score, proposing their economic offer and their parameters to satisfy the quantitative criteria imposed by the tender, representing their strategies. We obtained a Nash equilibrium framework for which we deduced the associated variational inequality formulation, as well as provided qualitative properties in terms of existence and uniqueness of the solutions. We also discussed the choice of normalization functions for calculating supplier offer scores, providing an example of such functions that satisfy the regularity properties required for the existence of solutions to alternative VI. We finally conducted illustrative numerical examples in which we analyzed the equilibrium solutions and we compared the normalization functions proposed in this chapter to well-known normalization functions commonly used.

This model can certainly be extended, for instance, considering the possibility that suppliers can establish coalitions and join consortia to win the contract, using the Strong Nash Equilibria Theory (see, for instance, [13]).

Acknowledgment

The research was partially supported by the research project "Programma ricerca di ateneo UNICT 2020-22 linea 2-OMNIA", University of Catania. This support is gratefully acknowledged.

References

[1] Autorità Nazionale Anticorruzione (ANAC). Linee Guida n. 2, di attuazione del D.lgs. 18 aprile 2016, n. 50, recanti "Offerta economicamente più vantaggiosa". http://www.anticorruzione.it/portal/rest/jcr/repository/collaboration/Digital%20Assets/anacd\ocs/Attivita/Atti/Delibere/2018/LineeGuida.n.2.OEPV.aggiornate.correttivo.pdf.

[2] A.A. Cournot, *Researches into the Mathematical Principles of the Theory of Wealth*, English translation. London, England: MacMillan, 1838.

[3] J.F. Nash, Equilibrium points in n-person games. *Proc. Nat. Acad. Sci.* USA **36**, 48–49, (1950).

[4] J.F. Nash, Noncooperative games, *Ann. Math.* **54**, 286–298, (1951).

[5] D. Gabay and H. Moulin, On the uniqueness and stability of Nash equilibria in noncooperative games. In: A. Bensoussan, P. Kleindorfer, and C.S. Tapiero (Eds.), *Applied Stochastic Control in Econometrics and Management Science* (pp. 271–294). (Amsterdam, The Netherlands: North-Holland, 1980).

[6] A. Nagurney, P. Daniele, and S. Shukla, A supply chain network game theory model of cybersecurity investments with nonlinear budget constraints. *Ann. Oper. Res.* **248**(1), 405–427, (2017).

[7] G. Colajanni, P. Daniele, and D. Sciacca, A projected dynamic system associated with a cybersecurity investment model with budget constraints and fixed demands, *J. Nonlinear Var. Anal.* **4**(1), 45–61, (2020). http://jnva.biemdas.com, https://doi.org/10.23952/jnva.4.2020.1.05.

[8] D. Kinderleher and G. Stampacchia, *Variational Inequalities and their Applications* (Academic Press: New York, 1980).

[9] N. Dimitri, G. Piga, and G. Spagnolo, *Handbook of Procurement* (CUP, 2006).

[10] P. Dupuis and A. Nagurney, Dynamical systems and variational inequalities, *Ann. Operations Res.* **44**, 9–42, (1993).

[11] P. Daniele and D. Sciacca, An optimization model for the management of green areas, *Intl. Trans. in Op. Res.* **28**, 3094–3116, (2021). Doi: 10.1111/itor.12987.

[12] J. Koshal, A. Nedic, and U.V. Shanbhag, Multiuser optimization, distributed algorithms and error analysis, *SIAM J. Optim.* **21**(3), 1046–1081, (2011).

[13] P. Daniele and L. Scrimali, Strong nash equilibria for cybersecurity investments with nonlinear budget constraints. In *Optimization and decision science: New trends in emerging complex real life problems* (AIRO springer series, 2018), pp. 199–207.

[14] T.H. Chen, An economic approach to public procurement, *J. Public Procurement* **8**(3), 407–430, (2008).

https://doi.org/10.1142/9789811261572_0008

Chapter 8

Perov-Type Results for Multivalued Mappings

Marija Cvetković*,§, Erdal Karapinar†,¶, Vladimir Rakočević*,**, and Seher Sultan Yeşilkaya‡,††

*Department of Mathematics, Faculty of Science and Mathematics, University of Niš, Niš, Serbia

†Department of Mathematics, Çankaya University, Etimesgut, Ankara, Turkey — Department of Medical Research, China Medical University Hospital, China Medical University, Taichung, Taiwan

‡Department of Mathematics Education, Tokat Gaziosmanpaşa University, Tokat, Turkey

§marijac@pmf.ni.ac.rs

¶erdalkarapinar@yahoo.com

**vrakoc@ptt.rs

††seher.yesilkaya@gop.edu.tr

Inspired by the results of Russian mathematician A. I. Perov from the 1960s, several authors have studied and extended these results by generalizing the contractive condition or changing the setting. In this chapter, we will present and analyze Perov-type results regarding multivalued operators and related applications.

1. Introduction

Russian mathematician Perov [1] in 1964 published a paper, in Russian, dealing with a Cauchy problem for a system of ordinary differential equations. In this paper, he defined a new type of generalized metric space (generalized metric space in the sense of Perov) and gave proof of a new kind of fixed point theorem. The Perov fixed point theorem is a generalization of a famous Banach fixed point theorem in a setting of a complete generalized metric space in the sense of Perov. It also gives convergence of an iterative sequence and is applicable in the area of Numerical Analysis. As there are numerous generalizations of Banach's result, it is important to mention that it was created as a tool in the area of differential equations and its area of application is much wider than in the case of Banach fixed point theorem and with a significantly better estimation as shown in Ref. [2]. The same concept was used once again in Perov's paper [3] in 1966 and then there were no major results on this topic till the 2000s. In the meantime, polish mathematician Czerwik [4] in 1976 published a similar result as a generalization of Edelstein's fixed point theorem [54]. In 1992, Zima [5], who also works in the area of differential equations, published a paper, quoting different works of Czerwik, which gave fixed point results on Banach space that could be related to the Perov fixed point theorem. Petruşel [6] did some research on Perov contractions for multivalued operators in 2005 that was followed by the results for Perov multivalued operators obtained by Petruşel and Filip in 2010 [7]. This led to several published papers on this topic [8–12]. Jurja [13] proved a version of Perov theorem for partially ordered generalized metric space. In 2014, Cvetković and Rakočević published a generalization of Perov fixed point theorem on cone metric spaces and this result obtained many extensions such as quasi-contraction, Fisher contraction, θ-contraction, F-contraction, coupled fixed point problem, common fixed point problem, etc. [14–30]. Many papers were published in 2010s citing Perov work, adjusting and generalizing that idea for multivalued operators, spaces endowed with a graph, ω-distance, etc., but will not be the main topic of this chapter. We will focus on Perov-type results for class multivalued mappings.

Three different frameworks: metric space, generalized metric space and generalized metric space equipped with ω-distance will be of interest in the case of multivalued mappings. We will collect some basic notations, definitions and properties of those kinds of spaces and related topics.
Let us recall some basic facts regarding metric spaces.

Definition 1. Let X be non-empty set and $d : X \times X \mapsto \mathbb{R}$ mapping such that

(d_1) $d(x,y) \geq 0$ and $d(x,y) = 0 \Leftrightarrow x = y$;
(d_2) $d(x,y) = d(y,x)$;
(d_3) $d(x,y) \leq d(x,z) + d(z,y)$.

Mapping d is a metric on X and (X,d) is called a metric space.

If $f : X \mapsto X$ is a mapping, then $x \in X$ is a fixed point of f if $f(x) = x$. Set of all fixed points of mapping f is denoted with $Fix(f)$. Banach in Ref. [31], published in 1922, gave a proof of the famous fixed point result regarding existence of a unique fixed point for a class of contractive mappings.

Definition 2. The mapping f on a metric space X is named contraction (contractive mapping) if there exists some constant $q \in (0, 1)$ such that

$$d(f(x), f(y)) \leq qd(x, y),\ x, y \in X.$$

The constant q is known as the contractive constant.

Clearly, every contraction is a non-expansive mapping. In the case of the self-mapping, for an arbitrary $x_0 \in X$, we define a sequence (x_n), $x_n = f(x_{n-1})$, $n \in \mathbb{N}$, for arbitrary $x_0 \in X$. It is called a sequence of successive approximations or an iterative sequence.

Theorem 1 ([31]). *Let (X, d) be a non-empty complete metric space with a contraction mapping $f : X \mapsto X$. Then f admits a unique fixed point in X and for any $x_0 \in X$, the iterative sequence (x_n) converges to the fixed point of f.*

Generalized metric space was introduced by Perov [1] and is defined in the following. Distances in the case of generalized metric spaces are in a set $\mathbb{R}^n$. Therefore, a self-mapping on a generalized metric spaces can satisfy contractive conditions similar to Banach's, but with a matrix A with non-negative entries instead of a constant q.

Let X be a non-empty set and $n \in \mathbb{N}$.

Definition 3 ([1]). A mapping $d : X \times X \mapsto \mathbb{R}^m$ is called a *vector-valued metric* on X if the following statements are satisfied for all $x, y, z \in X$.

(d_1) $d(x, y) \geq 0_n$ and $d(x, y) = 0_m \Leftrightarrow x = y$, where $0_m = (0, \ldots, 0) \in \mathbb{R}^m$;
(d_2) $d(x, y) = d(y, x)$;
(d_3) $d(x, y) \leq d(x, z) + d(z, y)$.

If $x = (x_1, \ldots, x_m)$, $y = (y_1, \ldots, y_m) \in \mathbb{R}^m$, then notation $x \leq y$ means $x_i \leq y_i$, $i = \overline{1, m}$.

Denote by $\mathcal{M}_{n,n}$ the set of all $n \times n$ matrices, by $\mathcal{M}_{n,n}(\mathbb{R}_+)$ the set of all $n \times n$ matrices with non-negative entries. We write O_n for the zero $n \times n$ matrix and I_n for the identity $n \times n$ matrix and further on we identify row and column vector in $\mathbb{R}^n$.

A matrix $A \in \mathcal{M}_{m,m}(\mathbb{R}_+)$ is said to be convergent to zero if $A^n \to O_m$, as $n \to \infty$ or, equivalently, if the matrix norm is less than 1.

Theorem 2 ([1,3]). *Let (X, d) be a complete generalized metric space, $f : X \mapsto X$, and $A \in \mathcal{M}_{m,m}(\mathbb{R}_+)$ a matrix convergent to zero, such that*

$$d(f(x), f(y)) \leq A(d(x, y)), \quad x, y \in X. \tag{1}$$

Then

(i) *f has a unique fixed point $x^* \in X$;*
(ii) *the sequence of successive approximations $x_n = f(x_{n-1}), n \in \mathbb{N}$, converges to x^* for any $x_0 \in X$;*
(iii) $d(x_n, x^*) \leq A^n(I_n - A)^{-1}(d(x_0, x_1)),\ n \in \mathbb{N}$;
(iv) *if $g : X \mapsto X$ satisfies the condition $d(f(x), g(x)) \leq c$ for all $x \in X$ and some $c \in \mathbb{R}^n$, then by considering the sequence $y_n = g^n(x_0), n \in \mathbb{N}$, we obtain*

$$d(y_n, x^*) \leq (I_n - A)^{-1}(c) + A^n(I_n - A)^{-1}(d(x_0, x_1)), \quad n \in \mathbb{N}.$$

This result has found its main application in solving differential and integral equations [1,3,32,33].

Note that a generalized metric space is just a special kind of normal cone metric space. Serbian mathematician Kurepa [34] presented the idea of pseudo metric and cone metric in 1934, but most authors in the fixed point theory cite Huang and Zhang's paper [35] from 2007 as a pioneer paper in the cone metric fixed point theory. All the definitions regarding property of convergence and continuity are, therefore, applicable from the cone metric space.

Definition 4. Let E be a real Banach space with a zero vector θ. A subset P of E is called a cone if:

(i) P is closed, non-empty and $P \neq \{\theta\}$;
(ii) $a, b \in \mathbb{R}$, $a, b \geq 0$, and $x, y \in P$ imply $ax + by \in P$;
(iii) $P \cap (-P) = \{\theta\}$.

Given a cone $P \subseteq E$, the partial ordering $\leq$ with respect to P is defined by $x \leq y$ if and only if $y - x \in P$. We write $x \prec y$ to indicate that $x \leq y$ but $x \neq y$, while $x \ll y$ denotes $y - x \in (\text{int})\, P$ where $\text{int}(P)$ is the interior of P.

The cone P in a real Banach space E is called normal if

$$\inf\{\|x + y\| \mid x, y \in P \text{ and } \|x\| = \|y\| = 1\} > 0,$$

or, equivalently, if there is a number $K > 0$ such that for all $x, y \in P$,

$$\theta \leq x \leq y \text{ implies } \|x\| \leq K \|y\| . \tag{2}$$

The least positive number satisfying (2) is called the normal constant of P. It has been shown that we can consider only case $K = 1$ for normal cone metric spaces. The cone P is called solid if $\operatorname{int}(P) \neq \emptyset$.

Introducing a concept of a cone in a real Banach space allows us to present a different type of pseudo metric related to defined partial ordering induced by observed cone.

Definition 5. Let X be a non-empty set, and let P be a cone on a real Banach space E. Suppose that the mapping $d : X \times X \mapsto E$ satisfies

(d_1) $\theta \leq d(x, y)$, for all $x, y \in X$ and $d(x, y) = \theta$ if and only if $x = y$;
(d_2) $d(x, y) = d(y, x)$, for all $x, y \in X$;
(d_3) $d(x, y) \leq d(x, z) + d(z, y)$, for all $x, y, z \in X$.

Then d is called a cone metric on X and (X, d) is a cone metric space.

Example 1. Defined partial ordering on $\mathbb{R}^n$ as in the definition of generalized metric in the sense of Perov determines a normal cone

$$P = \{x = (x_1, \ldots, x_n) \in \mathbb{R}^n \mid x_i \geq 0, i = \overline{1, n}\}$$

on $\mathbb{R}^n$, with the normal constant $K = 1$. Evidently, $A(P) \subseteq P$ if and only if $A \in \mathcal{M}_{n,n}(\mathbb{R}_+)$. It appears possible to adjust and probably broadly modify Perov's idea on a concept of cone metric space. Preferably, we will get some existence results. Nevertheless, forcing the transfer of contractive condition on cone metric space would be possible for some operator A instead of a matrix.

Definition 6. The sequence $(x_n) \subseteq X$ is convergent in X if there exists some $x \in X$ such that

$$(\forall\, c \gg \theta)(\exists\, n_0 \in \mathbb{N})\, n \geq n_0 \implies d(x_n, x) \ll c.$$

We say that a sequence $(x_n) \subseteq X$ converges to $x \in X$ and denote that with $\lim_{n\to\infty} x_n = x$ or $x_n \to x$, $n \to \infty$. Point x is called a limit of the sequence (x_n).

Definition 7. The sequence $(x_n) \subseteq X$ is a Cauchy sequence if

$$(\forall\, c \gg \theta)(\exists\, n_0 \in \mathbb{N})\, n, m \geq n_0 \implies d(x_n, x_m) \ll c.$$

Every convergent sequence is a Cauchy (fundamental) sequence, but the reverse does not hold. If any Cauchy sequence in a cone metric space (X, d) is convergent, then X is a complete cone metric space.

In the setting of normal cone metric space, emphasize that $\lim_{n\to\infty} x_n = x$ if and only if $\lim_{n\to\infty} \|x_n - x\| = 0$ and (x_n) is a Cauchy sequence if and only if $\lim_{n,m\to\infty} \|x_n - x_m\| = 0$ and policeman lemma holds despite that it is not the case in non-normal cone metric spaces.

Definition 8. Let (x_n) be a sequence in X and $x \in X$. If for every c in E with $0 \ll c$ there is n_0 such that for all $n > n_0$, $d(x_n, x) \ll c$, then (x_n) converges to x, and we denote this by $\lim_{n\to\infty} x_n = x$, or $x_n \to x$, $n \to \infty$.

If every Cauchy sequence is convergent in X, then X is called a complete cone metric space.

As proved in Ref. [35], if P is a normal cone, not related to if it is solid, a sequence $(x_n) \subseteq X$ converges to $x \in X$ if and only if $d(x_n, x) \to \theta$, $n \to \infty$. Similarly, $(x_n) \subseteq X$ is a Cauchy sequence if and only if $d(x_n, x_m) \to \theta$, $n, m \to \infty$. Also, if $\lim_{n\to\infty} x_n = x$ and $\lim_{n\to\infty} y_n = y$, then $d(x_n, y_n) \to d(x, y)$, $n \to \infty$. Let us emphasize that these equivalences do not hold if P is a non-normal cone.

Definition 9. The self-mapping $f : X \mapsto X$ is continuous on a generalized metric space (cone metric space) (X, d) if for any sequence $(x_n) \subseteq X$ converging to some $x^* \in X$, the sequence $(f(x_n))$ is convergent with a limit $f(x^*)$.

Definition 10. Assume that (X, d) is a generalized metric space (cone metric space). A mapping $f : X \to X$ is lower semicontinuous at $x \in X$ if for any $\epsilon > 0_m(\theta)$ there is n_0 in $\mathbb{N}$ such that

$$f(x) \le f(x_n) + \epsilon, \quad \text{for all} \quad n \ge n_0, \tag{3}$$

whenever $\{x_n\}$ is a sequence in X and $x_n \to x$.

Cvetković and Rakočević in Ref. [20] gave a generalization of Perov result for the class of cone metric spaces considering well-known property of bounded linear operators:

If the spectral radius of an operator $A \in \mathcal{B}(E)$ is less than 1, then the series $\sum_{n=0}^{\infty} A^n$ is absolutely convergent, $I - A$ is invertible in $\mathcal{B}(E)$ and

$$\sum_{n=0}^{\infty} A^n = (I - A)^{-1}.$$

This is, as mentioned, applicable if $A \in \mathcal{M}_{n,n}$.

Kada *et al.* [36] introduced ω-distance in 1996 and indicated that it is a more general concept than metric. They gave examples of ω-distance and improved Caristi's fixed point theorem [37], Ekeland's variationals principle [38] and the non-convex minimization theorem according to Takahashi [39].

Definition 11 ([36]). Let X be a metric space with metric d. Then a function $\omega : X \times X \to [0, \infty)$ is called a ω-distance on X if the following are satisfied:

(i) $\omega(x, z) \leq \omega(x, y) + \omega(y, z)$, for any $x, y, z \in X$;
(ii) For any $x \in X$, $\omega(x, \cdot) : X \to [0, \infty)$ is lower semi-continuous;
(iii) For any $\varepsilon > 0$, there exists $\delta > 0$ such that $\omega(z, x) \leq \delta$ and $\omega(z, y) \leq \delta$ imply $d(x, y) \leq \varepsilon$.

Example 2. If $(X, \| \cdot \|)$ is a normed space, then $\omega(x, y) = \|x\| + \|y\|$, $x, y \in X$, is a ω-distance on X. Obviously, every metric d is a ω-distance.

This concept has been extended on generalized metric space, but also on cone metric space overall by introducing the ω-cone distance.

Definition 12 ([40]). Let (X, d) be a cone metric space. Then a function $\omega : X \times X \to P$ is called a ω-cone distance on X if the following conditions are satisfied:

(w_1) $\omega(x, z) \leq \omega(x, y) + \omega(y, z)$, for any $x, y, z \in X$;
(w_2) For any $x \in X$, $\omega(x, \cdot) : X \to P$ is lower semi-continuous;
(w_3) For any ε in E with $\theta \ll \varepsilon$, there is δ in E with $\theta \ll \delta$, such that $\omega(z, x) \ll \delta$ and $\omega(z, y) \ll \delta$ imply $d(x, y) \ll \varepsilon$.

Remark 1. In the case of generalized ω-distance as a kind of ω-cone metric distance, we have 0_m in a role of θ and alphabetical partial ordering. So, if $\omega_i : X \times X \mapsto \mathbb{R}_+$ are ω-distances, $i = \overline{1, m}$, then $\omega : X \times X \mapsto \mathbb{R}_+$ defined as

$$\omega(x, y) = (\omega_1(x, y), \ldots, \omega_m(x, y)), \;\; x, y \in X$$

is a generalized ω-distance on X.

When talking about multivalued mappings we need to introduce several notations. Let X be a non-empty set and $\mathcal{P}(X)$ a partitive set.

$P(X) = \{Y \mid Y \subseteq X \,\text{and}\, Y \neq \emptyset\}$ is a set of all non-empty subsets of X.

$P_b(X) = \{Y \in P(X) \mid Y$ is a bounded set$\}$ and $P_{cl}(X) = \{Y \in P(X) \mid Y$ is a closed set$\}$ are, resp., families of all bounded (closed) non-empty subsets of X.

For a metric space (X, d), we may define a gap distance between sets in $P(X)$ such that, for any $A, B \in P(X)$,

$$D(A,B) = \inf\{d(a,b) \mid a \in A, b \in B\}.$$

In the case when $A = \{a\}$ for some $a \in X$, we can talk about a distance from the point a from the set B defined as $D(a,B) = \inf\{d(a,b) \mid b \in B\}$.

If $A, B \in P_b(X)$, then supremum of distances exists and

$$\delta(A,B) = \sup\{d(a,b) \mid a \in A, b \in B\}$$

is called a diameter functional.

Generalized excess functional $\rho : P(X) \times P(X) \mapsto [0, \infty]$ is defined as

$$\rho(A,B) = \sup\{D(a,B) \mid a \in A\}.$$

If $A, B \in P_{cl}(X)$, we may define a Pompeiu–Hausdorff generalized distance between the non-empty closed subsets A and B as

$$H(A,B) = \max\left\{\sup_{a\in A}\inf_{b\in B} d(a,b), \sup_{b\in B}\inf_{a\in A} d(a,b)\right\}.$$

Note that the generalized Pompeiu–Hausdorff functional can be observed as

$$H(A,B) = \max\{\rho(A,B), \rho(B,A)\}.$$

For a multivalued operator $T : X \mapsto P(Y)$, $G(T) = \{(x,y) \mid y \in T(x)\}$ is the graph of T. A set of fixed points of T is denoted by $Fix(T)$ (or sometimes as $F(T)$ or F_T) where $Fix(T) = \{x \in X \mid x \in T(x)\}$. Also, $SFix(T)$ is a set of all strict fixed points of T defined as

$$SFix(T) = \{x \in X \mid T(x) = \{x\}\}.$$

If $T : X \mapsto P(Y)$, then $f : X \mapsto Y$ is a selection for T if $f(x) \in T(x)$ for each $x \in X$.

2. Multivalued Perov-type Contractions and Generalizations

Bucur, Guran and Petruşel in Ref. [41] presented left and right multivalued A-contraction as a generalization of a Perov contraction. They gave several existence proofs and, as a corollary, existence and uniqueness in a single-valued case. The presented results were a foundation for several authors [7,

10,11,42] to extend many well-known results for multivalued operators to the generalized metric space and a new kind of a contraction introducing matrix in a contractive condition.

The evolution of macrosystems under uncertainty or lack of precision from biology, control theory, economics, etc. is often modeled by a semilinear inclusion system:

$$\begin{aligned} x_1 &\in T_1(x_1, x_2) \\ x_2 &\in T_1(x_1, x_2), \end{aligned} \tag{4}$$

where $T_i : X \times X \to P(X)$ for $i \in \{1, 2\}$ are multivalued operators on a Banach space X. The system (4) can be rephrased as a fixed-point problem

$$x \in Tx, \tag{5}$$

where $T = (T_1, T_2) : X \times X \longrightarrow P(X) \times P(X)$, $x = (x_1, x_2) \in X \times X$. Hence, it is of great interest to give fixed-point results for multivalued operators on a set endowed with vector-valued metrics or norms. Some benefits of estimations and precision when dealing with generalized metric space in the sense of Perov were already pointed out by Precup in Ref. [43] The following results are extensions of the theorems given by Perov [1], O'Regan *et al.* [44], Berinde and Berinde [45].

To solve the problem (4) we must prove some theoretical results regarding existence of a fixed point for multivalued operators.

Definition 13. Let $Y \subseteq X$ and $T : Y \mapsto P(X)$ be a multivalued operator. Then T is called a multivalued left A-contraction (Perov contraction) if there exists some $A \in \mathcal{M}_{m,m}(R_+)$ converging to the zero matrix such that for each $x, y \in Y$ and each $u \in Tx$ there exists some $v \in Ty$ such that

$$d(u, v) \leq Ad(x, y).$$

For a generalized metric space (X, d), $x_0 \in X$ and $r = (r_i) \in \mathbb{R}^m$, $r_i > 0$, $i = \overline{1, m}$, define an opened and closed ball, resp., as

$$B(x_0, r) = \{x \in X \mid d(x, x_0) < r\},$$

and

$$\overline{B}(x_0, r) = \{x \in X \mid d(x, x_0) \leq r\}.$$

Theorem 3. *Let (X, d) be a complete generalized metric space, $x_0 \in X, r = (r_i)_{i=1}^m \in \mathbb{R}_+^m$ with $r_i > 0$ for each $i \in \overline{1, m}$ and let $T : \overline{B}(x_0, r) \to P_{cl}(X)$ be multivalued left Perov contraction on $\overline{B}(x_0, r)$. We suppose that*

(i) *A is a matrix that converges toward zero;*
(ii) *if $u \in \mathbb{R}^m_+$ is such that $u(I-A)^{-1} \leq (I-A)^{-1}r$, then $u \leq r$;*
(iii) *there exists $x_1 \in T(x_0)$ such that $d(x_0, x_1)(I-A)^{-1} \leq r$.*

Then Fix $(T) \neq \emptyset$.

Proof. Let $x_1 \in T(x_0)$ be such that (iii) holds. By using the fact that T is a multivalued left Perov contraction, we get existence of $x_2 \in T(x_1)$ such that $d(x_1, x_2) \leq Ar$. Furthermore,

$$\begin{aligned} d(x_0, x_2)(I-A)^{-1} &\leq (d(x_0,x_1) + d(x_1,x_2))(I-A)^{-1} \\ &\leq Ir + Ar \\ &\leq (I-A)^[-1]r, \end{aligned}$$

so $x_2 \in \overline{B}(x_0, r)$.

Continuing in the same manner, we may construct the sequence $(x_n) \subseteq \overline{B}(x_0, r)$ such that, for any $n \in \mathbb{N}$, $x_{n+1} \in Tx_n$ and

$$d(x_n, x_{n+1})(I-A)^{-1} \leq A^n r.$$

Having in mind that, for any $n, m \in \mathbb{N}$, $m > n$,

$$\begin{aligned} d(x_n, x_m)(I-A)^{-1} &\leq \left(\sum_{i=n}^{m-1} d(x_i, x_{i+1}) \right) (I-A)^{-1} \\ &\leq \sum_{i=n}^{m-1} A^i r \\ &\leq A^n (I-A)^{-1} r, \end{aligned}$$

we may conclude that (x_n) is a Cauchy sequence in $\overline{B}(x_0, r)$ and since it is a closed subset of a complete generalized metric space, there exists some $x^* \in \overline{B}(x_0, r)$ such that $\lim_{n\to\infty} x_n = x^*$.

We shall prove that x^* is a fixed point of T. For any $n \in \mathbb{N}$ there exists some $u_n \in Tx^*$ such that

$$d(x_n, u_n) \leq Ad(x_{n-1}, x^*),$$

implying that $\lim_{n\to\infty} u_n = x^*$ since $d(x^*, u_n) \leq d(x^*, x_n) + d(x_n, u_n)$, $n \in \mathbb{N}$. However, Tx^* is a closed set so $x^* \in Tx^*$. □

Definition 14. Let (X, d) be a metric space. A mapping $T : X \mapsto P(X)$ is a multivalued weakly Picard operator (MWP) if for each $x \in X$ and $y \in Tx$ there exists a sequence (x_n) such that

(i) $x_0 = x, x_1 = y$;
(ii) $x_{n+1} \in Tx_n$, $n \in \mathbb{N}$;
(iii) (x_n) is a convergent sequence in X and its limit is a fixed point of T.

The sequence $(x_n) \in X$ fulfilling (i) and (ii) from the previous definition is called a sequence of successive approximations (iterative sequence) starting from $(x, y) \in G(T)$.

Obviously, every multivalued contraction $T : X \mapsto P_{cl}(X)$, where (X, d) is a complete metric space, is a MWP operator.

Corollary 1. *Let (X, d) be a complete generalized metric space and $T : X \mapsto P_{cl}(X)$ be a multivalued left Perov contraction. Then, T is a MWP operator.*

Definition 15. Let $Y \subseteq X$ and $T : Y \mapsto P(X)$ be a multivalued operator. Then, T is called a multivalued right A-contraction (Perov contraction) if $A \in \mathcal{M}_{m,m}(\mathbb{R}_+)$ is a matrix convergent to zero and for each $x, y \in Y$ and each $u \in Tx$ there exists $v \in Ty$ such that $d(u, v)^T \leq d(x, y)^T A$.

Theorem 4. *Let (X, d) be a complete generalized metric space, $x_0 \in X, r = (r_i)_{i=1}^m \in \mathbb{R}_+^m$ with $r_i > 0$ for each $i \in \overline{1, m}$ and let $T : \overline{B}(x_0, r) \to P_{cl}(X)$ be multivalued right A-contraction on $\overline{B}(x_0, r)$. If, $(I - A)^{-1} d(x_0, u) \leq r$ for each $u \in T(x_0)$, then:*

(i) *$\overline{B}(x_0, r)$ is invariant with respect to T;*
(ii) *T is an MWP operator on $\overline{B}(x_0, r)$.*

Proof. For some $x \in \overline{B}(x_0, r)$ let $v \in Tx$ be arbitrary. Since T is a multivalued left A-contraction, there exists $u \in T(x_0)$ such that $d(v, u) \leq d(x, x_0)A$ and

$$\begin{aligned} d(x_0, u) &\leq d(x_0, v) + d(v, u) \\ &\leq (I - A)r + Ad(x, x_0) \\ &\leq r. \end{aligned}$$

Hence, $\overline{B}(x_0, r)$ is invariant with respect to T.

Second part is obvious due to the Theorem 6. □

Theorem 5. *If (X, d) is a complete generalized metric space and $T : X \mapsto P_{cl}(X)$ is a multivalued left A-contraction on X, then T is a $(I - A)^{-1}$-MWP operator.*

Since any multivalued right A-contraction is a multivalued left A^T-contraction, we may state the following results:

Theorem 6. *Let (X, d) be a complete generalized metric space, $x_0 \in X, r = (r_i)_{i=1}^m \in \mathbb{R}_+^m$ with $r_i > 0$ for each $i \in \overline{1,m}$ and let $T : \overline{B}(x_0, r) \to P_{cl}(X)$ be multivalued right A-contraction on $\overline{B}(x_0, r)$. We suppose that*

(i) *if $u \in \mathbb{R}_+^m$ is such that $(I - A)^{-1}u \leq r(I - A)^{-1}$, then $u \leq r$;*
(ii) *there exists $x_1 \in T(x_0)$ such that $(I - A)^{-1}d(x_0, x_1) \leq r$.*

Then Fix $(T) \neq \emptyset$.

Proof. The proof goes analogously as in the case of the multivalued left A-contraction. □

Theorem 7. *Let (X, d) be a complete generalized metric space, $x_0 \in X, r = (r_i)_{i=1}^m \in \mathbb{R}_+^m$ with $r_i > 0$ for each $i \in \overline{1,m}$ and let $T : \overline{B}(x_0, r) \to P_{cl}(X)$ be multivalued right A-contraction on $\overline{B}(x_0, r)$. If $d(x_0, u)(I - A)^{-1} \leq r$ for each $u \in T(x_0)$, then:*

(i) *$\overline{B}(x_0, r)$ is invariant with respect to T;*
(ii) *T is an MWP operator on $\overline{B}(x_0, r)$.*

Theorem 8. *Let $(X, \|\cdot\|)$ be a Banach space and $T_1, T_2 : X \mapsto P_{cl}(X)$ be two multivalued operators. Suppose there exist $a_{ij} \in \mathbb{R}_+$, $i, = \overline{1,2}$ such that:*

(i) *For each $x = (x_1, x_2)$, $y = (y_1, y_2) \in X \times X$ and each $u_1 \in T_1(x_1, x_2)$ there exists $v_1 \in T_1(y_1, y_2)$ such that:*
$$\|u_1 - v_1\| \leq a_{11}\|x_1 - y - 1\| + a_{12}\|x_2 - y_2\|;$$
(ii) *For each $x = (x_1, x_2)$, $y = (y_1, y_2) \in X \times X$ and each $u_2 \in T_2(x_1, x_2)$ there exists $v_2 \in T_2(y_1, y_2)$ such that:*
$$\|u_2 - v_2\| \leq a_{21}\|x_1 - y - 1\| + a - 22\|x_2 - y_2\|.$$

In addition, assume that the matrix
$$A = \begin{bmatrix} a_{11} & a_{12} \\ a_{21} & a_{22} \end{bmatrix}$$
converges to the zero matrix. Then, the semilinear inclusion system
$$\begin{aligned} x_1 &\in T_1(x_1, x_2), \\ x_2 &\in T_1(x_1, x_2), \end{aligned} \tag{6}$$
has at least one solution in $X \times X$.

Proof. Denote with T an operator such that $T : X \times X \mapsto P_{cl}(X \times X)$ defined with

$$T(x, y) = (T_1(x, y), T_2(x, y)), \quad (x, y) \in X \times X.$$

Then (ii) along with (iii) forms a contractive condition for T. For any $x = (x_1, x_2)$, $y = (y_1, y_2) \in X \times X$ and $u = (u_1, u_2) \in Tx$, there exists some $v = (v_1, v_2) \in Ty$ such that

$$\|u - v\| \leq A\|x - y\|.$$

So the existence of the fixed point follows directly from Theorem 5. □

We will list, without the proof, several results on this topic with a remark that the concept of the proof is similar to already presented ideas in previous theorems.

Theorem 9. *Let (X, d) be a complete generalized metric space in the sense of Perov, $U \subseteq V \subseteq X$, such that U is an open set and V is closed in topology induced by generalized metric. Let $G : V \times [0, 1] \mapsto P_{cl}(X)$ be a multivalued operator with closed graph such that:*

(i) *For each $x \in V \setminus U$ and $t \in [0, 1]$, $x \notin G(x, t)$;*
(ii) *There exists a matrix $A \in \mathcal{M}_{m,m}(\mathbb{R}_+)$ convergent to the zero matrix such that for any $t \in [0, 1]$, $x, y \in Y$ and $u \in G(x, t)$, there exists $v \in G(y, t)$ such that $d(u, v) \leq Ad(x, y)$;*
(iii) *There exists a continuous increasing function $\phi : [0, 1] \mapsto \mathbb{R}^m$ such that for any $t, s \in [0, 1]$, $x \in V$ and $u \in G(x, t)$, there exists $v \in G(x, s)$ such that $d(u, v) \leq |\phi(t) - \phi(s)|$;*
(iv) *If $v, r \in \mathbb{R}^m_+$ are such that $v(I - A)^{-1} \leq (I - A)^{-1}r$, then $v \leq r$.*

Then $G(\cdot, 0)$ has a fixed point if and only if $G(x, 1)$ has a fixed point.

Theorem 10. *Let (X, d) be a complete generalized metric space in the sense of Perov and $T_1, T_2 : X \mapsto P_{cl}(X)$ be two multivalued operators such that*

(i) *T_i is a C_i-MWP operator, $i = \overline{1, 2}$;*
(ii) *There exists $r \in \mathbb{R}^m_+$ with $r_i > 0$, $i = \overline{1, m}$ such that $H(T_1x, T_2x) \leq r$, $x, y \in X$.*

Then

$$H(Fix(T_1), Fix(T_2)) \leq \max\{C_1r, C_2r\}.$$

Theorem 11. *Let (X, d) be a complete generalized metric space in the sense of Perov and $T : X \mapsto P_{cl}(X)$ be multivalued left A-contraction. If $SFix(T) \neq \emptyset$, then, for some $x^* \in X$,*

$$Fix(T) = SFix(T) = \{x^*\}.$$

Recall the definition of the well-posed fixed point problem.

Definition 16. Let (X, d) be a generalized metric space and $T : X \times X \mapsto P(X)$, a multivalued operator. The fixed-point problem $x = Tx$ is well-posed if

(i) $Fix(T) = \{x^*\}$;
(ii) For any sequence $(x_n) \subseteq X$ such that $\lim_{n\to\infty} D(x_n, Tx_n) = 0$, $\lim_{n\to\infty} d(x_n, x^*) = 0$.

Theorem 12. *Let (X, d) be a complete generalized metric space in the sense of Perov and $T : X \mapsto P_{cl}(X)$ be multivalued left A-contraction. If $SFix(T) \neq \emptyset$, then the fixed point problem is well-posed.*

Another question that arises regarding this topic is a fixed-point problem for a multivalued operator on a set endowed with two metrics as in the following theorem.

Theorem 13. *Let (X, d) be a complete generalized metric space in the sense of Perov, ρ, another generalized metric on X and $T : X \mapsto P_{cl}(X)$, a multivalued operator. Suppose that*

(i) *there exists $C \in \mathcal{M}_{m,m}(\mathbb{R}_+)$ such that*

$$d(x, y) \leq C\rho(x, y), \ \ x, y \in X;$$

(ii) *$T : (X, d) \mapsto (P(X, H_d))$ has a closed graph;*
(iii) *T is a multivalued left A-contraction on (X, ρ).*

Operator T is an MWP-operator with respect to d.

Proof. For arbitrary $x_0 \in X$ and $x_n \in Tx_{n-1}$, $n \in \mathbb{N}$, we have

$$\rho(x_n, x_{n+1}) \leq A^n \rho(x_0, x_1), \ \ n \in \mathbb{N}.$$

Hence, for $n, m \in \mathbb{N}$, $m > n$,

$$\begin{aligned}\rho(x_n, x_m) &\leq \sum_{i=n}^{m-1} \rho(x_i, x_{i+1}) \\ &\leq \sum_{i=n}^{m-1} A^i \rho(x_0, x_1) \\ &\leq A^n (I - A)^{-1} \rho(x_0, x_1).\end{aligned}$$

The sequence (x_n) is therefore a Cauchy sequence in (X, ρ) and so in (X, d). However, (X, d) is a complete generalized metric space, so there exists some $x^* \in X$ such that $\lim_{n\to\infty} d(x_n, x^*) = 0$. Due to the fact that T has a closed graph and $x_n \in Tx_{n-1}$, $n \in \mathbb{N}$, we get that x^* is a fixed point of T. □

Afterwards, Filip and Petrusel generalized the presented results from Ref. [41] in Ref. [7]. There are several single-valued results of this kind:

Theorem 14. *Let (X, d) be a complete generalized metric space, $x_0 \in X$, $r = (r_i)_{i=1}^m \in \mathbb{R}_+^m$ with $r_i > 0$ for each $i \in \overline{1, m}$ and let $f : \overline{B}(x_0, r) \to X$ have the property that there exist $A, B \in \mathcal{M}_{m,m}(\mathbb{R}_+)$ such that*

$$d(f(x), f(y)) \leq Ad(x, y) + Bd(y, f(x)), \tag{7}$$

for all $x, y \in \overline{B}(x_0, r)$. We suppose that

(i) *A is a matrix that converges toward zero;*
(ii) *if $u \in \mathbb{R}_+^m$ is such that $u(I - A)^{-1} \leq (I - A)^{-1} r$, then $u \leq r$;*
(iii) *$d(x_0, f(x_0))(I - A)^{-1} \leq r$.*

Then $Fix(f) \neq \emptyset$.

In addition, if the matrix $A + B$ converges to zero, then $\mathrm{Fix}(f) = \{x^*\}$.

Remark 2. By similitude to [46] a mapping $f : Y \subseteq X \to X$ satisfying the condition

$$d(f(x), f(y)) \leq Ad(x, y) + Bd(y, f(x)), \quad \forall x, y \in Y, \tag{8}$$

for some matrices $A, B \in M_{m,m}(\mathbb{R}_+)$, with A a matrix that converges toward zero, could be called an almost contraction of Perov type.

We have also a global version of Theorem 14 expressed by the following result:

Corollary 2. *Let (X, d) be a complete generalized metric space. Let $f : X \to X$ be a mapping having the property that there exist $A, B \in M_{m,m}(\mathbb{R}_+)$ such that*

$$d(f(x), f(y)) \leq Ad(x, y) + Bd(y, f(x)), \quad \forall x, y \in X. \tag{9}$$

If A is a matrix that converges toward zero, then

(i) $Fix(f) \neq \emptyset$;
(ii) *the sequence $(x_n)_{n\in\mathbb{N}}$ given by $x_n = f^n(x_0)$, $n \in \mathbb{N}_0$, converges toward a fixed point of f, for all $x_0 \in X$;*
(iii) *the estimation holds*

$$d(x_n, x^*) \leq A^n(I - A)^{-1} d(x_0, x_1), \tag{10}$$

where $x^* \in Fix(f)$.

In addition, if the matrix $A + B$ converges to the zero matrix, then $Fix(f) = \{x^*\}$.

Theorem 15. *Let $(X, |\cdot|)$ be a Banach space and let $f_1, f_2 : X \times X \to X$ be two operators. Suppose that there exist $a_{ij}, b_{ij} \in \mathbb{R}_+$, $i, j \in \{1, 2\}$ such that, for each $x = (x_1, x_2), y = (y_1, y_2) \in \mathrm{X} \times \mathrm{X}$, one has*

$$\begin{aligned}|f_1(x_1, x_2) - f_1(y_1, y_2)| \leq a_{11}|x_1 - y_1| + a_{12}|x_2 - y_2| \\ + b_{11}|x_1 - f_1(y_1, y_2)| + b_{12}|x_2 - f_2(y_1, y_2)| \\ |f_2(x_1, x_2) - f_2(y_1, y_2)| \leq a_{21}|x_1 - y_1| + a_{22}|x_2 - y_2| \\ + b_{21}|x_1 - f_1(y_1, y_2)| + b_{22}|x_2 - f_2(y_1, y_2)|.\end{aligned}$$

In addition, assume that the matrix $A = \begin{pmatrix} a_{11} a_{12} \\ a_{21} a_{22} \end{pmatrix}$ converges to the zero matrix. Then, the system

$$u_1 = f_1(u_1, u_2), \quad u_2 = f_1(u_1, u_2), \tag{11}$$

has at least one solution $x^* \in X \times X$. If, in addition, the matrix $A + B$ converges to the zero, then the above solution is unique.

As mentioned, we once again consider the case of a generalized metric space with two metrics.

Theorem 16. *Let X be a non-empty set and let d, ρ be two generalized metrics on X. Let $f : \mathrm{X} \to \mathrm{X}$ be an operator. We assume that*

(i) *there exists* $C \in M_{m,m}(\mathbb{R}_+)$ *such that* $d(f(x), f(y)) \leq \rho(x,y) \cdot C$;
(ii) (X,d) *is a complete generalized metric space*;
(iii) $f : (X,d) \to (X,d)$ *is continuous*;
(iv) *there exists* $A, B \in M_{m,m}(\mathbb{R}_+)$ *such that for all* $x, y \in X$ *one has*

$$\rho(f(x), f(y)) \leq A\rho(x,y) + B\rho(y, f(x)). \tag{12}$$

If the matrix A *converges toward zero, then* $Fix(f) \neq \emptyset$.

In addition, if the matrix $A+B$ *converges to zero, then* $Fix(f) = \{x^*\}$.

In the same manner, the authors of [7] considered existence of a fixed point for a multivalued operator of a same type.

Theorem 17. *Let* (X,d) *be a complete generalized metric space and let* $x_0 \in X, r = (r_i)_{i=1}^m \in \mathbb{R}_+^m$ *with* $r_i > 0$ *for each* $i \in \overline{1,m}$. *Consider* $T : \overline{B}(x_0, r) \to P_{cl}(X)$ *a multivalued operator. Assume that*

(i) *there exist* $A, B \in M_{m,m}(\mathbb{R}_+)$ *such that for all* $x, y \in \overline{B}(x_0, r)$ *and* $u \in Tx$ *there exists* $v \in Ty$ *with*

$$d(u,v) \leq Ad(x,y) + Bd(y,u); \tag{13}$$

(ii) *there exists* $x_1 \in T(x_0)$ *such that* $d(x_0, x_1)(I-A)^{-1} \leq r$;
(iii) *if* $u \in \mathbb{R}_+^m$ *is such that* $u(I-A)^{-1} \leq (I-A)^{-1}r$, *then* $u \leq r$.

If A *is a matrix convergent toward zero, then* $Fix(T) \neq \emptyset$.

Proof. In the same manner as in the proof of Theorem 3, for an arbitrary $x_0 \in X$, we construct a sequence $(x_n) \subseteq \overline{B}(x_0, r)$ such that $x_{n+1} \in Tx_n$ and $d(x_n, x_{n+1})(I-A)^{-1} \leq A^n r$, $n \in \mathbb{N}$. Let $n, m \in \mathbb{N}$ and $m > n$, then

$$\begin{aligned} d(x_n, x_m)(I-A)^{-1} &\leq \left(\sum_{i=n}^{m-1} d(x_i, x_{i+1}) \right) (I-A)^{-1} \\ &\leq \sum_{i=n}^{m-1} A^i r \\ &\leq A^n (I-A)^{-1} r, \end{aligned}$$

so (x_n) is a Cauchy sequence, and therefore convergent. Let $\lim_{n\to\infty} x_n = x^* \in \overline{B}(x_0, r)$. For any $n \in \mathbb{N}$ there exists some $u_n \in Tx^*$ such that

$$d(x_n, u_n) \leq Ad(x_{n-1}, x^*) + Bd(x^*, x_n),$$

and

$$d(x^*, u_n) \leq d(x^*, x_n)Ad(x_{n-1}, x^*) + Bd(x^*, x_n), \ n \in \mathbb{N}.$$

Hence, $\lim_{n\to\infty} u_n = x^*$ and $x^* \in Tx^*$ since Tx^* is a closed set. □

We have also a global variant for the Theorem 17 as follows:

Corollary 3. *Let (X, d) be a complete generalized metric space and $T : X \to P_{cl}(X)$ a multivalued operator. One supposes that there exist $A, B \in M_{m,m}(\mathbb{R}_+)$ such that for each $x, y \in X$ and all $u \in Tx$, there exists $v \in Ty$ with*

$$d(u, v) \leq Ad(x, y) + Bd(y, u). \tag{14}$$

If A is a matrix convergent toward zero, then $Fix(T) \neq \emptyset$.

Remark 3. By a similar approach to that given in Theorem 15, we can obtain an existence result for a system of operatorial inclusions of the following form:

$$\begin{aligned} x_1 &\in T_1(x_1, x_2), \\ x_2 &\in T_1(x_1, x_2), \end{aligned} \tag{15}$$

where $T_1, T_2 : X \times X \to P_{cl}(X)$ are multivalued operators satisfying a contractive type condition as in the previously presented example.

The case of a set X endowed with two metrics can also be discussed in the case of multivalued almost contraction.

Theorem 18. *Let (X, d) be a complete generalized metric space and ρ another generalized metric on X. Let $T : X \to P(X)$ be a multivalued operator. One assumes that*

(i) *there exists a matrix $C \in M_{m,m}(\mathbb{R}_+)$ such that $d(x, y) \leq \rho(x, y) \cdot C$, for all $x, y \in X$;*
(ii) *$T : (X, d) \to (P(X), H_d)$ has closed graph;*
(iii) *there exist $A, B \in M_{m,m}(\mathbb{R}_+)$ such that for all $x, y \in X$ and $u \in Tx$, there exists $v \in Ty$ with*

$$\rho(u, v) \leq A\rho(x, y) + B\rho(y, u). \tag{16}$$

If A is a matrix convergent toward zero, then $\text{Fix}(T) \neq \emptyset$.

Remark 4.

(i) Theorem 18 holds even if the assumption (iii) is replaced by (iii) there exist $A, B \in M_{m,m}(\mathbb{R}_+)$ such that for all $x, y \in X$ and $u \in Tx$, there exists $v \in Ty$ such that $\rho(u, v) \leq A\rho(x, y) + Bd(y, u)$.
(ii) Letting $p \to \infty$ in the estimation of $\rho(x_n, x_{n+p})$, presented in the proof of Theorem 18, we get

$$\rho(x_n, x^*) \leq A^n(I - A)^{-1}\rho(x_0, x_1). \tag{17}$$

Using the relation between the generalized metrics d and ρ, one has immediately

$$d\left(x_n, x^*\right) \leq CA^n(I-A)^{-1}\rho\left(x_0, x_1\right). \tag{18}$$

Theorem 19. *Let (X, d) be a complete generalized metric space and ρ another generalized metric on X. Let $x_0 \in X, r=(r_i)_{i=1}^m \in \mathbb{R}_+^m$ with $r_i>0$ for each $i \in \overline{1, m}$ and let $T: \overline{B}_\rho(x_0, r) \rightarrow P(X)$ be a multivalued operator. Suppose that*

(i) *there exists $C \in M_{m,m}(\mathbb{R}_+)$ such that $d(x, y) \leq C\rho(x, y)$, for all $x, y \in X$;*
(ii) *$T:\left(\tilde{B}_\rho(x_0, r), d\right) \rightarrow (P_b(X), H_d)$ has a closed graph;*
(iii) *there exist $A, B \in M_{m,m}(\mathbb{R}_+)$ such that A is a matrix that converges to zero and for all $x, y \in \overline{B}_\rho(x_0, r)$ and $u \in Tx$, there exists $v \in Ty$ such that*
$$\rho(u, v) \leq A\rho(x, y)+B\rho(y, u);$$
(iv) *if $u \in \mathbb{R}_+^m$ is such that $u(I-A)^{-1} \leq (I-A)^{-1}r$, then $u \leq r$;*
(v) *$\rho(x_0, x_1)(I-A)^{-1} \leq r$.*

Then $Fix(T) \neq \emptyset$.

A homotopy result for multivalued operators on a set endowed with a vector-valued metric is the following:

Theorem 20. *Let (X, d) be a generalized complete metric space in Perov sense, let U be an open subset of X, and let V be a closed subset of X, with $\mathrm{U} \subset \mathrm{V}$. Let $\mathrm{G}: \mathrm{V} \times [0,1] \rightarrow P(\mathrm{X})$ be a multivalued operator with closed graph (with respect to d), such that the following conditions are satisfied:*

(a) *$x \notin G(x, t)$, for each $x \in V \backslash U$ and each $t \in [0,1]$;*
(b) *there exist $A, B \in \mathcal{M}_{m,m}(\mathbb{R}_+)$ such that the matrix A is convergent to zero such that for each $t \in [0,1]$, for each $x, y \in X$ and all $u \in G(x, t))$ there exists $v \in G(y, t)$ with $d(u, v) \leq Ad(x, y)+Bd(y, u)$;*
(c) *there exists a continuous increasing function $\phi:[0,1] \rightarrow \mathbb{R}^m$ such that for all $t, s \in [0,1]$ each $x \in V$ and each $u \in G(x, t)$ there exists $v \in G(x, s)$ such that $d(u, v) \leq |\phi(t)-\phi(s)|$;*
(d) *if $v, r \in \mathbb{R}_+^m$ are such that $v \cdot (I-A)^{-1} \leq (I-A)^{-1} \cdot r$, then $v \leq r$.*

Then $G(\cdot, 0)$ has a fixed point if and only if $G(\cdot, 1)$ has a fixed point.

Remark 5. Usually, we take $V = \overline{U}$. Note that in this case, condition (a) becomes

$$(a')x \notin G(x,t), \text{ for each } x \in \partial U \text{ and each } t \in [0,1].$$

Remark 6. If $m = 1$, then we obtain several known results as those given by Berinde and Berinde [45], Precup [43], Petruşel and Rus [47], and Feng and Liu [9].

In 2008, Guran in Ref. [48] observed an MWP operators on a generalized metric space in the sense of Perov endowed with a generalized ω-distance.

Definition 17. If (X,d) is a generalized metric space and $T : X \mapsto P_{cl}(X)$ is a MWP operator, then $T^{\infty} : G(T) \mapsto P(Fix(T))$ is defined as follows:

$$T^{\infty}(x,y) = \{z \in Fix(T) \mid \text{is a sequence of successive approximations}$$
$$\text{starting from (x,y) and converging to z}\}.$$

Definition 18. Let (X,d) be a metric space and $T : X \mapsto P(X)$ be an MWP operator. Then T is called c-multivalued weakly Picard operator (c-MWP) if there exists a selection f^{∞} of T^{∞} such that

$$d(x, f^{\infty}(y,z)) \leq cd(y,z),\ x \in X,\ (y,z) \in G(T).$$

There are some important properties of ω-distance and its relation to metric that will transcend in the case of ω-cone distance, and, as a consequence, in the case of a generalized ω-distance as it proved in Ref. [40].

Lemma 1. *Let (X,d) be a metric space and let ω be a ω-cone distance on X. Let (x_n) and (y_n) are sequences in X, let (α_n) and $\{\beta_n\}$ be such that $\theta \leq \alpha_n$ along with $\theta \leq \beta_n$, be are sequences in E converging to θ, and let $x,y,z \in X$ be arbitrary. Then the following hold:*

(i) *If $\omega(x_n,y) \leq \alpha_n$ and $\omega(x_n,z) \leq \beta_n$ for any $n \in \mathbb{N}$, then $y = z$. In particular, if $\omega(x,y) = \theta$ and $\omega(x,z) = \theta$, then $y = z$;*
(ii) *If $\omega(x_n,y_n) \leq \alpha_n$ and $\omega(x_n,z) \leq \beta_n$ for any $n \in \mathbb{N}$, then $\{y_n\}$ converges to z;*
(iii) *If $\omega(x_n,x_m) \leq \alpha_n$ for any $n,m \in \mathbb{N}$ with $m > n$, then $\{x_n\}$ is a Cauchy sequence;*
(iv) *If $\omega(y,x_n) \leq \alpha_n$ for any $n \in \mathbb{N}$, then $\{x_n\}$ is a Cauchy sequence.*

Even though, Guran gave a proof of a similar, but more narrow, result for properties of generalized ω-distance in Ref. [10], it is abundant since it follows directly from Lemma presented in Ref. [40] with a comment that any generalized metric space (assume in the sense of Perov) is a normal cone metric space.

Theorem 21. *Let (X,d) be a complete generalized metric space, $\omega : X \times X \mapsto \mathbb{R}^m_+$ a generalized ω-distance on X and $T : X \mapsto P_{cl}(X)$ a multivalued operator. If there exists $A \in \mathcal{M}_{m,m}(\mathbb{R}_+)$ convergent to zero such that for any $x, y \in X$ and $u \in Tx$ there exists $v \in Ty$ such that*

$$\omega(u,v) \leq A(\omega(x,y)),$$

then T has a fixed point $x^* \in Tx^*$ such that $\omega(x^*,x^*) = 0$.

Proof. Observe that in this way we will prove that T is an MWP operator. For $x_0 \in X$ and $x_1 \in Tx_0$, choose $x_2 \in Tx_1$ such that $\omega(x_1,x_2) \leq A\omega(x_0,x_1)$ and, in a similar manner, $x_{n+1} \in Tx_n$ such that $\omega(x_n,x_{n+1}) \leq A\omega(x_{n-1},x_n)$ for any $n \in \mathbb{N}$. Thus,

$$\omega(x_n,x_{n+1}) \leq A^n\omega(x_0,x_1), \quad n \in \mathbb{N},$$

and, for $n, m \in \mathbb{N}$ such that $m > n$,

$$\begin{aligned}\omega(x_n,x_m) &\leq \sum_{i=n}^{m-1} \omega(x_i,x_{i+1}) \\ &\leq \sum_{i=n}^{m-1} A^i\omega(x_0,x_1) \\ &\leq A^n(I-A)^{-1}\omega(x_0,x_1).\end{aligned}$$

From the well-known properties of ω-distance previously presented, we have that (x_n) is a Cauchy sequence and so convergent. Denote the limit of the sequence (x_n) with x^*. Using that ω-distance is lower semi-continuous mapping, we get

$$\omega(x_n,x^*) \leq \lim\ inf_{m\to\infty}\omega(x_n,x_m) \leq A^n(I-A)^{-1}\omega(x_0,x_1).$$

There exists $u_n \in Tx^*$ such that

$$\omega(x_n,u_n) \leq A^n\omega(x_0,x_1).$$

Hence, $\lim_{n\to\infty} u_n = x^*$ and so $x^* \in Tx^*$. Existence of $y_1 \in Tx^*$ such that

$$\omega(x^*, y) \leq A\omega(x^*, x^*),$$

is followed by existence of $y_2 \in Ty_1$ such that

$$\omega(x^*, y_2) \leq A\omega(x, y_1) \leq A^2\omega(x^*, x^*).$$

Continuing in the same manner, we get the sequence (y_n) such that $y_n \in Ty_{n-1}$ and

$$\omega(x^*, y_n) \leq A^n\omega(x^*, x^*), \quad n \in \mathbb{N}.$$

As $n \to \infty$, $\lim_{n\to\infty} A^n\omega(x^*, x^*) \to 0$, so (y_n) is a Cauchy sequence and there exists some $y^* \in X$ such that $\lim_{n\to\infty} y_n = y^*$. Moreover,

$$0 \leq \omega(x^*, y^*) \leq \lim_{m\to\infty} \inf \omega(x^*, y_n) \leq \lim_{n\to\infty} A^n\omega(x^*, x^*).$$

Implying $\omega(x^*, y^*) = 0$ and since both $\lim_{n\to\infty} \omega(x_n, y^*)$ and $\lim_{n\to\infty} \omega(x_n, x^*)$ tend to zero, we have $x^* = z^*$ and $\omega(x^*, x^*) = 0$. □

The condition that Tx is a closed set for any $x \in X$ can be replaced in a following way without losing the conclusion that T is an MWP operator:

Theorem 22. *Let (X, d) be a complete generalized metric space, $\omega : X \times X \mapsto \mathbb{R}^m_+$, a generalized ω-distance on X and $T : X \mapsto P(X)$, a multivalued operator. Suppose that there exists $A \in \mathcal{M}_{m,m}(\mathbb{R}_+)$ convergent to zero such that for any $x, y \in X$ and $u \in Tx$ there exists $v \in Ty$ such that*

$$\omega(u, v) \leq A(\omega(x, y)),$$

and for every $z \in X$ such that $z \notin Tz$

$$\inf\{\omega(x, z) + \inf\{\omega(x, t) \mid t \in Tx\}\} > 0.$$

Then T is an MWP operator.

Proof. As in the proof of the previous theorem, observe the sequence (x_n) fulfilling

$$\omega(x_n, x_{n+1}) \leq A^n\omega(x_0, x_1) \text{ and } x_n \in Tx_{n-1,\, n\in\mathbb{N}}.$$

As $\omega(x_n, x_m) \leq A^n(I - A)^{-1}\omega(x_0, x_1)$, for $m > n$, it is a Cauchy sequence with a limit $x^* \in X$ and, since ω is a lower semi-continuous function, then

$$\omega(x_n, x^*) \leq \lim\ inf_{m\to\infty}\omega(x_n, x_m) \leq A^n(I - A)^{-1}\omega(x_0, x_1).$$

It remains to prove $x^* \in Tx^*$. Assume on the contrary, $x^* \notin Tx^*$. Thus,

$$\begin{aligned}0 &< \inf\{\omega(x_n, x^*) + \inf\{\omega(x,t) \mid t \in Tx\}\} \\ &\leq \inf\{\omega(x^*, x^*) + \omega(x_n, x_{n+1}) \mid n \in \mathbb{N}\} \\ &\leq \inf\{2A^n(I-A)^{-1}\omega(x_0, x_1) \mid n \in \mathbb{N}\} \\ &\leq 0,\end{aligned}$$

because $\lim_{n\to\infty} A^n(I-A)^{-1}\omega(x_0,x_1) = 0$. As it leads to a contradiction, the assumption is false, so $x^* \in Tx^*$ and T is an MWP operator. □

Considering results of [41], Guran in Ref. [10] gave some fixed-point results for a multivalued ω-left A-contraction and, analogously, for multivalued ω-right A-contraction.

Definition 19. Let $Y \subseteq X$ and $T : Y \mapsto P(X)$ be a multivalued operator. Then T is called a multivalued ω-left A-contraction if there exists some $A \in \mathcal{M}_{m,m}(R_+)$ converging to the zero matrix such that for each $x, y \in Y$ and each $u \in Tx$ there exists some $v \in Tx$ such that

$$\omega(u,v) \leq A\omega(x,y).$$

Theorem 23. *Let (X,d) be a complete generalized metric space, $x_0 \in X$, $r = (r_i)_{i=1}^m \in \mathbb{R}_+^m$ with $r_i > 0$ for each $i \in \overline{1,m}$ and let $T : \overline{B_\omega}(x_0, r) \to P_{cl}(X)$ be a multivalued ω-left A-contraction on $\overline{B_\omega}(x_0,r)$. We suppose that*

(i) *if $u \in \mathbb{R}_+^m$ is such that $u(I-A)^{-1} \leq (I-A)^{-1}r$, then $u \leq r$;*
(ii) *there exists $x_1 \in T(x_0)$ such that $\omega(x_0,x_1)(I-A)^{-1} \leq r$.*

Then $Fix(T) \neq \emptyset$.

Proof. Let $x_1 \in T(x_0)$ be such that (iii) holds. By using the fact that T is a multivalued left Perov contraction, we get existence of $x_2 \in T(x_1)$ such that $\omega(x_1, x_2) \leq Ar$. Furthermore,

$$\begin{aligned}\omega(x_0,x_2)(I-A)^{-1} &\leq (\omega(x_0,x_1) + \omega(x_1,x_2))(I-A)^{-1} \\ &\leq Ir + Ar \\ &\leq (I-A)^{-1}r,\end{aligned}$$

so $x_2 \in \overline{B_\omega}(x_0, r)$.

Continuing in the same manner, we may construct the sequence $(x_n) \subseteq \overline{B_\omega}(x_0, r)$ such that, for any $n \in \mathbb{N}$, $x_{n+1} \in Tx_n$, $\omega(x_0, x_n) \leq (I - A)^{-1}r$ and

$$\omega(x_n, x_{n+1})(I - A)^{-1} \leq A^n r.$$

Having in mind that, for any $n, m \in \mathbb{N}$, $m > n$,

$$\begin{aligned}\omega(x_n, x_m)(I - A)^{-1} &\leq \left(\sum_{i=n}^{m-1} \omega(x_i, x_{i+1})\right)(I - A)^{-1} \\ &\leq \sum_{i=n}^{m-1} A^i r \\ &\leq A^n (I - A)^{-1} r,\end{aligned}$$

we may conclude that (x_n) is a ω-Cauchy sequence in $\overline{B_\omega}(x_0, r)$ and, knowing some properties of ω-distance, it is a d-Cauchy sequence in a closed subset of a complete generalized metric space, so there exists some $x^* \in \overline{B_\omega}(x_0, r)$ such that $\lim_{n\to\infty} x_n = x^*$. Let us emphasize that the domain is d-closed due to the known correspondence between distance and ω-distance, both in standard or generalized case.

We shall prove that x^* is a fixed point of T. For any $n \in \mathbb{N}$ there exists some $u_n \in Tx^*$ such that

$$\omega(x_n, u_n) \leq A\omega(x_{n-1}, x^*) \leq \cdots \leq A^n (I - A)^{-1} r.$$

Accordingly, $\lim_{n\to\infty} u_n = x^*$ since

$$\begin{aligned}\omega(x^*, u_n) &\leq \omega(x^*, x_n) + \omega(x_n, u_n) \\ &\leq 2A^n (I - A)^{-1} r.\end{aligned}$$

Note that $u_n \in Tx^*$ further implies $x^* \in Tx^*$. However, Tx^* is a closed set so $x^* \in Tx^*$. □

Corollary 4. *Let (X, d) be a complete generalized metric space, $x_0 \in X, r = (r_i)_{i=1}^m \in \mathbb{R}_+^m$ with $r_i > 0$ for each $i \in \overline{1, m}$ and let $T : \overline{B_\omega}(x_0, r) \to P_{cl}(X)$ be a multivalued ω-left A-contraction on $\overline{B_\omega}(x_0, r)$. Then T is an MWP operator.*

Definition 20. Let $Y \subseteq X$ and $T : Y \mapsto P(X)$ be a multivalued operator. Then, T is called a multivalued ω-right A-contraction (Perov contraction) if $A \in \mathcal{M}_{m,m}(\mathbb{R}_+)$ is a matrix convergent to zero and for each $x, y \in Y$ and each $u \in Tx$ there exists $v \in Ty$ such that $\omega(u, v)^T \leq \omega(x, y)^T A$.

Since any multivalued right A-contraction is a multivalued left A^T-contraction, we may state the following results:

Theorem 24. *Let (X, d) be a complete generalized metric space, $x_0 \in X, r = (r_i)_{i=1}^m \in \mathbb{R}_+^m$ with $r_i > 0$ for each $i \in \overline{1,m}$ and let $T : \overline{B}_\omega(x_0, r) \to P_{cl}(X)$ be a multivalued ω-right A-contraction on $\overline{B}_\omega(x_0, r)$. We suppose that*

(i) *if $u \in \mathbb{R}_+^m$ is such that $(I - A)^{-1}u \le r(I - A)^{-1}$, then $u \le r$;*
(ii) *there exists $x_1 \in T(x_0)$ such that $(I - A)^{-1}\omega(x_0, x_1) \le r$.*

Then $Fix(T) \neq \emptyset$.

Proof. The proof goes analogously as in the case of the multivalued left A-contraction. □

We can state the global version of theorems regarding existence of a fixed point for a multivalued ω-right A-contraction and, resp., multivalued ω-left A-contraction.

Theorem 25. *Let (X, d) be a complete generalized metric space, $x_0 \in X, r = (r_i)_{i=1}^m \in \mathbb{R}_+^m$ with $r_i > 0$ for each $i \in \overline{1,m}$ and let $T : \overline{B}_\omega(x_0, r) \to P_{cl}(X)$ be a multivalued ω-right A-contraction on $\overline{B}_\omega(x_0, r)$. If $(I - A)^{-1}d(x_0, u) \le r$ for each $u \in T(x_0)$, then*

(i) *$\overline{B_\omega}(x_0, r)$ is invariant with respect to T;*
(ii) *T is an MWP operator on $\overline{B_\omega}(x_0, r)$.*

Proof. For some $x \in \overline{B}_\omega(x_0, r)$, let $v \in Tx$ be arbitrary. Since T is a multivalued ω-right A-contraction, there exists $u \in T(x_0)$ such that $\omega(v, u) \le \omega(x, x_0)A$ and

$$\begin{aligned}\omega(x_0, u) &\le \omega(x_0, v) + \omega(v, u)\\ &\le (I - A)r + A\omega(x, x_0)\\ &\le r.\end{aligned}$$

Hence, $\overline{B}_\omega(x_0, r)$ is invariant with respect to T. Second part is obvious due to the Theorem 24. □

Theorem 26. *If (X, d) is a complete generalized metric space and $T : X \mapsto P_{cl}(X)$ is a multivalued ω-left A-contraction, then T is a $(I - A)^{-1}$-MWP operator.*

Guran, Bota, Naseem, Mitrović, de la Sen and Radenović in Ref. [11] extended a contractive condition of Hardy–Rogers-type to generalized metric space in the sense of Perov and multivalued operator. The data dependence of the fixed point set, the well-posedness of the fixed point problem and the Ulam–Hyers stability have been also discussed.

They gave some definitions regarding Perov generalized space:

$\widetilde{D} : P(X) \times P(X) \to \mathbb{R}^m_+, \widetilde{D}(A,B) = (D_1(A,B),\ldots,D_m(A,B))^T$ for given $m \in \mathbb{N}-$ is the gap generalized functional.
$\widetilde{\rho} : P(X) \times P(X) \to \mathbb{R}^m_+ \cup \{+\infty\}, \widetilde{\rho}(A,B) = (\rho_1(A,B),\ldots,\rho_m(A,B))^T$, for given $m \in \mathbb{N}-$ the excess generalized functional.
$\widetilde{H} : P(X) \times P(X) \to \mathbb{R}^m_+ \cup \{+\infty\}, \widetilde{H}(A,B) = (H_1(A,B),\ldots,H_m(A,B)))^T$, for given $m \in \mathbb{N}$-the Pompeiu–Hausdorff generalized functional.

Obviously, D_i, ρ_i and H_i, for $i \in \{1,\ldots,m\}$ are pseudometrics. Recall that we made a decision not to make a difference between a row and column vector, so we may continue with already given notations in the Preliminaries section.

Lemma 2. *Let (X,d) be a generalized metric space in Perov's sense, $A, B \subseteq X$ and $q > 1$. Then for any $a \in A$ there exists $b \in B$ such that*

$$d(a,b) \le qH(A,B).$$

Lemma 3. *Let (X,d) be a generalized metric space in Perov's sense. Then $D(\{x\},A) = 0_{m\times 1}$ if and only if $x \in \bar{A}$.*

Lemma 4. *Let $A \in \mathcal{M}_{m,m}(\mathbb{R}_+)$ be a matrix that converges to zero. Then there exists $Q > 1$ such that for every $q \in (1,Q)$ we have that qA converges to zero.*

Let us give the definition of multivalued Hardy–Rogers-type operators on a generalized metric space in Perov's sense.

Definition 21. Let (X,d) be a generalized metric space in Perov's sense and $T : X \to P(X)$ be a given multivalued operator. If there exist $A, B, C \in \mathcal{M}_{m,m}(\mathbb{R}_+)$ such that

$$H(Tx,Ty) \le Ad(x,y) + B[D(x,Tx) + D(y,Ty)] \\ + C[D(x,Ty) + D(y,Tx)],$$

for all $x, y \in \mathbb{R}$, we say that T is a Hardy–Rogers-type operator.

Observe that the Hardy–Rogers operator on a generalized metric space in Perov's sense is an MWP operator.

Theorem 27. *Let (X, d) be a complete generalized metric space in Perov's sense, and $T : X \to P_{cl}(X)$ be a multivalued Hardy–Rogers-type operator. If there exist matrices $A, B, C \in \mathcal{M}_{m,m}(\mathbb{R}_+)$ such that*

(i) $I - q(B + C)$ *is non-singular and* $(I - q(B + C))^{-1} \in \mathcal{M}_{m,m}(\mathbb{R}_+)$, *for* $q \in (1, Q)$;
(ii) $M = (I - q(B + C))^{-1} q(A + B + C)$ *converges to* Θ.

Then T is a multivalued weakly Picard operator.

Proof. For some $x_0 \in X$ let $x_1 \in Tx_0$. If $x_1 = x_0$, x_0 is a fixed point of T. Otherwise,

$$\begin{aligned} d(x_1, x_2) &\le q(H)(Tx_0, Tx_1) \\ &\le qAd(x_0, x_1) + qB[D(x_0, TX_0) + D(x_1, Tx_1)] \\ &\quad + qC[D(x_0, Tx_1) + D(x_1, Tx_0)] \\ &\le qAd(x_0, x_1) + qB[d(x_0, x_1) + d(x_1, x_2)] + qCd(x_0, x_2) \\ &\le qAd(x_0, x_1) + qB[d(x_0, x_1) + d(x_1, x_2)] \\ &\quad + qC[d(x_0, x_1) + d(x_1, x_2) \\ &= q(A + B + C)d(x_0, x_1) + q(B + C)d(x_1, x_2), \end{aligned}$$

and

$$d(x_1, x_2) \le [I - q(B + C)]^{-1}(A + B + C)d(x_0, x_1).$$

Analogously,

$$d(x_n, x_{n+1}) \le \left[[I - q(B + C)]^{-1}(A + B + C)\right]^n d(x_0, x_1), \; n \in \mathbb{N}.$$

Denote with $M = [I - q(B + C)]^{-1}(A + B + C)$, then

$$d(x_n, x_{n+1}) \le M^n d(x_0, x_1), \; n \in \mathbb{N}.$$

Discussing on the Cauchiness of the sequence (x_n), for $m > n$, we have

$$\begin{aligned} d(x_n, x_m) &\le \sum_{i=n}^{m-1} M^i d(x_0, x_1) \\ &\le M^n \sum_{i=0}^{\infty} M^i d(x_0, x_1) \\ &= M^n (I - M)^{-1} d(x_0, x_1). \end{aligned}$$

There is some $x^* \in X$, the limit of sequence (x_n).

Since $D_i(x^*, Tx^*) \leq d(x^*, x_{n+1}) + D_i(x_{n+1}, Tx^*)$, $i = \overline{1,m}$, then

$$\begin{aligned} D(x^*, Tx^*) &\leq d(x^*, x_{n+1} + H(Tx_n, Tx^*)) \\ &\leq q(A+C)d(x_n, x^*) + qB[d(x_n, x_{n+1}) + D(x^*, Tx^*)] \\ &\quad + qCd(x^*, x_{n+1}). \end{aligned}$$

Letting $n \to \infty$, we have $D_i(x^*, Tx^*) = 0$, $i = \overline{1,m}$, and $x^* \in \overline{Tx^*}$, so $x^* \in Tx^*$. □

In this case, we may discuss the uniqueness of a fixed point.

Theorem 28. *Let (X, d) be a generalized metric space in Perov's sense and $T : X \to P_{cl}(X)$ be a multivalued Hardy–Rogers-type operator. If there exist matrices $A, B, C \in \mathcal{M}_{m,m}(\mathbb{R}_+)$ such that all the conditions of Theorem 27 are satisfied and, additionally, $I_m - q(A+2C)$ is non-singular and $[I - q(A+2C)]^{-1} \in \mathcal{M}_{m,m}(\mathbb{R}_+)$, $q \in (1, Q)$, then T has a unique fixed point x^*.*

Proof. Existence of the fixed point is guaranteed by the Theorem 27. Denote it with x^*. Assume that there is some $y \in Ty$ and estimate its distance from the already established fixed point x^*.

$$\begin{aligned} d(x^*, y) &\leq qH(Tx^*, Ty) \\ &\leq qAd(x^*, y) + qB[D(x^*, Tx^*) + D(y, Ty)] \\ &\quad + qC[D(x^*, Ty) + D(y, Tx^*)] \\ &\leq qAd(x^*, y) + 2qCd(x^*, y), \end{aligned}$$

leads to a conclusion that $y = x^*$ because of the additional request in the theorem statement saying that $I_m - q(A+2C)$ is nonsingular and $[I - q(A+2C)]^{-1} \in \mathcal{M}_{m,m}(\mathbb{R}_+)$. □

The following theorem is a consequences of Theorem 27.

Theorem 29. *Let (X, d) be a complete generalized metric space in Perov's sense, and $T : X \to P_{cl}(X)$ be a multivalued Hardy–Rogers-type operator. Suppose that all the hypothesis of Theorem 27 are fulfilled. Then the* *following statements are true:*

(i) $Fix(T) \neq \varnothing$;
(ii) *There exists a sequence* $(x_n)_{n\in\mathbb{N}} \in X$ *such that* $x_{n+1} \in T(x_n)$, *for all* $n \in \mathbb{N}$, *and converges to a fixed point of* T;
(iii) *For any* $n \in \mathbb{N}$, (x_n) *is a sequence of successive approximations defined in* (ii), $d(x_n, x^*) \leq (I - q(B+C))^{-1}[q(A+B+C)]^n d(x_0, x_1)$, *where* $x^* \in Fix(T)$.

In the case of Hardy–Rogers contractive condition, it is possible to discuss the common fixed-point problem and special cases related to the choice of matrices A, B and C.

Theorem 30. *Let* (X, d) *be a complete generalized metric space in Perov's sense and let* $T, G : X \to P_{cl}(X)$ *be two multivalued Hardy–Rogers-type operators. There exists the matrices* $A, B, C \in \mathcal{M}_{m,m}(\mathbb{R}_+)$ *such that*

(i) $I - q(B+C)$ *is non-singular and* $(I - q(B+C))^{-1} \in \mathcal{M}_{m,m}(\mathbb{R}_+)$, *for* $q \in (1, Q)$;
(ii) $I - q(A+2C)$ *is non-singular and* $[I - q(A+2C)]^{-1} \in \mathcal{M}_{m,m}(\mathbb{R}_+)$;
(iii) $M = (I - q(B+C))^{-1}q(A+B+C)$ *converges to* Θ.

Then T *and* G *have a common fixed point* $x^* \in X$ *and it is a unique common fixed point of* T *and* G.

Proof. For $x_0 \in X$, let us construct a sequence of successive approximations (x_n) in the following manner:

$$\begin{aligned} &x_{2n+1} \in Tx_{2n}, \\ &x_{2n+2} \in Gx_{2n+1}, \ n \in \mathbb{N}_0. \end{aligned}$$

Hence,

$$\begin{aligned} d(x_{2n}, x_{2n+1}) &\leq qH(Gx_{2n-1}, Tx_{2n}) \\ &\leq qAD(x_{2n-1}, x_{2n}) + qB[D(x_{2n}, Tx_{2n}) + d(x_{2n-1}, Gx_{2n-1})] \\ &\quad + qC[D(x_{2n-1}, Tx_{2n}) + D(x_{2n}, Gx_{2n-1})] \\ &\leq qAd(x_{2n-1}, x_{2n}) + qB[d(x_{2n}, x_{2n+1}) + d(x_{2n-1}, x_{2n})] \\ &\quad + qCd(x_{2n-1}, x_{2n+1}) \\ &\leq q(A+B+C)d(x_{2n-1}, x_{2n}) + q(B+C)d(x_{2n}, x_{2n+1}), \end{aligned}$$

and

$$\begin{aligned}d(x_{2n+1}, x_{2n+2}) &\leq qH(Gx_{2n}, Tx_{2n+1})\\&\leq qAD(x_{2n}, x_{2n+1}) + qB[D(x_{2n+1}, Gx_{2n+1}) + d(x_{2n}, Tx_{2n})]\\&\quad + qC[D(x_{2n}, Gx_{2n+1}) + D(x_{2n+1}, Tx_{2n})]\\&\leq qAd(x_{2n}, x_{2n+1}) + qB[d(x_{2n+1}, x_{2n+2}) + d(x_{2n}, x_{2n+1})]\\&\quad + qCd(x_{2n}, x_{2n+2})\\&\leq q(A + B + C)d(x_{2n}, x_{2n+1}) + q(B + C)d(x_{2n+1}, x_{2n+2}),\end{aligned}$$

imply that, for any $n \in \mathbb{N}$,

$$d(x_n, x_{n+1}) \leq (I - q(B + C))^{-1} q(A + B + C) d(x_{n-1}, x_n).$$

If we denote $(I - q(B + C))^{-1} q(A + B + C)$ with M, it follows that

$$d(x_n, x_{n+1}) \leq M^n d(x_0, x_1), \ \ n \in \mathbb{N}.$$

If $n, m \in \mathbb{N}$ and $m > n$, then

$$d(x_n, x_m) \leq< M^n (I - M)^{-1},$$

and $d(x_n, x_m) \to 0_m$ as $n, m \to \infty$ because the matrix M is converging to zero. Assume $\lim_{n\to\infty} x_n = x^* \in X$, then

$$\begin{aligned}D(x^*, Tx^*) &\leq qD(Tx^*, x_{2n+2}) + d(x_{2n+2}, x^*)\\&\leq H(Tx^*, Gx_{2n+1}) + d(x_{2n+2}, x^*)\\&\leq (I - q(B + C))^{-1} [qAd(x^*, x_{2n+1}) + qBd(x_{2n+1,x_{2n+2}})\\&\quad + qCd(x^*, x_{2n+2}) + qCd(x_{2n+1}, x^*) + d(x_{2n+2}, x^*)].\end{aligned}$$

As $n \to \infty$, we obtain $D(Tx^*, x^*) = 0_m$ and $D_i(Tx^*, x^*) = 0$, $i = \overline{1, m}$. Moreover, $x^* \in \overline{Tx^*}$ and x^* is a fixed point of T.

In a similar manner, estimating x_{2n+1} instead of x_{2n+2}, we get $x^* \in Gx^*$, so x^* is a common fixed point for T and G. □

Same authors considered the extension of Ulam–Hyers stability and well-posedness of fixed point inclusions for the case of multivalued operators on generalized metric space of Perov.

Recall that the stability problem has arisen from the famous question of Ulam posed in 1940: "Under what conditions does there exist an additive mapping near an approximately additive mapping?". It was further extended by Hyers in 1941 when he introduced that problem to the Banach space and partly extended by Rassias in 1978 ([49–51], etc.). This concept is often used in metric fixed point and has already got some generalizations like in cone metric space (and consequently in the case of a generalized metric space in a sense of Perov). We present the definition of Ulam–Hyers (Ulam–Hyers-Rassias) stability for a multivalued operator on a generalized metric space in a sense of Perov.

Definition 22. Let (X, d) be a generalized metric space in let Perov's sense and $T : X \to P(X)$ be an operator. The fixed point inclusion

$$x \in Tx \tag{19}$$

is Ulam–Hyers stable if there exists a real positive matrix $N \in \mathcal{M}_{m,m}(\mathbb{R}+)$ such that for each $\varepsilon > 0$ and each solution y^* of the equation

$$D(y, Ty) \leq \varepsilon I_{m\times 1}$$

there exists a solution x^* of Equation (19) fulfilling

$$d\left(y^*, x^*\right) \leq N\varepsilon I_{m\times 1}.$$

Note also that well-posedness problem means determining the fulfillment of an inequality:

$$D\left(x_n, T\left(x_n\right)\right) \to 0_m,\ n \to \infty, \implies x_n \to x^*,\ n \to \infty,$$

if (x_n) is a sequence of successive approximations and $x^* \in Fix(T)$.

Theorem 31. *Let (X, d) be a generalized metric space in Perov's sense and $T : X \to P_{cl}(X)$ be a multivalued Hardy–Rogers-type operator defined in (21). Then, for every non-singular matrix $I - q(A + B + 2C)$ such that $N = [I - q(A + B + 2C)]^{-1} \in \mathcal{M}_{m,m}(\mathbb{R}_+)$, for $q \in (1, Q)$, the fixed point equation (19) is Ulam–Hyers stable.*

Theorem 32. *Let (X, d) be a generalized metric space in Perov's sense and $T : X \to P_{cl}(X)$ be a multivalued Hardy–Rogers-type operator defined in Definition (21). Then, for every non-singular matrix $1 - q(A+2C)$ with $q \in (1, Q)$, such that the matrix $N = [I-q(A+2C)]^{-1}q(I+B+C) \in \mathcal{M}_{m,m}(\mathbb{R}_+)$ is a matrix convergent to Θ, and for every matrix $A, B, C \in \mathcal{M}_{m,m}(\mathbb{R}_+)$ the fixed-point equation (19) is well-posed.*

Theorem 33. *Let (X, d) be a generalized metric space in Perov's sense and $T, G : X \to P_{cl}(X)$ be a multivalued Hardy–Rogers-type operator defined in Definition (21). Then, if there exists a non-singular matrix $I - q(A + 2C)$ such that $(I - q(A + 2C))^{-1} \in \mathcal{M}_{m,m}(\mathbb{R}_+)$, with $q \in (1, Q)$, for every matrix $A, B, C \in \mathcal{M}_{m,m}(\mathbb{R}_+)$, then the fixed-point problem of T and G is well-posed.*

The following result studies the existence of a common fixed point in a case of multivalued operators of Perov type.

Theorem 34. *Let (X, d) be a generalized metric space in Perov's sense and $T_1, T_2 : X \to P_{cl}(X)$ be multivalued operators which satisfy the following conditions:*

(i) *For $A, B, C, M \in \mathcal{M}_{m,m}(\mathbb{R}_+)$ with $M = [I - q(B+C)]^{-1} q(A+B+C)$ a matrix convergent to Θ such that, for every $x, y \in X$ with $i \in \{1, 2\}$ and $q \in (1, Q)$, we have*

$$H(T_i x, T_i y) \le qAd(x, y) + qB[D(x, T_i x) + D(y, T_i y)] + qC[d(x, T_i y) + d(y, T_i x)];$$

(ii) *There exists $\eta > 0$ such that $H(T_1 x, T_2 x) \le (I - M)^{-1} \eta I_m$, for all $x \in X$.*

Then for $x_1^* \in T_1 x_1^*$ there exist $x_2^* \in T_2 x_2^*$ such that $d(x_1^*, x_2^*) \le (I - M)^{-1} \eta I_m$; (resp., for $x_2^* \in T_2 x_2^*$ there exist $x_1^* \in T_1 x_1^*$ such that $d(x_2^*, x_1^*) \le (I - M)^{-1} \eta I_m)$.

Ali and Kim in Ref. [52] extended the definition of generalized metric space in a sense of Czerwik Ref. [53] and gave the proof of some fixed-point theorems for multivalued Perov-type contraction in a new setting.

Definition 23. A mapping $d : X \times X \to \mathbb{R}^m$ is called a Czerwik vector-valued metric on X if there exists a matrix $S \in \mathcal{M}_{m,m}(R_+)$ such that $S = sI_m$, $s \ge 1$ and for each $x, y, z \in X$, the following conditions are satisfied:

(d_1) $d(x, y) \ge 0$ and $d(x, y) = 0 \Leftrightarrow x = y$;
(d_2) $d(x, y) = d(y, x)$;
(d_3) $d(x, y) \le S[d(x, z) + d(z, y)]$.

Then, a non-empty set X equipped with Czerwik vector-valued metric d is called Czerwik generalized metric space, denoted by (X, d, S).

Throughout this section, (X, d, S) is a Czerwik generalized metric space and $G = (V, E)$ is a directed graph such that the set V of its vertices coincides with X and the set E of its edges contains all loops, that is, $E \supseteq \{(x, x) \mid x \in V\}$.

Theorem 35. *Let (X, d, S) be a complete Czerwik generalized metric space endowed with the graph G. Let $T : X \to P_{cl}(X)$ be a multivalued mapping such that for each $(x, y) \in E$ and $u \in Tx$, there exists $v \in Ty$ satisfying the following inequality:*

$$d(u, v) \leq Ad(x, y) + Bd(y, u),$$

where $A, B \in M_{m,m}(\mathbb{R}_+)$. Further, assume that the following conditions hold:

(i) *The matrix SA converges to the zero matrix;*
(ii) *There exist $x_0 \in X$ and $x_1 \in Tx_0$ such that $(x_0, x_1) \in E$;*
(iii) *For each $u \in Tx$ and $v \in Ty$ with $d(u, v) \leq Ad(x, y)$, we have $(u, v) \in E$ whenever $(x, y) \in E$;*
(iv) *For each sequence (x_n) in X such that $x_n \to x$ and $(x_n, x_{n+1}) \in E$ for all $n \in \mathbb{N}$, we have $(x_n, x) \in E$ for all $n \in \mathbb{N}$.*

Then T has a fixed point.

Proof. As previously done several times, for x_0 and x_1 determined by (ii), we define a sequence of successive approximations such that $x_{n+1} \in Tx_n$ and $d(x_n, x_{n+1}) \leq A^n d(x_0, x_1)$, $n \in \mathbb{N}$. Indeed,

$$\begin{aligned} d(x_1, x_2) &\leq Ad(x_0, x_1) + Bd(x_1, x_1) \\ &\leq Ad(x_0, x_1). \end{aligned}$$

Assume that $d(x_n, x_{n+1}) \leq A^n d(x_0, x_1)$ and observe

$$\begin{aligned} d(x_{n+1}, x_{n+2}) &\leq Ad(x_n, x_{n+1}) + Bd(x_{n+1}, x_{n+1}) \\ &\leq A^{n+1} d(x_0, x_1). \end{aligned}$$

Using the principle of mathematical induction, we get that the inequality holds for any $n \in \mathbb{N}$, and letting $m > n$

$$\begin{aligned}
d(x_n, x_m) &\leq \sum_{i=n}^{m-1} S^i d(x_i, x_{i+1}) \\
&\leq \sum_{i=n}^{m-1} S^i A^i d(x_0, x_1) \\
&\leq S^n A^n \sum_{i=0}^{\infty} S^i A^i d(x_0, x_1) \\
&\leq S^n A^n (I - SA)^{-1}.
\end{aligned}$$

As we have assumed that SA converges to zero and we know that every matrix commutes with a diagonal matrix, it follows that (x_n) is a Cauchy sequence, thus convergent. If $\lim_{n\to\infty} x_n = x^*$, then $(x_n, x^*) \in E$ and there exists $u \in Tx^*$ satisfying

$$d(x_n, u) \leq Ad(x_{n-1}, x^*) + Bd(x^*, x_n),$$

and

$$\begin{aligned}
d(x^*, u) &\leq Sd(x^*, x_n) + Sd(x_n, u) \\
&\leq Sd(x^*, x_{n+1}) + SAd(x_{n-1}, x^*) + SBd(x^*, x_n)
\end{aligned}$$

yields, as $n \to \infty$, to $u = x^*$ and x^* is a fixed point of T. □

As a consequence, we have a single-valued case.

Corollary 5. *Let (X, d, S) be a complete Czerwik generalized metric space with the graph G. Let $T : X \to X$ be a mapping such that for each $(x, y) \in E$ we have*

$$d(Tx, Ty) \leq Ad(x, y) + Bd(y, Tx),$$

where $A, B \in M_{m,m}(\mathbb{R}_+)$. Further, assume that the following conditions hold:

(i) *The matrix SA converges to the zero matrix;*
(ii) *There exists $x_0 \in X$ such that $(x_0, Tx_0) \in E$;*
(iii) *For each $(x, y) \in E$, we have $(Tx, Ty) \in E$, provided $d(Tx, Ty) \leq Ad(x, y)$;*
(iv) *For each sequence (x_n) in X such that $x_n \to x$ and $(x_n, x_{n+1}) \in E$ for all $n \in \mathbb{N}$, we have $(x_n, x) \in E$ for all $n \in \mathbb{N}$.*

Then T has a fixed point.

By considering the graph $G = (V, E)$ as $V = X$ and $E = X \times X$, Corollary 5 reduces to the following result. Once again we can discuss on a case when a set is equipped with two metrics, in this case both Czerwik vector-valued metrics are dependant on the same matrix S.

Corollary 6. *Let (X, d, S) be a complete Czerwik generalized metric space. Let $T : X \to X$ be a mapping such that for each $x, y \in X$ we have*

$$d(Tx, Ty) \leq Ad(x, y) + Bd(y, Tx),$$

where $A, B \in M_{m,m}(\mathbb{R}_+)$. Also assume that the matrix SA converges to the zero matrix. Then T has a fixed point.

Theorem 36. *Let (X, d, S) be a complete Czerwik generalized metric space with the graph G and matrix S and ρ be another Czerwik generalized metric on X dependant on the same matrix S. Let $T : X \to P_{cl}(X)$ be a multivalued mapping such that for each $(x, y) \in E$ and $u \in Tx$ there exists $v \in Ty$ satisfying the following inequality:*

$$\rho(u, v) \leq A\rho(x, y) + B\rho(y, u)$$

where $A, B \in M_{n,n}(\mathbb{R}_+)$. Further, assume that the following conditions hold:

(i) *The matrix SA converges to the zero matrix;*
(ii) *There exist $x_0 \in X$ and $x_1 \in Tx_0$ such that $(x_0, x_1) \in E$;*
(iii) *For each $(x, y) \in E$, we have $(u, v) \in E$ provided $\rho(u, v) \leq A\rho(x, y)$, where $u \in Tx$ and $v \in Ty$;*
(iv) *There exists $C \in M_{m,m}(\mathbb{R}_+)$ such that $d(x, y) \leq C\rho(x, y)$, whenever, there exists a path between x and y, that is, we have a sequence $\{x_i\}_{i=0}^{p}$ such that $(x_i, x_{i+1}) \in E$ for each $i \in \overline{0, p-1}$ with $x_0 = x$ and $x_p = y$;*
(v) *$G(T) = \{(x, y) \in X \times X : y \in Tx\}$ is G-closed with respect to d, that is, if a sequence (x_n) is such that if $(x_n, x_{n+1}) \in E, (x_n, x_{n+1}) \in G(T)$; and $x_n \to x^*$, then $(x^*, x^*) \in \mathrm{Graph}(T)$.*

Then T has a fixed point.

Proof. As discussed in the case of one metric, we have a successive sequence (x_n) starting from x_0 and $x_1 \in Tx_0$ whose existence is determined by (ii) such that $x_{n+1} \in Tx_n$ and $\rho(x_n, x_{n+1}) \leq A^n\rho(x_0, x_1)$, $n \in \mathbb{N}$. It is easy to show that (x_n) is a Cauchy sequence in (X, ρ, S) by estimating

$$\rho(x_n, x_m) \leq (SA)^n(I - SA)^{-1}\rho(x_0, x_1),$$

and, moreover, in (X, d, S) since there exists a path between x_n and x_m for any $n, m \in \mathbb{N}$, $m > n$. Thus,

$$\begin{aligned} d(x_n, x_m) &\leq C\rho(x_n, x_m) \\ &\leq C(SA)^n(I - SA)^{-1}\rho(x_0, x_1). \end{aligned}$$

Completeness of (X, d, S) evidences the existence of a limit $x^* \in X$ and closedness of $G(T)$ implies that $x^* \in Tx^*$ is a fixed point of T. □

3. Conclusion

Perov's result on generalized metric space has shown a significant impact on fixed point theory and its applications. It was natural to extend these results to a case of multivalued operators and consider areas of application and its benefits. As presented in this chapter, there are many theoretical results on this topic, even including generalizations of Czerwik space and ω-distance. Several applications have been studied, as in the case of multivalued inclusion systems and differential equations. The question that arises is can we show, as in the case of multivalued operators, that Perov-type theorems give better estimations, faster convergence and have a wider area of applicability.

Acknowledgment

The authors wish to thank the referees for their careful reading of the manuscript and valuable suggestions. Supported by Grant No. 174025 of the Ministry of Education, Science and Technological Development of the Republic of Serbia.

Competing Interest

The authors declare that no competing interests exist.

References

[1] A.I. Perov, *On Cauchy problem for a system of ordinary differential equations* (in Russian), *Priblizhen. Metody Reshen. Difer. Uravn.* **2**, 115–134, (1964).

[2] R. Precup, *Methods in Nonlinear Integral Equations* (Springer, Netherlands, 2002).

[3] A.I. Perov and A.V. Kibenko, On a certain general method for investigation of boundary value problems, *Izv. Akad. Nauk SSSR Ser. Mat.* **30**, 249–264, (1966) (Russian).

[4] S. Czerwik, Generalization of Edelstein's fixed point theorem, *Demonstratio Mathematica* **9**(ii), 281–285, (1976).

[5] M. Zima, A certain fixed point theorem and its applications to integral-functional equations, *Bull. Austral. Math. Soc.* **46**, 179–186, (1992).

[6] G. Petruşel, Cyclic representations and periodic points, *Stud. Univ. Babes-Bolyai Math.* **50**, 107–112, (2005).

[7] A.D. Filip and A. Petruşel, Fixed point theorems on spaces endowed with vector-valued metrics, *Fixed Point Theory Appl.* **2010**, 281381, (2010)

[8] M. Abbas, T. Nazir, and V. Rakočević, Common fixed points results of multivalued Perov type contractions on cone metric spaces with a directed graph, *Bull. Belg. Math. Soc. Simon Stevin* **25**(iii), 331–354, (2018).

[9] Y. Feng and S. Liu, Fixed point theorems for multi-valued contractive mappings and multi-valued Caristi type mappings, *J. Math. Anal. Appl.* **317**, 103–112, (2006).

[10] L. Guran, Multivalued Perov-type theorems in generalized metric spaces, *Surv. Math. Appl.* **4**, 89–97, (2009).

[11] L. Guran, M-F. Bota, A. Naseem, Z.D. Mitrović, M. de la Sen, and S. Radenović, On some new multivalued results in the metric spaces of Perov's type, *Mathematics*, **8**(iii), 438, (2020).

[12] I.A. Rus, A. Petruşel, and A. Sıntămărian, Data dependence of the fixed point set of some multivalued weakly Picard operators, *Nonlinear Analy.: Theory, Methods Appl.* **52**(8), 1947–1959, (2003).

[13] N. Jurja, A Perov-type fixed point theorem in generalized ordered metric spaces, *Creative Math. Inf.* **17**, 137–140, (2008).

[14] M. Abbas, V. Rakočević, and A. Hussain, Proximal cyclic contraction of Perov type on regular cone metric spaces, *J. Adv. Math. Stud.* **9**, 65–71, (2016).

[15] M. Abbas, V. Rakočević, and A. Iqbal, Coincidence and common fixed points of Perov type generalized Ćirić–contraction mappings, *Mediterr. J. Math.* **13**, 3537–3555, (2016).

[16] M. Cvetković, *Fixed point theorems of Perov type*, PhD thesis, University of Niš, Niš, Serbia.

[17] M. Cvetković, On the equivalence between Perov fixed point theorem and Banach contraction principle, *Filomat* **31**(11), 3137–3146, (2017).

[18] M. Cvetković, Operatorial contractions on solid cone metric spaces, *J. Nonlinear Convex Anal.* **17**(7), 1399–1408, (2016).

[19] M. Cvetković and V. Rakočević, Quasi-contraction of Perov Type, *Appl. Math. Comput.* **235**, 712–722, (2014).

[20] M. Cvetković and V. Rakočević, Extensions of Perov theorem, *Carpathian J. Math.* **31**, 181–188, (2015).

[21] M. Cvetković and V. Rakočević, Fisher quasi-contraction of Perov type, *J. Nonlinear Convex. Anal.* **16**, 339–352, (2015).

[22] M. Cvetković and V. Rakočević, Common fixed point results for mappings of Perov type, *Math. Nach.* **288**, 1873–1890, (2015).

[23] M. Cvetković and V. Rakočević, Fixed point of mappings of Perov type for w-cone distance, *Bul. Cl. Sci. Math. Nat. Sci. Math.* **40**, 57–71, (2015).

[24] M. Cvetković, V. Rakočević, and B. E. Rhoades, Fixed point theorems for contractive mappings of Perov type, *Nonlinear Convex. Anal.* **16**, 2117–2127, (2015).

[25] Lj. Gajić, D. Ilić, and V. Rakočević, On Ćirić maps with a generalized contractive iterate at a point and Fisher's quasi-contractions in cone metric spaces, *Appl. Math. Comput.* **216**, 2240–2247, (2010).

[26] R.H. Haghi, V. Rakočević, S. Rezapour, and N. Shahzad, Best proximity results in regular cone metric spaces, *Rend. Circ. Mat. Palermo* **60**, 323–327, (2011).

[27] D. Ilić, M. Cvetković, Lj. Gajić, and V. Rakočević, Fixed points of sequence of Ćirić generalized contractions of Perov type, *Mediterr. J. Math.* **13**, 3921–3937, (2016).

[28] A. Petrusel, *Vector-valued metrics in fixed point theory*, Babes-Bolyai University, Cluj-Napoca, Faculty of Mathematics and Computer Science (2012).

[29] P.D. Proinov, A unified theory of cone metric spaces and its applications to the fixed point theory, *Fixed Point Theory Appl.* **2013**(103), (2013).

[30] Th. M. Rassias and L. Toth (Eds.), Topics in mathematical analysis and applications (Springer International Publishing, 2014).

[31] S. Banach, Sur les opérations dans les ensembles abstraits et leur applications aux équations intégrales, *Fund. Math.* **3**, 133–181, (1922).

[32] R. Precup, A. Viorel, Existence results for systems of nonlinear evolution equations, *Int. J. Pure Appl. Math.* **2**, 199–206, (2008).

[33] I.A. Rus and M-A. Şerban, Some existence results for systems of operatorial equations, *Bull. Math. Soc. Sci. Math. Roumanie* **57**, 101–108, (2014).

[34] Dj. Kurepa, Tableaux ramifies d'ensembles, Espaces pseudo-distancis, *C. R. Math. Acad. Sci. Paris* **198**, 1563–1565, (1934).

[35] L.G. Huang and X. Zhang, Cone metric spaces and fixed point theorems of contractive mappings, *J. Math. Anal. Appl.* **332**, 1468–1476, (2007).

[36] O. Kada, T. Suzuki, and W. Takahashi, Nonconvex minimization theorems and fixed point theorems in complete metric spaces, *Math. Jpon.* **44**, 381–591, (1996).

[37] J. Caristi, Fixed point theorems for mappings satisfying inwardness conditions, *Trans. Am. Math. Soc.* **215**, 241–251, (1976).

[38] I. Ekelend, Nonconvex minimization problems, *Bull. Am. Math. Soc.* **1**, 443–474, (1979).

[39] W. Takahashi, A convexity in metric space and nonexpansive mappings, *I. Kodai Math. Sem. Pep.* **22**, 142–149, (1970).

[40] Lj. Ćirić, H. Lakzian, and V. Rakočević, Fixed point theorems for w-cone distance contraction mappings in tvs-cone metric spaces, *Fixed Point Theory Appl.* **2012**(3), (2012).

[41] A. Bucur, L. Guran, and A. Petrusel, Fixed points for multivalued operators on a ser endowed with vector-valued metrics and applications, *Fixed Point Theory* **10**, 19–34, (2009).

[42] A.D. Filip, Perov's fixed point theorem for multivalued mappings in generalized Kasahara spaces, *Stud. Univ. Babes-Bolyai Math.* **56**, 19–28, (2011).

[43] R. Precup, The role of matrices that are convergent to zero in the study of semilinear operator systems, *Math. Comput. Modelling* **49**, 703–708, (2009).

[44] D. O'Regan, N. Shahzad, and R.P. Agarwal, Fixed point theory for generalized contractive maps on spaces with vector-valued metrics, *Fixed Point Theory Appl.* **6**, 143–149, (2007).

[45] M. Berinde and V. Berinde, On a general class of multivalued weakly Picard mappings, *Math. Anal. Appl.* **326**(ii), 772–782, (2007).

[46] V. Berinde and M. Pacurar, Fixed points and continuity of almost contractions, *Fixed Point Theory* **9**, 23–34, (2008).

[47] A. Petrusel and I.A. Rus, Fixed point theory for multivalued operators on a set with two metrics, *Fixed Point Theory* **8**, 97–104, (2007).

[48] L. Guran, A multivalued Perov-type theorem in generalized metric space, *Creative Math. Inf.* **17**(3), 412–419, (2008).

[49] D.H. Hyers, On the stability of the linear functional equation, *Proc. Natl. Acad. Sci. USA* **27**, 222–224, (1941).

[50] Th. M. Rassias, On the stability of the linear mapping in Banach spaces, *Proc. Amer. Math. Soc.* **72**, 297–300, (1978).

[51] S.M. Ulam, *A Collection of Mathematical Problems* (Interscience Publishers, New York, NY, USA, 1960). Reprinted as: *Problems in Modern Mathematics* (John Wiley & Sons, New York, NY, USA, 1964).

[52] M.U. Ali and J.K. Kim, An extension of vector-valued metric spaces and perov's fixed point theorem, *Nonlinear Anal. Convex Anal., RIMS Kokyuroku* **2114**, 12–20, (2019).

[53] S. Czerwik, Contraction mappings in *b*-metric spaces, *Acta Math. Inform. Univ. Ostrav.* **1**, 5–11, (1993).

[54] M. Edelstein, On fixed and periodic points under contractive mappings, *J. London Math. Soc.* **37**, 74–79, (1962).

https://doi.org/10.1142/9789811261572_0009

Chapter 9

Some Triple Integral Inequalities for Bounded Functions Defined on Three-Dimensional Bodies*

Silvestru Sever Dragomir

Mathematics, College of Engineering & Science Victoria University, PO Box 14428 Melbourne City, MC 8001, Australia

DST-NRF Centre of Excellence in the Mathematical and Statistical Sciences, School of Computer Science & Applied Mathematics, University of the Witwatersrand, Private Bag 3, Johannesburg 2050, South Africa

sever.dragomir@vu.edu.au

In this chapter, we provide some bounds for the absolute value of the quantity

$$\frac{1}{V(B)}\iiint_B f(x,y,z)\,dxdydz-\delta-\frac{1}{3V(B)}\times\iiint_B\left[(\alpha-x)\frac{\partial f(x,y,z)}{\partial x}+(\beta-y)\frac{\partial f(x,y,z)}{\partial y}+(\gamma-z)\frac{\partial f(x,y,z)}{\partial z}\right]dxdydz$$

for some choices of the parameters α, β, γ, δ and under the general assumption that B is a body in the three-dimensional space $\mathbb{R}^3$ and $f : B \to \mathbb{C}$ is differentiable on B. For this purpose, we use an identity obtained by the well-known *Gauss–Ostrogradsky* theorem for the divergence of a continuously differentiable vector field. An example for three-dimensional ball is also given.

---

*This chapter is dedicated to my granddaughters Audrey and Sienna.

1. Introduction

Recall the following inequalities of Hermite–Hadamard's type for convex functions defined on a ball $B(C, R)$, where $C = (a, b, c) \in \mathbb{R}^3$, $R > 0$ and

$$B(C,R) := \left\{ (x,y,z) \in \mathbb{R}^3 \middle| (x-a)^2 + (y-b)^2 + (z-c)^2 \le R^2 \right\}.$$

The following theorem holds [1].

Theorem 1. *Let $f : B(C,R) \to \mathbb{R}$ be a convex mapping on the ball $B(C, R)$. Then we have the inequality:*

$$f(a,b,c) \le \frac{1}{V(B(C,R))} \iiint_{B(C,R)} f(x,y,z)\, dxdydz$$
$$\le \frac{1}{\sigma(B(C,R))} \iint_{S(C,R)} f(x,y,z)\, dS, \tag{1}$$

where

$$S(C,R) := \left\{ (x,y,z) \in \mathbb{R}^3 \middle| (x-a)^2 + (y-b)^2 + (z-c)^2 = R^2 \right\}$$

and

$$V(B(C,R)) = \frac{4\pi R^3}{3}, \quad \sigma(B(C,R)) = 4\pi R^2.$$

If the assumption of convexity is dropped, then one can prove the following Ostrowski-type inequality for the center of the ball as well, see Ref. [2].

Theorem 2. *Assume that $f : B(C,R) \to \mathbb{C}$ is differentiable on $B(C,R)$. Then*

$$\left| f(a,b,c) - \frac{1}{V(B(C,R))} \iiint_{B(C,R)} f(x,y,z)\, dxdydz \right|$$
$$\le \frac{3}{8} R \left[\left\| \frac{\partial f}{\partial x} \right\|_{B(C,R),\infty} + \left\| \frac{\partial f}{\partial y} \right\|_{B(C,R),\infty} + \left\| \frac{\partial f}{\partial z} \right\|_{B(C,R),\infty} \right], \tag{2}$$

provided

$$\left\| \frac{\partial f}{\partial x} \right\|_{B(C,R),\infty} := \sup_{(x,y,z)\in B(C,R)} \left| \frac{\partial f(x,y,z)}{\partial x} \right| < \infty,$$

$$\left\| \frac{\partial f}{\partial y} \right\|_{B(C,R),\infty} := \sup_{(x,y,z)\in B(C,R)} \left| \frac{\partial f(x,y,z)}{\partial y} \right| < \infty,$$

and

$$\left\| \frac{\partial f}{\partial z} \right\|_{B(C,R),\infty} := \sup_{(x,y,z)\in B(C,R)} \left| \frac{\partial f(x,y,z)}{\partial y} \right| < \infty.$$

This fact can be furthermore generalized to the following Ostrowski-type inequality for any point in a convex body $B \subset \mathbb{R}^3$, see Ref. [2].

Theorem 3. *Assume that $f : B \to \mathbb{C}$ is differentiable on the convex body B and $(u, v, w) \in B$. If $V(B)$ is the volume of B, then*

$$\begin{aligned}
&\left| f(u,v,w) - \frac{1}{V(B)} \iiint_B f(x,y,z)\,dxdydz \right| \\
&\quad \le \frac{1}{V(B)} \iiint_B |x-u| \\
&\qquad \times \left(\int_0^1 \left| \frac{\partial f}{\partial x}[t(x,y,z)+(1-t)(u,v,w)] \right| dt \right) dxdydz \\
&\qquad + \frac{1}{V(B)} \iiint_B |y-v| \\
&\qquad \times \left(\int_0^1 \left| \frac{\partial f}{\partial y}[t(x,y,z)+(1-t)(u,v,w)] \right| dt \right) dxdydz \\
&\qquad + \frac{1}{V(B)} \iiint_B |z-w| \\
&\qquad \times \left(\int_0^1 \left| \frac{\partial f}{\partial y}[t(x,y,z)+(1-t)(u,v,w)] \right| dt \right) dxdydz \\
&\quad \le \left\| \frac{\partial f}{\partial x} \right\|_{B,\infty} \frac{1}{V(B)} \iiint_B |x-u|\,dxdydz \\
&\qquad + \left\| \frac{\partial f}{\partial y} \right\|_{B,\infty} \frac{1}{V(B)} \iiint_B |y-v|\,dxdydz \\
&\qquad + \left\| \frac{\partial f}{\partial z} \right\|_{B,\infty} \frac{1}{V(B)} \iiint_B |z-w|\,dxdydz, \qquad (3)
\end{aligned}$$

provided

$$\left\| \frac{\partial f}{\partial x} \right\|_{B,\infty}, \left\| \frac{\partial f}{\partial y} \right\|_{B,\infty}, \left\| \frac{\partial f}{\partial z} \right\|_{B,\infty} < \infty.$$

In particular,

$$\begin{aligned}
&\left| f(\overline{x_B}, \overline{y_B}, \overline{z_B}) - \frac{1}{V(B)} \iiint_B f(x,y,z)\,dxdydz \right| \\
&\quad \le \frac{1}{V(B)} \iiint_B |x-\overline{x_B}|
\end{aligned}$$

$$
\begin{aligned}
&\times \left(\int_0^1 \left| \frac{\partial f}{\partial x} \left[t(x,y,z) + (1-t)(\overline{x_B}, \overline{y_B}, \overline{z_B}) \right] \right| dt \right) dxdydz \\
&+ \frac{1}{V(B)} \iiint_B |y - \overline{y_B}| \\
&\times \left(\int_0^1 \left| \frac{\partial f}{\partial y} \left[t(x,y,z) + (1-t)(\overline{x_B}, \overline{y_B}, \overline{z_B}) \right] \right| dt \right) dxdydz \\
&+ \frac{1}{V(B)} \iiint_B |z - \overline{z_B}| \\
&\times \left(\int_0^1 \left| \frac{\partial f}{\partial y} \left[t(x,y,z) + (1-t)(\overline{x_B}, \overline{y_B}, \overline{z_B}) \right] \right| dt \right) dxdydz \\
\leq& \left\| \frac{\partial f}{\partial x} \right\|_{B,\infty} \frac{1}{V(B)} \iiint_B |x - \overline{x_B}| \, dxdydz \\
&+ \left\| \frac{\partial f}{\partial y} \right\|_{B,\infty} \frac{1}{V(B)} \iiint_B |y - \overline{y_B}| \, dxdydz \\
&+ \left\| \frac{\partial f}{\partial z} \right\|_{B,\infty} \frac{1}{V(B)} \iiint_B |z - \overline{z_B}| \, dxdydz, \qquad (4)
\end{aligned}
$$

where

$$
\overline{x_B} := \frac{1}{V(B)} \iiint_B x dxdydz, \quad \overline{y_B} := \frac{1}{V(B)} \iiint_B y dxdydz,
$$
$$
\overline{z_B} := \frac{1}{V(B)} \iiint_B z dxdydz,
$$

are the center of gravity coordinates for the convex body B.

For some Hermite–Hadamard-type inequalities for multiple integrals see Refs. [1,3–13]. For some Ostrowski-type inequalities see Refs. [2,14–26].

In this chapter, we provide some bounds for the absolute value of the quantity

$$
\begin{aligned}
&\frac{1}{V(B)} \iiint_B f(x,y,z) \, dxdydz - \delta \\
&\quad - \frac{1}{3V(B)} \iiint_B \left[(\alpha - x) \frac{\partial f(x,y,z)}{\partial x} + (\beta - y) \frac{\partial f(x,y,z)}{\partial y} \right. \\
&\qquad \left. + (\gamma - z) \frac{\partial f(x,y,z)}{\partial z} \right] dxdydz, \qquad (5)
\end{aligned}
$$

for certain choices of the parameters α, β, γ, δ and under the general assumption that B is a body in the three-dimensional space $\mathbb{R}^3$ and f :

$B \to \mathbb{C}$ is differentiable on B. For this purpose, we use an identity obtained via the well-known *Gauss–Ostrogradsky* theorem for the divergence of a continuously differentiable vector field. An example for three-dimensional balls is also given.

We need the following preparations.

2. Some Notations, Definitions and Preliminary Facts

Following Apostol [27], consider a surface described by the vector equation

$$r(u,v) = x(u,v)\vec{i} + y(u,v)\vec{j} + z(u,v)\vec{k}, \tag{6}$$

where $(u,v) \in [a,b] \times [c,d]$.

If x, y, z are differentiable on $[a,b] \times [c,d]$, we consider the two vectors

$$\frac{\partial r}{\partial u} = \frac{\partial x}{\partial u}\vec{i} + \frac{\partial y}{\partial u}\vec{j} + \frac{\partial z}{\partial u}\vec{k}$$

and

$$\frac{\partial r}{\partial v} = \frac{\partial x}{\partial v}\vec{i} + \frac{\partial y}{\partial v}\vec{j} + \frac{\partial z}{\partial v}\vec{k}.$$

The *cross-product* of these two vectors $\frac{\partial r}{\partial u} \times \frac{\partial r}{\partial v}$ will be referred to as the fundamental vector product of the representation r. Its components can be expressed as *Jacobian determinants*. In fact, we have [27, p. 420]

$$\begin{aligned}\frac{\partial r}{\partial u} \times \frac{\partial r}{\partial v} &= \begin{vmatrix} \frac{\partial y}{\partial u} & \frac{\partial z}{\partial u} \\ \frac{\partial y}{\partial v} & \frac{\partial z}{\partial v}\end{vmatrix}\vec{i} + \begin{vmatrix} \frac{\partial z}{\partial u} & \frac{\partial x}{\partial u} \\ \frac{\partial z}{\partial v} & \frac{\partial x}{\partial v}\end{vmatrix}\vec{j} + \begin{vmatrix} \frac{\partial x}{\partial u} & \frac{\partial y}{\partial u} \\ \frac{\partial x}{\partial v} & \frac{\partial y}{\partial v}\end{vmatrix}\vec{k} \\ &= \frac{\partial(y,z)}{\partial(u,v)}\vec{i} + \frac{\partial(z,x)}{\partial(u,v)}\vec{j} + \frac{\partial(x,y)}{\partial(u,v)}\vec{k}.\end{aligned} \tag{7}$$

Let $S = r(T)$ be a parametric surface described by a vector-valued function r defined on the box $T = [a,b] \times [c,d]$. The area of S denoted A_S is defined by the double integral [27, pp. 424–425]

$$\begin{aligned}A_S &= \int_a^b \int_c^d \left\| \frac{\partial r}{\partial u} \times \frac{\partial r}{\partial v} \right\| dudv \\ &= \int_a^b \int_c^d \sqrt{\left(\frac{\partial(y,z)}{\partial(u,v)}\right)^2 + \left(\frac{\partial(z,x)}{\partial(u,v)}\right)^2 + \left(\frac{\partial(x,y)}{\partial(u,v)}\right)^2} dudv.\end{aligned} \tag{8}$$

We define surface integrals in terms of a parametric representation for the surface. One can prove that under certain general conditions the value of the integral is independent of the representation.

Let $S = r(T)$ be a parametric surface described by a vector-valued differentiable function r defined on the box $T = [a, b] \times [c, d]$ and let $f : S \to \mathbb{C}$ be defined and bounded on S. The surface integral of f over S is defined by [27, p. 430]

$$\begin{aligned}\iint_S f dS &= \int_a^b \int_c^d f(x, y, z) \left\| \frac{\partial r}{\partial u} \times \frac{\partial r}{\partial v} \right\| dudv \\ &= \int_a^b \int_c^d f(x(u, v), y(u, v), z(u, v)) \\ &\quad \times \sqrt{\left(\frac{\partial (y, z)}{\partial (u, v)}\right)^2 + \left(\frac{\partial (z, x)}{\partial (u, v)}\right)^2 + \left(\frac{\partial (x, y)}{\partial (u, v)}\right)^2} dudv. \qquad (9)\end{aligned}$$

If $S = r(T)$ is a parametric surface, the fundamental vector product $N = \frac{\partial r}{\partial u} \times \frac{\partial r}{\partial v}$ is normal to S at each regular point of the surface. At each such point there are two unit normals, a unit normal n_1, which has the same direction as N, and a unit normal n_2, which has the opposite direction. Thus,

$$n_1 = \frac{N}{\|N\|} \quad \text{and} \quad n_2 = -n_1.$$

Let n be one of the two normals n_1 or n_2. Let also F be a vector field defined on S and assume that the surface integral,

$$\iint_S (F \cdot n) \, dS,$$

called the flux surface integral, exists. Here $F \cdot n$ is the dot or inner product.

We can write [27, p. 434]

$$\iint_S (F \cdot n) \, dS = \pm \int_a^b \int_c^d F(r(u, v)) \cdot \left(\frac{\partial r}{\partial u} \times \frac{\partial r}{\partial v}\right) dudv,$$

where the sign "+" is used if $n = n_1$ and the "−" sign is used if $n = n_2$.

If

$$F(x, y, z) = P(x, y, z) \overrightarrow{i} + Q(x, y, z) \overrightarrow{j} + R(x, y, z) \overrightarrow{k}$$

and

$$r(u, v) = x(u, v) \overrightarrow{i} + y(u, v) \overrightarrow{j} + z(u, v) \overrightarrow{k} \text{ where } (u, v) \in [a, b] \times [c, d],$$

then the flux surface integral for $n = n_1$ can be explicitly calculated as [27, p. 435]

$$\begin{aligned}\iint_S (F \cdot n)\, dS &= \int_a^b \int_c^d P(x(u,v), y(u,v), z(u,v)) \frac{\partial(y,z)}{\partial(u,v)} dudv \\ &\quad + \int_a^b \int_c^d Q(x(u,v), y(u,v), z(u,v)) \frac{\partial(z,x)}{\partial(u,v)} dudv \\ &\quad + \int_a^b \int_c^d R(x(u,v), y(u,v), z(u,v)) \frac{\partial(x,y)}{\partial(u,v)} dudv. \qquad (10)\end{aligned}$$

The sum of the double integrals on the right is often written more briefly as [27, p. 435]

$$\iint_S P(x,y,z)\, dy \wedge dz + \iint_S Q(x,y,z)\, dz \wedge dx + \iint_S R(x,y,z)\, dx \wedge dy.$$

Let $B \subset \mathbb{R}^3$ be a solid in three-space bounded by an orientable closed surface S, and let n be the unit outer normal to S. If F is a continuously differentiable vector field defined on B, we have the *Gauss–Ostrogradsky identity*

$$\iiint_B (\operatorname{div} F)\, dV = \iint_S (F \cdot n)\, dS. \qquad \text{(GO)}$$

If we express

$$F(x,y,z) = P(x,y,z)\overrightarrow{i} + Q(x,y,z)\overrightarrow{j} + R(x,y,z)\overrightarrow{k},$$

then (GO) can be written as

$$\begin{aligned}&\iiint_B \left(\frac{\partial P(x,y,z)}{\partial x} + \frac{\partial Q(x,y,z)}{\partial y} + \frac{\partial R(x,y,z)}{\partial z} \right) dxdydz \\ &= \iint_S P(x,y,z)\, dy \wedge dz + \iint_S Q(x,y,z)\, dz \wedge dx \\ &\quad + \iint_S R(x,y,z)\, dx \wedge dy. \qquad (11)\end{aligned}$$

By taking the real and imaginary part, we can extend the above inequality for complex-valued functions P, Q, R defined on B.

3. Some Identities of Interest

We have:

Lemma 1. *Let B be a solid in the three-dimensional space $\mathbb{R}^3$ bounded by an orientable closed surface S. If $f : B \to \mathbb{C}$ is a continuously differentiable function defined on an open set containing B, then we have the equality*

$$\begin{aligned}
&\frac{1}{V(B)} \iiint_B f(x,y,z)\, dxdydz - \delta \\
&= \frac{1}{3V(B)} \iiint_B \left[(\alpha - x) \frac{\partial f(x,y,z)}{\partial x} + (\beta - y) \frac{\partial f(x,y,z)}{\partial y} \right. \\
&\qquad \left. + (\gamma - z) \frac{\partial f(x,y,z)}{\partial z} \right] dxdydz \\
&\quad + \frac{1}{3V(B)} \left[\iint_S (x - \alpha) [f(x,y,z) - \delta]\, dy \wedge dz \right. \\
&\qquad + \iint_S (y - \beta) [f(x,y,z) - \delta]\, dz \wedge dx \\
&\qquad \left. + \iint_S (z - \gamma) [f(x,y,z) - \delta]\, dx \wedge dy \right], \qquad (12)
\end{aligned}$$

for all α, β, γ and δ complex numbers.

In particular, we have

$$\begin{aligned}
&\frac{1}{V(B)} \iiint_B f(x,y,z)\, dxdydz - \delta \\
&= \frac{1}{3V(B)} \iiint_B \left[(\overline{x_B} - x) \frac{\partial f(x,y,z)}{\partial x} + (\overline{y_B} - y) \frac{\partial f(x,y,z)}{\partial y} \right. \\
&\qquad \left. + (\overline{z_B} - z) \frac{\partial f(x,y,z)}{\partial z} \right] dxdydz \\
&\quad + \frac{1}{3V(B)} \left[\iint_S (x - \overline{x_B}) [f(x,y,z) - \delta]\, dy \wedge dz \right. \\
&\qquad + \iint_S (y - \overline{y_B}) [f(x,y,z) - \delta]\, dz \wedge dx \\
&\qquad \left. + \iint_S (z - \overline{z_B}) [f(x,y,z) - \delta]\, dx \wedge dy \right]. \qquad (13)
\end{aligned}$$

Proof. It would suffice to prove the equality (12) for $\delta = 0$ since the general case will follow by replacing f with $f - \delta$.

We have

$$\frac{\partial\left[(x-\alpha) f(x,y,z)\right]}{\partial x} = f(x,y,z) + (x-\alpha)\frac{\partial f(x,y,z)}{\partial x},$$

$$\frac{\partial\left[(y-\beta) f(x,y,z)\right]}{\partial y} = f(x,y,z) + (y-\beta)\frac{\partial f(x,y,z)}{\partial y}$$

and

$$\frac{\partial\left[(z-\gamma) f(x,y,z)\right]}{\partial z} = f(x,y,z) + (z-\gamma)\frac{\partial f(x,y,z)}{\partial z}.$$

By adding these three equalities, we get

$$\begin{aligned}\frac{\partial\left[(x-\alpha) f(x,y,z)\right]}{\partial x} &+ \frac{\partial\left[(y-\beta) f(x,y,z)\right]}{\partial y} \\ &+ \frac{\partial\left[(z-\gamma) f(x,y,z)\right]}{\partial z} = 3f(x,y,z) \\ &+ (x-\alpha)\frac{\partial f(x,y,z)}{\partial x} + (y-\beta)\frac{\partial f(x,y,z)}{\partial y} \\ &+ (z-\gamma)\frac{\partial f(x,y,z)}{\partial z}, \end{aligned} \tag{14}$$

for all $(x,y,z) \in B$.

Integrating this equality on B, we get

$$\begin{aligned}&\iiint_B \left(\frac{\partial\left[(x-\alpha) f(x,y,z)\right]}{\partial x} + \frac{\partial\left[(y-\beta) f(x,y,z)\right]}{\partial y}\right. \\ &\qquad \left. + \frac{\partial\left[(z-\gamma) f(x,y,z)\right]}{\partial z}\right) dxdydz \\ &= 3\iiint_B f(x,y,z)\, dxdydz \\ &\quad + \iiint_B \left[(x-\alpha)\frac{\partial f(x,y,z)}{\partial x} + (y-\beta)\frac{\partial f(x,y,z)}{\partial y}\right. \\ &\qquad \left. + (z-\gamma)\frac{\partial f(x,y,z)}{\partial z}\right] dxdydz. \end{aligned} \tag{15}$$

Applying the *Gauss–Ostrogradsky identity* (11) for the functions

$$P(x,y,z) = (x-\alpha) f(x,y,z), \quad Q(x,y,z) = (y-\beta) f(x,y,z)$$

and

$$R(x,y,z) = (z-\gamma) f(x,y,z)$$

we obtain

$$\begin{aligned}
&\iiint_B \left(\frac{\partial\left[(x-\alpha) f(x,y,z)\right]}{\partial x}+\frac{\partial\left[(y-\beta) f(x,y,z)\right]}{\partial y}\right.\\
&\qquad\left.+\frac{\partial\left[(z-\gamma) f(x,y,z)\right]}{\partial z}\right) dxdydz\\
&=\iint_S (x-\alpha) f(x,y,z)\, dy\wedge dz+\iint_S (y-\beta) f(x,y,z)\, dz\wedge dx\\
&\quad+\iint_S (z-\gamma) f(x,y,z)\, dx\wedge dy. \qquad (16)
\end{aligned}$$

By (15) and (16), we get

$$\begin{aligned}
&3\iiint_B f(x,y,z)\, dxdydz\\
&\quad+\iiint_B \left[(x-\alpha)\frac{\partial f(x,y,z)}{\partial x}+(y-\beta)\frac{\partial f(x,y,z)}{\partial y}\right.\\
&\qquad\left.+(z-\gamma)\frac{\partial f(x,y,z)}{\partial z}\right] dxdydz\\
&=\iint_S (x-\alpha) f(x,y,z)\, dy\wedge dz+\iint_S (y-\beta) f(x,y,z)\, dz\wedge dx\\
&\quad+\iint_S (z-\gamma) f(x,y,z)\, dx\wedge dy,
\end{aligned}$$

which is equivalent to

$$\begin{aligned}
&\iiint_B f(x,y,z)\, dxdydz\\
&\quad=\frac{1}{3}\iiint_B \left[(\alpha-x)\frac{\partial f(x,y,z)}{\partial x}+(\beta-y)\frac{\partial f(x,y,z)}{\partial y}\right.\\
&\qquad\left.+(\gamma-z)\frac{\partial f(x,y,z)}{\partial z}\right] dxdydz\\
&\quad+\frac{1}{3}\left[\iint_S (x-\alpha) f(x,y,z)\, dy\wedge dz\right.\\
&\qquad+\iint_S (y-\beta) f(x,y,z)\, dz\wedge dx\\
&\qquad\left.+\iint_S (z-\gamma) f(x,y,z)\, dx\wedge dy\right],
\end{aligned}$$

that, by division with $V(B)$, proves the claim. □

Remark 1. For a function f as in Lemma 1 above, we define the points

$$x_{B,\partial f} := \frac{\iiint_B x \frac{\partial f(x,y,z)}{\partial x} dxdydz}{\iiint_B \frac{\partial f(x,y,z)}{\partial x} dxdydz}, \; y_{B,\partial f} := \frac{\iiint_B y \frac{\partial f(x,y,z)}{\partial y} dxdydz}{\iiint_B \frac{\partial f(x,y,z)}{\partial y} dxdydz}$$

and

$$z_{B,\partial f} := \frac{\iiint_B z \frac{\partial f(x,y,z)}{\partial z} dxdydz}{\iiint_B \frac{\partial f(x,y,z)}{\partial z} dxdydz},$$

provided the denominators are not zero.

If we take $\alpha = x_{B,\partial f}$, $\beta = y_{B,\partial f}$ and $\gamma = z_{B,\partial f}$ in (12), then we get

$$\begin{aligned}
&\frac{1}{V(B)} \iiint_B f(x,y,z)\, dxdydz - \delta \\
&= \frac{1}{3V(B)} \left[\iint_S (x - x_{B,\partial f}) [f(x,y,z) - \delta]\, dy \wedge dz \right. \\
&\quad + \iint_S (y - \beta y_{B,\partial f}) [f(x,y,z) - \delta]\, dz \wedge dx \\
&\quad \left. + \iint_S (z - z_{B,\partial f}) [f(x,y,z) - \delta]\, dx \wedge dy \right], \qquad (17)
\end{aligned}$$

since, obviously,

$$\begin{aligned}
&\iiint_B \left[(x_{B,\partial f} - x) \frac{\partial f(x,y,z)}{\partial x} + (y_{B,\partial f} - y) \frac{\partial f(x,y,z)}{\partial y} \right. \\
&\quad \left. + (z_{B,\partial f} - z) \frac{\partial f(x,y,z)}{\partial z} \right] dxdydz = 0.
\end{aligned}$$

Remark 2. Let B be a solid in the three-dimensional space $\mathbb{R}^3$ bounded by an orientable closed surface S described by the vector equation

$$r(u,v) = x(u,v)\, \overrightarrow{i} + y(u,v)\, \overrightarrow{j} + z(u,v)\, \overrightarrow{k}, (u,v) \in [a,b] \times [c,d],$$

where $x(u,v)$, $y(u,v)$, $z(u,v)$ are differentiable.

From the equation (12), we get

$$\begin{aligned}
&\frac{1}{V(B)} \iiint_B f(x,y,z)\, dxdydz - \delta \\
&\quad - \frac{1}{3V(B)} \iiint_B \left[(\alpha - x) \frac{\partial f(x,y,z)}{\partial x} + (\beta - y) \frac{\partial f(x,y,z)}{\partial y} \right. \\
&\quad \left. + (\gamma - z) \frac{\partial f(x,y,z)}{\partial z} \right] dxdydz
\end{aligned}$$

$$
\begin{aligned}
&= \frac{1}{3V(B)} \left[\int_a^b \int_c^d (x(u,v) - \alpha) \left[f(x(u,v), y(u,v), z(u,v)) - \delta \right] \right. \\
&\qquad \times \frac{\partial(y,z)}{\partial(u,v)} du dv + \int_a^b \int_c^d (y(u,v) - \beta) \\
&\qquad \times \left[f(x(u,v), y(u,v), z(u,v)) - \delta \right] \frac{\partial(z,x)}{\partial(u,v)} du dv \\
&\qquad + \int_a^b \int_c^d (z(u,v) - \gamma) \left[f(x(u,v), y(u,v), z(u,v)) - \delta \right] \\
&\qquad \left. \times \frac{\partial(x,y)}{\partial(u,v)} du dv \right],
\end{aligned} \tag{18}
$$

for all α, β, γ and δ complex numbers, while from (13), we have

$$
\begin{aligned}
&\frac{1}{V(B)} \iiint_B f(x,y,z)\, dx dy dz - \delta \\
&\quad - \frac{1}{3V(B)} \iiint_B \left[(\overline{x_B} - x) \frac{\partial f(x,y,z)}{\partial x} + (\overline{y_B} - y) \frac{\partial f(x,y,z)}{\partial y} \right. \\
&\qquad \left. + (\overline{z_B} - z) \frac{\partial f(x,y,z)}{\partial z} \right] dx dy dz \\
&= \frac{1}{3V(B)} \left[\int_a^b \int_c^d (x(u,v) - \overline{x_B}) \left[f(x(u,v), y(u,v), z(u,v)) - \delta \right] \right. \\
&\qquad \times \frac{\partial(y,z)}{\partial(u,v)} du dv + \int_a^b \int_c^d (y(u,v) - \overline{y_B}) \\
&\qquad \times \left[f(x(u,v), y(u,v), z(u,v)) - \delta \right] \frac{\partial(z,x)}{\partial(u,v)} du dv \\
&\qquad + \int_a^b \int_c^d (z(u,v) - \overline{z_B}) \left[f(x(u,v), y(u,v), z(u,v)) - \delta \right] \\
&\qquad \left. \times \frac{\partial(x,y)}{\partial(u,v)} du dv \right],
\end{aligned} \tag{19}
$$

for all $\delta \in \mathbb{R}$.

From (17) we get

$$\begin{aligned}
&\frac{1}{V(B)} \iiint_B f(x,y,z)\,dxdydz - \delta \\
&= \frac{1}{3V(B)} \left[\int_a^b \int_c^d (x(u,v) - x_{B,\partial f}) \right. \\
&\qquad \times [f(x(u,v), y(u,v), z(u,v)) - \delta] \frac{\partial(y,z)}{\partial(u,v)} dudv \\
&\qquad + \int_a^b \int_c^d (y(u,v) - y_{B,\partial f}) \\
&\qquad \times [f(x(u,v), y(u,v), z(u,v)) - \delta] \frac{\partial(z,x)}{\partial(u,v)} dudv \\
&\qquad + \int_a^b \int_c^d (z(u,v) - z_{B,\partial f}) \\
&\qquad \left. \times [f(x(u,v), y(u,v), z(u,v)) - \delta] \frac{\partial(x,y)}{\partial(u,v)} dudv \right], \qquad (20)
\end{aligned}$$

for all $\delta \in \mathbb{R}$.

4. Inequalities for Bounded Functions

Let B be a solid in the three-dimensional space $\mathbb{R}^3$ bounded by an orientable closed surface S. Now, for ϕ, $\Phi \in \mathbb{C}$, define the sets of complex-valued functions

$$\begin{aligned}
\bar{U}_S(\phi,\Phi) := \Big\{ f : S \to \mathbb{C} \,|\, &\operatorname{Re}\left[(\Phi - f(x,y,z)) \left(\overline{f(x,y,z)} - \overline{\phi} \right) \right] \\
&\geq 0 \ \text{ for each } \ (x,y,z) \in S \Big\}
\end{aligned}$$

and

$$\begin{aligned}
\bar{\Delta}_S(\phi,\Phi) := \Big\{ f : S \to \mathbb{C} \,\Big|\, &\left| f(x,y,z) - \frac{\phi + \Phi}{2} \right| \\
&\leq \frac{1}{2} |\Phi - \phi| \ \text{for each } (x,y,z) \in S \Big\}.
\end{aligned}$$

The following representation result may be stated:

Proposition 1. *For any ϕ, $\Phi \in \mathbb{C}$, $\phi \neq \Phi$, we have that $\bar{U}_S(\phi,\Phi)$ and $\bar{\Delta}_S(\phi,\Phi)$ are non-empty, convex and closed sets and*

$$\bar{U}_S(\phi,\Phi) = \bar{\Delta}_S(\phi,\Phi). \tag{21}$$

Proof. We observe that for any $w \in \mathbb{C}$ we have the equivalence

$$\left|w - \frac{\phi+\Phi}{2}\right| \leq \frac{1}{2}|\Phi-\phi|$$

if and only if

$$\operatorname{Re}\left[(\Phi - w)(\overline{w} - \overline{\phi})\right] \geq 0.$$

This follows by the equality

$$\frac{1}{4}|\Phi-\phi|^2 - \left|w - \frac{\phi+\Phi}{2}\right|^2 = \operatorname{Re}\left[(\Phi - w)(\overline{w} - \overline{\phi})\right]$$

that holds for any $w \in \mathbb{C}$.

The equality (21) is thus a simple consequence of this fact. □

On making use of the complex numbers field properties, we can also state that:

Corollary 1. *For any ϕ, $\Phi \in \mathbb{C}$, $\phi \neq \Phi$, we have that*

$$\begin{aligned}\bar{U}_S(\phi,\Phi) = \{f : S \to \mathbb{C} \mid\ &(\operatorname{Re}\Phi - \operatorname{Re} f(x,y,z))(\operatorname{Re} f(x,y,z) - \operatorname{Re}\phi)\\ &+ (\operatorname{Im}\Phi - \operatorname{Im} f(x,y,z))(\operatorname{Im} f(x,y,z)\\ &- \operatorname{Im}\phi) \geq 0 \text{ for each } (x,y,z) \in S\}.\end{aligned} \tag{22}$$

Now, if we assume that $\operatorname{Re}(\Phi) \geq \operatorname{Re}(\phi)$ and $\operatorname{Im}(\Phi) \geq \operatorname{Im}(\phi)$, then we can define the following set of functions as well:

$$\begin{aligned}\bar{S}_S(\phi,\Phi) := \{f : S \to \mathbb{C} \mid\ &\operatorname{Re}(\Phi) \geq \operatorname{Re} f(x,y,z) \geq \operatorname{Re}(\phi)\\ &\text{and } \operatorname{Im}(\Phi) \geq \operatorname{Im} f(x,y,z) \geq \operatorname{Im}(\phi) \text{ for each } (x,y,z) \in S\}.\end{aligned} \tag{23}$$

One can easily observe that $\bar{S}_S(\phi,\Phi)$ is closed, convex and

$$\emptyset \neq \bar{S}_S(\phi,\Phi) \subseteq \bar{U}_S(\phi,\Phi). \tag{24}$$

Theorem 4. *Let B be a solid in the three-dimensional space $\mathbb{R}^3$ bounded by an orientable closed surface S described by the vector equation*

$$r(u,v) = x(u,v)\,\vec{i} + y(u,v)\,\vec{j} + z(u,v)\,\vec{k},\ \ (u,v) \in [a,b]\times[c,d],$$

where $x(u,v)$, $y(u,v)$, $z(u,v)$ *are differentiable. If* $f \in \bar{\Delta}_S(\phi,\Phi)$ *for some* ϕ, $\Phi \in \mathbb{C}$, $\phi \neq \Phi$, *then*

$$\begin{aligned}
&\left|\frac{1}{V(B)}\iiint_B f(x,y,z)\,dxdydz - \frac{\phi+\Phi}{2}\right.\\
&\qquad - \frac{1}{3V(B)}\iiint_B \left[(\alpha - x)\frac{\partial f(x,y,z)}{\partial x} + (\beta - y)\frac{\partial f(x,y,z)}{\partial y}\right.\\
&\qquad\qquad \left.\left. + (\gamma - z)\frac{\partial f(x,y,z)}{\partial z}\right] dxdydz\right|\\
&\quad \le \frac{1}{6V(B)}|\Phi - \phi|\, M(S,\alpha,\beta,\gamma), \qquad (25)
\end{aligned}$$

where

$$\begin{aligned}
M(S,\alpha,\beta,\gamma) := &\int_a^b\int_c^d |x(u,v) - \alpha|\left|\frac{\partial(y,z)}{\partial(u,v)}\right| dudv\\
&+ \int_a^b\int_c^d |y(u,v) - \beta|\left|\frac{\partial(z,x)}{\partial(u,v)}\right| dudv\\
&+ \int_a^b\int_c^d |z(u,v) - \gamma|\left|\frac{\partial(x,y)}{\partial(u,v)}\right| dudv.
\end{aligned}$$

Moreover, if we put $\Box := [a,b]\times[c,d]$, *then we have the bounds*

$$M(S,\alpha,\beta,\gamma) \le \begin{cases}
\left\|\frac{\partial(y,z)}{\partial(\cdot,\cdot)}\right\|_{\Box,\infty}\|x-\alpha\|_{\Box,1} + \left\|\frac{\partial(z,x)}{\partial(\cdot,\cdot)}\right\|_{\Box,\infty}\|y-\beta\|_{\Box,1}\\
\quad + \left\|\frac{\partial(x,y)}{\partial(\cdot,\cdot)}\right\|_{\Box,\infty}\|z-\gamma\|_{\Box,1},\\
\left\|\frac{\partial(y,z)}{\partial(\cdot,\cdot)}\right\|_{\Box,p}\|x-\alpha\|_{\Box,q} + \left\|\frac{\partial(z,x)}{\partial(\cdot,\cdot)}\right\|_{\Box,p}\|y-\beta\|_{\Box,q}\\
\quad + \left\|\frac{\partial(x,y)}{\partial(\cdot,\cdot)}\right\|_{\Box,p}\|z-\gamma\|_{\Box,q},\\
\left\|\frac{\partial(y,z)}{\partial(\cdot,\cdot)}\right\|_{\Box,1}\|x-\alpha\|_{\Box,\infty} + \left\|\frac{\partial(z,x)}{\partial(\cdot,\cdot)}\right\|_{\Box,p}\|y-\beta\|_{\Box,\infty}\\
\quad + \left\|\frac{\partial(x,y)}{\partial(\cdot,\cdot)}\right\|_{\Box,1}\|z-\gamma\|_{\Box,\infty}
\end{cases}. \qquad (26)$$

Proof. From (18) we have for $\delta = \frac{\phi+\Phi}{2}$ that

$$\begin{aligned}
&\left|\frac{1}{V(B)}\iiint_B f(x,y,z)\,dxdydz - \frac{\phi+\Phi}{2}\right. \\
&\quad - \frac{1}{3V(B)}\iiint_B \left[(\alpha - x)\frac{\partial f(x,y,z)}{\partial x} + (\beta - y)\frac{\partial f(x,y,z)}{\partial y}\right. \\
&\qquad\qquad \left.\left. + (\gamma - z)\frac{\partial f(x,y,z)}{\partial z}\right] dxdydz\right| \\
&\le \frac{1}{3V(B)}\left[\int_a^b\int_c^d |x(u,v)-\alpha|\left|f(x(u,v),y(u,v),z(u,v)) - \frac{\phi+\Phi}{2}\right|\right. \\
&\qquad \times \left|\frac{\partial(y,z)}{\partial(u,v)}\right| dudv + \int_a^b\int_c^d |y(u,v)-\beta|\,|f(x(u,v),y(u,v),z(u,v)) \\
&\qquad - \frac{\phi+\Phi}{2}\bigg|\left|\frac{\partial(z,x)}{\partial(u,v)}\right| dudv + \int_a^b\int_c^d |z(u,v)-\gamma| \\
&\qquad \times \left.\left|f(x(u,v),y(u,v),z(u,v)) - \frac{\phi+\Phi}{2}\right|\left|\frac{\partial(x,y)}{\partial(u,v)}\right| dudv\right] \\
&\le \frac{1}{6V(B)}|\Phi-\phi|\left[\int_a^b\int_c^d |x(u,v)-\alpha|\left|\frac{\partial(y,z)}{\partial(u,v)}\right| dudv\right. \\
&\qquad\qquad + \int_a^b\int_c^d |y(u,v)-\beta|\left|\frac{\partial(z,x)}{\partial(u,v)}\right| dudv \\
&\qquad\qquad \left.+ \int_a^b\int_c^d |z(u,v)-\gamma|\left|\frac{\partial(x,y)}{\partial(u,v)}\right| dudv\right] \\
&= \frac{1}{6V(B)}|\Phi-\phi|\,M(S,\alpha,\beta,\gamma)
\end{aligned}$$

which proves the inequality (25).

The bounds in (26) follow by Hölder's inequalities, for which we only mention

$$\begin{aligned}
&\int_a^b\int_c^d |x(u,v)-\alpha|\left|\frac{\partial(y,z)}{\partial(u,v)}\right| dudv \\
&\quad \le \begin{cases}
\sup_{(u,v)\in[a,b]\times[c,d]}\left|\frac{\partial(y,z)}{\partial(u,v)}\right| \int_a^b\int_c^d |x(u,v)-\alpha|\,dudv, \\
\left(\int_a^b\int_c^d |x(u,v)-\alpha|^q\,dudv\right)^{1/q}\left(\int_a^b\int_c^d \left|\frac{\partial(y,z)}{\partial(u,v)}\right|^p dudv\right)^{1/p} \\
\text{if } p,\ q>1 \text{ with } \frac{1}{p}+\frac{1}{q}=1, \\
\sup_{(u,v)\in[a,b]\times[c,d]}|x(u,v)-\alpha| \int_a^b\int_c^d \left|\frac{\partial(y,z)}{\partial(u,v)}\right| dudv
\end{cases}.
\end{aligned}$$

□

Corollary 2. *With the assumptions of Theorem 4, we have the inequality*

$$\begin{aligned}&\left|\frac{1}{V(B)}\iiint_B f(x,y,z)\,dxdydz-\frac{\phi+\Phi}{2}\right.\\&\quad-\frac{1}{3V(B)}\iiint_B\left[(\alpha-x)\frac{\partial f(x,y,z)}{\partial x}+(\beta-y)\frac{\partial f(x,y,z)}{\partial y}\right.\\&\qquad\left.\left.+(\gamma-z)\frac{\partial f(x,y,z)}{\partial z}\right]dxdydz\right|\\&\le\frac{1}{6V(B)}\left|\Phi-\phi\right|\iint_S\left(|x-\alpha|^2+|y-\beta|^2+|z-\gamma|^2\right)^{1/2}dS\\&\le\frac{A_S}{6V(B)}\left|\Phi-\phi\right|\sup_{(x,y,z)\in S}\left(|x-\alpha|^2+|y-\beta|^2+|z-\gamma|^2\right)^{1/2}.\end{aligned}\tag{27}$$

Proof. Using the discrete Cauchy–Bunyakovsky–Schwarz inequality, we have

$$\begin{aligned}&|x(u,v)-\alpha|\left|\frac{\partial(y,z)}{\partial(u,v)}\right|+|y(u,v)-\beta|\left|\frac{\partial(z,x)}{\partial(u,v)}\right|+|z(u,v)-\gamma|\left|\frac{\partial(x,y)}{\partial(u,v)}\right|\\&\le\left(|x(u,v)-\alpha|^2+|y(u,v)-\beta|^2+|z(u,v)-\gamma|^2\right)^{1/2}\\&\quad\times\left(\left|\frac{\partial(y,z)}{\partial(u,v)}\right|^2+\left|\frac{\partial(z,x)}{\partial(u,v)}\right|^2+\left|\frac{\partial(x,y)}{\partial(u,v)}\right|^2\right)^{1/2},\end{aligned}\tag{28}$$

for all $(u,v)\in[a,b]\times[c,d]$.

By taking the double integral over (u,v) on $[a,b]\times[c,d]$, we get

$$\begin{aligned}&M(S,\alpha,\beta,\gamma)\\&\quad\le\int_a^b\int_c^d\left(|x(u,v)-\alpha|^2+|y(u,v)-\beta|^2+|z(u,v)-\gamma|^2\right)^{1/2}\\&\qquad\times\left(\left|\frac{\partial(y,z)}{\partial(u,v)}\right|^2+\left|\frac{\partial(z,x)}{\partial(u,v)}\right|^2+\left|\frac{\partial(x,y)}{\partial(u,v)}\right|^2\right)^{1/2}dudv\\&\quad=\iint_S\left(|x-\alpha|^2+|y-\beta|^2+|z-\gamma|^2\right)^{1/2}dS,\end{aligned}$$

and by (25) we get the desired result (27). □

Remark 3. If $f \in \bar{\Delta}_S(\phi, \Phi)$ for some $\phi, \Phi \in \mathbb{C}$, $\phi \neq \Phi$, then by taking $(\alpha, \beta, \gamma) = (\overline{x_B}, \overline{y_B}, \overline{z_B})$ in Theorem 4, we get

$$\left| \frac{1}{V(B)} \iiint_B f(x,y,z)\, dxdydz - \frac{\phi + \Phi}{2} \right.$$
$$- \frac{1}{3V(B)} \iiint_B \left[(\overline{x_B} - x) \frac{\partial f(x,y,z)}{\partial x} + (\overline{y_B} - y) \frac{\partial f(x,y,z)}{\partial y} \right.$$
$$\left. \left. + (\overline{z_B} - z) \frac{\partial f(x,y,z)}{\partial z} \right] dxdydz \right|$$
$$\leq \frac{1}{6V(B)} |\Phi - \phi| M(\overline{x_B}, \overline{y_B}, \overline{z_B}), \tag{29}$$

where

$$M(\overline{x_B}, \overline{y_B}, \overline{z_B}) := \int_a^b \int_c^d |x(u,v) - \overline{x_B}| \left| \frac{\partial(y,z)}{\partial(u,v)} \right| dudv$$
$$+ \int_a^b \int_c^d |y(u,v) - \overline{y_B}| \left| \frac{\partial(z,x)}{\partial(u,v)} \right| dudv$$
$$+ \int_a^b \int_c^d |z(u,v) - \overline{z_B}| \left| \frac{\partial(x,y)}{\partial(u,v)} \right| dudv. \tag{30}$$

Moreover,

$$M(\overline{x_B}, \overline{y_B}, \overline{z_B}) \leq \begin{cases} \left\| \frac{\partial(y,z)}{\partial(\cdot,\cdot)} \right\|_{\square,\infty} \|x - \overline{x_B}\|_{\square,1} + \left\| \frac{\partial(z,x)}{\partial(\cdot,\cdot)} \right\|_{\square,\infty} \|y - \overline{y_B}\|_{\square,1} \\ \quad + \left\| \frac{\partial(x,y)}{\partial(\cdot,\cdot)} \right\|_{\square,\infty} \|z - \overline{z_B}\|_{\square,1}, \\ \left\| \frac{\partial(y,z)}{\partial(\cdot,\cdot)} \right\|_{\square,p} \|x - \overline{x_B}\|_{\square,q} + \left\| \frac{\partial(z,x)}{\partial(\cdot,\cdot)} \right\|_{\square,p} \|y - \overline{y_B}\|_{\square,q} \\ \quad + \left\| \frac{\partial(x,y)}{\partial(\cdot,\cdot)} \right\|_{\square,p} \|z - \overline{z_B}\|_{\square,q}, \\ \left\| \frac{\partial(y,z)}{\partial(\cdot,\cdot)} \right\|_{\square,1} \|x - \overline{x_B}\|_{\square,\infty} + \left\| \frac{\partial(z,x)}{\partial(\cdot,\cdot)} \right\|_{\square,p} \|y - \overline{y_B}\|_{\square,\infty} \\ \quad + \left\| \frac{\partial(x,y)}{\partial(\cdot,\cdot)} \right\|_{\square,1} \|z - \overline{z_B}\|_{\square,\infty} \end{cases}.$$

From (27) we also have

$$\left| \frac{1}{V(B)} \iiint_B f(x,y,z)\, dxdydz - \frac{\phi + \Phi}{2} \right.$$
$$- \frac{1}{3V(B)} \iiint_B \left[(\overline{x_B} - x) \frac{\partial f(x,y,z)}{\partial x} + (\overline{y_B} - y) \frac{\partial f(x,y,z)}{\partial y} \right.$$
$$\left. \left. + (\overline{z_B} - z) \frac{\partial f(x,y,z)}{\partial z} \right] dxdydz \right|$$

$$\leq \frac{1}{6V(B)} |\Phi - \phi| \iint_S \left(|x - \overline{x_B}|^2 + |y - \overline{y_B}|^2 + |z - \overline{z_B}|^2 \right)^{1/2} dS$$
$$\leq \frac{A_S}{6V(B)} |\Phi - \phi| \sup_{(x,y,z)\in S} \left(|x - \overline{x_B}|^2 + |y - \overline{y_B}|^2 + |z - \overline{z_B}|^2 \right)^{1/2}. \tag{31}$$

If $f \in \bar{\Delta}_S(\phi, \Phi)$ for some ϕ, $\Phi \in \mathbb{C}$, $\phi \neq \Phi$, then by taking $\alpha = x_{B,\partial f}$, $\beta = y_{B,\partial f}$ and $\gamma = z_{B,\partial f}$ in Theorem 4, we get

$$\left| \frac{1}{V(B)} \iiint_B f(x,y,z)\, dxdydz - \frac{\phi + \Phi}{2} \right|$$
$$\leq \frac{1}{6V(B)} |\Phi - \phi| \, M(S, x_{B,\partial f}, y_{B,\partial f}, z_{B,\partial f}), \tag{32}$$

where

$$\begin{aligned} M(S, x_{B,\partial f}, y_{B,\partial f}, z_{B,\partial f}) := &\int_a^b \int_c^d |x(u,v) - x_{B,\partial f}| \left| \frac{\partial(y,z)}{\partial(u,v)} \right| dudv \\ &+ \int_a^b \int_c^d |y(u,v) - y_{B,\partial f}| \left| \frac{\partial(z,x)}{\partial(u,v)} \right| dudv \\ &+ \int_a^b \int_c^d |z(u,v) - z_{B,\partial f}| \left| \frac{\partial(x,y)}{\partial(u,v)} \right| dudv. \end{aligned} \tag{33}$$

Moreover, we have the bounds

$$M(S, x_{B,\partial f}, y_{B,\partial f}, z_{B,\partial f})$$
$$\leq \begin{cases} \left\| \frac{\partial(y,z)}{\partial(\cdot,\cdot)} \right\|_{\square,\infty} \|x - x_{B,\partial f}\|_{\square,1} + \left\| \frac{\partial(z,x)}{\partial(\cdot,\cdot)} \right\|_{\square,\infty} \|y - y_{B,\partial f}\|_{\square,1} \\ \quad + \left\| \frac{\partial(x,y)}{\partial(\cdot,\cdot)} \right\|_{\square,\infty} \|z - z_{B,\partial f}\|_{\square,1}, \\ \left\| \frac{\partial(y,z)}{\partial(\cdot,\cdot)} \right\|_{\square,p} \|x - x_{B,\partial f}\|_{\square,q} + \left\| \frac{\partial(z,x)}{\partial(\cdot,\cdot)} \right\|_{\square,p} \|y - y_{B,\partial f}\|_{\square,q} \\ \quad + \left\| \frac{\partial(x,y)}{\partial(\cdot,\cdot)} \right\|_{\square,p} \|z - z_{B,\partial f}\|_{\square,q}, \\ \left\| \frac{\partial(y,z)}{\partial(\cdot,\cdot)} \right\|_{\square,1} \|x - x_{B,\partial f}\|_{\square,\infty} + \left\| \frac{\partial(z,x)}{\partial(\cdot,\cdot)} \right\|_{\square,p} \|y - y_{B,\partial f}\|_{\square,\infty} \\ \quad + \left\| \frac{\partial(x,y)}{\partial(\cdot,\cdot)} \right\|_{\square,1} \|z - z_{B,\partial f}\|_{\square,\infty}. \end{cases} \tag{34}$$

From (27) we also have

$$\begin{aligned}&\left|\frac{1}{V(B)}\iiint_B f(x,y,z)\,dxdydz-\frac{\phi+\Phi}{2}\right|\\&\quad\le\frac{1}{6V(B)}|\Phi-\phi|\\&\qquad\times\iint_S\left(|x-x_{B,\partial f}|^2+|y-y_{B,\partial f}|^2+|z-z_{B,\partial f}|^2\right)^{1/2}dS\\&\quad\le\frac{A_S}{6V(B)}|\Phi-\phi|\\&\qquad\times\sup_{(x,y,z)\in S}\left(|x-x_{B,\partial f}|^2+|y-y_{B,\partial f}|^2+|z-z_{B,\partial f}|^2\right)^{1/2}.\end{aligned}\tag{35}$$

5. Some Examples for Sphere

Consider the three-dimensional ball centered in $C=(a,b,c)$ and having the radius $R>0$,

$$B(C,R):=\left\{(x,y,z)\in\mathbb{R}^3\middle|\,(x-a)^2+(y-b)^2+(z-c)^2\le R^2\right\},$$

and the sphere

$$S(C,R):=\left\{(x,y,z)\in\mathbb{R}^3\middle|\,(x-a)^2+(y-b)^2+(z-c)^2=R^2\right\}.$$

Consider the parametrization of $B(C,R)$ and $S(C,R)$ given by

$$B(C,R):\begin{cases}x=r\cos\psi\cos\varphi+a\\y=r\cos\psi\sin\varphi+b\,;\ (r,\psi,\varphi)\in[0,R]\times\left[-\frac{\pi}{2},\frac{\pi}{2}\right]\times[0,2\pi],\\z=r\sin\psi+c\end{cases}$$

and

$$S(C,R):\begin{cases}x=R\cos\psi\cos\varphi+a\\y=R\cos\psi\sin\varphi+b\,;\ (\psi,\varphi)\in\left[-\frac{\pi}{2},\frac{\pi}{2}\right]\times[0,2\pi].\\z=R\sin\psi+c\end{cases}$$

By setting

$$A:=\begin{vmatrix}\frac{\partial y}{\partial\psi}&\frac{\partial z}{\partial\psi}\\\frac{\partial y}{\partial\varphi}&\frac{\partial z}{\partial\varphi}\end{vmatrix}=-R^2\cos^2\psi\cos\varphi,$$

$$B := \begin{vmatrix} \dfrac{\partial x}{\partial \psi} & \dfrac{\partial z}{\partial \psi} \\ \dfrac{\partial x}{\partial \varphi} & \dfrac{\partial z}{\partial \varphi} \end{vmatrix} = R^2 \cos^2 \psi \sin \varphi$$

and

$$C := \begin{vmatrix} \dfrac{\partial x}{\partial \psi} & \dfrac{\partial y}{\partial \psi} \\ \dfrac{\partial x}{\partial \varphi} & \dfrac{\partial y}{\partial \varphi} \end{vmatrix} = -R^2 \sin \psi \cos \psi,$$

we have that

$$A^2 + B^2 + C^2 = R^4 \cos^2 \psi \text{ for all } (\psi, \varphi) \in \left[-\frac{\pi}{2}, \frac{\pi}{2}\right] \times [0, 2\pi].$$

Obviously, $\overline{x_B} = a$, $\overline{y_B} = b$, $\overline{z_B} = c$ and

$$\begin{aligned} &\iint_S \left(|x - \overline{x_B}|^2 + |y - \overline{y_B}|^2 + |z - \overline{z_B}|^2\right)^{1/2} dS \\ &= \iint_S \left(|x - a|^2 + |y - b|^2 + |z - c|^2\right)^{1/2} dS \\ &= R^3 \int_{-\frac{\pi}{2}}^{\frac{\pi}{2}} \int_0^{2\pi} \cos \psi d\psi d\varphi = 4\pi R^3. \end{aligned}$$

Inequality (31) written for $B = B(C, R)$ and $S = S(C, R)$ becomes

$$\begin{aligned} &\left| \frac{1}{V(B(C,R))} \iiint_{B(C,R)} f(x,y,z)\, dxdydz - \frac{\phi + \Phi}{2} \right. \\ &- \frac{1}{3V(B(C,R))} \iiint_{B(C,R)} \left[(a - x) \frac{\partial f(x,y,z)}{\partial x} + (b - y) \frac{\partial f(x,y,z)}{\partial y} \right. \\ &\qquad \left.\left. + (c - z) \frac{\partial f(x,y,z)}{\partial z} \right] dxdydz \right| \le \frac{1}{2} |\Phi - \phi|, \end{aligned} \tag{36}$$

provided $f \in \bar{\Delta}_{S(C,R)}(\phi, \Phi)$ for some ϕ, $\Phi \in \mathbb{C}$, $\phi \ne \Phi$, where $V(B(C,R)) = \frac{4\pi R^3}{3}$.

References

[1] S.S. Dragomir, On Hadamard's inequality for the convex mappings defined on a ball in the space and applications, *Math. Ineq. & Appl.* **3**(2), 177–187, (2000).

[2] S.S. Dragomir, Ostrowski-type integral inequalities for multiple integral on general convex bodies, Preprint *RGMIA Res. Rep. Coll.* **22** (2019), Art. 50, 13 pp. [http://rgmia.org/papers/v22/v22a50.pdf].

[3] A. Barani, Hermite–Hadamard and Ostrowski type tnequalities on hemispheres, *Mediterr. J. Math.* **13**, 4253–4263, (2016).

[4] M. Bessenyei, The Hermite–Hadamard inequality on simplices, *Amer. Math. Monthly* **115**, 339–345, (2008).

[5] J. de la Cal and J. Cárcamo, Multidimensional Hermite-Hadamard inequalities and the convex order, *J. Math. Anal. Appl.* **324**(1), 248–261, (2006).

[6] S.S. Dragomir, On Hadamard's inequality on a disk, *J. Ineq. Pure & Appl. Math.* **1**(1), (2000), Art. 2. https://www.emis.de/journals/JIPAM/article95.html?sid=95.

[7] M. Matłoka, On Hadamard's inequality for h-convex function on a disk, *Appl. Math. Comput.* **235**, 118–123, (2014).

[8] F.-C. Mitroi and E. Symeonidis, The converse of the Hermite-Hadamard inequality on simplices, *Expo. Math.* **30**, 389–396, (2012).

[9] E. Neuman, Inequalities involving multivariate convex functions II, *Proc. Amer. Math. Soc.* **109**, 965–974, (1990).

[10] E. Neuman and J. Pečarić, Inequalities involving multivariate convex functions, *J. Math. Anal. Appl.* **137**, 541–549, (1989).

[11] S. Wasowicz and A. Witkowski, On some inequality of Hermite–Hadamard type, *Opusc. Math.* **32**(3), 591–600, (2012).

[12] F.-L. Wang, The generalizations to several-dimensions of the classical Hadamard's inequality, *Math. Pract. Theory* **36**(9), 370–373, (2006) (Chinese).

[13] F.-L. Wang, A family of mappings associated with Hadamard's inequality on a hypercube, *International Scholarly Research Network ISRN Mathematical Analysis* Vol. 2011, Article ID 594758, 9 pp. Doi: 10.5402/2011/594758.

[14] N.S. Barnett, F.C. Cîrstea, and S.S. Dragomir, Some inequalities for the integral mean of Hölder continuous functions defined on disks in a plane, in *Inequality Theory and Applications*, Vol. 2 (Chinju/Masan, 2001), pp. 7–18 (Nova Sci. Publ., Hauppauge, NY, 2003). Preprint *RGMIA Res. Rep. Coll.* **5** (2002), Nr. 1, Art. 7, 10 pp. https://rgmia.org/papers/v5n1/BCD.pdf.

[15] N.S. Barnett and S.S. Dragomir, An Ostrowski-type inequality for double integrals and applications for cubature formulae. *Soochow J. Math.* **27**(1), 1–10, (2001).

[16] N.S. Barnett, S.S. Dragomir, and C.E.M. Pearce, A quasi-trapezoid inequality for double integrals. *ANZIAM J.* **44**(3), 355–364, (2003).

[17] H. Budak and M.Z. Sarıkaya, An inequality of Ostrowski-Grüss type for double integrals. *Stud. Univ. Babeş-Bolyai Math.* **62**(2), 163–173, (2017).

[18] S.S. Dragomir, P. Cerone, N.S. Barnett, and J. Roumeliotis, An inequality of the Ostrowski-type for double integrals and applications for cubature formulae, *Tamsui Oxf. J. Math. Sci.* **16**(1), 1–16, (2000).

[19] S. Erden and M.Z. Sarikaya, On exponential Pompeiu's type inequalities for double integrals with applications to Ostrowski's inequality, *New Trends Math. Sci.* **4**(1), 256–267, (2016).

[20] G. Hanna, Some results for double integrals based on an Ostrowski-type inequality. In: *Ostrowski Type Inequalities and Applications in Numerical Integration* (Kluwer Acad. Publ., Dordrecht, 2002), pp. 331–371.

[21] G. Hanna, S.S. Dragomir, and P. Cerone, A general Ostrowski type inequality for double integrals, *Tamkang J. Math.* **33**(4), 319–333, (2002).

[22] Z. Liu, A sharp general Ostrowski type inequality for double integrals, *Tamsui Oxf. J. Inf. Math. Sci.* **28**(2), 217–226, (2012).

[23] M. Özdemir, E. Akdemir, and A.O.E. Set, A new Ostrowski type inequality for double integrals. *J. Inequal. Spec. Funct.* **2**(1), 27–34, (2011).

[24] B.G. Pachpatte, A new Ostrowski type inequality for double integrals, *Soochow J. Math.* **32**(2), 317–322, (2006).

[25] M.Z. Sarikaya, On the Ostrowski-type integral inequality for double integrals. *Demonstratio Math.* **45**(2), 533–540, (2012).

[26] M.Z. Sarikaya and H. Ogunmez, On the weighted Ostrowski-type integral inequality for double integrals, *Arab. J. Sci. Eng.* **36**(6), 1153–1160, (2011).

[27] T.M. Apostol, *Calculus Volume II, Multi Variable Calculus and Linear Algebra, with Applications to Differential Equations and Probability*, 2nd edn. (John Wiley & Sons, New York, London, Sydney, Toronto, 1969).

https://doi.org/10.1142/9789811261572_0010

Chapter 10

Generalized Ostrowski and Trapezoid Type-Rules for Approximating the Integral of Analytic Complex Functions on Paths from General Domains*

Silvestru Sever Dragomir

Mathematics, College of Engineering & Science Victoria University, PO Box 14428 Melbourne City, MC 8001, Australia
DST-NRF Centre of Excellence in the Mathematical and Statistical Sciences, School of Computer Science & Applied Mathematics, University of the Witwatersrand, Private Bag 3, Johannesburg 2050, South Africa.
sever.dragomir@vu.edu.au

In this chapter, we establish some generalized Ostrowski and trapezoid-type rules for approximating the integral of analytic complex functions on paths from general domains. Error bounds for these expansions in terms of p-norms, Hölder and Lipschitz constants are also provided. Examples for the complex logarithm and the complex exponential are given as well.

1. Introduction

Suppose γ is a *smooth path* parametrized by $z(t)$, $t \in [a,b]$ and f is a complex function which is continuous on γ. Put $z(a) = u$ and $z(b) = w$ with $u, w \in \mathbb{C}$. We define the integral of f on $\gamma_{u,w} = \gamma$ as

$$\int_\gamma f(z)\,dz = \int_{\gamma_{u,w}} f(z)\,dz := \int_a^b f(z(t))\,z'(t)\,dt.$$

We observe that the actual choice of parametrization of γ does not matter.

This definition immediately extends to paths that are *piecewise smooth.* Suppose γ is parametrized by $z(t)$, $t \in [a,b]$, which is differentiable on the

*Dedicated to my granddaughters Audrey and Sienna.

intervals $[a, c]$ and $[c, b]$, then assuming that f is continuous on γ, we define

$$\int_{\gamma_{u,w}} f(z)\,dz := \int_{\gamma_{u,v}} f(z)\,dz + \int_{\gamma_{v,w}} f(z)\,dz,$$

where $v := z(c)$. This can be extended for a finite number of intervals.

We also define the integral with respect to arc-length

$$\int_{\gamma_{u,w}} f(z)\,|dz| := \int_a^b f(z(t))\,|z'(t)|\,dt,$$

and the length of the curve γ is then

$$\ell(\gamma) = \int_{\gamma_{u,w}} |dz| = \int_a^b |z'(t)|\,dt.$$

Let f and g be holomorphic in G, an open domain and suppose $\gamma \subset G$ is a piecewise smooth path from $z(a) = u$ to $z(b) = w$. Then we have the *integration by parts formula*

$$\int_{\gamma_{u,w}} f(z)\,g'(z)\,dz = f(w)\,g(w) - f(u)\,g(u) - \int_{\gamma_{u,w}} f'(z)\,g(z)\,dz. \quad (1)$$

We recall also the *triangle inequality* for the complex integral, namely

$$\left|\int_\gamma f(z)\,dz\right| \le \int_\gamma |f(z)|\,|dz| \le \|f\|_{\gamma,\infty}\,\ell(\gamma), \quad (2)$$

where $\|f\|_{\gamma,\infty} := \sup_{z\in\gamma} |f(z)|$.

We also define the p-norm with $p \ge 1$ by

$$\|f\|_{\gamma,p} := \left(\int_\gamma |f(z)|^p\,|dz|\right)^{1/p}.$$

For $p = 1$, we have

$$\|f\|_{\gamma,1} := \int_\gamma |f(z)|\,|dz|.$$

If $p,\ q > 1$ with $\frac{1}{p} + \frac{1}{q} = 1$, then by Hölder's inequality we have

$$\|f\|_{\gamma,1} \le [\ell(\gamma)]^{1/q}\,\|f\|_{\gamma,p}.$$

In the recent paper [1] we obtained the following identity for the path integral of an analytic function defined on a convex domain:

Theorem 1. *Let $f : D \subseteq \mathbb{C} \to \mathbb{C}$ be an analytic function on the convex domain D and $x \in D$. Suppose $\gamma \subset D$ is a smooth path parametrized by $z(t)$, $t \in [a,b]$ with $z(a) = u$ and $z(b) = w$ where $u, w \in D$. Then we have the Ostrowski-type equality*

$$\int_\gamma f(z)\,dz = \sum_{k=0}^{n} \frac{1}{(k+1)!} f^{(k)}(x) \left[(w-x)^{k+1} + (-1)^k (x-u)^{k+1} \right] + R_n(x,\gamma), \tag{3}$$

for $n \geq 0$, where the remainder $R_n(x,\gamma)$ is given by

$$\begin{aligned} R_n(x,\gamma) &:= \frac{1}{n!} \int_\gamma (z-x)^{n+1} \left(\int_0^1 f^{(n+1)}[(1-s)x + sz](1-s)^n\,ds \right) dz \\ &= \frac{1}{n!} \int_0^1 \left(\int_\gamma (z-x)^{n+1} f^{(n+1)}[(1-s)x + sz]\,dz \right) (1-s)^n\,ds. \end{aligned} \tag{4}$$

We obtained among others the following simple error bound [1]:

Corollary 1. *Let $f : D \subseteq \mathbb{C} \to \mathbb{C}$ be an analytic function on the convex domain D and $x \in D$. Suppose $\gamma \subset D$ is a smooth path parametrized by $z(t)$, $t \in [a,b]$ with $z(a) = u$ and $z(b) = w$ where $u, w \in D$. If*

$$\left\| f^{(n+1)} \right\|_{D,\infty} := \sup_{z \in D} \left| f^{(n+1)}(z) \right| < \infty \quad \textit{for some } n \geq 0, \tag{5}$$

then we have the representation (3) *where the remainder $R_n(x,\gamma)$ satisfies the bound*

$$|R_n(x,\gamma)| \leq \frac{1}{(n+1)!} \left\| f^{(n+1)} \right\|_{D,\infty} \int_\gamma |z-x|^{n+1} |dz|. \tag{6}$$

The above results extend the inequalities for real valued functions of a real variable obtained in Refs. [2,3]. For similar results see Refs. [4–9].

2. Representation Results

We have the following identity for the integral on a path from a non-necessarily convex domain D as above:

Theorem 2. *Let $f : D \subseteq \mathbb{C} \to \mathbb{C}$ be an analytic function on the domain D and $x \in D$. Suppose $\gamma \subset D$ is a smooth path parametrized by $z(t)$, $t \in [a, b]$ with $z(a) = u$, $z(t) = x$ and $z(b) = w$ where $u, w \in D$. Then we have the equality*

$$\int_\gamma f(z)\,dz = \sum_{k=0}^{n-1} \frac{1}{(k+1)!} f^{(k)}(x) \left[(w-x)^{k+1} + (-1)^k (x-u)^{k+1} \right] + O_n(x,\gamma), \tag{7}$$

where the remainder $O_n(x,\gamma)$ is given by

$$O_n(x,\gamma) := \frac{(-1)^n}{n!} \int_\gamma K_n(x,z)\, f^{(n)}(z)\,dz, \tag{8}$$

and the kernel $K_n : \gamma \times \gamma \to \mathbb{C}$ is defined by

$$K_n(x,z) := \begin{cases} (z-u)^n & \text{if } z \in \gamma_{u,x} \\ (z-w)^n & \text{if } z \in \gamma_{x,w} \end{cases}, \quad x \in \gamma, \tag{9}$$

and n is a natural number, $n \geq 1$.

Proof. We prove the identity by induction over n. For $n = 1$, we have to prove the equality

$$\int_\gamma f(z)\,dz = (w-u)\, f(x) - \int_\gamma K_1(x,z)\, f'(z)\,dz, \tag{10}$$

where

$$K_1(x,z) := \begin{cases} z-u & \text{if } z \in \gamma_{u,x} \\ z-w & \text{if } z \in \gamma_{x,w} \end{cases}.$$

Integrating by parts, we have:

$$\begin{aligned} &\int_\gamma K_1(x,z)\, f'(z)\,dz \\ &= \int_{\gamma_{u,x}} (z-u)\, f'(z)\,dz + \int_{\gamma_{x,w}} (z-w)\, f'(z)\,dz \\ &= (z-u)\, f(z)\big|_u^x - \int_{\gamma_{u,x}} f(z)\,dz + (z-w)\, f(z)\big|_x^w - \int_{\gamma_{x,w}} f(z)\,dz \end{aligned}$$

$$= (x-u) f(x) + (w-x) f(x) - \int_\gamma f(z)\, dz$$

$$= (w-u) f(x) - \int_\gamma f(z)\, dz,$$

and the identity (10) is proved.

Assume that (7) holds for "n" and let us prove it for "$n+1$". That is, we have to prove the equality

$$\int_\gamma f(z)\, dz = \sum_{k=0}^{n} \left[\frac{(w-x)^{k+1} + (-1)^k (x-u)^{k+1}}{(k+1)!} \right] f^{(k)}(x) + \frac{(-1)^{n+1}}{(n+1)!} \int_\gamma K_{n+1}(x,z) f^{(n+1)}(z)\, dz. \quad (11)$$

We have, by using (9),

$$\frac{1}{(n+1)!} \int_\gamma K_{n+1}(x,z) f^{(n+1)}(z)\, dz = \int_{\gamma_{u,x}} \frac{(z-u)^{n+1}}{(n+1)!} f^{(n+1)}(z)\, dz + \int_{\gamma_{x,w}} \frac{(z-w)^{n+1}}{(n+1)!} f^{(n+1)}(z)\, dz,$$

and integrating by parts gives

$$\begin{aligned} \frac{1}{(n+1)!} \int_\gamma K_{n+1}(x,z) f^{(n+1)}(z)\, dz &= \left. \frac{(z-u)^{n+1}}{(n+1)!} f^{(n)}(z) \right|_u^x - \frac{1}{n!} \int_{\gamma_{u,x}} (z-u)^n f^{(n)}(z)\, dz \\ &\quad + \left. \frac{(z-w)^{n+1}}{(n+1)!} f^{(n)}(z) \right|_x^w - \frac{1}{n!} \int_{\gamma_{x,w}} (z-w)^n f^{(n)}(z)\, dz \\ &= \frac{(x-u)^{n+1} + (-1)^{n+2} (w-x)^{n+1}}{(n+1)!} f^{(n)}(x) \\ &\quad - \frac{1}{n!} \int_\gamma K_n(x,z) f^{(n)}(z)\, dz. \end{aligned}$$

That is

$$\begin{aligned}
&\frac{1}{n!}\int_{\gamma} K_n(x,z)\, f^{(n)}(z)\,dz \\
&\quad = \frac{(x-u)^{n+1} + (-1)^{n+2}(w-x)^{n+1}}{(n+1)!} f^{(n)}(x) \\
&\qquad - \frac{1}{(n+1)!}\int_{\gamma} K_{n+1}(x,z)\, f^{(n+1)}(z)\,dz \\
&\quad = \frac{(x-u)^{n+1} + (-1)^{n}(w-x)^{n+1}}{(n+1)!} f^{(n)}(x) \\
&\qquad - \frac{1}{(n+1)!}\int_{\gamma} K_{n+1}(x,z)\, f^{(n+1)}(z)\,dz.
\end{aligned}$$

By multiplying this with $(-1)^n$, we get

$$\begin{aligned}
&\frac{(-1)^n}{n!}\int_{\gamma} K_n(x,z)\, f^{(n)}(z)\,dz \\
&\quad = \frac{(w-x)^{n+1} + (-1)^{n}(x-u)^{n+1}}{(n+1)!} f^{(n)}(x) \\
&\qquad - \frac{(-1)^n}{(n+1)!}\int_{\gamma} K_{n+1}(x,z)\, f^{(n+1)}(z)\,dz. \qquad (12)
\end{aligned}$$

Now, using the mathematical induction hypothesis and (12), we get

$$\begin{aligned}
\int_{\gamma} f(z)\,dz &= \sum_{k=0}^{n-1}\left[\frac{(w-x)^{k+1} + (-1)^{k}(x-u)^{k+1}}{(k+1)!}\right] f^{(k)}(x) \\
&\quad + \frac{(w-x)^{n+1} + (-1)^{n}(x-u)^{n+1}}{(n+1)!} f^{(n)}(x) \\
&\quad - \frac{(-1)^n}{(n+1)!}\int_{\gamma} K_{n+1}(x,z)\, f^{(n+1)}(z)\,dz \\
&= \sum_{k=0}^{n}\left[\frac{(w-x)^{k+1} + (-1)^{k}(x-u)^{k+1}}{(k+1)!}\right] f^{(k)}(x) \\
&\quad + \frac{(-1)^{n+1}}{(n+1)!}\int_{\gamma} K_{n+1}(x,z)\, f^{(n+1)}(z)\,dz.
\end{aligned}$$

That is, identity (11) and the theorem is thus proved. □

Corollary 2. *With the assumptions of Theorem 2 and for λ_1, λ_2 complex numbers, we have the identity*

$$\int_\gamma f(z)\,dz = \sum_{k=0}^{n-1} \frac{1}{(k+1)!} f^{(k)}(x) \left[(w-x)^{k+1} + (-1)^k (x-u)^{k+1}\right]$$
$$+ \lambda_1 \frac{(-1)^n}{(n+1)!}(x-u)^{n+1} + \lambda_2 \frac{(-1)^{n+1}}{(n+1)!}(x-w)^{n+1}$$
$$+ O_n(x,\gamma,\lambda_1,\lambda_2), \tag{13}$$

where the remainder $O_n(x,\gamma,\lambda_1,\lambda_2)$ is given by

$$O_n(x,\gamma,\lambda_1,\lambda_2) := \frac{(-1)^n}{n!}\int_{\gamma_{u,x}} (z-u)^n \left[f^{(n)}(z) - \lambda_1\right] dz$$
$$+ \frac{(-1)^n}{n!}\int_{\gamma_{x,w}} (z-w)^n \left[f^{(n)}(z) - \lambda_2\right] dz. \tag{14}$$

In particular, for $\lambda_1 = \lambda_2 = \lambda$, we have

$$\int_\gamma f(z)\,dz = \sum_{k=0}^{n-1} \frac{1}{(k+1)!} f^{(k)}(x) \left[(w-x)^{k+1} + (-1)^k (x-u)^{k+1}\right]$$
$$+ \lambda \left[\frac{(-1)^n}{(n+1)!}(x-u)^{n+1} + \frac{(-1)^{n+1}}{(n+1)!}(x-w)^{n+1}\right]$$
$$+ O_n(x,\gamma,\lambda), \tag{15}$$

where

$$O_n(x,\gamma,\lambda) := \frac{(-1)^n}{n!}\int_{\gamma_{u,x}} (z-u)^n \left[f^{(n)}(z) - \lambda\right] dz$$
$$+ \frac{(-1)^n}{n!}\int_{\gamma_{x,w}} (z-w)^n \left[f^{(n)}(z) - \lambda\right] dz. \tag{16}$$

Proof. Observe that

$$\frac{(-1)^n}{n!}\int_{\gamma} K_n(x,z)\, f^{(n)}(z)\,dz$$

$$= \frac{(-1)^n}{n!}\int_{\gamma_{u,x}} (z-u)^n \left[f^{(n)}(z) - \lambda_1 + \lambda_1\right] dz$$

$$+ \frac{(-1)^n}{n!}\int_{\gamma_{x,w}} (z-w)^n \left[f^{(n)}(z) - \lambda_2 + \lambda_2\right] dz$$

$$= \frac{(-1)^n}{n!}\int_{\gamma_{u,x}} (z-u)^n \left[f^{(n)}(z) - \lambda_1\right] dz + \lambda_1 \frac{(-1)^n}{n!}\int_{\gamma_{u,x}} (z-u)^n\, dz$$

$$+ \frac{(-1)^n}{n!}\int_{\gamma_{x,w}} (z-w)^n \left[f^{(n)}(z) - \lambda_2\right] dz$$

$$+ \lambda_2 \frac{(-1)^n}{n!}\int_{\gamma_{x,w}} (z-w)^n\, dz$$

$$= \frac{(-1)^n}{n!}\int_{\gamma_{u,x}} (z-u)^n \left[f^{(n)}(z) - \lambda_1\right] dz + \lambda_1 \frac{(-1)^n}{(n+1)!}(x-u)^{n+1}$$

$$+ \frac{(-1)^n}{n!}\int_{\gamma_{x,w}} (z-w)^n \left[f^{(n)}(z) - \lambda_2\right] dz + \lambda_2 \frac{(-1)^{n+1}}{(n+1)!}(x-w)^{n+1},$$

and by (7) we then get (13). □

Corollary 3. *With the assumptions of Theorem 2 and for θ a complex number, we have the identity*

$$\int_{\gamma} f(z)\,dz = \sum_{k=0}^{n-1} \frac{1}{(k+1)!}\left[\theta(-1)^k f^{(k)}(w) + (1-\theta) f^{(k)}(u)\right](w-u)^{k+1}$$
$$+ T_n(\gamma,\theta), \tag{17}$$

where the remainder $T_n(x,\gamma,\theta)$ is given by

$$T_n(\gamma,\theta) := \frac{(-1)^n}{n!}\int_{\gamma} \left[\theta(z-u)^n + (1-\theta)(z-w)^n\right] f^{(n)}(z)\,dz. \tag{18}$$

In particular, for $\theta = \frac{1}{2}$, we have

$$\int_\gamma f(z)\,dz = \sum_{k=0}^{n-1} \frac{1}{(k+1)!} \left[\frac{(-1)^k f^{(k)}(w) + f^{(k)}(u)}{2} \right] \times (w-u)^{k+1} + T_n(\gamma), \tag{19}$$

where

$$T_n(\gamma) = \frac{(-1)^n}{2n!} \int_\gamma \left[(z-u)^n + (z-w)^n \right] f^{(n)}(z)\,dz. \tag{20}$$

Proof. From Theorem 2 for $x = w$, we have

$$\int_\gamma f(z)\,dz = \sum_{k=0}^{n-1} \frac{(-1)^k}{(k+1)!} f^{(k)}(w)(w-u)^{k+1} + \frac{(-1)^n}{n!} \int_\gamma (z-u)^n f^{(n)}(z)\,dz, \tag{21}$$

while for $x = u$ we have

$$\int_\gamma f(z)\,dz = \sum_{k=0}^{n-1} \frac{1}{(k+1)!} f^{(k)}(u)(w-u)^{k+1} + \frac{(-1)^n}{n!} \int_\gamma (z-w)^n f^{(n)}(z)\,dz. \tag{22}$$

If we multiply (21) by θ and (22) by $1-\theta$ and sum the obtained equalities, then we get

$$\begin{aligned}\int_\gamma f(z)\,dz &= \theta \sum_{k=0}^{n-1} \frac{(-1)^k}{(k+1)!} f^{(k)}(w)(w-u)^{k+1} \\ &\quad + \frac{(-1)^n}{n!} \theta \int_\gamma (z-u)^n f^{(n)}(z)\,dz \\ &\quad + (1-\theta) \sum_{k=0}^{n-1} \frac{1}{(k+1)!} f^{(k)}(u)(w-u)^{k+1} \\ &\quad + \frac{(-1)^n}{n!}(1-\theta) \int_\gamma (z-w)^n f^{(n)}(z)\,dz\end{aligned}$$

$$= \sum_{k=0}^{n-1} \frac{1}{(k+1)!} \left[\theta (-1)^k f^{(k)}(w) + (1-\theta) f^{(k)}(u)\right] (w-u)^{k+1}$$
$$+ \frac{(-1)^n}{n!} \int_\gamma \left[\theta (z-u)^n + (1-\theta)(z-w)^n\right] f^{(n)}(z)\, dz,$$

which proves the desired result (17). □

For the case of real variable functions, see Refs. [10,11].

Corollary 4. *With the assumptions of Theorem 2 and for θ and λ complex numbers, we have the identity*

$$\int_\gamma f(z)\, dz = \sum_{k=0}^{n-1} \frac{1}{(k+1)!} \left[\theta (-1)^k f^{(k)}(w) + (1-\theta) f^{(k)}(u)\right] (w-u)^{k+1}$$
$$+ \frac{(-1)^n}{(n+1)!} \lambda \left[\theta + (-1)^n (1-\theta)\right] (w-u)^{n+1} + T_n(\gamma, \theta, \lambda), \tag{23}$$

where the remainder $T_n(\gamma, \theta, \lambda)$ is given by

$$T_n(\gamma, \theta) := \frac{(-1)^n}{n!} \int_\gamma \left[\theta (z-u)^n + (1-\theta)(z-w)^n\right] \left[f^{(n)}(z) - \lambda\right] dz. \tag{24}$$

In particular, for $\theta = \frac{1}{2}$ we have

$$\int_\gamma f(z)\, dz = \sum_{k=0}^{n-1} \frac{1}{(k+1)!} \left[\frac{(-1)^k f^{(k)}(w) + f^{(k)}(u)}{2}\right] (w-u)^{k+1}$$
$$+ \frac{(-1)^n}{(n+1)!} \left[\frac{1 + (-1)^n}{2}\right] \lambda (w-u)^{n+1} + T_n(\gamma, \lambda), \tag{25}$$

where the remainder $T_n(\gamma, \lambda)$ is given by

$$T_n(\gamma, \lambda) := \frac{(-1)^n}{2n!} \int_\gamma \left[(z-u)^n + (z-w)^n\right] \left[f^{(n)}(z) - \lambda\right] dz. \tag{26}$$

3. *p*-Norm Error Bounds

We have the following:

Theorem 3. *Let $f : D \subseteq \mathbb{C} \to \mathbb{C}$ be an analytic function on the domain D and $x \in D$. Suppose $\gamma \subset D$ is a smooth path parametrized by $z(t)$, $t \in [a, b]$*

with $z(a) = u$, $z(t) = x$ and $z(b) = w$ where $u, w \in D$. Then we have the equality (7) and the remainder $O_n(x, \gamma)$ satisfies the bounds

$$|O_n(x,\gamma)| \leq B_n(x,\gamma), \tag{27}$$

where

$$B_n(x,\gamma) := \frac{1}{n!}\left[\int_{\gamma_{u,x}} |z-u|^n \left|f^{(n)}(z)\right| |dz| + \int_{\gamma_{x,w}} |z-w|^n \left|f^{(n)}(z)\right| |dz|\right].$$

Moreover, we have

$$B_n(x,\gamma) \leq \frac{1}{n!}\begin{cases} \left\|f^{(n)}\right\|_{\gamma_{u,x},\infty} \displaystyle\int_{\gamma_{u,x}} |z-u|^n |dz| \\ \left\|f^{(n)}\right\|_{\gamma_{u,x},p} \left(\displaystyle\int_{\gamma_{u,x}} |z-u|^{qn} |dz|\right)^{1/q}, \\ \text{where } p,\ q>1,\ \dfrac{1}{p}+\dfrac{1}{q}=1; \\ \left\|f^{(n)}\right\|_{\gamma_{u,x},1} \max_{z\in\gamma_{u,x}} |z-u|^n \end{cases}$$

$$+\ \frac{1}{n!}\begin{cases} \left\|f^{(n)}\right\|_{\gamma_{x,w},\infty} \displaystyle\int_{\gamma_{x,w}} |z-w|^n |dz| \\ \left\|f^{(n)}\right\|_{\gamma_{x,w},p} \left(\displaystyle\int_{\gamma_{x,w}} |z-w|^{qn} |dz|\right)^{1/q}, \\ \text{where } p,\ q>1,\ \dfrac{1}{p}+\dfrac{1}{q}=1; \\ \left\|f^{(n)}\right\|_{\gamma_{x,w},1} \max_{z\in\gamma_{x,w}} |z-w|^n \end{cases}$$

$$\leq \frac{1}{n!}\begin{cases} \left\|f^{(n)}\right\|_{\gamma_{u,w},\infty} \left[\displaystyle\int_{\gamma_{u,x}} |z-u|^n |dz| + \int_{\gamma_{x,w}} |z-w|^n |dz|\right], \\ \left\|f^{(n)}\right\|_{\gamma_{u,w},p} \left(\displaystyle\int_{\gamma_{u,x}} |z-u|^{qn} |dz| + \int_{\gamma_{x,w}} |z-w|^{qn} |dz|\right)^{1/q} \\ \text{where } p,\ q>1,\ \dfrac{1}{p}+\frac{1}{q}=1; \\ \left\|f^{(n)}\right\|_{\gamma_{u,w},1} \max\left\{\max_{z\in\gamma_{u,x}} |z-u|^n,\right. \\ \qquad \left.\max_{z\in\gamma_{x,w}} |z-w|^n\right\}. \end{cases} \tag{28}$$

Proof. We have

$$
\begin{aligned}
|O_n(x,\gamma)| &\le \left|\frac{(-1)^n}{n!}\int_{\gamma_{u,x}}(z-u)^n f^{(n)}(z)\,dz\right| \\
&\quad+\left|\frac{(-1)^n}{n!}\int_{\gamma_{x,w}}(z-w)^n f^{(n)}(z)\,dz\right| \\
&\le \frac{1}{n!}\int_{\gamma_{u,x}}|z-u|^n\left|f^{(n)}(z)\right||dz| \\
&\quad+\frac{1}{n!}\int_{\gamma_{x,w}}|z-w|^n\left|f^{(n)}(z)\right||dz|,
\end{aligned}
$$

which proves the inequality (27).

Using Hölder's integral inequality we have

$$
\begin{aligned}
&\int_{\gamma_{u,x}}|z-u|^n\left|f^{(n)}(z)\right||dz| \\
&\le \begin{cases}
\max_{z\in\gamma_{u,x}}\left|\left|f^{(n)}(z)\right|\right|\int_{\gamma_{u,x}}|z-u|^n|dz| \\
\left(\int_{\gamma_{u,x}}\left|f^{(n)}(z)\right|^p|dz|\right)^{1/p}\left(\int_{\gamma_{u,x}}|z-u|^{qn}|dz|\right)^{1/q}, \\
\text{where } p,\ q>1,\ \dfrac{1}{p}+\dfrac{1}{q}=1; \\
\int_{\gamma_{u,x}}\left|f^{(n)}(z)\right||dz|\max\limits_{z\in\gamma_{u,x}}|z-u|^n
\end{cases} \\
&= \begin{cases}
\left|\left|f^{(n)}\right|\right|_{\gamma_{u,x},\infty}\int_{\gamma_{u,x}}|z-u|^n|dz| \\
\left|\left|f^{(n)}\right|\right|_{\gamma_{u,x},p}\left(\int_{\gamma_{u,x}}|z-u|^{qn}|dz|\right)^{1/q}, \\
\text{where } p,\ q>1,\ \dfrac{1}{p}+\dfrac{1}{q}=1; \\
\left|\left|f^{(n)}\right|\right|_{\gamma_{u,x},1}\max_{z\in\gamma_{u,x}}|z-u|^n
\end{cases}
\end{aligned}
$$

and

$$\int_{\gamma_{x,w}} |z-w|^n \left|f^{(n)}(z)\right| |dz| \le \begin{cases} \left\|f^{(n)}\right\|_{\gamma_{x,w},\infty} \int_{\gamma_{x,w}} |z-w|^n |dz| \\ \left\|f^{(n)}\right\|_{\gamma_{x,w},p} \left(\int_{\gamma_{x,w}} |z-w|^{qn} |dz|\right)^{1/q}, \\ \text{where } p,\ q>1,\ \dfrac{1}{p}+\dfrac{1}{q}=1; \\ \left\|f^{(n)}\right\|_{\gamma_{x,w},1} \max_{z\in\gamma_{x,w}} |z-w|^n. \end{cases}$$

Using the elementary inequality

$$\alpha a+\beta b \le \begin{cases} \max\{\alpha,\beta\}(a+b) \\ (\alpha^p+\beta^p)^{1/p}(a^q+b^q)^{1/q}; \\ \text{where } p,\ q>1,\ \dfrac{1}{p}+\dfrac{1}{q}=1 \end{cases}$$

we have

$$\begin{aligned} &\left\|f^{(n)}\right\|_{\gamma_{u,x},\infty} \int_{\gamma_{u,x}} |z-u|^n |dz| + \left\|f^{(n)}\right\|_{\gamma_{x,w},\infty} \int_{\gamma_{x,w}} |z-w|^n |dz| \\ &\quad \le \max\left\{\left\|f^{(n)}\right\|_{\gamma_{u,x},\infty}, \left\|f^{(n)}\right\|_{\gamma_{x,w},\infty}\right\} \\ &\qquad \times \left[\int_{\gamma_{u,x}} |z-u|^n |dz| + \int_{\gamma_{x,w}} |z-w|^n |dz|\right] \\ &\quad = \left\|f^{(n)}\right\|_{\gamma_{u,w},\infty} \left[\int_{\gamma_{u,x}} |z-u|^n |dz| + \int_{\gamma_{x,w}} |z-w|^n |dz|\right], \end{aligned}$$

$$\begin{aligned} &\left\|f^{(n)}\right\|_{\gamma_{u,x},p} \left(\int_{\gamma_{u,x}} |z-u|^{qn} |dz|\right)^{1/q} \\ &\quad + \left\|f^{(n)}\right\|_{\gamma_{x,w},p} \left(\int_{\gamma_{x,w}} |z-w|^{qn} |dz|\right)^{1/q} \end{aligned}$$

$$\leq \left(\left\| f^{(n)} \right\|_{\gamma_{u,x},p}^{p} + \left\| f^{(n)} \right\|_{\gamma_{x,w},p}^{p} \right)^{1/p}$$
$$\times \left(\int_{\gamma_{u,x}} |z-u|^{qn} |dz| + \int_{\gamma_{x,w}} |z-w|^{qn} |dz| \right)^{1/q}$$
$$= \left\| f^{(n)} \right\|_{\gamma_{u,w},p} \left(\int_{\gamma_{u,x}} |z-u|^{qn} |dz| + \int_{\gamma_{x,w}} |z-w|^{qn} |dz| \right)^{1/q}$$

and

$$\left\| f^{(n)} \right\|_{\gamma_{u,x},1} \max_{z\in\gamma_{u,x}} |z-u|^{n} + \left\| f^{(n)} \right\|_{\gamma_{x,w},1} \max_{z\in\gamma_{x,w}} |z-w|^{n}$$
$$\leq \max\left\{ \max_{z\in\gamma_{u,x}} |z-u|^{n}, \max_{z\in\gamma_{x,w}} |z-w|^{n} \right\} \left[\left\| f^{(n)} \right\|_{\gamma_{u,x},1} + \left\| f^{(n)} \right\|_{\gamma_{x,w},1} \right]$$
$$= \left\| f^{(n)} \right\|_{\gamma_{u,w},1} \max\left\{ \max_{z\in\gamma_{u,x}} |z-u|^{n}, \max_{z\in\gamma_{x,w}} |z-w|^{n} \right\},$$

which proves the inequality (28). □

In a similar way, we can prove:

Theorem 4. *With the assumptions of Theorem 3 and for θ a complex number, we have the identity (17) and the remainder $T_n(x,\gamma,\theta)$ satisfies the bounds*

$$|T_n(\gamma,\theta)| \leq C_n(\gamma,\theta), \tag{29}$$

where

$$C_n(\gamma,\theta) := \frac{1}{n!}\left[|\theta| \int_{\gamma} |z-u|^{n} \left| f^{(n)}(z) \right| |dz| \right.$$
$$\left. + |1-\theta| \int_{\gamma} |z-w|^{n} \left| f^{(n)}(z) \right| |dz| \right]. \tag{30}$$

Moreover, we have

$$C_n(\gamma,\theta) \leq \frac{1}{n!}|\theta| \begin{cases} \|f^{(n)}\|_{\gamma,\infty} \int_{\gamma} |z-u|^{n} |dz| \\ \|f^{(n)}\|_{\gamma,p} \left(\int_{\gamma} |z-u|^{qn} |dz| \right)^{1/q} \\ \text{where } p,\ q>1,\ \frac{1}{p}+\frac{1}{q}=1; \\ \|f^{(n)}\|_{\gamma,1} \max_{z\in\gamma} |z-u|^{n} \end{cases}$$

$$
+\frac{1}{n!}|1-\theta|\begin{cases}\left\|f^{(n)}\right\|_{\gamma,\infty}\int_\gamma |z-w|^n\,|dz|\\ \left\|f^{(n)}\right\|_{\gamma,p}\left(\int_\gamma |z-w|^{qn}\,|dz|\right)^{1/q}\\ \text{where } p,\ q>1,\ \frac{1}{p}+\frac{1}{q}=1;\\ \left\|f^{(n)}\right\|_{\gamma,1}\max_{z\in\gamma}|z-w|^n\end{cases} \tag{31}
$$

$$
\leq \frac{1}{n!}\max\{|\theta|,|1-\theta|\}
$$

$$
\times\begin{cases}\left\|f^{(n)}\right\|_{\gamma,\infty}\left[\int_\gamma [|z-u|^n+|z-w|^n]\,|dz|\right]\\ \left\|f^{(n)}\right\|_{\gamma,p}\left[\left(\int_\gamma |z-u|^{qn}\,|dz|\right)^{1/q}+\left(\int_\gamma |z-w|^{qn}\,|dz|\right)^{1/q}\right]\\ \text{where } p,\ q>1,\ \frac{1}{p}+\frac{1}{q}=1;\\ \left\|f^{(n)}\right\|_{\gamma,1}[\max_{z\in\gamma}|z-u|^n+\max_{z\in\gamma}|z-w|^n].\end{cases} \tag{32}
$$

We observe that for $\theta=\frac{1}{2}$ we have the representation (19) and the remainder $T_n(\gamma)$ satisfies the inequalities

$$
|T_n(\gamma)|\leq\frac{1}{2n!}\int_\gamma [|z-u|^n+|z-w|^n]\left|f^{(n)}(z)\right||dz|
$$

$$
\leq\frac{1}{2n!}\begin{cases}\left\|f^{(n)}\right\|_{\gamma,\infty}\int_\gamma [|z-u|^n+|z-w|^n]\,|dz|\\ \left\|f^{(n)}\right\|_{\gamma,p}\left(\int_\gamma [|z-u|^n+|z-w|^n]^q\,|dz|\right)^{1/q}\\ \text{where } p,\ q>1,\ \frac{1}{p}+\frac{1}{q}=1;\\ \left\|f^{(n)}\right\|_{\gamma,1}\max_{z\in\gamma}[|z-u|^n+|z-w|^n].\end{cases} \tag{33}
$$

4. Error Bounds for Bounded Derivatives

Suppose $\gamma\subset\mathbb{C}$ is a piecewise smooth path parametrized by $z(t)$, $t\in\gamma$ from $z(a)=u$ to $z(b)=w$. Now, for ϕ, $\Phi\in\mathbb{C}$ and γ an interval of real numbers, define the sets of complex-valued functions

$$
\bar{U}_\gamma(\phi,\Phi):=\left\{f:\gamma\to\mathbb{C}\,|\,\mathrm{Re}\left[(\Phi-f(z))\left(\overline{f(z)}-\overline{\phi}\right)\right]\geq 0 \text{ for each } z\in\gamma\right\}
$$

and

$$\bar{\Delta}_{\gamma}(\phi,\Phi) := \left\{ f:\gamma\to\mathbb{C} \middle| \left| f(z) - \frac{\phi+\Phi}{2} \right| \le \frac{1}{2}|\Phi-\phi| \text{ for each } z\in\gamma \right\}.$$

The following representation result may be stated.

Proposition 1. *For any $\phi,\ \Phi\in\mathbb{C}$, $\phi\neq\Phi$, we have that $\bar{U}_{\gamma}(\phi,\Phi)$ and $\bar{\Delta}_{\gamma}(\phi,\Phi)$ are non-empty, convex and closed sets and*

$$\bar{U}_{\gamma}(\phi,\Phi) = \bar{\Delta}_{\gamma}(\phi,\Phi). \tag{34}$$

Proof. We observe that for any $w\in\mathbb{C}$ we have the equivalence

$$\left| w - \frac{\phi+\Phi}{2} \right| \le \frac{1}{2}|\Phi-\phi|,$$

if and only if

$$\operatorname{Re}\left[(\Phi-w)\left(\overline{w}-\overline{\phi}\right)\right] \ge 0.$$

This follows by the equality

$$\frac{1}{4}|\Phi-\phi|^2 - \left| w - \frac{\phi+\Phi}{2} \right|^2 = \operatorname{Re}\left[(\Phi-w)\left(\overline{w}-\overline{\phi}\right)\right],$$

that holds for any $w\in\mathbb{C}$.

The equality (34) is thus a simple consequence of this fact. □

On making use of the complex numbers field properties, we can also state the following:

Corollary 5. *For any $\phi,\ \Phi\in\mathbb{C}$, $\phi\neq\Phi$, we have that*

$$\begin{aligned}\bar{U}_{\gamma}(\phi,\Phi) = \{ f:\gamma\to\mathbb{C} |\, &(\operatorname{Re}\Phi - \operatorname{Re} f(z))(\operatorname{Re} f(z) - \operatorname{Re}\phi) \\ &+ (\operatorname{Im}\Phi - \operatorname{Im} f(z))(\operatorname{Im} f(z) - \operatorname{Im}\phi) \ge 0 \textit{ for each } z\in\gamma \}.\end{aligned} \tag{35}$$

Now, if we assume that $\operatorname{Re}(\Phi)\ge\operatorname{Re}(\phi)$ and $\operatorname{Im}(\Phi)\ge\operatorname{Im}(\phi)$, then we can define the following set of functions as well:

$$\begin{aligned}\bar{S}_{\gamma}(\phi,\Phi) := \{ f:\gamma\to\mathbb{C} |\, &\operatorname{Re}(\Phi)\ge\operatorname{Re} f(z)\ge\operatorname{Re}(\phi) \\ &\text{and } \operatorname{Im}(\Phi)\ge\operatorname{Im} f(z)\ge\operatorname{Im}(\phi) \text{ for each } z\in\gamma \}.\end{aligned} \tag{36}$$

One can easily observe that $\bar{S}_\gamma(\phi,\Phi)$ is closed, convex and

$$\emptyset \neq \bar{S}_\gamma(\phi,\Phi) \subseteq \bar{U}_\gamma(\phi,\Phi). \tag{37}$$

We have the following result:

Theorem 5. *Let $f : D \subseteq \mathbb{C} \to \mathbb{C}$ be an analytic function on the domain D and $x \in D$. Suppose $\gamma \subset D$ is a smooth path parametrized by $z(t)$, $t \in [a,b]$ with $z(a) = u$, $z(t) = x$ and $z(b) = w$ where u, $w \in D$. If $f^{(n)} \in \bar{\Delta}_\gamma(\phi_n,\Phi_n)$ for some ϕ_n, $\Phi_n \in \mathbb{C}$, $\phi_n \neq \Phi_n$, then we have the equality*

$$\begin{aligned}\int_\gamma f(z)\,dz &= \sum_{k=0}^{n-1} \frac{1}{(k+1)!} f^{(k)}(x)\left[(w-x)^{k+1} + (-1)^k (x-u)^{k+1}\right]\\ &\quad + \frac{\phi_n+\Phi_n}{2}\frac{(-1)^n}{(n+1)!}\left[(x-u)^{n+1} + (-1)^n (w-x)^{n+1}\right]\\ &\quad + O_n(x,\gamma,\phi_n,\Phi_n)\end{aligned} \tag{38}$$

and the remainder satisfies the bound

$$\begin{aligned}&|O_n(x,\gamma,\phi_n,\Phi_n)|\\ &\quad \leq \frac{1}{2n!}|\Phi_n-\phi_n|\left[\int_{\gamma_{u,x}}|z-u|^n|dz| + \int_{\gamma_{x,w}}|z-w|^n|dz|\right].\end{aligned} \tag{39}$$

Proof. From the equality (15) we have for $\lambda = \frac{\phi_n+\Phi_n}{2}$ that

$$\begin{aligned}\int_\gamma f(z)\,dz &= \sum_{k=0}^{n-1} \frac{1}{(k+1)!} f^{(k)}(x)\left[(w-x)^{k+1} + (-1)^k (x-u)^{k+1}\right]\\ &\quad + \frac{\phi_n+\Phi_n}{2}\frac{(-1)^n}{(n+1)!}\left[(x-u)^{n+1} + (-1)^n (w-x)^{n+1}\right]\\ &\quad + O_n(x,\gamma,\phi_n,\Phi_n),\end{aligned} \tag{40}$$

where

$$\begin{aligned}O_n(x,\gamma,\phi_n,\Phi_n) &= \frac{(-1)^n}{n!}\int_{\gamma_{u,x}}(z-u)^n\left(f^{(n)}(z) - \frac{\phi_n+\Phi_n}{2}\right)dz\\ &\quad + \frac{(-1)^n}{n!}\int_{\gamma_{x,w}}(z-w)^n\left(f^{(n)}(z) - \frac{\phi_n+\Phi_n}{2}\right)dz.\end{aligned} \tag{41}$$

Taking the modulus in (41) and taking into account that $f^{(n)} \in \bar{\Delta}_\gamma(\phi_n, \Phi_n)$, we have

$$\begin{aligned}
|O_n(x,\gamma,\phi_n,\Phi_n)| &\le \frac{1}{n!}\left|\int_{\gamma_{u,x}} (z-u)^n \left(f^{(n)}(z) - \frac{\phi_n+\Phi_n}{2}\right) dz\right| \\
&\quad + \frac{1}{n!}\left|\int_{\gamma_{x,w}} (z-w)^n \left(f^{(n)}(z) - \frac{\phi_n+\Phi_n}{2}\right) dz\right| \\
&\le \frac{1}{n!}\int_{\gamma_{u,x}} \left|(z-u)^n \left(f^{(n)}(z) - \frac{\phi_n+\Phi_n}{2}\right)\right| |dz| \\
&\quad + \frac{1}{n!}\int_{\gamma_{x,w}} \left|(z-w)^n \left(f^{(n)}(z) - \frac{\phi_n+\Phi_n}{2}\right)\right| |dz| \\
&= \frac{1}{n!}\int_{\gamma_{u,x}} |z-u|^n \left|f^{(n)}(z) - \frac{\phi_n+\Phi_n}{2}\right| |dz| \\
&\quad + \frac{1}{n!}\int_{\gamma_{x,w}} |z-w|^n \left|f^{(n)}(z) - \frac{\phi_n+\Phi_n}{2}\right| |dz| \\
&\le \frac{1}{2n!}|\Phi_n - \phi_n| \left[\int_{\gamma_{u,x}} |z-u|^n |dz| \right.\\
&\qquad \left. + \int_{\gamma_{x,w}} |z-w|^n |dz|\right],
\end{aligned}$$

which proves the desired inequality (39). □

We also have the following:

Theorem 6. *With the assumption of Theorem* 5 *and for* $\theta \in \mathbb{C}$, *we have*

$$\begin{aligned}
\int_\gamma f(z)\,dz &= \sum_{k=0}^{n-1} \frac{1}{(k+1)!}\left[\theta(-1)^k f^{(k)}(w) + (1-\theta) f^{(k)}(u)\right](w-u)^{k+1} \\
&\quad + \frac{\phi_n+\Phi_n}{2}\frac{(-1)^n}{(n+1)!}\left[\theta + (-1)^n(1-\theta)\right](w-u)^{n+1} \\
&\quad + T_n(\gamma,\theta,\phi_n,\Phi_n) \qquad (42)
\end{aligned}$$

and the remainder $T_n(\gamma,\theta,\phi_n,\Phi_n)$ *satisfies the bound*

$$\begin{aligned}
&|T_n(\gamma,\theta,\phi_n,\Phi_n)| \\
&\quad \le \frac{1}{2n!}|\Phi_n - \phi_n| \int_\gamma \left|\theta(z-u)^n + (1-\theta)(z-w)^n\right| |dz|
\end{aligned}$$

$$\leq \frac{1}{2n!} |\Phi_n - \phi_n| \left[|\theta| \int_\gamma |z-u|^n |dz| + |1-\theta| \int_\gamma |z-w|^n |dz| \right]$$
$$\leq \frac{1}{2n!} |\Phi_n - \phi_n| \max\{|\theta|, |1-\theta|\} \left[\int_\gamma [|z-u|^n + |z-w|^n] |dz| \right]. \quad (43)$$

In particular, for $\theta = \frac{1}{2}$ we get

$$\int_\gamma f(z)\, dz = \sum_{k=0}^{n-1} \frac{1}{(k+1)!} \left[\frac{(-1)^k f^{(k)}(w) + f^{(k)}(u)}{2} \right] (w-u)^{k+1}$$
$$+ \frac{\phi_n + \Phi_n}{2} \frac{(-1)^n}{(n+1)!} \left[\frac{1+(-1)^n}{2} \right] (w-u)^{n+1} + T_n(\gamma, \phi_n, \Phi_n) \quad (44)$$

and the remainder $T_n(\gamma, \phi_n, \Phi_n)$ satisfies the bounds

$$|T_n(\gamma, \phi_n, \Phi_n)| \leq \frac{1}{4n!} |\Phi_n - \phi_n| \int_\gamma |(z-u)^n + (z-w)^n| |dz|$$
$$\leq \frac{1}{4n!} |\Phi_n - \phi_n| \left[\int_\gamma [|z-u|^n + |z-w|^n] |dz| \right].$$

Proof. From the equality (23) for $\lambda = \frac{\phi_n + \Phi_n}{2}$, we have

$$\int_\gamma f(z)\, dz = \sum_{k=0}^{n-1} \frac{1}{(k+1)!} \left[\theta (-1)^k f^{(k)}(w) + (1-\theta) f^{(k)}(u) \right] (w-u)^{k+1}$$
$$+ \frac{(-1)^n}{(n+1)!} \frac{\phi_n + \Phi_n}{2} [\theta + (-1)^n (1-\theta)] (w-u)^{n+1}$$
$$+ T_n(\gamma, \theta, \phi_n, \Phi_n), \quad (45)$$

and the remainder $T_n(\gamma, \theta, \phi_n, \Phi_n)$ is given by

$$T_n(\gamma, \theta, \phi_n, \Phi_n) = \frac{(-1)^n}{n!} \int_\gamma [\theta (z-u)^n + (1-\theta)(z-w)^n]$$
$$\times \left(f^{(n)}(z) - \frac{\phi_n + \Phi_n}{2} \right) dz. \quad (46)$$

Taking the modulus in (46) and taking into account that $f^{(n)} \in \bar{\Delta}_\gamma(\phi_n, \Phi_n)$, we have

$$\begin{aligned}
&|T_n(\gamma,\theta,\phi_n,\Phi_n)| \\
&\le \frac{1}{n!}\int_\gamma |\theta(z-u)^n+(1-\theta)(z-w)^n|\left|f^{(n)}(z)-\frac{\phi_n+\Phi_n}{2}\right||dz| \\
&\le \frac{1}{2n!}|\Phi_n-\phi_n|\int_\gamma |\theta(z-u)^n+(1-\theta)(z-w)^n||dz|, \qquad (47)
\end{aligned}$$

which proves the first inequality in (43). The rest is obvious. □

5. Bounds for Hölder's Continuous Derivatives

A function $g:\gamma\subset D\subseteq\mathbb{C}\to\mathbb{C}\to C$ is *Hölder continuous* on γ with the constant $H>0$ and $r\in(0,1]$ if

$$|f(z)-f(w)|\le H|z-w|^r,$$

for all z, $w\in\gamma$.

Theorem 7. *Let $f:D\subseteq\mathbb{C}\to\mathbb{C}$ be an analytic function on the domain D and $x\in D$. Suppose $\gamma\subset D$ is a smooth path parametrized by $z(t)$, $t\in[a,b]$ with $z(a)=u$, $z(t)=x$ and $z(b)=w$ where u, $w\in D$. If $f^{(n)}$ is Hölder continuous on γ with the constant $H_n>0$ and $r\in(0,1]$,*

$$\begin{aligned}
\int_\gamma f(z)\,dz &= \sum_{k=0}^{n-1}\frac{1}{(k+1)!}f^{(k)}(x)\left[(w-x)^{k+1}+(-1)^k(x-u)^{k+1}\right] \\
&\quad+\frac{(-1)^n}{(n+1)!}\left[f^{(n)}(u)(x-u)^{n+1}+f^{(n)}(w)(-1)^n(w-x)^{n+1}\right] \\
&\quad+O_n(x,\gamma), \qquad (48)
\end{aligned}$$

where the remainder $O_n(x,\gamma)$ satisfies the bound

$$|O_n(x,\gamma)|\le\frac{1}{n!}H_n\left[\int_{\gamma_{u,x}}|z-u|^{n+r}|dz|+\int_{\gamma_{x,w}}|z-w|^{n+r}|dz|\right]. \qquad (49)$$

In particular, if $f^{(n)}$ is Lipschitzian on γ with the constant $L_n>0$, then we have the error bound

$$|O_n(x,\gamma)|\le\frac{1}{n!}L_n\left[\int_{\gamma_{u,x}}|z-u|^{n+1}|dz|+\int_{\gamma_{x,w}}|z-w|^{n+1}|dz|\right]. \qquad (50)$$

Proof. From the identity (13) we have for $\lambda_1 = f^{(n)}(u)$ and $\lambda_2 = f^{(n)}(w)$

$$\begin{aligned}\int_\gamma f(z)\,dz &= \sum_{k=0}^{n-1} \frac{1}{(k+1)!} f^{(k)}(x) \left[(w-x)^{k+1} + (-1)^k (x-u)^{k+1} \right] \\ &\quad + f^{(n)}(u) \frac{(-1)^n}{(n+1)!} (x-u)^{n+1} + f^{(n)}(w) \frac{(-1)^{n+1}}{(n+1)!} (x-w)^{n+1} \\ &\quad + O_n(x,\gamma), \qquad (51)\end{aligned}$$

with the remainder $O_n(x,\gamma)$ given by

$$\begin{aligned}O_n(x,\gamma) &= \frac{(-1)^n}{n!} \int_{\gamma_{u,x}} (z-u)^n \left[f^{(n)}(z) - f^{(n)}(u) \right] dz \\ &\quad + \frac{(-1)^n}{n!} \int_{\gamma_{x,w}} (z-w)^n \left[f^{(n)}(z) - f^{(n)}(w) \right] dz. \qquad (52)\end{aligned}$$

Taking the modulus in (52) and using the Hölder continuity, we have

$$\begin{aligned}|O_n(x,\gamma)| &\le \frac{1}{n!} \int_{\gamma_{u,x}} \left| (z-u)^n \left[f^{(n)}(z) - f^{(n)}(u) \right] \right| |dz| \\ &\quad + \frac{1}{n!} \int_{\gamma_{x,w}} \left| (z-w)^n \left[f^{(n)}(z) - f^{(n)}(w) \right] \right| |dz| \\ \text{53pt]} &= \frac{1}{n!} \int_{\gamma_{u,x}} |z-u|^n \left| f^{(n)}(z) - f^{(n)}(u) \right| |dz| \\ &\quad + \frac{1}{n!} \int_{\gamma_{x,w}} |z-w|^n \left| f^{(n)}(z) - f^{(n)}(w) \right| |dz| \\ &\le \frac{1}{n!} H_n \left[\int_{\gamma_{u,x}} |z-u|^n |z-u|^r |dz| \right. \\ &\quad \left. + \int_{\gamma_{x,w}} |z-w|^n |z-w|^r |dz| \right] \\ &= \frac{1}{n!} H_n \left[\int_{\gamma_{u,x}} |z-u|^{n+r} |dz| + \int_{\gamma_{x,w}} |z-w|^{n+r} |dz| \right],\end{aligned}$$

which proves the desired result (49). □

6. Examples for Logarithm and Exponential

Consider the function $f(z) = \operatorname{Log}(z)$ where $\operatorname{Log}(z) = \ln|z| + i \operatorname{Arg}(z)$ and $\operatorname{Arg}(z)$ is such that $-\pi < \operatorname{Arg}(z) \leq \pi$. Log is called the *principal branch* of the complex logarithmic function. The function f is analytic on all of $\mathbb{C}_\ell$ and

$$f^{(k)}(z) = \frac{(-1)^{k-1}(k-1)!}{z^k},\ k \geq 1,\ z \in \mathbb{C}_\ell.$$

Suppose $\gamma \subset \mathbb{C}_\ell$ is a *smooth path* parametrized by $z(t)$, $t \in [a,b]$ with $z(a) = u$ and $z(b) = w$ where u, $w \in \mathbb{C}_\ell$. Then

$$\begin{aligned}\int_\gamma f(z)\,dz = \int_{\gamma_{u,w}} f(z)\,dz &= \int_{\gamma_{u,w}} \operatorname{Log}(z)\,dz \\ &= z\operatorname{Log}(z)\big|_u^w - \int_{\gamma_{u,w}} (\operatorname{Log}(z))'\,z dz \\ &= w\operatorname{Log}(w) - u\operatorname{Log}(u) - \int_{\gamma_{u,w}} dz \\ &= w\operatorname{Log}(w) - u\operatorname{Log}(u) - (w-u),\end{aligned}$$

where u, $w \in \mathbb{C}_\ell$.

Consider the function $f(z) = \frac{1}{z}$, $z \in \mathbb{C}\setminus\{0\}$. Then

$$f^{(k)}(z) = \frac{(-1)^k k!}{z^{k+1}} \text{ for } k \geq 0,\ z \in \mathbb{C}\setminus\{0\},$$

and suppose $\gamma \subset \mathbb{C}_\ell := \mathbb{C}\setminus\{x+iy : x \leq 0,\ y = 0\}$ is a *smooth path* parametrized by $z(t)$, $t \in [a,b]$ with $z(a) = u$ and $z(b) = w$ where u, $w \in \mathbb{C}_\ell$. Then

$$\int_\gamma f(z)\,dz = \int_{\gamma_{u,w}} f(z)\,dz = \int_{\gamma_{u,w}} \frac{dz}{z} = \operatorname{Log}(w) - \operatorname{Log}(u),$$

for u, $w \in \mathbb{C}_\ell$.

Consider the function $f(z) = \exp(z)$, $z \in \mathbb{C}$. Then

$$f^{(k)}(z) = \exp(z) \quad \text{for} \quad k \geq 0,\ z \in \mathbb{C},$$

and suppose $\gamma \subset \mathbb{C}$ is a *smooth path* parametrized by $z(t)$, $t \in [a,b]$ with $z(a) = u$ and $z(b) = w$ where u, $w \in \mathbb{C}$. Then

$$\int_\gamma f(z)\,dz = \int_{\gamma_{u,w}} f(z)\,dz = \int_{\gamma_{u,w}} \exp(z)\,dz = \exp(w) - \exp(u).$$

We have by the equality (7) that

$$\begin{aligned}\int_\gamma f(z)\,dz &= f(x)(w-u) \\ &\quad+\sum_{k=1}^{n-1}\frac{1}{(k+1)!}f^{(k)}(x)\left[(w-x)^{k+1}+(-1)^k(x-u)^{k+1}\right] \\ &\quad+\frac{(-1)^n}{n!}\left[\int_{\gamma_{u,x}}(z-u)^n\ f^{(n)}(z)\,dz\right. \\ &\qquad\left.+\int_{\gamma_{x,w}}(z-w)^n f^{(n)}(z)\,dz\right], \qquad (53)\end{aligned}$$

for $n\geq 2$.

Suppose $\gamma\subset\mathbb{C}_\ell$ is a *smooth path* parametrized by $z(t)$, $t\in[a,b]$ with $z(a)=u$, $z(t)=x$ and $z(b)=w$ where u, x, $w\in\mathbb{C}_\ell$, then by writing the equality (53) for the function $f(z)=\frac{1}{z}$, we get the identity

$$\begin{aligned}&\operatorname{Log}(w)-\operatorname{Log}(u) \\ &\quad=\frac{w-u}{x}+\sum_{k=1}^{n-1}\frac{(-1)^k}{(k+1)}\left[\left(\frac{w-x}{x}\right)^{k+1}+(-1)^k\left(\frac{x-u}{x}\right)^{k+1}\right] \\ &\qquad+\int_{\gamma_{u,x}}\frac{(z-u)^n}{z^{n+1}}dz+\int_{\gamma_{x,w}}\frac{(z-w)^n}{z^{n+1}}dz, \qquad (54)\end{aligned}$$

for $n\geq 2$.

If we write the equality (53) for the function $f(z)=\operatorname{Log}(z)$, then we get the identity

$$\begin{aligned}&w\operatorname{Log}(w)-u\operatorname{Log}(u)-(w-u) \\ &\quad=\operatorname{Log}(x)(w-u)+x\sum_{k=1}^{n-1}\frac{(-1)^{k-1}}{(k+1)k} \\ &\qquad\times\left[\left(\frac{w-x}{x}\right)^{k+1}+(-1)^k\left(\frac{x-u}{x}\right)^{k+1}\right] \\ &\qquad\quad-\frac{1}{n}\left[\int_{\gamma_{u,x}}\left(\frac{z-u}{z}\right)^n dz+\int_{\gamma_{x,w}}\left(\frac{z-w}{z}\right)^n dz\right], \qquad (55)\end{aligned}$$

for $n\geq 2$.

Suppose $\gamma \subset \mathbb{C}$ is a *smooth path* parametrized by $z(t)$, $t \in [a,b]$ with $z(a) = u$, $z(t) = x$ and $z(b) = w$ where u, x, $w \in \mathbb{C}$. If we write the equality (53) for the function $f(z) = \exp z$, then we get

$$\begin{aligned}
&\exp(w) - \exp(u) \\
&= (w-u)\exp(x) + \exp(x)\sum_{k=1}^{n-1}\frac{1}{(k+1)!} \\
&\times \left[(w-x)^{k+1} + (-1)^k (x-u)^{k+1}\right] + \frac{(-1)^n}{n!} \\
&\times \left[\int_{\gamma_{u,x}} (z-u)^n \exp z dz + \int_{\gamma_{x,w}} (z-w)^n \exp z dz\right] \qquad (56)
\end{aligned}$$

for $n \geq 2$.

From the identity (19), we get

$$\begin{aligned}
\int_\gamma f(z)\,dz &= \frac{f(w)+f(u)}{2}(w-u) \\
&+ \sum_{k=1}^{n-1}\frac{1}{(k+1)!}\left[\frac{(-1)^k f^{(k)}(w) + f^{(k)}(u)}{2}\right](w-u)^{k+1} \\
&+ \frac{(-1)^n}{2n!}\int_\gamma \left[(z-u)^n + (z-w)^n\right] f^{(n)}(z)\,dz, \qquad (57)
\end{aligned}$$

for $n \geq 2$.

Suppose $\gamma \subset \mathbb{C}_\ell$ is a *smooth path* parametrized by $z(t)$, $t \in [a,b]$ with $z(a) = u$ and $z(b) = w$ where u, $w \in \mathbb{C}_\ell$, then by writing the equality (57) for the function $f(z) = \frac{1}{z}$, we get the identity

$$\begin{aligned}
\operatorname{Log}(w) - \operatorname{Log}(u) &= \frac{w+u}{2uw}(w-u) \\
&+ \sum_{k=1}^{n-1}\frac{1}{(k+1)}\left[\frac{u^{k+1} + (-1)^k w^{k+1}}{2u^{k+1}w^{k+1}}\right](w-u)^{k+1} \\
&+ \frac{1}{2}\int_\gamma \frac{(z-u)^n + (z-w)^n}{z^{n+1}}dz, \qquad (58)
\end{aligned}$$

for $n \geq 2$.

By writing the equality (57) for the function $f(z) = \mathrm{Log}(z)$, we get the identity

$$\begin{aligned} &w \,\mathrm{Log}\,(w) - u \,\mathrm{Log}\,(u) - (w-u) \\ &= \frac{\mathrm{Log}\,(w) + \mathrm{Log}\,(u)}{2}(w-u) \\ &\quad + \sum_{k=1}^{n-1} \frac{(-1)^{k-1}}{(k+1)\,k}\left[\frac{(-1)^k u^k + w^k}{2u^k w^k}\right](w-u)^{k+1} \\ &\quad - \frac{1}{2n}\int_\gamma \frac{(z-u)^n + (z-w)^n}{z^n}dz, \end{aligned} \tag{59}$$

for $n \geq 2$.

Suppose $\gamma \subset \mathbb{C}$ is a *smooth path* parametrized by $z(t)$, $t \in [a,b]$ with $z(a) = u$ and $z(b) = w$ where u, $w \in \mathbb{C}$. If we write the equality (57) for the function $f(z) = \exp z$, then we get

$$\begin{aligned} \exp(w) - \exp(u) &= \frac{\exp(w) + \exp(u)}{2}(w-u) \\ &\quad + \sum_{k=1}^{n-1} \frac{1}{(k+1)!}\left[\frac{(-1)^k \exp(w) + \exp(u)}{2}\right](w-u)^{k+1} \\ &\quad + \frac{(-1)^n}{2n!}\int_\gamma [(z-u)^n + (z-w)^n]\exp(z)\,dz, \end{aligned} \tag{60}$$

for $n \geq 2$.

From the equality (54), we get

$$\begin{aligned} &\left|\mathrm{Log}\,(w) - \mathrm{Log}\,(u) - \frac{w-u}{x}\right. \\ &\quad \left. - \sum_{k=1}^{n-1} \frac{(-1)^k}{(k+1)}\left[\left(\frac{w-x}{x}\right)^{k+1} + (-1)^k\left(\frac{x-u}{x}\right)^{k+1}\right]\right| \\ &\leq \int_{\gamma_{u,x}} \frac{|z-u|^n}{|z|^{n+1}}dz + \int_{\gamma_{x,w}} \frac{|z-w|^n}{|z|^{n+1}}dz, \end{aligned} \tag{61}$$

where u, x, $w \in \mathbb{C}_\ell$.

If $d_{u,x} := \inf_{z \in \gamma_{u,x}} |z|$ and $d_{x,w} := \inf_{z \in \gamma_{x,w}} |z|$ are positive and finite, then by (61) we get

$$\left| \operatorname{Log}(w) - \operatorname{Log}(u) - \frac{w-u}{x} \right.$$
$$\left. - \sum_{k=1}^{n-1} \frac{(-1)^k}{(k+1)} \left[\left(\frac{w-x}{x} \right)^{k+1} + (-1)^k \left(\frac{x-u}{x} \right)^{k+1} \right] \right|$$
$$\leq \frac{1}{d_{u,x}^{n+1}} \int_{\gamma_{u,x}} |z-u|^n \, dz + \frac{1}{d_{x,w}^{n+1}} \int_{\gamma_{x,w}} |z-w|^n \, dz. \quad (62)$$

If $d_{u,w} := \inf_{z \in \gamma_{u,w}} |z| \in (0, \infty)$, then we also have

$$\left| \operatorname{Log}(w) - \operatorname{Log}(u) - \frac{w-u}{x} \right.$$
$$\left. - \sum_{k=1}^{n-1} \frac{(-1)^k}{(k+1)} \left[\left(\frac{w-x}{x} \right)^{k+1} + (-1)^k \left(\frac{x-u}{x} \right)^{k+1} \right] \right|$$
$$\leq \frac{1}{d_{u,w}^{n+1}} \left[\int_{\gamma_{u,x}} |z-u|^n \, dz + \int_{\gamma_{x,w}} |z-w|^n \, dz \right]. \quad (63)$$

From the equality (55), we get

$$|w \operatorname{Log}(w) - u \operatorname{Log}(u) - (w-u)$$
$$- \operatorname{Log}(x)(w-u) - x \sum_{k=1}^{n-1} \frac{(-1)^{k-1}}{(k+1)k}$$
$$\times \left. \left[\left(\frac{w-x}{x} \right)^{k+1} + (-1)^k \left(\frac{x-u}{x} \right)^{k+1} \right] \right|$$
$$\leq \frac{1}{n} \left[\int_{\gamma_{u,x}} \left| \frac{z-u}{z} \right|^n dz + \int_{\gamma_{x,w}} \left| \frac{z-w}{z} \right|^n dz \right], \quad (64)$$

where u, x, $w \in \mathbb{C}_\ell$.

If $d_{u,x} := \inf_{z \in \gamma_{u,x}} |z|$ and $d_{x,w} := \inf_{z \in \gamma_{x,w}} |z|$ are positive and finite, then by (64) we get

$$\left| w \operatorname{Log}(w) - u \operatorname{Log}(u) - (w-u) \right.$$

$$- \operatorname{Log}(x)(w-u) - x \sum_{k=1}^{n-1} \frac{(-1)^{k-1}}{(k+1)k}$$

$$\left. \times \left[\left(\frac{w-x}{x} \right)^{k+1} + (-1)^k \left(\frac{x-u}{x} \right)^{k+1} \right] \right|$$

$$\leq \frac{1}{n} \left[\frac{1}{d_{u,x}^n} \int_{\gamma_{u,x}} |z-u|^n \, dz + \frac{1}{d_{x,w}^n} \int_{\gamma_{x,w}} |z-w|^n \, dz \right]. \tag{65}$$

If $d_{u,w} := \inf_{z \in \gamma_{u,w}} |z| \in (0, \infty)$, then we also have

$$\left| w \operatorname{Log}(w) - u \operatorname{Log}(u) - (w-u) \right.$$

$$- \operatorname{Log}(x)(w-u) - x \sum_{k=1}^{n-1} \frac{(-1)^{k-1}}{(k+1)k}$$

$$\left. \times \left[\left(\frac{w-x}{x} \right)^{k+1} + (-1)^k \left(\frac{x-u}{x} \right)^{k+1} \right] \right|$$

$$\leq \frac{1}{n d_{u,w}^n} \left[\int_{\gamma_{u,x}} |z-u|^n \, dz + \int_{\gamma_{x,w}} |z-w|^n \, dz \right]. \tag{66}$$

Similar inequalities may be stated by the use of the equalities (58) and (59) and we omit the details.

References

[1] S.S. Dragomir, Approximating the integral of analytic complex functions on paths from convex domains in terms of generalized Ostrowski and Trapezoid type rules, Preprint *RGMIA Res. Rep. Coll.* **22**, Art., (2019).

[2] P. Cerone, S.S. Dragomir, and J. Roumeliotis, Some Ostrowski type inequalities for n-time differentiable mappings and applications, *Demonstratio Math.* **32**(4), 697–712, (1999).

[3] P. Cerone and S.S. Dragomir, Midpoint-type rules from an inequalities point of view. *Handbook of Analytic-Computational Methods in Applied Mathematics*, pp. 135–200 (Chapman & Hall/CRC, Boca Raton, FL, 2000).

[4] K.M. Awan, J. Pečarić, and A. Vukelić, Harmonic polynomials and generalizations of Ostrowski-Grüss type inequality and Taylor formula, *J. Math. Inequal.* **9**(1), 297–319, (2015).

[5] H. Budak, F. Usta, and M.Z. Sarıkaya, New upper bounds of Ostrowski type integral inequalities utilizing Taylor expansion, *Hacet. J. Math. Stat.* **47**(3), 567–578, (2018).
[6] P. Cerone, S.S. Dragomir, and E. Kikianty, Jensen-Ostrowski inequalities and integration schemes via the Darboux expansion, Reprinted in *Ukrainian Math. J.* **69**(8), 1306–1327, (2018). *Ukraï n. Mat. Zh.* **69**(8), 1123–1140, (2017).
[7] S.S. Dragomir, Ostrowski type inequalities for Lebesgue integral: A survey of recent results, *Aust. J. Math. Anal. Appl.* **14**(1), Art. 1, 283, (2017).
[8] B. Meftah, Some new Ostrowski's inequalities for functions whose nth derivatives are logarithmically convex. *Ann. Math. Sil.* **32**(1), 275–284, (2018).
[9] A. Qayyum, A.R. Kashif, M. Shoaib, and I. Faye, Derivation and applications of inequalities of Ostrowski type for n-times differentiable mappings for cumulative distribution function and some quadrature rules, *J. Nonlinear Sci. Appl.* **9**(4), 1844–1857, (2016).
[10] P. Cerone and S.S. Dragomir, Trapezoidal-type rules from an inequalities point of view. *Handbook of Analytic-Computational Methods in Applied Mathematics*, pp. 65–134 (Chapman & Hall/CRC, Boca Raton, FL, 2000).
[11] P. Cerone, S.S. Dragomir, J. Roumeliotis, and J. Šunde, A new generalization of the trapezoid formula for n-time differentiable mappings and applications, *Demonstratio Math.* **33**(4), 719–736, (2000).

https://doi.org/10.1142/9789811261572_0011

Chapter 11

Hyperstability of Ternary Jordan Homomorphisms on Unital Ternary C^*-Algebras

Madjid Eshaghi Gordji* and Vahid Keshavarz†

Department of Mathematics, Semnan University,
P.O. Box 35195-363, Semnan, Iran
**m.eshaghi@semnan.ac.ir*
†v.keshavarz68@yahoo.com

In this chapter, we introduce the concept of ternary Jordan homomorphism on ternary C^*-algebras. Moreover, we investigate the hyperstability between unital ternary C^*-algebras for ternary Jordan homomorphisms.

1. Introduction and Preliminaries

A C^*-ternary algebra is a complex Banach space, equipped with a ternary product $(x, y, z) \to [x, y, z]$ of A^3 into A, which is C-linear in the outer variables, conjugate $\mathbb{C}$-linear in the middle variable, and associative in the sense that

$$\Big[x, y, [z, u, v]\Big] = \Big[x, [y, z, u]v\Big] = \Big[[x, y, z], u, v\Big]$$

and satisfies

$$\|[x, y, z]\| \leq \|x\|.\|y\|.\|z\|, \quad \|[x, x, x]\| = \|x\|^3,$$

(see Ref. [1]). Ternary structures and their extensions, known as n-ary algebras, have many applications in mathematical physics and photonics, such as the quark model and Nambu mechanics [2,3].

Today, many physical systems can be modeled as linear systems. The principle of additivity has various applications in physics and especially in calculating the internal energy in thermodynamics, as well as with regards to the meaning of the superposition principle (cf. [4–7]).

Consider the functional equation $\Im_1(f) = \Im_2(f)$ $(\Im)$ in a certain general setting. A mapping g is an approximate solution of $(\Im)$ if $\Im_1(g)$ and $\Im_2(g)$ are close, in some sense. The Ulam stability problem asks whether or not there is a true solution of $(\Im)$ near g. A functional equation is hyperstable if every approximate solution of the equation is an exact solution of it. Initially, the stability problem was formulated by Ulam [8] in 1940 and subsequently, one year later, Hyers [9] established the positive answer to the question of Ulam on Banach spaces. In 1978, the Hyers–Ulam Theorem was generalized by Rassias [10] and a new notion was created with respect to the stability problem, that is now known as the Hyers–Ulam–Rassias stability. In recent years, the problem of stability has been considered in a broad variety of settings by many mathematicians (cf. [11–33,42]). The reader is also referred to the book [34] for a broad collection of open problems which have attracted the interest of mathematicians worldwide over a very long period of time.

Definition 1 ([35]). A $\mathbb{C}$-linear mapping between ternary algebras $\mathfrak{A}$ and $\mathfrak{B}$; i.e. $H : \mathfrak{A} \to \mathfrak{B}$, is called

(1) ternary homomorphism if

$$H([x, y, z]) = [H(x), H(y), H(z)],$$

(2) ternary Jordan homomorphism if

$$H([x, x, x]) = [H(x), H(x), H(x)],$$

for all $x, y, z \in \mathfrak{A}$.

Throughout this chapter, let $\mathfrak{A}$ and $\mathfrak{B}$ be two unital ternary C^*-algebras with unit element e and unit element $e_{\mathfrak{B}}$, respectively Let $U(\mathfrak{A})$ be the set of unitary elements in $\mathfrak{A}$, $\mathfrak{A}_{sa} = \{x \in \mathfrak{A} |\ x^* = x\}$, and $I_1(\mathfrak{A}_{sa}) = \{u \in \mathfrak{A}_{sa} |\ \|u\| = 1, u \in Inv(\mathfrak{A})\}$.

We investigate the hyperstability of ternary Jordan homomorphisms between unital ternary C^*-algebras. Note that $\mathfrak{A}$ is a unital ternary C^*-algebra of real rank zero, if the set of invertible self-adjoint elements is dense in the set of self-adjoint elements [36,37]. We denote the algebraic center of $\mathfrak{A}$ by $Z(\mathfrak{A})$.

2. Main Results

In this section, let $\mathbb{T}^1_{1/n_0}$ be the set of all complex numbers $e^{i\theta}$, where $0 \leq \theta \leq \frac{2\pi}{n_0}$ and $\mathfrak{A}, \mathfrak{B}$ be two C^*-ternary Banach algebras.

Lemma 1. *Suppose that h is a linear mapping between $\mathfrak{A}$ and $\mathfrak{B}$. Then, the following relations are equivalent:*

$$h([x,x,x]) = [h(x), h(x), h(x)], \tag{1}$$

for all $x \in \mathfrak{A}$,

$$\begin{aligned} &h([x_1,x_2,x_3] + [x_2,x_3,x_1] + [x_3,x_1,x_2]) \\ &\quad = [h(x_1),h(x_2),h(x_3)] + [h(x_2),h(x_3),h(x_1)] + [h(x_3),h(x_1),h(x_2)], \end{aligned} \tag{2}$$

for all $x_1, x_2, x_3 \in \mathfrak{A}$.

Proof. In (1), substitute x by $x_1 + x_2 + x_3$. As a result, we derive that

$$\begin{aligned} &h[(x_1+x_2+x_3),(x_1+x_2+x_3),(x_1+x_2+x_3)] \\ &\quad = [h(x_1+x_2+x_3), h(x_1+x_2+x_3), h(x_1+x_2+x_3)], \end{aligned}$$

Thus, we have

$$\begin{aligned} &h[(x_1+x_2+x_3),(x_1+x_2+x_3),(x_1+x_2+x_3)] \\ &\quad = h([x_1,x_1,x_1] + [x_1,x_2,x_1] + [x_1,x_3,x_1] + [x_2,x_1,x_1] \\ &\qquad + [x_2,x_2,x_1] + [x_2,x_3,x_1] + [x_3,x_1,x_1] + [x_3,x_2,x_1] \\ &\qquad + [x_3,x_3,x_1] + [x_1,x_1,x_2] + [x_1,x_2,x_2] + [x_1,x_3,x_2] \\ &\qquad + [x_2,x_1,x_2] + [x_2,x_2,x_2] + [x_2,x_3,x_2] + [x_3,x_1,x_2] \\ &\qquad + [x_3,x_2,x_2] + [x_3,x_3,x_2] + [x_1,x_1,x_3] + [x_1,x_2,x_3] \\ &\qquad + [x_1,x_3,x_3] + [x_2,x_1,x_3] + [x_2,x_2,x_3] + [x_1,x_3,x_3] \\ &\qquad + [x_3,x_1,x_3] + [x_3,x_2,x_3] + [x_3,x_3,x_3]) \end{aligned}$$

$$
\begin{aligned}
&= h([x_1,x_1,x_1]) + h([x_1,x_2,x_1]) + h([x_1,x_3,x_1]) \\
&\quad + h([x_2,x_1,x_1]) + h([x_2,x_2,x_1]) + h([x_2,x_3,x_1]) \\
&\quad + h([x_3,x_1,x_1]) + h([x_3,x_2,x_1]) + h([x_3,x_1,x_1]) \\
&\quad + h([x_1,x_1,x_2]) + h([x_1,x_2,x_2]) + h([x_1,x_3,x_2]) \\
&\quad + h([x_2,x_1,x_2]) + h([x_2,x_2,x_2]) + h([x_2,x_3,x_2]) \\
&\quad + h([x_3,x_1,x_2]) + h([x_3,x_2,x_2]) + h([x_3,x_3,x_2]) \\
&\quad + h([x_1,x_1,x_3]) + h([x_1,x_2,x_3]) + h([x_1,x_3,x_3]) \\
&\quad + h([x_2,x_1,x_3]) + h([x_2,x_2,x_3]) + h([x_2,x_3,x_3]) \\
&\quad + h([x_3,x_1,x_3]) + h([x_3,x_2,x_3]) \\
&\quad + h([x_3,x_3,x_3]), \quad \forall x_1,x_2,x_3 \in \mathfrak{A}.
\end{aligned}
$$

On the other hand, for the right-hand side of the equation, we have

$$
\begin{aligned}
&[h(x_1+x_2+x_3), h(x_1+x_2+x_3), h(x_1+x_2+x_3)] \\
&\quad = [(h(x_1)+h(x_2)+h(x_3)), (h(x_1)+h(x_2) \\
&\qquad + h(x_3)), (h(x_1)+h(x_2)+f(x_3))] \\
&\quad = [h(x_1),h(x_1),h(x_1)] + [h(x_1),h(x_1),h(x_2)] + [h(x_1),h(x_1),h(x_3)] \\
&\qquad + [h(x_1),h(x_2),h(x_1)] + [h(x_1),h(x_2),h(x_2)] + [h(x_1),h(x_2),h(x_3)] \\
&\qquad + [h(x_1),h(x_3),h(x_1)] + [h(x_1),h(x_3),h(x_2)] + [h(x_1),h(x_3),h(x_3)] \\
&\qquad + [h(x_1),h(x_1),h(x_1)] + [h(x_2),h(x_1),h(x_2)] + [h(x_2),h(x_1),h(x_3)] \\
&\qquad + [h(x_2),h(x_2),h(x_1)] + [h(x_2),h(x_2),h(x_2)] + [h(x_2),h(x_2),h(x_3)] \\
&\qquad + [h(x_2),h(x_3),h(x_1)] + [h(x_2),h(x_3),h(x_2)] + [h(x_2),h(x_3),h(x_3)] \\
&\qquad + [h(x_3),h(x_1),h(x_1)] + [h(x_3),h(x_1),h(x_2)] + [h(x_3),h(x_1),h(x_3)] \\
&\qquad + [h(x_3),h(x_2),h(x_1)] + [h(x_3),h(x_2),h(x_2)] + [h(x_3),h(x_2),h(x_3)] \\
&\qquad + [h(x_3),h(x_3),h(x_1)] + [h(x_3),h(x_3),h(x_2)] + [h(x_3),h(x_3),h(x_3)],
\end{aligned}
$$

for all $x_1, x_2, x_3 \in \mathfrak{A}$. It follows that,

$$
\begin{aligned}
&h([x_1,x_2,x_3] + [x_2,x_3,x_1] + [x_3,x_1,x_2]) \\
&\quad = [h(x_1),h(x_2),h(x_3)] + [h(x_2),h(x_3),h(x_1)] + [h(x_3),h(x_1),h(x_2)]
\end{aligned}
$$

for all $x_1, x_2, x_3 \in \mathfrak{A}$. Hence, (2) holds. For the converse, replace x_1, x_2, x_3 with x_1 in (2). Then we have

$$h([x,x,x]) = [h(x), h(x), h(x)],$$

for all $x, x_1, x_2, x_3 \in \mathfrak{A}$.

We proved that the two relations (1) and (2) are equivalent. □

We need Theorem (1) in our main results.

Theorem 1 ([40]). *Let $n_0 \in \mathbb{N}$ be a fixed positive integer, and X and Y be linear spaces and $f : X \to Y$ be an additive function. Then $f \forall x \in X$ and*

$$\mu \in T^1_{\frac{1}{n_0}} = \left\{ e^{i\theta};\ 0 \leq \theta \leq \frac{2\pi}{n_0} \right\}$$

is linear if and only if $f(\mu x) = \mu f(x)$.

We use [38] in order to obtain the following theorem.

Theorem 2. *Suppose f is a mapping between $\mathfrak{A}$ and $\mathfrak{B}$ such that $f(0) = 0$ and*

$$\begin{aligned} &f([3^n u 3^n v y] + [3^n v y 3^n u] + [y 3^n u 3^n v]) \\ &\quad = ([f(3^n u) f(3^n v) f(y)] + [f(3^n v) f(y) f(3^n u)] + [f(y) f(3^n u) f(3^n v)]), \end{aligned} \tag{3}$$

for all $u, v \in I_1(\mathfrak{A}_{sa})$, all $y \in \mathfrak{A}$, and all $n = 0, 1, 2, \ldots$. Assume as well that there exists a function $\phi : (\mathfrak{A} - \{0\})^2 \to [0, \infty)$ such that

$$\widetilde{\phi}(x,y) = \sum_{n=0}^{\infty} 3^{-n} \phi(3^n x, 3^n y) < \infty \quad \text{for all } x, y \in \mathfrak{A} - \{0\},$$

and that

$$\left\| 2f\left(\frac{\mu x + \mu y}{2}\right) - \mu f(x) - \mu f(y) \right\| \leq \phi(x,y), \tag{4}$$

for all $\mu \in T_{\frac{1}{n_0}}$ and all $x, y \in \mathfrak{A}$. If

$$\lim_{n\to\infty} \frac{f(3^n e)}{3^n} \in I_1(\mathfrak{B}_{sa}) \bigcap Z(\mathfrak{B}),$$

then the mapping $f : \mathfrak{A} \to \mathfrak{B}$ is a ternary Jordan homomorphism.

Proof. Referring to [39, Theorem 1], there exists a unique additive mapping $h : \mathfrak{A} \to \mathfrak{B}$ such that

$$\|f(x) - h(x)\| \leq \frac{1}{3}\Big(\widetilde{\phi}(x, -x) + \widetilde{\phi}(-x, 3x)\Big), \quad \forall x \in \mathfrak{A} - \{0\}. \tag{5}$$

And

$$h(x) = \lim_{n\to\infty} \frac{f(3^n x)}{3^n}, \quad \forall x \in \mathfrak{A}.$$

For all $\mu \in T^1_{\frac{1}{n_0}}$, by Theorem 1 f is a $\mathbb{C}$-linear. By the proof of Theorem 3.3 of [40], we have that h is $\mathbb{C}$-linear. Using (3) we derive that

$$\begin{aligned}
&h([uvy] + [vyu] + [yuv])\\
&\quad = \lim_{n\to\infty} f\left(\frac{[3^n u 3^n vy]}{9^n} + \frac{[3^n vy 3^n u]}{9^n} + \frac{[y 3^n u 3^n v]}{9^n}\right)\\
&\quad = \lim_{n\to\infty} \left(\frac{[f(3^n u) f(3^n v) f(y)]}{9^n} + \frac{[f(3^n v) f(y) f(3^n u)]}{9^n}\right.\\
&\qquad\qquad \left. + \frac{[f(y) f(3^n u) f(3^n v)]}{9^n}\right)\\
&\quad = \lim_{n\to\infty} \left(\frac{[f(3^n u) f(3^n v) f(y)] + [f(y) f(3^n u) f(3^n v)] + [f(3^n v) f(y) f(3^n u)]}{9^n}\right)\\
&\quad = ([h(u)h(v)f(y)] + [h(v)f(y)h(u)] + [f(y)h(u)h(v)])\\
&\qquad \forall u, v \in U(\mathfrak{A}), \quad \forall y \in \mathfrak{A}.
\end{aligned} \tag{6}$$

Since h is additive, then by (6), we have

$$\begin{aligned}
&3^n h([uvy] + [vyu] + [yuv])\\
&\quad = h([uv(3^n y)] + [v(3^n y)u] + [(3^n y)uv])\\
&\quad = ([h(u)h(v)f(3^n y)] + [h(v)f(3^n y)h(u)] + [f(3^n y)h(u)h(v)]),
\end{aligned}$$

for all $u, v \in U(\mathfrak{A})$ and $y \in \mathfrak{A}$. Hence,

$$\begin{aligned}
&h([uvy] + [vyu] + [yuv])\\
&\quad = \lim_{n\to\infty} \left(\left[h(u)h(v)\frac{f(3^n y)}{3^n}\right] + \left[h(v)\frac{f(3^n y)}{3^n}h(u)\right] + \left[\frac{f(3^n y)}{3^n}h(u)h(v)\right]\right)\\
&\quad = ([h(u)h(v)h(y)] + [h(v)h(y)h(u)] + [h(y)h(u)h(v)]),
\end{aligned} \tag{7}$$

for all $u, v \in U(\mathfrak{A})$ and $y \in \mathfrak{A}$. Assuming that

$$h(e) = \lim_{n\to\infty} \frac{f(3^n e)}{3^n} \in U(\mathfrak{B}),$$

it follows by (6) and (7) that

$$\begin{aligned}
&([h(e)h(e)h(y)] + [h(e)h(y)h(e)] + [h(y)h(e)h(e)]) \\
&\quad = (h([eey] + [eye] + [yee])) \\
&\quad = ([h(e)h(e)f(y)] + [h(e)f(y)h(e)] + [f(y)h(e)h(e)]), \quad \forall y \in \mathfrak{A}.
\end{aligned}$$

Since $h(e)$ belongs to $I_1(\mathfrak{B}_{sa})$, we have

$$\begin{aligned}
3h(y) &= ([e_{\mathfrak{B}} e_{\mathfrak{B}} h(y)] + [e_{\mathfrak{B}} h(y) e_{\mathfrak{B}}] + [h(y) e_{\mathfrak{B}} e_{\mathfrak{B}}]) \\
&= [[h(e)^{-1} e_{\mathfrak{B}} h(e)] e_{\mathfrak{B}} h(y)] + [[h(e)^{-1} e_{\mathfrak{B}} h(e)] h(y) e_{\mathfrak{B}}] \\
&\quad + [h(y)[h(e) e_{\mathfrak{B}} h(e)^{-1}] e_{\mathfrak{B}}] \\
&= [h(e)^{-1}[e_{\mathfrak{B}} h(e) e_{\mathfrak{B}}] h(y)] + [h(e)^{-1} e_{\mathfrak{B}} [h(e) h(y) e_{\mathfrak{B}}]] \\
&\quad + [h(y) h(e) [e_{\mathfrak{B}} h(e)^{-1} e_{\mathfrak{B}}]] \\
&= [h(e)^{-1}[e_{\mathfrak{B}} e_{\mathfrak{B}} h(e)] h(y)] + [h(e)^{-1} e_{\mathfrak{B}} [h(e) h(y) [h(e)^{-1} e_{\mathfrak{B}} h(e)]]] \\
&\quad + [h(y) h(e) [[h(e) e_{\mathfrak{B}} h(e)^{-1}] h(e)^{-1} e_{\mathfrak{B}}]] \\
&= [h(e)^{-1} e_{\mathfrak{B}} [e_{\mathfrak{B}} h(e) h(y)]] + [h(e)^{-1} e_{\mathfrak{B}} [h(e) h(y) [h(e)^{-1} e_{\mathfrak{B}} h(e)]]] \\
&\quad + [h(y) h(e) [h(e) [e_{\mathfrak{B}} h(e)^{-1} h(e)^{-1}] e_{\mathfrak{B}}]] \\
&= [h(e)^{-1} e_{\mathfrak{B}} [[h(e)^{-1} e_{\mathfrak{B}} h(e)] h(e) h(y)]] + [h(e)^{-1} e_{\mathfrak{B}} [h(e) h(y) [h(e)^{-1} h(e) e_{\mathfrak{B}}]]] \\
&\quad + [[h(y) h(e) h(e)] [e_{\mathfrak{B}} h(e)^{-1} h(e)^{-1}] e_{\mathfrak{B}}] \\
&= [h(e)^{-1} e_{\mathfrak{B}} [h(e)^{-1} e_{\mathfrak{B}} [h(e) h(e) h(y)]]] + [h(e)^{-1} e_{\mathfrak{B}} [h(e) h(y) [h(e) h(e)^{-1} e_{\mathfrak{B}}]]] \\
&\quad + [[f(y) h(e) h(e)] [e_{\mathfrak{B}} h(e)^{-1} h(e)^{-1}] e_{\mathfrak{B}}] \\
&= [h(e)^{-1} e_{\mathfrak{B}} [h(e)^{-1} e_{\mathfrak{B}} [h(e) h(e) f(y)]]] + [h(e)^{-1} e_{\mathfrak{B}} [h(e) h(y) h(e)] h(e)^{-1} e_{\mathfrak{B}}]] \\
&\quad + [[f(y) h(e) [h(e) e_{\mathfrak{B}} h(e)^{-1}] h(e)^{-1}] e_{\mathfrak{B}}] \\
&= [h(e)^{-1} e_{\mathfrak{B}} [h(e)^{-1} e_{\mathfrak{B}} h(e)] h(e) f(y)]] + [h(e)^{-1} e_{\mathfrak{B}} [h(e) f(y) h(e)] h(e)^{-1} e_{\mathfrak{B}}]] \\
&\quad + [[f(y) h(e) [e_{\mathfrak{B}} h(e)^{-1} e_{\mathfrak{B}}]] \\
&= [h(e)^{-1}[e_{\mathfrak{B}} e_{\mathfrak{B}} h(e)] f(y)] + [h(e)^{-1} e_{\mathfrak{B}} [h(e) f(y) [h(e) h(e)^{-1} e_{\mathfrak{B}}]]] \\
&\quad + [f(y)[h(e) e_{\mathfrak{B}} h(e)^{-1}] e_{\mathfrak{B}}]
\end{aligned}$$

$$
\begin{aligned}
&= [h(e)^{-1}e_{\mathfrak{B}}[e_{\mathfrak{B}}h(e)f(y)]] + [h(e)^{-1}e_{\mathfrak{B}}[h(e)f(y)[h(e)e_{\mathfrak{B}}h(e)^{-1}]]] \\
&\quad + [f(y)e_{\mathfrak{B}}e_{\mathfrak{B}}] \\
&= [h(e)^{-1}[e_{\mathfrak{B}}e_{\mathfrak{B}}h(e)]f(y)] + [[h(e)^{-1}e_{\mathfrak{B}}h(e)]f(y)e_{\mathfrak{B}}] + f(y) \\
&= [h(e)^{-1}e_{\mathfrak{B}}h(e)]e_{\mathfrak{B}}f(y)] + [e_{\mathfrak{B}}f(y)e_{\mathfrak{B}}] + f(y) \\
&= [e_{\mathfrak{B}}e_{\mathfrak{B}}f(y)] + f(y) + f(y) \\
&= f(y) + f(y) + f(y) = 3f(y), \quad \forall y \in \mathfrak{A}. \qquad (8)
\end{aligned}
$$

It follows from (8) that $h(y) = f(y)$.

We will show that f is a ternary Jordan homomorphism. For every $a, b \in \mathfrak{A}$, we define $a\Diamond b := [aeb]$. Then $\Diamond$ from $\mathfrak{A} \times \mathfrak{A}$ into $\mathfrak{A}$ is a binary product for which $(\mathfrak{A}, \Diamond)$ may be considered as a (binary) C^*-algebra. Additionally, we have that

$$a \in U(\mathfrak{A}, [\ ]) \quad \text{if and only if} \quad a \in U((\mathfrak{A}, \Diamond)) \quad \forall a \in \mathfrak{A}.$$

Now, suppose that $a, b \in \mathfrak{A}$. Referring to [41, Theorem 4.1.7], a, b are finite linear combinations of unitary elements, i.e.

$$a = \sum_{i=1}^{n} c_i u_i, b = \sum_{j=1}^{m} d_j v_j c_i, d_j \in C, u_i, v_j \in U(\mathfrak{A})).$$

It follows by (7) that

$$
\begin{aligned}
&f([aby] + [bya] + [yab]) \\
&\quad = h([aby] + [bya] + [yab]) \\
&\quad = h\left(\left[\sum_{i=1}^{n}\sum_{j=1}^{m} c_i u_i d_j v_j y\right] + \left[\sum_{j=1}^{m} d_j v_j y \sum_{i=1}^{n} c_i u_i\right]\right. \\
&\qquad\qquad \left. + \left[y \sum_{i=1}^{n}\sum_{j=1}^{m} c_i u_i d_j v_j\right]\right) \\
&\quad = h\left(\left[\sum_{i=1}^{n}\sum_{j=1}^{m} c_i d_j u_i v_j y\right] + \left[\sum_{j=1}^{m} d_j v_j y \sum_{i=1}^{n} c_i u_i\right]\right. \\
&\qquad\qquad \left. + \left[y \sum_{i=1}^{n}\sum_{j=1}^{m} c_i d_j u_i v_j\right]\right)
\end{aligned}
$$

$$= \left(\left(\sum_{i=1}^{n} \sum_{j=1}^{m} c_i d_j h[u_i v_j y] \right) + \left(\sum_{j=1}^{m} d_j h[v_j y u_i] \sum_{i=1}^{n} c_i \right) \right.$$

$$\left. + (h[y u_i v_j] \sum_{i=1}^{n} \sum_{j=1}^{m} c_i d_j \right)$$

$$= \left(\left(\sum_{i=1}^{n} \sum_{j=1}^{m} c_i d_j [h(u_i) h(v_j) h(y)] \right) + \left(\sum_{j=1}^{m} d_j [h(v_j) h(y) h(u_i)] \sum_{i=1}^{n} c_i \right) \right.$$

$$\left. + \left([h(y) h(u_i) h(v_j)] \sum_{i=1}^{n} \sum_{j=1}^{m} c_i d_j \right) \right)$$

$$= \left(\left[\sum_{i=1}^{n} \sum_{j=1}^{m} c_i d_j h(u_i) h(v_j) h(y) \right] + \left[\sum_{j=1}^{m} d_j h(v_j) h(y) \sum_{i=1}^{n} c_i h(u_i) \right] \right.$$

$$\left. + \left[h(y) \sum_{i=1}^{n} \sum_{j=1}^{m} c_i d_j h(u_i) h(v_j) \right] \right)$$

$$= \left(\left[h \left(\sum_{i=1}^{n} c_i u_i \right) h \left(\sum_{j=1}^{m} d_j v_j \right) h(y) \right] \right.$$

$$+ \left[h \left(\sum_{j=1}^{m} d_j v_j \right) h(y) h \left(\sum_{i=1}^{n} c_i u_i \right) \right]$$

$$\left. + \left[h(y) h \left(\sum_{i=1}^{n} c_i u_i \right) h \left(\sum_{j=1}^{m} d_j v_j \right) \right] \right)$$

$$= ([h(a)h(b)h(y)] + [h(b)h(y)h(a)] + [h(y)h(a)h(b)]), \quad \forall y \in \mathfrak{A}.$$

This completes the proof of the theorem. □

Corollary 1. *Let $p \in (0,1), \theta \in [0,\infty)$ be real numbers. Let $f : \mathfrak{A} \to \mathfrak{B}$ be a mapping such that $f(0) = 0$ and that*

$$f([3^n u 3^n v y] + [3^n v y 3^n u] + [y 3^n u 3^n v])$$
$$= ([f(3^n u) f(3^n v) f(y)] + [f(3^n v) f(y) f(3^n u)] + [f(y) f(3^n u) f(3^n v)]),$$

for all $u, v \in U(\mathfrak{A})$, *all* $y \in \mathfrak{A}$, *and all* $n = 0, 1, 2, \ldots$. *Suppose that*

$$\left\| 2f\left(\frac{\mu x + \mu y}{2}\right) - \mu f(x) - \mu f(y) \right\| \leq \theta(\|x\|^p + \|y\|^p),$$

for all $\mu \in T_{\frac{1}{n_0}}$ *and all* $x, y \in \mathfrak{A}$. *If* $lim_{n\to\infty} \frac{f(3^n e)}{3^n} \in I_1(B_{sa})$, *then the mapping* $f : \mathfrak{A} \to \mathfrak{B}$ *is a ternary Jordan homomorphism.*

After here, let $\mathfrak{A}$ be a unital C^*-algebra of real rank zero.

Now, we investigate between unital C^*-algebras continuous ternary Jordan homomorphisms.

Theorem 3. *Let* $f : \mathfrak{A} \to \mathfrak{B}$ *be a continuous mapping such that* $f(0) = 0$ *and that*

$$\begin{aligned} &f([3^n u 3^n v y] + [3^n v y 3^n u] + [y 3^n u 3^n v]) \\ &\quad = ([f(3^n u) f(3^n v) f(y)] + [f(3^n v) f(y) f(3^n u)] + [f(y) f(3^n u) f(3^n v)]), \end{aligned} \tag{9}$$

for all $y \in \mathfrak{A}$, *all* $u, v \in I_1(\mathfrak{A}_{sa})$, *and all* $n = 0, 1, 2, \ldots$. *Suppose that there exists a function* $\phi : (\mathfrak{A} - \{0\})^2 \to [0, \infty)$ *satisfying* (4) *and* $\widetilde{\phi}(x, y) < \infty$ $\forall x, y \in \mathfrak{A} - \{0\}$. *If*

$$\lim_{n\to\infty} \frac{f(3^n e)}{3^n} \in I_1(\mathfrak{B}_{sa}),$$

then the mapping $f : \mathfrak{A} \to \mathfrak{B}$ *is a ternary Jordan homomorphism.*

Proof. By proof of Theorem 2, there exists a unique $\mathbb{C}$-linear mapping $h : \mathfrak{A} \to \mathfrak{B}$ satisfying (5). It follows from (9) that

$$\begin{aligned} &h([3^n u 3^n v y] + [3^n v y 3^n u] + [y 3^n u 3^n v]) \\ &\quad = \lim_{n\to\infty} f\left(\frac{[3^n u 3^n v y]}{9^n} + \frac{[3^n v y 3^n u]}{9^n} + \frac{[y 3^n u 3^n v]}{9^n}\right) \\ &\quad = \lim_{n\to\infty} \left(\frac{[f(3^n u) f(3^n v) f(y)] + [f(y) f(3^n u) f(3^n v)] + [f(3^n v) f(y) f(3^n u)]}{9^n}\right) \\ &\quad = ([h(u) h(v) f(y)] + [h(v) f(y) h(u)] + [f(y) h(u) h(v)]), \end{aligned} \tag{10}$$

for all $y \in \mathfrak{A}$ and all $u, v \in I_1(\mathfrak{A}_{sa})$. By additivity of h and (10), we obtain that

$$\begin{aligned} 3^n h([uvy] + [vyu] + [yuv]) &= h([uv(3^n y)] + [v(3^n y)u] + [(3^n y)uv]) \\ &= ([h(u)h(v)f(3^n y)] + [h(v)f(3^n y)h(u)] \\ &\quad + [f(3^n y)h(u)h(v)]), \end{aligned}$$

for all $y \in \mathfrak{A}$ and all $u, v \in I_1(\mathfrak{A}_{sa})$. Hence,

$$h([uvy]+[vyu]+[yuv]) \lim_{n\to\infty}\left(\left[h(u)h(v)\frac{f(3^n y)}{3^n}\right]+\left[h(v)\frac{f(3^n y)}{3^n}h(u)\right]\right.$$
$$\left.+\left[\frac{f(3^n y)}{3^n}h(u)h(v)\right]\right), \tag{11}$$

for all $y \in \mathfrak{A}$ and all $u, v \in I_1(\mathfrak{A}_{sa})$.

Assuming, we have

$$h(e) = \lim_{n\to\infty}\frac{f(3^n e)}{3^n} \in U(B).$$

Similar to the proof of Theorem 2, it follows from (10) and (11) that $h = f$ on $\mathfrak{A}$. So h is continuous. On the other hand, $\mathfrak{A}$ is real rank zero. One can easily show that $I_1(A_{sa})$ is dense in $\{x \in A_{sa} : \|x\| = 1\}$. Let $u, v \in \{x \in A_{sa} : \|x\| = 1\}$. There are $\{t_n\}, \{z_n\}$ in $I_1(\mathfrak{A}_{sa})$ such that $\lim_{n\to\infty} t_n = u$, $\lim_{n\to\infty} z_n = v$. Since h is continuous, it follows from (11) that

$$\begin{aligned}
&h([uvy]+[vyu]+[yuv])\\
&\quad= h\Big(\lim_{n\to\infty}((t_n z_n y)+(z_n y t_n)+(y t_n z_n))\Big)\\
&\quad= \lim_{n\to\infty} h\Big([t_n z_n y]+[z_n y t_n]+[y t_n z_n]\Big)\\
&\quad= \lim_{n\to\infty}\Big([h(t_n)h(z_n)h(y)]+[h(z_n)h(y)h(t_n)]+[h(y)h(t_n)h(z_n)]\Big)\\
&\quad= ([h(u)h(v)h(y)]+[h(v)h(y)h(u)]+[h(y)h(u)h(v)]), \quad \forall y \in \mathfrak{A}. \qquad (12)
\end{aligned}$$

New, let $a, b \in \mathfrak{A}$. Then we have $a = a_1 + ia_2, b = b_1 + ib_2$, where $a_1 := \frac{a+a^*}{2}, b_1 := \frac{b+b^*}{2}$ and $a_2 := \frac{a-a^*}{2i}, b_2 := \frac{b-b^*}{2i}$ are self-adjoint. First $a_2 = b_2 = 0, a_1, b_1 \neq 0$. Since h is $\mathbb{C}$-linear, it follows from (12) that

$$\begin{aligned}
&f([aby]+[bya]+[yab])\\
&\quad= h([aby]+[bya]+[yab])\\
&\quad= h([a_1 b_1 y]+[b_1 y a_1]+[y a_1 b_1])\\
&\quad= h\left(\left(\|a_1\|\|b_1\|\left[\frac{a_1}{\|a_1\|}\frac{b_1}{\|b_1\|}y\right]\right)+\left(\|b_1\|\left[\frac{b_1}{\|b_1\|}y\frac{a_1}{\|a_1\|}\right]\|a_1\|\right)\right.\\
&\qquad\left.+\left(\left[y\frac{a_1}{\|a_1\|}\frac{b_1}{\|b_1\|}\right]\|a_1\|\|b_1\|\right)\right)
\end{aligned}$$

$$\begin{aligned}
&= \left(\|a_1\|\|b_1\| h\left(\left[\frac{a_1}{\|a_1\|}\frac{b_1}{\|b_1\|}y\right]\right) + \|b_1\| h\left(\left[\frac{b_1}{\|b_1\|}y\frac{a_1}{\|a_1\|}\right]\right)\|a_1\| \right.\\
&\quad \left. + h\left(\left[y\frac{a_1}{\|a_1\|}\frac{b_1}{\|b_1\|}\right]\right)\|a_1\|\|b_1\| \right)\\
&= \left(\|a_1\|\|b_1\| \left[h\left(\frac{a_1}{\|a_1\|}\right) h\left(\frac{b_1}{\|b_1\|}\right) h(y)\right] \right.\\
&\quad + \|b_1\| \left[h\left(\frac{b_1}{\|b_1\|}\right) h(y) h\left(\frac{a_1}{\|a_1\|}\right)\right] \|a_1\|\\
&\quad \left. + \left[h(y) h\left(\frac{a_1}{\|a_1\|}\right) h\left(\frac{b_1}{\|b_1\|}\right)\right] \|a_1\|\|b_1\| \right)\\
&= \left(\left[h\left(\|a_1\|\frac{a_1}{\|a_1\|}\right) h\left(\|b_1\|\frac{b_1}{\|b_1\|}\right) h(y)\right] \right.\\
&\quad + \left[h\left(\|b_1\|\frac{b_1}{\|b_1\|}\right) h(y) h\left(\|a_1\|\frac{a_1}{\|a_1\|}\right)\right]\\
&\quad \left. + \left[h(y) h\left(\|a_1\|\frac{a_1}{\|a_1\|}\right) h\left(\|b_1\|\frac{b_1}{\|b_1\|}\right)\right] \right)\\
&= ([h(a_1)h(b_1)h(y)] + [h(b_1)h(y)h(a_1)] + [h(y)h(a_1)h(b_1)])\\
&= ([f(a)f(b)f(y)] + [f(b)f(y)f(a)] + [f(y)f(a)f(b)]), \quad \forall y \in \mathfrak{A}.
\end{aligned}$$

New, consider $a_1 = b_1 = 0, a_2, b_2 \neq 0$. Since h is $\mathbb{C}$-linear, it follows from (12) that

$$\begin{aligned}
&f([aby] + [bya] + [yab])\\
&\quad = h([aby] + [bya] + [yab])\\
&\quad = h([ia_2ib_2y] + [ib_2yia_2] + [yia_2ib_2])\\
&\quad = h\left(- \left(\|a_2\|\|b_2\| \left[\frac{a_2}{\|a_2\|}\frac{b_2}{\|b_2\|}y\right]\right) - \left(\|b_2\| \left[\frac{b_2}{\|b_2\|}y\frac{a_2}{\|a_2\|}\right] \|a_2\| \right) \right.\\
&\qquad \left. - \left(\left[y\frac{a_2}{\|a_2\|}\frac{b_2}{\|b_2\|}\right] \|a_2\|\|b_2\| \right) \right)\\
&\quad = \left(-\|a_2\|\|b_2\| h\left(\left[\frac{a_2}{\|a_2\|}\frac{b_2}{\|b_2\|}y\right]\right) - \|b_2\| h\left(\left[\frac{b_2}{\|b_2\|}y\frac{a_2}{\|a_2\|}\right]\right)\|a_2\| \right.\\
&\qquad \left. - h\left(\left[y\frac{a_2}{\|a_2\|}\frac{b_2}{\|b_2\|}\right]\right)\|a_2\|\|b_2\| \right)
\end{aligned}$$

$$
\begin{aligned}
&= \left(-\|a_2\|\|b_2\| \left[h\left(\frac{a_2}{\|a_2\|}\right) h\left(\frac{b_2}{\|b_2\|}\right) h(y)\right]\right. \\
&\quad - \|b_2\| \left[h\left(\frac{b_2}{\|b_2\|}\right) h(y) h\left(\frac{a_2}{\|a_2\|}\right)\right] \|a_2\| \\
&\quad \left. - \left[h(y) h\left(\frac{a_2}{\|a_2\|}\right) h\left(\frac{b_2}{\|b_2\|}\right)\right] \|a_2\|\|b_2\|\right) \\
&= \left(\left[h\left(i\|a_2\|\frac{a_2}{\|a_2\|}\right) h\left(i\|b_2\|\frac{b_2}{\|b_2\|}\right) h(y)\right]\right. \\
&\quad + \left[h\left(i\|b_2\|\frac{b_2}{\|b_2\|}\right) h(y) h\left(i\|a_2\|\frac{a_2}{\|a_2\|}\right)\right] \\
&\quad \left. + \left[h(y) h\left(i\|a_2\|\frac{a_2}{\|a_2\|}\right) h\left(i\|b_2\|\frac{b_2}{\|b_2\|}\right)\right]\right) \\
&= ([h(ia_2)h(ib_2)h(y)] + [h(ib_2)h(y)h(ia_2)] + [h(y)h(ia_2)h(ib_2)]) \\
&= ([f(a)f(b)f(y)] + [f(b)f(y)f(a)] + [f(y)f(a)f(b)]), \quad \forall y \in \mathfrak{A}.
\end{aligned}
$$

Suppose $a_2 = b_1 = 0, a_2, b_2 \neq 0$. Then by (12), we have

$$
\begin{aligned}
& f([aby] + [bya] + [yab]) \\
&\quad = h([aby] + [bya] + [yab]) \\
&\quad = h([a_1 ib_2 y] + [ib_2 y a_1] + [y a_1 ib_2]) \\
&\quad = h\left(\left(i\|a_1\|\|b_2\| \left[\frac{a_1}{\|a_1\|}\frac{b_2}{\|b_2\|} y\right]\right) + \left(i\|b_2\| \left[\frac{b_2}{\|b_2\|} y \frac{a_1}{\|a_1\|}\right] \|a_1\|\right]\right) \\
&\qquad \left. + \left(\left[y \frac{a_1}{\|a_1\|}\frac{b_2}{\|b_2\|}\right] i\|a_1\|\|b_2\|\right)\right) \\
&\quad = \left(\|a_1\|\|b_1\| h\left(\left[\frac{a_1}{\|a_1\|}\frac{b_1}{\|b_1\|} y\right]\right) + \|b_1\| h\left(\left[\frac{b_1}{\|b_1\|} y \frac{a_1}{\|a_1\|}\right]\right) \|a_1\|\right. \\
&\qquad \left. + h\left(\left[y \frac{a_1}{\|a_1\|}\frac{b_1}{\|b_1\|}\right]\right) i\|a_1\|\|b_1\|\right) \\
&\quad = \left(i\|a_1\|\|b_2\| \left[h\left(\frac{a_1}{\|a_1\|}\right) h\left(\frac{b_2}{\|b_2\|}\right) h(y)\right]\right. \\
&\qquad + i\|b_2\| \left[h\left(\frac{b_2}{\|b_2\|}\right) h(y) h\left(\frac{a_1}{\|a_1\|}\right)\right] \|a_1\|
\end{aligned}
$$

$$
\begin{aligned}
&\quad + \left[h(y)h\left(\frac{a_1}{\|a_1\|}\right)h\left(\frac{b_2}{\|b_2\|}\right)\right] i\|a_1\|\|b_2\|\Bigg)\\
&= \Bigg(\left[h\left(\|a_1\|\frac{a_1}{\|a_1\|}\right)h\left(i\|b_2\|\frac{b_2}{\|b_2\|}\right)h(y)\right]\\
&\quad + \left[h\left(i\|b_2\|\frac{b_2}{\|b_2\|}\right)h(y)h\left(\|a_1\|\frac{a_1}{\|a_1\|}\right)\right]\\
&\quad + \left[h(y)h\left(\|a_1\|\frac{a_1}{\|a_1\|}\right)h\left(i\|b_2\|\frac{b_2}{\|b_2\|}\right)\right]\Bigg)\\
&= ([h(a_1)h(ib_2)h(y)] + [h(ib_2)h(y)h(a_1)] + [h(y)h(a_1)h(2b_i)])\\
&= ([f(a)f(b)f(y)] + [f(b)f(y)f(a)] + [f(y)f(a)f(b)]), \quad \forall y \in \mathfrak{A}.
\end{aligned}
$$

Similarly, we can show that

$$
f\Big([aby]+[bya]+[yab]\Big) = \Big([f(a)f(b)f(y)]+[f(b)f(y)f(a)]+[f(y)f(a)f(b)]\Big),
$$

if $a_1 = b_2 = 0, a_2, b_1 \neq 0$. In the case that $b_2 = 0, a_1, a_2, b_1 \neq 0$, we have

$$
\begin{aligned}
&f\Big([aby] + [bya] + [yab]\Big) = h\Big([aby] + [bya] + [yab]\Big)\\
&\quad = h\Big([(a_1 + ia_2)b_1y] + [b_1y(a_1 + ia_2)] + [y(a_1 + ia_2)b_1]\Big)\\
&\quad = h\Big([a_1b_1y] + [b_1ya_1] + [ya_1b_1]\Big) + ih\Big([a_2b_1y] + [b_1ya_2] + [ya_2b_1]\Big)\\
&\quad = h\Bigg(\left(\|a_1\|\|b_1\|\left[\frac{a_1}{\|a_1\|}\frac{b_1}{\|b_1\|}y\right]\right) + \left(\|b_1\|\left[\frac{b_1}{\|b_1\|}y\frac{a_1}{\|a_1\|}\right]\|a_1\|]\right)\\
&\qquad + \left(\left[y\frac{a_1}{\|a_1\|}\frac{b_1}{\|b_1\|}\right]\|a_1\|\|b_1\|\right)\Bigg)\\
&\quad + ih\Bigg(\left(\|a_2\|\|b_1\|\left[\frac{a_2}{\|a_2\|}\frac{b_1}{\|b_1\|}y\right]\right) + \left(\|b_1\|\left[\frac{b_1}{\|b_1\|}y\frac{a_2}{\|a_{12}\|}\right]\|a_2\|]\right)\\
&\qquad + \left(\left[y\frac{a_2}{\|a_2\|}\frac{b_1}{\|b_1\|}\right]\|a_2\|\|b_1\|\right)\Bigg)\\
&\quad = \Bigg(\|a_1\|\|b_1\|h\left(\left[\frac{a_1}{\|a_1\|}\frac{b_1}{\|b_1\|}y\right]\right) + \|b_1\|h\left(\left[\frac{b_1}{\|b_1\|}y\frac{a_1}{\|a_1\|}\right]\right)\|a_1\|\\
&\qquad + h\left(\left[y\frac{a_1}{\|a_1\|}\frac{b_1}{\|b_1\|}\right]\right)\|a_1\|\|b_1\|\Bigg)
\end{aligned}
$$

$$+\left(i\|a_2\|\|b_1\|h\left(\left[\frac{a_2}{\|a_2\|}\frac{b_1}{\|b_1\|}y\right]\right)+i\|b_1\|h\left(\left[\frac{b_1}{\|b_1\|}y\frac{a_2}{\|a_2\|}\right]\right)\|a_2\|\right.$$

$$\left.+h\left(\left[y\frac{a_2}{\|a_2\|}\frac{b_1}{\|b_1\|}\right]\right)i\|a_2\|\|b_1\|\right)$$

$$=\left(\|a_1\|\|b_1\|\left[h\left(\frac{a_1}{\|a_1\|}\right)h\left(\frac{b_1}{\|b_1\|}\right)h(y)\right]\right.$$

$$+\|b_1\|\left[h\left(\frac{b_1}{\|b_1\|}\right)h(y)h\left(\frac{a_1}{\|a_1\|}\right)\right]\|a_1\|$$

$$\left.+\left[h(y)h\left(\frac{a_1}{\|a_1\|}\right)h\left(\frac{b_1}{\|b_1\|}\right)\right]\|a_1\|\|b_1\|\right)$$

$$+\left(i\|a_2\|\|b_1\|\left[h\left(\frac{a_2}{\|a_2\|}\right)h\left(\frac{b_1}{\|b_1\|}\right)h(y)\right]\right.$$

$$+i\|b_1\|\left[h\left(\frac{b_1}{\|b_1\|}\right)h(y)h\left(\frac{a_2}{\|a_2\|}\right)\right]\|a_2\|$$

$$\left.+\left[h(y)h\left(\frac{a_2}{\|a_2\|}\right)h\left(\frac{b_1}{\|b_1\|}\right)\right]i\|a_2\|\|b_2\|\right)$$

$$=\left(\left[h\left(\|a_1\|\frac{a_1}{\|a_1\|}\right)h\left(\|b_1\|\frac{b_1}{\|b_1\|}\right)h(y)\right]\right.$$

$$+\left[h\left(\|b_1\|\frac{b_1}{\|b_1\|}\right)h(y)h\left(\|a_1\|\frac{a_1}{\|a_1\|}\right)\right]$$

$$\left.+\left[h(y)h\left(\|a_1\|\frac{a_1}{\|a_1\|}\right)h\left(\|b_1\|\frac{b_1}{\|b_1\|}\right)\right]\right)$$

$$+\left(i\left[h\left(\|a_2\|\frac{a_2}{\|a_2\|}\right)h\left(\|b_1\|\frac{b_1}{\|b_1\|}\right)h(y)\right]\right.$$

$$+i\left[h\left(\|b_1\|\frac{b_1}{\|b_1\|}\right)h(y)h\left(\|a_2\|\frac{a_2}{\|a_2\|}\right)\right]$$

$$\left.+i\left[h(y)h\left(\|a_2\|\frac{a_2}{\|a_2\|}\right)h\left(\|b_1\|\frac{b_1}{\|b_1\|}\right)\right]\right)$$

$$
\begin{aligned}
&= \Big([h(a_1)h(b_1)h(y)] + [h(b_1)h(y)h(a_1)] + [h(y)h(a_1)h(b_1)]\Big)\\
&\quad + \Big(i[h(a_2)h(b_1)h(y)] + i[h(b_1)h(y)h(a_2)] + i[h(y)h(a_2)h(b_1)]\Big)\\
&= \Big([h(a_1+ia_2)h(b_1)h(y)] + [h(b_1)h(y)h(a_1+ia_2)]\\
&\quad + [h(y)h(a_1+ia_2)h(b_1)]\Big)\\
&= \Big([f(a)f(b)f(y)] + [f(b)f(y)f(a)] + [f(y)f(a)f(b)]\Big), \quad \forall y \in \mathfrak{A}.
\end{aligned}
$$

By above proof, we can denote that

$$
\begin{aligned}
&f\Big([aby] + [bya] + [yab]\Big)\\
&\quad = \Big([f(a)f(b)f(y)] + [f(b)f(y)f(a)] + [f(y)f(a)f(b)]\Big),
\end{aligned}
$$

for all $y \in \mathfrak{A}$, if $a_2 = 0, a_1, b_1 b_2 \neq 0$.

Now consider $a_1 = 0, a_2, b_1, b_2 \neq 0$. Then by (12), we have

$$
\begin{aligned}
f\Big([aby] &+ [bya] + [yab]\Big) = h\Big([aby] + [bya] + [yab]\Big)\\
&= h\Big([ia_2(b_1+ib_2)y] + [(b_1+ib_2)yia_2] + [yia_2(b_1+ib_2)]\Big)\\
&= h\Big([ia_2b_1y] + [b_1yia_2] + [yia_2b_1]\Big) - h\Big([a_2b_2y] + [b_2ya_2] + [ya_2b_2]\Big)\\
&= h\left(\left(i\|a_2\|\|b_1\|\left[\frac{a_2}{\|a_2\|}\frac{b_1}{\|b_1\|}y\right]\right) + \left(\|b_1\|\left[\frac{b_1}{\|b_1\|}y\frac{a_2}{\|a_2\|}\right]i\|a_2\|]\right)\right.\\
&\qquad \left. + \left(\left[y\frac{a_2}{\|a_2\|}\frac{b_1}{\|b_1\|}\right]i\|a_2\|\|b_1\|\right)\right)\\
&\quad - h\left(\left(\|a_2\|\|b_2\|\left[\frac{a_2}{\|a_2\|}\frac{b_2}{\|b_2\|}y\right]\right) + \left(\|b_2\|\left[\frac{b_2}{\|b_2\|}y\frac{a_2}{\|a_2\|}\right]\|a_2\|]\right)\right.\\
&\qquad \left. + \left(\left[y\frac{a_2}{\|a_2\|}\frac{b_2}{\|b_2\|}\right]\|a_2\|\|b_2\|\right)\right)\\
&= \left(i\|a_2\|\|b_1\|h\left(\left[\frac{a_2}{\|a_2\|}\frac{b_1}{\|b_1\|}y\right]\right) + \|b_1\|h\left(\left[\frac{b_1}{\|b_1\|}y\frac{a_2}{\|a_2\|}\right]\right)i\|a_2\|\right.
\end{aligned}
$$

$$
\begin{aligned}
&+ h\left(\left[y\frac{a_2}{\|a_2\|}\frac{b_1}{\|b_1\|}\right]\right)i\|a_2\|\|b_1\|\Bigg) \\
&- \Bigg(\|a_2\|\|b_2\|h\left(\left[\frac{a_2}{\|a_2\|}\frac{b_2}{\|b_2\|}y\right]\right) + \|b_2\|h\left(\left[\frac{b_2}{\|b_2\|}y\frac{a_2}{\|a_2\|}\right]\right)\|a_2\| \\
&\quad + h\left(\left[y\frac{a_2}{\|a_2\|}\frac{b_2}{\|b_2\|}\right]\right)\|a_2\|\|b_2\|\Bigg) \\
&= \Bigg(i\|a_2\|\|b_1\|\left[h\left(\frac{a_2}{\|a_2\|}\right)h\left(\frac{b_1}{\|b_1\|}\right)h(y)\right] \\
&\quad + \|b_1\|\left[h\left(\frac{b_1}{\|b_1\|}\right)h(y)h\left(\frac{a_2}{\|a_2\|}\right)\right]i\|a_2\| \\
&\quad + \left[h(y)h\left(\frac{a_2}{\|a_2\|}\right)h\left(\frac{b_1}{\|b_1\|}\right)\right]i\|a_2\|\|b_1\|\Bigg) \\
&- \Bigg(\|a_2\|\|b_2\|\left[h\left(\frac{a_2}{\|a_2\|}\right)h\left(\frac{b_2}{\|b_2\|}\right)h(y)\right] \\
&\quad + \|b_2\|\left[h\left(\frac{b_2}{\|b_2\|}\right)h(y)h\left(\frac{a_2}{\|a_2\|}\right)\right]\|a_2\| \\
&\quad + \left[h(y)h\left(\frac{a_2}{\|a_2\|}\right)h\left(\frac{b_2}{\|b_2\|}\right)\right]\|a_2\|\|b_2\|\Bigg) \\
&= \Bigg(\left[h\left(i\|a_2\|\frac{a_2}{\|a_2\|}\right)h\left(\|b_1\|\frac{b_1}{\|b_1\|}\right)h(y)\right] \\
&\quad + \left[h\left(\|b_1\|\frac{b_1}{\|b_1\|}\right)h(y)ih\left(\|a_2\|\frac{a_2}{\|a_2\|}\right)\right] \\
&\quad + \left[h(y)h\left(i\|a_2\|\frac{a_2}{\|a_2\|}\right)h\left(\|b_1\|\frac{b_1}{\|b_1\|}\right)\right]\Bigg) \\
&+ \Bigg(\left[ih\left(\|a_2\|\frac{a_2}{\|a_2\|}\right)ih\left(\|b_2\|\frac{b_2}{\|b_2\|}\right)h(y)\right]
\end{aligned}
$$

$$+\left[ih\left(\|b_2\|\frac{b_2}{\|b_2\|}\right)h(y)ih\left(\|a_2\|\frac{a_2}{\|a_2\|}\right)\right]$$

$$+\left[h(y)ih\left(\|a_2\|\frac{a_2}{\|a_2\|}\right)ih\left(\|b_2\|\frac{b_2}{\|b_2\|}\right)\right]\Bigg)$$

$$=\Big([ih(a_2)h(b_1)h(y)]+[h(b_1)h(y)ih(a_2)]+[h(y)ih(a_2)h(b_1)]\Big)$$

$$+\Big([ih(a_2)ih(b_2)h(y)]+[ih(b_2)h(y)ih(a_2)]+[h(y)ih(a_2)ih(b_2)]\Big)$$

$$=h\Big([ia_2(b_1+ib_2)y]+[(b_1+ib_2)yia_2]+[yia_2(b_1+ib_2)]\Big)$$

$$=\Big([f(a)f(b)f(y)]+[f(b)f(y)f(a)]+[f(y)f(a)f(b)]\Big),\quad \forall y\in\mathfrak{A}.$$

Also, by the same reasoning, we can denote that

$$f\Big([aby]+[bya]+[yab]\Big)$$
$$=\Big([f(a)f(b)f(y)]+[f(b)f(y)f(a)]+[f(y)f(a)f(b)]\Big),$$

for all $y\in\mathfrak{A}$ if $b_1=0, a_1, a_2, b_2\neq 0$.

Finally, consider that $b_1, a_1, a_2, b_2\neq 0$. Then by (12), we have

$$f\Big([aby]+[bya]+[yab]\Big)=h\Big([aby]+[bya]+[yab]\Big)$$

$$=h\Big([(a_1+ia_2)(b_1+ib_2)y]+[(b_1+ib_2)y(a_1+ia_2)]$$

$$+[y(a_1+ia_2)(b_1+ib_2)]\Big)$$

$$=h\Big([a_1b_1y]+[b_1ya_1]+[ya_1b_1]\Big)+h\Big([a_1ib_2y]+[ib_2ya_1]+[ya_1ib_2]\Big)$$

$$+h\Big([ia_2b_1y]+[b_1yia_2]+[yia_2b_1]\Big)-h\Big([a_2b_2y]+[b_2ya_2]+[ya_2b_2]\Big)$$

$$=h\Bigg(\left(\|a_1\|\|b_1\|\left[\frac{a_1}{\|a_1\|}\frac{b_1}{\|b_1\|}y\right]\right)+\left(\|b_1\|\left[\frac{b_1}{\|b_1\|}y\frac{a_1}{\|a_1\|}\right]\|a_1\|]\right)$$

$$+\left(\left[y\frac{a_1}{\|a_1\|}\frac{b_1}{\|b_1\|}\right]\|a_1\|\|b_1\|\right)\Bigg)$$

$$+ih\Bigg(\left(\|a_1\|\|b_2\|\left[\frac{a_1}{\|a_1\|}\frac{b_2}{\|b_2\|}y\right]\right)+\left(\|b_2\|\left[\frac{b_2}{\|b_2\|}y\frac{a_1}{\|a_1\|}\right]\|a_1\|]\right)$$

$$+\left(\left[y\frac{a_1}{\|a_1\|}\frac{b_2}{\|b_2\|}\right]\|a_1\|\|b_2\|\right)\Bigg)$$

$$+ ih\left(\left(\|a_2\|\|b_1\|\left[\frac{a_2}{\|a_2\|}\frac{b_1}{\|b_1\|}y\right]\right) + \left(\|b_1\|\left[\frac{b_1}{\|b_1\|}y\frac{a_2}{\|a_2\|}\right]\|a_2\|\right]\right)$$

$$+ \left(\left[y\frac{a_2}{\|a_2\|}\frac{b_1}{\|b_1\|}\right]\|a_2\|\|b_1\|\right)\right)$$

$$- h\left(\left(\|a_2\|\|b_2\|\left[\frac{a_2}{\|a_2\|}\frac{b_2}{\|b_2\|}y\right]\right) + \left(\|b_2\|\left[\frac{b_2}{\|b_2\|}y\frac{a_2}{\|a_2\|}\right]\|a_2\|\right]\right)$$

$$+ \left(\left[y\frac{a_2}{\|a_2\|}\frac{b_2}{\|b_2\|}\right]\|a_2\|\|b_2\|\right)\right)$$

$$= \left(\|a_1\|\|b_1\|h\left(\left[\frac{a_1}{\|a_1\|}\frac{b_1}{\|b_1\|}y\right]\right) + \|b_1\|h\left(\left[\frac{b_1}{\|b_1\|}y\frac{a_1}{\|a_1\|}\right]\right)\|a_1\|\right.$$

$$\left. + h\left(\left[y\frac{a_1}{\|a_1\|}\frac{b_1}{\|b_1\|}\right]\right)\|a_1\|\|b_1\|\right)$$

$$+ i\left(\|a_1\|\|b_2\|h\left(\left[\frac{a_1}{\|a_1\|}\frac{b_2}{\|b_2\|}y\right]\right) + \|b_2\|h\left(\left[\frac{b_2}{\|b_2\|}y\frac{a_1}{\|a_1\|}\right]\right)\|a_1\|\right.$$

$$\left. + h\left(\left[y\frac{a_1}{\|a_1\|}\frac{b_2}{\|b_2\|}\right]\right)\|a_2\|\|b_1\|\right)$$

$$+ i\left(\|a_2\|\|b_1\|h\left(\left[\frac{a_2}{\|a_2\|}\frac{b_1}{\|b_1\|}y\right]\right) + \|b_1\|h\left(\left[\frac{b_1}{\|b_1\|}y\frac{a_2}{\|a_2\|}\right]\right)\|a_2\|\right.$$

$$\left. + h\left(\left[y\frac{a_2}{\|a_2\|}\frac{b_1}{\|b_1\|}\right]\right)\|a_2\|\|b_1\|\right)$$

$$- \left(\|a_2\|\|b_2\|h\left(\left[\frac{a_2}{\|a_2\|}\frac{b_2}{\|b_2\|}y\right]\right) + \|b_2\|h\left(\left[\frac{b_2}{\|b_2\|}y\frac{a_2}{\|a_2\|}\right]\right)\|a_2\|\right.$$

$$\left. + h\left(\left[y\frac{a_2}{\|a_2\|}\frac{b_2}{\|b_2\|}\right]\right)\|a_2\|\|b_2\|\right)$$

$$= \left(\|a_1\|\|b_1\|\left[h\left(\frac{a_1}{\|a_1\|}\right)h\left(\frac{b_1}{\|b_1\|}\right)h(y)\right]\right.$$

$$+ \|b_1\|\left[h\left(\frac{b_1}{\|b_1\|}\right)h(y)h\left(\frac{a_1}{\|a_1\|}\right)\right]\|a_1\|$$

$$+\left[h(y)h\left(\frac{a_1}{\|a_1\|}\right)h\left(\frac{b_1}{\|b_1\|}\right)\right]\|a_1\|\|b_1\|\Bigg)$$
$$+i\Bigg(\|a_1\|\|b_2\|\left[h\left(\frac{a_1}{\|a_1\|}\right)h\left(\frac{b_2}{\|b_2\|}\right)h(y)\right]$$
$$+\|b_2\|\left[h\left(\frac{b_2}{\|b_2\|}\right)h(y)h\left(\frac{a_1}{\|a_1\|}\right)\right]\|a_1\|$$
$$+\left[h(y)h\left(\frac{a_1}{\|a_1\|}\right)h\left(\frac{b_2}{\|b_2\|}\right)\right]\|a_1\|\|b_2\|\Bigg)$$
$$+i\Bigg(\|a_2\|\|b_1\|\left[h\left(\frac{a_2}{\|a_2\|}\right)h\left(\frac{b_1}{\|b_1\|}\right)h(y)\right]$$
$$+\|b_1\|\left[h\left(\frac{b_1}{\|b_1\|}\right)h(y)h\left(\frac{a_2}{\|a_2\|}\right)\right]\|a_2\|$$
$$+\left[h(y)h\left(\frac{a_2}{\|a_2\|}\right)h\left(\frac{b_1}{\|b_1\|}\right)\right]\|a_2\|\|b_1\|\Bigg)$$
$$-\Bigg(\|a_2\|\|b_2\|\left[h\left(\frac{a_2}{\|a_2\|}\right)h\left(\frac{b_2}{\|b_2\|}\right)h(y)\right]$$
$$+\|b_2\|\left[h\left(\frac{b_2}{\|b_2\|}\right)h(y)h\left(\frac{a_2}{\|a_2\|}\right)\right]\|a_2\|$$
$$+\left[h(y)h\left(\frac{a_2}{\|a_2\|}\right)h\left(\frac{b_2}{\|b_2\|}\right)\right]\|a_2\|\|b_2\|\Bigg)$$
$$=\Bigg(\left[h\left(\|a_1\|\frac{a_1}{\|a_1\|}\right)h\left(\|b_1\|\frac{b_1}{\|b_1\|}\right)h(y)\right]$$
$$+\left[h\left(\|b_1\|\frac{b_1}{\|b_1\|}\right)h(y)h\left(\|a_1\|\frac{a_1}{\|a_1\|}\right)\right]$$
$$+\left[h(y)h\left(\|a_1\|\frac{a_1}{\|a_1\|}\right)h\left(\|b_1\|\frac{b_1}{\|b_1\|}\right)\right]\Bigg)$$

$$
\begin{aligned}
&+\left(\left[h\left(\|a_1\|\frac{a_1}{\|a_1\|}\right)ih\left(\|b_2\|\frac{b_2}{\|b_2\|}\right)h(y)\right]\right.\\
&\qquad+\left[h\left(i\|b_2\|\frac{b_2}{\|b_2\|}\right)h(y)h\left(\|a_1\|\frac{a_1}{\|a_1\|}\right)\right]\\
&\qquad\left.+\left[h(y)h\left(\|a_1\|\frac{a_1}{\|a_1\|}\right)ih\left(\|b_2\|\frac{b_2}{\|b_2\|}\right)\right]\right)\\
&+\left(\left[ih\left(\|a_2\|\frac{a_2}{\|a_2\|}\right)h\left(\|b_1\|\frac{b_1}{\|b_1\|}\right)h(y)\right]\right.\\
&\qquad+\left[h\left(\|b_1\|\frac{b_1}{\|b_1\|}\right)h(y)ih\left(\|a_2\|\frac{a_2}{\|a_2\|}\right)\right]\\
&\qquad\left.+\left[h(y)ih\left(\|a_2\|\frac{a_2}{\|a_2\|}\right)h\left(\|b_1\|\frac{b_1}{\|b_1\|}\right)\right]\right)\\
&+\left(\left[ih\left(\|a_2\|\frac{a_2}{\|a_2\|}\right)ih\left(\|b_2\|\frac{b_2}{\|b_2\|}\right)h(y)\right]\right.\\
&\qquad+\left[ih\left(\|b_2\|\frac{b_2}{\|b_2\|}\right)h(y)ih\left(\|a_2\|\frac{a_2}{\|a_2\|}\right)\right]\\
&\qquad\left.+\left[h(y)ih\left(\|a_2\|\frac{a_2}{\|a_2\|}\right)h\left(\|b_2\|\frac{b_2}{\|b_2\|}\right)\right]\right)\\
&=\Big([h(a_1)h(b_1)h(y)]+[h(b_1)h(y)h(a_1)]+[h(y)h(a_1)h(b_1)]\Big)\\
&+\Big([h(a_1)ih(b_2)h(y)]+[ih(b_2)h(y)h(a_1)]+[h(y)h(a_1)ih(b_2)]\Big)\\
&+\Big([ih(a_2)h(b_1)h(y)]+[h(b_1)h(y)ih(a_2)]+[h(y)ih(a_2)h(b_1)]\Big)\\
&+\Big([ih(a_2)ih(b_2)h(y)]+[ih(b_2)h(y)ih(a_2)]+[h(y)ih(a_2)ih(b_2)]\Big)\\
&=\Big([h(a_1+ia_2)h(b_1+ib_2)h(y)]+[h(b_1+ib_2)h(y)h(a_1+ia_2)]\\
&\qquad+[h(y)h(a_1+ia_2)h(b_1+ib_2)]\Big)\\
&=\Big([f(a)f(b)f(y)]+[f(b)f(y)f(a)]+[f(y)f(a)f(b)]\Big),\quad\forall y\in\mathfrak{A}.
\end{aligned}
$$

Hence, $f([aby]+[bya]+[yab]) = ([f(a)f(b)f(y)]+[f(b)f(y)f(a)]+[f(y)f(a)f(b)])$ for all $a, b, y \in \mathfrak{A}$ and f is a ternary Jordan homomorphism. □

Corollary 2. *Let $p \in (0,1), \theta \in [0,\infty)$ be real numbers. Let $f : \mathfrak{A} \to \mathfrak{B}$ be a mapping such that $f(0) = 0$ and that*

$$\begin{aligned} &f([3^n u 3^n v y] + [3^n v y 3^n u] + [y 3^n u 3^n v]) \\ &\quad = ([f(3^n u)f(3^n v)f(y)] + [f(3^n v)f(y)f(3^n u)] + [f(y)f(3^n u)f(3^n v)]), \end{aligned} \tag{13}$$

for all $y \in \mathfrak{A}$,, all $u, v \in I_1(\mathfrak{A}_{sa})$ and all $n = 0, 1, 2, \ldots$. Suppose that

$$\left\| 2f\left(\frac{\mu x + \mu y}{2}\right) - \mu f(x) - \mu f(y) \right\| \leq \theta(\|x\|^p + \|y\|^p),$$

for all $x, y \in \mathfrak{A}$. and all $\mu \in T_{\frac{1}{n_0}}$ If $lim_{n\to\infty} \frac{f(3^n e)}{3^n} \in U(B)$, then the mapping $f : \mathfrak{A} \to \mathfrak{B}$ is a ternary Jordan homomorphism.

References

[1] M. Amyari and M.S. Moslehian, Approximately ternary semigroup homomorphisms, *Lett. Math. Phys.* **77**, 1–9, (2006).

[2] R. Kerner, *Ternary Algebraic Structures and their Applications in Physics*, Pierre et Marie Curie University, Paris; 2000. http://arxiv.org/list/math-ph/0011 Ternary algebraic structures and their applications in physics, Proc. BTLP, 23rd International Conference on Group Theoretical Methods in Physics, Dubna, Russia, (2000).

[3] Y. Nambu, Generalized Hamiltonian mechanics, *Phys. Rev.* **7**, 2405–2412, (1973).

[4] I.R.U. Churchill, M. Elhamdadi, M. Green, and A. Makhlouf, Ternary and n-ary f-distributive structures, *Open Math.* **16**, 32–45, (2018).

[5] S. Mabrouk, O. Ncib, and S. Silvestrov, Generalized Derivations and Rota-Baxter operators of n-ary Hom-Nambu superalgebras, *Adv. Appl. Clifford Algebras* **31**, 32 pp. (2021).

[6] A. Nongmanee and S. Leeratanavalee, Ternary menger algebras: A generalization of ternary semigroups, *Mathematics*, **9**, 14 pp. (2021).

[7] C. Park, C*-ternary homomorphisms, C* Ternary Derivations, JB*-triple homomorphisms and JB*-triple derivations, *Int. J. Geom. Methods Mod. Phys.* **10**, 11 pp. (2013).

[8] S.M. Ulam, *Problems in Modern Mathematics*, Chapter VI, Science ed. (Wiley, New York, 1940).

[9] D.H. Hyers, On the stability of the linear functional equation, *Proc. Natl. Acad. Sci. U. S. A.* **27**, 222–224, (1941).

[10] M.Th. Rassias, On the stability of the linear mapping in Banach spaces, *Proc. Amer. Math. Soc.* **72**, 297–300, (1978).

[11] Z. Abbasbeygi, A. Bodaghi, and A. Gharibkhajeh, On an equation characterizing multi-quartic mappings and its stability, *Int. J. Nonlinear Anal. Appl.* **13**, 991–1002, (2022).

[12] M.R. Abdollahpour and M.Th. Rassias, Hyers-Ulam stability of hypergeometric differential equations, *Aequationes Math.* **93**(4), 691–698, (2019).

[13] M.R. Abdollahpour, R. Aghayaria, and M.Th. Rassias, Hyers-Ulam stability of associated Laguerre differential equations in a subclass of analytic functions, *J. Math. Anal. Appl.* **437**, 605–612, (2016).

[14] N. Ansari, M.H. Hooshmand, M.E. Gordji, and K. Jahedi, Stability of fuzzy orthogonally $*$-n-derivation in orthogonally fuzzy C*-algebras, *Int. J. Nonlinear Anal. Appl.* **12**, 533–540, (2021).

[15] A. Bodaghi, Th.M. Rassias, and A. Zivari-Kazempour, A fixed point approach to the stability of additive-quadratic-quartic functional equations, *Int. J. Nonlinear Anal. Appl.* **11**, 17–28, (2020).

[16] E. Elqorachi and M.Th. Rassias, Generalized Hyers-Ulam stability of trigonometric functional equations, *Mathematics*, **6**(5), 83, (2018). https://doi.org/10.3390/math6050083.

[17] M.E. Gordji, V. Keshavarz, C. Park, and S.Y. Jang, Ulam-Hyers stability of 3-Jordan homomorphisms in C*-ternary algebras, *J. Comput. Anal. Appl.* **22**, 573–578, (2017).

[18] V. Gupta and M.Th. Rassias, *Moments of Linear Positive Operators and Approximation* (Springer, 2019).

[19] V. Gupta and M.Th. Rassias, *Computation and Approximation* (Springer, 2021).

[20] S. Jahedi and V. Keshavarz, Approximate generalized additive-quadratic functional equations on ternary Banach, *J. Math. Extens.* **16**, 1–11, (2022).

[21] S.-M. Jung and M.Th. Rassias, A linear functional equation of third order associated to the Fibonacci numbers, *Abstract Appl. Anal.* **2014**, Article ID 137468, (2014).

[22] S.-M. Jung, D. Popa, and M.Th. Rassias, On the stability of the linear functional equation in a single variable on complete metric groups, *J. Global Optimizat.* **59**, 165–171, (2014).

[23] S.-M. Jung, C. Mortici, and M.Th. Rassias, On a functional equation of trigonometric type, *Appl. Math. Comput.* **252**, 294–303, (2015).

[24] Y.-H. Lee, S.-M. Jung, and M.Th. Rassias, On an n-dimensional mixed type additive and quadratic functional equation, *Appl. Math. Comput.* **228**, 13–16, (2014).

[25] Y.-H. Lee, S.-M. Jung, and M.Th. Rassias, Uniqueness theorems on functional inequalities concerning cubic-quadratic-additive equation, *J. Math. Inequalit.* **12**(1), 43–61, (2018).

[26] G.V. Milovanović and M.Th. Rassias, (Eds.), *Analytic Number Theory, Approximation Theory and Special Functions* (Springer, 2014).

[27] H.L. Montgomery, A. Nikeghbali, and M.Th. Rassias, (Eds.), *Exploring the Riemann Zeta Function: 190 years from Riemann's birth, (Preface by Freeman J. Dyson)* (Springer, 2017).

[28] C. Mortici, S.-M. Jung, and M.Th. Rassias, On the stability of a functional equation associated with the Fibonacci numbers, *Abstract Appl. Anal.* **2014**, Article ID 546046, 6 pp, (2014).
[29] C. Park and M.Th. Rassias, Additive functional equations and partial multipliers in C*-algebras, *Revista de la Real Academia de Ciencias Exactas, Serie A. Matemáticas*, **113**(3), 2261–2275, (2019).
[30] C. Pomerance and M.Th. Rassias, (Eds.), *Analytic Number Theory* (Springer, 2015).
[31] M.Th. Rassias, Solution of a functional equation problem of Steven Butler, *Octogon Math. Mag.* **12**, 152–153, (2004).
[32] M.Th. Rassias, *Problem-Solving and Selected Topics in Number Theory* (Springer, 2011).
[33] M.Th. Rassias, *Goldbach's Problem: Selected Topics* (Springer, 2017).
[34] J.F. Nash and M.Th. Rassias, (Eds.), *Open Problems in Mathematics* (Springer, 2016).
[35] J. Brzdek, W. Fechner, M.S. Moslehian, and J. Sikorska, Recent derivations of the conditional stability of the homomorphism equation, *Banach J. Math. Anal.* **9**, 278–326, (2015).
[36] L. Brown and G. Pedersen, C^*-algebras of real rank zero, *J. Funct. Anal.* **99**, 131–149, (1991).
[37] M.E. Gordji and Th.M. Rassias, Ternary homomorphisms between unital ternary C^*-algebras, *Proc. Romanian Acad. Sci.* **12**, 189–196, (2011).
[38] C. Park, D.-H. Boo, and J.-S. An, Homomorphisms between C^*-algebras and linear derivations on C^*-algebras, *J. Math. Anal. Appl.* **337**, 1415–1424, (2008).
[39] K. Jun and Y. Lee, A generalization of the Hyers-Ulam-Rassias stability of Jensens equation, *J. Math. Anal. Appl.* **238**, 305–315, (1999).
[40] M.E. Gordji and A. Fazeli, Stability and superstability of homomorphisms on C^*-ternary algebras, *An. St. Univ. Ovidius Constanta* **20**, 173–188, (2012).
[41] R.V. Kadison and J.R. Ringrose, *Fundamentals of the Theory of Operator Algebras, Elementary Theory* (Academic Press, New York, 1983).
[42] V. Keshavarz, S. Jahedi, and M.E. Gordji, Ulam-Hyers stability of C^*-ternary 3-Jordan derivations, *South East Aisian Bull. Math.* **45**, 55–64, (2021).

https://doi.org/10.1142/9789811261572_0012

Chapter 12

Analytic and Numerical Solutions to Nonlinear Partial Differential Equations in Biomechanics

Anastasios C. Felias, Konstantina C. Kyriakoudi, Kyriaki N. Mpiraki, and Michail A. Xenos*

Department of Mathematics, University of Ioannina, Ioannina, 45110, Greece

*mxenos@uoi.gr

The study of exact solutions to nonlinear equations is an active field of both, pure and applied mathematics. Plenty of the most interesting features of physical systems are hidden in their nonlinear behavior and can only be studied with appropriate methods designed to tackle nonlinearity. Therefore, seeking for suitable solving methods, exact, semi-exact or numerical, is an active task in branches of applied and computational mathematics. Complex phenomena in notable scientific fields, especially in physics, such as fluid and plasma dynamics, optical fibers, solid state physics, as well as in cardiac hemodynamics, can be efficiently mathematically modeled in terms of the Korteweg–de Vries (KdV), modified KdV (mKdV), Burgers and Korteweg–de Vries–Burgers (KdV–B) equations. In this review chapter, analytical solutions are sought for each of the aforementioned equations, by means of traveling wave and similarity transforms. Especially, for the KdV equation, one- and two-soliton solutions are derived. The Lax pairs are introduced and the Inverse Scattering Transform (IST) is discussed for the KdV equation. The Miura Transform is presented, connecting the KdV and mKdV equations. The Cole–Hopf Transform is described, converting the viscous Burgers equation to the linear Heat transport equation. Weak solution formulation is presented for the inviscid Burgers equation. Additionally, the Rankine–Hugoniot condition is discussed in the implementation of the method of characteristics. Semi-exact solutions are obtained through the Homotopy Analysis Method. Numerical solutions are derived by means of spectral Fourier analysis and are evolved in time, using the fourth-order explicit Runge–Kutta method. Qualitative analysis is performed for the inviscid Burgers equation, and conservation laws in general, and phase plane trajectories are obtained for the KdV–B equation.

1. Introduction

In recent years, much attention from a rather diverse group of scientists, including physicists, engineers and applied mathematicians, has been attracted to two contrasting research fields: (a) dynamical systems, most popularly associated with the study of chaos, and (b) integrable (or non-integrable) systems associated, among other things, with the study of solitary waves (solitons).

Recent advancements, concerning propagation of undular bores in shallow water [1,2], liquid flow containing gas bubbles [3], fluid flow in elastic tubes [4], crystal lattice theory, nonlinear circuit theory and turbulence [5–7], are all based on the study of the so-called Korteweg–de Vries–Burgers equation (KdVB) [8]

$$\frac{\partial u}{\partial t} + \gamma u \frac{\partial u}{\partial x} - \alpha \frac{\partial^2 u}{\partial x^2} + \beta \frac{\partial^3 u}{\partial x^3} = 0,$$
$$u = u(x,t), \ (x,t) \in \mathbb{R} \times (0,\infty), \ \alpha, \beta \in \mathbb{R}, \gamma \neq 0. \tag{1}$$

Equation (1) is non-integrable in the sense that its spectral problem is non-existent [9]. Multiplication of t, x and u by proper constants can be used, making the coefficients of any of the equation's four terms equal to any given non-zero constant. Thus, we will focus on the special cases in the following:

(1) $(\alpha = 0, \beta > 0)$, namely the Korteweg–de Vries (KdV) equation [10,11].
(2) $(\alpha > 0, \beta = 0)$, namely the viscous Burgers equation [12].
(3) $(\alpha = \beta = 0)$, namely the inviscid Burgers equation [12].

The modified KdV equation [10,13],

$$\frac{\partial u}{\partial t} + \gamma u^2 \frac{\partial u}{\partial x} + \beta \frac{\partial^3 u}{\partial x^3} = 0,$$
$$u = u(x,t), \ (x,t) \in \mathbb{R} \times (0,\infty), \ \beta > 0, \ \gamma \neq 0, \tag{2}$$

will also be studied. The main purpose of this review chapter is to introduce the main analytical and numerical approaches for solving nonlinear equations of Biomechanics.

2. The Korteweg–de Vries Equation

It is widely accepted that a plethora of physical phenomena, such as shallow-water waves with weakly nonlinear restoring forces, long internal waves in a density-stratified ocean, ion acoustic waves in a plasma, acoustic waves on a crystal lattice, as well as the pulse wave in cardiac dynamics, can be modeled, after some scaling, by the KdV equation [11,14–16],

$$\begin{aligned}&\frac{\partial u}{\partial t}+\gamma u\frac{\partial u}{\partial x}+\beta\frac{\partial^3 u}{\partial x^3}=0,\\ &u=u(x,t),\ \ (x,t)\in\mathbb{R}\times(0,\infty),\ \ \beta>0,\ \ \gamma\neq 0.\end{aligned}\tag{3}$$

The coefficients β and γ in the general form of the KdV equation can be freely changed through scaling/reflection transformations on the variables u, x and t. A conventional choice is $\beta = 1$ and $\gamma = \pm 6$, which eliminates awkward numerical factors in the expressions for soliton solutions.

The KdV equation is the simplest conservative one-dimensional wave equation with weak advective nonlinearity and dispersion. It is nonlinear, due to the uu_x term, dispersive partial differential equation for a function $u(x,t)$ of two dimensionless real variables, x and t, which are proportional to space and time, resp [17]. Dispersion, in general, corresponds to an initial wave, $u(x,0)$, broadening in space as time progresses.

In the case of $\gamma = -6$, the term -$u(x,t)$ corresponds to the vertical displacement of the water from the equilibrium at the location x at time t. Replacing u by -u amounts to replacing -6 by 6. Additionally, by scaling x, t and u, for example, by multiplying them with positive constants, it is possible to change the constants in front of each of the three terms on the left-hand side of (3) at will. Historically, the KdV equation was first introduced by Boussinesq (1877) [18] and rediscovered by Korteweg and de Vries (1895) [19].

The KdV model is particularly notable as the prototypical example of an exactly solvable mathematical model, that is, a nonlinear partial differential equation whose solutions can be exactly and precisely specified. The mathematical theory behind the KdV equation is a topic of active research.

2.1. *Derivation through a Hamiltonian density*

A natural derivation of the KdV equation, (3), arises when considering the Euler–Lagrange equation for the Lagrangian density, defined as [20],

$$L:=\frac{1}{2}\frac{\partial v}{\partial x}\frac{\partial v}{\partial t}+\left(\frac{\partial v}{\partial x}\right)^3-\frac{1}{2}\left(\frac{\partial^2 v}{\partial x^2}\right)^2.\tag{4}$$

Since the Lagrangian, (4), contains second-order derivatives, the Euler–Lagrange equation of motion for this field is

$$\partial_{ii}\left(\frac{\partial L}{\partial(\partial_{ii}v)}\right)-\partial_{i}\left(\frac{\partial L}{\partial(\partial_{i}v)}\right)+\frac{\partial L}{\partial v}=0, \tag{5}$$

with ∂ being a derivative with respect to the i component.

Now, a summation over i is implied, with (5) reading,

$$\begin{aligned}\partial_{tt}\left(\frac{\partial L}{\partial(\partial_{tt}v)}\right)+\partial_{xx}\left(\frac{\partial L}{\partial(\partial_{xx}v)}\right)-\partial_{t}\left(\frac{\partial L}{\partial(\partial_{t}v)}\right)\\ -\partial_{x}\left(\frac{\partial L}{\partial(\partial_{x}v)}\right)+\frac{\partial L}{\partial v}=0.\end{aligned} \tag{6}$$

At this stage, proceed evaluating the five terms of (6), by plugging in (4), to get

$$\begin{cases}\partial_{tt}\left(\dfrac{\partial L}{\partial(\partial_{tt}v)}\right)=0\\ \partial_{xx}\left(\dfrac{\partial L}{\partial(\partial_{xx}v)}\right)=\partial_{xx}\left(-\partial_{xx}v\right)\\ \partial_{t}\left(\dfrac{\partial L}{\partial(\partial_{t}v)}\right)=\partial_{t}\left(\dfrac{1}{2}\partial_{x}v\right)\\ \partial_{x}\left(\dfrac{\partial L}{\partial(\partial_{x}v)}\right)=\partial_{x}\left(\dfrac{1}{2}\partial_{t}v+3\left(\partial_{x}v\right)^{2}\right)\\ \dfrac{\partial L}{\partial v}=0.\end{cases}$$

Setting

$$u=\partial_{x}v,$$

we may simplify the above terms as follows:

$$\begin{cases}\partial_{tt}\left(\dfrac{\partial L}{\partial(\partial_{tt}v)}\right)=0\\ \partial_{xx}\left(-\partial_{xx}v\right)=-\partial_{xxx}u\\ \partial_{t}\left(\dfrac{1}{2}\partial_{x}v\right)=\dfrac{1}{2}\partial_{t}u\\ \partial_{x}\left(\dfrac{1}{2}\partial_{t}v+3\left(\partial_{x}v\right)^{2}\right)=\dfrac{1}{2}\partial_{t}u+3\partial_{x}\left(u^{2}\right)=\dfrac{1}{2}\partial_{t}u+6u\partial_{x}u\\ \dfrac{\partial L}{\partial v}=0.\end{cases}$$

Finally, after replacing the above terms into (6), we obtain

$$u_t + 6u\partial_x u + \partial_{xxx} u = 0,$$

which is exactly the KdV equation, for $\gamma = 6$.

2.2. *Conservation laws and motion integrals*

Firstly, we define the local conservation laws.

Definition 1 ([21]). A local conservation law is any partial differential equation (PDE) of the form

$$D_t + F_x = 0, \tag{7}$$

with D and F representing the local conserved density and its flux, resp.

Every (local) conservation law implies, under appropriate boundary conditions, the conservation of an integral of D,

$$\frac{d}{dt}\int_a^b D(x)dx + F(b) - F(a) = 0. \tag{8}$$

Therefore, in the case of F either vanishing or taking the same values at the boundary, the integral of D is conserved. Similarly, if F satisfies periodic boundary conditions on some finite interval.

Regarding the KdV equation, let us consider the boundary-value problem [10],

$$\begin{cases} u_t + \gamma u u_x + \beta u_{xxx} = 0, \quad u = u(x,t),\ (x,t) \in \mathbb{R} \times (0,\infty),\ \gamma = \pm 6, \\ \lim\limits_{|x|\to\infty} \dfrac{\partial^n u(x,t)}{\partial x^n} = 0, \quad n \in \mathbb{N}. \end{cases} \tag{9}$$

Our first task is showing that the equation conserves the mass functional,

$$m[u] := \int_{-\infty}^{\infty} u\, dx.$$

Note that (9) can be rewritten as the conservation law,

$$u_t + \left(\gamma\frac{u^2}{2} + \beta u_{xx}\right)_x = 0. \tag{10}$$

Now, integration of (10) over $\mathbb{R}$ for $t > 0$ yields

$$\begin{aligned} 0 &= \int_{-\infty}^{\infty} u_t \, dx + \left(\gamma \frac{u^2}{2} + \beta u_{xx} \right) \Big|_{-\infty}^{\infty} \\ &= \int_{-\infty}^{\infty} u_t \, dx \\ &= \frac{d}{dt} \int_{-\infty}^{\infty} u \, dx, \end{aligned} \tag{11}$$

implying that the quantity $\int_{-\infty}^{\infty} u \, dx$ is conserved in time. The latter may physically translate to the area of water under the free surface of the fluid not changing with time.

In a similar fashion, we shall also show that the equation conserves the momentum functional,

$$mom[u] := \int_{-\infty}^{\infty} u^2 dx.$$

We begin by multiplying the KdV equation by $2u$, getting

$$\left(u^2\right)_t + 2\gamma u^2 u_x + 2\beta u u_{xxx} = 0. \tag{12}$$

Observe that the second term of (12) is clearly an x derivative,

$$2\gamma u^2 u_x = \left(\frac{2\gamma}{3} u^3 \right)_x . \tag{13}$$

Now, regarding the third term, note that it may be rewritten as

$$\begin{aligned} 2\beta u u_{xxx} &= 2\beta u u_{xxx} + 2\beta u_x u_{xx} - 2\beta u_x u_{xx} \\ &= (2\beta u u_{xx})_x - \left(\beta u_x^2\right)_x \\ &= \left(2\beta u u_{xx} - \beta u_x^2\right)_x . \end{aligned} \tag{14}$$

From (13) and (14), we deduce that (12) can be rewritten as the conservation law,

$$\left(u^2\right)_t + \left(\frac{2\gamma}{3} u^3 + 2\beta u u_{xx} - \beta u_x^2 \right)_x = 0. \tag{15}$$

The latter implies that under the vanishing boundary conditions of (9), the momentum $\int_{-\infty}^{\infty} u^2 dx$ is conserved in time.

The KdV equation admits infinitely many conserved quantities, namely integrals of motion [22]. Explicitly, they can be given as

$$\int_{-\infty}^{\infty} P_{2n-1}\left(u, u_x, u_{xx}, \ldots\right) dx. \tag{16}$$

with the polynomials P_n defined recursively by

$$\begin{cases} P_1 = u \\ P_n = -\dfrac{dP_{n-1}}{dx} + \displaystyle\sum_{i=1}^{n-2} P_i P_{n-1-i}, \quad n \geq 2. \end{cases} \tag{17}$$

Namely, the first few integrals of motion are

(1) The mass $\int_{-\infty}^{\infty} u\, dx$.
(2) The momentum $\int_{-\infty}^{\infty} u^2 dx$.
(3) The energy $\int_{-\infty}^{\infty} \left(2u^3 - u_x^2\right) dx$.

Only the odd-numbered terms P_{2n+1} result in non-trivial integrals of motion [23].

2.3. *Analytic traveling wave solutions*

As we have previously showed, the area of water under the free surface of the fluid is not changing over time. The above key-property of the equation motivates into seeking wave-form solutions, vanishing at both infinities, maintaining their shape as they travel to the right at a constant phase speed λ. These waves are called solitary waves, being localized disturbances, meaning that the wave profile decays at infinity [24,25].

We introduce the traveling wave transform,

$$\begin{cases} u(x,t) = u(\zeta), \\ \zeta = x - \lambda t, \quad \lambda > 0. \end{cases} \tag{18}$$

The simplest classes of exact solutions to a given PDE are those obtained from a traveling-wave transform. Despite their very special nature, such solutions are understood to play a significant role in the evolution of a large class of initial profiles for each of the equations studied throughout the chapter.

Traveling wave solutions approaching different constant states at $\pm\infty$, as we will discuss in the viscous Burgers equation later on, are called wavefront solutions. They correspond to a traveling wave moving into and from constant states. If these two states are equal, as we now consider in the KdV case, then we say the wave front is a pulse [16,24,26].

Direct substitution into (9), using chain differentiation, leads to the boundary-value problem,

$$\begin{cases} -\lambda u'(\zeta) + \gamma u u'(\zeta) + \beta u'''(\zeta) = 0, \quad u = u(\zeta),\ \zeta \in \mathbb{R}, \\ \lim\limits_{|\zeta|\to\infty} \dfrac{d^n u(\zeta)}{d\zeta^n} = 0, \quad n \in \mathbb{N}. \end{cases} \tag{19}$$

Integration of (19) with respect to ζ, while applying the infinity conditions, leads to the autonomous second-order equation,

$$-\lambda u + \frac{\gamma}{2}u^2 + \beta u'' = 0, \tag{20}$$

where we may note the absence of the first derivative of u.

Multiplying (20) by u', we obtain

$$-\lambda u u' + \frac{\gamma}{2}u^2 u' + \beta u'' u' = 0,$$

which can be further integrated to give

$$-\lambda u^2 + \frac{\gamma}{3}u^3 + \beta\,(u')^2 = 0. \tag{21}$$

Now, (21) is a first-order, separable equation. Rearrangement of its terms gives

$$\begin{aligned} u' &= -\frac{1}{\sqrt{\beta}}u\sqrt{\lambda - \frac{\gamma}{3}u} \\ &\Leftrightarrow \frac{du}{d\zeta} = -\frac{1}{\sqrt{\beta}}u\sqrt{\lambda - \frac{\gamma}{3}u} \\ &\Leftrightarrow \zeta = -\sqrt{\beta}\int \frac{du}{u\sqrt{\lambda - \frac{\gamma}{3}u}} + c \\ u &= \frac{3\lambda}{\gamma}sech^2(\theta) \Leftrightarrow \zeta = \frac{2\sqrt{\beta}}{\sqrt{\lambda}}\int d\theta + c \\ &\Leftrightarrow \theta = \frac{\sqrt{\lambda}}{2\sqrt{\beta}}(\zeta - c), \end{aligned}$$

implying that

$$u(\zeta) = \frac{3\lambda}{\gamma} sech^2 \left(\frac{\sqrt{\lambda}}{2\sqrt{\beta}} (\zeta - c) \right), \tag{22}$$

with c being a real arbitrary constant of integration.

Now, recalling (18), equation (22) finally reads

$$u(x,t) = \frac{3\lambda}{\gamma} sech^2 \left(\frac{\sqrt{\lambda}}{2\sqrt{\beta}} (x - \lambda t - c) \right), \quad c \in \mathbb{R}. \tag{23}$$

The above final equation describes a right-moving solitary wave of KdV, being a self-reinforcing wave packet that maintains its shape while it propagates at a constant velocity. The width of the wave, defined by $\frac{2\sqrt{\beta}}{\sqrt{\lambda}}$, increases as β increases, meaning that the wave disperses.

In the case that the solitary wave retains its shape and speed after interacting with other waves of the same type, we say that the solitary wave is a soliton. Solitons, caused by a cancellation of nonlinear and dispersive effects in the medium, are waves of stable and steady form, although internal oscillations may occur, exhibiting unique characteristics when colliding with other solitary waves [24]. Moreover, solitons seem to be almost unaffected in shape by passing through each other, though this could result in a change in their position. These are evidently waves behaving like particles [24,25].

The soliton phenomenon was first described in 1834 by John Scott Russell, observing a solitary wave in the Union Canal in Scotland. He reproduced the phenomenon in a wave tank and named it the "Wave of Translation" [27].

In applications, if a pulse or signal travels as a soliton, then the information contained in the pulse can be carried over long distances with no distortion or loss of intensity. Recent developments in cardiac dynamics, found on an increasing number of studies, focus on describing the cardiac pulse as a soliton, due to the features those two seem to share. The pulsatility synchronization of the smooth arterial muscle allows the consideration of solitary profiles in cardiac hemodynamics [16,28–30].

In Fig. 1, a right-moving soliton of the KdV equation is demonstrated, maintaining its shape in time, as expected by (11).

2.4. *Multi-soliton solutions*

A special ansatz is applied to find one- and two-soliton solutions to the KdV equation. The ansatz is based on the Cole–Hopf transformation [31]

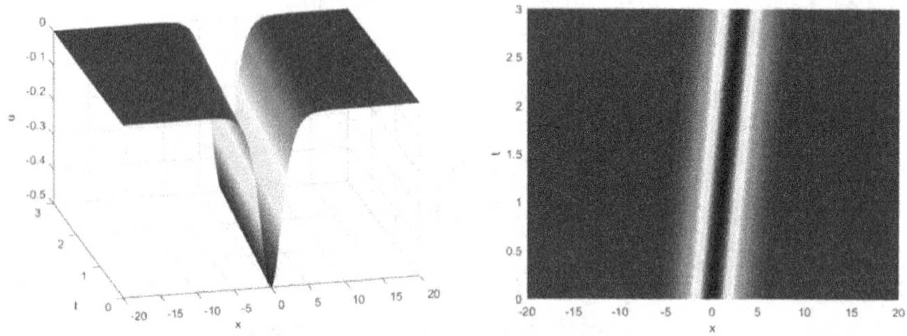

Fig. 1: A traveling wave solution, namely a right-moving soliton, of the KdV equation, for the values of $\lambda = \beta = 1$, $\gamma = -6$ and $c = 0$, with $(x, t) \in [-20, 20] \times [0, 3]$.

and is related to Hirota's bilinear method [32]. One may conjecture that all completely integrable nonlinear evolution equations can be put into bilinear form and will admit N-soliton solutions for any value of integer N. However, the fact that an equation admits a bilinear representation does not guarantee the existence of multi-soliton solutions of any order.

Consider a PDE,

$$F(u, u_t, u_{tt}, u_x, u_{tx}, u_{xx}, \ldots) = 0, \tag{24}$$

with F being a polynomial with respect to the unknown function $u = u(x, t)$ and its partial derivatives. In this setting, apply the Cole–Hopf transformation in the form [33],

$$u = u(x, t) = A\left(\log(f(x, t))_{xx},\right. \tag{25}$$

with A being some constant to be determined and $f(x, t)$ representing the new unknown function, called the auxiliary function. Most of the time, this function has the form of a traveling wave, that is,

$$f(x, t) = v(\zeta), \;\; \zeta = \kappa x - \omega t - c, \;\; c = \text{arbitrary constant}, \tag{26}$$

for some constants κ and ω.

Here $v = v(\zeta)$ is an unknown function. At this stage, substituting (26) into (24) yields a nonlinear ordinary differential equation,

$$G(v(\zeta), v'(\zeta), v''(\zeta), \ldots) = 0. \tag{27}$$

Exact solutions to (27), if any, are called one-soliton solutions. There are other types of solutions, the so-called two-soliton solutions, having the form (25) with

$$f(x,t) = v(\zeta_1, \zeta_2), \quad \zeta_1 = \kappa_1 x - \omega_1 t - c_1, \quad \zeta_2 = \kappa_2 x - \omega_2 t - c_2, \tag{28}$$

with $\kappa_1 \neq \kappa_2$ and c_1 and c_2 being arbitrary constants ($\kappa_1, \kappa_2, \omega_1, \omega_2$ are some non-zero constants).

Now, regarding the one-soliton solution to KdV, we shall use the ansatz

$$v_1(\zeta) = 1 + e^{\zeta}, \tag{29}$$

with ζ defined by (26), whereas for the two-soliton solution the ansatz

$$v_2(\zeta_1, \zeta_2) = 1 + e^{\zeta_1} + e^{\zeta_2} + B\, e^{\zeta_1 + \zeta_2} \tag{30}$$

is used with ζ_1 and ζ_2 defined by (28), with B being an additional constant to be determined.

Replace (29) into (25) and the latter to the KdV equation, (3), for $\beta = 1$ and $\gamma = 6$, to obtain a polynomial equation in the unknown quantity e^{ζ}. Equating the coefficients of different powers of e^{ζ} to zero yields a system of polynomial equations. The corresponding system is

$$\begin{cases} -A\kappa^2(\kappa^3 - \omega) = 0, \\ A\kappa^2(\kappa^3 - \omega) = 0, \\ -A\kappa^2(-11\kappa^3 + 6A\kappa^3 - \omega) = 0, \\ A\kappa^2(-11\kappa^3 + 6A\kappa^3 - \omega) = 0, \end{cases} \tag{31}$$

with the solution

$$\begin{cases} A = 2, \\ \omega = \kappa^3. \end{cases} \tag{32}$$

Therefore, the respective one-soliton solution is

$$u_1(x,t) = \frac{\kappa^2}{2} sech^2\left(\frac{\kappa}{2}\left(x - \kappa^2 t - c_0\right)\right). \tag{33}$$

Note that (33) and (23) are identical if setting

$$\kappa^2 = \lambda > 0.$$

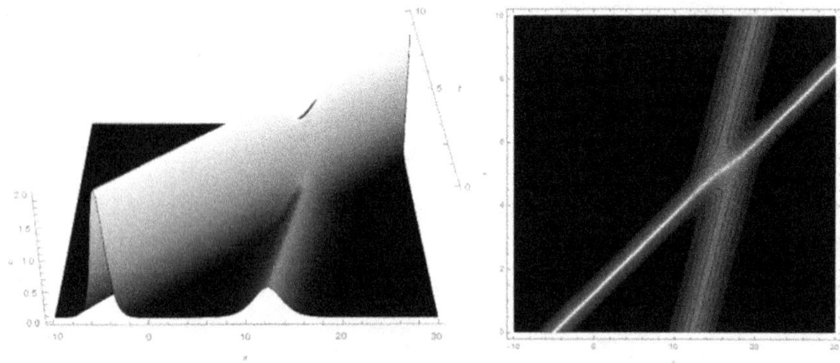

Fig. 2: A right-moving two-soliton solution to the KdV equation, exhibiting a soliton collision, for the values of $\beta = 1$, $\gamma = 6$, $\kappa_1 = 2$, $\kappa_2 = 1$, $c_1 = -10$ and $c_2 = 10$ with $(x,t) \in [-10, 30] \times [0, 10]$.

Following the same procedure for v_2, we get

$$\begin{cases} A = 2, \\ \omega_1 = \kappa_1^3, \\ \omega_2 = \kappa_2^3, \\ B = \left(\dfrac{\kappa_1 - \kappa_2}{\kappa_1 + \kappa_2}\right)^2. \end{cases} \tag{34}$$

Finally, we obtain a two-soliton solution to KdV from (34), (30) and (25).

In Fig. 2, we demonstrate a set of figures, consisting of a two-soliton solution to the KdV equation and the unique feature of wave overlap that is exhibited. Figure 3 the latter demonstrates that solitons are evidently waves behaving like particles [24,25].

2.5. *Self-similar solutions and the Painlevé property*

Let's again consider the KdV equation as follows:

$$u_t + \gamma u u_x + u_{xxx} = 0, \quad u = u(x,t), \ (x,t) \in \mathbb{R} \times (0, \infty), \ \gamma \pm 6. \tag{35}$$

the Current goal is to seek for self-similar solutions, namely solutions being similar to themselves, if the independent and dependent variables are appropriately scaled. Self-similar solutions are sought whenever the problem lacks a characteristic length or time scale, and are frequently used in the study of PDEs, particularly in fluid dynamics [34,35]. Introduce the

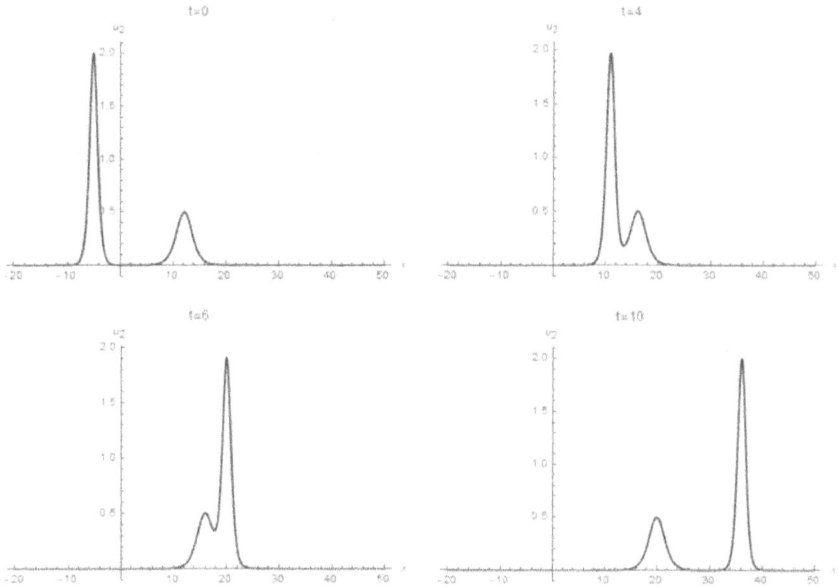

Fig. 3: Time evolution of a right-moving two-soliton solution to the KdV equation, for the values of $\beta = 1$, $\gamma = 6$, $\kappa_1 = 2$, $\kappa_2 = 1$, $c_1 = -10$ and $c_2 = 10$ with $(x, t) \in [-20, 50] \times \{0, 4, 6, 10\}$.

similarity transform,

$$\begin{cases} u(x,t) = t^m f(h), \\ h = xt^n, \end{cases} \tag{36}$$

where m and n shall be determined guaranteeing that the original PDE transforms to an ODE for f and h. The latter will be called the similarity equation.

Substitution of (36) into (35), with the use of chain differentiation, provides with

$$mf(h) + nhf'(h) + \gamma t^{m+n+1} f(h)f'(h) + t^{3n+1} f'''(h) = 0. \tag{37}$$

At this stage we require

$$\begin{cases} m + n + 1 = 0, \\ 3n + 1 = 0, \end{cases}$$

implying that $m = -\frac{2}{3}$ and $n = -\frac{1}{3}$. The latter leads to the ODE

$$f'''(h) + f'(h)\left(\gamma f(h) - \frac{h}{3}\right) - \frac{2}{3}f(h) = 0. \tag{38}$$

This ordinary differential equation can be shown to have the so-called Painlevé property.

Definition 2. An ordinary differential equation has the Painlevé property if it admits no movable singular points. A movable singular point is a point where the solution becomes singular, whose location depends on the arbitrary constants of integration [36].

For instance, the equation

$$y'(x) = y^2(x)$$

has the solution

$$y(x) = (C - x)^{-1},$$

which has a singular point whose location depends on the arbitrary constant of integration, C. Therefore, this equation does not have the Painlevé property.

There is a conjecture [37], that PDEs reducible to ODEs having the Painlevé property are integrable, that is, they admit soliton solutions and are solvable by the inverse scattering transform, which we will discuss later. Although we can't solve the ODE (38) in general, the act of deriving it can provide with useful information about the original PDE.

2.6. *The Lax pair and the zero curvature equation*

The term "Lax pair" refers to a set of two operators that, if they exist, indicate that a corresponding evolution equation,

$$F(x, t, u, \ldots) = 0,$$

is integrable. They represent a pair of differential operators having a characteristic feature whereby they yield a nonlinear evolution equation when they commute. Lax pairs were introduced by Peter Lax to discuss solitons in continuous media [38].

Now, a Lax pair shall consist of the Lax operator L, which is self-adjoint and may depend upon x, u_x, u_{xx} and so on, but not explicitly upon t, and the operator M that together with L, acting on a fixed Hilbert space, represents a given partial differential equation, namely Lax equation,

$$L_t = [M, L] := ML - LM, \tag{39}$$

with $[M, L]$ denoting the commutator of the two operators. Operators M and L can be either scalar or matrix operators.

Since the existence of a Lax pair indicates that the corresponding evolution equation is integrable, or equivalently exactly solvable, finding Lax pairs is a way of discovering new integrable evolution equations [39,40].

Additionally, in the case that a suitable Lax pair can be found for a particular nonlinear evolution equation, then it is possible that they can be used to solve the associated Cauchy problem using a method such as the Inverse Scattering Transform (IST) [25]. However, the latter is a difficult process and additional discussion can be found in Refs. [24,25,41–43].

Now, let's provide some deeper analysis behind both the Lax pairs and the Lax equation. Given a linear operator L, which may depend upon the function $u(x,t)$, the spatial variable x and spatial derivatives u_x, u_{xx} and so on, but not explicitly upon the time variable t, such that,

$$L\psi = \lambda\psi, \quad \psi = \psi(x,t), \tag{40}$$

the idea is to find another operator, M, whereby

$$\psi_t = M\psi. \tag{41}$$

More details about this technique can be found in Ref. [39]. Taking the time derivative of (40) gives

$$\begin{aligned} &L_t\psi + L\psi_t = \lambda_t\psi + \lambda\psi_t \\ &\overset{(40),(41)}{\Leftrightarrow} L_t\psi + LM\psi = \lambda_t\psi + ML\psi \\ &\Leftrightarrow (L_t + LM - ML)\,\psi = \lambda_t\psi. \end{aligned}$$

Therefore, solving for non-trivial eigenfunctions $\psi(x,t)$ demands

$$L_t + [L, M] = 0, \tag{42}$$

which will be true if and only if $\lambda_t = 0$, namely the "isospectrality" condition holds.

We may now conclude that a Lax pair, formed by (40) and (41), has the following properties.

(1) The eigenvalues λ are independent of time, $\lambda_t = 0$.
(2) The quantity $\psi_t - M\psi$ must remain a solution of (40).

(3) The compatibility relationship $L_t + (LM - ML) = 0$ must be true. This, together with property (1), implies that the operator L must be self-adjoint.

In the setting of the KdV equation, (35), consider the following Lax pair [43]:

$$\begin{cases} L = \partial_{xx} - u, \quad \text{(Sturm–Liouville operator)} \\ M = -4\partial_{xxx} + 6u\partial_x + 3u_x, \end{cases} \tag{43}$$

with all derivatives acting on all objects to the right. This specific operator L, introduced in (43), transforms (40) into the Schrödinger eigenvalue equation,

$$\Psi_{xx} - u\Psi = \lambda\Psi, \tag{44}$$

an equation that will be proven very useful as we progress.

Now, a straight substitution of (43) into the Lax equation, (39), yields

$$L_t = -u_t = [-4\partial_{xxx} + 6u\partial_x + 3u_x, \partial xx - u], \tag{45}$$

where the right-hand side of (45) is equal to

$$(ML - LM) = [M, L].$$

Now, expanding each element on the rhs of (45) separately, we get

$$\begin{cases} LM = -4\partial_x^5 + 10u\partial_{xxx} + 15u_x\partial_{xx} + \left(12u_{xx} - 6u^2\right)\partial_x + (3u_{xxx} - 3uu_x), \\ ML = -4\partial_x^5 + 10u\partial_{xxx} + 15u_x\partial_{xx} + \left(12u_{xx} - 6u^2\right)\partial_x + (4u_{xxx} - 9uu_x). \end{cases} \tag{46}$$

Thus, (46) implies that

$$(ML - LM) = (u_{xxx} - 6uu_x). \tag{47}$$

Therefore, on negating and rearranging (45), we finally arrive at

$$u_t - 6uu_x + u_{xxx} = 0,$$

which is exactly the familiar form of the KdV equation.

Another approach, avoiding the necessity of considering higher order Lax operators, is the *AKNS* method (Ablowitz *et al.*), which is a matrix formalism for Lax pairs [44]. The following system is introduced:

$$\begin{cases} D_x\Psi = X\Psi, \\ D_t\Psi = T\Psi, \end{cases} \tag{48}$$

where X and T correspond to operators L and M, resp., and Ψ is an auxiliary vector function. The compatibility condition for (48) shall be

$$\begin{aligned} 0 &= [D_t, D_x]\Psi \\ &= D_t(X\Psi) - D_x(T\Psi) \\ &= (D_tX)\Psi - XD_t\Psi - (D_xT)\Psi - TD_x\Psi, \end{aligned}$$

which may be rewritten, more compactly, as

$$(D_tX - D_xT + [X,T])\,\Psi = 0. \tag{49}$$

Equation (49) is known as the matrix Lax equation, with

$$[X,T] := XT - TX,$$

being defined as the matrix commutator. As a consequence of geometrical considerations, equation (49) is also known as the zero-curvature equation. For a more detailed discussion of this method, we refer to Refs. [10,45].

We demonstrate the application of the matrix Lax pair method to the KdV equation, (35). Consider the following matrix Lax pair:

$$\begin{cases} X = \begin{bmatrix} 0 & 1 \\ \lambda - \dfrac{\gamma}{6}u & 0, \end{bmatrix}, \\ T = \begin{bmatrix} au_x & -4\lambda - \dfrac{\gamma}{3}u \\ -4\lambda^2 + \dfrac{\lambda\gamma}{3}u + \dfrac{\gamma^2}{18}u^2 + \dfrac{\gamma}{6}u_{xx} & -\dfrac{\gamma}{6}u_x \end{bmatrix}, \end{cases} \tag{50}$$

with X and T being the matrices equivalent of the Lax operator L, and M, resp.

Inserting (50) into the zero curvature equation, (49), we are granted with,

$$\begin{bmatrix} 0 & 0 \\ -\dfrac{\gamma}{6}(u_t + \gamma uu_x + u_{xxx}) & 0 \end{bmatrix} = \begin{bmatrix} 0 & 0 \\ 0 & 0 \end{bmatrix}.$$

The latter implies that the compatibility condition should be

$$u_t + \gamma uu_x + u_{xxx} = 0,$$

which is the KdV equation, (35).

Therefore, we have just shown that the KdV equation can be thought of as the compatibility condition for the matrix Lax pair (50) [46]. The key idea of Lax pairs is that any equation that can be cast into such a framework for other operators L and M has automatically many of the features of the KdV equation, including an infinite number of local conservation laws.

2.7. *The inverse scattering transform*

Arguably, the biggest breakthrough in mathematical physics over the last 40 years, the landmark discovery of the Inverse Scattering Transform by Gardner, Green, Kruskal and Miura in 1967 [25] brought the KdV equation to the forefront of mathematical physics. It played a major role in the phenomenal development of the latter, involving multiple disciplines of science as well as several branches of mathematics.

The importance of the KdV arose in 1965, when Zabusky and Kruskal were able to explain the Fermi–Pasta–Ulam puzzle in terms of solitary-wave solutions to the KdV [47]. In their analysis of numerical solutions to the KdV, Zabusky and Kruskal observed solitary-wave pulses, named pulses solitons because of their particle-like behavior, and observed that such pulses interact with each other nonlinearly, yet come out of their interaction virtually unaffected in size or shape. Such unusual nonlinear interactions among soliton solutions to the KdV created a lot of excitement, but at that time no one knew how to solve such a nonlinear PDE, except numerically.

In their celebrated paper of 1967, Gardner, Greene, Kruskal and Miura presented a method, now known as the inverse scattering transform, to solve the initial-value problem for the KdV, assuming that the initial value $u(x, 0)$ approaches a constant sufficiently rapidly as $x \to \pm\infty$ [25]. There is no loss of generality in choosing that constant as zero.

Additionally, they showed that $u(x,t)$ can be obtained from $u(x, 0)$ with the help of the solution to the inverse scattering problem for the 1-D Schrödinger equation, (44), with the time-evolved scattering data. Moreover, it was explained that soliton solutions to the KdV corresponded to the case of zero reflection coefficient in the scattering data. They observed from various numerical studies of the KdV that, for large t, $u(x, t)$ in general consists of a finite train of solitons traveling in the positive x direction and an oscillatory train spreading in the opposite direction.

The method is a nonlinear analogue, and in some sense a generalization of the Fourier Transform, which itself is applied to solve many linear PDEs. The origin of the method is found in the key idea of recovering the time evolution of a potential from the time evolution of its scattering data. Inverse scattering refers to the problem of recovering a potential from its scattering matrix, as opposed to the direct scattering problem of finding the scattering matrix from the potential.

The Inverse Scattering Transform (IST) may be applied to many exactly solvable mathematical models, that is to say completely integrable infinite dimensional systems. Integrability means that there is a change of variables, action-angle variables, such that the evolution equation in the new variables is equivalent to a linear flow at constant speed [25,44]. A characteristic of solutions obtained by the inverse scattering method is the existence of solitons, solutions resembling both particles and waves, which have no analogue for linear PDEs.

At this stage, we shall briefly describe the steps followed in the solution process of the KdV initial value problem,

$$\begin{cases} u_t - 6uu_x + u_{xxx} = 0, \quad u = u(x,t), \quad (x,t) \in \mathbb{R} \times (0,\infty), \\ u(x,0) = f(x), \end{cases} \tag{51}$$

assuming that the initial value $u(x,0)$ vanishes rapidly as $x \to \pm\infty$.

To solve the initial value problem (51), we associate to our equation the Schrödinger eigenvalue equation,

$$\psi_{xx} - u\psi = \lambda\psi. \tag{52}$$

Here, ψ is an unknown function of x and t, and u is the solution of the KdV equation that is unknown except at $t = 0$. The constant λ is an eigenvalue.

A slight rearrangement of (52) reads

$$u = \frac{1}{\psi}\psi_{xx} - \lambda. \tag{53}$$

Inserting (53) into the KdV equation, (51), and integrating with respect to x, we are led to

$$\psi_t + \psi_{xxx} - 3(u-\lambda)\psi_x = C_1\psi + C_2\psi \int \frac{1}{\psi^2}dx. \tag{54}$$

The latter linear equation is known as a Gelfand–Levitan–Marchenko (GLM) equation.

The Inverse Scattering Method is summarized in the following steps:

(1) Determine the nonlinear PDE. This is usually accomplished by analyzing the physics of the situation being studied.
(2) Employ forward scattering. This consists in finding the Lax pair corresponding to the equation under study.

(3) Determine the time evolution of the eigenfunctions associated to each eigenvalue λ, the norming constants, and the reflection coefficient, all three comprising the so-called scattering data. This time evolution is given by a system of linear ordinary differential equations which can be solved.

(4) Perform the inverse scattering procedure by solving the Gelfand–Levitan–Marchenko integral equation, a linear integral equation, to obtain the final solution of the original nonlinear PDE [48,49]. All the scattering data are required in order to do this. If the reflection coefficient is zero, the process becomes much easier. This step works if the operator L, introduced in the previous section, is a differential or difference operator of order two, but not necessarily for higher orders. In all cases, however, the inverse scattering problem is reducible to a Riemann—Hilbert factorization problem [39,49].

2.8. *A spectral computational approach*

Aiming to a numerical solution of (9), instead of solving (20) exactly, we take the Fourier transform of the equation, getting

$$F\left\{-\lambda u + \frac{\gamma}{2}u^2 + u''\right\} = 0. \tag{55}$$

At this point, let's recall some essential facts on the Fourier Transform.

Definition 3. We define the Fourier and Inverse Fourier Transform of a function, say u, in the sense that the following symbols make sense, being [26,44],

$$\begin{cases} F\left\{u(x)\right\} = \hat{u}(\kappa) = \displaystyle\int_{-\infty}^{\infty} \mathrm{e}^{-i\kappa x} u(x) dx, \\ F^{-1}\left\{\hat{u}(\kappa)\right\} = u(x) = \dfrac{1}{2\pi}\displaystyle\int_{-\infty}^{\infty} \mathrm{e}^{i\kappa x}\hat{u}(\kappa) d\kappa. \end{cases} \tag{56}$$

Integrating by parts, regarding the nth derivative of f and its Fourier Transform, the following results hold:

$$\begin{cases} \dfrac{d^n u}{dx^n} = (i^n) F^{-1}\left\{\kappa^n \hat{u}(\kappa)\right\}, \\ \dfrac{d^n \hat{u}}{d\kappa^n} = (-i)^n F\left\{x^n u(x)\right\}. \end{cases} \tag{57}$$

Being an integral transform, the Fourier Transform preserves both linearity and homogeneity. Using the first property of (57), equation (55) reads

$$\begin{aligned} 0 &= -\lambda\hat{u} + \frac{\gamma}{2}F\left\{u^2\right\} - \kappa^2\hat{u} \\ &= (-\lambda - \kappa^2)\hat{u} + \frac{\gamma}{2}F\left\{u^2\right\} \\ \overset{\lambda \geq 0}{\Leftrightarrow} \hat{u} &= \frac{\gamma F\left\{u^2\right\}}{2(\lambda + \kappa^2)}. \end{aligned} \tag{58}$$

In order to construct a solution that its amplitude does not grow indefinitely nor tends to zero with each iteration, we introduce $v(x)$ such that

$$\begin{aligned} u &= cv \\ \Leftrightarrow \hat{u} &= c\hat{v}, \end{aligned} \tag{59}$$

where $c \neq 0$ is to be determined [50]. Substitution of (59) into (58), gives

$$\hat{v} = \frac{\gamma c F\left\{v^2\right\}}{2(\lambda + \kappa^2)}. \tag{60}$$

Additionally, multiplying (60) by $\hat{v}^*$, which is the conjugate of $\hat{v}$ and integrating over the frequency domain, we obtain

$$c = \frac{2}{\gamma}\frac{\int_{-\infty}^{\infty}(\lambda + \kappa^2)|\hat{v}|^2 d\kappa}{\int_{-\infty}^{\infty} F\left\{v^2\right\}\hat{v}^* d\kappa}. \tag{61}$$

Now, equations (60) and (61) form a fixed point iteration scheme, whose solution for c and v can be expressed as

$$\begin{cases} \widehat{v_{n+1}} = \dfrac{\gamma c_n F\left\{v_n^2\right\}}{2(\lambda + \kappa^2)}, \\ c_n = \dfrac{2}{\gamma}\dfrac{\int_{-\infty}^{\infty}(\lambda + \kappa^2)|\hat{v_n}|^2 d\kappa}{\int_{-\infty}^{\infty} F\left\{v_n^2\right\}\hat{v_n}^* d\kappa} \end{cases} \tag{62}$$

For the initial guesses, c_0 and v_0, we propose

$$c_0 = 1, \quad v_0(x) = \mathrm{e}^{-x^2}.$$

Next goal is to evolve our spectral solution in time, using the explicit fourth-order Runge–Kutta method. A proper rearrangement of the terms of our equation,

$$u_t + \gamma u u_x + u_{xxx} = 0$$

yields

$$\frac{\partial u}{\partial t} = -\gamma u \frac{\partial u}{\partial x} - \frac{\partial^3 u}{\partial x^3}. \tag{63}$$

At this stage, by means of the Inverse Fourier Transform, F^{-1}, equation (63) can be rewritten in the form [44,51],

$$\frac{\partial u}{\partial t} = f(t, u), \tag{64}$$

where a substitution is made on the x-partial derivatives, namely,

$$\begin{cases} \dfrac{\partial u}{\partial x} = iF^{-1}\{\kappa \hat{u}\}, \\ \dfrac{\partial^3 u}{\partial x^3} = -iF^{-1}\{\kappa^3 \hat{u}\}. \end{cases}$$

Equation (64) is suitable for applying the fourth-order explicit Runge–Kutta method. In Fig. 4, a comparison is made between the time-evolved spectral solution and the right-moving soliton of the KdV equation. It is worth noting that the two almost perfectly coincide.

2.9. *The modified KdV (mKdV) equation*

Consider the modified KdV (mKdV) boundary-value problem [10,13],

$$\begin{cases} u_t + \gamma u^2 u_x + u_{xxx} = 0, \quad u = u(x,t), \quad (x,t) \in \mathbb{R} \times (0,\infty), \quad \gamma = \pm 6, \\ \lim\limits_{|x|\to\infty} \dfrac{\partial^n u(x,t)}{\partial x^n} = 0, \quad n \in \mathbb{N}. \end{cases} \tag{65}$$

Same as the KdV equation, (65) can be rewritten as the conservation law [21],

$$u_t + \left(\gamma \frac{u^3}{3} + u_{xx}\right)_x = 0. \tag{66}$$

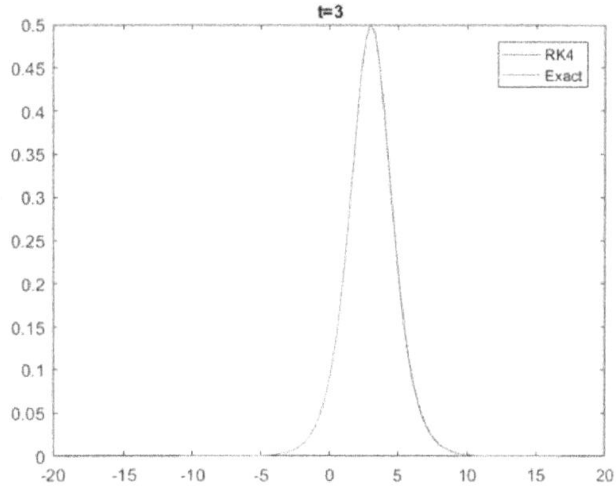

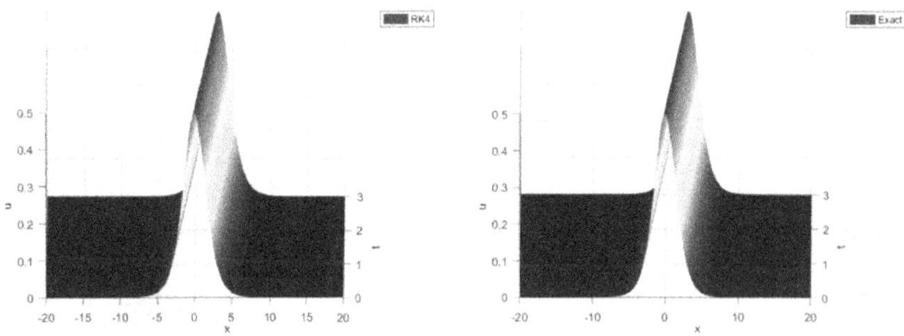

Fig. 4: The RK4 solution (left) and the right-moving soliton (right) of the KdV equation, for the values of $\lambda = 1$ and $\gamma = 6$, with $(x, t) \in [-20, 20] \times [0, 3]$.

Through (66), it is implied that the mass quantity $\int_{-\infty}^{\infty} u\, dx$ is conserved in time.

Therefore, it is again reasonable seeking for traveling wave solutions of (65). By applying the traveling wave transform (18), we are led to the second-order autonomous ODE,

$$-\lambda u + \frac{\gamma}{3}u^3 + u'' = 0. \tag{67}$$

Again, multiplying by u', we get

$$-\lambda uu' + \frac{\gamma}{3}u^3u' + u''u' = 0,$$

which can be integrated to get

$$-\lambda u^2 + \frac{\gamma}{6}u^4 + (u')^2 = 0. \tag{68}$$

Now, (68) is a first-order, separable equation. Following similar steps as those for the KdV equation, we arrive at a right-moving soliton of the mKdV equation, of the form

$$u(x,t) = \sqrt{\frac{6\lambda}{\gamma}} sech\left(\sqrt{\lambda}(x - \lambda t - c)\right), \quad c \in \mathbb{R}. \tag{69}$$

Regarding a computational approach, the spectral method can be applied in this case as well. In a similar manner as in the KdV equation, the iterative scheme shall now be

$$\begin{cases} \hat{v_{n+1}} = \dfrac{\gamma c_n^2 F\left\{v_n^3\right\}}{3(\lambda+\kappa^2)}, \\ c_n^2 = \dfrac{3}{\gamma}\dfrac{\int_{-\infty}^{\infty}(\lambda+\kappa^2)|\hat{v_n}|^2 d\kappa}{\int_{-\infty}^{\infty} F\left\{v_n^3\right\}\hat{v_n}^* d\kappa}. \end{cases} \tag{70}$$

Note that, since (65) is symmetrical under $v \mapsto -v$, it suffices to consider the positive square root for c_n. For the initial guesses, c_0 and v_0, again, we select $c_0 = 1$ and $v_0 = \mathrm{e}^{-x^2}$.

Time-evolution of the spectral solution, using the explicit fourth-order Runge–Kutta method, is performed as in the KdV equation.

The concept underlying the spectral method was to transform the underlying equation governing the soliton into Fourier space and determine a nonlinear non-local integral equation coupled to an algebraic equation. The coupling prevents the numerical scheme from diverging. The nonlinear mode is then determined from a convergent fixed-point iteration scheme. The spectral method is an iterative method which in each iteration adjusts the ratio between the dispersive and the nonlinear parts of the equation so that a "balance" between the two effects is maintained. It is a method of spectral accuracy, regarding the spatial variables, converging rapidly to a solution of the required accuracy [50,52,53].

2.10. *The Miura transform*

The Miura Transform is known as the transformation between the KdV and the mKdV equations. The KdV equation, in a setting convenient for our purpose, reads

$$KdV(u) := u_t - 6uu_x + u_{xxx} = 0, \tag{71}$$

while its modified counterpart, the mKdV equation, equals

$$mKdV(v) := v_t - 6v^2 v_x + v_{xxx} = 0. \tag{72}$$

Assume $v \neq 0$ satisfies the mKdV equation (72).

Definition 4. The Miura Transform of v is defined as

$$u_\pm(x,t) := v(x,t)^2 \pm v_x(x,t). \tag{73}$$

The transform is formally the same as Riccati's differential equation of a variable x if u is assumed to be a known function.

Our task will be to prove that (73) solves the KdV equation, (71). A straight substitution of (73) into (71) yields

$$\begin{aligned}&\left(v^2 \pm v_x\right)_t - 6\left(v^2 \pm v_x\right)\left(v^2 \pm v_x\right)_x + \left(v^2 \pm v_x\right)_{xxx}\\ &\quad = 2vv_t \pm v_{tx} - 12v^3 v_x \mp 6v^2 v_{xx} \mp 12vv_x^2 \pm v_{xxxx} + 2vv_{xxx}.\end{aligned} \tag{74}$$

Multiplying (72) by $2v$, we get

$$2vv_t - 12v^3 v_x + 2vv_{xxx} = 0. \tag{75}$$

Differentiating (72) with respect to x, we obtain

$$v_{tx} - 12vv_x^2 - 6v^2 v_{xx} + v_{xxxx} = 0. \tag{76}$$

Combining (74) with (75) and (76), we are left with

$$\left(v_x + v^2\right)_t - 6\left(v_x + v^2\right)\left(v_x + v^2\right)_x + \left(v_x + v^2\right)_{xxx} = 0.$$

The latter implies that $u_\pm$ defined as the Miura Transform of v indeed satisfies the KdV equation. Furthermore, explicit calculations by Miura showed the validity of the identity

$$KdV(u_\pm) = (2v \pm \partial_x) mKdV(v). \tag{77}$$

The latter identity enables to transfer solutions of the mKdV equation to solutions of the KdV equation. The Miura transformation was also quite prominently used in the construction of an infinite series of conservation laws for the KdV equation [22,40].

The Miura transformation, when applied to the Lax pair for the KdV equation, given by (43), yields the following Lax pair:

$$\begin{cases} L = \partial_{xx} - \left(v_x + v^2\right), \\ \quad = -4\partial_{xxx} + 6\left(v_x + v^2\right)\partial_x + 3\partial_x\left(v_x + v^2\right). \end{cases} \tag{78}$$

Substituting (78) into the Lax equation, (39), and negating, we get

$$v_{tx} + 2vv_t + v_{xxxx} + 2vv_{xxx} - 6v^2 v_{xx} - 12v\,(v_x)^2 - 12v^3 v_x = 0,$$

which can be further simplified into

$$(2v + \partial_x)\left(v_t - 6v^2 v_x + v_{xxx}\right) = 0. \tag{79}$$

At this stage, due to the nontrivial kernel of $(2v+\partial_x)$, it is not immediately clear how to reverse the procedure so as to transfer solutions of the KdV equation to solutions of the mKdV equation. Introduce the first-order differential expression

$$\widetilde{P}(u) = 2u\partial_x - u_x. \tag{80}$$

to derive

$$mKdV(v) = \partial_x\left(\frac{1}{\psi}(\psi_t - \widetilde{P}(u\pm)\psi)\right), \tag{81}$$

with

$$\begin{cases} v = \dfrac{\psi_x}{\psi}, & \psi > 0, \\ u_\pm = v^2 \pm v_x. \end{cases} \tag{82}$$

Next, assume that $u = u(x,t)$ solves the KdV equation $KdV(u) = 0$, and $\psi > 0$ satisfies

$$\begin{cases} \psi_t = \widetilde{P}(u)\psi, \\ -\psi_{xx} + u\psi = 0. \end{cases} \tag{83}$$

Then, we immediately deduce that v solves the mKdV equation $mKdV(v) = 0$, and hence, the Miura transformation has been "inverted" [54,55].

The KdV equation, (71), and the mKdV equation, (72), are just the first nonlinear evolution equations in a countably infinite hierarchy of such equations, namely the (m)KdV hierarchy. The considerations performed throughout the current section can extend to the entire hierarchy of these equations [54,56,57].

2.11. *Homotopy analysis method*

The Homotopy Analysis Method (HAM) is a semi-analytic technique to solve nonlinear ordinary and partial differential equations. The HAM employs the concept of homotopy from topology to generate a convergent series solution for nonlinear systems. This is enabled by utilizing a homotopy-Maclaurin series to deal with the nonlinearities in the system.

The method was first devised in 1992 by Liao Shijun of Shanghai Jiaotong University in his PhD dissertation [58] and further modified [59] in 1997 to introduce a non-zero auxiliary parameter, referred to as the convergence-control parameter, $\hbar$, to construct a homotopy on a differential system in general form [60]. The convergence-control parameter is a non-physical variable that provides a simple way to verify and enforce convergence of a solution series.

The HAM is distinguished from various other analytical methods in four important aspects. First, it is a series expansion method that is not directly dependent on small or large physical parameters. Thus, it is applicable for not only weakly but also strongly nonlinear problems, going beyond some of the inherent limitations of the standard perturbation methods. Second, the HAM is a unified method for the Lyapunov artificial small parameter method, the delta expansion method, the Adomian decomposition method [61] and the homotopy perturbation method [62,63]. The greater generality of the method often allows for strong convergence of the solution over larger spatial and parameter domains. Third, the HAM gives excellent flexibility in the expression of the solution and how the solution is explicitly obtained. It provides great freedom to choose the basis functions of the desired solution and the corresponding auxiliary linear operator of the homotopy. Finally, unlike the other analytic approximation techniques, the HAM provides a simple way to ensure the convergence of the solution series.

The method is able to combine with other techniques employed in nonlinear differential equations such as spectral methods [64] and Padé approximants. It may further be combined with computational methods, such as the boundary element method, to allow the linear method to solve nonlinear systems. Different from the numerical technique of homotopy continuation, the homotopy analysis method is an analytic approximation method as opposed to a discrete computational method. Furthermore, HAM uses the homotopy parameter only on a theoretical level to demonstrate that a nonlinear system may be split into an infinite set of linear systems which are solved analytically, while the continuation methods require solving a discrete linear system as the homotopy parameter is varied to solve the nonlinear system [65].

In the last 20 years, HAM has been applied to solve a growing number of nonlinear ordinary/partial differential equations in science, finance and engineering [65–68]. For example, multiple steady-state resonant waves in deep and finite water depth [69] were found with the wave resonance criterion of arbitrary number of traveling gravity waves; this agreed with Phillips's criterion for four waves with small amplitude. Further, a unified wave model applied with the HAM [70] admits not only the traditional smooth progressive periodic/solitary waves, but also the progressive solitary waves with peaked crest in finite water depth. This model shows peaked solitary waves are consistent solutions along with the known smooth ones. Additionally, the HAM has been applied to many other nonlinear problems such as nonlinear heat transfer [71], the limit cycle of nonlinear dynamic systems [72], the American put option [73], the exact Navier–Stokes equation [74], the option pricing under stochastic volatility [75], the electrohydrodynamic flows [76], the Poisson–Boltzmann equation for semiconductor devices [77], study of boundary layer equations [68], and others.

We consider the following differential equation:

$$N\left[u(\tau)\right] = 0, \tag{84}$$

where N is a nonlinear operator, τ denotes the independent variable and $u(\tau)$ is an unknown function, resp. For simplicity, we ignore all boundary or initial conditions, which can be treated in a similar way.

By means of generalizing the traditional homotopy method, Liao [60], constructs the so-called zero-order deformation equation,

$$(1-p)L\left[\phi(\tau;p) - u_0(\tau)\right] = p\hbar H(\tau) N\left[\phi(\tau;p)\right], \tag{85}$$

where $p \in [0,1]$ is the embedding parameter, $\hbar \neq 0$ is a non-zero auxiliary parameter, $H(\tau) \neq 0$ is an auxiliary function, L is an auxiliary linear operator, $u_0(\tau)$ is an initial guess of $u(\tau)$ and $u(\tau;p)$ is an unknown function, resp. It is crucial that one has great freedom in making auxiliary choices in HAM. Obviously, when $p = 0$ and $p = 1$, it holds that

$$\phi(\tau;0) = u_0(\tau), \quad \phi(\tau;1) = u(\tau). \tag{86}$$

Thus, as p increases from 0 to 1, the solution $u(\tau;p)$ continuously varies from the initial guess $u_0(\tau)$ to the solution $u(\tau)$.

Expanding $u(\tau;p)$ in Taylor series with respect to p, we obtain

$$\phi(\tau;p) = u_0(\tau) + \sum_{m=1}^{\infty} u_m(\tau)p^m, \tag{87}$$

where

$$u_m(\tau) = \frac{1}{m!}\frac{\partial^m \phi(\tau;p)}{\partial p^m}\bigg|_{p=0}. \tag{88}$$

If the auxiliary linear operator, the initial guess, the auxiliary parameter $\hbar$ and the auxiliary function are properly chosen, then the series in (87) converges at $p = 1$, giving

$$u(\tau) = u_0(\tau) + \sum_{m=1}^{\infty} u_m(\tau), \tag{89}$$

which must be among the solutions of the original nonlinear equation, as proven by Liao [60].

As $\hbar = -1$ and $H(\tau) = 1$, equation (85) becomes

$$(1-p)L\left[\phi(\tau;p) - u_0(\tau)\right] + pN\left[\phi(\tau;p)\right] = 0, \tag{90}$$

which is used mostly in the homotopy perturbation method [78], where the solution is directly obtained, without the use of Taylor series.

Differentiating equation (85) m times with respect to the embedding parameter p, taking $p = 0$ and finally dividing by $m!$, we obtain the so-called mth-order deformation equation

$$L\left[u_m(\tau) - \chi_m u_{m-1}(\tau)\right] = \hbar H(\tau) R_m(u_{m-1}), \tag{91}$$

where,

$$R_m(u_{m-1}) = \frac{1}{(m-1)!}\frac{\partial^{m-1} N\left[\phi(\tau;p)\right]}{\partial p^{m-1}}\bigg|_{p=0}, \tag{92}$$

and

$$\chi_m = \begin{cases} 0, & m \le 1, \\ 1, & m > 1. \end{cases} \tag{93}$$

It should be emphasized that $u_m(\tau)$ for $m \ge 1$ is fully governed by the linear equation (91) under the linear boundary conditions that arise from the original problem, which can easily be solved by a symbolic computation software, such as Mathematica.

For the convergence of the above method, a reliable reference is Liao's work [60]. Should equation (84) admit a unique solution, then the described method produces this unique solution. Otherwise, it gives a solution among many other possible solutions.

In this section, the Homotopy analysis method is applied to the KdV equation, equation (9). Consider a bell-shaped initial condition,

$$u(x,0) = \frac{\mathrm{e}^{-x^2}}{10}. \tag{94}$$

Let's consider the linear operator [79,80],

$$L\left[u\right] = \frac{\partial u}{\partial t}, \tag{95}$$

satisfying

$$L\left[u\right] = 0 \Leftrightarrow u = u(x). \tag{96}$$

We now proceed in constructing the R_m term, for $m \geq 1$, having

$$R_m(u_{m-1}) = \frac{1}{(m-1)!} \left.\frac{\partial^{m-1} N\left[u\right]}{\partial p^{m-1}}\right|_{p=0}, \quad N\left[u\right] = \frac{\partial u}{\partial t} + \gamma u \frac{\partial u}{\partial x} + \beta \frac{\partial^3 u}{\partial x^3}, \tag{97}$$

which gives,

$$R_m(u_{m-1}) = \frac{\partial u_{m-1}}{\partial t} + \gamma \left(\sum_{k=0}^{m-1} \frac{\partial u_k}{\partial x} u_{m-1-k} \right) + \beta \frac{\partial^3 u_{m-1}}{\partial x^3}. \tag{98}$$

Then, the mth order deformation equation is obtained as

$$\begin{cases} L\left[u_m - \chi_m u_{m-1}\right] = \hbar R_m(u_{m-1}), \\ u_m(x,0) = \begin{cases} u(x,0), & m = 1 \\ 0, & \text{otherwise.} \end{cases} \end{cases} \tag{99}$$

Using equations (95) and (99) can be rewritten as

$$\begin{cases} \dfrac{\partial u_m}{\partial t} - \chi_m \dfrac{\partial u_{m-1}}{\partial t} = \hbar \left(\dfrac{\partial u_{m-1}}{\partial t} + \gamma \left(\displaystyle\sum_{k=0}^{m-1} \dfrac{\partial u_k}{\partial x} u_{m-1-k} \right) \right. \\ \qquad\qquad\qquad\qquad \left. + \beta \dfrac{\partial^3 u_{m-1}}{\partial x^3} \right), \\ u_m(x,0) = \begin{cases} u(x,0), & m = 1 \\ 0, & \text{otherwise.} \end{cases} \end{cases} \tag{100}$$

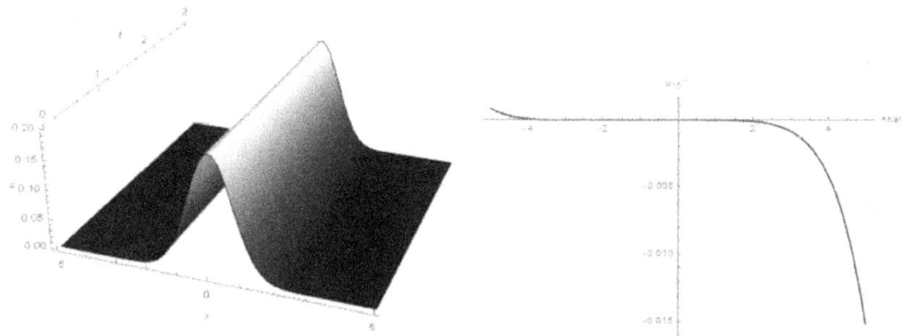

Fig. 5: The Homotopy analysis solution of 15th-order for the values of $\beta = 0.01$, $\gamma = 0.2$ and $\hbar = 0.01$, with $(x,t) \in [-5,5] \times [0,3]$ and the corresponding $\hbar$-curve ensuring convergence for $\hbar \in [-4,2]$.

The latter is recursively solved for u_m by utilizing Mathematica, where, respecting the $\hbar$-curve analysis, $\hbar = 0.01$ is chosen. A 15th-order approximation reveals an error, $||\delta||_\infty = 3.4 \times 10^{-3}$. In Fig. 5, both the $\hbar$-curve and the HAM-obtained solution are demonstrated, exhibiting a solitary waveform.

3. The Viscous Burgers Equation

The effects of diffusion and nonlinear advection are combined in the viscous Burgers equation, most commonly studied in the form [81],

$$u_t + \gamma u u_x - \alpha u_{xx} = 0, \quad u = u(x,t), \quad (x,t) \in \mathbb{R} \times (0,\infty), \quad \alpha > 0, \quad \gamma = 1. \tag{101}$$

Initially introduced by Harry Bateman in 1915 [82,83] and later studied by Johannes Martinus Burgers in 1948 [81], the equation models plenty of phenomena in fluid mechanics (relating to the Navier–Stokes momentum equation with negligible pressure term) [84–86], gas dynamics [87], and traffic flow [88].

There are four parameters in the Burgers equation, u, x, t and α. In a system consisting of a moving viscous fluid with one spatial (x) and one temporal (t) dimension, like a thin ideal pipe with fluid running through it, the Burgers equation describes the velocity of the fluid at each location along the pipe as time progresses. The terms of the equation represent the following quantities:

(1) x: spatial coordinate.
(2) t: temporal coordinate (time).

(3) $u(x,t)$: velocity of the fluid at the indicated spatial and temporal coordinates.
(4) α: viscosity of fluid.

The viscosity is a constant physical property of the fluid, and the other parameters represent the dynamics contingent on that viscosity.

3.1. *Derivation from the Navier–Stokes equations*

We may obtain the Burgers equation as a simplified version of the Navier–Stokes equations. For a Newtonian incompressible fluid, the Navier–Stokes equations, in vector form, read [84–86,89]

$$\rho\left(\frac{\partial \mathbf{u}}{\partial t} + \mathbf{u}\cdot\nabla\mathbf{u}\right) = -\nabla p + \mu\nabla^2\mathbf{u} + \mathbf{F}. \tag{102}$$

Here, ρ is the fluid density, $\mathbf{u}$ is the velocity vector field, p is the pressure, μ is the viscosity, and $\mathbf{F}$ is an external force field. Considering that the effect of pressure drop is negligible, $\nabla p = 0$, we obtain the Burgers equation,

$$\rho\left(\frac{\partial \mathbf{u}}{\partial t} + \mathbf{u}\cdot\nabla\mathbf{u}\right) = \mu\nabla^2\mathbf{u} + \mathbf{F}. \tag{103}$$

The latter simplifies further, assuming the external force field term being zero and taking advantage of the fact that ρ is a constant for an incompressible fluid. Now we are enabled to define a new constant, the kinematic viscosity,

$$\nu = \frac{\mu}{\rho}. \tag{104}$$

Combining (103) with (104), we may write Burgers equation as

$$\frac{\partial \mathbf{u}}{\partial t} + \mathbf{u}\cdot\nabla\mathbf{u} = \nu\nabla^2\mathbf{u}. \tag{105}$$

In the following sections, we consider Burgers equation with non-zero viscosity, but in the special case of one space dimension ($n = 1$). In that case, (105) becomes

$$\frac{\partial u}{\partial t} + u\frac{\partial u}{\partial x} = \nu\frac{\partial^2 u}{\partial x^2}, \tag{106}$$

being exactly, the previously introduced, Burgers equation.

3.2. *Analytic traveling wave solutions*

At this section, as in the early stages of the KdV analysis, we shall initially seek for special solutions. The most general initial-value problem shall be dealt with later on. Consider the viscous Burgers equation,

$$\begin{cases} u_t + uu_x - \alpha u_{xx} = 0, \ u = u(x,t), \ \alpha > 0, \\ \lim\limits_{x\to-\infty} u(x,t) = u_1, \ \lim\limits_{x\to\infty} u(x,t) = u_2, \ u_1 > u_2, \text{ being two constant values} \end{cases} \tag{107}$$

and apply the traveling wave transform (18). Application of the chain rule leads to

$$-\lambda u'(\zeta) + u(\zeta)u'(\zeta) - \alpha u''(\zeta) = 0, \quad u = u(\zeta). \tag{108}$$

Integrating gives

$$\begin{aligned} &-\lambda u + \frac{1}{2}u^2 - \alpha u' = C \\ &\Leftrightarrow u' = \frac{1}{2\alpha}\left(u^2 - 2\lambda u - 2C\right), \end{aligned} \tag{109}$$

where C is a constant of integration. Clearly, (109) suggests that u_1 and u_2 are the roots of the quadratic equation,

$$u^2 - 2\lambda u - 2C = 0. \tag{110}$$

Thus, λ and C shall be determined from the sum and the product of the roots u_1 and u_2 of (110), and therefore,

$$\begin{cases} \lambda = \dfrac{u_1 + u_2}{2}, \\ C = -\dfrac{u_1 u_2}{2}. \end{cases} \tag{111}$$

The latter implies that the wave speed λ is the average of the two speeds of asymptotic states at infinity. At this stage, (109) can be rewritten as

$$u' = \frac{1}{2\alpha}(u - u_1)(u - u_2), \tag{112}$$

being a first-order separable equation for u. Now, integration of (112), taking the integration constant to 0, leads to

$$u(\zeta) = \frac{u_1 + u_2 \mathrm{e}^{\frac{u_1 - u_2}{2\alpha}\zeta}}{1 + \mathrm{e}^{\frac{u_1 - u_2}{2\alpha}\zeta}}. \tag{113}$$

Another useful expression for u can be obtained from (113) in the form

$$\frac{(u_1+u_2)}{2} - \frac{(u_1-u_2)}{2}\tanh\left(\frac{(u_1-u_2)}{4\alpha}\zeta\right). \tag{114}$$

Finally, in terms of x and t our traveling shock wave solution shall be

$$u(x,t) = \frac{(u_1+u_2)}{2} - \frac{(u_1-u_2)}{2}\tanh\left(\frac{(u_1-u_2)}{4\alpha}\left(x - \frac{(u_1+u_2)}{2}t\right)\right). \tag{115}$$

As $u_1 > u_2$, the wave profile $u(\zeta)$ decreases monotonically with ζ from the constant value u_1, as $\zeta \to -\infty$, to the constant value u_2, as $\zeta \to -\infty$.

At $\zeta = 0$, note that

$$u = \frac{u_1+u_2}{2}.$$

The shape of the waveform (114) is significantly affected by the diffusion coefficient α. This means that the presence of diffusion processes prevents the gradual distortion of the wave profile, and so, it does not break, as seen on Fig. 6. On the other hand, if the diffusion term is absent ($\alpha \to 0^+$), the solution becomes discontinuous, with

$$\lim_{\alpha\to 0+} u(x,t) = \begin{cases} u_1, & x < \dfrac{(u_1+u_2)}{2}t, \\ u_2, & x > \dfrac{(u_1+u_2)}{2}t. \end{cases} \tag{116}$$

The advantage of the above analysis is that the convection and diffusion terms in the Burgers equation exhibit opposite effects.

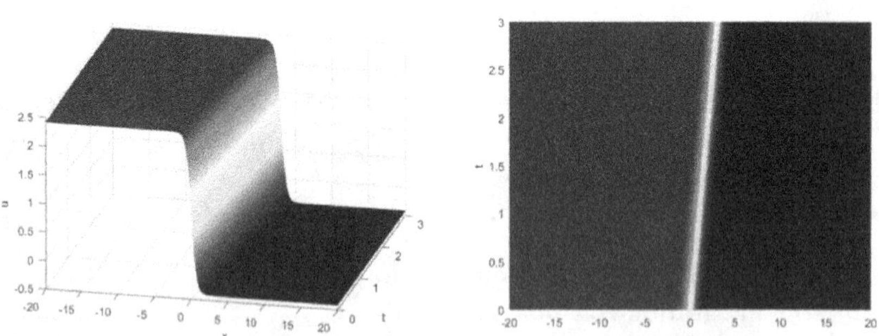

Fig. 6: A traveling wave solution of the viscous Burgers equation, for the values of $\alpha = 0.5$, $u_1 = 1+\sqrt{2}$ and $u_2 = 1-\sqrt{2}$, with $(x,t) \in [-20,20] \times [0,3]$.

3.3. Self-similar solutions

Following the same reasoning as in the KdV equation, the non-trivial self-similar solutions shall be given by

$$\begin{cases} u(x,t) = t^m f(h), \\ mf(h) + nhf'(h) + t^{m+n+1} f(h)f'(h) - \alpha t^{2n+1} f''(h) = 0, \quad h = xt^n, \end{cases} \tag{117}$$

reducing to an ODE, for $f(h)$, provided,

$$m = n = -\frac{1}{2}. \tag{118}$$

Thus, the equation under study shall be

$$\alpha f''(h) + \frac{1}{2}\left(f(h) + hf'(h)\right) - f(h)f'(h) = 0. \tag{119}$$

Integration of the latter with respect to h, respecting vanishing infinity conditions, yields

$$\begin{aligned} &\alpha f'(h) + \frac{1}{2}\left(hf(h) - f^2(h)\right) = 0, \\ &\overset{\alpha \gtrless 0}{\Leftrightarrow} f'(h) = -\frac{h}{2\alpha} f(h) + \frac{1}{2\alpha} f^2(h). \end{aligned} \tag{120}$$

Now, (120) is a first-order Bernoulli equation. Seeking for non-trivial solutions, we start by dividing both sides of (120) by $f^2(h)$, obtaining

$$f^{-2}(h)f'(h) = -\frac{h}{2\alpha} f^{-1}(h) + \frac{1}{2\alpha}. \tag{121}$$

Next, setting

$$v(h) = f^{-1}(h), \tag{122}$$

(121) reads

$$v'(h) = \frac{h}{2\alpha} v(h) - \frac{1}{2\alpha}. \tag{123}$$

At this stage, note that (123) is a first-order linear non-homogeneous equation for $v(h)$, with its general solution being

$$\begin{aligned} v(h) &= \mathrm{e}^{\frac{1}{2\alpha}\int h dh}\left(C - \frac{1}{2\alpha}\int \mathrm{e}^{-\frac{1}{2\alpha}\int h dh} dh\right) \\ &= \mathrm{e}^{\frac{h^2}{4\alpha}}\left(C - \frac{1}{2\alpha}\int \mathrm{e}^{-\frac{h^2}{4\alpha}} dh\right) \\ &= \mathrm{e}^{\frac{h^2}{4\alpha}}\left(C - \frac{\sqrt{\pi}}{2\sqrt{\alpha}} erf\left(\frac{h}{2\sqrt{\alpha}}\right)\right). \end{aligned} \tag{124}$$

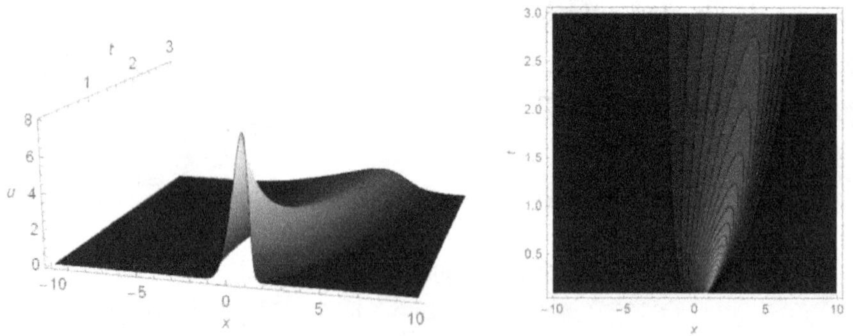

Fig. 7: A self-similar solution of the viscous Burgers equation, for the values of $\alpha = 0.8$ and $C = 1$, with $(x, t) \in [-10, 10] \times [0.1, 3]$.

Reverting (122), we obtain

$$f(h) = \frac{e^{-\frac{h^2}{4\alpha}}}{C - \frac{\sqrt{\pi}}{2\sqrt{\alpha}} erf\left(\frac{h}{2\sqrt{\alpha}}\right)}. \tag{125}$$

Combining (125) with (118) and (117), we are finally led to

$$u(x,t) = \frac{e^{-\frac{x^2}{4\alpha t}}}{\sqrt{t}\left(C - \frac{\sqrt{\pi}}{2\sqrt{\alpha}} erf\left(\frac{x}{2\sqrt{\alpha}\sqrt{t}}\right)\right)}. \tag{126}$$

In Fig. 7 a self-similar solution of the viscous Burgers equation is exhibited.

3.4. *The Cole–Hopf transform*

We now turn the focus on the general initial-value problem,

$$\begin{cases} u_t + uu_x - \alpha u_{xx} = 0, \ u = u(x,t), \ (x,t) \in \mathbb{R} \times (0, \infty), \ \alpha > 0, \\ u(x, 0) = g(x), \ x \in \mathbb{R}. \end{cases} \tag{127}$$

Setting

$$\begin{cases} u := \int_{-\infty}^{x} u(y,t)\, dy, \\ h(x) := \int_{-\infty}^{x} g(y)\, dy, \end{cases} \tag{128}$$

we are led to the initial-value problem

$$\begin{cases} w_t + \dfrac{w_x^2}{2} - \alpha w_{xx} = 0, \ (x,t) \in \mathbb{R} \times (0,\infty), \\ w(x,0) = h(x), \ x \in \mathbb{R}. \end{cases} \tag{129}$$

Assume, momentarily, that w is a smooth solution of (129), and set

$$v = \phi(w), \tag{130}$$

where ϕ shall be determined in a way that v solves a linear equation.

Chain differentiation of (130) provides with

$$\begin{cases} v_t = \phi'(w)w_t, \\ v_{xx} = \phi''(w)w_x^2 + \phi'(w)w_{xx}. \end{cases} \tag{131}$$

Now, as a consequence, (129) implies

$$\begin{aligned} v_t &= \phi'(w)\left(\alpha w_{xx} - \frac{w_x^2}{2}\right) \\ &= \alpha v_{xx} - \left(\alpha\phi''(w) + \frac{\phi'(w)}{2}\right) w_x^2 \\ &= \alpha v_{xx}, \end{aligned}$$

provided we choose ϕ such that

$$\alpha\phi'' + \frac{\phi'}{2} = 0.$$

The latter is solved if setting

$$\phi = \mathrm{e}^{-\frac{z}{2\alpha}}.$$

Therefore, we note that, in case w solves (129), then

$$v = \mathrm{e}^{-\frac{w}{2\alpha}} \tag{132}$$

solves the initial value problem for the Heat transfer equation, with conductivity α, as follows:

$$\begin{cases} v_t - \alpha v_{xx} = 0, \ (x,t) \in \mathbb{R} \times (0,\infty), \\ v(x,0) = \mathrm{e}^{-\frac{g(x)}{2\alpha}}. \end{cases} \tag{133}$$

Definition 5. The formula (132) is known as the Cole–Hopf transform.

At this point, the linear problem (133) possesses a unique bounded solution,

$$v(x,t) = \frac{1}{\sqrt{4\pi\alpha t}} \int_{-\infty}^{\infty} \mathrm{e}^{-\frac{(x-y)^2}{4\alpha t} - \frac{g(y)}{2\alpha}}, \quad (x,t) \in \mathbb{R} \times (0,\infty). \tag{134}$$

Now, since (132) may be rewritten as

$$w = -2\alpha \log v,$$

we obtain thereby the explicit formula for w,

$$w(x,t) = -2\alpha \log \frac{1}{\sqrt{4\pi\alpha t}} \int_{-\infty}^{\infty} \mathrm{e}^{-\frac{(x-y)^2}{4\alpha t} - \frac{g(y)}{2\alpha}}, \quad (x,t) \in \mathbb{R} \times (0,\infty). \tag{135}$$

Finally, regarding the initial unknown function, u, (128) along with differentiation of (135) gives

$$u(x,t) = \frac{\int_{-\infty}^{\infty} \frac{x-y}{t} \mathrm{e}^{-\frac{(x-y)^2}{4\alpha t} - \frac{h(y)}{2\alpha}}}{\int_{-\infty}^{\infty} \mathrm{e}^{-\frac{(x-y)^2}{4\alpha t} - \frac{h(y)}{2\alpha}}}. \tag{136}$$

3.5. *Homotopy analysis method*

Following the same procedure as in the KdV equation, the initial guess,

$$u(x,0) = \frac{\mathrm{e}^{-x}}{1+\mathrm{e}^{-x}}, \tag{137}$$

is proposed.

A seventh-order approximation reveals an error, $||\delta||_\infty = 1.8 \times 10^{-5}$. In Fig. 8, the HAM obtained solution is demonstrated, exhibiting a shock waveform.

4. The Inviscid Burgers Equation

A central issue in the study of nonlinear evolution equations is that solutions may exist locally, yet not globally in time. This is caused by a phenomenon called "blow-up". One of the simplest nonlinear PDEs, exhibiting blow-up,

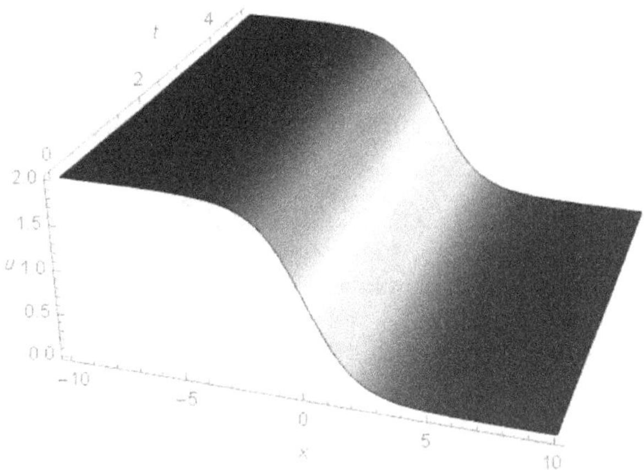

Fig. 8: A Homotopy analysis solution of seventh-order for the values of $\alpha = 0.2$, $\gamma = 1$ and $\hbar = 0.01$, with $(x,t) \in [-10, 10] \times [0, 5]$.

is the inviscid Burgers equation, [21,31,90]

$$\begin{cases} u_t + \gamma \left(\dfrac{u^2}{2}\right)_x = 0, \; u = u(x,t) \in \mathbb{R}, \; (x,t) \in \mathbb{R} \times (0, \infty), \\ u(x,0) = g(x), \; (x,t) \in \mathbb{R} \times \{t = 0\}. \end{cases} \tag{138}$$

Without loss of generality, assume $\gamma = 1$ with g denoting the initial-time data, which for the moment is assumed to be smooth.

We may derive the inviscid Burgers equation by considering free fluid (that is, non-interacting) particles where at the point $x \in \mathbb{R}$ at time t the velocity of the particles is $u(x,t)$ [84,89]. As time passes, the velocity of each particle does not change. However, the particles move, so it is the convective derivative of u that vanishes,

$$0 = \frac{Du}{Dt} = \frac{\partial u}{\partial t} + u \frac{\partial u}{\partial x}. \tag{139}$$

4.1. *The method of characteristics*

We shall solve (138) using the method of characteristics.

More specifically, let Γ be a smooth curve on the boundary $\mathbb{R} \times \{t = 0\}$ such that

$$\Gamma := (x(s), t(s))$$

and

$$z(s) = u(x(s), t(s))$$

give the values of u along the curve Γ. Then, using chain differentiation, we obtain

$$\frac{dz(s)}{ds} = \frac{du(x(s), t(s))}{ds} = u_x \frac{dx}{ds} + u_t \frac{dt}{ds}.$$

Thus, our characteristic equations will be given by

$$\begin{cases} \dfrac{dt}{ds} = 1, & t(r, 0) = 0, \\ \dfrac{dx}{ds} = z, & x(r, 0) = r, \\ \dfrac{dz}{ds} = 0, & z(r, 0) = g(r). \end{cases}$$

We conclude that the characteristic curves are

$$\begin{cases} t = s, \\ x = g(r)s + r, \\ z = g(r). \end{cases} \tag{140}$$

Noting that $\frac{dz}{ds} = 0$, we conclude that u is constant along the projected characteristic curves,

$$x = g(r)t + r, \tag{141}$$

with r being the x-intercept of the characteristic curve. Now, (141) defines $r = r(x, t)$ implicitly as a function of x and t. Therefore, regarding u, it will hold the following:

$$u(x, t) = u(r, 0) = g(r). \tag{142}$$

Differentiating (141) with respect to x and t, we get

$$\begin{cases} 1 = (1 + tg'(r))\dfrac{\partial r}{\partial x}, \\ 0 = g(r) + (1 + tg'(r))\dfrac{\partial r}{\partial t}. \end{cases}$$

Again, differentiating (142) with respect to x and t, we get

$$\begin{cases} \dfrac{\partial u}{\partial x} = g'(r)\dfrac{\partial r}{\partial x}, \\ \dfrac{\partial u}{\partial t} = g'(r)\dfrac{\partial r}{\partial t}. \end{cases}$$

Eliminating r_x and r_t from the above expressions gives

$$\begin{cases} \dfrac{\partial u}{\partial x} = \dfrac{g'(r)}{1+tg'(r)}, \\ \dfrac{\partial u}{\partial t} = -\dfrac{g'(r)g(r)}{1+tg'(r)}. \end{cases} \tag{143}$$

It is now clear that (138) is satisfied only if

$$1 + tg'(r) \neq 0.$$

The solution (142) also satisfies the initial condition at $t = 0$, since $r = x$, and the solution (142) is unique. Summarizing, we have established the following Theorem.

Theorem 1 (Existence and Uniqueness for the inviscid Burgers initial-value problem (IVP)). *The nonlinear IVP,*

$$\begin{cases} u_t + \left(\dfrac{u^2}{2}\right)_x = 0, \quad u = u(x,t) \in \mathbb{R}, \quad (x,t) \in \mathbb{R} \times (0,\infty), \\ u(x,0) = g(x), \quad (x,t) \in \mathbb{R} \times \{t = 0\}, \end{cases}$$

has a unique solution provided g is a $\mathbb{C}^1$ initial-data function, satisfying

$$1 + tg'(r) \neq 0. \tag{144}$$

The solution is given in the parametric form

$$\begin{cases} u(x,t) = g(r), \\ x = g(r)t + r. \end{cases} \tag{145}$$

4.2. *Breaking time*

We have seen that the solution, a differentiable function $u(x,t)$, of the nonlinear initial value problem, (138), exists provided (144) holds. However, for smooth initial data, this condition is always satisfied for sufficiently small time t. It follows from (143) that both u_x and u_t tend to infinity as $1 + tg'(r) \to 0$. The latter implies that the solution develops a singularity (discontinuity) when $1 + tg'(r) = 0$. We consider a point $(x,t) = (r,0)$ so that this condition is satisfied on the characteristics through the point $(r,0)$ at a time t such that

$$t = -\frac{1}{g'(r)}, \tag{146}$$

which is positive provided $g'(r) < 0$. Hence, the solution ceases to exist for all time if the initial data are such that $g'(r) < 0$ for some value of r. The time t^* at which this happens for the first time is called the breaking time. As a consequence, the earliest breaking time is

$$t^* = min\left\{-\frac{1}{g'(x)}\right\} \quad \text{when } g'(x) < 0. \tag{147}$$

It is instructive to compare the solution, (145), of the quasilinear PDE in (138) with the solution

$$u(x,t) = g(x - ct), \tag{148}$$

of the transport equation,

$$\begin{cases} u_t + cu_x = 0, \quad u = u(x,t) \in \mathbb{R}, \quad (x,t) \in \mathbb{R} \times (0,\infty), \\ u(x,0) = g(x), \quad (x,t) \in \mathbb{R} \times \{t = 0\}. \end{cases} \tag{149}$$

In the case of the linear transport equation, the solution represents a steady translation of the initial wave profile along the x-axis with speed c, and without change of shape or scale. In the (x,t) plane, where the solution represents a propagating wave, the function $u(x,t)$ is said to define the wave profile at time t.

On the other hand, in the quasilinear case (inviscid Burgers equation) the speed of translation of the wave depends on u, so different parts of the wave will move with different speeds, causing it to distort as it propagates. It is this distortion that can lead to the non-uniqueness of the solution in the quasilinear case.

A physical example of this phenomena is found in the theory of shallow water waves, where the speed of propagation of a surface element of water is proportional to the square root of the depth. This has the effect that in shallow water the crest of the wave moves faster than the trough, leading to wave breaking close to the shore line [91,92].

4.3. *Self-similar solutions*

Consider the equation,

$$u_t + uu_x = 0, \quad u = u(x,t), \quad (x,t) \in \mathbb{R} \times (0,\infty). \tag{150}$$

Now, apply to (150) a uniform dilatation of space and time as follows:

$$(x,t) \mapsto (\kappa x, \kappa t), \tag{151}$$

to get

$$\kappa u_t + u\left(\kappa u_x\right) = \kappa\left(u_t + uu_x\right) = 0.$$

The latter implies our equation's invariance by (151). Applying the exact same steps as before, the equation under study shall be

$$mf(h) + nhf'(h) + t^{m+n+1}f(h)f'(h) = 0, \tag{152}$$

reducing to an ODE provided

$$m + n + 1 = 0.$$

For various scaling parameters n, we can obtain analytical solutions. Due to the problem's invariance by uniform dilatation of space and time, a natural scaling is $m = 0$ and $n = -1$. For the above m and n, we obtain the following equation:

$$(f(h) - h)f'(h) = 0, \tag{153}$$

having non-trivial solutions,

$$f(h) = h. \tag{154}$$

From (154), we conclude that the self-similar solution to (150) is

$$u(x,t) = \frac{x}{t}. \tag{155}$$

A self-similar solution of this form is called a rarefaction wave. These type of wave-solutions shall be further analyzed in the process.

4.4. *Weak solution formulation*

Consider the first-order, quasilinear initial-value problem for scalar conservation laws in one space dimension,

$$\begin{cases} u_t + F(u)_x = 0 & \text{in} \quad \mathbb{R}\times(0,\infty), \\ u = g & \text{on} \quad \mathbb{R}\times\{t=0\}. \end{cases} \tag{156}$$

Here, $F, g : \mathbb{R} \to \mathbb{R}$ are given and $u : \mathbb{R}\times[0,\infty) \to \mathbb{R}$ is the unknown function, $u = u(x,t)$.

The case where $F(u) = \frac{u^2}{2}$ reduces (156) to the inviscid Burgers equation (138). The method of characteristics demonstrates that there does not exist in general a smooth solution of (156) for all times $t > 0$, since the characteristic lines may cross. Therefore, we must devise a way of interpreting a less regular function u to somehow "solve" this initial-value problem, beyond the breaking time, if any.

Observe that, if u is temporarily assumed to be smooth, we can rewrite as follows, so that the resulting expressions do not involve derivatives of u. The idea is to multiply (156) by a smooth function v and by applying integration by parts to transfer derivatives to v. Note that this is the "standard" way followed to obtain the form of a weak solution.

At this point we remind ourselves of the notion of compactness and compact support in the Euclidean space.

Definition 6. A subset $U \subset \mathbb{R}^n$ is compact if and only if it is both closed and bounded.

Definition 7. A function $v : \mathbb{R}^n \to \mathbb{R}$ has compact support if and only if $v = 0$ outside of some compact set.

At this stage, assume,

$$v : \mathbb{R} \times [0, \infty) \to \mathbb{R}, \quad \text{is smooth with compact support.} \tag{157}$$

We call v a test function. Now multiply (156) by v and integrate by parts, using both Fubini's Theorem and the fact that u, v both have compact support, to obtain

$$\begin{aligned}
0 &= \int_0^\infty \int_{-\infty}^\infty [u_t + F(u)_x] v \, dxdt \\
&= \int_{-\infty}^\infty \int_0^\infty u_t v \; dtdx + \int_0^\infty \int_{-\infty}^\infty F(u)_x v \, dxdt \\
&= \int_{-\infty}^\infty \left(uv \bigg|_{t=0}^{t=\infty} - \int_0^\infty uv_t \; dt \right) dx \\
&\quad + \int_0^\infty \left(F(u)v \bigg|_{x=-\infty}^{x=\infty} - \int_{-\infty}^\infty F(u)v_x \, dx \right) dt \\
&= \int_{-\infty}^\infty -uv|_{t=0} \, dx - \int_{-\infty}^\infty \int_0^\infty uv_t \; dtdx - \int_0^\infty \int_{-\infty}^\infty F(u)v_x \, dxdt \\
&= -\int_{-\infty}^\infty u(x,0)v(x,0) \, dx - \int_0^\infty \int_{-\infty}^\infty uv_t \, dxdt - \int_0^\infty \int_{-\infty}^\infty F(u)v_x \, dxdt \\
&= -\int_{-\infty}^\infty u(x,0)v(x,0) \, dx - \int_0^\infty \int_{-\infty}^\infty [uv_t + F(u)v_x] \, dxdt.
\end{aligned}$$

In view of the initial condition $u = g$ on $\mathbb{R} \times \{t = 0\}$, we thereby obtain the identity

$$\int_0^\infty \int_{-\infty}^\infty [uv_t + F(u)v_x]\, dxdt + \int_{-\infty}^\infty g(x)v(x,0)\, dx = 0, \tag{158}$$

for all test functions $v \in C_c^\infty(\mathbb{R} \times [0,\infty))$ [21,31].

The above equality was derived with u assumed to be a smooth solution of (156), yet the resulting formula, (158), has meaning even if u is only bounded. However, the above reasoning shows that a strong solution for (156) is always a weak solution, as well.

Definition 8. A function $u \in \mathbb{L}^\infty(\mathbb{R} \times (0,\infty))$ is an integral solution of (156), provided equality (158) holds for each test function v satisfying (157) [31].

Assume now we have an integral solution of (156). We wish to deduce some information about this solution by studying (158). As mentioned above, the notion of weak solution allows for solutions u that can be discontinuous. Yet, weak solutions u have some restrictions on types of discontinuities. Thus, our choices, even though being more now, are again limited.

Consider u as a weak solution of (156) such that u is discontinuous across some curve $x = \xi(t)$, but both u and its first derivatives are uniformly continuous on either side of the curve. We observe that on either side of the suggested curve, everything behaves well, meaning the weak solution there is a strong solution, since we can differentiate (158) and go backwards to (156).

Let $u^-(x,t)$ be the limit of u approaching (x,t) from the left and let $u^+(x,t)$ be the limit of u approaching (x,t) from the right. The claim is that the curve $x = \xi(t)$ cannot be arbitrary, but rather there is a relation between $x = \xi(t)$, u^- and u^+ Fig. 9 explains this state. The latter relation is of fundamental importance in the study of conservation laws.

Theorem 2 (Rankine–Hugoniot condition [10,21,31,90]). *If u is a weak solution of (156) such that u is discontinuous across the curve $x = \xi(t)$ but both u and its first derivatives are uniformly continuous on either side of $x = \xi(t)$, then u must satisfy the condition*

$$\frac{F(u^-) - F(u^+)}{u^- - u^+} = \xi'(t), \tag{159}$$

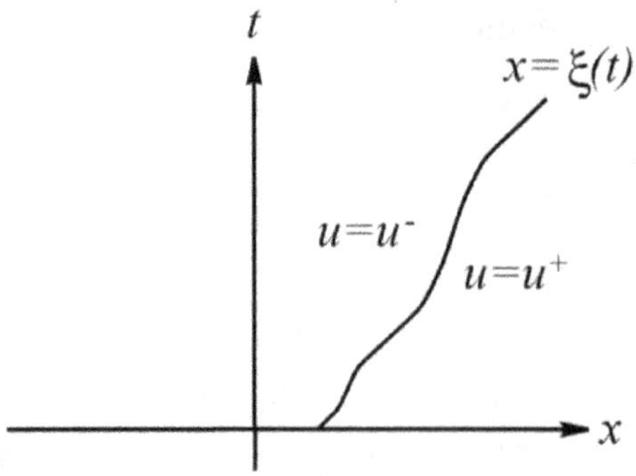

Fig. 9: The aforementioned state for $u(x,t)$.

across the curve of discontinuity, where $u^-(x,t)$ *is the limit of* u *approaching* (x,t) *from the left and* $u^+(x,t)$ *is the limit of* u *approaching* (x,t) *from the right, where* $F(u)$ *was defined in* (156).

Definition 9. The condition (159) is called the Rankine–Hugoniot jump condition along the shock curve $x = \xi(t)$.

For shorthand notation, define

$$\begin{cases} [u] = u^- - u^+, \\ [F(u)] = F[u^-] - F[u^+], \\ \sigma = \xi'(t). \end{cases}$$

In that case, (159) is equivalently rewritten as

$$\sigma = \frac{[F(u)]}{[u]}. \tag{160}$$

Definition 10. The quantities $[u]$ and $[F(u)]$ are called jumps of u and $F(u)$ across the discontinuity curve and σ, the speed of the curve of discontinuity.

Remark 1. Observe that the speed σ and the values u^-, u^+, $F(u^-)$ and $F(u^+)$ will generally vary along the shock curve. The point is that although these quantities may change, the expressions $\sigma[u]$ and $[F(u)]$ must always exactly balance, satisfying the Rankine–Hugoniot condition.

4.5. *Shock wave solutions*

In order to apply and further understand our latest result (159), we consider Burgers equation (138), where the initial condition satisfies,

$$g(x) = \begin{cases} 1, & x \leq 0, \\ 1 - x, & 0 \leq x \leq 1, \\ 0, & x \geq 1. \end{cases} \tag{161}$$

We shall solve this equation using the method of characteristics. Let Γ be a smooth curve on the boundary $\mathbb{R} \times \{t = 0\}$ such that

$$\Gamma := (x(s), t(s))$$

and

$$z(s) = u(x(s), t(s))$$

give the values of u along the curve Γ. Then, by the chain rule we have

$$\frac{dz(s)}{ds} = \frac{du(x(s), t(s))}{ds} = u_x \frac{dx}{ds} + u_t \frac{dt}{ds}.$$

Thus, the characteristic equations will be given by

$$\begin{cases} \dfrac{dt}{ds} = 1, & t(r, 0) = 0, \\ \dfrac{dx}{ds} = z, & x(r, 0) = r, \\ \dfrac{dz}{ds} = 0, & z(r, 0) = g(r). \end{cases}$$

We see that the characteristic curves are

$$\begin{cases} t = s, \\ x = g(r)s + r, \\ z = g(r). \end{cases} \tag{162}$$

From these solutions, we arrive at an implicit solution for (138) as $u = g(x - ut)$.

Noting that $\frac{dz}{ds} = 0$, we conclude that u is constant along the projected characteristic curves, $x = g(r)t + r$. We parameterize Γ by r such that $\Gamma = \{(r, 0)\}$.

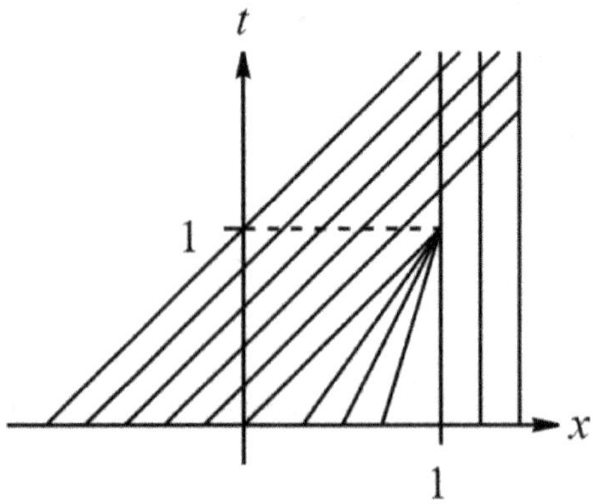

Fig. 10: The related, crossing, characteristic curves.

We see that for $r \leq 0$, we have $g(r) = 1$ and, therefore, these projected characteristic curves are given by

$$x = t + r \quad \text{for } -\infty < r \leq 0, \quad \text{and} \quad u(x,t) = 1,$$

along these curves.

For $0 \leq r \leq 1$, we have $g(r) = 1 - r$ and therefore, these projected characteristic curves are given by

$$x = (1-r)t + r \quad \text{for } 0 \leq r \leq 1 \quad \text{and} \quad u(x,t) = z(r,s) = 1 - r = \frac{1-x}{1-t},$$

along these curves.

Finally, for $r \geq 1$, we have $g(r) = 0$ and, therefore, the projected characteristic curves are now given by

$$x = r \quad \text{for } r \geq 1 \quad \text{and} \quad u = 0,$$

along these curves.

For $t \leq 1$, the solution is defined as

$$u(x,t) = \begin{cases} 1, & x \leq t,\ 0 \leq t \leq 1, \\ \dfrac{1-x}{1-t}, & t \leq x \leq 1,\ 0 \leq t \leq 1, \\ 0, & x \geq 1,\ 0 \leq t \leq 1. \end{cases} \tag{163}$$

However, note that the curves intersect at $t = 1$ Fig. 10 summarizes the results. Beyond that time t, the different projected characteristics demand for the solution u to satisfy different conditions. This cannot happen.

Thus, we no longer have a classical solution. Instead, let's look for a weak solution of (138) for $t \geq 1$ which satisfies (161).

From Theorem (2), a weak solution must satisfy the Rankine–Hugoniot jump condition discussed above. That is,

$$[F(u)] = \sigma[u],$$

or more specifically,

$$\frac{(u^-)^2}{2} - \frac{(u^+)^2}{2} = \xi'(t)[u^- - u^+].$$

Note that the initial data for $x < 0$ requires $u = 1$, while the initial data for $x > 1$ demands $u = 0$ for $t \geq 1$. Now, let's try to make a compromise by defining a curve $x = \xi(t)$ such that $u = 1$ to the left of the curve and $u = 0$ to the right of the curve.

In other words, let

$$u^- = 1 \quad \text{and} \quad u^+ = 0.$$

Now the curve $x = \xi(t)$ is defined by the Rankine–Hugoniot jump condition. In particular, we have

$$\xi'(t) = \frac{1}{2}.$$

In addition, we want the curve $x = \xi(t)$ to contain the point $(x, t) = (1, 1)$. The latter assumption specifies the curve. Thus, the curve must be given by

$$x - 1 = \frac{t-1}{2} \Leftrightarrow x = \frac{t+1}{2}.$$

Therefore, for $t \geq 1$, let

$$u(x,t) = \begin{cases} 1, & x < \dfrac{t+1}{2}, \\ 0, & x > \dfrac{t+1}{2}. \end{cases} \tag{164}$$

Now u defined by (164) is a classical solution of (138) on either side of the curve $x(t) = \frac{t+1}{2}$, satisfying the Rankine–Hugoniot jump condition along the curve of discontinuity Fig. 11 explains the application of the Rankine–Hugoniot condition. Therefore, $u(x,t)$ defined by (164) is a weak solution of (138) for $t \geq 1$. To summarize, the solution regarding (138) with initial condition (161) is given by (162) for $t \leq 1$ and (164) for $t \geq 1$.

Finally, we depict the obtained solution in Fig. 12. This solution shows a shock-like behavior, as expected by Fig. 13.

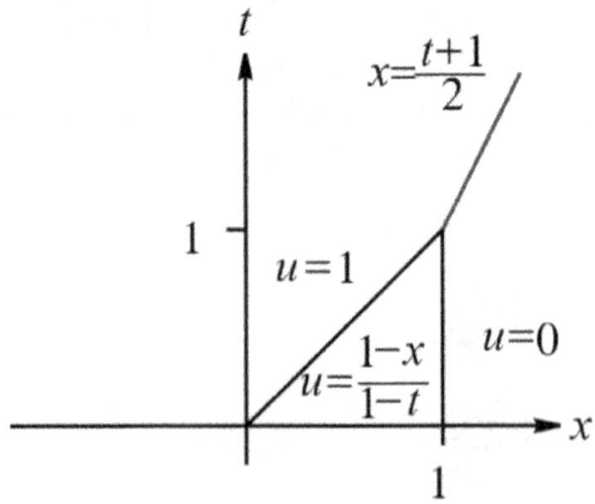

Fig. 11: Application of the Rankine–Hugoniot condition for $u(x,t)$.

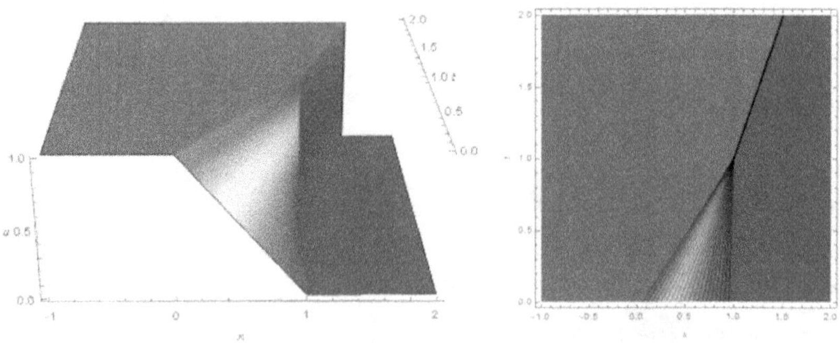

Fig. 12: The shock solution $u(x,t)$.

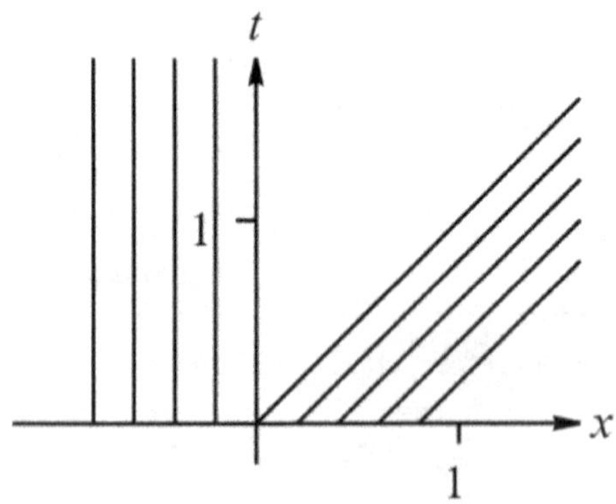

Fig. 13: The related, non-crossing characteristic curves, revealing a region of information loss.

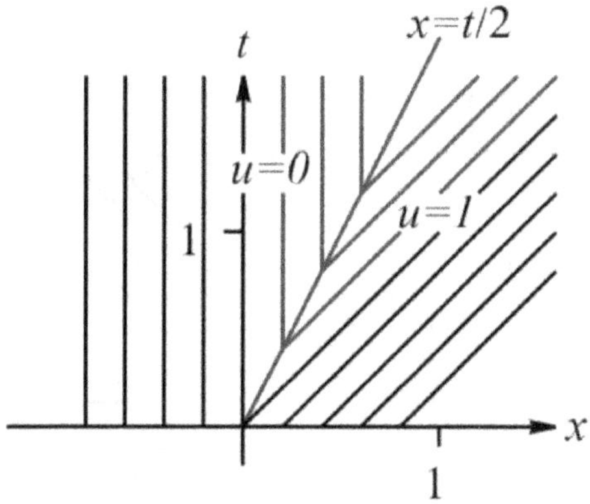

Fig. 14: u_1 as a possible, allowed, filling of the information gap.

4.6. *Rarefaction wave solutions*

We consider Burgers equation (138), this time with the initial condition

$$g(x) = \begin{cases} 0, & x < 0, \\ 1, & x > 0. \end{cases} \tag{165}$$

Studying the characteristics, we deduce that u should be constant along the projected characteristic curves, $x = g(r)t + r$. Here, if $r < 0$, then $g(r) = 0$ and, therefore, $x = r$. If $r > 0$, then $g(r) = 1$ and, therefore, $x = t + r$. Consequently, we have no crossing of characteristics. However, we still have a problem. In fact, what is interesting here is that we have a region on which we do not have enough information! The question that arises now is in which way should we define the solution in this region?

One possible way is to mimic our previous work by letting

$$u_1(x,t) = \begin{cases} 0, & x < \dfrac{t}{2}, \\ 1, & x > \dfrac{t}{2}. \end{cases} \tag{166}$$

Clearly, $u_1(x,t)$ is a classical solution on either side of the curve of discontinuity $x = \frac{t}{2}$. In addition, from previous work, it is easy to see that $u_1(x,t)$ satisfies the Rankine–Hugoniot jump condition along the curve of discontinuity. Therefore, $u_1(x,t)$ is a weak solution of (138) satisfying (165), as seen on Fig. 14.

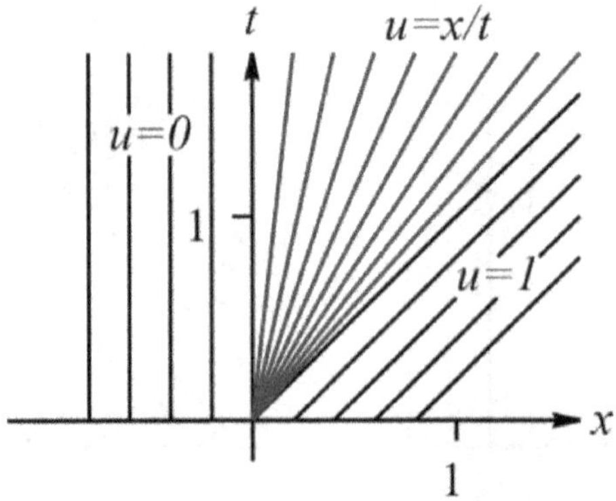

Fig. 15: u_2 as another possible, allowed, filling of the information gap.

Another possibility is letting

$$u_2(x,t) = \begin{cases} 0, & x \leq 0, \\ \dfrac{x}{t}, & 0 \leq x \leq t, \\ 1, & x \geq t. \end{cases} \tag{167}$$

Note that $u_2(x,t)$ is a continuous solution of (138), satisfying (165), as seen on Fig. 15. Concluding, we have found two different solutions. Is it possible, however, that one solution is more physically realistic than the other? If so, we would like to consider that as the "real" solution. To get there, in the following subsection, we introduce the notion of an entropy condition. Solutions of quasilinear equations of the form of (156) which satisfy this entropy condition are considered more physically realistic, and, thus, when looking for solutions of (156), we only allow for solutions which satisfy this extra condition. As we will describe, solution $u_2(x,t)$ is considered to be the more physically realistic solution and we consider that as the "real" solution.

Definition 11. This type of solution, u_2, which "fans" the wedge $0 < x < t$ is called a rarefaction wave solution [21,31,93,94].

4.7. *The nature of rarefaction waves*

A rarefaction wave is also called a relief wave an unloading wave and a Taylor wave. The rarefaction wave is the progression of fluid particles being accelerated away from a compressed or shocked zone. It travels in the direction opposite to the acceleration of the particles. This is opposite to a shock wave, where the fluid particles are accelerated in the direction of the shock [93,94].

Unlike a shock wave, which is nearly a discontinuity and is very steady, a rarefaction wave is spread out in space, and continues to spread with time. This is because the expansion of the high-density material to a lower density one takes time. In other words, the velocity of a rarefaction wave is dependent on time.

A common rarefaction wave is the area of low relative pressure following a shock wave. Rarefaction waves expand with time, much like sea waves spread out as they reach a beach. In most cases rarefaction waves keep the same overall profile at all times throughout the wave's movement: it is a self-similar expansion. Each part of the wave travels at the local speed of sound, in the local medium. This expansion behavior is in contrast to the behavior of pressure increases, which gets narrower with time, until they steepen into shock waves.

A natural example of rarefaction occurs in the layers of Earth's atmosphere. Because the atmosphere has mass, most atmospheric matter is nearer to the Earth due to the Earth's gravitation. Therefore, air at higher layers of the atmosphere is less dense, or rarefied, relative to air at lower layers. Thus, rarefaction can refer either to a reduction in density over space at a single point of time, or a reduction of density over time for one particular area. Rarefaction can be easily observed by compressing a spring and releasing it [95].

4.8. *The entropy condition*

Previously, we have seen that integral solutions are not in general unique. The current purpose is finding a criterion ensuring uniqueness.

In this matter, consider the conservation law in the form of (156). This equation can also be written equivalently as,

$$u_t + F'(u)u_x = 0. \tag{168}$$

The characteristic equations, associated with (168), are given by

$$\begin{cases} \dfrac{dt}{ds} = 1, \\ \dfrac{dx}{ds} = F'(z), \\ \dfrac{dz}{ds} = 0. \end{cases}$$

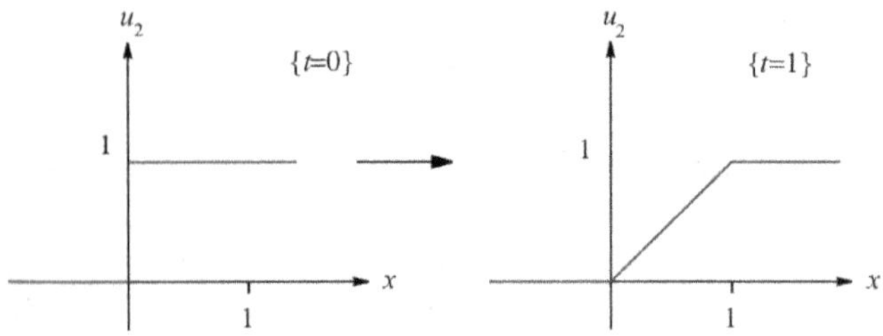

Fig. 16: Time evolution of $u_2(x,t)$.

From these equations, we have that the speed of a solution u, $\frac{du}{dt}$ is given by

$$\frac{du}{dt} = F'(u).$$

Particularly, for Burgers equation, the speed of a solution u is given by

$$\frac{du}{dt} = u.$$

This points to taller waves moving faster than shorter waves. As a result, in (138), we expect the part of the wave to the left to overtake the part of the wave to the right (and cause the wave to break). This resulted in the curve of discontinuity. In ((138), (165)), however, for the initial data, the wave is higher to the right.

Consequently, we expect the right part of the wave to move faster. Physically, we do not want to allow for solution $u_1(x,t)$. Instead, we accept $u_2(x,t)$ as a physically more realistic solution. We depict the formation of u_2 in Fig. 16.

Now, let's make these ideas more precise. In particular, for an equation of the form (168), we only allow for a curve of discontinuity in the solution $u(x,t)$ if the wave to the left is moving faster than the wave to the right. That means, we only allow for a curve of discontinuity between u^- and u^+ if

$$F'(u^-) > \sigma > F'(u^+). \tag{169}$$

Definition 12. The condition (169) is called the entropy condition, from a rough analogy with the thermodynamic principle stating that entropy cannot decrease as time goes forward [21,31,96].

Definition 13. A curve of discontinuity is called a shock curve for a solution u, if the curve satisfies the Rankine–Hugoniot jump condition and the entropy condition for that solution u [31].

Therefore, to eliminate the physically less realistic solutions, we only "accept" solutions u for which curves of discontinuity in the solution are shock curves. We state this more precisely as follows.

Consider the initial-value problem,

$$\begin{cases} u_t + F(u)_x = 0 \quad \text{in} \quad \mathbb{R} \times (0, \infty), \\ u = g \quad \text{on} \quad \mathbb{R} \times \{t = 0\}. \end{cases} \tag{170}$$

Definition 14. A function u is a weak, admissible solution of (170) only if u is a weak solution such that any curve of discontinuity for u is a shock curve [21,31,96].

In ((138), (165)), possibility one, for

$$\begin{cases} u^- = 0, \\ u^+ = 1, \\ \sigma = \xi'(t) = \dfrac{1}{2}, \end{cases}$$

we had,

$$F'(u^-) = u^- = 0 \leq \frac{1}{2} \leq 1 = u^+ = F'(u^-).$$

Therefore, u_1 fails to satisfy the entropy condition along the curve of discontinuity $x = \frac{t}{2}$. Consequently, $x = \frac{t}{2}$ is not a shock curve, and, therefore, u_1 is not an admissible solution. Solution u_2, however, is a continuous solution. Therefore, we accept this solution as the physically relevant one.

We now turn the focus to initial-value problems of the form (170) when F has a particular structure.

Definition 15 ([31]). A function F is called uniformly convex if

$$F'' \geq \theta > 0 \quad \text{for } \theta > 0.$$

More specifically, this points to F' being strictly increasing. If F' is strictly increasing, then u satisfies the entropy condition (169) if and only if

$$u^- > u^+,$$

on any curve of discontinuity.

That is, for F uniformly convex, u will be a weak, admissible solution to

$$u_t + F(u)_x = 0,$$

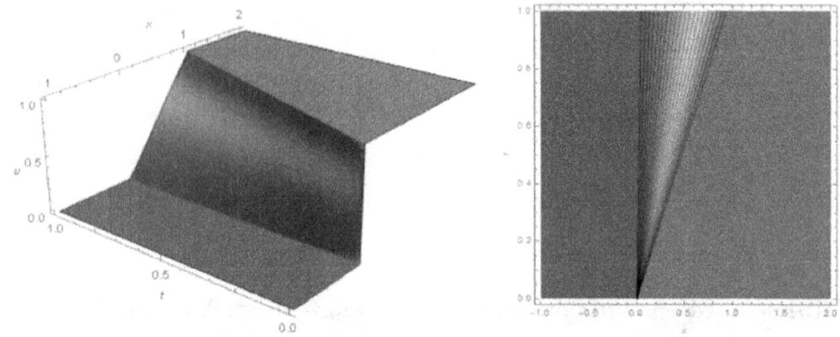

Fig. 17: The rarefaction wave solution $u_2(x, t)$.

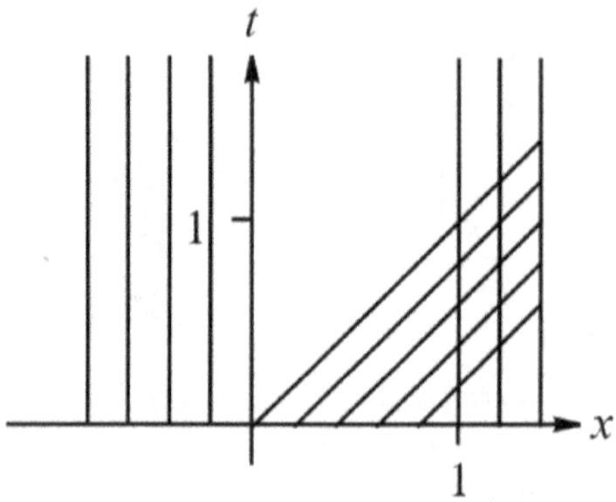

Fig. 18: The related, crossing, characteristic curves, revealing a region of information loss.

if and only if u satisfies the Rankine–Hugoniot condition, (169), as well as

$$u^- > u^+,$$

along any curves of discontinuity. In Fig. 17, we present the rarefaction wave solution $u_2(x, t)$, behaving as predicted by Fig. 18.

4.9. *Riemann's problem*

Let's deal with the initial-value problem (141) with g, such that

$$g(x) = \begin{cases} 1 & 0 < x < 1, \\ 0 & \text{elsewhere.} \end{cases} \tag{171}$$

We deduce that the characteristic equations are

$$\begin{cases} \dfrac{dt}{ds} = 1 & t(r,0) = 0, \\ \dfrac{dx}{ds} = z & x(r,0) = r, \\ \dfrac{dz}{ds} = 0 & z(r,0) = g(r). \end{cases}$$

The solution is constant across the projected characteristic curves given by

$$x(r,t) = g(r)t + r.$$

We consider the following cases:

(1) $r < 0$, $g(r) = 0$ implies

$$\begin{cases} x = r, \\ u = 0, \end{cases}$$

along such curves.

(2) $0 < r < 1$, $g(r) = 1$ leads to

$$\begin{cases} x = t + r, \\ u = 1, \end{cases}$$

along such curves.

(3) $r > 1$, $g(r) = 0$ gives

$$\begin{cases} x = r, \\ u = 0, \end{cases}$$

along such curves.

We are hence led to a rarefaction wave, between $x = 0$ and $x = t$, and a formed shock, due to the intersection of the lines $x = t + r$, for $0 < r < 1$, and $x = 1$.

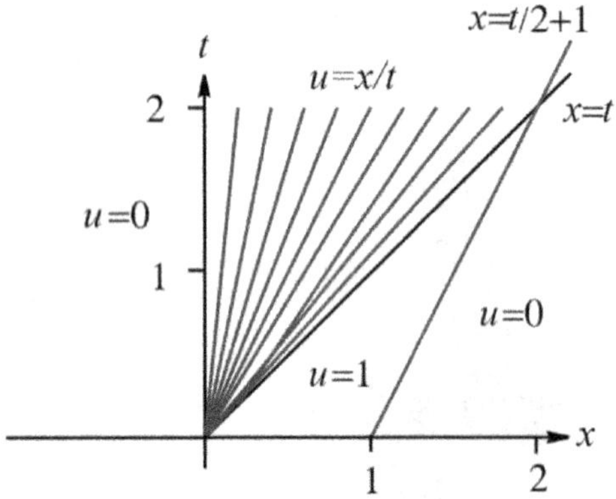

Fig. 19: The proper, entropy-respecting filling of the information gap.

Because of the Rankine–Hugoniot jump condition, the shock speed, σ, satisfies

$$\begin{aligned}\sigma &= \frac{[F(u)]}{[u]}\\ &= \frac{F(u^-) - F(u^+)}{u^- - u^+}\\ &= \frac{\frac{(u^-)^2}{2} - \frac{(u^+)^2}{2}}{u^- - u^+}\\ &= \frac{\frac{1}{2} - 0}{1 - 0}\\ &= \frac{1}{2}.\end{aligned}$$

Therefore, the shock curve emanating from $(1, 0)$ shall be $x - 1 = \frac{t}{2}$ or $x = 1 + \frac{t}{2}$, implying that the weak solution u will take on the values shown in Fig. 19.

Focusing at $t = 2$, however, the rarefaction wave hits the shock curve $x = 1 + \frac{t}{2}$. We demand the jump along the shock to obey the Rankine–Hugoniot jump condition. To the left of the jump, we have $u^- = \frac{x}{t}$ and to the right, we have $u^+ = 0$. The Rankine–Hugoniot jump condition reads

$$\begin{aligned}\sigma &= \frac{[F(u)]}{[u]}\\ &= \frac{F(u^-) - F(u^+)}{u^- - u^+}\end{aligned}$$

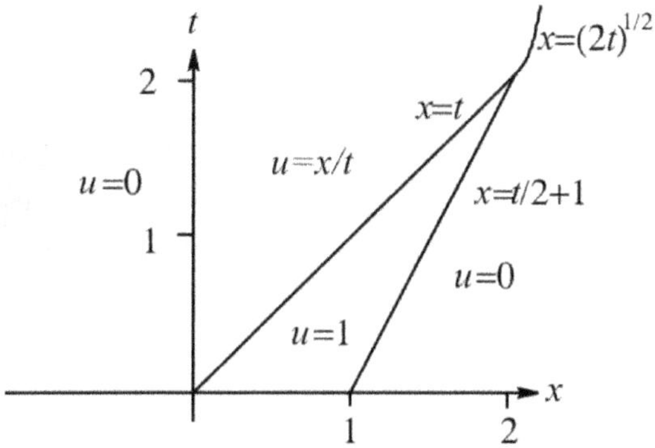

Fig. 20: Application of the Rankine–Hugoniot condition to $u(x,t)$.

$$= \frac{\frac{(u^-)^2}{2} - \frac{(u^+)^2}{2}}{u^- - u^+}$$

$$= \frac{\frac{1}{2}\left(\frac{x}{t}\right)^2 - 0}{\left(\frac{x}{t}\right) - 0}$$

$$= \frac{x}{2t}.$$

Thus, we get a new shock curve emanating from the point $(2,2)$ with speed $\sigma = \xi'(t) = \frac{x}{2t}$. This curve is given as $x(t) = \sqrt{2t}$, as seen on Fig. 20. In summary, we have

$$u(x,t) = \begin{cases} 0 & x < 0, \\ \dfrac{x}{t} & 0 < x < t, \\ 1 & t < x < 1 + \dfrac{t}{2}, \\ 0 & x > 1 + \dfrac{t}{2}, \end{cases} \quad t \leq 2 \tag{172}$$

and

$$u(x,t) = \begin{cases} 0 & x < 0, \\ \dfrac{x}{t} & 0 < x < \sqrt{2t}, \\ 0 & x > \sqrt{2t}, \end{cases} \quad t \geq 2. \tag{173}$$

In Fig. 21, we depict the formation of our obtained solution.

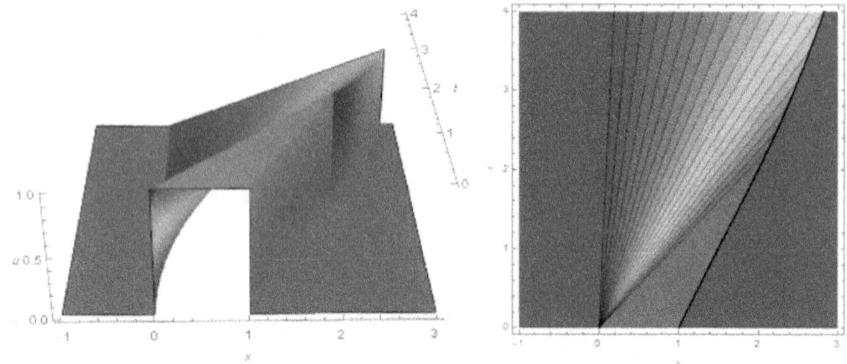

Fig. 21: The solution $u(x,t)$, showing both shock and rarefaction wave characteristics.

Note that $|u| \to 0$ as t tends to infinity. More precisely, using the solution formula derived above, we observe that

$$\begin{cases} |u(x,t)| \leq 1, & \text{for } 0 \leq t \leq 2, \\ |u(x,t)| \leq \dfrac{x}{t} \leq \dfrac{\sqrt{2t}}{t} = \dfrac{\sqrt{2}}{\sqrt{t}}, & t \geq 2, \end{cases}$$

implying that

$$|u(x,t)| \leq \frac{\sqrt{2}}{\sqrt{t}}, \quad \text{for any } t > 0, \ x \in \mathbb{R}.$$

The latter shows that u vanishes like $\frac{1}{\sqrt{t}}$ as t gets arbitrary large. This key-property leads to the next section.

4.10. *Long-time asymptotics*

Consider the most general conservation law (156). We consider the following assumptions.

(1) $g(x)$ is a bounded and integrable initial-data function.
(2) F is smooth and uniformly convex, such that $F(0) = 0$.

We present the following result on the qualitative way the quantity $|u|$ decays to zero as time becomes sufficiently large.

Theorem 3 (Asymptotics in $\mathbb{L}^\infty$ norm [10,21,31]). *Consider an initial-value problem of the form of* (156), *with F and g obeying the* *hypothesis indicated. In that case, the solution of* (156) *satisfies the following decay estimate. There exists some constant* $C > 0$, *such that*

$$|u(x,t)| \leq \frac{C}{\sqrt{t}}, \quad \textit{for all } t > 0, \ x \in \mathbb{R}.$$

Even more specifically, solutions of (156), in general, can be proven to decay to an N-wave [31].

Definition 16 ([21,31]). Given constants $p, q \geq 0$, $d > 0$ and σ, an N-wave is defined as a function of the form

$$N(x,t) = \begin{cases} \frac{1}{d}\left(\frac{x}{t} - \sigma\right), & \text{for } -\sqrt{pdt} < x - \sigma t < \sqrt{pdt} \\ 0, & \text{elsewhere.} \end{cases} \tag{174}$$

We now focus on an N-wave of a specific form, given by

$$\begin{cases} p = -2 \min\limits_{y\in\mathbb{R}} \displaystyle\int_{-\infty}^{y} g(x)\,dx, \\ q = 2 \max\limits_{y\in\mathbb{R}} \displaystyle\int_{y}^{\infty} g(x)\,dx, \\ d = F''(0) > 0, \\ \sigma = F'(0), \end{cases} \tag{175}$$

which will be useful in introducing the following result on the shape u evolves into in $\mathbb{L}_1$.

Theorem 4 (Asymptotics in $\mathbb{L}_1$ norm [21,31]). *Assuming an initial-value problem of the form* (156) *where F is smooth and uniformly convex, g has compact support, p, q, d and σ are defined in* (175) *and $N(x,t)$ is defined in* (174), *there exists a constant $C > 0$ such that the solution u of* (156) *satisfies*

$$\int_{-\infty}^{\infty} |u(x,t) - N(x,t)|\,dx \leq \frac{C}{\sqrt{t}}, \quad \textit{for all } t > 0. \tag{176}$$

The two aforementioned Theorems are central tools in the qualitative analysis of laws (156), with the inviscid Burgers equation being one of these conservation laws.

5. The Korteweg–de Vries–Burgers Equation

In this section, an emphasis is given to the theoretical and numerical analysis of the Korteweg–de Vries–Burgers (KdV–B) equation [30]

$$u_t + \gamma u u_x - \alpha u_{xx} + \beta u_{xxx} = 0, \quad u = u(x,t), \ (x,t) \in \mathbb{R} \times (0,\infty). \quad (177)$$

The latter represents a marriage of the KdV equation (35), when $\alpha = 0$, and the viscous Burgers equation (101), when $\beta = 0$. Therefore, it is intuitive to hope for solutions to (177) to exhibit both solitary and shock profile characteristics.

Applications of (177) range from undular bores in shallow water [1,2], liquid flow containing gas bubbles [3], fluid flow in elastic tubes [4], crystal lattice theory, nonlinear circuit theory and turbulence [5–7], to cardiac hemodynamics [30,97,98].

5.1. *Phase plane analysis*

The traveling wave transform, (18), reduces the KdV–B equation to

$$\begin{cases} \dfrac{du}{d\zeta} = v, \\ \dfrac{dv}{d\zeta} = -\dfrac{u}{\beta}\left(\dfrac{\gamma}{2}u - \lambda\right) + \dfrac{\alpha}{\beta}v. \end{cases} \quad (178)$$

The phase plane of the latter is thoroughly analyzed on [30], exhibiting the following results:

(1) $(0,0)$ is invariably a saddle point.
(2) For $\alpha \geq 2\sqrt{\lambda\beta}$, the point $(\frac{2\lambda}{\gamma}, 0)$ is a source (unstable node).
(3) For $0 < \alpha < 2\sqrt{\lambda\beta}$, the point $(\frac{2\lambda}{\gamma}, 0)$ is a foci (unstable node).
(4) For $\alpha = 0$, the point $(\frac{2\lambda}{\gamma}, 0)$ is a center (stable node).

The geometric nature of the above points is summarized in Refs. [30,99].

In what follows, the phase plane trajectories of the KdV–B equation are plotted, in the steady state where diffusion is absent ($\alpha = 0$), as seen on Fig. 22. Moreover, Fig. 23 depicts the different waveforms of the numerical solutions for each case, with $u_1(\zeta)$, $u_2(\zeta)$ and $u_3(\zeta)$ representing the non-diffusive stable case, diffusion-dominated unstable case and dispersion-dominated unstable case, resp.

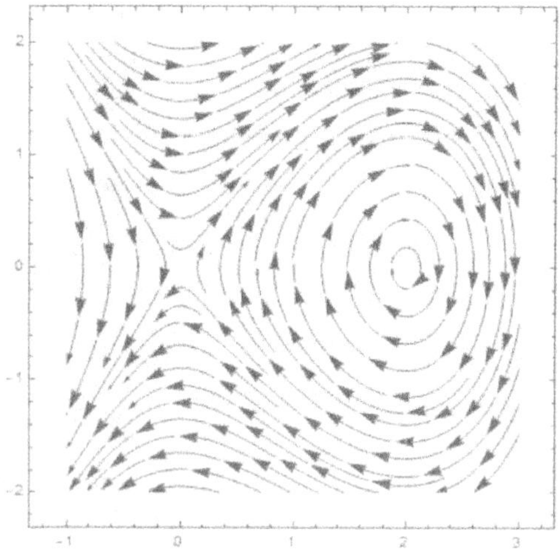

Fig. 22: The non-diffusive state of the KdV–B equation, for the values of $\alpha = 0$, $\beta = 0.5$ and $\gamma = \lambda = 1$, with $(u, v) \in [-1, 3] \times [-2, 2]$, revealing a saddle point at $(0, 0)$ and a central point at $(2, 0)$.

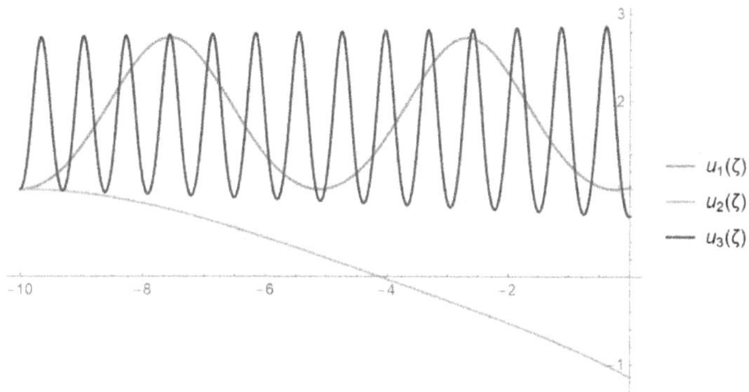

Fig. 23: Non-diffusive (u_1), diffusion-dominated (u_2) and dispersion-dominated (u_3) numerical solutions of the KdV–B equation, for $\zeta \in [-10, 0]$.

5.2. *Hyperbolic methods for traveling wave solutions of KdV–B*

Since the late 1980s, various methods for seeking explicit exact solutions to the KdV–B equation have been independently proposed by many mathematicians, engineers and physicists [100–103].

The Cauchy problem for the KdV–B equation was thoroughly investigated [104]. Both existence and uniqueness of bounded traveling wave solutions, tending to constant states at plus and minus infinity, were proven. Notable results regarding the 2-D KdV–B equation can be found on Ref. [105].

In the present study, in order to derive traveling wave solutions to the KdV–B equation, the hyperbolic tangent method [5,30,106–109] is followed. Consider the traveling wave transform of u,

$$\begin{cases} u(x,t) = u(\zeta), \\ \zeta = \mu(x - \lambda t), \quad \mu > 0, \quad \lambda \neq 0. \end{cases} \tag{179}$$

Here, μ represents the wave number and λ the, perhaps unknown, velocity of the traveling wave. Recall that the wave number μ is inversely proportional to the width of the wave [39,44].

Now, (179) transforms the KdV–B equation, (177), into the ODE for $u(\zeta)$,

$$-\lambda u'(\zeta) + \gamma u(\zeta)u'(\zeta) - \alpha\mu u''(\zeta) + \beta\mu^2 u'''(\zeta) = 0. \tag{180}$$

Integration of (180) with respect to ζ, yields

$$-\lambda u(\zeta) + \frac{\gamma}{2}u^2(\zeta) - \alpha\mu u'(\zeta) + \beta\mu^2 u''(\zeta) = C. \tag{181}$$

The concept behind the hyperbolic tangent, or Tanh, method lies on the key property of the tanh function derivatives all being written in terms of the tanh function itself. The latter transforms the differential equation to a polynomial equation for successive powers of the tanh function. Introducing the new variable,

$$y = \tanh\zeta,$$

the solution(s) we are looking for shall be written as a finite power series in y,

$$u(y) = \sum_{n=0}^{N} a_n y^n, \tag{182}$$

limiting them to solitary- and shock-wave profiles.

Determination of N (highest order of y) lies on the following balancing procedure. At least two terms proportional to y^N must appear after substituting ansatz (182) into (181). As a result, we definitely require $a_{N+1} = 0$ and $a_N \neq 0$ for a particular N. It turns out that N equals 1 or 2 in most cases. This balance (and thus N) is obtained by comparing the behavior of y^N in the highest derivative against its counterpart within the nonlinear term(s).

As soon as N is determined in this way, substitution of (182) into (181) produces an algebraic system for a_n, $n = 0, 1, \ldots, N$. Depending on the problem under study, the wave number μ remains fixed or undetermined, whereas the velocity λ of the traveling wave is always a function of μ.

In the present study, $N = 2$, hence we propose a solution of the form

$$u(\zeta) = a_0 + a_1 \tanh\zeta + a_2 \tanh^2\zeta. \tag{183}$$

Introducing (183) into (181), using the identity,

$$sech^2\zeta = 1 - \tanh^2\zeta \tag{184}$$

and setting the like powers of equal to zero, we obtain the following set of algebraic equations:

$$\begin{cases} \dfrac{a_0^2\gamma}{2} - a_0\lambda - a_1\alpha\mu + 2a_2\beta\mu^2 = C, \\ a_0a_1\gamma - a_1\lambda - 2a_2\alpha\mu - 2a_1\beta\mu^2 = 0, \\ \dfrac{a_1^2\gamma}{2} + a_0a_2\gamma - a_2\lambda + a_1\alpha\mu - 8a_2\beta\mu^2 = 0, \\ a_1a_2\gamma + 2a_2\alpha\mu + 2a_1\beta\mu^2 = 0, \\ \dfrac{a_2^2\gamma}{2} + 6a_2\beta\mu^2 = 0. \end{cases} \tag{185}$$

The latter admits the following solutions:

$$\begin{cases} \lambda = \pm\dfrac{\sqrt{2}\sqrt{18\alpha^4 - 625C\beta^2\gamma}}{25\beta}, \\ \mu = \dfrac{\alpha}{10\beta}, \\ a_0 = \dfrac{3\alpha^2 + 25\beta\lambda}{25\beta\gamma}, \\ a_1 = -\dfrac{6\alpha^2}{25\beta\gamma}, \\ a_2 = -\dfrac{3\alpha^2}{25\beta\gamma}. \end{cases} \tag{186}$$

Therefore, combining (186) with (183), while using (184), we obtain

$$u(\zeta) = \frac{\lambda}{\gamma} + \frac{3\alpha^2\left(sech^2\zeta - 2\tanh\zeta\right)}{25\beta\gamma}. \tag{187}$$

Note that the arbitrary constant C is related to the value of the wave at infinity. Demanding that the wave amplitude at infinity is such that the constant C vanishes, (186) gives

$$\lambda = \pm\frac{6\alpha^2}{25\beta}, \tag{188}$$

and in that case,

$$\begin{aligned} u(\zeta) &= \pm\frac{6\alpha^2}{25\beta\gamma} + \frac{3\alpha^2\left(sech^2\zeta - 2\tanh\zeta\right)}{25\beta\gamma} \\ &= \begin{cases} \dfrac{3\alpha^2\left(2 + sech^2\zeta - 2\tanh\zeta\right)}{25\beta\gamma} := u_1(\zeta), \\ -\dfrac{3\alpha^2\left(1+\tanh^2\zeta\right)}{25\beta\gamma} := u_2(\zeta). \end{cases} \end{aligned} \tag{189}$$

Finally, using (179), (189) leads to

$$u(x,t) = \begin{cases} \dfrac{3\alpha^2\left(2 + sech^2\left[\frac{\alpha}{10\beta}\left(x - \frac{6\alpha^2}{25\beta}t\right)\right] - 2\tanh\left[\frac{\alpha}{10\beta}\left(x - \frac{6\alpha^2}{25\beta}t\right)\right]\right)}{25\beta\gamma} \\ := u_1(x,t), \\ -\dfrac{3\alpha^2\left(1 + \tanh^2\left[\frac{\alpha}{10\beta}\left(x + \frac{6\alpha^2}{25\beta}t\right)\right]\right)}{25\beta\gamma} := u_2(x,t), \end{cases} \tag{190}$$

with u_1 and u_2 representing a right- and left-moving traveling wave of the KdV–B equation, resp.

In Fig. 24, we demonstrate right- and left-moving traveling waves of the KdV–B equation for two distinct times. Note that, indeed, u_1 propagates to the right whereas u_2 propagates to the left. No collision of the waves is observed.

5.3. *Homotopy analysis method*

Choosing, as an initial solution

$$u(x,0) = \frac{e^{-x}}{1+e^{-x}}, \tag{191}$$

a sixth-order approximation, reveals an error, $||\delta||_\infty = 9.3\times10^{-6}$. In Fig. 25, the HAM obtained solution is demonstrated, exhibiting a shock waveform.

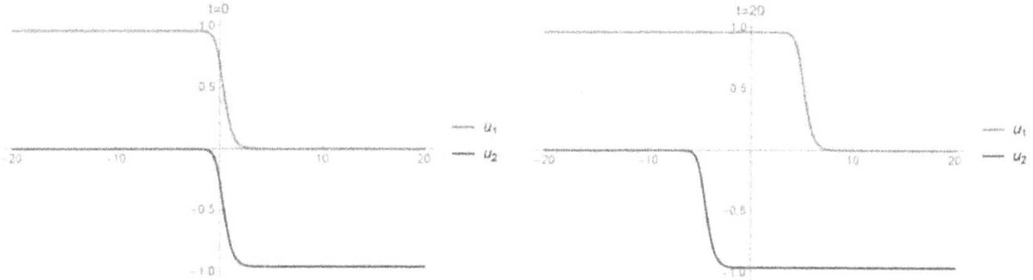

Fig. 24: A right- (u_1) and left-moving (u_2) traveling wave solution of the KdV–B equation, for the values of $\alpha = 0.1$, $\beta = 0.01$ and $\gamma = 0.5$, with $x \in [-20, 20]$ and $t \in \{0, 20\}$.

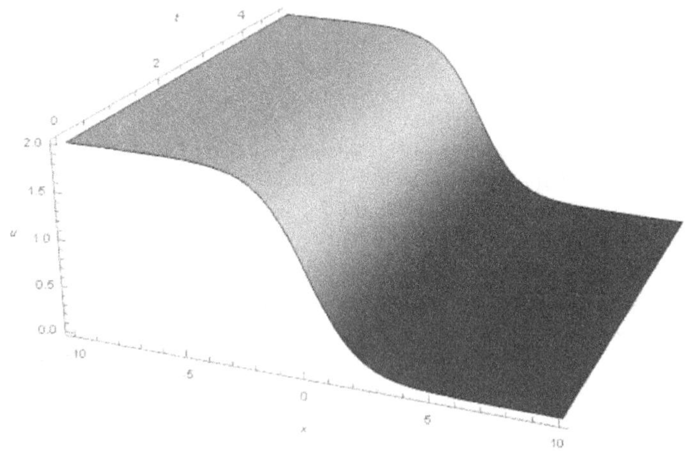

Fig. 25: A Homotopy analysis solution of the sixth-order for the values of $\alpha = 0.1$, $\beta = 0.01$, $\gamma = 0.5$ and $\hbar = 0.01$, with $(x, t) \in [-10, 10] \times [0, 5]$.

6. Conclusion

Plenty of the most interesting features of physical systems are hidden in their nonlinear behavior, and can only be studied with appropriate methods designed to tackle nonlinear problems. The study of exact solutions to nonlinear equations is an active field of both pure and applied mathematics. Complex phenomena in notable scientific fields, especially in physics, such as fluid and plasma dynamics, optical fibers, solid state physics, as well as in cardiac hemodynamics, can be mathematically modeled in terms of the Korteweg–de Vries (KdV), modified KdV (mKdV), Burgers and Korteweg–de Vries–Burgers (KdV–B) equations. A mixture of analytic, semi-analytic and approximate methods was implemented for the behavior of solutions to these equations. Semi-exact solutions are obtained through the Homotopy method for each of the proposed equations. Numerical solutions are derived by means of spectral Fourier analysis and are evolved in time, using the fourth-order explicit Runge–Kutta method. Additionally, qualitative analysis is performed for the inviscid Burgers equation, and conservation laws in general. Phase plane trajectories are obtained for the KdV–B equation. This analysis provides vital information on the connection and applicability of these fundamental equations to mathematical physics and cardiac hemodynamics.

References

[1] D. Benney, Long waves on liquid films, *J. Math. Phys.* **45**(1-4), 150–155, (1966).

[2] R. Johnson, Shallow water waves on a viscous fluid? The undular bore, *Phys. Fluids* **15**(10), 1693–1699, (1972).

[3] L.v. Wijngaarden, One-dimensional flow of liquids containing small gas bubbles, *Ann. Rev. Fluid Mech.* **4**(1), 369–396, (1972).

[4] T. Kawahara, Weak nonlinear magneto-acoustic waves in a cold plasma in the presence of effective electron-ion collisions, *J. Phys. Soc. Japan* **28**(5), 1321–1329, (1970).

[5] G. Gao, A theory of interaction between dissipation and dispersion of turbulence, *SSSMP* **28**, 616–627, (1985).

[6] S.D. Liu and S.K. Liu, Kdv-burgers equation modelling of turbulence, *Sci. China Series A-Math., Phys., Astronomy Tech. Sci.* **35**(5), 576–586, (1992).

[7] M. Wadati, Wave propagation in nonlinear lattice. *J. Phys. Soc. Japan* **38**(3), 673–680, (1975).

[8] J. Canosa and J. Gazdag, The Korteweg-de Vries–Burgers equation, *J. Comput. Phys.* **23**(4), 393–403, (1977).

[9] F. Feudel and H. Steudel, Non-existence of prolongation structure for the Korteweg-de Vries–Burgers equation, *Phys. Lett. A* **107**(1), 5–8, (1985).

[10] M.J. Ablowitz, *Nonlinear Dispersive Waves: Asymptotic Analysis and Solitons*, Vol. 47 (Cambridge University Press, 2011).

[11] D.J. Korteweg and G. De Vries Xli. on the change of form of long waves advancing in a rectangular canal, and on a new type of long stationary waves, *London, Edinburgh, and Dublin Philosophical Magazine and J. Sci.* **39**(240), 422–443, (1895).

[12] J. Burgers, *Correlation Problems in a One-dimensional Model of Turbulence II* (North-Holland Publishing, 1950).

[13] M. Wadati, The modified Korteweg-de Vries equation, *J. Phys. Soc. Japan* **34**(5), 1289–1296, (1973).

[14] E. Crépeau and M. Sorine, A reduced model of pulsatile flow in an arterial compartment, *Chaos, Solitons & Fractals* **34**(2), 594–605, (2007).

[15] T.P. Horikis and D.J. Frantzeskakis, On the nls to KdV connection, *Rom. J. Phys.* **59**, 195–203, (2014).
[16] A. Mangel, M. Fahim, and C. Van Breemen, Control of vascular contractility by the cardiac pacemaker, *Science* **215**(4540), 1627–1629, (1982).
[17] A.C. Newell, Solitons in mathematics and physics, *SIAM* (1985).
[18] J. Boussinesq, *Essai sur la théorie des eaux courantes, Impr. nationale* (1877).
[19] O. Darrigol, *Worlds of Flow: A History of Hydrodynamics from the Bernoullis to Prandtl* (Oxford University Press, 2005).
[20] Y. Nutku, Hamiltonian formulation of the kdv equation, *J. Math. Phys.* **25**(6), 2007–2008, (1984).
[21] P.D. Lax, Hyperbolic systems of conservation laws and the mathematical theory of shock waves, *SIAM* (1973).
[22] R.M. Miura, Korteweg-de Vries equation and generalizations. I. A remarkable explicit nonlinear transformation, *J. Math. Phys.* **9**(8), 1202–1204, (1968).
[23] M.W. Dingemans, *Water Wave Propagation Over Uneven Bottoms: Linear Wave Propagation*, Vol. 13 (World Scientific, 1997).
[24] M.J. Ablowitz and H. Segur, Solitons and the inverse scattering transform, *SIAM* (1981).
[25] C.S. Gardner, J.M. Greene, M.D. Kruskal, and R.M. Miura, Method for solving the Korteweg-de Vries equation, *Phys. Rev. Lett.* **19**(19), 1095, (1967).
[26] J.D. Logan, *Applied Mathematics* (John Wiley & Sons, 2013).
[27] J.S. Russell, *The Wave of Translation in the Oceans of Water, Air, and Ether* (Trübner & Company, 1885).
[28] M.R. Alfonso, L.J. Cymberknop, W. Legnani, F. Pessana, and R.L. Armentano, Conceptual model of arterial tree based on solitons by compartments, In: *36th Annual International Conference of the IEEE Engineering in Medicine and Biology Society*, pp. 3224–3227 (IEEE, 2014).
[29] E. Middlemas and J. Knisley, Soliton solutions of a variation of the nonlinear schrödinger equation, In: *Topics from the 8th Annual UNCG Regional Mathematics and Statistics Conference*, pp. 39–53 (Springer, 2013).
[30] M.A. Xenos and A.C. Felias, Nonlinear dynamics of the KdV-b equation and its biomedical applications. In: *Nonlinear Analysis, Differential Equations, and Applications*, pp. 765–793 (Springer, 2021).
[31] L.C. Evans, Partial differential equations, *Graduate Stud. Math.* **19**(2), (1998).
[32] R. Hirota, Exact solution of the Korteweg–de Vries equation for multiple collisions of solitons, *Phys. Rev. Lett.* **27**(18), 1192, (1971).
[33] A. Salas, W. Lozano, and L. Vallejo, One and two soliton solutions for the KdV equation via mathematica 7, *Int. J. Appl. Math. (IJAM)* **23**, 1075–1080, (2010).
[34] G.I. Barenblatt, *Scaling, Self-Similarity, and Intermediate Asymptotics: Dimensional Analysis and Intermediate Asymptotics*, Vol. 14 (Cambridge University Press, 1996).

[35] J. Gratton, Similarity and self similarity in fluid dynamics, *Fundam. Cosm. Phys.* **15**, 1–106, (1991).

[36] J. Weiss, M. Tabor, and G. Carnevale, The painlevé property for partial differential equations, *J. Math. Phys.* **24**(3), 522–526, (1983).

[37] M.J. Ablowitz, A. Ramani, and H. Segur, A connection between nonlinear evolution equations and ordinary differential equations of p-type, *J. Math. Phys.* **21**(4), 715–721, (1980).

[38] P.D. Lax, Integrals of nonlinear equations of evolution and solitary waves, *Commun. Pure Appl. Math.* **21**(5), 467–490, (1968).

[39] M.J. Ablowitz, M. Ablowitz, P. Clarkson, and P.A. Clarkson, *Solitons, Nonlinear Evolution Equations and Inverse Scattering*, Vol. 149 (Cambridge University Press, 1991).

[40] P.G. Drazin and R.S. Johnson, *Solitons: An Introduction*, Vol. 2 (Cambridge University Press, 1989).

[41] E. Infeld and G. Rowlands, *Nonlinear Waves, Solitons and Chaos* (Cambridge University Press, 2000).

[42] R.S. Johnson, *A Modern Introduction to the Mathematical Theory of Water Waves*, Vol. 19 (Cambridge University Press, 1997).

[43] A.D. Polyanin and A.V. Manzhirov, *Handbook of Mathematics for Engineers and Scientists* (Chapman and Hall/CRC, 2006).

[44] M.J. Ablowitz, D.J. Kaup, A.C. Newell, and H. Segur, The inverse scattering transform-fourier analysis for nonlinear problems, *Studies Appl. Math.* **53**(4), 249–315, (1974).

[45] M. Hickman, W. Hereman, J. Larue, and Ü. Göktaş, Scaling invariant lax pairs of nonlinear evolution equations, *Appl. Anal.* **91**(2), 381–402, (2012).

[46] G.W. Griffiths, *Lax Pairs* (2012).

[47] N.J. Zabusky and M.D. Kruskal, Interaction of "solitons" in a collisionless plasma and the recurrence of initial states, *Phys. Rev. Lett.* **15**(6), 240, (1965).

[48] I.M. Gel'fand and B.M. Levitan, On the determination of a differential equation from its spectral function, *Izvestiya Rossiiskoi Akademii Nauk. Seriya Matematicheskaya*, **15**(4), 309–360, (1951).

[49] V.A. Marchenko, *Sturm-Liouville Operators and Applications*, Vol. 373 (American Mathematical Soc., 2011).

[50] M. Ablowitz and T. Horikis, Solitons and spectral renormalization methods in nonlinear optics, *Eur. Phys. J. Special Topics* **173**(1), 147–166, (2009).

[51] L.N. Trefethen, *Spectral Methods in Matlab, Volume 10 of Software, Environments, and Tools*, (Society for Industrial and Applied Mathematics (SIAM), Philadelphia, 2000).

[52] M.J. Ablowitz and Z.H. Musslimani, Spectral renormalization method for computing self-localized solutions to nonlinear systems, *Optics Lett.* **30**(16), 2140–2142, (2005).

[53] E. Tzirtzilakis, M. Xenos, V. Marinakis, and T. Bountis, Interactions and stability of solitary waves in shallow water, *Chaos, Solitons & Fract.* **14**(1), 87–95, (2002).

[54] F. Gesztesy and H. Holden, The cole-hopf and miura transformations revisited. In: *Mathematical Physics and Stochastic Analysis: Essays in Honour of Ludwig Streit*, pp. 198–214 (World Scientific, 2000).

[55] F. Gesztesy and B. Simon, Constructing solutions of the mkdv-equation, *J. Funct. Anal.* **89**(1), 53–60, (1990).

[56] F. Gesztesy, Some applications of commutation methods, In: *Schrödinger Operators*, pp. 93–117. (Springer, 1989).

[57] F. Gesztesy, W. Schweiger, and B. Simon, Commutation methods applied to the mkdv-equation, *Trans. Am. Math. Soc.* **324**(2), 465–525, (1991).

[58] S.J. Liao, The proposed homotopy analysis technique for the solution of nonlinear problems, Ph.D. Thesis, Shanghai Jiao Tong University Shanghai (1992).

[59] S. Liao, An explicit, totally analytical approximate solution for blasius equation, *Int. J. Non-Linear Mech* **34**, 759–778, (1999).

[60] S. Liao, *Beyond Perturbation: Introduction to the Homotopy Analysis Method* (CRC Press, 2003).

[61] G. Adomian, Solving frontier problems of physics: The decomposition method, with a preface by yves cherruault, *Fundamental Theories of Physics*, Vol. 1 (Kluwer Academic Publishers Group, Dordrecht, 1994).

[62] S. Liang and D.J. Jeffrey, Comparison of homotopy analysis method and homotopy perturbation method through an evolution equation, *Commun. Nonlinear Sci. Numer. Simul.* **14**(12), 4057–4064, (2009).

[63] M. Sajid and T. Hayat, Comparison of ham and hpm methods in nonlinear heat conduction and convection equations, *Nonlinear Analy: Real World Appl.* **9**(5), 2296–2301, (2008).

[64] S. Motsa, P. Sibanda, F. Awad, and S. Shateyi, A new spectral-homotopy analysis method for the mhd Jeffery–Hamel problem, *Computers & Fluids* **39**(7), 1219–1225, (2010).

[65] E.N. Petropoulou and M.A. Xenos, Qualitative, approximate and numerical approaches for the solution of nonlinear differential equations, In: *Applications of Nonlinear Analysis*, pp. 611–664 (Springer, 2018).

[66] S. Liao, *Homotopy Analysis Method in Nonlinear Differential Equations* (Springer, 2012).

[67] K. Vajravelu and R. Van Gorder, *Nonlinear Flow Phenomena and Homotopy Analysis* (Springer, 2013).

[68] M.A. Xenos, E.N. Petropoulou, A. Siokis, and U. Mahabaleshwar, Solving the nonlinear boundary layer flow equations with pressure gradient and radiation, *Symmetry* **12**(5), 710, (2020).

[69] D. Xu, Z. Lin, S. Liao, and M. Stiassnie, On the steady-state fully resonant progressive waves in water of finite depth, *J. Fluid Mech.* **710**, 379–418, (2012).

[70] S. Liao, Do peaked solitary water waves indeed exist? *Commun. Nonlinear Sci. Numer. Simul.* **19**(6), 1792–1821, (2014).

[71] S. Abbasbandy, The application of homotopy analysis method to nonlinear equations arising in heat transfer, *Phys. Lett. A* **360**(1), 109–113, (2006).

[72] Y. Chen and J. Liu, Uniformly valid solution of limit cycle of the duffing–van der pol equation, *Mech. Res. Communications* **36**(7), 845–850, (2009).
[73] S.P. Zhu, An exact and explicit solution for the valuation of american put options, *Quant. Finance* **6**(3), 229–242, (2006).
[74] M. Turkyilmazoglu, Purely analytic solutions of the compressible boundary layer flow due to a porous rotating disk with heat transfer, *Phys. Fluids* **21**(10), 106104, (2009).
[75] S.H. Park and J.H. Kim, Homotopy analysis method for option pricing under stochastic volatility, *Appl. Math. Lett.* **24**(10), 1740–1744, (2011).
[76] A. Mastroberardino, Homotopy analysis method applied to electrohydrodynamic flow, *Commun. Nonlinear Sci. Numer. Simul.* **16**(7), 2730–2736, (2011).
[77] C.J. Nassar, J.F. Revelli, and R.J. Bowman, Application of the homotopy analysis method to the Poisson–Boltzmann equation for semiconductor devices, *Commun. Nonlinear Sci. Numer. Simul.* **16**(6), 2501–2512, (2011).
[78] J.H. He, A coupling method of a homotopy technique and a perturbation technique for non-linear problems, *Int. J. Non-linear Mech.* **35**(1), 37–43, (2000).
[79] H. Jafari and M. Firoozjaee, Homotopy analysis method for solving KdV equations, *Surveys Math. Appl.* **5**, 89–98, (2010).
[80] M. Nazari, F. Salah, X. Abdul Aziz, and M. Nilashi, Approximate analytic solution for the KdV and Burger equations with the homotopy analysis method, *J. Appl. Math.* **2012**, (2012).
[81] J.M. Burgers, A mathematical model illustrating the theory of turbulence, In: *Advances in Applied Mechanics*, Vol. 1, pp. 171–199 (Elsevier, 1948).
[82] H. Bateman, Some recent researches on the motion of fluids, *Monthly Weather Rev.* **43**(4), 163–170, (1915).
[83] G.B. Whitham, *Linear and Nonlinear Waves*, Vol. 42 (John Wiley & Sons, 2011).
[84] C.K. Batchelor and G. Batchelor, *An Introduction to Fluid Dynamics* (Cambridge University Press, 2000).
[85] A.J. Chorin, J.E. Marsden, and J.E. Marsden, *A Mathematical Introduction to Fluid Mechanics*, Vol. 168 (Springer, 1990).
[86] H.K. Versteeg and W. Malalasekera, *An Introduction to Computational Fluid Dynamics: The Finite Volume Method* (Pearson Education, 2007).
[87] L. Yang and X. Pu, Derivation of the Burgers' equation from the gas dynamics, *Commun. Math. Sci.* **14**(3), 671–682, (2016).
[88] T. Nagatani, H. Emmerich, and K. Nakanishi, Burgers equation for kinetic clustering in traffic flow, *Physica A: Statistical Mech. Appl.* **255**(1–2), 158–162, (1998).
[89] H. Schlichting and K. Gersten, *Boundary-layer Theory* (Springer, 2016).
[90] W.A. Strauss, *Partial Differential Equations: An Introduction* (John Wiley & Sons, 2007).
[91] C. Cao, D.D. Holm, and E.S. Titi, Traveling wave solutions for a class of one-dimensional nonlinear shallow water wave models, *J. Dynam. Diff. Eq.* **16**(1), 167–178, (2004).

[92] G.A. El, M.A. Hoefer, and M. Shearer, Expansion shock waves in regularized shallow-water theory, *Proc. Royal Soc. A: Math. Phys. Eng. Sci.* **472**(2189), 20160141, (2016).

[93] A. Matsumura and K. Nishihara, Global stability of the rarefaction wave of a one-dimensional model system for compressible viscous gas, *Commun. Math. Physi.* **144**(2), 325–335, (1992).

[94] Z.X. Zhang, *Rock Fracture and Blasting: Theory and Applications* (Butterworth-Heinemann, 2016).

[95] O. Ivashnev, M. Ivashneva, and N. Smirnov, Rarefaction waves in nonequilibrium-boiling fluid flows, *Fluid Dynamics* **35**(4), 485–495, (2000).

[96] T.P. Liu, The entropy condition and the admissibility of shocks, *J. Math. Anal. Appl.* **53**(1), 78–88, (1976).

[97] N. Antar and H. Demiray, Weakly nonlinear waves in a prestressed thin elastic tube containing a viscous fluid, *Int. J. Eng. Sci.* **37**(14), 1859–1876, (1999).

[98] H. Erbay, S. Erbay, and S. Dost, Wave propagation in fluid filled nonlinear viscoelastic tubes, *Acta Mechanica* **95**(1-4), 87–102, (1992).

[99] G. Nagy, Ordinary Differential Equations, Online notes (2020).

[100] H. Demiray, A note on the exact travelling wave solution to the KdV–Burgers equation, *Wave Motion* **38**(4), 367–369, (2003).

[101] A. Jeffrey and M. Mohamad, Exact solutions to the KdV-Burgers' equation, *Wave Motion* **14**(4), 369–375, (1991).

[102] M. Wang, Exact solutions for a compound KdV-Burgers equation, *Phys. Lett. A* **213**(5–6), 279–287, (1996).

[103] S. Xiong, An analytic solution of Burgers–KdV equation, *Chinese Sci. Bull.* **34**(14), 1158–1162, (1989).

[104] J. Bona and M. Schonbek, Travelling-wave solutions to the Korteweg-de Vries–Burgers equation, *Proc. Royal Soc. Edinburgh Sect A* **101**, 207–226, (1985).

[105] Z. Feng, Travelling wave solutions and proper solutions to the two-dimensional Burgers–Korteweg-de Vries equation, *J. Physics A: Math. General* **36**(33), 8817, (2003).

[106] R. Hirota, Direct method of finding exact solutions of nonlinear evolution equations, In: *Bäcklund Transformations, the Inverse Scattering Method, Solitons, and their Applications*, pp. 40–68 (Springer, 1976).

[107] W. Malfliet, Solitary wave solutions of nonlinear wave equations, *Am. J. Phys.* **60**(7), 650–654, (1992).

[108] W. Malfliet and W. Hereman, The tanh method: I exact solutions of nonlinear evolution and wave equations, *Phys. Scripta* **54**, 563–568, (1996).

[109] A.M. Wazwaz, The tanh–coth and the sech methods for exact solutions of the Jaulent–Miodek equation, *Phys. Lett. A* **366**(1-2), 85–90, (2007).

https://doi.org/10.1142/9789811261572_0013

Chapter 13

Localized Cerami Condition and a Deformation Theorem

Lucas Fresse*,‡ and Viorica V. Motreanu†,§

*Université de Lorraine, Institut Élie Cartan
54506 Vandoeuvre-lès-Nancy, France,
†Lycée Varoquaux, 10 rue Jean Moulin
54510 Tomblaine, France
‡lucas.fresse@univ-lorraine.fr
§vmotreanu@gmail.com

We establish a deformation theorem and a general minimax principle for locally Lipschitz maps subject to a localized compactness-type condition which is weaker than the classical Cerami condition.

1. Introduction

Let $(X, \|\cdot\|)$ be a Banach space. The classical Cerami condition ((C)-condition for short) expresses a compactness-type property of a C^1-functional $f : X \to \mathbb{R}$. Namely, f is said to satisfy the (C)-condition at level $c \in \mathbb{R}$ if every sequence $(x_n) \subset X$ such that

$$f(x_n) \to c, \quad (1 + \|x_n\|)f'(x_n) \to 0$$

has a convergent subsequence. Under the Cerami condition, one has the so-called first deformation theorem (see Ref. [1]), and then minimax principles such as the mountain pass theorem (see Refs. [2,3]), which can be obtained as consequences of the deformation theorem.

In this chapter, we establish a deformation theorem (Theorem 5) and a general minimax principle (Theorem 12) under a condition weaker than the classical Cerami condition as it is required only for sequences localized within a subset Z; see Definition 1. Moreover, our condition is expressed in terms of a Lipschitz map $\varphi : Z \to [1, +\infty)$, following the approach already considered in Ref. [4] (but where the condition is global).

Our setting is also more general than the setting presented above since we are dealing with locally Lipschitz functionals instead of C^1-functionals. We refer to [5–7] for deformation and minimax theorems for non-differentiable functionals; see also [8, §4–5] and references therein.

The rest of the text is organized as follows. In Section 2, we give a short review on critical point theory for locally Lipschitz functionals and we define our localized Cerami condition. In Section 3, we prove a deformation theorem. In Section 4, we derive a general minimax principle.

2. Localized Cerami Condition

We first review some basic facts on the critical point theory for locally Lipschitz maps $f : X \to \mathbb{R}$. By $\partial f(x)$, we denote the subdifferential of f at $x \in X$, defined by

$$\partial f(x) = \{x^* \in X^* : \forall h \in X,\ \langle x^*, h\rangle \le f^0(x; h)\},$$

where

$$f^0(x;h) := \limsup_{\substack{x' \to x \\ t \to 0^+}} \frac{f(x' + th) - f(x')}{t}$$

stands for the generalized directional derivative of f at x in the direction h. We summarize the properties of $\partial f(x)$ (see Ref. [9]) as follows:

- For all $x \in X$, the set $\partial f(x)$ is non-empty, convex and w*-compact.
- In particular, $\lambda_f(x) := \inf_{x^* \in \partial f(x)} \|x^*\|_*$ is attained.
- The map $\lambda_f : X \to [0, +\infty)$ is lower semicontinuous.
- The multimap $\partial f : X \to 2^{X^*}$ is upper semicontinuous from X endowed with the norm topology to X^* endowed with the w*-topology, i.e.

$$C \subset X^* \text{ is w}^*\text{-closed} \quad \Longrightarrow \quad \{x \in X : \partial f(x) \cap C \neq \emptyset\} \text{ is closed.}$$

- *Chain rule.* If $u : [0,1] \to X$ is of class C^1, then $f \circ u$ is differentiable almost everywhere, and for almost every $t \in [0,1]$ we have

$$(f \circ u)'(t) \le \max\{\langle x^*, u'(t)\rangle : x^* \in \partial f(u(t))\}.$$

- If f is of class C^1, then $\partial f(x) = \{f'(x)\}$.

We say that x is a *critical point* of f if $0 \in \partial f(x)$, which equivalently means that $\lambda_f(x) = 0$. Moreover, we denote by $K_f \subset X$ the set of critical points of f and, for $c \in \mathbb{R}$, we denote by

$$K_f^c := \{x \in K_f : f(x) = c\}$$

the subset of critical points at level c. Note that K_f and K_f^c are closed subsets of X (due to the lower semicontinuity of λ_f).

We arrive at the definition of a localized Cerami condition.

Definition 1. For a subset $Z \subset X$, a globally Lipschitz map $\varphi : Z \to [1, +\infty)$, and a value $c \in \mathbb{R}$, we say that f satisfies the $(\varphi - \mathrm{C})_{Z,c}$-condition if every sequence (x_n) in Z such that

$$f(x_n) \to c \quad \text{and} \quad \varphi(x_n)\lambda_f(x_n) \to 0 \tag{1}$$

has a convergent subsequence.

Remark 2. If $Z = X$, then we recover the $(\varphi - \mathrm{C})_c$-condition introduced in Ref. [4]. If $\varphi(x) = 1 + \|x\|$ for all $x \in X$, then we get the condition $(\mathrm{C})_{Z,c}$ considered in Ref. [10, §5.2]. The classical Cerami (resp. Palais–Smale) condition is retrieved by taking $Z = X$ and $\varphi(x) = 1 + \|x\|$ for all $x \in X$ (resp. $\varphi \equiv 1$).

Remark 3. If the set Z is relatively compact, or more generally $Z \cap f^{-1}([a, b])$ is relatively compact for some $a < c < b$, then condition $(\varphi - \mathrm{C})_{Z,c}$ is verified.

Lemma 4. *Assume that f satisfies the $(\varphi - \mathrm{C})_{Z,c}$-condition. Then the set $K_f^c \cap Z$ is relatively compact.*

Proof. Any sequence (x_n) in $K_f^c \cap Z$ is such that $x_n \in Z$, $f(x_n) = c$, and $\lambda_f(x_n) = 0$ for all n, hence it admits a converging subsequence by virtue of the $(\varphi - \mathrm{C})_{Z,c}$-condition. □

3. Deformation Theorem

In this section, we show a deformation theorem under the localized Cerami condition introduced in Definition 1. Hereafter, whenever $A, B \subset X$ are two subsets, we write

$$\mathrm{dist}(A, B) := \inf_{(x,y) \in A \times B} \|x - y\|,$$

and $\operatorname{dist}(x, B) := \operatorname{dist}(\{x\}, B)$ if $x \in X$. Given $f : X \to \mathbb{R}$ and $a \in \mathbb{R}$, we denote

$$f^a := \{x \in X : f(x) \leq a\}.$$

Theorem 5. *Let $f : X \to \mathbb{R}$ be a locally Lipschitz map. Let $Z \subset X$ be a subset whose complement $S := X \setminus Z$ is bounded, and assume that f satisfies the $(\varphi - \mathrm{C})_{Z,c}$-condition with respect to $c \in \mathbb{R}$ and a Lipschitz map $\varphi : Z \to [1, +\infty)$.*

Then, for every open subset $U \subset X$ with

$$K_f^c \cup S \subset U, \quad \operatorname{dist}(S, X \setminus U) > 0,$$

every $\varepsilon_0 \in (0, +\infty)$ and every $\theta \in (0, +\infty)$, there exist $\varepsilon \in (0, \varepsilon_0)$ and a continuous map $h : [0, 1] \times X \to X$ such that, for every $(t, x) \in [0, 1] \times X$, we have the following:

(a) $x \in Z \Leftrightarrow h(t, x) \in Z$.
(b) $\|h(t, x) - x\| \leq \theta\varphi(x)t$ *if* $x \in Z$; $h(t, x) = x$ *if* $x \in S$.
(c) $f(h(t, x)) \leq f(x)$.
(d) $h(t, x) \neq x \Rightarrow f(h(t, x)) < f(x)$.
(e) $|f(x) - c| \geq \varepsilon_0 \Rightarrow h(t, x) = x$.
(f) $h(1, f^{c+\varepsilon}) \subset f^{c-\varepsilon} \cup U$ *and* $h(1, f^{c+\varepsilon} \setminus U) \subset f^{c-\varepsilon}$.

Remark 6. Beyond the fact that this deformation theorem is stated under a condition $(\varphi - \mathrm{C})_{Z,c}$ which is weaker than the classical Cerami condition (since it is localized within the set Z), Theorem 5 is also a refined version of the usual first deformation theorem since it permits the choice of a bounded set S which will not be modified by the deformation (which is expressed by the second property in Theorem 5 (b)). This fact plays a key role in the proof of the minimax principle obtained in Theorem 12.

Proof of Theorem 5. Let $\hat{\varphi} : X \to [0, +\infty)$ be the map such that $\hat{\varphi}|_Z = \varphi$ and $\hat{\varphi}|_S \equiv 0$. For any subset $A \subset X$, we write

$$A_\delta := \{x \in X : \operatorname{dist}(x, A) < \delta\},$$

understanding that A_δ is empty whenever A is empty.

Claim 7. $\operatorname{dist}(K_f^c \cup S, X \setminus U) > 0$.

We have indeed

$$\operatorname{dist}(K_f^c \cup S, X \setminus U) = \min\{\operatorname{dist}(K_f^c \cap Z, X \setminus U), \operatorname{dist}(S, X \setminus U)\},$$

since $K_f^c \cup S = (K_f^c \cap Z) \cup S$. By Lemma 4, the set $\overline{K_f^c \cap Z} \subset K_f^c \subset U$ is compact. Since U is open, this implies that $\operatorname{dist}(K_f^c \cap Z, X \setminus U) > 0$. Moreover, we have $\operatorname{dist}(S, X \setminus U) > 0$ by assumption. Claim 7 ensues.

On the basis of Claim 7, we can take $\delta > 0$ such that

$$(K_f^c \cup S)_{3\delta} \subset U.$$

Claim 8. *There are $\varepsilon_1 \in (0, \varepsilon_0)$ and $m \in (0, +\infty)$ such that*

$$\Big(|f(x) - c| \leq \varepsilon_1 \quad \textit{and} \quad \operatorname{dist}(x, K_f^c \cup S) \geq \delta\Big) \quad \Longrightarrow \quad \varphi(x)\lambda_f(x) > m.$$

Indeed, if this is not true, then we can find a sequence $(x_n) \subset X$ such that

$$f(x_n) \to c, \quad \varphi(x_n)\lambda_f(x_n) \to 0, \quad \text{and} \quad \operatorname{dist}(x_n, K_f^c \cup S) \geq \delta \text{ for all } n. \tag{2}$$

Note that we have $x_n \in Z = X \setminus S$ for all n. Then, in view of the first two parts of (2), and by virtue of the $(\varphi-\mathrm{C})_{Z,c}$ condition, (x_n) should converge along a subsequence to an element in K_f^c. But this is incompatible with the last part of (2). This establishes Claim 8.

With ε_1 taken from Claim 8, we set

$$\begin{aligned} A &:= \{x \in X : |f(x) - c| \leq \varepsilon_1 \text{ and } x \notin (K_f^c \cup S)_\delta\}, \\ A_0 &:= \left\{x \in X : |f(x) - c| \leq \frac{\varepsilon_1}{2} \text{ and } x \notin (K_f^c \cup S)_{2\delta}\right\}, \\ B &:= \left\{x \in X : |f(x) - c| \geq \varepsilon_1 \text{ or } x \in \overline{(K_f^c \cup S)_\delta}\right\}, \end{aligned}$$

which are closed subsets of X such that

$$X = A \cup B, \quad A_0 \subset \operatorname{int}(A), \quad A_0 \cap B = \emptyset.$$

Let

$$\gamma : X \to [0, +\infty), \quad x \mapsto \frac{\operatorname{dist}(x, B)}{\operatorname{dist}(x, A_0) + \operatorname{dist}(x, B)},$$

which is a locally Lipschitz map such that

$$\gamma|_B \equiv 0, \quad \gamma|_{A_0} \equiv 1.$$

Since γ is zero on a neighborhood of $S = X \setminus Z$, the map $x \mapsto \gamma(x)\hat{\varphi}(x)$ is also locally Lipschitz.

Claim 9. *There is a locally Lipschitz map* $V : X \to X$, *with* $V|_B = 0$, *such that*

$$\|V(x)\| \le \gamma(x)\hat{\varphi}(x) \text{ and } \langle x^*, V(x)\rangle \ge m\gamma(x) \quad \text{for all } x \in X,\ x^* \in \partial f(x).$$

We show Claim 9. Given any $x \in A$, we first construct an element $v = v(x) \in X$ with $\|v\| = 1$ and a radius $r = r(x) \in (0, +\infty)$ such that

$$\forall y \in B(x,r) \subset Z,\ \forall y^* \in \partial f(y),\ \varphi(y)\langle y^*, v\rangle > m. \tag{3}$$

Here, $B(x,r) := \{y \in X : \|y - x\| < r\}$ stands for the open ball in X. We will use the notation $B^*(\phi, s) := \{\psi \in X^* : \|\psi - \phi\|_* < s\}$ for open balls in X^*. In order to construct v, r satisfying (3), we first note that $B^*(0, \lambda_f(x))$ and $\partial f(x)$ are convex subsets of X^*, such that $B^*(0, \lambda_f(x)) \cap \partial f(x) = \emptyset$ and, moreover, $B^*(0, \lambda_f(x))$ has internal points. These facts enable us to apply the separation theorem (for the w*-topology) which yields $v \in X$, $v \neq 0$, such that

$$\sup_{z^* \in B^*(0,\lambda_f(x))} \langle z^*, v\rangle \le \inf_{x^* \in \partial f(x)} \langle x^*, v\rangle. \tag{4}$$

Moreover, up to normalizing it, we may assume that $\|v\| = 1$. Note also that

$$\sup_{z^* \in B^*(0,\lambda_f(x))} \langle z^*, v\rangle = \lambda_f(x). \tag{5}$$

Indeed, the inequality $\le$ is clear, knowing that $\|v\| = 1$. The inequality $\ge$ comes as a consequence of Hahn–Banach theorem (see Ref. [11, Corollary 1.3]). Since $x \in A$, by (4), (5) and Claim 8, we get

$$\forall x^* \in \partial f(x),\ \varphi(x)\langle x^*, v\rangle \ge \varphi(x)\lambda_f(x) > m.$$

Now since φ is continuous and since the multimap ∂f is upper semicontinuous from X endowed with the norm topology to X^* endowed with the w*-topology, we can find an open ball $B(x,r) \subset Z$ such that (3) is valid.

The collection $\{B(x, r(x)) : x \in A\}$ is an open covering of A. By the paracompacity of X, we can extract a locally finite covering $\{B_i := B(x_i, r_i) : i \in I\}$ of A, with $r_i := r(x_i)$, and let $v_i := v(x_i)$. Then we define

$$V : X \to X,\ x \mapsto \hat{\varphi}(x)\gamma(x) \sum_{i \in I} \frac{\operatorname{dist}(x, X \setminus B_i)}{\operatorname{dist}(x, A) + \sum_{j \in I} \operatorname{dist}(x, X \setminus B_j)} v_i.$$

The map V so obtained is locally Lipschitz (as composed by product/sum of locally Lipschitz maps). For all $x \in A$, we have

$$\|V(x)\| \leq \gamma(x)\varphi(x) \sum_{i\in I} \frac{\operatorname{dist}(x, X\setminus B_i)}{\sum_{j\in I}\operatorname{dist}(x, X\setminus B_j)} \|v_i\| = \gamma(x)\varphi(x),$$

since $\|v_i\| = 1$ for all i. Moreover, letting $I_0 := \{i \in I : x \in B_i\}$, for all $x^* \in \partial f(x)$ we have also

$$\langle x^*, V(x)\rangle = \gamma(x) \sum_{i\in I_0} \frac{\operatorname{dist}(x, X\setminus B_i)}{\sum_{j\in I_0}\operatorname{dist}(x, X\setminus B_j)} \varphi(x)\langle x^*, v_i\rangle \geq \gamma(x)m,$$

where the last inequality is obtained by applying (3). For all $x \in B$, we have $\gamma(x) = 0$, so $V(x) = 0$, and the claimed assertions are still valid in this case. This establishes Claim 9.

Let $\kappa > 0$ be the Lipschitz constant for φ. We choose $\eta \in (0, +\infty)$ such that $e^{\eta\kappa} - 1 \leq \kappa\theta$. Given $x \in X$, we consider the Cauchy problem

$$\begin{cases} u'(t) = -\eta V(u(t)) \ \text{ in } [0,1], \\ u(0) = x. \end{cases} \tag{6}$$

By Claim 9, the map V is locally Lipschitz and of sublinear growth, hence problem (6) admits a unique global solution $h(\cdot, x) := u_x \in C^1([0,1], X)$. The so-obtained map $h : [0,1] \times X \to X$ is therefore the flow (localized within $[0,1]$) of problem (6); in particular it is continuous. It remains to check that h fulfills the conditions (a)–(f) of the theorem.

Since $X \setminus Z \subset B$, (a) and the second part of (b) are easy consequences of the following property:

$$\exists t_0 \in [0,1],\ h(t_0, x) \in B \quad \Longrightarrow \quad \forall t \in [0,1],\ h(t,x) = x, \tag{7}$$

which itself can be shown as follows: if $x_0 := h(t_0, x) \in B$, then $V(x_0) = 0$ and so the unique solution of the Cauchy problem

$$\begin{cases} u'(t) = -\eta V(u(t)) \ \text{ in } [0,1], \\ u(t_0) = x_0 \end{cases}$$

is the constant function $u \equiv x_0$. Since this unique solution is also $h(\cdot, x)$, we get $h(t,x) = x_0 = x$ for all $t \in [0,1]$ (using that $h(0,x) = x$). This shows (7).

For $x \in A$, by integrating (6) (and invoking Claim 9), we have

$$\begin{aligned}\|h(t,x)-x\| &\le \eta \int_0^t \|V(h(s,x))\|\,ds \le \eta \int_0^t \varphi(h(s,x))\,ds\\ &= \eta \int_0^t (\varphi(h(s,x))-\varphi(x))\,ds + \eta\varphi(x)t\\ &\le \eta\kappa \int_0^t \|h(s,x)-x\|\,ds + \eta\varphi(x)t.\end{aligned}$$

By Gronwall's inequality, we obtain

$$\begin{aligned}\|h(t,x)-x\| &\le \eta\varphi(x)t + \int_0^t \eta^2\kappa\varphi(x)se^{\eta\kappa(t-s)}\,ds = \varphi(x)(e^{\eta\kappa t}-1)\kappa^{-1}\\ &\le \varphi(x)t(e^{\eta\kappa}-1)\kappa^{-1} \le \theta\varphi(x)t,\end{aligned}$$

for all $t \in [0,1]$, all $x \in A$ (recall that η has been chosen so that $e^{\eta\kappa} - 1 \le \kappa\theta$). This combined with (7) finishes the proof of part (b) of the theorem.

Since f is locally Lipschitz and $h(\cdot, x)$ is of class C^1, by the chain rule, the map $t \mapsto f(h(t,x))$ is differentiable a.e. in $[0,1]$ and we have

$$\begin{aligned}\frac{d}{dt}f(h(t,x)) &\le \max\{\langle x^*, \frac{d}{dt}h(t,x)\rangle : x^* \in \partial f(h(t,x))\}\\ &= -\eta\max\{\langle x^*, V(h(t,x))\rangle : x^* \in \partial f(h(t,x))\}\\ &\le -\eta m\gamma(h(t,x))\\ &\le 0, \end{aligned}\tag{8}$$

for a.a. $t \in [0,1]$. Since $x = h(0,x)$, this implies that

$$f(h(t,x)) - f(x) \le 0 \quad \text{for all } t \in [0,1], \text{ all } x \in X.$$

Whence part (c) of the theorem.

Let $(t,x) \in [0,1] \times X$ be such that $f(h(t,x)) = f(x)$. In view of (8), we must have $\gamma(h(s,x)) = 0$ for all $s \in [0,t]$. In particular, this means that $\gamma(x) = \gamma(h(0,x)) = 0$, hence $V(x) = 0$. Therefore, the constant function $u \equiv x$ is a solution of the problem (6). By uniqueness of the solution of (6), we must have $h(s,x) = x$ for all $s \in [0,1]$, and so $h(t,x) = x$. This establishes part (d) of the theorem.

If $x \in X$ is such that $|f(x) - c| \ge \varepsilon_0$, then we have in particular $|f(x) - c| > \varepsilon_1$ and so $x \in B$. Invoking again (7), we must have $h(t,x) = x$ for all $t \in [0,1]$. Whence part (e) of the theorem.

It remains to show the claim in part (f). Since S and $K_f^c \cap Z$ are bounded (see Lemma 4), we can find $R > 0$ such that

$$(K_f^c \cup S)_{3\delta} \subset B(0, R). \tag{9}$$

Then let $\varepsilon \in (0, \varepsilon_0)$ be small enough so that

$$2\varepsilon < \varepsilon_1, \; 2\varepsilon \leq m\eta, \text{ and } \; 2\theta \left(\sup_{B(0,R)} \hat{\varphi} \right) \varepsilon \leq m\eta\delta. \tag{10}$$

Let us show part (f) of the theorem for the chosen ε. Arguing indirectly, suppose that there is $x \in X$ with

$$f(x) \leq c + \varepsilon, \; f(h(1, x)) > c - \varepsilon, \text{ and } (x \notin U \quad \text{or} \quad h(1, x) \notin U). \tag{11}$$

By part (c) of the theorem and (8), we have

$$c - \varepsilon < f(h(1, x)) \leq f(h(t, x)) \leq f(x) \leq c + \varepsilon \; \text{ for all } t \in [0, 1]. \tag{12}$$

In particular, this implies that

$$|f(h(t, x)) - c| \leq \varepsilon < \frac{\varepsilon_1}{2} \quad \text{for all } t \in [0, 1].$$

We have that

$$\exists t \in [0, 1], \; h(t, x) \in (S \cup K_f^c)_{2\delta}. \tag{13}$$

Indeed, supposing that (13) does not hold, we have $h(t, x) \in A_0$ and so $\gamma(h(t, x)) = 1$ for all $t \in [0, 1]$. Then (8) together with (12) implies that

$$\eta m \leq f(x) - f(h(1, x)) < 2\varepsilon,$$

which is impossible in view of (10). Hence, (13) is verified. Note also that

$$\exists s \in [0, 1], \; h(s, x) \notin (S \cup K_f^c)_{3\delta},$$

because of (11) (recalling that $(S \cup K_f^c)_{3\delta} \subset U$). Combining this with (13), we can find $t_1, t_2 \in [0, 1]$ such that

$$\begin{aligned} 2\delta = \operatorname{dist}(h(t_1, x), S \cup K_f^c) &< \operatorname{dist}(h(t, x), S \cup K_f^c) \\ &< \operatorname{dist}(h(t_2, x), S \cup K_f^c) = 3\delta, \end{aligned}$$

for all $t \in (t_1', t_2')$ with $t_1' := \min\{t_1, t_2\}$, $t_2' := \max\{t_1, t_2\}$. We get in particular $h(t, x) \in A_0$ so $\gamma(h(t, x)) = 1$ whenever $t \in (t_1', t_2')$. Whence

$$\eta m(t_2' - t_1') \leq f(h(t_1', x)) - f(h(t_2', x)) < 2\varepsilon$$

(by (8) and (12)). We have $h(t_1' + t, x) = h(t, h(t_1', x))$ for all $t \in (0, 1 - t_1')$, which holds because both functions are solutions of the same Cauchy problem

$$\begin{cases} u'(t) = -\eta V(u(t)) & \text{in } [0,1], \\ u(0) = h(t_1', x). \end{cases}$$

In view of part (b) of the theorem and (9), (10), we obtain

$$\delta \leq \|h(t_2', x) - h(t_1', x))\| \leq \theta\varphi(h(t_1', x))(t_2' - t_1') < \theta \left(\sup_{B(0,R)} \hat{\varphi} \right) \frac{2\varepsilon}{\eta m} \leq \delta,$$

a contradiction. This proves part (f) and completes the proof of the theorem. □

4. Minimax Principle

In this section, as an application of Theorem 5, we obtain a general minimax principle. It is based on the notion of linking that we recall in the following definition.

Definition 10. If $E_0 \subset E$ and D are non-empty subsets of X and $\gamma_0 : E_0 \to X$ is a continuous map, such that

(a) $\gamma_0(E_0) \cap D = \emptyset$,
(b) $\gamma(E) \cap D \neq \emptyset$ whenever $\gamma : E \to X$ is a continuous map with $\gamma|_{E_0} = \gamma_0$,

then we say that the pair (E_0, E) is linking with D via γ_0.

Example 11. Let $x, y \in X$, $x \neq y$. The pair (E_0, E) is linking with D in the following situation:

$$E = \{tx + (1-t)y : t \in [0,1]\} \supset E_0 = \{x, y\}, \quad \gamma_0 = \mathrm{id}_{E_0},$$
$$D = \partial B(x, r) \quad \text{with } r \in (0, \|y - x\|).$$

Theorem 12. *Assume that (E_0, E) is linking with D via γ_0. Let $f : X \to \mathbb{R}$ be a locally Lipschitz map such that*

$$a := \sup_{\gamma_0(E_0)} f \leq \inf_D f =: b \tag{14}$$

and

$$c := \inf_{\gamma \in \Gamma} \sup_{x \in E} f(\gamma(x)) \in \mathbb{R} \quad \textit{with} \quad \Gamma = \{\gamma : E \to X \textit{ continuous} : \gamma|_{E_0} = \gamma_0\},$$

so that $c \geq b$ (see Definition 10). Moreover, we assume that f satisfies the $(\varphi - \mathrm{C})_{X \setminus S, c}$-condition for some bounded set S such that

- $\mathrm{dist}(S, \{x \in X : f(x) \geq d\}) > 0$ *for some* $d \in (b, c)$, *in the case where* $c > b$, *or*
- $\gamma_0(E_0) \subset S$ *and* $\mathrm{dist}(S, D) > 0$, *in the case where* $c = b$,

and with respect to some Lipschitz map $\varphi : X \setminus S \to [1, +\infty)$.

Then $K_f^c \neq \emptyset$. *Moreover, if* $c = b$, *then* $K_f^c \cap \overline{D} \neq \emptyset$.

Remark 13.

(a) In the case where $c > b$, the bounded set S can be empty (provided that f satisfies the global $(\varphi - \mathrm{C})_{X,c}$-condition).

(b) In the case where $c = b$, as a necessary condition for applying the theorem, we must have $\gamma_0(E_0)$ bounded and $\mathrm{dist}(\gamma_0(E_0), D) > 0$. Conversely, if these conditions are fulfilled and if f satisfies the global $(\varphi - \mathrm{C})_{X,c}$-condition, then the conclusion of the theorem is valid (one can take $S = \gamma_0(E_0)$).

Proof. First we deal with the case where $c > b$. Arguing by contradiction, assume that $K_f^c = \emptyset$. We apply Theorem 5 with $\varepsilon_0 \in (0, c - d)$ and $U = \{x \in X : f(x) < d\}$, which yields $\varepsilon \in (0, \varepsilon_0)$ and a continuous map $h : [0, 1] \times X \to X$ satisfying conditions (a)–(f) of the theorem. Let $\gamma \in \Gamma$ be such that

$$f(\gamma(x)) \leq c + \varepsilon \quad \text{for all } x \in E.$$

We claim that

$$\gamma_1 := h(1, \gamma(\cdot)) \in \Gamma. \tag{15}$$

For all $x \in E_0$, we have indeed $f(\gamma(x)) < c - \varepsilon_0$, hence $\gamma_1(x) = \gamma(x) = \gamma_0(x)$ (by Theorem 5 (e)), whence $\gamma_1|_{E_0} = \gamma_0$, which is the required property for having (15). Also, we have $f(\gamma_1(x)) \leq c - \varepsilon$ for all $x \in E$ (by Theorem 5 (f)), but this contradicts the definition of c.

Next, we assume that $c = b$. Arguing by contradiction, we assume that $K_f^c \cap \overline{D} = \emptyset$. Then we can apply Theorem 5 with any $\varepsilon_0 \in (0, +\infty)$ and $U := X \setminus \overline{D}$, and we get $\varepsilon \in (0, \varepsilon_0)$ and a continuous map $h : [0, 1] \times X \to X$ satisfying conditions (a)–(f) in the theorem. As above we choose $\gamma \in \Gamma$ such that $f(\gamma(x)) \leq c + \varepsilon$ for all $x \in E$, and we define $\gamma_1 := h(1, \gamma(\cdot))$. As before, we can see that $\gamma_1 \in \Gamma$, in other words

$$\gamma_1(x) = \gamma(x) = \gamma_0(x) \quad \text{for all } x \in E_0,$$

this time this is due to Theorem 5 (b) and the fact that $\gamma(E_0) \subset S$. Moreover, by Theorem 5 (f), for all $x \in E$ we have

$$f(\gamma_1(x)) \leq c - \varepsilon < \inf_D f \quad \text{or} \quad \gamma_1(x) \in U = X \setminus \overline{D}.$$

This implies that $\gamma_1(E) \cap D = \emptyset$, which contradicts the assumption of linking. The proof of the theorem is complete. □

Remark 14. It can be noted that in the proof of Theorem 12 above, we do not distinguish between the case $a < b$ and the limit case $a = b$. This unified treatment is possible due to the form of the deformation result on which we rely (Theorem 5). The key point is that in the application of Theorem 5, we can specify a bounded set S that will not be modified by the deformation (see Theorem 5 (b)).

References

[1] P. Bartolo, V. Benci, and D. Fortunato, Abstract critical point theorems and applications to some nonlinear problems with "strong" resonance at infinity, *Nonlinear Anal.* **7**(9), 981–1012, (1983).

[2] A. Ambrosetti and P.H. Rabinowitz, Dual variational methods in critical point theory and applications, *J. Funct. Anal.* **14**, 349–381, (1973).

[3] P.H. Rabinowitz, *Minimax Methods in Critical Point Theory with Applications to Differential Equations.* (American Mathematical Society, Providence, RI, 1986).

[4] A. Kristály, V.V. Motreanu, and Cs. Varga, A minimax principle with a general Palais-Smale condition, *Commun. Appl. Anal.* **9**(2), 285–297, (2005).

[5] K.C. Chang, Variational methods for nondifferentiable functionals and their applications to partial differential equations, *J. Math. Anal. Appl.* **80**(1), 102–129, (1981).

[6] S.A. Marano and D. Motreanu, A deformation theorem and some critical point results for non-differentiable functions, *Topol. Methods Nonlinear Anal.* **22**(1), 139–158, (2003).

[7] D. Motreanu and P.D. Panagiotopoulos, *Minimax Theorems and Qualitative Properties of the Solutions of Hemivariational Inequalities* (Kluwer Academic Publishers, Dordrecht, 1999).

[8] N. Costea, A. Kristály, and Cs. Varga, *Variational and Monotonicity Methods in Nonsmooth Analysis* (Birkhäuser, Cham, 2021).

[9] F.H. Clarke, *Optimization and Nonsmooth Analysis* (John Wiley & Sons, Inc., New York, 1983).

[10] D. Motreanu, V.V. Motreanu, and N. Papageorgiou, *Topological and Variational Methods with Applications to Nonlinear Boundary Value Problems* (Springer, New York, 2014).

[11] H. Brezis, *Functional Analysis, Sobolev Spaces and Partial Differential Equations* (Springer, New York, 2011).

https://doi.org/10.1142/9789811261572_0014

Chapter 14

A Two-phase Problem Related to Phase-change Material

D. Goeleven* and R. Oujja†

University of La Réunion, PIMENT EA4518,
97715 Saint-Denis Messag cedex 9 La Réunion.
**daniel.goeleven@univ-reunion.fr*
†*rachid.oujja@univ-reunion.fr*

A semi-discretized scheme is applied to a two-phase free boundary problem issued from the thermal behavior in phase-change materials. The numerical method is based on backward Euler scheme. *A priori* estimates for the solution of the semi-discretized problem are stated and convergence to the solution of the continuous problem is studied. The existence of a solution of the semi-discretized method yields to a sequence of second-kind variational inequalities. A duality algorithm is then applied for iterative approximations of the solution.

1. Introduction

Thermal energy storage systems containing phase-change materials (PCM) have been widely recognized as one of the most advanced technical means of improving the energy efficiency and environmental impact of buildings. Heat transfers in PCM can be described using partial differential equations for solid–liquid boundary formation when the phase boundary can move with time. This phase-change process is quite complex due to its highly nonlinear nature. Both the solid and the liquid phases are present and the material has different thermo-physical characteristics for each phase (see Ref. [1]).

An analytical solution for phase transition was first found in 1861 by Franz Neumann, who introduced it in his lecture notes (see Ref. [2]). Later in 1889, this physical phenomenon was studied by the Slovenian physicist Josef Stefan for the case of water freezing [3]. More analytical solutions for the phase-change processes in PCMs are discussed in Refs. [4–10]. Nowadays, the solution of the Stefan problem is widely considered as one of the mostly used analytical solutions for one-dimensional solid–liquid phase transition [11]. However, some of the referenced analytical solutions may provide useful guidance but may not be applicable for actual PCM applications in buildings.

We consider here the case of a building roof equipped with PCM, and represented by the two-dimensional domain $\Omega = [l_1, l_2] \times [0, 1]$ in the $x = (x_1, x_2)$ coordinates. The transient local two-dimensional equation of the energy balance under enthalpy reads

$$\frac{\partial \chi(x,t)}{\partial t} = \nabla.(\lambda(x,t)\nabla T(x,t)),\ t \in [0, \mathcal{T}],\ x \in \Omega, \tag{1}$$

where T represents the temperature, λ is the thermal conductivity of the roof and χ is the enthalpy defined by

$$\chi \in H(T) = \begin{cases} \rho_s c_s (T - T_m) & \text{if} \quad T < T_m, \\ [0, \rho_l L_m] & \text{if} \quad T = T_m, \\ \rho_l c_l (T - T_m) + \rho_l L_m & \text{if} \quad T > T_m. \end{cases}$$

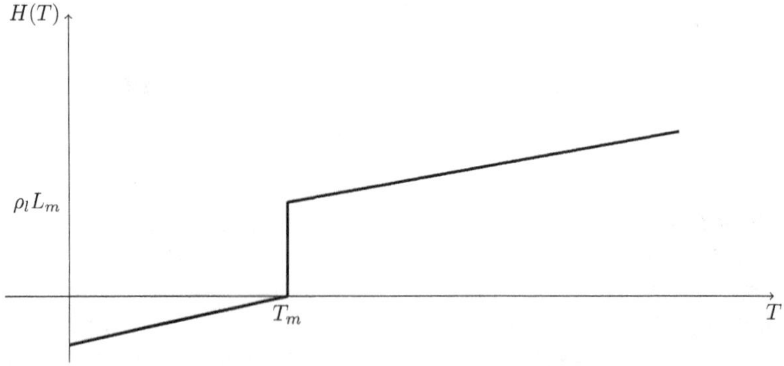

Here c_i $(i = s, l)$ represent the heat capacity, ρ_i $(i = s, l)$, the volume density, T_m the melt temperature and L_m, the latent heat of the roof.

Equation (1) is completed with the initial condition

$$\chi(x, 0) = \chi_0(x) \quad \text{a.e.}\, x \in \Omega \tag{2}$$

and the boundary conditions

$$(\forall t \in [0, \mathcal{T}], \forall x_2 \in [0,1]) : T(l_1, x_2, t) = T_G, \; T(l_2, x_2, t) = T_D, \tag{3}$$

$$(\forall t \in [0, \mathcal{T}], \forall x_1 \in [l_1, l_2]) : \frac{\partial T}{\partial x_2}(x_1, 0, t) = \frac{\partial T}{\partial x_2}(x_1, 1, t) = 0, \tag{4}$$

where T_G and T_D are given values of boundary temperature, and χ_0 is the initial enthalpy function in Ω.

Let $Q_\mathcal{T} = \Omega \times [0, \mathcal{T}]$ be the time-dependent domain. Let V and W be the spaces defined by

$$\begin{aligned} V &= \{z \in H^1(\Omega) : z(l_1, .) = z(l_2, .) = 0\}, \\ W &= \{\xi \in H^1(Q_\mathcal{T}) : \forall t \in [0, \mathcal{T}], \; \xi(., t) \in V \text{ and } \forall x \in \Omega, \\ &\qquad \xi(x, 0) = \xi(x, \mathcal{T}) = 0\}. \end{aligned}$$

Let $\overline{T} \in L^2(\Omega)$ be a function such that $\overline{T}(l_1, x_2) = T_G$ and $\overline{T}(l_2, x_2) = T_D$ $(\forall x_2 \in [0, 1])$. We say that (T, χ) is a weak solution of the boundary value problem (1)–(4) if the following properties are fulfilled:

$$T \in \overline{T} + L^2(0, \mathcal{T}; V), \quad \chi \in H^1(0, \mathcal{T}; V'), \tag{5}$$

$$\int_{Q_\mathcal{T}} \chi \frac{\partial \varphi}{\partial t} \, dx dt = \int_{Q_\mathcal{T}} \lambda \nabla T . \nabla \varphi \, dx dt \quad \forall \varphi \in W, \tag{6}$$

$$\chi \in H(T), \tag{7}$$

$$\chi(., 0) = \chi_0 \text{ a.e. in } \Omega. \tag{8}$$

We suppose that

$$\lambda \in C^1(0, \mathcal{T}; L^\infty(\Omega)). \tag{9}$$

We suppose also that there exist two constants $\lambda_m > 0$ and λ_M such that

$$\lambda_m \leq \lambda(x, t) \leq \lambda_M, \forall (x, t) \in Q_\mathcal{T}. \tag{10}$$

2. A Semi-discretized Problem

In order to give an approximation solution of problem (6)–(8), we use a semi-implicit time discretization formula defined by the backward Euler scheme applied to equation (1) which is written in the following weak formulation: For $n = 1, 2, ..., N$, find (T^n, χ^n) such that

$$T^n - \overline{T} \in V, \quad \chi^n \in V', \tag{11}$$

$$\chi^n \in H(T^n), \tag{12}$$

$$\int_\Omega \lambda^n \nabla T^n . \nabla \varphi \, dx + \int_\Omega \frac{\chi^n - \chi^{n-1}}{k} \varphi dx = 0 \quad \forall \varphi \in V, \tag{13}$$

$$\chi^0(x) = \chi_0(x), \quad \text{a.e. in } \Omega. \tag{14}$$

where N is a positive integer, $k = \mathcal{T}/N$, $t_n = kn$, $\lambda^n(x) = \lambda(x, t_n)$, $n = 0, 1, ..., N$.

The set-valued graph H is the subdifferential of the function $\Phi : \mathbb{R} \to \mathbb{R} \cup \{+\infty\}$ defined by

$$\Phi(T) = \begin{cases} \frac{1}{2}\rho_s c_s (T - T_m)^2, & \text{if } T \leq T_m \\ \frac{1}{2}\rho_l c_l (T - T_m)^2 + \rho_l L_m (T - T_m), & \text{if } T \geq T_m. \end{cases} \tag{15}$$

Proposition 1. *Let (T^n, χ^n) be a solution of problem (11)–(14). Then $w^n = T^n - \overline{T}$ is a solution of the following variational inequality problem:*

Find $w^n \in V$ such that (16)

$$k \int_\Omega \lambda^n \nabla w^n . \nabla(\varphi - w^n) \, dx + J(\varphi) - J(w^n) \geq \langle f, \varphi - w^n \rangle,$$

$$\forall \varphi \in V, \tag{17}$$

where

$$\langle f, \varphi \rangle = \langle \chi^{n-1}, \varphi \rangle - k \int_\Omega \lambda^n \nabla \overline{T} . \nabla \varphi \, dx, \quad \forall \varphi \in V$$

and J is the functional defined by

$$J(v) = \int_\Omega \Phi(v + \overline{T})(x) dx, \quad \forall v \in L^2(\Omega). \tag{18}$$

Proof. Let (T^n, χ^n) be a solution of (11)–(14). Taking $\varphi - w^n$ as a test function in (13) yields

$$k \int_\Omega \lambda^n \nabla w^n . \nabla(\varphi - w^n) \, dx + \int_\Omega \chi^n (\varphi - w^n) dx = \langle f, \varphi - w^n \rangle,$$

$$\forall \varphi \in V. \tag{19}$$

From the definition of J, we have

$$J(\varphi) - J(w^n) = \int_\Omega (\Phi(\varphi + \overline{T})(x) - \Phi(T^n)(x))dx,$$

and from the inclusion $\chi^n \in H(T^n) = \partial\Phi(T^n)$, we get

$$\Phi(y) - \Phi(T^n(x)) \geq \chi^n(y - T^n(x)), \quad \forall y \in \mathbb{R} \text{ and } a.e.\, x \in \Omega.$$

Thus,

$$\Phi((\varphi + \overline{T})(x)) - \Phi(T^n(x)) \geq \chi^n((\varphi + \overline{T})(x) - T^n(x)) \; a.e.\, x \in \Omega,$$

and using (19), we get

$$J(\varphi) - J(w^n) \geq \langle f, \varphi - w^n \rangle - k \int_\Omega \lambda^n \nabla w^n . \nabla(\varphi - w^n)\, dx,$$

which is the desired variational inequality. □

Lemma 1. *Let J and Φ be as defined above. We have $\forall \varphi \in L^2(\Omega)$: $\partial J(\varphi) \subset L^2(\Omega)$. Moreover, for all w in $L^2(\Omega)$,*

$$\chi \in \partial J(w) \iff \chi \in H(w + \overline{T}).$$

Proof. The mapping Φ is convex and continuous on $\mathbb{R}$, then, by linearity, J is a convex continuous operator from $L^2(\Omega)$ in $\mathbb{R}$. Thus, $\partial J(\varphi)$ is well defined in $L^2(\Omega)$ [12]. Let $w \in L^2(\Omega)$ and $\chi \in L^2(\Omega)$ such that $\chi \in H(w+\overline{T})$. Then we have

$$\forall v \in L^2(\Omega),\ \Phi(v + \overline{T}) - \Phi(w + \overline{T}) \geq \chi(v(x) - w(x)) \; a.e.x \in \Omega.$$

Integrating on Ω, we obtain

$$J(v) - J(w) \geq \int_\Omega \chi(x)(v(x) - w(x))dx,\ \forall v \in L^2(\Omega).$$

Thus, $\chi \in \partial J(w)$.

If $\chi \in \partial J(w)$, then, $J(v) - J(w) \geq \int_\Omega \chi(x)(v(x) - w(x))dx$, $\forall v \in L^2(\Omega)$. For a given point x_0 in Ω, we build a sequence of open sets $(S_i)_{i\in\mathbb{N}^*}$ such that $S_{i+1} \subset S_i$, $|S_i| \to 0$ (where $|S_i|$ denotes the Lebesgue measure of S_i) and $\cap_i S_i = \{x_0\}$. Let us set the following for $a \in \mathbb{R}$ and $i \in \mathbb{N}$ fixed:

$$v_i(x) = \begin{cases} w(x) & \text{if } x \notin S_i, \\ a - \bar{T}(x) & \text{if } x \in S_i. \end{cases} \tag{20}$$

we have

$$J(v_i) - J(w) = \int_\Omega (\Phi(v_i(x) + \overline{T}(x)) - \Phi(w(x) + \overline{T}(x)))dx$$
$$= \int_{S_i} (\Phi(a) - \Phi(w(x) + \overline{T}(x))dx.$$

And thus,

$$\int_{S_i} (\Phi(a) - \Phi(w(x) + \overline{T}(x))dx \geq \int_{S_i} \chi(v_i(x) - w(x))dx.$$

Therefore,

$$\frac{1}{|S_i|}\int_{S_i} (\Phi(a) - \Phi(w(x) + \overline{T}(x))dx \geq \frac{1}{|S_i|}\int_{S_i} \chi(v_i(x) - w(x))dx.$$

Passing to the limit $i \to +\infty$ and using Lebesgue theorem, we obtain

$$(\Phi(a) - \Phi(w(x_0) + \overline{T}(x_0)) \geq \chi(x_0)(a - (w(x_0) + \overline{T}(x_0)), \quad \forall a \in \mathbb{R}.$$

Thus, $\chi(x_0) \in \partial\Phi(w(x_0) + \overline{T}(x_0))$, a.e. $x_0 \in \Omega$. □

Proposition 2. *Let w^n be the unique solution of problem* (16)–(17), *then there exists a unique χ^n such that $(w^n + \overline{T}, \chi^n)$ is a solution of* (11)–(14).

Proof. Let $w^n \in V$ be a solution of problem (16)–(17). The linear form

$$u \in V \mapsto \langle f, u\rangle - k\int_\Omega \lambda^n \nabla w^n.\nabla u\, dx$$

is a continuous mapping from V into $\mathbb{R}$. Therefore, from the density of V in $L^2(\Omega)$ and the Riez theorem, there exists a unique $\chi^n \in L^2(\Omega)$ such that

$$\forall u \in V,\ \langle f, u\rangle - k\int_\Omega \lambda^n \nabla w^n.\nabla u\, dx = (\chi^n, u)_{L^2(\Omega)}.$$

We have shown that for all w^n solution of (16)–(17), there exists a unique $\chi^n \in L^2(\Omega)$ such that for all $\varphi \in V$,

$$k\int_\Omega \lambda^n \nabla w^n.\nabla(\varphi - w^n)\, dx + \int_\Omega \chi^n(\varphi - w^n)dx = \langle f, \varphi - w^n\rangle. \qquad (21)$$

A comparison of (17) and (21) yields

$$J(\varphi) - J(w^n) \geq \int_\Omega \chi^n(\varphi - w^n)dx,\ \forall\varphi \in V.$$

From the density of V in $L^2(\Omega)$ and the continuity of J on $L^2(\Omega)$, the previous inequality is still valid on $L^2(\Omega)$, and thus $\chi^n \in \partial J(w^n)$. Taking into account lemma 1, $\chi^n \in H(w^n + \overline{T})$ and thus $(T^n = w^n + \overline{T}, \chi^n)$ is a solution of (11)–(14). □

Theorem 1. *For $n = 1, ..., N$, there exists a unique solution (T^n, χ^n) of the semi-discretized problem* (11)–(14).

Proof. From Propositions 1 and 2, solving the sequence of problems (11)–(14) is equivalent to solving a sequence of a variational inequalities of the second-kind (16)–(17) for $n = 1$ to N. It is easy to show that the function J, the bilinear form defined by $a(u, v) = \int_\Omega \lambda^n \nabla u.\nabla v$ and the function f defined in Proposition 1 satisfy the classical hypotheses of variational inequalities of the second-kind ([12]). □

3. *A Priori* Estimates

Proposition 3. *Assuming that there exists T^0 such that $T^0 - \overline{T} \in V$ satisfying $\chi^0 \in H(T^0)$, there exist constants C_1 and C_2 such that for all n in $\mathbb{N}$*

$$\|T^n\|_{H^1(\Omega)} \leq C_1, \tag{22}$$

$$\|\frac{\chi^n - \chi^{n-1}}{k}\|_{V'} \leq C_2. \tag{23}$$

Proof. If we take $\varphi = T^n - T^{n-1}$ in (13), we obtain

$$k \int_\Omega \lambda^n \nabla T^n.\nabla(T^n - T^{n-1})\, dx + \int_\Omega (\chi^n - \chi^{n-1})(T^n - T^{n-1})\, dx = 0.$$

Since the operator H is monotone, we obtain

$$\int_\Omega (\chi^n - \chi^{n-1})(T^n - T^{n-1})\, dx \geq 0.$$

And by using the identity $\dfrac{a^2 - b^2}{2} \leq a^2 - ab$ for $a, b \in \mathbb{R}$, we get

$$\frac{k}{2} \int_\Omega \lambda^n |\nabla T^n|^2\, dx - \frac{k}{2} \int_\Omega \lambda^n |\nabla T^{n-1}|^2\, dx \leq 0.$$

Thus,

$$\int_\Omega \lambda^n |\nabla T^n|^2\, dx - \int_\Omega \lambda^{n-1} |\nabla T^{n-1}|^2\, dx \le \int_\Omega (\lambda^n - \lambda^{n-1}) |\nabla T^{n-1}|^2\, dx. \quad (24)$$

Now, summing up to some integer $j \le N$ and using (10) yields

$$\begin{aligned}\int_\Omega \lambda^j |\nabla T^j|^2\, dx - \int_\Omega \lambda^0 |\nabla T^0|^2\, dx \\ \le \sum_{n=1}^{j} \int_\Omega \frac{|\lambda^n - \lambda^{n-1}|}{\lambda_m} \lambda^{n-1} |\nabla T^{n-1}|^2\, dx. \quad (25)\end{aligned}$$

From hypothesis (9), we have for a.e. $x \in \Omega$,

$$|\lambda^n(x) - \lambda^{n-1}(x)| \le \|\lambda^n - \lambda^{n-1}\|_{L^\infty(\Omega)} \le k \sup_{t\in[0,\mathcal{T}]} \|\lambda'(t)\|_{L^\infty(\Omega)}.$$

By substituting this last inequality in (25), we get

$$\int_\Omega \lambda^j |\nabla T^j|^2\, dx - \int_\Omega \lambda^0 |\nabla T^0|^2\, dx \le c_0 \sum_{n=0}^{j-1} \int_\Omega k \lambda^n |\nabla T^n|^2\, dx. \quad (26)$$

with $c_0 = \dfrac{1}{\lambda_m} \sup_{t\in[0,\mathcal{T}]} \|\lambda'(t)\|_{L^\infty(\Omega)}$. Then we obtain for $j = 1, ..., N$ the inequality

$$\int_\Omega \lambda^j |\nabla T^j|^2\, dx \le \int_\Omega \lambda^0 |\nabla T^0|^2\, dx + c_0 \sum_{n=0}^{j-1} \int_\Omega k \lambda^n |\nabla T^n|^2\, dx. \quad (27)$$

By applying this inequality for $j = 1$, we obtain

$$\begin{aligned}\int_\Omega \lambda^1 |\nabla T^1|^2\, dx &\le \int_\Omega \lambda^0 |\nabla T^0|^2\, dx + c_0 k \int_\Omega \lambda^0 |\nabla T^0|^2\, dx \\ &\le (1 + c_0 k) \int_\Omega \lambda^0 |\nabla T^0|^2\, dx.\end{aligned}$$

By induction and by using simple calculations, we get

$$\int_\Omega \lambda^n |\nabla T^n|^2\, dx \le (1 + c_0 k)^n \int_\Omega \lambda^0 |\nabla T^0|^2\, dx.$$

For a time-step k small enough, there exists a constant C such that

$$(1 + c_0 k)^n \le (1 + c_0 k)^N \le C e^{c_0 \mathcal{T}}.$$

Thus,

$$\int_\Omega \lambda_m |\nabla T^n|^2\, dx \le \int_\Omega \lambda^n |\nabla T^n|^2\, dx \le Ce^{c_0\mathcal{T}} \int_\Omega \lambda_M |\nabla T^0|^2\, dx,$$

then

$$\|\nabla T^n\|^2_{L^2(\Omega)} \le \frac{Ce^{c_0\mathcal{T}}\lambda_M}{\lambda_m}\|\nabla T^0\|^2_{L^2(\Omega)}. \tag{28}$$

Now from Poincaré inequality, there exists a constant $C(\Omega)$, depending on the domain Ω, such that

$$\|(T^n - \bar{T})\|_{L^2(\Omega)} \le C(\Omega)\|\nabla(T^n - \bar{T})\|_{L^2(\Omega)}.$$

It follows that

$$\|T^n\|_{L^2(\Omega)} \le \|\bar{T}\|_{L^2(\Omega)} + C(\Omega)(\|\nabla T^n\|_{L^2(\Omega)} + \|\nabla \bar{T}\|_{L^2(\Omega)}). \tag{29}$$

From (28) and (29), we get (22) with $C_1 = C_1(c_0, \lambda_m, \lambda_M, C, \mathcal{T}, T^0, \bar{T}, C(\Omega))$.

If we consider $\frac{\chi^n - \chi^{n-1}}{k}$ as an element of $V' \subset H^{-1}(\Omega)$, ($H^{-1}(\Omega)$ is the dual of the Sobolev space $H^1_0(\Omega)$), we have from the definition of the norm in $H^{-1}(\Omega)$,

$$\left\|\frac{\chi^n - \chi^{n-1}}{k}\right\|_{H^{-1}(\Omega)} = \sup_{\|\varphi\|_{H^1(\Omega)}\le 1}\left|\int_\Omega \left(\frac{\chi^n - \chi^{n-1}}{k}\right)\varphi\, dx\right|.$$

We have from (13)

$$\begin{aligned}\left|\int_\Omega \left(\frac{\chi^n - \chi^{n-1}}{k}\right)\varphi\, dx\right| &= \left|\int_\Omega \lambda^n \nabla T^n.\nabla\varphi\, dx\right| \\ &\le \lambda_M \|\nabla T^n\|_{L^2(\Omega)}\|\varphi\|_{H^1(\Omega)}, \quad \forall \varphi \in V\end{aligned}$$

Thus,

$$\left\|\frac{\chi^n - \chi^{n-1}}{k}\right\|_{H^{-1}(\Omega)} \le \lambda_M \|\nabla T^n\|_{L^2(\Omega)}$$

Finally, we deduce (23) from (22). □

4. Convergence and Existence Results

Let χ_k, T_k and λ_k be defined on $Q_\mathcal{T}$ by $\chi_k(t,x) = \chi^n(x)$, $T_k(t,x) = T^n(x)$ and $\lambda_k(t,x) = \lambda^n(x)$ for $t \in](n-1)k, nk[$ and for all $n = 1, 2, \ldots N$. We set also $\chi_k(0,x) = \chi^0(x)$. We define the linear function $\widetilde{\chi}$ by

$$\widetilde{\chi}_k(t,x) = \frac{\chi^n(x) - \chi^{n-1}(x)}{k}(t-(n-1)k) + \chi^{n-1}(x),\ t \in [(n-1)k, nk].$$

We obtain from equation (13) that

$$\int_{Q_T} \frac{\partial \widetilde{\chi}_k}{\partial t}\varphi\, dxdt + \int_{Q_T} \lambda_k \nabla T_k . \nabla \varphi\, dxdt = 0, \quad \forall \varphi \in W. \tag{30}$$

Proposition 4. *There exist constants D_1 and D_2 such that*

$$\|T_k\|_{L^2(0,\mathcal{T};H^1(\Omega))} \leq D_1, \tag{31}$$

$$\|\widetilde{\chi}_k\|_{H^1(0,\mathcal{T};V')} \leq D_2. \tag{32}$$

Proof. By the definition of T_k, we have

$$\|T_k\|^2_{L^2(0,\mathcal{T};H^1(\Omega))} = \sum_{n=1}^{N} k\|T^n\|^2_{H^1(\Omega)},$$

and from (22), we obtain the estimate (31).

From (13), we have for all $\varphi \in V$ and $t \in](n-1)k, nk[$

$$\left\langle \frac{\partial \widetilde{\chi}_k}{\partial t}, \varphi \right\rangle_{V' \times V} \leq \lambda_M \|T^n\|_{H^1(\Omega)} \|\varphi\|_V,$$

thus,

$$\left\| \frac{\partial \widetilde{\chi}_k}{\partial t} \right\|_{V'} \leq \lambda_M \|T^n\|_{H^1(\Omega)}.$$

By summing on n up to N, we obtain

$$\int_0^{\mathcal{T}} \left\| \frac{\partial \widetilde{\chi}}{\partial t} \right\|^2_{V'} dt \leq \lambda_M^2 \int_0^{\mathcal{T}} \|T^n\|^2_{H^1(\Omega)} dt.$$

Using estimate (31), we obtain (32) with $D_2 = \lambda_M^2 D_1^2$. □

Lemma 2. *There exists $\chi \in H^1(0,\mathcal{T};V')$ such that, up to a subsequence, $\widetilde{\chi}_k$ and χ_k satisfy*

$$\widetilde{\chi}_k \underset{k\to 0}{\to} \chi \text{ weakly in } H^1(0,\mathcal{T};V'), \tag{33}$$

$$\chi_k \underset{k\to 0}{\to} \chi \text{ weakly in } L^2(0,\mathcal{T};V') \text{ and } L^2(Q_{\mathcal{T}}). \tag{34}$$

Proof. From (32), there exists $\chi \in H^1(0,\mathcal{T};V')$ such that, up to a subsequence, $\widetilde{\chi}_k \to \chi$ weakly in $H^1(0,\mathcal{T};V')$ when $k \to 0$, we have for all $n = 1, 2, ..., N$ and for all $t \in](n-1)k, nk[$

$$\widetilde{\chi}_k(t,x) - \chi_k(t,x) = \frac{t-nk}{k}(\chi^n(x) - \chi^{n-1}(x)).$$

From (23) and after simple calculations, we obtain

$$\|\widetilde{\chi}_k - \chi_k\|_{L^2(0,\mathcal{T},V')} \leq Ck,$$

where $C = C(C_2, \mathcal{T})$ is a constant. Then $\widetilde{\chi}_k$ and χ_k converge to the same limit in $L^2(0,\mathcal{T};V')$ when $k \to 0$. From (33) and the compact injection of $H^1(0,\mathcal{T};V')$ in $L^2(0,\mathcal{T};V')$, we deduce that $\widetilde{\chi}_k$ converges weakly in $L^2(0,\mathcal{T};V')$ when $k \to 0$. From the density of $L^2(0,\mathcal{T};V)$ in $L^2(Q_\mathcal{T})$ we deduce $L^2(Q_\mathcal{T})$-weak convergence. Thus, we have (34). □

Proposition 5. *There exist $T \in \overline{T} + L^2(0,\mathcal{T};V)$ and $\chi \in H^1(0,\mathcal{T};V')$ such that up to a subsequence we have*

$$T_k \underset{k\to 0}{\to} T \text{ weakly in } L^2(0,\mathcal{T};H^1(\Omega)), \tag{35}$$

$$\chi_k \underset{k\to 0}{\to} \chi \text{ weakly in } L^2(0,\mathcal{T};V') \text{ and } L^2(Q_\mathcal{T}). \tag{36}$$

In addition, (T,χ) satisfy in $L^2(0,\mathcal{T};V')$

$$\nabla.(\lambda\nabla T) = \frac{\partial\chi}{\partial t}. \tag{37}$$

Proof. From (31), the sequence (T_k) is uniformly bounded in $L^2(0,\mathcal{T};H^1(\Omega))$ and thus there exists $T \in L^2(0,\mathcal{T};H^1(\Omega))$ such that, up to a subsequence,

$$T_k \underset{k\to 0}{\to} T \text{ weakly in } L^2(0,\mathcal{T};H^1(\Omega)).$$

Since $T_k \in \overline{T} + L^2(0,\mathcal{T};V)$, T lies in the same space. The weak formulation (30) can be rewritten in $L^2(0,\mathcal{T};V')$ as

$$\nabla.(\lambda_k\nabla T_k) = \frac{\partial\widetilde{\chi_k}}{\partial t}. \tag{38}$$

From the regularity of λ, $\lambda_k \to \lambda$ in $L^2(Q_\mathcal{T})$ when $k \to 0$ and from (35), we deduce

$$\nabla.(\lambda_k\nabla T_k) \underset{k\to 0}{\to} \nabla.(\lambda\nabla T) \text{ in } L^2(0,\mathcal{T};V').$$

The convergence (33) yields

$$\frac{\partial \widetilde{\chi_k}}{\partial t} \underset{k\to 0}{\to} \frac{\partial \widetilde{\chi}}{\partial t} \text{ weakly in } L^2(0,\mathcal{T};V').$$

Finally, passing to the limit when $k \to 0$ in (38) and using the previous results, we obtain (37). □

Proposition 6. *Let (T,χ) be the limit functions defined in Proposition* 5, *then $\chi \in H(T)$.*

Proof. Let Φ be the function defined in (15). Since $\partial\Phi = H$, it is enough to prove that

$$\forall v \in L^2(Q_\mathcal{T}),\ \int_{Q_\mathcal{T}} (\Phi(v) - \Phi(T))\, dx\, dt \geq \int_{Q_\mathcal{T}} \chi(v-T)\, dx\, dt. \tag{39}$$

Since $C^0(0,\mathcal{T};L^2(\Omega))$ is dense in $L^2(0,\mathcal{T},H^1(\Omega))$, it is enough to prove (39) for all $v \in C^0(0,\mathcal{T};L^2(\Omega))$. Let $v \in C^0(0,\mathcal{T};L^2(\Omega))$ and define the sequence v_k by $v_k(t,x) = v(nk,x)$, for $t \in](n-1)k, nk[$. We have $v_k \to v$ when $k \to 0$ in $L^2(Q_\mathcal{T})$. Using also the convexity and the lower semicontinuous of Φ, we obtain

$$\begin{aligned}
\int_{Q_\mathcal{T}} (\Phi(v) - \Phi(T))\, dx\, dt &\geq \varliminf_{k\to 0} \int_{Q_\mathcal{T}} (\Phi(v_k) - \Phi(T_k))\, dx\, dt \\
&= \varliminf_{k\to 0} k \sum_{n=1}^{N} \int_\Omega (\Phi(v^n) - \Phi(T^n))\, dx \\
&\geq \varliminf_{k\to 0} k \sum_{n=1}^{N} \int_\Omega \chi^n (v^n - T^n)\, dx \ (since\, \chi^n \in H(T^n)) \\
&= \varliminf_{k\to 0} \int_{Q_\mathcal{T}} \chi_k (v_k - T_k)\, dx\, dt.
\end{aligned}$$

It remains to prove that $\lim_{k\to 0} \int_{Q_\mathcal{T}} \chi_k(v_k - T_k)\, dx\, dt = \int_{Q_\mathcal{T}} \chi(v-T)\, dx\, dt$. We have

$$\int_{Q_\mathcal{T}} \chi_k(v_k - T_k)\, dx\, dt = \int_{Q_\mathcal{T}} \chi_k(v_k - \overline{T})\, dx\, dt - \int_{Q_\mathcal{T}} \chi_k(T_k - \overline{T})\, dx\, dt.$$

From the $L^2(Q_T)$-weak convergence of χ_k and the $L^2(Q_T)$-strong convergence of v_k, the first term, converges to $\int_{Q_T} \chi(v - \overline{T})\, dx\, dt$. For the second term, we have

$$\begin{aligned}
&\left| \int_{Q_T} (\chi_k(T_k - \overline{T}) - \chi(T - \overline{T}))\, dx\, dt \right| \\
&\quad = \left| \int_{Q_T} (\chi_k - \chi)(T_k - \overline{T})\, dx\, dt \right. \\
&\qquad \left. + \int_{Q_T} \chi(T_k - T)\, dx\, dt \right| \\
&\quad \leq \|\chi_k - \chi\|_{L^2(0,T;V')} \cdot \|T_k - \overline{T}\|_{L^2(0,T;V)} \\
&\qquad + \left| \int_{Q_T} \chi(T_k - T) \right| dx\, dt.
\end{aligned}$$

The proof is completed in using Proposition 5. □

5. Iterative Approximation

Numerical approach of (11)–(14) can be placed in the frame of duality methods for variational inequalities (see Ref. [13]) based on classical results for monotone maps (see Refs. [14,15]). For convenience, we give first a brief introduction to the monotonous maximal operator theory. Let G be a maximal monotone multivalued map on a Hilbert space H, and β, a non-negative parameter. It can be proved that for all $f \in H$ there exists a unique $y \in H$ such that $f \in (I + \beta G)(y)$. The single-valued map $J_\beta^G = (I + \beta G)^{-1}$, i.e. the resolvent of G, is a well-defined and contraction map on H ([14]).

The map $G_\beta = \frac{I - J_\beta^G}{\beta}$ is the Moreau–Yosida approximation of G. This map is maximal monotone, single-valued and $\frac{1}{\beta}$-Lipschitz continuous. Our approximation is based on the following property:

Proposition 7.

$$\forall u, y \in H : \; u \in G(y) \iff u = G_\beta(y + \beta u). \tag{40}$$

Proof. Let be $u = G_\beta(x)$. Then

$$\begin{aligned}
u = \frac{I - J_\beta^G}{\beta}(x) &\iff \beta u = x - J_\beta^G(x) \\
&\iff J_\beta^G(x) = x - \beta u
\end{aligned}$$

$$\iff x \in (I + \beta G)(x - \beta u) = x - \beta u + \beta G(x - \beta u)$$
$$\iff \beta u \in \beta G(x - \beta u)$$
$$\iff u \in G(x - \beta u),$$

and by taking $x - \beta u = y$, we get $u \in G(y) \iff u = G_\beta(y + \beta u)$. □

Now, returning to problem (11)–(14), we first introduce the perturbed operator $L = H - \omega I$ for some $0 \le \omega << 1$. And for a parameter $\beta > 0$ such that $\beta\omega < 1$ we consider the map $I + \beta L$ defined by

$$(I + \beta L)(T) = \begin{cases} (1 - \beta\omega)T + \beta\rho_s c_s(T - T_m) & \text{for} \quad T < T_m, \\ [(1 - \beta\omega)T_m, (1 - \beta\omega)T_m + \beta\rho_l L_m] & \text{for} \quad T = T_m, \\ (1 - \beta\omega)T + \beta\rho_l c_l(T - T_m) + \beta\rho_l L_m & \text{for} \quad T > T_m. \end{cases}$$

It is easy to see that for all $f \in \mathbb{R}$ there exists a unique $y \in \mathbb{R}$ such that $f \in (I + \beta L)(y)$. The resolvent of L is defined by

$$J_\beta^L(y) = \begin{cases} \dfrac{y + \beta\rho_s c_s T_m}{1 - \beta\omega + \beta\rho_s c_s} \text{ for } y < (1 - \beta\omega)T_m, \\ T_m \text{ for } (1 - \beta\omega)T_m < y < 1 - \beta\omega)T_m + \beta\rho_l L_m, \\ \dfrac{y + \beta\rho_l c_l T_m - \beta\rho_l L_m}{1 - \beta\omega + \beta\rho_l c_l} \text{ for } y > (1 - \beta\omega)T_m + \beta\rho_l L_m, \end{cases}$$

and its Moreau–Yosida approximation is given by $L_\beta = \dfrac{I - J_\beta^L}{\beta}$.

From (40), we have the following equivalence:

$$u \in L(y) \iff u = L_\beta(y + \beta u), \quad \text{for } u, y \in \mathbb{R}. \tag{41}$$

By introducing the perturbed operator $L = H - \omega I$ in problem (11)–(14) and taking into account the equivalence (41), it leads to the following:

$$\text{Find } T^n \in \overline{T} + V \text{ and } \gamma^n \in V' \text{such that } \gamma^n = L_\beta(T^n + \beta\gamma^n), \tag{42}$$

$$k \int_\Omega \lambda^n \nabla T^n . \nabla\varphi \, dx + \omega \int_\Omega T^n \varphi dx + \int_\Omega \gamma^n \varphi dx$$
$$= \int_\Omega \chi^{n-1} \varphi dx, \; \forall \varphi \in V, \tag{43}$$

$$\gamma^0(x) = \chi_0(x) - \omega T^0(x) \quad a.e. \; x \in \Omega. \tag{44}$$

To solve (42)–(44), we consider the following method:

(0) Start with some arbitrary value of the multiplier γ_0^n.
(1) For γ_j^n known, compute T_j^n solution to

$$k\int_\Omega \lambda^n \nabla T_j^n . \nabla\varphi\, dx + \omega\int_\Omega T_j^n \varphi dx + \int_\Omega \gamma_j^n \varphi\, dx = \int_\Omega \chi^{n-1}\varphi\, dx, \quad \forall\varphi \in V. \tag{45}$$

(2) Update multiplier γ_j^n as

$$\gamma_{j+1}^n = L_\beta(T_j^n + \beta\gamma_j^n). \tag{46}$$

(3) Go to (1) until stop criterion is reached.

Theorem 2. *If parameters β and ω satisfy the condition $\frac{1}{2} \leq \beta\omega < 1$, then we have the convergence*

$$\lim_{j\to\infty} \|T_j^n - T^n\| = 0.$$

Proof. The mapping L_β is $\frac{1}{\beta}$-Lipschitz and thus

$$\begin{aligned}
\|\gamma^n - \gamma_{j+1}^n\|^2 &= \|L_\beta(T^n + \beta\gamma^n) - L_\beta(T_j^n + \beta\gamma_j^n)\|^2 \\
&\leq \frac{1}{\beta^2}\|(T^n + \beta\gamma^n) - (T_j^n + \beta\gamma_j^n)\|^2 \\
&= \frac{1}{\beta^2}\|(T^n - T_j^n) + \beta(\gamma^n - \gamma_j^n)\|^2 \\
&= \frac{1}{\beta^2}\|T^n - T_j^n\|^2 + \frac{2}{\beta}(T^n - T_j^n, \gamma^n - \gamma_j^n) + \|\gamma^n - \gamma_j^n\|^2.
\end{aligned}$$

Therefore,

$$\|\gamma^n - \gamma_j^n\|^2 - \|\gamma^n - \gamma_{j+1}^n\|^2 \geq -\frac{1}{\beta^2}\|T^n - T_j^n\|^2 - \frac{2}{\beta}(T^n - T_j^n, \gamma^n - \gamma_j^n). \tag{47}$$

Now by subtracting (45) from (43), we obtain

$$\begin{aligned}
&k\int_\Omega \lambda^n \nabla(T^n - T_j^n).\nabla\varphi\, dx + \omega\int_\Omega (T^n - T_j^n)\varphi dx \\
&\quad + \int_\Omega (\gamma^n - \gamma_j^n)\varphi dx = 0,\ \forall\varphi \in V
\end{aligned}$$

and if we take $\varphi = T^n - T_j^n$ as a test function, it yields $\omega\|T^n - T_j^n\|^2 \leq -(\gamma^n - \gamma_j^n, T^n - T_j^n)$. By substituting this inequality in (47), we get

$$\|\gamma^n - \gamma_j^n\|^2 - \|\gamma^n - \gamma_{j+1}^n\|^2 \geq -\frac{1}{\beta^2}\|T^n - T_j^n\|^2 + \frac{2\omega}{\beta}\|T^n - T_j^n\|^2$$

$$= \frac{1}{\beta}(2\omega - \frac{1}{\beta})\|T^n - T_j^n\|^2.$$

Now under the condition $\beta\omega \geq \frac{1}{2}$, we have

$$\|\gamma^n - \gamma_j^n\|^2 - \|\gamma^n - \gamma_{j+1}^n\|^2 \geq 0.$$

The sequence $(\|\gamma_j^n - \gamma^n\|^2)_{j\geq 0}$ is then decreasing and positive. It results that $\lim_{j\to\infty} \|\gamma^n - \gamma_j^n\|^2 = 0$, and finally,

$$\lim_{j\to\infty} \|T^n - T_j^n\|^2 = 0.$$

□

References

[1] P. Lamberg, R. Lehtiniemi, and A.M. Henell, Numerical and experimental investigation of melting and freezing processes in phase change material storage, *Int. J. Therm. Sci.* **43**, 277–287, (2000).

[2] M. Brillouin, Sur Quelques Problèmes non Résolus de la Physique Mathématique Classique, *Propagation de la Fusion. Annales de l'institut Henri Poincaré*, **1**(3), 285–308, (1930).

[3] J. Stefan, Uber Einige Probleme der Theorie der Warmeleitung, *Sitzungsber Wiener Akad Math Naturwiss Abt* **98**, 473–484, (1889).

[4] H.S. Carslaw and J.C. Jaeger, *Conduction of Heat in Solids* (Clarendon Press, Oxford, 1959).

[5] J. Crank, *Free and Moving Boundary Problems.* (Clarendon Press, Oxford, 1984).

[6] J. Douglas, A uniqueness theorem for the solution of a Stefan problem, *Proc. Am. Math. Soc.* **8**, 402–408, (1957).

[7] G.W. Evans, A note on the existence of a solution to a problem of Stefan, *Q. Appl. Math.* **9**, 185–193, (1951).

[8] A. Friedman, *Partial Differential Equations of Parabolic Type* (Prentice-Hall, New Jersey, 1964).

[9] J.M. Hill, *One-dimensional Stefan Problems: An Introduction* (Longman Scientific Technical, Harlow, 1987).

[10] L. I. Rubinstein, *The Stefan Problem* (Amer. Math. Soc. (Translated from Russian), 1971).

[11] W. Ogoh and D. Groulx, *Stefan's problem: Validation of a One-dimensional Solid-liquid Phase Change Heat Transfer Process* (Proceedings of the COMSOL conference, Boston, USA, 2010).

[12] I. Ekeland and R. Teman, *Analyse Convexe et Problèmes Variationels* (Gautier-Villars, Paris, 1984).

[13] A. Bermudez and C. Moreno, Duality methods for solving variational inequalities, *Comput. Math. Appl.* **7**, 43–58, (1981).

[14] H. Brezis, *Opérateurs Maximaux Monotones et Semi-groupes De Contractions Dans les Espaces de Hilbert* (North-Holland, Amsterdam 1973).

[15] A. Pazy, *Semigroups of Non-linear Contractions in Hilbert Spaces* (Problems in Nonlinear Analysis (C.I.M.E. Ed. Cremonese, Roma, 1971)).

https://doi.org/10.1142/9789811261572_0015

Chapter 15

Motive of the Representation Varieties of Torus Knots for Low Rank Affine Groups

Ángel González-Prieto*,§,¶, Marina Logares†,||, and Vicente Muñoz‡,**

*Departamento de Matemáticas, Facultad de Ciencias, Universidad Autónoma de Madrid. C. Francisco Tomás y Valiente, 7, 28049 Madrid, Spain, and Instituto de Ciencias Matemáticas (CSIC-UAM-UC3M-UCM), C. Nicolás Cabrera 15, 28049 Madrid, Spain

†Facultad de Ciencias Matemáticas, Universidad Complutense de Madrid, Plaza Ciencias 3, 28040 Madrid, Spain

‡Departamento de Álgebra, Geometría y Topología, Facultad de Ciencias, Universidad de Málaga, Campus de Teatinos s/n, 29071 Málaga, Spain

§angel.gonzalezprieto@uam.es

¶angelgonzalezprieto@ucm.es

||mlogares@ucm.es

**vicente.munoz@ucm.es

We compute the motive of the variety of representations of the torus knot of type (m, n) into the affine groups $\mathrm{AGL}_1(\mathbb{C})$ and $\mathrm{AGL}_2(\mathbb{C})$. For this, we stratify the varieties and show that the motives lie in the subring generated by the Lefschetz motive $q = [\mathbb{C}]$.

1. Introduction

Since the foundational work of Culler and Shalen [1], the varieties of $\mathrm{SL}_2(\mathbb{C})$-characters have been extensively studied. Given a manifold M, the variety of representations of $\pi_1(M)$ into $\mathrm{SL}_2(\mathbb{C})$ and the variety of characters of such representations both contain information on the topology of M. It is especially interesting for three-dimensional manifolds, where the fundamental group and the geometrical properties of the manifold are strongly related. This can be used to study knots $K \subset S^3$, by analyzing the $\mathrm{SL}_2(\mathbb{C})$-character variety of the fundamental group of the knot complement $S^3 - K$ (these are called *knot groups*).

For a very different reason, the case of fundamental groups of surfaces has also been extensively analyzed [2–6], in this situation focusing more on geometrical properties of the moduli space in itself (cf. non-abelian Hodge theory).

Much less is known of the character varieties for other groups. The character varieties for $\mathrm{SL}_3(\mathbb{C})$ for free groups have been described in Ref. [7]. In the case of 3-manifolds, little has been done. In this chapter, we focus on the case of the torus knots $K_{m,n}$ for coprime m, n, which are the first family of knots where the computations are rather feasible. The case of $\mathrm{SL}_2(\mathbb{C})$-character varieties of torus knots was carried out in Refs. [8,9]. For $\mathrm{SL}_3(\mathbb{C})$, it has been carried out by Muñoz and Porti in Refs. [10]. The case of $\mathrm{SL}_4(\mathbb{C})$ has been computed by two of the authors of the current chapter through a computer-assisted proof in Ref. [11].

The group $\mathrm{SL}_r(\mathbb{C})$ is reductive, which allows to use Geometric Invariant Theory (GIT) to define the moduli of representations, the so-called *character variety*. In Ref. [12], we started the analysis of character varieties for the first non-reductive groups, notably computing by three different methods (geometric, arithmetic and through a Topological Quantum Field Theory) the motive of the variety of representations for a surface group into the rank one affine group $\mathrm{AGL}_1(\mathbb{C})$.

In the current work, we study the variety of representations of the torus knot $K_{m,n}$ into the affine groups $\mathrm{AGL}_1(\mathbb{C})$ and $\mathrm{AGL}_2(\mathbb{C})$. We prove the following result:

Theorem 1. *Let $m, n \in \mathbb{N}$ with $\gcd(m,n) = 1$. The motives of the $\mathrm{AGL}_1(\mathbb{C})$ and $\mathrm{AGL}_2(\mathbb{C})$-representation variety of the (m,n)-torus knot in the Grothendieck ring of complex algebraic varieties are*

$$\begin{aligned}
\left[\mathfrak{X}_{m,n}(\mathrm{AGL}_1(\mathbb{C}))\right] &= (mn - m - n + 2)(q^2 - q). \\
\left[\mathfrak{X}_{m,n}(\mathrm{AGL}_2(\mathbb{C}))\right] &= q^2 + q^6 - 2q^4 + q^3 + \left[\mathfrak{X}^{\mathrm{irr}}_{m,n}(\mathrm{GL}_2(\mathbb{C}))\right] q^2 \\
&\quad + (m-1)(n-1) \\
&\quad \times \left((q^4 - 3q^3 + 2q^2) \frac{(m-2)(n-2)q + mn - 4}{4} \right. \\
&\qquad \left. + (q+1)^2(q-1)q^3 \right)
\end{aligned}$$

$$+ (m-1)(n-1)(mn-m-n)\left(\frac{1}{2}(q^2-1)-(q-1)\right)$$
$$\times (q^2+q)q^2.$$

Here, $q = [\mathbb{C}] \in \mathrm{K}\mathbf{Var}_{\mathbb{C}}$ *denotes the Lefschetz motive, and* $\left[\mathfrak{X}^{\mathrm{irr}}_{m,n}(\mathrm{GL}_2(\mathbb{C}))\right]$

$$= \begin{cases} (q^3-q)\frac{1}{4}(m-1)(n-1)(q-2)(q-1), & m,n \text{ both odd,} \\ (q^3-q)\left(\frac{1}{4}(n-2)(m-1)(q-2)+\frac{1}{2}(m-1)(q-1)\right)(q-1), & m \text{ odd, } n \text{ even,} \\ (q^3-q)\left(\frac{1}{4}(n-1)(m-2)(q-2)+\frac{1}{2}(n-1)(q-1)\right)(q-1), & m \text{ even, } n \text{ odd.} \end{cases}$$

2. Basic Notions

2.1. *Representation varieties of torus knots*

Let Γ be a finitely presented group, and let G be a complex algebraic group. A *representation* of Γ in G is a homomorphism $\rho : \Gamma \to G$. Consider a presentation $\Gamma = \langle x_1, \ldots, x_k \,|\, r_1, \ldots, r_s \rangle$. Then ρ is completely determined by the k-tuple $(A_1, \ldots, A_k) = (\rho(x_1), \ldots, \rho(x_k))$ subject to the relations $r_j(A_1, \ldots, A_k) = \mathrm{Id}$, $1 \le j \le s$. The *representation variety* is

$$\mathfrak{X}_\Gamma(G) = \mathrm{Hom}\,(\Gamma, G)$$
$$= \{(A_1, \ldots, A_k) \in G^k \,|\, r_j(A_1, \ldots, A_k) = \mathrm{Id},\ 1 \le j \le s\} \subset G^k.$$

Therefore, $\mathfrak{X}_\Gamma(G)$ is an affine algebraic set.

Suppose in addition that G is a linear group, say $G \subset \mathrm{GL}_r(\mathbb{C})$. A representation ρ is *reducible* if there exists some proper subspace $V \subset \mathbb{C}^r$ such that for all $g \in G$ we have $\rho(g)(V) \subset V$; otherwise ρ is *irreducible*. This distinction induces a natural stratification of the representation variety into its irreducible and reducible parts $\mathfrak{X}_\Gamma(G) = \mathfrak{X}^{\mathrm{irr}}_\Gamma(G) \sqcup \mathfrak{X}^{\mathrm{red}}_\Gamma(G)$.

Let $T^2 = S^1 \times S^1$ be the 2-torus and consider the standard embedding $T^2 \subset S^3$. Let m, n be a pair of coprime positive integers. Identifying T^2 with the quotient $\mathbb{R}^2/\mathbb{Z}^2$, the image of the straight line $y = \frac{m}{n}x$ in T^2 defines the *torus knot* of type (m,n), which we shall denote as $K_{m,n} \subset S^3$ (see Ref. [13], Chapter 3). For a knot $K \subset S^3$, we denote by Γ_K the fundamental group of the exterior $S^3 - K$ of the knot. It is known that

$$\Gamma_{m,n} = \Gamma_{K_{m,n}} \cong \langle x, y \,|\, x^n = y^m \rangle .$$

Therefore, the variety of representations of the torus knot of type (m,n) is described as

$$\mathfrak{X}_{m,n}(G) = \mathfrak{X}_{\Gamma_{m,n}}(G) = \{(A,B) \in G^2 \mid A^n = B^m\}.$$

In this work, we shall focus on the case $G = \mathrm{AGL}_r(\mathbb{C})$, the group of affine automorphisms of the complex r-dimensional affine space.

2.2. *The Grothendieck ring of algebraic varieties*

Take the category of complex algebraic varieties with regular morphisms $\mathbf{Var}_{\mathbb{C}}$. We can construct its Grothendieck group, $\mathrm{K}\mathbf{Var}_{\mathbb{C}}$, as the abelian group generated by isomorphism classes of algebraic varieties with the relation that $[X] = [Y] + [U]$ if $X = Y \sqcup U$, with $Y \subset X$ a closed subvariety. The cartesian product of varieties also provides $\mathrm{K}\mathbf{Var}_{\mathbb{C}}$ with a ring structure, as $[X]\cdot[Y] = [X \times Y]$. The elements of $\mathrm{K}\mathbf{Var}_{\mathbb{C}}$ are usually referred to as virtual classes. A very important element of $\mathrm{K}\mathbf{Var}_{\mathbb{C}}$ is the class of the affine line, $q = [\mathbb{C}]$, the so-called [14, Section *Lefschetz motive*].

Virtual classes are well-behaved with respect to two typical geometric situations that we will encounter in the upcoming sections. A proof of the following facts can be found for instance in Ref. [14, Section 4.1].

- Let $E \to B$ be a regular morphism that is a locally trivial bundle in the Zariski topology with fiber F. In this situation, we have that in $\mathrm{K}\mathbf{Var}_{\mathbb{C}}$,
$$[E] = [F]\cdot[B].$$
- Suppose that X is an algebraic variety with an action of $\mathbb{Z}_2$. Setting $[X]^+ = [X/\mathbb{Z}_2]$ and $[X]^- = [X] - [X]^+$, we have the formula
$$[X \times Y]^+ = [X]^+[Y]^+ + [X]^-[Y]^-, \tag{1}$$
for two varieties X, Y with $\mathbb{Z}_2$-actions.

Example 1. Consider the fibration $\mathbb{C}^2 - \mathbb{C} \to \mathrm{GL}_2(\mathbb{C}) \to \mathbb{C}^2 - \{(0,0)\}$, $f \mapsto f(1,0)$. It is locally trivial in the Zariski topology, and therefore $[\mathrm{GL}_2(\mathbb{C})] = [\mathbb{C}^2 - \mathbb{C}]\cdot[\mathbb{C}^2 - \{(0,0)\}] = (q^2-q)(q^2-1) = q^4 - q^3 - q^2 + q$. Analogously, the quotient map defines a locally trivial fibration $\mathbb{C}^* = \mathbb{C} - \{0\} \to \mathrm{GL}_2(\mathbb{C}) \to \mathrm{PGL}_2(\mathbb{C})$, so $[\mathrm{PGL}_2(\mathbb{C})] = q^3 - q$.

We have the following computation that we will need later.

Lemma 1. *Let $\mathbb{Z}_2$ act on $\mathbb{C}^2$ by exchange of coordinates. Then $[(\mathbb{C}^*)^2 - \Delta]^+ = (q-1)^2$, $[(\mathbb{C}^*)^2 - \Delta]^- = -q+1$, where Δ denotes the diagonal.*

Also let $X = \mathrm{GL}_2(\mathbb{C})/\mathrm{GL}_1(\mathbb{C}) \times \mathrm{GL}_1(\mathbb{C})$, and $\mathbb{Z}_2$ acting by exchange of columns in $\mathrm{GL}_2(\mathbb{C})$. Then $[X]^+ = q^2$ and $[X]^- = q$.

Proof. The quotient $\mathbb{C}^2/\mathbb{Z}_2$ is parametrized by $s = x + y$, $p = xy$, where (x, y) are the coordinates of $\mathbb{C}^2$. Then $((\mathbb{C}^*)^2 - \Delta)/\mathbb{Z}_2$ is given by the equations $p \neq 0$, $4p \neq s^2$. Therefore, $[(\mathbb{C}^*)^2 - \Delta]^+ = [((\mathbb{C}^*)^2 - \Delta)/\mathbb{Z}_2] = q^2 - q - (q-1) = (q-1)^2$, and $[(\mathbb{C}^*)^2 - \Delta]^- = [(\mathbb{C}^*)^2 - \Delta] - [(\mathbb{C}^*)^2 - \Delta]^+ = (q-1)^2 - (q-1) - (q-1)^2 = -q + 1$.

For the second case, note that $X = \mathbb{P}^1 \times \mathbb{P}^1 - \Delta$, and $\mathbb{Z}_2$ acts by exchange of coordinates. The whole quotient is $(\mathbb{P}^1 \times \mathbb{P}^1)/\mathbb{Z}_2 = \mathrm{Sym}^2(\mathbb{P}^1) = \mathbb{P}^2$. The diagonal goes down to a smooth conic (the completion of $4p = s^2$), hence, $[X]^+ = [X/\mathbb{Z}_2] = [(\mathbb{P}^1 \times \mathbb{P}^1 - \Delta)/\mathbb{Z}_2] = q^2 + q + 1 - (q+1) = q^2$. Also $[X] = (q+1)^2 - (q+1) = q^2 + q$, hence $[X]^- = q$. □

3. $\mathbf{AGL_1(\mathbb{C})}$-representation Varieties of Torus Knots

In this section, we shall compute the motive of the $\mathrm{AGL}_1(\mathbb{C})$-representation variety of the (m, n)-torus knot by describing it explicitly. Suppose that we have an element $(A, B) \in \mathfrak{X}_{m,n}(\mathrm{AGL}_1(\mathbb{C}))$ with matrices of the form

$$A = \begin{pmatrix} 1 & 0 \\ \alpha & a_0 \end{pmatrix}, \quad B = \begin{pmatrix} 1 & 0 \\ \beta & b_0 \end{pmatrix}.$$

A straightforward computation shows that

$$A^n = \begin{pmatrix} 1 & 0 \\ (1 + a_0 + \ldots + a_0^{n-1})\alpha & a_0^n \end{pmatrix},$$

$$B^m = \begin{pmatrix} 1 & 0 \\ (1 + b_0 + \ldots + b_0^{m-1})\beta & b_0^m \end{pmatrix}.$$

Note that, since $\gcd(m, n) = 1$, for any pair $(a_0, b_0) \in \mathbb{C}^2$ with $a_0^n = b_0^m$ and $a_0, b_0 \neq 0$, there exists a unique $t \in \mathbb{C}^* = \mathbb{C} - \{0\}$ such that $t^m = a_0$ and $t^n = b_0$. This means that the representation variety can be explicitly described as

$$\mathfrak{X}_{m,n}(\mathrm{AGL}_1(\mathbb{C})) = \left\{(t, \alpha, \beta) \in \mathbb{C}^* \times \mathbb{C}^2 \mid \Phi_n(t^m)\alpha = \Phi_m(t^n)\beta\right\},$$

where Φ_l is the polynomial

$$\Phi_l(x) = 1 + x + \cdots + x^{l-1} = \frac{x^l - 1}{x - 1} \in \mathbb{C}[x].$$

Written in a more geometric fashion, the morphism $(t, \alpha, \beta) \mapsto t$ defines a regular map

$$\mathfrak{X}_{m,n}(\mathrm{AGL}_1(\mathbb{C})) \longrightarrow \mathbb{C}^*. \tag{2}$$

The fiber over $t \in \mathbb{C}^*$ is the annihilator of the vector $(\Phi_n(t^m), \Phi_m(t^n)) \in \mathbb{C}^2$ (in other words, the orthogonal complement with respect to the standard euclidean metric). This annihilator is $\mathbb{C}$ if $(\Phi_n(t^m), \Phi_m(t^n)) \neq (0,0)$ and is $\mathbb{C}^2$ otherwise.

Denote by μ_l the group of l-th roots of units. Recall that the roots of the polynomial Φ_l are the elements of $\mu_l^* = \mu_l - \{1\}$. Hence, $(\Phi_n(t^m), \Phi_m(t^n)) = (0,0)$ if and only if

$$t \in \Omega_{m,n} = \mu_{mn} - (\mu_m \cup \mu_n).$$

The number of elements of $\Omega_{m,n}$ is

$$|\Omega_{m,n}| = mn - m - n + 1 = (m-1)(n-1).$$

The space (2) decomposes into the two Zariski locally trivial fibrations

$$\mathbb{C} \longrightarrow \mathfrak{X}^{(1)}_{m,n}(\mathrm{AGL}_1(\mathbb{C})) \longrightarrow \mathbb{C}^* - \Omega_{m,n},$$
$$\mathbb{C}^2 \longrightarrow \mathfrak{X}^{(2)}_{m,n}(\mathrm{AGL}_1(\mathbb{C})) \longrightarrow \Omega_{m,n},$$

with $\mathfrak{X}_{m,n}(\mathrm{AGL}_1(\mathbb{C})) = \mathfrak{X}^{(1)}_{m,n}(\mathrm{AGL}_1(\mathbb{C})) \sqcup \mathfrak{X}^{(2)}_{m,n}(\mathrm{AGL}_1(\mathbb{C}))$. This implies that the motive of the whole representation variety is

$$\begin{aligned}
[\mathfrak{X}_{m,n}(\mathrm{AGL}_1(\mathbb{C}))] &= \left[\mathfrak{X}^{(1)}_{m,n}(\mathrm{AGL}_1(\mathbb{C}))\right] + \left[\mathfrak{X}^{(2)}_{m,n}(\mathrm{AGL}_1(\mathbb{C}))\right] \\
&= [\mathbb{C}^* - \Omega_{m,n}]\,[\mathbb{C}] + [\Omega_{m,n}]\,[\mathbb{C}^2] \\
&= (q - 1 - |\Omega_{m,n}|)q + |\Omega_{m,n}|q^2 \\
&= (mn - m - n + 2)(q^2 - q).
\end{aligned}$$

This proves the first assertion of Theorem 1.

4. $\mathrm{AGL}_2(\mathbb{C})$-representation Varieties of Torus Knots

In this section, we compute the motive of the $\mathrm{AGL}_2(\mathbb{C})$-representation variety of the (m,n)-torus knot. Suppose that we have an element $(A,B) \in \mathfrak{X}_{m,n}(\mathrm{AGL}_2(\mathbb{C}))$ with matrices of the form

$$A = \begin{pmatrix} 1 & 0 \\ \alpha & A_0 \end{pmatrix}, \quad B = \begin{pmatrix} 1 & 0 \\ \beta & B_0 \end{pmatrix}.$$

Note that in this setting $A_0, B_0 \in \mathrm{GL}_2(\mathbb{C})$, while $\alpha, \beta \in \mathbb{C}^2$. Computing the powers we obtain

$$A^n = \begin{pmatrix} 1 & 0 \\ \Phi_n(A_0)\alpha & A_0^n \end{pmatrix}, \quad B^m = \begin{pmatrix} 1 & 0 \\ \Phi_m(B_0)\beta & B_0^m \end{pmatrix}.$$

Therefore, the $\mathrm{AGL}_2(\mathbb{C})$-representation variety is explicitly given by

$$\mathfrak{X}_{m,n}(\mathrm{AGL}_2(\mathbb{C})) = \left\{ (A_0, B_0, \alpha, \beta) \in \mathrm{GL}_2(\mathbb{C})^2 \times \mathbb{C}^2 \;\middle|\; \begin{array}{c} A_0^n = B_0^m \\ \Phi_n(A_0)\alpha = \Phi_m(B_0)\beta \end{array} \right\}, \tag{3}$$

In particular, these conditions imply that $(A_0, B_0) \in \mathfrak{X}_{m,n}(\mathrm{GL}_2(\mathbb{C}))$. Let us decompose

$$\mathfrak{X}_{m,n}(\mathrm{AGL}_2(\mathbb{C})) = \mathfrak{X}_{m,n}^{\mathrm{irr}}(\mathrm{AGL}_2(\mathbb{C})) \sqcup \mathfrak{X}_{m,n}^{\mathrm{red}}(\mathrm{AGL}_2(\mathbb{C})),$$

where $\mathfrak{X}_{m,n}^{\mathrm{irr}}(\mathrm{AGL}_2(\mathbb{C}))$ (resp. $\mathfrak{X}_{m,n}^{\mathrm{red}}(\mathrm{AGL}_2(\mathbb{C}))$) are the representations (A, B) with (A_0, B_0) an irreducible (resp. reducible) representation of $\mathfrak{X}_{m,n}(\mathrm{GL}_2(\mathbb{C}))$.

Remark 1. Beware of the notation: the superscripts refer to the reducibility/irreducibility of the vectorial part of the representation, not to the representation itself.

4.1. *The irreducible stratum*

First of all, let us analyze the case where (A_0, B_0) is an irreducible representation. In that case, the eigenvalues are restricted as the following result shows.

Lemma 2. *Let $\rho = (A_0, B_0) \in \mathfrak{X}_{m,n}^{\mathrm{irr}}(\mathrm{GL}_r(\mathbb{C}))$ be an irreducible representation. Then $A_0^n = B_0^m = \omega\,\mathrm{Id}$, for some $\omega \in \mathbb{C}^*$.*

Proof. Note that A_0^n is a linear map that is equivariant with respect to the representation ρ. By Schur's lemma, this implies that A_0^n must be a multiple of the identity, say $A_0^n = \omega\,\mathrm{Id}$ and, since $B_0^m = A_0^n$, the result follows. □

Corollary 1. *Let $\rho = (A_0, B_0) \in \mathfrak{X}_{m,n}^{\mathrm{irr}}(\mathrm{GL}_r(\mathbb{C}))$ be an irreducible representation and let $\lambda_1, \ldots, \lambda_r$ and $\eta_1, \ldots, \eta_r$ be the eigenvalues of A_0 and B_0, resp. Then A_0 and B_0 are diagonalizable and $\lambda_1^n = \cdots = \lambda_r^n = \eta_1^m = \cdots = \eta_r^m$.*

In order to analyze the conditions of (3), observe that $(A, B) \mapsto (A_0, B_0)$ defines a morphism

$$\mathfrak{X}_{m,n}^{\mathrm{irr}}(\mathrm{AGL}_2(\mathbb{C})) \longrightarrow \mathfrak{X}_{m,n}^{\mathrm{irr}}(\mathrm{GL}_2(\mathbb{C})). \tag{4}$$

The fiber of this morphism at (A_0, B_0) is the kernel of the map

$$\Lambda : \mathbb{C}^2 \times \mathbb{C}^2 \to \mathbb{C}^2, \quad \Lambda(\alpha, \beta) = \Phi_n(A_0)\alpha - \Phi_m(B_0)\beta. \tag{5}$$

The following appears in Ref. [10, Proposition 7.3]. Recall from Example 1 that $[\mathrm{PGL}_2(\mathbb{C})] = q^3 - q$.

Proposition 1. *For the torus knot of type (m, n), we have the following:*

- *If m, n are both odd, then* $[\mathfrak{X}^{\mathrm{irr}}_{m,n}(\mathrm{GL}_2(\mathbb{C}))] = [\mathrm{PGL}_2(\mathbb{C})]\frac{1}{4}(m-1)(n-1)(q-2)(q-1)$.
- *If n is even and m is odd, then* $[\mathfrak{X}^{\mathrm{irr}}_{m,n}(\mathrm{GL}_2(\mathbb{C}))] = [\mathrm{PGL}_2(\mathbb{C})]\left(\frac{1}{4}(n-2)(m-1)(q-2) + \frac{1}{2}(m-1)(q-1)\right)(q-1)$.
- *If m is even and n is odd, then* $[\mathfrak{X}^{\mathrm{irr}}_{m,n}(\mathrm{GL}_2(\mathbb{C}))] = [\mathrm{PGL}_2(\mathbb{C})]\left(\frac{1}{4}(n-1)(m-2)(q-2) + \frac{1}{2}(n-1)(q-1)\right)(q-1)$.

To understand the kernel of (5), we use the following lemma:

Lemma 3. *Let A be a diagonalizable matrix and let $p(x) \in \mathbb{C}[x]$, a polynomial. Then, the dimension of the kernel of the matrix $p(A)$ is the number of eigenvalues of A that are roots of $p(x)$.*

Proof. Write $A = QDQ^{-1}$ with $D = \mathrm{diag}(\lambda_1, \ldots, \lambda_r)$ a diagonal matrix. Then $p(A) = Qp(D)Q^{-1}$ and, since $p(D) = \mathrm{diag}(p(\lambda_1), \ldots, p(\lambda_r))$, the dimension of its kernel is the number of eigenvalues that are also roots of p. □

Using the previous lemma for $r = 2$, we get that the dimension of the kernel of $\Phi_n(A_0)$ is the number of eigenvalues of A_0 that belong to μ_n^*, and analogously for $\Phi_m(B_0)$. Let λ_1, λ_2 be the eigenvalues of A_0 and η_1, η_2, the eigenvalues of B_0. Recall that $\lambda_1 \neq \lambda_2$ and $\eta_1 \neq \eta_2$ since otherwise (A_0, B_0) is not irreducible. Then, we have the following options:

(1) Case $\lambda_1, \lambda_2 \in \mu_n^*$ and $\eta_1, \eta_2 \in \mu_m^*$. In this situation, $\Lambda \equiv 0$ so $\mathrm{Ker}\,\Lambda = \mathbb{C}^4$. Hence, if we denote by $\mathfrak{X}^{\mathrm{irr},(1)}_{m,n}(\mathrm{AGL}_2(\mathbb{C}))$ and $\mathfrak{X}^{\mathrm{irr},(1)}_{m,n}(\mathrm{GL}_2(\mathbb{C}))$, the corresponding strata in (4) of the total and base space, resp., we have that

$$\left[\mathfrak{X}^{\mathrm{irr},(1)}_{m,n}(\mathrm{AGL}_2(\mathbb{C}))\right] = \left[\mathfrak{X}^{\mathrm{irr},(1)}_{m,n}(\mathrm{GL}_2(\mathbb{C}))\right][\mathbb{C}^4].$$

To get the motive of $\mathfrak{X}^{\mathrm{irr},(1)}_{m,n}(\mathrm{GL}_2(\mathbb{C}))$, the eigenvalues define a fibration

$$\mathfrak{X}^{\mathrm{irr},(1)}_{m,n}(\mathrm{GL}_2(\mathbb{C})) \longrightarrow ((\mu_n^*)^2 - \Delta)/\mathbb{Z}_2 \times ((\mu_m^*)^2 - \Delta)/\mathbb{Z}_2, \tag{6}$$

where Δ is the diagonal and $\mathbb{Z}_2$ acts by permutation of the entries. The fiber of this map is the collection of representations $(A_0, B_0) \in \mathfrak{X}^{\mathrm{irr}}_{m,n}(\mathrm{GL}_2)$ with fixed eigenvalues, denoted by $\mathfrak{X}^{\mathrm{irr}}_{m,n}(\mathrm{GL}_2(\mathbb{C}))_0$. An element of $\mathfrak{X}^{\mathrm{irr}}_{m,n}(\mathrm{GL}_2(\mathbb{C}))_0$ is completely determined by the two pairs of eigenspaces of (A_0, B_0) up to conjugation. Since the representation (A_0, B_0) must be irreducible, these eigenspaces must be pairwise distinct. Hence, this variety is $\mathfrak{X}^{\mathrm{irr}}_{m,n}(\mathrm{GL}_2(\mathbb{C}))_0 = (\mathbb{P}^1)^4 - \Delta_c$, where $\Delta_c \subset (\mathbb{P}^1)^4$ is the "coarse diagonal" of tuples with two repeated entries. There is a free and closed action of $\mathrm{PGL}_2(\mathbb{C})$ on $(\mathbb{P}^1)^4$ with a quotient

$$\frac{(\mathbb{P}^1)^4 - \Delta_c}{\mathrm{PGL}_2(\mathbb{C})} = \mathbb{P}^1 - \{0, 1, \infty\}.$$

To see this, note that there is a $\mathrm{PGL}_2(\mathbb{C})$-equivariant map that sends the first three entries to $0, 1, \infty \in \mathbb{P}^1$, resp., so the orbit is completely determined by the image of the fourth point under this map. Hence, $[\mathfrak{X}^{\mathrm{irr}}_{m,n}(\mathrm{GL}_2(\mathbb{C}))_0] = [\mathbb{P}^1 - \{0,1,\infty\}]\,[\mathrm{PGL}_2(\mathbb{C})] = (q-2)(q^3-q)$.

Coming back to the fibration (6), we have that the basis is a set of $\binom{n-1}{2}\binom{m-1}{2} = \frac{(n-1)(n-2)(m-1)(m-2)}{4}$ points, so

$$\left[\mathfrak{X}^{\mathrm{irr},(1)}_{m,n}(\mathrm{GL}_2(\mathbb{C}))\right] = \frac{(n-1)(n-2)(m-1)(m-2)}{4}(q-2)(q^3-q),$$

and thus,

$$\left[\mathfrak{X}^{\mathrm{irr},(1)}_{m,n}(\mathrm{AGL}_2(\mathbb{C}))\right] = \frac{(n-1)(n-2)(m-1)(m-2)}{4}(q^5-2q^4)(q^3-q).$$

(2) Case $\lambda_1, \lambda_2 \in \mu_n^*$, $\eta_1 \in \mu_m^*$ and $\eta_2 = 1$. In this case, $\mathrm{Ker}\,\Lambda = \mathbb{C}^3$ and the base space is made of $\binom{n-1}{2}(m-1)$ copies of $\mathfrak{X}^{\mathrm{irr}}_{m,n}(\mathrm{GL}_2(\mathbb{C}))_0$. Hence, this stratum contributes

$$\begin{aligned}\left[\mathfrak{X}^{\mathrm{irr},(2)}_{m,n}(\mathrm{AGL}_2(\mathbb{C}))\right] &= \frac{(n-1)(n-2)(m-1)}{2}\,[\mathbb{P}^1 - \{0,1,\infty\}] \\ &\quad \times [\mathrm{PGL}_2(\mathbb{C})]\,[\mathbb{C}^3] \\ &= \frac{(n-1)(n-2)(m-1)}{2}(q^4-2q^3)(q^3-q).\end{aligned}$$

(3) Case $\lambda_1 \in \mu_n^*$, $\lambda_2 = 1$ and $\eta_1, \eta_2 \in \mu_m^*$. This is analogous to the previous stratum and contributes

$$\begin{aligned}\left[\mathfrak{X}^{\mathrm{irr},(3)}_{m,n}(\mathrm{AGL}_2(\mathbb{C}))\right] &= \frac{(m-1)(n-1)(m-2)}{2}\,[\mathbb{P}^1 - \{0,1,\infty\}] \\ &\quad \times [\mathrm{PGL}_2(\mathbb{C})]\,[\mathbb{C}^3] \\ &= \frac{(m-1)(n-1)(m-2)}{2}(q^4-2q^3)(q^3-q).\end{aligned}$$

(4) Case $\lambda_1 \in \mu_n^*$, $\lambda_2 = 1$ and $\eta_1 \in \mu_m^*$, $\eta_2 = 1$. Now, $\operatorname{Ker}\Lambda = \mathbb{C}^2$ and this stratum contributes

$$\left[\mathfrak{X}_{m,n}^{\mathrm{irr},(4)}(\mathrm{AGL}_2(\mathbb{C}))\right] = (m-1)(n-1)\left[\mathbb{P}^1 - \{0,1,\infty\}\right][\mathrm{PGL}_2(\mathbb{C})][\mathbb{C}^2]$$
$$= (m-1)(n-1)(q^3-2q^2)(q^3-q).$$

(5) Case $\lambda_1 \notin \mu_n^*, \lambda_2 \notin \mu_n^*, \eta_1 \notin \mu_m^*$ and $\eta_2 \notin \mu_m^*$. Recall that by Corollary 1, these conditions are all equivalent. In this situation, Λ is surjective so $\operatorname{Ker}\Lambda = \mathbb{C}^2$. The motive $\left[\mathfrak{X}_{m,n}^{\mathrm{irr}}(\mathrm{GL}_2(\mathbb{C}))\right]$ is given in Proposition 1. To this space, we have to remove the orbits corresponding to the forbidden eigenvalues, which are

$$\begin{aligned}\ell_{m,n} &= \frac{(n-1)(n-2)(m-1)(m-2)}{4} + \frac{(n-1)(n-2)(m-1)}{2}\\ &\quad + \frac{(m-1)(n-1)(m-2)}{2} + (m-1)(n-1)\\ &= \frac{1}{4}mn(m-1)(n-1)\end{aligned}$$

copies of $[\mathfrak{X}_{m,n}^{\mathrm{irr}}(\mathrm{GL}_2(\mathbb{C}))_0] = [\mathbb{P}^1 - \{0,1,\infty\}][\mathrm{PGL}_2(\mathbb{C})]$. Hence, this stratum contributes

$$\begin{aligned}\left[\mathfrak{X}_{m,n}^{\mathrm{irr},(5)}(\mathrm{AGL}_2(\mathbb{C}))\right] &= \left(\left[\mathfrak{X}_{m,n}^{\mathrm{irr}}(\mathrm{GL}_2(\mathbb{C}))\right] - \ell_{m,n}(q-2)(q^3-q)\right)\left[\mathbb{C}^2\right]\\ &= \left[\mathfrak{X}_{m,n}^{\mathrm{irr}}(\mathrm{GL}_2(\mathbb{C}))\right]q^2 - \frac{1}{4}mn(m-1)(n-1)\\ &\quad \times (q^3-2q^2)(q^3-q).\end{aligned}$$

Adding up all the contributions, we get

$$\begin{aligned}\left[\mathfrak{X}_{m,n}^{\mathrm{irr}}(\mathrm{AGL}_2(\mathbb{C}))\right] &= \sum_{k=1}^{5}\left[\mathfrak{X}_{m,n}^{\mathrm{irr},(k)}(\mathrm{AGL}_2(\mathbb{C}))\right] = \left[\mathfrak{X}_{m,n}^{\mathrm{irr}}(\mathrm{GL}_2(\mathbb{C}))\right]q^2\\ &\quad + \frac{(m-1)(n-1)(q^3-2q^2)(q-1)(q^3-q)}{4}\\ &\quad \times ((m-2)(n-2)q + mn - 4).\end{aligned}$$

4.2. *The reducible stratum*

In this section, we shall consider the case in which $(A_0, B_0) \in \mathfrak{X}_{m,n}^{\mathrm{red}}(\mathrm{GL}_2(\mathbb{C}))$ is a reducible representation. After a change in the basis, since $A_0^n = B_0^m$, we can suppose that (A_0, B_0) has exactly one of the following three forms:

$$\text{(A)} \left(\begin{pmatrix} t_1^m & 0 \\ 0 & t_2^m \end{pmatrix}, \begin{pmatrix} t_1^n & 0 \\ 0 & t_2^n \end{pmatrix} \right), \text{(B)} \left(\begin{pmatrix} t^m & 0 \\ 0 & t^m \end{pmatrix}, \begin{pmatrix} t^n & 0 \\ 0 & t^n \end{pmatrix} \right),$$

$$\text{(C)} \left(\begin{pmatrix} t^m & 0 \\ x & t^m \end{pmatrix}, \begin{pmatrix} t^n & 0 \\ y & t^n \end{pmatrix} \right),$$

with $t_1, t_2, t \in \mathbb{C}^*$, $x, y \in \mathbb{C}$ and satisfying $t_1 \neq t_2$ and $(x, y) \neq (0, 0)$.

Restricting to the representations of each stratum $S = \text{(A)}, \text{(B)}, \text{(C)}$, we have a morphism

$$\mathfrak{X}^S_{m,n}(\mathrm{AGL}_2(\mathbb{C})) \longrightarrow \mathfrak{X}^S_{m,n}(\mathrm{GL}_2(\mathbb{C})), \tag{7}$$

whose fiber is the kernel of the linear map (5).

4.2.1. *Case (A)*

In this case, as for the irreducible part of Section 4.1, the kernel of Λ depends on whether t_1, t_2 are roots of the polynomial Φ_l. In this case, the base space is

$$\mathfrak{X}^{(\mathrm{A})}_{m,n}(\mathrm{GL}_2(\mathbb{C})) = \left(((\mathbb{C}^*)^2 - \Delta) \times \frac{\mathrm{GL}_2(\mathbb{C})}{\mathrm{GL}_1(\mathbb{C}) \times \mathrm{GL}_1(\mathbb{C})} \right) / \mathbb{Z}_2,$$

with the action of $\mathbb{Z}_2$ given by exchange of eigenvalues and eigenvectors. Using Lemma 1 and equation (1), we have

$$\begin{aligned} \left[\mathfrak{X}^{(\mathrm{A})}_{m,n}(\mathrm{GL}_2)\right] &= [(\mathbb{C}^*)^2 - \Delta]^+ \left[\frac{\mathrm{GL}_2(\mathbb{C})}{\mathrm{GL}_1(\mathbb{C}) \times \mathrm{GL}_1(\mathbb{C})}\right]^+ \\ &\quad + [(\mathbb{C}^*)^2 - \Delta]^- \left[\frac{\mathrm{GL}_2(\mathbb{C})}{\mathrm{GL}_1(\mathbb{C}) \times \mathrm{GL}_1(\mathbb{C})}\right]^- \\ &= q^2(q-1)^2 - q(q-1). \end{aligned}$$

On the other hand, if we fix the eigenvalues of (A_0, B_0) as in Section 4.1, the corresponding fiber $\mathfrak{X}^{(\mathrm{A})}_{m,n}(\mathrm{GL}_2(\mathbb{C}))_0$ is

$$\left[\mathfrak{X}^{(\mathrm{A})}_{m,n}(\mathrm{GL}_2(\mathbb{C}))_0\right] = \left[\frac{\mathrm{GL}_2(\mathbb{C})}{\mathrm{GL}_1(\mathbb{C}) \times \mathrm{GL}_1(\mathbb{C})}\right] = q^2 + q.$$

As in Section 3, set $\Omega_{m,n} = \mu_{mn} - (\mu_m \cup \mu_n)$ for those $t \in \mathbb{C}^*$ such that $\Phi_n(t^m) = 0$ and $\Phi_m(t^n) = 0$. With this information at hand, we compute for each stratum the following:

(1) Case $t_1, t_2 \in \Omega_{m,n}$. In this situation, $\Lambda \equiv 0$ so $\operatorname{Ker}\Lambda = \mathbb{C}^4$. The eigenvalues yield a fibration

$$\mathfrak{X}_{m,n}^{(A),(1)}(\mathrm{GL}_2(\mathbb{C})) \longrightarrow \left(\Omega_{m,n}^2 - \Delta\right)/\mathbb{Z}_2,$$

whose fiber is $\mathfrak{X}_{m,n}^{(A)}(\mathrm{GL}_2(\mathbb{C}))_0$. Observe that $\left(\Omega_{m,n}^2 - \Delta\right)/\mathbb{Z}_2$ is a finite set of $(m-1)(n-1)((m-1)(n-1)-1)/2$ points, so we have

$$\begin{aligned}\left[\mathfrak{X}_{m,n}^{(A),(1)}(\mathrm{AGL}_2(\mathbb{C}))\right] &= \left[\mathfrak{X}_{m,n}^{(A),(1)}(\mathrm{GL}_2(\mathbb{C}))\right][\mathbb{C}^4] \\ &= \left[\mathfrak{X}_{m,n}^{(A)}(\mathrm{GL}_2(\mathbb{C}))_0\right][\mathbb{C}^4]\left[\left(\Omega_{m,n}^2 - \Delta\right)/\mathbb{Z}_2\right] \\ &= \frac{(m-1)(n-1)(mn-m-n)}{2} q^4(q^2+q).\end{aligned}$$

(2) Case $t_1 \in \Omega_{m,n}$ but $t_2 \notin \Omega_{m,n}$ (or vice versa, the order is not important here). Now, we have a locally trivial fibration

$$\mathfrak{X}_{m,n}^{(A),(2)}(\mathrm{GL}_2(\mathbb{C})) \longrightarrow \Omega_{m,n} \times \left(\mathbb{C}^* - \Omega_{m,n}\right),$$

with fiber $\mathfrak{X}_{m,n}^{(A)}(\mathrm{GL}_2(\mathbb{C}))_0$. The kernel of Λ is $\mathbb{C}^3$, so this stratum contributes

$$\left[\mathfrak{X}_{m,n}^{(A),(2)}(\mathrm{AGL}_2(\mathbb{C}))\right] = (m-1)(n-1)(q-mn+n+m-2)q^3(q^2+q).$$

(3) Case $t_1, t_2 \notin \Omega_{m,n}$. The kernel is now $\mathbb{C}^2$ and we have a fibration

$$\mathfrak{X}_{m,n}^{(A),(3)}(\mathrm{GL}_2(\mathbb{C})) \longrightarrow B,$$

where the motive of the base space B is

$$\begin{aligned}[B] &= [(\mathbb{C}^*)^2 - \Delta]^+ - \left[\Omega_{m,n}^2 - \Delta\right]^+ - [\Omega_{m,n}]\left(q - 1 - [\Omega_{m,n}]\right) \\ &= (q-1)^2 - \frac{(m-1)(n-1)(mn-m-n)}{2} \\ &\quad - (m-1)(n-1)(q-mn+n+m-2) \\ &= q^2 - (mn-m-n+3)q - \frac{1}{4}(m-1)(n-1)(mn-8).\end{aligned}$$

Therefore, this space contributes

$$\left[\mathfrak{X}_{m,n}^{(\mathrm{A}),(3)}(\mathrm{AGL}_2(\mathbb{C}))\right] = \left[\mathfrak{X}_{m,n}^{(\mathrm{A}),(3)}(\mathrm{GL}_2(\mathbb{C}))\right][\mathbb{C}^2]$$
$$= q^2(q^2+q)\left(q^2 - (mn-m-n+3)q\right.$$
$$\left. + (m-1)(n-1)(mn-8)/4\right).$$

Adding up all the contributions, we get that

$$\left[\mathfrak{X}_{m,n}^{(\mathrm{A})}(\mathrm{AGL}_2(\mathbb{C}))\right] = (q^2+q)q^2\Bigg(\frac{(m-1)(n-1)(mn-m-n)}{2}$$
$$\times(q^2-1) + (m-1)(n-1)$$
$$\times(q-mn+n+m-2)(q-1) + (q-1)^2\Bigg).$$

4.2.2. *Case (B)*

In this setting, this situation is simpler. Observe that the adjoint action of $\mathrm{GL}_2(\mathbb{C})$ on the vectorial part is trivial, so the corresponding $\mathrm{GL}_2(\mathbb{C})$-representation variety is just

$$\mathfrak{X}_{m,n}^{(\mathrm{B})}(\mathrm{GL}_2(\mathbb{C})) = \mathbb{C}^*.$$

Analogously, the variety with fixed eigenvalues, $\mathfrak{X}_{m,n}^{(\mathrm{B})}(\mathrm{GL}_2(\mathbb{C}))_0$ is just a point. With these, we obtain the following:

(1) If $t \in \Omega_{m,n}$, then $\operatorname{Ker}\Lambda = \mathbb{C}^4$. We have a fibration

$$\mathfrak{X}_{m,n}^{(\mathrm{B}),(1)}(\mathrm{GL}_2(\mathbb{C})) \longrightarrow \Omega_{m,n},$$

whose fiber is $\mathfrak{X}_{m,n}^{(\mathrm{B})}(\mathrm{GL}_2(\mathbb{C}))_0$. Hence, this stratum contributes

$$\left[\mathfrak{X}_{m,n}^{(\mathrm{B}),(1)}(\mathrm{AGL}_2(\mathbb{C}))\right] = \left[\mathfrak{X}_{m,n}^{(\mathrm{B}),(1)}(\mathrm{GL}_2(\mathbb{C}))\right][\mathbb{C}^4] = (m-1)(n-1)q^4.$$

(2) If $t \notin \Omega_{m,n}$, then $\operatorname{Ker}\Lambda = \mathbb{C}^2$. We have a fibration

$$\mathfrak{X}_{m,n}^{(\mathrm{B}),(2)}(\mathrm{GL}_2(\mathbb{C})) \longrightarrow \mathbb{C}^* - \Omega_{m,n}.$$

Thus, the contribution of this stratum is

$$\left[\mathfrak{X}_{m,n}^{(\mathrm{B}),(2)}(\mathrm{AGL}_2(\mathbb{C}))\right] = (q-1-(m-1)(n-1))q^2.$$

The total contribution is

$$\left[\mathfrak{X}_{m,n}^{(\mathrm{B})}(\mathrm{AGL}_2(\mathbb{C}))\right] = (m-1)(n-1)(q^4-q^2) + (q-1)q^2.$$

4.2.3. *Case (C)*

In this case, an extra calculation must be done to control the off-diagonal entry. If (A_0, B_0) has the form

$$\left(\begin{pmatrix} t^m & 0 \\ x & t^m \end{pmatrix}, \begin{pmatrix} t^n & 0 \\ y & t^n \end{pmatrix}\right),$$

then the condition $A_0^n = B_0^m$ reads as

$$\begin{pmatrix} t^{mn} & 0 \\ nt^{m(n-1)}x & t^{mn} \end{pmatrix} = \begin{pmatrix} t^{mn} & 0 \\ mt^{n(m-1)}y & t^{mn} \end{pmatrix}.$$

The later conditions reduce to $nt^{m(n-1)}x = mt^{n(m-1)}y$ and, since $t \neq 0$, this means that (x, y) should lie in a line minus $(0, 0)$. The stabilizer of a Jordan-type matrix in $\mathrm{GL}_2(\mathbb{C})$ is the subgroup $U = (\mathbb{C}^*)^2 \times \mathbb{C} \subset \mathrm{GL}_2(\mathbb{C})$ of upper triangular matrices. Hence, the corresponding $\mathrm{GL}_2(\mathbb{C})$-representation variety is

$$\mathfrak{X}_{m,n}^{(\mathrm{C})}(\mathrm{GL}_2(\mathbb{C})) = (\mathbb{C}^*)^2 \times \mathrm{GL}_2(\mathbb{C})/U.$$

In particular, $[\mathfrak{X}_{m,n}^{(\mathrm{C})}(\mathrm{GL}_2(\mathbb{C}))] = (q-1)^2(q^4 - q^3 - q^2 + q)/q(q-1)^2 = (q-1)^2(q+1)$. Moreover, if we fix the eigenvalues, we get that $[\mathfrak{X}_{m,n}^{(\mathrm{C})}(\mathrm{GL}_2(\mathbb{C}))_0] = [\mathbb{C}^* \times \mathrm{GL}_2(\mathbb{C})/U] = (q-1)(q+1)$.

To analyze the condition $\Phi_n(A_0) = \Phi_m(B_0)$, a straightforward computation reduces it to

$$\begin{pmatrix} \Phi_n(t^m) & 0 \\ x\sum_{i=1}^{n-1} it^{m(i-1)} & \Phi_n(t^m) \end{pmatrix} = \begin{pmatrix} \Phi_m(t^n) & 0 \\ y\sum_{i=1}^{m-1} it^{n(i-1)} & \Phi_m(t^n) \end{pmatrix}.$$

The off-diagonal entries can be recognized as $x\Phi_n'(t^m)$ and $y\Phi_m'(t^n)$, resp., where $\Phi_l'(x)$ denotes the formal derivative of $\Phi_l(x)$. Since Φ_l has no repeated roots, we have that $\Phi_n'(t^m)$ and $\Phi_n(t^m)$ (resp., $\Phi_m'(t^n)$ and $\Phi_m(t^n)$) cannot vanish simultaneously. Therefore, stratifying according to the kernel of Λ, we get the following two possibilities:

(1) If $t \in \Omega_{m,n}$, then $\operatorname{Ker}\Lambda = \mathbb{C}^3$. We have a fibration

$$\mathfrak{X}_{m,n}^{(\mathrm{C}),(1)}(\mathrm{GL}_2(\mathbb{C})) \longrightarrow \Omega_{m,n},$$

whose fiber is $\mathfrak{X}_{m,n}^{(\mathrm{C})}(\mathrm{GL}_2(\mathbb{C}))_0$. Hence, this stratum contributes

$$\begin{aligned}\left[\mathfrak{X}_{m,n}^{(\mathrm{C}),(1)}(\mathrm{AGL}_2(\mathbb{C}))\right] &= \left[\mathfrak{X}_{m,n}^{(\mathrm{C}),(1)}(\mathrm{GL}_2(\mathbb{C}))\right][\mathbb{C}^3] \\ &= (m-1)(n-1)q^3(q-1)(q+1).\end{aligned}$$

(2) If $t \in \mathbb{C}^* - \Omega_{m,n}$, then $\operatorname{Ker}\Lambda = \mathbb{C}^2$. The fibration we get is now

$$\mathfrak{X}_{m,n}^{(\mathrm{C}),(2)}(\mathrm{GL}_2(\mathbb{C})) \longrightarrow \mathbb{C}^* - \Omega_{m,n}.$$

Therefore, this stratum contributes

$$\begin{aligned}\left[\mathfrak{X}_{m,n}^{(\mathrm{C}),(2)}(\mathrm{AGL}_2(\mathbb{C}))\right] &= \left[\mathfrak{X}_{m,n}^{(\mathrm{C}),(2)}(\mathrm{GL}_2(\mathbb{C}))\right][\mathbb{C}^2] \\ &= \Big[\mathfrak{X}_{m,n}^{(\mathrm{C})}(\mathrm{GL}_2(\mathbb{C})) - (m-1)(n-1)\mathfrak{X}_{m,n}^{(\mathrm{C}),(2)} \\ &\quad \times (\mathrm{GL}_2(\mathbb{C}))_0\Big][\mathbb{C}^2] \\ &= \big((q-1)^2(q+1) - (m-1)(n-1)(q-1) \\ &\quad \times (q+1)\big)q^2.\end{aligned}$$

Adding up all the contributions, we get that

$$\left[\mathfrak{X}_{m,n}^{(\mathrm{C})}(\mathrm{AGL}_2(\mathbb{C}))\right] = (q-1)^2(q+1)q^2 + (m-1)(n-1)(q-1)(q+1)(q^3-q^2).$$

Putting the results of Sections 4.1, 4.2.1, 4.2.2 and 4.2.3 together, we prove the second formula in Theorem 1.

5. Character Varieties of Torus Knots

As we have said in Section 2.1, the G-representation variety of a (m,n)-torus knot $\mathfrak{X}_{m,n}(G)$ parameterizes all the representations $\rho : \pi_1(\mathbb{R}^3 - K_{m,n}) \to G$. However, this space does not take into account the fact that two representations might be isomorphic. To remove this redundancy, consider the adjoint action of G on $\mathfrak{X}_{m,n}(G)$ given by $(P \cdot \rho)(\gamma) = P\rho(\gamma)P^{-1}$ for $P \in G$, $\rho \in \mathfrak{X}_{m,n}(G)$ and $\gamma \in \pi_1(\mathbb{R}^3 - K_{m,n})$.

Ideally, we would like to take the quotient space $\mathfrak{X}_{m,n}(G)/G$ as the moduli space of isomorphism classes of representations. However, typically this orbit space is not an algebraic variety, and we need to consider instead the Geometric Invariant Theory (GIT) quotient [15]

$$\mathcal{R}_{m,n}(G) = \mathfrak{X}_{m,n}(G) /\!\!/ G,$$

usually known as the *character variety*. Roughly speaking, the character variety is obtained by collapsing those orbits of isomorphism classes of representations of the representation variety whose Zariski closures intersect. This collapsing can be justified intuitively since those orbits are indistinguishable from the point of their structure sheaf.

In the case that G is affine (so that $\mathfrak{X}_{m,n}(G)$ is also an affine variety), there is a very simple description of the GIT quotient. Let $\mathcal{O}(\mathfrak{X}_{m,n}(G))$ be the ring of regular functions on $\mathfrak{X}_{m,n}(G)$ (the global sections of its structure sheaf). The action of G on $\mathfrak{X}_{m,n}(G)$ induces an action on $\mathcal{O}(\mathfrak{X}_{m,n}(G))$. Set $\mathcal{O}(\mathfrak{X}_{m,n}(G))^G$ for the collection of G-invariant functions. By Nagata's theorem [16], if G is a reductive group, then this is a finitely generated algebra so we can take as the GIT quotient the algebraic variety

$$\mathcal{R}_{m,n}(G) = \mathfrak{X}_{m,n}(G) /\!\!/ G = \mathrm{Spec}\left(\mathcal{O}(\mathfrak{X}_{m,n}(G))^G\right).$$

This is the construction of character varieties that is customarily developed in the literature for the classical groups $G = \mathrm{GL}_r(\mathbb{C}), \mathrm{SL}_r(\mathbb{C})$. However, the affine case $G = \mathrm{AGL}_r(\mathbb{C})$ is problematic since $\mathrm{AGL}_r(\mathbb{C})$ is not a reductive group. Roughly speaking, the underlying reason is that we have for $\mathrm{AGL}_r(\mathbb{C}) = \mathbb{C}^r \rtimes \mathrm{GL}_r(\mathbb{C})$ a description as a semi-direct product and the factor $\mathbb{C}^r$ is the canonical example of a non-reductive group.

For this reason, it is not guaranteed by Nagata's theorem that $\mathcal{O}(\mathfrak{X}_{m,n}(\mathrm{AGL}_r(\mathbb{C})))^{\mathrm{AGL}_r(\mathbb{C})}$ is a finitely generated algebra so the GIT quotient may not be defined as an algebraic variety. However, in this situation we have the following result.

Proposition 2. *For any $r \geq 1$, we have that*

$$\mathcal{O}(\mathfrak{X}_{m,n}(\mathrm{AGL}_r(\mathbb{C})))^{\mathrm{AGL}_r(\mathbb{C})} = \mathcal{O}(\mathfrak{X}_{m,n}(\mathrm{GL}_r(\mathbb{C})))^{\mathrm{GL}_r(\mathbb{C})}.$$

Proof. We shall explode the natural description of $\mathfrak{X}_{m,n}(\mathrm{GL}_r(\mathbb{C}))$ as a subvariety of the whole representation variety $\mathfrak{X}_{m,n}(\mathrm{AGL}_r(\mathbb{C}))$. By restriction, there is a natural homomorphism

$$\varphi : \mathcal{O}(\mathfrak{X}_{m,n}(\mathrm{AGL}_r(\mathbb{C})))^{\mathrm{AGL}_r(\mathbb{C})} \longrightarrow \mathcal{O}(\mathfrak{X}_{m,n}(\mathrm{GL}_r(\mathbb{C})))^{\mathrm{GL}_r(\mathbb{C})}.$$

Note that the action of $\mathrm{AGL}_r(\mathbb{C})$ on the subvariety $\mathfrak{X}_{m,n}(\mathrm{GL}_r(\mathbb{C}))$ agrees with the $\mathrm{GL}_r(\mathbb{C})$-action. Hence, given an invariant function $f \in \mathcal{O}(\mathfrak{X}_{m,n}(\mathrm{GL}_r(\mathbb{C})))^{\mathrm{GL}_r(\mathbb{C})}$, we can consider the lifting $\tilde{f} \in \mathcal{O}(\mathfrak{X}_{m,n}(\mathrm{AGL}_r(\mathbb{C})))^{\mathrm{AGL}_r(\mathbb{C})}$ given by $\tilde{f}(A,B) = f(A_0,B_0)$ where (A_0,B_0) is the vectorial part of the representation $(A,B) \in \mathfrak{X}_{m,n}(\mathrm{AGL}_r(\mathbb{C}))$. The map $f \mapsto \tilde{f}$ gives a right inverse to φ.

To show that this morphism is also a left inverse, let $(A, B) \in \mathfrak{X}_{m,n}(\mathrm{AGL}_r(\mathbb{C}))$, say

$$(A, B) = \left(\begin{pmatrix} 1 & 0 \\ \alpha & A_0 \end{pmatrix}, \begin{pmatrix} 1 & 0 \\ \beta & B_0 \end{pmatrix} \right),$$

with $A_0, B_0 \in \mathrm{GL}_r(\mathbb{C})$ and $\alpha, \beta \in \mathbb{C}^r$. Consider the homothety

$$P = \begin{pmatrix} 1 & 0 \\ 0 & \lambda \,\mathrm{Id} \end{pmatrix} \in \mathrm{AGL}_r(\mathbb{C}).$$

Then, we have that

$$P \cdot (A, B) = \left(\begin{pmatrix} 1 & 0 \\ \lambda\alpha & A_0 \end{pmatrix}, \begin{pmatrix} 1 & 0 \\ \lambda\beta & B_0 \end{pmatrix} \right).$$

By letting $\lambda \to 0$, this implies that the Zariski closure of the orbit contains the representation

$$\left(\begin{pmatrix} 1 & 0 \\ 0 & A_0 \end{pmatrix}, \begin{pmatrix} 1 & 0 \\ 0 & B_0 \end{pmatrix} \right) \in \mathfrak{X}_{m,n}(\mathrm{GL}_r(\mathbb{C})).$$

Now, observe that any $\mathrm{AGL}_r(\mathbb{C})$-invariant function $f : \mathfrak{X}_{m,n}(\mathrm{AGL}_r(\mathbb{C})) \to \mathbb{C}$ must take the same value on the closure of an orbit, so for any $(A, B) \in \mathfrak{X}_{m,n}(\mathrm{AGL}_r(\mathbb{C}))$ we have that $f(A, B) = f(A_0, B_0)$. In particular, this shows that $f \mapsto \tilde{f}$ is also a left inverse of φ, so φ is an isomorphism. □

Remark 2. In fact, there is nothing special in considering torus knots in the previous proof. Exactly the same argument actually proves that we have $\mathcal{O}(\mathfrak{X}_\Gamma(\mathrm{AGL}_r(\mathbb{C})))^{\mathrm{AGL}_r(\mathbb{C})} = \mathcal{O}(\mathfrak{X}_\Gamma(\mathrm{AGL}_r(\mathbb{C})))^{\mathrm{AGL}_r(\mathbb{C})}$ for the representation variety of representations $\rho : \Gamma \to \mathrm{AGL}_r(\mathbb{C})$ for any finitely presented group Γ.

In particular, the previous proof shows that $\mathcal{O}(\mathfrak{X}_{m,n}(\mathrm{AGL}_r(\mathbb{C})))^{\mathrm{AGL}_r(\mathbb{C})}$ is a finitely generated algebra, so we can harmlessly define the $\mathrm{AGL}_r(\mathbb{C})$-character variety and it satisfies

$$\mathcal{R}_{m,n}(\mathrm{AGL}_r(\mathbb{C})) = \mathcal{R}_{m,n}(\mathrm{GL}_r(\mathbb{C})).$$

The motive of the $\mathrm{GL}_r(\mathbb{C})$-character variety has been previously computed in the literature for low rank r, for instance in Ref. [9] for $r = 2$ (cf. Proposition 1) and in Ref. [10] for $r = 3$.

Acknowledgments

The first author was partially supported by MICINN (Spain) grant PID2019-106493RB-I00. The third author was partially supported by MINECO (Spain) grant PGC2018-095448-B-I00.

References

[1] M. Culler and P.B. Shalen, Varieties of group representations and splitting of 3-manifolds, *Ann. of Math. (2)* **117**, 109–146, (1983).

[2] T. Hausel and M. Thaddeus, Mirror symmetry, Langlands duality, and the Hitchin system, *Invent. Math.* **153**, 197–229, (2003).

[3] N.J. Hitchin, The self-duality equations on a Riemann surface, *Proc. London Math. Soc. (3)* **55**, 59–126, (1987).

[4] M. Logares and V. Muñoz, P.E. Newstead, Hodge polynomials of $\mathrm{SL}(2,\mathbb{C})$-character varieties for curves of small genus, *Rev. Mat. Complut.* **26**, 635–703, (2013).

[5] J. Martínez and V. Muñoz, E-polynomials of the $\mathrm{SL}(2,\mathbb{C})$-character varieties of surface groups, *Int. Math. Research Notices* **2016**, 926–961, (2016).

[6] J. Martínez and V. Muñoz, E-polynomial of the $\mathrm{SL}(2,\mathbb{C})$-character variety of a complex curve of genus 3, *Osaka J. Math.* **53**, 645–681, (2016).

[7] S. Lawton and V. Muñoz, E-polynomial of the $\mathrm{SL}(3,\mathbb{C})$-character variety of free groups, *Pacific J. Math.* **282**, 173–202, (2016).

[8] J. Martín-Morales and A.M. Oller-Marcén, Combinatorial aspects of the character variety of a family of one-relator groups, *Topology Appl.* **156**, 2376–2389, (2009).

[9] V. Muñoz, The $\mathrm{SL}(2,\mathbb{C})$-character varieties of torus knots, *Rev. Mat. Complut.* **22**, 489–497, (2009).

[10] V. Muñoz and J. Porti, Geometry of the $\mathrm{SL}(3,\mathbb{C})$-character variety of torus knots, *Algebr. Geom. Topol.* **16**, 397–426, (2016).

[11] A. González-Prieto and V. Muñoz, Motive of the SL_4-character variety of torus knots, arxiv.org/abs/2006.01810.

[12] A. González-Prieto, M. Logares, and V. Muñoz, *Representation variety for the rank one affine group.* In: I.N. Parasidis, E. Providas and Th.M. Rassias (Eds.), *Mathematical Analysis in Interdisciplinary Research* (Springer, to appear).

[13] D. Rolfsen, *Knots and links*, Mathematics Lecture Series **7** (Publish or Perish, 1990).

[14] Á. González-Prieto, Pseudo-quotients of algebraic actions and their application to character varieties, arxiv.org/abs/1807.08540v4.

[15] P.E. Newstead, *Introduction to moduli problems and orbit spaces*, Tata Institute of Fundamental Research Lectures on Mathematics and Physics **51** (TIFR, 1978).

[16] M. Nagata, Invariants of a group in an affine ring, *J. Math. Kyoto Univ.* **3**, 369–377, (1963/1964).

https://doi.org/10.1142/9789811261572_0016

Chapter 16

Quaternionic Fractional Wavelet Transform

Bivek Gupta[*,‡], Amit K. Verma[*,§], and Carlo Cattani[†,¶]

[*]*Department of Mathematics, Indian Institute of Technology Patna, Bihta, 801103, (BR) India*

[†]*Engineering School, DEIM, University of Tuscia, 01100 Viterbo (IT)*

[‡]*bivekgupta040792@gmail.com*

[§]*akverma@iitp.ac.in*

[¶]*cattani@unitus.it*

In this chapter, we define the quaternionic fractional Fourier transform and study some of its properties like Riemann–Lebesgue lemma, Parseval's formula and convolution theorem. We also define the fractional wavelet transform of the quaternionic valued function and obtain its basic properties, the inner product relation, inversion formula and the range of the transform. We derive Heisenberg's uncertainty principle for the quaternionic fractional Fourier transform, which enables us to derive also Heisenberg's uncertainty principle for the quaternionic fractional wavelet transform.

1. Introduction

In 1843, W.R. Hamilton introduced the quaternion algebra so that a quaternion is usually denoted by $\mathbb{H}$ in his honor. Because of the non-commutativity of the quaternion multiplication, the Fourier transform of the quaternion-valued function on $\mathbb{R}^2$ can be classified into various types, *viz.*, right-sided, left-sided and double-sided Fourier transform [1–3]. Based on the definitions of quaternion Fourier transform, the Gabor transform and wavelet transform [4–7] have been extended to the space of quaternion-valued function. Analogously, the fractional Fourier transform [8–10] and linear canonical transform for the complex-valued function on $\mathbb{R}^2$ have been extended to the quaternion-valued function [11], so that the quaternionic windowed fractional Fourier transform [12] and quaternionic linear canonical wavelet transform [13] have been obtained and intensively studied.

He *et al.* [14] extended the Fourier transform of a complex-valued function in $\mathbb{R}$ to the quaternion-valued function on $\mathbb{R}$. They studied also the theory of wavelet transform for the quaternion-valued function defined on $\mathbb{R}$. Akila *et al.* [15] defined a natural convolution of quaternion-valued function and studied the associated convolution theorem. They also defined the wavelet transform and proved Parseval's identity without any additional condition. Based on the convolution, some integral transforms like Stockwell transform, shearlet transform and ridgelet transform have been extended to the quaternion-valued function defined on $\mathbb{R}$ [16–18]. Roopkumar [19] studied the quaternionic one-dimensional fractional Fourier transform together with the corresponding properties like the inversion formula, Parseval's theorem, convolution and product theorem. Using the properties of the quaternionic fractional Fourier transform, Roopkumar [20] derived the inner product relation and the inversion formula of the quaternion fractional wavelet transform. Saima *et al.* [21] extended the properties of quaternionic one-dimensional fractional Fourier transform, in Ref. [19], in the context of linear canonical transform.

Luchko *et al.* [22], introduced a new fractional Fourier transform (FrFT) in the Lizorkin space, and discussed many important results involving fractional derivatives (see also [23,24]). In Refs. [25,26], the authors studied the theory of fractional wavelet transform (FrWT), associated with the FrFT already given in Refs. [23,24]. They studied the properties like inner product relation, inversion formula, etc. and also the corresponding multiresolution analysis (MRA). In Ref. [27] Verma *et al.* complemented this theory by studying the FrWT on some more general functional spaces like Hardy space [28] and Morrey space [29,30].

In the following, we will extend the theory of FrFT and FrWT to the quaternion-valued functions. In particular, we will define the one-dimensional fractional Fourier transform for the quaternion-valued functions and discuss the corresponding properties such as Parseval's theorem, Riemann–Lebesgue lemma and the convolution theorem where the convolution in the quaternionic setting is given in Ref. [15]. We will extend the one-dimensional fractional wavelet transform [27] to the quaternion valued functions. We will give the basic properties along with the inner product relation and the inversion formula for the proposed quaternionic fractional wavelet transform (QFrWT). Heisenberg's uncertainty principle for the QFrWT [31], will also be given with the help of the quaternionic fractional Fourier transform (QFrFT). This transform can be used for the solution of many types of quaternionic differential equations.

The organization of this chapter is as follows: In Section 2, we recall some basic definitions and results. In Section 3, we give the definition of the QFrFT together with some associated results like Riemann–Lebesgue lemma and Parseval's identity. We also derive the convolution theorem of this transform. In Section 4, the QFrWT is studied together with its basis properties like linearity, translation, parity, scaling, inner product relation and the inversion formula. In Section 5, we derive the uncertainty principle for the QFrFT and we extend Heisenberg's uncertainty principle to the QFrWT. Conclusion and future perspectives are discussed in Section 6.

2. Preliminaries

2.1. *Quaternion algebra* $\mathbb{H}$

The field of real and complex numbers are, resp., denoted by $\mathbb{R}$ and $\mathbb{C}$. Let $\mathbb{H} = \{z_1 + jz_2 : z_1, z_2 \in \mathbb{C}\}$, where j is an imaginary number other than i, satisfying $j^2 = -1$ and $jz = \bar{z}j$, for all $z \in \mathbb{C}$, and $\bar{z}$ is the conjugate of z. The set $\mathbb{H}$ of quaternion form a division ring with respect to the addition and multiplication defined, resp., as follows:

$$(z_1 + jz_2) + (w_1 + jw_2) = (z_1 + w_1) + j(z_2 + w_2),$$
$$(z_1 + jz_2)(w_1 + jw_2) = (z_1 w_1 - \bar{z_2} w_2) + j(\bar{z_1} w_2 + z_2 w_1).$$

The conjugate and absolute value of the quaternion $z = z_1 + jz_2 \in \mathbb{H}$ are given, resp., by

$$z^c = \bar{z_1} - jz_2 \text{ and } |z| = \sqrt{|z_1|^2 + |z_2|^2},$$

where $|z_k|$ is the absolute value of the complex number z_k, $k = 1, 2$. We list some well-known properties of the conjugate and absolute value of the quaternions.

$$(z^c)^c = z, \quad (z+w)^c = z^c + w^c, \quad (zw)^c = w^c z^c, \quad |z|^2 = zz^c,$$
$$|zw| = |z||w|, \quad \forall z, w \in \mathbb{H}.$$

Let $L^p(\mathbb{R},\mathbb{H}) = \{f_1 + jf_2 : f_1, f_2 \in L^p(\mathbb{R},\mathbb{C})\}$, where $L^p(\mathbb{R},\mathbb{C})$ is the Banach space of all complex-valued functions f defined on $\mathbb{R}$, satisfying $\int_{\mathbb{R}} |f|^p dx < \infty$, and $p = 1, 2$. Then $L^p(\mathbb{R},\mathbb{H})$ is a Banach space with respect to the norm given by $\|f\|_{L^p(\mathbb{R},\mathbb{H})} = (\int_{\mathbb{R}} |f|^p dx)^{\frac{1}{p}}$.

Moreover, $L^2(\mathbb{R},\mathbb{H})$ is a quaternion left Hilbert space, where the inner product inducing the norm

$$\|f\|_{L^2(\mathbb{R},\mathbb{H})} = \left(\int_{\mathbb{R}} |f|^2 dx\right)^{\frac{1}{2}} = \left(\int_{\mathbb{R}} \left(|f_1(x)|^2 + |f_2(x)|^2\right) dx\right)^{\frac{1}{2}}$$

is given by

$$\langle f, g\rangle_{L^2(\mathbb{R},\mathbb{H})} = \int_{\mathbb{R}} f(x)(g(x))^c dx, \; f, g \in L^2(\mathbb{R},\mathbb{H}),$$

where integral of a quaternion-valued function $f = f_1 + jf_2$ defined on $\mathbb{R}$ is given by

$$\int_{\mathbb{R}} f(x)dx = \int_{\mathbb{R}} f_1(x)dx + j\int_{\mathbb{R}} f_2(x)dx,$$

whenever the integral exists.

2.2. *Fractional wavelet transform*

In this subsection, we recall basic theory of fractional wavelet transform for complex-valued functions [27]. For a function $f \in L^2(\mathbb{R},\mathbb{C})$, the fractional wavelet transform $W_\psi^\theta f$ of f with respect to $\psi \in L^1(\mathbb{R},\mathbb{C}) \cap L^2(\mathbb{R},\mathbb{C})$ is defined by

$$(W_\psi^\theta f)(b, a) = \int_{\mathbb{R}} f(t)\overline{\psi_{a,b,\theta}(t)}dt,$$

where

$$\psi_{a,b,\theta}(t) = \frac{1}{|a|^{\frac{1}{2\theta}}}\psi\left(\frac{t-b}{(\operatorname{sgn} a)|a|^{\frac{1}{\theta}}}\right). \tag{1}$$

Let $f, g \in L^2(\mathbb{R},\mathbb{C})$ and $\psi \in L^1(\mathbb{R},\mathbb{C}) \cap L^2(\mathbb{R},\mathbb{C})$ be such that $0 < C_{\psi,\theta} = \int_{\mathbb{R}} \frac{|(\mathfrak{F}_\theta\psi)(u)|^2}{|u|} du < \infty$, then the inner product relation for the fractional wavelet transform is given by

$$\int_{\mathbb{R}}\int_{\mathbb{R}} \left(W_\psi^\theta f\right)(b,a)\overline{\left(W_\psi^\theta g\right)(b,a)}\frac{dbda}{|a|^{\frac{1}{\theta}+1}} = C_{\psi,\theta}\langle f, g\rangle_{L^2(\mathbb{R})},$$

and the inversion formula for the fractional wavelet transform is given by

$$f(t) = \frac{1}{C_{\psi,\theta}}\int_{\mathbb{R}}\int_{\mathbb{R}} \psi_{a,b,\theta}(t)\left(W_\psi^\theta f\right)(b,a)\frac{dbda}{|a|^{\frac{1}{\theta}+1}}.$$

3. Quaternionic Fractional Fourier Transform

In this section, we give a definition of quaternionic fractional Fourier transform and study its properties like Riemann–Lebesgue lemma and Parseval's identity. We also obtain the convolution theorem associated with the QFrFT.

Definition 1. Let $f = f_1 + jf_2 \in L^2(\mathbb{R}, \mathbb{H})$ and $0 < \theta \leq 1$, then the quaternionic fractional Fourier transform (QFrFT) of f is defined by $\mathfrak{F}_\theta f = \mathfrak{F}_\theta f_1 + j\mathfrak{F}_\theta f_2$, where $\mathfrak{F}_\theta$ for $g \in L^2(\mathbb{R}, \mathbb{C})$ is defined by [25]

$$(\mathfrak{F}_\theta g)(\xi) = \int_{\mathbb{R}} g(t)e^{-i(\operatorname{sgn}\xi)|\xi|^{\frac{1}{\theta}}t}dt, \quad \xi \in \mathbb{R}. \tag{2}$$

We now prove the following:

Theorem 1. *The QFrFT $\mathfrak{F}_\theta$ is $\mathbb{H}-$linear on $L^1(\mathbb{R}, \mathbb{H})$.*

Proof. Let $f = f_1 + jf_2, g = g_1 + jg_2 \in L^1(\mathbb{R}, \mathbb{H})$. Since $\mathfrak{F}_\theta$ is linear on $L^1(\mathbb{R}, \mathbb{C})$, we have

$$\begin{aligned}\mathfrak{F}_\theta(f+g) &= \mathfrak{F}_\theta(f_1+g_1) + j\mathfrak{F}_\theta(f_2+g_2)\\ &= (\mathfrak{F}_\theta f_1 + \mathfrak{F}_\theta g_1) + j(\mathfrak{F}_\theta f_2 + \mathfrak{F}_\theta g_2)\\ &= (\mathfrak{F}_\theta f_1 + j\mathfrak{F}_\theta f_2) + (\mathfrak{F}_\theta g_1 + j\mathfrak{F}_\theta g_2)\\ &= \mathfrak{F}_\theta f + \mathfrak{F}_\theta g.\end{aligned}$$

Again, let $z = z_1 + jz_2 \in \mathbb{H}$. Since

$$\mathfrak{F}_\theta(jf) = \mathfrak{F}_\theta(jf_1 - f_2) = j\mathfrak{F}_\theta f_1 - \mathfrak{F}_\theta f_2 = j(\mathfrak{F}_\theta f_1 + j\mathfrak{F}_\theta f_2) = j\mathfrak{F}_\theta f,$$

we have

$$\mathfrak{F}_\theta(zf) = \mathfrak{F}_\theta(z_1 f) + \mathfrak{F}_\theta(jz_2 f) = z_1\mathfrak{F}_\theta f + jz_2\mathfrak{F}_\theta f = z\mathfrak{F}_\theta f.$$

This completes the proof. □

Theorem 2 (Riemann–Lebesgue lemma). *Let* $f \in L^1(\mathbb{R}, \mathbb{H})$, *then*

$$\lim_{|\xi|\to\infty} |(\mathfrak{F}_\theta f)(\xi)| = 0.$$

Proof. We first prove the Riemann–Lebesgue lemma for the function $\phi \in L^1(\mathbb{R}, \mathbb{C})$.

We have

$$e^{i(\operatorname{sgn}\xi)|\xi|^{\frac{1}{\theta}}y}(\mathfrak{F}_\theta\phi)(\xi) = \int_{\mathbb{R}} \phi(t+y)e^{-i(\operatorname{sgn}\xi)|\xi|^{\frac{1}{\theta}}t}dt. \tag{3}$$

Fix $\xi \neq 0$ and let $y_\xi = \frac{\pi}{(\operatorname{sgn}\xi)|\xi|^{\frac{1}{\theta}}}$. Then $e^{i(\operatorname{sgn}\xi)|\xi|^{\frac{1}{\theta}}y_\xi} = -1$ and hence from equation (3), it follows that

$$(\mathfrak{F}_\theta\phi)(\xi) = -\int_{\mathbb{R}} \phi(t+y_\xi)e^{-i(\operatorname{sgn}\xi)|\xi|^{\frac{1}{\theta}}t}dt. \tag{4}$$

From the Definition 1 and equation (4), we obtain

$$2(\mathfrak{F}_\theta\phi)(\xi) = \int_{\mathbb{R}} (\phi(t) - \phi(t+y_\xi))e^{-i(\operatorname{sgn}\xi)|\xi|^{\frac{1}{\theta}}}dt,$$

which implies

$$|(\mathfrak{F}_\theta\phi)(\xi)| \leq \frac{1}{2}\int_{\mathbb{R}} |\phi(t) - \phi(t+y_\xi)|dt. \tag{5}$$

Now, $|y_\xi| \to 0$ as $|\xi| \to \infty$, since $|y_\xi| = \frac{\pi}{|\xi|^{\frac{1}{\theta}}}$. Therefore, from equation (5), we have

$$\lim_{|\xi|\to\infty} |(\mathfrak{F}_\theta\phi)(\xi)| \leq \lim_{y\to 0} \|\phi(\cdot) - \phi(\cdot - y_\xi)\|_{L^1(\mathbb{R})}. \tag{6}$$

Using continuity of translation in $L^1(\mathbb{R})$, in equation (6), we have $\lim_{|\xi|\to\infty} |(\mathfrak{F}_\theta\phi)(\xi)| = 0$.

Let $f = f_1 + jf_2 \in L^1(\mathbb{R}, \mathbb{H})$, then we have $\mathfrak{F}_\theta f = \mathfrak{F}_\theta f_1 + j\mathfrak{F}_\theta f_2$. So,

$$|(\mathfrak{F}_\theta f)(\xi)| = \sqrt{|(\mathfrak{F}_\theta f_1)(\xi)|^2 + |(\mathfrak{F}_\theta f_2)(\xi)|^2}. \tag{7}$$

Since $f_k \in L^1(\mathbb{R}, \mathbb{C})$, $k = 1, 2$ and Riemann–Lebesgue lemma holds for the function on $L^1(\mathbb{R}, \mathbb{C})$,

i.e.

$$\lim_{|\xi|\to\infty} |(\mathfrak{F}_\theta f_k)(\xi)| = 0, \quad k = 1, 2. \tag{8}$$

Using equation (8) in equation (7), we get

$$\lim_{|\xi|\to\infty} |(\mathfrak{F}_\theta f)(\xi)| = 0.$$

This completes the proof. □

Theorem 3 (Parseval's formula). *Let $f, g \in L^2(\mathbb{R}, \mathbb{H})$, then*

$$\langle f, g\rangle_{L^2(\mathbb{R},\mathbb{H})} = \frac{1}{2\pi\theta}\left\langle |\cdot|^{\frac{1}{\theta}-1}(\mathfrak{F}_\theta f)(\cdot), (\mathfrak{F}_\theta g)(\cdot)\right\rangle_{L^2(\mathbb{R},\mathbb{H})}.$$

Proof. Let $f = f_1 + jf_2$ and $g = g_1 + jg_2$, then

$$\begin{aligned}
&\langle f, g\rangle_{L^2(\mathbb{R},\mathbb{H})}\\
&\quad= \int_{\mathbb{R}} f(x)(g(x))^c dx\\
&\quad= \int_{\mathbb{R}} (f_1(x) + jf_2(x))\,(g_1(x) + jg_2(x))^c\, dx\\
&\quad= \int_{\mathbb{R}} \left(f_1(x)\overline{g_1(x)} + \overline{f_2(x)}g_2(x)\right)\\
&\qquad + j\left(f_2(x)\overline{g_1(x)} - \overline{f_1(x)}g_2(x)\right) dx\\
&\quad= \frac{1}{2\pi\theta}\int_{\mathbb{R}} |\xi|^{\frac{1}{\theta}-1}\left((\mathfrak{F}_\theta f_1)(\xi)\overline{(\mathfrak{F}_\theta g_1)(\xi)} + \overline{(\mathfrak{F}_\theta f_2)(\xi)}(\mathfrak{F}_\theta g_2)(\xi)\right) d\xi\\
&\qquad + j\frac{1}{2\pi\theta}\int_{\mathbb{R}} |\xi|^{\frac{1}{\theta}-1}\Big((\mathfrak{F}_\theta f_2)(\xi)\overline{(\mathfrak{F}_\theta g_1)(\xi)}\\
&\qquad\qquad - \overline{(\mathfrak{F}_\theta f_1)(\xi)}(\mathfrak{F}_\theta g_2)(\xi)\Big)\, d\xi\\
&\quad= \frac{1}{2\pi\theta}\int_{\mathbb{R}} |\xi|^{\frac{1}{\theta}-1}\Big\{\left((\mathfrak{F}_\theta f_1)(\xi)\overline{(\mathfrak{F}_\theta g_1)(\xi)} + \overline{(\mathfrak{F}_\theta f_2)(\xi)}(\mathfrak{F}_\theta g_2)(\xi)\right)\\
&\qquad\qquad + j\left((\mathfrak{F}_\theta f_2)(\xi)\overline{(\mathfrak{F}_\theta g_1)(\xi)} - \overline{(\mathfrak{F}_\theta f_1)(\xi)}(\mathfrak{F}_\theta g_2)(\xi)\right)\Big\} d\xi
\end{aligned}$$

$$= \frac{1}{2\pi\theta} \int_{\mathbb{R}} |\xi|^{\frac{1}{\theta}-1} \left\{ (\mathfrak{F}_\theta f_1)(\xi) + j(\mathfrak{F}_\theta f_2)(\xi) \right\}$$
$$\times \left\{ (\mathfrak{F}_\theta g_1)(\xi) + j(\mathfrak{F}_\theta g_2)(\xi) \right\}^c d\xi$$
$$= \frac{1}{2\pi\theta} \int_{\mathbb{R}} |\xi|^{\frac{1}{\theta}-1} (\mathfrak{F}_\theta f)(\xi) \left((\mathfrak{F}_\theta g)(\xi) \right)^c d\xi$$
$$= \frac{1}{2\pi\theta} \left\langle |\cdot|^{\frac{1}{\theta}-1} (\mathfrak{F}_\theta f)(\cdot), (\mathfrak{F}_\theta g)(\cdot) \right\rangle_{L^2(\mathbb{R},\mathbb{H})}.$$

This completes the proof. □

Remark 1. If in Theorem 3, we replace $f_2 = g_2 = 0$, then Parseval's formula reduces to the one already obtained in Ref. [25].

The following natural definition of convolution of quaternion-valued function was defined by Akila *et al.* [15] to study the convolution theorem for the Fourier transform on quaternion-valued functions.

Definition 2. Let f and g be two quaternion-valued functions of $\mathbb{R}$, then the convolution of f and g is defined by

$$f \otimes g = (f_1 \star g_1)(x) - (\check{\overline{f_2}} \star g_2)(x) + j\left((\check{\overline{f_1}} \star g_2)(x) + (f_2 \star g_1)(x) \right),$$

provided the convolutions on the right exist, which for two complex-valued functions u and v are defined on $\mathbb{R}$ by

$$(u \star v)(x) = \int_{\mathbb{R}} u(y) v(x-y) dy.$$

Theorem 4 (Convolution theorem). *Let $f \in L^2(\mathbb{R},\mathbb{H})$ and $g \in L^1(\mathbb{R},\mathbb{H})$, then*

$$\mathfrak{F}_\theta(f \otimes g) = (\mathfrak{F}_\theta f)(\mathfrak{F}_\theta g).$$

Proof. Let $f = f_1 + jf_2$ and $g = g_1 + jg_2$, then

$$(\mathfrak{F}_\theta(f \otimes g))(\xi)$$
$$= (\mathfrak{F}_\theta(f_1 \star g_1))(\xi) - \left(\mathfrak{F}_\theta(\check{\overline{f_2}} \star g_2) \right)(\xi)$$
$$+ j\left(\left(\mathfrak{F}_\theta(\check{\overline{f_1}} \star g_2) \right)(\xi) + (\mathfrak{F}_\theta(f_2 \star g_1))(\xi) \right)$$

$$= (\mathfrak{F}_\theta f_1)(\xi)(\mathfrak{F}_\theta g_1)(\xi) - \left(\mathfrak{F}_\theta \check{\overline{f_2}}\right)(\xi)(\mathfrak{F}_\theta g_2)(\xi)$$

$$+ \left(\left(\mathfrak{F}_\theta \check{\overline{f_1}}\right)(\xi)(\mathfrak{F}_\theta g_2)(\xi) + (\mathfrak{F}_\theta f_2)(\xi)(\mathfrak{F}_\theta g_1)(\xi)\right)$$

$$= (\mathfrak{F}_\theta f_1)(\xi)(\mathfrak{F}_\theta g_1)(\xi) - \overline{(\mathfrak{F}_\theta f_2)(\xi)}(\mathfrak{F}_\theta g_2)(\xi)$$

$$+ j\left(\overline{(\mathfrak{F}_\theta f_1)(\xi)}(\mathfrak{F}_\theta g_2)(\xi) + (\mathfrak{F}_\theta f_2)(\xi)(\mathfrak{F}_\theta g_1)(\xi)\right)$$

$$= ((\mathfrak{F}_\theta f_1)(\xi) + j(\mathfrak{F}_\theta f_2)(\xi))\,((\mathfrak{F}_\theta g_1)(\xi) + j(\mathfrak{F}_\theta g_2)(\xi))$$

$$= (\mathfrak{F}_\theta f)(\xi)(\mathfrak{F}_\theta g)(\xi).$$

This proves the theorem. □

Definition 3. For a function $f = f_1 + jf_2 \in L^2(\mathbb{R}, \mathbb{H})$, we define the unary operator U by

$$U(f) = \check{\overline{f_1}} - jf_2.$$

Lemma 1. *If* $f \in L^2(\mathbb{R}, \mathbb{H})$, *then* $\mathfrak{F}_\theta\left(U(f)\right) = (\mathfrak{F}_\theta f)^c$.

Proof. Let $f = f_1 + jf_2 \in L^2(\mathbb{R}, \mathbb{H})$, then

$$\begin{aligned}\mathfrak{F}_\theta\left(U(f)\right) &= \mathfrak{F}_\theta\left(\check{\overline{f_1}} - jf_2\right)\\ &= \mathfrak{F}_\theta(\check{\overline{f_1}}) - j\mathfrak{F}_\theta(f_2)\\ &= \overline{\mathfrak{F}_\theta(f_1)} - j\mathfrak{F}_\theta(f_2)\\ &= (\mathfrak{F}_\theta f)^c.\end{aligned}$$

This proves the lemma. □

4. Quaternionic Fractional Wavelet Transform

Definition 4. Let $\psi = \psi_1 + j\psi_2 \in L^1(\mathbb{R}, \mathbb{H}) \cap L^2(\mathbb{R}, \mathbb{H})$, then ψ is called a wavelet if

$$C_{\psi,\theta} := \int_{\mathbb{R}} \frac{|(\mathfrak{F}_\theta \psi)(u)|^2}{|u|} du < \infty.$$

Now, we give the definition of the quaternionic fractional wavelet transform.

Definition 5. Let ψ be a wavelet and $f \in L^2(\mathbb{R}, \mathbb{H})$, then the quaternionic fractional wavelet transform (QFrWT) is defined by

$$\left(W_\psi^\theta f\right)(b, a) = \left(f \otimes \left(U(\psi_{a,\theta})\right)\right)(b), \quad a, b \in \mathbb{R},$$

where $\psi_{a,\theta}(t) = \frac{1}{|a|^{\frac{1}{2\theta}}}\psi\left(\frac{t}{(\operatorname{sgn} a)|a|^{\frac{1}{\theta}}}\right)$, $t \in \mathbb{R}$, and $\otimes$ and U denote the convolution and the unary operator given in Definitions 2 and 3, resp.

Before we go to the main result of this section, we give the following lemma that will help us to study the basic properties of the transform like linearity, translation, parity and scaling.

Lemma 2. *Let ψ be a wavelet and $f \in L^2(\mathbb{R}, \mathbb{H})$, then*

$$W^\theta_\psi f = W^\theta_{\psi_1} f_1 + W^\theta_{\check{\psi_2}} \check{\overline{f_2}} + j\left(W^\theta_{\psi_1} f_2 - W^\theta_{\check{\psi_2}} \check{\overline{f_1}}\right),$$

where $f = f_1 + jf_2$ and $\psi = \psi_1 + j\psi_2$.

Proof. We have, from the definition of the wavelet transform,

$$\begin{aligned}
(W^\theta_\psi f)(b,a) &= \left(f \otimes (U(\psi_{a,\theta}))\right)(b) \\
&= \left((f_1 + jf_2) \otimes \left(\overline{(\check{\psi_1})_{a,\theta}} - j(\psi_2)_{a,\theta}\right)\right)(b) \\
&= \left(f_1 \star \overline{(\check{\psi_1})_{a,\theta}}\right)(b) + \left(\check{\overline{f_1}} \star (\psi_2)_{a,\theta}\right)(b) \\
&\quad + j\left(\left(f_2 \star \overline{(\check{\psi_1})_{a,\theta}}\right)(b) - \left(\check{\overline{f_1}} \star (\psi_2)_{a,\theta}\right)(b)\right) \\
&= \int_{\mathbb{R}} f_1(t) \frac{1}{|a|^{\frac{1}{2\theta}}} \overline{\psi_1\left(\frac{t-b}{(\operatorname{sgn} a)|a|^{\frac{1}{\theta}}}\right)} dt \\
&\quad + \int_{\mathbb{R}} \check{\overline{f_2(t)}} \frac{1}{|a|^{\frac{1}{2\theta}}} \overline{\check{\psi_2}\left(\frac{t-b}{(\operatorname{sgn} a)|a|^{\frac{1}{\theta}}}\right)} \\
&\quad + j\left\{\int_{\mathbb{R}} f_2(t) \frac{1}{|a|^{\frac{1}{2\theta}}} \overline{\psi_1\left(\frac{t-b}{(\operatorname{sgn} a)|a|^{\frac{1}{\theta}}}\right)} dt\right. \\
&\qquad \left. - \int_{\mathbb{R}} \check{\overline{f_1(t)}} \frac{1}{|a|^{\frac{1}{2\theta}}} \overline{\check{\psi_2}\left(\frac{t-b}{(\operatorname{sgn} a)|a|^{\frac{1}{\theta}}}\right)}\right\} \\
&= (W^\theta_{\psi_1} f_1)(b,a) + \left(W^\theta_{\check{\psi_2}} \check{\overline{f_2}}\right)(b,a) \\
&\quad + j\left\{\left(W^\theta_{\psi_1} f_2\right)(b,a) - \left(W^\theta_{\check{\psi_2}} \check{\overline{f_1}}\right)(b,a)\right\}.
\end{aligned}$$

Therefore,

$$W_{\psi}^{\theta}f = W_{\psi_1}^{\theta}f_1 + W_{\check{\overline{\psi_2}}}^{\theta}\check{\overline{f_2}} + j\left(W_{\psi_1}^{\theta}f_2 - W_{\check{\overline{\psi_2}}}^{\theta}\check{\overline{f_1}}\right). \qquad \square$$

Remark 2. If in Lemma 2 we replace $f_2 = \psi_2 = 0$, the Definition 5 coincides with the definition of fractional wavelet transform of complex-valued functions in Ref. [27].

The following theorem gives the basic properties of the quaternionic fractional wavelet transform.

Theorem 5. *Let ψ, ϕ be wavelets and $f, g \in L^2(\mathbb{R}, \mathbb{H})$. Also let $\lambda > 0$ and $\alpha, \beta \in \mathbb{H}$, then*

(*i*) *Linearity*:
 (a) $W_{\psi}^{\theta}(\alpha f + \beta g) = \alpha W_{\psi}^{\theta}f + \beta W_{\psi}^{\theta}g$
 (b) $W_{\psi+\phi}^{\theta}f = W_{\psi}^{\theta}f + W_{\phi}^{\theta}f$
 (c) $W_{c\psi}^{\theta}f = cW_{\psi}^{\theta}f,\ c \in \mathbb{R}$.

(*ii*) *Translation*:
 $(W_{\psi}^{\theta}(\tau_y f))(b,a) = (W_{\psi_1}^{\theta}(f_1 + jf_2))(b-y,a) - j\left(W_{\check{\overline{\psi_2}}}^{\theta}(\check{\overline{f_1}} + j\check{\overline{f_1}})\right)(b+y,a)$, *where* $\tau_y f(t) = f(t-y)$.

(*iii*) *Parity*: $(W_{\psi}^{\theta}\check{f})(b,a) = (W_{\psi}^{\theta}f)(-b,-a)$.

(*iv*) *Scaling*: $(W_{\psi}^{\theta}f_{\lambda})(b,a) = (W_{\psi}^{\theta}f)(\lambda b, \lambda^{\theta}a)$, *where* $(f_{\lambda})(t) = \sqrt{\lambda}f(\lambda t)$.

Proof. We present the proof of the linearity first.
$i(a)$ We have

$$\begin{aligned}\left(W_{\psi}^{\theta}(\alpha f + \beta g)\right)(b,a) &= \left((\alpha f + \beta g) \otimes (U(\psi_{a,\theta}))\right)(b)\\ &= \left(\alpha f \otimes (U(\psi_{a,\theta}))\right)(b) + \left(\beta g \otimes (U(\psi_{a,\theta}))\right)(b)\\ &= \alpha\left(f \otimes (U(\psi_{a,\theta}))\right)(b) + \beta\left(g \otimes (U(\psi_{a,\theta}))\right)(b).\end{aligned}$$

Therefore,

$$W_{\psi}^{\theta}(\alpha f + \beta g) = \alpha W_{\psi}^{\theta}f + \beta W_{\psi}^{\theta}g.$$

$i(b)$ We have

$$\left(W_{\psi+\phi}^{\theta}f\right)(b,a) = \left(f \otimes (U((\psi+\phi)_{a,\theta}))\right)(b).$$

Using $(\psi+\phi)_{a,\theta}(t) = \psi_{a,\theta}(t)+\phi_{a,\theta}(t)$ and $U((\psi+\phi)_{a,\theta}) = U(\psi_{a,\theta})+U(\phi_{a,\theta})$, we get

$$\begin{aligned}\left(W^{\theta}_{\psi+\phi}f\right)(b,a) &= \left(f \otimes (U(\psi_{a,\theta}) + U(\phi_{a,\theta}))\right)(b)\\ &= \left(f \otimes (U(\psi_{a,\theta})\right)(b) + \left(f \otimes (U(\phi_{a,\theta})\right)(b).\end{aligned}$$

Therefore, we have

$$W^{\theta}_{\psi+\phi}f = W^{\theta}_{\psi}f + W^{\theta}_{\phi}f.$$

$i(c)$ We have

$$\left(W^{\theta}_{c\psi}f\right)(b,a) = \left(f \otimes (U((c\psi)_{a,\theta}))\right)(b).$$

Using $(c\psi)_{a,\theta}(t) = c\psi_{a,\theta}(t)$ and $U((c\psi)_{a,\theta}) = cU(\psi_{a,\theta})$, we have

$$\begin{aligned}\left(W^{\theta}_{c\psi}f\right)(b,a) &= \left(f \otimes (U(c\psi_{a,\theta}))\right)(b)\\ &= \left(f \otimes (cU(\psi_{a,\theta}))\right)(b)\\ &= c\left(f \otimes (U(\psi_{a,\theta}))\right)(b), \text{ since } c \in \mathbb{R}.\end{aligned}$$

Therefore,

$$W^{\theta}_{c\psi}f = cW^{\theta}_{\psi}f, \ c \in \mathbb{R}.$$

(ii) From Lemma 2, we have

$$\begin{aligned}&(W^{\theta}_{\psi}(\tau_y f))(b,a)\\ &= (W^{\theta}_{\psi_1}(\tau_y f_1))(b,a) + \left(W^{\theta}_{\check{\overline{\psi_2}}}\overline{(\check{\tau_y f_2})}\right)(b,a)\\ &\quad + j\left\{\left(W^{\theta}_{\psi_1}(\tau_y f_2)\right)(b,a) - \left(W^{\theta}_{\check{\overline{\psi_2}}}\overline{(\check{\tau_y f_1})}\right)(b,a)\right\}.\end{aligned} \tag{9}$$

Now for $m = 1, 2$, we have

$$\begin{aligned}(W^{\theta}_{\psi_1}(\tau_y f_1))(b,a) &= \int_{\mathbb{R}} (\tau_y f_m)(t)\frac{1}{|a|^{\frac{1}{2\theta}}}\overline{\psi_1\left(\frac{t-b}{(\operatorname{sgn} a)|a|^{\frac{1}{\theta}}}\right)}dt\\ &= \left(W^{\theta}_{\psi_1}f_m\right)(b-y,a).\end{aligned} \tag{10}$$

Also,

$$\left(W^{\theta}_{\check{\overline{\psi_1}}}\overline{(\check{\tau_y f_m})}\right)(b,a) = \int_{\mathbb{R}} \overline{(\check{\tau_y f_m})}(t)\frac{1}{|a|^{\frac{1}{\theta}}}\overline{\check{\overline{\psi_2}}\left(\frac{t-b}{(\operatorname{sgn} a)|a|^{\frac{1}{\theta}}}\right)}dt$$

$$= \int_{\mathbb{R}} \overline{(\tau_y f_m)}(t) \frac{1}{|a|^{\frac{1}{\theta}}} \overline{\check{\psi}_2 \left(\frac{-t-b}{(\operatorname{sgn} a)|a|^{\frac{1}{\theta}}} \right)} dt$$

$$= \int_{\mathbb{R}} \overline{\check{f}_m}(t) \frac{1}{|a|^{\frac{1}{\theta}}} \overline{\check{\psi}_2 \left(\frac{t-(b+y)}{(\operatorname{sgn} a)|a|^{\frac{1}{\theta}}} \right)} dt.$$

Thus, it follows that

$$\left(W^{\theta}_{\check{\psi}_1} \overline{(\tau_y \check{f}_m)} \right)(b,a) = \left(W^{\theta}_{\check{\psi}_1} \overline{\check{f}_m} \right)(b+y,a). \tag{11}$$

Using equations (10), (11) in equation (9), we have

$$\begin{aligned}
&(W^{\theta}_{\psi}(\tau_y f))(b,a) \\
&= (W^{\theta}_{\psi_1} f_1)(b-y,a) + \left(W^{\theta}_{\check{\psi}_2} \overline{\check{f}_2} \right)(b+y,a) \\
&\quad + j \left\{ \left(W^{\theta}_{\psi_1} f_2 \right)(b-y,a) - \left(W^{\theta}_{\check{\psi}_2} \overline{\check{f}_1} \right)(b+y,a) \right\} \\
&= (W^{\theta}_{\psi_1} f_1)(b-y,a) + j \left(W^{\theta}_{\psi_1} f_2 \right)(b-y,a) \\
&\quad - j \left\{ \left(W^{\theta}_{\check{\psi}_2} \overline{\check{f}_1} \right)(b+y,a) + j \left(W^{\theta}_{\check{\psi}_2} \overline{\check{f}_2} \right)(b+y,a) \right\}.
\end{aligned}$$

Thus, we have

$$(W^{\theta}_{\psi}(\tau_y f))(b,a) = (W^{\theta}_{\psi_1}(f_1 + j f_2))(b-y,a) - j \left(W^{\theta}_{\check{\psi}_2} (\overline{\check{f}_1} + j \overline{\check{f}_1}) \right)(b+y,a).$$

(iii) Using Lemma 2, we get

$$\begin{aligned}
(W^{\theta}_{\psi} \check{f})(b,a) &= (W^{\theta}_{\psi_1} \check{f}_1)(b,a) + \left(W^{\theta}_{\check{\psi}_2} \overline{\check{\check{f}}_2} \right)(b,a) \\
&\quad + j \left\{ \left(W^{\theta}_{\psi_1} \check{f}_2 \right)(b,a) - \left(W^{\theta}_{\check{\psi}_2} \overline{\check{\check{f}}_1} \right)(b,a) \right\}.
\end{aligned} \tag{12}$$

For $m = 1, 2$, it can be shown that

$$(W^{\theta}_{\psi_1} \check{f}_m)(b,a) = (W^{\theta}_{\psi_1} f_m)(-b,-a) \tag{13}$$

and

$$\left(W^{\theta}_{\check{\psi}_2}\check{\overline{f_m}}\right)(b,a) = \left(W^{\theta}_{\check{\psi}_2}\check{\overline{f_m}}\right)(-b,-a). \tag{14}$$

Using (13), (14) in (12), we get

$$\begin{aligned}(W^{\theta}_{\psi}\check{f})(b,a) &= (W^{\theta}_{\psi_1}f_1)(-b,-a) + \left(W^{\theta}_{\check{\psi}_2}\check{\overline{f_2}}\right)(-b,-a)\\ &\quad + j\left\{\left(W^{\theta}_{\psi_1}f_2\right)(-b,-a) - \left(W^{\theta}_{\check{\psi}_2}\check{\overline{f_1}}\right)(-b,-a)\right\},\end{aligned}$$

i.e.

$$(W^{\theta}_{\psi}\check{f})(b,a) = (W^{\theta}_{\psi}f)(-b,-a).$$

(iv) Using Lemma 2, we get

$$\begin{aligned}(W^{\theta}_{\psi}f_{\lambda})(b,a) &= (W^{\theta}_{\psi_1}(f_1)_{\lambda})(b,a) + \left(W^{\theta}_{\check{\psi}_2}\check{\overline{(f_2)_{\lambda}}}\right)(b,a)\\ &\quad + j\left\{\left(W^{\theta}_{\psi_1}(f_2)_{\lambda}\right)(b,a) - \left(W^{\theta}_{\check{\psi}_2}\check{\overline{(f_1)_{\lambda}}}\right)(b,a)\right\}.\end{aligned} \tag{15}$$

For $m = 1, 2$, it can be shown that

$$(W^{\theta}_{\psi_1}(f_m)_{\lambda})(b,a) = (W^{\theta}_{\psi_1}f_m)(\lambda b, \lambda^{\theta}a) \tag{16}$$

and

$$\left(W^{\theta}_{\check{\psi}_2}\check{\overline{(f_m)_{\lambda}}}\right)(b,a) = \left(W^{\theta}_{\check{\psi}_2}\check{\overline{f_m}}\right)(\lambda b, \lambda^{\theta}a). \tag{17}$$

Using (16), (17) in (15), we get

$$\begin{aligned}(W^{\theta}_{\psi}f_{\lambda})(b,a) &= (W^{\theta}_{\psi_1}f_1)(\lambda b, \lambda^{\theta}a) + \left(W^{\theta}_{\check{\psi}_2}\check{\overline{f_2}}\right)(\lambda b, \lambda^{\theta}a)\\ &\quad + j\left\{\left(W^{\theta}_{\psi_1}f_2\right)(\lambda b, \lambda^{\theta}a) - \left(W^{\theta}_{\check{\psi}_2}\check{\overline{f_1}}\right)(\lambda b, \lambda^{\theta}a)\right\},\end{aligned}$$

i.e.

$$(W^{\theta}_{\psi}f_{\lambda})(b,a) = (W^{\theta}_{\psi}f)(\lambda b, \lambda^{\theta}a).$$

This completes the proof of Theorem 5. □

Theorem 6 (Inner product relation). *If $f, g \in L^2(\mathbb{R}, \mathbb{H})$, then*

$$\left\langle W_\psi^\theta f, W_\psi^\theta g \right\rangle_{L^2\left(\mathbb{R}\times\mathbb{R},\mathbb{H};\frac{dbda}{|a|^{\frac{1}{\theta}+1}}\right)} = C_{\psi,\theta}\langle f, g\rangle_{L^2(\mathbb{R},\mathbb{H})}.$$

Proof. We have

$$\begin{aligned}
&\langle W_\psi^\theta f, W_\psi^\theta g\rangle_{L^2\left(\mathbb{R}\times\mathbb{R},\mathbb{H};\frac{dbda}{|a|^{\frac{1}{\theta}+1}}\right)} \\
&= \int_{\mathbb{R}}\int_{\mathbb{R}} \left(W_\psi^\theta f\right)(b,a)\left(\left(W_\psi^\theta g\right)(b,a)\right)^c \frac{dbda}{|a|^{\frac{1}{\theta}+1}} \\
&= \int_{\mathbb{R}}\int_{\mathbb{R}} \left(f \otimes (U(\psi_{a,\theta}))\right)(b)\left(\left(g \otimes (U(\psi_{a,\theta}))\right)(b)\right)^c \frac{dbda}{|a|^{\frac{1}{\theta}+1}} \\
&= \frac{1}{2\pi\theta}\int_{\mathbb{R}}\int_{\mathbb{R}} |\xi|^{\frac{1}{\theta}-1}\left(\mathfrak{F}_\theta\left(f \otimes (U(\psi_{a,\theta}))\right)\right)(\xi) \\
&\quad \times \left(\left(\mathfrak{F}_\theta\left(g \otimes (U(\psi_{a,\theta}))\right)\right)(\xi)\right)^c d\xi \frac{da}{|a|^{\frac{1}{\theta}+1}} \\
&= \frac{1}{2\pi\theta}\int_{\mathbb{R}}\int_{\mathbb{R}} |\xi|^{\frac{1}{\theta}-1}(\mathfrak{F}_\theta f)(\xi)\left(\mathfrak{F}_\theta(U(\psi_{a,\theta}))(\xi)\right) \\
&\quad \times \left((\mathfrak{F}_\theta g)(\xi)\left(\mathfrak{F}_\theta(U(\psi_{a,\theta}))(\xi)\right)\right)^c d\xi \frac{da}{|a|^{\frac{1}{\theta}+1}} \\
&= \frac{1}{2\pi\theta}\int_{\mathbb{R}}\int_{\mathbb{R}} |\xi|^{\frac{1}{\theta}-1}(\mathfrak{F}_\theta f)(\xi)\left(\mathfrak{F}_\theta(U(\psi_{a,\theta}))(\xi)\right) \\
&\quad \times \left(\left(\mathfrak{F}_\theta(U(\psi_{a,\theta}))(\xi)\right)\right)^c \left((\mathfrak{F}_\theta g)(\xi)\right)^c d\xi \frac{da}{|a|^{\frac{1}{\theta}+1}} \\
&= \frac{1}{2\pi\theta}\int_{\mathbb{R}}\int_{\mathbb{R}} |\xi|^{\frac{1}{\theta}-1}(\mathfrak{F}_\theta f)(\xi)\left(\mathfrak{F}_\theta(\psi_{a,\theta})(\xi)\right)^c \\
&\quad \times \left(\left(\mathfrak{F}_\theta(\psi_{a,\theta})(\xi)\right)^c\right)^c \left((\mathfrak{F}_\theta g)(\xi)\right)^c d\xi \frac{da}{|a|^{\frac{1}{\theta}+1}} \\
&= \frac{1}{2\pi\theta}\int_{\mathbb{R}} |\xi|^{\frac{1}{\theta}-1}(\mathfrak{F}_\theta f)(\xi)\left((\mathfrak{F}_\theta g)(\xi)\right)^c \\
&\quad \times \left\{\int_{\mathbb{R}} |\left(\mathfrak{F}_\theta(\psi_{a,\theta})\right)(\xi)|^2 \frac{da}{|a|^{\frac{1}{\theta}+1}}\right\} d\xi. \qquad (18)
\end{aligned}$$

Since

$$\begin{aligned}\int_{\mathbb{R}} |\left(\mathfrak{F}_\theta(\psi_{a,\theta})\right)(\xi)|^2 \frac{da}{|a|^{\frac{1}{\theta}+1}} &= \int_{\mathbb{R}} \Big(|\left(\mathfrak{F}_\theta((\psi_1)_{a,\theta})\right)(\xi)|^2 \\ &\qquad + |\left(\mathfrak{F}_\theta((\psi_2)_{a,\theta})\right)(\xi)|^2 \Big) \frac{da}{|a|^{\frac{1}{\theta}+1}} \\ &= \int_{\mathbb{R}} \left(\mathfrak{F}_\theta(\psi_1)\right)(a\xi) \frac{da}{|a|} + \int_{\mathbb{R}} \left(\mathfrak{F}_\theta(\psi_2)\right)(a\xi) \frac{da}{|a|} \\ &= \int_{\mathbb{R}} \left|\left(\mathfrak{F}_\theta(\psi_1)\right)(u)\right|^2 \frac{du}{|u|} + \int_{\mathbb{R}} \left|\left(\mathfrak{F}_\theta(\psi_2)\right)(u)\right|^2 \frac{du}{|u|} \\ &= \int_{\mathbb{R}} \left|\left(\mathfrak{F}_\theta\psi\right)(u)\right|^2 \frac{du}{|u|} \\ &= C_{\psi,\theta}. \end{aligned} \tag{19}$$

Therefore, from equations (18) and (19), we get

$$\begin{aligned}\left\langle W_\psi^\theta f, W_\psi^\theta g\right\rangle_{L^2\left(\mathbb{R}\times\mathbb{R},\mathbb{H};\frac{dbda}{|a|^{\frac{1}{\theta}+1}}\right)} &= \frac{1}{2\pi\theta} C_{\psi,\theta} \int_{\mathbb{R}} |\xi|^{\frac{1}{\theta}-1} (\mathfrak{F}_\theta f)(\xi) \left((\mathfrak{F}_\theta g)(\xi)\right)^c d\xi \\ &= C_{\psi,\theta} \langle f, g\rangle_{L^2(\mathbb{R},\mathbb{H})}.\end{aligned}$$

This completes the proof. □

Theorem 7 (Inversion formula). *Let $f \in L^2(\mathbb{R},\mathbb{H})$, then*

$$f(t) = \frac{1}{C_{\psi,\theta}} \int_{\mathbb{R}} \left(\left((W_\psi^\theta f)(\cdot, a) \right) \otimes (\psi_{a,\theta}) \right)(t) \frac{da}{|a|^{\frac{1}{\theta}+1}} \quad in \quad L^2(\mathbb{R},\mathbb{H}).$$

Proof. We have

$$\begin{aligned}&\langle f, g\rangle_{L^2(\mathbb{R},\mathbb{H})} \\ &= \frac{1}{C_{\psi,\theta}} \left\langle W_\psi^\theta f, W_\psi^\theta \right\rangle_{L^2\left(\mathbb{R}\times\mathbb{R},\mathbb{H};\frac{dbda}{|a|^{\frac{1}{\theta}+1}}\right)} \\ &= \frac{1}{2\pi\theta C_{\psi,\theta}} \int_{\mathbb{R}} \int_{\mathbb{R}} |\xi|^{\frac{1}{\theta}-1} \left(\mathfrak{F}_\theta \left(f \otimes (U(\psi_{a,\theta})) \right) \right)(\xi) \\ &\quad \times \left(\left(\mathfrak{F}_\theta \left(g \otimes (U(\psi_{a,\theta})) \right) \right)(\xi) \right)^c d\xi \frac{da}{|a|^{\frac{1}{\theta}+1}}.\end{aligned} \tag{20}$$

Since

$$\begin{aligned}
&\Big(\mathfrak{F}_\theta(f\otimes(U(\psi_{a,\theta})))\Big)(\xi)\left(\Big(\mathfrak{F}_\theta\left(g\otimes(U(\psi_{a,\theta}))\right)\Big)(\xi)\right)^c\\
&=\Big(\mathfrak{F}_\theta\left(f\otimes(U(\psi_{a,\theta}))\right)\Big)(\xi)\left(\left(\mathfrak{F}_\theta(U(\psi_{a,\theta}))(\xi)\right)\right)^c\left((\mathfrak{F}_\theta g)(\xi)\right)^c\\
&=\Big(\mathfrak{F}_\theta\left(f\otimes(U(\psi_{a,\theta}))\right)\Big)(\xi)\left(\left(\mathfrak{F}_\theta(\psi_{a,\theta})(\xi)\right)^c\right)^c\left((\mathfrak{F}_\theta g)(\xi)\right)^c\\
&=\Big(\mathfrak{F}_\theta\left(f\otimes(U(\psi_{a,\theta}))\right)\Big)(\xi)\left(\mathfrak{F}_\theta(\psi_{a,\theta})(\xi)\right)\left((\mathfrak{F}_\theta g)(\xi)\right)^c\\
&=\Big(\mathfrak{F}_\theta\left(f\otimes(U(\psi_{a,\theta}))\right)\Big)(\xi)\left(\mathfrak{F}_\theta(\psi_{a,\theta})(\xi)\right)\left((\mathfrak{F}_\theta g)(\xi)\right)^c\\
&=\Big(\mathfrak{F}_\theta\Big(\left(f\otimes(U(\psi_{a,\theta}))\otimes(\psi_{a,\theta}\right)\Big)\Big)(\xi)\left((\mathfrak{F}_\theta g)(\xi)\right)^c. \qquad (21)
\end{aligned}$$

Using equation (21) in equation (20), we obtain

$$\begin{aligned}
\langle f,g\rangle_{L^2(\mathbb{R},\mathbb{H})}&=\frac{1}{2\pi\theta C_{\psi,\theta}}\int_{\mathbb{R}}\int_{\mathbb{R}}|\xi|^{\frac{1}{\theta}-1}\Big(\mathfrak{F}_\theta\Big(\left(f\otimes(U(\psi_{a,\theta}))\otimes(\psi_{a,\theta}\right)\Big)\Big)(\xi)\\
&\quad\times\left((\mathfrak{F}_\theta g)(\xi)\right)^c d\xi\frac{da}{|a|^{\frac{1}{\theta}+1}}\\
&=\frac{1}{C_{\psi,\theta}}\int_{\mathbb{R}}\int_{\mathbb{R}}\Big(\left(f\otimes(U(\psi_{a,\theta}))\otimes(\psi_{a,\theta}\right)\Big)(b)(g(b))^c db\frac{da}{|a|^{\frac{1}{\theta}+1}}\\
&=\frac{1}{C_{\psi,\theta}}\int_{\mathbb{R}}\int_{\mathbb{R}}\Big(\Big((W_\psi^\theta f)(\cdot,a)\Big)\otimes(\psi_{a,\theta})\Big)(b)(g(b))^c db\frac{da}{|a|^{\frac{1}{\theta}+1}}\\
&=\left\langle\frac{1}{C_{\psi,\theta}}\int_{\mathbb{R}}\Big(\Big((W_\psi^\theta f)(\cdot,a)\Big)\otimes(\psi_{a,\theta})\Big)(b)\frac{da}{|a|^{\frac{1}{\theta}+1}},g(b)\right\rangle_{L^2(\mathbb{R},\mathbb{H})}. \qquad (22)
\end{aligned}$$

From equation (22), it follows that

$$f(t)=\frac{1}{C_{\psi,\theta}}\int_{\mathbb{R}}\Big(\Big((W_\psi^\theta f)(\cdot,a)\Big)\otimes(\psi_{a,\theta})\Big)(t)\frac{da}{|a|^{\frac{1}{\theta}+1}}\quad\text{in } L^2(\mathbb{R},\mathbb{H}).$$

This completes the proof of the theorem. □

Now, we state a lemma, which will be used in characterizing the range of the CFrWT.

Lemma 3. *Let $f \in L^2(\mathbb{R}, \mathbb{H})$, and $\psi = \psi_1 + j\psi_2$ be a wavelet, then*

$$f(t) = \frac{1}{C_{\psi,\theta}} \int_{\mathbb{R}} \int_{\mathbb{R}} \Big[(W^{\theta}_{\psi_1} f)(b,a)(\psi_1)_{a,b,\theta}(t) + (W^{\theta}_{\psi_2} f)(b,a)(\psi_2)_{a,b,\theta}(t) \Big] \times \frac{dbda}{|a|^{\frac{1}{\theta}}+1},$$

where $(\psi_k)_{a,b,\theta}(t) = \frac{1}{|a|^{\frac{1}{2\theta}}} \psi_k\big(\frac{t-b}{(\operatorname{sgn} a)|a|^{\frac{1}{\theta}}}\big)$, $k = 1, 2$.

Proof. Let $f = f_1 + jf_2$, using the definition of $\otimes$, it can be shown that

$$\begin{aligned}\Big(f \otimes (U\psi_{a,\theta})\Big) \otimes (\psi_{a,\theta}) = {} & \Big(f_1 \star \overline{\check{(\psi_1)}_{a,\theta}}\Big) \star (\psi_1)_{a,\theta} \\ & + \Big(f_1 \star \overline{\check{(\psi_2)}_{a,\theta}}\Big) \star (\psi_2)_{a,\theta} \\ & + j\Big\{\Big(f_1 \star \overline{\check{(\psi_1)}_{a,\theta}}\Big) \star (\psi_1)_{a,\theta} \\ & + \Big(f_1 \star \overline{\check{(\psi_2)}_{a,\theta}}\Big) \star (\psi_2)_{a,\theta}\Big\}. \end{aligned} \tag{23}$$

Using Theorem 7 in equation (23), we get

$$\begin{aligned} f(t) = \frac{1}{C_{\psi,\theta}} \Bigg[& \int_{\mathbb{R}} \Big(\Big(f_1 \star \overline{\check{(\psi_1)}_{a,\theta}}\Big) \star (\psi_1)_{a,\theta}\Big)(t) \frac{da}{|a|^{\frac{1}{\theta}+1}} \\ & + \int_{\mathbb{R}} \Big(\Big(f_1 \star \overline{\check{(\psi_2)}_{a,\theta}}\Big) \star (\psi_2)_{a,\theta}\Big)(t) \frac{da}{|a|^{\frac{1}{\theta}+1}} \\ & + j\Big\{\int_{\mathbb{R}} \Big(\Big(f_2 \star \overline{\check{(\psi_1)}_{a,\theta}}\Big) \star (\psi_1)_{a,\theta}\Big)(t) \frac{da}{|a|^{\frac{1}{\theta}+1}} \\ & + \int_{\mathbb{R}} \Big(\Big(f_2 \star \overline{\check{(\psi_2)}_{a,\theta}}\Big) \star (\psi_2)_{a,\theta}\Big)(t) \frac{da}{|a|^{\frac{1}{\theta}+1}}\Big\}\Bigg]. \end{aligned} \tag{24}$$

Now, for $r \in \{1,2\}$ and $k \in \{1,2\}$, we have

$$\begin{aligned} & \int_{\mathbb{R}} \Big(\Big(f_r \star \overline{\check{(\psi_k)}_{a,\theta}}\Big) \star (\psi_k)_{a,\theta}\Big)(t) \frac{da}{|a|^{\frac{1}{\theta}+1}} \\ & = \int_{\mathbb{R}} \int_{\mathbb{R}} \Big(f_r \star \overline{\check{(\psi_k)}_{a,\theta}}\Big)(b)\,(\psi_k)_{a,\theta}(t-b)db \frac{da}{|a|^{\frac{1}{\theta}+1}} \\ & = \int_{\mathbb{R}} \int_{\mathbb{R}} \Big(W^{\theta}_{\psi_k} f_r\Big)(b,a)\,(\psi_k)_{a,b,\theta}(t) \frac{dbda}{|a|^{\frac{1}{\theta}+1}}. \end{aligned} \tag{25}$$

Using equation (25) in equation (24), we get

$$f(t) = \frac{1}{C_{\psi,\theta}} \int_{\mathbb{R}} \int_{\mathbb{R}} \left[\left(W^{\theta}_{\psi_k} f_r \right) (b,a) \left(\psi_k \right)_{a,b,\theta} (t) \right.$$
$$\left. + \left(W^{\theta}_{\psi_k} f_r \right) (b,a) \left(\psi_k \right)_{a,b,\theta} (t) \right] \frac{dbda}{|a|^{\frac{1}{\theta}+1}}.$$

This completes the proof. □

Now, we characterize the range of the CFrWT, where the wavelet involved in the transform is complex-valued.

Theorem 8 (Characterization of the range). *Let* $F \in L^2\big(\mathbb{R} \times \mathbb{R}, \mathbb{H}, \frac{dbda}{|a|^{\frac{1}{\theta}+1}}\big)$ *and* ψ *be a complex-valued wavelet, then* $F \in W^{\theta}_{\psi}\left(L^2(\mathbb{R}, \mathbb{H})\right)$ *if and only if*

$$F(b_0, a_0) = \int_{\mathbb{R}} \int_{\mathbb{R}} F(b,a) K_{\psi,\theta}(b_0, a_0; b, a) \frac{dbda}{|a|^{\frac{1}{\theta}+1}}, \quad (b_0, a_0) \in \mathbb{R} \times \mathbb{R}, \quad (26)$$

where $K_{\psi,\theta}(b_0, a_0; b, a) = \frac{1}{C_{\psi,\theta}} \int_{\mathbb{R}} \psi_{a,b,\theta}(t) \overline{\psi_{a_0,b_0,\theta}(t)} dt$, *where* $\psi_{a,b,\theta}(t)$ *is given by equation* (1).

Proof. Let us assume that $F \in W^{\theta}_{\psi}\big(L^2(\mathbb{R}, \mathbb{H})\big)$. Then there exists $f \in L^2(\mathbb{R}, \mathbb{H})$ such that $W^{\theta}_{\psi} f = F$. Now,

$$\begin{aligned} F(b_0, a_0) &= \left(W^{\theta}_{\psi} f \right) (b_0, a_0) \\ &= \left(f \otimes \overline{\check{\psi}_{a_0,\theta}} \right) (b_0) \\ &= \int_{\mathbb{R}} f_1(t) \overline{\check{\psi}_{a_0,\theta}}(b_0 - t) dt + j \int_{\mathbb{R}} f_2(t) \overline{\check{\psi}_{a_0,\theta}}(b_0 - t) dt \\ &= \int_{\mathbb{R}} f_1(t) \overline{\psi_{a_0,b_0,\theta}(t)} dt + j \int_{\mathbb{R}} f_2(t) \overline{\psi_{a_0,b_0,\theta}(t)} dt \\ &= \int_{\mathbb{R}} f(t) \overline{\psi_{a_0,b_0,\theta}(t)} dt. \end{aligned} \quad (27)$$

Using Lemma 3 in equation (27), we get

$$\begin{aligned} F(b_0, a_0) &= \int_{\mathbb{R}} \left\{ \left(W_\psi^\theta f\right)(b,a)\psi_{a,b,\theta}(t)\frac{dbda}{|a|^{\frac{1}{\theta}+1}} \right\} \overline{\psi_{a_0,b_0,\theta}(t)}dt \\ &= \int_{\mathbb{R}}\int_{\mathbb{R}} \left(W_\psi^\theta f\right)(b,a) \left\{ \frac{1}{C_{\psi,\theta}} \int_{\mathbb{R}} \psi_{a,b,\theta}(t)\overline{\psi_{a_0,b_0,\theta}(t)}dt \right\} \frac{dbda}{|a|^{\frac{1}{\theta}+1}} \\ &= \int_{\mathbb{R}}\int_{\mathbb{R}} F(b,a)K_{\psi,\theta}(b_0,a_0;b,a)\frac{dbda}{|a|^{\frac{1}{\theta}+1}}. \end{aligned}$$

Conversely, for the given $F \in L^2(\mathbb{R}\times\mathbb{R}, \mathbb{H}; \frac{dbda}{|a|^{\frac{1}{\theta}+1}})$, let equation (26) hold. Define

$$f(t) = \frac{1}{C_{\psi,\theta}} \int_{\mathbb{R}}\int_{\mathbb{R}} F(b,a)\psi_{a,b,\theta}(t)\frac{dbda}{|a|^{\frac{1}{\theta}+1}}. \tag{28}$$

Using equation (28) and Fubini's theorem, we get

$$\begin{aligned} \|f\|^2_{L^2(\mathbb{R},\mathbb{H})} &= \int_{\mathbb{R}} f(t)\left(f(t)\right)^c dt \\ &= \int_{\mathbb{R}} \left\{ \frac{1}{C_{\psi,\theta}} \int_{\mathbb{R}}\int_{\mathbb{R}} F(b,a)\psi_{a,b,\theta}(t)\frac{dbda}{|a|^{\frac{1}{\theta}+1}} \right\} \\ &\quad \times \left\{ \frac{1}{C_{\psi,\theta}} \int_{\mathbb{R}}\int_{\mathbb{R}} \overline{\psi_{a',b',\theta}(t)}(F(b',a'))^c \frac{db'da'}{|a'|^{\frac{1}{\theta}+1}} \right\} dt \\ &= \frac{1}{C_{\psi,\theta}} \int_{\mathbb{R}}\int_{\mathbb{R}} \left\{ \int_{\mathbb{R}}\int_{\mathbb{R}} F(b,a)K_{\psi,\theta}(b',a';b,a)\frac{dbda}{|a|^{\frac{1}{\theta}+1}} \right\} \\ &\quad \times (F(b',a'))^c \frac{db'da'}{|a'|^{\frac{1}{\theta}+1}} \\ &= \frac{1}{C_{\psi,\theta}} \int_{\mathbb{R}}\int_{\mathbb{R}} F(b',a')(F(b',a'))^c \frac{db'da'}{|a'|^{\frac{1}{\theta}+1}} \\ &= \frac{1}{C_{\psi,\theta}} \|F\|^2_{L^2\left(\mathbb{R}\times\mathbb{R},\mathbb{H},\frac{dbda}{|a|^{\frac{1}{\theta}+1}}\right)} < \infty. \end{aligned}$$

Therefore, $f \in L^2(\mathbb{R}, \mathbb{H})$. Now,

$$\begin{aligned}\left(W_\psi^\theta f\right)(b,a) &= \int_{\mathbb{R}} f(t)\overline{\psi_{a,b,\theta}(t)}dt \\ &= \int_{\mathbb{R}} \left\{ \frac{1}{C_{\psi,\theta}} \int_{\mathbb{R}} \int_{\mathbb{R}} F(b',a')\psi_{a',b',\theta}(t) \frac{db'da'}{|a'|^{\frac{1}{\theta}+1}} \right\} \\ &\quad \times \overline{\psi_{a,b,\theta}}dt, \text{ using equation (28)} \\ &= \int_{\mathbb{R}} \int_{\mathbb{R}} F(b',a')K_{\psi,\theta}(b,a;b',a') \frac{db'da'}{|a'|^{\frac{1}{\theta}+1}} \\ &= F(b,a), \text{ using equation (26).}\end{aligned}$$

This completes the proof. □

5. Uncertainty Principle for QFrWT

Heisenberg's uncertainty principle can be shortly summarized as a principle that gives the information about both the signal and its Fourier transform. According to this principle, the signal cannot be highly localized in both in time and in frequency domain. Wilczok [31] introduced a new class of uncertainty principle that compares the localization of a function with the localization of its wavelet transform, similar to Heisenberg's uncertainty principle governing the localization of the complex-valued function and the corresponding Fourier transform. This section is devoted to the Heisenberg-type uncertainty principle for the fractional Fourier transform and the fractional wavelet transform of the quaternion-valued function in $\mathcal{S}(\mathbb{R}, \mathbb{H})$, where the functions in $\mathcal{S}(\mathbb{R}, \mathbb{H})$ are of the form $f_1 + jf_2$, where f_k, $k = 1, 2$ is in the Schwartz class $\mathcal{S}(\mathbb{R}, \mathbb{C})$.

The following theorem gives the Heisenberg-type uncertainty principle for the QFrFT.

Theorem 9 (Uncertainty principle for the QFrFT). *Let $f \in \mathcal{S}(\mathbb{R}, \mathbb{H})$, then the following inequality holds:*

$$\left(\int_{\mathbb{R}} |x|^2 |f(x)|^2 dx\right) \left(\int_{\mathbb{R}} |\xi|^{\frac{3}{\theta}-1} |(\mathfrak{F}_\theta f)(\xi)|^2\right) \geq \frac{\pi\theta}{4} \|f\|^4_{L^2(\mathbb{R},\mathbb{H})}.$$

Proof. Let $\phi \in \mathcal{S}(\mathbb{R}, \mathbb{C})$, then

$$\begin{aligned}\|\phi\|^2_{L^2(\mathbb{R},\mathbb{C})} &= \int_{\mathbb{R}} |\phi(x)|^2 dx \\ &= -\int_{\mathbb{R}} x \frac{d}{dx} |\phi(x)|^2 dx.\end{aligned}$$

Now

$$\begin{aligned}\frac{d}{dx}|\phi(x)|^2 &= \frac{d}{dx}\left\{\phi(x)\overline{\phi(x)}\right\} \\ &= 2Re\left(\phi'(x)\overline{\phi(x)}\right).\end{aligned}$$

So, we have

$$\begin{aligned}\|\phi\|^2_{L^2(\mathbb{R},\mathbb{C})} &\leq \int_{\mathbb{R}} |x| \left| Re\left(\phi'(x)\overline{\phi(x)}\right)\right| dx \\ &\leq 2\int_{\mathbb{R}} |x||\phi'(x)||\phi(x)|dx, \text{ since } |Re(z)| \leq |z|,\ z \in \mathbb{C} \\ &\leq 2\left(\int_{\mathbb{R}} |x|^2|\phi(x)|^2 dx\right)^{\frac{1}{2}} \left(\int_{\mathbb{R}} |\phi'(x)|^2\right)^{\frac{1}{2}}, \\ &\quad \text{using Holder's inequality.}\end{aligned}$$

Using Parseval's formula [25], Theorem 1, we get

$$\|\phi\|^2_{L^2(\mathbb{R},\mathbb{C})} \leq \frac{2}{\sqrt{2\pi\theta}} \left(\int_{\mathbb{R}} |x|^2|\phi(x)|^2 dx\right)^{\frac{1}{2}} \left(\int_{\mathbb{R}} |\xi|^{\frac{1}{\theta}-1}|(\mathfrak{F}_\theta\phi')(\xi)|^2\right)^{\frac{1}{2}}. \quad (29)$$

Now

$$\begin{aligned}(\mathfrak{F}_\theta\phi')(\xi) &= \int_{\mathbb{R}} \phi'(x)e^{-ix(\operatorname{sgn}\xi)|\xi|^{\frac{1}{\theta}}} dx \\ &= \int_{\mathbb{R}} i(\operatorname{sgn}\xi)|\xi|^{\frac{1}{\theta}}\phi(x)e^{-ix(\operatorname{sgn}\xi)|\xi|^{\frac{1}{\theta}}} dx, \\ &\quad \text{(using integration by parts)} \\ &= i(\operatorname{sgn}\xi)|\xi|^{\frac{1}{\theta}}(\mathfrak{F}_\theta\phi)(\xi). \quad (30)\end{aligned}$$

Using equation (30) in equation (29), we get

$$\|\phi\|^2_{L^2(\mathbb{R},\mathbb{C})} \leq \sqrt{\frac{2}{\pi\theta}} \left(\int_{\mathbb{R}} |x|^2 |\phi(x)|^2 dx\right)^{\frac{1}{2}} \left(\int_{\mathbb{R}} |\xi|^{\frac{1}{\theta}-1} |\xi|^{\frac{2}{\theta}} |(\mathfrak{F}_\theta \phi)(\xi)|^2\right)^{\frac{1}{2}}.$$

Therefore, the uncertainty principle for the function $\phi \in \mathcal{S}(\mathbb{R},\mathbb{C})$ is given by

$$\left(\int_{\mathbb{R}} |x|^2 |\phi(x)|^2 dx\right) \left(\int_{\mathbb{R}} |\xi|^{\frac{3}{\theta}-1} |(\mathfrak{F}_\theta \phi)(\xi)|^2\right) \geq \frac{\pi\theta}{2} \|\phi\|^4_{L^2(\mathbb{R},\mathbb{C})}. \tag{31}$$

This inequality holds for the function in $\phi \in \mathcal{S}(\mathbb{R},\mathbb{C})$.

Let $f \in \mathcal{S}(\mathbb{R},\mathbb{H})$ and $f(x) = f_1(x) + j f_2(x)$, we have

$$|f(x)|^2 = |f_1(x)|^2 + |f_2(x)|^2$$

and

$$|(\mathfrak{F}_\theta f_1)(\xi)|^2 = |(\mathfrak{F}_\theta f_2)(\xi)|^2 + |(\mathfrak{F}_\theta f)(\xi)|^2.$$

So,

$$\begin{aligned}
&\left(\int_{\mathbb{R}} |x|^2 |f(x)|^2 dx\right) \left(\int_{\mathbb{R}} |\xi|^{\frac{3}{\theta}-1} |(\mathfrak{F}_\theta f)(\xi)|^2\right) \\
&= \left(\int_{\mathbb{R}} |x|^2 |f_1(x)|^2 dx + \int_{\mathbb{R}} |x|^2 |f_2(x)|^2 dx\right) \\
&\quad \times \left(\int_{\mathbb{R}} |\xi|^{\frac{3}{\theta}-1} |(\mathfrak{F}_\theta f_1)(\xi)|^2 d\xi + \int_{\mathbb{R}} |\xi|^{\frac{3}{\theta}-1} |(\mathfrak{F}_\theta f_2)(\xi)|^2 d\xi\right) \\
&\geq \left(\int_{\mathbb{R}} |x|^2 |f_1(x)|^2 dx\right) \left(\int_{\mathbb{R}} |\xi|^{\frac{3}{\theta}-1} |(\mathfrak{F}_\theta f_1)(\xi)|^2 d\xi\right) \\
&\quad + \left(\int_{\mathbb{R}} |x|^2 |f_2(x)|^2 dx\right) \left(\int_{\mathbb{R}} |\xi|^{\frac{3}{\theta}-1} |(\mathfrak{F}_\theta f_2)(\xi)|^2 d\xi\right).
\end{aligned}$$

Using the inequality (31) for the complex-valued functions f_1 and f_2, we get

$$\begin{aligned}
&\left(\int_{\mathbb{R}} |x|^2 |f(x)|^2 dx\right) \left(\int_{\mathbb{R}} |\xi|^{\frac{3}{\theta}-1} |(\mathfrak{F}_\theta f)(\xi)|^2 d\xi\right) \\
&\quad \geq \frac{\pi\theta}{2} \left\{ \left(\int_{\mathbb{R}} |f_1(x)|^2 dx\right)^{\frac{1}{2}} + \left(\int_{\mathbb{R}} |f_2(x)|^2 dx\right)^{\frac{1}{2}} \right\}.
\end{aligned} \tag{32}$$

Let $r = \int_{\mathbb{R}} |f_1(x)|^2 dx$ and $s = \int_{\mathbb{R}} |f_2(x)|^2 dx$. Then from equation (32), we get

$$\left(\int_{\mathbb{R}} |x|^2 |f(x)|^2 dx\right) \left(\int_{\mathbb{R}} |\xi|^{\frac{3}{\theta}-1} |(\mathfrak{F}_\theta f)(\xi)|^2\right) \geq \frac{\pi\theta}{2}(r^2 + s^2)$$
$$\geq \frac{\pi\theta}{4}(r+s)^2, \text{ using } \frac{r+s}{2} \geq (rs)^{\frac{1}{2}}.$$

Thus, we have

$$\left(\int_{\mathbb{R}} |x|^2 |f(x)|^2 dx\right) \left(\int_{\mathbb{R}} |\xi|^{\frac{3}{\theta}-1} |(\mathfrak{F}_\theta f)(\xi)|^2\right)$$
$$\geq \frac{\pi\theta}{4}\left(\int_{\mathbb{R}} |f_1(x)|^2 dx + \int_{\mathbb{R}} |f_2(x)|^2 dx\right)^2$$
$$= \frac{\pi\theta}{4}\left(\int_{\mathbb{R}} |f(x)|^2 dx\right)^2.$$

i.e.

$$\left(\int_{\mathbb{R}} |x|^2 |f(x)|^2 dx\right) \left(\int_{\mathbb{R}} |\xi|^{\frac{3}{\theta}-1} |(\mathfrak{F}_\theta f)(\xi)|^2\right) \geq \frac{\pi\theta}{4} \|f\|^4_{L^2(\mathbb{R},\mathbb{H})}.$$

This completes the proof. □

Now, we derive Heisenberg's uncertainty principle for the quaternionic fractional wavelet transform.

Theorem 10 (Heisenberg's uncertainty principle for QFrWT). *Let the wavelet* ψ *and function* f *be in* $\mathcal{S}(\mathbb{R}, \mathbb{H})$, *then the following inequality holds*:

$$\left(\int_{\mathbb{R}}\int_{\mathbb{R}} |b|^2 |(W^\theta_\psi f)(b,a)|^2 \frac{dbda}{|a|^{\frac{1}{\theta}+1}}\right) \left(\int_{\mathbb{R}} |\xi|^{\frac{3}{\theta}-1} |(\mathfrak{F}_\theta f)(\xi)|^2 d\xi\right)$$
$$\geq \frac{\pi\theta}{4} C_{\psi,\theta} \|f\|^4_{L^2(\mathbb{R},\mathbb{H})}.$$

Proof. Replacing $f(\cdot)$ by $(W^\theta_\psi f)(\cdot, a)$, we get

$$\left(\int_{\mathbb{R}} |b|^2 |(W^\theta_\psi f)(b,a)|^2 db\right) \left(\int_{\mathbb{R}} |\xi|^{\frac{3}{\theta}-1} |(\mathfrak{F}_\theta (W^\theta_\psi f)(\cdot,a))(\xi) d\xi|^2\right)$$
$$\geq \frac{\pi\theta}{4} \|(W^\theta_\psi f)(\cdot,a)\|^4_{L^2(\mathbb{R},\mathbb{H})}.$$

Taking square root on both sides and integrating with respect to $\frac{da}{|a|^{\frac{1}{\theta}+1}}$, we get

$$\begin{aligned}
&\int_{\mathbb{R}} \left(\int_{\mathbb{R}} |b|^2 |(W_\psi^\theta f)(b,a)|^2 db \right)^{\frac{1}{2}} \left(\int_{\mathbb{R}} |\xi|^{\frac{3}{\theta}-1} |(\mathfrak{F}_\theta (W_\psi^\theta f)(\cdot,a))(\xi)|^2 d\xi \right)^{\frac{1}{2}} \\
&\quad \times \frac{da}{|a|^{\frac{1}{\theta}+1}} \\
&\quad \geq \sqrt{\frac{\pi\theta}{4}} \int_{\mathbb{R}} \|(W_\psi^\theta f)(\cdot,a)\|^2_{L^2(\mathbb{R},\mathbb{H})} \frac{da}{|a|^{\frac{1}{\theta}+1}} \\
&\quad = \sqrt{\frac{\pi\theta}{4}} \int_{\mathbb{R}} \int_{\mathbb{R}} |(W_\psi^\theta f)(b,a)|^2 \frac{dadb}{|a|^{\frac{1}{\theta}+1}} \\
&\quad = \sqrt{\frac{\pi\theta}{4}} C_{\psi,\theta} \|f\|^2_{L^2(\mathbb{R},\mathbb{H})},
\end{aligned} \tag{33}$$

using Theorem 6. Using Holder's inequality in (33), we get

$$\begin{aligned}
&\left(\int_{\mathbb{R}} \int_{\mathbb{R}} |b|^2 |(W_\psi^\theta f)(b,a)|^2 \frac{dbda}{|a|^{\frac{1}{\theta}+1}} \right) \\
&\quad \times \left(\int_{\mathbb{R}} \int_{\mathbb{R}} |\xi|^{\frac{3}{\theta}-1} |(\mathfrak{F}_\theta (W_\psi^\theta f)(\cdot,a))(\xi)|^2 \frac{d\xi da}{|a|^{\frac{1}{\theta}+1}} \right) \\
&\quad \geq \frac{\pi\theta}{4} C_{\psi,\theta}^2 \|f\|^4_{L^2(\mathbb{R},\mathbb{H})}.
\end{aligned} \tag{34}$$

Now, we see that

$$\begin{aligned}
&\int_{\mathbb{R}} \int_{\mathbb{R}} |\xi|^{\frac{3}{\theta}-1} |(\mathfrak{F}_\theta (W_\psi^\theta f)(\cdot,a))(\xi)|^2 \frac{d\xi da}{|a|^{\frac{1}{\theta}+1}} \\
&\quad = \int_{\mathbb{R}} \int_{\mathbb{R}} |\xi|^{\frac{3}{\theta}-1} |(\mathfrak{F}_\theta f)(\xi)|^2 |(\mathfrak{F}_\theta \psi_{a,\theta})(\xi)|^2 \frac{d\xi da}{|a|^{\frac{1}{\theta}+1}} \\
&\quad = \int_{\mathbb{R}} |\xi|^{\frac{3}{\theta}-1} |(\mathfrak{F}_\theta f)(\xi)|^2 \left\{ \int_{\mathbb{R}} |(\mathfrak{F}_\theta \psi_{a,\theta})(\xi)|^2 \frac{da}{|a|^{\frac{1}{\theta}+1}} \right\} d\xi \\
&\quad = C_{\psi,\theta} \int_{\mathbb{R}} |\xi|^{\frac{3}{\theta}-1} |(\mathfrak{F}_\theta f)(\xi)|^2 d\xi.
\end{aligned} \tag{35}$$

Using equation (35) in equation (34), we get

$$\left(\int_{\mathbb{R}}\int_{\mathbb{R}}|b|^2|(W_\psi^\theta f)(b,a)|^2\frac{dbda}{|a|^{\frac{1}{\theta}+1}}\right)\left(\int_{\mathbb{R}}|\xi|^{\frac{3}{\theta}-1}|(\mathfrak{F}_\theta f)(\xi)|^2d\xi\right)$$
$$\geq\frac{\pi\theta}{4}C_{\psi,\theta}\|f\|^4_{L^2(\mathbb{R},\mathbb{H})}.$$

This completes the proof. □

6. Conclusion

In this chapter, we have studied the properties of Fractional Fourier transform like Riemann–Lebesgue lemma, Parseval's identity and the convolution theorem for the quaternion-valued functions. We have defined a quaternionic fractional wavelet transform and studied its basic properties like linearity, translation, parity and scaling. We have also obtained the inner product relation and inversion formula for the transform along with the characterization of its range, where the wavelet involved in the transform is complex-valued. We discussed the uncertainty principle for the QFrWT for the functions in $\mathcal{S}(\mathbb{R},\mathbb{H})$.

Acknowledgment

This work is partially supported by UGC File No.16-9 (June 2017)/ 2018(NET/CSIR), New Delhi, India.

References

[1] M. Bahri, R. Ashino, and R. Vaillancourt, Continuous quaternion Fourier and wavelet transforms, *Int. J. Wavelets, Multiresolution Inf. Proc.* **12**(4), 1460003, (2014).
[2] M. Bahri, E. Hitzer, A. Hayashi, and R. Ashino, An uncertainty principle for quaternion Fourier transform, *Comp. Math. Appl.* **56**(9), 2398–2410, (2008).
[3] T. Ell, Quaternion-Fourier transforms for analysis of two-dimensional linear time-invariant partial differential systems, In: *Proceedings of 32nd IEEE Conference on Decision and Control.* IEEE, pp. 1830–1841 (IEEE, 1993).
[4] M. Bahri, E. Hitzer, R. Ashino, and R. Vaillancourt, Windowed Fourier transform of two-dimensional quaternionic signals, *Appl. Math. Comput.* **216**(8), 2366–2379, (2010).
[5] L. Akila and R. Roopkumar, Multidimensional quaternionic Gabor transforms, *Adv. Appl. Clifford Algebras* **26**(3), 985–1011, (2016).

[6] M. Bahri and R. Ashino, Uncertainty principles related to quaternionic windowed Fourier transform, *Int. J. Wavelets, Multiresolution Inf. Proc.* **18**(3), 2050015, (2020).
[7] M. Bahri, R. Ashino, and R. Vaillancourt, Two-dimensional quaternion wavelet transform, *Appl. Math. Comp.* **218**(1), 10–21, (2011).
[8] V. Namias, The fractional order Fourier transform and its application to quantum mechanics, *IMA J. Appl. Math.* **25**(3), 241–265, (1980).
[9] L. Almeida, The fractional Fourier transform and time-frequency representations, *IEEE Trans. Signal Proc.* **42**(11), 3084–3091, (1994).
[10] A. Verma and B. Gupta, A note on continuous fractional wavelet transform in $\mathbb{R}^n$, *Int. J. Wavelets, Multiresolution Inf. Proc.* 2150050, (2021).
[11] K. Kou, J. Ou, and J. Morais, Uncertainty principles associated with quaternionic linear canonical transforms, *Math. Meth. Appl. Sci.* **39**(10), 2722–2736, (2016).
[12] W. Gao and B. Li, Quaternion windowed linear canonical transform of two-dimensional signals, *Adv. Appl. Clifford Algebras* **30**(1), 1–18, (2020).
[13] F. Shah, A. Teali, and A. Tantary, Linear canonical wavelet transform in quaternion domains, *Adv. Appl. Clifford Algebras* **31**(3), 1–24, (2021).
[14] J. He and B. Yu, Continuous wavelet transforms on the space $L^2(\mathbb{R}, \mathbb{H}; dx)$, *Appl. Math. Letts.* **17**(1), 111–121, (2004).
[15] L. Akila and R. Roopkumar, A natural convolution of quaternion valued functions and its applications, *Appl. Math. Comp.* **242**, 633–642, (2014).
[16] L. Akila and R. Roopkumar, Quaternionic Stockwell transform, *Int. Transf. Special Funct.* **27**(6), 484–504, (2016).
[17] L. Akila and R. Roopkumar, Ridgelet transform for quarternion-valued functions, *Int. J. Wavelets, Multiresolution Inf. Proc.* **14**(1), 1650006, (2016b).
[18] F. Shah and A. Tantary, Quaternionic Shearlet transform, *Optik* **175**, 115–125, (2018).
[19] R. Roopkumar, Quaternionic one-dimensional fractional Fourier transform, *Optik* **127**(24), 11657–11661, (2016).
[20] R. Roopkumar, Quaternionic fractional wavelet transform, *J. Anal.* **26**(2), 313–322, (2018).
[21] S. Saima and B. Li, Quaternionic one-dimensional linear canonical transform, *Optik* **244**, 166914, (2021).
[22] Y. Luchko, H. Martinez, and J. Trujillo, Fractional Fourier transform and some of its applications, *Fract. Calc. Appl. Anal* **11**(4), 457–470, (2008).
[23] A. Kilbas, Y. Luchko, H. Martinez, and J. Trujillo, Fractional Fourier transform in the framework of fractional calculus operators, *Integral Transf. Special Funct.* **21**(10), 779–795, (2010).
[24] H. Srivastava, S. Upadhyay, and K. Khatterwani, A family of pseudo-differential operators on the Schwartz space associated with the fractional Fourier transform, *Russian J. Math. Phys.* **24**(4), 534–543, (2017).
[25] H. Srivastava, K. Khatterwani, and S. Upadhyay, A certain family of fractional wavelet transformations, *Math. Meth. Appl. Sci.* **42**(9), 3103–3122, (2019).

[26] S. Upadhyay and K. Khatterwani, Continuous fractional wavelet transform, *J. Int. Acad. Phys. Sci.* **21**(1), 55–61, (2018).

[27] A. Verma and B. Gupta, Certain properties of continuous fractional wavelet transform on Hardy space and Morrey space, *Opuscula Mathematica* **41**(5), 701–723, (2021).

[28] N. Chuong and D. Duong, Boundedness of the wavelet integral operator on weighted function spaces, *Russian J. Math. Phys.* **20**(3), 268–275, (2013).

[29] A. Almeida and S. Samko, Approximation in Morrey spaces, *J. Funct. Anal.* **272**(6), 2392–2411, (2017).

[30] C. Morrey, On the solutions of quasi-linear elliptic partial differential equations, *Trans. Am. Math. Soc.* **43**(1), 126–166, (1938).

[31] E. Wilczok, New uncertainty principles for the continuous Gabor transform and the continuous wavelet transform, *Documenta Mathematica* **5**, 201–226, (2000).

https://doi.org/10.1142/9789811261572_0017

Chapter 17

From Variational Inequalities to Singular Integral Operator Theory — A Note on the Lions–Stampacchia Theorem*

Joachim Gwinner

Institut für Mathematik und Rechneranwendung,
Fakultät für Luft- und Raumfahrttechnik,
Universität der Bundeswehr München,
85577 Neubiberg/München
joachim.gwinner@unibw-muenchen.de

In this chapter, we are concerned with the famous Lions–Stampacchia existence theorem for variational inequalities (VIs). We provide a partial converse result and show that unique solvability of VIs on arbitrary closed convex sets implies that the linear operator associated to the VI is strictly monotone. If the operator is self-adjoint, then it follows that the operator is coercive. Here we present this converse result in the setting of reflexive Banach spaces. The converse result applies to singular integral operator theory. Here we demonstrate this approach by a study of the simple-layer potential operator for the Laplace equation in three dimensions. More complicated singular integral operators in mathematical physics can similarly be treated.

1. Introduction

The starting point of this chapter is the celebrated Lions–Stampacchia Theorem [1] for linear variational inequalities (VIs) on closed convex sets (possibly unbounded) in a Hilbert space that demands for solvability the coerciveness of the bilinear form, or equivalently, the coerciveness of the corresponding linear operator that is associated to the VI. Note that coerciveness automatically implies monotonicity for the linear operator.

*Dedicated to Professor W.L. Wendland on the occasion of his 85th birthday.

Here vice versa, we show that strict monotonicity is necessary for unique solvability, moreover if the operator is self-adjoint, then even coerciveness is necessary. Thus, we extend a recent partial converse result to existence in Ref. [2] in various aspects. The analysis of the Lions–Stampacchia Theorem in VI theory toward converse results is inspired by the paper of Ernst and Théra [3]. They showed in a Hilbert space setting that solvability of linear VIs on non-empty bounded closed convex sets is equivalent to topological pseudomonotonicity (in the sense of Brézis) of the associated linear operator. For a broad comparison among various existence theorems for VIs including necessary conditions for solvability, we also refer to [4].

Whereas the partial converse result in Ref. [2] was limited to the Hilbert space case, here the more general setting of reflexive Banach spaces allows to give applications in operator theory. In particular, this furnishes an elegant way to show coerciveness of singular integral operators [5] which occur in mathematical physics. We modify the reasoning of Nedelec [6], focus here on the simple case of the inhomogeneous Dirichlet problem for the Laplace equation in three dimensions, and show the coerciveness of the simple-layer potential operator in the relevant function spaces. To conclude this introduction, let us point out that by a similar reasoning one can show the coerciveness of the hypersingular boundary integral operator (= normal derivative of the double-layer potential operator) for the Laplace equation in $\mathbb{R}^d$ $(d = 2, 3)$ and the coerciveness of more involved integral operators in mathematical physics. In particular, let us mention the Poincaré–Steklov operator (Dirichlet to Neumann map) that plays a dominant part in the mathematical treatment and numerical analysis in the boundary integral approach, resp. in boundary element methods for the solution of boundary value, transmission and contact problems without and with friction in classic linear elasticity and mathematical physics; here we can refer to [7,8] and the references therein.

2. Solvability of Variational Inequalities

In what follows, E is a real reflexive Banach space, E^*, its dual and $\langle \cdot, \cdot \rangle = \langle \cdot, \cdot \rangle_{E^* \times E}$, its associated dual form. Let $\mathcal{L}(E, E^*)$ denote the class of linear continuous operators from E to E^*. We start off by the

Theorem 1 (Lions–Stampacchia theorem). *Suppose* $A \in \mathcal{L}(E, E^*)$ *satisfies the coerciveness condition*

$$(CC) \quad \exists \alpha > 0 : \langle Av, v \rangle \geq \alpha \|v\|^2, \quad \forall v \in E.$$

Then for any $f \in E^*$ and any closed, convex non-void subset K of E there exists a unique solution to the variational inequality $VI(A, f, K)$:
Find $u \in K$ such that

$$\langle Au,\ v-u\rangle \ \geq\ \langle f,\ v-u\rangle, \quad \forall v \in K.$$

Note that (CC) implies that A is monotone, that is, $\langle Av, v\rangle \geq 0$ for any $v \in E$; moreover, strict monotonicity, that is, $\langle Av, v\rangle > 0$ for any $v \neq 0$.

We can here give a partial converse result to the existence theorem above showing that the coerciveness condition (CC) is indeed necessary. More precisely, we first show strict monotonicity and thus drop the monotonicity assumption in Proposition 2.1 in Ref. [2]. Then if the operator A is moreover self-adjoint, i.e. coincides with its adjoint $A^* \in \mathcal{L}(E, E^*)$, then we show that even (CC) is necessary.

Theorem 2. *Suppose that for any $f \in E^*$ and any closed convex non-void subset K of E, $VI(A, f, K)$ admits a unique solution. Then A is strictly monotone. Moreover, if A is self-adjoint, then A satisfies (CC).*

Proof. We establish the result in several steps. We first show monotonicity by an indirect argument. Suppose there exists $\tilde{x} \in E\backslash\{0\}$ such that $\langle A\tilde{x}, \tilde{x}\rangle < 0$. Then define

$$\tilde{K} = \{\lambda\tilde{x} \in E : \lambda \geq 1\} = [1, \infty)\tilde{x},$$

clearly convex and closed. Hence, there exists $\hat{x} \in \tilde{K}$ that solves $VI(A, 0, \tilde{K})$. This gives $\hat{x} = \hat{\lambda}\tilde{x}$ for some $\hat{\lambda} \geq 1$ and

$$\langle A\hat{\lambda}\tilde{x}, (\lambda - \hat{\lambda})\tilde{x}\rangle = \hat{\lambda}(\lambda - \hat{\lambda})\langle A\tilde{x}, \tilde{x}\rangle \geq 0$$

for all $\lambda \geq 1$. With $\langle A\tilde{x}, \tilde{x}\rangle < 0$ and $\lambda := 2\hat{\lambda}$, we obtain $\hat{\lambda}^2 \leq 0$ and arrive at a contradiction.

Next, we show strict monotonicity of A by an indirect argument. Suppose there exists some $\hat{x} \neq 0$ such that $\langle A\hat{x}, \hat{x}\rangle = 0$. Now define

$$\hat{K} = \{\lambda\hat{x} \in E :\ 0 \leq \lambda \leq 1\} = [0, 1]\hat{x},$$

clearly convex and closed. Let $y \in \hat{K}, y = \lambda\hat{x}$. Then

$$\langle A\hat{x}, y - \hat{x}\rangle = (\lambda - 1)\langle A\hat{x}, \hat{x}\rangle = 0.$$

Thus, $\hat{x}$ solves $VI(A, 0, \hat{K})$. However, $0 \in \hat{K}$ trivially solves this VI, too. This contradicts the assumed unique solvability.

Take $K = E$, then $VI(A, f, E)$ is equivalent to the linear operator equation $Au = f$. Hence, unique solvability implies that A is bijective. Therefore, by Banach's inverse mapping theorem, $A^{-1} \in \mathcal{L}(E^*, E)$.

Finally, we treat the self-adjoint case. We have to modify the argument in Ref. [2], since the argument given there is limited to the Hilbert space case.

We claim that the corresponding bilinear form $a(u,v) = \langle Au, v\rangle$ satisfies

$$\inf_{\|u\|=1} \sup\{|a(u,v)| : \|v\| = 1\} \geq \frac{1}{\|A^{-1}\|} > 0.$$

Indeed, for any $u \in E, u = A^{-1}Au$, hence

$$\|u\| \leq \|A^{-1}\| \; \|Au\| = C^{-1}\|Au\|,$$

with $C := \|A^{-1}\|^{-1} > 0$. This gives, in particular, $\|Au\| \geq C$ for all u with $\|u\| = 1$. Note that for $Au \in E^*$,

$$\|Au\| = \sup_{\|v\|=1} |\langle Au, v\rangle|.$$

Therefore, we obtain

$$\sup_{\|v\|=1} |a(u,v)| = \sup_{\|v\|=1} |\langle Au, v\rangle| \geq C, \; \forall\, u \text{ with } \|u\| = 1,$$

and thus the claim above.

With $A = A^*$, hence,

$$a(u,v) = \langle Au, v\rangle = \langle Av, u\rangle = a(v,u),$$

the bilinear form a is symmetric, moreover non-negative by the shown monotonicity of A. Therefore, the bilinear form satisfies the Cauchy–Schwarz inequality and, hence,

$$\begin{aligned} a(v,v) &\geq \sup_{w\neq 0} \frac{|a(v,w)|^2}{a(w,w)} \\ &\geq \frac{1}{\|A\|} \sup_{w\neq 0} \left[\frac{|a(v,w)|}{\|w\|}\right]^2 \\ &\geq \frac{1}{\|A\|} \frac{1}{\|A^{-1}\|^2} \|v\|^2, \end{aligned}$$

which proves the assertion. □

To clarify the self-adjoint case and the Hilbert space case, we give the following remarks.

Remark 1. In the self-adjoint case, E can be renormed to a Hilbert space, provided A satisfies (CC). Then the associated bilinear form a is a scalar product, $\|x\|_a := a(x,x)^{1/2}$ is equivalent to the original norm, and the Lions–Stampacchia Theorem is equivalent to the projection theorem for convex closed sets.

Remark 2. In the case of a Hilbert space E, we can even drop the assumption of continuity of the linear operator A, since then by virtue of a result of Albrecht and Neumann [9, Korollar 4], a dissipative (= monotone) linear operator is automatically continuous.

3. Coerciveness of Simple-Layer Potential Operator for the Laplace Equation

In this section, we use the well-known representation formula in potential theory, which itself follows from Green's second formula (see Refs. [7, p. 2, 5, p. 220 ff.] for the Laplacian, [7, p. 131] for the general case of a uniformly strongly elliptic differential operator) and we use the Lax–Milgram theorem, which is a corollary of the Lions–Stampacchia theorem, here Theorem 1. Moreover, we apply Theorem 2, more precisely its second part, to derive the coerciveness of the simple-layer potential operator for the Laplace equation in three dimensions, which is the simplest singular boundary integral operator (in $\mathbb{R}^3$). By this approach, we can dispense with Fredholm theory as used in Ref. [10]. Instead, our arguments slightly modify the reasoning of Nedelec [6] in the following steps.

Consider the interior Dirichlet boundary value problem for the Laplacian on a bounded, sufficiently smooth domain $\Omega \subset \mathbb{R}^3$ and the exterior Dirichlet boundary value problem for the Laplacian on the outer domain $\Omega' := \mathbb{R}^3 \setminus \overline{\Omega}$:

$$\Delta u = 0 \text{ in } \Omega, \quad u|_\Gamma = u_0 \text{ in } \Gamma, \tag{1}$$

$$\Delta u = 0 \text{ in } \Omega', \quad u|_\Gamma = u_0 \text{ in } \Gamma, \tag{2}$$

for some given $u_0 \in H^{1/2}(\Gamma)$ on the joint boundary $\Gamma = \overline{\Omega} \cap \overline{\Omega'}$. Note that in a classic treatment, unique solvability of the exterior boundary value problem requires an additional condition at infinity; see, e.g. [11, p. 352 ff.] for a detailed exposition of these dimension-dependent conditions.

By contrast, we employ functional analytic methods and in the first step we show unique solvability of the above problems in the appropriate function spaces, namely in the standard Sobolev space $H^1(\Omega)$ for the interior problem, resp., in

$$W^1(\Omega') := \left\{u \mid (1+|x|^2)^{\frac{|\alpha|-1}{2}} D^\alpha u \in L^2(\Omega') \text{ for } 0 \le |\alpha| \le 1\right\}.$$

for the exterior problem. Thus, the weight $\frac{1}{(1+|x|^2)^{\frac{1}{2}}}$ controls the behavior of a function in $W^1(\Omega')$ at infinity. Therefore (see Ref. [12, Theorem 1, Chapter XI, §4] the seminorm

$$|u|_{1,\Omega'} = \left(\sum_{k=l}^{3} \int_{\Omega'} \left|\frac{\partial u_k}{\partial x_k}(x)\right|^2 dx\right)^{\frac{1}{2}}$$

is equivalent to the intrinsic norm $\|u\|_{1,\Omega'}$ given by

$$\|u\|^2_{1,\Omega'} = \int_{\Omega'} \frac{|u(x)|^2}{1+|x|^2}\, dx + |u|^2_{1,\Omega'}.$$

For the interior problem, by extension obtain $Ru_0 \in H^1(\Omega)$ such that $Ru_0|_\Gamma = u_0$. Then the problem (1) is equivalent to the Poisson boundary value problem (BVP)

$$-\Delta w = \Delta Ru_0 \text{ in } \Omega, w|_\Gamma = 0,$$

for $w := u - Ru_0$. Using Green's first formula, the latter writes in variational form as follows: Find $w \in H^1_0(\Omega) :=$ closure of $\mathcal{D}(\Omega)$ in $H^1(\Omega)$ such that for all $v \in H^1_0(\Omega)$,

$$a(w,v) = \langle f, v\rangle\,,$$

where

$$\langle f, v\rangle := \int_\Omega \nabla Ru_0 \cdot \nabla v\, dx,$$

$$a(v,w) := \int_\Omega \nabla v \cdot \nabla w\, dx$$

is a continuous linear form, resp., continuous bilinear form that is coercive by virtue of the Poincaré inequality. Thus, by the Lax–Milgram theorem there exists a unique solution $w \in H^1_0(\Omega)$ to the latter BVP and hence a unique solution $u \in H^1(\Omega)$ to the interior problem (1). The same procedure applies to the exterior problem (2).

Next, we use that $x \mapsto (4\pi|x|)^{-1}$ is the fundamental solution of the Laplace equation in $\mathbb{R}^3$ and hence by Green's second formula (for details see Refs. [5, p. 220 ff., 8, Section 2.4]), the solution u of (1) and of (2) can be represented by

$$u(y) = \frac{1}{4\pi} \int_\Gamma \left(\frac{\partial u}{\partial n}|_{\text{int }\Gamma} - \frac{\partial u}{\partial n}|_{\text{ext }\Gamma} \right) \frac{1}{|x-y|}\, d\gamma(x), \quad y \in \mathbb{R}^3 \setminus \Gamma.$$

Since the above weakly singular integral is continuous when sending y to the boundary, we obtain the boundary integral equation of the first kind as follows:

$$u_0(y) = \frac{1}{4\pi} \int_\Gamma \frac{q(x)}{|x-y|}\, d\gamma(x), \quad y \in \Gamma, \tag{3}$$

where

$$q := \left[\frac{\partial u}{\partial n}\right] := \frac{\partial u}{\partial n}|_{\text{int }\Gamma} - \frac{\partial u}{\partial n}|_{\text{ext }\Gamma}$$

is the jump of the normal derivative. This leads to the following result in the Gelfand triple $H^{1/2}(\Gamma) \subset L^2(\Gamma) \subset H^{-1/2}(\Gamma) = [H^{1/2}(\Gamma)]^*$ of reflexive separable Banach spaces with continuous and dense imbeddings.

Theorem 3. *Let $u_0 \in H^{1/2}(\Gamma)$. Then the integral equation* (3) *admits a unique solution $q \in H^{-1/2}(\Gamma)$. Moreover, the integral equation* (3) *writes in variational form as follows: Find $q \in H^{-1/2}(\Gamma)$ such that for all $\tilde{q} \in H^{-1/2}(\Gamma)$,*

$$b(q, \tilde{q}) = \langle u_0, \tilde{q} \rangle\,,$$

where the continuous bilinear form

$$b(q, \tilde{q}) := \frac{1}{4\pi} \int_\Gamma \int_\Gamma \frac{q(x)\; \tilde{q}(y)}{|x-y|}\, d\gamma(x)\, d\gamma(y)$$

is coercive on the space $H^{-1/2}(\Gamma)$. Thus, the corresponding boundary single-layer potential operator is coercive from $H^{-1/2}(\Gamma)$ to $H^{1/2}(\Gamma)$.

Proof. From the first Green's formula we have for $u \in H^1(\Omega)$ with $\Delta u = 0$,

$$\int_\Omega \nabla u \cdot \nabla v\, dx = \int_\Gamma \frac{\partial u}{\partial n}\bigg|_{\text{int }\Gamma} v\, d\gamma. \tag{4}$$

Here the left integral defines a continuous linear form on $H^1(\Omega)$, hence, by the trace theorem, the right integral gives a continuous linear form on $H^{1/2}(\Gamma)$. Thus, $\frac{\partial u}{\partial n}|_{\text{int }\Gamma}$ is defined in the dual $H^{-1/2}(\Gamma)$. In the same way — note the opposite orientation of the normal — we have for $u \in W^1(\Omega')$ with

$\Delta u = 0,$

$$\int_{\Omega'} \nabla u \cdot \nabla v \, dx = - \int_{\Gamma} \frac{\partial u}{\partial n}\bigg|_{\text{ext } \Gamma} v \, d\gamma, \tag{5}$$

for any $v \in W^1(\Omega')$. Thus also $\frac{\partial u}{\partial n}|_{\text{ext } \Gamma}$ is defined in the dual $H^{-1/2}(\Gamma)$. Therefore, for any given $u_0 \in H^{1/2}(\Gamma)$, we obtain the variational solution of the above integral equation (3) by the formula $q = [\frac{\partial u}{\partial n}]$ in $H^{-1/2}(\Gamma)$. Furthermore, addition of (4) and (5) leads to

$$\int_{\mathbb{R}^3} \nabla u \cdot \nabla v \, dx = - \int_{\Gamma} q \, v \, d\gamma, \tag{6}$$

for any $v \in W^1(\mathbb{R}^3)$, where the latter integral is understood as the duality between $H^{1/2}(\Gamma)$ and $H^{-1/2}(\Gamma)$.

Vice versa, with q given in $H^{-1/2}(\Gamma)$, (6) is a variational problem in the space $W^1(\mathbb{R}^3)$, where the left integral defines a continuous and coercive bilinear form and by the trace theorem, the right integral defines a continuous linear form on this space $W^1(\mathbb{R}^3)$. Hence, by the Lax–Milgram theorem, the solution u of the variational problem (6) is uniquely defined. Its trace gives us u_0 back in $H^{1/2}(\Gamma)$. Thus, we have shown that the mapping $q \mapsto u_0$ is an isomorphism from $H^{-1/2}(\Gamma)$ onto $H^{1/2}(\Gamma)$. Since the bilinear form b introduced in the theorem is symmetric, the second part of Theorem 2 applies to conclude the proof. □

References

[1] J.L. Lions and G. Stampacchia, Variational inequalities, *Comm. Pure Appl. Math.* **20**, 493–519, (1967).

[2] J. Gwinner and N. Ocharova, From solvability and approximation of variational inequalities to solution of nondifferentiable optimization problems in contact mechanics, *Optimization* **64**, 1683–1702, (2015).

[3] E. Ernst and M. Théra, A converse to the Lions-Stampacchia theorem, *ESAIM COCV.* **15**, 810–817, (2009).

[4] A. Maugeri and F. Raciti, On existence theorems for monotone and non-monotone variational inequalities, *J. Convex Anal.* **16**, 899–911, (2009).

[5] S.G. Mikhlin, *Multidimensional Singular Integrals and Integral Equations* (Pergamon Press, 1965).

[6] J.C. Nédélec, *Approximation des Équations Intégrales en Mécanique et en Physique* (Ecole Polytechnique, Palaiseau, France, 1977).

[7] G.C. Hsiao and W.L. Wendland, *Boundary Integral Equations* (Springer, Berlin, 2008).

[8] J. Gwinner and E.P. Stephan, *Advanced Boundary Element Methods — Treatment of Boundary Value, Transmission and Contact Problems* (Springer, Berlin, 2018).

[9] E. Albrecht and M. Neumann, Über die Stetigkeit von dissipativen linearen Operatoren, *Arch. Math.* **31**, 74–88, (1978).

[10] W. McLean, *Strongly Elliptic Systems and Boundary Integral Equations* (Cambridge University Press, 2000).

[11] R. Dautray and J.-L. Lions, Mathematical analysis and numerical methods for science and technology. In: *Physical Origins and Classical Methods*, Vol. 1. With the collaboration of P. Bénilan, M. Cessenat, A. Gervat, A. Kavenoky, and H. Lanchon (Springer, Berlin, 1990).

[12] R. Dautray and J.-L. Lions, Mathematical analysis and numerical methods for science and technology. In: *Integral Equations and Numerical Methods*, Vol. 4. With the collaboration of M. Artola, P. Bénilan, M. Bernadou, M. Cessenat, J.-C. Nédélec, J. Planchard and B. Scheurer (Springer, Berlin, 1990).

https://doi.org/10.1142/9789811261572_0018

Chapter 18

Graph Pursuit Games and New Algorithms for Nash Equilibrium Computation

Athanasios Kehagias* and Michael Koutsmanis†

Department of Electrical and Computer Engineering,
Faculty of Engineering,
Aristotle University of Thessaloniki,
Greece
**kehagiat@ece.auth.gr*
†mkoutsman@ece.auth.gr

In this chapter, we introduce a family of *N-player graph pursuit games* (GPG) and some algorithms for the computation of GPG *Nash Equilibria* (NE). We establish theoretical properties of both the GPGs and the associated algorithms and also evaluate the algorithms by extensive numerical experimentation.

GPGs are games played by N players on a graph. The players take turns in moving from vertex to neighboring vertex (i.e. along the edges); a player can be either a *pursuer* or an *evader* (or both). A pursuer's goal is to land on a vertex which contains an evader and thus effect a *capture*; an evader's goal is to avoid capture.

In the analysis of an N-player stochastic game, the main goals are to establish the existence of NE and compute one or more of these. Classical algorithms, such as *Value Iteration* (VI), have been developed for the solution of two-player games but cannot be applied to N-player games. Hence, in this chapter we present two extensions of VI, appropriate for N-player GPGs.

The first extension is the "basic" *Multi-Value Iteration* (MVI) algorithm. This is a deterministic algorithm which, *when convergent*, will provably produce one (always the same) NE of a given GPG. The second extension is *Multi-Start MVI* (MS–MVI), a simple modification in which the basic MVI is run multiple times, with a vertex label permutation applied before each run. Numerical experiments indicate that MS–MVI improves the convergence behavior of the basic MVI algorithm and can obtain multiple NE.

1. Introduction

In this chapter, we introduce a family of *N-player graph pursuit games* (GPG) and some algorithms for the computation of GPG *Nash Equilibria* (NE). We establish theoretical properties of both the GPGs and the associated algorithms and also evaluate the algorithms by extensive numerical experimentation.

GPGs are games played by N players on a graph. The players take turns in moving from vertex to neighboring vertex (i.e. along the edges); a player can be either a *pursuer* or an *evader* (or both). A pursuer's goal is to land on a vertex which contains an evader and thus effect a *capture*; an evader's goal is to avoid capture. The GPG family consists of variants of the above theme, obtained by different specifications of the pursuer/evader relationship between players. We provide a framework in which every GPG can be formulated as an *N-player discounted stochastic game of perfect information* [1]; a particular game is obtained by specifying the *payoff functions* of the N players.

The inspiration for the study of GPGs comes from the classical *Cops and Robbers* (CR), an extensively studied *two*-player game [2,3]. The extension to general two-player pursuit games (*Generalized Cops and Robber Games* or GCR Games) has been presented in Ref. [4]. Special cases of N-player pursuit games have been previously presented in Refs. [5–7].

In the analysis of an N-player stochastic game, the main goals are to establish the existence of NE and compute one or more of these. For *two*-player *zero-sum* games, these goals can be achieved by the classic *Value Iteration* (VI) algorithm [1]; but no general algorithm exists for N-player games. Hence, the need arises for more sophisticated algorithms. The problem is similar to global optimization, where one or more global optima must be selected from a multiplicity of local optima (with the Nash equilibrium taking the role of a global optimum). Indeed, there is a long and fruitful connection between game theory and global optimization. Global optimization methods have often been used for the solution of N-player games, including linear and nonlinear programming [8–11], integer programming [12], evolutionary computation [13], Bayesian optimization [14], computational intelligence [15], etc. The reverse direction has also appeared in the literature, i.e. utilizing game theoretic concepts to achieve global optimization [16–18].

In this chapter, we combine the VI algorithm with ideas from the above papers, to synthesize two extensions of VI, appropriate for N-player GPGs.

(1) The first extension is the *Multi-Value Iteration* (MVI) algorithm. This is a deterministic algorithm which, for certain classes of GPG N-player games, can compute an NE. More specifically, MVI is a deterministic algorithm which, *when convergent*, will provably produce one (always the same) NE of a given GPG. The algorithm works reasonably well but has two drawbacks. For a given GPG:
  (a) convergence is not guaranteed and
  (b) the computed NE will always be the same.
(2) To address the above drawbacks, we introduce the *Multi-Start MVI* (MS–MVI) algorithm. This consists in running the "basic" MVI multiple times, with a vertex label permutation applied before each run. For small graphs, MS–MVI can utilize *all possible* vertex permutations; for larger graphs, one can use a computationally viable number of *randomly generated* permutations. Numerical experimentation gives strong evidence for the following advantages of MS–MVI over deterministic MVI:
  (a) Convergence behavior is much better, i.e. over multiple reruns the algorithm may converge for at least some vertex permutations. Recall that for every convergent run, the algorithm produces an NE.
  (b) Over multiple reruns the algorithm may compute several distinct NE.

This chapter is organized as follows. In Section 2, we present preliminary notations and definitions. The GPG family is defined rigorously in Section 3. In Section 4, we establish the existence of NE for every GPG. In Section 5, we present the MVI and establish some of its properties; then we present the MS–MVI modification. Section 6 is devoted to the evaluation of the algorithm by numerical experiments. Finally, we present conclusions and future research directions in Section 7.

2. Preliminaries

The following are standard mathematical notations:

(1) The cardinality of set A is denoted by $|A|$.
(2) The set of elements of set A which are not elements of set B is denoted by $A \backslash B$.
(3) The set of natural numbers is $\mathbb{N} = \{1, 2, 3, \ldots\}$ and $\mathbb{N}_0 = \{0, 1, 2, 3, \ldots\}$.

(4) For any $M \in \mathbb{N}$, we define $[M] = \{1, 2, \ldots, M\}$.
(5) A permutation of an ordered set is, intuitively speaking, a reordering of its elements.

The following are standard graph theoretic concepts.

(1) A *graph*, denoted as G, is a pair $G = (V, E)$, where the set of *vertices* is

$$V = \{x_1, \ldots, x_N\},$$

and the set of *edges*, is E, where

$$E \subseteq \{\{x, y\} : \text{with } x, y \in V \text{ and } x \neq y\}.$$

In what follows, we will always take (without loss of generality) $V = [N]$.
(2) Given a graph $G = (V, E)$, for any $x \in V$, the *neighborhood* of x is

$$N(x) = \{y : \{x, y\} \in E\},$$

and the *closed neighborhood* is

$$N[x] = N(x) \cup \{x\}.$$

(3) Given a graph $G = (V, E)$, the *degree* of a vertex $x \in V$ is $D(x) = |N(x)|$.
(4) A *path in* $G = (V, E)$ *between vertices* $x, y \in V$ is a sequence $v_0, v_1, \ldots, v_K \in V$ such that

$$v_0 = x, \qquad \forall k \in \{1, 2, \ldots, K\} : v_k \in N(v_{k-1}), \qquad v_K = y.$$

(5) Given a graph, the *graph distance* between $x, y \in V$ is the length of the shortest path in G between x and y and is denoted by $d_G(x, y)$ or simply by $d(x, y)$.
(6) The graph $G = (V, E)$ is called *connected* iff every pair of vertices x, y is connected by at least one path or, equivalently, iff $d_G(x, y) < \infty$. Otherwise the graph is called *disconnected*.

3. The GPG Family

3.1. *Definitions*

A *Graph Pursuit Game* (GPG) is played between N players; the *player set* will be denoted by either $\mathcal{P} = \{P_1, \ldots, P_N\}$ or, for simplicity, by $\mathcal{P} = [N]$.

(1) The game proceeds at discrete *turns* (or time steps) $t \in \{0, 1, 2, \ldots\}$) on a graph $G = (V, E)$.

(2) At the t-th turn, the n-th player is located at $x_t^n \in V$.
(3) At the t-th turn, a single player can (but is not obliged to) change his location by moving to a vertex adjacent to his current one; all other players must remain at their locations.
(4) The game ends when a *capture* takes place (what constitutes a capture will be rigorously defined later).

A *game position* or *game state* has the form $s = (x^1, x^2, \ldots, x^N, i)$ where $x^n \in V$ is the position (vertex) of the n-th player and $i \in [N]$ is the number of the player who has the next move. The set of *non-terminal* states is

$$S = \{(x^1, x^2, \ldots, x^N, i) : (x^1, x^2, \ldots, x^N) \in V \times V \times \cdots \times V \text{ and } i \in [N]\}.$$

We introduce an additional *terminal state* τ. Hence, the full state set is

$$\overline{S} = S \cup \{\tau\}.$$

We define S^n to be the set of states in which P_n has the next move:

$$\text{for each } n \in [N] : S^n = \{s : s = (x^1, x^2, \ldots, x^N, n) \in S\},$$

Hence, the set of non-terminal states can be partitioned as follows:

$$S = S^1 \cup S^2 \cup \cdots \cup S^N.$$

Captures play a central role in GPGs. Informally, a capture occurs iff for certain pairs $(n, m) \in [N] \times [N]$ (which are defined for each game), P_n and P_m are located in the same vertex. The above is an approximate description; a more precise definition will be given when *payoff functions* are defined. At any rate, we can partition the set of non-terminal states as follows:

$$S = S_{nc} \cup S_c \text{ with } S_{nc} \cap S_c = \emptyset,$$

where S_{nc} (resp. S_c) is the set of *non-capture* (resp. *capture*) states. When the game state is $s = (x^1, x^2, \ldots, x^N, m)$, the *$n$-th player's action set* is

$$A^n(s) = \begin{cases} N[x^n] & \text{for } s \in S^n \cap S_{nc}, \\ \{x^n\} & \text{for } s \in S^m \cap S_{nc} \text{ with } n \neq m, \\ \{\lambda\} & \text{for } s \in S_c \text{ (where } \lambda \text{ is the } \textit{null} \text{ move)}, \\ \{\lambda\} & \text{for } s = \tau. \end{cases}$$

The *full action set* is

$$A = \cup_{s \in S} \cup_{n \in \mathcal{P}} A^n(s).$$

The *successor function* $Suc : \mathcal{P} \to \mathcal{P}$ represents the sequence by which the players are moving. For example, suppose we have $\mathcal{P} = \{P_1, P_2, P_3\}$ (three players) and

$$Suc(P_1) = P_3, \qquad Suc(P_2) = P_1, \qquad Suc(P_3) = P_2.$$

Then the players move in the sequence

$$\cdots \to P_1 \to P_3 \to P_2 \to P_1 \to \cdots$$

(the player who has the first move is specified by the initial state). The movement rules of the game are completely specified by the successor function, the action set (for each state) and the capture set.

(1) When the game is in a non-capture state $s = (x^1, \ldots, x^N, i) \in S_{nc}$, the only player who can change his position is P_i; his possible actions are those of the closed neighborhood of x_t^i. In other words, the set of possible actions is

$$N\left[x_t^i\right] = \{y^1, \ldots, y^K\}.$$

If P_i chooses action y^k, then the game moves to the next state $s' = (z^1, \ldots, z^N, j)$ where

$$z^i = y^k \text{ and } \forall m \neq i : z^m = x^m$$

and j, the next player to move, is determined by the successor function.

(2) When the game is in a capture state $s = (x^1, \ldots, x^N, i) \in S_c$, it will necessarily move to the terminal state τ.

(3) When the game is in the terminal state τ, it will remain there for all subsequent rounds.

The game always terminates immediately after a capture, in the sense that a capture state always leads to the terminal state.

The above ideas can be expressed more compactly through the use of a *transition function* $\mathbf{T} : \overline{S} \times A \to \overline{S}$, which specifies the state to which the game moves when the player having the move chooses action a. Note especially that

$$\forall a, \forall s \in S_c \cup \{\tau\} : \mathbf{T}(s, a) = \tau.$$

According to the above, the game starts at some preassigned state $s_0 = (x_0^1, x_0^2, x_0^3, i_0)$ and at the t-th turn ($t \in \mathbb{N}$) is in the state $s_t = (x_t^1, x_t^2, \ldots, x_t^N, i_t)$. This results in a *game history* $\mathbf{s} = s_0 s_1 s_2 \ldots$ with $s_i \in \bar{S}$ for all $i \in \mathbb{N}_0$. In other words, we assume each play of the game lasts an infinite number of turns (but it "effectively" terminates when the terminal state τ is entered). We define the following history sets:

(1) Histories of length $k : H_k = \{\mathbf{s} = s_0 s_1 ... s_k\}$.
(2) Histories of finite length: $H_* = \cup_{k=1}^{\infty} H_k$.
(3) Histories of infinite length: $H_\infty = \{\mathbf{s} = s_0 s_1 ... s_k ...\}$.

For a given game (with given capture set S_c) and given history $s_0 s_1 s_2 ...$, the *capture time* is defined to be

$$T_C (s_0 s_1 s_2 ...) = \min \{t : s_t \in S_c\}.$$

If no capture takes place, the *capture time* is $T_C (s_0 s_1 s_2 ...) = \infty$. When a particular history is understood from the context, we will often write simply T_C. While in our formulation a GPG always lasts an infinite number of turns, if $T_C < \infty$, then $s_t = \tau$ for every $t > T_C$; hence, the game *effectively* ends at T_C.

Note that, at every turn $t \in \mathbb{N}$, for every player except one, the *action set* is a singleton. This, in addition to the fact that all players are aware of all previously executed moves, means that GPG is a *perfect information game.*

A *deterministic strategy* (also known as a *pure strategy*) is a function σ^n which assigns a move to each finite-length history:

$$\sigma^n : H_* \to V.$$

At the start of the game, P_n selects a σ^n, which determines all his subsequent moves. We will only consider *admissible*[a] deterministic strategies. As will be seen, since GPGs are games of perfect information, the player loses nothing by using only deterministic strategies. A strategy σ^n is called *positional* if the next move depends only on the current state of the game (but not on previous states or current time):

$$\sigma^n (s_0 s_1 ... s_t) = \sigma^n (s_t).$$

We write $\sigma = (\sigma^1, \ldots, \sigma^N)$ to denote the strategy vector or *strategy profile* of all players. We also write $\sigma^{-n} = (\sigma^j)_{j \in [N] \setminus \{n\}}$; for instance, if $\sigma = (\sigma^1, \sigma^2, \sigma^3)$, then $\sigma^{-1} = (\sigma^2, \sigma^3)$.

To complete the description of a GPG, we must specify the players' *payoff functions.* The *total payoff* function of the n-th player ($n \in [N]$) has the form

$$Q^n (s_0, \sigma) = \sum_{t=0}^{\infty} \gamma^t q^n (s_t),$$

[a]That is, they never produce moves outside the player's action set.

where: q^n is the *turn payoff* (it depends on s_t, the game state at time t) which is assumed to be bounded:

$$\exists M : \forall n \in [N], \forall s \in S : |q^n(s)| \leq M,$$

and $\gamma \in (0,1)$ is the *discount factor*. Specific payoff functions $(q^n)_{n=1}^N$ will be introduced presently; for each such we obtain a specific GPG. The general rule is that

$$(\forall n : q^n(s) \neq 0) \Rightarrow s \in S_c,$$

i.e. the capture states are the ones associated with non-zero stage payoffs.

We summarize all of the above as follows: A *Graph Pursuit Game* (GPG) is a tuple $(G, \overline{S}, S_{nc}, \mathbf{q}, Suc, \gamma, s_0)$ where

(1) $G = (V, E)$ is the *graph* on which the game is played.
(2) $\overline{S} = (V \times \cdots \times V \times \{1, \ldots, N\}) \cup \{\tau\}$ is the *state set* (this implicitly also defines the number of players).
(3) $S_{nc} \subseteq S$ is the set of *non-capture states* (this also defines implicitly the set of *capture states* $S_c = S \backslash S_{nc}$).
(4) $\mathbf{q} = \{q^1, \ldots, q^N\}$ are the *turn payoffs* for the N players.
(5) Suc is the *successor function*, which determines the sequence by which the players move.
(6) $\gamma \in (0,1)$ is the *discount factor*.
(7) s_0 is the *initial state*.

We will also often refer to the GPG *family*

$$(G, \overline{S}, S_{nc}, \mathbf{q}, Suc, \gamma) = \{(G, \overline{S}, S_{nc}, \mathbf{q}, Suc, \gamma, s_0) : s_0 \in S\}.$$

This is the *set* of all GPGs which share the same rules but start from a different initial condition. From each family $(G, \overline{S}, S_{nc}, \mathbf{q}, Suc, \gamma)$ we can obtain a specific game $(G, \overline{S}, S_{nc}, \mathbf{q}, Suc, \gamma, s_0)$ by specifying the initial state s_0.

It is worth emphasizing that a GPG is a type of stochastic game [1] (i.e. a game which is played in stages, with a one-shot game being played at each stage) with the following characteristics:

(1) Sequential player moves.
(2) Deterministic state transitions and payoffs.
(3) Perfect information and recall.
(4) Non-zero payoffs only at "preterminal" (i.e. capture) states.

It is also worth noting that, while we call a GPG a "stochastic" game, it actually evolves completely deterministically.

3.2. *Examples*

Let us now introduce some particular *3-player* GPG families $(G, \overline{S}, S_{nc}, \mathbf{q}, Suc, \gamma)$. As mentioned, we obtain a particular GPG games class by specifying S_c and $\mathbf{q}$. In each of the following examples, the *total* payoff function is, of course,

$$\forall n \in [3] : Q^n (s_0, s_1, ...) = \sum_{t=0}^{\infty} \gamma^t q^n (s_t).$$

3-Player Linear Pursuits. In this game P_1 pursues P_2 and P_2 pursues P_3; the game ends when either P_1 captures P_2 or P_2 captures P_3 or both. The capture set is $S_c = \widetilde{S}^1 \cup \widetilde{S}^2 \cup \widetilde{S}^{12}$ where

$$\begin{aligned}
\widetilde{S}^1 &= \{s : (x, x, y, p), x \neq y \in V\} \text{ (P_1 captures P_2 and P_2 does not capture P_3)},\\
\widetilde{S}^2 &= \{s : (y, x, x, p), x \neq y \in V\} \text{ (P_1 does not capture P_2 and P_2 captures P_3)},\\
\widetilde{S}^{12} &= \{s : (x, x, x, p), x \in V\} \quad \text{(P_1 captures P_2 and P_2 captures P_3)}.
\end{aligned}$$

For all players n and states s, we have $q^n (s) = 0$, except in the following cases:

(1) $q^1 (s) = 1$ iff $s \in \widetilde{S}^1 \cup \widetilde{S}^{12}$, i.e. P_1 is rewarded when he captures P_2.
(2) $q^3 (s) = -1$ iff $s \in \widetilde{S}^2 \cup \widetilde{S}^{12}$, i.e. P_3 is punished when he is captured by P_2.
(3) The case of P_2 is a little more complicated.
    (a) $q^2 (s) = -1$ iff $s \in \widetilde{S}^1$, i.e. P_2 is punished when he is captured by P_1 *and has not simultaneously captured* P_3.
    (b) $q^2 (s) = 1$ iff $s \in \widetilde{S}^2$, i.e. P_2 is rewarded when he captures P_3 *and is not simultaneously captured by* P_1.

Note that $q^2 (s) = 0$ iff $s \in \widetilde{S}^{12}$, i.e. when P_2 simultaneously captures P_3 and is captured by P_1, he is neither rewarded nor punished (or: he receives one unit of reward and one of punishment, which cancel out).

Modified 3-Player Linear Pursuit. This game is almost identical to the previously defined 3-Player Linear Pursuit. The only difference is that the game does not terminate when it enters a state $s \in \widetilde{S}^{12}$. In other words, if all players are located in the same vertex, no capture takes place, all players receive zero payoff and the game continues.

3-Player Cyclic Pursuit. In this game, P_1 pursues P_2, P_2 pursues P_3 and P_3 pursues P_1; the game ends when either P_1 captures P_2, P_2 captures P_3 or P_3 captures P_1. The capture set is $S_c = \widetilde{S}^1 \cup \widetilde{S}^2 \cup \widetilde{S}^3 \cup \widetilde{S}^{123}$ where

$$
\begin{aligned}
\widetilde{S}^1 &= \{s : (x,x,y,p), x \neq y \in V\} \text{ (P_1 captures P_2 only)},\\
\widetilde{S}^2 &= \{s : (y,x,x,p), x \neq y \in V\} \text{ (P_2 captures P_3 only)},\\
\widetilde{S}^3 &= \{s : (x,y,x,p), x \neq y \in V\} \text{ (P_3 captures P_1 only)},\\
\widetilde{S}^{123} &= \{s : (x,x,x,p), x \in V\} \quad \text{(P_1 captures P_2, P_2 captures}\\
&\qquad \text{P_3 and P_3 captures P_1)}
\end{aligned}
$$

For all players n and states s, we have $q^n(s) = 0$, except in the following cases (the semantics are easily understood):

(1) $q^1(s) = 1$ iff $s \in \widetilde{S}^1$ and $q^1(s) = -1$ iff $s \in \widetilde{S}^3$.
(2) $q^2(s) = 1$ iff $s \in \widetilde{S}^2$ and $q^2(s) = -1$ iff $s \in \widetilde{S}^1$.
(3) $q^3(s) = 1$ iff $s \in \widetilde{S}^3$ and $q^3(s) = -1$ iff $s \in \widetilde{S}^2$.

Modified 3-Player Cyclic Pursuit: This game is almost identical to the previously defined 3-Player Cyclic Pursuit. The only difference is that the game does not terminate when it enters a state $s \in \widetilde{S}^{123}$. In other words, if all players are located in the same vertex, no capture takes place (and such a state yields zero payoff to all players).

Two Selfish Cops and one Adversarial Robber (SCAR): In this game P_1 and P_2 (the cops, also called C_1 and C_2) pursue P_3 (the robber, also called R) who tries to evade them. Unlike the classic CR game, in SCAR the capturing cop receives a higher payoff than the non-capturing one; each cop is selfish and wants to maximize his own payoff, i.e. he wants to be the one who effects the capture. We define the capture set as in terms of the following sets:

$$
\begin{aligned}
\widetilde{S}^1 &= \left\{s : s = (x^1, x^2, x^3, n) \text{ with } x^1 = x^3,\ x^2 \neq x^3\right\}:\\
&\quad R \text{ is captured } \textit{only} \text{ by } C_1;\\
\widetilde{S}^2 &= \left\{s : s = (x^1, x^2, x^3, n) \text{ with } x^1 \neq x^3,\ x^2 = x^3\right\}:\\
&\quad R \text{ is captured } \textit{only} \text{ by } C_2;\\
\widetilde{S}^{12} &= \left\{s : s = (x^1, x^2, x^3, n) \text{ with } x^1 = x^2 = x^3\right\}:\\
&\quad R \text{ is captured by } C_1 \textit{ and } C_2.
\end{aligned}
$$

and accordingly the capture set is

$$S_c = \widetilde{S}^1 \cup \widetilde{S}^2 \cup \widetilde{S}^{12}.$$

To define the turn payoffs, we introduce an additional fixed constant $\varepsilon \in \left[0, \frac{1}{2}\right]$. For every n and s, $q^n(s)$ is zero, except in the following cases:

(1) For $n \in \{1, 2\}$, P_n is rewarded when either he or the other cop captures P_3; but P_n's payoff is greater in the first case:
   (a) $q^n(s) = 1 - \varepsilon$ iff $s \in \widetilde{S}^n$,
   (b) $q^n(s) = \varepsilon$ iff $s \in \widetilde{S}^m$ with $n \neq m$,
   (c) $q^n(s) = \frac{1}{2}$ iff $s \in \widetilde{S}^{12}$.

(2) P_3 is punished when he is captured by P_1 or P_2: $q^3(s) = -1$ if $s \in S_c$.

It is worth noting that each cop could delay the capture in order to effect it when this is profitable.

N-Player Games. Each of the above 3-Player GPGs can be generalized, in a rather obvious manner, to an N-Player GPG.

4. Nash Equilibria of GPGs

In this chapter, we will prove that all GPGs have both positional and non-positional NE. In what follows, we will use the following conventions. First, we always assume the play sequence

$$\cdots P_1 \to P_2 \to \cdots \to P_N \to P_1 \to \cdots .$$

Secondly, we assume S, S_{nc}, $\mathbf{q}$, Suc to be known from the context and, omitting them from the notation, we will denote a specific GPG as $\Gamma_N(G, \gamma | s_0)$ or (also omitting γ) as $\Gamma_N(G | s_0)$.

We first establish that every GPG has an NE in *deterministic positional* strategies. The proof is based on a theorem by Fink [19], which establishes the existence of an NE in *probabilistic* (more specifically *behavioral*) *positional* strategies; the proof presented here shows that in turn-based (and hence perfect information) games, deterministic strategies can be used without loss to the players. $\Gamma_N(G|s_0)$ designates any N-player GPG.

Theorem 1. *For every graph G, every $N \geq 2$ and every initial state $s_0 \in S$, the game $\Gamma_N(G|s_0)$ admits a profile of deterministic positional strategies $\widehat{\sigma} = \left(\widehat{\sigma}^1, \widehat{\sigma}^2, \ldots, \widehat{\sigma}^N\right)$ such that*

$$\forall n \in [N], \forall s_0 \in S, \forall \sigma^n : Q^n\left(s_0, \widehat{\sigma}^n, \widehat{\sigma}^{-n}\right) \geq Q^n\left(s_0, \sigma^n, \widehat{\sigma}^{-n}\right). \quad (1)$$

For every s and n, let $u^n(s) = Q^n(s, \widehat{\sigma})$. Then the following equations are satisfied:

$$\forall n, \forall s \in S^n : \widehat{\sigma}^n(s) = \arg \max_{a^n \in A^n(s)} [q^n(s) + \gamma u^n(\mathbf{T}(s, a^n))], \tag{2}$$

$$\forall n, m, \forall s \in S^n : u^m(s) = q^m(s) + \gamma u^m(\mathbf{T}(s, \widehat{\sigma}^n(s))). \tag{3}$$

Proof. Fink has proved in Ref. [19] that every N-player discounted stochastic game has a positional NE in *probabilistic* strategies; this result holds for the general game (i.e. with *concurrent* moves and probabilistic strategies and state transitions). According to [19], at equilibrium the following equations must be satisfied for all m and s:

$$\mathrm{u}^m(s) = \max_{\mathbf{p}^m(s)} \sum_{a^1 \in A^1(s)} \ldots \sum_{a^N \in A^N(s)} p^1_{a^1}(s) \ldots p^N_{a^N}(s) A(\mathrm{u}^m(s')), \tag{4}$$

where

$$A(\mathrm{u}^m(s')) = \left[q^m(s) + \gamma \sum_{s'} \Pi(s'|s, a^1, a^2, \ldots, a^N) \mathrm{u}^m(s') \right].$$

In the above, we have modified Fink's notation to fit our own.

(1) $\mathrm{u}^m(s)$ is the expected value of $u^m(s)$.
(2) $p^m_{a^m}(s)$ is the probability that, given the current game state is s, the m-th player plays action a^m.
(3) $\mathbf{p}^m(s) = (p^m_{a^m}(s))_{a^m \in A^m(s)}$ is the vector of all such probabilities (one probability per available action).
(4) $\Pi(s'|s, a^1, a^2, \ldots, a^N)$ is the probability that, given the current state is s and the player actions are $a^1, a^2, \ldots, a^N$, the next state is s'.

Choose any n and any $s \in S^n$. For all $m \neq n$, P_m has a single move, i.e. $A^m(s) = \{a^m\}$, and so $p^m_{a_m}(s) = 1$. Also, since transitions are deterministic,

$$\sum_{s'} \Pi(s'|s, a^1, a^2, \ldots, a^N) \mathrm{u}^n(s') = \mathrm{u}^n(\mathbf{T}(s, a^n)).$$

Hence, for $m = n$, (4) becomes

$$\mathrm{u}^n(s) = \max_{\mathbf{p}^n(s)} \sum_{a^n \in A^n(s)} p^n_{a^n}(s) [q^n(s) + \gamma \mathrm{u}^n(\mathbf{T}(s, a^n))]. \tag{5}$$

Furthermore, let us define $\widehat{\sigma}^n(s)$ (for the specific s and n) by

$$\widehat{\sigma}^n(s) = \arg\max_{a^n \in A^n(s)} [q^n(s) + \gamma \mathbf{u}^n(\mathbf{T}(s, a^n))]. \tag{6}$$

If (5) is satisfied by more than one a^n, we set $\widehat{\sigma}^n(s)$ to one of these arbitrarily. Then, to maximize the sum in (5) the n-th player can set $p^n_{\widehat{\sigma}^n(s)}(s) = 1$ and $p^n_a(s) = 0$ for all $a \neq \widehat{\sigma}^n(s)$. This is true for all states and all players (i.e. every player can, without loss, use deterministic strategies); hence, $\mathbf{u}^n(s) = u^n(s)$ and (5) becomes

$$u^n(s) = \max_{a^n \in A^n(s)} [q^n(s) + \gamma u^n(\mathbf{T}(s, a^n))] = q^n(s) + \gamma u^n(\mathbf{T}(s, \widehat{\sigma}^n(s))). \tag{7}$$

For $m \neq n$, the mth player has no choice of action and (5) becomes

$$u^m(s) = q^m(s) + \gamma u^m(\mathbf{T}(s, \widehat{\sigma}^n(s))). \tag{8}$$

We recognize that (6)–(8) are (2)–(3). Also, (6) defines $\widehat{\sigma}^n(s)$ for every n and s and the required deterministic positional strategies are $\widehat{\sigma} = (\widehat{\sigma}^1, \widehat{\sigma}^2, \ldots, \widehat{\sigma}^N)$. □

Note that the initial state s_0 plays no special role in the system (2)–(3). In other words, using the notation $u(s) = (u^1(s), u^2(s), \ldots, u^N(s))$ and $\mathbf{u} = (u(s))_{s \in S}$, we see that $\mathbf{u}$ and $\widehat{\sigma}$ are the same for every starting position s_0 and every game $\Gamma_N(G|s_0)$ (when N, G and γ are fixed). Fink's proof requires that, for every n, the total payoff is $Q^n(s_0, \sigma) = \sum_{t=0}^{\infty} \gamma^t q^n(s_t)$; but does not place any restrictions (except boundedness) on q^n; hence, the theorem applies to all GPG games.

We will next show that every GPG also has *non-positional* deterministic NE. For the sake of simplicity we will restrict our analysis to 3-player GPGs, denoted as $\Gamma(G|s_0)$; but the results can be immediately generalized to N-player games.

To prove the desired result, we introduce a family of *auxiliary games* and *threat strategies*. For every $n \in [3]$, we define the game $\widetilde{\Gamma}^n(G|s_0)$ played on G (and starting at s_0) by P_n against a player P_{-n} who controls the remaining two entities ("tokens"). For example, in $\widetilde{\Gamma}^1(G|s_0)$, P_1 plays against P_{-1} who controls P_2 and P_3. The $\widetilde{\Gamma}^n(G|s_0)$ elements (e.g. movement sequence, states, action sets, capturing conditions, etc.) are the same as in $\Gamma(G|s_0)$. P_n uses a strategy σ^n and P_{-n} uses a strategy profile σ^{-n}; these form a strategy profile $\sigma = (\sigma^1, \sigma^2, \sigma^3)$ (which can also be used in $\Gamma(G|s_0)$). The payoffs to P_n and P_{-n} in $\widetilde{\Gamma}^n(G|s_0)$ are

$$\widetilde{Q}^n(s_0,\sigma) = Q^n(s_0,\sigma) = \sum_{t=0}^{\infty} \gamma^t q^n(s_t) \quad \text{and} \quad \widetilde{Q}^{-n}(s_0,\sigma) = -\widetilde{Q}^n(s_0,\sigma).$$

Since the capture rules of $\widetilde{\Gamma}^n(G|s_0)$ are those of $\Gamma(G|s_0)$, P_{-n} can use one of his tokens to capture the other. For instance, in $\widetilde{\Gamma}^1(G|s_0)$, P_{-1} can use P_2 to capture P_3. Note that in this case P_1 receives zero payoff (since he did not capture) and P_{-1} also receives zero payoff, since $\widetilde{\Gamma}^1(G|s_0)$ is a zero-sum game.

Since $\widetilde{\Gamma}^n(G|s_0)$ is a two-player *zero-sum* discounted stochastic game, the next lemma follows from Ref. [1, Theorem 4.3.2].

Lemma 1. *For every n, G and s_0, the game $\widetilde{\Gamma}^n(G|s_0)$ has a value and the players have optimal deterministic positional strategies.*

Furthermore, the value and optimal strategies can be computed by Shapley's value-iteration algorithm [1]. Let us denote by $\widehat{\phi}_n^n$ (resp. $\widehat{\phi}_n^{-n}$) the optimal strategy of P_n (resp. P_{-n}) in $\widetilde{\Gamma}^n(G|s_0)$. For example, in $\widetilde{\Gamma}^1(G|s_0)$, P_1 has the optimal strategy $\widehat{\phi}_1^1$ and P_{-1} has the optimal strategy $\widehat{\phi}_1^{-1} = (\widehat{\phi}_1^2, \widehat{\phi}_1^3)$. In fact, the same $\widehat{\phi}_n^m$s (for fixed n and any $m \in [3]$) are optimal in $\widetilde{\Gamma}_3^n(G|s_0)$ for every initial position s_0.

We return to $\Gamma(G|s_0)$, and for each P_n we introduce the *threat strategy* $\widehat{\pi}^n$ defined as follows:

(1) As long as every player P_m (with $m \neq n$) follows $\widehat{\phi}_m^m$, P_n follows $\widehat{\phi}_n^n$.
(2) As soon as some player P_m (with $m \neq n$) deviates from $\widehat{\phi}_m^m$, P_n switches to $\widehat{\phi}_m^n$ and uses it for the rest of the game.[b]

Note that the $\widehat{\pi}^n$ strategies are *not* positional. In particular, the action of a player at time t may be influenced by the action (deviation) performed by another player at time $t-2$. However, as we will now prove, $(\widehat{\pi}^1, \widehat{\pi}^2, \widehat{\pi}^3)$ is a (non-positional) NE in $\Gamma(G|s_0)$.

[b]Since $\Gamma(G|s_0)$ is a perfect information game, the deviation will be detected immediately.

Theorem 2. *For every G, s_0 and γ, we have*

$$\forall n \in \{1,2,3\}, \forall \pi^n : Q^n(s, \widehat{\pi}^1, \widehat{\pi}^2, \widehat{\pi}^3) \geq Q^n(s, \pi^n, \widehat{\pi}^{-n}). \tag{9}$$

Proof. We choose some initial state s_0 and fix it for the rest of the proof. Now let us prove (9) for the case $n = 1$. In other words, we will show that

$$\forall \pi^1 : Q^1(s_0, \widehat{\pi}^1, \widehat{\pi}^2, \widehat{\pi}^3) \geq Q^1(s_0, \pi^1, \widehat{\pi}^2, \widehat{\pi}^3). \tag{10}$$

We take any π^1 and let

the history produced by $(\widehat{\pi}^1, \widehat{\pi}^2, \widehat{\pi}^3)$ be $\widehat{\mathbf{s}} = \widehat{s}_0\widehat{s}_1\widehat{s}_2\ldots,$
the history produced by $(\pi^1, \widehat{\pi}^2, \widehat{\pi}^3)$ be $\widetilde{\mathbf{s}} = \widetilde{s}_0\widetilde{s}_1\widetilde{s}_2\ldots,$

(where $\widehat{s}_0 = \widetilde{s}_0 = s_0$). We define T_1 as the earliest time in which π^1 and $\widehat{\pi}^1$ produce different states:

$$T_1 = \min\{t : \widetilde{s}_t \neq \widehat{s}_t\},$$

If $T_1 = \infty$, then $\widetilde{\mathbf{s}} = \widehat{\mathbf{s}}$ and

$$Q^1(s, \widehat{\pi}^1, \widehat{\pi}^2, \widehat{\pi}^3) = Q^1(s, \pi^1, \widehat{\pi}^2, \widehat{\pi}^3). \tag{11}$$

If $T_1 < \infty$, on the other hand, then $\widetilde{s}_t = \widehat{s}_t$ for every $t < T_1$ and we have

$$\begin{aligned} Q^1(s, \widehat{\pi}^1, \widehat{\pi}^2, \widehat{\pi}^3) &= \sum_{t=0}^{T_1-2} \gamma^t q^1(\widehat{s}_t) + \sum_{t=T_1-1}^{\infty} \gamma^t q^1(\widehat{s}_t) \\ &= \sum_{t=0}^{T_1-2} \gamma^t q^1(\widetilde{s}_t) + \sum_{t=T_1-1}^{\infty} \gamma^t q^1(\widehat{s}_t), \end{aligned} \tag{12}$$

$$Q^1(s, \pi^1, \widehat{\pi}^2, \widehat{\pi}^3) = \sum_{t=0}^{T_1-2} \gamma^t q^1(\widetilde{s}_t) + \sum_{t=T_1-1}^{\infty} \gamma^t q^1(\widetilde{s}_t). \tag{13}$$

We define $s^* = \widehat{s}_{T_1-1} = \widetilde{s}_{T_1-1}$ and proceed to compare the sums in (12) and (13).

First consider $\sum_{t=T_1-1}^{\infty} \gamma^t q^1(\widehat{s}_t)$. The history $\widehat{\mathbf{s}} = \widehat{s}_0\widehat{s}_1\widehat{s}_2\ldots$ is produced by $(\widehat{\phi}_1^1, \widehat{\phi}_2^2, \widehat{\phi}_3^3)$ and, since the $\widehat{\phi}_n^n$s are positional strategies, we have

$$\sum_{t=T_1-1}^{\infty} \gamma^t q^1(\widehat{s}_t) = \gamma^{T_1-1} \sum_{t=0}^{\infty} \gamma^t q^1(\widehat{s}_{T_1-1+t}) = \gamma^{T_1-1} \widetilde{Q}^1\left(s^*, \widehat{\phi}_1^1, \widehat{\phi}_2^2, \widehat{\phi}_3^3\right), \tag{14}$$

i.e. up to the multiplicative constant γ^{T_1-1}, the sum in (14) is the payoff to P_1 in $\widetilde{\Gamma}^1(G|s^*)$, under the strategies $\widehat{\phi}_1^1, (\widehat{\phi}_2^2, \widehat{\phi}_3^3)$. Since $\widetilde{\Gamma}^1(G|s^*)$ is a zero-sum game in which the optimal response to $\widehat{\phi}_1^1$ is $(\widehat{\phi}_1^2, \widehat{\phi}_1^3)$, we have

$$\gamma^{T_1-1}\widetilde{Q}^1\left(s^*, \widehat{\phi}_1^1, \widehat{\phi}_2^2, \widehat{\phi}_3^3\right) \geq \gamma^{T_1-1}\widetilde{Q}^1\left(s^*, \widehat{\phi}_1^1, \widehat{\phi}_1^2, \widehat{\phi}_1^3\right). \tag{15}$$

Next consider $\sum_{t=T_1-1}^{\infty}\gamma^t q^1(\widetilde{s}_t)$. The history $\widetilde{\mathbf{s}} = \widetilde{s}_0\widetilde{s}_1\widetilde{s}_2...$ is produced by $(\pi^1, \widehat{\phi}_1^2, \widehat{\phi}_1^3)$ and, since π^1 is not necessarily positional, $\widetilde{s}_{T_1}\widetilde{s}_{T_1+1}\widetilde{s}_{T_1+2}...$ may depend on $\widetilde{s}_0\widetilde{s}_1...\widetilde{s}_{T_1-2}$. However, we can introduce a strategy ρ^1 which will in general depend on $\widetilde{s}_0\widetilde{s}_1...\widetilde{s}_{T_1-2}$ (hence, it is not positional) and will produce the same history $\widetilde{s}_{T_1}\widetilde{s}_{T_1+1}\widetilde{s}_{T_1+2}...$ as σ^1. Then, since in $\widetilde{\Gamma}^1(G|s^*)$ the optimal response to $(\widehat{\phi}_1^2, \widehat{\phi}_1^3)$ is $\widehat{\phi}_1^1$, we have

$$\gamma^{T_1-1}\widetilde{Q}^1\left(s^*, \widehat{\phi}_1^1, \widehat{\phi}_1^2, \widehat{\phi}_1^3\right) \geq \gamma^{T_1-1}\widetilde{Q}^1\left(s^*, \rho^1, \widehat{\phi}_1^2, \widehat{\phi}_1^3\right) = \sum_{t=T_1-1}^{\infty}\gamma^t q^1(\widetilde{s}_t). \tag{16}$$

Combining (12)–(16), we have

$$\begin{aligned}
Q^1\left(s_0, \widehat{\pi}^1, \widehat{\pi}^2, \widehat{\pi}^3\right) &= \sum_{t=0}^{T_1-2}\gamma^t q^1(\widetilde{s}_t) + \gamma^{T_1-1}\widetilde{Q}^1\left(s^*, \widehat{\phi}_1^1, \widehat{\phi}_2^2, \widehat{\phi}_3^3\right) \\
&\geq \sum_{t=0}^{T_1-2}\gamma^t q^1(\widetilde{s}_t) + \gamma^{T_1-1}\widetilde{Q}^1\left(s^*, \widehat{\phi}_1^1, \widehat{\phi}_1^2, \widehat{\phi}_1^3\right) \\
&\geq \sum_{t=0}^{T_1-2}\gamma^t q^1(\widetilde{s}_t) + \gamma^{T_1-1}\widetilde{Q}^1\left(s^*, \rho^1, \widehat{\phi}_1^2, \widehat{\phi}_1^3\right) \\
&= Q^1\left(s, \pi^1, \widehat{\pi}^2, \widehat{\pi}^3\right).
\end{aligned}$$

and we have proved (10), which is (9) for $n = 1$. The proof for the cases $n = 2$ and $n = 3$ is similar and hence omitted. □

5. Algorithms for NE Computation

5.1. *The basic multi-value iteration (MVI) algorithm*

The goal of the MVI algorithm is to compute a positional deterministic NE $\widehat{\sigma} = \left(\widehat{\sigma}^1, \widehat{\sigma}^2, \widehat{\sigma}^3\right)$ and the corresponding NE payoff vectors

$$Q(s_0, \widehat{\sigma}) = \left(Q^1(s_0, \widehat{\sigma}), Q^2(s_0, \widehat{\sigma}), Q^3(s_0, \widehat{\sigma})\right)$$

for the GPG game $\Gamma_3(G|s_0, \gamma)$ and all initial states s_0. Here is a description of the algorithm (which is presented in pseudocode in the sequel) for three-player games with "natural" successor function; generalizations to N-players and general succession is straightforward.

(1) The algorithm is implemented as a function, which will form the basic building block of more sophisticated implementation, to be presented in the sequel.

(2) The inputs to the function are G, S_c, q, γ, which give a full description of the specific GPG to which MVI is applied. In addition, we provide $I_{\max}$, which is the maximum number of iterations for which MVI will be run; by letting $I_{\max} = \infty$, we let the algorithm run ad infinitum, *unless* the break condition of lines 32–34 holds (this will be further discussed a little later).

(3) Lines 2–12 perform the algorithm initialization.

  (a) In line 3, we set vertex set V equal to the vertex set of the given graph G.

  (b) In what follows, note that the correspondence between GPG states and algorithm variables is $s \leftrightarrow (n_1, n_2, n_3, p)$.

  (c) In the loop of lines 4–12, all initial payoffs are set to zero, except for the ones which correspond to capture states $s = (n_1, n_2, n_3, p) \in S_c$.

(4) The main part of the algorithm is in lines 13–37, which are repeated until (a) either the maximum number of iterations $I_{\max}$ is exceeded or (b) both Q and $\widehat{\sigma}$ remain unchanged in two successive iterations (lines 32–34).

(5) In every iteration the algorithm:

  (a) loops through all non-capture states (lines 17 and 18),

  (b) assigns to each player optimal next move from current state (lines 19–21),

  (c) updates accordingly the payoffs.

(6) The algorithm may terminate in two ways.

  (a) Either the no-change condition (lines 32–34) is satisfied, in which case the function returns the obtained $[Q, \widehat{\sigma}]$ pair and breaks out of the loop. In this case we say that the algorithm has *converged* or *reached equilibrium.* As we will show in the sequel, in this case the returned $\widehat{\sigma}$ is a positional deterministic NE of the GPG and Q is the vector of corresponding payoffs to the players.

  (b) Or the maximum number of iterations $I_{\max}$ is exceeded, in which case the function does not return an output.

To further elucidate the rationale of the algorithm, consider the following:

(1) In the *zero-th* iteration we assign to the capture states the appropriate payoff vector. Capture states correspond to games of length 0.
(2) In the *first* iteration of the algorithm, for each non-capture state $s = (x^1, x^2, x^3, n) \in S_{nc}$, player P_n chooses a move which maximizes his payoff $Q^n(s)$ (lines 17–19). *This also determines the payoff of the other players*, namely $Q^{-n}(s)$. Note that if, at the end of the first iteration, for some s there exists an m such that $Q^m(s) \neq 0$, then, under optimal play, s results in a capturing game of length 1.
(3) In the *second* iteration, for each state $s = (x^1, x^2, x^3, n)$ player P^n chooses a move which maximizes his payoff, *under the assumption that the next player has also played optimally* (actually, in accordance to the results of the first iteration and *provided that the results of the first iteration are not modified at a later iteration*). Again, P^n's choice also determines the payoffs of the other players. Note that if, at the end of the second iteration, for some s there exists an m such that $Q^m(s) \neq 0$ (*which can only happen if P^n can move the game into a state $s' = (y^1, y^2, y^3, n')$) such that $Q^m(s') \neq 0$*), then, under optimal play, s results in a capturing game of length 2.
(4) The algorithm proceeds in the same manner, determining the payoffs for games of progressively greater length. However, these payoffs may be revised at a later iteration.

We now give the pseudocode description of the MVI function.

Multi-Value Iteration (MVI)

1: **function** $[Q, \widehat{\sigma}]$=MVI$(G, S_c, q, \gamma, I_{\max})$
2: **//Initialization**
3: $V = V(G)$
4: **for all**$(n_1, n_2, n_3, p, n) \in V \times V \times V \times \{1,2,3\} \times \{1,2,3\}$ **do**
5: $\quad Q^n(n_1, n_2, n_3, p) = 0$
6: $\quad \widehat{\sigma}^n(n_1, n_2, n_3, p) = 0$
7: $\quad$ **if** $(n_1, n_2, n_3, p) \in S_c$ **then**
8: $\quad\quad$ **for all** $n \in \{1,2,3\}$ **do**
9: $\quad\quad\quad Q^n(n_1, n_2, n_3, p) = q^n(n_1, n_2, n_3, p)$
10: $\quad\quad$ **end for**
11: $\quad$ **end if**
12: **end for**
13: **//Main**
14: **for** $i = 0$ **to** $I_{\max}$ **do**
15: $\quad Q_{new} = Q$
16: $\quad \widehat{\sigma}_{new} = \widehat{\sigma}$

Multi-Value Iteration (MVI)

    for all $(n_1, n_2, n_3) \in V \times V \times V$ **do**
      if $(n_1, n_2, n_3, p) \in S_{nc}$ **then**
        $\widehat{n}_1 = \arg\max_{m \in N[n_1]} \gamma Q^1 (m, n_2, n_3, 2)$
        $\widehat{n}_2 = \arg\max_{m \in N[n_2]} \gamma Q^2 (n_1, m, n_3, 3)$
        $\widehat{n}_3 = \arg\max_{m \in N[n_3]} \gamma Q^3 (n_1, n_2, m, 1)$
        $\widehat{\sigma}^1 (n_1, n_2, n_3) = \widehat{n}_1$
        $\widehat{\sigma}^2 (n_1, n_2, n_3) = \widehat{n}_2$
        $\widehat{\sigma}^3 (n_1, n_2, n_3) = \widehat{n}_3$
        for all $k \in \{1, 2, 3\}$ **do**
          $Q^k_{new} (n_1, n_2, n_3, 1) = \gamma Q^k (\widehat{n}_1, n_2, n_3, 2)$
          $Q^k_{new} (n_1, n_2, n_3, 2) = \gamma Q^k (n_1, \widehat{n}_2, n_3, 3)$
          $Q^k_{new} (n_1, n_2, n_3, 3) = \gamma Q^k (n_1, n_2, \widehat{n}_3, 1)$
        end for
      end if
    end for
    if $Q_{new} = Q$ & $\widehat{\sigma}_{new} = \widehat{\sigma}$ **then**
      return $[Q, \widehat{\sigma}]$
      break
    end if
    $Q = Q_{new}$
    $\widehat{\sigma} = \widehat{\sigma}_{new}$
  end for
  end function

We claim that: for every graph G, *if MVI converges* (i.e. if at some iteration produces no further changes to the payoffs of any state), then it has computed a positional deterministic equilibrium $\widehat{\sigma}$ and the corresponding equilibrium payoff vectors for a three-player GPG game $\Gamma (G|s_0, \gamma)$, for all initial states s_0. This is the subject of the following.

Theorem 3. *For every* G, γ, s_0*, if MVI converges, then the output* $\widehat{\sigma}$ *is a deterministic positional equilibrium of* $\Gamma (G|s_0, \gamma)$, i.e. *the following holds*:

$$\forall n : \forall \sigma^n : \forall s_0 : Q^n \left(s_0, \widehat{\sigma}^n, \widehat{\sigma}^{-n}\right) \geq Q^n \left(s_0, \sigma^n, \widehat{\sigma}^{-n}\right). \tag{17}$$

Proof. We have $\widehat{\sigma} = \left(\widehat{\sigma}^1, \widehat{\sigma}^2, \widehat{\sigma}^3\right)$. Let us choose some n and some deterministic positional strategy σ^n and let $\sigma = (\sigma^n, \widehat{\sigma}^{-n})$. Then (9) can be rewritten as

$$\forall n : \forall \sigma^n : \forall s_0 = \left(x^1, x^2, x^3, p\right) : Q^n (s_0, \widehat{\sigma}) \geq Q^n (s_0, \sigma). \tag{18}$$

For the sake of concreteness, we will only prove the case $n = p = 1$ (the remaining cases are proved similarly). In this case, $\sigma = (\sigma^1, \widehat{\sigma}^2, \widehat{\sigma}^3)$; so we will prove

$$\forall \sigma^1 : \forall s_0 = \left(x^1, x^2, x^3, 1\right) : Q^1\left(s_0, \widehat{\sigma}\right) \geq Q^1\left(s_0, \sigma\right).$$

Let us first define the state sequences $s_0 s_1 s_2 ...$ and $s_0 \widehat{s}_1 \widehat{s}_2 ...$ as follows:

$$\begin{array}{l} s_1 = \sigma\left(s_0\right), \; s_2 = \sigma\left(s_1\right), \; s_3 = \sigma\left(s_2\right), \; s_4 = \sigma\left(s_3\right), \; ... \\ \widehat{s}_1 = \widehat{\sigma}\left(s_0\right), \; \widehat{s}_2 = \widehat{\sigma}\left(s_1\right), \; \widehat{s}_3 = \widehat{\sigma}\left(s_2\right), \; \widehat{s}_4 = \widehat{\sigma}\left(s_3\right), \; ... \end{array}$$

Note that $s_0 s_1 s_2 ...$ is the history produced by (s_0, σ) and

$$s_2 = \widehat{s}_2, \quad s_3 = \widehat{s}_3, \quad s_5 = \widehat{s}_5, \quad s_6 = s_6, \quad s_8 = s_8, \quad \ldots$$

Assuming that the algorithm has converged, the resulting $\widehat{\sigma}$ always chooses maximizing successor states. Hence, we have the following sequence of inequalities.

$$\begin{array}{l} Q^1\left(s_0, \widehat{\sigma}\right) = q^1\left(s_0\right) + \gamma Q^1\left(\widehat{s}_1, \widehat{\sigma}\right) \geq q^1\left(s_0\right) + \gamma Q^1\left(s_1, \widehat{\sigma}\right) \\ Q^1\left(s_1, \widehat{\sigma}\right) = q^1\left(s_1\right) + \gamma Q^1\left(\widehat{s}_2, \widehat{\sigma}\right) = q^1\left(s_1\right) + \gamma Q^1\left(s_2, \widehat{\sigma}\right) \\ \qquad \Rightarrow Q^1\left(s_0, \widehat{\sigma}\right) \geq \sum_{t=0}^{1} \gamma^t q^1\left(s_t\right) + \gamma^2 Q^1\left(s_2, \widehat{\sigma}\right) \\ Q^1\left(s_2, \widehat{\sigma}\right) = q^1\left(s_2\right) + \gamma Q^1\left(\widehat{s}_3, \widehat{\sigma}\right) = q^1\left(s_2\right) + \gamma Q^1\left(s_3, \widehat{\sigma}\right) \\ \qquad \Rightarrow Q^1\left(s_0, \widehat{\sigma}\right) \geq \sum_{t=0}^{2} \gamma^t q^1\left(s_t\right) + \gamma^3 Q^1\left(s_3, \widehat{\sigma}\right) \\ Q^1\left(s_3, \widehat{\sigma}\right) = q^1\left(s_3\right) + \gamma Q^1\left(\widehat{s}_4, \widehat{\sigma}\right) \geq q^1\left(s_3\right) + \gamma Q^1\left(s_4, \widehat{\sigma}\right) \\ \qquad \Rightarrow Q^1\left(s_0, \widehat{\sigma}\right) \geq \sum_{t=0}^{3} \gamma^t q^1\left(s_t\right) + \gamma^4 Q^1\left(s_4, \widehat{\sigma}\right), \\ \qquad \ldots \end{array}$$

Continuing in this way, we get

$$\forall T : Q^1\left(s_0, \widehat{\sigma}\right) \geq \sum_{t=0}^{T-1} \gamma^t q^1\left(s_t\right) + \gamma^T Q^1\left(s_T, \widehat{\sigma}\right),$$

and, taking the limit as $T \to \infty$,

$$Q^1(s_0, \widehat{\sigma}) \geq \sum_{t=0}^{\infty} \gamma^t q^1(s_t) + \lim_{T\to\infty} \left(\gamma^T Q^1(s_T, \widehat{\sigma})\right) = \sum_{t=0}^{\infty} \gamma^t q^1(s_t) + 0$$
$$= Q^1(s_0 s_1 s_2 ...) = Q^1(s_0, \sigma)$$

which is the required result. □

In conclusion, when the MVI algorithm is run on a specific game $\Gamma(G, \gamma|s_0)$, there exist only two possible outcomes.

(1) The algorithm will not converge (will not terminate before exceeding the maximum number of iterations) and will not output a strategy profile.
(2) The algorithm will converge and will output a strategy profile $\widehat{\sigma}$ which, by Theorem 3, is a positional deterministic NE of $\Gamma(G, \gamma|s_0)$.

5.2. *Multi-start multi-value iteration algorithm*

Multi-Start Multi-Value Iteration (MS–MVI) addresses two problems of the "basic" MVI algorithm presented above. These problems follow from the concluding remarks of Section 5.1, which can be rephrased as follows: when the MVI algorithm is run on a specific game $\Gamma(G, \gamma|s_0)$, there exist only two possible outcomes.

(1) The algorithm will not converge and will not output a $\widehat{\sigma}$.
(2) When the algorithm converges, it will always output *the same* NE $\widehat{\sigma}$ (since MVI is a deterministic algorithm).

Now, the search for a positional deterministic NE $\widehat{\sigma}$ of $\Gamma(G, \gamma|s_0)$ is essentially the search for a solution of the system of equations (3). This system will in general have more (sometimes many more) than one solutions. Hence, we would like a procedure which *with high probability will find as many solutions as possible.*

It turns out that these requirements can be satisfied with slight modifications of the "basic" MVI algorithm. Apparently, a factor which influences the convergence of MVI is the order of the payoff updates (lines 17–31 of the pseudocode); this will be seen in the experiments presented in Section 6. Obviously, one way to exploit this is to run MVI multiple times, with different update orders. Hence, we have initially experimented

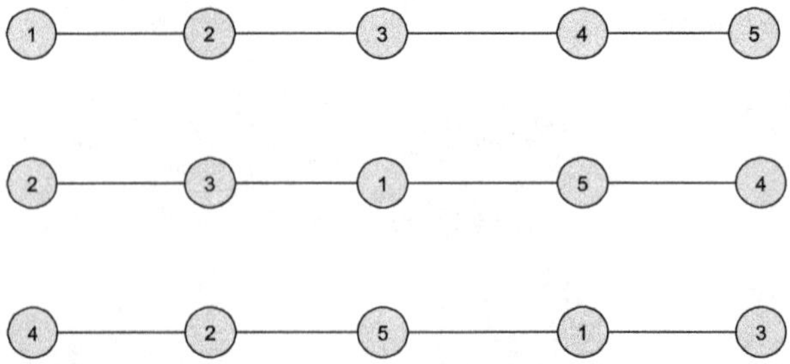

Fig. 1: Permuted paths.

with using a *randomized* update order in MVI. This has resulted in two improvements:

(1) The algorithm converges to a $\widehat{\sigma}$ more often than when a fixed update order is used.
(2) On different runs, the algorithm converges to different $\widehat{\sigma}$s.

Further numerical experiments revealed that it is not necessary to randomize the update order in *every* iteration of the MVI algorithm. Instead, it suffices to *randomly choose a fixed update order*. Then, each run of the algorithm has a non-zero convergence probability and different runs have a non-zero probability of producing different $\widehat{\sigma}$.

A final simplification follows from the observation that, rather than changing the update order, we can keep it fixed and instead permute the vertex labels (since the labels are simply the numbers $1, ..., N$, a vertex permutation is essentially equivalent to a permutation of the update order).

To better understand this, suppose that we want to solve a given GPG played on the 5-vertices path illustrated at the top of Fig. 1. Then we can apply MVI (with fixed update order) to the same GPG played on any of the paths of Fig. 1 (or any other isomorphic graph). This is equivalent to a changed update order and any solution (NE) for an isomorphic graph can be easily transformed to one for the original graph.

When we are dealing with a "small" graph, i.e. one with a small number N of vertices, we can use *every* permutation of vertex labels. This results in the following algorithm, presented in pseudocode:

MS–MVI with Full Vertex Permutations

function $[\mathbf{Q}, \mathbf{S}]=$ MVIAllVertexPerms$(G, S_c, q, \gamma, I_{\max})$
$\mathcal{G}=$AllVertexPerms(G)
$k=0$
for all $G' \in \mathcal{G}$ **do**
  $k \leftarrow k+1$
  $[Q, \widehat{\sigma}]=$MVI$(G', S_c, q, \gamma, I_{\max})$
  $\mathbf{S}_k=$InvPermS$(\widehat{\sigma}, G, G')$
  $\mathbf{Q}_k=$InvPermQ(Q, G, G')
end for
return [**Q**,**S**]
end function

The above function returns a list $[\mathbf{Q}, \mathbf{S}]$ of NE strategy profiles and corresponding payoffs. The function AllVertexPerms(G) returns all graphs resulting from vertex label permutations of G and the functions InvPermS$(\widehat{\sigma}, G, G')$, InvPermQ$(Q, G, G')$ perform the inverse label permutations, so that the returned strategies and payoffs correspond to the original graph labeling.

If N is relatively large (say, $N > 10$), it is not practical to generate all (i.e. $N!$) possible permutations. In this case, we can use a manageable number of *randomly generated permutations* and we have the following algorithm (The function RandVertexPerm(G) returns a graph resulting from a randomly chosen vertex label permutation of G):

MS–MVI with Random Vertex Permutations

function $[\mathbf{Q}, \mathbf{S}]=$MVIRandVertexPerms$(G, S_c, q, \gamma, I_{\max}, K)$
for $k=0$ **to** K **do**
  $G'=$RandVertexPerm(G)
  $[Q, \widehat{\sigma}]=$MVI$(G', S_c, q, \gamma, I_{\max})$
  $\mathbf{S}_k=$InvPermS$(\widehat{\sigma}, G, G')$
  $\mathbf{Q}_k=$InvPermQ(Q, G, G')
end for
return [**Q**,**S**]
end function

We refer to both of the above algorithms as *Multi-Start Multi-Value Iteration* (MS–MVI). The analogy to heuristic multi-start search procedures for global optimization is obvious and we can invoke the usual justification: "Multi-start procedures were originally conceived as a way to exploit a local or neighborhood search procedure, by simply applying it from multiple random initial solutions. Modern multi-start methods usually incorporate a powerful form of diversification in the generation of solutions to help overcome local optimality" [20].

5.3. *Regarding the comparison of MVI and MS–MVI*

Section 6 is devoted to experimental evaluation of the MVI and MS–MVI algorithms. Before the actual experiments, let us present some considerations which will put the experimental results in perspective.

Given a particular GPG game $\Gamma(G|s_0)$, our goal is to find as many of its NE as possible. It must be emphasized that we can expect most GPGs to have a large number of NE. To illustrate this, we present an example. Suppose that the Modified three-Player Linear Pursuit game $\Gamma(G|s_0)$ is played on a six-vertices path. Now consider two instances of the game: the first instance is $\Gamma(G|s_0')$ with $s_0' = (4,6,1,1)$ and the second is $\Gamma(G|s_0'')$ with $s_0'' = (4,4,4,1)$. These initial states appear in Fig. 2.

The solution of the game $\Gamma(G|s_0')$ is "essentially" always the same: P_1 will head toward P_2, who will stay in place and will be eventually captured. There are many NE which will capture this behavior. The important thing is that P_3 can use *any* strategy by which he stays "behind" P_1; for example, P_3 can stay in place, or move toward P_1, etc. Each such strategy can be part of an NE profile (i.e. P_3 has no incentive to change such a strategy, as long as the other two players follow their abovementioned strategies).[c] Hence, in this case we have a multitude of NE which, however, essentially result in the same outcome.

The situation is different for the game $\Gamma(G|s_0'')$. Here there exist *qualitatively different* classes of NE profiles. We leave as an exercise to the reader to check that each of the following situations yields an NE.[d]

[c]To be precise, the behaviors we have described above are "strategy fragments", i.e. they are the parts of the functions σ^n (for $n \in [3]$) which describe the player moves for the game states which appear in the above particular situations; while a full strategy describes a player's move for *every* possible game state.

[d]In proving that each of the above is an NE, keep in mind that the Modified three-Player Linear Pursuit game does not terminate when all players are located in the same vertex.

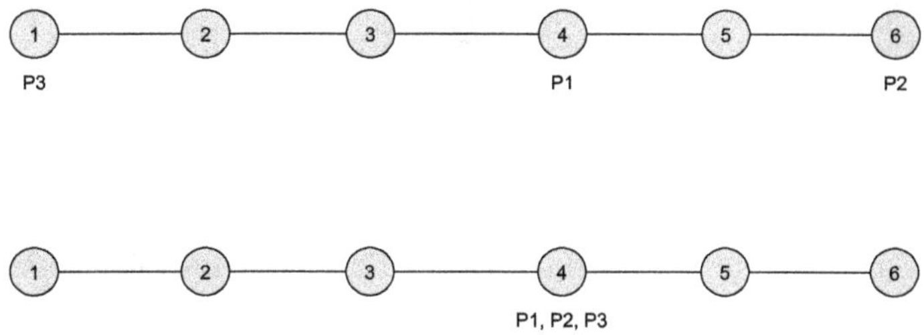

Fig. 2: The modified three-player linear pursuit game played from different initial conditions.

(1) P_1 captures P_2 in 4 turns.
(2) P_2 captures P_3 in 1 turn.
(3) All players stay in place and no capture ever takes place.

While many NE strategy profiles correspond to each of the above situations, the situations themselves are different in important respects, namely in the capture time and the capturing player.

The above considerations are important in the analysis of experimental results. Namely, we expect that the randomized MS–MVI algorithm will produce a large number of different NE; however, many of these will correspond to the same "situation" (i.e. same capture time and capturing player). This motivates us to introduce the *outcome function* $O_G(s_0, \sigma)$, which, for a given GPG game and graph, is defined as follows:

$$O_G(s_0, \sigma) = (T_c, \mathbf{P}_c). \tag{19}$$

In (19), T_c is the *capture time* and $\mathbf{P}_c$ is the *capturing players set* (note that in certain situations we can have more than one capturing players). Both of these are completely determined assuming that the initial state is s_0 and the players use strategy profile $\sigma = (\sigma^1, \sigma^2, \sigma^3)$; in case no capture takes place, we let $T_c = \infty$ and $P_c = \lambda$. In other words,

$$O_G : S \times \Sigma \to (N_0 \cup \{\infty\}) \times \wp(\{P_1, P_2, P_3, \lambda\}),$$

where $\wp(\{P_1, P_2, P_3, \lambda\})$ is the power set (set of all subsets) of $\{P_1, P_2, P_3, \lambda\}$. The dependence on the particular game played is omitted from the notation and assumed to be known from the context.

So, for example, the three possible outcomes described for the second game of Fig. 2 can be described thus: there exist strategy profiles $\sigma', \sigma'', \sigma'''$

such that

$$O_G((4,4,4,1),\sigma') = (4, P_1),$$
$$O_G((4,4,4,1),\sigma'') = (1, P_2),$$
$$O_G((4,4,4,1),\sigma''') = (\infty, \lambda).$$

We understand that there exist many different (but not *substantially different*) σ' which yield the same outcome $(4, P_1)$; and the same is true for σ'' and $(1, P_2)$ as well as for σ''' and (∞, λ).

6. Experiments

6.1. *Introductory remarks*

In this section, we will present numerical experiments to evaluate the performance of "basic" MVI and MS–MVI. The results are arranged by game type: Section 6.2 presents Linear Pursuit and Modified Linear Pursuit, Section 6.3 presents Cyclic Pursuit and Modified Cyclic Pursuit and Section 6.4 presents SCAR. All reported experiments have been performed on a Windows 8 personal computer with Intel Core i5-4590s CPU running at 3 GHz and using an 8 GB RAM. The results are presented in tabular form, where each table has the following structure:

(1) In the first column, we list the specific initial condition used and whether results of the corresponding row concern MVI or MS–MVI.
(2) In the second column, we indicate whether the algorithm converged or not.
(3) If the algorithm has converged, then it has produced an NE which may be capturing or not. In the third column, we list the capturing players (there may exist more than one). If the NE is non-capturing, we write 0, to indicate that no player effected a capture.
(4) In the fourth column, we list the capturing time (i.e. the duration of the game), which can be infinity (if the NE is non-capturing) or N/A (if the algorithm did not converge).
(5) In the sixth column, we list the proportion (between zero and one) of the computed NE which correspond to the specifics of the current row.
    (a) The first row corresponds to the MVI algorithm, which is run only once. Hence, in this row the proportion is always one.
    (b) The remaining rows correspond to various outputs of the MS–MVI algorithm, which can produce different NE for different runs; hence, the final column indicates the proportions with which the different

NE appear. To be more precise, the proportions correspond to different $O_G(s_0, \sigma)$ outcomes; as already discussed, many different NE can result in the same outcome.

6.2. *Three-Player Linear and Modified Linear Pursuit*

6.2.1. *Paths*

We first present results for Linear Pursuit and Modified Linear Pursuit played on the five-vertices path illustrated in Fig. 3.

We apply MVI to the Linear Pursuit game, played from various initial conditions. The algorithm always reaches equilibrium (and terminates) in an average of 0.0335 seconds, a quite short time. Since the graph has five vertices, there is a total of $5! = 120$ possible vertex label permutations and it is computationally feasible to run MS–MVI with all possible vertex permutations (we expect a total runtime of around 3 to 4 seconds). The results are listed in Table 1.

The results presented in Table 1 are actually quite simple. We give results for three initial states: $(1, 3, 5, 1)$, $(1, 5, 3, 1)$, $(3, 1, 5, 1)$. The average computation time (for one initial state, and all vertex label permutations) is 4.4458 seconds. For each initial state, all runs of the MS–MVI converge to the same NE as the basic MVI run.

In Table 2, we present results for the Modified Linear Pursuit game, played on the same graph. Similarly to Table 1, we see that MS–MVI

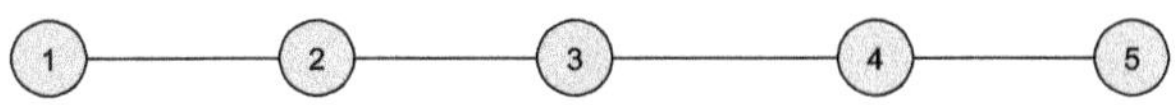

Fig. 3: A five-vertices path.

Table 1: Linear pursuit game on five-vertices path.

	Conv.	Capt. players	Game duration	Prop.
$s_0 = (1, 3, 5, 1)$				
MVI	Yes	2	5	1.0000
MS–MVI	Yes	2	5	1.0000
$s_0 = (1, 5, 3, 1)$				
MVI	Yes	1	10	1.0000
MS–MVI	Yes	1	10	1.0000
$s_0 = (3, 1, 5, 1)$				
MVI	Yes	1	4	1.0000
MS–MVI	Yes	1	4	1.0000

Table 2: Modified linear pursuit game on five-vertices path.

	Conv.	Capt. players	Game duration	Prop.
$s_0 = (1, 3, 5, 1)$				
MVI	Yes	2	5	1.0000
MS–MVI	Yes	2	5	1.0000
$s_0 = (1, 5, 3, 1)$				
MVI	Yes	1	10	1.0000
MS–MVI	Yes	0	∞	0.3750
	Yes	1	10	0.6250
$s_0 = (3, 3, 3, 1)$				
MVI	Yes	1	4	1.0000
MS–MVI	Yes	0	∞	0.3333
	Yes	1	4	0.6667

always converges. However, in Table 2 we see that for certain initial states (namely, for $(1, 5, 3, 1)$ and $(3, 3, 3, 1)$) MS–MVI can compute more NE than basic MVI. In particular, for $(3, 3, 3, 1)$, 33% of the MS–MVI runs compute the"non-obvious" non-capturing NE.

We observe that the most common result of the algorithm MS–MVI for the Modified Linear Pursuit game is the same as in the Linear Pursuit game and this holds for every graph tested.

6.2.2. *Trees*

We continue our analysis with experiments on games played on the tree presented in Fig. 4.

We apply the algorithms to the Linear Pursuit game and both of them converge. In contrast to paths, we do computations only for a few (namely, 100) out of 14! possible vertex label permutations, since experiments on the full set of permutations would take up too much time. The results for the initial states $(14, 10, 2, 1)$, $(14, 7, 2, 1)$, $(14, 2, 10, 1)$ and $(2, 14, 10, 1)$ are shown in Table 3 and can be produced in average time of 2 minutes and 17.329 seconds.

We repeated the procedure for the Modified Linear Pursuit game. The results for the same initial states are presented in Table 4 and can be produced in 2 minutes and 18.815 seconds average time. As in the case of paths, here too, for some initial states, the Modified Linear Pursuit game produces more than one NE.

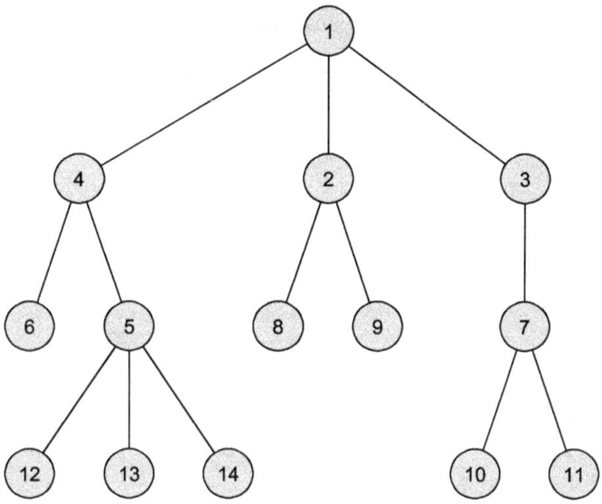

Fig. 4: A tree.

Table 3: Linear pursuit game on a tree.

	Conv.	Capt. players	Game duration	Prop.
$s_0 = (14, 10, 2, 1)$				
MVI	Yes	1	16	1.0000
MS–MVI	Yes	1	16	1.0000
$s_0 = (14, 7, 2, 1)$				
MVI	Yes	1	16	1.0000
MS–MVI	Yes	1	16	1.0000
$s_0 = (14, 2, 10, 1)$				
MVI	Yes	2	11	1.0000
MS–MVI	Yes	2	11	1.0000
$s_0 = (2, 14, 10, 1)$				
MVI	Yes	1	10	1.0000
MS–MVI	Yes	1	10	1.0000

6.2.3. *Cycles*

We continue our analysis with experiments on games played on the six-vertices cycle illustrated in Fig. 5.

In this case, both Linear and Modified Linear Pursuit give the same results presented in Table 4. Furthermore, the MS–MVI algorithm does not yield any advantage, i.e. it always computes the same NE. These results are not surprising: each evader can avoid his pursuer forever, hence, essentially

Table 4: Modified linear pursuit game on a tree.

	Conv.	Capt. players	Game duration	Prop.
$s_0 = (14, 10, 2, 1)$				
MVI	Yes	0	∞	1.0000
MS–MVI	Yes	0	∞	0.1400
	Yes	1	16	0.8600
$s_0 = (14, 7, 2, 1)$				
MVI	Yes	0	∞	1.0000
MS–MVI	Yes	0	∞	0.0900
	Yes	1	16	0.8500
	Yes	2	11	0.0600
$s_0 = (14, 2, 10, 1)$				
MVI	Yes	2	11	1.0000
MS–MVI	Yes	2	11	1.0000
$s_0 = (2, 14, 10, 1)$				
MVI	Yes	1	10	1.0000
MS–MVI	Yes	1	10	1.0000

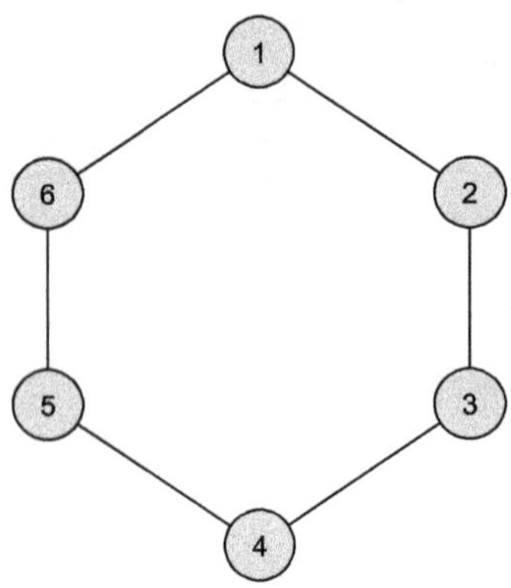

Fig. 5: Six-vertices cycle.

the only NE is the one presented above, which could be computed from theoretical analysis, rather than computationally.

6.2.4. *Cycle-rays*

In Fig. 6, we perform experiments on games played on a "cycle-ray" graph, which consists of a cycle and some paths (rays) starting from the cycle's vertices.

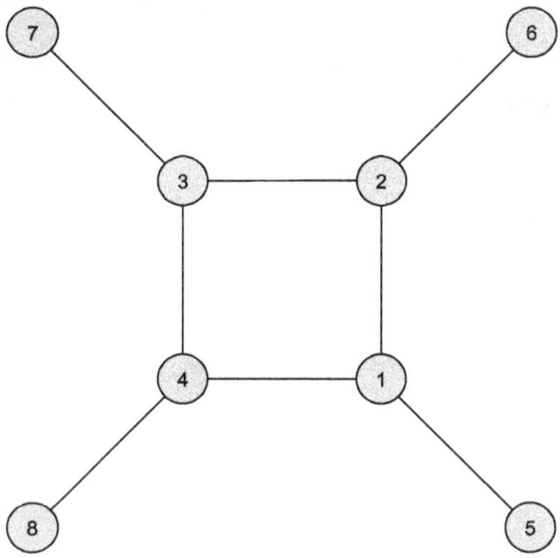

Fig. 6: A cycle-ray graph consisting of a four-vertices cycle and a one-vertex ray.

Table 5: Linear and modified linear pursuit game on a six-vertices cycle: Experiment results.

	Conv.	Capt. players	Game duration	Prop.
$s_0 = (2, 6, 2, 1)$				
MVI	Yes	0	∞	1.0000
MS–MVI	Yes	0	∞	1.0000

Table 6: Linear and modified linear pursuit game on a cycle-ray.

	Conv.	Capt. players	Game duration	Prop.
$s_0 = (1, 3, 5, 1)$				
MVI	Yes	0	∞	1.0000
MS–MVI	Yes	0	∞	0.8400
	Yes	2	8	0.1600

We apply MVI and MS–MVI to the above graph for Linear Pursuit game and the algorithms converge. The total computation time for 100 randomly chosen permutations is 14.143 seconds. In the same way, we conduct experiments for the Modified Linear Pursuit game with a total computation time of 14.283 seconds. The results for both games are qualitatively the same and are presented in Table 6. As can be seen, both Linear Pursuit games result in more than one NE, when the algorithm MS–MVI is applied to them for the cycle-ray graph. This is the only example of a graph from those we have tested in which the Linear Pursuit game possesses more than one NE.

6.2.5. *Gavenciak graphs*

We conclude our study of Linear and Modified Linear Pursuit with some experiments on *Gavenciak graphs* (these have been introduced in Ref. [21] and have the property of being cop-win graphs with maximum capture-time). In Fig. 7, we present the smallest (seven-vertices) Gavenciak graph.

In Table 7, we present the outcome of the experiments on the seven-vertices Gavenciak graph for the Linear Pursuit game. The average computation time (for 100 randomly chosen permutations) is 17.46 seconds.

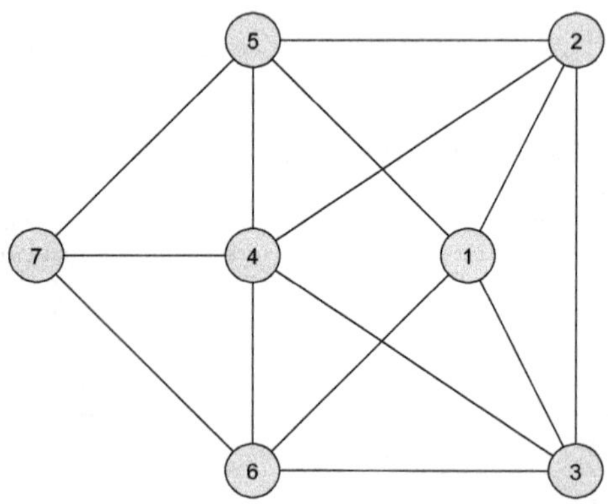

Fig. 7: Seven-vertices Gavenciak graph.

Table 7: Linear pursuit game on a seven-vertices Gavenciak graph.

	Conv.	Capt. players	Game duration	Prop.
$s_0 = (2, 6, 2, 1)$				
MVI	Yes	1	7	1.0000
MS–MVI	Yes	1	7	1.0000
$s_0 = (2, 6, 2, 2)$				
MVI	Yes	1	9	1.0000
MS–MVI	Yes	1	9	1.0000
$s_0 = (2, 6, 2, 3)$				
MVI	Yes	1	8	1.0000
MS–MVI	Yes	1	8	1.0000

Table 8: Modified linear pursuit game on seven-vertices Gavenciak graph.

	Conv.	Capt. players	Game duration	Prop.
$s_0 = (2, 6, 2, 1)$				
MVI	Yes	1	7	1.0000
MS–MVI	No	N/A	N/A	0.0100
	Yes	0	∞	0.0900
	Yes	1	7	0.9000
$s_0 = (2, 6, 2, 2)$				
MVI	Yes	0	∞	1.0000
MS–MVI	No	N/A	N/A	0.0100
	Yes	0	∞	0.7900
	Yes	1	9	0.1900
	Yes	1	15	0.0100
$s_0 = (2, 6, 2, 3)$				
MVI	Yes	0	∞	1.0000
MS–MVI	No	N/A	N/A	0.0300
	Yes	0	∞	0.5600
	Yes	1	8	0.4100

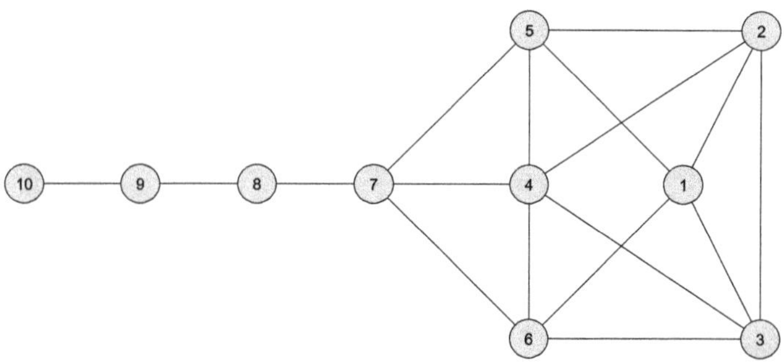

Fig. 8: Ten-vertices Gavenciak graph.

In Table 8, we present the results of applying MVI and MS–MVI to Modified Linear Pursuit played on the graph of Fig. 7.

While MS–MVI does not give any advantage over basic MVI in the Linear Pursuit game, the situation is different in the Modified Linear Pursuit game, where the MS–MVI algorithm reveals the existence of several qualitatively different NE; there exist NE in which the same player captures his evader in different capture times.

We perform similar experiments using the ten-vertices Gavenciak graph presented in Fig. 8. In Table 9, we present the outcome of the experiments

Table 9: Linear pursuit game on a ten-vertices Gavenciak graph.

	Conv.	Capt. players	Game duration	Prop.
$s_0 = (2, 6, 2, 1)$				
MVI	Yes	1	16	1.0000
MS–MVI	No	N/A	N/A	0.4200
	Yes	1	16	0.5800
$s_0 = (2, 6, 2, 2)$				
MVI	Yes	1	18	1.0000
MS–MVI	No	N/A	N/A	0.4300
	Yes	1	18	0.5700
$s_0 = (2, 6, 2, 3)$				
MVI	Yes	1	17	1.0000
MS–MVI	No	N/A	N/A	0.4400
	Yes	1	17	0.5600

Table 10: Modified linear pursuit game on 10-vertices Gavenciak graph.

	Conv.	Capt. players	Game duration	Prop.
$s_0 = (2, 6, 2, 1)$				
MVI	Yes	1	16	1.0000
MS–MVI	No	N/A	N/A	0.4900
	Yes	0	∞	0.1800
	Yes	1	16	0.3300
$s_0 = (2, 6, 2, 2)$				
MVI	Yes	0	∞	1.0000
MS–MVI	No	N/A	N/A	0.4300
	Yes	0	∞	0.5200
	Yes	1	18	0.0500
$s_0 = (2, 6, 2, 3)$				
MVI	Yes	0	∞	1.0000
MS–MVI	No	N/A	N/A	0.4500
	Yes	0	∞	0.3100
	Yes	1	17	0.2400

on the ten-vertices Gavenciak graph for the Linear Pursuit game. The average computation time is 4 minutes and 48.19 seconds. In Table 10, we present the results of applying MVI and MS–MVI to Modified Linear Pursuit, for the ten-vertices Gavenciak graph.

We observe that for the seven-vertices Gavenciak graph, MS–MVI always converges when it is applied to the Linear Pursuit game, while for the ten-vertices graph, it converges only for some permutations; MVI and MS–MVI may have different response to graphs of the same family.

6.3. *Three-player Cyclic and Modified Cyclic Pursuit*

6.3.1. *Paths*

We perform several experiments on the graph of the six-vertices path of Fig. 9.

Since there is a total of $6! = 720$ possible vertex label permutations, which would result in rather excessive computation time, we work instead with 100 randomly chosen permutations. The results are identical for the Cyclic Pursuit and Modified Cyclic Pursuit games and are presented in Table 11.

6.3.2. *Trees*

For the tree of Fig. 4, results appear in Table 12 for Linear, and in Table 13 for Modified Linear, game.

Regarding the capturing player, there exists an NE for each possible case of $P_c \in \{P_1, P_2, P_3, \lambda\}$ for the initial states $(14, 8, 7, 1)$ and $(6, 8, 7, 1)$. However, even MS–MVI does not produce each one in the experiments of Tables 12 and 13. The most obvious explanation is that the MS–MVI algorithm did not run for a sufficiently large set of permutations (only 100 from 14! cases have been tested). Probably, the proportion of permutations leading to the NE, which have not been produced, is too small; hence, we claim that for a larger testing set they would have been produced. Although

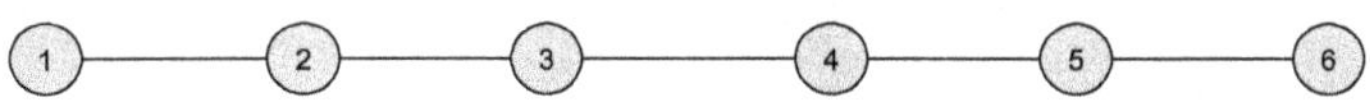

Fig. 9: Six-vertices path.

Table 11: Cyclic and modified cyclic pursuit game on six-vertices path.

	Conv.	Capt. players	Game duration	Prop.
$s_0 = (1, 3, 5, 1)$				
MVI	Yes	2	9	1.0000
MS–MVI	Yes	2	9	1.0000
$s_0 = (1, 5, 3, 1)$				
MVI	Yes	4	7	1.0000
MS–MVI	Yes	4	7	1.0000
$s_0 = (3, 1, 5, 1)$				
MVI	Yes	1	5	1.0000
MS–MVI	Yes	1	5	1.0000

Table 12: Cyclic pursuit game on a tree.

	Conv.	Capt. players	Game duration	Prop.
$s_0 = (14, 2, 10, 1)$				
MVI	Yes	2	11	1.0000
MS–MVI	No	N/A	N/A	0.9400
	Yes	2	11	0.0600
$s_0 = (5, 2, 10, 1)$				
MVI	Yes	1	10	1.0000
MS–MVI	No	N/A	N/A	0.9600
	Yes	1	10	0.0400
$s_0 = (14, 8, 7, 1)$				
MVI	Yes	0	∞	1.0000
MS–MVI	No	N/A	N/A	0.9500
	Yes	1	13	0.0400
	Yes	2	14	0.0100
$s_0 = (14, 8, 3, 1)$				
MVI	Yes	0	∞	1.0000
MS–MVI	No	N/A	N/A	0.9700
	Yes	2	14	0.0300
$s_0 = (6, 8, 7, 1)$				
MVI	Yes	0	∞	1.0000
MS–MVI	No	N/A	N/A	0.9700
	Yes	0	∞	0.0100
	Yes	1	10	0.0200

for the majority of permutations the algorithm does not converge (hence, we are not sure if it produces an NE for every s_0), even in those cases when it is halted, it produces the NE missing from the Tables 12 and 13.

We can also observe that Cyclic game and Modified Cyclic game have the same outcome and the only difference exists in the initial states of the form (x, x, x, p); the results for this initial state can be seen in Table 13 for $s_0 = (1, 1, 1, 1)$.

6.3.3. *Cycles*

In Table 14, we present results of experiments conducted on cycles for Modified Linear Pursuit game only, since, apart from initial states of the form $s_0 = (x, x, x, p)$, Linear and Modified Linear Pursuit game are identical. The average computation time is 2.92 seconds.

Since MS–MVI always converges for cycles when applied to Cyclic and Modified Cyclic game, the proportion of Table 14 for $s_0 = (x, x, x, p)$ is more representative than the respective proportion of Table 13.

Table 13: Modified cyclic pursuit game on a tree.

	Conv.	Capt. players	Game duration	Prop.
$s_0 = (14, 2, 10, 1)$				
MVI	Yes	2	11	1.0000
MS–MVI	No	N/A	N/A	0.9600
	Yes	2	11	0.0400
$s_0 = (5, 2, 10, 1)$				
MVI	Yes	1	10	1.0000
MS–MVI	No	N/A	N/A	0.9600
	Yes	1	10	0.0400
$s_0 = (14, 8, 7, 1)$				
MVI	Yes	0	∞	1.0000
MS–MVI	No	N/A	N/A	0.9500
	Yes	0	∞	0.0100
	Yes	1	13	0.0400
$s_0 = (14, 8, 3, 1)$				
MVI	Yes	0	∞	1.0000
MS–MVI	No	N/A	N/A	0.9300
	Yes	0	∞	0.0100
	Yes	2	14	0.0600
$s_0 = (6, 8, 7, 1)$				
MVI	Yes	0	∞	1.0000
MS–MVI	No	N/A	N/A	0.9200
	Yes	0	∞	0.0400
	Yes	1	10	0.0400
$s_0 = (1, 1, 1, 1)$				
MVI	Yes	0	∞	1.0000
MS–MVI	No	N/A	N/A	0.9500
	Yes	2	1	0.0500

Table 14: Modified cyclic pursuit game on six-vertices cycle.

	Conv.	Capt. players	Game duration	Prop.
$s_0 = (1, 3, 5, 1)$				
MVI	Yes	0	∞	1.0000
MS–MVI	Yes	0	∞	1.0000
$s_0 = (1, 1, 1, 1)$				
MVI	Yes	0	∞	1.0000
MS–MVI	Yes	0	∞	0.3500
	Yes	2	1	0.6500

6.3.4. *Cycle-rays*

In Table 15, we present the outcome of the experiments on the cycle-ray of Fig. 6 for the Linear and Modified Linear Pursuit game, since their results

Table 15: Cyclic and modified cyclic pursuit game on a cycle-ray.

	Conv.	Capt. players	Game duration	Prop.
$s_0 = (4, 2, 5, 1)$				
MVI	Yes	2	5	1.0000
MS–MVI	Yes	0	∞	0.3400
	Yes	2	5	0.6600

Table 16: Cyclic and modified cyclic pursuit game on seven-vertices Gavenciak graph.

	Conv.	Capt. players	Game duration	Prop.
$s_0 = (7, 2, 6, 1)$				
MVI	Yes	0	∞	1.0000
MS–MVI	No	N/A	N/A	0.7500
	Yes	0	∞	0.2500

are qualitatively the same. The average computation time is 13.54 seconds. Note that a similar initial position exists also in trees. However, unlike the case of trees, here the algorithm always converges. A possible explanation could be that for trees, any case of capturing player would be possible, whereas for cycle-rays it could only be two of them (especially for the case of Table 15, they are P_2 or λ).

6.3.5. *Gavenciak graphs*

We have performed several experiments (for both Cyclic Pursuit and Modified Cyclic Pursuit) on the seven-vertices and ten-vertices Gavenciak graphs. The results are qualitatively the same in all cases, so we only present, in Table 16, the results for Cyclic Pursuit on the seven-vertices graph. The computation time is 2 minutes and 8.36 seconds. Note the high proportion of non-convergent runs of the MS–MVI algorithm.

6.4. *Two selfish cops and an adversarial robber*

The game of two Selfish cops and an adversarial robber (henceforth called SCAR) is a particularly interesting GPG. It is characterized by two parameters: γ, which is the discount factor, and ε, which is the proportion of the capture reward received by the non-capturing cop. What makes the game interesting is the interplay of γ and ε: for certain combinations of their values a cop may prefer to delay capture so that he captures the robber himself (instead of helping the other cop to effect an earlier capture). Roughly, we will have early capture when $\gamma < \frac{\varepsilon}{1-\varepsilon}$ and deferred capture when $\gamma > \frac{\varepsilon}{1-\varepsilon}$. Examples of this behavior (which will lead to different NE) will be seen in the subsequent experiments.

6.4.1. *Paths*

We start with experiments on the six-vertices path. In Table 17, we present results on SCAR played with $\gamma = 0.2$ and $\varepsilon = 0.2$ and in Table 18 for $\gamma = 0.99$ and $\varepsilon = 0.2$.

Consider the case with $s_0 = (1, 5, 3, 1)$. In Table 17, the second cop (P_2) "facilitates" capture by the first cop (P_1) in four moves. On the other hand, in Table 18, P_2 plays in such a manner that capture is delayed (takes place in five moves) but is effected by himself rather than by P_1. The reason is

Table 17: SCAR ($\gamma = 0.2$ and $\varepsilon = 0.2$) on six-vertices path.

	Conv.	Capt. players	Game duration	Prop.
$s_0 = (1, 3, 5, 1)$				
MVI	Yes	2	8	1.0000
MS–MVI	Yes	2	8	1.0000
$s_0 = (1, 5, 3, 1)$				
MVI	Yes	1	4	1.0000
MS–MVI	Yes	1	4	1.0000
$s_0 = (3, 1, 5, 1)$				
MVI	Yes	1	7	1.0000
MS–MVI	Yes	1	7	1.0000

Table 18: SCAR ($\gamma = 0.99$ and $\varepsilon = 0.2$) on six-vertices path.

	Conv.	Capt. players	Game duration	Prop.
$s_0 = (1, 3, 5, 1)$				
MVI	Yes	2	8	1.0000
MS–MVI	Yes	2	8	1.0000
$s_0 = (1, 5, 3, 1)$				
MVI	Yes	2	5	1.0000
MS–MVI	Yes	2	5	1.0000
$s_0 = (3, 1, 5, 1)$				
MVI	Yes	1	7	1.0000
MS–MVI	Yes	1	7	1.0000

that the cost for delaying capture is more than offset by the lower discount factor γ. However, both MVI and MS–MVI are able to compute the correct NE in every case.

6.4.2. *Trees*

We next present results on SCAR for the tree of Fig. 4. Here we present only the cases in which all players lie on different branches of the tree, since every other case has already been discussed in paths. The results are the same for each value of γ and ε, therefore the results are summed up in Table 19 and the computation time needed was 2 minutes and 10.28 seconds.

6.4.3. *Cycles*

We next present experiments on cycles for SCAR when $\gamma < \frac{\varepsilon}{1-\varepsilon}$ (Table 20) and Modified Linear Pursuit game when $\gamma > \frac{\varepsilon}{1-\varepsilon}$ (Table 21). The computation time is 16.6 s in the first case and 1 m and 19.9 s in the latter.

The outcome of MVI and MS–MVI for cycles are identical to those for paths. In fact, SCAR is the only game of those we have tested that has a

Table 19: SCAR on a tree.

	Conv.	Capt. players	Game duration	Prop.
$s_0 = (14, 2, 10, 1)$				
MVI	Yes	2	11	1.0000
MS–MVI	Yes	2	11	1.0000
$s_0 = (14, 10, 2, 1)$				
MVI	Yes	1	13	1.0000
MS–MVI	Yes	1	13	1.0000
$s_0 = (2, 10, 14, 1)$				
MVI	Yes	1	10	1.0000
MS–MVI	Yes	1	10	1.0000

Table 20: SCAR ($\gamma = 0.2$ and $\varepsilon = 0.2$) on six-vertices cycle.

	Conv.	Capt. players	Game duration	Prop.
$s_0 = (1, 5, 3, 1)$				
MVI	Yes	1	4	1.0000
MS–MVI	Yes	1	4	1.0000

Table 21: SCAR ($\gamma = 0.99$ and $\varepsilon = 0.2$) on six-vertices cycle.

	Conv.	Capt. players	Game duration	Prop.
$s_0 = (1, 5, 3, 1)$				
MVI	Yes	2	5	1.0000
MS–MVI	Yes	2	5	1.0000

Table 22: SCAR ($\gamma = 0.2$ and $\varepsilon = 0.2$) on cycle-rays.

	Conv.	Capt. players	Game duration	Prop.
$s_0 = (1, 3, 5, 1)$				
MVI	No	N/A	N/A	1.0000
MS–MVI	No	N/A	N/A	1.0000

Table 23: SCAR ($\gamma = 0.2$ and $\varepsilon = 0.2$) on cycle-rays.

	Conv.	Capt. players	Game duration	Prop.
$s_0 = (5, 7, 4, 1)$				
MVI	Yes	1	7	1.0000
MS–MVI	Yes	1	7	1.0000

capturing player different than λ. However, even MS–MVI cannot produce all possible NE as can be seen; this, together with the fact that all results of SCAR have a proportion equal to one when MS–MVI is applied to all graphs that have been tested, reinforces the conjecture that it can compute only one NE for each initial state for SCAR.

6.4.4. *Cycle-rays*

Finally, we present experiments conducted on cycle-rays (Tables 22 and 23). When $\gamma > \frac{\varepsilon}{1-\varepsilon}$, MS–MVI does not converge for any permutation; this establishes the remarkable property that MS–MVI cannot converge when applied to any graph for at least some permutations, since the whole set of 8! permutations have been tested (the computation time was about 1 h and 50 m).

Unlike the previous case, when $\gamma < \frac{\varepsilon}{1-\varepsilon}$, the algorithms always converge and the computation time for these experiments is 16.548 s. This case of γ and ε values seems to have the same behavior as the Cops and Robbers game with two cops and one robber for all graphs that have been tested.

7. Conclusion

We have introduced a family of *N-player graph pursuit games* (GPG) and shown that they can be seen as a special case of stochastic games. We have proved that every GPG possesses at least one-positional and one non-positional NE. We have also introduced (a) the basic Multi-Value Iteration (MVI), which generalizes the Value Iteration algorithm to N-player games; (b) a multi-start variant (MS–MVI) of MVI. We have evaluated these algorithms by numerical experiments which indicate that, for every graph and every GPG tested (with the single exception of SCAR played on cycle-rays) MS–MVI will compute several NE of a GPG. Consequently, despite the fact that MS–MVI is not guaranteed to compute *all* NE of a given GPG and graph combination, it is a powerful tool for the analysis of GPGs, which constitute a novel generalization of two-player pursuit games on graphs.

References

[1] J. Filar and K. Vrieze, *Competitive Markov Decision Processes* (Springer Science & Business Media, 2012).

[2] R. Nowakowski and P. Winkler, Vertex-to-vertex pursuit in a graph, *Discrete Math.* **43**(2–3), 235–239, (1983).

[3] A. Bonato and R. Nowakowski, *The Game of Cops and Robbers on Graphs* (American Mathematical Soc., 2011).

[4] A. Bonato and G. MacGillivray, Characterizations and algorithms for generalized Cops and Robbers games, *Contrib. Discret. Math.* **12**(1), (2017).

[5] A. Kehagias, Generalized cops and robbers: A multi-player pursuit game on graphs, *Dyn. Games Appl.* **9**(4), 1076–1099, (2019).

[6] G. Konstantinidis and A. Kehagias, Selfish cops and active robber: Multi-player pursuit evasion on graphs, *Theoret. Comput. Sci.* **780**, 84–102, (2019).

[7] G. Konstantinidis and A. Kehagias, On positionality of trigger strategies Nash equilibria in SCAR, *Theoret. Comput. Sci.* **845**, 144–158, (2020).

[8] S. Batbileg and R. Enkhbat, Global optimization approach to game theory, *Mongolian Math. Soc.* **14**, 2–11, (2010).

[9] S. Batbileg and R. Enkhbat, Global optimization approach to nonzero sum n-person game, *Int. J. Adv. Model. Optimizat.* **13**(1), 59–66, (2011).

[10] R. Matulevicius, Search for dynamic equilibrium in duel problems by global optimization, *Informatica* **13**(1), 73–88, (2002).

[11] J. Mockus, Walras competition model, an example of global optimization, *Informatica* **15**(4), 525–550, (2004).

[12] T. Sandholm, G. Andrew, and C. Vincent, Mixed-integer programming methods for finding Nash equilibria, *AAAI.* (2005).

[13] R.I. Lung and D. Dumitrescu, Computing Nash equilibria by means of evolutionary computation, *Int. J. of Comput. Commun. Control* **3**(suppl. issue), 364–368, (2008).
[14] V. Picheny, B. Mickael, and H. Abderrahmane, A Bayesian optimization approach to find Nash equilibria, *J. Global Optimizat.* **73**(1), 171–192, (2019).
[15] N.G. Pavlidis, E.P. Kostantinos, and N.V. Michael, Computing Nash equilibria through computational intelligence methods, *J. Comput. Appl. Math.* **175**(1), 113–136, (2005).
[16] D. Cheng and Z. Liu, Optimization via game theoretic control, *Nat. Sci. Rev.* **7**(7), 1120–1122, (2020).
[17] S. Xu and H. Chen, Nash game based efficient global optimization for large-scale design problems, *J. Global Optimizat.* **71**(2), 361–381, (2018).
[18] G. Yang, Game theory-inspired evolutionary algorithm for global optimization, *Algorithms* **10**(4), 111, (2017).
[19] A.M. Fink, Equilibrium in a stochastic n-person game, *J. Sci. Hiroshima University*, series AI (Mathematics) **28**(1), 89–93, (1964).
[20] R. Martí, Multi-start methods. In: *Handbook of Metaheuristics*, pp. 355–368 (Springer, Boston, MA, 2003).
[21] T. Gavenciak, *Hry na grafech* (Games on graphs), Master's Thesis, Department of Applied Mathematics, Charles University, Prague, 2007.
[22] A. Quilliot, A short note about pursuit games played on a graph with a given genus, *J. Combinat. Theory Series B* **38**(1), 89–92, (1985).
[23] M. Takahashi, Equilibrium points of stochastic non-cooperative n-person games, *J. Sci. Hiroshima University*, Series AI (Mathematics) **28**(1), 95–99, (1964).
[24] F. Thuijsman and T.E.S. Raghavan, Perfect information stochastic games and related classes, *Int. J. Game Theory* **26**(3), 403–408, (1997).

https://doi.org/10.1142/9789811261572_0019

Chapter 19

Geometry in Quantitative Methods and Applications

Christos Kitsos* and Stavros Fatouros†
School of Engineering, University of West Attica
Agiou Spiridonos Street, Egaleo 12243, Greece
**xkitsos@uniwa.gr*
†*fatouros@uniwa.gr*

The target of this chapter is to investigate the geometrical line of thought adopted to proceed in quantitative methods in application. Both Euclidean and non-Euclidean geometries were applied in statistical methods and not only typical examples from Biology but Industrial Methods are also discussed. The Pythagoras Theorem was the attracting point, at the beginning but even elegant methods for affine or projective geometry were eventually adopted, offering solutions to a number of qualitative applications.

1. Introduction

There are a number of ideas, in different scientific fields, where the underlying line of thought comes from geometry. Physics is the most popular field for geometrical applications [1,2].

In linear programming, the sense of a convex set is essential. Even the well-known Newton–Raphson iteration is based on the iteration of appropriate tangents, moving toward the limit of the method, the solution of an equation.

Geometry covers fields such as biology, statistics, econometrics. An example from Statistics is to consider the vector of observations $x \in R^n$

and the mean value as follows:

$$\bar{x} = \frac{1}{n}\sum_{i=1}^{n} x_i = \frac{1}{n}\mathbf{1x'}, \tag{1}$$

with $\mathbf{1} = (1, 1, ..., 1)$, when the inner product is considered. Vectors $\bar{x}_+ = (\bar{x}, \bar{x}, ..., \bar{x})$ and $x - \bar{x}$ are orthogonal as $\bar{x}_+(x - \bar{x})' = 0$, thus from Pythagoras's theorem it holds

$$\|x\|^2 = \|\bar{x}_+\|^2 + \|x - \bar{x}_+\|^2,$$

and

$$\|x - \bar{x}_+\| = \sum_{i=1}^{n} (x_i - \bar{x})^2 = (n-1)s^2 = ns_*^2, \tag{2}$$

with s^2 being the unbiased and s_*^2 the biased variance. The well-known statistical t-criterion is defined as

$$t = \frac{\|\bar{x}_+\|}{\|x - \bar{x}_+\|} = \sqrt{n-1}\cot(\bar{x}_+, x). \tag{3}$$

In this geometrical context, statistical ideas are discussed in Section 2, while in Section 4 we refer to the use of curves in biology.

Euclidean and non-Euclidean geometry offers a variety of applications in science. Non-Euclidean applications are well known in physics; we shall focus on non-Euclidean applications in Statistics in Section 3.

Under Rassias's point of view, on the major trends in Mathematics, [3], unification is the first in a list of four directions of activity. Under this line of thought, we believe, that we have invariance under a linear transformation with relation to affine spaces. That is the reason why non-Euclidean geometries are discussed in Section 4.

It is impossible to discuss the minimum projection, to a hyperplane, with no mention of the least square method and the projection through the linear algebra, so useful to multivariable statistics, [4–6], see Section 3. Since the hexagon is the "best" area for beans and a number of elementary curves have been noticed in nature, new optimal curves are introduced and fitted to automobile surfaces.

There are a lot of geometrical results that are applied to physics, while statistics adopted and extended the sense of entropy [7]. Moreover, in quantum mechanics either Fermi-Dirac statistics or Bose–Einstein statistics can be applied, see Appendix A. Therefore, physics fed statistics with new ideas and developments, in the sense that Lie groups provided a new line of thought in physics [2]. The Fermi-Dirac statistics in contrast to Bose–Einstein statistics are only applied in the case were the particles under investigation are not limited to single occupancy of the same state; particles that do not obey the Pauli excursion principle.

2. Euclidean Geometry and Statistics

We recall that an orthogonal coordinate system in n-space is the set $\{u_1, u_2, \ldots, u_n\}$ where $(u_i \cdot u_j) = 0$ when $i \neq j$ for all $i, j = 1, 2, \ldots, n$.

The projection of a given vector α onto a unit vector u is the vector $(\alpha \cdot u)u$ while the projection of $x \in \mathbb{R}^n$ onto $K \subseteq \mathbb{R}^n$, the subspace spanned by the orthogonal unit vectors $u_1, u_2, \ldots, u_n$, is $(y \cdot u_1)\, u_1 + \cdots + (y \cdot u_k)\, u_k$.

The distance, i.e. the l_2-norm, from the mean vector y of the observations in $\mathbb{R}^n$ is given by

$$\frac{1}{n-1} \|y - \bar{y}\|^2 = s^2,$$

known as the unbiased sample variance. This simple notation gives rise to the geometrical relation between F and t tests as

$$F = \frac{\|\bar{y}\|^2}{s^2} = \frac{n\bar{y}^2}{s^2} = \left(\frac{\bar{y}}{s/\sqrt{n}}\right) = t^2. \tag{4}$$

Under this line of thought evaluating distances, [8] in a one-way ANOVA the total corrected sum of squares (TSS), the sum of squares of treatments (SST) and the sum of squares of errors (SSE) obey a Pythagorean breakup, see Ref. [9]

$$TSS = SST + SSE,$$

as

$$TSS = \|y - \bar{y}_{..}\|^2,\ SSE = \|y - \bar{y}_{1.}\|^2,\ SST = \|\bar{y}_{1.} - \bar{y}_{..}\|.$$

An $n \times n$ *Latin Square* can be analyzed under the following decomposition where the corrected observation vector is the sum of treatment, row and column vectors as follows:

$$y - \bar{y}_{...} = (\bar{y}_{i..} - \bar{y}_{...}) + (\bar{y}_{.j.} - \bar{y}_{...}) + (\bar{y}_{..k} - \bar{y}_{...}) + error,$$

with $i, j, k \in \{1, 2, \ldots, n\}$, [10].

The Pythagorean decomposition is still valid as follows:

$$TSS = SSR + SSC + SST + SSE, \tag{5}$$

where
SS of Rows, $SSR = \|\bar{y}_{.j.} - \bar{y}_{...}\|^2$, SS of Columns, $SSR = \|\bar{y}_{..k} - \bar{y}_{...}\|^2$, SS of Treatments, $SST = \|\bar{y}_{i..} - \bar{y}_{...}\|^2$, Total SS $TSS = \|y - \bar{y}_{...}\|^2$.

Consider the simple regression model, i.e. to fit the straight line to the dataset as follows:

$$y = \beta_0 + \beta_1 x + e,$$

$x, y, e \in R^{n\times 1}$, where y is the response, x, the input variable and e, the stochastic error of a zero-mean distribution with σ^2 variance. If we consider an inference, we assume that $E(e) = 0$ and $E(e^2) = \sigma^2$ and the distribution the errors follow is the Normal Gaussian Distribution. Fitting the model, we obtain the response $\hat{y}$ and the residual $\hat{e}$ in terms that

$$\text{Observed Value} = \text{Fitted Value} + \text{Residual}.$$

That is $y = \hat{y} + \hat{e}$ where the fitted value is $\hat{y} = \hat{b}_0 + \hat{b}_1 x = \hat{b}_0 z + \hat{b}_1 x$ introducing the vector z of 1s, which is another input variable. $z = (1, 1, \ldots, 1)$ so that $z \cdot x = n\bar{x}$.

Considering that vector x can always be translated so that $\bar{x} = 0$, therefore $z \perp x$, and that explains the frequent motivation to translate the vector x onto a zero-mean one. The orthogonal vectors x, y form a plane $E \subseteq \mathbb{R}^2$ and the vector of observations y is given as in Fig. 1, [11,12].

Due to the theorem of the *three perpendiculars*, see Fig. 1, we have

$$e \perp E \Rightarrow e \perp \hat{y}\,,\ \hat{y}_0 \perp \varepsilon_0\,,\ \hat{y}_1 \perp \varepsilon_1.$$

Moreover, due to the projections, we obtain

$$\hat{y} = \hat{y}_0 + \hat{y}_1,$$

or

$$\hat{y} = \frac{y \cdot z}{\|z\|^2} z + \frac{y \cdot x_1}{\|zx_1\|^2} x_1 = \hat{\beta}_0 z + \hat{\beta}_1 x_1. \tag{6}$$

As $y \cdot z = \sum_{i=1}^n y_i$, $\|z\|^2 = n$, $y \cdot x_1 = \sum_{i=1}^n x_i y_i$ and $\|x_i\|^2 = \sum_{i=1}^n x_i^2$, we obtain the usual, since the era of Gauss, estimates for Ordinary Least Squares (OLS)

$$\hat{b}_0 = \bar{y}, \hat{b}_1 = \frac{1}{\sum_{i=1}^n x_i^2} \sum_{i=1}^n x_i y_i.$$

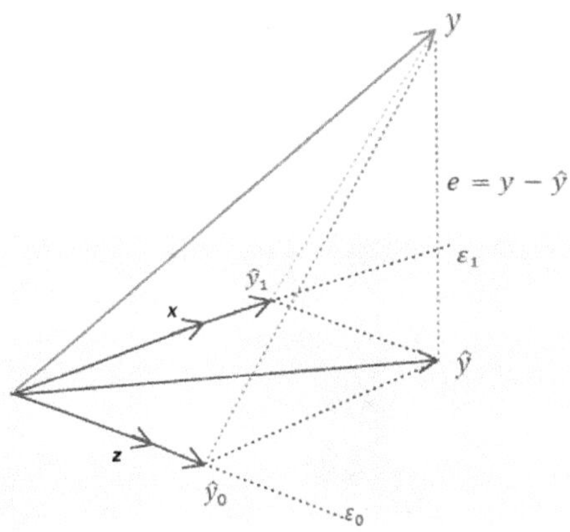

Fig. 1: OLS as the "three perpendiculars problem".

Therefore, in Fig. 1 $\hat{y}_0 = \bar{y}$, thus $\|y - \hat{y}_0\|^2$ presents the total variation and $\|y - \hat{y}\|^2$, the unexplained residual variation. On the ε_1-axis as $\|\beta_1 x\|^2 = \beta_1^2 \| x\|^2$, the explained variation has been evaluated and from the Pythagorean Theorem we obtain

$$\|y - \hat{y}_0\|^2 = \beta_1^2 \|x\|^2 + \|y - \hat{y}\|^2.$$

The above discussion provides food for thought for the general linear regression problem. Consider the General Linear Model (GLM) of the form

$$Y + X\beta + e, \tag{7}$$

with $X \in R^{n\times p}$, $Y \in R^{n\times 1}$, $\beta \in R^{p\times 1}$ and the stochastic error term $e \in R^{n\times 1}$.

When inference is asked, the usual normal assumption is imposed, $e \sim N(0, \sigma^2 I)$, and the errors are from the normal distribution with zero-mean and variance σ^2. There are experimental design problems where the elements of $X = (x_{ij})$, $i = 1, \ldots, n$, $j = 1, \ldots, p$ are either 0 or 1. In this case, X is known as the design matrix, Y is known as the response vector and β is the vector of the unknown parameters estimated under the optimality criterion

$$\text{minimize } e'e = \|Y - X\beta\|^2 = \|Y - \tau\|^2, \tag{8}$$

with $\tau = X\beta \in R[X]$ the range space of X. As τ varies it achieves its minimum value $\hat{\tau}$ when

$$(Y - \hat{\tau}) \perp R[X] \rightarrow X'(Y - \hat{\tau}) = 0 \rightarrow X'\hat{\tau} = X'Y.$$

Recall that τ is the unique orthogonal projection of Y onto $R[X]$ and as for the columns of X, we need to assume that they are linearly independent (otherwise another approach is needed). There exists a unique vector $\hat{\beta}$ so that $\hat{\tau} = X\hat{\beta}$. thus we obtain the so-called normal equation, [13]

$$X'X\hat{\beta} = X'Y. \tag{9}$$

As we need $\det(XX') \neq 0$, we assume that $rank(X) = p$ so that XX' is positive definite and therefore non-singular. Therefore, under the imposed assumptions

$$\hat{\beta} = (X'X)^{-1}X'Y.$$

By denoting by $\hat{y}$ the fitted regression $X\hat{\beta}$, it is easy to see that for the estimate of e, say ε, known as residuals, we have

$$\varepsilon = Y - \hat{Y} = (I_n - P)Y\ ,\ P = (X'X)^{-1}X',$$

and

$$\min \varepsilon'\varepsilon = Y'Y - \hat{\beta}'(X'X)\beta = \text{RSS}, \tag{10}$$

known as *Residual Sum of Squares* in literature [14].

It is easy to see that P and $I_n - P$ are symmetric and idempontent, therefore,

$$rank(I_n - P) = tr(I_n - P) = n - tr(P) = n - p.$$

Given the assumption that the errors are modified to $e \sim N(0, \sigma^2 V)$, that is the variance–covariance matrix $\Sigma = \sigma^2 V$ permits the errors to be correlated, matrix V is positive definite and it can be proved that the Generalized Least Square estimate is

$$\hat{\beta}_* = (X'V^{-1}X)^{-1}. \tag{11}$$

When V is a diagonal matrix, we refer to weight least squares.

It was Kruskal [15] who considered the range space of V and VX and concluded the nice result

Proposition 1. *The ordinary least square estimate $\hat{\beta}$ and the generalized least square estimate are identical if and only if*

$$R[X] = R[VX].$$

The choice of matrix X in the GLM gives rise to the optimal design theory, originated by De la Gaza [16]. When the target is to minimize the volume of the ellipsoidal of the parameters confidence intervals,

$$\min Var(\hat{\beta}) \Leftrightarrow \max \det(X'X) \Leftrightarrow \max \prod_{i=1}^{p} \lambda_i,$$

where λ_i, $i = 1, 2, ..., p$ are the eigenvalues of the matrix $X'X$. In this case, we refer to D-optimality.

Another criterion can be to minimize the total variance,

$$\min tr[(X'X)^{-1}] \Leftrightarrow \min \sum_{i=1}^{p} \frac{1}{\lambda_i},$$

where we refer to A-optimality.

Moreover, maximizing $\det(X'X)$ over some region $U \subseteq R^p$ is equal to the minimization of the maximum variance of the future response $\hat{Y} = x'\hat{\beta}$, for $x \in U$, known as G-optimality. This is the so-called equivalence theorem [17].

Another fundamental alphabetical optimal design criterion is the c-optimality, which was extensively applied to the calibration problem providing a particular geometrical interpretation [18]. In a three-dimensional problem, Geometry provides an illustrative approach to the linear optimal design criteria.

The multivariate observations offer an illustrative geometrical interpretation, [4] while upper bounds for the greatest eigenvalue have been evaluated. Moreover, the orthogonal transformation of the notation is essential to factor analysis as any linear transformation keeps the optimality criteria invariant.

The idea of invariant transformation will be discussed in Section 4. As for the nonlinear experimental designs see Ref. [19].

Orthogonal transformations are very useful in statistics. The most well-known example for the given random vector $(x_1, x_2, ..., x_n)$ coming from a multivariable normal distribution is the orthogonal transformation

$$Y_i = \frac{1}{\sqrt{i}\sqrt{i-1}} \left(\sum_{j=1}^{i-1} x_j - (i-1)x_i \right),$$

which provides the normal distribution observations $(y_1, y_2, \ldots, y_n)$ with

$$y_1 = \sqrt{\bar{x}}, \ \sum_{i=2}^{n} y_i^2 = \sum_{i=1}^{n} (x_i - \bar{x})^2 = (n-1)s^2.$$

3. Statistics and Non-Euclidean Geometries

Affine geometry can be considered as a study which lies between Euclidean geometry and projective geometry, [20,21]. Affine Geometry can be, roughly speaking, an Euclidean one with no metric (distance, angle, surface, etc). We can also think affine geometry as an Euclidean geometry with no congruence. Moreover, projective geometry might generate an affine by the designation of a particular line or plane to present the points at infinity. The parallel postulate is essential in affine geometry and it can be the basis for Euclidean development when perpendicular lines are defined. Moreover affine Geometry is the study of geometrical properties through the group, G, of affine transformations. Under this rough presentation an affine transformation is a projective transformation that does not permute finite points with points at infinity [22,23]. We shall focus mainly on affine geometry and the applications in statistics, [24,25], and to a lesser extend to projective geometry.

We shall recall that the elementary, provided by Euler, formula for a polyhedron is

$$V - E + F = 2, \tag{12}$$

with F the number of faces, V the number of vertices, E the number of edges, can be considered as the beginning of Topology. We shall not go into details but a short description of a geometrical approach [26] is given in Appendix B.

Let S^2 be the unit sphere. In spherical Geometry, the first non-Euclidean, the distance between two points $p_1, p_2 \in S^2$, $dS^2(p_1, p_2) \in [0, \pi]$ is defined as

$$\cos[ds^2(p_1, p_2)] = p_1 \cdot p_2. \tag{13}$$

Moreover, for any $w_1, w_2 \in \mathbb{C}$, $dS^2(w_1, w_2)$ due to stereographic projection of S^2 to w_1, w_2, we have

$$\tan\left[\frac{1}{2} dS^2(w_1, w_2)\right] = \frac{|w_1 - w_2|}{|w_1 + \bar{w_1} w_2|}. \tag{14}$$

An interesting observation is that in spherical geometry similar triangulars are equals.

In projective geometry, the distance is defined in the half-plane

$$E_+ = \{(x, y) \in \mathbb{R}^2, y > 0\} = \{z \in \mathbb{C}, Imz > 0\}.$$

For any two points p_1, $p_2 \in H$, the distance is

$$dH = 2\tanh^{-1} \frac{|p_2 - p_1|}{|p_2 - \bar{p_1}|}. \tag{15}$$

The use of tanh is the reason we refer to hyperbolic geometry. In hyperbolic geometry, the similar triangulars are equal, as happens in spherical geometry, too.

Let us consider the unit disk

$$D = \{z \in \mathbb{C} : |z| < 1\}.$$

Let us consider the Poincaré function

$$P : \mathbb{C} \setminus \{-i\} \to \mathbb{C} \setminus \{-1\} : z \to p(z) = \frac{z - i}{z + i}, \tag{16}$$

while

$$p^{-1}(z) = \frac{z + i}{i(z - i)}.$$

Therefore, P is a function of H to the $D \cup R_A$ where R_A is the real axis on the boundary of D, equivalent to the unit circle of $\mathbb{C}$, $|z| = 1$. This disk, known as Poincare disk, DP, forms another line of perspective in hyperbolic geometry in which the distance of two points p_1, $p_2 \in DP$ is now

$$dDP(p_1, p_2) = 2\tanh^{-1} \frac{|p_2 - p_1|}{|1 - \bar{p_1}p_2|}. \tag{17}$$

The Klein disk, DK, is defined as

$$K(z) = \frac{2z}{|z|^2 + 1}.$$

The Beltrami–Klein model is the development of a non-Euclidean geometry where the distance of two points p_1, p_2 is defined as

$$dDBK(p_1, p_2) = \frac{1}{2} \ln \frac{|p_1 - q_1||p_2 - q_2|}{|p_1 - p_2||q_1 - q_2|}, \tag{18}$$

where q_1, q_2 are the points where the line through p_1, p_2 intersects with the circle C, associated with Beltrami–Klein geometry.

The fraction

$$(p_1, p_2; q_1, q_2) = \frac{|p_1 - q_1||p_2 - q_2|}{|p_1 - p_2||q_1 - q_2|} \tag{19}$$

is known as double ratio. Double ratios are known from the early Euclidean geometry, see Appendix C.

Beltrami–Klein geometry tries to offer a bridge between a non-Euclidean geometry and a projective geometry in the sense that the hyperbolic lines of the Beltrami–Klein geometry are Euclidean linear parts lying in DK.

In the early steps of statistics, Euclidean geometry and especially Pythagoras's theorem played an important role in the development of Experimental Design [27] among others. Geometrical aspects were adopted to calibration problems [18]. The Analytic Geometry [28] was the first approach, see also Section 1. The application of Affine and projective geometry occurred latter. We refer to this connection in the sequence.

The sense of invariance is fundamental in statistics, [29,30], and the following proposition plays an important role.

Proposition 2. *The set of the affine transformations G as*

$$G = \left\{ g = \begin{pmatrix} 1 & 0 & \dots & 0 & 0 \\ 0 & 1 & \dots & 0 & 0 \\ \vdots & & & & \vdots \\ 0 & 0 & \dots & 1 & 0 \\ \beta_1 & \beta_2 & \dots & \beta_p & \sigma \end{pmatrix} = \begin{pmatrix} I_p & O^t \\ \beta & \sigma \end{pmatrix}, \; g \in R^{(p+1)\times(p+1)} \right\}, \tag{20}$$

with $I_p = diag(1, 1, \dots, 1) \in R^{p\times p}$, $O = (0, 0, \dots, 0) \in R^{1\times p}$, $\beta = (\beta_1 \beta_2, \dots, \beta_p) \in R^{1\times p}$, $\sigma \in R^+$ *forms a group under matrix multiplication.*

Proof. Considering two elements $g, g' \in G$, with

$$g' = \begin{pmatrix} I_p & O^t \\ \beta' & \sigma' \end{pmatrix}, \; \beta' = (\beta'_1 \beta'_2, \dots, \beta'_p) \in \mathbb{R}^{1\times p}.$$

Then for the g, g' can be easily verified that

$$g' = \begin{pmatrix} I_p & O^t \\ \gamma & \sigma\sigma' \end{pmatrix} := h,$$

with $\gamma = (\gamma_1, \gamma_2, \dots, \gamma_p) \in \mathbb{R}^{1\times p}$ and $\gamma_r = \beta_p + \beta'_p \sigma$, $r = 1, 2, \dots, p$. Therefore, $h \in \mathbb{R}^{1\times p}$. The unit matrix I can be defined as

$$I = \begin{pmatrix} I_p & O^t \\ O & 1 \end{pmatrix} = diag(1, 1, \dots, 1) \in R^{1\times(p+1)}.$$

For given g, the inverse $g^{-1} \in G$ can be defined as

$$g^{-1} = \begin{pmatrix} 1 & 0 & \dots & 0 & 0 \\ 0 & 1 & \dots & 0 & 0 \\ \vdots & & & & \vdots \\ 0 & 0 & \dots & 1 & 0 \\ -\sigma\beta_1 & -\sigma\beta_2 & \dots & -\sigma\beta_p & \sigma \end{pmatrix}.$$

Note that from the definition of the product

$$g^{-1}g = gg^{-1} = \begin{pmatrix} I_p & O^t \\ \gamma & \sigma\sigma' \end{pmatrix} \in R^{(p+1)\times(p+1)},$$

with $\gamma_i = \beta - i + (-\sigma(-1))\beta_i = 1$, with $i = 1, 2, \dots, p$. Therefore, $g^{-1}g = gg^{-1} = I$, I the unit element of the defined group.

Moreover, it can be easily verified that given elements g_1, g_2, $g_3 \in G$ with

$$g_i = \begin{pmatrix} I_p & O^t \\ \beta_j & \sigma_j \end{pmatrix},$$

with $\beta_j = (\beta_{j1}, \beta_{j2}, \dots\beta_{jp})$, $j = 1, 2, 3$, then

$$g_1(g_2g_3) = (g_1g_2)g_3.$$

This is true as

$$\beta_{1i} + (\beta_{2i} + \beta_{3i}\sigma_2)\sigma_1 = \beta_{1i} + \beta_{2i}\sigma_1 + \beta_{3i}\sigma_2\sigma_1 = (\beta_{1i} + \beta_{2i}\sigma_1) + \beta_{3i}\sigma_2\sigma_1$$

and

$$\sigma_1(\sigma_2\sigma_3) = (\sigma_1\sigma_2)\sigma_3.$$ □

Based on Proposition 2, Ref. [24] developed a compact theoretical development on invariance and although the affine geometry point of view was behind it, he never mentioned it, as Ref. [25] did.

Adopting the Fraser structural of inference [31] provided a method for choosing the optimal degree of a response polynomial using the min-max criterion. The idea of tolerance region was adopted, while [32] proved that either classical or affine tolerance region remain the same under linear transformation. Therefore, a function of their volume offers the chance to minimize and provide optimal design points, useful in "sensitive" applications [33]. The affine transformation can be adopted to the binary response problem, trough a "canonical form", i.e. under an affine transformation of the data [26]. The result is to gain calculations as in nonlinear experimental design, where computations are always a problem [34]. But gaining calculations is not always possible despite the affine transformation. Consider the tolerance regions [30] in environmental economics, where the polluted place X cannot be transformed to another area in a distance h, although the transformation $X + h$ is affine.

Similar problem can appear in biological problems. In quantitative methods, the Experimental Design Theory plays an important role, not only in statistics, and needs a special consideration. In the early steps, statistics were developed due to experimental design theory, solving agricultural problems at the beginning, chemical later [35], or economical problems [36] and eventually developed into the Optimal Design Theory. We follow the existing general notation for both lines of thought, Design and Geometry, either projective or affine. We consider that the Euclidean approach to design theory is considered known [9].

A $t-(v,k,\lambda)$ block design is the pair $(P, B) := D$, with P being a set of v points, B, a collection of k-subsets of P known as blocks. The value λ, known as the index of the design, refers to the imposed condition that every t-subset of P must be contained in exactly λ blocks. In principle, $t \geq 2$, $0 < k < v$.

The number of blocks is

$$b = v(B) = \lambda \binom{v}{t} \binom{k}{t}^{-1}, \tag{21}$$

while every point appears in exactly r blocks with

$$r = \lambda \binom{v-1}{t-1} \binom{k-1}{t-1}^{-1}. \tag{22}$$

The Fano plane is an example of a 2-(7,3,1) design under a finite geometry. From the above evaluations, we obtain for this particular design $b = 7, r = 3$. The set of points are partitioned due to the block consideration and therefore the blocks offer parallel classes. Such a partition is called resolution in design theory, [37]. In such a case, despite the fact that different resolutions can exist, the design D is considered as resolvable. A non-resolvable design D can possess parallel classes.

The most well-known design is the balanced incomplete block design (BIBD) which is reached when $t = 2$, while Kitsos [33] adopted the Experimental Design Theory to estimate "small" percentiles. For spacial cases on ROCCD, see Ref. [38].

Let the projective geometry of dimension m over F_q be $PG(m, q)$. The projective geometry design $PG_t(m, q)$, is the design whose points are the points of $PG(m,q)$ and whose blocks are the t-dimensional projective subspaces of $PG(m,q)$.

Let us now consider the finite geometry designs constructed from affine geometry of dimension m over F_q, $AG(m,q)$ say. The points of $AG(m,q)$ are the vectors of F_q^m and the t-dimensional affine subspaces of the geometry are the t-dimensional vector subspaces of F_q^m and their cosets. Under this notation, in a similar way with the projective geometry designs, the affine geometry design $AG_t(m,q)$ is the design whose points are the points of $AG(m,q)$ and whose blocks are the t-dimensional affine subspaces of $AG(m,q)$.

4. Curves and Surfaces in Action

The smoothing problem is in practice roughly described here, the quantity q is estimated by a set of n observations y_i, $i = 1, 2, ..., n$ following the form $y_i = q + e_1$, where e_i are random variables known as errors. Thus, estimation of q depends on the approximation by a q^* so that $\|q - q^*\|$ is minimum. The criterion usually imposed to obtain q^* is based on the L_p norm

$$\min \|y_i - q\|_p = \|y_i - q^*\|_p. \tag{23}$$

Quantity q might be a number, a polynomial of degree $n-1$, a linear or a nonlinear function. For $p = 2$, the L_2 norm introduced by Legendre [39] is assumed and we refer to the "Least Square" problem, see Section 2. In the case of a polynomial, the method introduced by Gauss and Laplace with the L_∞ norm [40] could not be performed due to calculations. The approximation was efficient after the application of well-known curves due to conic sections.

As it is mentioned by [41], "Apollonius of Perga, who was born in around 262 BC in Perga, Greece, and died in about 190 BC in Alexandria, wrote a book called *Conics*, where terms such as parabola, ellipse, and hyperbola were introduced. It was J. Kepler (1571–1630) who showed that the orbits of planets around the sun form ellipses". See also [26]. So as it was also pointed out, it took almost 16 centuries to find an application for the ellipse. The fact that an ellipse in $\mathbb{R}^2$,

$$El = \left\{(x,y) : \left(\frac{x}{a}\right)^2 + \left(\frac{y}{b}\right)^2 = 1\right\}, \tag{24}$$

can be transformed with an affine transformation to the cycle, $Cyc = \{(x,y) : x^2 + y^2 = 1\}$, attracts only mathematical interest. As an ellipse can be not only an orbit but also a confidence or tolerance interval in statistics, there is no interest to transform it to a cycle. What's more the simple models $y = \beta x + e$ and $y = \beta_0 + \beta_1 x + e$ are certainly related through an affine transformation, but the economical, statistical background for these models is essentially different. In practice, we need to return to the transformed model and this implies complications related to the nature of the scientific field covering the model. Mathematical modeling reflects the nature of the study and there is no other way to look for generalizations.

Typical examples are the affine tolerance regions [26] and the canonical form of the generalized Logit model [26]. The former might explain why tolerance regions in a p_2 station, in distance h from a station p_1, are certainly related in an affine geometry point of view. The implications relate to the environment around p_1 and p_2, provided that in both places the cause of the pollution is the same.

As for the simple, one-hit, Logit model, known in Ca problems, and the generalized, multistage, Logit, the biological insight for both is completely different, where one model describes an affine transformation of the other [26].

Therefore, the fact can be closer to the real life situation that there is no affine transformation between ellipse, hyperbola, parabola. The orbits are well-defined by Nature!

The conic sections appeal not only in astronomy but in Nature in general, since the pioneering works of [42,43], through the rolling curves principle. If a conic section is rolling on an axis l, the focus graph shows a curve related to l and the conic section itself when it is rotated in the space around l graphs on the rotation surface.

Maxwell [42], provides a number of examples where even the spiral of Archimedes is considered; if we roll a parabola with parameter 4α on the spiral of Archimedes $r = \alpha\theta$, then the pole will trace the axis of a parabola. When a rolling curve is a cycle $r = \rho(1 + \cos\theta)$, then the traced curve is a Cardioid $r = \rho(1 + \cos\theta)$.

Rolling curves play an important role on the development and description of the main characteristics in real life organisms as has been well-known since the compact and pioneering work of Thompson [44], who offers an excellent approach for the Nature and Growth, the transformation and the evolution in general, adopting mathematical ideas. We emphasize the use of geometry in biology as we are very familiar with its use in physics, architecture, engineering, etc.

The problem of choosing the appropriate curve describing an economical phenomenon and the uncertainty which implies such a decision had been extensively discussed by [36].

The use of Geometry for the General Linear Model (GLM) has been discussed by Herr [14] who actually offers a review of the application of OLS to statistics. For the nonlinear models, [19] refers to both on fitting the model problems and the optimal experimental design point of view, while [72] discuss a number of approaches to tackle a nonlinear problem. The optimization adopting the minimal distance approach is not only reached by OLS but also with splines. Both OLS and splines are based on the general approach under the $L2$-norm; $\min \|f(x) - p(x)\|_2$, with a given function $f(x)$ and the optimum polynomial $p(x)$ approximating $f(x)$ either by OLS or by (usually) cubic splines. In practice the problem is that the spline approach needs a number of "smooth" assumptions, which are not in most cases available.

Through Probability theory, the Receiver Operating Characteristic (ROC) curves can be constructed without an interesting geometric approach, but it is a very important tool when the performance of a machine learning model is requested. Therefore, it can be evaluated with a number of available statistical packages. The main interest focuses on TPR and TNR, while TPR provides the sensitivity S_e and TNR the specificity S_p. See Appendix B for S_p and [45] for an introduction to ROC analysis.

Moving from curves to surfaces, the GLM can offer a presentation of the second order linear model. The full quadratic response surface model is

$$y = \beta_0 + \sum_{i=1}^{k} \beta_i x_i + \sum_{i=1}^{k} \beta_{ii} x_i^2 + \sum_{i=1}^{k-1} \sum_{\substack{j=i+1 \\ i<j}}^{k} \beta_{ij} x_i x_j + e, \tag{25}$$

where there are k input variables $x_1, x_2, ..., x_k$, $p = (k+1)(k+2)/2$ unknown coefficients β and the errors are independent with mean $\mu = 0$ and variance σ^2. The design consists of $n \geq p$ points and $X = [x_{j1}, x_{j2}, ..., x_{jk}]$. The matrix of moments is

$$M = \frac{1}{n} X'X, \tag{26}$$

and the prediction variance at a given point x is

$$Var[\hat{y}(x)] = \frac{\sigma^2}{n} f(x) M^{-1} f'(x), \tag{27}$$

with

$$f(x) = [1, x_1, \dots, x_k, x_1^2, \dots, x_k^2, x_1, x_2, \dots, x_k - 1, x_k]. \tag{28}$$

Proposition 3. *For the fitted response surface ds :*

$$\hat{y} = \hat{\beta}_0 + x'\hat{\beta} + x'Bx, \tag{29}$$

with $x = (x_1, x_2, ..., x_k)'$, $\hat{\beta} = (\hat{\beta}_1 \hat{\beta}_2, ..., \hat{\beta}_k)$ *and*

$$B = (\hat{\beta}_{i,j}) = \begin{pmatrix} \hat{\beta}_{11} & \frac{1}{2}\hat{\beta}_{12} & \cdots & \frac{1}{2}\hat{\beta}_{1k} \\ \frac{1}{2}\hat{\beta}_{12} & \hat{\beta}_{22} & \cdots & \frac{1}{2}\hat{\beta}_{2k} \\ \vdots & & \ddots & \\ \frac{1}{2}\hat{\beta}_{1k} & \frac{1}{2}\hat{\beta}_{2k} & & \hat{\beta}_{kk} \end{pmatrix}. \tag{30}$$

Interest is focused on the corresponding stationary points of the surface

$$x_0 = -\frac{1}{2}B^{-1}\hat{\beta}\,,\ y(\hat{x}_0) = \hat{\beta}_0 + \frac{1}{2}x_0'\hat{\beta}. \tag{31}$$

The response surface method (RSM) concerns the minimum or maximum of the response. Through the transformation

$$w = T(x) = x - x_0,$$

the transformed surface is

$$\hat{y} = y_0 + \sum_{i=1}^{k} \lambda_i w_i^2, \tag{32}$$

with λ_i the eigenvalues of the matrix B.

Therefore, if $\lambda_i > 0\,,\ i = 1, 2, ...,$ x_0 corresponds to minimum otherwise to maximum, and when some of the λ_i-s are some of positive and some negative, to saddle point.

Certain conditions have been imposed to obtain designs for which the standardized variance of the predicted response $\hat{y}$ is roughly the same throughout the experimental region, therefore,

$$Var(x) = \frac{n}{\sigma^2} Var(\hat{y}(x)) \tag{33}$$

remains constant on spheres and as Fisher's information I is the inverse of the variance, it is demanded to work actually on Poincaré disk. Recall Section 3, if we impose $\delta < 1$,

$$I = I(\delta),\ \delta = \|x\|_2^{1/2}, \tag{34}$$

see Refs. [37,46,47] for the available software. Under this geometrical restriction, a number of optimal experimental designs are constructed.

When the response surface is "large" (an airplane or boat surface), the B-splines offer solutions to estimate different surfaces [48]. Splines offer a solution to "larger" areas than the response surfaces but still, in real life geographical areas, interpolation is difficult. Kriging method offers a solution to even larger surfaces, between regions in a city or country [49,50], among others. Both B-splines and Kriging are based on a well-constructed software offering graphs, appropriate for applications which are beyond the target of the present chapter.

When we refer to surfaces by revolution of a given curve $c(t)$ in the plane, the easiest way to define cylindrical coordinates is

$$S(t, q) = (r, \theta, z) = (c_x(t), s, c_y(t)). \tag{35}$$

When the $c(t)$ curve is a circle, the torus is an example with several applications. Let L be the distance from the center of the tube to the outer extent of the torus and let r be the radius of the tube, then interesting results can be obtained. From Pappus's centroid theorem the surface A and the volume V of this solid are

$$A = 4\pi^2 Lr, \ V = (\pi r^2)(2\pi L). \tag{36}$$

Furthermore, the topology of a torus is interesting but beyond the present discussion. But it is an interesting point that the curvatures K_1 and K_2 are such that always $K_1 > 0$ and therefore a point of torus depends on K_2, as follows:

$$K_2 > 0, \text{ the point is elliptic,}$$
$$K_2 = 0, \text{ the point is parabolic,}$$
$$K_2 < 0, \text{ the point is hyperbolic.}$$

An extension of splines, adopting computer aided design (CAD) and finite element models (FEM), are the Bezier surfaces, first applied on the car industry in 1960s. It took about 50 years to realize that the Bernstein polynomials, see Appendix E, which were introduced in 1962, can be useful in practice. A Bezier curve is a parametric curve with the equation

$$B(t) = (1 - t)^3 b_1 + 3(1 - t)^2 b_2 + 3(1 - t)t^2 b_3 + t^3 b_4. \tag{37}$$

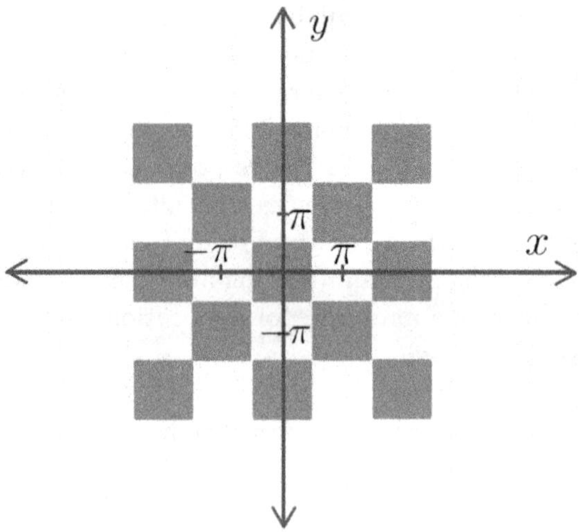

Fig. 2: Checkboard pattern, see Ref. [73].

The parameter $t \in [0, 1]$ is called the time along the curve. A Bezier path is a set of Bezier curves which connect in a chain at their end points to form a more complex curve, in the line of thought of knots in splines.

A set of such closed paths, as each one path forms an area, produces the Bezier contour [51,52], based on Bernstein polynomials. Therefore, a bridge between geometry and computer science is developing and this is computational geometry, playing a significant role in recent applications. Therefore, Bezier curves have helped develop either bicubic or biquadratic Bezier patches.

In architecture, geometry is adopted following the thought of Monge (1746–1868), see Ref. [21] for the extended approach of Ref. [53]. Even today, architecture adopts minimal surfaces [43,54] for constructions. A minimal surface (extending the idea of minimal distance, related to geodesics) is defined as a surface of the smallest area spanned by a given curve $\gamma(t)$ in plane. A surface $S \subseteq \mathbb{R}^3$ is minimal if and only if

(1) The mean curvature is equal to zero at all points.
(2) For every point $p \in S \subseteq \mathbb{R}$ there exists a neighborhood N_p bounded by a simple closed curve, with the minimum area $A(N_p)$ of all surfaces having the same boundary.
(3) It is a critical point of the area functional for all compactly supported variations.

(4) It can be expressed locally by $u = u(x, y)$ and provides the least area across a given closed curve by the Lagrange partial differential equation

$$(1 + u_x^2)u_{yy} - 2u_x u_y u_{xy} + (1 + u_y^2)u_{xx} = 0. \tag{38}$$

A trivial example is the plane, but in constructions the checkerboard pattern is

$$z = \log\left(\frac{\cos y}{\cos x}\right), \tag{39}$$

with $(\cos y/\cos x) > 0$, $(x, y) \in (-\frac{\pi}{2}, \frac{\pi}{2})^2$. See Refs. [55,73,74] for the Scherk surface as above, Fig. 2.

5. Discussion

To approximate a surface is very important and covers a great number of applications, mainly in engineering. In recent years there have been an increasing number of references where the medical problem of identifying a tumor related to fractals, see Refs. [56–58]. The main reasons are: (i) Fractal Geometry provides a framework to the study of "irregular sets"; (ii) the breast tissues structure obeys the number of statistical parameters that can be measured.

In this chapter, we would like to emphasize that at least as far as the affine geometry is equivalent to the three fundamental ideas of (i) Riemann surface; (ii) meromorphic function; and (iii) holomorphic differential, it escapes from the Euclidean geometry, ignores metric notions and approaches mathematical analysis. But this is difficult to be applied in biological studies. At the first stage, the shape was investigated, but after the shape is known, a special development is needed [59]. New techniques have been developed since [44], the OLS for Econometrics appeared, a new development since [60] see Refs. [61,62], and fractals appear in applications in Cancer [57,58,63]. All of which seems elementary now but was then pioneering work, [64–66] we moved to more elegant mathematical methods in quantitative applications. The next step was to work vice versa, apply the geometrical insight of tolerance regions for decision-making [67]. But geometry remains there to support, either in obvious ways or invisible.

But Euclidean spaces are affine spaces in the sense that the associated vector space is equipped with a positive-definite quadratic form $q(x)$ so that the inner product is

$$< x, y >= \frac{1}{2}(q(x + y) - q(x) - q(y)). \tag{40}$$

Therefore, geometry either as Euclidean or Non-Euclidean, affine or projective, offers solutions to quantitative methods in a number of fields that try to adopt the current geometrical development: from rolling curves to fractals, from minimum distance to minimal surfaces, and always remains the "right hand" in Physics and Architecture.

Appendix A. Statistics and Physics

Consider the $(BC)_1$ model of a simple random sampling r times from a population size n without replacement. Let $(BC)_\infty$ be the model that represents random sampling r times from a population size n with replacement. Let

$$\Omega = \{(x_1, ..., x_r) : \ x_i \text{ be the number of the cell where the } i\text{-th ball is placed}\},$$

$$\Omega_0 = \{(b_1, ..., b_n) : \ b_j, \text{ the number of balls in } j \text{ cell}\}.$$

Then with $V(A)$ being the cardinal number of the set A, under $(BC)_1$:

$$V(\Omega_1) = (n)_r,$$

$$V(\Omega_0) = \binom{n}{k}.$$

Under $(BC)_\infty$:

$$V(\Omega_1) = n^r,$$

$$V(\Omega_0) = \binom{n+r-1}{r}.$$

Moreover, the classification of Maxwell–Boltzman statistics, known as the classical case, the Bose–Einstein statistics for the quantum mechanical case and the Fermi statistics can be represented as in the following table, based on the above compact discussion.

	Ω_1	Ω_0
$(BC)_\infty$	Maxwell–Boltzman Statistics	Bose–Einstein Statistics
$(BC)_1$		Fermi–Dirac Statistics

As far as the distribution of the mean number of the particles, $< u_i >$, in state i, (the balls in cells), it is evaluated as

$$< u_i >= \frac{1}{\exp(\beta(\varepsilon_i - \mu)) - \gamma}, \ \beta = \frac{1}{kT},$$

with

$$\gamma = 0\,, \text{ Maxwell–Boltzman,}$$

$$\gamma = 1\,, \text{ Bose–Einstein,}$$

$$\gamma = -1\,, \text{ Fermi–Dirac,}$$

where T is the temperature, ε_i, the energy of the particle in state i, k, a known constant and μ, an adjusted factor of the chemical potential.

Appendix B. Geometry and Topology

Let $x_0, x_1, ..., x_p$, $p \geq 0$ be the linear independent points of $\mathbb{R}^n$. Then the point set

$$S_p = \left\{ \sum_{i=0}^{p} w_i x_i\,,\ \sum_{i=1}^{p} w_i = 1\,,\ w_i \geq 0 i = 1, 2, ..., p \right\} \tag{B.1}$$

is called an affine simplex. The point

$$C_p = \frac{1}{p+1} \sum_{i=0}^{p} x_i \tag{B.2}$$

is called the barycenter of S_p, while the points x_i are called the vertices of S_p. Any non-empty subset of vertices spans an affine simplex, $S_q,\ 0 \leq q \leq p$, called face of S_p. There are

$$\binom{p+1}{q+1}$$

proper faces with $\dim S_q = q \in [0, p]$.

Every affine simplex is a convex hull of its vertices. Considering x_0 a vertex of S_p, then S_p is the cone with x_0 as vertex and the face of S_p opposite x_0 as base.

Let $z \in S_p$. Then z is an interior point of S_p and therefore defines exactly one face of S_p with $\min \dim S_p$ that contains x. For the simplex S_p as in (B.2), consider the union of $(p-1)$ faces, known as geometric boundary $\dot{S}_p$. Moreover, it holds the following:

Let S_p be an affine p-simplex and let d be its diameter, $d = diam(S_p)$. For every simplex of the normal subdivision S'_p of S_p, then

$$d' = diam(S_p) \leq d\frac{p}{p+1}.$$

Appendix C. Double Ratios

Consider the points A_1, A_2, A_3, A_4 on the line l_A and B_1, B_2, B_3, B_4 on the line l_B. Then it holds

$$(A_2, A_3; A_1, A_4) = (B_2, B_3; B_1, B_4),$$

with

$$(A_2, A_3; A_1, A_4) = \frac{(A_2 - A_1)(A_3 - A_4)}{(A_2 - A_4)(A_3 - A_1)} = \frac{|A_2 - A_1||A_3 - A_4|}{|A_2 - A_4||A_3 - A_1|}.$$

The general approach to harmonic points has been considered under Geometric Algebra, in Artin (1988), which offers a new look on Pappus's and Desargues's theorem. It is emphasized that the cross or double ratio [74], is that quantity which is preserved from projective transformations (see Fig. C.1). Regarding Euclidean distance, it is the projective equivalent as far as rigid transformations are concerned.

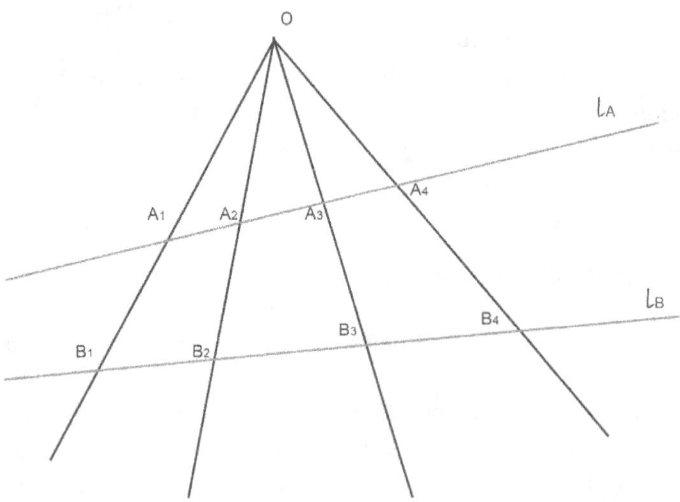

Fig. C.1: Double ratios.

Moreover, while the above evaluation of $(A_1, A_3; A_1, A_4)$ is a Euclidean one in the sense that $|A_i - A_j| = d(A_i, A_j),\ i \neq j,\ i, j = 1, 2, 3, 4$ with $d(\cdot)$ being the Euclidean distance, the affine definition is an extension.

For the affine lines $A_i = [KA_i, 1],\ i = 1, 2, 3, 4,$

$$(A_2, A_3; A_1, A_4) = \frac{(KA_2 - KA_1)(KA_3 - KA_4)}{(KA_3 - KA_1)(KA_2 - KA_4)}, \tag{C.1}$$

therefore,

$$(\infty, A_3; A_1, A_4) = \frac{(KA_3 - KA_4)}{(KA_3 - KA_1)},$$

while

$$(A_2, 1; 0, \infty) = A_2.$$

Note that a double ratio remains invariant under a Mobius transformation.

Appendix D. ROC Curves

Consider the following table:

Real Situation	Results of Test	
	Positive, $Y = 1$	Negative, $Y = 0$
Present, $D = 1$	TPR	FNR
Absent, $D = 0$	FPR	TNR

where:
TPR: True Positive Ratio FPR: False Positive Ratio.
FNR: False Negative Ratio TNR: True Negative Ratio.

The percentage of the positive results in the set of positive situations is the TPR, while on the negative situation, it is the FNR. The percentage of the negative results is expressed by FNR in the set of positive situation, while by TNR in the set of the negative situation. Then we define the ROC curve

$$ROC(\cdot) = \{(FPR(\gamma), TPR(\gamma),\ \gamma \in \mathbb{R}) = \{(t, ROC(t), t \in [0, 1])\}\}. \tag{D.1}$$

The ROC curve is an increasing function in the square $[0,1]^2$. Moreover, comparing two tests X, Y when $ROC_x(t) > ROC_y(t)$, $t \in [0,1]$, the decision is that test X provides a better diagnosis than Y.

Appendix E. Bernstein Polynomial

The nth degree polynomial of Bernstein is defined as

$$\beta_{v,n}(x) = \binom{n}{v} x^n (1-x)^{n-v}, \quad v = 0, 1, \ldots, n. \tag{E.1}$$

The Bernstein (basis) polynomial as above forms a vector space and a linear combination such as

$$B_n(x) = \sum_{v=0}^{n} b_n \beta_{v,n}(x), \tag{E.2}$$

forming an nth degree polynomial where the b_n coefficients are known as the Bezier coefficients.

Appendix F. Main Definitions

Let φ be a smooth non-singular function from a parameter domain Ω to a Euclidean space $\mathbb{R}^3$, $\varphi : \Omega \longrightarrow \mathbb{R}^3$, and let $< \cdot, \cdot >$ be the standard scalar product. As φ is non-singular, the differential $d\varphi$ has maximal rank 2. The Riemannian metric of V, W is defined

$$g(V, W) =< d\varphi \cdot V, d\varphi \cdot W > . \tag{F.1}$$

Therefore, for the curve $C : [a, b] \longrightarrow \Omega$, the length $L(\cdot)$ of $\varphi \circ c$ is

$$L(C_L) = \int_a^b [g(c'(t), c'(t))]^{\frac{1}{2}}. \tag{F.2}$$

Based on the above, a parametrization is conformal or isothermal if it preserves the angles between tangent vectors

$$< d\varphi \cdot V, d\varphi \cdot W >= k^2 < V, W > \tag{F.3}$$

with k known as stretch factor of φ. In other words, the Riemannian metric is, pointwise, proportional to the Euclidean scalar product defined in Ω.

Now recall that the stereographic projection maps the unit sphere to the complex plane. Under the stereographic projection, as below the North Pole is mapped to ∞, the South Pole is mapped to zero and the equator to a unit circle. The inverse σ^{-1} of a stereographic projection σ is a most well-known example of conformal parametrization

$$\sigma^{-1} : \mathbb{C} \longrightarrow \mathbb{R}^3 \; : \; z = x + iy \longrightarrow \sigma^{-1}(z), \tag{F.4}$$

with

$$\sigma^{-1}(z) = \frac{1}{1+|z|^2}(2Re(z), 2Im(z), |z|^2 - 1). \tag{F.5}$$

Now for every point $p \in \Omega$, provided the coordinate vectors

$$U = \frac{\partial}{\partial u}, \quad V = \frac{\partial}{\partial v},$$

the vectors $U_1 = d\varphi(p) \cdot U$, $V_1 = d\varphi(p) \cdot V$ span the tangent plane $T \subseteq \mathbb{R}^2$ at $\phi(p)$. The perpendicular unit vector $N(p)$ to T, $N(p) \perp T$, known as normal vector, is defined through the Gauss function

$$N : \Omega \longrightarrow \mathbb{R}^2, \; p \longrightarrow N(p), \tag{F.6}$$

with

$$N(p) = \left.\frac{d\varphi \cdot U \times d\varphi \cdot V}{|d\varphi \cdot U||d\varphi \cdot V|}\right|_p = \left.\frac{|U_1 \times V_1|}{|U_1||V_1|}\right|_p. \tag{F.7}$$

The sense of curvature is defined through the identification of the tangent space of the surface at $\varphi(p)$, with the tangent space of the unit sphere at $N(p)$, by an affine (parallel) translation in the sense that require

$$d\varphi \cdot S \cdot V = dN \cdot V, \tag{F.8}$$

with S being the shape operator, or Weingarten function. The eigenvalues of the Weingarten function are known as principle curvatures, which are of two kinds: Mean curvature, $K_M = tr(S)$ and Gauss curvature, $K_G = \det(S)$.

Acknowledgments

C.K. would like to express his sincere gratitude to the University of Alberta, Portugal, for offering him the chance to teach the PhD course *Topicos de Estatica Mathematica*.

References

[1] Y. Choquet-Bruhat, C. DeWitt-Morette, and M. Dillard-Bleik, *Analysis, Manifolds and Physics* (Gulf Professional Publishing, 1977).

[2] B.F. Schutz, *Geometrical methods of Mathematical Physics* (Cambridge University Press, Cambridge, UK, 1980).

[3] T.M. Rassias, On some major trends in mathematics, *Approximation and Computation*, pp. 45–49 (Springer, New York, 2010).

[4] D.F. Morrison, *Multivariate Statistical Methods* (McGraw-Hill, New York, 1976).

[5] W.T. Anderson, *An Introduction to Multivariate Statistical Analysis* (Wiley, London, UK, 1958).

[6] T.D. Wickens, *The Geometry of Multivariate Statistics* (Psychology Press, 1995).

[7] C. Kitsos, Inequalities for the Fisher's Information Measures. In: T.M. Rassias (Ed.), *Handbook of Functional Inequalities*, pp. 281–313 (Springer, 2014).

[8] C.R. Rao, On the use and interpretation of distance functions in statistics, *Bull. Int. Statistical Institute* **34**, 90–100, (1954).

[9] D. Saville and G.R. Wood, *Statistical Methods: The Geometric Approach* (Springer-Verlag, New York, 1991).

[10] W. Mendenhall, *Introduction to Linear Models and the Design and Analysis of Experiments* (Wadsworth Publishing Company, 1968).

[11] J.R. Wannacott and H.T. Wannacott, *Econometrics* (Wiley, New York, 1979).

[12] J. Durbin and M.G. Kendall, The geometry of estimation, *Biometrika* **38**(1/2), 150–158, (1951).

[13] A.N. Kolmogorov, On the proof of the method of least squares, *Uspekhi Mat. Nauk*, **1**(11) 57–70, (1946).

[14] D. Herr, On the History of the use of geometry in the general linear model, *Am Stat.* **34**(1), 43–47, (1980).

[15] W. Kruskal, The geometry of generalized inverses, *J. R. Stat. Soc., B: Stat. Methodol.* **37**(2), 272–283, (1975).

[16] A. De la Garza, Spacing of information in polynomial regression, *Ann. Math. Stat.* 123–130, (1954).

[17] J. Kiefer and J. Wolfowitz, The equivalence of two extremum problems, *Canad. J. Math.* **12**, 363–366, (1960).

[18] C. Kitsos, The simple linear calibration problem as an optimal experimental design, *Commun. Statistics-Theory Meth.* **31**(7), 1167–1177, (2002).

[19] I. Ford, D.M. Titterington, and C. Kitsos, Recent advances in nonlinear experimental design, *Technometrics* **31**(1), 49–60, (1989).

[20] J.V. Poncelet, Traité des propriétés projectives des figures (Paris France, 1862).

[21] G. Monge, *Application de l' Analyse a la Geometrie* (Ecole Imperiale Polytechnique, Paris, 1807).

[22] P. Bryant, Geometry, statistics, probability: Variations on a common theme, *Am. Stat.* **98**, 38–48, (1984).

[23] N. Prakash, *Differential Geometry: An Integrated Approach* (Tata McGraw-Hill, New Delhi, India, 1981).

[24] D.A.S. Fraser, *The Structure of Inference* (Wiley, London, UK, 1968).

[25] M.K. Murray and J.W. Rice, *Differential Geometry and Statistics* (Chapman & Hall, London, UK, 1993).

[26] C. Kitsos, Invariant canonical form for the multiple logistic regression, *Math. Eng. Sci. Aerosp.* **2**(3), 267–275, (2011).
[27] T. Botts, Finite planes and Latin squares, *Math. Teacher* **54**(5), 300–306, (1961).
[28] J.G. Garnier, *Élémens de Géometrie Analytique* (Paris, France, 1808).
[29] C. Kitsos and T. Toulias, Confidence and tolerance regions for the signal process, *Recent Patents Signal Proc. (Discontinued)* **2**(2), 149–155, (2012).
[30] C.P. Kitsos and T.L. Toulias, Adopting tolerance regions in environmental Economics, *J. Adv. Math. Comput. Sci.* **34**(6), 1–12, (2019).
[31] R.R.W. Ellerton, P. Kitsos, and S. Rinco, Choosing the optimal order of a response polynomial — Structural approach with minimax criterion, *Commun. Stat. Theory. Meth.* **15**(1), 129–136, (1986).
[32] C.H. Müller and C.P. Kitsos, Optimal design criteria based on tolerance regions. In: *mODa 7—Advances in Model-Oriented Design and Analysis*, pp. 107–115 (Physica, Heidelberg, 2004).
[33] C. Kitsos, Optimal designs for estimating the percentiles of the risk in multistage models of carcinogenesis, *Biom. J.* **41**, 33–43, (1999).
[34] C. Kitsos and A. Oliveira, *On the Computational Methods in Non-linear Design of Experiments.* In: N. Daras and T. Rassias (Eds.), *Computational Mathematics and Variational Analysis, Optimization and Its Applications*, Vol. 159 (Springer, 2020).
[35] P. Kitsos and K. Kolovos, An optimal calibration design for pH meters, *Instrum. Sci. Technol.* **38**(6), 436–447, (2010).
[36] G. Halkos and C. Kitsos, Uncertainty in environmental economics: The problem of entropy and model choice, *Econom. Anal. Policy* **60**, 127–140, (2018).
[37] A.I. Khuri and J.A. Cornell, *Response Surfaces: Designs and Analyses*, (CRC Press, 1996).
[38] C.P. Kitsos, On the design points for a rotatable orthogonal central composite design, *REVSTAT-Statistical J.* **17**(1), 25–34, (2019).
[39] A.M. Legendre, *Nouvelles Méthodes pour la Détermination des Orbites des Comètes* (Courcier, Paris, France, 1806).
[40] S.P. Laplace, *Traité de Mécanique Céleste* (1799).
[41] T. Sasaki, Affine geometry, projective geometry, and non-euclidean geometry. In: *Mathematics: Concepts, and Foundations*, Vol. I, (2011).
[42] J.C. Maxwell, On the theory of rolling curves. *Earth Environ. Sci. Trans. Royal Soc. Edinburgh* **16**(5), 519–540, (1849).
[43] K.J. Wittemore, Minimal surfaces applicable to surfaces of revolution, *Ann. Math. Second Series* **19**(1), 1–20, (1917).
[44] D'A. Thompson, *On Growth and Form.* (Cambridge University Press, 1917).
[45] T. Fawcett, An introduction to ROC analysis, *Pattern Recognition Lett.* **27**(8), 861–874, (2006).
[46] R.H. Myers and D.C. *Montgomery, Response Surface Methodology: Process and Product in Optimization Using Designed Experiments* (Wiley, New York, 1995).

[47] P. Kitsos, *Computational Statistics* (in Greek) (New Tech. Pub., Athens, Greece, 1996).

[48] A. Entezari and D. Van De Ville, Practical box splines for reconstruction on the body centered cubic lattice, *IEEE Trans. Visualizat. Comput. Graphics* **14**(2), 313–328 (2008).

[49] C. Kitsos and P. Iliopoulou, (2020). Distance measures in spatial statistics (Greek). In: K. Kalambokides, G. Korres, N. Soulakelis, and N. Fidias (Eds.), *Social Studies and Geography Theory, Methods and Techniques in Spatial Analysis*, pp. 96–108.

[50] P. Iliopoulou and C. Kitsos, (2019) Statistics in Geography: Spatial Analysis. In: G. Korres, H. Kourliourgos, and A. Kokkinos (Eds.), *Recent Work in Statistics and Geography: Theory and Policies*, pp. 212–224, 43–47, (1980).

[51] J. Hoschek and N. Wissel, Optimal approximate conversion of spline curves and spline approximation of offset curves, *Comput. Aided Des.* **20**(8), 475–483, (1988).

[52] R.T. Farouki and V.T. Rajan, Algorithms for polynomials in Bernstein form, *Comput. Aided Geom. Des.* **5**(1), 1–26, (1988).

[53] G. Lefkaditis and G. Exarchakos, *Descriptive Geometry* (Greek) (Anelixi, Athens, Greece, 2018).

[54] D'A. Delaunay, Sur la surface de révolution dont la courbure moyenne est constante, *J. de math. Pures et Appliquées 1re Série* **6**, 309–314, (1841).

[55] S.L. Velimirovic, G. Radivojevic, S.M. Stankovic, and D. Kostic, Minimal surfaces for architectural constructions Facta universitatis – series. *Architecture Civil Eng.* **6**(1), 89–96, (2008).

[56] J.W. Baish and R.J. Jain, Fractals and cancer, *Cancer Res.* **60**, 3683–3688, (2000).

[57] M.E. Dokukin, N.V. Guz, C.D. Woodworth, and I. Sokolov, Emergence of fractal geometry on the surface of human cervical epithelial cells during progression towards cancer, *New J. Phys.* **17**, (2015).

[58] M. Stehlík, P. Hermann, and S. Giebel, J.P. Schenk, Multifractal analysis on cancer risk. In: T. Oliveira C. Kitsos C., A. Oliveira, and L. Grilo (Eds.), *Recent Studies on Risk Analysis and Statistical Modeling. Contributions to Statistics* (Springer, Cham. 2018).

[59] C.P. Kitsos, D.M. Titterington, and B. Torsney, An optimal design problem in rhythmometry, *Biometrics* **44**(3), 657–671, (1988).

[60] A. Aitken, On least squares and linear combination of observations, *Proce. Royal Soc. Edinburgh* **55**, 42–48, (1936).

[61] T. Amemiya, *Advanced Econometrics* (Cambridge: Harvard University Press, 1985).

[62] G. Halkos, *Econometrics: Theory and practice: Instructions in using Eviews, Minitab, SPSS and Excel* (Gutenberg, Athens Greece, 2007).

[63] B.B. Mandelbrot, *The Fractal Geometry of Nature* (W.H. Freeman, New York, 1982).

[64] T. Dwight, The range of variation of the human shoulder-blade, *Am. Naturalist* **21**(7), 627–638, (1887).

[65] T. Cook, Spirals in nature and art, *Nature* **68**, 296, (1903).

[66] T. Cook, *The Curves of Life* (Constable and Company LTD, London, UK, 1914).

[67] V. Zarikas, V. Gikas, and P. Kitsos, Evaluation of the optimal design "cosinor model" for enhancing the potential of robotic theodolite kinematic observations, *Measurement* **43**(10), 1416–1424, (2010).

[68] L.E. George and K.H. Sager, Breast cancer diagnosis using multi-fractal dimension spectra, *IEEE International Conference on Signal Processing and Communications*, Dubai, UAE, pp. 592–595 (November 2007).

[69] C. Kitsos, *The Geometry of Greeks* (New Tech. Pub., Athens Greece, 2021).

[70] W. Kruskal, When are Gauss-Markov and least squares estimators identical? A coordinate-free approach, *Ann. Math. Statistics* **39**(1), 70–75, (1968).

[71] P. Phillips, Geometry of the Equivalence of OLS and GLS in the Linear Model. *J. Econ. Theory* **8**(1), 158–159, (1992).

[72] T. Toulias, and C. Kitsos, Estimation aspects of the Michaelis–Menten model, *REVSTAT–Statistical J.* **14**(2), 101–118, (2016).

[73] A. Treibergs, USAC Colloquium. *Minimal Surfaces: Nonparametric Theory* (University of Utah, 2016).

[74] A. Pressley, *Elementary Differential Geometry*, 2nd Edition (Springer-Verlag London Limited, 2010).

https://doi.org/10.1142/9789811261572_0020

Chapter 20

Norm Inequalities for Fractional Laplace-Type Integral Operators in Banach Spaces

Jichang Kuang

Department of Mathematics, Hunan Normal University, Changsha, 410081, P.R. China

jckuang@163.com

This chapter introduces the new fractional Laplace-type integral operators in Banach spaces. It contains the generalized Laplace transform, the generalized Stieltjes transform and the Hankel transform, etc. the corresponding new operator norm inequalities are obtained. As applications, a large number of known and new results have been obtained by proper choice of kernel. They are significant improvement and generalizations of many famous results.

1. Introduction

It is well known that fractional integral operator is one of the important operators in harmonic analysis with background of partial differential equations. In the recent past, different versions of fractional integral operators have been developed which are useful in the study of different classes of differential and integral equations. These fractional integral operators act as ready tools to study the classes of differential and integral equations. Fractional integral operators were utilized extensively to study the classical inequalities. A generalization of classical inequalities by means of fractional integral operators is considered as an interesting subject area. Many mathematicians have established a variety of inequalities for those fractional integral and derivative operators, some of which have turned out to be useful in analyzing solutions of certain fractional integral and differential equations. Hence, fractional integral inequalities are very important in the theory and applications of differential equations. See e.g. [1–18].

First, we recall the following definitions and some related results.

Definition 1 (cf. [1,4]). Let $f \in L[a,b]$, then Riemann–Liouville fractional integrals of f of order $\alpha > 0$ with $a \geq 0$ are defined by

$$T_1(f,x) = \frac{1}{\Gamma(\alpha)} \int_a^x (x-t)^{\alpha-1} f(t)dt, \quad x > a, \tag{1}$$

and

$$T_2(f,x) = \frac{1}{\Gamma(\alpha)} \int_x^b (t-x)^{\alpha-1} f(t)dt, \quad x < b, \tag{2}$$

resp., where

$$\Gamma(\alpha) = \int_0^\infty t^{\alpha-1} e^{-t} dt, \tag{3}$$

is the Gamma function and when $\alpha = 0$, $T_1(f,x) = T_2(f,x) = f(x)$.

We can generalize the region from $[a,b]$ to $[0,\infty)$ by defining a function $f_1 : [0,\infty) \to \mathbb{R}^1$ as follows:

$$f_1(t) = \begin{cases} f(t), & t \in [a,b] \\ 0, & t \in [0,a) \bigcup (b,\infty), \end{cases}$$

and we can still define f_1 as f. Let $f \in L[0,\infty), \alpha > 0, x > 0$, then α — Riemann–Liouville fractional integrals of f are defined by

$$T_3(f,x) = \frac{1}{\Gamma(\alpha)} \int_0^x (x-t)^{\alpha-1} f(t)dt, \tag{4}$$

and

$$T_4(f,x) = \frac{1}{\Gamma(\alpha)} \int_x^\infty (t-x)^{\alpha-1} f(t)dt, \tag{5}$$

resp. (5) is also called the α-Weyl integral of f. (cf. [1]).

Definition 2 (cf. [3]). Let $f \in L[a,b]$, then Riemann–Liouville k — fractional integrals of f of order $\alpha > 0$ with $a \geq 0$ are defined by

$$T_5(f,x) = \frac{1}{k\Gamma_k(\alpha)} \int_a^x (x-t)^{(\alpha/k)-1} f(t)dt, \quad x > a, \tag{6}$$

and

$$T_6(f,x) = \frac{1}{k\Gamma_k(\alpha)} \int_x^b (t-x)^{(\alpha/k)-1} f(t)dt, \quad x < b, \tag{7}$$

resp., where

$$\Gamma_k(\alpha) = \int_0^\infty t^{\alpha-1} e^{-(\frac{t^k}{k})} dt \quad (\alpha > 0) \tag{8}$$

is the k — Gamma function. Also, $\Gamma(x) = \lim_{k\to 1}\Gamma_k(x), \Gamma_k(\alpha) = k^{(\alpha/k)-1}\Gamma(\frac{\alpha}{k})$ and $\Gamma_k(\alpha+k) = \alpha\Gamma_k(\alpha)$.

It is well known that the Mellin transform of f is defined by

$$M(f,\alpha) = \int_0^\infty t^{\alpha-1} f(t)dt, \ (\alpha > 0).$$

Therefore, the Mellin transform of the exponential function $\exp\{-\frac{t^k}{k}\}$ is the k — Gamma function.

Definition 3 (cf. [6]). Let $f \in L[a,b], g : [a,b] \to (0,\infty)$ be an increasing function, and $g' \in C[a,b], \alpha > 0$. Then g — Riemann–Liouville fractional integrals of f with respect to the function g on $[a,b]$ are defined by

$$T_7(f,x) = \frac{1}{\Gamma(\alpha)} \int_a^x g'(t)[g(x)-g(t)]^{\alpha-1} f(t)dt, \quad x > a, \tag{9}$$

and

$$T_8(f,x) = \frac{1}{\Gamma(\alpha)} \int_x^b g'(t)[g(t)-g(x)]^{\alpha-1} f(t)dt, \quad x < b, \tag{10}$$

resp.

In 2020, Kuang [12] introduced the new notion of the generalized fractional integral operators and fractional area balance operators as follows:

Definition 4 ([12]). Let $f \in L[a,b]$, $g : [a,b] \to (0,\infty)$ be an increasing function, and $g \in AC[a,b], k, c, \alpha > 0$, $a \geq 0$. Then the generalized fractional integral operator T_9 with respect to the function g on $[a,b]$ is defined by

$$T_9(f,x) = \frac{c}{k\Gamma_k(\alpha)} \int_a^b g'(t)|g(x)-g(t)|^{(\alpha/k)-1} f(t)dt, \tag{11}$$

where $\Gamma_k(\alpha)$ is defined by (8)

Let

$$T_{10}(f,x)=\frac{c}{k\Gamma_k(\alpha)}\int_a^x g'(t)[g(x)-g(t)]^{(\alpha/k)-1}f(t)dt,\quad x>a, \tag{12}$$

and

$$T_{11}(f,x)=\frac{c}{k\Gamma_k(\alpha)}\int_x^b g'(t)[g(t)-g(x)]^{(\alpha/k)-1}f(t)dt,\quad x<b, \tag{13}$$

then

$$T_9(f,x)=T_{10}(f,x)+T_{11}(f,x). \tag{14}$$

Note indicate that for suitable and appropriate choice of the parameters and function, one can obtain various new and old results. For example:

If $c=k=1$ in (14), then (14) reduces to

$$T_9(f,x)=T_7(f,x)+T_8(f,x). \tag{15}$$

Hence, Definition 4 unified and generalized many known and new classes of fractional integral operators.

It is noted that all the fractional integral operators above are established for functions of one variable. We naturally ask, how do we generalize these results to functions of several variables?

In 2014, Sarikaya [13] gives the definitions for Riemann–Liouville fractional integrals of two variable functions as follows:

Definition 5 ([13–15]). Let $Q=[a,b]\times[c,d]=\bigcup_{k=1}^4 Q_k$, where $Q_1=[a,x]\times[c,y]$, $Q_2=[a,x]\times[y,d]$, $Q_3=[x,b]\times[c,y]$, $Q_4=[x,b]\times[y,d]$. The Riemann–Liouville fractional integrals $I_k(1\le k\le 4)$ are defined by

$$I_1(f;x,y)=\frac{1}{\Gamma(\alpha)\Gamma(\beta)}\int_{Q_1}(x-t)^{\alpha-1}(y-s)^{\beta-1}f(t,s)dsdt,$$

$$I_2(f;x,y)=\frac{1}{\Gamma(\alpha)\Gamma(\beta)}\int_{Q_2}(x-t)^{\alpha-1}(s-y)^{\beta-1}f(t,s)dsdt,$$

$$I_3(f;x,y)=\frac{1}{\Gamma(\alpha)\Gamma(\beta)}\int_{Q_3}(t-x)^{\alpha-1}(y-s)^{\beta-1}f(t,s)dsdt,$$

and

$$I_4(f;x,y)=\frac{1}{\Gamma(\alpha)\Gamma(\beta)}\int_{Q_4}(t-x)^{\alpha-1}(s-y)^{\beta-1}f(t,s)dsdt.$$

If the Q in the above Definition 5 is applied to the n — dimensional interval $Q = \bigcup_{k=1}^{n} Q_k$ in $\mathbb{R}_+^n$, where $Q_k = [a_k, b_k]$, then the problem becomes very complicated when n is large.

In 2021, Kuang [16] introduced the new notion of the generalized fractional integral operators and fractional area balance operators in Banach spaces.

Throughout this chapter, we write

$$E_n = \left\{ x = (x_1, x_2, \ldots, x_n) : x_k \geq 0, 1 \leq k \leq n, \|x\| \right.$$
$$\left. = \left(\sum_{k=1}^{n} |x_k|^\beta \right)^{1/\beta}, \beta > 0 \right\}.$$

When $1 \leq \beta < \infty, E_n$ is a Banach space. In particular, when $\beta = 2, E_n$ is an n-dimensional Euclidean space $\mathbb{R}_+^n$. When $\beta = 1, \|x\| = \sum_{k=1}^{n} |x_k|$ is a Cartesian norm. Let

$$D_1 = \{y = (y_1, y_2, \ldots, y_n) : y_k \geq 0, 1 \leq k \leq n, 0 \leq \|y\| < \|x\|, x \in E_n\},$$
$$D_2 = \{y = (y_1, y_2, \cdots, y_n) : y_k \geq 0, 1 \leq k \leq n, \|x\| < \|y\| < \infty, x \in E_n\}.$$

Then $E_n = D_1 \bigcup D_2$. The norm of operator $T : L^p(E_n) \to L_\omega^p(E_n)$ $(1 \leq p < \infty)$ is defined by

$$\|T\| = \sup_{f \neq 0} \frac{\|Tf\|_{p,\omega}}{\|f\|_p}.$$

Definition 6 ([16]). Let $f \in L(E_n)$. Then the generalized conformable fractional integral operator T_{12} is defined by

$$T_{12}(f, x) = \frac{(r+s)^{1-(\alpha/k)}}{k\Gamma_k(\alpha)} \int_{E_n} |\|x\|^{r+s} - \|y\|^{r+s}|^{(\alpha/k)-1} \|y\|^{r+s-1} f(y) dy, \tag{16}$$

where $\Gamma_k(\alpha)$ is defined by (8).

Let

$$T_{13}(f, x) = \frac{(r+s)^{1-(\alpha/k)}}{k\Gamma_k(\alpha)} \int_{D_1} [\|x\|^{r+s} - \|y\|^{r+s}]^{(\alpha/k)-1} \|y\|^{r+s-1} f(y) dy, \tag{17}$$

and

$$T_{14}(f,x) = \frac{(r+s)^{1-(\alpha/k)}}{k\Gamma_k(\alpha)} \int_{D_2} [\|y\|^{r+s} - \|x\|^{r+s}]^{(\alpha/k)-1} \|y\|^{r+s-1} f(y)dy. \tag{18}$$

Thus,

$$T_{12}(f,x) = T_{13}(f,x) + T_{14}(f,x). \tag{19}$$

The generalized conformable fractional area balance operator is defined by

$$T_{15}(f,x) = T_{14}(f,x) - T_{13}(f,x). \tag{20}$$

Definition 7 ([16]). Let $f \in L(E_n)$, $g : (0,\infty) \to (0,\infty)$ be an increasing function, and $g \in AC(0,\infty)$, $k, c, \alpha > 0, a \geq 0$. Then the generalized fractional integral operator T_{16} with respect to the function g on E_n is defined by

$$T_{16}(f,x) = \frac{c}{k\Gamma_k(\alpha)} \int_{E_n} g'(\|y\|)|g(\|x\|) - g(\|y\|)|^{(\alpha/k)-1} f(y)dy, \tag{21}$$

where $\Gamma_k(\alpha)$ is defined by (8). Let

$$T_{17}(f,x) = \frac{c}{k\Gamma_k(\alpha)} \int_{D_1} g'(\|y\|)[g(\|x\|) - g(\|y\|)]^{(\alpha/k)-1} f(y)dy, \tag{22}$$

and

$$T_{18}(f,x) = \frac{c}{k\Gamma_k(\alpha)} \int_{D_2} g'(\|y\|)[g(\|y\|) - g(\|x\|)]^{(\alpha/k)-1} f(y)dy. \tag{23}$$

Thus,

$$T_{16}(f,x) = T_{17}(f,x) + T_{18}(f,x). \tag{24}$$

The fractional area balance operator with respect to the function g on E_n is defined by

$$T_{19}(f,x) = T_{18}(f,x) - T_{17}(f,x). \tag{25}$$

Given a function f on $(0,\infty)$ such that $e^{-\alpha y}|f(y)|$ is integrable over the interval $(0,\infty)$ for some real α, we define $F(z)$ as

$$F(z) = \int_0^\infty e^{-zy} f(y)dy, \tag{26}$$

where we require that $\mathrm{Re}(z) > \alpha$ so that the integral in (26) converges. F is called the (one-sided) Laplace transform of f. We consider only one-sided

Laplace transforms with the real parameter x, that is,

$$F(x) = \int_0^\infty e^{-xy} f(y)dy, \quad x > \alpha, \tag{27}$$

as these play the most important role in the solution of initial and boundary value problems for partial differential equations (see Refs. [19–21]). In fact, the tools we shall use for solving Cauchy and initial and boundary value problems are integral transforms. Specifically, we shall consider the Fourier transform, the Fourier sine and cosine transforms, the Hankel transform, and the Laplace transform.

The aim of this chapter is to introduce new fractional Laplace-type integral operators in the Banach space, which includes R_+^n as the special case. In Section 2, we define fractional Laplace-type integral operators in the Banach space. In Section 3, the corresponding norm inequalities are established. In Section 4, the proof of the main result is given. In Section 5, some applications are also given. They are significant improvements and generalizations of many known and new classes of fractional integral operators.

2. Fractional Laplace-Type Integral Operators

Definition 8. Let the radial kernel $K(\|x\|^{\lambda_1} \cdot \|y\|^{\lambda_2})$ be a non-negative measurable function defined on $E_n \times E_n, \lambda_1, \lambda_2 > 0$, then the fractional Laplace-type integral operators of f are defined by

$$T(f,x) = \frac{c}{k\Gamma_k(\alpha)} \int_{E_n} (K(\|x\|^{\lambda_1} \cdot \|y\|^{\lambda_2}))^{((\alpha/k)-1)} f(y)dy, \tag{28}$$

where $k, c, \alpha > 0$, $\Gamma_k(\alpha)$ is defined by (8).

If $c = k = 1$, $\alpha = 2, E_n = R_+^n$ in (28), then (28) reduces to the wider class of integral operators introduced by Kuang [22]:

$$T_{20}(f,x) = \int_{R_+^n} K(\|x\|_2^{\lambda_1} \cdot \|y\|_2^{\lambda_2}) f(y)dy. \tag{29}$$

where $x \in R_+^n = \{x = (x_1, x_2, \ldots, x_n) : x_k \geq 0, 1 \leq k \leq n\}$, $\|x\|_2 = (\sum_{k=1}^n |x_k|^2)^{1/2}, \lambda_1 \times \lambda_2 \neq 0$, and obtained the operator norm inequalities for T_{20} defined by (29) on the multiple weighted Orlicz spaces. When $n=1$, (29) reduces to

$$T_{21}(f,x) = \int_0^\infty K(x^{\lambda_1} y^{\lambda_2}) f(y)dy, \quad x \in (0,\infty). \tag{30}$$

If $\lambda_1 = \lambda_2 = 1$, then (30) reduces to the generalized Laplace transform of f, it is introduced by Hardy [4] as follows:

$$T_{22}(f, x) = \int_0^\infty K(xy)f(y)dy, \quad x \in (0, \infty). \tag{31}$$

If $K(xy) = (1 + xy)^{-\lambda_3}$ in (31), then

$$T_{23}(f, x) = x^{\lambda_3} \int_0^\infty \frac{1}{(1 + xy)^{\lambda_3}} f(y)dy \tag{32}$$

is called the generalized Stieltjes transform of f. If $K(xy) = J_\alpha(xy)(xy)^{1/2}$ in (31), then

$$T_{24}(f, x) = \int_0^\infty J_\alpha(xy)(xy)^{1/2} f(y)dy \tag{33}$$

is called the Hankel transform of f, where $J_\alpha(t)$ is the Bessel function of the first kind of order α, that is,

$$J_\alpha(t) = \left(\frac{t}{2}\right)^\alpha \sum_{k=0}^\infty \frac{(-1)^k}{k!\Gamma(\alpha + k + 1)} \left(\frac{t}{2}\right)^{2k} \quad \text{(see Ref. [23])}.$$

We use the following standard notations:

$$\|f\|_{p,\omega} = \left(\int_{E_n} |f(x)|^p \omega(x)dx\right)^{1/p},$$

$$L^p(\omega) = \{f : f \text{ is measurable and } \|f\|_{p,\omega} < \infty\},$$

where, ω is a non-negative measurable function on E_n.

If $\omega(x) \equiv 1$, we will denote $L^p(\omega)$ by $L^p(E_n)$, and $\|f\|_{p,1}$ by $\|f\|_p$.

$$B(u, v) = \int_0^1 x^{u-1}(1 - x)^{v-1}dx \ (u, v > 0)$$

is the Beta function.

3. Main Results

Our main result read as follows.

Theorem 1. *Let* $1 < p, q < \infty, \frac{1}{p} + \frac{1}{q} \le 1$, $0 < \frac{\lambda}{n} = 2 - \frac{1}{p} - \frac{1}{q}$, $\omega(x) = \|x\|^{n(\frac{\lambda_1}{\lambda_2}(p-1)-1)}$, $x \in E_n$, *and the radial kernel* $K(\|x\|^{\lambda_1} \cdot \|y\|^{\lambda_2})$ *be a non-negative measurable function defined on* $E_n \times E_n, \lambda_1, \lambda_2 > 0$.

(i) *If*

$$c_1 = \frac{(\Gamma(1/\beta))^n}{\beta^{n-1}\Gamma(n/\beta)} \int_0^\infty \{K(u)\}^{(\frac{\alpha}{k}-1)\frac{np}{\lambda}(1-\frac{1}{q})} u^{\frac{n^2}{\lambda\lambda_2}(p-1)(1-\frac{1}{q})-1} du < \infty, \tag{34}$$

and

$$c_2 = \frac{(\Gamma(1/\beta))^n}{\beta^{n-1}\Gamma(n/\beta)} \int_0^\infty \{K(u)\}^{(\frac{\alpha}{k}-1)\frac{n}{\lambda}} u^{\frac{n}{\lambda_2}[1-\frac{n}{\lambda}(1-\frac{1}{q})]-1} du < \infty, \tag{35}$$

then the integral operator T is defined by (28): $T: L^p(E_n) \to L^p(\omega)$, which exists as a bounded operator and

$$\|Tf\|_{p,\omega} \leq \frac{c}{k\Gamma_k(\alpha)} \left(\frac{c_1}{\lambda_1}\right)^{1/p} \left(\frac{c_2}{\lambda_2}\right)^{1-(1/p)} \|f\|_p.$$

This implies that

$$\|T\| = \sup_{f\neq 0} \frac{\|Tf\|_{p,\omega}}{\|f\|_p} \leq \frac{c}{k\Gamma_k(\alpha)} \left(\frac{c_1}{\lambda_1}\right)^{1/p} \left(\frac{c_2}{\lambda_2}\right)^{1-(1/p)}. \tag{36}$$

(ii) *If $0 < \lambda_1 < \lambda_2$ and*

$$c_3 = \frac{(\Gamma(1/\beta))^n}{\beta^{n-1}\Gamma(n/\beta)} \int_0^\infty \{K(u)\}^{(\frac{\alpha}{k}-1)} u^{\frac{n}{\lambda_2}(1-\frac{1}{p})-1} du < \infty, \tag{37}$$

then

$$\|T\| \geq \frac{cc_3}{k\Gamma_k(\alpha)\lambda_2}. \tag{38}$$

In the conjugate case: $\lambda = n$, $\frac{1}{p} + \frac{1}{q} = 1$, then by (34),(35) and (37), we get

$$c_0 = c_1 = c_2 = c_3 = \frac{(\Gamma(1/\beta)^n)}{\beta^{n-1}\Gamma(n/\beta)} \int_0^\infty (K(u))^{\frac{\alpha}{k}-1} u^{\frac{n}{q\lambda_2}-1} du. \tag{39}$$

Thus, by Theorem 1, we get

Corollary 1. *Let $1 < p < \infty$, $\frac{1}{p} + \frac{1}{q} = 1$, $\omega(x) = \|x\|^{n(\frac{\lambda_1}{\lambda_2}(p-1)-1)}$, $x \in E_n$, and the radial kernel $K(\|x\|^{\lambda_1} \cdot \|y\|^{\lambda_2})$ be a non-negative measurable function defined on $E_n \times E_n, \lambda_1, \lambda_2 > 0$. If*

$$c_0 = \frac{(\Gamma(1/\beta))^n}{\beta^{n-1}\Gamma(n/\beta)} \int_0^\infty (K(u))^{\frac{\alpha}{k}-1} u^{\frac{n}{q\lambda_2}-1} du < \infty, \tag{40}$$

then the integral operator T is defined by (28): $T : L^p(E_n) \to L^p(\omega)$, which exists as a bounded operator and

$$\|Tf\|_{p,\omega} \leq \frac{cc_0}{k\Gamma_k(\alpha)\lambda_1^{1/p}\lambda_2^{1/q}}\|f\|_p.$$

This implies that

$$\|T\| = \sup_{f\neq 0}\frac{\|Tf\|_{p,\omega}}{\|f\|_p} \leq \frac{cc_0}{k\Gamma_k(\alpha)\lambda_1^{1/p}\lambda_2^{1/q}}.$$

If $0 < \lambda_1 \leq \lambda_2$, then

$$\|T\| \geq \frac{cc_0}{k\Gamma_k(\alpha)\lambda_2}.$$

Corollary 2. *Under the same conditions as those of Corollary 1, if $\lambda_1 = \lambda_2 = \lambda_0$, then*

$$\|Tf\|_{p,\omega} \leq c_4\|f\|_p,$$

where

$$c_4 = \|T\| = \frac{cc_0}{k\Gamma_k(\alpha)\lambda_0} \tag{41}$$

is the best possible.

In particular, for $\lambda_0 = n = k = 1, \alpha = 2$, we get the following:

Corollary 3. *Under the same conditions as those of Corollary 2, if $\lambda_0 = n = k = 1, \alpha = 2$, then the integral operator T_{22} is defined by (31): $T_{22} : L^p(0,\infty) \to L^p(\omega)$, which exists as a bounded operator and*

$$\|T_{22}f\|_{p,\omega} \leq c_4\|f\|_p,$$

where $\omega(x) = x^{p-2}$ and

$$c_4 = \|T_{22}\| = \int_0^\infty K(u)u^{(1/q)-1}du = \int_0^\infty K(u)u^{-(1/p)}du$$

is the best possible.

Hence, Corollary 3 reduces to Theorem 2 in Ref. [24].

Remark 1. If $\lambda_1\lambda_2 < 0$, such as, $\lambda_1 < 0, \lambda_2 > 0$, let $\mu_1 = -\lambda_1 > 0$, we can define a new fractional integral operator as follows:

$$T_0(f,x) = \frac{c}{k\Gamma_k(\alpha)}\int_{E_n}(K(\|x\|^{\mu_1} - \|y\|^{\lambda_2}))^{(\alpha/k)-1}f(y)dy,$$

with different properties, which we will discuss in another paper.

4. Proof of Theorem 1

We require the following Lemmas to prove our main results.

Lemma 1 ([11]). *If $a_k, b_k, p_k > 0, 1 \leq k \leq n$, f is a measurable function on $(0, \infty)$, then*

$$\int_{B(r_1,r_2)} f\left(\sum_{k=1}^{n}\left(\frac{x_k}{a_k}\right)^{b_k}\right) x_1^{p_1-1}\cdots x_n^{p_n-1}dx_1\cdots dx_n$$

$$= \frac{\prod_{k=1}^{n} a_k^{p_k}}{\prod_{k=1}^{n} b_k}\cdot\frac{\prod_{k=1}^{n}\Gamma(\frac{p_k}{b_k})}{\Gamma(\sum_{k=1}^{n}\frac{p_k}{b_k})}\int_{r_1}^{r_2} f(t)t^{(\sum_{k=1}^{n}\frac{p_k}{b_k}-1)}dt,$$

where $B(r_1, r_2) = \{x \in E_n : 0 \leq r_1 \leq \|x\| \leq r_2\}$.

We get the following Lemma 2 by taking $a_k = 1, b_k = \beta > 0$, $p_k = 1, 1 \leq k \leq n$, $r_1 = 0, r_2 = \infty$ in Lemma 1.

Lemma 2. *Let f be a measurable function on $(0, \infty)$, then*

$$\int_{E_n} f(\|x\|)dx = \frac{(\Gamma(1/\beta))^n}{\beta^n\Gamma(n/\beta)}\int_0^{\infty} f(t^{1/\beta})t^{(n/\beta)-1}dt. \tag{42}$$

Lemma 3. *Let $f \in L^p(\omega), g \in L^q(E_n)$, $1 < p < \infty, \frac{1}{p}+\frac{1}{q} = 1, \omega$ be a non-negative measurable function on E_n, then*

$$\|f\|_{p,\omega} = \sup\left\{\left|\int_{E_n} fg\omega^{1/p}d\mu\right| : \|g\|_q \leq 1\right\}. \tag{43}$$

Proof. This is an immediate consequence of the Hölder inequality with weight (see Ref. [25]). □

Proof of Theorem 1. Let $p_1 = \frac{p}{p-1}, q_1 = \frac{q}{q-1}$, it follows that

$$\frac{1}{p_1}+\frac{1}{q_1}+\left(1-\frac{\lambda}{n}\right) = 1, \quad \frac{p}{q_1}+p\left(1-\frac{\lambda}{n}\right) = 1.$$

By Hölder's inequality, we get

$$\begin{aligned}
T(f,x) &= \frac{c}{k\Gamma_k(\alpha)}\int_{E_n}(K(\|x\|^{\lambda_1}\cdot\|y\|^{\lambda_2}))^{\frac{\alpha}{k}-1}f(y)dy \\
&= \frac{c}{k\Gamma_k(\alpha)}\int_{E_n}\{\|y\|^{(\frac{n^2}{\lambda p_1})}[K(\|x\|^{\lambda_1}\cdot\|y\|^{\lambda_2})]^{(\frac{\alpha}{k}-1)\frac{n}{\lambda}}f^p(y)\}^{1/q_1} \\
&\quad\times\left\{\|y\|^{-(\frac{n^2}{\lambda q_1})}[K(\|x\|^{\lambda_1}\cdot\|y\|^{\lambda_2})]^{(\frac{\alpha}{k}-1)\frac{n}{\lambda}}\right\}^{1/p_1}\{f(y)\}^{p(1-\frac{\lambda}{n})}dy \\
&\le \frac{c}{k\Gamma_k(\alpha)}\left\{\int_{E_n}\|y\|^{(\frac{n^2}{\lambda p_1})}[K(\|x\|^{\lambda_1}\cdot\|y\|^{\lambda_2})]^{(\frac{\alpha}{k}-1)\frac{n}{\lambda}}|f(y)|^p dy\right\}^{1/q_1} \\
&\quad\times\left\{\int_{E_n}\|y\|^{-(\frac{n^2}{\lambda q_1})}[K(\|x\|^{\lambda_1}\cdot\|y\|^{\lambda_2})]^{(\frac{\alpha}{k}-1)\frac{n}{\lambda}}dy\right\}^{1/p_1}\|f\|_p^{p(1-\frac{\lambda}{n})} \\
&= \frac{c}{k\Gamma_k(\alpha)}I_1^{1/q_1}\times I_2^{1/p_1}\times\|f\|_p^{p(1-\frac{\lambda}{n})}. \qquad (44)
\end{aligned}$$

In I_2, by using Lemma 2 and letting $u=\|x\|^{\lambda_1}t^{\lambda_2/\beta}$, as well as using (35), we get

$$\begin{aligned}
I_2 &= \int_{E_n}\|y\|^{-(\frac{n^2}{\lambda q_1})}[K(\|x\|^{\lambda_1}\cdot\|y\|^{\lambda_2})]^{(\frac{\alpha}{k}-1)\frac{n}{\lambda}}dy \\
&= \frac{(\Gamma(1/\beta))^n}{\beta^n\Gamma(n/\beta)}\int_0^\infty t^{-(\frac{n^2}{\lambda\beta q_1})}\left[K\left(t^{\frac{\lambda_2}{\beta}}\|x\|^{\lambda_1}\right)\right]^{(\frac{\alpha}{k}-1)\frac{n}{\lambda}}\times t^{\frac{n}{\beta}-1}dt \\
&= \frac{(\Gamma(1/\beta))^n}{\lambda_2\beta^{n-1}\Gamma(n/\beta)}\|x\|^{\frac{n\lambda_1}{\lambda_2}(\frac{n}{\lambda q_1}-1)}\int_0^\infty\{K(u)\}^{(\frac{\alpha}{k}-1)\frac{n}{\lambda}}u^{\frac{n}{\lambda_2}(1-\frac{n}{\lambda q_1})-1}du \\
&= \frac{c_2}{\lambda_2}\|x\|^{\frac{n\lambda_1}{\lambda_2}(\frac{n}{\lambda q_1}-1)}. \qquad (45)
\end{aligned}$$

Note that $\frac{1}{p}+\frac{1}{q}\le 1$ implies that $\frac{p}{q_1}\ge 1$, thus, by (44), (45), (42), (34) and Minkowski's inequality for integrals:

$$\left\{\int_X\left(\int_Y|f(x,y)|dy\right)^p\omega(x)dx\right\}^{1/p}\le\int_Y\left\{\int_X|f(x,y)|^p\omega(x)dx\right\}^{1/p}dy,$$
$$1\le p<\infty,$$

and letting $v = \|y\|^{\lambda_2} t^{(\lambda_1/\beta)}$, we conclude that

$$\begin{aligned}
\|Tf\|_{p,\omega} &= \left(\int_{E_n} |T(f,x)|^p \omega(x) dx \right)^{1/p} \\
&\le \frac{c}{k\Gamma_k(\alpha)} \left(\int_{E_n} I_1^{p/q_1} I_2^{p/p_1} \|f\|_p^{p^2(1-\frac{\lambda}{n})} \omega(x) dx \right)^{1/p} \\
&= \frac{c}{k\Gamma_k(\alpha)} \left(\frac{c_2}{\lambda_2} \right)^{1/p_1} \|f\|_p^{p(1-\frac{\lambda}{n})} \Bigg\{ \int_{E_n} \|x\|^{\frac{\lambda_1 np}{\lambda_2 p_1}(\frac{n}{\lambda q_1}-1)+n(\frac{\lambda_1}{\lambda_2}(p-1)-1)} \\
&\quad \times \left(\int_{E_n} \|y\|^{\frac{n^2}{p_1\lambda}} [K(\|x\|^{\lambda_1}\cdot\|y\|^{\lambda_2})]^{(\frac{\alpha}{k}-1)\frac{n}{\lambda}} |f(y)|^p dy \right)^{\frac{p}{q_1}} dx \Bigg\}^{1/p} \\
&\le \frac{c}{k\Gamma_k(\alpha)} \left(\frac{c_2}{\lambda_2} \right)^{1/p_1} \|f\|_p^{p(1-\frac{\lambda}{n})} \\
&\quad \times \Bigg\{ \int_{E_n} \|y\|^{\frac{n^2}{p_1\lambda}} |f(y)|^p \Bigg(\int_{E_n} \|x\|^{(\frac{\lambda_1 np}{p_1\lambda_2})(\frac{n}{q_1\lambda}-1)+n(\frac{\lambda_1}{\lambda_2}(p-1)-1)} \\
&\qquad\qquad \times [K(\|x\|^{\lambda_1}\cdot\|y\|^{\lambda_2})]^{(\frac{\alpha}{k}-1)\frac{np}{q_1\lambda}} dx \Bigg)^{\frac{q_1}{p}} dy \Bigg\}^{\frac{1}{q_1}} \\
&= \frac{c}{k\Gamma_k(\alpha)} \left(\frac{c_2}{\lambda_2} \right)^{1/p_1} \|f\|_p^{p(1-\frac{\lambda}{n})} \\
&\quad \times \Bigg\{ \int_{E_n} \|y\|^{\frac{n^2}{p_1\lambda}} |f(y)|^p \Bigg(\frac{(\Gamma(1/\beta))^n}{\beta^n \Gamma(n/\beta)} \int_0^\infty t^{(\frac{\lambda_1 np}{p_1\beta\lambda_2})(\frac{n}{q_1\lambda}-1)+\frac{n}{\beta}(\frac{\lambda_1}{\lambda_2}(p-1)-1)} \\
&\qquad\qquad \times [K(t^{(\lambda_1/\beta)}\|y\|^{\lambda_2})]^{(\frac{\alpha}{k}-1)\frac{np}{q_1\lambda}} t^{\frac{n}{\beta}-1} dt \Bigg)^{\frac{q_1}{p}} dy \Bigg\}^{1/q_1} \\
&= \frac{c}{k\Gamma_k(\alpha)} \left(\frac{c_2}{\lambda_2} \right)^{1/p_1} \|f\|_p^{p(1-\frac{\lambda}{n})} \times \left(\frac{c_1}{\lambda_1} \right)^{1/p} \|f\|_p^{p/q_1} \\
&= \frac{c}{k\Gamma_k(\alpha)} \left(\frac{c_2}{\lambda_2} \right)^{1/p_1} \left(\frac{c_1}{\lambda_1} \right)^{1/p} \|f\|_p.
\end{aligned}$$

Thus,

$$\|Tf\|_{p,\omega} \le \frac{c}{k\Gamma_k(\alpha)} \left(\frac{c_1}{\lambda_1} \right)^{1/p} \left(\frac{c_2}{\lambda_2} \right)^{1-(1/p)} \|f\|_p. \tag{46}$$

This implies that

$$\|T\| = \sup_{f\neq 0} \frac{\|Tf\|_{p,\omega}}{\|f\|_p} \le \frac{c}{k\Gamma_k(\alpha)} \left(\frac{c_1}{\lambda_1} \right)^{1/p} \left(\frac{c_2}{\lambda_2} \right)^{1-(1/p)}. \tag{47}$$

To prove the reversed inequality (38), setting f_ε and g_ε be as follows:

$$f_\varepsilon(x) = \|x\|^{-(n/p)+\varepsilon}\varphi_B(x), \tag{48}$$

$$g_\varepsilon(x) = (p\varepsilon)^{1/p_1}\left\{\frac{\beta^{n-1}\Gamma(n/\beta)}{(\Gamma(1/\beta))^n}\right\}^{1/p_1}\|x\|^{-(\frac{n}{p_1})+(\frac{\lambda_1}{\lambda_2}-p)\varepsilon}\varphi_{B^c}(x), \tag{49}$$

where $0 < \varepsilon < \alpha/p$, $B = B(0,1) = \{x \in E_n : \|x\| < 1\}$, φ_{B^c} is the characteristic function of the set $B^c = \{x \in E_n : \|x\| \geq 1\}$, that is,

$$\varphi_{B^c}(x) = \begin{cases} 1, \ x \in B^c, \\ 0, \ x \in B. \end{cases}$$

Thus, we get

$$\|f_\varepsilon\|_p = \left(\frac{(\Gamma(1/\beta))^n}{p\varepsilon\beta^{n-1}\Gamma(n/\beta)}\right)^{1/p}, \tag{50}$$

$$\|g_\varepsilon\|_{p_1}^{p_1} = \frac{p-1}{p-\frac{\lambda_1}{\lambda_2}} \leq 1. \tag{51}$$

Using Lemma 3 and the Fubini Theorem, we get

$$\begin{aligned}
\|Tf_\varepsilon\|_{p,\omega} &\geq \int_{E_n} T(f_\varepsilon, x)g_\varepsilon(x)\{\omega(x)\}^{1/p}dx \\
&= \frac{c}{k\Gamma_k(\alpha)}\int_{E_n}\int_{E_n} K\left(\|x\|^{\lambda_1}\cdot\|y\|^{\lambda_2}\right)^{\frac{\alpha}{k}-1} \\
&\quad \times f_\varepsilon(y)g_\varepsilon(x)\left(\|x\|^{n(\frac{\lambda_1}{\lambda_2}(p-1)-1)}\right)^{1/p} dydx \\
&= \frac{c}{k\Gamma_k(\alpha)}(p\varepsilon)^{1/p_1}\left\{\frac{\beta^{n-1}\Gamma(n/\beta)}{(\Gamma(1/\beta))^n}\right\}^{1/p_1} \\
&\quad \times \int_{B^c}\left\{\int_B K(\|x\|^{\lambda_1}\cdot\|y\|^{\lambda_2})^{\frac{\alpha}{k}-1}\|y\|^{-(n/p)+\varepsilon}dy\right\} \\
&\quad \times \|x\|^{\frac{\lambda_1}{\lambda_2}(\frac{n}{p_1}+\varepsilon)-p\varepsilon-n}dx.
\end{aligned} \tag{52}$$

Letting $u = t^{\lambda_2/\beta}\|x\|^{\lambda_1}$ and using (42), we have

$$\begin{aligned}
&\int_B K(\|x\|^{\lambda_1}\cdot\|y\|^{\lambda_2})^{(\frac{\alpha}{k}-1)}\|y\|^{-(n/p)+\varepsilon}dy \\
&= \frac{(\Gamma(1/\beta))^n}{\beta^n\Gamma(n/\beta)}\int_0^1 K(t^{\lambda_2/\beta}\|x\|^{\lambda_1})t^{-(\frac{n}{p\beta})+\frac{\varepsilon}{\beta}+\frac{n}{\beta}-1}dt \\
&= \frac{(\Gamma(1/\beta))^n}{\beta^{n-1}\Gamma(n/\beta)\lambda_2}\|x\|^{-\frac{\lambda_1}{\lambda_2}(\frac{n}{p_1}+\varepsilon)}\int_0^{\|x\|^{\lambda_1}} K(u)^{(\frac{\alpha}{k}-1)}u^{\frac{1}{\lambda_2}((\frac{n}{p_1})+\varepsilon)-1}du.
\end{aligned} \tag{53}$$

We insert (53) into (52) and use Fubini's theorem to obtain

$$\begin{aligned}
\|Tf_\varepsilon\|_{p,\omega} &\geq \frac{c}{k\Gamma_k(\alpha)}(p\varepsilon)^{1/p_1}\left\{\frac{(\Gamma(1/\beta))^n}{\beta^{n-1}\Gamma(n/\beta)}\right\}^{1/p}\times\frac{1}{\lambda_2}\\
&\quad\times\int_{B^c}\|x\|^{-p\varepsilon-n}\left(\int_0^{\|x\|^{\lambda_1}}K(u)u^{\frac{1}{\lambda_2}(\frac{n}{p_1}+\varepsilon)-1}du\right)dx\\
&\geq \frac{c}{k\Gamma_k(\alpha)}(p\varepsilon)^{1/p_1}\left\{\frac{(\Gamma(1/\beta))^n}{\beta^{n-1}\Gamma(n/\beta)}\right\}^{1/p}\times\frac{1}{\lambda_2}\\
&\quad\times\int_0^{\infty}K(u)u^{\frac{1}{\lambda_2}(\frac{n}{p_1}+\varepsilon)-1}\left(\int_{\beta(u)}^{\infty}\|x\|^{-p\varepsilon-n}dx\right)du\\
&= \frac{c}{k\Gamma_k(\alpha)}(p\varepsilon)^{1/p_1}\left\{\frac{(\Gamma(1/\beta))^n}{\beta^{n-1}\Gamma(n/\beta)}\right\}^{(1/p)+1}\times\frac{1}{\beta\lambda_2}\\
&\quad\times\int_0^{\infty}K(u)u^{\frac{1}{\lambda_2}(\frac{n}{p_1}+\varepsilon)-1}\left(\int_{G(u)}^{\infty}t^{-(p\varepsilon)/\beta-1}dt\right)du\\
&= \frac{c}{k\Gamma_k(\alpha)}(p\varepsilon)^{-(1/p)}\left\{\frac{(\Gamma(1/\beta))^n}{\beta^{n-1}\Gamma(n/\beta)}\right\}^{(1/p)+1}\times\frac{1}{\lambda_2}\\
&\quad\times\int_0^{\infty}K(u)u^{\frac{1}{\lambda_2}(\frac{n}{p_1}+\varepsilon)-1}(G(u))^{-(p\varepsilon)/\beta}du, \qquad (54)
\end{aligned}$$

where $G(u)=\max\{1,u^{1/\lambda_1}\}$. Thus, we get

$$\begin{aligned}
\|T\| &= \sup_{f\neq 0}\frac{\|Tf\|_{p,\omega}}{\|f\|_p}\geq\frac{\|Tf_\varepsilon\|_{p,\omega}}{\|f_\varepsilon\|_p}\\
&\geq\frac{c(\Gamma(1/\beta))^n}{k\Gamma_k(\alpha)\beta^{n-1}\Gamma(n/\beta)\lambda_2}\int_0^{\infty}K(u)u^{\frac{1}{\lambda_2}(\frac{n}{p_1}+\varepsilon)-1}(G(u))^{-(p\varepsilon)/\beta}du.
\end{aligned} \tag{55}$$

By letting $\varepsilon\to 0^+$ in (55) and using the Fatou lemma, we get

$$\|T\|\geq\frac{c(\Gamma(1/\beta))^n}{k\Gamma_k(\alpha)\beta^{n-1}\Gamma(n/\beta)\lambda_2}\int_0^{\infty}K(u)u^{\frac{n}{p_1\lambda_2}-1}du=\frac{cc_3}{k\Gamma_k(\alpha)\lambda_2}. \tag{56}$$

The proof is complete. □

5. Some Applications

As applications, a large number of known and new results have been obtained by proper choice of kernel K. In this section, we present some interesting model applications which display the importance of our results. Also, these examples are of fundamental importance in analysis.

In what follows, without loss of generality, we may assume $0 < \lambda_1 \leq \lambda_2$, thus under the same conditions as those of Theorem 1, we have

$$\frac{cc_3}{k\Gamma_k(\alpha)\lambda_2} \leq \|T\| \leq \frac{c}{k\Gamma_k(\alpha)}\left(\frac{c_1}{\lambda_1}\right)^{1/p}\left(\frac{c_2}{\lambda_2}\right)^{1-(1/p)}. \tag{57}$$

In the conjugate case ($\lambda = n$), we get

$$\frac{cc_0}{k\Gamma_k(\alpha)\lambda_2} \leq \|T\| \leq \frac{cc_0}{k\Gamma_k(\alpha)\lambda_1^{1/p}\lambda_2^{1/q}}. \tag{58}$$

If $\lambda_1 = \lambda_2 = \lambda_0$, then

$$\|T\| = \frac{cc_0}{k\Gamma_k(\alpha)\lambda_0} \tag{59}$$

where the constants c_1, c_2, c_3 and c_0 are defined by (34), (35), (37) and (40), resp.

Example 1. If $K(\|x\|^{\lambda_1} \cdot \|y\|^{\lambda_2}) = \exp\{-(\|\|x\|^{\lambda_1} \cdot \|y\|^{\lambda_2})^{\lambda_3}\}, \lambda_3 > 0$ in Theorem 1, then the operator T_{25} is defined by

$$T_{25}(f,x) = \frac{c}{k\Gamma_k(\alpha)}\int_{E_n}\left\{\exp[-(\|\|x\|^{\lambda_1} \cdot \|y\|^{\lambda_2})^{\lambda_3}]\right\}^{(\alpha/k)-1} f(y)dy. \tag{60}$$

Thus, by (57) and (58), we have

$$\begin{aligned} c_1 &= \frac{(\Gamma(1/\beta))^n}{\beta^{n-1}\Gamma(n/\beta)}\int_0^\infty \{\exp(-u^{\lambda_3})\}^{(\frac{\alpha}{k}-1)\frac{np}{\lambda}(1-\frac{1}{q})}u^{\frac{n^2}{\lambda\lambda_2}(p-1)(1-\frac{1}{q})-1}du, \\ &= \frac{(\Gamma(1/\beta))^n}{\lambda_3\beta^{n-1}\Gamma(n/\beta)} \times \left(\frac{kq\lambda}{np(\alpha-k)(q-1)}\right)^{\frac{n^2}{\lambda\lambda_2\lambda_3}(p-1)(1-\frac{1}{q})} \\ &\quad \times\Gamma\left(\frac{n^2}{\lambda\lambda_2\lambda_3}(p-1)\left(1-\frac{1}{q}\right)\right). \end{aligned} \tag{61}$$

$$\begin{aligned} c_2 &= \frac{(\Gamma(1/\beta))^n}{\beta^{n-1}\Gamma(n/\beta)}\int_0^\infty \{\exp(-u^{\lambda_3})\}^{(\frac{\alpha}{k}-1)\frac{n}{\lambda}} u^{\frac{n}{\lambda_2}[1-\frac{n}{\lambda}(1-\frac{1}{q})]-1}du \\ &= \frac{(\Gamma(1/\beta))^n}{\lambda_3\beta^{n-1}\Gamma(n/\beta)}\left(\frac{k\lambda}{n(\alpha-k)}\right)^{\frac{n}{\lambda_2\lambda_3}(1-\frac{n}{\lambda}(1-\frac{1}{q}))} \\ &\quad \times\Gamma\left(\frac{n}{\lambda_2\lambda_3}\left(1-\frac{n}{\lambda}\left(1-\frac{1}{q}\right)\right)\right). \end{aligned} \tag{62}$$

$$c_3 = \frac{(\Gamma(1/\beta))^n}{\beta^{n-1}\Gamma(n/\beta)} \int_0^\infty \{\exp(-u^{\lambda_3})\}^{\frac{\alpha}{k}-1} u^{\frac{n}{\lambda_2}(1-\frac{1}{p})-1} du$$

$$= \frac{(\Gamma(1/\beta))^n}{\lambda_3\beta^{n-1}\Gamma(n/\beta)} \left(\frac{k}{\alpha-k}\right)^{\frac{n}{\lambda_2\lambda_3}(1-\frac{1}{p})} \Gamma\left(\frac{n}{\lambda_2\lambda_3}\left(1-\frac{1}{p}\right)\right). \tag{63}$$

$$c_0 = \frac{(\Gamma(1/\beta))^n}{\lambda_3\beta^{n-1}\Gamma(n/\beta)} \left(\frac{k}{\alpha-k}\right)^{\frac{n}{q\lambda_2\lambda_3}} \Gamma\left(\frac{n}{q\lambda_2\lambda_3}\right). \tag{64}$$

In particular, if $\lambda_1 = \lambda_2 = \lambda_0$, then by Corollary 2, we get

$$\|T_{25}\| = \frac{c(\Gamma(1/\beta))^n}{k\Gamma_k(\alpha)\lambda_0\lambda_3\beta^{n-1}\Gamma(n/\beta)} \left(\frac{k}{\alpha-k}\right)^{\frac{n}{q\lambda_2\lambda_3}} \Gamma\left(\frac{n}{q\lambda_2\lambda_3}\right). \tag{65}$$

If $\lambda_0 = \lambda_3 = 1, \beta = 2$, then T_{25} reduces to

$$T_{25}(f,x) = \frac{c}{k\Gamma_k(\alpha)} \int_{\mathbb{R}_+^n} \{\exp[-(\|x\|_2 \cdot \|y\|_2)]\}^{(\alpha/k)-1} f(y) dy, \tag{66}$$

thus by (65), we get

$$\|T_{25}\| = \frac{c\pi^{n/2}}{k\Gamma_k(\alpha)2^{n-1}\Gamma(n/2)} \left(\frac{k}{\alpha-k}\right)^{n/q} \Gamma\left(\frac{n}{q}\right). \tag{67}$$

If $k = 1, \alpha = 2, c = 1$ in (66), then T_{25} reduces to the n — dimensional Laplace transform of f, thus is,

$$T_{25}(f,x) = \int_{R_+^n} \exp\{-(\|x\|_2 \cdot \|y\|_2)\} f(y) dy, \tag{68}$$

thus by (67), we get

$$\|T_{25}\| = \frac{\pi^{n/2}}{2^{n-1}\Gamma(n/2)} \Gamma\left(\frac{n}{q}\right). \tag{69}$$

If $n = 1$ in (68), then T_{25} reduces to the one-sided Laplace transform of f, thus by (69), we have

$$\|T_{25}\| = \Gamma\left(\frac{1}{q}\right) = \int_0^\infty u^{\frac{1}{q}-1} e^{-u} du.$$

That is, it reduces to Theorem 2 in Ref. [24].

Example 2. If

$$K(\|x\|^{\lambda_1} \cdot \|y\|^{\lambda_2}) = \left(\frac{2}{\pi}\right)^{n/\beta} \sin(\lambda_3\|x\|^{\lambda_1} \cdot \|y\|^{\lambda_2})(\lambda_1 > 0, \lambda_3 > 0, \lambda_2 > n/q)$$

in Theorem 1, then the operator T_{26} is defined by

$$T_{26}(f,x)=\frac{c}{k\Gamma_k(\alpha)}\left(\frac{2}{\pi}\right)^{n/\beta}\int_{E_n}\{\sin(\lambda_3\|x\|^{\lambda_1}\cdot\|y\|^{\lambda_2})\}^{(\alpha/k)-1}f(y)dy. \tag{70}$$

By (40), we have

$$c_0=\frac{(\Gamma(1/\beta))^n}{\beta^{n-1}\Gamma(n/\beta)}\left(\frac{2}{\pi}\right)^{n/\beta}\int_0^\infty\frac{(\sin(\lambda_3 u))^{(\alpha/k)-1}}{u^{1-\frac{n}{q\lambda_2}}}du. \tag{71}$$

By Corollary 2, we have

$$\|T_{26}\|=\frac{c(\Gamma(1/\beta))^n}{k\Gamma_k(\alpha)\lambda_0\beta^{n-1}\Gamma(n/\beta)}\left(\frac{2}{\pi}\right)^{n/\beta}\int_0^\infty\frac{(\sin(\lambda_3 u))^{(\alpha/k)-1}}{u^{1-\frac{n}{q\lambda_2}}}du. \tag{72}$$

In particular, if $k=c=1,\alpha=2$ in (72), we get

$$\|T_{26}\|=\frac{(2/\pi)^{(n/\beta)-1}\lambda_3^{-(\frac{n}{q\lambda_2})}(\Gamma(1/\beta))^n}{\lambda_0\beta^{n-1}\Gamma\left(\frac{n}{\beta}\right)\Gamma(1-\frac{n}{q\lambda_2})\cos\left(\frac{n\pi}{2q\lambda_2}\right)}. \tag{73}$$

If $\lambda_1=\lambda_2=\lambda_0$ in (73), then

$$\|T_{26}\|=\frac{(2/\pi)^{(n/\beta)-1}\lambda_3^{-(\frac{n}{q\lambda_0})}(\Gamma(1/\beta))^n}{\lambda_0\beta^{n-1}\Gamma(\frac{n}{\beta})\Gamma(1-\frac{n}{q\lambda_0})\cos(\frac{n\pi}{2q\lambda_0})}. \tag{74}$$

When $\beta=2,\lambda_3=\lambda_0=1,T_{26}$ is called the n-dimensional Fourier sine transform of f, then by (74), we get

$$\|T_{26}\|=\frac{\pi}{2^{n/2}\Gamma(\frac{n}{2})\Gamma(1-\frac{n}{q})\cos(\frac{n\pi}{2q})}.$$

If $n=1$, then T_{26} reduces to the Fourier sine transform of f in Ref. [21] as follows:

$$T_{26}(f,x)=\sqrt{\frac{2}{\pi}}\int_0^\infty\sin(xy)f(y)dy.$$

Thus, we have

$$\|T_{26}\|=\frac{(\pi/2)^{1/2}}{\Gamma(1/p)\cos(\frac{\pi}{2q})}.$$

Example 3. If

$$K(\|x\|^{\lambda_1} \cdot \|y\|^{\lambda_2}) = \left(\frac{2}{\pi}\right)^{n/\beta} \cos(\lambda_3\|x\|^{\lambda_1} \cdot \|y\|^{\lambda_2})(\lambda_1 > 0, \lambda_3 > 0, \lambda_2 > n/q)$$

in Theorem 1, then the operator T_{27} in defined by

$$T_{27}(f, x) = \frac{c}{k\Gamma_k(\alpha)} \left(\frac{2}{\pi}\right)^{n/\beta} \int_{E_n} \{\cos(\lambda_3\|x\|^{\lambda_1} \cdot \|y\|^{\lambda_2})\}^{(\alpha/k)-1} f(y)dy. \tag{75}$$

By (40), we have

$$c_0 = \frac{(\Gamma(1/\beta))^n}{\beta^{n-1}\Gamma(n/\beta)} \left(\frac{2}{\pi}\right)^{n/\beta} \int_0^\infty \frac{\{\cos(\lambda_3 u)\}^{(\alpha/k)-1}}{u^{1-\frac{n}{q\lambda_2}}} du. \tag{76}$$

If $k = 1, \alpha = 2$ in (76), then

$$c_0 = \frac{(2/\pi)^{(n/\beta)-1} \lambda_3^{-(\frac{n}{q\lambda_2})} (\Gamma(1/\beta))^n}{\beta^{n-1}\Gamma\left(\frac{n}{\beta}\right)\Gamma\left(1 - \frac{n}{q\lambda_2}\right)\sin\left(\frac{n\pi}{2q\lambda_2}\right)}. \tag{77}$$

In particular, if $\lambda_1 = \lambda_2 = \lambda_0$ in (77), then by Corollary 2, we get

$$\|T_{27}\| = \frac{c(2/\pi)^{(n/\beta)-1} \lambda_3^{-(\frac{n}{q\lambda_0})} (\Gamma(1/\beta))^n}{\lambda_0\beta^{n-1}\Gamma\left(\frac{n}{\beta}\right)\Gamma\left(1 - \frac{n}{q\lambda_0}\right)\sin\left(\frac{n\pi}{2q\lambda_0}\right)}. \tag{78}$$

When $\beta = 2, \lambda_3 = \lambda_0 = c = 1, T_{27}$ is called the n-dimensional Fourier cosine transform of f, then by (78), we get

$$\|T_{27}\| = \frac{\pi}{2^{n/2}\Gamma\left(\frac{n}{2}\right)\Gamma\left(1 - \frac{n}{q}\right)\sin\left(\frac{n\pi}{2q}\right)}.$$

If $n = 1$, then T_{27} reduces to the Fourier cosine transform of f in Ref. [21]:

$$T_{27}(f, x) = \sqrt{\frac{2}{\pi}} \int_0^\infty \cos(xy) f(y)dy.$$

Thus, we have

$$\|T_{27}\| = \frac{(\pi/2)^{1/2}}{\Gamma(1/p)\sin\left(\frac{\pi}{2q}\right)}.$$

Remark 2. Because the Fourier transform of f can be decomposed into Fourier sine and cosine transform of f, thus, the corresponding operator norm can be derived from Examples 2 and 3.

Example 4. If $K(\|x\|^{\lambda_1} \cdot \|y\|^{\lambda_2}) = \frac{1}{1+(\|x\|^{\lambda_1} \cdot \|y\|^{\lambda_2})^{\lambda_3}}$ in Theorem 1, then the operator T_{28} is defined by

$$T_{28}(f,x) = \frac{c}{k\Gamma_k(\alpha)} \int_{E_n} \left\{ \frac{1}{1+(\|x\|^{\lambda_1} \cdot \|y\|^{\lambda_2})^{\lambda_3}} \right\}^{(\alpha/k)-1} f(y)dy. \tag{79}$$

By (57) and (58), if

$$\frac{n}{\lambda_2\lambda_3} < \frac{p}{p-1}\left(\frac{\alpha}{k}-1\right),$$

then

$$\begin{aligned} c_1 &= \frac{(\Gamma(1/\beta))^n}{\beta^{n-1}\Gamma(n/\beta)} \int_0^\infty \left\{ \frac{1}{1+u^{\lambda_3}} \right\}^{(\frac{\alpha}{k}-1)\frac{np}{\lambda}(1-\frac{1}{q})} u^{\frac{n^2}{\lambda\lambda_2}(p-1)(1-\frac{1}{q})-1} du \\ &= \frac{(\Gamma(1/\beta))^n}{\lambda_3\beta^{n-1}\Gamma(n/\beta)} \\ &\quad \times B\left(\frac{n^2}{\lambda\lambda_2\lambda_3}(p-1)\left(1-\frac{1}{q}\right), \frac{n}{\lambda}\left(1-\frac{1}{q}\right) \right. \\ &\qquad \left. \times \left(p\left(\frac{\alpha}{k}-1\right) - \frac{n}{\lambda_2\lambda_3}(p-1) \right) \right), \end{aligned}$$

and

$$\begin{aligned} c_3 &= \frac{(\Gamma(1/\beta))^n}{\beta^{n-1}\Gamma(n/\beta)} \int_0^\infty \left\{ \frac{1}{1+u^{\lambda_3}} \right\}^{\frac{\alpha}{k}-1} u^{\frac{n}{\lambda_2}(1-\frac{1}{p})-1} du \\ &= \frac{(\Gamma(1/\beta))^n}{\lambda_3\beta^{n-1}\Gamma(n/\beta)} B\left(\frac{n}{\lambda_2\lambda_3}\left(1-\frac{1}{p}\right), \left(\frac{\alpha}{k}-1\right) - \frac{n}{\lambda_2\lambda_3}\left(1-\frac{1}{p}\right) \right). \end{aligned}$$

If $0 < 1 - \frac{n}{\lambda}(1-\frac{1}{q}) < \frac{\lambda_2\lambda_3}{\lambda}\left(\frac{\alpha}{k}-1\right)$, then

$$\begin{aligned} c_2 &= \frac{(\Gamma(1/\beta))^n}{\beta^{n-1}\Gamma(n/\beta)} \int_0^\infty \left\{ \frac{1}{1+u^{\lambda_3}} \right\}^{(\frac{\alpha}{k}-1)\frac{n}{\lambda}} u^{\frac{n}{\lambda_2}[1-\frac{n}{\lambda}(1-\frac{1}{q})]-1} du \\ &= \frac{(\Gamma(1/\beta))^n}{\lambda_3\beta^{n-1}\Gamma(n/\beta)} \\ &\quad \times B\left(\frac{n}{\lambda_2\lambda_3}\left(1-\frac{n}{\lambda}\left(1-\frac{1}{q}\right)\right), \left(\frac{\alpha}{k}-1\right)\frac{n}{\lambda} \right. \\ &\qquad \left. -\frac{n}{\lambda_2\lambda_3}\left(1-\frac{n}{\lambda}\left(1-\frac{1}{q}\right)\right) \right). \end{aligned}$$

In the conjugate case ($\lambda = n$), if $\frac{n}{q\lambda_2\lambda_3} < \frac{\alpha}{k} - 1$, then

$$c_0 = \frac{(\Gamma(1/\beta))^n}{\lambda_3\beta^{n-1}\Gamma(n/\beta)} B\left(\frac{n}{q\lambda_2\lambda_3}, \left(\frac{\alpha}{k} - 1\right) - \frac{n}{q\lambda_2\lambda_3}\right).$$

In particular, if $\lambda_1 = \lambda_2 = \lambda_0$, then by Corollary 2, we get

$$\|T_{28}\| = \frac{c(\Gamma(1/\beta))^n}{\lambda_0\lambda_3\beta^{n-1}\Gamma(n/\beta)} B\left(\frac{n}{q\lambda_0\lambda_3}, \left(\frac{\alpha}{k} - 1\right) - \frac{n}{q\lambda_0\lambda_3}\right). \tag{80}$$

Example 5. If

$$K(\|x\|^{\lambda_1} \cdot \|y\|^{\lambda_2}) = \frac{1}{\{1 + (\|x\|^{\lambda_1} \cdot \|y\|^{\lambda_2})\}^{\lambda_3}}$$

in Theorem 1, then the operator T_{29} is defined by

$$T_{29}(f, x) = \frac{c}{k\Gamma_k(\alpha)} \int_{E_n} \left\{\frac{1}{1 + (\|\|x\|^{\lambda_1} \cdot \|y\|^{\lambda_2})}\right\}^{\lambda_3(\frac{\alpha}{k} - 1)} f(y)dy. \tag{81}$$

If

$$\frac{n}{\lambda_2\lambda_3} < \frac{p}{p-1}\left(\frac{\alpha}{k} - 1\right), \text{ and } 0 < 1 - \frac{n}{\lambda}\left(1 - \frac{1}{q}\right) < \frac{\lambda_2\lambda_3}{\lambda}\left(\frac{\alpha}{k} - 1\right),$$

then by (57) and (58), we get

$$\begin{aligned} c_1 &= \frac{(\Gamma(1/\beta))^n}{\beta^{n-1}\Gamma(n/\beta)} \int_0^\infty \left(\frac{1}{1+u}\right)^{(\frac{\alpha}{k}-1)\frac{np\lambda_3}{\lambda}(1-\frac{1}{q})} u^{\frac{n^2}{\lambda\lambda_2}(p-1)(1-\frac{1}{q})-1} du \\ &= \frac{(\Gamma(1/\beta))^n}{\beta^{n-1}\Gamma(n/\beta)} \\ &\quad \times B\left(\frac{n^2}{\lambda\lambda_2}(p-1)\left(1 - \frac{1}{q}\right), \frac{n}{\lambda}\left(1 - \frac{1}{q}\right)\left(p\lambda_3\left(\frac{\alpha}{k} - 1\right) - \frac{n}{\lambda_2}(p-1)\right)\right). \end{aligned}$$

$$\begin{aligned} c_2 &= \frac{(\Gamma(1/\beta))^n}{\beta^{n-1}\Gamma(n/\beta)} \int_0^\infty \left\{\frac{1}{1+u}\right\}^{(\frac{\alpha}{k}-1)\frac{n\lambda_3}{\lambda}} u^{\frac{n}{\lambda_2}[1-\frac{n}{\lambda}(1-\frac{1}{q})]-1} du \\ &= \frac{(\Gamma(1/\beta))^n}{\beta^{n-1}\Gamma(n/\beta)} \\ &\quad \times B\left(\frac{n}{\lambda_2}\left(1 - \frac{n}{\lambda}\left(1 - \frac{1}{q}\right)\right), \frac{n\lambda_3}{\lambda}\left(\frac{\alpha}{k} - 1\right) - \frac{n}{\lambda_2}\left(1 - \frac{n}{\lambda}\left(1 - \frac{1}{q}\right)\right)\right). \end{aligned}$$

$$\begin{aligned} c_3 &= \frac{(\Gamma(1/\beta))^n}{\beta^{n-1}\Gamma(n/\beta)} \int_0^\infty \left(\frac{1}{1+u}\right)^{(\frac{\alpha}{k}-1)\lambda_3} u^{\frac{n}{\lambda_2}(1-\frac{1}{p})-1} du \\ &= \frac{(\Gamma(1/\beta))^n}{\beta^{n-1}\Gamma(n/\beta)} B\left(\frac{n}{\lambda_2}\left(1 - \frac{1}{p}\right), \left(\frac{\alpha}{k} - 1\right)\lambda_3 - \frac{n}{\lambda_2}\left(1 - \frac{1}{p}\right)\right). \end{aligned}$$

In the conjugate case ($\lambda = n$), if $\left(\frac{\alpha}{k} - 1\right)\lambda_3 > \frac{n}{\lambda_2 q}$, then

$$c_0 = \frac{(\Gamma(1/\beta))^n}{\beta^{n-1}\Gamma(n/\beta)} B\left(\frac{n}{\lambda_2 q}, \left(\frac{\alpha}{k} - 1\right)\lambda_3 - \frac{n}{\lambda_2 q}\right).$$

In particular, if $\lambda_1 = \lambda_2 = \lambda_0$, then by Corollary 2, we get

$$\|T_{29}\| = \frac{c(\Gamma(1/\beta))^n}{\lambda_0 \beta^{n-1}\Gamma(n/\beta)} B\left(\frac{n}{\lambda_0 q}, \left(\frac{\alpha}{k} - 1\right)\lambda_3 - \frac{n}{\lambda_0 q}\right). \tag{82}$$

Example 6. If $K(xy) = J_\beta(xy)(xy)^{1/2}$ in Theorem 1, where $J_\beta(t)$ is the Bessel function of the first kind of order β, then the operator T_{30} is defined by

$$T_{30}(f, x) = \frac{c}{k\Gamma_k(\alpha)} \int_0^\infty (J_\beta(xy)(xy)^{1/2})^{(\alpha/k)-1} f(y)dy. \tag{83}$$

By (40), we have

$$c_0 = \int_0^\infty (J_\beta(u)u^{1/2})^{(\alpha/k)-1} u^{(n/q)-1} du. \tag{84}$$

Then by Corollary 2, we get

$$\|T_{30}\| = \frac{c}{k\Gamma_k(\alpha)} \int_0^\infty (J_\beta(u)u^{1/2})^{(\alpha/k)-1} u^{(n/q)-1} du. \tag{85}$$

If $k = 1, \alpha = 2, c = 1$ in (83), then T_{30} reduces to T_{24} in (33). Then, by (85), we have

$$\|T_{24}\| = \int_0^\infty (J_\beta(u)u^{1/2}) u^{(n/q)-1} du.$$

Remark 3. Defining other forms of the kernel K, we can obtain new results of interest.

References

[1] J.C. Kuang *Applied Inequalities*, 5th edu. (Shanggdong Science and Technology Press, Jinan, 2021), (in Chinese).

[2] A.A. Kilbas, H.M. Srivgstava, and J.J. Trujillo, *Theory and Applications of Fractional Differential Equations*, Vol. 204 (North-Holland Mathematics Studies, Elsevier, New York, 2006).

[3] S. Mubeen and S. Iqbal, Grüss type integral inequalities for generalized Riemann–Liouville k-fractional integral, *J. Inequal. Appl.* **109**, (2016).

[4] M.Z. Sarikaya, Z. Dahmani, M.E. Kiris, and F. Ahmad, (k, s)-Riemann–Liouville fractional integral and applications, *Hacet. J. Math. Stat.* **45**(1) 77–89, (2016).

[5] G. Abbas, A.K. Khuram, G. Farid, and A. Ur Rehman, Generalizations of some fractional integral inequalities via generalized Mittag — Leffler function, *J. Inequal. Appl.* **121**, (2017).
[6] Y. Zhao, S. Haiwei, X. Weicheng, and C. Zhongwei, Hermite–Hadamard type inequalities involving ψ-Riemann–Liouville k-fractional integrals via s-convex functions, *J. Inequal. Appl.* **128**, (2020).
[7] J.U. Khan, M.A. Khan, Generalized conformable fractional integral operators, *J. Comput. Appl. Math.* **346**, 378–389, (2019).
[8] S. Rashid, A.O. Akdemir, K.S. Nisar, T. Abdeljawad, and G. Rahman, New generalized reverse Minkowski and related integral inequalities involving generalized fractional conformable integrals, *J. Inequali. Appl.* **177**, (2020).
[9] B. Çelik, Gürbüz, Ç. Mustafa, M.E. Özdemir, and S. Erhan, On integral inequalities related to the weighted and the extended Chebyshev functionals involving different fractional operators, *J. Inqual. Appl.* **246**, (2020).
[10] I. Iscan, Jensen–Mercer inequality for GA — Convex functions and some related inequalities, *J. Inequal. Appl.* **212**, (2020).
[11] J.C. Kuang, Norm inequalities for generalized Laplace transforms. In: A. Raigorodskii and M. Th. Rassias (Eds.), *Trigonometric Sums and Their Applications* (Springer, 2020).
[12] J.C. Kuang, Some new inequalities for fractional integral operators. In: Th. M. Rassias (Ed.), *Approximation and Computation in Science and Engineering* (Springer, 2021).
[13] M.Z. Sarikaya, On the Hermite–Hadamard — Type inequalities for co-ordinated convex function via fractional integral, *Integral Transforms Spec. Funct.* **25**(2), 134–147, (2014).
[14] S. Erden, H. Budak, M.Z. Sarikaya, S. Iftikhar, and P. Kumann, Fractional Ostrowski type inequalities for bounded functions, *J. Inequali. Appl.* **123**, (2020).
[15] S. Erden, H. Budak, and M.Z. Sarikaya, Fractional Ostrowski type inequalities for functions of bounded variation with two variable, *Miskolc Math. Notes* **21**(1), 171–188, (2020).
[16] J.C. Kuang, Norm inequalities for generalized fractional integral operators in Banach spaces. In: Th. M. Rassias (Ed.), *Mathematical Analysis, Optimization, Approximation and Applications* (World Scientific Publishing Company Pte, Ltd, 2021).
[17] M.Z. Sarikaya and H. Yildirim, On generalization of the Riesz potential. *Indian J. Math. Math. Sci.* **3**, 231–235, (2007).
[18] A. Kashuri and Th.M. Rassias, Fractional trapezium-type inequalities for strongly exponentially generalized preinvex functions with applications, *Appl. Anal. Discrete Math.* **14**(3), 560–578, (2020).
[19] D.V. Widder, *The Laplace Transform* (Princeton University Press, 1972).
[20] K.B. Wolff, *Integral Transforms in Science and Engineering* (Pleum, 1979).
[21] E. Zauderer, *Partial Differential Equations of Applied Mathematics* (A Wiley-Interscience Publication, John Wiley & Sons. Inc. 1983).

[22] J.C. Kuang, Generalized Laplace transform inequalities in multiple Orlicz spaces. In: *Computation, Cryptography, and Network Security*, Chapter 13 (Springer, 2015).
[23] N. Bleistein, and R.A. Handelsman, *Asymptotic Expansions of Integrals* (Dover Publications, Inc., New York, 1986).
[24] G.H. Hardy, The constants of certain inequalities, *J. London Math. Soc.* **8**, 114–119, (1933).
[25] J.C. Kuang, *Real and Functional Analysis (Continuation)(1–2 Volume)* (Beijing: Higher Education Press, 2015, in Chinese).

https://doi.org/10.1142/9789811261572_0021

Chapter 21

Hyers–Ulam–Rassias Stability of Functional Equations in G-Normed Spaces

Jung Rye Lee*,§, Choonkil Park†,¶, and Themistocles M. Rassias‡,‖

*Department of Data Science, Daejin University, Kyunggi 11159, Korea

†Department of Mathematics, Hanyang University, Seoul 04763, Korea

‡Department of Mathematics, National Technical University of Athens, Zografou Campus, 15780 Athens, Greece

§jrlee@daejin.ac.kr

¶baak@hanyang.ac.kr

‖trassias@math.ntua.gr

In this chapter, we introduce functional equations in G-normed spaces and we prove the Hyers–Ulam stability of the Cauchy additive functional equation and of the quadratic functional equation in complete G-normed spaces.

1. Introduction and Preliminaries

The stability problem of functional equations was originated from a question of Ulam [1] concerning the stability of group homomorphisms. Hyers [2] gave a first affirmative partial answer to the question of Ulam for Banach spaces. Hyers's Theorem was generalized by Aoki [3] for additive mappings and by Rassias [4] for linear mappings by considering an unbounded Cauchy difference. A generalization of the Rassias theorem was obtained by Găvruta [5] by replacing the unbounded Cauchy difference by a general control function in the spirit of Rassias's approach.

The functional equation

$$f(x+y) + f(x-y) = 2f(x) + 2f(y)$$

is called a *quadratic functional equation*. In particular, every solution of the quadratic functional equation is said to be a *quadratic mapping*. The Hyers–Ulam stability problem for the quadratic functional equation was proved by Skof [6] for mappings $f : X \to Y$, where X is a normed space and Y is a Banach space. Cholewa [7] noticed that the theorem of Skof is still true if the relevant domain X is replaced by an Abelian group. Czerwik [8] proved the Hyers–Ulam stability of the quadratic functional equation. The stability problems of several functional equations have been extensively investigated by a number of authors and there are many interesting results concerning this problem [9–27].

Definition 1 ([28]). Let X be a vector space. A function $G : X^3 \to [0, \infty)$ is called a *G-metric* if the following conditions are satisfied:

(1) $G(x, y, z) = 0$ if $x = y = z$.
(2) $G(x, x, z) > 0$ for all $x, z \in X$ with $x \neq z$.
(3) $G(x, x, z) \leq G(x, y, z)$ for all $x, y, z \in X$ with $y \neq z$.
(4) $G(x, y, z) = G(p(x), p(y), p(z))$, where p is a permutation of x, y, z.
(5) $G(x, y, z) \leq G(x, w, w) + G(w, y, z)$ for all $x, y, z, w \in X$.

The pair (X, G) is called a *G-metric space.*

Definition 2 ([28]). Let (X, G) be a G-metric space.

(1) A sequence $\{x_n\}$ in X is said to be a *G-Cauchy sequence* if, for each $\varepsilon > 0$, there exists an integer N such that, for all $m, n, l \geq N$,

$$G(x_m, x_n, x_l) < \varepsilon.$$

(2) A sequence $\{x_n\}$ in X is said to be *G-convergent to a point* x if, for each $\varepsilon > 0$, there exists an integer N such that, for all $m, n \geq N$,

$$G(x_m, x_n, x) < \varepsilon.$$

A G-metric space (X, G) is called *complete* if every G-Cauchy sequence is G-convergent.

Example 1 ([28]). Let (X, d) be a metric space. Then $G : X^3 \to [0, \infty)$, defined by

$$G(x, y, z) = \max\{d(x, y), d(y, z), d(x, z)\}, \quad x, y, z \in X,$$

is a G-metric.

One can define the following:

Definition 3. Let X be a vector space over a field $F = \mathbb{R}$ or $\mathbb{C}$. A function $\|\cdot,\cdot\| : X^2 \to [0,\infty)$ is called a *G-norm* if the following conditions are satisfied:

(1) $\|x, y\| = 0$ if and only if $x = y = 0$.
(2) $\|x - y, 0\| \leq \|x - w, 0\| + \|w - y, 0\|$ for all $x, y, w \in X$.
(3) $\|x - y, x - z\| \leq \|x - w, x - w\| + \|w - y, w - z\|$ for all $x, y, z, w \in X$.
(4) $\|\lambda x, \lambda y\| = |\lambda|\ \|x, y\|$ for all $x, y \in X$ and all $\lambda \in F$.

The pair $(X, \|\cdot,\cdot\|)$ is called a *G-normed space.*

Definition 4. Let $(X, \|\cdot,\cdot\|)$ be a G-normed space.

(1) A sequence $\{x_n\}$ in X is said to be a *G-Cauchy sequence* if, for each $\varepsilon > 0$, there exists an integer N such that, for all $m, l \geq N$,

$$\|x_l - x_m, x_l - x_m\| < \varepsilon \quad \text{and} \quad \|x_l - x_m, 0\| < \varepsilon.$$

(2) A sequence $\{x_n\}$ in X is said to be *G-convergent to a point* $x \in X$ if, for each $\varepsilon > 0$, there exists an integer N such that, for all $m \geq N$,

$$\|x - x_m, x - x_m\| < \varepsilon \quad \text{and} \quad \|x - x_m, 0\| < \varepsilon.$$

We will denote x by $G\text{-}\lim_{n\to\infty} x_n$.

A G-normed space $(X, \|\cdot,\cdot\|)$ is called *complete* if every G-Cauchy sequence is G-convergent.

It is easy to show that if there exists a G-limit $x \in X$ of a sequence $\{x_n\}$ in X, then the G-limit is *unique.*

Example 2. Let $(X, \|\cdot\|)$ be a normed space. It is easy to show that $\|\cdot,\cdot\| : X^2 \to [0,\infty)$, defined by

$$\|x, y\| = \max\{\|x\|, \|y\|\}, \quad x, y \in X,$$

is a G-norm.

In this chapter, we prove the Hyers–Ulam stability of the Cauchy additive functional equation and of the quadratic functional equation in complete G-normed spaces.

Throughout this chapter, let X be a G-normed space and let Y be a complete G-normed space.

2. Stability of the Cauchy Additive Functional Equation in G-normed Spaces

For the given mapping $f : X \to Y$, we define the difference operator $Df : X \times X \to Y$ by

$$Df(x,y) := f(x+y) - f(x) - f(y),$$

for all $x, y \in X$.

We prove the Hyers–Ulam stability of the Cauchy additive functional equation in G-normed spaces.

Theorem 1. *Let θ be a positive real number and p a real number with $0 < p < 1$. Suppose that $f : X \to Y$ is a mapping such that*

$$\|Df(x,y), Df(z,w)\| \leq \theta(\|x,x\|^p + \|y,y\|^p + \|z,z\|^p + \|w,w\|^p), \quad (1)$$

for all $x, y, z, w \in X$. Then there exists a unique Cauchy additive mapping $A : X \to Y$ such that

$$\|f(x) - A(x), f(x) - A(x)\| \leq \frac{4\theta}{2-2^p}\|x,x\|^p,$$

$$\|f(x) - A(x), 0\| \leq \frac{2\theta}{2-2^p}\|x,x\|^p, \quad (2)$$

for all $x \in X$.

Proof. Setting $x = y = z = w = 0$ in (1), we obtain

$$\| - f(0), -f(0)\| \leq 0,$$

for all $x \in X$. So $f(0) = 0$.

Setting $x = y = z = w$ in (1), we obtain

$$\|f(2x) - 2f(x), f(2x) - 2f(x)\| \leq 4\theta\|x,x\|^p, \quad (3)$$

for all $x \in X$. So

$$\left\|f(x) - \frac{1}{2}f(2x), f(x) - \frac{1}{2}f(2x)\right\| \leq 2\theta\|x,x\|^p,$$

for all $x \in X$. Since $\|x-y, x-z\| \leq \|x-w, x-w\| + \|w-y, w-z\|$ for all $x, y, z, w \in X$,

$$\|x_l - x_m, x_l - x_m\| \leq \|x_l - x_n, x_l - x_n\| + \|x_n - x_m, x_n - x_m\|,$$

for all $x_l, x_m, x_n \in X$. Hence,

$$\left\| \frac{1}{2^l} f(2^l x) - \frac{1}{2^m} f(2^m x), \frac{1}{2^l} f(2^l x) - \frac{1}{2^m} f(2^m x) \right\| \leq 2\theta \sum_{j=l}^{m-1} \frac{1}{2^j} \|2^j x, 2^j x\|^p$$

$$= 2\theta \sum_{j=l}^{m-1} \frac{2^{pj}}{2^j} \|x, x\|^p, \quad (4)$$

for all non-negative integers m and l with $m > l$ and all $x \in X$.

Setting $x = y$ and $z = w = 0$ in (1), we obtain

$$\|f(2x) - 2f(x), 0\| \leq 2\theta \|x, x\|^p, \quad (5)$$

for all $x \in X$. So

$$\left\| f(x) - \frac{1}{2} f(2x), 0 \right\| \leq \theta \|x, x\|^p,$$

for all $x \in X$. Since $\|x - y, 0\| \leq \|x - w, 0\| + \|w - y, 0\|$ for all $x, y, w \in X$,

$$\left\| \frac{1}{2^l} f(2^l x) - \frac{1}{2^m} f(2^m x), 0 \right\| \leq \theta \sum_{j=l}^{m-1} \frac{1}{2^j} \|2^j x, 2^j x\|^p = \theta \sum_{j=l}^{m-1} \frac{2^{pj}}{2^j} \|x, x\|^p, \quad (6)$$

for all non-negative integers m and l with $m > l$ and all $x \in X$.

It follows from (4) and (6) that the sequence $\{\frac{1}{2^j} f(2^j x)\}$ is a G-Cauchy sequence for all $x \in X$. Since Y is complete, the sequence $\{\frac{1}{2^j} f(2^j x)\}$ is G-convergent. Thus, one can define the mapping $A : X \to Y$ by

$$A(x) := G\text{-}\lim_{j \to \infty} \frac{1}{2^j} f(2^j x),$$

for all $x \in X$.

By (1),

$$\|DA(x, y), 0\| = \lim_{n \to \infty} \frac{1}{2^n} \|Df(2^n x, 2^n y), 0\|$$

$$\leq \lim_{n \to \infty} \frac{2^{pn}}{2^n} \theta(\|x, x\|^p + \|y, y\|^p) = 0,$$

for all $x, y \in X$. So $DA(x, y) = 0$. Thus, the mapping $A : X \to Y$ is a Cauchy additive mapping. Moreover, letting $l = 0$ and passing the limit $m \to \infty$ in (4) and (6), we get the inequality (2).

Now, let $T : X \to Y$ be another Cauchy additive mapping satisfying (2). Then

$$\begin{aligned}\|A(x) - T(x), 0\| &= \frac{1}{2^n}\|A(2^n x) - T(2^n x), 0\| \\ &\le \frac{1}{2^n}\left(\|A(2^n x) - f(2^n x), 0\| + \|T(2^n x) - f(2^n x), 0\|\right) \\ &\le \frac{4 \cdot 2^{pn}}{2^n(2-2^p)}\theta\|x, x\|^p,\end{aligned}$$

which tends to zero as $n \to \infty$ for all $x \in X$. So we can conclude that $A(x) = T(x)$ for all $x \in X$. This proves the uniqueness of A. □

Corollary 1. *Let θ be a positive real number. Suppose that $f : X \to Y$ is a mapping with $f(0) = 0$ such that*

$$\|Df(x, y), Df(z, w)\| \le \theta, \tag{7}$$

for all $x, y, z, w \in X$. Then there exists a unique Cauchy additive mapping $A : X \to Y$ such that

$$\|f(x) - A(x), f(x) - A(x)\| \le \theta \quad \text{and} \quad \|f(x) - A(x), 0\| \le \theta,$$

for all $x \in X$.

Proof. Setting $x = y = z = w$ in (7), we obtain

$$\|f(2x) - 2f(x), f(2x) - 2f(x)\| \le \theta$$

for all $x \in X$. So

$$\left\|f(x) - \frac{1}{2}f(2x), f(x) - \frac{1}{2}f(2x)\right\| \le \frac{\theta}{2},$$

for all $x \in X$.

Setting $x = y$ and $z = w = 0$ in (7), we obtain

$$\|f(2x) - 2f(x), 0\| \le \theta,$$

for all $x \in X$. So

$$\left\|f(x) - \frac{1}{2}f(2x), 0\right\| \le \frac{\theta}{2},$$

for all $x \in X$.

The rest of the proof is similar to the proof of Theorem 1. □

Similarly, one can obtain the following:

Theorem 2. *Let θ be a positive real number and p a real number with $p > 1$. Suppose that $f : X \to Y$ is a mapping satisfying* (1). *Then there exists a unique Cauchy additive mapping $A : X \to Y$ such that*

$$\|f(x) - A(x), f(x) - A(x)\| \leq \frac{4\theta}{2^p - 2}\|x, x\|^p \quad \text{and} \quad \|f(x) - A(x), 0\| \leq \frac{2\theta}{2^p - 2}\|x, x\|^p,$$

for all $x \in X$.

3. Stability of the Quadratic Functional Equation in G-normed Spaces

For the given mapping $f : X \to Y$, we define the difference operator $Cf : X \times X \to Y$ by

$$Cf(x, y) := f(x + y) + f(x - y) - 2f(x) - 2f(y),$$

for all $x, y \in X$.

We prove the Hyers–Ulam stability of the quadratic functional equation in G-normed spaces.

Theorem 3. *Let θ be a positive real number and p a real number with $0 < p < 2$. Suppose that $f : X \to Y$ is a mapping such that*

$$\|Cf(x, y), Cf(z, w)\| \leq \theta(\|x, x\|^p + \|y, y\|^p + \|z, z\|^p + \|w, w\|^p), \tag{8}$$

for all $x, y, z, w \in X$. Then there exists a unique quadratic mapping $Q : X \to Y$ such that

$$\|f(x) - Q(x), f(x) - Q(x)\| \leq \frac{4\theta}{4 - 2^p}\|x, x\|^p,$$

$$\|f(x) - Q(x), 0\| \leq \frac{2\theta}{4 - 2^p}\|x, x\|^p,$$

for all $x \in X$.

Proof. Setting $x = y = z = w = 0$ in (8), we obtain

$$\| - 2f(0), -2f(0)\| \leq 0,$$

for all $x \in X$. So $f(0) = 0$.

Setting $x = y = z = w$ in (8), we obtain

$$\|f(2x) - 4f(x), f(2x) - 4f(x)\| \leq 4\theta\|x, x\|^p,$$

for all $x \in X$. So

$$\left\|f(x)-\frac{1}{4}f(2x), f(x)-\frac{1}{4}f(2x)\right\| \le \theta\|x,x\|^p,$$

for all $x \in X$.

Setting $x = y$ and $z = w = 0$ in (8), we obtain

$$\|f(2x)-4f(x),0\| \le 2\theta\|x,x\|^p,$$

for all $x \in X$. So

$$\left\|f(x)-\frac{1}{4}f(2x),0\right\| \le \frac{\theta}{2}\|x,x\|^p,$$

for all $x \in X$.

The rest of the proof is similar to the proof of Theorem 1. □

Corollary 2. *Let θ be a positive real number. Suppose that $f : X \to Y$ is a mapping with $f(0) = 0$ such that*

$$\|Cf(x,y), Cf(z,w)\| \le \theta, \tag{9}$$

for all $x, y, z, w \in X$. Then there exists a unique quadratic mapping $Q : X \to Y$ such that

$$\|f(x)-Q(x), f(x)-Q(x)\| \le \frac{\theta}{3} \quad \text{and} \quad \|f(x)-Q(x),0\| \le \frac{\theta}{3},$$

for all $x \in X$.

Proof. Setting $x = y = z = w$ in (9), we obtain

$$\|f(2x)-4f(x), f(2x)-4f(x)\| \le \theta,$$

for all $x \in X$. So

$$\left\|f(x)-\frac{1}{4}f(2x), f(x)-\frac{1}{2}f(2x)\right\| \le \frac{\theta}{4},$$

for all $x \in X$.

Setting $x = y$ and $z = w = 0$ in (9), we obtain

$$\|f(2x)-4f(x),0\| \le \theta,$$

for all $x \in X$. So

$$\left\|f(x)-\frac{1}{4}f(2x),0\right\| \le \frac{\theta}{4},$$

for all $x \in X$.

The rest of the proof is similar to the proof of Theorem 1. □

Similarly, one can obtain the following:

Theorem 4. *Let θ be a positive real number and p a real number with $p > 2$. Suppose that $f : X \to Y$ is a mapping satisfying* (8). *Then there exists a unique quadratic mapping $Q : X \to Y$ such that*

$$\|f(x) - Q(x), f(x) - Q(x)\| \leq \frac{4\theta}{2^p - 4}\|x, x\|^p,$$
$$\|f(x) - Q(x), 0\| \leq \frac{2\theta}{2^p - 4}\|x, x\|^p,$$

for all $x \in X$.

References

[1] S.M. Ulam, *A Collection of the Mathematical Problems* (Interscience Publ., New York, 1960).

[2] D.H. Hyers, On the stability of the linear functional equation, *Proc. Nat. Acad. Sci. USA* **27**, 222–224, (1941).

[3] T. Aoki, On the stability of the linear transformation in Banach spaces, *J. Math. Soc. Japan* **2**, 64–66, (1950).

[4] Th.M. Rassias, On the stability of the linear mapping in Banach spaces, *Proc. Am. Math. Soc.* **72**, 297–300, (1978).

[5] P. Găvruta, A generalization of the Hyers–Ulam–Rassias stability of approximately additive mappings, *J. Math. Anal. Appl.* **184**, 431–436, (1994).

[6] F. Skof, Propriet locali e approssimazione di operatori, *Rend. Sem. Mat. Fis. Milano* **53**, 113–129, (1983).

[7] P.W. Cholewa, Remarks on the stability of functional equations, *Aequationes Math.* **27**, 76–86, (1984).

[8] S. Czerwik, On the stability of the quadratic mapping in normed spaces, *Abh. Math. Sem. Univ. Hamburg* **62**, 59–64, (1992).

[9] M.R. Abdollahpour, R. Aghayaria, and M.Th. Rassias, Hyers–Ulam stability of associated Laguerre differential equations in a subclass of analytic functions, *J. Math. Anal. Appl.* **437**, 605–612, (2016).

[10] M.R. Abdollahpour and M.Th. Rassias, Hyers–Ulam stability of hypergeometric differential equations, *Aequationes Math.* **93**, 691–698, (2019).

[11] J. Aczel and J. Dhombres, *Functional Equations in Several Variables* (Cambridge University Press, Cambridge, 1989).

[12] D.G. Bourgin, Multiplicative transformations, *Proc. Nat. Acad. Sci. USA* **36**, 564–570, (1950).

[13] D.G. Bourgin, Classes of transformations and bordering transformations, *Bull. Am. Math. Soc.* **57**, 223–237, (1951).

[14] Z. Gajda, On stability of additive mappings, *Int. J. Math. Math. Sci.* **14**, 431–434, (1991).

[15] S. Czerwik, *Functional Equations and Inequalities in Several Variables* (World Scientific, Singapore, 2002).

[16] A. Gilányi, Hyers-Ulam stability of monomial functional equations on a general domain, *Proc. Nat. Acad. Sci. USA* **96**, 10588–10590, (1999).
[17] D.H. Hyers, G. Isac, and Th.M. Rassias, *Stability of Functional Equations in Several Variables* (Birkhäuser, Basel, 1998).
[18] D.H. Hyers and Th.M. Rassias, Approximate homomorphisms, *Aequationes Math.* **44**, 125–153, (1992).
[19] S. Jung, Hyers–Ulam–Rassias, *Stability of Functional Equations in Nonlinear Analysis* (Springer, New York, 2011).
[20] S. Jung, C. Mortici, and M.Th. Rassias, On a functional equation of trigonometric type, *Appl. Math. Comput.* **252**, 294–303, (2015).
[21] S. Jung, D. Popa and M. Th. Rassias, On the stability of the linear functional equation in a single variable on complete metric groups, *J. Global Optim.* **59**, 165–171, (2014).
[22] S. Jung and M.Th. Rassias, A linear functional equation of third order associated to the Fibonacci numbers, *Abstr. Appl. Anal.* **2014**, Article ID 137468 (2014).
[23] Pl. Kannappan, *Functional Equations and Inequalities with Applications* (Springer, New York, 2009).
[24] Y. Lee, S. Jung, and M.Th. Rassias, On an n-dimensional mixed type additive and quadratic functional equation, *Appl. Math. Comput.* **228**, 13–16, (2014).
[25] Y. Lee, S. Jung, and M. Th. Rassias, Uniqueness theorems on functional inequalities concerning cubic-quadratic-additive equation, *J. Math. Inequal.* **12**, 43–61, (2018).
[26] C. Mortici, S. Jung, and M. Th. Rassias, On the stability of a functional equation associated with the Fibonacci numbers, *Abstr. Appl. Anal.* **2014**, Article ID 546046 (2014).
[27] C. Park and M. Th. Rassias, Additive functional equations and partial multipliers in C^*-algebras, *Rev. R. Acad. Cien. Exactas Fís. Nat. Ser. A Mat. RACSAM* **113**, 2261–2275, (2019).
[28] Z. Mustafa and B. Sims, A new approach to generalized metric spaces, *J. Nonlinear Convex Anal.* **7**, 289–297, (2006).

https://doi.org/10.1142/9789811261572_0022

Chapter 22

Wavelet Detrended Fluctuation Analysis: Review and Extension to Mixed Cases

Anouar Ben Mabrouk[*,†,‡,||], Mohamed Essaied Hamrita[†,§,**], and Carlo Cattani[¶,††]

[*] *Department of Mathematics, Higher Institute of Applied Mathematics and Computer Science, Street of Assad Ibn Alfourat, 3100 Kairouan, Tunisia*

[†] *Laboratory of Algebra, Number Theory and Nonlinear Analysis, LR18ES15, Department of Mathematics, Faculty of Sciences, 5000 Monastir, Tunisia*

[‡] *Department of Mathematics, Faculty of Sciences, University of Tabuk, Kingdom of Saudi Arabia*

[§] *Department of Quantitative Methods, Higher Institute of Business, University of Sousse, 4054 Sousse, Tunisia*

[¶] *Engineering School (DEIM), Tuscia University, Viterbo, Italy*

[||] *anouar.benmabrouk@fsm.rnu.tn*

[**] *mohamed_essaied_hamrita@ihecso-sousse.tn*

[††] *cattani@unitus.it*

Detrended fluctuation analysis considers single statistical series and studied possible fluctuations in time by estimating scaling exponents, dynamics, etc. However, in many cases it is necessary to consider simultaneously many series and study their behavior once. This leads to the extension of detrended fluctuation analysis to mixed detrended fluctuation analysis. No study has been developed on simultaneously many detrended fluctuation analyses or detrended fluctuation analysis of simultaneously many series. Inspired from mixed multifractal analysis developed in Ref. [1] and next extended in Refs. [2–4], we developed in the present chapter a new type of mixed detrended fluctuation analysis based on wavelets. The originality consists in exploiting wavelet analysis which splits a series into detail components, and applying thus a type of mixed wavelet detrended multifractal analysis to the detail components of finitely many statistical series simultaneously.

1. Introduction

Detrended fluctuation analysis has been proved to be a powerful tool in many cases of studies such as time series, physico-financial series, natural series, DNA, proteins, etc. Detrended fluctuation analysis considers a single statistical series relative to some data and studied its fluctuation in time by estimating scaling exponents, dynamics, etc. In some cases, it is related and/or somehow ameliorated by the inclusion of fractal/multifractal analysis in the study leading to fractal/multifractal detrended fluctuation analysis. Besides, multifractal analysis is related strongly to wavelets. This leads to the extension of detrended fluctuation analysis to wavelet detrended fluctuation analysis and to wavelet fractal/multifractal detrended fluctuation analysis.

In Ref. [5], a clinical study on patients suffering from Congestive Heart Failure has been conducted based on detrended fluctuation analysis of cardiac rhythm variability. It is shown effectively that such an analysis is an efficient tool for diagnosing heart failure. On the same subject of heart rate variability diagnosis, in Ref. [6], a classification of certain diseases using correlation dimension and detrended fluctuation analysis has been developed on ECG signals related to heart rate variability. In another subject out of bio-signal studies, detrended multifractal fluctuation analysis has been applied for physico-economic signals. In Ref. [7], daily records of international crude oil prices have been studied. The findings reveals that re-scaled range Hurst analysis may be a good factor to show persistence and long-memory in the crude oil market.

In Ref. [8], multifractal fluctuation analysis has been conducted among other techniques for statistical analysis of DNA sequences. By applying wavelet theory the authors showed that bias in the DNA walk can be removed and thus the existence of power-law correlations with specific scale invariance properties can be revealed accurately. In Ref. [9], a comparison study has been developed between wavelet leaders and multifractal detrended fluctuation analysis for the estimation of the multifractal structure in natural series. In Ref. [10], detrended fluctuation analysis has been applied to quantify fractal-like auto-correlation properties of some biomedical signals. Selected examples of application in cardiology, neurology have been reviewed revealing that detrended fluctuation analysis may be an efficient tool to show the complexity of signals. In Ref. [11], the scaling exponents of the EEG dynamics are obtained by using detrended fluctuation analysis. It is shown that the mean scaling exponents of EEG are related to the so-called rapid eye movement and waken stage, and gradually increased from stage 1 to stage 2, 3 and 4. In Ref. [12], the fractal scaling properties of human sleep EEG dynamics have been Quantified and next each normal sleep stage has been compared with sleep apnea one. Using detrended fluctuation analysis, the fractal scaling exponents due to power-law correlations have been estimated.

In Ref. [13], p-leader multifractal analysis has been developed leading to a p-leader multifractal formalism by conjecturing an explicit and universal closed-form correction to the classical version permitting an accurate estimation of scaling exponents based on wavelet cascades. In Ref. [14], medical images fractal analysis has been conducted based on the computation of fractal dimension of the irregular regions based on the computation of the intensity difference scaling aspect. In Ref. [15], starting from the evidence idea that DNA sequences in genes containing non-coding regions are correlated in a remarkably long range, the authors proposed a solution for the non-stationary feature of DNA sequence of base pairs by applying detrended fluctuation analysis based algorithms.

In Ref. [16], the complexity of surface electromyography signals have been studied by using detrended fluctuation analysis. The experimental results of mean and standard deviation have shown that the scaling exponents issued from detrended fluctuation analysis have significant difference values in various hand motions. By applying a cluster-to-cluster distance and scatter plot between scaling exponents of hand movements the authors proved that detrended fluctuation analysis is suitable for surface electromyography feature extraction.

In Ref. [17], scaling properties of electrocardiogram recordings of healthy subjects and heart failure patients have been estimated by using detrended fluctuation analysis. The findings have shown that intra-beat dynamics of healthy subject are less correlated than for heart failure dynamics. In Ref. [18], a multifractal study of hydrophobicity scale of amino acids and the 6-letter model of protein have been developed. The main focuses are the study of the relationship between the primary structure and the secondary structural classification of proteins. In the same subject, the authors in Ref. [19] developed a wavelet multifractal study of hydrophobicity scale of amino acids to localize transmembrane segments.

In Ref. [20], spectral analysis of heart rate variability has been developed based on detrended fluctuation analysis on a sample of pregnant women. It is discovered that late pregnant women have elevated global scaling exponent, elevated short-term scaling exponent and lower heart rate variability measures in the low and high frequency ranges than those of the healthy controls and 3 months after delivery. Based on the findings, the authors suggested that the global and short-term detrended fluctuation scaling exponents may be an efficient independent measure of heart rate variability in late pregnancy. As it is noticed in this literature review, all existing studies have been based on a single detrended fluctuation analysis type. No study has been developed on simultaneously many detrended fluctuation analyses or detrended fluctuation analysis of simultaneously many signals resulting in a mixed detrended fluctuation analysis, which is the main subject of the present work. Indeed, as for the case of fractal/multifractal analysis, a mixed type of such an analysis has been already developed and extended from functional framework to measures. Mixed multifractal analysis based on wavelets had been firstly developed in Ref. [1] and next extended to other cases such as Ref. [2–4].

More about the use of DFA and its extensions may be found in [41,42,47, 52] for the analysis of energy market, [45,48] for the agriculture market, [50, 51,54,56] on exchange rates, currencies, and other stock markets, and [43, 44,46,49,53,55] for general signals, time series, software, and mathematical methods.

In the present work, inspired from these references, we aim to develop a mixed type of detrended fluctuation analysis and to conduct some concrete applications to show their efficiency. As it is mentioned above, detrended fluctuation analysis considers single statistical series and studies the type or the structure of eventual fluctuations by estimating scaling exponents, dynamics, etc. However, in many situations such as financial markets, cross-correlation analysis is well known and informs us that simultaneous behavior of markets and/or their components is of great effect and interest. This leads to the extension of the detrended fluctuation analysis to mixed detrended fluctuation analysis. To the best of our knowledge, there is no study that has been developed on simultaneously many detrended fluctuation analyses or detrended fluctuation analysis of simultaneously many series. This motivates us to develop in the present chapter a new type of mixed detrended fluctuation analysis based on wavelets. The original idea consists in exploiting wavelet analysis, which splits the series into detail components, applying thus a type of mixed wavelet detrended multifractal analysis to the detail components of finitely many statistical series simultaneously. We thus combine the classical fluctuation analysis, the detrended variant, the multifractal structure of series and the wavelet decomposition to obtain a hybrid model.

In the next section, a brief review of detrended fluctuation analysis will be developed. Next, Section 3 will be devoted to the review of multifractal detrended fluctuation analysis as a first extension of detrended fluctuation analysis. Section 4 will be subject of wavelet multifractal detrended fluctuation analysis as a second step ahead for ameliorating the detrended fluctuation analysis and to include wavelets in the study. Recall that wavelets since their appearance have been proved to be the most efficient tool to study fluctuations, dynamics, scalings for signals. Section 5 is devoted to the introduction of our main ideas of extension to the mixed detrended fluctuation analysis. We will distinguish three extensions: the mixed detrended fluctuation analysis, mixed multifractal detrended fluctuation analysis and mixed wavelet multifractal detrended fluctuation analysis. Section 6 will be concerned about the development of some empirical studies.

2. The Detrended Fluctuation Analysis

Let $N \in \mathbb{N}$ be a fixed integer, $T > 0$, $I = [0, T]$ and $I_N = \{t_i,\ i = 1, 2, \ldots, N\}$ the set of integers in the interval I. Let next $X_i = X_{t_i}$, $i = 1, 2, \ldots, N$, be a time series (also called signal). The integrated series associated to X_t will be denoted by IX_t and is defined by

$$IX_t = \sum_{i=1}^{t} (X_i - \overline{X}), \tag{1}$$

where $\overline{X}$ is the mean value of the series X_t. Let next $n \in \mathbb{N}$ be fixed and $I_1, I_2, \ldots, I_n$ be a subdivision of the time interval I_N. In each segment I_k, we consider the least squares fit of the series IX_t and denote it by IX_t^k. The overage fluctuation of X_t around the trend is defined by

$$AF_{X_t}(n) = \sqrt{\frac{1}{N} \sum_{i=1}^{N} (IX_i - IX_i^n)^2}.$$

The detrended fluctuation analysis (DFA) consists in evaluating the power-law dependence of the function $AF_{X_t}(n)$ on n. When $AF_{X_t}(n) \sim n^{\alpha}$, the series is self-fluctuating with scaling exponent α.

3. The Multifractal Detrended Fluctuation Analysis

The multifractal fluctuation analysis is essentially based on the computation of two characterizing indices of the time series. One parameter is known as the Hölder exponent which quantifies the local regularity of the signal and thus detects the fluctuation (see Refs. [21,22]). A second parameter is called the multifractal spectrum of the analyzed signal and quantifies the so-called multifractality of it. It somehow classifies the data into sub-signals or subsets with the same Hölder exponent and computes the fractal dimension of these subsets by means of the Hausdorff dimension. This leads to some correspondence (function) between the set of Hölder exponents and the one of the Hausdorff dimensions (see Refs. [21,22]).

Let X_t be a time series as above, and t_0 be a given time position. The multifractal analysis of X_t near t_0 starts by computing the so-called Hölder exponent at t_0. The regularity of X_t near t_0 is measured by means of its Hölder exponent $H_X(t_0)$. This is the maximum exponent γ satisfying the estimation

$$|X_t - P(t - t_0)| = o(|t - t_0)|^\gamma), \tag{2}$$

near t_0 for some polynomial $P(t - t_0)$ of degree at most $[\gamma]$, where the brackets designate the integer part. This means that the Hölder exponent $H_X(t_0)$ of X_t at t_0 is

$$H_X(t_0) = \sup\{\gamma \geq 0\,;\ (2)\ \text{holds}\}.$$

The Holder exponent has an important role in understanding and modeling many time series such as the financial one. It permits essentially to decompose the whole domain of the series into disjoint special species, known as the singularity sets. For $\gamma > 0$, the γ-level singularity set is defined by

$$E(\gamma) = \{t\,;\ H_X(t) = \gamma\}.$$

The spectrum of singularities of the series X_t in the fractal dimension of $E(\gamma)$ in the Hausdorff sense, which will be denoted by $Spect_X(\gamma)$. Recall that the Hausdorff measure is defined on $\mathbb{R}$ by

$$\mathcal{H}^s(E) = \liminf_{\delta \downarrow 0} \left\{ \sum_i (2r_i)^s \right\},$$

for any set $E \subset \mathbb{R}$, and where the lower bound above is taken on all the intervals centered in E, with radius at most δ, and covering E. The Hausdorff dimension of E is

$$dimE = \inf\{s \in \mathbb{R}\ \mathcal{H}^s(E) = 0\}.$$

The spectrum of singularity of the time series X_t is simply the function $\gamma \mapsto D_X(\gamma) = dimE(\gamma)$.

The main problem in fractal analysis is the direct computation of such a spectrum from the mathematical definition due to the Hausdorff measure. For this reason, researchers proposed different methods to compute the spectrum. One of them states that

$$D_X(\gamma) = \inf_p(\gamma p - \eta_X(p) + 1), \tag{3}$$

where $\eta_X(p)$ is Besov's exponent, and is due to Arneodo *et al.* [23] as explained in the following.

Let $m \in \mathbb{N}$ and f be a real-valued function on $\mathbb{R}$. The m-iterated difference of f is defined for x and h in $\mathbb{R}$ by

$$\Delta_h^m f(x) = \sum_{j=0}^{m}(-1)^{m-j}\binom{m}{j} f(x+jh).$$

Next for $s > 0$, the integer parameter m will be fixed such that $m-1 \leq s < m$, $(s = m-1+\sigma$ with $0 \leq \sigma < 1)$ and Besov's space $B_p^{s,q}$ $(1 \leq p, q \leq \infty)$ will be

$$B_p^{s,q}(\mathbb{R}) = \left\{ f \in L^p(\mathbb{R}),\ \int_{\mathbb{R}} \frac{\|\Delta_h^m f\|_p^q}{|h|^{1+sq}} < \infty \right\}.$$

In fact, the parameter p reflects the original norm or the original functional space that contains f. The index s concerns the regularity and finally the index q is somehow a correction of such a regularity. For example, when $q = \infty$, we obtain Besov's space

$$B_p^{s,\infty}(\mathbb{R}) = \left\{ f \in L^p(\mathbb{R}),\ \sup_{h\in\mathbb{R}} \frac{\|\Delta_h^m f\|_p}{|h|^s} < \infty \right\}.$$

Besov's exponent of a function f is

$$\eta_F(p) = \sup\left\{ s,\ F \in B_p^{s/p,\infty}(\mathbb{R}) \right\}. \tag{4}$$

In general, we compute the L^p-modulus of continuity magnitude order

$$\zeta_F(p) = \liminf_{|h|\to 0} \frac{\log \|\Delta_h^1 F\|_p}{\log |h|^{1/p}}. \tag{5}$$

For $s > 0$, we denote $\sigma = s - [s]$ where $[s]$ is the integer part of s and for $p \geq 1$, we denote $H^{s,p}(\mathbb{R})$ the Nikol'skij space

$$H^{s,p}(\mathbb{R}) = \left\{F \in L^p(\mathbb{R});\ \|\Delta_h^1 F^{([s])}\|_p \leq |h|^\sigma;\ |h| <<< 1\right\}.$$

It is well known that

$$H^{s+\varepsilon,p}(\mathbb{R}) \hookrightarrow B_p^{s,\infty}(\mathbb{R}) \hookrightarrow H^{s-\varepsilon,p}(\mathbb{R})\,\forall\,\varepsilon > 0.$$

Therefore, $\xi_F(p) = \eta_F(p)$ where $\xi_F(p)$ is the Nikol'skij exponent of F, defined by

$$\xi_F(p) = \sup\left\{s; F \in H^{s/p,p}(\mathbb{R})\right\}.$$

On the other hand, it is well known in functional spaces theory that whenever $\zeta(p) < p$, we have

$$\zeta_F(p) = \xi_F(p).$$

Indeed, if $s < \zeta_F(p)$, there exists $r > 0$ such that

$$\|\Delta_h^1 F^{([s])}\|_p < |h|^{s/p},\ \forall\,|h| < r.$$

Hence,

$$\|\Delta_h^1 F\|_p < |h|^{\widetilde{\sigma}},$$

where $\widetilde{\sigma} = \frac{s}{p} - [\frac{s}{p}] = \frac{s}{p}$ because $s < p$. As a result, equation (5) yields that $s \leq \xi(p)$ and then $\zeta_F(p) \leq \xi_F(p)$. Conversely, let s be such that $p > s > \xi_F(p)$. Then $F \in H^{s/p,p}(\mathbb{R})$. Hence,

$$\|\Delta_h^1 F\|_p \leq C|h|^{\widetilde{\sigma}},$$

where $\sigma = \frac{s}{p} - [\frac{s}{p}] = \frac{s}{p}$ and $|\gamma| = 0$. Consequently, the previous inequality becomes

$$\|\Delta_h^1 F\|_p \leq C|h|^{s/p}.$$

Therefore, $s \leq \zeta_F(p)$ and thus $\xi_F(p) \leq \zeta_F(p)$.

For more backgrounds on these spaces, associated exponents, embedding, and also wavelet characterization, the readers may refer to [21,24–30].

In practice, we decompose the domain of the signal X_t into cells (dyadic for example) $C_{j,k} = [\frac{k}{2^j}, \frac{k+1}{2^j})$, $j,k \in \mathbb{N}$. For t fixed, let $I_j(t)$ be a j-level dyadic cell that contains it and assume that for some measure μ on the set of cells we have the scaling property

$$\mu(2I_j(t)) \sim 2^{-jH_X(t)}$$

where $\alpha_X(t)$ is the Hölder exponent of X_t at the point t. This permits to define a structure function

$$S_X(p,j) = 2^{-j} \sum_{k,\mu(I_{jk})>0} (\mu(2I_{jk}))^p. \tag{6}$$

This function shall satisfy the estimation

$$S_X(p,j) \sim 2^{-j(1-D_X(\gamma)+\gamma p)},$$

whenever $t \in E(\gamma)$. When $j \to +\infty$ (and thus $I_j(t) \to \{t\}$, which explains somehow the pointwise behavior), the smallest exponent will characterize the principal contribution. For this we may expect that Besov's exponent is

$$\eta_X(p) = \inf_\gamma (1 - D_X(\gamma) + \gamma p),$$

which gives the equation (3) when η_X is concave.

The multifractal detrended fluctuation analysis consists in replacing the fit IX_t^k with the best polynomials, which will be denoted here PIX_t^k and next computing the new detrended time series $Y_k(t)$ for each segment $s = 1, \ldots, n$ defined by

$$Y_k(t) = IX_t - PIX_t^k.$$

The new fluctuation analysis starts by computing for each $k = 1, \ldots, n$ the quantity

$$F_k(s) = \sqrt{\frac{1}{k} \sum_{i=1}^{k} [Y_k((s-1)k+i)]^2}.$$

It represents somehow the variance of the detrended time series Y_k associated to each segment I_k. Next, averaging on all the segments we obtain the fluctuation function

$$F(s) = \sqrt{\frac{1}{2n} \sum_{i=1}^{2n} [F_s(i)]^2}.$$

This quantity may be generalized to obtain the p-th fluctuation function

$$F(p,s) = \left\{ \frac{1}{2n} \sum_{i=1}^{2n} [F_s(i)]^p \right\}^{1/p}.$$

The multifractal detrended fluctuation analysis seeks the exponent $h(p)$ satisfying

$$F(p,s) \sim s^{h(p)}.$$

One standard idea relating the new detrended fluctuation analysis may be discovered in the Parisi–Frisch work [31] where the measure μ is evaluated by

$$\mu(I_{jk}) = \left| X\left(\frac{k+1}{2^j}\right) - X\left(\frac{k}{2^j}\right) \right|, \tag{7}$$

or sometimes by

$$\mu(I_{jk}) = 2^j \left| \int_{k/2^j}^{(k+1)/2^j} (X(t) - X(t+1))dt \right|. \tag{8}$$

For more details on these facts, we refer to [9,15,32] and the references therein.

4. The Wavelet Multifractal Detrended Fluctuation Analysis

Wavelets have been since their appearance a powerful tool in multifractal analysis. Indeed, wavelets are characterized by their cancellation, localization and regularity properties, which permit to characterize pointwise singularities of functions by the decay of their wavelet transforms near them. Then, *a priori*, the wavelet analysis is the best adaptable tool in the study of multifractal signals. In the case where some self-similarity appears in the signal, the wavelet coefficients inherit in fact a type of self-similarity from the original function. The estimations of these coefficients permitted the researchers to justify the multifractal formalism due to Frisch–Parisi and were revisited next by Arneodo and his collaborators with wavelets [1,8,31]. If $X : \mathbb{R} \to \mathbb{R}$ satisfies $X(t) = \lambda X(rt)$ for all t and for some λ and r, then its wavelet transform at a scale $a > 0$ and a position $b \in \mathbb{R}^m$ relatively to any analyzing wavelet satisfies

$$C_{a,b}(X) = \lambda C_{ra,rb}(X) \;\; \forall\, a > 0 \text{ and } \forall\, b \in \mathbb{R}.$$

This allows the estimation of the size of the wavelet transform $C_{a,b}(X)$ everywhere. When X is periodic with period T, the wavelet coefficient is also periodic with respect to the second variable of position b. In other words,

$$C_{a,b}(X) = C_{a,b+T}(X) \;\; \forall\, a > 0 \text{ and } \forall\, b \in \mathbb{R}.$$

Using wavelet transforms, the equation (8) becomes

$$\mu(I_{jk}) = |C_{j,k}| \tag{9}$$

where $C_{j,k}$ is the wavelet coefficient relatively to a discrete grid $a = 2^{-j}$ and $b = k2^{-j}$. The structure function $S_X(p, j)$ becomes

$$S_X(p,j) = 2^{-j} \sum_k |C_{j,k}|^p \,. \tag{10}$$

However, the equation (9) above does not yield a true measure because of the fact that it is not necessarily increasing. To overcome this obstacle, a more sophisticated formulation has been introduced by considering instead the wavelet leaders

$$W_{j,k} = \sup\{|C_{a,b}|\,;\ I_{a,b} \subset MI_{j,k}\}, \tag{11}$$

where $M > 0$ is an appropriate scaling factor that allows to dilate the interval $I_{j,k}$ in the sense that $MI_{j,k}$ and $I_{j,k}$ have the same center and $|MI_{j,k}| = M\,|I_{j,k}|$. The wavelet leaders structure function is defined by

$$S_X(p,j) = 2^{-j} \sum_k |W_{j,k}|^p \,. \tag{12}$$

To handle the multifractal structure and thus compute the multifractal spectrum, we seek a function η_X that reflects the power-law behavior of $S_X(p, j)$. Let

$$\eta_X(p) = \liminf_{j \to +\infty} \frac{\log(S_X(p,j))}{\log 2^j}.$$

The multifractal formalism is evaluated in the same way as (3), where in this case $\eta_X(p)$ is evaluated in a more general context of wavelet theory [23]. We have precisely,

$$\eta_F(p) = \liminf_{a \to 0} \frac{\log \displaystyle\int |C_{a,b}(F)|^p db}{\log a}.$$

This means that Besov's space itself may be redefined via wavelet theory as

$$B_p^{s/p,\infty}(\mathbb{R}) = \left\{F;\ \int |C_{a,b}(F)|^p db \leq Ca^s;\ a << 1\right\}.$$

5. Extension to Mixed Cases

A mixed model is generally composed of factors with simultaneous effects, some may be fixed and others with random effects. Almost all the known methods for linear models in statistics have been reviewed and extended to mixed linear models such as estimations, tests, forecasting, etc. See Refs. [33–39].

In this section, we propose to develop a new context of mixed analysis related to the different fluctuation analyses described above. The idea consists in considering finitely many statistical series and introducing one single controller that may describe best their simultaneous behavior.

5.1. *Mixed fluctuation analysis*

The first step of the extension to mixed cases starts with the fluctuation analysis as it consists of the first step in the whole topic. For the sake of simplicity, we will restrict to the two-dimensional case. The results may be extended in a natural way for any finite number of time series.

Consider in this case a vector-valued time series $X_t = (Y_t, Z_t)$ on an interval I_N as in Section 2. We associate here also an integrated series IX_t to X_t as in equation (1), with $(X_i - \overline{X}) = (Y_i - \overline{Y}, Z_i - \overline{Z})$ in the present case. Consider as in Section 2 the subdivision I_1, $I_2, \ldots, I_n$ of I_N. In each segment I_k, we consider the least squares fit of the series IX_t and denote it by $IX_t^k = (IY_t^k, IZ_t^k)$. The mixed overage fluctuation of X_t around the trend is defined by

$$AF_{X_t}(n) = \sqrt{\frac{1}{N}\sum_{i=1}^{N}(IY_i - IY_i^n)^2(IZ_i - IZ_i^n)^2}.$$

It consists of a type of cross-correlation analysis that permits to evaluate the simultaneous fluctuation analysis of the two series Y_t and Z_t. The mixed Detrended Fluctuation Analysis (DFA) consists in evaluating the power-law dependence of the function $AF_{X_t}(n)$ on n. When $AF_{X_t}(n) \sim n^\alpha$, the series is self-fluctuating with mixed scaling exponent α.

5.2. *The mixed multifractal detrended fluctuation analysis*

The motivation behind the present extension is essentially due to some recent work dealing with a mixed multifractal analysis of functions and measures developed in Refs. [1–4].

In this section, we will investigate a different variant of the Holder exponent and the associated spectrum. For a point t_0, the mixed Hölder exponent $H_X(t_0)$ of $X_t = (Y_t, Z_t)$ at t_0 will be the vector

$$H_X(t_0) = (H_Y(t_0), H_Z(t_0)).$$

For $\gamma = (\gamma_1, \gamma_2) \in \mathbb{R}_+^2$, the mixed γ-level singularities set is defined by

$$E(\gamma) = \{t;\ H_X(t) = \gamma\}.$$

The mixed spectrum of singularities of X_t is the fractal dimension of this set denoted as previously $D_X(\gamma)$, and evaluated by a quietly modified formula as

$$D_X(\gamma) = \inf_p(<\gamma, p> -\eta_X(p) + 1),$$

where $\eta_X(p)$ is the mixed Besov exponent.

In the present context of mixed multifractal detrended analysis, we seek a function $\eta_X(p_1, p_2)$ that depends on two parameters p_1 and p_2 that guarantee that two functions f and g are simultaneously in one mixed Besov space.

In practice, we decompose the domain of the signal X_t into cells (dyadic for example) $C_{j,k} = [\frac{k}{2^j}, \frac{k+1}{2^j})$, $j, k \in \mathbb{N}$. Pour t fixed, let $I_j(t)$ be j-level dyadic cell that contains it. Next, consider a vector $\mu = (\mu_1, \mu_2)$ of measures for which a mixed scaling property holds for these cells such as

$$\mu_1(2I_j(t)) \sim 2^{-jH_Y(t)} \quad \text{and} \quad \mu_2(2I_j(t)) \sim 2^{-jH_Z(t)}.$$

The measure μ will act on the cells in the sense that for $p = (p_1, p_2)$,

$$\begin{aligned}(\mu(2I_j(t)))^p &= (\mu_1(2I_j(t)))^{p_1}(\mu_2(2I_j(t)))^{p_2} \sim 2^{-j(p_1H_Y(t)+p_2H_Z(t))}\\ &= 2^{-j<p,H_X(t)>}.\end{aligned}$$

This permits to define a mixed structure function

$$S_X(p, j) = 2^{-j} \sum_{k,\mu(I_{jk})>0} (\mu_1(2I_{jk}))^{p_1}(\mu_2(2I_{jk}))^{p_2}. \tag{13}$$

We seek that this function satisfy the estimation

$$S_X(p, j) \sim 2^{-j(1-D_X(\gamma)+<\gamma,p>)},$$

whenever $t \in E(\gamma)$. When $j \to +\infty$, we obtain as in the single case the smallest exponent that will characterize the principal contribution. Consequently, a suitable definition of the mixed Besov exponent may be

$$\eta_X(p) = \inf_\gamma(1 - D_X(\gamma)+ <\gamma, p>),$$

and that

$$D_X(\gamma) = \inf_p(1 - \eta_X(p)+ <\gamma, p>).$$

So, here also for the mixed multifractal detrended fluctuation analysis we replace the fit IX_t^k with the best polynomials PIX_t^k and we next compute a new detrended time series $\widetilde{X}_k(t)$ for each segment $s = 1, \ldots, n$ defined by

$$\widetilde{X}_k(t) = IX_t - PIX_t^k.$$

The new mixed fluctuation analysis starts by computing for each $k = 1, \ldots, n$ the quantity

$$F_k(s) = \sqrt{\frac{1}{k} \sum_{i=1}^{k} \left[\widehat{Y}_k((s-1)k+i) \widehat{Z}_k((s-1)k+i) \right]^2}.$$

It represents somehow a cross-correlation associated to each segment I_k. Next, averaging on all the segments, we obtain the fluctuation function

$$F(s) = \sqrt{\frac{1}{2n} \sum_{i=1}^{2n} [F_s(i)]^2}.$$

This quantity may be generalized to obtain the p-th fluctuation function

$$F(p,s) = \left\{ \frac{1}{2n} \sum_{i=1}^{2n} [F_s(i)]^p \right\}^{1/p}.$$

The multifractal detrended fluctuation analysis seeks the exponent $h(p)$ satisfying

$$F(p,s) \sim s^{h(p)}.$$

5.3. *The mixed wavelet multifractal detrended fluctuation analysis*

Mixed wavelet multifractal analysis has been firstly developed in Refs. [1,2] where simultaneous behaviors of functions are studied. Mixed spectra are introduced and conjectures of mixed formalism have been put and proved in some cases, especially when the presence of scaling laws and self-similar laws exist in the studied functions. This is motivating as these laws mark the majority of financial time series. Such a mixed analysis has been generating a great interest and thus proved to be powerful in describing the local simultaneous behaviors of signals, especially fractal-like ones.

For the vector-valued time series $X_t = (Y_t, Z_t)$ and $\alpha = (\alpha_1, \alpha_2) \in \mathbb{R}^2$, to the mixed Hölder spectrum $D_X(\alpha)$, one computes firstly for $p = (p_1, p_2) \in \mathbb{R}_+^2$ the structure function

$$\Gamma_X(a,p) = \int_{\mathbb{R}} |C_{a,b}(Y)|^{p_1} \, |C_{a,b}(Z)|^{p_2} \, db,$$

and evaluates its scaling low. Whenever the order of magnitude of $\Gamma_X(a,p)$, is $a^{\eta_X(p)}$, we expect that

$$D_X(\alpha) = \inf_{q \in (0,\infty)^2} (<\alpha, q> - \eta_X(q) + 1),$$

where $<,>$ denotes the usual inner product in $\mathbb{R}^2$ and we call it the mixed multifractal formalism for functions. This means that the mixed Besov exponent η_X may be defined as

$$\eta_X(p) = \liminf_{a \to 0} \frac{\log \Gamma_X(a,p)}{\log a}.$$

This induces immediately an equivalent definition that looks like a natural extension of the single case as

$$\eta_X(p) = \sup \left\{ s; \int_{\mathbb{R}} |C_{a,b}(Y)|^{p_1} |C_{a,b}(Z)|^{p_2}\, db \leq C a^s \right\}.$$

We call this function the mixed scaling function of the vector X. Now, as for the single case, we may show that near the Hölder exponent α of X, we have, in a small ball of size a, $|C_{a,b}(Y)| \sim a^{\alpha_1}$ and $|C_{a,b}(Z)| \sim a^{\alpha_2}$. From the definition of $D_X(\alpha)$, there are about $a^{-D_X(\alpha)}$ such balls, each of volume a, so that the contribution to the integral above is $a^{<\alpha,p>-D_X(\alpha)+1}$. The real order of magnitude of the integral should be given by the largest contribution, which yields $\eta_X(p)$ as the lower bound.

Using wavelet transform, the relation (9) should be replaced by a mixed extension as

$$\mu_X(I_{jk}) = |C_{j,k}(Y)|\, |C_{j,k}(Z)|\,, \tag{14}$$

where $C_{j,k}$ is the wavelet coefficient relatively to a discrete grid $a = 2^{-j}$ and $b = k2^{-j}$. The structure function $S_X(p,j)$ becomes

$$S_X(p,j) = 2^{-j} \sum_k |C_{j,k}(Y)|^{p_1} |C_{j,k}(Z)|^{p_2}\,. \tag{15}$$

However, as for the single case, the definition (14) above is not a true measure because of the fact that it is not necessarily increasing. The sophisticated formulation with the wavelet leaders may be then

$$S_X(p,j) = 2^{-j} \sum_k |W_{j,k}(Y)|^{p_1} |W_{j,k}(Z)|^{p_2}\,. \tag{16}$$

Hence, the wanted exponent η to handle the multifractal structure and thus compute the multifractal spectrum may be defined by

$$\eta_X(p) = \liminf_{j \to +\infty} \frac{\log(S_X(p,j))}{\log 2^j}.$$

The multifractal formalism is evaluated in a way that

$$D_X(\gamma) = \inf_p (1 - \eta_X(p) + <\gamma, p>).$$

6. Some Empirical Tests

To our knowledge, only a few papers have dealt with detrended fluctuation analysis of biological series such as DNA series and proteins ones. Some interesting results were obtained in the analysis. See for example [5,6,8–12, 14,15,17,18,20,38]. In this chapter, we focus instead on a special empirical example issued from financial trading.

In the first part, we analyze price–volume multifractal cross-correlation (MF-CC) between price change and volume change of Bitcoin (Fig. 1). The period chosen dates from February 1, 2015 to February 2, 2021 to investigate the multifractal behavior of price change and volume change series. Then, we used the multifractal detrended cross-correlation analysis method (MF-DCC) and multifractal detrended fluctuation analysis (MFDFA).

We denote p_t the closing price of the index on day t. In the present chapter, the method applied to the natural logarithmic returns of the index is defined by

$$r_t = \ln\left(\frac{p_t}{p_{t-1}}\right). \tag{17}$$

We have 2202 logarithm daily returns and daily volume changes of Bitcoin series.

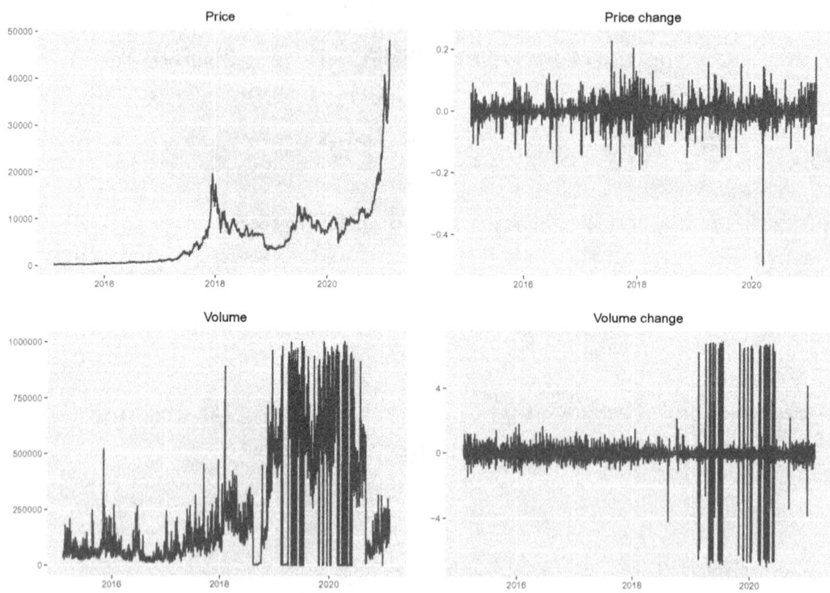

Fig. 1: Price, price change, volume and volume change of Bitcoin.

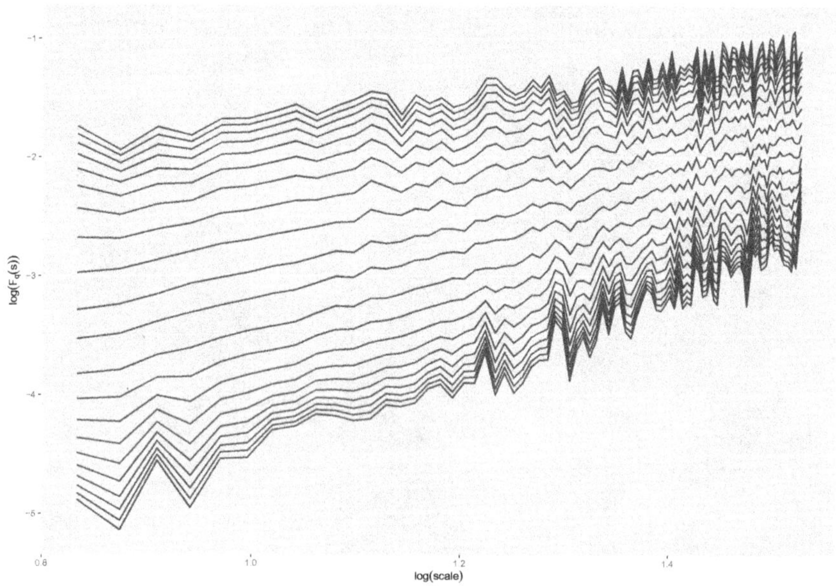

Fig. 2: Logarithmic fluctuation functions $F_q(s)$ vs. logarithmic time scales s of Bitcoin price–volume change.

Firstly, we use multifractal detrended fluctuation analysis algorithm, which is an MF-DCC method, to evaluate cross-correlation of price change and volume change of Bitcoin.

Figure 2 shows a correlated power law of price–volume multifractal cross-correlation of Bitcoin. Function of fluctuations $F_q(s)$ increases for large values of s, as a power law, which means that memory effect varies in function of scales s and moments q.

Using multifractal detrended fluctuation cross-correlation and MFDFA algorithms, we calculate the generalized Hurst exponent, then we deduce the Holder exponent and singularity spectrum. We present in Fig. 3 the plotting of $H_{XY}(q)$ of price change and volume change.

In Fig. 4, Holder exponent presents a nonlinear curve, which confirms the existence of multifractal cross-correlations between price and volume. Using MF-DFA, we find that both volume change series and price change series have a nonlinear curve.

$$\rho_{DCCA}(s) = \frac{F^2_{DCCA}(s)}{F_{DFA,x}(s)F_{DFA,y}(s)}.$$

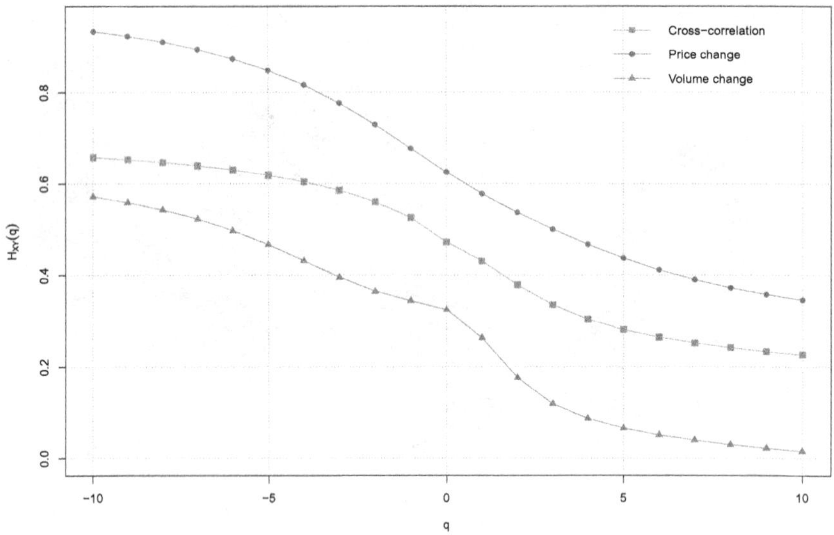

Fig. 3: Generalized Hurst exponent for different values of q.

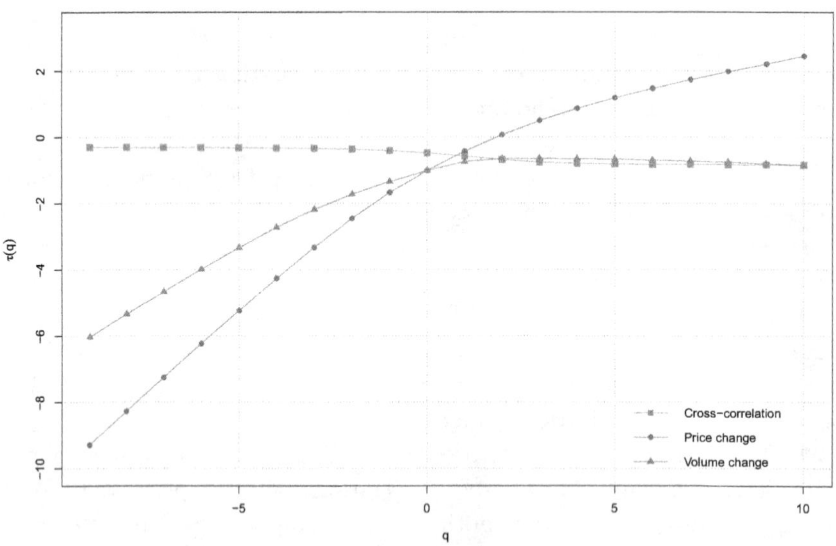

Fig. 4: Holder exponent for different values of q.

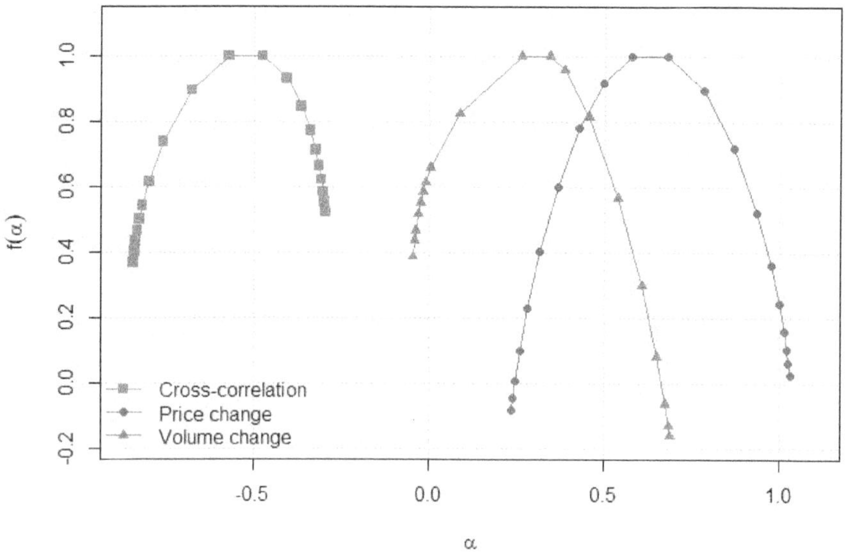

Fig. 5: The multifractal spectrum of singularities.

In the remaining part, we conduct a mixed wavelet multifractal detrended analysis of the well-known U.S. SP500 and the French CAC40 financial series. Recall that these indices are one of the best representatives of the economic situations in their countries. The SP500 is the stock market index composed of 500 large companies listed on stock exchanges. Similarly, the CAC40 is the most significant index representing the French market. Fig. 5 illustrates the multifractal spectra of the three series, the cross-correlation, the price change and the volume change. Fig. 6 illustrates the cross-correlation coefficient due to the price change and the volume change of the bitcoin series. In Figs. 7 and 8, we provided the wavelet decomposition of the SP500 series and the CAC40 series, resp., during the period from February 6, 2017 to March 2, 2021. We get a size $1024 = 2^12$ for both series. As previously, we applied a logarithmic return to both series as in equation (17). A decomposition level is fixed to $J = 3$, and the Daubechies wavelet Db8 is applied for the analysis. The figures show clearly the fluctuation character of both series.

Next, a single multifractal fluctuation analysis has been applied for both series separately, to show and confirm the multifractal fluctuation character of the series. Figure 9 illustrates the spectra of singularities of both series, and shows clearly the multifractal structure of both. To finish,

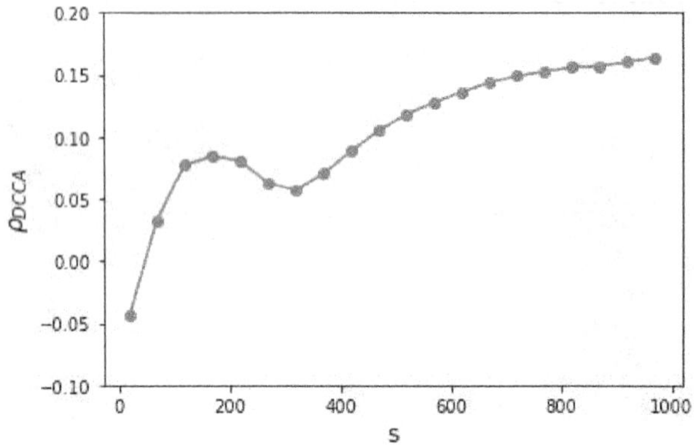

Fig. 6: DCCA cross-correlation coefficient between price change and volume change of Bitcoin.

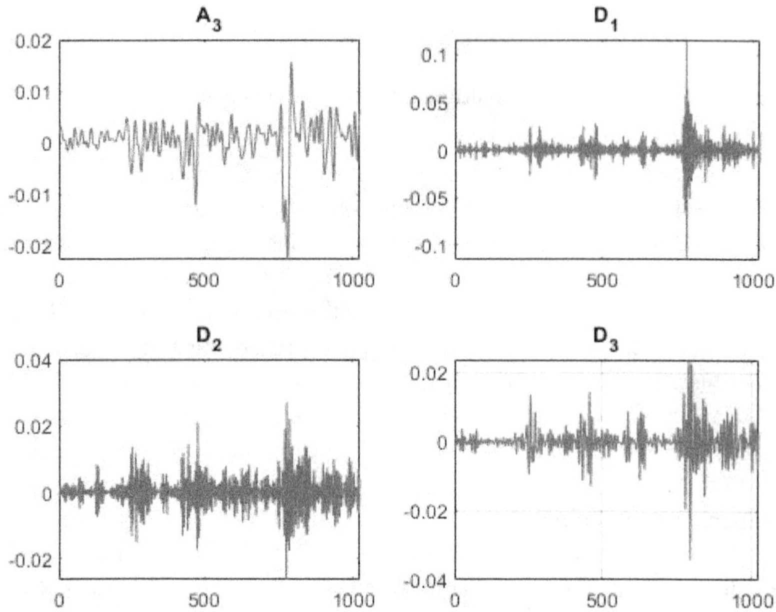

Fig. 7: The Db8-wavelet decomposition of the SP500 series at level 3.

we applied the well-known wavelet leaders method to compute the mixed wavelet multifractal spectrum of the vector-valued series (SP500,CAC40). This shows the role of the mixed analysis in detecting the fluctuation aspect of the series simultaneously. Figure 10 illustrates the result.

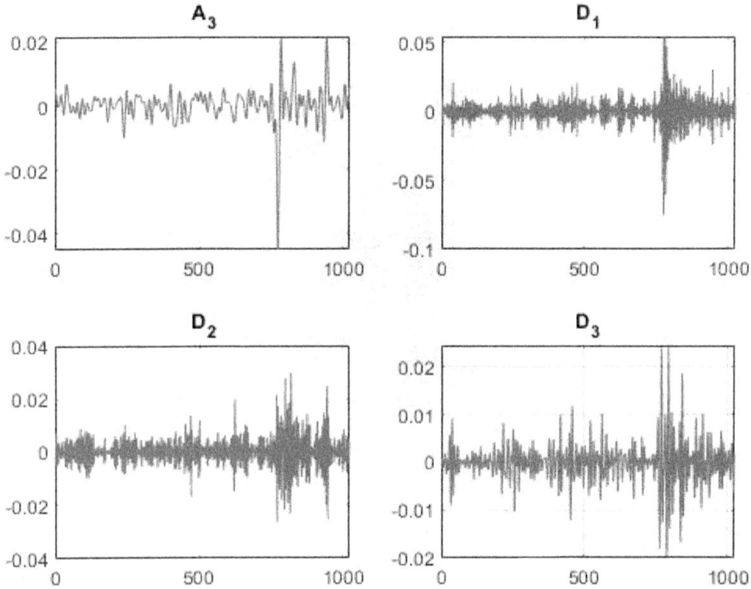

Fig. 8: The Db8-wavelet decomposition of the CAC40 series at level 3.

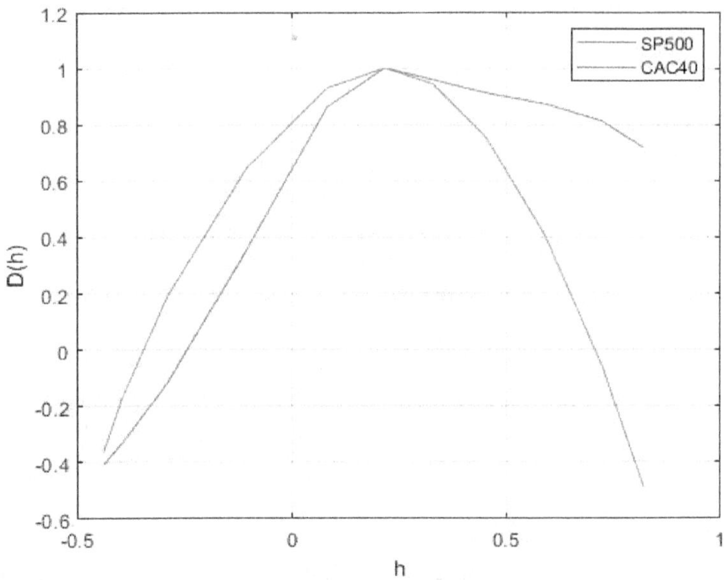

Fig. 9: The multifractal spectra of both series SP500 and CAC40.

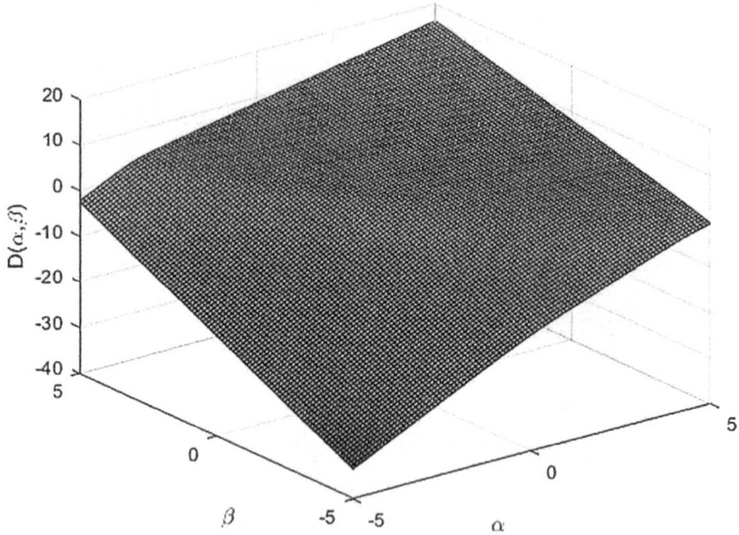

Fig. 10: The mixed wavelet multifractal spectrum of the series (SP500,CAC40).

7. Conclusion

In this chapter, detrended fluctuation analysis has been reviewed starting from the classical variant of single time/statistical series. Next, multifractal detrended fluctuation analysis has been recalled with necessary details and ideas. The natural extension of multifractal variant is the wavelet multifractal detrended fluctuation analysis which is based on the fact that wavelets since their discovery have been strongly related to fractals, especially in studying scaling laws hidden there. As a consequence, a new mixed detrended fluctuation analysis has been introduced in the present work extending all the existing variants. The originality consists in exploiting wavelet analysis, which splits the series into detail components, applying thus a type of mixed wavelet detrended multifractal analysis to the detail components of finitely many statistical series simultaneously. Future applications of the theoretical concepts developed here will be of interest, especially in simultaneous behaviors of markets, environmental factors, social indices, etc.

References

[1] A. Ben Mabrouk, M. Ben Slimane, and J. Aouidi, A wavelet multifractal formalism for simultaneous singularities of functions, *Int. J. Wavelets, Multiresoluti. Infor. Proc.* **12**(1), (2014).

[2] M. Ben Slimane, A. Ben Mabrouk, and J. Aouidi, Mixed multifractal analysis for functions: General upper bound and optimal results for vectors of self-similar or quasi-self-similar of functions and their superpositions, *Fractals*, **24**(4), 12 pp. (2016).

[3] M. Menceur, A. Ben Mabrouk, and K. Betina, The multifractal formalism for measures, review and extension to mixed cases, *Anal. Theory Appl.* **32**(1) 77–106, (2016).

[4] M. Menceur and A. Ben Mabrouk, A joint multifractal analysis of vector valued non Gibbs measures, *Chaos, Solitons Fract.* **126**, 203–217, (2019).

[5] P.A. Absil, R. Sepulchre, A. Bilge, and P. Gerard, Nonlinear analysis of cardiac rhythm fluctuations using DFA method, *Phys. A* **272**, 235–244, (1999).

[6] R.U. Acharya, C.M. Lim, and P. Joseph, Heart rate variability analysis using correlation dimension and detrended fluctuation analysis, *ITBM-RBM* 333–339, (2002).

[7] J. Alvarez-Ramirez, M. Cisneros, C. Ibarra-Valdez, and A. Soriano, Multifractal Hurst analysis of crude oil prices, *Phys. A* **313**, 651–670, (2002).

[8] A. Arneodo, C. Vaillant, B. Audit, Y. d'Aubenton-Carafa, and C. Thermes, La transformation en ondelettes continue: un microscope mathématique adapté à l'étude des propriétés d'invariance d'échelle et de corrélations à longue portée des séquences d'ADN. In: *19th GRETSI Symposium on Signal and Image Processing*, Vol. III, pp. 1–10, (Paris, 2003).

[9] A. Figliola, E. Serrano, G. Paccosi and M. Rosenblatt, About the effectiveness of different methods for the estimation of the multifractal spectrum of natural series, *Int. J. Bifurcation Chaos* **20**(2), 331–339, (2010).

[10] A.K. Golinska, Detrended fluctuation analysis in biomedical signal processing: Selected examples, *Studis in Logic, Grammar and Rhetoric*, **29**(42), 107–115, (2012).

[11] J.M. Lee, D.J. Kim, and I.Y. Kim, Detrended fluctuation analysis of EEG in sleep apnea using MIT/BIH polysomnography data, *Comput. Biol. Med.* **32** 37–47, (2002).

[12] J.M. Lee, D.J. Kim, and I.Y. Kim, Nonlinear analysis of human sleep EEG using detrended fluctuation analysis, *Med. Eng. Phys.* **26**, 773–776, (2004).

[13] R. Leonarduzzi, H. Wendt, P. Abry, S. Jaffard, and C. Melot, Finite resolution effects in p-leader multifractal analysis. arXiv:1612.01430.

[14] E. Oczeretko, M. Borowska, A. Kitlas, A. Borusiewicz, and M. Sobolewska-Siemieniuk, Fractal analysis of medical images in the irregular regions of interest. In: *Proceedings of the 8th IEEE International Conference on Bioinformatics and Bioengineering*, BIBE 2008, October 8–10, 2008, Athens, Greece, pp. 1–6.

[15] C.K. Peng, S.V. Buldyrev, A.L. Goldberger, S. Havlin, R.N. Mantegna, M. Simons, and H.E. Stanley, Statistical properties of DNA sequences, *Phys. A* **221**, 180–192, (1995).

[16] A. Phinyomark, M. Phothisonothai, C. Limsakul, and P. Phukpattaranont, Detrended fluctuation analysis of electromyography signal to identify hand movement, In: *The 2nd Biomedical Engineering International Conference* (BMEiCON) pp. 324–329, (2009).

[17] E. Rodriguez, J.C. Echeverria, and J. Alvarez-Ramirez, Detrended fluctuation analysis of heart intrabeat dynamics, *Phys. A* **384**, 429–438, (2007).

[18] J.-Y Yang, Z.-G Yu, and V. Anh, Clustering structures of large proteins using multifractal analyses based on a 6-letter model and hydrophobicity scale of amino acids, *Chaos, Solitons Fract.* **40**, 607–620, (2009).

[19] M.M. Ibrahim Mahmoud, A. Ben Mabrouk, and M.H.A. Hashim, Wavelet multifractal models for transmembrane proteins series. *Int. J. Wavelets Multires Infor. Proc.* **14**(6), 1650044, 36 pp, (2016).

[20] R.G. Yeh, J.S. Shieh, G.Y. Chen, and C.D. Kuo, Detrended fluctuation analysis of short-term heart rate variability in late pregnant women, *Auton. Neurosci.: Basic Clin.* **150**, 122–126, (2009).

[21] S. Arfaoui, A. Ben Mabrouk, and C. Carlo, *Wavelet Analysis, Basic Concepts and Applications* (CRC Press, Taylor & Francis Group, 2021), ISBN: 9780367562182.

[22] C. Azizieh, *Modélisation des series financieres par un modele multifractal* (Memoire d'actuaire, Universite Libre de Bruxelle, 2002).

[23] A. Arneodo, E. Bacry, and J. F. Muzy, Singularity spectrum of fractal signals from wavelet analysis: Exact results, *J. Statist. Phys.* **70**, 635–674, (1993).

[24] V.I. Bogachev, E.D. Kosov, and S.N. Popova, A new approach to Nikolskii–Besov classes. arXiv:1707.06477v1 [math.FA] July 20, 2017.

[25] M. Bownik and K.-P. Ho, Atomic and molecular decompositions of anisotropic triebel-lizorkin spaces, *Trans. Am. Math. Soc.* **358**(4), 1469–1510, (2005).

[26] F. Enriquez, A. Montes, and J. Perez, A characterization of structural Nikol'skii-Besov spaces using fractional derivatives, *Bol. Mat.* **17**(1), 77–98, (2010).

[27] G. Garrigos and A. Tabacco, Wavelet decompositions of anisotropic Besov spaces, *Mathematische Nachrichten* **239–240**(1), 80–102, (2002).

[28] N. Nikoloski, *Hardy Space* (Cambridge University Press, January 2019), Doi: 10.1017/9781316882108.

[29] H. Triebel, *Theory of Function Spaces. Monographs in Mathematic* (Birkhäuser Basel, 2006).

[30] H. Triebel, *Function Spaces and Wavelets on Domains* (European Mathematical Society, 2008).

[31] U. Frisch and G. Parisi, On the singularity structure of fully developed turbulence, In: R. Benzi, M. Ghill, and G. Parisi (Eds.), *Turbulence and Predictability in Geophysical Fluid Dynamics and Climate Dynamics* (North-Holland, New York), pp. 84–88.

[32] J.W. Kantelhardt, S.A. Zschiegner, E. Koscielny-Bunde, S. Havlin, A. Bunde and H.E. Stanley, Multifractal detrended fluctuation analysis of nonstationary time series, *Phys. A* **316**, 87–114, (2002).

[33] R. Benzi, L. Biferale, A. Crisanti, G. Paladin, M. Vergassolad, and A. Vulpiani, A random process for the construction of multiaffine fields, *Phys. D* **65**, 352–358, (1993).

[34] S.M. Miller and F.S. Rusek, Fiscal structures and economic growth: International evidence, *Econ. Inq.* **35**(3), 603–613, (1997).

[35] J.C. Pinheiro and D.M. *Bates, Mixed-effect Models in S and S-Plus* (Springer, New York, 2000).

[36] S.R. Searle, *Matrix Algebra Useful for Statistics* (Wiley, New York, 1982).

[37] S.R. Searle, Extending some results and proofs for the singular linear model. *Linear Algebra Appl.*, 4th Special Issue on Linear Algebra and Statistics, **210**, 139–151, (1994).

[38] S.R. Searle, On mixed models, REML and BLUP, In: *Biometrics Unit and Statistics Center*, Cornell University, Ithaca, NY, U.S.A. BU-1256-M, 15 pp (August 1994).

[39] G. Verbeke and G. Molenberghs, *Linear Mixed Models for Longitudinal Data.* (Springer Series in Statistics, 2000).

[40] J. Alvarez-Ramirez and R. Escarela-Perez, Time-dependent correlations in electricity markets, *Energy Economics* **32**(2), 269–277, (2010).

[41] J. Alvarez-Ramirez, R. Escarela-Perez, G. Espinoza-Perez, and R. Urrea, Dynamics of electricity market correlations, *Phys. A* **388**, 2173–2188, (2009).

[42] J. Alvarez-Ramirez, J. Alvarez and E. Rodriguez, Short-term predictability of crude oil markets: A detrended fluctuation analysis approach, *Energy Economics* **30**(5), 2645–2656, (2008).

[43] J.-P. Antoine, R. Murenzi, and P. Vandergheynst, Two-dimensional directional wavelets in image processing, *Int. J. Imaging Syst. Technol.* **7**(3), 152–165, (1996).

[44] S. Arfaoui, I. Rezgui, and A. Ben Mabrouk, *Wavelet Analysis on the Sphere: Spheroidal Wavelets. Walter de Gruyter* (March 20, 2017), ISBN-10: 311048109X, ISBN-13: 978-3110481099.

[45] S.-P. Chen and L.-Y. He, Multifractal spectrum analysis of nonlinear dynamical mechanisms in China's agricultural futures markets, *Phys. A* **389**, 1434–1444, (2010).

[46] V.D.A. Corino, F. Ziglio, and F. Lombardini, Detrended fluctuation analysis of atrial signal during adrengenic activation in atrial fibrillation, *Comput. Cardiol.* **33**, 141–144, (2006).

[47] L.-Y. He and S.-P. Chen, Are crude oil markets multifractal? Evidence from MF-DFA and MF-SSA perspectives, *Phys. A* **389**, 3218–3229, (2010).

[48] L.-Y. He and S.-P. Chen, Are developed and emerging agricultural futures markets multifractal? A comparative perspective, *Phys. A* **389**, 3828–3836, (2010).

[49] E.A.F. Ihlen, Introduction to multifractal detrended fluctuation analysis in matlab, *Front. Physiol.* **3**(141), 1–18, (2012).

[50] P. Norouzzadeh and G.R. Jafari, Application of multifractal measures to Tehran price index, *Phys. A* **356**, 609–627, (2005).
[51] P. Norouzzadeh and B. Rahmani, A multifractal detrended fluctuation description of Iranian rial–US dollar exchange rate, *Phys. A* **367**, 328–336, (2006).
[52] P. Norouzzadeh, W. Dullaert, and B. Rahmani, Anti-correlation and multifractal features of Spain electricity spot market, *Physica A: Statistical Mech. Appl.* **380**, 333–342, (2007).
[53] L. Olsen, Mixed generalized dimensions of self-similar measures, *J. Math. Anal. Appl.* **306**, 516–539, (2005).
[54] S.A.R. Rizvi, G. Dewandaru, O.I. Bacha and M. Masih, An analysis of stock market efficiency: Developed vs Islamic stock markets using MF-DFA, *Physica A: Statistical Mech. Appl.* **407**, 86–99, (2014).
[55] H.E. Stanley, L.A. Amaral, P. Gopikrishnan, P.Ch. Ivanov, T.H. Keitt, and V. Plerou, Scale invariance and universality: Organizing principles in complex systems, *Phys. A* **281**, 60–68, (2000).
[56] D. Stosic, D. Stosic, T. Stosic, and H.E. Stanley, Multifractal properties of price change and volume change of stock market indices, *Phys. A* **428**, 46–51, (2015).

https://doi.org/10.1142/9789811261572_0023

Chapter 23

Stability of Some Functional Equations on Restricted Domains

Abbas Najati*,‡, Mohammad B. Moghimi*,§, Batool Noori*, and Themistocles M. Rassias†,¶

*Department of Mathematics, Faculty of Sciences
University of Mohaghegh Ardabili, Ardabil, Iran
†Department of Mathematics, National Technical University of Athens
Zografou Campus, 15780 Athens, Greece
‡a.nejati@yahoo.com,
§mbfmoghimi@yahoo.com, noori.batool@yahoo.com
¶trassias@math.ntua.gr

In this work, we investigate the Hyers–Ulam stability for some functional equations on restricted domains. As a consequence, we obtain asymptotic behaviors of these functional equations.

1. Introduction

Ulam [1] in 1940 presented an intriguing and famous lecture that triggered the study of stability problems for various functional equations. He discussed a number of important unsolved mathematical problems. Among them, a question regarding the stability of homomorphisms between groups seemed too abstract for anyone to attain any conclusion. In fact, he asked the following question regarding the stability of homomorphisms:

Let $(G_1, *)$ *be a group and let* $(G_2, \diamond)$ *be a metric group with a metric* d. *Given* $\varepsilon > 0$, *does there exist a* $\delta > 0$ *such that if a function* $f : G_1 \to G_2$ *satisfies the inequality* $d(f(x*y), f(x)\diamond f(y)) < \delta$ *for all* $x, y \in G_1$; *then there is a homomorphism* $h : G_1 \to G_2$ *with* $d(f(x), h(x)) < \varepsilon$ *for all* $x \in G_1$?

If the answer is affirmative, the functional equation of homomorphisms is called stable. In 1941, Hyers [2] was able to give a partial solution to the Ulam question, which was the first significant step forward and a step toward further solutions in this area. He was the first mathematician to present the result concerning the stability of functional equations. He masterly answered the question of Ulam for the case where G_1 and G_2 are assumed to be Banach spaces. Later, the result of Hyers was significantly generalized by Aoki [3] and Rassias [4] (see also [5]). During the last decades, several stability problems of functional equations have been investigated by several mathematicians. A large list of references concerning the stability of functional equations can be found in Refs. [6–16].

It will also be interesting to study the stability problems of the additive Cauchy equation on a restricted domain. More precisely, the goal is whether there is a true additive function in the neighborhood of a function f which only satisfies $\|f(x+y)-f(x)-f(y)\| \leqslant \varepsilon$ in a restricted domain. Skof [17] proved the following theorem and applied the result to the study of an asymptotic behavior of additive functions.

Theorem 1. *Let E be a Banach space, and let $d>0$ be a given constant. Suppose a function $f:\mathbb{R}\to E$ satisfies the inequality*

$$\|f(x+y)-f(x)-f(y)\| \leqslant \varepsilon, \quad |x|+|y|>d$$

for some $\varepsilon \geqslant 0$. Then there exists a unique additive function $A:\mathbb{R}\to E$ such that

$$\|f(x)-A(x)\| \leqslant 9\varepsilon, \quad x\in\mathbb{R}.$$

Using this theorem, Skof [17] investigated an interesting asymptotic behavior of additive functions, as we see in the following theorem.

Theorem 2. *Let X and $\mathcal{Y}$ be a normed space and a Banach space, resp. Suppose z is a fixed point of X. For a function $f:X\to\mathcal{Y}$, the following two conditions are equivalent:*

(i) $f(x+y)-f(x)-f(y)\to z$ *as* $\|x\|+\|y\|\to\infty$.
(ii) $f(x+y)-f(x)-f(y)=z$ *for all* $x,y\in X$.

Among the normed linear spaces, inner product spaces play an important role. In an inner product space E the parallelogram law is an algebraic identity, i.e.

$$\|x+y\|^2+\|x-y\|^2=2\|x\|^2+2\|y\|^2, \quad x,y\in E.$$

This translates into a functional equation well-known as the quadratic functional equation

$$f(x+y)+f(x-y)=2f(x)+2f(y), \quad x,y\in X,$$

where X is a linear space. Most mathematicians may be interested in the study of the quadratic functional equation since the quadratic functions are applied to almost every field of mathematics.

Skof [17] was the first person who proved the Hyers–Ulam stability of the quadratic functional equation for the functions $f : X \to \mathcal{Y}$, where X is a normed space and $\mathcal{Y}$ is a Banach space. In 1998, Jung [13] investigated the Hyers–Ulam stability of the quadratic functional equation on the unbounded restricted domains. For more detailed information on the stability of the Cauchy and quadratic functional equations, we can refer to [18–26].

2. An Additive Functional Inequality

Theorem 3. *Let $f : X \to Y$ and $f(0) = 0$. If*

$$\|f(x)+f(y)+f(z)\| \leqslant \left\| rf\left(\frac{x+y+z}{r}\right)\right\|, \quad \|x\|+\|y\|+\|z\| \geqslant d. \tag{1}$$

Then f is additive.

Proof. Letting $z=-x-y$ in (1), we get

$$f(x)+f(y)+f(-x-y)=0, \quad \|x\|+\|y\|\geqslant d. \tag{2}$$

Therefore,

$$f(-x)=-f(x), \quad \|x\|\geqslant d. \tag{3}$$

It follows from (2) and (3) that

$$f(x+y)=f(x)+f(y), \quad \|x+y\|\geqslant d. \tag{4}$$

Let $x\in X$ and choose $y\in X$ such that $\|y\|\geqslant d+\|x\|$. It is clear that $\|x+y\|\geqslant d$. So (3) and (4) imply that

$$f(-x)-f(y)=f(-x)+f(-y)=f(-x-y)=-f(x+y)=-f(x)-f(y).$$

Hence, $f(-x)=-f(x)$. This means f is odd. We now show f is additive. Let $x,y\in X\setminus\{0\}$ such that $x+y\neq 0$. We can choose a positive integer n

such that $\min\{\|nx\|, \|ny\|, \|n(x+y)\|\} \geqslant d$. Then (4) yields

$$\begin{aligned} f(x+y)+f(n(x+y)) &= f((n+1)(x+y)) \\ &= f(nx+x)+f(ny+y) \\ &= [f(nx)+f(x)]+[f(ny)+f(y)] \\ &= [f(x)+f(y)]+[f(nx)+f(ny)] \\ &= [f(x)+f(y)]+f(n(x+y)). \end{aligned}$$

Hence,

$$f(x+y) = f(x)+f(y), \quad x, y, x+y \neq 0. \tag{5}$$

Since f is odd, we infer that (5) holds true for all $x, y \in X$. This means f is additive. □

Theorem 4. *Let X be a linear normed space and $\mathcal{Y}$ a Banach space. Suppose that $f : X \to \mathcal{Y}$ satisfies*

$$\|f(x)+f(y)+f(z)\| \leqslant \left\|3f\left(\frac{x+y+z}{3}\right)\right\| + \varepsilon, \quad \|x\|+\|y\|+\|z\| \geqslant d \tag{6}$$

for some $d > 0$. Then there exists a unique additive function $A : X \to \mathcal{Y}$ such that

$$\|f(x)-A(x)\| \leqslant 18\|f(0)\| + 6\varepsilon, \quad x \in X. \tag{7}$$

Proof. Letting $z = -x-y$ in (6), we get

$$\|f(x)+f(y)+f(-x-y)\| \leqslant 3\|f(0)\| + \varepsilon, \quad \|x\|+\|y\| \geqslant d. \tag{8}$$

Therefore,

$$\|f(x)+f(-x)\| \leqslant 3\|f(0)\| + \varepsilon, \quad \|x\| \geqslant d. \tag{9}$$

By inequalities (8) and (9), we obtain

$$\|f(x+y)-f(x)-f(y)\| \leqslant 6\|f(0)\| + 2\varepsilon, \quad \|x+y\| \geqslant d. \tag{10}$$

Putting $y = x$ in (10), we get $\|f(2x)-2f(x)\| \leqslant 6\|f(0)\|+2\varepsilon$ for all $\|x\| \geqslant d$. Then

$$\left\|\frac{f(2^{n+1}x)}{2^{n+1}} - \frac{f(2^m x)}{2^m}\right\| \leqslant \sum_{k=m}^{n} \frac{3\|f(0)\|+\varepsilon}{2^k}, \quad \|x\| \geqslant d. \tag{11}$$

This implies that the sequence $\{\frac{f(2^n x)}{2^n}\}_n$ is Cauchy for all $x \in X$. Define $A : X \to \mathcal{Y}$ by

$$A(x) := \lim_{n\to\infty} \frac{f(2^n x)}{2^n}, \quad x \in X.$$

In view of the definition of A, (10) implies that $A(x+y) = A(x) + A(y)$ for all $x, y \in X$ with $\|x+y\| \geqslant d$. Hence, according to the argument given in proof of Theorem 3, it follows that A is additive. Letting $m = 0$ and allowing n tending to infinity in (11), we get

$$\|f(x) - A(x)\| \leqslant 6\|f(0)\| + 2\varepsilon, \quad \|x\| \geqslant d. \tag{12}$$

To extend (12) to the whole X, let $x \in X$ and choose $y \in X$ such that $\|y\| \geqslant d + \|x\|$. Then $\|x+y\| \geqslant d$, and (12) yields

$$\|f(y) - A(y)\| \leqslant 6\|f(0)\| + 2\varepsilon$$

$$\text{and} \quad \|f(x+y) - A(x+y)\| \leqslant 6\|f(0)\| + 2\varepsilon.$$

Using these inequalities together with (10), we obtain

$$\|A(x+y) - A(y) - f(x)\| \leqslant 18\|f(0)\| + 6\varepsilon.$$

Since A is additive, we get (7). The uniqueness of A follows easily from (7). □

Corollary 1. *Let X and Y be linear normed spaces. Suppose that $f : X \to Y$ satisfies the asymptotic behavior*

$$\lim_{\|x\|+\|y\|+\|z\|\to\infty} \left| \left\| 3f\left(\frac{x+y+z}{3}\right) \right\| - \|f(x) + f(y) + f(z)\| \right| = 0. \tag{13}$$

Then f is affine.

Proof. Define $g : X \to Y$ by $g := f - f(0)$. Then (13) yields

$$\lim_{\|x\|+\|y\|+\|z\|\to\infty} \left| \left\| 3g\left(\frac{x+y+z}{3}\right) \right\| - \|g(x) + g(y) + g(z)\| \right| = 0 \tag{14}$$

Let $\varepsilon > 0$ be an arbitrary real number. By (14) there exists $d_\varepsilon > 0$ such that

$$\|g(x) + g(y) + g(z)\| \leqslant \left\| 3g\left(\frac{x+y+z}{3}\right) \right\| + \varepsilon, \quad \|x\| + \|y\| + \|z\| \geqslant d_\varepsilon.$$

Let $\mathcal{Y}$ be the completion of Y. In view of $g(0) = 0$, by Theorem 4 there exists a unique additive function $A_\varepsilon : X \to \mathcal{Y}$ such that

$$\|g(x) - A_\varepsilon(x)\| \leqslant 6\varepsilon, \quad x \in X.$$

Then

$$\begin{aligned}\|g(x+y) - g(x) - g(y)\| &\leqslant \|g(x+y) - A_\varepsilon(x+y)\| + \|g(x) - A_\varepsilon(x)\| \\ &\quad + \|g(y) - A_\varepsilon(y)\| \leqslant 18\varepsilon, \quad x, y \in X.\end{aligned}$$

Since ε is arbitrary, we get g is additive. Hence, f is affine and this completes the proof. □

3. Fréchet's Result

First, let us recall a result from Fréchet [27].

Theorem 5 (Fréchet's Result). *If a continuous function $g : \mathbb{R} \to \mathbb{R}$ satisfies*

$$\sum_{j=0}^{n} \binom{n}{j} (-1)^{n-j} g(x + jy) = 0, \quad x, y \in \mathbb{R}, \tag{†}$$

then g is a polynomial of degree $< n$.

In Ref. [15, Theorem 7.20], the general solution of (†) was obtained for $n = 3$ without assuming any regularity condition on g, where g is a function between two linear spaces. Indeed, we have the following results:

Lemma 1. *Let $g : X \to Y$ be an odd function between linear spaces X and Y. If*

$$g(x + 3y) - 3g(x + 2y) + 3g(x + y) - g(x) = 0, \quad x, y \in X,$$

then g is additive.

Lemma 2. *Let $g : X \to Y$ be an even function between linear spaces X and Y. If*

$$g(x + 3y) - 3g(x + 2y) + 3g(x + y) - g(x) = 0, \quad x, y \in X,$$

then $g - g(0)$ is quadratic.

Theorem 6. *Suppose $g : X \to Y$ satisfies*

$$g(x + 3y) - 3g(x + 2y) + 3g(x + y) - g(x) = 0, \quad x, y \in X.$$

Then g has the form

$$g(x) = g(0) + A(x) + Q(x), \quad x \in X,$$

where $A : X \to Y$ is additive and $Q : X \to Y$ is quadratic.

Now, we prove the Hyers–Ulam stability of (†) (for $n = 3$) on restricted domains.

In this section, X, Y are normed linear spaces and $\mathcal{Y}$ is a Banach space.

Theorem 7. *Suppose that $\varepsilon \geqslant 0$ and $g : X \to Y$ is an odd function that satisfies*

$$\|g(x + 3y) - 3g(x + 2y) + 3g(x + y) - g(x)\| \leqslant \varepsilon, \quad \|x + y\| \geqslant d, \tag{15}$$

for some $d > 0$. Then there exists a unique additive function $A : X \to \mathcal{Y}$ such that

$$\|A(x) - g(x)\| \leqslant 8\varepsilon, \quad x \in X. \tag{16}$$

Proof. Letting $x = -2y$ in (15), we get

$$\|g(2y) - 2g(y)\| \leqslant \varepsilon, \quad \|y\| \geqslant d.$$

Then

$$\left\| \frac{g(2^{n+1}y)}{2^{n+1}} - \frac{g(2^m y)}{2^m} \right\| \leqslant \sum_{k=m}^{n} \frac{\varepsilon}{2^{k+1}}, \quad \|y\| \geqslant d. \tag{17}$$

This implies that the sequence $\{\frac{g(2^n y)}{2^n}\}_n$ is Cauchy for all $y \in X$. Define $A : X \to \mathcal{Y}$ by

$$A(y) := \lim_{n\to\infty} \frac{g(2^n y)}{2^n}, \quad y \in X.$$

Using the definition of A, (15) yields

$$A(x+3y) - 3A(x+2y) + 3A(x+y) - A(x) = 0, \quad x+y \neq 0.$$

Since g is odd, we infer that A is odd and then the equation above holds true for all $x, y \in X$. Hence, A is additive by Lemma 1. If $m = 0$ and n goes to infinity, then (17) yields

$$\|g(y) - A(y)\| \leqslant \varepsilon, \quad \|y\| \geqslant d. \tag{18}$$

To extend (18) to the whole X, let $x \in X$ and choose $y \in X$ such that $\|y\| \geqslant d + \|x\|$. Then $\min\{\|x+y\|, \|x+2y\|, \|x+3y\|\} \geqslant d$, and (18) yields

$$\|A(x+3y) - g(x+3y)\| \leqslant \varepsilon,$$
$$\|3g(x+2y) - 3A(x+2y)\| \leqslant 3\varepsilon,$$
$$\|3A(x+y) - 3g(x+y)\| \leqslant 3\varepsilon.$$

Using these inequalities together with (15), we obtain

$$\|A(x+3y) - 3A(x+2y) + 3A(x+y) - g(x)\| \leqslant 8\varepsilon.$$

Since A is additive, we get

$$\|A(x) - g(x)\| \leqslant 8\varepsilon, \quad x \in X.$$

The uniqueness of A follows easily from (16). □

Theorem 8. *Suppose that* $\varepsilon \geqslant 0$ *and* $g : X \to Y$ *is an even function that satisfies* (15) *for some* $d > 0$. *Then there exists a unique quadratic function* $Q : X \to \mathcal{Y}$ *such that*

$$\|Q(x) - g(x) - g(0)\| \leqslant \frac{10\varepsilon}{3}, \quad x \in X. \tag{19}$$

Proof. Letting $x = -2y$ in (15) and using the evenness of g, we get

$$\|g(2y) - 4g(y) + 3g(0)\| \leqslant \varepsilon, \quad \|y\| \geqslant d.$$

Then

$$\left\| \frac{g(2^{n+1}y)}{4^{n+1}} - \frac{g(2^m y)}{4^m} + 3\sum_{k=m}^{n} \frac{g(0)}{4^{k+1}} \right\| \leqslant \sum_{k=m}^{n} \frac{\varepsilon}{4^{k+1}}, \quad \|y\| \geqslant d. \tag{20}$$

This implies that the sequence $\{\frac{g(2^n y)}{4^n}\}_n$ is Cauchy for all $y \in X$. Define $Q : X \to \mathcal{Y}$ by

$$Q(y) := \lim_{n\to\infty} \frac{g(2^n y)}{4^n}, \quad y \in X.$$

By using the definition of Q, (15) yields

$$Q(x+3y) - 3Q(x+2y) + 3Q(x+y) - Q(x) = 0, \quad x + y \neq 0.$$

Since $Q(0) = 0$, $Q(2y) = 4Q(y)$ and Q is even, we infer that the equation above holds true for all $x, y \in X$. Hence, Q is quadratic by Lemma 2. If $m = 0$ and n goes to infinity, then (20) yields

$$\|g(y) - Q(y) + g(0)\| \leqslant \frac{\varepsilon}{3}, \quad \|y\| \geqslant d. \tag{21}$$

To extend (21) to the whole X, let $x \in X$ and choose $y \in X$ such that $\|y\| \geqslant d + \|x\|$. Then $\min\{\|x+y\|, \|x+2y\|, \|x+3y\|\} \geqslant d$, and (21) yields

$$\|Q(x+3y) - g(x+3y) - g(0)\| \leqslant \frac{\varepsilon}{3},$$

$$\|3g(x+2y) - 3Q(x+2y) + 3g(0)\| \leqslant \varepsilon,$$

$$\|3Q(x+y) - 3g(x+y) - 3g(0)\| \leqslant \varepsilon.$$

Using these inequalities together with (15), we obtain

$$\|Q(x+3y) - 3Q(x+2y) + 3Q(x+y) - g(x) - g(0)\| \leqslant \frac{10\varepsilon}{3}.$$

Since Q satisfies $Q(x+3y) - 3Q(x+2y) + 3Q(x+y) - Q(x) = 0$, we get

$$\|Q(x) - g(x) - g(0)\| \leqslant \frac{10\varepsilon}{3}, \quad x \in X.$$

The uniqueness of Q follows easily from (19). □

Theorem 9. *Suppose that $\varepsilon \geqslant 0$ and $g : X \to \mathcal{Y}$ is a function that satisfies (15) for some $d > 0$. Then there exist a unique additive $A : X \to \mathcal{Y}$ and a unique quadratic function $Q : X \to \mathcal{Y}$ such that*

$$\|A(x) + Q(x) - g(x) - g(0)\| \leqslant \frac{34\varepsilon}{3}, \quad x \in X.$$

Proof. We know every function $f : X \to Y$ can be written as $f(x) = f_e(x) + f_o(x)$ for all $x \in X$, where $f_e(x) = \frac{f(x)+f(-x)}{2}$ is called the even part of f and $f_o(x) = \frac{f(x)-f(-x)}{2}$ is called the odd part of f. It is clear that f_e is even and f_o is odd.

It is easy to see that g_e and g_o satisfy (15). By Theorems 7 and 8, there exist a unique additive $A : X \to \mathcal{Y}$ and a unique quadratic function $Q : X \to \mathcal{Y}$ such that

$$\|Q(x) - g_e(x) - g(0)\| \leqslant \frac{10\varepsilon}{3} \quad \text{and} \quad \|A(x) - g_o(x)\| \leqslant 8\varepsilon, \quad x \in X.$$

Then

$$\|A(x) + Q(x) - g(x) - g(0)\| \leqslant \frac{34\varepsilon}{3}, \quad x \in X. \qquad \square$$

Remark 1. Since

$$\{(x, y) \in X \times X : \|x + y\| \geqslant d\} \subseteq \{(x, y) \in X \times X : \|x\| + \|y\| \geqslant d\},$$

the above results remain valid if the condition $\|x+y\| \geqslant d$ in (15) is replaced by $\|x\| + \|y\| \geqslant d$.

Now, we can prove the following corollary concerning an asymptotic property of the mixed functional equation

$$f(x + 3y) - 3f(x + 2y) + 3f(x + y) - f(x) = 0.$$

Corollary 2. *Suppose that $f : X \to Y$ satisfies one of the following asymptotic behaviors:*

(i) $\lim_{\|x\|+\|y\|\to\infty} [f(x + 3y) - 3f(x + 2y) + 3f(x + y) - f(x)] = 0.$
(ii) $\lim_{\|x+y\|\to\infty} [f(x + 3y) - 3f(x + 2y) + 3f(x + y) - f(x)] = 0.$

Then f has the form $f = f(0) + A + Q$, where $A : X \to Y$ is additive and $Q : X \to Y$ is quadratic.

4. A Mixed Functional Equation

In 2010, Najati and Zamani [22] considered the following mixed type functional equation:

$$\sum_{i=1}^{k} f\left(2x_i + \sum_{j=1, j\neq i}^{k} x_j\right) = (k + 2)f\left(\sum_{i=1}^{k} x_j\right) + \sum_{i=1}^{k} f(-x_i), \quad (k \geqslant 2). \tag{✠}$$

It is easy to show that the function $f : \mathbb{R} \to \mathbb{R}$ given by $f(x) = ax + bx^2$ is a solution of the functional equation (✠), where a, b are arbitrary constants. They established the general solution of the functional equation (✠), and then proved the generalized Hyers–Ulam stability of the functional equation (✠) in Banach modules over a unital Banach algebra. For $k = 2$ in (✠), we obtain the functional equation

$$f(2x + y) + f(x + 2y) = 4f(x + y) + f(-x) + f(-y). \tag{22}$$

In this section, we prove the Hyers–Ulam stability of (22) on restricted domains.

We start with the following results from Ref. [22].

Lemma 3. *Let $f : X \to Y$ be an odd function between linear spaces X and Y. If f satisfies* (22), *then f is additive.*

Lemma 4. *Let $f : X \to Y$ be an even function between linear spaces X and Y. If f satisfies* (22), *then f is quadratic.*

Theorem 10. *Suppose $f : X \to Y$ satisfies* (22). *Then f has the form*

$$f(x) = A(x) + Q(x), \quad x \in X$$

where $A : X \to Y$ is additive and $Q : X \to Y$ is quadratic.

Now, we prove the Hyers–Ulam stability of (22) on restricted domains.

In this section X, Y are normed linear spaces and $\mathcal{Y}$ is a Banach space.

Theorem 11. *Suppose that $\varepsilon \geqslant 0$ and $f : X \to Y$ is an odd function that satisfies*

$$\|f(2x + y) + f(x + 2y) - 4f(x + y) - f(-x) - f(-y)\| \leqslant \varepsilon, \quad \|x + y\| \geqslant d \tag{23}$$

for some $d > 0$. Then there exists a unique additive function $A : X \to \mathcal{Y}$ such that

$$\|A(x) - f(x)\| \leqslant 8\varepsilon, \quad x \in X. \tag{24}$$

Proof. Letting $y = 0$ in (23), we get

$$\|f(2x) - 2f(x)\| \leqslant \varepsilon, \quad \|x\| \geqslant d.$$

Then

$$\left\| \frac{f(2^{n+1}x)}{2^{n+1}} - \frac{f(2^m x)}{2^m} \right\| \leqslant \sum_{k=m}^{n} \frac{\varepsilon}{2^{k+1}}, \quad \|x\| \geqslant d. \tag{25}$$

This implies that the sequence $\{\frac{f(2^n x)}{2^n}\}_n$ is Cauchy for all $x \in X$. Define $A : X \to \mathcal{Y}$ by

$$A(x) := \lim_{n\to\infty} \frac{f(2^n x)}{2^n}, \quad x \in X.$$

Using the definition of A, (23) yields

$$A(2x + y) + A(x + 2y) = 4A(x + y) + A(-x) + A(-y), \quad x + y \neq 0.$$

Since f is odd, we infer that A is odd and then the equation above holds true for all $x, y \in X$. Hence, A is additive by Lemma 3. If $m = 0$ and n goes to infinity, then (25) yields

$$\|f(x) - A(x)\| \leqslant \varepsilon, \quad \|x\| \geqslant d. \tag{26}$$

To extend (26) to the whole X, let $x \in X$ and choose $y \in X$ such that $\|y\| \geqslant d + 2\|x\|$. Then $\min\{\|2x + y\|, \|x + 2y\|, \|x + y\|\} \geqslant d$, and (26) yields

$$\|A(2x + y) - f(2x + y)\| \leqslant \varepsilon,$$
$$\|A(x + 2y) - f(x + 2y)\| \leqslant \varepsilon,$$
$$\|4f(x + y) - 4A(x + y)\| \leqslant 4\varepsilon,$$
$$\|f(-y) - A(-y)\| \leqslant \varepsilon.$$

Using these inequalities together with (23), we obtain

$$\|A(2x + y) + A(x + 2y) - 4A(x + y) - A(-y) - f(-x)\| \leqslant 8\varepsilon.$$

Since A is additive, we get

$$\|f(x) - A(x)\| \leqslant 8\varepsilon, \quad x \in X.$$

The uniqueness of A follows easily from (24). □

Theorem 12. *Suppose that $\varepsilon \geqslant 0$ and $f : X \to Y$ is an even function satisfies (23) for some $d > 0$. Then there exists a unique quadratic function $Q : X \to \mathcal{Y}$ such that*

$$\|Q(x) - f(x) + f(0)\| \leqslant \frac{10\varepsilon}{3}, \quad x \in X. \tag{27}$$

Proof. Letting $y = 0$ in (23) and using the evenness of g, we get

$$\|f(2x) - 4f(x) - f(0)\| \leqslant \varepsilon, \quad \|x\| \geqslant d.$$

Then

$$\left\| \frac{f(2^{n+1}x)}{4^{n+1}} - \frac{f(2^m x)}{4^m} - \sum_{k=m}^{n} \frac{f(0)}{4^{k+1}} \right\| \leqslant \sum_{k=m}^{n} \frac{\varepsilon}{4^{k+1}}, \quad \|x\| \geqslant d. \tag{28}$$

This implies that the sequence $\{\frac{f(2^n x)}{4^n}\}_n$ is Cauchy for all $x \in X$. Define $Q : X \to \mathcal{Y}$ by

$$Q(x) := \lim_{n\to\infty} \frac{f(2^n x)}{4^n}, \quad x \in X.$$

By using the definition of Q, (23) yields

$$Q(2x+y) + Q(x+2y) - 4Q(x+y) - Q(-x) - Q(-y) = 0, \quad x+y \neq 0.$$

Since $Q(0) = 0$ and Q is even, we infer that the equation above holds true for all $x, y \in X$. Hence, Q is quadratic by Lemma 4. If $m = 0$ and n goes to infinity, then (28) yields

$$\left\| Q(x) - f(x) - \frac{f(0)}{3} \right\| \leqslant \frac{\varepsilon}{3}, \quad \|x\| \geqslant d. \tag{29}$$

To extend (29) to the whole X, let $x \in X$ and choose $y \in X$ such that $\|y\| \geqslant d + 2\|x\|$. Then $\min\{\|2x+y\|, \|x+2y\|, \|x+y\|\} \geqslant d$, and (29) yields

$$\left\| Q(2x+y) - f(2x+y) - \frac{f(0)}{3} \right\| \leqslant \frac{\varepsilon}{3},$$

$$\left\| Q(x+2y) - f(x+2y) - \frac{f(0)}{3} \right\| \leqslant \frac{\varepsilon}{3},$$

$$\left\| 4f(x+y) - 4Q(x+y) + \frac{4f(0)}{3} \right\| \leqslant \frac{4\varepsilon}{3},$$

$$\left\| f(-y) - Q(-y) + \frac{f(0)}{3} \right\| \leqslant \frac{\varepsilon}{3}.$$

Using these inequalities together with (23), we obtain

$$\|Q(2x+y) + Q(x+2y) - 4Q(x+y) - Q(-y) - f(-x) + f(0)\| \leqslant \frac{10\varepsilon}{3}.$$

Since Q is even and satisfies $Q(2x+y) + Q(x+2y) - 4Q(x+y) - Q(-x) - Q(-y) = 0$, we get

$$\|Q(x) - f(x) + f(0)\| \leqslant \frac{10\varepsilon}{3}, \quad x \in X.$$

The uniqueness of Q follows easily from (27). □

Theorem 13. *Suppose that $\varepsilon \geqslant 0$ and $f : X \to \mathcal{Y}$ is a function that satisfies (23) for some $d > 0$. Then there exist a unique additive $A : X \to \mathcal{Y}$ and a unique quadratic function $Q : X \to \mathcal{Y}$ such that*

$$\|A(x) + Q(x) - f(x) + f(0)\| \leqslant \frac{34\varepsilon}{3}, \quad x \in X.$$

Proof. Let f_e and f_o be the even and odd parts of f, resp. It is easy to see that f_e and f_o satisfy (23). By Theorems 11 and 12, there exist a unique additive $A : X \to \mathcal{Y}$ and a unique quadratic function $Q : X \to \mathcal{Y}$ such that

$$\|Q(x) - f_e(x) + f(0)\| \leqslant \frac{10\varepsilon}{3} \quad \text{and} \quad \|A(x) - f_o(x)\| \leqslant 8\varepsilon, \quad x \in X.$$

Then

$$\|A(x) + Q(x) - f(x) + f(0)\| \leqslant \frac{34\varepsilon}{3}, \quad x \in X. \qquad \square$$

Remark 2. Since

$$\{(x, y) \in X \times X : \ \|x + y\| \geqslant d\} \subseteq \{(x, y) \in X \times X : \ \|x\| + \|y\| \geqslant d\},$$

the above results remain valid if the condition $\|x+y\| \geqslant d$ in (23) is replaced by $\|x\| + \|y\| \geqslant d$.

Now, we can prove the following corollary concerning an asymptotic property of the mixed functional equation

$$f(2x + y) + f(x + 2y) = 4f(x + y) + f(-x) + f(-y).$$

Corollary 3. *Suppose that $f : X \to Y$ satisfies one of the following asymptotic behaviors*

(i) $\lim_{\|x\| + \|y\| \to \infty}[f(2x + y) + f(x + 2y) - 4f(x + y) - f(-x) - f(-y)] = 0.$
(ii) $\lim_{\|x+y\| \to \infty}[f(2x + y) + f(x + 2y) - 4f(x + y) - f(-x) - f(-y)] = 0.$

Then f has the form $f = f(0) + A + Q$. where $A : X \to Y$ is additive and $Q : X \to Y$ is quadratic.

References

[1] S.M. Ulam, *A Collection of Mathematical Problems* (Interscience, New York, 1960).

[2] D.H. Hyers, On the stability functional equation, *Proc. Natl. Acad. Sci. USA* **27**, 222–224, (1941).

[3] T. Aoki, On the stability of the linear transformation in Banach spaces, *J. Math. Soc. Jpn.* **2**, 64–66, (1950).

[4] Th.M. Rassias, On the stability of the linear mapping in Banach spaces, *Proc. Amer. Math. Soc.* **72** 297–300, (1978).

[5] D.G. Bourgin, Classes of transformations and bordering transformations, *Bull. Amer. Math. Soc.* **57**, 223–237, (1951).

[6] M.R. Abdollahpour and M.Th. Rassias, Hyers–Ulam stability of hypergeometric differential equations, *Aequationes Math.* **93**(4), 691–698, (2019).
[7] M.R. Abdollahpour and A. Najati, Stability of linear differential equations of third order, *Appl. Math. Lett.* **24**(11), 1827–1830, (2011).
[8] J. Aczél and J. Dhombres, *Functional Equations in Several Variables* (Cambridge University Press, Cambridge, 1989).
[9] S. Czerwik, On the stability of the quadratic mapping in normed spaces, *Abh. Math. Sem. Univ. Hamburg* **62**, 59–64, (1992).
[10] S. Czerwik, *Functional Equations and Inequalities in Several Variables* (World Scientific Publishing Company, New Jersey, Hong Kong, Singapore and London, 2002).
[11] H. Haruki and Th.M. Rassias, New generalizations of Jensen's functional equation, *Proc. Amer. Math. Soc.* **123**, 495–503, (1995).
[12] D.H. Hyers, G. Isac and Th.M. Rassias, *Stability of Functional Equations in Several Variables* (Birkhäuser, Basel, 1998).
[13] S.-M. Jung, On the Hyers–Ulam stability of the functional equations that have the quadratic property, *J. Math. Anal. Appl.* **222**, 126–137, (1998).
[14] S.-M. Jung, *Hyers–Ulam–Rassias Stability of Functional Equations in Nonlinear Analysis* (Springer, 2011).
[15] Pl. Kannappan, *Functional Equations and Inequalities with Applications* (Springer, New York, 2009).
[16] Y.-H. Lee, S.-M. Jung and M.Th. Rassias, On an n-dimensional mixed type additive and quadratic functional equation, *Appl. Math. Comput.* **228**, 13–16, (2014).
[17] F. Skof, Proprietá locali e approssimazione di operatori, *Rend. Sem. Mat. Fis. Milano* **53**, 113–129, (1983).
[18] M.B. Moghimi, A. Najati, and C. Park, A functional inequality in restricted domains of Banach modules, *Adv. Difference Equ.* Art. ID 973709, 14, (2009).
[19] D. Molaei and A. Najati, Hyperstability of the general linear equation on restricted domains, *Acta Math. Hungar.* **149**(1), 238–253, (2016).
[20] A. Najati and S.-Mo. Jung, Approximately quadratic mappings on restricted domains, *J. Inequal. Appl.* Art. ID 503458, 10, (2010).
[21] A. Najati and Th. M. Rassias, Stability of the Pexiderized Cauchy and Jensen's equations on restricted domains, *Commun. Math. Anal.* **8**(2), 125–135, (2010).
[22] A. Najati and G.Z. Eskandani, A fixed point method to the generalized stability of a mixed additive and quadratic functional equation in Banach modules, *J. Difference Equ. Appl.* **16**(7), 773–788, (2010).
[23] C. Park, B. Noori, M.B. Moghimi, A. Najati, and J.M. Rassias, Local stability of mappings on multi-normed spaces, *Adv. Difference Equ.* **395**, 14, (2020).
[24] J. Senasukh and S. Saejung, On the hyperstability of the Drygas functional equation on a restricted domain, *Bull. Aust. Math. Soc.* **102**(1), 126–137, (2020).

[25] J.M. Rassias, On the Ulam stability of mixed type mappings on restricted domains, *J. Math. Anal. Appl.* **276**(2), 747–762, (2002).
[26] J.M. Rassias and M.J. Rassias, On the Ulam stability of Jensen and Jensen type mappings on restricted domains, *J. Math. Anal. Appl.* **281**(2), 516–524, (2003).
[27] M. Fréchet, Une définition fonctionelle des polynômes, *Nouv. Ann*, **49**, 145–162, (1909).

https://doi.org/10.1142/9789811261572_0024

Chapter 24

System of General Variational Inclusions

Muhammad Aslam Noor[*,‡], Khalida Inayat Noor[*,§], and Michael Th. Rassias[†,¶]

[*]*Department of Mathematics, COMSATS University Islamabad, Park Road, Islamabad, Pakistan*
[†]*Institute of Mathematics, University of Zurich, 8057 Zurich, Switzerland*
[‡]*noormaslam@gmail.com*
[§]*khalidan@gmail.com*
[¶]*michail.rassias@math.uzh.ch*

In this chapter, we introduce and consider a new system of variational inclusions involving three different operators, which is called the system of general variational inequalities. We prove that the system of general variational inequalities is equivalent to the fixed-point problem using the resolvent operator technique. This equivalent formulation is used to suggest and analyze some new explicit iterative methods for solving general variational inclusions. We also study the convergence analysis of the new iterative method under certain mild conditions. Since this new system includes the general variational inequalities involving the two operators, variational inequalities and related optimization problems as special cases, results obtained in this chapter continue to hold for these problems. Our results can be viewed as a refinement and improvement of the previously known results for variational inclusions.

1. Introduction

Multivalued variational inclusions, which were introduced and studied by Noor [1,2] are being used as mathematical programming models to study wide class of equilibrium problems arising in industry, finance, economics, transportation, optimization, operations research and engineering sciences in a unified and general frame work. It has been shown that variational inclusions contain variational inequalities. which were introduced and studied by Stampacchia [3]. For the formulation, numerical methods, generalizations, dynamical systems, sensitivity analysis and other aspects of variational inequalities and related optimization problems, see Refs. [1–46] and the references therein. Variational inclusions can provide new insights regarding problems being studied and can stimulate new and innovative ideas for problem solving. These activities have motivated to generalize and extend the variational inequalities and related optimization problems in several directions using new and novel techniques.

Inspired and motivated by research going on in this area, we introduce and consider a new system of general variational inclusions involving three different nonlinear operators. This class of system includes the system of general variational inequalities involving three operators and the classical variational inequalities as special cases. It is shown that odd order and non-symmetric boundary value problems can be studied in the framework of general variational inequality. Using the resolvent operator technique, we prove that the system of general variational inequalities are equivalent to the fixed-point problem. This alternative equivalent formulation is used to suggest and analyze some iterative methods for solving the system of general variational inclusions. We also prove the convergence of the proposed iterative methods under weaker conditions. Since the new system of variational inclusions includes the system of variational inequalities and related optimization problems as special cases, results proved in this chapter continue to hold for these problems. Our result can be viewed as a refinement and improvement of the previous results in this field. The comparison of these methods with other methods is a subject of future research.

2. Preliminaries

Let H be a real Hilbert space whose inner product and norm are denoted by $\langle \cdot, \cdot \rangle$ and $\|.\|$, resp. Let K be a closed and convex set in H. Let $T_1, T_2, A, g : H \times H \to H$ be nonlinear different operators and let $\varphi : H \longrightarrow R \cup \{+\infty\}$ be a continuous function.

We now consider the problem of finding $x^*, y^* \in H$ such that

$$\left.\begin{array}{ll} 0 \in \rho T_1(y^*, x^*) + \rho A(g(x^*)) - g(y^*) + g(x^*), & \rho > 0 \\ 0 \in \eta T_2(x^*, y^*) + \eta A(g(y^*)) + g(y^*) - g(x^*), & \eta > 0 \end{array}\right\}, \tag{1}$$

which is called the system of general variational inclusions. We now discuss some special cases of the problem (1), which appear to new ones.

Special cases

(I) If $T_1 = T_2 = T$, the problem (1) is equivalent to finding $x^*, y^* \in H$ such that

$$\left.\begin{array}{ll} 0 \in \rho T(y^*, x^*) + \rho A(g(x^*)) - g(y^*) + g(x^*), & \rho > 0 \\ 0 \in \eta T(x^*, y^*) + \eta A(g(y^*)) + g(y^*) - g(x^*), & \eta > 0 \end{array}\right\}, \tag{2}$$

(II) If T_1, T_2 are univariate operators, then problem (1) is equivalent to finding $x^*, y^* \in H$ such that

$$\left.\begin{array}{ll} 0 \in \rho T_1(y^*) + \rho A(g(x^*)) - g(y^*) + g(x^*), & \rho > 0 \\ 0 \in \eta T_2(x^*) + \eta A(g(y^*)) + g(y^*) - g(x^*), & \eta > 0 \end{array}\right\}, \tag{3}$$

which appears to be a new one.

(III) If $T_1 = T_2 = T, \quad \rho = \eta, \quad x = x^* = y^*$, then problem (1) is equivalent to finding $x \in H$ such that

$$0 \in \rho T(x) + g(x) - g(x) + \rho A(g(x)), \tag{4}$$

which is known as the variational inclusion problem or finding the zero of the sum of two (more) monotone operators. It is well known that a wide class of linear and nonlinear problems can be studied via variational inclusion problems.

(IV) If $A(.) = \partial\varphi(.)$, the subdifferential of a proper, convex and lower-semicontinuous function, then the system of variational inclusions (1) is equivalent to finding $x^*, y^* \in H$ such that

$$\left.\begin{array}{ll} 0 \in \rho T_1(y^*, x^*) + \rho\partial\varphi(g(x^*)) - g(y^*) + g(x^*), & \rho > 0 \\ 0 \in \eta T_2(x^*, y^*) + \eta\partial\varphi(g(y^*)) + g(y^*) - g(x^*), & \eta > 0 \end{array}\right\}, \tag{5}$$

or equivalently the problem of finding $x^*, y^* \in H$ such that

$$\left.\begin{array}{lll} \langle \rho T_1(y^*, x^*) + g(x^*) - g(y^*), g(x) - g(x^*)\rangle \geq \rho\varphi(g(x^*)) - \rho\varphi(g(x)), & \forall x \in H, & \rho > 0 \\ \langle \eta T_2(x^*, y^*) + g(y^*) - g(x^*), g(x) - g(y^*)\rangle \geq \rho\varphi(g(y^*)) - \rho\varphi(g(x)), & \forall x \in H, & \eta > 0 \end{array}\right\}, \tag{6}$$

which is called the system of nonlinear mixed variational inequalities involving two different nonlinear operators, which has been introduced and considered by Noor [20,29,30].

(V) If $T_1 = T_2 = T$, then problem (6) reduces to the following system of mixed variational inequalities of finding $x^*, y^* \in H$ such that

$$\left.\begin{array}{ll}\langle \rho T(y^*,x^*) + g(x^*) - g(y^*), g(x) - g(x^*)\rangle \geq \rho\varphi(g(x^*)) - \rho\varphi(g(x)), & \forall x \in H, \quad \rho > 0\\ \langle \eta T(x^*,y^*) + g(y^*) - g(x^*), g(x) - g(y^*)\rangle \geq \rho\varphi(g(y^*)) - \rho\varphi(g(x)), & \forall x \in H, \quad \eta > 0\end{array}\right\}. \tag{7}$$

(VI) If $T_1, T_2 : H \to H$ are univariate mappings, then problem (6) reduces to finding $x^*, y^* \in H$ such that

$$\left.\begin{array}{ll}\langle \rho T_1(y^*) + g(x^*) - g(y^*), g(x) - g(x^*)\rangle \geq \rho\varphi(g(x^*)) - \rho\varphi(g(x)), & \forall x \in H, \quad \rho > 0\\ \langle \eta T_2(x^*) + g(y^*) - g(x^*), g(x) - g(y^*)\rangle \geq \rho\varphi(g(y^*)) - \rho\varphi(g(x)), & \forall x \in H, \quad \eta > 0\end{array}\right\}, \tag{8}$$

which appears to be a new one.

(VII) If $T_1 = T_1 = T$ is a univariate operator and $\rho = \eta$, $x = x^* = y^*$, then problem (6) is equivalent to finding $x \in H$ such that

$$\langle Tx, g(y) - g(x)\rangle \geq \varphi(g(x)) - \varphi(g(y)), \quad \forall y \in H, \tag{9}$$

which is known as the mixed general variational inequality or general variational inequality of the second type. For the applications and numerical methods for solving the mixed general variational inequalities, see Noor [20,28–30,33] and Noor *et al.* [34,36].

(VIII) If φ is an indicator function of a closed convex set K in H, then problem (6) is equivalent to finding $x^*, y^* \in K$ such that

$$\left.\begin{array}{ll}\langle \rho T_1(y^*,x^*) + g(x^*) - g(y^*), g(x) - g(x^*)\rangle \geq 0, & \forall x \in K, \quad \rho > 0\\ \langle \eta T_2(x^*,y^*) + g(y^*) - g(x^*), g(x) - g(y^*)\rangle \geq 0, & \forall x \in K, \quad \eta > 0\end{array}\right\} \tag{10}$$

is called the system of variational inequalities, and it has been considered and studied by Huang and Noor Refs. [12].

(IX) If $T_1 = T_2 = T$, then the problem (10) is equivalent to the following system of variational inequalities (SVI) of finding $x^*, y^* \in K$ such that

$$\left.\begin{array}{ll}\langle \rho T(y^*,x^*) + g(x^*) - g(y^*), g(x) - g(x^*)\rangle \geq 0, & \forall x \in K, \quad \rho > 0\\ \langle \eta T(x^*,y^*) + g(y^*) - g(x^*), g(x) - g(y^*)\rangle \geq 0, & \forall x \in K, \quad \eta > 0\end{array}\right\}, \tag{11}$$

which has been considered and studied in Refs. [12,34].

(X) If $T_1, T_2 :\longrightarrow H$ are univariate operators, then problem (10) reduces to finding $x^*, y^* \in K$ such that

$$\left.\begin{aligned}\langle \rho T_1(y^*) + g(x^*) - g(y^*), g(x) - g(x^*)\rangle \geq 0, \quad \forall x \in K, \quad \rho > 0\\ \langle \eta T_2(x^*) + g(y^*) - g(x^*), g(x) - g(y^*)\rangle \geq 0, \quad \forall x \in K, \quad \eta > 0\end{aligned}\right\}, \tag{12}$$

which has been considered and studied in Ref. [12].

(XI) If $\varphi(.)$ is the indicator function of a closed convex set K, then problem (9) is equivalent to finding $x^* \in K$ such that

$$\langle Tx^*, g(x) - g(x^*)\rangle \geq 0, \quad \forall x \in K, \tag{13}$$

which is known as the general variational inequality introduced and studied by Noor [15–17,22,24]. For the formulation of numerical methods, error bounds, dynamical system and other aspects of the general variational inequalities, see Refs. [1,2,15–33]. and the references therein

It is known [9,44] that if the operator is not symmetric and non-positive, then it can be made symmetric and positive with respect to an arbitrary operator. For the sake of completeness and to convey the idea, we include all the details.

Definition 1. [6,32] An operator $T : H \longrightarrow H$ with respect to an arbitrary operator g is said to be

(a) *g-symmetric*, if and only if,

$$\langle Tu, g(v)\rangle = \langle g(u), Tv\rangle, \quad \forall u, v \in H.$$

(b) g-positive, if and only if,

$$\langle Tu, g(u)\rangle \geq 0, \quad \forall u, v \in H.$$

(c) *g-coercive g-elliptic*, if there exists a constant $\alpha > 0$ such that

$$\langle Tu, g(u)\rangle \geq \alpha \|g(u)\|^2, \quad \forall u, v \in H.$$

Note that g-coercivity implies g-positivity, but the converse is not true. It is also worth mentioning that there are operators which are not g-symmetric but g-positive. On the other hand, there are g-positive, but not g-symmetric operators. Furthermore, it is well known [9,44] that, for a linear operator T, there exists an inverse operator T^{-1} operator on $R(T)$ with $\overline{R(T)} = H$, then one can find an infinite set of auxiliary operators g such that the operator T is both g-symmetric and g-positive. We now show that the third-order boundary value problems can be studied via problem (11).

Example 1. Consider the third-order absolute boundary value problem of finding u such that

$$\left.\begin{array}{ll} -u'''(x(\geq f(x) & \text{on } \Omega=[a,b] \\ u(x)\geq \phi(x) & \text{on } \Omega=[a,b] \\ -u'''(x)-f(x)[u-\phi(x)]=0 & \text{on } \Omega=[a,b] \\ u(a)=0, \quad u'(a)=0, \quad u'(b)=0, & \end{array}\right\}, \tag{14}$$

where $f(x)$ is a continuous function and $\phi(x)$ is the cost (obstacle) function. The problem (14) can be studied in the framework of the general variational inequality (11). To do so, let

$$H_0^2[a,b]=\{u\in H, u(a)=0, \quad u'(a)=0, \quad u'(b)=0\}$$

be a Hilbert space, see Ref. [9]. One can easily show that the energy functional associated with (11) is

$$\begin{aligned} I[v] &= -\int_a^b \frac{d^3v}{dx^3}vdx - 2\int_a^b f\frac{dv}{dx}dx, \quad \forall \frac{dv}{dx}\in H_0^2[a,b] \\ &= \int_a^b \left(\frac{d^2v}{dx^2}\right)^2 - 2\int_a^b f\frac{dv}{dx}vdx \\ &= \langle Lv, g(v)\rangle - 2\langle f, g(v)\rangle. \end{aligned}$$

where

$$\langle Lu, g(v)\rangle = -\int_a^b \frac{d^3u}{dx^3}\frac{dv}{dx}dx = \int_a^b \left(\frac{d^2u}{dx^2}\right)\left(\frac{d^2v}{dx^2}\right)dx, \tag{15}$$

and

$$\langle f, g(v)\rangle = \int_a^b f\frac{dv}{dx}dx,$$

where $g=\frac{d}{dx}$ is a linear operator. It is clear that the operator L defined by (15) is linear, g-symmetric, g-positive and g is a linear operator.

Thus, the minimum of the functional $I[v]$ defined on the Hilbert space $H_0^1[a,b]$ can be characterized by the inequality (11). This shows that the third-order absolute boundary value problems can be studied in the framework of the general variational inequality (11).

(XII) If $g=I$, the identity operator, then problem (13) is equivalent to finding $u\in K$, such that

$$\langle Tx^*, x-x^*\rangle \geq 0, \quad \forall x\in K, \tag{16}$$

is called the classical variational inequality, introduced and studied by Stampacchia [3] in potential theory. It has been shown that a wide class of unrelated problems, which arise in various branches of pure and applied sciences, can be studied in the unified and general framework of variational inequalities. Clearly, the system of general variational inclusions (1) is more general and includes several classes of variational inequalities and related optimization problems as special cases. For the recent applications, numerical methods and formulations of variational inequalities and variational inclusions and for more details, see Refs. [1–46]. and the references therein.

3. Iterative Methods

In this section, we suggest some explicit iterative algorithms for solving the system of general variational inclusion (1). First of all, we establish the equivalence between the system of variational inclusions and fixed point problems. For this purpose, we recall the following well-known result.

Definition 2. For any maximal operator T, the resolvent operator associated with T, for any $\rho > 0$, is defined as

$$J_T(u) = (I + \rho T)^{-1}(u), \quad \forall u \in H.$$

It is well known that an operator T is maximal monotone, if and only if, its resolvent operator J_T is defined everywhere. It is single-valued and non-expansive, that is,

$$\|J_A u - J_A v\| \leq \|u - v\|, \quad \forall u, v \in H.$$

If $\varphi(.)$ is a proper, convex and lower-semicontinuous function, then its subdifferential $\partial\varphi(.)$ is a maximal monotone operator. In this case, we can define the resolvent operator

$$J_\varphi(u) = (I + \rho\partial\varphi)^{-1}(u), \quad \forall u \in H,$$

associated with the subdifferential $\partial\varphi(.)$. The resolvent operator J_φ has the following useful characterization.

Lemma 1 ([5]). *For a given $z \in H$, $u \in H$ satisfies the inequality*

$$\langle u - z, v - u\rangle + \rho\varphi(v) - \rho\varphi(u) \geq 0, \quad \forall v \in H,$$

if and only if

$$u = J_\varphi(z),$$

where $J_\varphi = (I + \rho\partial\varphi)^{-1}$ is the resolvent operator. It is well-known that the resolvent operator J_φ is non-expansive.

We now show that the system of general variational inclusions (1) is equivalent to the fixed-point problem and this is the motivation of our next result.

Lemma 2. *If the operator A is maximal monotone, then $(x^*, y^*) \in H$ is a solution of the system of general variational inclusions* (1), *if and only if, $x^*, y^* \in H$ satisfies*

$$g(x^*) = J_A[g(y^*) - \rho T_1(y^*, x^*)], \tag{17}$$

$$g(y^*) = J_A[g(x^*) - \eta T_2(x^*, y^*)]. \tag{18}$$

Proof. Let $(x^*, y^*) \in H$ be a solution of (1). Then

$$\left.\begin{aligned} g(y^*) - \rho T_1(y^*, x^*) &\in (I + \rho A)(g(x^*)) \\ g(x^*) - \eta T_2(x^*, y^*) &\in (I + \eta A)(g(y^*), \end{aligned}\right\},$$

which implies that

$$g(x^*) = J_A[g(y^*) - \rho T_1(y^*, x^*)],$$

$$g(y^*) = J_A[g(x^*) - \eta T_2(x^*, y^*)],$$

the required result. □

Lemma 2 shows the system of general variational inequalities to the fixed-point problems. This alternative equivalence formulation enables us to suggest the following explicit iterative method for solving the system of general variational inclusions (1):

Algorithm 1. For initial points $x_0, y_0 \in H$, compute the sequence $\{x_n\}$ and $\{y_n\}$ by

$$g(x_{n+1}) = (1 - a_n)g(x_n) + a_n J_A[g(y_n) - \rho T_1(y_n, x_n)],$$

$$g(y_{n+1}) = J_A[g(x_{n+1}) - \eta T_2(x_{n+1}, y_n)],$$

where $a_n \in [0, 1]$ for all $n \geq 0$.

If $T_1, T_2 : H \to H$ are univariate mappings and $g = I$, the identity operator, then Algorithm 1 reduces to the following iterative method for solving the system of variational inclusion:

Algorithm 2. For initial points $x_0, y_0 \in K$ compute the sequence $\{x_n\}$ and $\{y_n\}$ by

$$x_{n+1} = (1 - a_n)x_n + a_n J_A[y_n - \rho T_1(y_n)], \tag{19}$$

$$y_{n+1} = J_A[x_{n+1} - \eta T_2(x_{n+1})], \tag{20}$$

where $a_n \in [0, 1]$ for all $n \geq 0$ satisfies some suitable conditions.

If $T_1 = T_2 = T$, then Algorithm 1 reduces to the following method for solving (2):

Algorithm 3. For initial points $x_0, y_0 \in H$, compute the sequence $\{x_n\}$ and $\{y_n\}$ by

$$g(x_{n+1}) = (1 - a_n)g(x_n) + a_n J_A[g(y_n) - \rho T(y_n, x_n)],$$
$$g(y_{n+1}) = J_A[g(x_{n+1}) - \eta T(x_{n+1}, y_n)],$$

where $a_n \in [0, 1]$ for all $n \geq 0$.

If $A(.) = \partial\varphi(.)$, the subdifferential of a proper, convex and lower semicontinuous function φ, then $J_A = J_\varphi$. In this case, Algorithms 1–3 collapse to the following iterative methods for solving the system of mixed general variational inequalities (6)–(**??**8):

Algorithm 4. For initial points $x_0, y_0 \in H$, compute the sequence $\{x_n\}$ and $\{y_n\}$ by

$$g(x_{n+1}) = (1 - a_n)g(x_n) + a_n J_\varphi[g(y_n) - \rho T_1(y_n, x_n)],$$
$$g(y_{n+1}) = J_\varphi[g(x_{n+1}) - \eta T_2(x_{n+1}, y_n)],$$

where $a_n \in [0, 1]$ for all $n \geq 0$.

If $T_1, T_2 : H \to H$ are univariate mappings, then Algorithm 3.4 reduces to the following iterative method for solving the system of mixed general variational inequalities (8):

Algorithm 5. For initial points $x_0, y_0 \in K$ compute the sequence $\{x_n\}$ and $\{y_n\}$ by

$$g(x_{n+1}) = (1 - a_n)g(x_n) + a_n J_\varphi[g(y_n) - \rho T_1(y_n)],$$
$$g(y_{n+1}) = J_\varphi[g(x_{n+1}) - \eta T_2(x_{n+1})],$$

where $a_n \in [0, 1]$ for all $n \geq 0$ satisfies some suitable conditions.

If $T_1 = T_2 = T$, then Algorithm 4 reduces to the following method for solving the system of general variational inequalities (7):

Algorithm 6. For initial points $x_0, y_0 \in H$, compute the sequence $\{x_n\}$ and $\{y_n\}$ by

$$g(x_{n+1}) = (1 - a_n)g(x_n) + a_n J_\varphi[y_n - \rho T(y_n, x_n)],$$
$$g(y_{n+1}) = J_\varphi[g(x_{n+1}) - \eta T(x_{n+1}, y_n)],$$

where $a_n \in [0, 1]$ for all $n \geq 0$.

If φ is the indicator function of a closed convex set K in H, then $J_\varphi = P_K$, the projection of H onto the closed convex set K. Then Algorithms 4–6 collapse to the following iterative projection method for solving systems of general variational inequalities (10)–(12):

Algorithm 7. For initial points $x_0, y_0 \in K$, compute the sequence $\{x_n\}$ and $\{y_n\}$ by

$$g(x_{n+1}) = (1-a_n)g(x_n) + a_n P_K[g(y_n) - \rho T_1(y_n, x_n)],$$

$$g(y_{n+1}) = P_K[g(x_{n+1}) - \eta T_2(x_{n+1}, y_n)],$$

where $a_n \in [0,1]$ for all $n \geq 0$.

If $T_1, T_2 : K \to H$ are univariate mappings, then Algorithm 7 reduces to the following iterative method for solving the system of general variational inequalities (12):

Algorithm 8. For initial points $x_0, y_0 \in K$, compute the sequence $\{x_n\}$ and $\{y_n\}$ by

$$x_{n+1} = (1-a_n)x_n + a_n P_K[y_n - \rho T_1(y_n)],$$

$$y_{n+1} = P_K[x_{n+1} - \eta T_2(x_{n+1})].$$

If $T_1 = T_2 = T$, then Algorithm 7 is reduced to the following:

Algorithm 9. For initial points $x_0, y_0 \in K$, compute the sequence $\{x_n\}$ and $\{y_n\}$ by

$$g(x_{n+1}) = (1-a_n)g(x_n) + a_n P_K[y_n - \rho T(y_n, x_n)],$$
$$g(y_{n+1}) = P_K[g(x_{n+1}) - \eta T(x_{n+1}, y_n)],$$

where $a_n \in [0,1]$ for all $n \geq 0$.

We again use Lemma 1 to suggest another method for solving the system of general variational inclusions (1):

Algorithm 10. For initial points $x_0, y_0 \in H$, compute the sequence $\{x_n\}$ and $\{y_n\}$ by

$$g(x_{n+1}) = (1-a_n)g(x_n) + a_n J_A[g(y_{n+1}) - \rho T_1(y_{n+1}, x_n)],$$
$$g(y_{n+1}) = J_A[g(x_n) - \eta T_2(x_n, y_n)],$$

where $a_n \in [0,1]$ for all $n \geq 0$. Algorithm 10 is called the Gauss–Seidel-type algorithms for a class of solving systems of general Variational Inequalities.

If $g = I$, then Algorithm 10 reduces to the following method for solving the system of variational inclusions:

Algorithm 11. For initial points $x_0, y_0 \in H$, compute the sequence $\{x_n\}$ and $\{y_n\}$ by

$$\begin{aligned} x_{n+1} &= (1 - a_n)x_n + a_n J_A[y_{n+1} - \rho T_1(y_{n+1}, x_n)], \\ y_{n+1} &= J_A[x_n - \eta T_2(x_n, y_n)], \end{aligned}$$

where $a_n \in [0, 1]$ for all $n \geq 0$. Using the technique of Noor *et al.* [34], one can investigate the convergence analysis of Algorithm 11.

Remark 1. For appropriate and suitable choices of the operator T_1, T_2, g, and the spaces, one can suggest several new iterative methods for solving the system of general variational inclusions. This shows that Algorithms 1 and 10 are quite flexible and general and include various known and new algorithms for solving general variational inequalities and related optimization problems as special cases.

Definition 3. A mapping $T : H \to H$ is called r-strongly monotone, if there exists a constant $r > 0$, such that

$$\langle Tx - Ty, x - y\rangle \geq r||x - y||^2, \quad \forall x, y \in H.$$

Definition 4. A mapping $T : H \to H$ is called relaxed γ-cocoercive, if there exists a constant $\gamma > 0$, such that

$$\langle Tx - Ty, x - y\rangle \geq -\gamma||Tx - Ty||^2, \quad \forall x, y \in H.$$

Definition 5. A mapping $T : H \to H$ is called relaxed (γ, r)-cocoercive, if there exist constants $\gamma > 0, r > 0$, such that

$$\langle Tx - Ty, x - y\rangle \geq -\gamma||Tx - Ty||^2 + r||x - y||^2, \quad \forall x, y \in H.$$

The class of relaxed (γ, r)-cocoercive mappings is more general than the class of strongly monotone mappings.

Definition 6. A mapping $T : H \to H$ is called μ-Lipschitzian, if there exists a constant $\mu > 0$, such that

$$||Tx - Ty|| \leq \mu||x - y||, \quad \forall x, y \in H.$$

Lemma 3. *Suppose $\{\delta_n\}_{n=0}^{\infty}$ is a non-negative sequence satisfying the following inequality:*

$$\delta_{n+1} \leq (1-\lambda_n)\delta_n + \sigma_n, \quad \forall \quad n \geq 0,$$

with $\lambda_n \in [0,1]$, $\sum_{n=0}^{\infty} \lambda_n = \infty$, and $\sigma_n = o(\lambda_n)$. Then $\lim_{n\to\infty} \delta_n = 0$.

4. Convergence Analysis

In this section, we consider the convergence criteria of Algorithm 2 under some suitable mild conditions and this is the main motivation of this chapter. In a similar way, one can consider the convergence analysis of other algorithms.

Theorem 1. *Let (x^*, y^*) be the solution of (1). If $T_1 : H \to H$ is relaxed (γ_1, r_1)-cocoercive and μ_1-Lipschitzian, and $T_2 : H \times H \to H$ is relaxed (γ_2, r_2)-cocoercive and μ_2-Lipschitzian with conditions*

$$0 < \rho < \min\{2(r_1 - \gamma_1\mu_1^2)/\mu_1^2,\ 2(r_2 - \gamma_2\mu_2^2)/\mu_2^2\}, \quad r_1 > \gamma_1\mu_1^2, \tag{21}$$

$$0 < \eta < \min\{2(r_1 - \gamma_1\mu_1^2)/\mu_1^2,\ 2(r_2 - \gamma_2\mu_2^2)/\mu_2^2\}, \quad r_2 > \gamma_2\mu_2^2, \tag{22}$$

and $a_n \in [0,1]$, $\sum_{n=0}^{\infty} a_n = \infty$, then for arbitrarily initial points $x_0, y_0 \in H$, x_n and y_n obtained from Algorithm 2 converge strongly to x^* and y^*, resp.

Proof. From (17), (19), and the non-expansive property of the resolvent operator J_A, we have

$$\begin{aligned}
&||x_{n+1} - x^*|| \\
&\quad = ||(1-a_n)x_n + a_n J_A[y_n - \rho T_1(y_n)] - (1-a_n)x^* \\
&\qquad - a_n J_A[y^* - \rho T_1(y^*)]|| \\
&\quad \leq (1-a_n)||x_n - x^*|| + a_n||J_A[y_n - \rho T_1(y_n)] - J_A[y^* - \rho T_1(y^*)]|| \\
&\quad \leq (1-a_n)||x_n - x^*|| + a_n||[y_n - \rho T_1(y_n)] - [y^* - \rho T_1(y^*)]|| \\
&\quad = (1-a_n)||x_n - x^*|| + a_n||y_n - y^* - \rho[T_1(y_n) - T_1(y^*)]||. \qquad (23)
\end{aligned}$$

From the relaxed (γ_1, r_1)-cocoercive and μ_1-Lipschitzian definition of T_1, we have

$$
\begin{aligned}
&||y_n - y^* - \rho[T_1(y_n) - T_1(y^*)]||^2 \\
&= ||y_n - y^*||^2 - 2\rho\langle T_1(y_n) - T_1(y^*), y_n - y^*\rangle + \rho^2||T_1(y_n) - T_1(y^*)||^2 \\
&\le ||y_n - y^*||^2 - 2\rho[-\gamma_1||T_1(y_n) - T_1(y^*)||^2 + r_1||y_n - y^*||^2] \\
&\quad + \rho^2||T_1(y_n) - T_1(y^*)||^2 \\
&\le ||y_n - y^*||^2 + 2\rho\gamma_1\mu_1^2||y_n - y^*||^2 - 2\rho r_1||y_n - y^*||^2 \\
&\quad + \rho^2\mu_1^2||y_n - y^*||^2 \\
&= [1 + 2\rho\gamma_1\mu_1^2 - 2\rho r_1 + \rho^2\mu_1^2]||y_n - y^*||^2.
\end{aligned}
\tag{24}
$$

Set $\theta_1 = [1 + 2\rho\gamma_1\mu_1^2 - 2\rho r_1 + \rho^2\mu_1^2]^{1/2}$. It is clear from the condition (21) that $0 \le \theta_1 < 1$. Hence, from (24), it follows that

$$||y_n - y^* - \rho[T_1(y_n, x_n) - T_1(y^*, x^*)]|| \le \theta_1||y_n - y^*||. \tag{25}$$

Similarly, from the relaxed (γ_2, r_2)-cocoercive and μ_2-Lipschitzian definition of T_2, we obtain

$$
\begin{aligned}
&||x_{n+1} - x^* - \eta[T_2(x_{n+1}) - T_2(x^*)]||^2 \\
&= ||x_{n+1} - x^*||^2 - 2\eta\langle T_2(x_{n+1}) - T_2(x^*), x_{n+1} - x^*\rangle \\
&\quad + \eta^2||T_2(x_{n+1}) - T_2(x^*)||^2 \\
&\le ||x_{n+1} - x^*||^2 - 2\eta[-\gamma_2||T_2(x_{n+1}) - T_2(x^*)||^2 + r_2||x_{n+1} - x^*||^2] \\
&\quad + \eta^2||T_2(x_{n+1}) - T_2(x^*)||^2 \\
&= ||x_{n+1} - x^*||^2 + 2\eta\gamma_2||T_2(x_{n+1}) - T_2(x^*)||^2 - 2\eta r_2||x_{n+1} - x^*||^2 \\
&\quad + \eta^2||T_2(x_{n+1}) - T_2(x^*)||^2 \\
&\le ||x_{n+1} - x^*||^2 + 2\eta\gamma_2\mu_2^2||x_{n+1} - x^*||^2 - 2\eta r_2||x_{n+1} - x^*||^2 \\
&\quad + \eta^2\mu_2^2||x_{n+1} - x^*||^2 \\
&= [1 + 2\eta\gamma_2\mu_2^2 - 2\eta r_2 + \eta^2\mu_2^2]||x_{n+1} - x^*||^2.
\end{aligned}
\tag{26}
$$

Set $\theta_2 = [1 + 2\eta\gamma_2\mu_2^2 - 2\eta r_2 + \eta^2\mu_2^2]^{1/2}$. It is clear from the condition (22) that $0 \le \theta_2 < 1$. Then from (26), it follows that

$$
\begin{aligned}
&||x_{n+1} - x^* - \eta[T_2(x_{n+1}, y_n) - T_2(x^*, y^*)]|| \\
&\le \theta_2||x_{n+1} - x^*||.
\end{aligned}
\tag{27}
$$

Hence, from (23), (25), (27), and the non-expansive property of the resolvent operator J_A, we have

$$\begin{aligned}
&||y_{n+1} - y^*|| \\
&= ||J_A[x_{n+1} - \eta T_2(x_{n+1}, y_n)] - J_A[x^* - \eta T_2(x^*, y^*)]|| \\
&\leq ||[x_{n+1} - \eta T_2(x_{n+1}, y_n)] - [x^* - \eta T_2(x^*, y^*)]|| \\
&= ||x_{n+1} - x^* - \eta[T_2(x_{n+1}, y_n) - T_2(x^*, y^*)]|| \\
&\leq \theta_2 ||x_{n+1} - x^*||, \qquad (28)
\end{aligned}$$

which implies that for all $n \geq 1$,

$$||y_n - y^*|| \leq \theta_2 ||x_n - x^*||. \qquad (29)$$

Then from (23), (28) and (29), we obtain that

$$\begin{aligned}
&||x_{n+1} - x^*|| \\
&\leq (1 - a_n)||x_n - x^*|| + a_n||y_n - y^* - \rho[T_1(y_n, x_n) - T_1(y^*, x^*)]|| \\
&\leq (1 - a_n)||x_n - x^*|| + a_n\theta_1||y_n - y^*|| \\
&\leq (1 - a_n)||x_n - x^*|| + a_n\theta_1 \cdot \theta_2||x_n - x^*|| \\
&= [1 - a_n(1 - \theta_1\theta_2)]||x_n - x^*||.
\end{aligned}$$

Since the constant$(1 - \theta_1\theta_2) \in (0, 1]$, and $\sum_{n=0}^{\infty} a_n(1 - \theta_1\theta_2) = \infty$, from Lemma 3, we have $\lim_{n\to\infty} ||x_n - x^*|| = 0$. Hence, the result $\lim_{n\to\infty} ||y_n - y^*|| = 0$ is from (29). □

If $\varphi(.)$ is the indicator function of a closed convex set K in H, then $J_\varphi = P_K$, and $g = I$, the identity operator, then the projection of H onto K. Consequently, Theorem 4.1 reduces to the result for solving systems of variational inequalities (10)–(12), which is mainly due to Huang and Noor [12].

5. Conclusion

In this chapter, we have introduced a new system of variational inclusions involving three different operators. Using the resolvent operator technique, we have established the equivalence between the system of variational inclusions and the fixed-point problems. We have used this alternative equivalent formulation to suggest and analyze a number of iterative resolvent methods for solving this new system. Convergence analysis is also considered. Several special cases of our results are also discussed. It is interesting to compare the idea and technique of this chapter with other techniques. These research areas do not appear to have developed to an extent that it provides a complete framework for studying these problems. These problems can be very useful in practice. It is worth mentioning that the concept of fuzzy mappings can be extended for system of general variational inclusions (1) by using the technique of Noor [34]. Much more work is needed in all these areas to develop a sound basis for applications in engineering and physical sciences. The interested reader is advised to explore this field further and discover new and interesting applications of the system of variational inclusions and related optimization problems.

6. Acknowledgments

The authors wish to express their deepest gratitude to their teachers, students, colleagues, collaborators and friends, who have directly or indirectly contributed in the preparation of this chapter.

References

[1] M.A. Noor, Generalized set-valued variational inclusions and resolvent equations, *J. Math. Anal. Appl.* **228**, 206–220, (1998).
[2] M.A. Noor, Three-step iterative algorithms for multivalued quasi variational inclusions, *J. Math. Anal. Appl.* **255**, 589–604, (2001).
[3] G. Stampacchia, Formes bilineaires coercivities sur les ensembles convexes, C.R. *Acad. Sci. Paris* **258**, 4413–4416, (1964).
[4] A. Bnouhachem, M.A. Noor, and M. Khalfaoui, Modified descent-projection method for solving variational inequalities, *Appl. Math. Comput.* **190**, 1691–1700, (2007).
[5] H. Brezis, *Operateurs Maximaux Monotone et Semigroupes de Contractions dans les Espace d'Hilbert* (North-Holland, Amsterdam, Holland, 1973).
[6] S.S. Chang, H.W.J. Lee, and C.K. Chan, Generalized system for relaxed cocoercive variational inequalities in Hilbert spaces, *Appl. Math. Lett.* **20**, 329–334, (2006).
[7] X.P. Ding, Perturbed proximal point algorithms for generalized quasi variational inclusions, *J. Math. Anal. Appl.* **210**, 88–101, (1997).
[8] F. Giannessi and A. Maugeri, *Variational Inequalities and Network Equilibrium Problems* (Plenum Press, New York, 1995).
[9] V.M. Filippov, *Variational Principles for Nonpotential Operators* (Amer. Math. Soc. Providence, Rode Island, USA, 1989).

[10] F. Giannessi, A. Maugeri and P.M. Pardalos, *Equilibrium Problems, Nonsmooth Optimization and Variational Inequalities Problems* (Kluwer Academic Publishers, Dordrecht Holland, 2001).

[11] R. Glowinski, J.L. Lions and R. Tremolieres, *Numerical Analysis of Variational Inequalities* (North-Holland, Amsterdam, Holland 1981).

[12] Z. Huang and M.A. Noor, An explicit projection method for a system of nonlinear variational inequalities with different (γ, r)-cocoercive mappings, *Appl. Math. Comput.* **190**, 356–361, (2007).

[13] J.L. Lions and G. Stampacchia, Variational inequalities, *Commun. Pure Appl. Math.* **20**, 493–512, (1967).

[14] M.A. Noor, *On Variational Inequalities*, Ph.D. Thesis, Brunel University, London, UK, 1975.

[15] M.A. Noor, General variational inequalities, *Appl. Math. Lett.* **1**, 119–121, (1988).

[16] M.A, Noor, Wiener-Hopf equations and variational inequalities, *J. Optim. Theory Appl.* **79**, 197–206, (1993).

[17] M.A. Noor, Variational inequalities in physical oceanography. In: M. Rahman (Ed.), *Ocean Wave Engineering*, pp. 201–226 (Computational Mechanics Publications, Southampton, UK, 1994).

[18] M.A. Noor, Some recent advances in variational inequalities, Part I, basic concepts, *New Zealand J. Math.* **26**, 53–80, (1997).

[19] M.A. Noor, Some recent advances in variational inequalities, Part II, other concepts, *New Zealand J. Math.* **26**, 229–255, (1997).

[20] M.A. Noor, Some algorithms for general monotone mixed variational inequalities, *Math. Comput. Model.* **29**, 1–9, (1999).

[21] M.A. Noor, Variational inequalities for fuzzy mappings, III, *Fuzzy Sets and Systems* **110**, 101–108, (2000).

[22] M.A. Noor, New approximation schemes for general variational inequalities, *J. Math. Anal. Appl.* **251**, 217–229, (2000).

[23] M.A. Noor, New extragradient-type methods for general variational inequalities, *J. Math. Anal. Appl.* **277**, 379–395, (2003).

[24] M.A. Noor, Some developments in general variational inequalities, *Appl. Math. Comput.* **152**, 199–277, (2004).

[25] M.A. Noor, Mixed quasi variational inequalities, *Appl. Math. Comput.* **146**, 553–578, (2003).

[26] M.A. Noor, Fundamentals of mixed quasi variational inequalities, *Inter. J. Pure Appl. Math.* **15**, 137–258, (2004).

[27] M.A. Noor, Fundamentals of equilibrium problems, *Math. Inequal. Appl.* **9**, 529–566, (2006).

[28] M.A. Noor, Merit functions for general variational inequalities, *J. Math. Anal. Appl.* **316**, 736–752, (2006).

[29] M.A. Noor, On iterative methods for solving a system of mixed variational inequalities, *Appl. Anal.* **87**, 99–108, (2008).

[30] M.A. Noor, Differentiable nonconvex functions and general variational inequalities, *Appl. Math. Comput.* **199**, 623–630, (2008).

[31] M.A. Noor, Extended general variational inequalities, *Appl. Math. Lett.* **22**, 182–186, (2009).
[32] M.A. Noor, On a class of general variational inequalities, *J. Adv. Math. Stud.* **1**, 75–86, (2008).
[33] M.A. Noor, On a system of general mixed variational inequalities, *Optim. Lett.* **3**, 437–457, (2009).
[34] M.A, Noor, A.G. Khan, K.I. Noor, and A, Pervez, Gauss-Seidel type algorithms for a class of variational inequalities, *Filomat* **32**(2), 395–407, (2018).
[35] M.A. Noor and K.I. Noor, Sensitivity analysis of quasi variational inclusions, *J. Math. Anal. Appl.* **236**, 290–299, (1999).
[36] M.A. Noor and K.I. Noor, Projection algorithms for solving system of general variational inequalities, *Nonl. Anal.* **70**, 2700–2706, (2009).
[37] M.A. Noor, K.I. Noor and H. Yaqoob, On general mixed variational inequalities, *Acta Appl. Math.* **110**, 227–246, (2010). https://doi.org/10.1007/s10440-008-9402-4.
[38] M.A. Noor and K.I. Noor, Self-adaptive projection algorithms for general variational inequalities, *Appl. Math. Comput.* **151**, 659–670, (2004).
[39] M.A. Noor, K.I. Noor, and Th.M. Rassias, Some aspects of variational inequalities, *J. Comput. Appl. Math.* **47**, 285–312, (1993).
[40] M.A. Noor, K.I. Noor and M.Th. Rassias, New trends in general variational inequalities, *Acta Math. Applicandae* **170**(1), 981–1046, (2020).
[41] M.A. Noor, K.I. Noor and T.M. Rassias, Set-valued resolvent equations and mixed variational inequalities, *J. Math. Anal. Appl.* **220**, 741–759, (1998).
[42] M. Patriksson, *Nonlinear Programming and Variational Inequalities: A Unified Approach* (Kluwer Academic Publishers, Dordrecht, 1998).
[43] S.M. Robinson, Generalized equations and their solutions, *Math. Program. Stud.* **10**, 128–141, (1979).
[44] E. Tonti, Variational formulation for every nonlinear problem, *Intern. J. Eng. Sci.* **22**(11–12), 1343–1371, (1984).
[45] L.U. Uko, Strongly non-linear generalized equations, *J. Math. Anal. Appl.* **220**, 65–76, (1998).
[46] X.L. Weng, Fixed point iteration for local strictly pseudocontractive mappings, *Proc. Amer. Math. Soc.* **113**, 727–731, (1991).

https://doi.org/10.1142/9789811261572_0025

Chapter 25

Analytical Solution of nth-Order Volterra Integro-Differential Equations of Convolution Type with Non-local Conditions

E. Providas*,† and I.N. Parasidis*,‡

*Department of Environmental Sciences, University of Thessaly, Gaiopolis Campus, 415 00 Larissa, Greece

†providas@uth.gr

‡paras@uth.gr

We present a method for the analytical exact solution of boundary value problems for nth-order Volterra integro-differential equations of the second kind with convolution-type kernels and multipoint and integral conditions. Existence and uniqueness criteria are established and a formula for computing the solution is derived. The method is based on the Laplace transform.

1. Introduction

Integro-differential equations arise in many fields in sciences and engineering. For instance, Fredholm integro-differential equations are utilized to describe the response of plates and shells in the theory of elasticity [1,2]; Volterra integro-differential equations are employed in modeling population growth [3], heat flow in materials with memory [4,5], elastic liquids [6,7] and viscoelasticity [8]; Fredholm and Volterra integro-differential equations are used in modeling neural networks [9,10]. This is why integro-differential equations have been studied heavily by many researchers [11]. Volterra integro-differential equations have also been investigated intensively and numerous mainly numerical methods have been developed for their solution [12–16].

Non-local boundary value problems for integro-differential equations, however, have not received much attention in the past. These kinds of problems not only have theoretical interest, but many phenomena and processes are also accurately described in this way [17]. An existence and uniqueness analysis for the solution of such problems was given by Agarwal [18]. Recently, the authors in a series of papers have developed direct techniques for the exact solution of certain class of boundary value problems for Fredholm integro-differential equations with multipoint and integral conditions [19–23].

Two-point boundary value problems for Volterra integro-differential equations are solved numerically by the modified form of Adomian decomposition method and the variational iteration method by Wazwaz [24] and Wazwaz and Khuri [25], resp. Existence and uniqueness results of solutions to the second-order Volterra integro-differential equations with non-local and boundary conditions where given recently by Shikhare *et al.* [26]. Nakagiri [27] investigated the solution of a boundary control problem for the first-order Volterra integro-differential equation with nonlocal terms both in state and in boundary conditions. Setia *et al.* [28] proposed a numerical method to solve Fredholm–Volterra fractional integro-differential equations with non-local boundary conditions by using Chebyshev wavelets of the second kind.

An important class of integro-differential equations with many applications in life sciences, mechanics, image processing and others are the Volterra integro-differential equations of the convolution type. Initial value problems for this kind of equations with constant coefficients can be solved efficiently by the powerful method of Laplace transform [29–31]. A numerical method for the solution of linear and nonlinear initial value problems for convolution-type Volterra integro-differential equations was presented recently by Katsikadelis [32], where more information can be found on the subject. Motivated by these works, we propose here an analytical technique for examining the existence and uniqueness and constructing the solution of nth-order Volterra integro-differential equations of the second kind involving convolution integrals and non-local boundary conditions. The method is based on the Laplace transform combined by a direct matrix method developed by the authors for the solution of boundary value problems [19–23].

In particular, this chapter is devoted to the exact solution of the nth-order linear Volterra integro-differential equation of the second kind which assumes the form

$$\sum_{i=0}^{n} a_i u^{(n-i)}(x) - \sum_{i=0}^{n} \int_0^x k_i(x-t) u^{(n-i)}(t)dt = f(x), \tag{1}$$

subject to multipoint and integral conditions

$$u^{(i-1)}(0) - \sum_{j=1}^{m} \mu_{ij} \Psi_j(u) = c_i, \quad i = 1, 2, \ldots, n, \tag{2}$$

where $b \in (0, +\infty)$, $0 \leq x \leq b$, $u(x) \in C^n[0, b]$ is the unknown function, $u^{(n)}(x)$ denotes the nth derivative of $u(x)$, a_i, $i = 0, 1, \ldots, n$, are real constants and $a_0 \neq 0$, $f(x) \in C[0, b]$ is the input function, $k_i(x) \in C[0, b]$, $i = 0, 1, \ldots, n$, are kernels of convolution type, $\mu_{ij} \in \mathbb{R}$, $i = 1, 2, \ldots, n$, $j = 1, 2, \ldots, m$, Ψ_j, $j = 1, 2, \ldots, m$, are linear bounded functionals defined on $C[0, b]$ involving values at the fixed points $0 \leq \xi_\ell \leq b$, $\ell = 1, 2, \ldots, m_1$ and definite integrals of $u^{(i-1)}(x)$, $i = 1, 2, \ldots, n$, and c_i, $i = 1, 2, \ldots, n$, are real constants.

The rest of the chapter is organized as follows. In Section 2, some notations and terminology are introduced and the problem is formulated in a convenient operator form. Section 3 is devoted to the study of the problem and the presentation of the solution method. In Section 4, two examples problems are solved. Finally, in Section 5, some conclusions are given.

2. Formulation of the Problem

Let $A : C[0, b] \to C[0, b]$ be the nth-order linear differential operator

$$Au = \sum_{i=0}^{n} a_i u^{(n-i)}(x), \tag{3}$$

where $n \in \mathbb{N}$, a_i, $i = 0, 1, \ldots, n$, are real constants, $a_0 \neq 0$, $u(x) \in C^n[0, b]$, and $u^{(i)}(x) = \frac{d^i u}{dx^i}$, $i = 1, 2, \ldots, n$. Let $K : C[0, b] \to C[0, b]$ be the linear Volterra integral operator of the type

$$Ku = \sum_{i=0}^{n} \int_0^x k_i(x-t) u^{(n-i)}(t)dt, \tag{4}$$

where the kernels $k_i(x) \in C[0,b]$, $i = 0, 1, \ldots, n$. Let the $n \times 1$ vectors $\vec{u}(x)$ with elements the function $u(x)$ and its derivatives $u^{(i)}(x)$, $i = 1, 2, \ldots, n-1$, $\Psi(u)$, whose elements are the functionals $\{\Psi_j\}$, and $\vec{c}$ whose elements are real constants be defined by

$$\vec{u}(x) = \begin{pmatrix} u(x) \\ u'(x) \\ \vdots \\ u^{(n-1)}(x) \end{pmatrix}, \quad \Psi(u) = \begin{pmatrix} \Psi_1(u) \\ \Psi_2(u) \\ \vdots \\ \Psi_m(u) \end{pmatrix}, \quad c = \begin{pmatrix} c_1 \\ c_2 \\ \vdots \\ c_n \end{pmatrix}, \tag{5}$$

and the $n \times m$ matrix

$$M = \begin{pmatrix} \mu_{11} & \mu_{12} & \cdots & \mu_{1m} \\ \mu_{21} & \mu_{22} & \cdots & \mu_{2m} \\ \vdots & \vdots & \ddots & \vdots \\ \mu_{n1} & \mu_{n2} & \cdots & \mu_{nm} \end{pmatrix}. \tag{6}$$

Then by means of (3)–(6), the non-local boundary value problem (1), (2) is written in the symbolic form

$$Au(x) - Ku(x) = f(x), \quad x \geq 0, \tag{7}$$

$$\vec{u}(0) - M\Psi(u) = \vec{c}. \tag{8}$$

It is noted that if the vector $\vec{h} = (h_1(x),\ h_2(x), \ldots, h_n(x))$, then by $\Psi(\vec{h})$ we mean the $m \times n$ matrix

$$\Psi(\vec{h}) = \begin{pmatrix} \Psi_1(h_1) & \cdots & \Psi_1(h_n) \\ \vdots & \ddots & \vdots \\ \Psi_m(h_1) & \cdots & \Psi_m(h_n) \end{pmatrix},$$

where the element $\Psi_i(h_j)$ is the value of the functional Ψ_i on the element h_j. It is easy to show that for an $n \times m$ constant matrix C,

$$\Psi(\vec{h}C) = \Psi(\vec{h})C.$$

We recall that in a Banach space X, a linear operator $T : X \to X$ is injective if for every $u_1, u_2 \in D(T)$, $u_1 \neq u_2$ implies $Tu_1 \neq Tu_2$. The operator T is surjective if $R(T) = X$. If T is both injective and surjective, then the operator is called bijective and there exists the inverse operator $T^{-1} : X \to X$ defined by $T^{-1}f = u$ if and only if $Tu = f$ for each $f \in X$. The operator T is said to be *correct* if it is bijective and the inverse operator T^{-1} is bounded on X. The problem $Tu = f$ is correct if the operator T is correct.

3. Main Results

We assume that the unknown function $u(x)$, the kernel functions $k_i(x)$, $i = 0, 1, \ldots, n$, and the input function $f(x)$ are of exponential order γ and that their corresponding Laplace transforms are $U(s) = \mathcal{L}\{u\}$, $K_i(s) = \mathcal{L}\{k_i\}$, $i = 0, 1, \ldots, n$, and $F(s) = \mathcal{L}\{f\}$ for $s > \gamma$.

Then the Laplace transform of Au in (3) is as follows:

$$\begin{aligned}
\mathcal{L}\{Au\} &= \sum_{i=0}^{n} a_i \mathcal{L}\{u^{(n-i)}(x)\} \\
&= a_0 \left(s^n U(s) - \sum_{\ell=0}^{n-1} s^{n-1-\ell} u^{(\ell)}(0) \right) \\
&\quad + a_1 \left(s^{n-1} U(s) - \sum_{\ell=0}^{n-2} s^{n-2-\ell} u^{(\ell)}(0) \right) \\
&\quad \cdots \\
&\quad + a_{n-1} \left(sU(s) - u(0) \right) \\
&\quad + a_n U(s),
\end{aligned}$$

or after expanding and collecting like terms,

$$\begin{aligned}
\mathcal{L}\{Au\} = U(s) &\sum_{\ell=0}^{n} a_\ell s^{n-\ell} \\
&- u(0) \sum_{\ell=0}^{n-1} a_\ell s^{n-1-\ell} \\
&\cdots \\
&- u^{(n-2)}(0)(a_0 s + a_1) \\
&- u^{(n-1)}(0) a_0.
\end{aligned} \tag{9}$$

Similarly, the Laplace transform of Ku in (4) by using first the convolution theorem assumes the form

$$\begin{aligned}
\mathcal{L}\{Ku\} &= \sum_{i=0}^{n} \mathcal{L}\left\{ \int_a^x k_i(x-t) u^{(n-i)}(t) dt \right\} \\
&= \sum_{i=0}^{n} \mathcal{L}\{k_i(x)\} \mathcal{L}\{u^{(n-i)}(x)\} \\
&= \sum_{i=0}^{n} K_i(s) \mathcal{L}\{u^{(n-i)}(t)\},
\end{aligned}$$

and then by repeating the operations as above becomes

$$\begin{aligned}\mathcal{L}\{Ku\} = U(s)\sum_{\ell=0}^{n} K_\ell(s)s^{n-\ell} \\ -u(0)\sum_{\ell=0}^{n-1} K_\ell(s)s^{n-1-\ell} \\ \dots \\ -u^{(n-2)}(0)(K_0(s)s + K_1(s)) \\ -u^{(n-1)}(0)K_0(s).\end{aligned} \tag{10}$$

Taking the Laplace transform of the Volterra integro-differential equation (7) and using (9) and (10), we get

$$\begin{aligned}\mathcal{L}\{Au - Ku\} &= U(s)\sum_{\ell=0}^{n} \left[a_\ell - K_\ell(s)\right] s^{n-\ell} \\ &\quad -u(0)\sum_{\ell=0}^{n-1} \left[a_\ell - K_\ell(s)\right] s^{n-1-\ell} \\ &\quad \dots \\ &\quad -u^{(n-2)}(0)\left[(a_0 - K_0(s))s + a_1 - K_1(s)\right] \\ &\quad -u^{(n-1)}(0)\left[a_0 - K_0(s)\right] \\ &= \mathcal{L}\{f\} = F(s).\end{aligned}$$

Solving for $U(s)$, we obtain

$$\begin{aligned}U(s) &= u(0)\frac{\sum_{\ell=0}^{n-1}\left[a_\ell - K_\ell(s)\right] s^{n-1-\ell}}{\sum_{\ell=0}^{n}\left[a_\ell - K_\ell(s)\right] s^{n-\ell}} \\ &\quad + u'(0)\frac{\sum_{\ell=0}^{n-2}\left[a_\ell - K_\ell(s)\right] s^{n-2-\ell}}{\sum_{\ell=0}^{n}\left[a_\ell - K_\ell(s)\right] s^{n-\ell}} \\ &\quad \dots \\ &\quad + u^{(n-2)}(0)\frac{\left[a_0 - K_0(s)\right] s + a_1 - K_1(s)}{\sum_{\ell=0}^{n}\left[a_\ell - K_\ell(s)\right] s^{n-\ell}} \\ &\quad + u^{(n-1)}(0)\frac{a_0 - K_0(s)}{\sum_{\ell=0}^{n}\left[a_\ell - K_\ell(s)\right] s^{n-\ell}} \\ &\quad + \frac{F(s)}{\sum_{\ell=0}^{n}\left[a_\ell - K_\ell(s)\right] s^{n-\ell}},\end{aligned}$$

or

$$
\begin{aligned}
U(s) = u(0) &\left\{ \frac{[a_0 - K_0(s)]\, s^{n-1}}{p(s)} + \frac{\sum_{\ell=1}^{n-1} [a_\ell - K_\ell(s)]\, s^{n-1-\ell}}{p(s)} \right\} \\
&+ u'(0) \frac{\sum_{\ell=0}^{n-2} [a_\ell - K_\ell(s)]\, s^{n-2-\ell}}{p(s)} \\
&\cdots \\
&+ u^{(n-2)}(0) \frac{[a_0 - K_0(s)]\, s + a_1 - K_1(s)}{p(s)} \\
&+ u^{(n-1)}(0) \frac{a_0 - K_0(s)}{p(s)} + \frac{F(s)}{p(s)},
\end{aligned} \tag{11}
$$

where $p(s) = \sum_{\ell=0}^{n} [a_\ell - K_\ell(s)]\, s^{n-\ell}$.

Application of the inverse Laplace transform on (11) yields

$$
\begin{aligned}
u(x) = u(0)\mathcal{L}^{-1} &\left\{ \frac{[a_0 - K_0(s)]\, s^{n-1}}{p(s)} + \frac{\sum_{\ell=1}^{n-1} [a_\ell - K_\ell(s)]\, s^{n-1-\ell}}{p(s)} \right\} \\
&+ u'(0)\mathcal{L}^{-1} \left\{ \frac{\sum_{\ell=0}^{n-2} [a_\ell - K_\ell(s)]\, s^{n-2-\ell}}{p(s)} \right\} \\
&\cdots \\
&+ u^{(n-2)}(0)\mathcal{L}^{-1} \left\{ \frac{[a_0 - K_0(s)]\, s + a_1 - K_1(s)}{p(s)} \right\} \\
&+ u^{(n-1)}(0)\mathcal{L}^{-1} \left\{ \frac{a_0 - K_0(s)}{p(s)} \right\} + \mathcal{L}^{-1} \left\{ \frac{F(s)}{p(s)} \right\}.
\end{aligned} \tag{12}
$$

Let us assume that there exists a function $\phi(x)$ such that

$$
\phi(x) = \mathcal{L}^{-1} \left\{ \frac{s^{n-1}}{p(s)} \right\} = \mathcal{L}^{-1} \left\{ \frac{s^{n-1}}{\sum_{\ell=0}^{n} [a_\ell - K_\ell(s)]\, s^{n-\ell}} \right\}. \tag{13}
$$

Then by utilizing the convolution theorem and for $i = 1, 2, \ldots, n$, we have

$$\begin{aligned}\mathcal{L}^{-1}\left\{\frac{s^{n-i-\ell}}{p(s)}\right\} &= \mathcal{L}^{-1}\left\{\frac{1}{s^{\ell+i-1}}\frac{s^{n-1}}{p(s)}\right\}\\ &= \mathcal{L}^{-1}\left\{\frac{1}{s^{\ell+i-1}}\right\} * \mathcal{L}^{-1}\left\{\frac{s^{n-1}}{p(s)}\right\}\\ &= \left(\phi * \frac{x^{\ell+i-2}}{(\ell+i-2)!}\right)(x),\\ &\ell = 0, 1, \ldots, n-i,\ (i,\ell) \neq (1,0). \end{aligned} \tag{14}$$

As a consequence,

$$\begin{aligned}\mathcal{L}^{-1}\left\{\frac{[a_0 - K_0(s)]\, s^{n-1}}{p(s)}\right\} &= \mathcal{L}^{-1}\left\{a_0\frac{s^{n-1}}{p(s)} - \frac{K_0(s)s^{n-1}}{p(s)}\right\}\\ &= a_0\mathcal{L}^{-1}\left\{\frac{s^{n-1}}{p(s)}\right\} - \mathcal{L}^{-1}\left\{K_0(s)\frac{s^{n-1}}{p(s)}\right\}\\ &= a_0\phi(x) - (k_0 * \phi)(x), \end{aligned} \tag{15}$$

and for $i = 1, 2, \ldots, n$, and $\ell = 0, 1, \ldots, n-i$, $(i,\ell) \neq (1,0)$,

$$\begin{aligned}\mathcal{L}^{-1}\left\{\frac{[a_\ell - K_\ell(s)]\, s^{n-i-\ell}}{p(s)}\right\} &= \mathcal{L}^{-1}\left\{a_\ell\frac{s^{n-i-\ell}}{p(s)} - \frac{K_\ell(s)s^{n-i-\ell}}{p(s)}\right\}\\ &= a_\ell\mathcal{L}^{-1}\left\{\frac{s^{n-i-\ell}}{p(s)}\right\} - \mathcal{L}^{-1}\left\{K_\ell(s)\frac{s^{n-i-\ell}}{p(s)}\right\}\\ &= a_\ell\left(\phi * \frac{x^{\ell+i-2}}{(\ell+i-2)!}\right)(x)\\ &\quad - \left(k_\ell * \left(\phi * \frac{x^{\ell+i-2}}{(\ell+i-2)!}\right)\right)(x). \end{aligned} \tag{16}$$

Also,

$$\begin{aligned}\mathcal{L}^{-1}\left\{\frac{F(s)}{p(s)}\right\} &= \mathcal{L}^{-1}\{F(s)\} * \mathcal{L}^{-1}\left\{\frac{1}{p(s)}\right\} = (f * \phi)(x), \quad n = 1,\\ \mathcal{L}^{-1}\left\{\frac{F(s)}{p(s)}\right\} &= \mathcal{L}^{-1}\{F(s)\} * \mathcal{L}^{-1}\left\{\frac{s^{n-1}}{p(s)}\frac{1}{s^{n-1}}\right\}\\ &= \left(f * \left(\phi * \frac{x^{n-2}}{(n-2)!}\right)\right)(x), \quad n > 1. \end{aligned} \tag{17}$$

Using now the relations (15)–(17), equation (12) is carried into

$$\begin{aligned}
u(x) = {} & u(0)\left[a_0\phi(x) - (k_0 * \phi)(x)\right] \\
& + u(0)\sum_{\ell=1}^{n-1}\left[a_\ell\left(\phi * \frac{x^{\ell-1}}{(\ell-1)!}\right)(x) - \left(k_\ell * \left(\phi * \frac{x^{\ell-1}}{(\ell-1)!}\right)\right)(x)\right] \\
& + u'(0)\sum_{\ell=0}^{n-2}\left[a_\ell\left(\phi * \frac{x^{\ell}}{\ell!}\right)(x) - \left(k_\ell * \left(\phi * \frac{x^{\ell}}{\ell!}\right)\right)(x)\right] \\
& \dots \\
& + u^{(n-2)}(0)\left[a_0\left(\phi * \frac{x^{n-3}}{(n-3)!}\right)(x) - \left(k_0 * \left(\phi * \frac{x^{n-3}}{(n-3)!}\right)\right)(x)\right. \\
& \qquad \left. + a_1\left(\phi * \frac{x^{n-2}}{(n-2)!}\right)(x) - \left(k_1 * \left(\phi * \frac{x^{n-2}}{(n-2)!}\right)\right)(x)\right] \\
& + u^{(n-1)}(0)\left[a_0\left(\phi * \frac{x^{n-2}}{(n-2)!}\right)(x) - \left(k_0 * \left(\phi * \frac{x^{n-2}}{(n-2)!}\right)\right)(x)\right] \\
& + \hat{f}(x),
\end{aligned} \tag{18}$$

where we have put

$$\hat{f}(x) = \begin{cases} (f * \phi)(x), & n = 1, \\ \left(f * \left(\phi * \frac{x^{n-2}}{(n-2)!}\right)\right)(x), & n > 1. \end{cases} \tag{19}$$

Moreover, by letting

$$\begin{aligned}
h_1(x) = {} & a_0\phi(x) - (k_0 * \phi)(x) \\
& + \sum_{\ell=1}^{n-1}\left[a_\ell\left(\phi * \frac{x^{\ell-1}}{(\ell-1)!}\right)(x) - \left(k_\ell * \left(\phi * \frac{x^{\ell-1}}{(\ell-1)!}\right)\right)(x)\right], \\
h_2(x) = {} & \sum_{\ell=0}^{n-2}\left[a_\ell\left(\phi * \frac{x^{\ell}}{\ell!}\right)(x) - \left(k_\ell * \left(\phi * \frac{x^{\ell}}{\ell!}\right)\right)(x)\right], \\
& \dots \\
h_{n-1}(x) = {} & a_0\left(\phi * \frac{x^{n-3}}{(n-3)!}\right)(x) - \left(k_0 * \left(\phi * \frac{x^{n-3}}{(n-3)!}\right)\right)(x) \\
& + a_1\left(\phi * \frac{x^{n-2}}{(n-2)!}\right)(x) - \left(k_1 * \left(\phi * \frac{x^{n-2}}{(n-2)!}\right)\right)(x), \\
h_n(x) = {} & a_0\left(\phi * \frac{x^{n-2}}{(n-2)!}\right)(x) - \left(k_0 * \left(\phi * \frac{x^{n-2}}{(n-2)!}\right)\right)(x),
\end{aligned}$$

and the vector

$$\vec{h} = \left(h_1(x)\ h_2(x)\ \cdots\ h_n(x) \right), \tag{20}$$

equation (18) may be written compactly as

$$u(x) = \vec{h}\vec{u}(0) + \hat{f}. \tag{21}$$

We now state the theorem for constructing the solution of initial value problems for nth-order linear Volterra integro-differential equations of convolution type.

Theorem 1. *Let the functions $u(x)$, $k_i(x)$, $i = 0, 1, \ldots, n$, and $f(x)$ be of exponential order γ with their Laplace transforms denoted by $U(s) = \mathcal{L}\{u\}$, $K_i(s) = \mathcal{L}\{k_i\}$, $i = 0, 1, \ldots, n$, and $F(s) = \mathcal{L}\{f\}$, resp., for $s > \gamma$. Let the operator $A : C[0,b] \to C[0,b]$, $b > 0$, be an nth-order linear differential operator of the type* (3) *and $K : C[0,b] \to C[0,b]$ be a linear Volterra operator as in* (4)*. Let $\vec{u}(0) = \mathrm{col}(u(0), u'(0), \ldots, u^{n-1}(0))$ and the column vector $\vec{c}$ of n be arbitrary constants. Let there exist a function $\phi(x)$ as in* (13)*. Then the unique solution of the initial value problem*

$$Au(x) - Ku(x) = f(x), \quad x \geq 0, \tag{22}$$

$$\vec{u}(0) = \vec{c}, \tag{23}$$

is given by

$$u(x) = \vec{h}\vec{c} + \hat{f}, \tag{24}$$

where $\hat{f}$ is given in (19) *and the vector of functions $\vec{h}$ is defined in* (20).

Next, we state and prove our main theorem for solving non-local boundary value problems for nth-order linear Volterra integro-differential equations with convolution-type kernels.

Theorem 2. *Let the functions $u(x)$, $k_i(x)$, $i = 0, 1, \ldots, n$, and $f(x)$ be of exponential order γ with their Laplace transforms denoted by $U(s) = \mathcal{L}\{u\}$, $K_i(s) = \mathcal{L}\{k_i\}$, $i = 0, 1, \ldots, n$, and $F(s) = \mathcal{L}\{f\}$, resp., for $s > \gamma$. Let the operator $A : C[0,b] \to C[0,b]$, $b > 0$, be an nth-order linear differential operator of the type* (3) *and $K : C[0,b] \to C[0,b]$ be a linear Volterra operator as in* (4)*. Let the vectors $\vec{u}$, Ψ and $\vec{c}$ be as defined in* (5)*, and the matrix M as in* (6)*. Let there exist a function $\phi(x)$ as in* (13)*. Then the non-local boundary value problem*

$$Au(x) - Ku(x) = f(x), \quad x \geq 0, \tag{25}$$

$$\vec{u}(0) - M\Psi(u) = \vec{c}, \tag{26}$$

is correct if

$$\det W = \det[I_n - M\Psi(\vec{h})] \neq 0, \tag{27}$$

and its unique solution is given by

$$u(x) = \vec{h}[I_n - M\Psi(\vec{h})]^{-1}\left(M\Psi(\hat{f}) + \vec{c}\right) + \hat{f}, \tag{28}$$

where $\hat{f}$ is given in (19) *and the vector of functions $\vec{h}$ is defined in* (20).

Proof. By applying the Laplace transform on Volterra integro-differential equation (25), we have shown that its solution may be obtained by the formula (21), viz.

$$u(x) = \vec{h}\vec{u}(0) + \hat{f}, \tag{29}$$

where $\vec{u}(0)$ has to be evaluated by considering some initial or boundary conditions.

Acting by the functional vector Ψ on both sides of (29), we get

$$\Psi(u) = \Psi\left(\vec{h}\vec{u}(0)\right) + \Psi(\hat{f}) = \Psi(\vec{h})\vec{u}(0) + \Psi(\hat{f}).$$

Substituting into (26), we obtain

$$\vec{u}(0) - M\Psi(u) = \vec{u}(0) - M\left[\Psi(\vec{h})\vec{u}(0) + \Psi(\hat{f})\right] = \vec{c},$$

and after rearranging terms

$$\left[I_n - M\Psi(\vec{h})\right]\vec{u}(0) = M\Psi(\hat{f}) + \vec{c}, \tag{30}$$

where I_n denotes the $n \times n$ identity matrix.

If condition (27) holds true, then equation (30) can be solved uniquely with respect to $\vec{u}(0)$ to acquire

$$\vec{u}(0) = \left[I_n - M\Psi(\vec{h})\right]^{-1}\left(M\Psi(\hat{f}) + \vec{c}\right). \tag{31}$$

Substitution then of (31) into (29) yields the solution in (28).

Finally, since equation (28) holds for every $f(x) \in C[0, b]$ and the elements of $\vec{h}$, the function $\hat{f}$ and the functionals $\{\Psi_i\}$ are bounded, it is implied that the operator $A - K$ is correct and hence the problem (25), (26) is correct. □

4. Examples

In this section, we solve two illustrative examples to explain the application of the method and to demonstrate its effectiveness and ease of use.

Example 1. Consider the second-order Volterra integro-differential equation of convolution type

$$\begin{aligned} a_0u''(x) + a_1u'(x) + a_2u(x) - \int_0^x k_0(x-t)u''(t)dt \\ - \int_0^x k_1(x-t)u'(t)dt - \int_0^x k_2(x-t)u(t)dt = f(x), \end{aligned} \tag{32}$$

with initial conditions $u(0) = 0$ and $u'(0) = 1$. Let

$$\begin{aligned} a_0 &= a_1 = a_2 = 1, \\ k_0(x) &= k_1(x) = k_2(x) = \sin x, \\ f(x) &= \frac{2\cos x - x\sin x}{2}, \end{aligned} \tag{33}$$

so that the exact solution is $u(x) = \sin x$.

We will construct its unique solution by applying Theorem 1 in the following three steps.

Step 1: Comparing (32), (33) with (22), (23), it is natural to take

$$\begin{aligned} n &= 2, \\ Au &= u''(x) + u'(x) + u(x), \\ Ku &= \int_0^x k_0(x-t)u''(t)dt - \int_0^x k_1(x-t)u'(t)dt \\ &\quad - \int_0^x k_2(x-t)u(t)dt, \\ \vec{u} &= \begin{pmatrix} u \\ u' \end{pmatrix}, \quad \vec{c} = \begin{pmatrix} 0 \\ 1 \end{pmatrix}. \end{aligned} \tag{34}$$

Step 2: We find the function $\phi(x)$ in (13), viz.

$$\phi(x) = \mathcal{L}^{-1}\left\{\frac{s^2+1}{s(s^2+s+1)}\right\} = 1 - \frac{2\sqrt{3}e^{-\frac{x}{2}}\sin(\frac{\sqrt{3}x}{2})}{3}. \tag{35}$$

Then, using (20) we compute

$$\begin{aligned}
h_1(x) &= a_0\phi(x) - (k_0 * \phi)(x) + a_1(\phi * 1)(x) - (k_1 * (\phi * 1))(x) \\
&= \phi(x) - \int_0^x \sin(x-t)\phi(t)dt + \int_0^x \phi(x-t)dt \\
&\quad - \int_0^x \sin(x-t) \int_0^t \phi(t-\tau)d\tau dt, \\
h_2(x) &= a_0(\phi * 1)(x) - (k_0 * (\phi * 1))(x) \\
&= \int_0^x \phi(x-t)dt - \int_0^x \sin(x-t) \int_0^t \phi(t-\tau)d\tau dt, \\
\vec{h} &= \begin{pmatrix} h_1 & h_2 \end{pmatrix}, \\
\hat{f}(x) &= (f * (\phi * 1))(x) \\
&= \int_0^x f(x-t) \int_0^t \phi(t-\tau)d\tau dt. \qquad (36)
\end{aligned}$$

Step 3: Substituting (34)–(36) into the formula (24) and after performing the vector product and adding, we obtain the solution to the initial value problem (32), (33) in closed form.

Example 2. Let the second-order Volterra integro-differential equation

$$u''(x) - \int_0^x \cos(x-t)u'(t)dt = x - 1, \qquad (37)$$

with the boundary conditions

$$u(0) = 6u(1) - 2u'(1), \quad u'(0) = 12u(1) + 8u'(1). \qquad (38)$$

According to Theorem 2 for solving boundary value problems, we proceed as follows.

Step 1: We formulate the problem (37), (38) in the symbolic form (25), (26) by taking

$$\begin{aligned}
n &= 2, \\
Au &= u''(x), \quad a_0 = 1,\ a_1 = 0,\ a_2 = 0, \\
Ku &= \int_0^x k_1(x-t)u'(t)dt, \quad k_0(x) = 0, \quad k_1(x) = \cos(x), \quad k_2(x) = 0, \\
f(x) &= x - 1,
\end{aligned}$$

$$\vec{u} = \begin{pmatrix} u \\ u' \end{pmatrix}, \quad M = \begin{pmatrix} 6 & -2 \\ 12 & 8 \end{pmatrix}, \quad \Psi(u) = \begin{pmatrix} u(1) \\ u'(1) \end{pmatrix}, \quad \vec{c} = \begin{pmatrix} 0 \\ 0 \end{pmatrix}. \tag{39}$$

Step 2: We check the existence and uniqueness of the solution. First, from (13) we find

$$\phi(x) = \mathcal{L}^{-1}\left\{\frac{s}{s^2 - sK_1(s)}\right\} = \mathcal{L}^{-1}\left\{\frac{s}{s^2 - \frac{s^2}{s^2+1}}\right\} = 1 + \frac{x^2}{2}. \tag{40}$$

Then, by using (20) and (39), we compute

$$\begin{aligned}
h_1(x) &= a_0\phi(x) - (k_0 * \phi)(x) + a_1(\phi * 1)(x) - (k_1 * (\phi * 1))(x) \\
&= \phi(x) - (k_1 * (\phi * 1))(x) \\
&= \phi(x) - \int_0^x \cos(x-t) \int_0^t \phi(t-\tau)d\tau dt, \\
h_2(x) &= a_0(\phi * 1)(x) - (k_0 * (\phi * 1))(x) \\
&= (\phi * 1)(x) \\
&= \int_0^x \phi(x-t)dt, \\
\vec{h} &= \begin{pmatrix} h_1 & h_2 \end{pmatrix}, \\
\Psi(\vec{h}) &= \begin{pmatrix} \Psi_1(h_1) & \Psi_1(h_2) \\ \Psi_2(h_1) & \Psi_2(h_2) \end{pmatrix} = \begin{pmatrix} h_1(1) & h_2(1) \\ h_1'(1) & h_2'(1) \end{pmatrix},
\end{aligned} \tag{41}$$

and set up the matrix

$$W = I_2 - M\Psi(\vec{h}). \tag{42}$$

Since $\det W = 77 \neq 0$, the problem (37), (38) admits exactly one solution.

Step 3: We further calculate

$$\begin{aligned}
\hat{f}(x) &= (f * (\phi * 1))(x) \\
&= \int_0^x f(x-t) \int_0^t \phi(t-\tau)d\tau dt, \\
\Psi(\hat{f}) &= \begin{pmatrix} \Psi_1(\hat{f}) \\ \Psi_2(\hat{f}) \end{pmatrix} = \begin{pmatrix} \hat{f}(1) \\ \hat{f}'(1) \end{pmatrix}.
\end{aligned} \tag{43}$$

Substituting (39)–(43) into (28), we obtain the solution of the problem (37), (38) as follows:

$$u(x) = \frac{77x^5 - 385x^4 + 2252x^3 - 4620x^2 + 4272x - 1662}{9240}.$$

5. Conclusion

A method for constructing the analytical exact solution of a class of nth-order Volterra integro-differential equations of the second kind with convolution integrals and non-local boundary conditions was presented. Existence and uniqueness criteria were established and a ready-to-use solution formula was derived by using the Laplace transform and a formalism of a direct matrix method. The matrix procedure can be implemented easily to any software for symbolic computations. The technique requires the construction of a function $\phi(x)$ via the inverse Laplace transform and the analytical evaluation of the integrals involved. The solution of the two test problems showed the efficiency and the ease of use of the technique.

References

[1] B.N. Fradlin and F.A. Tsykunov, On the development of the method of integro-differential equations in the theory of plates and shells, *Soviet Appl. Mech.* **3**, 8–12, (1967). https://doi.org/10.1007/BF00886224.

[2] G.V. Kostin and V.V. Saurin, Integro-differential approach to solving problems of linear elasticity theory, *Dokl. Phys.* **50**, 535–538, (2005). https://doi.org/10.1134/1.2123305.

[3] V. Volterra, *Theory of Functionals of Integral and Integro-Differential Equations* (Dover, New York, 1959).

[4] R.C. MacCamy, An integro-differential equation with application in the heat flow, *Quarterly of Appl. Math.* **35**(1), 1–19, (1977). http://www.jstor.org/stable/43636851.

[5] G. Seifert, A temperature equation for rigid heat conductor with memory, *Quart. Appl. Math.* **38**(2), 246–252, (1980). http://www.jstor.org/stable/43637032.

[6] P. Markowich and M. Renardy, The numerical solution of a class of quasilinear parabolic volterra equations arising in polymer rheology, *SIAM J. Numer. Anal.* **20**(5), 890–908, (1983). http://www.jstor.org/stable/2157105.

[7] H. Engler, A matrix Volterra integrodifferential equation occurring in polymer rheology, *Pacific J. Math.* **149**, 25–60, (1991).

[8] D.A. Zakora, Abstract linear Volterra second-order integro-differential equations, *Eurasian Math. J.* **7**(2), 75–91, (2016).

[9] X. Lou and B. Cui, Passivity analysis of integro-differential neural networks with time-varying delays, *Neurocomputing* **70**, 1071–1078, (2007). https://doi.org/10.1016/j.neucom.2006.09.007.

[10] Y. Guo, Global asymptotic stability analysis for integro-differential systems modeling neural networks with delays, *Z. Angew. Math. Phys.* **61**, 971–978, (2010). https://doi.org/10.1007/s00033-009-0057-4.

[11] A.M. Wazwaz, Volterra integro-differential equations. In: *Linear and Nonlinear Integral Equations* (Springer, Berlin, Heidelberg, 2011). https://doi.org/10.1007/978-3-642-21449-3_5.

[12] H. Brunner, A survey of recent advances in the numerical treatment of Volterra integral and integro-differential equations, *J. Comput. Appl. Math.* **8**(3), 213–229, (1982). https://doi.org/10.1016/0771-050X(82)90044-4.

[13] T. Tang, A note on collocation methods for Volterra integro-differential equations with weakly singular kernels, *IMA J. Numer. Anal.* **13**, 93–99, (1993).

[14] H. Chen and C. Zhang, Block boundary value methods for solving Volterra integral and integro-differential equations, *J. Comput. Appl. Math.* **236**(11), 2822–2837, (2012). https://doi.org/10.1016/j.cam.2012.01.018.

[15] A. Cardone, D. Conte, R. D'Ambrosio, and B. Paternoster, Collocation methods for Volterra integral and integro-differential equations: A review, *Axioms* **45**(7), (2018). https://doi.org/10.3390/axioms7030045.

[16] D. Rani and V. Mishra, Solutions of Volterra integral and integro-differential equations using modified Laplace Adomian decomposition method, *J. Appl. Math., Stat. Inform.* **15**(1), 5–18, (2019). https://doi.org/10.2478/jamsi-2019-0001.

[17] C. Thaiprayoon, W. Sudsutad, and S.K. Ntouyas, Mixed nonlocal boundary value problem for implicit fractional integro-differential equations via ψ-Hilfer fractional derivative, *Adv. Differ. Equ.* **2021**, 50, (2021). https://doi.org/10.1186/s13662-021-03214-1.

[18] R.P. Agarwal, Boundary value problems for higher order integro-differential equations, *Nonlinear Anal.* **7**(3), 259–270, (1983).

[19] N.N. Vassiliev, I.N. Parasidis, and E. Providas, Exact solution method for Fredholm integro-differential equations with multipoint and integral boundary conditions. Part 1. Extension method, *Infor. Control Syst.* **6**, 14–23, (2018). https://doi.org/10.31799/1684-8853-2018-6-14-23.

[20] M.M. Baiburin and E. Providas, Exact solution to systems of linear first-order integro-differential equations with multipoint and integral conditions, In: T. Rassias and P. Pardalos, (Eds.), *Mathematical Analysis and Applications*, Vol. 154, Springer Optimization and Its Applications, pp. 1–16 (Springer, Cham, 2019). https://doi.org/10.1007/978-3-030-31339-5_1.

[21] I.N. Parasidis and E. Providas, Exact solutions to problems with perturbed differential and boundary operators, In: T.M. Rassias and V.A. Zagrebnov (Eds.), *Analysis and Operator Theory*, Vol. 146, pp. 301–317 (Springer, Cham, 2019). https://doi.org/10.1007/978-3-030-12661-2_14.

[22] I.N. Parasidis and E. Providas, Factorization method for solving nonlocal boundary value problems in Banach space, *Bull. Karaganda University: Math. Series* **3**(103), 76–85, (2021).

[23] E. Providas and I.N. Parasidis, A procedure for factoring and solving nonlocal boundary value problems for a type of linear integro-differential equations, *Algorithms* **14**(12), 346, (2021). https://doi.org/10.3390/a14120346.

[24] A.M. Wazwaz, A reliable algorithm for solving boundary value problems for higher-order integro-differential equations, *Appl. Math. Comput.* **118**(2–3), 327–342, (2001). https://doi.org/10.1016/S0096-3003(99)00225-8.

[25] A.M. Wazwaz and S.A. Khuri, The variational iteration method for solving the Volterra integro-differential forms of the Lane-Emden and the Emden-Fowler problems with initial and boundary value conditions, *Open Eng.* **5**(1), (2015). https://doi.org/10.1515/eng-2015-0006.

[26] P.U. Shikhare, K.D. Kucche and J.V.C. Sousa, Analysis of Volterra integrodifferential equations with nonlocal and boundary conditions via Picard operator, arXiv: 1908.08224, (2019).

[27] S. Nakagiri, Deformation formulas and boundary control problems of first-order Volterra integro-differential equations with nonlocal boundary conditions, *IMA J. Math. Control Infor.* **30**(3), 345–377, (2013). Doi: 10.1093/imamci/dns026.

[28] A. Setia, Y. Liu, and A.S. Vatsala, Numerical solution of Fredholm-Volterra fractional integro-differential equations with nonlocal boundary conditions, *J. Fract. Calculus Appl.* **5**(2), 155–165, (2014). http://fcag-egypt.com/Journals/JFCA/.

[29] A. Lunardi, Laplace transform methods in integrodifferential equations, *J. Integral Eq.* **10**(1/3), 185–211, (1985).

[30] H. Rezaei, S.M Jung, and T.M. Rassias, Laplace transform and Hyers–Ulam stability of linear differential equations, *J. Math. Anal. Appl.* **403**(1), 244–251, (2013). https://doi.org/10.1016/j.jmaa.2013.02.034.

[31] D. Inoan and D. Marian, Semi-Hyers–Ulam–Rassias stability of a Volterra integro-differential equation of order I with a convolution type Kernel via Laplace transform, *Symmetry*, **13**(11), 2181, (2021). https://doi.org/10.3390/sym13112181.

[32] J.T. Katsikadelis, Numerical solution of integrodifferential equations with convolution integrals, *Arch. Appl. Mech.* **89**, 2019–2032, (2019). https://doi.org/10.1007/s00419-019-01557-6.

https://doi.org/10.1142/9789811261572_0026

Chapter 26

Some Families of Finite Sums Associated with Interpolation Functions for many Classes of Special Numbers and Polynomials and Hypergeometric Series

Yilmaz Simsek

Department of Mathematics, Faculty of Science
University of Akdeniz TR-07058 Antalya, Turkey
ysimsek@akdeniz.edu.tr

We aim at presenting and surveying a systematic investigation of many families of finite sums which are associated with interpolation functions for many classes of special numbers and polynomials, hypergeometric series, the polygamma functions, the digamma functions, the generalized harmonic functions, the generalized harmonic numbers, the harmonic numbers, the combinatorial numbers and sums, and also the Dedekind sums. We gave new classes of Apostol-type numbers and polynomials with their generating functions, which are constructed by applications of the p-adic integrals including the Volkenborn integral and the fermionic integral. With the aid of these generating functions and their functional equations, we gave many identities and formulas which include the combinatorial numbers and polynomials, the Bernoulli numbers and polynomials, the Euler numbers and polynomials, the Apostol–Bernoulli numbers and polynomials, the Apostol–Euler numbers and polynomials, the Stirling numbers of the second kind, the combinatorial numbers, the generalized Eulerian type numbers, the Eulerian polynomials, the Fubini numbers, the Dobinski numbers. Moreover, we use partial zeta functions of interpolation functions for these numbers. These interpolation functions are associated with zeta-type functions such as the Riemann zeta function, the Hurwitz zeta function, the polygamma functions and the digamma functions. Finally, we mention that for a reasonably detailed histrorical account of the finite sums, interpolation functions, and special combinatotial numbers and polynomials, we refer the interested reader to our works which are given in Refs. [1–4]. Because we survey especially the results of these works.

The contents of this chapter are organized as follows: In Section 1, we give the main motivation of this chapter, and also some well-known definitions, notations, formulas and relations which are needed to give the results of this chapter.

In Section 2, Apostol-type new classes of numbers and polynomials are given. Some properties of these numbers are presented.

In Section 3, using p-adic integrals, two new families of special combinatorial numbers and polynomials with their generating functions are given. Some properties of these numbers and polynomials are presented.

In Section 4, generating functions for new classes of combinatorial numbers $y_{8,n}(\lambda; a)$ and polynomials $y_{8,n}(x, \lambda; a)$ are given with aid of the p-adic integral. Some properties of these numbers and polynomials are presented. Interpolation functions for the numbers $y_{8,n}(\lambda; a)$ are introduced. Partial Zeta-type functions for this interpolation function are presented.

In Section 5, generating functions for new classes of combinatorial numbers $y_{9,n}(\lambda; a)$ and polynomials $y_{9,n}(x, \lambda; a)$ are given with the aid of the p-adic integral. Some properties of these numbers and polynomials are presented. Interpolation functions for the numbers $y_{9,n}(\lambda; a)$ are introduced. Partial Zeta-type function for this interpolation function is presented.

In Section 6, many new finite sums arising from the numbers $y(n, \lambda)$ are given. Some interesting formulas for these sums are presented.

In Section 7, finite sums involving powers of binomial coefficients and combinatorial numbers are given.

In Section 8, finite sums derived from the Dedekind sums and the numbers $y(n, \lambda)$ are given.

In Section 9, finite sums derived from decomposition of the multiple Hurwitz zeta functions with the help of the numbers $y(n, \lambda)$ are given.

1. Introduction

We observe in the literature that numbers and their applications are the most used in mathematics and science. It is also known that there are various scientific studies about on the numbers almost every day. In addition to the applications of numbers and especially to special functions, the discovery of special numbers, may be starting with Bernoulli in 1500, have various applications in many areas of mathematics and other sciences. Today, generating functions, which have many applications in mathematics, in statistics, in probability, in quantum physics, in mathematical physics and in engineering and other sciences, have begun to play a leading role in the study of special numbers and polynomials. Generating functions are used not only for the study of many properties of special numbers, but also for special polynomials, differential equations, moment generating functions and other important subjects. They play a vital role in finite sums, which are just as important as generating functions for studying special numbers and special polynomials. After human beings started counting, they began of understand the importance of numbers, involving their summations, subtractions, multiplications and divisions, and also began to investigate their properties. Besides counting, finite sums and infinite sums (also known as series) of numbers quickly entered their field of interest. Therefore, in every developing period of science, finite sums have always been a source of need. At the same time, finite sums continued to be the focus of attention. Today, finite sums are one of the most used topics in mathematics, in mathematical models, mathematical software, and other vital research areas used in solving real-world problems.

Generating functions for special numbers and polynomials and finite sums, which are briefly mentioned and summarized in the above, form the basis and focus of this chapter. Therefore, with reference to the results of some articles written by the author in recent years, various novel and interesting new and some known formulas for (special) finite sums and special numbers and polynomials with their interpolation functions are presented. Their relations with some (special) finite sums, especially hypergeometric series, the Newton–Mercator series which is known as the Taylor series for the logarithm function, and their relations with other special functions are also discussed and interpreted in detail.

This section has been shaped by focusing on the following special sum (or numbers) and other related special numbers, special polynomials, specifically hypergeometric series, (special) finite sums, which emerged in the interpolation function of the special numbers $y_{8,k}(\lambda; a)$, which was published in 2020 by the author [2]:

$$y(n,\lambda) = \sum_{j=0}^{n} \frac{(-1)^n}{(j+1)\lambda^{j+1}(\lambda-1)^{n+1-j}}. \tag{1}$$

For $a, b \in \mathbb{R}$ with $a < b$ and $\lambda = \frac{x-a}{b-a}$, in Ref. [5], we modified the equation (1) as follows:

$$y\left(n, \frac{x-a}{b-a}\right) = \sum_{j=0}^{n} \frac{(-1)^{j-1}(b-a)^{n+2}}{(j+1)(x-a)^{j+1}(b-x)^{n+1-j}}, \tag{2}$$

Using (2), we also gave the following finite sum:

$$\sum_{j=0}^{n} \frac{(-1)^{j}(b-x)^{n+2}}{(j+1)(x-a)^{j+1}(b-x)^{n+1-j}} = \int_0^{\frac{b-x}{x-a}} \frac{(-1)^n y^{n+1} + 1}{y+1} dy,$$

(*cf.* [5]).

By using the equations (1) and (2), the author gave various interesting formulas and relations, see, for detail [2,3,5], and others. On the other hand, with the aid of the equations (1) and (2), it is still possible to discover many new formulas and relations involving special numbers and polynomials, finite sums, integral formulas and other special functions.

The sum given by equation (1) is discussed in detail in the following sections. There are many interesting and useful formulas related to this topic. At the same time, in the following sections, the zeta-type function from which this special sum arises and some of its properties are discussed. In addition to these, some new results of this function are also presented. Herewith, in the following sections, we study and survey relations among generating functions under integral transforms and derivative operators, many zeta-type functions and Dirichlet-type series, which are one of the most popular areas of analytic number theory and other mathematical analysis areas. Zeta-type functions involving Dirichlet-type series have been used for various kinds of interesting problems associated with models for solving some real-world problems (*cf.* [1–137]).

To give the results of this chapter, the following well-known definitions, notations, formulas, relations are needed, which will be used throughout the chapter.

Let $\mathbb{N}$, $\mathbb{Z}$ and $\mathbb{C}$ denote the set of natural numbers, the set of integer numbers and the set of complex numbers, resp. And also it is assumed that $\mathbb{N}_0 = \mathbb{N} \cup \{0\}$, $\mathbb{Z}^- = \{-1, -2, -3, \ldots\}$, $\mathbb{Z}_0 = \mathbb{Z}^- \cup \{0\}$.

We also tacitly suppose that for $z \in \mathbb{C}$, $\log z$ denotes the principal branch of the many-valued function $\text{Im}(\log z)$ with the imaginary part $\log z$ constrained by

$$-\pi < \text{Im}(\log z) \leq \pi.$$

Thus,

$$\log e = 1,$$

is considered throughout this chapter. We also assume that

$$0^n = \begin{cases} 1, & n = 0 \\ 0, & n \in \mathbb{N}. \end{cases}$$

Pochhammer's symbol for the rising factorial is given by the following notation:

$$(\lambda)_v = \frac{\Gamma(\lambda + v)}{\Gamma(\lambda)} = \lambda(\lambda+1)\cdots(\lambda+v-1),$$

and

$$(\lambda)_0 = 1,$$

for $\lambda \neq 1$, where $v \in \mathbb{N}$, $\lambda \in \mathbb{C}$ and $\Gamma(\lambda)$ denotes the gamma function, which is an important special function in mathematics.

$$\binom{z}{v} = \frac{z(z-1)\cdots(z-v+1)}{v!} = \frac{(z)^{\underline{v}}}{v!} \quad (v \in \mathbb{N}, z \in \mathbb{C}),$$

and

$$\binom{z}{0} = 1.$$

Observe that

$$(-\lambda)_v = (-1)^v (\lambda)^{\underline{v}}.$$

The following references can be given for briefly the above notations (*cf.* [1–137]).

Let k be an indeterminate.

The Bernoulli numbers of order k, $B_n^{(k)}$, and the Bernoulli polynomials of order k, $B_n^{(k)}(x)$, are, resp., defined by means of the following generating function:

$$F_{bs}(t; k) = \left(\frac{t}{e^t - 1}\right)^k = \sum_{n=0}^{\infty} B_n^{(k)} \frac{t^n}{n!}$$

and

$$F_{bp}(t, x; k) = e^{xt} F_{bs}(t; k) = \sum_{n=0}^{\infty} B_n^{(k)}(x) \frac{t^n}{n!},$$

where $|t| < 2\pi$. By using the function $F_{bs}(t;k)$, for $k \in \mathbb{N}$, we have

$$t^k = \sum_{n=0}^{\infty}\sum_{j=0}^{k}(-1)^{k-j}\binom{k}{j}j^n\frac{t^n}{n!}\sum_{n=0}^{\infty}B_n^{(k)}\frac{t^n}{n!}.$$

Therefore,

$$t^k = \sum_{n=0}^{\infty}\sum_{l=0}^{n}\binom{n}{l}\sum_{j=0}^{k}(-1)^{k-j}\binom{k}{j}j^l B_{n-l}^{(k)}\frac{t^n}{n!}.$$

Hence, we have

$$\sum_{l=0}^{n}\binom{n}{l}\sum_{j=0}^{k}(-1)^{k-j}\binom{k}{j}j^l B_{n-l}^{(k)} = \begin{cases} 1, & \text{if } n = k, \\ 0, & \text{if } n \neq k, \end{cases}$$

(*cf.* [2,3,18,24,34,43,44,50,54,63–66,72,81,92–95,110–113,117–119,121,124, 129]; and cited references therein).

By using the following functional equation:

$$F_{bp}(t,x;k) = e^{xt}F_{bs}(t;k),$$

we have

$$B_n^{(k)}(x) = \sum_{l=0}^{n}\binom{n}{l}x^{n-l}B_l^{(k)}.$$

Here, we note that

$$B_n^{(0)}(x) = x^n.$$

Since

$$F_{bp}(t,0;k) = F_{bs}(t;k),$$

we have

$$B_n^{(k)} = B_n^{(k)}(0).$$

Here, we note that in the literature, there are various computational formulae for the numbers $B_n^{(k)}$.

The function $F_{bs}(t;1)$ gives us the generating function for the Bernoulli numbers, B_n. That is

$$B_n = B_n^{(1)}.$$

(*cf.* [2,3,18,24,34,43,44,50,54,63–66,72,81,92–95,110–113,117–119,121,124, 129]; and cited references therein).

Now, we recall in the following another type of Bernoulli numbers and polynomials which were defined by Apostol [9].

The Apostol–Bernoulli numbers $\mathcal{B}_n(\lambda)$ are defined by means of the following generating function:

$$F_{as}(t;\lambda) = \frac{t}{\lambda e^t - 1} = \sum_{n=0}^{\infty} \mathcal{B}_n(\lambda) \frac{t^n}{n!}.$$

By using the function $F_{as}(t;\lambda)$, we have

$$t = \sum_{n=0}^{\infty} \left(\lambda \sum_{l=0}^{n} \binom{n}{l} \mathcal{B}_l(\lambda) - \mathcal{B}_n(\lambda) \right) \frac{t^n}{n!}.$$

Therefore,

$$\lambda \sum_{l=0}^{n} \binom{n}{l} \mathcal{B}_l(\lambda) \frac{t^n}{n!} - \mathcal{B}_n(\lambda) = \begin{cases} 1, & \text{if } n = 1 \\ 0, & \text{if } n \neq 1, \end{cases}$$

where

$$\mathcal{B}_0(\lambda) = 0.$$

The Apostol–Bernoulli polynomials $\mathcal{B}_n(x;\lambda)$ are defined by means of the following generating function:

$$F_{ap}(t,x;\lambda) = F_{as}(t;\lambda)e^{xt} = \sum_{n=0}^{\infty} \mathcal{B}_n(x;\lambda) \frac{t^n}{n!}, \tag{3}$$

where λ is an arbitrary (real or complex) parameter and $|t| < 2\pi$ when $\lambda = 1$ and $|t| < |\log \lambda|$ when $\lambda \neq 1$ (*cf.* [9,76,125,129]).

Using (3), we have

$$\sum_{n=0}^{\infty} x^n \frac{t^{n+1}}{n!} = \lambda \sum_{n=0}^{\infty} \sum_{j=0}^{n} \binom{n}{j} \mathcal{B}_j(x;\lambda) \frac{t^n}{n!} - \sum_{n=0}^{\infty} \mathcal{B}_n(x;\lambda) \frac{t^n}{n!}.$$

Comparing the coefficients of $\frac{t^n}{n!}$ on both sides of the above equation, we arrive at the following well-known relation:

$$\mathcal{B}_n(x;\lambda) = \lambda \sum_{j=0}^{n} \binom{n}{j} \mathcal{B}_j(x;\lambda) - nx^{n-1}.$$

Moreover, for $x = 0$, we have

$$F_{ap}(t,0;\lambda) = F_{as}(t;\lambda).$$

That is

$$\mathcal{B}_n(0;\lambda) = \mathcal{B}_n(\lambda),$$

(*cf.* [9,76,125,129]). For details about the Apostol–Bernoulli polynomials and numbers, see Ref. [9].

The Apostol–Euler numbers $\mathcal{E}_n(x;\lambda)$ are defined by means of the following generating function:

$$F_{es}(t;\lambda) = \frac{2}{\lambda e^t + 1} = \sum_{n=0}^{\infty} \mathcal{E}_n(\lambda) \frac{t^n}{n!}.$$

By using the function $F_{es}(t;\lambda)$, we have

$$2 = \sum_{n=0}^{\infty} \left(\lambda \sum_{l=0}^{n} \binom{n}{l} \mathcal{E}_l(\lambda) + \mathcal{E}_n(\lambda) \right) \frac{t^n}{n!}.$$

Therefore,

$$\lambda \sum_{l=0}^{n} \binom{n}{l} \mathcal{E}_l(\lambda) \frac{t^n}{n!} + \mathcal{E}_n(\lambda) = \begin{cases} 2, & \text{if } n = 0 \\ 0, & \text{if } n > 0, \end{cases}$$

where

$$\mathcal{E}_n(\lambda) = \frac{2}{\lambda + 1}.$$

The Apostol–Euler polynomials, $\mathcal{E}_n(x;\lambda)$, are defined by the following generating function:

$$F_{ep}(t,x;\lambda) = F_{es}(t;\lambda)e^{xt} = \sum_{n=0}^{\infty} \mathcal{E}_n(x;\lambda) \frac{t^n}{n!}, \tag{4}$$

where λ is an arbitrary (real or complex) parameter and $|t| < \pi$ when $\lambda = 1$ and $|t| < |\log(-\lambda)|$ when $\lambda \neq 1$ (*cf.* [76,125,129]).

Using (4), we have

$$2\sum_{n=0}^{\infty} x^n \frac{t^n}{n!} = \lambda \sum_{n=0}^{\infty} \sum_{j=0}^{n} \binom{n}{j} \mathcal{E}_j(x;\lambda) \frac{t^n}{n!} + \sum_{n=0}^{\infty} \mathcal{E}_n(x;\lambda) \frac{t^n}{n!}.$$

Comparing the coefficients of $\frac{t^n}{n!}$ on both sides of the above equation, we arrive at the following well-known relation:

$$\mathcal{E}_n(x;\lambda) = 2x^n - \lambda \sum_{j=0}^{n} \binom{n}{j} \mathcal{E}_j(x;\lambda).$$

Moreover, for $x = 0$, we have

$$F_{ep}(t,0;\lambda) = F_{es}(t;\lambda).$$

That is

$$\mathcal{E}_n(0;\lambda) = \mathcal{E}_n(\lambda),$$

(*cf.* [76,125,129]).

When $\lambda = 1$ in (4), we have

$$\mathcal{E}_n(1) = E_n,$$

where E_n denotes the Euler numbers (which are also the so-called Euler numbers of the first kind (*cf.* [7,129]; and references therein).

A known relation between the Apostol–Euler numbers, $\mathcal{E}_n(\lambda)$ and the Apostol–Bernoulli numbers, $\mathcal{B}_n(\lambda)$ is given by

$$\mathcal{E}_n(\lambda) = -\frac{2}{n+1}\mathcal{B}_{n+1}(-\lambda), \tag{5}$$

(*cf.* [121, Eq. (1.28)], [129]).

Let $a, b, c \in \mathbb{R}^+ \ (a \neq b), x \in \mathbb{R}, \lambda \in \mathbb{C}$ and $u \in \mathbb{C} \setminus \{\lambda\}$. The generalized Eulerian-type polynomials, $\mathcal{H}_n(x; u; a, b, c; \lambda)$, are defined by means of the following generating function:

$$F_\lambda(t, x; u, a, b, c; \lambda) = \frac{(a^t - u)\, c^{xt}}{\lambda b^t - u} = \sum_{n=0}^{\infty} \mathcal{H}_n(x; u; a, b, c; \lambda)\frac{t^n}{n!}, \tag{6}$$

where $|t| < \frac{2\pi}{|\log b|}$ when $\lambda = u$; $\left|t \log b + \log\left(\frac{\lambda}{u}\right)\right| < 2\pi$ when $\lambda \neq u$ (*cf.* [106, Eq. (13)]; and also see Refs. [67,107,117]).

$$F_\lambda(t, x; u, a, b, c; \lambda) = \frac{(a^t - u)}{ut \ln b} F_{ap}\left(t \ln b, x\frac{\ln c}{\ln b}; \frac{\lambda}{u}\right).$$

By using the above functional equation, we have

$$\sum_{n=0}^{\infty} n\mathcal{H}_{n-1}(x; u; a, b, c; \lambda)\frac{t^n}{n!} = \frac{1}{u \ln b}\sum_{n=0}^{\infty}\sum_{j=0}^{n}\binom{n}{j}(\ln a)^j \mathcal{B}_{n-j}\left(x\frac{\ln c}{\ln b}; \frac{\lambda}{u}\right)\frac{t^n}{n!}$$
$$-\sum_{n=0}^{\infty}(\ln b)^{n-1}\mathcal{B}_n\left(x\frac{\ln c}{\ln b}; \frac{\lambda}{u}\right)\frac{t^n}{n!}.$$

Comparing the coefficients of $\frac{t^n}{n!}$ on both sides of the above equation, we arrive at the following theorem:

Theorem 1.

$$\begin{aligned} &n\mathcal{H}_{n-1}(x; u; a, b, c; \lambda) \\ &= \frac{1}{u \ln b}\sum_{j=0}^{n}\binom{n}{j}(\ln a)^j \mathcal{B}_{n-j}\left(x\frac{\ln c}{\ln b}; \frac{\lambda}{u}\right) \\ &\quad - (\ln b)^{n-1}\mathcal{B}_n\left(x\frac{\ln c}{\ln b}; \frac{\lambda}{u}\right). \end{aligned} \tag{7}$$

When $x = 0$ in (8), we have the generalized Eulerian-type numbers:

$$\mathcal{H}_n(0; u; a, b, c; \lambda) = \mathcal{H}_n(u; a, b, c; \lambda),$$

which are defined by means of the following generating function:

$$\frac{a^t - u}{\lambda b^t - u} = \sum_{n=0}^{\infty} \mathcal{H}_n(u; a, b, c; \lambda) \frac{t^n}{n!}, \tag{8}$$

(*cf.* [106]; and also see Refs. [67,107,117]).

In the special case when $\lambda = a = 1$ and $b = c = e$, the generalized Eulerian-type polynomials are reduced to the Eulerian polynomials (or Frobenius–Euler polynomials) which are defined by means of the following generating function:

$$\frac{1-u}{e^t - u} e^{xt} = \sum_{n=0}^{\infty} H_n(x; u) \frac{t^n}{n!}, \tag{9}$$

(*cf.* [19,21–23,59,67,105]).

Setting $a = 1$ and $b = c = e$ in (7), we have

$$\begin{aligned} n\mathcal{H}_{n-1}(x; u; 1, e, e; 1) &= H_n(x; u) \\ &= \frac{1-u}{u} \mathcal{B}_n\left(x; \frac{1}{u}\right), \end{aligned}$$

(*cf.* [104]).

Putting $x = 0$ in (9), we have the following Eulerian numbers:

$$H_n(0; u) = H_n(u),$$

(*cf.* [19,21–23,59,67,105]).

When $u = -1$ and $x = 0$, equation (9) gives us

$$H_n(0; -1) = H_n(-1) = E_n,$$

(*cf.* [19,21–23,59,67,105]).

The Stirling numbers of the first kind, denoted by $S_1(n, k)$ and also in general represented with the symbol $s(n, k)$, are defined by the following generating function:

$$\frac{(\log(1+z))^k}{k!} = \sum_{n=0}^{\infty} S_1(n, k) \frac{z^n}{n!}, \tag{10}$$

with $S_1(n, k) = 0$ if $k > n$, and $k \in \mathbb{N}_0$; the numbers $S_1(n, k)$ are also given by

$$(x)^{\underline{n}} = \sum_{v=0}^{n} S_1(n, v) x^v.$$

(*cf.* [2,3,18,24,34,37,43,44,50,54,63–66,72,81,92–95,110–113,117–119,121, 124,129]; and cited references therein).

Observe that the signs of the (signed) Stirling numbers of the first kind are given as

$$(-1)^{n-k}S_1(n,k).$$

The Stirling numbers of the second kind, denoted by $S_2(n,k)$ and also in general represented with the symbol $S(n,k)$, are defined by the following generating function:

$$\frac{(e^t-1)^k}{k!}=\sum_{n=0}^{\infty}S_2(n,k)\frac{t^n}{n!}, \tag{11}$$

(*cf.* [7,129]; and references therein).

By using (11), an explicit formula for the numbers $S_2(n,k)$ is given by

$$S_2(n,k)=\frac{1}{k!}\sum_{j=0}^{k}(-1)^{k-j}\binom{k}{j}j^n, \tag{12}$$

with $k>n$, $S_2(n,k)=0$ (*cf.* [7,129]).

The Stirling numbers of the second kind are also given by the following generating function including the falling factorial:

$$x^n=\sum_{v=0}^{n}S_2(n,v)(x)^{\underline{v}}, \tag{13}$$

(*cf.* [7,129]; and references therein).

The harmonic numbers, denoted by $f(n)$ in Euler's works and also in general represented with the symbol h_n, H_n and $H(n)$, are defined by means of the following generating function:

$$\sum_{n=1}^{\infty}h_nz^n=\frac{\log(1-z)}{z-1}, \tag{14}$$

where $|z|<1$ (*cf.* [18,34,43,92,95,124,129]).

By using (14), we have

$$\sum_{n=1}^{\infty}(-1)^{n+1}h_nz^n+\sum_{n=1}^{\infty}(-1)^{n+1}h_nz^{n-1}=\log(1+z).$$

By combining equation (13) in Ref. [3], we have

$$\sum_{n=1}^{\infty}(-1)^{n+1}h_nz^n+\sum_{n=0}^{\infty}(-1)^nh_{n+1}z^n=\sum_{n=0}^{\infty}B_n\frac{(\log(1+z))^n}{n!},$$

Therefore,

$$\sum_{m=0}^{\infty}(-1)^{m+1}h_m z^m+\sum_{m=1}^{\infty}(-1)^m h_{m+1}z^m=\sum_{m=0}^{\infty}\sum_{n=0}^{m}B_nS_1(m,n)\frac{z^m}{m!},$$

such that we here use the fact that $S_1(m,n)=0$ if $n>m$. Now, equating the coefficients of z^m on both sides of the above equation, we arrive at the following theorem:

Theorem 2. *Let* $m\in\mathbb{N}$. *Then we have*

$$\sum_{n=0}^{m}B_nS_1(m,n)=(-1)^m m!(h_{m+1}-h_m). \tag{15}$$

By using (15), we have the following well-known result:

$$\sum_{n=0}^{m}B_nS_1(m,n)=(-1)^m\frac{m!}{m+1}, \tag{16}$$

(*cf.* [3,24, p. 117, 50,92, p. 45, Exercise 19(b)]).

In Refs. [3,18,95,124], the harmonic numbers are also given by the following integral formula:

$$h_n=\int_0^1\frac{1-x^n}{1-x}dx, \tag{17}$$

and

$$h_n=-n\int_0^1 x^{n-1}\log(1-x)dx=-n\int_0^1(1-x)^{n-1}\log(x)dx, \tag{18}$$

where $h_0=0$.

The generalized harmonic numbers (or the generalized harmonic functions) are defined by the following formula: $h_0(x)=0$,

$$h_m(x)=\sum_{n=1}^{m}\frac{1}{n+x},$$

where x is an indeterminate and $m\in\mathbb{N}$ (*cf.* [33]).

It is clear that

$$h_m=h_m(0).$$

The alternating harmonic numbers, denoted by $\mathfrak{h}_n$, are defined by means of the following generating function:

$$F_2(z)=\frac{\log(1+z)}{z-1}=\sum_{n=1}^{\infty}\mathfrak{h}_n z^n, \tag{19}$$

where $|z|<1$ (*cf.* [3,34,124]).

The Fubini numbers, $w_g(n)$, are defined by means of the following generating function:

$$\frac{1}{2-e^t} = \sum_{n=0}^{\infty} w_g(n)\frac{t^n}{n!}, \tag{20}$$

(*cf.* [34,42,43,45], [46]).

By using (20), and assuming that $|1-e^t|<1$, we have

$$\begin{aligned}\sum_{m=0}^{\infty} w_g(m)\frac{t^m}{m!} &= \sum_{n=0}^{\infty}(-1)^n(1-e^t)^n \\ &= \sum_{m=0}^{\infty}\sum_{n=0}^{m} n!S_2(m,n)\frac{t^m}{m!}.\end{aligned}$$

Comparing the coefficients of $\frac{t^m}{m!}$ on both sides of the above equation, we arrive at the following well-known result:

$$w_g(n) = \sum_{j=0}^{n} j!S_2(n,j), \tag{21}$$

(*cf.* [34,42,45,46]).

The Dobinski numbers, $D(n)$, are defined by means of the following generating function:

$$F_d(t) = \frac{1}{e^{e^t-1}} = \sum_{n=0}^{\infty} D(n)\frac{t^n}{n!}, \tag{22}$$

(*cf.* [43, Eq. (3.15)]).

The Dobinski numbers are related to the exponential numbers (or the Bell numbers) and other combinatorial numbers. The exponential numbers, which not only occur often in probability, but also are associated with that of the Poisson–Charlier polynomial, are defined by

$$F_e(t) = e^{e^t-1} = \sum_{n=0}^{\infty} B(n)\frac{t^n}{n!},$$

(*cf.* [34,58,94,108]). Using the above generating functions, one has the following well-known results:

$$B(n) = \sum_{j=0}^{n} S_2(n,j),$$

(*cf.* [34,58,94,108]). Since

$$F_e(t)F_d(t) = 1,$$

for $n \in \mathbb{N}$, we have

$$\sum_{j=0}^{n} \binom{n}{j} B(n-j)D(j) = 0.$$

In Refs. [115,116], we defined the λ-Apostol–Daehee numbers of higher order. Generating function for the λ-Apostol–Daehee numbers, $\mathfrak{D}_n(\lambda)$, is given by

$$\frac{\log \lambda + \log(1+\lambda t)}{\lambda(1+\lambda t) - 1} = \sum_{n=0}^{\infty} \mathfrak{D}_n(\lambda) \frac{t^n}{n!}, \tag{23}$$

(*cf.* [115,116]). Recently, the λ-Apostol–Daehee numbers have also been studied by many authors, such as [28,64,66,115,116,122].

By using (23), we have

$$\begin{aligned}
\mathfrak{D}_0(\lambda) &= \frac{\log \lambda}{\lambda - 1}, \\
\mathfrak{D}_1(\lambda) &= -\frac{\lambda^2 \log \lambda}{(\lambda-1)^2} + \frac{\lambda}{\lambda - 1}, \\
\mathfrak{D}_2(\lambda) &= \frac{2\lambda^4 \log \lambda}{(\lambda-1)^3} + \frac{\lambda^2(1-3\lambda)}{(\lambda-1)^2}, \\
\mathfrak{D}_3(\lambda) &= -\frac{6\lambda^6 \log \lambda}{(\lambda-1)^4} + \frac{\lambda^3(11\lambda^2 - 7\lambda + 2)}{(\lambda-1)^3},
\end{aligned}$$

and so on (*cf.* [66,115,116,122]).

For $n \in \mathbb{N}$, combining equation (23) with the following well-known series:

$$\log(1+t) = \sum_{n=1}^{\infty} (-1)^{n+1} \frac{t^n}{n},$$

$|t| < 1$, we have

$$\begin{aligned}
&\mathfrak{D}_0(\lambda) + \sum_{n=1}^{\infty} \mathfrak{D}_n(\lambda) \frac{t^n}{n!} \\
&\quad = \frac{\log \lambda}{\lambda - 1} + \frac{\log \lambda}{\lambda - 1} \sum_{n=1}^{\infty} (-1)^n \left(\frac{\lambda^2}{\lambda - 1} t\right)^n + \frac{1}{\lambda - 1} \sum_{n=1}^{\infty} (-1)^{n+1} \lambda^n \frac{t^n}{n} \\
&\qquad + \frac{1}{\lambda - 1} \sum_{n=1}^{\infty} (-1)^{n+1} \lambda^n \frac{t^n}{n} \sum_{n=1}^{\infty} (-1)^n \left(\frac{\lambda^2}{\lambda - 1} t\right)^n,
\end{aligned}$$

assuming that $|\lambda t| < 1$. By applying the Cauchy multiplication rule to the right-hand side of the above equation for two series products, after some elementary calculation, we arrive at the following explicit formula for the numbers $\mathfrak{D}_n(\lambda)$:

For $n = 0$, we have

$$\mathfrak{D}_0(\lambda) = \frac{\log \lambda}{\lambda - 1}$$

and for $n \geq 1$, we have

$$\mathfrak{D}_n(\lambda) = (-1)^{n+1} n! \left(\frac{\lambda^n}{n(\lambda-1)} - \frac{\lambda^{2n} \log \lambda}{(\lambda-1)^{n+1}} + \sum_{k=1}^{n} \frac{\lambda^{n+k}}{(n-k)(\lambda-1)^{k+1}} \right).$$

There are many different explicit formulae for the numbers $\mathfrak{D}_n(\lambda)$, for details, see (*cf.* [66,115,116,122]).

Kucukoglu and Simsek [66] gave the following explicit formula for the numbers $\mathfrak{D}_n(\lambda)$:

$$\mathfrak{D}_n(\lambda) = n!(-1)^n \left(\frac{\lambda^2}{\lambda-1} \right)^n \left(\frac{\log \lambda}{\lambda-1} - \frac{1}{\lambda} \sum_{k=0}^{n-1} \frac{1}{k+1} \left(\frac{\lambda-1}{\lambda} \right)^k \right). \quad (24)$$

When $\lambda = 1$ in (23), we have

$$\mathfrak{D}_n(1) = D_n$$

where D_n denotes the Daehee numbers (*cf.* [36,50]; and see also [28,66,115, 116,122]).

By using (23), one has a computation formula for the Daehee numbers D_n as follows:

$$D_n = (-1)^n \frac{n!}{n+1} \quad (25)$$

(*cf.* [50,91]; and see also [121]).

The Changhee numbers, Ch_n, are defined by the following generating function:

$$\frac{2}{2+t} = \sum_{n=0}^{\infty} Ch_n \frac{t^n}{n!} \quad (26)$$

(*cf.* [52,60]).

By using (26), one has a computation formula for the Changhee numbers Ch_n as follows:

$$Ch_n = (-1)^n \frac{n!}{2^n} \tag{27}$$

(*cf.* [52], see also [60,121]).

The numbers $Y_n(u;a)$ and the polynomials $Y_n(x,u;a)$ are defined by means of the following generating functions, resp.:

$$G_Y(t,u,a) = \frac{1}{a^t - u} = \sum_{n=0}^{\infty} Y_n(u;a)\frac{t^n}{n!}, \tag{28}$$

and

$$H_Y(x,t,u,a) = G_Y(t,u,a)a^{xt} = \sum_{n=0}^{\infty} Y_n(x,u;a)\frac{t^n}{n!}, \tag{29}$$

where $a \geq 1$; $u \neq 0$, $u \neq 1$ and $\left|t\log a + \log\left(\frac{1}{u}\right)\right| < 2\pi$ (*cf.* [106, Eq. (37), 117]).

With the aid of Eq. (28), we get

$$\sum_{n=0}^{\infty} Y_n(u;a)\frac{t^{n+1}}{n!} = \frac{t}{u\left(\frac{1}{u}e^{t\ln a} - 1\right)}$$

Combining (3) with the above equation, we obtain

$$Y_{n-1}(u;a) = \frac{(\ln a)^{n-1}}{un}\mathcal{B}_n\left(\frac{1}{u}\right),$$

where $n \in \mathbb{N}$ (*cf.* [2,117]).

By using (28), we also have

$$\sum_{n=0}^{\infty} Y_n(u;a)\frac{t^n}{n!} = \frac{1-u}{(1-u)\left(e^{t\ln a} - u\right)}.$$

Combining the above equation with (9) for $x = 0$, we obtain

$$Y_n(u;a) = \frac{(\ln a)^n}{1-u}H_n(x;u),$$

where $n \in \mathbb{N}_0$ (*cf.* [2]).

Setting $a = 1$ in (28), we have

$$Y_0(u;1) = \frac{1}{1-u},$$

(*cf.* [106, p. 22]). Putting $a = 1$ into (29), we also have

$$Y_0(x,u;1) = \frac{1}{1-u}.$$

For details about these numbers and polynomials, see also the following references [7,106,117]:

When we substitute $x = 0$ into (29), we have

$$\mathrm{Y}_n(0, u; a) = \mathrm{Y}_n(u; a).$$

Combining (29) with (28), we have

$$\mathrm{Y}_n(x, u; a) = \sum_{j=0}^{n} \binom{n}{j} x^{n-j} \mathrm{Y}_j(u; a),$$

(*cf.* [106, Eq. (37), 117).

Generalized hypergeometric function ${}_pF_p$ is defined by

$${}_pF_q\left[\begin{matrix}\alpha_1, ..., \alpha_p \\ \beta_1, ..., \beta_q\end{matrix}; z\right] = \sum_{m=0}^{\infty} \left(\frac{\prod_{j=1}^{p} (\alpha_j)_m}{\prod_{j=1}^{q} (\beta_j)_m}\right) \frac{z^m}{m!}.$$

The above series converges for all z if $p < q+1$, and for $|z| < 1$ if $p = q+1$. For this series, one can assume that all parameters have general values, real or complex, except for the β_j, $j = 1, 2, ..., q$ none of which is equal to zero or to a negative integer.

Some special hypergeometric functions are given as follows:

$${}_0F_0(z) = e^z,$$

$${}_2F_1(\alpha_1; \alpha_2; \beta_1; z) = \sum_{k=0}^{\infty} \frac{(\alpha_1)_k (\alpha_2)_k}{(\beta_1)_k} \frac{z^k}{k!}$$

and

$${}_1F_0\left[\begin{matrix} b \\ - \end{matrix}; x\right] = \frac{1}{(1-x)^b},$$

(*cf.* [6,16,32,63,73,79,88,90,129,130,132,133,137]).

We also need the following formulae for operations on power series:

The formula for division of the following power series $\sum_{k=0}^{\infty} a_k x^k$ and $\sum_{k=0}^{\infty} b_k x^k$ is given by

$$\frac{\sum_{v=0}^{\infty} b_v x^v}{\sum_{v=0}^{\infty} a_v x^v} = \frac{1}{a_0} \sum_{v=0}^{\infty} c_v x^v,$$

where

$$0 = c_n + \frac{1}{a_0} \sum_{k=0}^{n} c_{n-k} a_k - b_n,$$

(*cf.* [138, p. 17]). The above relation is also given by

$$c_n = \frac{(-1)^n}{a_0^n}\begin{vmatrix} a_1b_0 - a_0b_1 & a_0 & 0 & \cdots & 0 \\ a_2b_0 - a_0b_2 & a_1 & a_0 & \cdots & 0 \\ a_3b_0 - a_0b_3 & a_2 & a_1 & \cdots & 0 \\ \vdots & \vdots & \vdots & \ddots & \vdots \\ a_{n-1}b_0 - a_0b_{n-1} & a_{n-2} & a_{n-3} & \cdots & a_0 \\ a_nb_0 - a_0b_n & a_{n-1} & a_{n-2} & \cdots & a_1 \end{vmatrix},$$

(*cf.* [138, p. 17]).

The formula for power series raised to powers is as follows:

$$\left(\sum_{v=0}^{\infty} a_v x^v\right)^n = \sum_{v=0}^{\infty} c_v x^v,$$

where $n \in \mathbb{N}$,

$$(a_0)^n = c_0,$$

and for $v \in \mathbb{N}$,

$$c_v = \frac{1}{va_0}\sum_{j=0}^{v}(jn - v + j)\, a_j c_{v-j},$$

(*cf.* [138, p. 17]).

The formulas for the substitution of one series into another are as follows:

$$\sum_{v=1}^{\infty} b_v y^v = \sum_{v=1}^{\infty} c_v x^v,$$

$$y = \sum_{v=1}^{\infty} a_v x^v,$$

where

$$\begin{aligned} c_1 &= a_1b_1 \\ c_2 &= a_2b_1 + a_1^2b_2 \\ c_3 &= a_3b_1 + 2a_1b_2b_2 + a_1^3b_3 \\ c_4 &= a_4b_1 + a_2^2b_2 + 2a_1a_3b_2 + 3a_1^2a_2b_3 + a_1^4b_4, \end{aligned}$$

so on (*cf.* [138, p. 17]).

The formulas for the multiplication of power series are as follows:

$$\sum_{v=0}^{\infty} a_v x^v \sum_{k=0}^{\infty} b_v x^v = \sum_{v=0}^{\infty} c_v x^v,$$

where $v \in \mathbb{N}_0$,

$$c_v = \sum_{j=0}^{v} a_j b_{v-j},$$

(*cf.* [138, p. 18]).

2. Some Classes of Apostol-Type Numbers and Polynomials

In Ref. [2], we defined some classes of Apostol-type numbers and polynomials by means of the following generating functions. These numbers and polynomials denote by $\mathcal{Y}_n(u; a, \lambda)$ and $\mathcal{Y}_n(x, u; a, \lambda)$, resp.

$$g_Y(t, u, a, \lambda) = \frac{1}{\lambda a^t - u} = \sum_{n=0}^{\infty} \mathcal{Y}_n(u; a, \lambda) \frac{t^n}{n!}, \tag{30}$$

and

$$h_Y(x, t, u, a, \lambda) = g_Y(t, u, a, \lambda) a^{xt} = \sum_{n=0}^{\infty} \mathcal{Y}_n(x, u; a, \lambda) \frac{t^n}{n!}, \tag{31}$$

where $a \geq 1$; $u \neq 0, 1$.

These numbers $\mathcal{Y}_n(u; a, \lambda)$ are related to the many well-known numbers and polynomials which are given as follows (*cf.* for detail, see Ref. [2]):

Substituting $a = 1$ into (30), we have

$$\mathcal{Y}_0(u; a, \lambda) = \frac{1}{\lambda - u}.$$

We also note that

$$\mathcal{Y}_n(u; a, \lambda) = \frac{1}{\lambda} \mathrm{Y}_n\left(\frac{u}{\lambda}; a\right). \tag{32}$$

When $a = e$, for $n \in \mathbb{N}$, we also obtain

$$\mathcal{Y}_{n-1}(u; e, \lambda) = \frac{1}{nu} \mathcal{B}_n\left(\frac{u}{\lambda}\right). \tag{33}$$

Combining (33) with (5), we get

$$\mathcal{Y}_n(u; e, \lambda) = -\frac{1}{2u} \mathcal{E}_n\left(-\frac{u}{\lambda}\right),$$

$$\mathcal{Y}_n(u; e, \lambda) = \frac{1}{\lambda - u} H_n\left(\frac{u}{\lambda}\right).$$

Substituting $u = -1$ and $\lambda = 1$ into the above equation, we have

$$\mathcal{Y}_n(-1; e, 1) = \frac{1}{2} E_n.$$

Theorem 3 (*cf.* [2]). *Let*

$$\mathcal{Y}_0(u;a,\lambda)=\frac{1}{\lambda-u}.$$

For $n\geq 1$, we have

$$\mathcal{Y}_n(u;a,\lambda)=\frac{\lambda}{u}\sum_{v=0}^{n}\binom{n}{v}\mathcal{Y}_v(u;a,\lambda)(\log a)^{n-v}. \tag{34}$$

Using equation (34), for $n=1,2,3,\ldots$, some few values of the numbers $\mathcal{Y}_n(u;a,\lambda)$ are found as follows:

$$\mathcal{Y}_1(u;a,\lambda)=-\frac{\lambda\log a}{(\lambda-u)^2},$$
$$\mathcal{Y}_2(u;a,\lambda)=\frac{\lambda(\log a)^2}{(\lambda-u)^3}(u+2-\lambda),$$

and so on (*cf.* [2], see also [141]).

Theorem 4 (*cf.* [2]). *Let $n\in\mathbb{N}_0$. Then we have*

$$\mathcal{Y}_n(x,u;a,\lambda)=\sum_{v=0}^{n}\binom{n}{v}\mathcal{Y}_v(u;a,\lambda)(x\log a)^{n-v}. \tag{35}$$

Using the equations (34) and (35), for $n=1,2,3,\ldots$, some few values of the polynomials $\mathcal{Y}_n(x,u;a,\lambda)$ are found as follows:

$$\mathcal{Y}_0(x,u;a,\lambda)=\frac{1}{\lambda-u},$$
$$\mathcal{Y}_1(x,u;a,\lambda)=\frac{\log a}{\lambda-u}x-\frac{\lambda\log a}{(\lambda-u)^2},$$
$$\mathcal{Y}_2(x,u;a,\lambda)=\frac{(\log a)^2}{\lambda-u}x^2-\frac{2\lambda(\log a)^2}{(\lambda-u)^2}x+\frac{\lambda(\log a)^2}{(\lambda-u)^3}(u+2-\lambda),$$

and so on (*cf.* [2]).

3. Two New Families of Special Combinatorial Numbers and Polynomials Derived from p-adic Integrals

In this section, we introduce two new families of special combinatorial numbers and polynomials derived from p-adic integrals which are briefly given as follows:

Let $\mathbb{Z}_p$ denote the set of p-adic integers. Let $\mathbb{K}$ be a field with a complete valuation. Let $f \in C^1(\mathbb{Z}_p \to \mathbb{K})$ be a set of continuous derivative functions.

The Volkenborn integral (bosonic p-adic integral) of the function f on $\mathbb{Z}_p$ is given by

$$\int_{\mathbb{Z}_p} f(x)\, d\mu_1(x) = \lim_{N\to\infty} \frac{1}{p^N} \sum_{x=0}^{p^N-1} f(x), \tag{36}$$

where $x \in \mathbb{Z}_p$ and

$$\mu_1(x) = \frac{1}{p^N},$$

(*cf.* [53,54,96,121]; and the references cited therein).

By using (36), the Bernoulli numbers B_n are also given by

$$\int_{\mathbb{Z}_p} y^n d\mu_1(y) = B_n, \tag{37}$$

$y \in \mathbb{Z}_p$ (*cf.* [53,54,96,121]; and the references cited therein).

With the help of (36), Kim *et al.* [50] gave the following formula:

$$\int_{\mathbb{Z}_p} (y)^{\underline{n}}\, d\mu_1(y) = \frac{(-1)^n n!}{n+1} = D_n, \tag{38}$$

where $y \in \mathbb{Z}_p$.

Recently, by using (36) and (38), many properties and applications of the function $(x)^{\underline{n}}$ were given by Simsek [121].

The fermionic p-adic integral of function f on $\mathbb{Z}_p$ is given by

$$\int_{\mathbb{Z}_p} f(x)\, d\mu_{-1}(x) = \lim_{N\to\infty} \sum_{x=0}^{p^N-1} (-1)^x f(x), \tag{39}$$

where $x \in \mathbb{Z}_p$ and

$$\mu_{-1}(x) = (-1)^x,$$

(*cf.* [55–57,59,121] and the references therein).

With the help of equation (39), the Euler numbers E_n are also given by the fermionic p-adic integral of the function x^n as follows:

$$\int_{\mathbb{Z}_p} x^n d\mu_{-1}(x) = E_n, \tag{40}$$

(*cf.* [55–57,59,121] and the references therein).

Using (39), Kim *et al.* [52] gave the following formula:

$$\int_{\mathbb{Z}_p} (y)^{\underline{n}} \, d\mu_{-1}(y) = \frac{(-1)^n \, n!}{2^n} = Ch_n, \tag{41}$$

where $y \in \mathbb{Z}_p$.

Recently, by using (39) and (41), many properties and applications of the function $(y)^{\underline{n}}$ were also given by the author [121].

3.1. *Construction of generating function for combinatorial numbers denoted by* $y_{8,n}(\lambda; a)$

By applying the Volkenborn integral to the following function:

$$f(t, x; \lambda, a) = \left(\lambda + a^t\right)^x, \qquad (\lambda, x, t \in \mathbb{Z}_p), \tag{42}$$

we construct the following function which is used to define the generating function for the numbers $y_{8,n}(\lambda; a)$:

$$\int_{\mathbb{Z}_p} \left(\lambda + a^t\right)^x d\mu_1(x) = \frac{\log\left(\lambda + a^t\right)}{a^t + \lambda - 1}, \tag{43}$$

(*cf.* [2]). By applying Mahler's theorem, (proved by Mahler (1958) [96]) to (43), we have many interesting results since continuous p-adic-valued function f on $\mathbb{Z}_p$ can be written as in terms of polynomials. By using Binomial theorem (43), we get

$$\sum_{m=0}^{\infty} \left(\frac{a^t}{\lambda}\right)^m \int_{\mathbb{Z}_p} \binom{x}{m} \lambda^x d\mu_1(x) = \frac{\log\left(\lambda + a^t\right)}{a^t + \lambda - 1}. \tag{44}$$

With the aid of the equation (43), we set

$$D_1(n; \lambda) = \int_{\mathbb{Z}_p} \binom{x}{m} \lambda^x d\mu_1(x),$$

and

$$\sum_{n=0}^{\infty} D_1(n; \lambda) \left(\frac{a^t}{\lambda}\right)^n = \frac{\log\left(\lambda + a^t\right)}{a^t + \lambda - 1}. \tag{45}$$

Substituting $a^t = u$ into (45), and combining with the following known relation (*cf.* [139]):

$$\log(1 + u) = -\log 2 + 2 \sum_{n=1}^{\infty} \frac{(-1)^{n+1}}{n} T_n(u), \quad (-1 < u < 1), \tag{46}$$

we obtain the following result:

$$\sum_{n=0}^{\infty} D_1(n;\lambda)\left(\frac{u}{\lambda}\right)^n = \log\left(\frac{\lambda}{2}\right)\sum_{n=0}^{\infty}\frac{(-1)^n u^n}{(\lambda-1)^{n+1}} + \sum_{n=0}^{\infty}\frac{(-1)^n u^n}{(\lambda-1)^{n+1}}\sum_{n=1}^{\infty}\frac{(-1)^{n+1}}{n}T_n(u),$$

where $T_n(u)$ denotes the Chebyshev polynomial of the first kind, which is given by the following explicit formula:

$$\begin{aligned} T_n(u) &= \sum_{k=0}^{\left[\frac{n}{2}\right]}\binom{n}{2k}\left(u^2-1\right)^k u^{n-2k} \\ &= {}_2F_1\left[\begin{matrix}-n, n \\ \frac{1}{2}\end{matrix}; \frac{1-u}{2}\right], \end{aligned}$$

where $n \in \mathbb{N}$ (*cf.* [139]).

3.2. *Construction of generating function for combinatorial numbers denoted by $y_{9,n}(\lambda; a)$*

Applying the fermionic integral to equation (42) on $\mathbb{Z}_p$, we [2] gave the following generating function for the numbers $y_{9,n}(\lambda; a)$:

$$\int_{\mathbb{Z}_p}\left(\lambda + a^t\right)^x d\mu_{-1}(x) = \frac{2}{a^t+\lambda}. \tag{47}$$

Using (47), we have

$$\sum_{m=0}^{\infty}\left(\frac{a^t}{\lambda}\right)^m \int_{\mathbb{Z}_p}\binom{x}{m}\lambda^x d\mu_{-1}(x) = \frac{2}{a^t+\lambda},$$

(*cf.* [2]).

Combining the above equation with (41) for $\lambda = 1$, we have

$$\sum_{m=0}^{\infty} Ch_m \frac{a^{tm}}{m!} = \frac{2}{a^t+1},$$

(*cf.* [2]).

4. Generating Functions for New Classes of Combinatorial Numbers $y_{8,n}(\lambda; a)$ and Polynomials $y_{8,n}(x, \lambda; a)$ Derived from the p-adic Integral (43)

In Ref. [2], with aid of equation (43), we defined the combinatorial numbers $y_{8,n}(\lambda; a)$ and the combinatorial polynomials $y_{8,n}(x, \lambda; a)$, resp., as follows:

$$K_1(t; a, \lambda) := \frac{\log(\lambda + a^t)}{a^t + \lambda - 1} = \sum_{n=0}^{\infty} y_{8,n}(\lambda; a) \frac{t^n}{n!}, \tag{48}$$

and

$$\begin{aligned} K_2(t, x; a, \lambda) &:= a^{xt} K_1(t; a, \lambda) \\ &= \sum_{n=0}^{\infty} y_{8,n}(x, \lambda; a) \frac{t^n}{n!}. \end{aligned} \tag{49}$$

For functions $K_1(t; a, \lambda)$ and $K_2(t, x; a, \lambda)$, when $\lambda \neq 0$, we assume that

$$\left| \frac{a^t}{\lambda} \right| < 1,$$

and

$$\left| t \log a + \log\left(\frac{1}{\lambda - 1} \right) \right| < \pi,$$

(*cf.* [2, Eqs. (4.1) and (4.2), p. 47]).

By applying the following well-known Newton–Mercator series:

$$\ln(1+z) = \sum_{j=0}^{\infty} \frac{(-1)^j}{j+1} z^{j+1}, \quad (|z| < 1),$$

to (48) for $\lambda = 1$, we get

$$\sum_{j=0}^{\infty} \frac{(-1)^j}{j+1} e^{jt \log a} = \sum_{n=0}^{\infty} y_{8,n}(1; a) \frac{t^n}{n!}.$$

Therefore,

$$\sum_{n=0}^{\infty} \sum_{j=0}^{\infty} \frac{(-1)^j}{j+1} \frac{(j \log a)^n t^n}{n!} = \sum_{n=0}^{\infty} y_{8,n}(\lambda; a) \frac{t^n}{n!}.$$

Comparing the coefficients of $\frac{t^n}{n!}$ on both sides of the above equation, we arrive at the following theorem:

Theorem 5. *Let $n \in \mathbb{N}_0$. Then we have*

$$y_{8,n}(1;a) = \sum_{j=0}^{\infty} \frac{(-1)^j}{j+1} (j \log a)^n. \tag{50}$$

Substituting $a = e$ into (50), we arrive at the following result:

$$y_{8,n}(1;e) = \sum_{j=0}^{\infty} \frac{(-1)^j}{j+1} j^n.$$

When $\lambda = 0$, using (48), we have

$$\frac{t \log a}{e^{t \log a} - 1} = \sum_{n=0}^{\infty} y_{8,n}(0;a) \frac{t^n}{n!}.$$

Combining the above function with (3), we get

$$y_{8,n}(0;a) = (\log a)^n B_n.$$

By using (48), we have

$$\sum_{n=0}^{\infty} y_{8,n}(\lambda;a) \frac{t^n}{n!} = \frac{2}{\lambda - 1} \sum_{n=0}^{\infty} (\log a)^n \mathcal{E}_n \left(\frac{1}{\lambda - 1}\right) \frac{t^n}{n!} \times \left(\log(\lambda) + \log\left(1 + \frac{1}{\lambda} e^{t \log a}\right)\right).$$

Therefore,

$$\sum_{m=0}^{\infty} y_{8,m}(\lambda;a) \frac{t^m}{m!} = \frac{2 \log(\lambda)}{\lambda - 1} \sum_{m=0}^{\infty} (\log a)^m \mathcal{E}_m \left(\frac{1}{\lambda - 1}\right) \frac{t^m}{m!} + \frac{2}{\lambda - 1} \sum_{m=0}^{\infty} \sum_{j=0}^{m} (\log a)^{m-j} \mathcal{E}_{m-j} \left(\frac{1}{\lambda - 1}\right) \times \sum_{n=0}^{\infty} \frac{(-1)^n (n \log a)^j}{(n+1)\lambda^n} \frac{t^m}{m!}.$$

Comparing the coefficients of $\frac{t^m}{m!}$ on both sides of the above equation, we have the following theorem:

Theorem 6 (*cf.* **[2, Eqs. (4.1) and (4.2), p. 47]).** *Let* $m \in \mathbb{N}_0$ *and* $\lambda \neq 1$. *Then we have*

$$y_{8,m}(\lambda;a) = \frac{2\log(\lambda)(\log a)^m}{\lambda-1}\mathcal{E}_m\left(\frac{1}{\lambda-1}\right) + \frac{2(\log a)^m}{\lambda-1}\sum_{j=0}^{m}\mathcal{E}_{m-j}\left(\frac{1}{\lambda-1}\right)\sum_{n=0}^{\infty}\frac{(-1)^n n^j}{(n+1)\lambda^n}.$$

By using (48) and (4), we have

$$\sum_{m=0}^{\infty} y_{8,m}(\lambda;a)\frac{t^m}{m!} = \frac{2\log(\lambda)}{\lambda-1}\sum_{m=0}^{\infty}(\log a)^m\mathcal{E}_m\left(\frac{1}{\lambda-1}\right)\frac{t^m}{m!} + \sum_{n=0}^{\infty}\frac{(-1)^n}{n+1}\frac{1}{\lambda^n}\sum_{m=0}^{\infty}(\log a)^m\mathcal{E}_m\left(\frac{1}{\lambda-1}\right)\frac{n^m t^m}{m!}.$$

Comparing the coefficients of $\frac{t^m}{m!}$ on both sides of the above equation, we have the following theorem:

Theorem 7 (*cf.* **[2, Eqs. (4.1) and (4.2), p. 47]).** *Let* $m \in \mathbb{N}_0$ *and* $\lambda \neq 1$. *Then we have*

$$\sum_{n=0}^{\infty}\frac{(-1)^n}{n+1}\frac{n^m}{\lambda^n} = \frac{(\lambda-1)\,y_{8,m}(\lambda;a)}{2(\log a)^m\mathcal{E}_m\left(\frac{1}{\lambda-1}\right)} - \log(\lambda). \tag{51}$$

Combining (51) with (5), we obtain the following theorem:

Theorem 8. *Let* $m \in \mathbb{N}_0$ *and* $\lambda \neq 1$. *Then we have*

$$\sum_{n=0}^{\infty}\frac{(-1)^n}{n+1}\frac{n^m}{\lambda^n} = \frac{(1-\lambda)(m+1)\,y_{8,m}(\lambda;a)}{4(\log a)^m\mathcal{B}_{m+1}\left(\frac{1}{1-\lambda}\right)} - \log(\lambda).$$

The author has recently defined many new types of combinatorial numbers and polynomials. He gave some notations for these numbers and polynomials. For instance, in order to distinguish them from each other, these polynomials are given by the following notations:

$$y_{j,n}(x;\lambda,q),\ j=1,2,\ldots,9,\ \text{and also } Y_n(x;\lambda).$$

For instance, for the numbers $y_{9,n}(\lambda;a)$ the number 9 is only used for index representation for these polynomials (*cf.* [2,103,123]).

Here, by using these generating functions, we give not only fundamental properties of these polynomials and numbers, but also new identities and formulas including these numbers and polynomials, the Daehee numbers, the Stirling numbers of the second kind, the Apostol–Bernoulli numbers, the Apostol–Euler numbers, the λ-Apostol–Daehee numbers and the numbers $\mathrm{Y}_n(\lambda; a)$. In addition, we introduce a presumably new zeta-type function which interpolates the numbers $y_{8,m}(1; a)$ at negative integers.

Substituting $\lambda = 1$ into (45), we have the generating function for the numbers $\mathcal{D}_m := y_{8,n}(1; a)$:

$$\sum_{m=0}^{\infty} \mathcal{D}_m \frac{a^{tm}}{m!} = \frac{\log(1 + a^t)}{a^t},$$

(*cf.* [2, p. 47]).

Substituting $\lambda = 1$ and $u = a^t$ into (45), we have

$$\sum_{n=0}^{\infty} \left(D_n \frac{u^n}{n!} - \frac{y_{8,n}(1; a)}{n!} \left(\frac{\log u}{\log a} \right)^n \right) = 0.$$

By using (48) and (49), for $x = 0$, we have

$$y_{8,n}(0, \lambda; a) = y_{8,n}(\lambda; a).$$

Using the Eqs. (48) and (49), we have the following theorem:

Theorem 9 (*cf.* [2, Eqs. (4.1) and (4.2), p. 47]). *Let $n \in \mathbb{N}_0$. Then, we have*

$$y_{8,n}(x, \lambda; a) = \sum_{j=0}^{n} \binom{n}{j} (x \log a)^{n-j} y_{8,j}(\lambda; a). \tag{52}$$

Setting $a = 1$ in (48), we have

$$K_1(t; 1, \lambda) = \frac{\log(\lambda + 1)}{\lambda} = \sum_{n=0}^{\infty} y_{8,n}(\lambda; 1) \frac{t^n}{n!}. \tag{53}$$

From the previous equation, the following relations are derived:

$$y_{8,0}(\lambda; 1) = \frac{\log(\lambda + 1)}{\lambda}, \tag{54}$$

and for $n \geq 1$,

$$y_{8,n}(\lambda; 1) = 0.$$

By combining (53) with (46), for $-1 < \lambda < 1$, we obtain

$$\sum_{n=1}^{\infty} \frac{(-1)^{n+1}}{n} {}_2F_1 \left[\begin{matrix} -n, n \\ \frac{1}{2} \end{matrix} ; \frac{1-\lambda}{2} \right] = \frac{\lambda y_{8,0}(\lambda;1) + \log 2}{2}.$$

Moreover, combining (54) with (23), we also obtain

$$y_{8,0}(\lambda;1) = \frac{\log(\lambda+1)}{\lambda} = \sum_{n=0}^{\infty} D_n \frac{\lambda^n}{n!}.$$

Thus, we see that $y_{8,0}(\lambda;1)$ gives us the generating function for the Daehee numbers.

By combining (28) with (48), we get the following functional equation:

$$G_Y(t, 1-\lambda, a) \log\left(\lambda + a^t\right) = \sum_{n=0}^{\infty} y_{8,n}(\lambda;a) \frac{t^n}{n!}. \tag{55}$$

By using (55), for $\left|\frac{a^t}{\lambda}\right| < 1$, we arrive at the following result (*cf.* [2, Eqs. (4.1) and (4.2), p. 47]):

Theorem 10. *Let $\lambda \neq 0, 1$ and $m \in \mathbb{N}_0$. Then, we have*

$$y_{8,m}(\lambda;a) = \mathrm{Y}_m(1-\lambda;a)\log(\lambda) + \sum_{k=0}^{m} \binom{m}{k} \mathrm{Y}_{m-k}(1-\lambda;a) \times \sum_{n=0}^{k} \sum_{j=0}^{n+1} \binom{n+1}{j} (-1)^n \frac{j! S_2(k,j) (\log a)^k}{(n+1)\lambda^{n+1}}. \tag{56}$$

Using (56) and (25), we get the following theorem:

Corollary 1 (*cf.* [2, Eqs. (4.1) and (4.2), p. 47]). *Let $\lambda \neq 0, 1$ and $m \in \mathbb{N}_0$. Then we have*

$$y_{8,m}(\lambda;a) - \mathrm{Y}_m(1-\lambda;a)\log(\lambda) = \sum_{k=0}^{m} \binom{m}{k} \mathrm{Y}_{m-k}(1-\lambda;a) \sum_{n=0}^{k} (n+1) \sum_{j=0}^{n+1} \frac{S_2(k,j) (\log a)^k D_n}{\lambda^{n+1}(n+1-j)!}. \tag{57}$$

Using (57) and (32), we get the following result:

Corollary 2 (*cf.* [2, Eqs. (4.1) and (4.2), p. 47]). *Let $\lambda \neq 0,1$ and $m \in \mathbb{N}_0$. Then we have*

$$y_{8,m}(\lambda;a) - \lambda\mathcal{Y}_m\left(\lambda-\lambda^2;a,\lambda\right)\log(\lambda)$$
$$= \sum_{k=0}^{m}\binom{m}{k}\lambda\mathcal{Y}_{m-k}\left(\lambda-\lambda^2;a,\lambda\right)\sum_{n=0}^{k}(n+1)$$
$$\times\sum_{j=0}^{n+1}\frac{S_2(k,j)(\log a)^k D_n}{\lambda^{n+1}(n+1-j)!}. \tag{58}$$

Substituting $a=e$ into (48), we have

$$K_1(t;e,\lambda) = \frac{\log\left(\lambda+e^t\right)}{e^t+\lambda-1} = \sum_{n=0}^{\infty} y_{8,n}(\lambda;e)\frac{t^n}{n!}, \tag{59}$$

(*cf.* [2, Eqs. (4.1) and (4.2), p. 47]).

Using (59), (4) and (11), we have

$$\sum_{n=0}^{\infty} y_{8,n}(\lambda;e)\frac{t^n}{n!} = \frac{\log(\lambda)}{2(\lambda-1)}\sum_{n=0}^{\infty}\mathcal{E}_n\left(\frac{1}{\lambda-1}\right)\frac{t^n}{n!} + \frac{1}{2(\lambda-1)}\sum_{n=0}^{\infty}\frac{t^n}{n!}\sum_{k=0}^{n}\binom{n}{k}$$
$$\times\sum_{m=0}^{k}\sum_{j=0}^{m+1}\frac{(-1)^m j!\binom{m+1}{j}}{(m+1)\lambda^{m+1}}\mathcal{E}_{n-k}\left(\frac{1}{\lambda-1}\right)S_2(k,j).$$

Comparing the coefficients of $\frac{t^n}{n!}$ on both sides of the above equation, we get the following theorem:

Theorem 11 (*cf.* [2, Eqs. (4.1) and (4.2), p. 47]). *Let $\lambda \neq 0,1$ and $n \in \mathbb{N}_0$. Then, we have*

$$y_{8,n}(\lambda;e) = \frac{\mathfrak{D}_0(\lambda)}{2}\mathcal{E}_n\left(\frac{1}{\lambda-1}\right) + \frac{1}{2(\lambda-1)}\sum_{k=0}^{n}\binom{n}{k}$$
$$\times\sum_{m=0}^{k}\sum_{j=0}^{m+1}\frac{(-1)^m j!\binom{m+1}{j}}{(m+1)\lambda^{m+1}}\mathcal{E}_{n-k}\left(\frac{1}{\lambda-1}\right)S_2(k,j). \tag{60}$$

Using (60) and (5), we get the following theorem:

Theorem 12 (*cf.* [2, Eqs. (4.1) and (4.2), p. 47]). *Let* $\lambda \neq 0,1$ *and* $n \in \mathbb{N}_0$. *Then, we have*

$$y_{8,n}(\lambda;e) = \frac{\log\lambda}{(n+1)(1-\lambda)}\mathcal{B}_{n+1}\left(\frac{1}{1-\lambda}\right) - \frac{1}{(\lambda-1)}\sum_{k=0}^{n}\binom{n}{k}\sum_{m=0}^{k}\frac{(-1)^m}{(m+1)\lambda^{m+1}} \times \sum_{j=0}^{m+1} j!\binom{m+1}{j}\frac{\mathcal{B}_{n-k+1}\left(\frac{1}{1-\lambda}\right)S_2(k,j)}{n-k+1}.$$

By using the above theorem and (5), we also get

$$y_{8,n}(\lambda;e) = \frac{\log\lambda}{\lambda-1}\mathcal{E}_n\left(\frac{1}{\lambda-1}\right) - \frac{1}{(\lambda-1)}\sum_{k=0}^{n}\binom{n}{k}\sum_{m=0}^{k}\frac{(-1)^m}{(m+1)\lambda^{m+1}} \times \sum_{j=0}^{m+1} j!\binom{m+1}{j}\frac{\mathcal{B}_{n-k+1}\left(\frac{1}{1-\lambda}\right)S_2(k,j)}{n-k+1}.$$

Substituting $\lambda = 1$ into (48), and using (23), we have

$$\sum_{m=0}^{\infty} y_{8,m}(1;a)\frac{t^m}{m!} = \sum_{n=0}^{\infty} D_n\frac{a^{nt}}{n!}.$$

Therefore,

$$\sum_{m=0}^{\infty} y_{8,m}(1;a)\frac{t^m}{m!} = \sum_{m=0}^{\infty}\sum_{n=0}^{\infty}\frac{D_n}{n!}(n\log a)^m\frac{t^m}{m!}.$$

Comparing the coefficients of $\frac{t^m}{m!}$ on both sides of the above equation, we arrive at the following theorem:

Theorem 13 (*cf.* [2, Eqs. (4.1) and (4.2), p. 47]). *Let* $m \in \mathbb{N}_0$. *Then we have*

$$y_{8,m}(1;a) = \sum_{n=0}^{\infty}\frac{D_n}{n!}(n\log a)^m. \tag{61}$$

Theorem 14 (*cf.* [2, Eqs. (4.1) and (4.2), p. 47]). *Let $m \in \mathbb{N}_0$. Then we have*

$$y_{8,m}(1;a) = \sum_{n=0}^{\infty} \frac{D_n}{n!} \sum_{j=0}^{m} \binom{n}{j} j! S_2(m,j) (\log a)^m .$$

4.1. *Interpolation function for the numbers $y_{8,n}(\lambda; a)$*

In Ref. [2, Eqs. (4.1) and (4.2), p. 47], we defined interpolation function for the numbers $y_{8,n}(\lambda; a)$. This interpolation function gives us a new type of zeta functions. It is well-known that number theory involves various kinds of zetas, starting with Riemann's — necessary ingredient in the study of the distribution of prime numbers, the properties of Bernoulli numbers and the Hurwitz zeta function involving the Riemann zeta function and the Bernoulli polynomials at negative integers with aid of analytic continuation. Here, we will find some properties of the interpolation function for the numbers $y_{8,n}(\lambda; a)$.

Here, using the following derivative operator in the following equation:

$$\frac{\partial^k}{\partial t^k} f(t)|_{t=0}, \tag{62}$$

to the generating function for the numbers $y_{8,n}(\lambda; a)$, the author gave unification of zeta type function which interpolates the numbers $y_{8,n}(\lambda; a)$ at negative integers.

By applying the derivative operator in Eq. (62) to Eq. (75), we have

$$y_{8,k}(\lambda; a) = \frac{\partial^k}{\partial t^k} \left\{ \frac{\log(\lambda + a^t)}{a^t + \lambda - 1} \right\} |_{t=0},$$

(*cf.* [2, Eqs. (4.1) and (4.2), p. 47]).

Assuming that $|\frac{a^t}{\lambda}| < 1$ $(\lambda \neq 0, 1)$ to guarantee the convergence range of the following power series, the above equation reduces to the following relation:

$$\begin{aligned} y_{8,k}(\lambda; a) = \frac{\partial^k}{\partial t^k} \Bigg\{ & \sum_{n=0}^{\infty} (-1)^n \frac{a^{tn} \log \lambda}{(\lambda - 1)^{n+1}} \\ & + \sum_{n=0}^{\infty} \sum_{j=0}^{n} \frac{(-1)^n a^{t(n+1)}}{(j+1)\lambda^{j+1} (\lambda - 1)^{n+1-j}} \Bigg\} |_{t=0}. \end{aligned}$$

(*cf.* [2, Eqs. (4.1) and (4.2), p. 47]). After some elementary calculations, the above equation reduces to

$$y_{8,k}(\lambda;a) = (\log\lambda)\sum_{n=0}^{\infty}(-1)^n\frac{n^k(\log a)^k}{(\lambda-1)^{n+1}}+\sum_{n=0}^{\infty}\sum_{j=0}^{n}\frac{(-1)^n(n+1)^k(\log a)^k}{(j+1)\lambda^{j+1}(\lambda-1)^{n+1-j}}. \tag{63}$$

(*cf.* [2, Eqs. (4.1) and (4.2), p. 47]).

With the help of analytic continuation technique applied to Lerch-type zeta function, using (63), we arrive at the following definition of the interpolation function for the number $y_{8,k}(\lambda;a)$:

Definition 1 (*cf.* [2, Eqs. (4.1) and (4.2), p. 47]). Let $a \geq 1$. For $\lambda \in \mathbb{C}\setminus\{0,1\}$ ($|\frac{1}{\lambda-1}| < 1$) and $s \in \mathbb{C}$, a unification of zeta-type function $\mathcal{Z}_1(s;a,\lambda)$ is defined by

$$\mathcal{Z}_1(s;a,\lambda) = \frac{\log\lambda}{(\log a)^s}\sum_{n=1}^{\infty}\frac{(-1)^n}{n^s(\lambda-1)^{n+1}} + \frac{1}{(\log a)^s}\sum_{n=0}^{\infty}\left(\sum_{j=0}^{n}\frac{(-1)^n}{(j+1)\lambda^{j+1}(\lambda-1)^{n+1-j}}\right)\frac{1}{(n+1)^s}, \tag{64}$$

where $\lambda \in \mathbb{C}\setminus\{0,1\}$ ($|\frac{1}{\lambda-1}| < 1$; $\mathrm{Re}(s) > 1$).

Substituting $a = e$ into (64), we have

$$\mathcal{Z}_1(s;e,\lambda) = \log\lambda\sum_{n=1}^{\infty}\frac{(-1)^n}{n^s(\lambda-1)^{n+1}} + \sum_{n=0}^{\infty}\sum_{j=0}^{n} \times\frac{(-1)^n}{(j+1)\lambda^{j+1}(\lambda-1)^{n+1-j}(n+1)^s}.$$

Putting $\lambda = 2$ and $\mathrm{Re}(s) > 1$, we have

$$\mathcal{Z}_1(s;e,2) = \log 2\zeta_E(s) + \frac{1}{2}\sum_{n=0}^{\infty}\sum_{j=0}^{n}\frac{(-1)^n}{(j+1)2^j(n+1)^s}, \tag{65}$$

where $\zeta_E(s)$ denotes the alternating Riemann zeta function:

$$\zeta_E(s) = \sum_{n=1}^{\infty}\frac{(-1)^n}{n^s}.$$

Thus, we have

$$\mathcal{Z}_1(s;e,2) = \log 2\zeta_E(s) + \frac{1}{2}\sum_{n=0}^{\infty}\frac{(-1)^n}{(n+1)^s}\sum_{j=0}^{n}\frac{1}{(j+1)2^j}$$

$$= \log 2\zeta_E(s) + \frac{1}{2}\sum_{n=0}^{\infty}\frac{(-1)^n}{(n+1)^s}\sum_{j=0}^{n}\frac{1}{\frac{d}{dx}\left\{x^{j+1}\right\}|_{x=2}}.$$

Combining the following equation:

$$\mathcal{Z}_1(s;e,2) = \log 2\zeta_E(s) + \sum_{n=0}^{\infty}\frac{(-1)^n}{(n+1)^s}\sum_{j=0}^{n}\frac{1}{(j+1)2^{j+1}},$$

with the following well-known identity, for $n \in \mathbb{N}_0$,

$$\sum_{j=1}^{n}\frac{1}{j2^j} = h_n - 2^{-n}\sum_{j=0}^{n}\binom{n}{j}h_j, \tag{66}$$

(*cf.* [30, Eq. (1.25)]), we arrive at the following theorem which includes the alternating Riemann zeta function, Dirichlet series and harmonic numbers:

Theorem 15. *Let $n \in \mathbb{N}_0$. Then we have*

$$\mathcal{Z}_1(s;e,2) = \log 2\zeta_E(s) + \sum_{n=0}^{\infty}\frac{(-1)^n h_{n+1}}{(n+1)^s}$$

$$-\sum_{n=0}^{\infty}\frac{(-1)^n}{2^{n+1}(n+1)^s}\sum_{j=0}^{n+1}\binom{n+1}{j}h_j.$$

Note that the function $\mathcal{Z}_1(s;a,\lambda)$ also has the following property: The function $\mathcal{Z}_1(s;a,\lambda)$ is analytic continuation, except $s = 1$ and $\lambda = 1$ in whole complex plane, therefore, combining (63) with (64) at negative integers, we get the following theorem which shows that the function $\mathcal{Z}_1(s;a,\lambda)$ interpolates the numbers $y_{8,k}(\lambda;a)$ at negative integer.

Theorem 16 (*cf.* [2, Eqs. (4.1) and (4.2), p. 47]). *Let $\lambda \in \mathbb{C}$ ($|\frac{1}{\lambda}| < 1$) and $k \in \mathbb{N}$. Then, we have*

$$\mathcal{Z}_1(-k;a,\lambda) = y_{8,k}(\lambda;a). \tag{67}$$

The polylogarithm function is defined by

$$Li_s(z) = \sum_{n=1}^{\infty}\frac{z^n}{n^s}. \tag{68}$$

The definition of the $Li_s(z)$ is valid for arbitrary complex order s and for all complex arguments z with $|z| < 1$; this definition can be extended to $|z| \geq 1$ by the process of analytic continuation.

For $\text{Re}(s) > 1$, and $z = 1$, we have

$$Li_s(1) = \zeta(s) = \sum_{n=1}^{\infty} \frac{1}{n^s}.$$

The special case $s = 1$ involves the ordinary natural logarithm,

$$Li_1(z) = -\log(1-z),$$

$$Li_1\left(\frac{1}{2}\right) = \log(2),$$

while the special cases $s = 2$ and $s = 3$ are called the dilogarithm (also referred to as Spence's function) and trilogarithm, resp. For $s = m$ with $|z| \leq 1$; $m \in \mathbb{N} \setminus \{1\}$, one has

$$Li_m(z) = \sum_{n=1}^{\infty} \frac{z^n}{n^m}$$

(*cf.* [130]).

In Ref. [2, Eqs. (4.1) and (4.2), p. 47], we defined the following finite sum which is called the numbers represented by $y(n, \lambda)$:

$$y(n,\lambda) = \sum_{j=0}^{n} \frac{(-1)^n}{(j+1)\lambda^{j+1}(\lambda-1)^{n+1-j}}.$$

Substituting the above finite sum into (64), we get the relation between the function $\mathcal{Z}_1(s; a, \lambda)$ and the polylogarithm function as follows:

$$\mathcal{Z}_1(s;a,\lambda) = \frac{1}{(\log a)^s}\left(\frac{\log\lambda}{(\lambda-1)}Li_s\left(\frac{1}{1-\lambda}\right) + \sum_{n=0}^{\infty}\frac{y(n,\lambda)}{(n+1)^s}\right). \tag{69}$$

Substituting $s = -k$ into (69) and using (67), we get

$$\sum_{n=0}^{\infty} y(n,\lambda)(n+1)^k = \frac{y_{8,k}(\lambda;a)}{(\log a)^k} - \frac{\log\lambda}{\lambda-1}Li_{-k}\left(\frac{1}{1-\lambda}\right). \tag{70}$$

Combining equation (70) with the following well-known formula:

$$Li_{-k}(z) = \left(z\frac{\partial}{\partial z}\right)^k\left(\frac{z}{1-z}\right) = \sum_{j=0}^{k} j!S_2(k+1,j+1)\left(\frac{z}{1-z}\right)^{j+1}, \tag{71}$$

we get

$$\sum_{n=0}^{\infty} y(n,\lambda)(n+1)^k = \frac{y_{8,k}(\lambda;a)}{(\log a)^k} - \frac{\log\lambda}{\lambda-1}\left(\frac{1}{1-\lambda}\frac{\partial}{\partial\lambda}\right)^k\left(\frac{\frac{1}{1-\lambda}}{1-\frac{1}{1-\lambda}}\right)$$

$$= \frac{y_{8,k}(\lambda;a)}{(\log a)^k} - \frac{\log\lambda}{\lambda-1}\left(\frac{1}{1-\lambda}\frac{\partial}{\partial\lambda}\right)^k\left(\frac{1}{-\lambda}\right),$$

we arrive at the following theorem:

Theorem 17.

$$\sum_{n=0}^{\infty} y(n,\lambda)(n+1)^k = \frac{y_{8,k}(\lambda;a)}{(\log a)^k} + \frac{\log\lambda}{\lambda-1}\sum_{j=0}^{k}(-1)^j\frac{j!S_2(k+1,j+1)}{\lambda^{j+1}}.$$

Assuming that $|a^t| < 1$. Combining (61) with (25), we have

$$\sum_{m=0}^{\infty} y_{8,m}(1;a)\frac{t^m}{m!} = \sum_{n=0}^{\infty}(-1)^n\frac{a^{tn}}{n+1}.$$

By substituting Taylor series of the function a^{tn} into the above equation, we get

$$\sum_{m=0}^{\infty} y_{8,m}(1;a)\frac{t^m}{m!} = \sum_{n=0}^{\infty}(-1)^n\frac{1}{n+1}\sum_{m=0}^{\infty}(n\log a)^m\frac{t^m}{m!}.$$

Comparing the coefficients of $\frac{t^m}{m!}$ on both sides of the above equation, the following theorem is gotten:

Theorem 18 (*cf.* [2, Eqs. (4.1) and (4.2), p. 47]). *Let $m \in \mathbb{N}_0$. Then we have*

$$y_{8,m}(1;a) = (\log a)^m \sum_{n=0}^{\infty}(-1)^n\frac{n^m}{n+1}. \tag{72}$$

By using (72), we have

$$y_{8,m}(1;a) = (\log a)^m \sum_{j=0}^{m}\binom{m}{j}\sum_{n=0}^{\infty}(-1)^n(n+1)^{j-1}$$

$$= (\log a)^m \sum_{j=0}^{m}\binom{m}{j}\zeta_E(1-j)$$

$$= (\log a)^m \sum_{j=0}^{m}\binom{m}{j}E_j.$$

By substituting $m = 0$ into (72), we get

$$y_{8,0}(1;a) = Li_1\left(\frac{1}{2}\right) = \log(2),$$

(*cf.* [2, Eqs. (4.1) and (4.2), p. 47]).

With the aid of (72), the following ***zeta-type function*** $\mathcal{Z}_2(s,a)$, which interpolates the numbers $y_{8,m}(1;a)$ at negative integers, is defined:

Definition 2 (*cf.* [2, Eqs. (4.1) and (4.2), p. 47]). Let $s \in \mathbb{C}$ with $\mathrm{Re}(s) > 0$. Let $a \in (1,\infty)$. We define

$$\mathcal{Z}_2(s,a) = \frac{1}{(\log a)^s}\sum_{n=1}^{\infty}\frac{(-1)^n}{(n+1)\,n^s}. \tag{73}$$

Putting $s = -m$, $(m \in \mathbb{N})$, in (73), using (72), we have the following result:

Theorem 19 (*cf.* [2, Eqs. (4.1) and (4.2), p. 47]). *Let $m \in \mathbb{N}$. Then, we have*

$$\mathcal{Z}_2(-m,a) = y_{8,m}(1;a).$$

4.2. *Partial zeta-type function*

Here, we define partial zeta-type function for the unification of zeta-type function $\mathcal{Z}_1(s;a,\lambda)$.

Let s be a complex variable. Let d be an integer. Let v be an odd integer with $0 < d < v$. Then we define partial zeta-type function for the unification of zeta-type function $\mathcal{Z}_1(s;a,\lambda)$ as follows:

Definition 3. Let $a \geq 1$. For $\lambda \in \mathbb{C}\setminus\{0,1\}$ $(|\frac{1}{\lambda-1}| < 1)$ and $s \in \mathbb{C}$, define partial zeta-type function for unification of zeta-type function $\mathcal{H}_1(s,b;v;a,\lambda)$ as

$$\begin{aligned}
&\mathcal{H}_1(s,b;v;a,\lambda)\\
&\quad = \frac{\log\lambda}{(\log a)^s}\sum_{n\equiv b(\mathrm{mod}\,v),n>0}^{\infty}\frac{(-1)^n}{n^s(\lambda-1)^{n+1}} + \frac{1}{(\log a)^s}\\
&\qquad \times \sum_{n\equiv a(\mathrm{mod}\,v),n>0}^{\infty}\left(\sum_{j=0}^{n}\frac{(-1)^n}{(j+1)\lambda^{j+1}(\lambda-1)^{n+1-j}}\right)\frac{1}{(n+1)^s},
\end{aligned} \tag{74}$$

where $\lambda \in \mathbb{C}\setminus\{0,1\}$ $(|\frac{1}{\lambda-1}| < 1;\ \mathrm{Re}(s) > 1)$.

Substituting $n = b + mv$, we get

$$\mathcal{H}_1(s,b;v;a,\lambda) = \frac{\log\lambda}{(\log a)^s}\sum_{m=0}^{\infty}\frac{(-1)^{b+mv}}{(b+mv)^s(\lambda-1)^{b+mv+1}}$$
$$+\frac{1}{(\log a)^s}\sum_{m=0}^{\infty}\left(\sum_{j=0}^{b+mv}\frac{(-1)^{b+mv}}{(j+1)\lambda^{j+1}(\lambda-1)^{b+mv+1-j}}\right)$$
$$\times\frac{1}{(b+mv+1)^s}.$$

Therefore,

$$\mathcal{H}_1(s,b;v;a,\lambda)$$
$$=\frac{\log\lambda(-1)^b}{(\log a)^s v^s(\lambda-1)^b}\sum_{m=0}^{\infty}\frac{(-1)^m}{\left(\frac{b}{v}+m\right)^s(\lambda-1)^{mv+1}}$$
$$+\frac{(-1)^b}{(\log a)^s v^s(\lambda-1)^b}\sum_{m=0}^{\infty}\left(\sum_{j=0}^{b+mv}\frac{(-1)^m}{(j+1)\lambda^{j+1}(\lambda-1)^{mv+1-j}}\right)$$
$$\times\frac{1}{\left(\frac{b+1}{v}+m\right)^s}.$$

5. Generating Functions for New Classes of Combinatorial Numbers $y_{9,n}(\lambda;a)$ and Polynomials $y_{9,n}(x,\lambda;a)$ Derived from the p-adic Integral (47)

In this section, by the help of (47), we construct generating functions for the numbers $y_{9,n}(\lambda;a)$ and the polynomials $y_{9,n}(x,\lambda;a)$, resp., as follows:

$$Y_1(t;a,\lambda) := \frac{2}{a^t+\lambda} = \sum_{n=0}^{\infty} y_{9,n}(\lambda;a)\frac{t^n}{n!}, \tag{75}$$

and

$$Y_2(t,x;a,\lambda) := a^{tx}Y_1(t;a,\lambda)$$
$$= \sum_{n=0}^{\infty} y_{9,n}(x,\lambda;a)\frac{t^n}{n!}. \tag{76}$$

For generating functions $Y_1(t;a,\lambda)$ and $Y_2(t,x;a,\lambda)$, when $\lambda \neq 0$, we assume that

$$\left|t\log a + \log\left(\frac{1}{\lambda}\right)\right| < \pi.$$

When $\lambda = 0$, using (75), we have

$$y_{9,n}(0;a) = (-1)^n 2 \log^n a.$$

Using (76), we get

$$\frac{2e^{tx\log a}}{\lambda\left(\frac{1}{\lambda}e^{t\log a}+1\right)} = \sum_{n=0}^{\infty} y_{9,n}(x,\lambda;a)\frac{t^n}{n!}.$$

Combining the above function with (4), after some elementary calculations, we get the following relation between the polynomials $\mathcal{E}_n(x;\lambda)$ and $y_{9,n}(x,\lambda;a)$:

Theorem 20 (*cf.* [2, Eqs. (4.1) and (4.2), p. 47]).

$$y_{9,n}(x,\lambda;a) = \frac{(\log a)^n}{\lambda}\mathcal{E}_n\left(x;\frac{1}{\lambda}\right). \tag{77}$$

Here, we note that due to the relation in equation (77), all the constraints given in Refs. [127–129] also apply to the generating functions given in Eqs. (75) and (76).

By using these generating functions, we give not only fundamental properties of these polynomials and numbers, but also new identities and formulas including these numbers and polynomials, the generalized Eulerian-type numbers, the Euler numbers, the Fubini numbers, the Stirling numbers, the Dobinski numbers, the numbers $\mathrm{Y}_n(\lambda;a)$.

By using (75) and (76), we have

$$y_{9,n}(0,\lambda;a) = y_{9,n}(\lambda;a).$$

From equations (75) and (76), we have the following theorem:

Theorem 21 (*cf.* [2, Eqs. (4.1) and (4.2), p. 47]). *Let $n \in \mathbb{N}_0$. Then we have*

$$y_{9,n}(x,\lambda;a) = \sum_{j=0}^{n}\binom{n}{j}(x\log a)^{n-j}\, y_{9,j}(\lambda;a). \tag{78}$$

Setting $a = e$ and $\lambda = -2$ in (75) yields the following equation:

$$\frac{2}{e^t-2} = \sum_{n=0}^{\infty} y_{9,n}(-2;e)\frac{t^n}{n!}. \tag{79}$$

Combing the above equation with (20), we have

$$\sum_{n=0}^{\infty} y_{9,n}(-2;e)\frac{t^n}{n!} = -2\sum_{n=0}^{\infty} w_g(n)\frac{t^n}{n!}.$$

Comparing the coefficients of $\frac{t^n}{n!}$ on both sides of the above equation, we have the following corollary:

Corollary 3 (*cf.* [2, Eqs. (4.1) and (4.2), p. 47]). *Let* $n \in \mathbb{N}_0$. *Then, we have*

$$y_{9,n}(-2;e) = -2w_g(n). \tag{80}$$

By using (79), we have

$$-\sum_{n=0}^{\infty} n! \frac{(e^t-1)^n}{n!} = \sum_{m=0}^{\infty} y_{9,m}(-2;e) \frac{t^m}{m!}.$$

Therefore,

$$-\sum_{m=0}^{\infty}\sum_{n=0}^{\infty} n! S_2(m,n) \frac{t^m}{m!} = \sum_{m=0}^{\infty} y_{9,m}(-2;e) \frac{t^m}{m!}.$$

Comparing the coefficients of $\frac{t^n}{n!}$ on both sides of the above equation, we have the following corollary:

Corollary 4. *Let* $n \in \mathbb{N}_0$. *Then, we have*

$$y_{9,n}(-2;e) = -2\sum_{j=0}^{n} j! S_2(n,j). \tag{81}$$

Observe that by aid of (21) and (80), we also arrive at equation (81) (*cf.* [2, Eqs. (4.1) and (4.2), p. 47]).

Substituting $a^t = e^{e^t-1}$ and $\lambda = 0$ into (75), we have

$$\frac{2}{e^{e^t-1}} = \sum_{m=0}^{\infty} y_{9,m}(\lambda;e) \frac{(e^t-1)^m}{m!}.$$

Combining the above equation with (11) and (22), we obtain

$$2\sum_{n=0}^{\infty} D(n)\frac{t^n}{n!} = \sum_{m=0}^{\infty} y_{9,m}(\lambda;e) \sum_{n=0}^{\infty} S_2(n,m)\frac{t^n}{n!}.$$

Thus,

$$2\sum_{n=0}^{\infty} D(n)\frac{t^n}{n!} = \sum_{n=0}^{\infty}\sum_{m=0}^{n} y_{9,m}(\lambda;e) S_2(n,m)\frac{t^n}{n!}.$$

Comparing the coefficients of $\frac{t^n}{n!}$ on both sides of the above equation, we arrive at the following theorem:

Theorem 22. *Let $n \in \mathbb{N}_0$. Then we have*

$$D(n) = \frac{1}{2}\sum_{m=0}^{n} y_{9,m}(\lambda; e) S_2(n,m).$$

By replacing λ by $-\lambda$ in (75), we get

$$\frac{2}{a^t - \lambda} = \sum_{n=0}^{\infty} y_{9,n}(-\lambda; a)\frac{t^n}{n!}.$$

Combining the above equation with (28), we have

$$2\sum_{n=0}^{\infty} \mathrm{Y}_n(\lambda; a)\frac{t^n}{n!} = \sum_{n=0}^{\infty} y_{9,n}(-\lambda; a)\frac{t^n}{n!}.$$

Comparing the coefficients of $\frac{t^n}{n!}$ on both sides of the above equation, we have the following corollary:

Corollary 5 (*cf.* [2, Eqs. (4.1) and (4.2), p. 47]). *Let $n \in \mathbb{N}_0$. Then, we have*

$$\mathrm{Y}_n(\lambda; a) = \frac{y_{9,n}(-\lambda; a)}{2}.$$

Setting $a = 1$, $x = 0$ and $\lambda = 1$ in (8) yields the following equation:

$$\frac{1-u}{b^t - u} = \sum_{n=0}^{\infty} \mathcal{H}_n(u; 1, b, c; 1)\frac{t^n}{n!}.$$

Combining the above equation with (75), we get

$$\sum_{n=0}^{\infty} \mathcal{H}_n(u; 1, b, c; 1)\frac{t^n}{n!} = \frac{1-u}{2}\sum_{n=0}^{\infty} y_{9,n}(-u; b)\frac{t^n}{n!}.$$

Comparing the coefficients of $\frac{t^n}{n!}$ on both sides of the above equation, we arrive at the following result:

Corollary 6 (*cf.* [2, Eqs. (4.1) and (4.2), p. 47]). *Let $n \in \mathbb{N}_0$. Then we have*

$$y_{9,n}(-u; b) = \frac{2}{1-u}\mathcal{H}_n(u; 1, b, c; 1). \tag{82}$$

Substituting $b = e$ into (82), we get

$$y_{9,n}(-u; e) = \frac{2}{1-u}H_n(u).$$

When $u = -1$ in the above equation, we arrive at the following corollary:

Corollary 7. *Let $n \in \mathbb{N}_0$. Then we have*

$$y_{9,n}(1; e) = E_n.$$

5.1. *Interpolation function for the numbers $y_{9,n}(\lambda; a)$*

By applying the derivative operator in equations (62) to the generating function for the numbers $y_{9,n}(\lambda; a)$, we [2, Eqs. (4.1) and (4.2), p. 47] defined by unification of zeta-type function $\mathcal{Z}_3(s; a, \lambda)$. The function $\mathcal{Z}_3(s; a, \lambda)$ interpolates the numbers $y_{9,n}(\lambda; a)$ at negative integers.

Definition 4 (*cf.* [2, Eqs. (4.1) and (4.2), p. 47]). Let $a \geq 1$. For

$$\lambda \in \mathbb{C}; \left(\left|\frac{1}{\lambda}\right| < 1\right) \text{ and } s \in \mathbb{C},$$

a unification of zeta-type function $\mathcal{Z}_3(s; a, \lambda)$ is defined by

$$\mathcal{Z}_3(s; a, \lambda) = \frac{2}{(\log a)^s} \sum_{n=1}^{\infty} \frac{(-1)^n}{n^s \lambda^{n+1}}, \tag{83}$$

where $\mathrm{Re}(s) > 1$.

Combining (83) with (68), we get

$$\mathcal{Z}_3(s; a, \lambda) = \frac{2}{\lambda(\log a)^s} Li_s\left(-\frac{1}{\lambda}\right). \tag{84}$$

The function $\mathcal{Z}_3(s; a, \lambda)$ has analytic continuation, except $s = 1$ and $\lambda = 1$ in the whole complex plane. By using (83) we have the following result:

Theorem 23 (*cf.* [2, Eqs. (4.1) and (4.2), p. 47]). *Let $\lambda \in \mathbb{C}$ $\left(\left|\frac{1}{\lambda}\right| < 1\right)$ and $k \in \mathbb{N}$. Then we have*

$$\mathcal{Z}_3(-k; a, \lambda) = y_{9,k}(\lambda; a). \tag{85}$$

For $s = -k$ $(k \in \mathbb{N})$, combining (84) with (85), we get

$$y_{9,k}(\lambda; a) = \frac{2 \log^k a}{\lambda} Li_{-k}\left(-\frac{1}{\lambda}\right).$$

Substituting (71) into the above equation, we obtain

$$y_{9,k}(\lambda;a)=\frac{2\log^k a}{\lambda}\sum_{j=0}^{k}(-1)^{j+1}\frac{j!S_2(k+1,j+1)}{(\lambda+1)^{j+1}}.$$

Corollary 8 (*cf.* [2, Eqs. (4.1) and (4.2), p. 47]). *Let* $\lambda \in \mathbb{C}$ $\left(\left|\frac{1}{\lambda}\right|<1\right)$ *and* $k\in\mathbb{N}$. *Then we have*

$$\sum_{n=1}^{\infty}\frac{(-1)^n n^k}{\lambda^n}=\frac{\lambda y_{9,k}(\lambda;a)}{2(\log a)^k}.$$

Remark 1. In Refs. [83,126], Srivastava *et al.* studied and investigated many properties of the following unification of the Riemann-type zeta functions: For $\beta\in\mathbb{C}$ $(|\beta|<1)$,

$$\zeta_\beta(s;v,c,d)=\left(-\frac{1}{2}\right)^{v-1}\sum_{n=1}^{\infty}\frac{\beta^{dn}}{c^{d(n+1)}n^s},\tag{86}$$

where $v\in\mathbb{N}_0$ and c and d are positive real numbers. Substituting $s=-m$ into the above equation, we have

$$\zeta_\beta(-m;v,c,d)=\frac{(-1)^v m!}{(m+v)!}y_{m+v,\beta}(v,c,d),$$

where the numbers $y_{m+v,\beta}(v,c,d)$ are a unification of the Bernoulli, Euler and Genocchi numbers defined by Ozden [82] by means of the following generating function:

$$\frac{2^{v-1}t^v}{\beta^b e^{tx}-a^b}=\sum_{m=m}^{\infty}y_{m,\beta}(v,c,d)\frac{t^m}{m!}.$$

In Ref. [61], Kim *et al.* applied the derivative operator $\frac{d^k}{dt^k}$ to the λ-Bernoulli numbers, the constructed interpolation of these numbers (*cf.* [2, Eqs. (4.1) and (4.2), p. 47]).

By combining (83) with (86), we get the following result:

Corollary 9 (*cf.* [2, Eqs. (4.1) and (4.2), p. 47]). *Under the conditions given for the above Eqs.* (83) *and* (86), *we have*

$$\mathcal{Z}_3(s;a,\lambda)=\frac{2}{\lambda(\log a)^s}\zeta_{-\frac{1}{\lambda}}(s;1,1,1)$$

Remark 2. Recently, many authors have studied the unification of the Bernoulli, Euler and Genocchi numbers $y_{m,\beta}(v,c,d)$ with their interpolation function (*cf.* [2, Eqs. (4.1) and (4.2), 11, p. 47, 12,28,48,70,71,75,82,84,114, 125,133]).

Partial zeta-type function for the unification of zeta-type function $\mathcal{Z}_3(s; a, \lambda)$:

Let s be a complex variable. Let d be an integer. Let v be an odd integer with $0 < d < v$. Then we define partial zeta-type function for the unification of zeta-type function $\mathcal{Z}_3(s; a, \lambda)$ as follows:

Definition 5. Let $a \geq 1$. For $\lambda \in \mathbb{C} \setminus \{0, 1\}$ ($|\frac{1}{\lambda-1}| < 1$) and $s \in \mathbb{C}$, define partial zeta-type function for unification of zeta-type function $\mathcal{H}_3(s, b; v; a, \lambda)$ defined by

$$\mathcal{H}_3(s, b; v; a, \lambda) = \frac{2}{(\log a)^s} \sum_{n \equiv b (\operatorname{mod} v), n>0}^{\infty} \frac{(-1)^n}{n^s \lambda^{n+1}}. \tag{87}$$

By using (87), we get

$$\begin{aligned}
\mathcal{H}_3(s, b; v; a, \lambda) &= \frac{2}{(\log a)^s} \sum_{n=0}^{\infty} \frac{(-1)^{b+nv}}{(b+nv)^s \lambda^{b+nv+1}} \\
&= \frac{2(-1)^b}{\lambda^{b+1} (v \log a)^s} \sum_{n=0}^{\infty} \frac{(-1)^{nv}}{\left(\frac{b}{v} + n\right)^s \lambda^{nv}}, \\
&= \frac{2(-1)^b}{\lambda^{b+1} (v \log a)^s} \Phi \left(\frac{b}{v}, \frac{-1}{\lambda^v}, s \right),
\end{aligned}$$

where $\Phi\left(\frac{b}{v}, \frac{-1}{\lambda^v}, s\right)$ denotes the Hurwitz–Lerch zeta function which interpolates the Apostol–Bernoulli numbers due to the following relation:

$$\Phi(\lambda, m, 0) = \sum_{n=0}^{\infty} \lambda^n n^m = -\frac{\mathcal{B}_{m+1}(\lambda)}{m+1},$$

(*cf.* [129]; and also see the references cited therein).

Thus, setting $s = -m$, $m \in \mathbb{N}$, we have

$$\begin{aligned}
\mathcal{H}_3(-m, b; v; a, \lambda) &= \frac{2(-1)^b (v \log a)^m}{\lambda^{b+1}} \Phi \left(\frac{b}{v}, \frac{-1}{\lambda^v}, -m \right) \\
&= \frac{2(-1)^{b+1} (v \log a)^m}{\lambda^{b+1} (m+1)} \mathcal{B}_{m+1} \left(\frac{b}{v}, \frac{-1}{\lambda^v} \right).
\end{aligned}$$

6. Finite Sums Arising from (64)

In this section, we gave some properties of the numbers $y(n, \lambda)$, which are raised from equation (64).

Substituting $\lambda = 2$ into (1), we get

$$y(n) := y(n,2) = \sum_{j=0}^{n} \frac{(-1)^n}{(j+1)2^{j+1}}, \tag{88}$$

(*cf.* [2]).

Combining (88) with (66), we obtain the following formula:

Theorem 24. *Let $n \in \mathbb{N}_0$. Then we have*

$$y(n,2) = (-1)^n h_{n+1} + \left(-\frac{1}{2}\right)^{n+1} \sum_{j=0}^{n+1} \binom{n+1}{j} h_j.$$

In Ref. [2], we gave the following **open problems**:

(1) One of the first questions that comes to mind is what is the generating function for the numbers $y(n)$ and the numbers $y(n,\lambda)$.
(2) Some of the other questions are what are the special families of numbers the numbers $y(n)$ are related to.
(3) What are the combinational applications of the numbers $y(n)$.
(4) Can we find a special arithmetic function representing this family of numbers?

We can partially solve the second question as follows:

By (24), Kucukoglu and Simsek [66] also gave the following novel finite combinatorial sum in terms of the numbers $\mathfrak{D}_n(\lambda)$:

$$\sum_{k=0}^{n-1} \frac{1}{k+1} \left(\frac{\lambda-1}{\lambda}\right)^k = \frac{(-1)^{n+1} \lambda \mathfrak{D}_n(\lambda)}{n!} \left(\frac{\lambda-1}{\lambda^2}\right)^n + \frac{\lambda \log \lambda}{\lambda - 1}, \tag{89}$$

(*cf.* [66]).

Combining (1) and (89) shows that there exists a relationship between the numbers $y(n,\lambda)$ and λ-Apostol–Daehee numbers. In general, response to the second question is are there any other relationships of the numbers $y(n,\lambda)$ with other well-known numbers and functions?

Even substituting $\lambda = 2$ into (89), and replacing n by $n+1$, we get

$$\sum_{k=0}^{n} \frac{1}{(k+1)\,2^k} = \frac{(-1)^n\, \mathfrak{D}_{n+1}(2)}{2^{2n-1}\,(n+1)!} + 2\log 2, \tag{90}$$

which, by (88), yields that there exists a relationship between the numbers $y(n)$ and the λ-Apostol–Daehee numbers $\mathfrak{D}_n(\lambda)$ as in the following form:

$$y\left(n\right)=\frac{\mathfrak{D}_{n+1}(2)}{2^{2n}\left(n+1\right)!}+\left(-1\right)^n\log 2,$$

(*cf.* [2]).

Recently, reciprocals of binomial coefficients, combinatorial sums, have been studied in many different areas (*cf.* [13,34,110,112,131], see also [142]).

With the help of the beta function and the gamma function, Sury *et al.* [131, Eq. (3)] gave the following combinatorial sum:

$$\frac{1}{n+1}\sum_{j=0}^{n}\frac{\lambda^j}{\binom{n}{j}}=\sum_{j=0}^{n}\frac{\lambda^{n+1}+\lambda^{n-j}}{(j+1)(1+\lambda)^{n+1-j}},$$

(*cf.* [2]).

The left-hand side of the above type sum has been recently studied by many mathematicians such as Mansour [77], Simsek [110–112], and Sury *et al.* [131]; and also the references cited therein.

Substituting $\lambda=1$ into the above equation, we obtain

$$y(n,-1)=\frac{1}{2\left(n+1\right)}\sum_{j=0}^{n}\frac{1}{\binom{n}{j}}\tag{91}$$

(*cf.* [2], see also [142]).

The left-hand side is the analogue of our alternating combinatorial sum in Eq. (88).

Remark 3. In Refs. [110–112] and also [142], the author showed that the finite sums, containing reciprocals of binomial coefficients, are also related to the Beta-type polynomials and the Bernstein basis functions. The readers may refer to the aforementioned papers in order to see these relationships.

Some known results for the numbers $y(n,\lambda)$ are as follows:

Theorem 25 (*cf.* [3]). *Let $n\in\mathbb{N}_0$. Then we have*

$$y(n,\lambda)=\frac{-1}{\lambda}\sum_{j=0}^{n}\sum_{k=0}^{j}\binom{n+1}{j}\frac{B_kS_1(j,k)}{j!B_j^{n+1}\left(\lambda\right)},\tag{92}$$

where $B_j^n\left(\lambda\right)$ denote the Bernstein basis function which are defined by

$$B_j^n\left(\lambda\right)=\binom{n}{j}\lambda^j(1-\lambda)^{n-j},$$

where $j\in\{0,1,2,\ldots,n\}$ (cf. [72]).

Theorem 26 (cf. [3]). *Let $n \in \mathbb{N}_0$. Then we have*

$$y(n,\lambda) = \frac{(-1)^n}{(\lambda-1)^{n+2}} \int_0^{\frac{\lambda-1}{\lambda}} \frac{1-x^{n+1}}{1-x} dx.$$

Theorem 27 (cf. [3]). *Let $n \in \mathbb{N}_0$. Then we have*

$$y\left(n,\frac{1}{2}\right) = 2^{n+2}\left(h_{\left[\frac{n}{2}\right]} - h_n + \frac{(-1)^{n+1}}{n+1}\right), \tag{93}$$

where $[x]$ denotes the integer part of x.

Theorem 28 (cf. [3]). *Let $n \in \mathbb{N}_0$. Then we have*

$$y\left(n,\frac{1}{2}\right) = \frac{1}{2^{n+2}} \int_0^1 \frac{1+(-1)^n x^{n+1}}{1+x} dx. \tag{94}$$

Theorem 29 (cf. [3]). *Let $n \in \mathbb{N}_0$. Then we have*

$$\int_0^1 \frac{1-x^{\left[\frac{n}{2}\right]}}{1-x} dx - \int_0^1 \frac{1-x^n}{1-x} dx = \frac{y\left(n,\frac{1}{2}\right)}{2^{n+2}} + \frac{(-1)^n}{n+1}. \tag{95}$$

Theorem 30 (cf. [3]). *Let $n \in \mathbb{N}_0$. Then we have*

$$y\left(n,\frac{1+\sqrt{5}}{2}\right) = \sum_{j=0}^{n} \frac{(-1)^{j+1}}{(j+1)\left(\frac{1+\sqrt{5}}{2}F_{j+1}+F_j\right)\left(\frac{1-\sqrt{5}}{2}F_{n-j+1}+F_{n-j}\right)}, \tag{96}$$

where F_n denotes the Fibonacci numbers.

Theorem 31 (cf. [3]). *Let $n \in \mathbb{N}_0$. Then we have*

$$y\left(n,\frac{1}{2}\right) = 2^{n+2}\left(n\int_0^1 (1-x)^{n-1}\log(x)dx - \left[\frac{n}{2}\right] \times \int_0^1 (1-x)^{\left[\frac{n}{2}\right]-1}\log(x)dx\right) + \frac{(-1)^{n+1}2^{n+2}}{n+1}.$$

In Ref. [3], we gave the following relation among the numbers $y\left(n,\frac{1}{2}\right)$, the Digamma function and the Euler constant:

Euler's constant (or Euler–Mascheroni) constant is given by

$$\gamma = \lim_{m\to\infty}\left(-\log(m) + \sum_{j=1}^{m}\frac{1}{j}\right),$$

and the Psi (or Digamma) function

$$\psi(z) = \frac{d}{dz} \{\log \Gamma(z)\},$$

where $\Gamma(z)$ denotes the Euler gamma function, which is defined by

$$\Gamma(z) = \int_0^\infty t^{z-1} e^{-t} dt,$$

where $z = x + iy$ with $x > 0$. For $z = n \in \mathbb{N}$,

$$\Gamma(n+1) = n!$$

(*cf.* [18,95,129]).

Sofo [124] gave the following formula:

$$h_n = \gamma + \psi(n+1), \tag{97}$$

where $h_0 = 0$.

Combining (93) with (97), we arrive at the following theorem:

Theorem 32 (*cf.* [3]). *Let $n \in \mathbb{N}_0$. Then we have*

$$y\left(n, \frac{1}{2}\right) = 2^{n+2} \left(\frac{(-1)^{n+1}}{n+1} + h_{\left[\frac{n}{2}\right]} - \gamma - \psi(n+1)\right).$$

Combining (92) with (16) and (25), we also arrive at the following result:

Corollary 10 (*cf.* [3]). *Let $n \in \mathbb{N}_0$. Then we have*

$$y(n, \lambda) = \frac{-1}{\lambda} \sum_{j=0}^{n} \binom{n+1}{j} \frac{D_j}{j! B_j^{n+1}(\lambda)}. \tag{98}$$

Theorem 33 (*cf.* [3]). *Let $n \in \mathbb{N}_0$. Then we have*

$$\sum_{v=0}^{n} \frac{S_1(v,1)}{v!} - \sum_{v=0}^{\left[\frac{n}{2}\right]} \frac{S_1(v,1)}{v!} + \sum_{j=0}^{n} B_j S_1(n,j) = 2^{-n-2} y\left(n, \frac{1}{2}\right). \tag{99}$$

Proof. Combining (14) with (10), we obtain

$$\sum_{n=1}^{\infty} H_n z^n = -\sum_{n=0}^{\infty} S_1(n,1) \frac{z^n}{n!} \sum_{n=0}^{\infty} z^n.$$

Therefore,

$$\sum_{n=1}^{\infty} h_n z^n = -\sum_{n=0}^{\infty}\sum_{v=0}^{n} S_1(v,1)\frac{z^n}{v!}.$$

We have the following well-known formula:

$$h_n = -\sum_{v=0}^{n} \frac{S_1(v,1)}{v!}. \tag{100}$$

Combining (100) with (93) and (16) yields the desired result. □

By combining the following well-known formula:

$$S_1(v,1) = (-1)^{v+1}(v-1)!, \tag{101}$$

(*cf.* [129, p. 76]), with (99), we have

$$-\sum_{v=1}^{n}\frac{(-1)^v}{v} + \sum_{v=1}^{\left[\frac{n}{2}\right]}\frac{(-1)^v}{v!} + \sum_{j=0}^{n} B_j S_1(n,j) = 2^{-n-2} y\left(n, \frac{1}{2}\right).$$

Combining the above equation with (19), we arrive at the following result:

Corollary 11. *Let $n \in \mathbb{N}$. Then we have*

$$y\left(n,\frac{1}{2}\right) = 2^{n+2}\left(\mathfrak{h}_{\left[\frac{n}{2}\right]} - \mathfrak{h}_n + \sum_{j=0}^{n} B_j S_1(n,j)\right),$$

(*cf.* [3, *Corollary* 2]).

7. Finite Sums of Powers of Binomial Coefficients

Finite sums of powers of binomial coefficients have been studied for long years. Because these sums have been used by many mathematicians, statisticians, physicists, engineers and other researchers, many different manuscripts have been published associated with these sums with different methods (*cf.* [1]; see also the references cited in each of these earlier works).

For n, $k \in \mathbb{N}$, Golombek [40] gave the following interesting finite sums involving binomial coefficients:

$$B(n,k) = \sum_{j=1}^{k}\binom{k}{j} j^n = \frac{d^n}{dt^n}\left\{\left(e^t+1\right)^k\right\}|_{t=0}. \tag{102}$$

For $n \in \mathbb{N}$, Golombek and Marburg [41] also gave the following formula for finite sums of powers of binomial coefficients:

$$\frac{d^m}{dt^m}\left\{\sum_{k=0}^{n}\binom{n}{k}^2 e^{tk}\right\}|_{t=0} = \sum_{k=1}^{n}\binom{n}{k}^2 k^m. \tag{103}$$

By the aid of the hypergeometric function, we [1] defined the following finite sums involving powers of binomial coefficients with their generating functions. That is, for n, $p \in \mathbb{N}$, and $\lambda \in \mathbb{R}$ (or $\mathbb{C}$), we define the following finite sums:

$$F_{y6}(t, n; \lambda, p) = \frac{1}{n!}\sum_{k=0}^{n}\binom{n}{k}^p \lambda^k e^{tk} \tag{104}$$

and

$$P(x; m, n; \lambda, p) = \sum_{j=0}^{n}\binom{n}{j}^p \lambda^j (x+j)^m \tag{105}$$

(*cf.* [1]).

Taking v times derivative of (105), with respect to x, we obtain

$$\begin{aligned}\frac{d^v}{dx^v}\{P(x; m, n; \lambda, p)\} &= \sum_{j=0}^{n}\binom{n}{j}^p \binom{m}{v} v!\lambda^j (x+j)^{m-v}\\ &= \sum_{j=0}^{n}\binom{n}{j}^p (m)_{\underline{v}} \lambda^j (x+j)^{m-v}.\end{aligned}$$

In Ref. [1], we know that there are two types of finite sums of powers of binomial coefficients, which are given by the equations (104) and (105). That is, for $m, p \in \mathbb{N}_0$, these type of sums are given as follows:

$$\sum_{j=0}^{n}\binom{n}{j}^p \lambda^j j^m,$$

and

$$\sum_{j=0}^{n}\binom{n}{j}^p j^m.$$

We constructed the following generating functions for a new family of combinatorial numbers $y_6(n,k;\lambda,p)$ involving finite sums of powers of binomial coefficients:

$$F_{y_6}(t,n;\lambda,p) = \frac{1}{n!}\, {}_pF_{p-1}\left[\begin{matrix} -n,-n,...,-n \\ 1,1,...,1 \end{matrix}; (-1)^p \lambda e^t\right]$$
$$= \sum_{m=0}^{\infty} y_6(m,n;\lambda,p)\frac{t^m}{m!}, \tag{106}$$

where $n,p \in \mathbb{N}$ and $\lambda \in \mathbb{R}$ (or $\mathbb{C}$) (*cf.* [1]).

By using (106), we derive (104).

Remark 4. Substituting $\lambda = 1$ and $p = 2$ into (104), we have

$$\sum_{k=0}^{n}\binom{n}{k}^2 e^{tk} = n!F_{y_6}(t,n;1,2)$$

(*cf.* [41]).

Substituting Taylor series for e^{tk} into (104), we obtain the following explicit representation for the numbers $y_6(m,n;\lambda,p)$ generated by equation (106):

Theorem 34 (*cf.* [1]). *Let $n,m,p \in \mathbb{N}_0$. Then we have*

$$y_6(m,n;\lambda,p) = \frac{1}{n!}\sum_{k=0}^{n}\binom{n}{k}^p k^m \lambda^k. \tag{107}$$

Taking m times derivative of (106), with respect to t, we also get the following formula for the numbers $y_6(m,n;\lambda,p)$:

Corollary 12 (*cf.* [1]). *Let $m \in \mathbb{N}$. Then we have*

$$y_6(m,n;\lambda,p) = \frac{\partial^m}{\partial t^m}\left\{F_{y_6}(t,m,n;\lambda,p)\right\}\Big|_{t=0}. \tag{108}$$

Remark 5. Upon setting $\lambda = 1$ in (107), we have

$$n!y_6(m,n;1,p) = M_{m,p}(n) = \sum_{k=0}^{n}\binom{n}{k}^p k^m$$

(*cf.* [1,80, p. 159, Eq. (5.1.1)]; see also [35,41]).

Remark 6. Substituting $\lambda = 1$ and $p = 2$ into (107) and (108), we have the following well-known results, resp.:

$$n!y_6(m,n;1,2) = \frac{\partial^m}{\partial t^m}\left\{F_{y_6}(t,m,n;1,2)\right\}\Big|_{t=0},$$

(*cf.* [1,41]).

By using equation (107), the numbers $y_6(0,n;\lambda,p)$ are represented by the following hypergeometric function:

Corollary 13 (*cf.* [1]).

$$y_6(0,n;\lambda,p) = \frac{1}{n!}\ {}_pF_{p-1}\left[\begin{matrix}-n,-n,\ldots,-n\\ 1,1,\ldots,1\end{matrix};(-1)^p\lambda\right]. \tag{109}$$

Remark 7. Setting $\lambda = 1$ and $p = 2$ in (109), we have

$$n!y_6(0,n;1,2) = \sum_{k=0}^{n}\binom{n}{k}^2 = {}_2F_1\left[\begin{matrix}-n,-n\\ 1\end{matrix};1\right].$$

The above finite sum is obviously a particular case of the Chu–Vandermonde identity (*cf.* [63, p. 37]). Setting $\lambda = 1$ and $p = 3$ and replacing n by $2n$ in (109), we have

$$\begin{aligned}(2n)!y_6(0,2n;-1,3) &= {}_3F_2\left[\begin{matrix}-2n,-2n,-2n\\ 1,1\end{matrix};1\right]\\ &= (-1)^n\frac{(3n)!}{(n!)^3},\end{aligned}$$

which is a special case of Dixon's identity (*cf.* [1,63, pp. 37–38, p. 64]).

7.1. *Generalization of the Franel numbers*

Generalized Franel numbers $F_p(m,n;\lambda)$ are defined by

$$F_p(m,n;\lambda) = n!y_6(m,n;\lambda,p).$$

Here, the numbers $F_p(m,n;\lambda)$ are so-called *generalized p-th*-order Franel numbers.

Here we note that, in Ref. [1], we gave many properties involving generating functions, computation formula, recurrence relations for the generalized Franel numbers. We gave relations among the generalized Franel numbers, the Bernoulli numbers, the Stirling numbers of the first kind, the

Catalan numbers, the Daehee numbers, and the numbers $y_6(m, n; \lambda, p)$. For instance, using definition of the generalized Franel numbers, one has

$$F_2(3, n; 1) = n! y_6(3, n; 1, 2) = n^2(n+1)\binom{2n-3}{n-1},$$

(*cf.* [41]), and also

$$n! y_6(0, n; 1, 3) = {}_3F_2\left[\begin{matrix} -n, -n, -n \\ 1, 1 \end{matrix}; -1\right]$$

and

$$n! y_6(0, n; 1, 4) = {}_4F_3\left[\begin{matrix} -n, -n, -n, -n \\ 1, 1, 1 \end{matrix}; 1\right],$$

where the numbers $n! y_6(0, n; 1, 3)$ and $n! y_6(0, n; 1, 4)$ denote the Franel numbers. (*cf.* [1]).

Substituting $m = 0$, $\lambda = 1$ and $p = 2$ into (107), we obtain

$$y_6(0, n; 1, 2) = \frac{n+1}{n!} C_n,$$

where C_n denotes the Catalan numbers. We also gave the following interesting finite sum:

$$y_6(0, n; 1, 2) = \frac{(-1)^n C_n}{\sum_{k=0}^{n} s(n, k) B_k} \tag{110}$$

(*cf.* [1]).

Substituting $\lambda = -1$ into (107), we introduce generalized alterne Franel numbers of the order p as follows:

$$y_6(m, n; -1, p) = \frac{1}{n!} \sum_{k=0}^{n} (-1)^k \binom{n}{k}^p k^m. \tag{111}$$

Substituting $m = 0$ into the above equation, we have

$$y_6(0, n; -1, p) = \frac{1}{n!} \sum_{k=0}^{n} (-1)^k \binom{n}{k}^p. \tag{112}$$

Remark 8. Setting $\lambda = 1$ in (107), we have

$$L_j(n) = n! y_6(0, n; 1, j),$$

(*cf.* [80]). Moll gave in Ref. [80] a recurrence relation for the numbers $L_j(n)$.

Remark 9. In Ref. [1], we gave the following alterne Franel numbers of order j:

$$n!y_6(0,n;-1,j).$$

Remark 10. For $n \in \mathbb{N}_0$. Setting $p=2$ and $p=3$ into (112), we have the following well-known identities, resp.:

$$n!y_6(0,n;-1,2)=\begin{cases} 0, & n \text{ odd positive integer} \\ \frac{(-1)^{\frac{n}{2}} n!}{\left(\left(\frac{n}{2}\right)!\right)^2}, & \text{otherwise,} \end{cases}$$

$$n!y_6(0,n;-1,2)=\frac{\sqrt{\pi}2^n}{\Gamma\left(\frac{2+n}{2}\right)\Gamma\left(\frac{1-n}{2}\right)}, \tag{113}$$

where

$$\Gamma\left(n+\frac{1}{2}\right)=\frac{(2n)!\sqrt{\pi}}{2^{2n}(2n)!}$$

and

$$n!y_6(0,n;-1,3)=\begin{cases} 0, & n \text{ odd} \\ \frac{(-1)^{\frac{n}{2}}\left(\frac{3n}{2}\right)!}{\left(\left(\frac{n}{2}\right)!\right)^3}, & \text{otherwise} \end{cases}$$

(*cf.* [1,63, p. 11]).

We next give an open problem related to the numbers mentioned above:

Open problem:
How can we find relations among the numbers $y(n,\lambda)$, the generalized Franel numbers $F_p(m,n;\lambda)$, the numbers $y_6(m,n;\lambda,p)$?

8. Special Finite Sums Involving the Generalized Harmonic Numbers and Harmonic Functions

We [4] defined some special finite sums involving the generalized harmonic numbers and the numbers $y(n,\lambda)$.

Let $a_1,a_2,\ldots,a_v$ be indeterminate and $m,v\in\mathbb{N}$. We [4] defined a sum $\mathfrak{s}_{m,s}(a_1,a_2,\ldots,a_{v-1};a_v)$ as follows:

$$\mathfrak{s}_{m,s}(a_1,a_2,\ldots,a_{v-1};a_v,v)=\sum_{j=1}^{v}\sum_{k=1}^{m}\frac{1}{(a_j+k)^s}, \tag{114}$$

$(v,m\in\mathbb{N};\ s\in\mathbb{C};\ a_j\in\mathbb{C}\backslash\mathbb{Z}^-;j\in\{1,2,3,\ldots,v\})$.

The sum $\mathfrak{s}_{m,s}(a_1, a_2, \ldots a_{v-1}; a_v, v)$ has the following properties:

Substituting $a_1 = z - 1$ and $a_2 = \cdots = a_{v-1} = a_v = 0$ into (114), we obtain

$$\mathfrak{s}_{m,s}(z-1, 0, \ldots, 0; 0, v) = v h_m^{(s)}(z-1)$$

(*cf.* [4]).

Putting $v = 1$ in the above equation, we obtain the following well-known formula for the generalized harmonic numbers $h_m^{(\kappa)}(z)$, which are also denoted by $h_m(z;\kappa)$ (*cf.* [4]):

$$\mathfrak{s}_{m,\kappa}(z-1;1) = \sum_{k=1}^{m} \frac{1}{(z+k-1)^{\kappa}} = h_m^{(\kappa)}(z), \quad \left(m \in \mathbb{N};\ \kappa \in \mathbb{C};\ z \in \mathbb{C}\backslash\mathbb{Z}_0^-\right),$$

which, for $s = \kappa$, is recorded by Rassias and Srivastava [135, Eq. (1.11)].

Substituting $a_1 = a_2 = \cdots = a_{v-1} = a_v = 0$ into (114), we obtain

$$\mathfrak{s}_{m,s}(0, 0, \ldots, 0; 0, v) = v h_m^{(s)}$$

(*cf.* [4]).

For $v = 1$, we have

$$\mathfrak{s}_{m,s}(0;1) = h_m^{(s)}.$$

For $v = s = 1$, we also have

$$\mathfrak{s}_{m,1}(0;1) = h_m.$$

Substituting $a_j = b_j - 1$, $j \in \{1, 2, 3, \ldots, v\}$ into (114), we obtain

$$\begin{aligned}
&\mathfrak{s}_{m,d+1}(b_1 - 1, b_2 - 1, \ldots, b_{v-1} - 1; b_v - 1, v) \\
&= \sum_{j=1}^{v} (\zeta(d+1, b_j) - \zeta(d+1, b_j + m))
\end{aligned}$$

(*cf.* [4]).

The sum $\mathfrak{s}_{m,s}(a_1, a_2, \ldots, a_{v-1}; a_v, v)$ is related to the generalized harmonic functions $h_m^{(s)}(z)$. That is

$$\mathfrak{s}_{m,s}(a_1, a_2, \ldots, a_{v-1}; a_v, v) = \sum_{j=1}^{v} h_m^{(s)}(a_j) \tag{115}$$

(*cf.* [4]).

Theorem 35 (*cf.* [4]). *Symmetric property for the sum* $\mathfrak{s}_{m,s}(a_1, a_2, \ldots, a_{v-1}; a_v, v)$ *is given as follows:*

$$\mathfrak{s}_{m,s}(a_2, a_3, \ldots, a_v; a_1, v) = \mathfrak{s}_{m,s}(a_1, a_3, \ldots, a_v; a_2, v)$$
$$= \cdots = \mathfrak{s}_{m,s}(a_1, a_2, \ldots, a_{v-1}; a_v, v).$$

Theorem 36 (*cf.* [4]). *Reciprocity law for the sum* $\mathfrak{s}_{m,s}(a_1, a_2, \ldots, a_{v-1}; a_v, v)$ *is given as follows:*

$$\begin{aligned}&\mathfrak{s}_{m,d+1}(a_2 - 1, a_3 - 1, \ldots, a_v - 1; a_1 - 1, v)\\&\quad +\mathfrak{s}_{m,d+1}(a_1 - 1, a - 1, \ldots, a_v - 1; a_2 - 1, v)\\&\quad +\cdots + \mathfrak{s}_{m,d+1}(a_1 - 1, a_2 - 1, \ldots, a_{v-1} - 1; a_v - 1, v)\\&\quad = \sum_{j=1}^{v} \frac{\psi^{(d)}(a_j - 1 + m) - \psi^{(d)}(a_j - 1)}{(-1)^d d!}.\end{aligned} \tag{116}$$

In order to give proof of the assertion of Theorem 36, and to obtain new formulas, the following well-known formula involving the polygamma functions $\psi^{(d)}(z)$ $(d \in \mathbb{N})$, the function $H_m^{(s)}(z)$ (*cf.* [130, p. 22, Eq. (20)]; see also [30, Eq. (1.8), 135]), is needed:

$$\psi^{(d)}(z + m) - \psi^{(d)}(z) = (-1)^d d! h_m^{(d+1)}(z - 1), \tag{117}$$

where $m, d \in \mathbb{N}_0$.

Using (117), (114) and (115), for $s = d + 1$ and $a_j = b_j - 1$; $j \in \{1, 2, 3, \ldots, v\}$, we have

$$\mathfrak{s}_{m,d+1}(b_1 - 1, b_2 - 1, \ldots, b_{v-1} - 1; b_v - 1, v) = \sum_{j=1}^{v} h_m^{(d+1)}(b_j - 1),$$

and

$$\begin{aligned}&\mathfrak{s}_{m,d+1}(b_1 - 1, b_2 - 1, \ldots, b_{v-1} - 1; b_v - 1, v)\\&\quad = \sum_{j=1}^{v} \frac{\psi^{(d)}(b_j - 1 + m) - \psi^{(d)}(b_j - 1)}{(-1)^d d!}\end{aligned} \tag{118}$$

(*cf.* [4]).

Substituting $s = 1$ into (114), we get

$$\psi^{(d)}(1 + m) = \psi^{(d)}(1) + (-1)^d d! h_m^{(d+1)}(0).$$

Combining the above equation with the following known result (*cf.* [135, Eq. (4.8)]):

$$\psi^{(d)}(1) = (-1)^{d-1} d! \zeta(p+1), \ (d \in \mathbb{N}).$$

Thus, we get

$$\psi^{(d)}(1+m) = (-1)^{d-1} d! \left(H_m^{(d+1)} - \zeta(d+1) \right).$$

or

$$\psi^{(d)}(1+m) = (-1)^{d-1} d! \left(h_m^{(d+1)} - \sum_{k=d}^{\infty} \frac{(-1)^{k-d}}{k.k!} S_1(k,d) \right).$$

For $d = 1$, the following relations reduce to the following result:

$$\psi^{(1)}(1+m) = h_m^{(2)} - \sum_{k=d}^{\infty} \frac{(-1)^{k-1}}{k.k!} S_1(k,1).$$

Combining the above equation with (101), we have

$$\begin{aligned} \psi^{(1)}(1+m) &= h_m^{(2)} - \sum_{k=1}^{\infty} \frac{1}{k^2} \\ &= h_m^{(2)} - \frac{\pi^2}{6}. \end{aligned}$$

Substituting $d = l-1$, $l \in \mathbb{N}$, and $a_1 = a_2 = \cdots = a_v = 2$ into (116), we obtain

$$\begin{aligned} \mathfrak{s}_{m,l-1}(1,1,\ldots,1;1,v) &= \psi^{(l)}(1) - \psi^{(l)}(1+m) \\ &= (-1)^{l-1} l! \left(2\zeta(l+1) - h_m^{(l+1)} \right). \end{aligned}$$

When $l = 1$, we have

$$\mathfrak{s}_{m,0}(1,1,\ldots,1;1,v) = \frac{\pi^2}{3} - h_m^{(2)}.$$

Similarly, we also have

$$\sum_{m=1}^{n} \mathfrak{s}_{m,1}(1,1,\ldots,1;1,v) = v \sum_{m=1}^{n} h_m.$$

where $n \in \mathbb{N}$. Combining the above equation with the following known formula:

$$\sum_{m=1}^{n} h_m = (n+1)h_n - n,$$

(*cf.* [26, Eq. (1.28)]), we have the following summation formula:

$$\sum_{m=1}^{n} \mathfrak{s}_{m,1}(1,1,\ldots,1;1,v) = v(n+1)h_n - vn.$$

A relation between the sum $\mathfrak{s}_{m,1}(a_1,a_2,\ldots,a_{v-1};a_v,v)$ and generalized harmonic functions of order b is given by

$$\mathfrak{s}_{m,b}(a_1,a_2,\ldots,a_{v-1};a_v,v) = \sum_{j=1}^{v} h_m^{(b)}(a_j)$$

(*cf.* [4]).

For $b=1$, one has

$$\mathfrak{s}_{m,1}(a_1,a_2,\ldots,a_{v-1};a_v,v) = \sum_{j=1}^{v} h_m(a_j)$$

(*cf.* [4]).

For $v=3$ and $b=1$, we (*cf.* [4]) gave known relations between the sum $\mathfrak{s}_m(a_1,a_2;a_3,3)$ and Gauss's hypergeometric series as follows:

$$\mathfrak{s}_{m,1}(a_1-1,a_2-1;0,3) - 3h_m = \sum_{j=1}^{m} \frac{1}{a_1-1+j} + \sum_{j=1}^{m} \frac{1}{a_1-1+j} - 2\sum_{j=1}^{m} \frac{1}{j},$$

or

$$\mathfrak{s}_{m,1}(a_1-1,a_2-1;0,3) - 3h_m = h_m(a_1-1) + h_m(a_2-1) - 2h_m.$$

Observe that the right-hand side of the above equation is derived from the following equation, which was proved by Beukers [17, p. 25]:

$$\log z\ {}_2F_1\left[\begin{matrix} a_1, a_2 \\ 1 \end{matrix}; z\right] + \sum_{m=1}^{\infty} \frac{(a_1)_m (a_2)_m}{(m!)^2} z^m \left(h_m(a_1-1) + h_m(a_2-1) - 2h_m\right).$$

A relation between the sum $\mathfrak{s}_m(a_1,a_2,\ldots,a_{v-1};0,v)$ and the numbers $y\left(n,\frac{1}{2}\right)$ is given by (*cf.* [4]):

$$\mathfrak{s}_{m,1}(a_1,a_2,\ldots,a_{v-1};0,v) = \sum_{j=1}^{v-1} h_m(a_j) + 2^{n+2}\left(h_{\left[\frac{m}{2}\right]} + \frac{(-1)^{m+1}}{m+1}\right)$$
$$-2^{-m-2} y\left(m,\frac{1}{2}\right),$$

where $[x]$ denotes the integer part of x and $y\left(m, \frac{1}{2}\right)$ is related to equation (93), which was proved in Ref. [3, Theorem 3].

Theorem 37 (*cf.* [4]). *Let $n \in \mathbb{N}_0$. Then, we have*

$$y(n,2) = (-1)^n h_n + \frac{(-1)^n}{(n+1)2^{n+1}} + (-1)^{n+1} 2^{-n} \sum_{j=0}^{n} \binom{n}{j} h_j. \tag{119}$$

Proof. By using (88), we have

$$(-1)^n y(n,2) - \frac{1}{(n+1)2^{n+1}} = \sum_{j=1}^{n} \frac{1}{j 2^j}.$$

Combining the above equation with (66), we obtain

$$h_n - 2^{-n} \sum_{j=0}^{n} \binom{n}{j} h_j = (-1)^n y(n,2) - \frac{1}{(n+1)2^{n+1}},$$

which proves the assertion (119) of Theorem 37. □

Recall that

$$\frac{d}{d\lambda} \left\{ (\lambda)_v \right\} = (\lambda)_v \left(\psi(\lambda + v) - \psi(\lambda) \right), \tag{120}$$

and

$$\psi(s+m) - \psi(s) = h_m(s-1) \tag{121}$$

(*cf.* [26]).

By combining the equation (120) with the equation (114), for $d = 0$ and $j \in \{1, 2, \ldots, v\}$, we get

$$\begin{aligned} \sum_{j=1}^{v} \frac{1}{(a_j)_m} \frac{d}{d\lambda} \left\{ (\lambda)_m \right\} \Big|_{\lambda = a_j} &= \sum_{j=1}^{v} h_m(a_j - 1) \\ &= \mathfrak{s}_{m,1}(a_1 - 1, a_2 - 1, \ldots, a_{v-1} - 1; a_v - 1, v), \end{aligned}$$

Thus, we get the following theorem:

Theorem 38 (*cf.* [4]). *Let $n \in \mathbb{N}_0$. Then we have*

$$\begin{aligned} &\mathfrak{s}_{m,1}(a_1 - 1, a_2 - 1, \ldots, a_{v-1} - 1; a_v - 1, v) \\ &= \sum_{j=1}^{v} \frac{1}{(a_j)_m} \frac{d}{d\lambda} \left\{ (\lambda)_m \right\} \big|_{\lambda = a_j}. \end{aligned}$$

Combining the following well-known formula:

$$\frac{d}{d\lambda}\left\{(\lambda)_m\right\}|_{\lambda=-m} = -m!h_m,$$

(*cf.* [26]) with the equation (93), we arrive at the following theorem:

Theorem 39 (*cf.* [4]). *Let $m \in \mathbb{N}$. Then we have*

$$y\left(m, \frac{1}{2}\right) = 2^{m+2}\left(h_{\left[\frac{m}{2}\right]} + \frac{(-1)^{m+1}}{m+1} + \frac{1}{m!}\frac{d}{d\lambda}\left\{(\lambda)_m\right\}|_{\lambda=-m}\right).$$

In Ref. [135, Eqs. (1.5) and (1.7)], Rassias and Srivastava gave the following partial derivative formulas for Gauss's hypergeometric series:

$$\frac{\partial}{\partial a}\left\{\sum_{m=1}^{\infty}\frac{(a)_m(b)_m}{(c)_m m!}z^m\right\} = \sum_{m=1}^{\infty}\frac{(a)_m(b)_m}{(c)_m m!}\left(\psi(m+a)-\psi(a)\right)z^m, \tag{122}$$

$\left(|z|<1; |z|=0 \text{ when } \mathrm{Re}\{c-a-b\}>0; c \notin \mathbb{Z}_0^-\right)$

$$\frac{\partial}{\partial b}\left\{\sum_{m=1}^{\infty}\frac{(a)_m(b)_m}{(c)_m m!}z^m\right\} = \sum_{m=1}^{\infty}\frac{(a)_m(b)_m}{(c)_m m!}\left(\psi(m+b)-\psi(b)\right)z^m, \tag{123}$$

$\left(|z|<1; |z|=0 \text{ when } \mathrm{Re}\{c-a-b\}>0; c \notin \mathbb{Z}_0^-\right)$

$$\frac{\partial}{\partial c}\left\{\sum_{m=1}^{\infty}\frac{(a)_m(b)_m}{(c)_m m!}z^m\right\} = \sum_{m=1}^{\infty}\frac{(a)_m(b)_m}{(c)_m m!}\left(\psi(m+c)-\psi(c)\right)z^m, \tag{124}$$

$\left(|z|<1; |z|=0 \text{ when } \mathrm{Re}\{c-a-b\}>0; c \notin \mathbb{Z}_0^-\right)$.

Combining the equations from (122) to (124) with (121), we have the following results:

$$\frac{\partial}{\partial a}\left\{\sum_{m=1}^{\infty}\frac{(a)_m(b)_m}{(c)_m m!}z^m\right\} = \sum_{m=1}^{\infty}\frac{(a)_m(b)_m}{(c)_m m!}h_m(a-1)z^m,$$

$\left(|z|<1; |z|=0 \text{ when } \mathrm{Re}\{c-a-b\}>0; c \notin \mathbb{Z}_0^-\right)$

$$\frac{\partial}{\partial b}\left\{\sum_{m=1}^{\infty}\frac{(a)_m(b)_m}{(c)_m m!}z^m\right\} = \sum_{m=1}^{\infty}\frac{(a)_m(b)_m}{(c)_m m!}h_m(b-1)z^m,$$

$\left(|z| < 1; |z| = 0 \text{ when} \operatorname{Re}\{c-a-b\} > 0; c \notin \mathbb{Z}_0^-\right)$

$$\frac{\partial}{\partial c}\left\{\sum_{m=1}^{\infty} \frac{(a)_m (b)_m}{(c)_m\, m!} z^m\right\} = \sum_{m=1}^{\infty} \frac{(a)_m (b)_m}{(c)_m\, m!} h_m(c-1) z^m,$$

$\left(|z| < 1; |z| = 0 \text{ when } \operatorname{Re}\{c-a-b\} > 0; c \notin \mathbb{Z}_0^-\right).$

$$\begin{aligned}&\frac{\partial}{\partial a}\left\{\sum_{m=1}^{\infty} \frac{(a)_m (b)_m}{(c)_m\, m!} z^m\right\} + \frac{\partial}{\partial b}\left\{\sum_{m=1}^{\infty} \frac{(a)_m (b)_m}{(c)_m\, m!} z^m\right\} \\ &\quad + \frac{\partial}{\partial c}\left\{\sum_{m=1}^{\infty} \frac{(a)_m (b)_m}{(c)_m\, m!} z^m\right\} \\ &= \sum_{m=1}^{\infty} \frac{(a)_m (b)_m}{(c)_m\, m!} \left(h_m(a-1) + h_m(b-1) + h_m(c-1)\right) z^m.\end{aligned}$$

Combining the above equation with (114), we arrive at the following theorem:

Theorem 40 (*cf.* [4]).

$$\begin{aligned}&\frac{\partial}{\partial a}\left\{\sum_{m=1}^{\infty} \frac{(a)_m (b)_m}{(c)_m\, m!} z^m\right\} + \frac{\partial}{\partial b}\left\{\sum_{m=1}^{\infty} \frac{(a)_m (b)_m}{(c)_m\, m!} z^m\right\} \\ &\quad + \frac{\partial}{\partial c}\left\{\sum_{m=1}^{\infty} \frac{(a)_m (b)_m}{(c)_m\, m!} z^m\right\} \\ &= \sum_{m=1}^{\infty} \frac{(a)_m (b)_m}{(c)_m\, m!} \mathfrak{s}_{m,1}(a-1, b-1; c-1, 3) z^m,\end{aligned}$$

where

$$\mathfrak{s}_{m,1}(a-1, b-1; c-1, 3) = h_m(a-1) + h_m(b-1) + h_m(c-1).$$

9. Finite Sums Derived from the Dedekind Sums and the Sum $y(n, \lambda)$

In various kinds of applications of elliptic modular functions to number theory and in analysis, in number theory, in combinatorics, in q-series, in theory of the Weierstrass elliptic functions, in modular forms, in Kronecker limit formula, and in theory of cryptography, the Dedekind eta function plays a central role. In 1877, this function, which was introduced by Dedekind, was defined by

$$\eta(\tau) = e^{\frac{\pi i \tau}{12}} \prod_{m=1}^{\infty} \left(1 - e^{2\pi i m \tau}\right),$$

where $\tau \in H$ denotes the upper half-plane (*cf.* [10,89]). The infinite product has the form $\prod_{n=1}^{\infty}(1-x^n)$ where $x = e^{2\pi i \tau}$. If $\tau \in H$, then $\mid x \mid< 1$, so the product $\prod_{n=1}^{\infty}(1-x^n)$ converges absolutely and is non-zero. Moreover, since the convergence is uniform on compact subsets of H, the function $\eta(\tau)$ is analytic on H (*cf.* [16,89,100,103]).

The behavior of the function $\log \eta(z)$ under the modular group, denoted by $\Gamma(1)$, which is given by

$$\Gamma(1) = \left\{ A = \begin{bmatrix} a & b \\ c & d \end{bmatrix} : ad - bc = 1, a, b, c, d \in \mathbb{Z} \right\},$$

where $Az = \frac{az+b}{cz+d}$, is given as follows:

$$\log \eta(Az) = \log \eta(z) + \frac{\pi i(a+d)}{12c} - \pi i \left(s(d,c) - \frac{1}{4} \right) + \frac{1}{2}\log(cz+d), \quad (125)$$

where $z \in\in H$ and $s(d,c)$ denote the Dedekind sum which defined by

$$s(h,k) = \sum_{j \bmod k} \left(\left(\frac{j}{k} \right) \right) \left(\left(\frac{hj}{k} \right) \right),$$

where $h, k \in \mathbb{Z}$ and $k > 0$, $(h,k) = 1$ and

$$((x)) = \begin{cases} x - [x] - \frac{1}{2} & x \notin \mathbb{Z} \\ 0 & x \in \mathbb{Z} \end{cases}$$

(*cf.* [16,78,89]).

Then the well-known reciprocity theorem for the sum $s(h,k)$ is given by

$$s(h,k) + s(k,h) = -\frac{1}{4} + \frac{1}{12}\left(\frac{h}{k} + \frac{k}{h} + \frac{1}{hk} \right), \quad (126)$$

where h and k are coprime positive integers (*cf.* [89]).

By setting

$$h(c,d) = s(d,c) + s(c,d),$$

and

$$f(c,d) = -\frac{1}{4} + \frac{d}{12c} + \frac{c}{12d} + \frac{1}{12dc},$$

we next give a brief explanation of the notations written under the sigma summation symbol by the following examples:

The following sum runs over coprime integers:

$$\sum_{\substack{c=1\\(c,d)=1}}^{m} h(c,d).$$

We can exemplify the above sum as follows:

$$\begin{aligned}\sum_{\substack{c=1\\(c,3)=1}}^{12} h(c,3) &= h(1,3)+h(2,3)+h(4,3)+h(5,3)\\ &\quad +h(7,3)+h(8,3)+h(10,3)+h(11,3).\end{aligned}$$

Similarly, the following sum runs over non-coprime integers:

$$\sum_{\substack{c=1\\(c,d)>1}}^{m} f(c,d).$$

The above sum can be exemplified as follows:

$$\sum_{\substack{c=1\\(c,3)>1}}^{12} f(c,3) = f(3,3)+f(6,3)+f(9,3)+f(12,3).$$

If we sum both sides of (126) from $c=1$ to m, we get

$$\sum_{\substack{c=1\\(c,d)=1}}^{m} h(c,d) + \sum_{\substack{c=1\\(c,d)>1}}^{m} f(c,d) = \frac{m(m+1)-6dm+2\left(d^2+1\right)h_m}{24d}, \tag{127}$$

where $m,d \in \mathbb{N}$.

If we sum both sides of (126) from $d=1$ to m, we obtain

$$\sum_{\substack{d=1\\(d,c)=1}}^{m} h(c,d) + \sum_{\substack{d=1\\(d,c)>1}}^{m} f(c,d) = \frac{m(m+1)-6cm+2\left(c^2+1\right)h_m}{24c}, \tag{128}$$

where $m,c \in \mathbb{N}$.

By combining (127) with (93), and then by making some calculations, we arrive at the following theorem:

Theorem 41 (*cf.* [4]). *Let c and d be coprime positive integers. Let* $m \in \mathbb{N}$*. Then we have*

$$y\left(m, \frac{1}{2}\right) = 2^{m+2}\left(h_{\left[\frac{m}{2}\right]} + \frac{(-1)^{m+1}}{m+1} + \frac{m(m+1) - 6dm}{d^2+1}\right) - 2^{m+4}\frac{3d}{d^2+1}\left(\sum_{\substack{d=1\\(d,c)=1}}^{m} h(c,d) + \sum_{\substack{d=1\\(d,c)>1}}^{m} f(c,d)\right). \tag{129}$$

Remark 11. For other applications associated with the numbers $y\left(m, \frac{1}{2}\right)$ and the Dedekind-type sums, the reader may glance at the recent paper [4] of the author.

10. Finite Sums Derived from Decomposition of the Multiple Hurwitz Zeta Functions with the Aid of the Numbers $y(n, \lambda)$

In this section, we give finite sums derived from decomposition of the multiple Hurwitz zeta functions in terms of the Bernoulli polynomials of higher order with the aid of the numbers $y(n, \lambda)$.

Putting $\lambda = e^{-t}$ in (1), we have the following finite sum involving generating function for the Bernoulli polynomials of order $n+1-j$:

$$y\left(n, \frac{1}{e^t}\right) = \sum_{j=0}^{n} \frac{(-1)^{j-1}}{(j+1)} \frac{e^{t(n+2)}}{(e^t-1)^{n+1-j}}, \tag{130}$$

(*cf.* [3, Eq. (28)]).

By combining (130) and (3), we have

$$y\left(n, \frac{1}{e^t}\right) = \sum_{m=0}^{\infty}\sum_{j=0}^{n} (-1)^{j-1} \frac{B_{m+n+1-j}^{(n+1-j)}(n+2)}{(j+1)\binom{m+n+1-j}{n+1-j}(n+1-j)!} \frac{t^m}{m!} \tag{131}$$

(*cf.* [3, Eq. (29)]).

Using (130), we obtain

$$y\left(n,\frac{1}{e^t}\right)=\sum_{m=0}^{\infty}\sum_{j=0}^{n}\frac{(-1)^{j-1}t^m\,(n+2)^{n+1-j}\,B_0^{(n+1-j)}}{(j+1)\binom{m+n+1-j}{n+1-j}(n+1-j)!m!}$$
$$+\sum_{m=0}^{\infty}\sum_{j=0}^{n}\frac{(-1)^{j-1}t^m\sum_{l=1}^{m+n+1-j}\binom{m+n+1-j}{l}(n+2)^{n+1-j-l}}{(j+1)\binom{m+n+1-j}{n+1-j}(n+1-j)!m!}$$
$$\times\left(\left(1-\frac{l}{n-j}\right)B_l^{(n-j)}-B_{l-1}^{(n-j)}\right), \tag{132}$$

where we used the following well-known formula for the Bernoulli numbers of order $n+1-j$, for $l, n+1-j\in\mathbb{N}\setminus\{1\}$:

$$B_l^{(n+1-j)}=\left(1-\frac{l}{n-j}\right)B_l^{(n-j)}-B_{l-1}^{(n-j)},$$

with the aid of this formula

$$B_l^{(l+1)}=(-1)^l\,l!$$

(*cf.* [20]).

Using (130), we also have

$$y\left(n,\frac{1}{e^t}\right)=\sum_{j=0}^{n}\frac{(-1)^n}{(j+1)}\sum_{v=0}^{\infty}\sum_{m=0}^{\infty}\binom{v+n-j}{v}(v+n+2)^m\frac{t^m}{m!}. \tag{133}$$

(*cf.* [3, Eq. (30)]).

Combining (132) and (133), we have the following finite sum:

Theorem 42. *Let $m,n\in N_0$. Then we have*

$$\sum_{j=0}^{n}\frac{(-1)^n}{(j+1)}\sum_{v=0}^{\infty}\sum_{m=0}^{\infty}\binom{v+n-j}{v}(v+n+2)^m$$
$$=\sum_{j=0}^{n}\frac{(-1)^{j-1}\,(n+2)^{n+1-j}\,B_0^{(n+1-j)}}{(j+1)\binom{m+n+1-j}{n+1-j}(n+1-j)!}$$
$$+\sum_{j=0}^{n}\frac{(-1)^{j-1}\sum_{l=1}^{m+n+1-j}\binom{m+n+1-j}{l}(n+2)^{n+1-j-l}\left(\left(1-\frac{l}{n-j}\right)B_l^{(n-j)}-B_{l-1}^{(n-j)}\right)}{(j+1)\binom{m+n+1-j}{n+1-j}(n+1-j)!}.$$

Theorem 43 (*cf.* [2]). *Let $m, n \in N_0$. Then we have*

$$\sum_{j=0}^{n} \frac{1}{j+1} \left((-1)^n \sum_{v=0}^{\infty} \binom{v+n-j}{v} (v+n+2)^m + \frac{(-1)^j B_{m+n+1-j}^{(n+1-j)}(n+2)}{\binom{m+n+1-j}{n+1-j}(n+1-j)!} \right) = 0. \tag{134}$$

We [3] gave a relation between $y\left(n, \frac{1}{e^t}\right)$ and the d-ple (multiple) Hurwitz zeta functions or the Hurwitz zeta function of order $d \in \mathbb{N}$, which is defined by

$$\zeta_d(s, x) = \sum_{v=0}^{\infty} \binom{v+d-1}{v} \frac{1}{(x+v)^s},$$

where $\mathrm{Re}(s) > d$, when $d = 1$, we have the Hurwitz zeta function

$$\zeta(s, x) = \zeta_1(s, x) = \sum_{v=0}^{\infty} \frac{1}{(x+v)^s},$$

(*cf.* [15,25,49,68,98,99,127,129,130]), as follows:

$$y\left(n, \frac{1}{e^t}\right) = \sum_{j=0}^{n} \frac{(-1)^n}{(j+1)} \zeta_{n+1-j}(-m, n+2) \frac{t^m}{m!}. \tag{135}$$

Since

$$\zeta_d(-m, x) = \frac{(-1)^d m! B_{m+d}^{(d)}(x)}{(d+m)!}, \tag{136}$$

where $m \in \mathbb{N}_0$ (*cf.* [15,25,49,68,98,99,127,129,130]), we arrive at the following theorem:

Theorem 44 (*cf.* [3, Eq. (31)]). *Let $m, n \in \mathbb{N}$. Then we have*

$$\sum_{j=0}^{n} \frac{1}{j+1} \zeta_{n+1-j}(-m, n+2) = \sum_{j=0}^{n} \frac{(-1)^{-n+j+1}}{j+1} \frac{B_{m+n+1-j}^{(n+1-j)}(n+2)}{\binom{m+n+1-j}{n+1-j}(n+1-j)!}. \tag{137}$$

Using (137), we have the novel decomposition of the function $\zeta_d(s, x)$ (*cf.* [3, Eq. (31)]). That is,

$$\begin{aligned}&\zeta_{n+1}(-m,n+2)+\frac{1}{2}\zeta_n(-m,n+2)+\frac{1}{3}\zeta_{n-1}(-m,n+2)\\&\quad+\cdots+\frac{1}{n+1}\zeta_1(-m,n+2)\\&=\sum_{j=0}^{n}(-1)^{-n+j+1}\sum_{l=1}^{m+n+1-j}\frac{\binom{m+n+1-j}{l}}{\binom{m+n+1-j}{n+1-j}}\frac{(n+2)^{n+1-j-l}}{(j+1)(n+1-j)!}B_l^{(n+1-j)}.\end{aligned}$$

Putting $n=0$ in (137), we have the following well-known special value of the above decomposition:

$$\zeta(-m,2)=-\frac{B_{m+1}(2)}{m+1}.$$

References

[1] Y. Simsek, Generating functions for finite sums involving higher powers of binomial coefficients: Analysis of hypergeometric functions including new families of polynomials and numbers, *J. Math. Anal. Appl.* **477**, 1328–1352, (2019).

[2] Y. Simsek, Interpolation functions for new classes special numbers and polynomials via applications of p-adic integrals and derivative operator, *Montes Taurus J. Pure Appl. Math.* **3**(1), 38–61, (2021).

[3] Y. Simsek, New integral formulas and identities involving special numbers and functions derived from certain class of special combinatorial sums, *RACSAM* **115**(66), (2021). https://doi.org/10.1007/s13398-021-01006-6.

[4] Y. Simsek, Some classes of finite sums related to the generalized harmonic functions and special numbers and polynomials, *Montes Taurus J. Pure Appl. Math.* **4**(3), 61–79, (2022). Article ID: MTJPAM-D-21-00002.

[5] Y. Simsek, Miscellaneous formulae for the certain class of combinatorial sums and special numbers, *Bulletin T.CLIV de l'Académie serbe des sciences et des arts — 2021 Classe des Sciences mathématiques et naturelles Sciences mathématiques* **46**, 151–167, (2021).

[6] M. Ali, M. Ghayasuddin, and T.K. Pogany, Integrals with two–variable generating function in the integrand, *Montes Taurus J. Pure Appl. Math.* **3**(3), 95–103, (2021). Article ID: MTJPAM-D-20-00048.

[7] M. Alkan and Y. Simsek, Generating function for q-Eulerian polynomials and their decomposition and applications, *Fixed Point Theory Appl.* **2013**(72), 1–14, (2013).

[8] H. Alzer and J. Choi, The Riemann zeta function and classes of infinite series, *Appl. Anal. Discrete Math.* **11**, 386–398, (2017).

[9] T.M. Apostol, On the Lerch zeta function, *Pacific J. Math.* **1**, 161–167, (1951).

[10] T.M. Apostol, *Modular Functions and Dirichlet Series in Number Theory* (Springer-Verlag, New York, 1976).

[11] A.A. Aygunes, An integral formula generated by Hurwitz-Lerch zeta function with order 1, *Montes Taurus J. Pure Appl. Math.* **3**(3), 12–16, (2021). Article ID: MTJPAM-D-20-00029.

[12] A.A. Aygunes and Y. Simsek, Unification of multiple Lerch-zeta type functions, *Adv. Studies Contemp. Math.* **21**, 367–373, (2011).

[13] T.-T. Bai and Q.-M. Luo, A Simple proof of a binomial identity with applications, *Montes Taurus J. Pure Appl. Math.* **1**(2), 13–20, (2019). Article ID: MTJPAM-D-19-00008.

[14] A. Bayad and Y. Simsek, Values of twisted Barnes zeta functions at negative integers, *Russ. J. Math. Phys.* **139**(20), 129–137, (2013).

[15] A. Bayad and Y. Simsek, Note on the Hurwitz Zeta Function of Higher Order, *AIP Conf. Proc.* **1389**, 389–391 (2011). Doi: 10.1063/1.3636744.

[16] A. Bayad, Jacobi forms in two variables: Multiple elliptic Dedekind sums, the Kummer-Von Staudt Clausen congruences for elliptic Bernoulli functions and values of Hecke L-functions, *Montes Taurus J. Pure Appl. Math.* **1**(2), 58–129, (2019). Article ID: MTJPAM-D-19-00009.

[17] F. Beukers, Chapter 2: Gauss' Hypergeometric Function, In: *Arithmetic and Geometry Around Hypergeometric Functions* (R.P. Holzapfel, M. Uludag, and M. Yoshida), Lecture Notes of a CIMPA Summer School held at Galatasaray University, Istanbul, Part of the Progress in Mathematics book series (PM, Vol. 260), pp. 23–42 (Birkhäuser Verlag Basel/Switzerland, 2007).

[18] R.E. Bradley and C.E. Sandifer, *Leonhard Euler Life, Work and Legacy.* (Elsevier Science, Amsterdam, 2007).

[19] L. Carlitz, Eulerian numbers and polynomials, *Math. Mag.* **32**, 247–260, (1959).

[20] L. Carlitz, Some theorems on Bernoulli numbers of higher order, *Pac. Math.* **2**(2), 127–139, (1952).

[21] L. Carlitz, Generating functions, *Fibonacci Q.* **7**, 359–393, (1969).

[22] L. Carlitz, Some numbers related to the Stirling numbers of the first and second kind, *Publ. Elektroteh. Fak. Univ. Beogr., Mat.* **544–576**, 49–55, (1976).

[23] L. Carlitz, A note on the multiplication formulas for the Bernoulli and Euler polynomials, *Proc. Am. Math. Soc.* **4**, 184–188, (1953).

[24] C.A. Charalambides, *Combinatorial Methods in Discrete Distributions* (Wiley-Interscience, Hoboken, New Jersey, 2005).

[25] J. Choi, Explicit formulas for Bernoulli polynomials of order n, *Indian J. Pure Appl. Math.* **27**, 667–674, (1996).

[26] J. Choi, Certain summation formulas involving harmonic numbers and generalized harmonic numbers, *App. Math. Comput.* **218**(3), 734–740, (2011). Doi: 10.1016/j.amc.2011.01.062.

[27] J. Choi, Remark on the Hurwitz-Lerch zeta function, *Fixed Point Theory and Appl.* **2013**(70), (2013).

[28] J. Choi, Note on Apostol–Daehee polynomials and numbers, *Far East J. Math. Sci.* **101**(8), 1845–1857, (2017).

[29] J. Choi and H.M. Srivastava, Certain families of series associated with the Hurwitz–Lerch zeta function, *Appl. Math. Comput.* **170**(1), 399–409, (2005).

[30] J. Choi and H.M. Srivastava, Some summation formulas involving harmonic numbers and generalized harmonic numbers, *Math. Comput. Model.* **54**(9–10), 2220–2234, (2011).

[31] J. Choi and H.M. Srivastava, The multiple Hurwitz–zeta function and the multiple Hurwitz-Euler eta function, *Taiwanese J. Math.* **15**(2), 501–522, (2011).

[32] J. Choi and A.K. Rathie, General summation formulas for the Kampe de Feriet function, *Montes Taurus J. Pure Appl. Math.* **1**(1), 107–128, (2019). Article ID: MTJPAM-D-19-00004.

[33] W. Chu and L. De Donno, Hypergeometric series and harmonic number identities, *Adv. Appl. Math.* **34**, 123–137, (2005).

[34] L. Comtet, *Advanced Combinatorics.* (D. Reidel Publication Company, Dordrecht-Holland/ Boston-U.S.A., 1974).

[35] T.W. Cusick, Recurrences for sums of powers of binomials, *J. Combinatorials Theory Series A* **52**, 77–83, (1989).

[36] B.S. El-Desouky and A. Mustafa, New results and matrix representation for Daehee and Bernoulli numbers and polynomials, *Appl. Math. Sci.* **9**(73), 3593–3610, (2015). arXiv:1412.82 59v1.

[37] G.B. Djordjevic and G.V. Milovanovic, *Special Classes of Polynomials* (University of Nis, Faculty of Technology Leskovac, 2014).

[38] J. Franel, On a question of Laisant, *L'intermédiaire des Mathématiciens* **1**, 45–47, (1894).

[39] J. Franel, On a question of J. Franel, *L'intermédiaire des Mathématiciens* **2**, 33–35, (1895).

[40] R. Golombek, Aufgabe 1088, *El. Math.* **49**, 126–127, (1994).

[41] R. Golombek and D. Marburg, Aufgabe 1088, Summen mit Quadraten von Binomialkoeffizienten, *El. Math.* **50**, 125–131, (1995).

[42] I.J. Good, The number of ordering of n candidates when ties are permitted, *Fibonacci Quart.* **13**, 11–18, (1975).

[43] H.W. Gould, Combinatorial Numbers and Associated Identities: Table 1: Stirling Numbers, (Edited and Compiled by Jocelyn Quaintance, 2010), https://math.wvu.edu/~hgould/Vol.7.PDF.

[44] D. Gun and Y. Simsek, Some new identities and inequalities for Bernoulli polynomials and numbers of higher order related to the Stirling and Catalan numbers, *RACSAM* **114**(167), 1–12, (2020).

[45] N. Kilar and Y. Simsek, A new family of Fubini numbers and polynomials associated with Apostol–Bernoulli numbers and polynomials, *J. Korean Math. Soc.* **54**(5), 1605–1621, (2017).

[46] N. Kilar and Y. Simsek, Identities and relations for Fubini type numbers and polynomials via generating functions and p-adic integral approach, *Publ. Inst. Math., Nouv. Sér.* **106**(120), 113–123, (2019).

[47] N. Kilar and Y. Simsek, Formulas and relations of special numbers and polynomials arising from functional equations of generating functions, *Montes Taurus J. Pure Appl. Math.* **3**(1), 106–123, (2021). Article ID: MTJPAM-D-20-00035.

[48] D. Kim, H. Ozden Ayna, Y. Simsek, and A. Yardimci, New families of special numbers and polynomials arising from applications of p-adic q-integrals, *Adv. Diff. Equ.* **2017**(207). 1–11, (2017). Doi: 10.1186/s13662-017-1273-4.

[49] M.-S. Kim, A note on sums of products of Bernoulli numbers, *Appl. Math. Lett.* **24**, 55–61 (2011).

[50] D.S. Kim and T. Kim, Daehee numbers and polynomials, *Appl. Math. Sci. (Ruse)* **7**(120), 5969–5976, (2013).

[51] D.S. Kim, T. Kim, S.-H. Lee and J.-J. Seo, A note on the lambda-Daehee polynomials, *Int. J. Math. Anal,* **7**(62), 3069–3080, (2013).

[52] D.S. Kim, T. Kim, and J. Seo, A note on Changhee numbers and polynomials, *Adv. Stud. Theor. Phys.* **7**, 993–1003, (2013).

[53] T. Kim, On a q-analogue of the p-adic log gamma functions and related integrals, *J. Number Theory* **76**, 320–329, 1999.

[54] T. Kim, q-Volkenborn integration, *Russ. J. Math. Phys.* **19**, 288–299, (2002).

[55] T. Kim, A note on q-Volkenborn integration, *Proc. Jangjeon Math. Soc.* **8**(1), 13–17, (2005).

[56] T. Kim, On the analogs of Euler numbers and polynomials associated with p-adic q-integral on $\mathbb{Z}_p$ at $q = -1$, *J. Math. Anal. Appl.* **331**(2), 779–792, (2007).

[57] T. Kim, q-Euler numbers and polynomials associated with p-adic q-integrals, *J. Nonlinear Math. Phys.* **14**(1), 15–27, (2007).

[58] T. Kim and D.S. Kim, A note on central Bell numbers and polynomials, *Russian J. Math. Phy.* **27**, 76–81, (2020).

[59] T. Kim, M.S. Kim, and L.C. Jang, New q-Euler numbers and polynomials associated with p-adic q-integrals, *Adv. Stud. Contemp. Math.* **15**, 140–153, (2007).

[60] T. Kim, D.V. Dolgy, D.S. Kim, and J.J. Seo, Differential equations for Changhee polynomials and their applications, *J. Nonlinear Sci. Appl.* **9**, 2857–2864, (2016).

[61] T. Kim, S.H. Rim, Y. Simsek, and D. Kim, On the analogs of Bernoulli and Euler numbers, related identities and zeta and L-functions, *J. Korean Math. Soc.* **45**, 435–453, (2008).

[62] T. Kim, D.S. Kim, and J. Kwon, Analogues of Faulhaber's formula for poly-Bernoulli and type 2 poly-Bernoulli polynomials, *Montes Taurus J. Pure Appl. Math.* **3**(1), 1–6, (2021). Article ID: MTJPAM-D-20-00033.

[63] W. Koepf, Hypergeometric summation, In: *An Algorithmic Approach to Summation and Special Function Identities*, 2nd edn. (Springer-Verlag, London, 2014).

[64] I. Kucukoglu, Implementation of computation formulas for certain classes of Apostol-type polynomials and some properties associated with these

polynomials, *Commun. Fac. Sci. Univ. Ank. Ser. A1 Math. Stat.* **70**(1), 426–442, (2021).

[65] I. Kucukoglu, B. Simsek, and Y. Simsek, An approach to negative hypergeometric distribution by generating function for special numbers and polynomials, *Turk. J. Math.* **43**, 2337–2353, (2019).

[66] I. Kucukoglu and Y. Simsek, On a family of special numbers and polynomials associated with Apostol-type numbers and polynomials and combinatorial numbers, *Appl. Anal. Discrete Math.* **13**, 478–494, (2019).

[67] I. Kucukoglu and Y. Simsek, Identities and relations on the q-Apostol type Frobenius-Euler numbers and polynomials, *J. Korean Math. Soc.* **56**(1), 265–284, (2019).

[68] I. Kucukoğlu, Y. Simsek, and H.M. Srivastava, A new family of Lerch-type zeta functions interpolating a certain class of higher-order Apostol-type numbers and Apostol-type polynomials, *Quaest. Math.* **42**(4), 465–478 (2019). Doi: 10.2989/16073606.2018.1459925.

[69] I. Kucukoglu and Y. Simsek, New formulas and numbers arising from analyzing combinatorial numbers and polynomials, *Montes Taurus J. Pure Appl. Math.* **3**(3), 238–259, (2021). Article ID: MTJPAM-D-20-00038.

[70] B. Kurt and Y. Simsek, Notes on generalization of the Bernoulli type polynomials, *Appl. Math. Comput.* **218**, 906–911, (2011).

[71] B. Kurt and Y. Simsek, On the generalized Apostol-type Frobenius-Euler polynomials, *Adv. Differ. Equ.* **1** (2013). https://doi.org/10.1186/1687-1847-2013-1.

[72] G.G. Lorentz, *Bernstein Polynomials* (Chelsea, New York, 1986).

[73] Y.L. Luke, The Special Functions and Their Approximations, Vol. 1. (Academic Press, New York, 1969).

[74] Q.-M. Luo, Apostol-Euler polynomials of higher order and Gaussian hypergeometric functions, *Taiwanese J. Math.* **10**, 917–925, (2006).

[75] Q.-M. Luo and H.M. Srivastava, Some generalizations of the Apostol–Genocchi polynomials and the Stirling numbers of the second kind, *Appl. Math. Comput.* **217**, 5702–5728, (2011).

[76] Q.-M. Luo and H.M. Srivastava, Some generalizations of the Apostol-Bernoulli and Apostol–Euler polynomials, *J. Math. Anal. Appl.* **308**, 290–302, (2005).

[77] T. Mansour, Combinatoral identities and inverse binomial coefficients, *Adv. Appl. Math.* **28**, 196–202, (2002).

[78] G.V. Milovanovic and Y. Simsek, Dedekind and Hardy type sums and trigonometric sums induced by quadrature formulas. In: A.M. M.Th. Rassias, (Eds.), *Trigonometric Sums and Their Applications* eBook ISBN 978-3-030-37904-9, pp. 183–228 (Springer Nature Publishing Group Switzerland, Basel, 2020).

[79] G.V. Milovanovic and A.K. Rathie, Four unified results for reducibility of Srivastava's triple hypergeometric series H_B, *Montes Taurus J. Pure Appl. Math.* **3**(3), 155–164, (2021). Article ID: MTJPAM-D-20-00062.

[80] V.H. Moll, *Numbers and Functions: From a Classical-Experimental Mathematician's Point of View, Student Mathematical Library*, Vol. 65 (American Mathematical Society, Providence, Rhode Island, 2012).

[81] G. Ozdemir, Y. Simsek and G. V. Milovanović, Generating functions for special polynomials and numbers including Apostol-type and Humbert-type polynomials, *Mediterr. J. Math.* **14**, 1–17, (2017).
[82] H. Ozden, Unification of generating function of the Bernoulli, Euler and Genocchi numbers and polynomials, *AIP Conference Proceedings* **1281**, 1125, (2010). https://doi.org/10.1063/1.3497848.
[83] H. Ozden, Y. Simsek, and H.M. Srivastava, A unified presentation of the generating functions of the generalized Bernoulli, Euler and Genocchi polynomials, *Comput. Math. Appl.* **60**, 2779–2787, (2010).
[84] M.A. Özarslan, Unified Apostol–Bernoulli, Euler and Genocchi polynomials, *Comput. Math. Appl.* **62**, 2452–2462, (2011).
[85] J.-W. Park, On the λ-Daehee polynomials with q-parameter, *J. Comput. Anal. Appl.* **20**(1), 11–20, (2016).
[86] M.A. Perlstadt, Some recurrences for sums of powers of binomial coefficients, *J. Number Theory* **27**, 304–309, (1987).
[87] F. Qi and B.-N. Guo, Sums of infinite power series whose coefficients involve products of the Catalan–Qi numbers, *Montes Taurus J. Pure Appl. Math.* **1**(2), 1–12, (2019). Article ID: MTJPAM-D-19-00007.
[88] M.I. Qureshi and S.A. Dar, Some hypergeometric summation theorems and reduction formulas via Laplace transform method, *Montes Taurus J. Pure Appl. Math.* **3**(3), 182–199, (2021). Article ID: MTJPAM-D-20-00016.
[89] H. Rademacher and E. Grosswald, *Dedekind Sums, Carus Mathematical Monograph.* No. 16 (Mathematical Association of America, Washington D.C., 1972).
[90] E.D. Rainville, *Special Functions* (The Macmillan Company, New York, 1960).
[91] S.-H. Rim, T. Kim, and S.S. Pyo, Identities between harmonic, hyperharmonic and Daehee numbers, *J. Inequal Appl.* **1**, 168, (2018).
[92] J. Riordan, *Introduction to Combinatorial Analysis* (Dover Publications, 2002).
[93] S. Roman, *The Umbral Calculus* (Dover Publications, New York, 2005).
[94] G.-C. Rota, The number of partitions of a set, *American Math. Monthly* **71**(5), 498–504, (1964).
[95] C.E. Sandifer, *How Euler Did Even More* (The Mathematical Association of America, Washington, 2014).
[96] W.H. Schikhof, Ultrametric calculus: An introduction to p-adic analysis. In: *Cambridge Studies in Advanced Mathematics* 4 (Cambridge University Press, Cambridge, 1984).
[97] J.B. Seaborne, *Hypergeometric Functions and their Applications* (Springer-Verlag, New York, 1991).
[98] Y. Simsek, Multiple interpolation functions of higher order $(h;q)$-Bernoulli numbers, *AIP Conf. Proc.* **1048**, 486–489, (2008).
[99] Y. Simsek, q-Hardy Berndt type sums associated with q-Genocchi type zeta and q-l-functions, *Nonlinear Anal.* **71**, e377–e395, (2009).
[100] Y. Simsek, Relations between Theata-functions Hardy sums Eisenstein and Lambert series in the transformation formula $log\eta_{g,h}(z)$, *J. Number Theory* **99**, 338–360, (2003).

[101] Y. Simsek, Generalized Dedekind sums associated with the Abel sum and the Eisenstein and Lambert series, *Adv. Stud. Contemp. Math.* **9**, 125–137, (2004).

[102] Y. Simsek, Remarks on reciprocity laws of the Dedekind sums and Hardy sums, *Adv. Stud. Contemp. Math. (Kyungshang)* **12**(2), 125–137, 237–246. (2006).

[103] Y. Simsek, Special functions related to Dedekind-type DC-sums and their applications, *Russ. J. Math. Phys.* **17**(4), 495–508, (2010).

[104] Y. Simsek, Twisted p-adic (h, q)-L-functions, *Comput. Math. Appl.* **59**(6), 2097–2110 (2010).

[105] Y. Simsek, On twisted generalized Euler numbers, *Bull. Korean Math. Soc.* **41**, 299–306, (2004).

[106] Y. Simsek, Generating functions for generalized Stirling type numbers, Array type polynomials, Eulerian type polynomials and their applications, *Fixed Point Theory Appl.* **87**, 1–28, (2013).

[107] Y. Simsek, Generating functions for q-Apostol type Frobenius–Euler numbers and polynomials, *Axioms* **1**, 395–403, (2012). Doi: 10.3390/axioms1030395.

[108] Y. Simsek, On q-deformed Stirling numbers, *Int. J. Math. Comput.* **17**(2), 70–80, (2012).

[109] Y. Simsek, Special numbers on analytic functions, *Appl. Math.* **5**, 1091–1098, (2014).

[110] Y. Simsek, A new combinatorial approach to analysis: Bernstein basis functions, combinatorial identities and Catalan numbers, *Math. Meth. Appl. Sci.* **38**(14), 3007–3021, (2015).

[111] Y. Simsek, Beta-type polynomials and their generating functions, *Appl. Math. Comput.* **254**, 172–182, (2015).

[112] Y. Simsek, Combinatorial sums and binomial identities associated with the Beta-type polynomials, *Hacet. J. Math. Stat.* **47**(5), 1144–1155, (2018).

[113] Y. Simsek, Computation methods for combinatorial sums and Euler-type numbers related to new families of numbers, *Math. Meth. Appl. Sci.* **40**(7), 2347–2361, (2017).

[114] Y. Simsek, Analysis of the p-adic q-Volkenborn integrals: An approach to generalized Apostol-type special numbers and polynomials and their applications, *Cogent Math. Stat.* **2016**, 1269393, (2016). https://dx.doi.org/10.1080/23311835.2016.1269393.

[115] Y. Simsek, Apostol type Daehee numbers and polynomials, *Adv. Stud. Contemp. Math. (Kyungshang)* **26**(3), 555–566, (2016).

[116] Y. Simsek, Identities on the Changhee numbers and Apostol-type Daehee polynomials, *Adv. Stud. Contemp. Math. (Kyungshang)* **27**(2), 199–212, (2017).

[117] Y. Simsek, On generating functions for the special polynomials, *Filomat* **31**(1), 9–16, (2017).

[118] Y. Simsek, New families of special numbers for computing negative order Euler numbers and related numbers and polynomials, *Appl. Anal. Discrete Math.* **12**, 1–35, (2018). https://doi.org/10.2298/AADM1801001S.

[119] Y. Simsek, Combinatorial identities and sums for special numbers and polynomials, *Filomat* **32**(20), 6869–6877, (2018).

[120] Y. Simsek, Construction of some new families of Apostol-type numbers and polynomials via Dirichlet character and p-adic q-integrals, *Turk. J. Math.* **42**, 557–577, (2018).

[121] Y. Simsek, Explicit formulas for p-adic integrals: Approach to p-adic distributions and some families of special numbers and polynomials, *Montes Taurus J. Pure Appl. Math.* **1**(1), 1–76, (2019). Article ID: MTJPAM-D-19-00005.

[122] Y. Simsek and A. Yardimci, Applications on the Apostol-Daehee numbers and polynomials associated with special numbers, polynomials, and p-adic integrals, *Adv. Difference Equ.* **308**, (2016). https://dx.doi.org/10.1186/s13662-016-1041-x.

[123] Y. Simsek, T. Kim, D.W. Park, Y.S. Ro, L.J. Jang, and S.H. Rim, An explicit formula for the multiple Frobenius–Euler numbers and polynomials, *JP J. Algebra Number Theory Appl.* **4**, 519–529, (2004).

[124] A. Sofo, Quadratic alternating harmonic number sums, *J. Number Theory* **154**, 144–159, (2015).

[125] H.M. Srivastava, Some generalizations and basic (or q-) extensions of the Bernoulli, Euler and Genocchi polynomials, *Appl. Math. Inf. Sci.* **5**(3), 390–444, (2011).

[126] H.M. Srivastava, H. Ozden, I.N. Cangul, and Y. Simsek, A unified presentation of certain meromorphic functions related to the families of the partial zeta type functions and the L-functions, *Appl. Math. Comput.* **219**, 3903–3913, (2012).

[127] H.M. Srivastava, B. Kurt, and Y. Simsek, Some families of Genocchi type polynomials and their interpolation functions, *Integr. Transf. Spec. F.* **23**(12), 919–938, (2012).

[128] H.M. Srivastava, B. Kurt, and Y. Simsek, Corrigendum: Some families of Genocchi type polynomials and their interpolation functions, *Integr. Transf. Spec. F.* **23**(12), 939–940, (2012). Doi: 10.1080/10652469.2012.690950.

[129] H.M. Srivastava and J. Choi, *Zeta and q-Zeta Functions and Associated Series and Integrals* (Elsevier Science Publishers, Amsterdam, 2012).

[130] H.M. Srivastava and J. Choi, *Series Associated with the Zeta and Related Functions* (Kluwer Academic Publishers, Dordrecht, Boston and London, 2001).

[131] B. Sury, T. Wang, and F.-Z. Zhao, Some identities involving reciprocals of binomial coefficients. *J. Integer Sequences* **7**, (2004). Article 04.2.8.

[132] N.M. Temme, *Special Functions: An Introduction to the Classical Functions of Mathematical Physics* (John Wiley and Sons, New York, 1996).

[133] R. Tremblay and B.J. Fugere, Products of two restricted hypergeometric functions, *J. Math. Anal. Appl.* **198**(3), 844–852, (1996).

[134] R. Tremblay, S. Gaboury, and B.-J. Fugère, A new class of generalized Apostol–Bernoulli polynomials and some analogues of the Srivastava–Pintér addition theorem, *Appl. Math. Lett.* **24**, 1888–1893, (2011).

[135] T.M. Rassias and H.M. Srivastava, Some classes of infinite series associated with the Riemann Zeta and Polygamma functions and generalized harmonic numbers, *Appl. Math. Comput.* **131**, 734–740, (2011).

[136] A. Sofo and H.M. Srivastava, Identities for the harmonic numbers and binomial coefficients, *Ramanujan J.* **25**, 93–113, (2011).

[137] T. Usman, N. Khan, M. Saif, and J. Choi, A unified family of Apostol-Bernoulli based poly-Daehee polynomials, *Montes Taurus J. Pure Appl. Math.* **3**(3), 1–11, (2021). Article ID: MTJPAM-D-20-00009.

[138] D. Zwillinger, *Table of Integrals, Series, and Products* (Academic Press, 2007).

[139] https://en.wikipedia.org/wiki/Chebyshev_polynomials.

[140] F. Yalcin and Y. Simsek, A new class of symmetric beta type distributions constructed by means of symmetric Bernstein type basis functions, *Symmetry* **12**, 779, (2020). Doi: 10.3390/sym12050779.

[141] N. Kilar and Y. Simsek, Families of unified and modified presentation of Fubini numbers and polynomials, *Montes Taurus J. Pure Appl. Math.* **5**(1), 1–21, (2023).

[142] Y. Simsek, Derivation of computational formulas for certain class of finite sums: Approach to generating functions arising from p-adic integrals and special functions, **45**(16), 9520–9544, (2022). https://doi.org/10.1002/mma.8321.

https://doi.org/10.1142/9789811261572_0027

Chapter 27

Transitive Pseudometric Principles and Caristi–Kirk Theorems

Mihai Turinici
A. Myller Mathematical Seminar, A. I. Cuza University,
700506 Iaşi, Romania
mturi@uaic.ro

The transitive maximal principle in Turinici [An. Şt. Univ. Ovidius Constanţa (Mat.), 17 (2009), 231–246] is equivalent to the (Bernays–Tarski) Dependent Choice Principle; and as such, equivalent with Ekeland's Variational Principle [*Bull. Amer. Math. Soc.* (New Series), 1 (1979), 443–474]. Applications of the obtained facts to Caristi–Kirk fixed point theorems over KST-metric spaces are then given.

1. Introduction

Let X be a non-empty set; and $d : X \times X \to R_+$ be a *metric* on X; then (X, d) will be referred to as a *metric space*. The following condition will be considered:

(com) d is *complete*: each d-Cauchy sequence is d-convergent.

Further, let $\varphi : X \to R \cup \{\infty\}$ be a function; we call it *inf-proper*, when

(inf-pr-1) φ is proper $[\text{Dom}(\varphi) := \{x \in X; \varphi(x) < \infty\} \neq \emptyset]$,
(inf-pr-2) φ is bounded below $[\inf[\varphi(X)] > -\infty]$.

An extra condition to be used here is

(lsc) φ is d-lsc: $\liminf_n \varphi(x_n) \geq \varphi(x)$, whenever $x_n \xrightarrow{d} x$.

The following 1979 statement in Ekeland [1] (referred to as: Ekeland's Variational Principle; in short: (EVP)) is our starting point.

Theorem 1. *Assume that d is complete and $\varphi : X \to R \cup \{\infty\}$ is inf-proper, d-lsc. Then, for each $u \in \mathrm{Dom}(\varphi)$, there exists $v \in \mathrm{Dom}(\varphi)$, such that*

(11-*a*) $d(u,v) \leq \varphi(u) - \varphi(v)$ (*hence* $\varphi(u) \geq \varphi(v)$)
(11-*b*) $d(v,x) > \varphi(v) - \varphi(x)$, *for all* $x \in X \setminus \{v\}$.

This principle found some basic applications in control and optimization, generalized differential calculus, critical point theory and global analysis; see the 1997 monograph by Hyers *et al.* [2, Ch 5] for a survey of these. As a consequence, many extensions of (EVP) were proposed. For example, the (*abstract*) *order* one starts from the fact that, with respect to the Brøndsted quasi-order [3],

(bro) $(x, y \in X)$: $x \leq y$ iff $d(x,y) + \varphi(y) \leq \varphi(x)$,

the point $v \in X$ appearing in (11-b) is *maximal*; so that, (EVP) is nothing but a denumerable variant of the Zorn–Bourbaki maximality principle [4], [5]; its precise formulation is just the 1976 ordering principle due to Brezis and Browder [6] (in short: (BB)). Further, the *dimensional* way of extension refers to the ambient space (R) of $\varphi(X)$ being substituted by a (topological or not) vector space; an account of the results in this area is to be found in the 2003 monograph by Goepfert *et al.* [7, Ch 3]. Finally, the *metrical* one consists in the conditions imposed on the ambient metric d over X being relaxed; the basic result in this direction was obtained by Kada *et al.* [8].

Now, the natural question arising here is: are all these extensions effective? Some partial answers were stated in Altman [9], Anisiu [10], Tataru [11], and Turinici [12]; see also Bao and Khanh [13]. According to these, the dimensional and metrical extensions of (EVP) are obtainable from either (EVP) or (BB), via straightforward techniques. Concerning the question of (BB) (and its subsequent extensions) being reducible to (EVP), the basic tool for solving it is the Dependent Choice Principle (in short: (DC)) due — independently — to Bernays [14] and Tarski [15]. Precisely, note that by the developments in Cârjă *et al* [39, Ch 2, Sect 2.1], (DC) $\Longrightarrow$ (BB) $\Longrightarrow$ (EVP); moreover, as shown in Brunner [16], (EVP) $\Longrightarrow$ (DC). Hence, any maximal/variational result — (MP) say — with (DC) $\Longrightarrow$ (MP) $\Longrightarrow$ (EVP) is equivalent with both (DC) and (EVP); see Turinici [17] for details. Note that, this is the case with many extensions of (EVP) and/or (BB); in particular (*cf.* Turinici [18]), the conclusion is also true for the 1987 extension of (EVP) represented by the Smooth Variational Principle due to Borwein and Preiss [19].

Having these precise, it is our aim in the following to verify this assertion upon some refinements of the transitive maximal principle in Turinici [12]. As a by-product of this, some Caristi–Kirk fixed point theorems due to Alegre and Marin [20] are being derived. Further aspects will be delineated in a separate paper.

2. Dependent Choice Principles

Throughout this exposition, the axiomatic system in use is Zermelo–Fraenkel's (abbreviated: (ZF)), as described by Cohen [21, Ch 2]. The notations and basic facts to be considered in this system are more or less standard. Some important ones are described below.

(A) Let X be a non-empty set. By a *relation* over X, we mean any (non-empty) part $\mathcal{R}$ of $X \times X$; then, $(X, \mathcal{R})$ will be referred to as a *relational structure*. For simplicity, we sometimes write $(x, y) \in \mathcal{R}$ as $x\mathcal{R}y$. Note that $\mathcal{R}$ may be regarded as a mapping between X and $\exp[X]$ (=the class of all subsets in X). In fact, denote for $x \in X$:

$$X(x, \mathcal{R}) = \{y \in X; x\mathcal{R}y\} \text{ (the } section \text{ of } \mathcal{R} \text{ through } x);$$

then, the desired mapping representation is $[\mathcal{R}(x) = X(x, \mathcal{R}), x \in X]$.

A basic example of relational structure is to be constructed as follows. Let $N = \{0, 1, \ldots\}$ be the set of *natural* numbers, endowed with the usual addition and (partial) order; note that

$(N, \leq)$ is well ordered:
any (non-empty) subset of N has a first element.

Further, denote for $p, q \in N$, $p \leq q$,

$$N[p, q] = \{n \in N; p \leq n \leq q\},\ N]p, q[= \{n \in N; p < n < q\},$$
$$N[p, q[= \{n \in N; p \leq n < q\},\ N]p, q] = \{n \in N; p < n \leq q\};$$

as well as, for $r \in N$,

$$N[r, \infty[= \{n \in N; r \leq n\},\ N]r, \infty[= \{n \in N; r < n\}.$$

For each $r \geq 1$, $N[0, r[= N(r, >)$ is referred to as the *initial interval* (in N) induced by r. Any set P with $P \sim N$ (in the sense: there exists a bijection from P to N) will be referred to as *effectively denumerable*. In addition, given some natural number $n \geq 1$, any set Q with $Q \sim N(n, >)$ will be said to be *n-finite*; when n is generic here, we say that Q is *finite*. Finally, the (non-empty) set Y is called (at most) *denumerable* iff it is either effectively denumerable or finite.

Let X be a non-empty set. By a *sequence* in X, we mean any mapping $x : N \to X$, where $N = \{0, 1, \ldots\}$ is the set of *natural* numbers. For simplicity reasons, it will be useful to denote it as $(x(n); n \geq 0)$, or $(x_n; n \geq 0)$; moreover, when no confusion can arise, we further simplify this notation as $(x(n))$ or (x_n), resp. Also, any sequence $(y_n := x_{i(n)}; n \geq 0)$ with

$(i(n); n \geq 0)$ is *strictly ascending* (hence, $i(n) \to \infty$ as $n \to \infty$)

will be referred to as a *subsequence* of $(x_n; n \geq 0)$. Note that, under such a convention, the relation "subsequence of" is transitive; i.e.

(z_n) = subsequence of (y_n) and (y_n) = subsequence of (x_n)
imply (z_n) = subsequence of (x_n).

(B) Remember that, an outstanding part of (ZF) is the *Axiom of Choice* (abbreviated: (AC)); which, in a convenient manner, may be written as

(AC) For each couple (J, X) of non-empty sets and each function
$F : J \to \exp(X)$, there exists a (selective) function
$f : J \to X$, with $f(\nu) \in F(\nu)$, for each $\nu \in J$.

(Here, $\exp(X)$ stands for the class of all non-empty elements in $\exp[X]$). Sometimes, when the index set J is denumerable, the existence of such a selective function may be determined by using a weaker form of (AC), called: *Dependent Choice* principle (in short: (DC)). Some preliminaries are needed.

Call the relation $\mathcal{R}$ over X *proper* when

$(X(x, \mathcal{R}) =)\mathcal{R}(x)$ is non-empty, for each $x \in X$.

Then, $\mathcal{R}$ is to be viewed as a mapping between X and $\exp(X)$; and the couple $(X, \mathcal{R})$ will be referred to as a *proper relational structure*. For each natural number $k \geq 1$, call the map $F : N(k, >) \to X$ a *k-sequence*; if $k \geq 1$ is generic, we talk about a *finite* sequence. The following result, referred to as the *Finite Dependent Choice principle* (in short: (DC-fin)) is available in the strongly reduced Zermelo–Fraenkel system (ZF-AC). Given $a \in X$, let us say that the k-sequence $F : N(k, >) \to X$ (where $k \geq 2$) is *$(a, \mathcal{R})$-iterative*, provided

$F(0) = a$ and $F(i)\mathcal{R}F(i + 1)$, for all $i \in N(k - 1, >)$.

Proposition 1. *Let the relational structure $(X, \mathcal{R})$ be proper. Then, for each $k \geq 2$, the following property holds:*

(P(k)) *for each $a \in X$, there exists an $(a, \mathcal{R})$-iterative k-sequence.*

Proof. Clearly, $(P(2))$ is true; just take $b \in \mathcal{R}(a)$ and define $F : N(2, >) \to X$ as: $F(0) = a$, $F(1) = b$. Assume that $(P(k))$ is true, for some $k \geq 2$; we claim that $(P(k+1))$ is true as well. In fact, let $F : N(k, >) \to X$ be an $(a, \mathcal{R})$-iterative k-sequence, assured by hypothesis. As $\mathcal{R}$ is proper, $\mathcal{R}(F(k-1))$ is non-empty; let u be some element of it. The map $G : N(k+1, >) \to X$ introduced as

$$G(i) = F(i), i \in N(k, >); \; G(k) = u$$

is an $(a, \mathcal{R})$-iterative $(k+1)$-sequence; and then, we are done. □

Now, it is natural to see what happens when k "tends to infinity". Given $a \in X$, let us say that the sequence $(x_n; n \geq 0)$ in X is *$(a; \mathcal{R})$-iterative* provided

$$x_0 = a \text{ and } [x_n \mathcal{R} x_{n+1} \text{ (i.e. } x_{n+1} \in \mathcal{R}(x_n)), \forall n].$$

The following statement, referred to as *Dependent Choice principle* (in short: (DC)) naturally comes into this discussion.

Proposition 2. *Let the relational structure $(X, \mathcal{R})$ be proper. Then, for each $a \in X$ there is at least an $(a, \mathcal{R})$-iterative sequence in X.*

At a first glance, the Dependent Choice principle (DC) is obtainable in (ZF-AC) by means of a *limit* process upon (DC-fin). However — from a technical perspective — the underlying procedure does not work in (ZF-AC); whence (DC) is not obtainable from the axioms of our strongly reduced system. On the other hand, this principle — proposed, independently, by Bernays [14] and Tarski [15] — is deductible from (AC), but not conversely; *cf.* Wolk [22]. Denote, for simplicity

(ZF-AC+DC) (the reduced Zermelo–Fraenkel system) = the strongly reduced Zermelo–Fraenkel system (ZF-AC) completed with the Dependent Choice principle (DC).

According to the developments in Moskhovakis [23, Ch 8] and Schechter [24, Ch 6], the reduced system (ZF-AC+DC) is large enough so as to cover the "usual" mathematics; see also Moore [25, Appendix 2, Table 4]. This reduced system will be largely used along our present exposition devoted to (countable) maximal principles. So, determining various equivalents of (DC) may not be without profit.

Let $(\mathcal{R}_n; n \geq 0)$ be a sequence of relations on X. Given $a \in X$, let us say that the sequence $(x_n; n \geq 0)$ in X is $(a; (\mathcal{R}_n; n \geq 0))$-*iterative*, provided

$$x_0 = a \text{ and } [x_n \mathcal{R}_n x_{n+1} \text{ (i.e.: } x_{n+1} \in \mathcal{R}_n(x_n)), \forall n].$$

The following *Diagonal Dependent Choice* principle (in short: (DDC)) is available.

Proposition 3. *Let $(\mathcal{R}_n; n \geq 0)$ be a sequence of proper relations on X. Then, for each $a \in X$ there exists an $(a; (\mathcal{R}_n; n \geq 0))$-iterative sequence in X.*

Clearly, (DDC) includes (DC); to which it reduces when $(\mathcal{R}_n; n \geq 0)$ is constant. The reciprocal of this is also true. In fact, letting the premises of (DDC) hold, put $P = N \times X$; and let $\mathcal{S}$ be the relation over P introduced as

$$\mathcal{S}(i, x) = \{i + 1\} \times \mathcal{R}_i(x), \quad (i, x) \in P.$$

It will suffice applying (DC) to $(P, \mathcal{S})$ and $b := (0, a) \in P$ to get the conclusion in the statement; we do not give details.

Summing up, (DDC) is provable in (ZF-AC+DC). This is valid as well for its variant, referred to as: the *Selected Dependent Choice* principle (in short: (SDC)).

Proposition 4. *Let the map $F : N \to \exp(X)$ and the relation $\mathcal{R}$ over X fulfill*

$$(\forall n \in N) : \mathcal{R}(x) \cap F(n + 1) \neq \emptyset, \quad \forall x \in F(n).$$

Then, for each $a \in F(0)$, there exists a sequence $(x(n); n \geq 0)$ in X, with

$$x(0) = a, \ x(n) \in F(n), \ x(n + 1) \in \mathcal{R}(x(n)), \ \forall n.$$

As before, (SDC) $\Longrightarrow$ (DC) ($\Longleftrightarrow$ (DDC)); just take $(F(n) = X, n \geq 0)$. But, the reciprocal is also true, in the sense: (DDC) $\Longrightarrow$ (SDC). This follows from

Proof (Proposition 4). Let the premises of (SDC) be true. Define a sequence of relations $(\mathcal{R}_n; n \geq 0)$ over X as: for each $n \geq 0$,

$$\mathcal{R}_n(x) = \mathcal{R}(x) \cap F(n + 1), \ \text{ if } x \in F(n),$$
$$\mathcal{R}_n(x) = \{x\}, \text{ otherwise } (x \in X \setminus F(n)).$$

Clearly, $\mathcal{R}_n$ is proper for all $n \geq 0$. So, by (DDC), we have that, for the starting $a \in F(0)$, there exists an $(a, (R_n; n \geq 0))$-iterative sequence $(x(n); n \geq 0)$ in X. Combining with the very definition above gives the conclusion. □

In particular, when $\mathcal{R} = X \times X$, the regularity condition imposed in (SDC) holds. The corresponding variant of underlying statement is just the *Denumerable Axiom of Choice* [in short: (AC(N))]. Precisely, we have

Proposition 5. *Let $F : N \to \exp(X)$ be a function. Then, for each $a \in F(0)$ there exists a function $f : N \to X$ with $f(0) = a$ and $(f(n) \in F(n), \forall n)$.*

As a consequence of these, (DC) $\Longrightarrow$ (AC(N)) in (ZF-AC). A direct verification is obtainable by an application of (DC) to the relational structure $(Q, \mathcal{T})$, where

$$Q = N \times X,\ \mathcal{T}(n, x) = \{n+1\} \times F(n+1),\ \ n \in N,\ x \in X;$$

we do not give details. The reciprocal inclusion is not true; see, for instance, Moskhovakis [23, Ch 8, Sect 8.25].

3. Pseudometric Maximal Principles

Let M be a non-empty set. Call the subset P of M, *almost singleton* (in short: *asingleton*) provided $[p_1, p_2 \in P$ implies $p_1 = p_2]$; and *singleton* if, in addition, P is non-empty; note that in this case $P = \{p\}$, for some $p \in M$. Denote by $\mathcal{S}(M)$ the class of all sequences (x_n) in M. By a (sequential) *convergence structure* on M we mean, as in Kasahara [26], any part $\mathcal{C}$ of $\mathcal{S}(M) \times M$ with the property

(conv-1) $\mathcal{C}$ is hereditary:
$((x_n); x) \in \mathcal{C} \Longrightarrow ((y_n); x) \in \mathcal{C}$, for each subsequence (y_n) of (x_n).

For simplicity, the relation $((x_n); x) \in \mathcal{C}$ will be denoted $x_n \xrightarrow{\mathcal{C}} x$; and reads

(x_n), $\mathcal{C}$-converges to x; or: x is the $\mathcal{C}$-*limit* of (x_n);

if the set $\mathcal{C} - \lim_n(x_n)$ of all such x is not empty, we say that (x_n) is $\mathcal{C}$-*convergent*. The following conditions about this structure are to be optionally considered:

(conv-2) $\mathcal{C}$ is *reflexive*: for each $a \in M$,
the constant sequence $(x_n = a \geq 0)$ fulfills $((x_n); a) \in \mathcal{C}$
(conv-3) $\mathcal{C}$ is *separated*: $\mathcal{C} - \lim_n(x_n)$ is an asingleton,
for each sequence $(x_n; n \geq 0)$ in M.

Further, by a (sequential) *Cauchy structure* on M we mean any part $\mathcal{H}$ of $\mathcal{S}(M)$ with the property

(Cauchy-1) $\mathcal{H}$ is *hereditary*:
$(x_n) \in \mathcal{H} \Longrightarrow (y_n) \in \mathcal{H}$, for each subsequence (y_n) of (x_n).

Each element of $\mathcal{H}$ will be referred to as a $\mathcal{H}$-*Cauchy* sequence in M. As a rule, the optional condition about this structure is to be considered

(Cauchy-2) $\mathcal{H}$ is *reflexive*: for each $a \in M$,
the constant sequence $(x_n = a; n \geq 0)$ fulfills $(x_n) \in \mathcal{H}$.

Finally, the couple $(\mathcal{C}, \mathcal{H})$ is referred to as a *conv-Cauchy structure* on M. The optional conditions about this combined structure to be considered here are

(CC-1) $(\mathcal{C}, \mathcal{H})$ is *regular*: each $\mathcal{C}$-convergent sequence is $\mathcal{H}$-Cauchy,
(CC-2) $(\mathcal{C}, \mathcal{H})$ is *complete*: each $\mathcal{H}$-Cauchy sequence is $\mathcal{C}$-convergent.

In the following, a basic example of conv-Cauchy structure is given.

Let M be some non-empty set. By a *pseudometric* over M we shall mean any map $e : M \times M \to R_+$. Suppose that we fixed such an object; with, in addition,

(tri) e is *triangular*: $e(x, z) \leq e(x, y) + e(y, z)$, for all $x, y, z \in M$,

then, $e(.,.)$ is called a *triangular pseudometric* (in short: *t-pseudometric*) on M. We introduce an e-convergence and an e-Cauchy structure on X as follows. Given the sequence (x_n) in X and the point $x \in X$, we say that (x_n), *e-converges* to x (written as: $x_n \xrightarrow{e} x$) provided $e(x_n, x) \to 0$ as $n \to \infty$; i.e.

$$\forall \varepsilon > 0, \exists i = i(\varepsilon):\ n \geq i \Longrightarrow e(x_n, x) < \varepsilon.$$

This will also be referred to as: x is an *e-limit* of (x_n); the set of all such points will be denoted as $e - \lim_n(x_n)$ [or $\lim_n(x_n)$, when no confusion can arise]; when it is non-empty, we say that (x_n) is *e-convergent*. By this very definition, we have the hereditary property:

(conv-1) $x_n \xrightarrow{e} x$ implies $y_n \xrightarrow{d} x$,
for each subsequence $(y_n; n \geq 0)$ of $(x_n; n \geq 0)$;

hence, $(\xrightarrow{d})$ is a convergence structure on M. The following conditions about this structure are to be optionally considered:

(conv-2) $(\xrightarrow{e})$ is *reflexive*: for each $a \in M$,
the constant sequence $(x_n = a; n \geq 0)$ fulfills $x_n \xrightarrow{e} a$
(conv-3) $(\xrightarrow{e})$ is *separated*:
$\lim_n(x_n)$ is an asingleton, for each sequence (x_n) in M.

Note that the former of these holds whenever

(ref) e is *reflexive*: $e(x, x) = 0$, $\forall x \in M$.

Further, call the sequence (x_n), *e-Cauchy* when $e(x_n, x_m) \to 0$ as $n, m \to \infty$, $n < m$; i.e.

$$\forall \varepsilon > 0, \exists j = j(\varepsilon): \ j \leq n < m \Longrightarrow e(x_n, x_m) < \varepsilon.$$

Clearly, we have the hereditary property

(Cauchy-1) (x_n) is e-Cauchy implies (y_n) is e-Cauchy,
for each subsequence $(y_n; n \geq 0)$ of $(x_n; n \geq 0)$,

so that, $Cauchy(e)$ (=the class of all such sequences) is a Cauchy structure on M. As a rule, the following condition is to be optionally considered

(Cauchy-2) $Cauchy(e)$ is *reflexive*: for each $a \in M$,
the constant sequence $(x_n = a; n \geq 0)$ fulfills $(x_n) \in Cauchy(e)$,

for example, this holds whenever e is reflexive.

Now — according to the general setting — call the couple $((\xrightarrow{e}), Cauchy(e))$, a *conv-Cauchy structure* induced by e. The following optional conditions about this structure are to be considered:

(CC-1) e is *regular*: each e-convergent sequence in M is e-Cauchy,
(CC-2) e is *complete*: each e-Cauchy sequence in M is e-convergent;

note that the former of these holds if (in addition)

(sym) e is *symmetric*: $e(x, y) = e(y, x)$, $\forall x, y \in M$.

Returning to the general case, let $e : M \times M \to R_+$ be a triangular pseudometric on M. Further, let $\varphi : M \to R \cup \{\infty\}$ be some *inf-proper* function; i.e.

(inf-pr-1) $\mathrm{Dom}(\varphi) := \{x \in M; \varphi(x) \neq \infty\} \neq \emptyset$
(inf-pr-2) φ is bounded below: $\inf \varphi(M) > -\infty$.

In the following, some basic concepts attached to (e, φ) will be introduced.

(I) Let $\nabla := \nabla[e, \varphi]$ be the relation over M

$$(x, y \in M)\ x\nabla y \text{ iff } e(x, y) + \varphi(y) \leq \varphi(x).$$

The following properties are immediate (so, we do not give details):

(a-1) ∇ is transitive ($\forall x, y, z \in M$: $x\nabla y, y\nabla z \Longrightarrow x\nabla z$)
(a-2) (∇, φ) is decreasing: $x, y \in M$, $x\nabla y$ imply $\varphi(x) \geq \varphi(y)$.

Note that (in general)

∇ is not reflexive: $x\nabla x$ may be false for certain $x \in M$.

So, for the given $u \in M$, it is possible that

u is *singular* (modulo ∇): $M(u, \nabla)$ is empty.

The opposite to this alternative will be expressed as

$u \in M$ is *starting* (modulo ∇): $M(u, \nabla)$ is non-empty.

(II) Let $(\perp)$ be a relation over M, endowed with

(b-1) (transitivity) $\forall x, y, z \in M$: $x \perp y, y \perp z \Longrightarrow x \perp z$,
(b-2) (comparison property) $\forall x, y \in M$: $x \perp y$ implies $x\nabla y$,

we then say that $(e, \varphi; \perp, \nabla)$ is an *admissible system*. Note that, as a consequence

(b-3) $(\perp, \varphi)$ is decreasing: $x, y \in M$, $x \perp y$ imply $\varphi(x) \geq \varphi(y)$.

A basic example is the following one. Let $\Box := \Box(e, \varphi)$ be the relation over M

$$(x, y \in M)\text{: } x\Box y \text{ iff } x\nabla y,\ \varphi(x) > \varphi(y).$$

Note that by this very definition,

(c-1) $\Box$ is transitive [$\forall x, y, z \in M$: $x\Box y, y\Box z \Longrightarrow x\Box z$],
(c-2) $\Box$ is *coarser* than ∇ [$x\Box y$ implies $x\nabla y$],

whence, $(e, \varphi; \Box, \nabla)$ is an admissible system.

(III) Returning to the general case, let us introduce two new concepts. Given a sequence (x_n) in M and an element $u \in M$, define the properties

(asc) (x_n) is $\perp$-ascending: $x_n \perp x_{n+1}$, $\forall n$,
or, equivalently (by transitivity): $x_n \perp x_m$ if $n < m$
(bound) $(x_n) \perp u$ iff $x_n \perp u$, for all $n \geq 0$.

Having these precise, call $v \in \mathrm{Dom}(\varphi)$, *BB-variational* (modulo $\perp$) provided

$$v \perp x \Longrightarrow \varphi(v) = \varphi(x) \text{ [so, by comparison, } e(v,x) = 0].$$

Likewise, let us say that $v \in \mathrm{Dom}(\varphi)$ is *BB-variational* (modulo ∇), if

$$v \nabla x \Longrightarrow \varphi(v) = \varphi(x) \text{ [so, by definition, } e(v,x) = 0].$$

(The introduced terminologies are being related to the methods in Brezis and Browder [6]). Note that, as $\perp$ is coarser than ∇, we must have

($\forall v \in \mathrm{Dom}(\varphi)$):
BB-variational (modulo ∇) implies BB-variational (modulo $\perp$).

The reverse inclusion is not in general true.

Clearly, the BB-variational (modulo $\perp$) property is vacuously fulfilled, when

$v \in \mathrm{Dom}(\varphi)$ is *singular* (modulo $\perp$): $M(v, \perp)$ is empty.

So, its verification is to be needed only when

$v \in \mathrm{Dom}(\varphi)$ is *starting* (modulo $\perp$): $M(v, \perp) \neq \emptyset$.

In this case, the following characterization of these points is available.

Proposition 6. *Suppose that* $v \in \mathrm{Dom}(\varphi)$ *is starting* (*modulo* $\perp$). *Then,*

(31-1) v *is BB-variational* (*modulo* $\perp$) $\Longleftrightarrow \varphi(M(v, \perp)) = \{\varphi(v)\}$,
(31-2) In this case, $\forall x \in M{:}\, v \perp x$ implies $\varphi(v) = \varphi(x), e(v,x) = 0$.

Proof. (i) If $v \in \mathrm{Dom}(\varphi)$ is BB-variational (modulo $\perp$), then $\varphi(M(v, \perp)) = \{\varphi(v)\}$. (ii) Conversely, suppose that $\varphi(M(v, \perp)) = \{\varphi(v)\}$; and let $x \in M(v, \perp)$ be arbitrary fixed. From the admitted hypothesis, $\varphi(v) = \varphi(x)$. Moreover, by the comparison property,

$$x \in M(v, \nabla); \text{ whence, } 0 \leq e(v,x) \leq \varphi(v) - \varphi(x) = 0; \text{ so, } e(v,x) = 0.$$

Putting these together yields the desired facts. □

Some basic properties of such points are collected in

Proposition 7. *Suppose that $v \in \mathrm{Dom}(\varphi)$ is starting and BB-variational (modulo $\perp$). Then, the following are true*

(32-1) $e(v,x) \geq \varphi(v) - \varphi(x)$, *for all* $x \in M(v,\perp)$,
(32-2) $e(v,x) > \varphi(v) - \varphi(x)$, *for each* $x \in M(v,\perp)$ *with* $e(v,x) > 0$.

Proof. (i) Assume that the underlying conclusion would be false:

$$e(v,x) < \varphi(v) - \varphi(x), \text{ for some } x \in M(v,\perp).$$

This, along with the characterization of our concept, yields $\varphi(v) = \varphi(x)$ and $e(v,x) = 0$; in contradiction with the relation above.

(ii) As before, assume that the underlying conclusion would be false:

$$e(v,x) \leq \varphi(v) - \varphi(x), \text{ for some } x \in M(v,\perp) \text{ with } e(v,x) > 0.$$

This, along with the very definition of our concept, yields $\varphi(v) = \varphi(x)$; wherefrom (by the working condition) $0 < e(v,x) = 0$, contradiction; hence the claim. □

Under these preliminaries, we may now state a useful result involving our data (referred to as: pseudometric Ekeland Variational Principle; in short: (EVP-p)). The following specific condition will be essential for us:

(s-desc-com-perp) $(e,\varphi;\perp)$ is strictly descending complete:
for each $\perp$-ascending sequence (x_n) in $\mathrm{Dom}(\varphi)$ with
$(\varphi(x_n))$ = strictly descending, there exists $x \in M$ with
$x_n \xrightarrow{e} x$, $(x_n) \perp x$ and $(\varphi(x_n)) \geq \varphi(x)$ (hence, $\lim_n \varphi(x_n) \geq \varphi(x)$).

Theorem 2. *Let the general conditions upon $(e,\varphi;\nabla,\perp)$ be valid; and $(e,\varphi;\perp)$ be strictly descending complete. Then, for each starting (modulo $\perp$) point $u \in \mathrm{Dom}(\varphi)$ there exists an associated point $v \in \mathrm{Dom}(\varphi)$, with*

$(31-a)$ $u \perp v$; *hence,* $e(u,v) \leq \varphi(u) - \varphi(v)$
$(31-b)$ v *is BB-variational (modulo* $\perp$*).*

Proof. Let $u \in \mathrm{Dom}(\varphi)$ be as in the statement. As $(\perp)$=transitive,

$$M_u := M(u,\perp) \text{ is non-empty, hereditary: } v \in M_u \Longrightarrow M(v,\perp) \subseteq M_u.$$

If the following alternative holds

(alt-1) at least one $v \in M_u$ is singular (modulo $\perp$) (i.e. $M(v,\perp) = \emptyset$), we are done; so, without loss, one may assume that

(alt-2) each $v \in M_u$ is starting (modulo $\perp$) (i.e. $M(v, \perp) \neq \emptyset$).

Suppose by contradiction that

(alt-3) all $v \in M_u$ are not BB-variational (modulo $\perp$).

By a previous characterization of this property, we necessarily have

$(\forall v \in M_u)$: $M(v, \perp) \neq \emptyset$ and $\inf \varphi(M(v, \perp)) < \varphi(v)$;
hence, $\inf \varphi(M(v, \perp)) < (1/2)[\varphi(v) + \inf \varphi(M(v, \perp))] < \varphi(v)$.

This, by the very definition of infimum, tells us that

for each $v \in M_u$, there exists $y \in M_u$, with
$v \perp y$, $\varphi(y) < (1/2)[\varphi(v) + \inf \varphi(M(v, \perp))] < \varphi(v)$ (hence, $\varphi(y) < \varphi(v)$).

Let $\mathcal{R}$ denote the relation over M_u assured by these relations. Clearly, $\mathcal{R}$ is proper; so, by the Dependent Choice principle (DC), there must be a sequence $(u_n; n \geq 0)$ in M_u, such that

$(\forall n)$: $u_n \perp u_{n+1}$, $\varphi(u_{n+1}) < (1/2)[\varphi(u_n) + \inf \varphi(M(u_n, \perp))] < \varphi(u_n)$.

From the first half of this, (u_n) is a $\perp$-ascending sequence in M_u; in addition, by the second part of the same, $(\varphi(u_n))$ is strictly descending. Putting these together, it follows, via $(e, \varphi; \perp)$=descending complete, that there exists $v \in M$ with

$u_n \xrightarrow{e} v$, $(u_n) \perp v$, and $(\varphi(u_n)) \geq \varphi(v)$ (hence: $\lambda := \lim_n \varphi(u_n) \geq \varphi(v)$).

We now claim that existence of such an element yields a contradiction. In fact, the second part of this relation gives (via $(\perp)$=transitive)

$v \in M_u$; whence, v is an upper bound (modulo $\perp$) of (u_n) in M_u.

Combining with our starting assumption, $M(v, \perp) \neq \emptyset$; so, let x be an element of this set. By the previous comparison property

$\varphi(u_n) \geq \varphi(v) \geq \varphi(x), \forall n$; hence, $\lambda \geq \varphi(v) \geq \varphi(x)$,

by passing to limit as $n \to \infty$. Moreover, the initial choice of our sequence gives

$\varphi(u_{n+1}) \leq (1/2)[\varphi(u_n) + \varphi(x)], \forall n$; hence, $\lambda \leq \varphi(x) \leq \varphi(v)$, if we again pass to limit as $n \to \infty$. Combining these relations gives $\varphi(v) = \varphi(x)$; which [by the arbitrariness of $x \in M(v, \perp)$] tells us that v is BB-variational (modulo $\perp$); in contradiction with a previous hypothesis about the points of M_u. The proof is thereby complete. $\square$

An interesting completion of Theorem 2 is the following. Call $v \in \mathrm{Dom}(\varphi)$, *E-variational* (modulo $\perp$) provided

$v \perp x$ implies $v = x$; hence, $[\varphi(v) = \varphi(x)$ and $e(v, x) = 0]$,

where the last assertion follows by the comparison property. This is stronger than the concept of BB-variational (modulo $\perp$) element. To get a corresponding form of (EVP-p) involving such points we have to impose (in addition to the above)

(tr-suf) e is *transitively sufficient* $[e(z, x) = e(z, y) = 0 \Longrightarrow x = y]$.

The following statement (referred to as: strong pseudometric Ekeland Variational Principle; in short: (EVP-sp)) is now available.

Theorem 3. *Suppose (under the same general conditions) that $(e, \varphi; \perp)$ is strictly descending complete and e is transitively sufficient. Then, for each starting (modulo $\perp$) point $u \in \mathrm{Dom}(\varphi)$ there exists an associated point $w \in \mathrm{Dom}(\varphi)$, with*

$(32-a)$ *$u \perp w$; hence, $e(u, w) \le \varphi(u) - \varphi(w)$*
$(32-b)$ *w is E-variational (modulo $\perp$) :*
$w \perp x$ implies $w = x$ (hence, $\varphi(w) = \varphi(x)$ and $e(w, x) = 0$).

Proof. Let $u \in \mathrm{Dom}(\varphi)$ be taken as in the statement. By Theorem 2, we have promised a BB-variational (modulo $\perp$) element $v \in \mathrm{Dom}(\varphi)$ with $u \perp v$. If the obtained point v is E-variational (modulo $\perp$), we are done (with $w = v$); so, it remains the alternative of v fulfilling the opposite property:

$v \perp w$ for some $w \in M(v, \perp) \setminus \{v\}$; note that, additionally (by the choice of v), $\varphi(v) = \varphi(w)$ and $e(v, w) = 0$.

In this case, w is our desired element. Assume not:

$w \perp y$, for some $y \in M(w, \perp)$, $y \neq w$.

By the preceding relation (and transitivity of $\perp$)

$v \perp y$ (hence, $\varphi(v) = \varphi(y)$ and $e(v, y) = 0$).

This yields

$$\varphi(v) = \varphi(w) = \varphi(y) \text{ and } e(v, w) = e(v, y) = 0,$$

wherefrom (by the transitive sufficiency of e) $w = y$; contradiction. As a consequence, we must have

$$w \perp y \in M \text{ implies } w = y; \text{ hence, } [\varphi(v) = \varphi(y),\ e(w, y) = 0],$$

if we take the comparison property into account; and this ends the argument. □

Now, it is natural to ask under which circumstances is the strict descending complete condition upon $(e, \varphi; \perp)$ available. An appropriate answer to this is to be given in the particular case of

(perp=na) $(\perp)$ is identical with ∇; i.e. $x \perp y$ iff $e(x, y) + \varphi(y) \leq \varphi(y)$.

Precisely, let us consider the particular condition

(s-desc-com) (e, φ) is strictly descending complete:
for each e-Cauchy sequence (x_n) in $\mathrm{Dom}(\varphi)$ with
$(\varphi(x_n))$ = strictly descending, there exists $x \in M$ with
$x_n \xrightarrow{e} x$ and $(\varphi(x_n)) \geq \varphi(x)$ (hence, $\lim_n \varphi(x_n) \geq \varphi(x)$).

Proposition 8. *Suppose that (e, φ) is strictly descending complete. Then,*

$(e, \varphi; \nabla)$ is strictly descending complete:
for each ∇-ascending sequence (x_n) in $\mathrm{Dom}(\varphi)$ *with*
$(\varphi(x_n))$ = strictly descending, there exists $x \in M$ with
$x_n \xrightarrow{e} x$, $(x_n)\nabla x$ and $(\varphi(x_n)) \geq \varphi(x)$ (hence, $\lim_n \varphi(x_n) \geq \varphi(x)$).

Proof. Let (x_n) be a ∇-ascending sequence in $\mathrm{Dom}(\varphi)$ with $(\varphi(x_n))$ = strictly descending. From the ascending property

$$e(x_n, x_m) \leq \varphi(x_n) - \varphi(x_m), \text{ if } n < m.$$

The sequence $(\varphi(x_n); n \geq 0)$ is strictly descending and bounded from below; hence, a Cauchy one; and this, along with the working condition above, assures us that $(x_n; n \geq 0)$ is e-Cauchy in $\mathrm{Dom}(\varphi)$, with $(\varphi(x_n))$ = strictly descending. Combining with (e, φ) = strictly descending complete, it results that there exists $x \in M$ with

$$x_n \xrightarrow{e} x \text{ and } (\varphi(x_n)) \geq \varphi(x) \text{ (hence, } \lim_n \varphi(x_n) \geq \varphi(x)).$$

The only fact to be clarified is that of

$(x_n)\nabla x$; that is: $x_n\nabla x$, for all n.

To do this, fix some rank n. By the working condition

$$e(x_n, x_m) \le \varphi(x_n) - \varphi(x_m) \le \varphi(x_n) - \varphi(x),\ \forall m > n.$$

Combining with the triangular property of e, gives

$$e(x_n, x) \le e(x_n, x_m) + e(x_m, x) \le \varphi(x_n) - \varphi(x) + e(x_m, x),\ \forall m > n.$$

This, along with the choice of x, yields by a limit process (relative to m)

$$e(x_n, x) \le \varphi(x_n) - \varphi(x);\ \text{i.e.: } x_n\nabla x,$$

wherefrom (by the arbitrariness of n), x is an upper bound of (x_n) (modulo ∇), as claimed. □

Now, by simply combining this with Theorem 2, one gets the following practical statement involving these data (referred to as: narrow pseudometric Ekeland Variational Principle; in short: (EVP-n-p)).

Theorem 4. *Let the general conditions upon (e, φ) be accepted; and (e, φ) be strictly descending complete. Then, for each starting (modulo ∇) point $u \in \mathrm{Dom}(\varphi)$ there exists an associated point $v \in \mathrm{Dom}(\varphi)$, with*

(33-*a*) $u\nabla v$; *hence,* $e(u, v) \le \varphi(u) - \varphi(v)$
(33-*b*) v *is BB-variational (modulo* ∇*)* :
$v\nabla x \Longrightarrow \varphi(v) = \varphi(x)$ *(hence,* $e(v, x) = 0$*).*

Likewise, as an immediate application of Theorem 3, the following statement (referred to as: narrow strong pseudometric Ekeland Variational Principle; in short: (EVP-n-sp)) is now available.

Theorem 5. *Suppose (in addition to the same general conditions) that (e, φ) is strictly descending complete and e is transitively sufficient. Then, for each starting (modulo ∇) point $u \in \mathrm{Dom}(\varphi)$ there exists an associated point $w \in \mathrm{Dom}(\varphi)$, with*

(34-*a*) $u\nabla w$; *hence,* $e(u, w) \le \varphi(u) - \varphi(w)$
(34-*b*) w *is E-variational (modulo* ∇ *:)*
$w\nabla x$ *implies* $w = x$ *(hence,* $\varphi(w) = \varphi(x)$ *and* $e(w, x) = 0$*).*

Now, evidently, (e, φ) is strictly descending complete under

(desc-com) (e, φ) is descending complete:
for each e-Cauchy sequence (x_n) in $\mathrm{Dom}(\varphi)$

with $(\varphi(x_n))$=descending, there exists $x \in M$ with
$x_n \xrightarrow{e} x$ and $(\varphi(x_n)) \geq \varphi(x)$ (hence, $\lim_n \varphi(x_n) \geq \varphi(x)$).

For example, this is retainable whenever

(com) (e, φ) is complete:
each e-Cauchy sequence in Dom(φ) is e-convergent
(desc-lsc) φ is e-descending-lsc:
$(\varphi(x_n)) \geq \varphi(x)$ (hence, $\lim_n \varphi(x_n) \geq \varphi(x)$), whenever
$x_n \xrightarrow{e} x$ and $(\varphi(x_n))$=descending.

On the other hand, the transitive sufficiency of e holds whenever

(suf) e is *sufficient* $[e(x, y) = 0 \Longrightarrow x = y]$.

In this case, (EVP-n-sp) becomes the variational principle in Turinici [27]; see also Alegre *et al.* [28]. Note finally that, by a technique developed in Cârjă and Ursescu [29], all these results may be adapted to functions $\varphi : M \to R \cup \{-\infty\} \cup \{\infty\}$; some partial aspects of these may be found in Turinici [30].

4. Relative Aspects

In the following, some relative type versions of these results are to be stated.

Let M be a non-empty set; and $d : M \times M \to R_+$ be a (triangular or not) pseudometric over it. We introduce a d-convergence and a d-Cauchy structure on X as follows. Given the sequence (x_n) in M and the point $x \in M$, we say that (x_n), *d-converges* to x (written as: $x_n \xrightarrow{d} x$) if $d(x_n, x) \to 0$ as $n \to \infty$; i.e.

$\forall \varepsilon > 0$, $\exists i = i(\varepsilon)$: $n \geq i \Longrightarrow d(x_n, x) < \varepsilon$; or, equivalently:
$\forall \varepsilon > 0$, $\exists i = i(\varepsilon)$: $n \geq i \Longrightarrow d(x_n, x) \leq \varepsilon$.

This will be referred to as: x is a *d-limit* of (x_n); the set of all these will be denoted as $d - \lim_n(x_n)$, or $\lim_n(x_n)$, when d is understood; if it is non-empty, we say that (x_n) is *d-convergent*. By this definition, we have the hereditary property:

(conv-1) $x_n \xrightarrow{d} x$ implies $y_n \xrightarrow{d} x$,
for each subsequence $(y_n; n \geq 0)$ of $(x_n; n \geq 0)$;

hence, $(\xrightarrow{d})$ is a convergence structure on X. The following conditions about this structure are to be optionally considered:

(conv-2) $(\xrightarrow{d})$ is *reflexive*: for each $a \in M$,
the constant sequence $(x_n = a; n \geq 0)$ fulfills $x_n \xrightarrow{d} a$
(conv-3) $(\xrightarrow{d})$ is *separated*:
$d - \lim_n(x_n)$ is an asingleton, for each sequence (x_n) in M,

note that the former of these holds whenever d is reflexive.

Further, call the sequence (x_n), *d-Cauchy* when $d(x_n, x_m) \to 0$ as $n, m \to \infty$, $n < m$; i.e.

$\forall \varepsilon > 0$, $\exists j = j(\varepsilon)$: $j \leq n < m \Longrightarrow d(x_n, x_m) < \varepsilon$; or, equivalently,
$\forall \varepsilon > 0$, $\exists j = j(\varepsilon)$: $j \leq n < m \Longrightarrow d(x_n, x_m) \leq \varepsilon$.

As before, we have the hereditary property

(Cauchy-1) (x_n) is d-Cauchy implies (y_n) is d-Cauchy,
for each subsequence $(y_n; n \geq 0)$ of $(x_n; n \geq 0)$;

so that, *Cauchy*(d) (=the class of all such sequences) is a Cauchy structure on M. As before, the following condition is to be optionally considered:

(Cauchy-2) *Cauchy*(d) is *reflexive*: for each $a \in M$,
the constant sequence $(x_n = a; n \geq 0)$ fulfills $(x_n) \in$ *Cauchy*(d);

clearly, this holds whenever d is reflexive.

Now — according to the general setting — call the couple $((\xrightarrow{d})$, *Cauchy*$(d))$, a *conv-Cauchy structure* induced by d. The following optional conditions about this structure are to be considered

(CC-1) d is *regular*: each d-convergent sequence is d-Cauchy
(CC-2) d is *complete*: each d-Cauchy sequence is d-convergent;

note that the former of these holds if d is triangular and symmetric.

Having these precise, let $d : M \times M \to R_+$ be a general pseudometric over M; and $e : M \times M \to R_+$ be a triangular pseudometric over the same. The following concept is introduced:

(Cd-lsc-2) e is Cauchy d-lsc in the second variable: (y_n) is e-Cauchy and $y_n \xrightarrow{d} y$ imply $\liminf_n e(x, y_n) \geq e(x, y), \forall x \in M$.

Then, define

(C-s-sub) e is Cauchy semi-subordinated to d:
each e-Cauchy sequence admits a d-Cauchy subsequence

(C-sub) e is Cauchy subordinated to d:
each e-Cauchy sequence is d-Cauchy.

In this case, we say that

(kst-sm) e is *KST-semimetric* (modulo d), if (Cd-lsc-2)+(C-s-sub) hold;
(kst-m) e is *KST-metric* (modulo d), provided (Cd-lsc-2)+(C-sub) hold.

(The proposed terminology comes from the developments in Kada *et al.* [8]). Clearly, each KST-metric (modulo d) is a KST-semimetric (modulo d); the motivation of introducing it follows from the fact that the KST-metric (modulo d) property is assured for $e = d$, under mild conditions upon d. Precisely, we have

Proposition 9. *Suppose that the pseudometric d is triangular. Then, necessarily, d is KST-metric (modulo d).*

Proof. Clearly, d is Cauchy subordinated to itself; so, it remains to establish that d is Cauchy d-lsc in the second variable. Let $x, y \in M$ be fixed; and take a d-Cauchy sequence (y_n) in M with $y_n \xrightarrow{d} y$. By the triangular property,

$$d(x, y) \le d(x, y_n) + d(y_n, y), \forall n.$$

Passing to inferior limit, we get $d(x, y) \le \liminf_n d(x, y_n)$; whence, all is clear. □

Now, let $\varphi : M \to R \cup \{\infty\}$ be some *inf-proper* function; i.e.

(inf-pr-1) $\mathrm{Dom}(\varphi) := \{x \in M; \varphi(x) \neq \infty\} \neq \emptyset$
(inf-pr-2) φ is bounded below: $\inf \varphi(M) > -\infty$.

For each pseudometric $g : M \times M \to R_+$, define the property

(g, φ) is strictly descending complete: for each g-Cauchy sequence (x_n) in $\mathrm{Dom}(\varphi)$ with $(\varphi(x_n)) =$ strictly descending, there exists $x \in M$ with $x_n \xrightarrow{g} x$ and $(\varphi(x_n)) \ge \varphi(x)$ (hence, $\lim_n \varphi(x_n) \ge \varphi(x)$).

The following relative auxiliary fact will be useful for us.

Proposition 10. *Assume that the pseudometric $d(.,.)$ and the triangular pseudometric $e(.,.)$ over M are such that*

$e(.,.)$ is a KST-semimetric (modulo d).

Then, the following inclusion holds

(d,φ) is strictly descending complete, implies
(e,φ) is strictly descending complete.

Proof. Let (x_n) be some e-Cauchy sequence in $\mathrm{Dom}(\varphi)$, with $(\varphi(x_n)) =$ strictly descending. From the Cauchy semi-subordination property, there exists a subsequence $(y_n = x_{i(n)}; n \geq 0)$ of (x_n), such that (y_n) is d-Cauchy in $\mathrm{Dom}(\varphi)$; in addition, $(\varphi(y_n)) =$ strictly descending. By the strict descending completeness of (d,φ), there exists some $x \in M$ with

$$y_n \xrightarrow{d} x \text{ as } n \to \infty, \text{ and } (\varphi(y_n)) \geq \varphi(x); \text{ hence, } \lim_n \varphi(y_n) \geq \varphi(x).$$

We claim that this is our desired point for the strict descending completeness of (e,φ). In fact, let $\gamma > 0$ be arbitrary fixed. By the e-Cauchy property of (x_n), there exists some index $k = k(\gamma)$, so that

$$e(x_p, x_m) \leq \gamma, \text{ whenever } k \leq p < m.$$

Fix in the following $q > p$. As $(i(n); n \geq 0)$ is strictly ascending, we have

$$i(n) \geq n, \ \forall n; \text{ hence, } i(n) \geq q > p, \text{ for all } n \geq q.$$

Combining with the above relation, gives

$$e(x_p, y_n) \leq \gamma, \text{ whenever } k \leq p < q \leq n.$$

Passing to limit as $n \to \infty$ gives (taking into account the d-lsc property of $e(.,.)$ in its second variable)

$$e(x_p, x) \leq \gamma, \text{ for each } p \geq k;$$

and since $\gamma > 0$ was arbitrarily chosen, $x_n \xrightarrow{e} x$. This, along with $[\lim_n \varphi(x_n) = \lim_n \varphi(y_n) \geq \varphi(x)]$, completes the reasoning. □

Under these preliminaries, we may now state some relative type versions of our previous results. Let $d : M \times M \to R_+$ be a pseudometric over M; and $e : M \times M \to R_+$ be a triangular pseudometric over the same. Further, let $\varphi : M \to R \cup \{\infty\}$ be some inf-proper function (see above). By the obtained facts and narrow pseudometric Ekeland Variational Principle (EVP-n-p), one gets the following variational statement involving these data (referred to as: relative narrow pseudometric Ekeland Variational Principle; in short: (EVP-r-n-p)). Remember that we introduced a relation ∇ on M as

$$(x, y \in M)\text{: } x \nabla y \text{ iff } e(x,y) + \varphi(y) \leq \varphi(x).$$

Theorem 6. *Suppose that* $e(.,.)$ *is KST-semimetric (modulo* d*) and* (d, φ) *is strictly descending complete. Then, for each starting (modulo* ∇*) point* $u \in \mathrm{Dom}(\varphi)$ *there exists an associated point* $v \in \mathrm{Dom}(\varphi)$*, with*

$$(41-a) \quad e(u,v) \leq \varphi(u) - \varphi(v) \ (\textit{hence, } \varphi(u) \geq \varphi(v))$$
$$(41-b) \quad v \textit{ is BB-variational } (\textit{modulo } \nabla):$$
$$e(v,x) \leq \varphi(v) - \varphi(x) \Longrightarrow \varphi(v) = \varphi(x) \ (\textit{hence, } e(v,x) = 0).$$

Note that, by the developments above, one has (EVP-n-p) $\Longrightarrow$ (EVP-r-n-p). The reciprocal inclusion ((EVP-r-n-p) $\Longrightarrow$ (EVP-n-p)) is not in general true; but, when d is triangular, this holds (see above).

Now, as in the *absolute* setting, a problem to be posed is that of getting corresponding forms of the obtained result involving E-variational (modulo ∇) points. The appropriate answer to this is obtainable via the narrow strong pseudometric Ekeland Variational Principle, (EVP-n-sp). Precisely, the following variational statement involving these data (referred to as: relative narrow strong pseudometric Ekeland Variational Principle; in short: (EVP-r-n-sp)) is available. Remember that the triangular pseudometric $e : M \times M \to R_+$ is *transitively sufficient*, when

$$(\text{tr-suf}) \quad e(z,x) = e(z,y) = 0 \Longrightarrow x = y.$$

Theorem 7. *Suppose that* $e(.,.)$ *is a transitively sufficient KST-semimetric (modulo* d*) and* (d, φ) *is strictly descending complete. Then, for each starting (modulo* ∇*) point* $u \in \mathrm{Dom}(\varphi)$ *there exists an associated point* $w \in \mathrm{Dom}(\varphi)$*, with*

$$(42-a) \quad e(u,w) \leq \varphi(u) - \varphi(w) \ (\textit{hence, } \varphi(u) \geq \varphi(w))$$
$$(42-b) \quad w \textit{ is E-variational } (\textit{modulo } \nabla):$$
$$e(w,x) \leq \varphi(w) - \varphi(x) \Longrightarrow w = x \ (\textit{hence, } \varphi(w) = \varphi(x) \textit{ and } e(w,x) = 0).$$

As before, note that, by the developments above, one has (EVP-n-sp) $\Longrightarrow$ (EVP-r-n-sp). The reciprocal inclusion ((EVP-r-n-sp) $\Longrightarrow$ (EVP-n-sp)) is not in general true; but, when d is triangular, this holds (see above). Further aspects may be found in Turinici [12].

Finally, the following technical problem is useful in practice. Let $d : M \times M \to R_+$ be a general pseudometric over M; and $e : M \times M \to R_+$ be a triangular pseudometric over the same, subjected to

(Cd-lsc-2) e is Cauchy d-lsc in the second variable (see above).

We may ask under which extra conditions, $e(.,.)$ is a KST-semimetric (modulo d) or a KST-metric (modulo d). An appropriate answer to this is formulated in

Proposition 11. *Let the couple (d, e) be taken as before. Then,*

(43-1) *$e(.,.)$ is a KST-semimetric (modulo d), under*

e is chain-subordinated to d : for each $\varepsilon > 0$, there exists $\delta > 0$, such that $e(x, y) \leq \delta$, $e(y, z) \leq \delta \Longrightarrow d(x, z) \leq \varepsilon$

(43-2) *$e(.,.)$ is a KST-metric (modulo d) under either of the conditions*

(43-2-a) *e is left-subordinated to d : for each $\varepsilon > 0$, there exists $\delta > 0$, such that $e(x, y) \leq \delta$, $e(x, z) \leq \delta \Longrightarrow d(y, z) \leq \varepsilon$*
(43-2-b) *e is right-subordinated to d : for each $\varepsilon > 0$, there exists $\delta > 0$, such that $e(x, z) \leq \delta$, $e(y, z) \leq \delta \Longrightarrow d(x, y) \leq \varepsilon$.*

Proof. (i) Suppose that e is chain-subordinated to d; and let (x_n) be an e-Cauchy sequence in M. Given $\varepsilon > 0$, let $\delta > 0$ be the number attached to it, by the chain subordination property. By definition, there must be some index $i(\delta)$, such that

$$i(\delta) \leq n < m \Longrightarrow e(x_n, x_m) \leq \delta.$$

For each (n, m) like before, we thus have (via $i(\delta) \leq 2n < 2n + 1 < 2m$)

$$e(x_{2n}, x_{2n+1}) \leq \delta,\ e(x_{2n+1}, x_{2m}) \leq \delta,$$

so that, by the chain subordination of e,

$$d(x_{2n}, x_{2m}) \leq \varepsilon, \text{ for all } (n, m) \text{ with } i(\delta) \leq n < m.$$

In particular, this tells us that the subsequence $(y_n := x_{2n}; n \geq 0)$ is d-Cauchy (in M); and our claim follows.

(ii) Suppose that e is left-subordinated to d; and let (x_n) be an e-Cauchy sequence in M. Given $\varepsilon > 0$, let $\delta > 0$ be the number attached to it, via left subordination property. By definition, there must be some index $i := i(\delta)$, with

$$i \leq n < m \Longrightarrow e(x_n, x_m) \leq \delta.$$

Fix in the following $j > i$. For each couple (n, m) with $j \leq n < m$, we thus have

$e(x_i, x_n) \leq \delta$, $e(x_i, x_m) \leq \delta$; so that (by the left subordination of e)
$d(x_n, x_m) \leq \varepsilon$, for all such (n, m).

As a consequence, $(x_n; n \geq 0)$ is d-Cauchy (in M); and the claim follows.

(iii) Suppose that e is right-subordinated to d; and let (x_n) be an e-Cauchy sequence in M. Given $\varepsilon > 0$, let $\delta > 0$ be the number attached to it, via right subordination property. By definition, there exists an index $i := i(\delta)$, such that

$$i \leq n < m \Longrightarrow e(x_n, x_m) \leq \delta.$$

Given such a couple (n, m), let p be an index with $(n <)m < p$. Then,

$e(x_n, x_p) \leq \delta$, $e(x_m, x_p) \leq \delta$; so that (by the right subordination of e) $d(x_n, x_m) \leq \varepsilon$, for all (n, m) with $i \leq n < m$.

As a consequence, $(x_n; n \geq 0)$ is d-Cauchy (in M); and the claim follows. □

Concerning the imposed setting, it is natural to ask whether the triangular condition upon e is or not removable. Some (positive) answers to these may be found under the lines in Gajek and Zagrodny [31]; see also Turinici [32]. Further aspects will be delineated elsewhere.

5. Monotone Variational Principles

In the following, a basic application of the pseudometric statements above is given to *monotone* variational principles over almost metric spaces.

Let M be a non-empty set. By an *almost metric* over M, we mean any map $e : M \times M \to R_+$ endowed with the properties

(am-1) e is *triangular*: $e(x, z) \leq e(x, y) + e(y, z)$, $\forall x, y, z \in M$,
(am-2) e is *reflexive-sufficient*: $e(x, y) = 0$ iff $x = y$.

Note that e has all properties of a metric, except *symmetry*; this, among others, motivates our convention.

Fix in the following such an object $e(.,.)$; as well as a *quasi-order* [i.e. reflexive transitive relation] $(\leq)$ over M; then, $(M, e; \leq)$ will be referred to as a *quasi-ordered almost metric space*. Further, let the function $\varphi : M \to R \cup \{\infty\}$ be *inf-proper*; i.e.

(inf-pr-1) $\mathrm{Dom}(\varphi) := \{x \in M; \varphi(x) \neq \infty\} \neq \emptyset$,
(inf-pr-2) φ is bounded below: $\varphi_* := \inf \varphi(M) > -\infty$.

The following specific conditions about the triple $(e, \leq; \varphi)$ will be considered:

(sc-1) $(e, \leq; \varphi)$ is complete:
each $(\leq)$-ascending e-Cauchy sequence in $\mathrm{Dom}(\varphi)$ converges (in M)
(sc-2) $(e, \leq; \varphi)$ is self-closed: the e-limit of each ascending sequence in $\mathrm{Dom}(\varphi)$ is an upper bound of it (modulo $(\leq)$)
(sc-3) φ is $(e, \leq)$-lsc over $\mathrm{Dom}(\varphi)$: $\liminf_n \varphi(x_n) \geq \varphi(x)$,
whenever the sequence (x_n) in $\mathrm{Dom}(\varphi)$ is $(\leq)$-ascending and $x_n \xrightarrow{e} x$.

The following statement (referred to as: monotone Ekeland Variational Principle on almost metric spaces; in short: (EVP-mon-am)) is then available.

Theorem 8. *Let the conditions $(sc-1)-(sc-3)$ be in force. Then*
(51-a) *for each $u \in \mathrm{Dom}(\varphi)$ there exists $v \in \mathrm{Dom}(\varphi)$ with*

(51-*a*-1) $u \leq v$, $e(u, v) \leq \varphi(u) - \varphi(v)$ *(hence $\varphi(u) \geq \varphi(v)$)*
(51-*a*-2) $e(v, x) > \varphi(v) - \varphi(x)$, *for each $x \in M(v, \leq) \setminus \{v\}$*

(51-b) *if $u \in \mathrm{Dom}(\varphi)$, $\rho > 0$ fulfill $\varphi(u) - \varphi_* \leq \rho$, then* (51-*a*-1) *gives*

($\varphi(u) \geq \varphi(v)$ and) $u \leq v, e(u, v) \leq \rho$.

Proof. Let $\nabla := \nabla[e, \varphi]$ be the associated relation over M

$(x, y \in M)$: $x \nabla y$ iff $e(x, y) + \varphi(y) \leq \varphi(x)$.

Clearly, ∇ is reflexive and transitive — hence, a quasi-ordering — on M. Moreover,

∇ is antisymmetric on $\mathrm{Dom}(\varphi)$: $x, y \in \mathrm{Dom}(\varphi)$, $x \nabla y$, $y \nabla x$ imply $x = y$;

so that, ∇ appears as a (partial) order on $\mathrm{Dom}(\varphi)$. In fact, let $x, y \in \mathrm{Dom}(\varphi)$ be taken as before. By these conditions

$\varphi(x) \geq \varphi(y)$, $\varphi(y) \geq \varphi(x)$; hence, $\varphi(x) = \varphi(y)$,

and this yields (via $\varphi(x)$, $\varphi(y)$ being (finite) real numbers)

$e(x, y) = e(y, x) = 0$; hence $x = y$.

Further, let the relation $\perp$ on M be introduced as

$x \perp y$ iff $x \leq y$ and $x \nabla y$.

From the above properties, $\perp$ is reflexive and transitive — hence, a quasi-ordering — on M. Moreover,

$\perp$ is antisymmetric on $\mathrm{Dom}(\varphi)$: $x, y \in \mathrm{Dom}(\varphi)$, $x \perp y$, $y \perp x$ imply $x = y$,

so that, $\perp$ is (partial) order on $\mathrm{Dom}(\varphi)$; note that, as a consequence of this,

each $u \in \mathrm{Dom}(\varphi)$ is $\perp$-starting since $u \in M(u, \perp)$).

Finally, it is clear that [by the very definition of this relation]

$\perp$ is coarser than ∇: $x \perp y$ implies $x \nabla y$.

Under these preliminaries, we now claim that the strong pseudometric Ekeland Variational Principle (EVP-sp) is applicable here; and from this, we are done. The verification will necessitate a number of steps.

Step 1. Let us verify that the following condition holds:

(cond-1) $(e, \varphi; \perp)$ is strictly descending complete:
for each $\perp$-ascending sequence (x_n) in $\mathrm{Dom}(\varphi)$ with
$(\varphi(x_n))$ = strictly descending there exists $x \in M$ with
$x_n \xrightarrow{e} x$, $(x_n) \perp x$ and $(\varphi(x_n)) \geq \varphi(x)$ (hence, $\lim_n \varphi(x_n) \geq \varphi(x)$).

In fact, let the sequence (x_n) in $\mathrm{Dom}(\varphi)$ be as in the premise above. From the $\perp$-ascending property (and the choice of e),

(perp-asc) $x_n \leq x_m$, $e(x_n, x_m) \leq \varphi(x_n) - \varphi(x_m)$, whenever $n \leq m$.

The sequence $(\varphi(x_n))$ is descending and bounded from below; hence, a Cauchy one. This, along with (perp-asc), shows that (x_n) is an ascending (modulo $(\leq)$) e-Cauchy sequence in $\mathrm{Dom}(\varphi)$; wherefrom, as $(e, \leq; \varphi)$ is complete,

$x_n \xrightarrow{e} x$ as $n \to \infty$, for some $x \in M$.

Moreover, as $(e, \leq; \varphi)$ is self-closed, and φ is $(e, \leq)$-lsc over M, we must have

$(x_n) \leq x$, and $(\varphi(x_n)) \geq \varphi(x)$ (hence, $\lim_n \varphi(x_n) \geq \varphi(x)$).

Finally, let the rank $n \geq 0$ be arbitrary fixed. By the working condition (and the previous relations)

$$e(x_n, x_m) \leq \varphi(x_n) - \varphi(x_m) \leq \varphi(x_n) - \varphi(x), \ \forall m > n.$$

Combining with the triangular property of e, gives

$$e(x_n, x) \leq e(x_n, x_m) + e(x_m, x) \leq \varphi(x_n) - \varphi(x) + e(x_m, x), \ \forall m > n.$$

This, along with the choice of x, yields by a limit process (relative to m)

$$e(x_n, x) \leq \varphi(x_n) - \varphi(x); \text{ that is: } x_n \nabla x.$$

As $n \geq 0$ was arbitrarily chosen, we get (combining with a previous relation)

$$(x_n)\nabla x; \text{ hence, } (x_n) \perp x; \text{ and our claim follows.}$$

Step 2. Further, we have to verify that

(cond-2) e is *transitively sufficient* $[e(z, x) = e(z, y) = 0 \Longrightarrow x = y]$.

This is evident by the sufficiency of e; so, the assertion is proved.

Summing up, (EVP-sp) is indeed applicable here. This (by a previous observation about the $\perp$-starting points) tells us that, for each $u \in \mathrm{Dom}(\varphi)$ there exists an associated point $v \in \mathrm{Dom}(\varphi)$, with

(con-1) $u \perp v$; hence, $u \leq v$, $e(u, v) \leq \varphi(u) - \varphi(v)$,
(con-2) v is E-variational (modulo $\perp$): $v \perp x$ implies $v = x$.

The former of these is just conclusion (51-a-1). Moreover, the latter of these yields

$$v \leq x,\ e(v, x) \leq \varphi(v) - \varphi(x) \text{ imply } v = x,$$

and, from this, conclusion (51-a-2) is clear. Finally, the last part (51-b) is evident. The proof is complete. □

In particular, when

e is symmetric (wherefrom: $e(.,.)$ is a *metric* on M),

the corresponding version of (EVP-mon-am) is just the variational principle in Turinici [33].

A basic particular case of these developments corresponds to the choice

$(\leq) = M \times M$ (=the *trivial* quasi-order on M).

So, let $e(.,.)$ be an almost metric over M; the couple (M, e) will be referred to as an *almost metric space*. Further, let the function $\varphi : M \to R \cup \{\infty\}$ be taken according to

(inf-pro) φ is inf-proper (see above).

The following specific conditions about the couple (e, φ) will be considered (in our trivial quasi-order setting)

(t-sc-1) (e, φ) is complete:
each e-Cauchy sequence in $\mathrm{Dom}(\varphi)$ converges (in M)
(t-sc-2) φ is e-lsc over $\mathrm{Dom}(\varphi)$: $\liminf_n \varphi(x_n) \geq \varphi(x)$,
whenever the sequence (x_n) in $\mathrm{Dom}(\varphi)$ fulfills $x_n \stackrel{e}{\longrightarrow} x$.

The following statement (referred to as: Ekeland Variational Principle on almost metric spaces; in short: (EVP-am)) is then available.

Theorem 9. *Let the precise conditions be in force. Then*
(52-a) *for each* $u \in \mathrm{Dom}(\varphi)$ *there exists* $v \in \mathrm{Dom}(\varphi)$ *with*

(52-a-1) $e(u, v) \leq \varphi(u) - \varphi(v)$ *(hence,* $\varphi(u) \geq \varphi(v)$*)*
(52-a-2) $e(v, x) > \varphi(v) - \varphi(x)$, *for each* $x \in M \setminus \{v\}$.

(52-b) *if* $u \in \mathrm{Dom}(\varphi)$, $\rho > 0$ *fulfill* $\varphi(u) - \varphi_* \leq \rho$, *then* (52-$a$-1) *gives*

$(\varphi(u) \geq \varphi(v)$ *and)* $u \leq v, e(u, v) \leq \rho$.

As precise, this statement is obtainable from the preceding one. For technical reasons, it would be useful to establish that (EVP-am) is also deductible from the relative narrow strong pseudometric Ekeland Variational Principle (EVP-r-n-sp).

Proposition 12. *The following conclusions hold*

(EVP-am) is deductible from (EVP-r-n-sp) in (ZF-AC);
hence, (EVP-am) is deductible in (ZF-AC+DC).

Proof. Let $\nabla := \nabla[e, \varphi]$ be the associated relation over M

$(x, y \in M)$: $x \nabla y$ iff $e(x, y) + \varphi(y) \leq \varphi(x)$.

Clearly, ∇ is reflexive and transitive — hence, a quasi-ordering — on M. Moreover, ∇ is antisymmetric on $\mathrm{Dom}(\varphi)$; i.e.

$x, y \in \mathrm{Dom}(\varphi)$, $x \nabla y$, $y \nabla x$ imply $x = y$,

so that, ∇ is a (partial) order on $\mathrm{Dom}(\varphi)$; note that, as a consequence of this,

each $u \in \mathrm{Dom}(\varphi)$ is ∇-starting [since $u \in M(u, \nabla))$].

Part 1. Let us verify that the condition below holds

(condi-1) (e,φ) is strictly descending complete: for each e-Cauchy sequence (x_n) in $\mathrm{Dom}(\varphi)$ with $(\varphi(x_n))=$strictly descending, there exists $x\in M$ with $x_n \overset{e}{\longrightarrow} x$ and $(\varphi(x_n))\geq\varphi(x)$ (hence, $\lim_n\varphi(x_n)\geq\varphi(x)$).

In fact, let the sequence (x_n) in $\mathrm{Dom}(\varphi)$ be as in the premise above. As (e,φ) is complete, we must have

$$x_n \overset{e}{\longrightarrow} x \text{ as } n\to\infty, \text{ for some } x\in M.$$

Moreover, as φ is e-lsc over M, it results that

$$(\varphi(x_n))\geq\varphi(x) \text{ (hence, } \lim_n\varphi(x_n)\geq\varphi(x)),$$

and our claim follows.

Part 2. Further, we have to verify that

(condi-2) $e(.,.)$ is a transitively sufficient KST-semimetric (modulo e).

This is evident, by the choice of e; so, the assertion is proved.

Summing up, (EVP-r-n-sp) is indeed applicable here. This (by a previous observation about the ∇-starting points) tells us that, for each $u\in\mathrm{Dom}(\varphi)$ there exists an associated point $v\in\mathrm{Dom}(\varphi)$, with

(conc-1) $u\nabla v$; hence, $e(u,v)\leq\varphi(u)-\varphi(v)$,
(conc-2) v is E-variational (modulo ∇): $v\nabla x\Longrightarrow v=x$.

The former of these is just conclusion (52-a-1). Moreover, the latter of these yields

$e(v,x)\leq\varphi(v)-\varphi(x)$ imply $v=x$; whence, conclusion (52-a-2) is clear.

Finally, the last part (52-b) is evident. □

Clearly, the regularity condition (t-sc-2) holds under

(t-sc-2a) φ is e-lsc over M: $\liminf_n\varphi(x_n)\geq\varphi(x)$, whenever $x_n \overset{e}{\longrightarrow} x$.

Then, in the context of

e is symmetric (wherefrom: $e(.,.)$ is a metric on M)

(EVP-am) is just Ekeland's variational principle [1] (in short: (EVP)). Further aspects may be found in Kang and Park [34]; see also Hyers *et al.* [2, Ch 5].

Summing up, all variational results in this exposition are extensions of (EVP). Precisely, note that, by the developments above, we have the following chain of implications:

(DC) $\Longrightarrow$ (EVP-p) $\Longrightarrow$ (EVP-sp) $\Longrightarrow$ (EVP-mon-am) $\Longrightarrow$ (EVP-am)
(EVP-p) $\Longrightarrow$ (EVP-n-p) $\Longrightarrow$ (EVP-r-n-p) $\Longrightarrow$ (EVP-r-n-sp)
(EVP-sp) $\Longrightarrow$ (EVP-n-sp) $\Longrightarrow$ (EVP-r-n-sp) $\Longrightarrow$ (EVP-am) $\Longrightarrow$ (EVP).

Clearly, all these inclusions may be reversed as long as (EVP) $\Longrightarrow$ (DC) is retainable. An appropriate (positive) answer to this is to be given in the *strongly reduced* Zermelo–Fraenkel system (ZF-AC). Some preliminaries are needed.

Let X be a non-empty set; and $(\leq)$ be an order on it. We say that $(\leq)$ has the *inf-lattice* property, provided:

$$x \wedge y := \inf(x, y) \text{ exists, for all } x, y \in X.$$

Remember that $z \in X$ is a $(\leq)$-*maximal* element if $X(z, \leq) = \{z\}$; the class of all these points will be denoted as $\max(X, \leq)$. In this case, $(\leq)$ is called a *Zorn order* when $\max(X, \leq)$ is non-empty and *cofinal* in X [for each $u \in X$ there exists a $(\leq)$-maximal $v \in X$ with $u \leq v$]. Further aspects are to be described in a metric setting. Let $d : X \times X \to R_+$ be a metric over X; and $\varphi : X \to R_+$ be some function. Then, the natural choice for $(\leq)$ above is

$$x \leq_{(d,\varphi)} y \text{ iff } d(x, y) \leq \varphi(x) - \varphi(y),$$

referred to as the Brøndsted order [3] attached to (d, φ). Denote

$$X(x, \rho) = \{u \in X; d(x, u) < \rho\},\ x \in X,\ \rho > 0$$

[the open sphere with center x and radius ρ]. Call the ambient metric space (X, d), *discrete* when for each $x \in X$ there exists $\rho = \rho(x) > 0$ such that $X(x, \rho) = \{x\}$. Note that, under such an assumption, any function $\psi : X \to R$ is continuous over X. However, the *Lipschitz* property

$$|\psi(x) - \psi(y)| \leq Ld(x, y),\ x, y \in X, \text{ for some } L > 0,$$

cannot be assured, in general; this is also true for the *non-expansive* one $(L = 1)$.

Now, the following statement is a particular case of (EVP):

Theorem 10. *Let the metric space (X, d) and the function $\varphi : X \to R_+$ satisfy*

(53-*i*) (X, d) *is discrete bounded and complete*
(53-*ii*) $(\leq_{(d,\varphi)})$ *has the inf-lattice property*
(53-*iii*) φ *is d-non-expansive and* $\varphi(X)$ *is countable.*

Then, the Brøndsted order $(\leq_{(d,\varphi)})$ *is a Zorn one.*

We will term it as: the *discrete Lipschitz countable* version of (EVP) (in short: (EVP-dLc)). Clearly, (EVP) $\Longrightarrow$ (EVP-dLc). The remarkable fact to be added is that this last principle yields (DC); so, it completes the circle between all these.

Proposition 13. *We have (in the strongly reduced system (ZF-AC)) (EVP-dLc)* $\Longrightarrow$ *(DC). So (by the above), the variational principles (EVP-p), (EVP-sp), (EVP-n-p), (EVP-n-sp), (EVP-r-n-p), (EVP-r-n-sp), (EVP-mon-am), (EVP-am) and (EVP) are all equivalent with (DC); hence, mutually equivalent.*

For a complete proof, see Turinici [17].

Summing up, all variational/maximal principles in this exposition are nothing but logical equivalents of (EVP). Concerning this aspect, we may ask whether the abstract ordering principle in Qiu [35], the perturbed minimization principle in Deville and Ghoussoub [36], or the generalized Ekeland Variational Principle due to Farkas *et al.* [37] are also comprised in our list. The answer to this is affirmative; further aspects will be delineated elsewhere. Some other aspects of the described facts may be found in Turinici [18].

6. Functional Caristi–Kirk theorems

In the following, some applications of the developments above to Caristi–Kirk theorems are given.

Let M be some non-empty set; and $e : M \times M \to R_+$ be a pseudometric over it; with, in addition,

e is *triangular*: $e(x, z) \leq e(x, y) + e(y, z)$, for all $x, y, z \in M$.

We then say that $e(.,.)$ is a *triangular pseudometric* (in short: *t-pseudometric*) on M; and (M, e) is referred to as a *triangular pseudometric space* (in short: *t-pseudometric space*). Further, let $\varphi : M \to R \cup \{\infty\}$ be some *inf-proper* function:

(inf-pr-1) $\text{Dom}(\varphi) := \{x \in M; \varphi(x) \neq \infty\} \neq \emptyset$
(inf-pr-2) φ is bounded below: $\inf \varphi(M) > -\infty$.

As before, we attach to this couple the transitive relation $\nabla = \nabla[e, \varphi]$ over M

$(x, y \in M)$: $x\nabla y$ iff $e(x, y) + \varphi(y) \leq \varphi(x)$.

Note that, in general, ∇ is not reflexive. So, for $u \in M$, it is possible that

u is *singular* (modulo ∇): $M(u, \nabla)$ is empty.

The opposite to this alternative will be expressed as

$u \in M$ is *starting* (modulo ∇): $M(u, \nabla)$ is non-empty.

Finally, call $v \in \text{Dom}(\varphi)$,

BB-*variational* (modulo ∇), provided
$e(v, x) \leq \varphi(v) - \varphi(x) \Longrightarrow \varphi(v) = \varphi(x)$ [so, by definition, $e(v, x) = 0$]
E-*variational* (modulo ∇), provided
$x \in M$ and $v\nabla x$ imply $v = x$ (hence, $\varphi(v) = \varphi(x)$ and $e(v, x) = 0$).

Note that the implications above are vacuously fulfilled when $v \in \text{Dom}(\varphi)$ is singular (modulo ∇); so, the verification is to be considered only when $v \in \text{Dom}(\varphi)$ is starting (modulo ∇).

Having these precise, let $T : M \to M$ be a selfmap of M with

(pro) $x\nabla Tx$, for each $x \in M$; referred to as: T is ∇-*progressive.*

Note that, whenever $\varphi(x) = \infty$, this relation holds; hence, the convention above may be also written as

(pro-var) $e(x, Tx) \leq \varphi(x) - \varphi(Tx)$, for each $x \in \text{Dom}(\varphi)$.

Suppose that we fixed such a map; note that, by this very definition, each point of $\text{Dom}(\varphi)$ is starting (modulo ∇). Denote, for simplicity

$\text{Fix}(T; \varphi) := \{x \in \text{Dom}(\varphi); \varphi(x) = \varphi(Tx)\}$;

each point of this set will be referred to as φ-*fixed* under T. For both practical and theoretical reasons, we are interested to determine sufficient conditions under which such points exist. As the definition above suggests, these points are to be determined among the BB-variational (modulo ∇) points of M. The natural strategy is to apply the narrow pseudometric Ekeland Variational Principle (EVP-n-p), based on (s-desc-com) (e, φ) is strictly descending complete: for each e-Cauchy sequence (x_n) in $\mathrm{Dom}(\varphi)$ with $(\varphi(x_n))$=strictly descending, there exists $x \in M$ with $x_n \xrightarrow{e} x$ and $(\varphi(x_n)) \geq \varphi(x)$ (hence, $\lim_n \varphi(x_n) \geq \varphi(x)$).

Precisely, the following functional pseudometric Caristi–Kirk theorem (in short: (CK-p-f)) is available.

Theorem 11. *Suppose (in addition) that (e, φ) is strictly descending complete. Then, for each $u \in \mathrm{Dom}(\varphi)$ (which, as noted, is a starting (modulo ∇) point), there exists an associated point $v \in \mathrm{Dom}(\varphi)$, with*

(61-*a*) $e(u, v) \leq \varphi(u) - \varphi(v)$ *(hence, $\varphi(u) \geq \varphi(v)$)*
(61-*b*) *v is BB-variational (modulo ∇) :*
$e(v, x) \leq \varphi(v) - \varphi(x) \Longrightarrow \varphi(v) = \varphi(x)$, *and* $e(v, x) = 0$
(61-*c*) *v is φ-fixed under T (i.e.: $\varphi(v) = \varphi(Tv)$), and $e(v, Tv) = 0$.*

The proof is immediate by the very definition of BB-variational (modulo ∇) point, and the progressiveness of T (modulo (e, φ)); so, we do not give details.

A stronger variant of this result is to be stated under the following lines. Denote, for simplicity

$$\mathrm{Fix}(T) = \{x \in \mathrm{Dom}(\varphi); x = Tx\},$$

each point of this set will be referred to as *fixed under T*. As the definition suggests, these points are to be determined among the E-variational (modulo ∇) points of M. The natural strategy is to apply the narrow strong pseudometric Ekeland Variational Principle (EVP-n-sp), based on

(tr-suf) e is *transitively sufficient* $[e(z, x) = e(z, y) = 0 \Longrightarrow x = y]$.

Precisely, as a direct application of the quoted statement, the following pseudometric Caristi–Kirk theorem (in short: (CK-p)) is available.

Theorem 12. *Suppose (in addition) that (e, φ) is strictly descending complete, and e is transitively sufficient. Then, for each $u \in \mathrm{Dom}(\varphi)$ (which, as noted, is a starting (modulo ∇) point), there exists an associated point $w \in \mathrm{Dom}(\varphi)$, with*

(62-*a*) $e(u, w) \leq \varphi(u) - \varphi(w)$ *(hence, $\varphi(u) \geq \varphi(w)$)*
(62-*b*) *w is E-variational (modulo ∇) :*
$e(w, x) \leq \varphi(w) - \varphi(x) \Longrightarrow w = x,\ e(w, x) = 0.$
(62-*c*) *w is fixed under T (i.e. $w = Tw$), and $e(w, Tw) = 0$.*

A relative form of these results may be given along the following lines. Let $d : M \times M \to R_+$ be a general pseudometric over M; and $e : M \times M \to R_+$ be a triangular pseudometric over the same. Remember that $e(.,.)$ is a *KST-semimetric* (modulo d), provided

(kst-sm-1) e is Cauchy d-lsc in the second variable: (y_n) is e-Cauchy and $y_n \xrightarrow{d} y$ imply $\liminf_n e(x, y_n) \geq e(x, y), \forall x \in M$
(kst-sm-2) e is Cauchy semi-subordinated to d:
each e-Cauchy sequence admits a d-Cauchy subsequence.

Further, let $\varphi : M \to R \cup \{\infty\}$ be some *inf-proper* function; and the selfmap $T : M \to M$ be ∇*-progressive* (see above).

As a direct consequence of the relative narrow pseudometric Ekeland Vatiational Principle (EVP-r-n-p), the following relative functional pseudometric Caristi–Kirk result (in short: (CK-r-f-p)) is available.

Theorem 13. *Suppose that* $e(.,.)$ *is KST-semimetric* (*modulo* d) *and* (d, φ) *is strictly descending complete. Then, for each* $u \in \mathrm{Dom}(\varphi)$ (*which, as noted, is a starting* (*modulo* ∇) *point*) *there exists an associated point* $v \in \mathrm{Dom}(\varphi)$, *with*

(63-*a*) $e(u, v) \leq \varphi(u) - \varphi(v)$ (*hence,* $\varphi(u) \geq \varphi(v)$)
(63-*b*) v *is BB-variational* (*modulo* ∇) :
$e(v, x) \leq \varphi(v) - \varphi(x) \implies \varphi(v) = \varphi(x)$, *and* $e(v, x) = 0$.
(63-*c*) v *is* φ*-fixed under* T (*i.e.* $\varphi(v) = \varphi(Tv)$), *and* $e(v, Tv) = 0$.

As before, a natural problem to be posed is that of getting a corresponding form of this result involving E-variational (modulo ∇) points. The appropriate answer to this is obtainable by means of the relatively narrow strong pseudometric Ekeland variational principle (EVP-r-n-sp). Precisely, the following relative pseudometric Caristi–Kirk result involving these data (in short: (CK-r-p)) is available.

Theorem 14. *Suppose that* $e(.,.)$ *is a transitively sufficient KST-semimetric* (*modulo* d), *and* (d, φ) *is strictly descending complete. Then, for each* $u \in \mathrm{Dom}(\varphi)$ (*which, as noted, is a starting* (*modulo* (e, φ)) *point*) *there exists an associated point* $w \in \mathrm{Dom}(\varphi)$, *with*

(64-*a*) $e(u, w) \leq \varphi(u) - \varphi(w)$ (*hence,* $\varphi(u) \geq \varphi(w)$)
(64-*b*) w *is E-variational* (*modulo* ∇) :
$e(w, x) \leq \varphi(w) - \varphi(x) \implies w = x$ *and* $e(w, x) = 0$
(64-*c*) w *is fixed under* T (*i.e.:* $w = Tw$), *and* $e(w, Tw) = 0$.

In particular, the strict descending completeness of (d, φ) is fulfilled whenever

d is complete and φ is d-lsc.

In this case, under the extra assumptions

d is triangular and weakly sufficient ($d(x,y) = d(y,x) = 0$ imply $x = y$),

the relative functional pseudometric Caristi–Kirk result (CK-r-f-p) is just the statement in Alegre and Marin [20]; see also Turinici [27]. The corresponding versions of our relative pseudometric Caristi–Kirk result (CK-r-p) seem to be new. Further aspects are to be found in Cobzaş [38].

References

[1] I. Ekeland, Nonconvex minimization problems, *Bull. Amer. Math. Soc. (New Series)* **1**, 443–474, (1979).
[2] D.H. Hyers, G. Isac, and T.M. Rassias, *Topics in Nonlinear Analysis and Applications* (World Sci. Publ., Singapore, 1997).
[3] A. Brøndsted, Fixed points and partial orders, *Proc. Amer. Math. Soc.* **60**, 365–366, (1976).
[4] N. Bourbaki, Sur le théorème de Zorn, *Archiv Math.* **2**, 434–437, (1949/1950).
[5] M. Zorn, A remark on method in transfinite algebra, *Bull. Amer. Math. Soc.* **41**, 667–670, (1935).
[6] H. Brezis and F.E. Browder, A general principle on ordered sets in nonlinear functional analysis, *Advances Math.* **21**, 355–364, (1976).
[7] A. Goepfert, H. Riahi, C. Tammer, and C. Zălinescu, Variational methods in partially ordered spaces, *Canad. Math. Soc. Books Math.* Vol. 17. (Springer, New York, 2003).
[8] O. Kada, T. Suzuki, and W. Takahashi, Nonconvex minimization theorems and fixed point theorems in complete metric spaces, *Math. Japonica* **44**, 381–391, (1996).
[9] M. Altman, A generalization of the Brezis–Browder principle on ordered sets, *Nonlinear Analysis* **6**, 157–165, (1982).
[10] M.C. Anisiu, *On maximality principles related to Ekeland's theorem,* Seminar Funct. Analysis Numer. Meth. (Faculty Math. Research Seminars), Preprint No. 1, Babeş-Bolyai University, Cluj-Napoca (Romania), 1987.
[11] D. Tataru, Viscosity solutions of Hamilton-Jacobi equations with unbounded nonlinear terms, *J. Math. Anal. Appl.* **163**, 345–392, (1992).
[12] M. Turinici, Variational statements on KST-metric structures, *An. Şt. Univ. "Ovidius" Constanţa (Mat.)* **17**, 231–246, (2009).
[13] T.Q. Bao and P.Q. Khanh, Are several recent generalizations of Ekeland's variational principle more general than the original principle?, *Acta Math. Vietnamica* **28**, 345–350, (2003).

[14] P. Bernays, A system of axiomatic set theory: Part III. Infinity and enumerability analysis, *J. Symbolic Logic* **7**, 65–89, (1942).
[15] A. Tarski, Axiomatic and algebraic aspects of two theorems on sums of cardinals, *Fund. Math.* **35**, 79–104, (1948).
[16] N. Brunner, Topologische Maximalprinzipien, *Zeitschr. Math. Logik Grundl. Math.* **33**, 135–139, (1987).
[17] M. Turinici, Sequential maximality principles, In: T.M. Rassias and P.M. Pardalos, (Eds.), *Mathematics Without Boundaries*, pp. 515–548 (Springer, New York, 2014).
[18] M. Turinici, Variational principles in gauge spaces, In: T.M. Rassias, C.A. Floudas, and S. Butenko, (Eds.), *Optimization in Science and Engineering*, pp. 503–542 (Springer, New York, 2014).
[19] J.M. Borwein and D. Preiss, A smooth variational principle with applications to subdifferentiability and to differentiability of convex functions, *Trans. Amer. Math. Soc.* **303**, 517–527, (1987).
[20] C. Alegre and J. Marin, A Caristi fixed point theorem for complete quasi-metric spaces by using mw-distances, *Fixed Point Th.* **19**, 25–32, (2018).
[21] P.J. Cohen, *Set Theory and the Continuum Hypothesis.* (Benjamin, New York, 1966).
[22] E.S. Wolk, On the principle of dependent choices and some forms of Zorn's lemma, *Canad. Math. Bull.* **26**, 365–367, (1983).
[23] Y. Moskhovakis, *Notes on Set Theory* (Springer, New York, 2006).
[24] E. Schechter, *Handbook of Analysis and its Foundation* (Academic Press, New York, 1997).
[25] G.H. Moore, *Zermelo's Axiom of Choice: Its Origin, Development and Influence.* (Springer, New York, 1982).
[26] S. Kasahara, On some generalizations of the Banach contraction theorem, *Publ. Res. Inst. Math. Sci. Kyoto Univ.* **12**, 427–437, (1976).
[27] M. Turinici, Pseudometric versions of the Caristi–Kirk fixed point theorem, *Fixed Point Theory (Cluj-Napoca)* **5**, 147–161, (2004).
[28] C. Alegre, J. Marin, and S. Romaguera, A fixed point theorem for generalized contractions involving w-distances on complete quasi-metric spaces, *Fixed Point Th. Appl.* **2014**, 40, (2014).
[29] O. Cârjă and C. Ursescu, The characteristics method for a first order partial differential equation, *An. Şt. Univ. "A. I. Cuza" Iaşi (Sect I-a, Mat.)* **39**, 367–396, (1993).
[30] M. Turinici, Minimal points in product spaces, *An. Şt. Univ. "Ovidius" Constanţa (Ser. Math.)* **10**, 109–122, (2002).
[31] L. Gajek and D. Zagrodny, Countably orderable sets and their application in optimization, *Optimization* **26**, 287–301, (1992).
[32] M. Turinici, Relational Brezis-Browder principles, *Fixed Point Theory (Cluj-Napoca)* **7**, 111–126, (2006).
[33] M. Turinici, A monotone version of the variational Ekeland's principle, *An. Şt. Univ. "A. I. Cuza" Iaşi (S. I-a: Mat.)* **36**, 329–352, (1990).
[34] B.G. Kang and S. Park, On generalized ordering principles in nonlinear analysis, *Nonlinear Analysis* **14**, 159–165, (1990).

[35] J.-H. Qiu, A pre-order principle and set-valued Ekeland variational principle, *J. Math. Anal. Appl.* **419**, 904–937, (2014).

[36] R. Deville and N. Ghoussoub, Perturbed minimization principles and applications. In: W.B. Johnson and J. Lindenstrauss, (Eds.), *Handbook of the Geometry of Banach Spaces*, Vol. I, Ch. 10, pp. 399–435 (Elsevier Science B. V., 2001).

[37] C. Farkas, A.E. Molnár and S. Nagy, A generalized variational principle in b-metric spaces, *Le Matematiche* **69**, 205–221, (2014).

[38] S. Cobzaş, Ekeland, Takahashi, and Caristi principles in quasi-pseudometric spaces, *Arxiv, 1902-09743-v2*, 3 Mar 2019.

[39] O. Cârjă, M. Necula, and I. I. Vrabie, *Viability, Invariance and Applications*, North Holland Mathematics Studies, Vol. 207 (Elsevier B. V., Amsterdam, 2007).

https://doi.org/10.1142/9789811261572_0028

Chapter 28

Metrical Coercivity for Monotone Functionals

Mihai Turinici
A. Myller Mathematical Seminar, A. I. Cuza University, 700506 Iaşi, Romania
mturi@uaic.ro

A quasi-metric-type coercivity result is established for order non-smooth functionals fulfilling Palais-Smale conditions. The core of this approach is an asymptotic statement obtained via local versions of the monotone variational principle in Turinici [An. Şt. UAIC Iaşi, 36 (1990), 329–352].

1. Introduction

Let $(X, \|.\|)$ be a (real) Banach space. Given a (proper) functional $F : X \to R \cup \{\infty\}$, we say that it is *coercive*, provided

(coer) $F(u) \to \infty$, whenever $u \to \infty$ (in the sense: $\|u\| \to \infty$).

Sufficient conditions for such a property are obtainable in a differential setting, by means of the celebrated 1964 Palais–Smale condition [1]. A typical result in this direction is due to Caklovic *et al.* [2]; it states that, whenever

(G-lsc) F is Gateaux differentiable and lower semicontinuous (lsc)

the relation (coer) is deductible under a Palais-Smale requirement like

(PS) each sequence (v_n) in X with $(F(v_n)) =$ bounded and $F'(v_n) \to 0$ (in X^*) has a convergent (in X) subsequence.

(Here, $(X^*, \|.\|)$ is the *topological dual* of X). Note that (G-lsc) holds when $F \in C^1(X)$; hence, their statement includes Brezis–Nirenberg's [3]. An extension of it was obtained by Goeleven [4], under (G-lsc) substituted by the representation

(G-lsc-conv) $F = F_1 + F_2$,
where F_1 is Gateaux differentiable lsc and F_2 is (proper) convex lsc;

and the Palais–Smale (PS) condition being adapted to this decomposition. Further enlargements of (G-lsc-conv) were given by Motreanu and Motreanu [5], under the general requirement

(loc-L-conv) $F = F_1 + F_2$, where F_1 is locally Lipschitz (hence continuous), and F_2 is (proper) convex lsc;

and a Palais–Smale condition like in Motreanu and Panagiotopoulos [6, Ch 3]. Further coercivity statements may be found in Bae *et al.* [7] and Xu [8]; see also Turinici [9–12].

The basic tools of all these are, essentially, Ekeland's Variational Principle [13] (in short: (EVP)), and the Zhong Variational Principle [14] (in short: (ZVP)); requiring that F should be lsc (on X). So, we may ask whether this is removable; an appropriate answer is available in a quasi-order quasi-metrical context. As far as we know, the first result in this direction is the 2002 one due to Motreanu and Turinici [15], based on the monotone Ekeland Variational Principle in Turinici [16]; some related developments of this theory are to be found in Motreanu *et al.* [17] and Turinici [18,19]. It is our aim in the following to show that further refinements of this result to such structures are possible. The basic tool of these investigations is an asymptotic type result involving such functionals, obtained via "local" quasi-order quasi-metric versions of Ekeland Variational Principle comparable with the one in Turinici [20]. And the specific tool is a quasi-order quasi-metric version of the *slope* concept introduced by DeGiorgi *et al.* [21]. Note, finally, that multivalued and functional extensions of these results are available — under the described methods — by following the lines in Turinici [22,23]; further aspects will be delineated elsewhere.

2. Dependent Choice Principles

Throughout this exposition, the axiomatic system in use is Zermelo–Fraenkel's (abbreviated: (ZF)), as described by Cohen [24, Ch 2]. The notations and basic facts to be considered in this system are more or less standard. Some important ones are discussed in what follows.

(A) Let X be a non-empty set. By a *relation* over X, we mean any (non-empty) part $\mathcal{R}$ of $X \times X$; then, $(X, \mathcal{R})$ will be referred to as a *relational structure*. For simplicity, we sometimes write $(x, y) \in \mathcal{R}$ as $x\mathcal{R}y$. Note that $\mathcal{R}$ may be regarded as a mapping between X and $\exp[X]$ (=the class of all subsets in X). In fact, denote for $x \in X$:

$X(x, \mathcal{R}) = \{y \in X; x\mathcal{R}y\}$ (the *section* of $\mathcal{R}$ through x);

then, the desired mapping representation is $[\mathcal{R}(x) = X(x, \mathcal{R}),\ x \in X]$. Call $\mathcal{R}$, *proper* when $\mathcal{R}(x) \neq \emptyset$, for all $x \in X$. Note that, in such a case, $\mathcal{R}$ appears as a mapping between X and $\exp(X)$ (=the class of all non-empty parts in X); this will also be referred to as: $(X, \mathcal{R})$ is a *proper* relational structure. A basic example of such an object is $\mathcal{I} = \{(x, x); x \in X\}$ (the *identical relation* over X).

Given the relations $\mathcal{R}$, $\mathcal{S}$ over X, define their *product* $\mathcal{R} \circ \mathcal{S}$ as

$(x, z) \in \mathcal{R} \circ \mathcal{S}$, if there exists $y \in X$ with $(x, y) \in \mathcal{R}$, $(y, z) \in \mathcal{S}$.

Also, for each relation $\mathcal{R}$ in X, denote

$\mathcal{R}^{-1} = \{(x, y) \in X \times X; (y, x) \in \mathcal{R}\}$ (the *inverse* of $\mathcal{R}$).

Finally, given the relations $\mathcal{R}$ and $\mathcal{S}$ on X, let us say that $\mathcal{R}$ is *coarser* than $\mathcal{S}$ (or, equivalently: $\mathcal{S}$ is *finer* than $\mathcal{R}$), provided

$\mathcal{R} \subseteq \mathcal{S}$; i.e.: $x\mathcal{R}y$ implies $x\mathcal{S}y$.

Given a relation $\mathcal{R}$ on X, the following properties are to be discussed here:

(P1) $\mathcal{R}$ is *reflexive*: $\mathcal{I} \subseteq \mathcal{R}$
(P2) $\mathcal{R}$ is *irreflexive*: $\mathcal{R} \cap \mathcal{I} = \emptyset$
(P3) $\mathcal{R}$ is *transitive*: $\mathcal{R} \circ \mathcal{R} \subseteq \mathcal{R}$
(P4) $\mathcal{R}$ is *symmetric*: $\mathcal{R}^{-1} = \mathcal{R}$
(P5) $\mathcal{R}$ is *antisymmetric*: $\mathcal{R}^{-1} \cap \mathcal{R} \subseteq \mathcal{I}$.

This yields the classes of relations to be used; the following ones are important for our developments:

(C1) $\mathcal{R}$ is a *quasi-order* (reflexive and transitive)
(C2) $\mathcal{R}$ is a *strict order* (irreflexive and transitive)
(C3) $\mathcal{R}$ is an *equivalence* (reflexive, transitive, symmetric)

(C4) $\mathcal{R}$ is a *(partial) order* (reflexive, transitive, antisymmetric)
(C5) $\mathcal{R}$ is *trivial* (i.e.: $\mathcal{R} = X \times X$).

A basic example of relational structure is to be constructed as follows. Let $N = \{0, 1, ...\}$ be the set of *natural* numbers, endowed with the usual addition and (partial) order; note that

$(N, \leq)$ is well ordered: any (non-empty) subset of N has a first element.

Further, denote for $p, q \in N$, $p \leq q$,

$$N[p, q] = \{n \in N; p \leq n \leq q\},\ N]p, q[= \{n \in N; p < n < q\},$$
$$N[p, q[= \{n \in N; p \leq n < q\},\ N]p, q] = \{n \in N; p < n \leq q\},$$

as well as, for $r \in N$,

$$N[r, \infty[= \{n \in N; r \leq n\},\ N]r, \infty[= \{n \in N; r < n\}.$$

For each $r \geq 1$, $N[0, r[= N(r, >)$ is referred to as the *initial interval* (in N) induced by r. Any set P with $P \sim N$ (in the sense: there exists a bijection from P to N) will be referred to as *effectively denumerable.* In addition, given some natural number $n \geq 1$, any (non-empty) set Q with $Q \sim N(n, >)$ will be said to be *n-finite*; when n is generic here, we say that Q is *finite.* Finally, the (non-empty) set Y is called (at most) *denumerable* iff it is either effectively denumerable or finite.

Let X be a non-empty set. By a *sequence* in X, we mean any mapping $x : N \to X$, where $N = \{0, 1, ...\}$ is the set of *natural* numbers. For simplicity reasons, it will be useful to denote it as $(x(n); n \geq 0)$, or $(x_n; n \geq 0)$; moreover, when no confusion can arise, we further simplify this notation as $(x(n))$ or (x_n), respectively. Also, any sequence $(y_n := x_{i(n)}; n \geq 0)$ where

$(i(n); n \geq 0)$ is *strictly ascending* (hence, $i(n) \to \infty$ as $n \to \infty$)

will be referred to as a *subsequence* of $(x_n; n \geq 0)$. Note that, under such a convention, the relation "subsequence of" is transitive; i.e.

(z_n)=subsequence of (y_n) and (y_n)=subsequence of (x_n)
imply (z_n)=subsequence of (x_n).

(B) Remember that, an outstanding part of (ZF) is the *Axiom of Choice* (abbreviated: (AC)); which, in a convenient manner, may be written as

(AC) For each couple (J, X) of non-empty sets and each function
$F : J \to \exp(X)$, there exists a (selective) function
$f : J \to X$, with $f(\nu) \in F(\nu)$, for each $\nu \in J$.

Sometimes, when the index set J is denumerable, the existence of such a selective function may be determined by using a weaker form of (AC), called: *Dependent Choice* principle (in short: (DC)). Some preliminaries are needed. For each natural number $k \geq 1$, call the map $F : N(k, >) \to X$, a *k-sequence*; if $k \geq 1$ is generic, we talk about a *finite* sequence. The following result, referred to as the *Finite Dependent Choice property* (in short: (DC-fin)) is available in the *strongly reduced* Zermelo–Fraenkel system (ZF-AC). Given $a \in X$, let us say that the k-sequence $F : N(k, >) \to X$ (where $k \geq 2$) is $(a, \mathcal{R})$-*iterative*, provided

$$F(0) = a \text{ and } F(i)\mathcal{R}F(i+1), \text{ for all } i \in N(k-1, >).$$

Proposition 1. *Let the relational structure* $(X, \mathcal{R})$ *be proper. Then, for each* $k \geq 2$, *the following property holds*:

$(P(k))$ *for each* $a \in X$, *there exists an* $(a, \mathcal{R})$*-iterative* k*-sequence.*

Proof. Clearly, $(P(2))$ is true; just take $b \in \mathcal{R}(a)$ and define $F : N(2, >) \to X$ as: $F(0) = a$, $F(1) = b$. Assume that $(P(k))$ is true, for some $k \geq 2$; we claim that $(P(k+1))$ is true as well. In fact, let $F : N(k, >) \to X$ be an $(a, \mathcal{R})$-iterative k-sequence, assured by hypothesis. As $\mathcal{R}$ is proper, $\mathcal{R}(F(k-1))$ is non-empty; let u be some element of it. The map $G : N(k+1, >) \to X$ introduced as

$$G(i) = F(i), i \in N(k, >); \; G(k) = u$$

is an $(a, \mathcal{R})$-iterative $(k+1)$-sequence; and then, we are done. □

Now, it is natural to see what happens when k "tends to infinity". At a first glance, the following *Dependent Choice* principle (in short: (DC)) is obtainable in (ZF-AC) from this "limit" process. Given $a \in X$, let us say that the sequence $(x_n; n \geq 0)$ in X is $(a; \mathcal{R})$-*iterative*, provided

$$x_0 = a \text{ and } [x_n \mathcal{R} x_{n+1} \text{ (i.e.: } x_{n+1} \in \mathcal{R}(x_n)), \forall n].$$

Proposition 2. *Let the relational structure* $(X, \mathcal{R})$ *be proper. Then, for each* $a \in X$, *there is at least an* $(a, \mathcal{R})$*-iterative sequence in* X.

Formally, the "argument" involved here consists in the possibility of constructing an (infinite) sequence via finite sequences. This, ultimately, cannot be done under the precise context; so, the limit process in question does not work in our strongly reduced system (ZF-AC)]; whence, (DC) is not obtainable from its axioms. On the other hand, this principle —

proposed, independently, by Bernays [25] and Tarski [26] — is deductible from (AC), but not conversely; *cf.* Wolk [27]. Denote, for simplicity,

(ZF-AC+DC) (the *reduced Zermelo–Fraenkel system*) = the strongly reduced Zermelo–Fraenkel system (ZF-AC) completed with the Dependent Choice principle (DC).

According to the developments in Moskhovakis [28, Ch 8], and Schechter [29, Ch 6], the *reduced* system (ZF-AC+DC) it large enough so as to cover the "usual" mathematics; see also Moore [30, Appendix 2, Table 4]. So, determining various equivalents of (DC) may not be without profit.

(C) Let $(\mathcal{R}_n; n \geq 0)$ be a sequence of relations on X. Given $a \in X$, let us say that the sequence $(x_n; n \geq 0)$ in X is $(a; (\mathcal{R}_n; n \geq 0))$-*iterative*, provided

$$x_0 = a \text{ and } [x_n \mathcal{R}_n x_{n+1} \text{ (i.e.: } x_{n+1} \in \mathcal{R}_n(x_n)), \forall n].$$

The following *Diagonal Dependent Choice* principle (in short: (DDC)) is available.

Proposition 3. *Let $(\mathcal{R}_n; n \geq 0)$ be a sequence of proper relations on X. Then, for each $a \in X$, there exists an $(a; (\mathcal{R}_n; n \geq 0))$-iterative sequence in X.*

Clearly, (DDC) includes (DC); to which it reduces when $(\mathcal{R}_n; n \geq 0)$ is constant. The reciprocal of this is also true. In fact, letting the premises of (DDC) hold, put $P = N \times X$; and let $\mathcal{S}$ be the relation over P introduced as

$$\mathcal{S}(i, x) = \{i + 1\} \times \mathcal{R}_i(x), \quad (i, x) \in P.$$

It will suffice applying (DC) to $(P, \mathcal{S})$ and $b := (0, a) \in P$ to get the conclusion in the statement; we do not give details.

Summing up, (DDC) is provable in (ZF-AC+DC). This is valid as well for its variant, referred to as: the *Selected Dependent Choice* principle (in short: (SDC)).

Proposition 4. *Let the map $F : N \to \exp(X)$ and the relation $\mathcal{R}$ over X fulfill*

$$(\forall n \in N) : \mathcal{R}(x) \cap F(n + 1) \neq \emptyset, \quad \forall x \in F(n).$$

Then, for each $a \in F(0)$, there exists a sequence $(x(n); n \geq 0)$ in X, with

$$x(0) = a, \; x(n) \in F(n), \; x(n + 1) \in \mathcal{R}(x(n)), \; \forall n.$$

As before, (SDC) $\Longrightarrow$ (DC) ($\Longleftrightarrow$ (DDC)); just take $(F(n) = X, n \geq 0)$. But, the reciprocal is also true, in the sense: (DDC) $\Longrightarrow$ (SDC). This follows from

Proof. (Proposition 4). Let the premises of (SDC) be true. Define a sequence of relations $(\mathcal{R}_n; n \geq 0)$ over X as: for each $n \geq 0$,

$$\mathcal{R}_n(x) = \mathcal{R}(x) \cap F(n+1), \text{ if } x \in F(n),$$
$$\mathcal{R}_n(x) = \{x\}, \text{ otherwise } (x \in X \setminus F(n)).$$

Clearly, $\mathcal{R}_n$ is proper for all $n \geq 0$. So, by (DDC), we have that, for the starting $a \in F(0)$, there exists an $(a, (R_n; n \geq 0))$-iterative sequence $(x(n); n \geq 0)$ in X. Combining with the very definition above gives the conclusion. □

In particular, when $\mathcal{R} = X \times X$, the regularity condition imposed in (SDC) holds. The corresponding variant of the underlying statement is just (AC(N)) (=the *Denumerable Axiom of Choice*). Precisely, we have

Proposition 5. *Let $F : N \to \exp(X)$ be a function. Then, for each $a \in F(0)$ there exists a function $f : N \to X$ with $f(0) = a$ and $(f(n) \in F(n), \forall n)$.*

As a consequence of these, (DC) $\Longrightarrow$ (AC(N)) in (ZF-AC). A direct verification is obtainable by an application of (DC) to the relational structure $(Q, \mathcal{T})$, where

$$Q = N \times X, \ \mathcal{T}(n, x) = \{n+1\} \times F(n+1), \ \ n \in N, x \in X;$$

we do not give details. The reciprocal inclusion is not true; see, for instance, Moskhovakis [28, Ch 8, Sect 8.25].

3. Pseudometric Structures

In the following, some preliminary facts about convergent and Cauchy sequences in pseudometric spaces are being discussed.

(A) Let X be a non-empty set. By a *pseudometric* over X we mean any map $d : X \times X \to R_+$. For an easy reference, we list the conditions to be used:

(ref) d is *reflexive*: $d(x, x) = 0$, for each $x \in X$,
(suf) d is *sufficient*: $x, y \in X$, $d(x, y) = 0$ imply $x = y$,
(F-tri) d is *Frink triangular*:
$\forall \varepsilon > 0, \exists \delta > 0$, such that: $d(x, y), d(y, z) < \delta$ imply $d(x, z) < \varepsilon$,

(B-tri) d is *Bakhtin triangular*:
there exists $\beta \geq 1$, such that $d(x,z) \leq \beta(d(x,y)+d(y,z))$, $\forall x,y,z \in X$,
(tri) d is *triangular*: $d(x,z) \leq d(x,y) + d(y,z)$, $\forall x,y,z \in X$,
(sym) d is *symmetric*: $d(x,y) = d(y,x)$, for all $x,y \in X$.

For the moment, we only assume that the reflexive property holds. Then d is called a *r-pseudometric* on X; and (X,d) is referred to as a *r-pseudometric space.*

Define a d-convergence and a d-Cauchy structure on X as follows. Let us say that the sequence (x_n) in X, *d-converges* to $x \in X$ (and write: $x_n \xrightarrow{d} x$) iff $d(x_n,x) \to 0$ as $n \to \infty$; that is

$\forall \varepsilon > 0$, $\exists p = p(\varepsilon)$, $\forall n$: $(p \leq n \Longrightarrow d(x_n,x) < \varepsilon)$; or, equivalently:
$\forall \varepsilon > 0$, $\exists p = p(\varepsilon)$, $\forall n$: $(p \leq n \Longrightarrow d(x_n,x) \leq \varepsilon)$.

Then x is called a *d-limit* of (x_n); the set of all these will be denoted as $d - \lim_n(x_n)$ [or $\lim_n(x_n)$, when d is understood]; if such elements exist, we say that (x_n) is *d-convergent.* Clearly, the introduced convergence $(\xrightarrow{d})$ has the properties

(conve-1) ($(\xrightarrow{d})$ is *hereditary*)
if $x_n \xrightarrow{d} x$, then $y_n \xrightarrow{d} x$, for each subsequence (y_n) of (x_n)
(conve-2) ($(\xrightarrow{d})$ is *reflexive*) for each $u \in X$,
the constant sequence $(x_n = u; n \geq 0)$ fulfills $x_n \xrightarrow{d} u$,

so, it fulfills the general requirements in Kasahara [31].

Let us say that the sequence $(x_n; \geq 0)$ in X is *d-Cauchy,* when $d(x_m,x_n) \to 0$ as $m,n \to \infty$, $m < n$; i.e.

$\forall \varepsilon > 0$, $\exists q = q(\varepsilon)$, $\forall (m,n)$: $(q \leq m < n \Longrightarrow d(x_m,x_n) < \varepsilon)$;
or, equivalently:
$\forall \varepsilon > 0$, $\exists q = q(\varepsilon)$, $\forall (m,n)$: $(q \leq m < n \Longrightarrow d(x_m,x_n) \leq \varepsilon)$.

The class of all these will be indicated as $Cauchy(d)$; some basic properties of it are described as follows:

(Cauchy-1) ($Cauchy(d)$ is *hereditary*)
(x_n) is d-Cauchy implies (y_n) is d-Cauchy,
for each subsequence (y_n) of (x_n)
(Cauchy-2) ($Cauchy(d)$ is *reflexive*) for each $u \in X$,
the constant sequence $(x_n = u; n \geq 0)$ is d-Cauchy,

so, this concept fulfills the general requirements in Turinici [32].

Finally, call the couple $((\xrightarrow{d}), Cauchy(d))$, a *conv-Cauchy structure* induced by d. The following optional conditions about this structure are to be considered:

(conv-Cauchy-1) d is *regular*:
each d-convergent sequence in X is d-Cauchy
(conv-Cauchy-2) d is *complete*:
each d-Cauchy sequence in X is d-convergent.

Note that, by the ambient r-pseudometric setting, neither of these properties is attainable, in general; we do not give details.

(B) Let (X, d) be an r-pseudometric space. In the following, an adherence concept over $\exp[X]$ will be introduced, as well as its associated closure operator.

Given $Y \in \exp[X]$ and $w \in X$, let us define a couple of concepts according to the lines below:

(adhe-1) w is *d-sequentially-adherent* to Y when
$w \in d - \lim_n(y_n)$, for some sequence (y_n) of Y
(adhe-2) w is *d-spherically-adherent* to Y when
for each $\varepsilon > 0$, $Y(\varepsilon, w) := \{y \in Y; d(y, w) < \varepsilon\}$ is non-empty.

Proposition 6. *We have, in (ZF-AC+DC), for Y and w as before*

d-sequentially-adherent is equivalent with d-spherically adherent.

Proof. The left to right inclusion is clear, by the very definition of convergence. Concerning the right to left inclusion, let $(\varepsilon_n; n \geq 0)$ be a strictly descending sequence in R_+^0 with $\varepsilon_n \to 0$ as $n \to \infty$; and put $(G(n) = Y(\varepsilon_n, w); n \geq 0)$; this is a mapping from N to $\exp(X)$. By the Denumerable Axiom of Choice (AC(N)) (deductible in (ZF-AC+DC)), there exists a selective function $g : N \to X$ with $(g(n) \in G(n), \forall n)$. It is now clear that $(y_n = g(n); n \geq 0)$ is a sequence in Y with $y_n \xrightarrow{d} w$ as $n \to \infty$; so that (adhe-1) follows. □

By definition, the class of all *d-adherent* points $w \in X$ (in the sense of (adhe-1) or (adhe-2)) will be denoted as $\mathrm{cl}(Y)$; and referred to as the *closure* (or, *adherence*) of Y. The selfmap $Y \mapsto \mathrm{cl}(Y)$ of $\exp[X]$ is a semi-closure over X, in the sense

(cl-1) $\text{cl}(\emptyset) = \emptyset$, $\text{cl}(X) = X$,
(cl-2) $\text{cl}(U \cup V) = \text{cl}(U) \cup \text{cl}(V)$, for each $U, V \in \exp[X]$
(cl-3) $Y \subseteq \text{cl}(Y)$, for each $Y \in \exp[X]$.

Unfortunately, $Y \mapsto \text{cl}(Y)$ is not involutive; i.e.

(cl-inv) $\text{cl}(\text{cl}(Y)) = \text{cl}(Y)$, for each $Y \in \exp[X]$

is not in general true; so that, $Y \mapsto \text{cl}(Y)$ is not a closure over X as in Kuratowski [33, Ch I, Sect 4]. Further, let us say that $Y \in \exp[X]$ is *d-closed*, provided

the d-limit of each sequence in Y is included in Y.

It is not hard to see that

(cl-char) (for each $Y \in \exp[X]$): Y is d-closed iff $Y = \text{cl}(Y)$.

In fact, the right to left inclusion is clear, by definition. For the left to right inclusion, let $w \in \text{cl}(Y)$ be arbitrary fixed. By definition,

$w \in d - \lim_n(y_n)$, for some sequence (y_n) of Y,

and this, by the imposed condition, yields $w \in Y$. By the arbitrariness of our underlying point, one gets

$Y \supseteq \text{cl}(Y)$, hence, $Y = \text{cl}(Y)$,

proving our assertion.

Returning to the involutive question above, we may ask under which extra conditions upon our pseudometric is this property valid. The most general ones are

(r-s) d is *reflexive sufficient*: $d(x, y) = 0$ iff $x = y$
(F-tri) d is *Frink triangular*:
$\forall \varepsilon > 0$, $\exists \delta > 0$, such that $d(x, y), d(y, z) < \delta$, imply $d(x, z) < \varepsilon$.

We then say that d is a *Frink quasi-metric* on X; and (X, d) is referred to as a *Frink quasi-metric space*.

Proposition 7. *Let (X, d) be a Frink quasi-metric space. Then, the following properties are valid:*

(32-1) *the selfmap $Y \mapsto \text{cl}(Y)$ of $\exp[X]$ is involutive:*
$\text{cl}(\text{cl}(Y)) = \text{cl}(Y)$, for all $Y \in \exp[X]$
(32-2) *the selfmap $Y \mapsto \text{cl}(Y)$ of $\exp[X]$ is a closure operator over X, as in Kuratowski* [33, *Ch* I, *Sect* 4].

Proof. Clearly, it will suffice proving the former one; because the latter one follows at once by means of the stated properties of closure operator. The right to left inclusion is clear; so, it remains to establish the left to right inclusion. Let $w \in \mathrm{cl}(\mathrm{cl}(Y))$ be arbitrary fixed. Given $\varepsilon > 0$, let $\delta > 0$ be the number assured by the Frink triangular property. By the spherical definition of closure, there exists $v \in \mathrm{cl}(Y)$ such that $d(v, w) < \delta$. Further, again by the underlying definition, there exists $u \in Y$ with $d(u, v) < \delta$. This, via Frink triangular inequality, gives $d(u, w) < \varepsilon$. Combining with the arbitrariness of $\varepsilon > 0$, we derive $w \in \mathrm{cl}(Y)$; wherefrom (as w is arbitrarily chosen, too) $\mathrm{cl}(\mathrm{cl}(Y)) \subseteq \mathrm{cl}(Y)$; as desired. □

(C) Let X be a non-empty set. By a *quasi-metric* over X we shall mean any mapping $e : X \times X \to R_+$ with the properties

(qm-1) e is *reflexive sufficient*: $e(x, x) = 0$ iff $x = y$,
(qm-2) e is *triangular*: $e(x, z) \leq e(x, y) + e(y, z)$, $\forall x, y, z \in X$,

in this case, the couple (X, e) will be referred to as a *quasi-metric space*. Clearly, any quasi-metric is a Frink quasi-metric.

Technically speaking, a quasi-metric has all properties of a metric, except symmetry; hence, in particular, any metric is a quasi-metric. The converse is not in general true; as shown by

Example 1.
Put $X = \{1, 2, 3\}$; and define the mapping $e : X \times X \to R_+$ as

$$e(1,1) = e(2,2) = e(3,3) = 0;$$
$$e(1,2) = e(1,3) = 1,\ e(2,3) = 2;\ e(2,1) = e(3,1) = e(3,2) = 1.$$

Clearly, $e(.,.)$ is a quasi-metric on X. On the other hand, $e(.,.)$ is not a metric on X; because $e(2, 3) = 2$, $e(3, 2) = 1$.

Finally, note that, given a quasi-metric e on X, its *conjugate map*

$$(e^* : X \times X \to R_+) \colon e^*(x, y) = e(y, x),\ x, y \in X$$

is a quasi-metric too. On the other hand, the associated (to e and e^*) mapping $d : X \times X \to R_+$, given as

$$d(x, y) = \max\{e(x, y), e^*(x, y)\},\ x, y \in X \text{ (in short: } d = max(e, e^*)),$$

is a symmetric quasi-metric; hence, a (standard) metric on X.

The concept of quasi-metric seems to have a pretty long tradition in metrical spaces theory. For example, in his 2001 PhD Thesis, Hitzler [34,

Ch 1, Sect 1.2] introduced this notion as a useful tool for the topological study of logic semantics. Later, in a 2004 paper, Turinici [35] used the same concept [referred to as: *reflexive triangular sufficient pseudometric*] with the aim of establishing a Caristi–Kirk fixed point theorem over such structures. Further aspects may be found in Cobzaş [36].

Suppose that we introduced a quasi-metric $e(.,.)$ (on X). The natural e-convergence and e-Cauchy structures over X were already defined. It is our aim in the following to discuss a lot of specific properties appearing in this context.

(I) Let us say that e is *separated*, when

(qm-sep) $(\overset{e}{\longrightarrow})$ is *separated*:
$e - \lim_n(x_n)$ is an asingleton, for each sequence $(x_n; n \geq 0)$ in X.

In this case, note that — given the e-convergent sequence (x_n) — we must have

$\lim_n(x_n) = \{z\}$ (written as: $\lim_n(x_n) = z$), for some $z \in X$.

Remark 1. The separated property of a quasi-metric e includes its sufficiency. In fact, let $u, v \in X$ be such that $e(u, v) = 0$. The constant sequence $(x_n = u; n \geq 0)$ fulfills $x_n \overset{e}{\longrightarrow} u$, $x_n \overset{e}{\longrightarrow} v$; so that (by the separated property), $u = v$. On the other hand, when, in addition,

e is symmetric (hence, e is a metric on X)

the separated property holds. In fact, let (x_n) be a sequence in X with $x_n \overset{d}{\longrightarrow} u$, $x_n \overset{d}{\longrightarrow} v$. As d is triangular and symmetric,

$d(u, v) \leq d(x_n, u) + d(x_n, v)$, $\forall n$,

so, by a limit process, $d(u, v) = 0$; and this, via $d =$ sufficient, yields $u = v$.

(II) Let us say that e is *first variable continuous*, provided

(fv-cont) for each $y \in X$, the map $x \mapsto e(x, y)$
is (sequentially) continuous: $x_n \overset{e}{\longrightarrow} x$ implies $e(x_n, y) \to e(x, y)$.

For example, if $e =$ metric, this condition is fulfilled. In fact, a stronger property holds; as results from

Proposition 8. *Suppose that $d(.,.)$ is a metric on X. Then,*
(33-1) *The mapping $(x, y) \mapsto d(x, y)$ is d-Lipschitz, in the sense*

$|d(x, y) - d(u, v)| \leq d(x, u) + d(y, v)$, $\forall (x, y), (u, v) \in X \times X$.

(33-2) *As a consequence, this map is d-continuous; i.e.*

$x_n \xrightarrow{d} x$, $y_n \xrightarrow{d} y$ *imply* $d(x_n, y_n) \to d(x, y)$.

The proof is immediate, by the properties of $d(.,.)$; so, we do not give details.

Remark 2. Concerning the introduced convention, let us consider the following (apparently weaker) form of it

(fv-lsc) e is *first variable lsc*: for each $y \in X$, the map $x \mapsto e(x, y)$ is (sequentially) e-lsc: $x_n \xrightarrow{e} x$ implies $\liminf_n e(x_n, y) \geq e(x, y)$.

Note that, in many concrete circumstances to be considered, this variant works; so, it is legitimate to ask what is the reason for introducing its strong version. The answer to this is very simple: by the conditions imposed upon $e(.,.)$, the weaker form in question is equivalent with its initial (strong) counterpart. In fact, assume that e is first variable e-lsc; and fix some $y \in X$. Given the sequence (x_n) in X and the point $x \in X$ with $x_n \xrightarrow{e} x$, we have (by the triangular inequality)

$e(x_n, y) \leq e(x_n, x) + e(x, y)$, for all n;

wherefrom (as $e(x_n, x) \to 0$ as $n \to \infty$)

$\limsup_n e(x_n, y) \leq e(x, y)$ (as $n \to \infty$).

Combining with the working assumption about e, gives

$e(x, y) \leq \liminf_n e(x_n, y) \leq \limsup_n e(x_n, y) \leq e(x, y)$;
or, equivalently: $\lim_n e(x_n, y) = e(x, y)$;

which tells us that $e(.,.)$ is first variable continuous. Hence, passing from (fv-cont) to (fv-lsc) does not bring any generality relative to the precise framework.

Finally, we show by an example — patterned after Alsulami *et al.* [37] — that such extra conditions imposed upon quasi-metrics are effective.

Example 2. Let $X = R_+$; and $(g(x, y) = |x - y|; x, y \in X)$ stand for the usual metric over it. Further, let $\beta \in]0, 1[$ be fixed in the sequel. Define a mapping $e : X \times X \to R_+$ as

$(e(x, y) = \beta g(x, y)$, if $x < y)$, $(e(x, y) = g(x, y)$, if $x \geq y)$.

Clearly, $e(.,.)$ is a quasi-metric over X. However, it is not a metric; because, e.g.

$$e(1,2) = \beta < 1 = e(2,1).$$

Nevertheless, its convergence structure is equivalent with the one induced by the usual metric, in view of

$$\beta g(x,y) \leq e(x,y) \leq g(x,y),\ \forall x,y \in X.$$

In fact, as a direct consequence of this,

$$(\forall (x_n),\ \forall x)\colon x_n \xrightarrow{e} x \text{ iff } x_n \xrightarrow{g} x;$$

wherefrom, e is separated. Moreover, for each $y \in X$, the partial map (in $\mathcal{F}(X, R_+)$) $\varphi_y := e(., y)$ defined as

$$(\varphi_y(x) = \beta g(x,y), \text{ if } x < y),\ \ (\varphi_y(x) = g(x,y), \text{ if } x \geq y)$$

is continuous with respect to the usual convergence structure of X; hence (by the above), with respect to the convergence structure attached to e; so that, $e(.,.)$ is first variable (sequentially) continuous.

(D) In the following, a relative-type construction is proposed for obtaining a large class of quasi-metrics. Let the locally Riemann integrable function $b : R_+ \to R_+^0$ be given; and $B : R_+ \to R_+$ stand for its *primitive*:

$$B(t) = \int_0^t b(s)ds,\ t \in R_+.$$

Suppose in the following that

(norm) (b, B) is *normal*: b is decreasing and $B(\infty) = \infty$.

In particular, we have the (integral) representation

(int-rep) $\int_p^q b(\xi)d\xi = (q-p)\int_0^1 b(p+\tau(q-p))d\tau$, when $0 \leq p < q < \infty$.

Some basic facts involving this couple are being collected in

Theorem 1. *The following are valid*

(31-a) *B is a continuous order isomorphism of R_+; hence, so is B^{-1}*
(31-b) $b(s) \leq (B(s) - B(t))/(s-t) \leq b(t), \forall t, s \in R_+, t < s$
(31-c) *B is almost concave:*
$t \mapsto [B(t+s) - B(t)]$ *is decreasing on* R_+, $\forall s \in R_+$

(31-d) *B is concave*: $B(t+\lambda(s-t)) \geq B(t)+\lambda(B(s)-B(t))$, *for all $t, s \in R_+$ with $t < s$ and all $\lambda \in [0,1]$*

(31-e) *B is sub-additive (hence, B^{-1} is super-additive).*

The proof is immediate, by (norm) above; so, we do not give details. Note that the properties in (31-c) and (31-d) are equivalent to each other, under (31-a). This follows at once from the (non-differential) mean value theorem in Bantaş and Turinici [38].

Now, let X be some non-empty set; and $d : X \times X \to R_+$ be a quasi-metric over it. Further, let $\Gamma : X \to R_+$ be chosen as

(a-nexp) Γ is *almost d-non-expansive*: $\Gamma(x) - \Gamma(y) + d(x,y) \geq 0, \forall x, y \in X$.

Define a pseudometric $e := [d; B, \Gamma]$ over X as

(e-expli) $e(x,y) = B(\Gamma(x) + d(x,y)) - B(\Gamma(x))$, $x, y \in X$.

This may be viewed as an "explicit" formula; the "implicit" version of it is

(e-impli) $d(x,y) = B^{-1}(B(\Gamma(x)) + e(x,y)) - \Gamma(x)$, $x, y \in X$.

We will establish some properties of this map, useful in the sequel.

First, the "metrical" nature of $(x,y) \mapsto e(x,y)$ is of interest.

Theorem 2. *The mapping $(x,y) \mapsto e(x,y)$ is a quasi-metric over X.*

Proof. The reflexivity and sufficiency are clear by Theorem 1 (first part); so, it remains to establish the triangular property. Let $x, y, z \in X$ be arbitrary fixed. The triangular property of $d(.,.)$ yields [via Theorem 1 (first part)]

$$e(x,z) \leq B(\Gamma(x) + d(x,y) + d(y,z)) - B(\Gamma(x) + d(x,y)) + e(x,y).$$

On the other hand, the almost d-non-expansiveness of Γ gives

$$\Gamma(x) + d(x,y) \geq \Gamma(y); \text{ so (by Theorem 1 (third part))}$$
$$B(\Gamma(x) + d(x,y) + d(y,z)) - B(\Gamma(x) + d(x,y)) \leq e(y,z).$$

Combining with the previous relation yields our desired conclusion. □

By definition, e will be called the *Zhong metric* attached to d and the couple (B, Γ). The following properties of (d, e) are immediate (via Theorem 1):

Proposition 9. *Under the prescribed conventions,*

(34-1) $b(\Gamma(x)+d(x,y))d(x,y) \le e(x,y) \le b(\Gamma(x))d(x,y)$, $\forall x,y \in X$
(34-2) $e(x,y) \le B(d(x,y))$, $\forall x,y \in X$; *hence,* $x_n \xrightarrow{d} x$ *implies* $x_n \xrightarrow{e} x$.

A basic property of $e(.,.)$ to be checked is d-compatibility. Given the couple (d,e) of quasi-metrics over X, we say that e is *d-compatible* provided

(comp-1) each e-Cauchy sequence is d-Cauchy, too
(comp-2) $y \mapsto e(x,y)$ is d-lsc, for each $x \in X$.

Theorem 3. *The Zhong metric $e(.,.)$ is d-compatible (see above).*

Proof. (i) We firstly check (comp-2); which may be written as

$$(e(x,y_n) \le \lambda, \forall n) \text{ and } y_n \xrightarrow{d} y \text{ imply } e(x,y) \le \lambda.$$

So, let x, (y_n), λ and y be as in the premise of this relation. By Proposition 9, we have $y_n \xrightarrow{e} y$ as $n \to \infty$. Moreover (as e is triangular)

$$e(x,y) \le e(x,y_n) + e(y_n,y) \le \lambda + e(y_n,y), \text{ for all } n.$$

It will suffice passing to limit as $n \to \infty$ to get the desired conclusion.

(ii) Further, we claim that (comp-1) holds, too, in the sense:

(for each sequence) d-Cauchy $\Longleftrightarrow$ e-Cauchy.

The left to right implication is clear, via Proposition 9. For the right to left one, assume that (x_n) is an e-Cauchy sequence in X. As e = triangular,

$$e(x_i,x_j) \le \mu, \text{ for all } (i,j) \text{ with } i \le j, \text{ and some } \mu \ge 0.$$

This, along with the implicit definition of e, yields

$$d(x_0,x_i) = B^{-1}(B(\Gamma(x_0)) + e(x_0,x_i)) - \Gamma(x_0) \le$$
$$B^{-1}(B(\Gamma(x_0)) + \mu) - \Gamma(x_0), \ \forall i \ge 0;$$

wherefrom (by the choice of Γ)

$$\Gamma(x_i) \le \Gamma(x_0) + d(x_0,x_i) \le B^{-1}(B(\Gamma(x_0)) + \mu)$$
$$(\text{hence, } B(\Gamma(x_i)) \le B(\Gamma(x_0)) + \mu), \text{ for all } i \ge 0.$$

Putting these facts together yields (again via implicit definition of e)

$$\Gamma(x_i) + d(x_i,x_j) = B^{-1}(B(\Gamma(x_i)) + e(x_i,x_j)) \le$$
$$\nu := B^{-1}(B(\Gamma(x_0)) + 2\mu), \text{ for all } (i,j) \text{ with } i \le j.$$

And this, via Proposition 9, gives (for the same pairs (i,j))

$$e(x_i,x_j) \geq b(\Gamma(x_i)+d(x_i,x_j))d(x_i,x_j) \geq b(\nu)d(x_i,x_j).$$

But then, the d-Cauchy property of (x_n) is clear; and the proof is complete. □

Note finally that this relative construction $d \mapsto [d;B,\Gamma]$ of quasi-metrics may be viewed as an iterative process of generating such objects; and yields a lot of interesting problems. But, for the moment, this will suffice.

4. Local Maximality Principles

Let X be a non-empty set. By a *pseudometric* over X we mean any map $d: X\times X \to R_+$. For an easy reference, we list the conditions to be used in the sequel:

(ref) d is *reflexive*: $d(x,x)=0$, for each $x\in X$,
(suf) d is *sufficient*: $x,y\in X$, $d(x,y)=0$ imply $x=y$,
(F-tri) d is *Frink triangular*:
$\forall \varepsilon>0$, $\exists\delta>0$, such that: $d(x,y), d(y,z)<\delta$ imply $d(x,z)<\varepsilon$,
(B-tri) d is *Bakhtin triangular*:
there exists $\beta\geq 1$, such that $d(x,z)\leq \beta(d(x,y)+d(y,z))$, $\forall x,y,z\in X$,
(tri) d is *triangular*: $d(x,z)\leq d(x,y)+d(y,z)$, $\forall x,y,z\in X$,
(sym) d is *symmetric*: $d(x,y)=d(y,x)$, for all $x,y\in X$.

(A) For the moment, we only assume that the reflexive property holds. Then d will be called a *r-pseudometric* on X; and (X,d) will be referred to as a *r-pseudometric space*. Further, let $(\preceq)$ be a *quasi-order* (i.e. reflexive and transitive relation) over X; the triple $(X,d,\preceq)$ will be referred to as a *quasi-ordered r-pseudometric space*. Given $M\in \exp(X)$, call $z\in X$,

(max-1) $(d,\preceq)$-*maximal* over M, if $(u,v\in M,\ z\preceq u\preceq v) \Longrightarrow d(u,v)=0$,
(max-2) $(\preceq)$-*maximal* over M, if $(u\in M,\ z\preceq u) \Longrightarrow d(z,u)=0$.

Note that, under pretty general conditions upon d, these properties are somehow equivalent. Precisely, suppose that

(r-s) d is *reflexive sufficient*: $d(x,y)=0$ iff $x=y$,

we then say that d is a *rs-pseudometric* over X.

Proposition 10. *Let the quasi-ordered rs-pseudometric space* $(X, d, \preceq)$ *be given. Then the following equivalence is true*

for each $Y \in \exp(X)$ *and each* $z \in Y$:
z *is* $(d, \preceq)$*-maximal over* Y *iff* z *is* $(\preceq)$*-maximal over* Y.

Proof. The left to right inclusion follows via

$$z \preceq w \in Y \Longrightarrow z \preceq z \preceq w \text{ (and } z, w \in Y)$$
$$\Longrightarrow d(z, w) = 0 \Longrightarrow z = w \text{ (as } d \text{ is rs-pseudometric).}$$

On the other hand, the right to left inclusion is immediate, by the very definition of $(\preceq)$-maximal element, combined with

$$(z \preceq u \preceq v,\ u, v \in Y) \Longrightarrow (z \preceq u \in Y) \text{ and } (z \preceq v \in Y)$$
$$\Longrightarrow (z = u \text{ and } z = v) \Longrightarrow u = v \Longrightarrow (d(u, v) = 0).$$

This ends the argument. □

Returning to the r-pseudometric context, let us remember that the sequence (x_n) in X is called *d-Cauchy* when $d(x_m, x_n) \to 0$ as $m, n \to \infty$, $m < n$; that is

$\forall\varepsilon > 0, \exists n(\varepsilon)$, such that $n(\varepsilon) \le p \le q \Longrightarrow d(x_p, x_q) \le \varepsilon$;
or, equivalently (passing to $n(\varepsilon) + 1$)
$\forall\varepsilon > 0, \exists n(\varepsilon)$, such that $n(\varepsilon) < p < q \Longrightarrow d(x_p, x_q) \le \varepsilon$.

Further, let us say that (x_n) is *d-asymptotic*, if $\lim_n d(x_n, x_{n+1}) = 0$; that is

$\forall\varepsilon > 0, \exists n(\varepsilon)$, such that $n(\varepsilon) \le p \Longrightarrow d(x_p, x_{p+1}) \le \varepsilon$.

Then, for each $M \in \exp(X)$, let us consider the global conditions

(C-reg) $(M, d, \preceq)$ is *Cauchy regular*:
each $(\preceq)$-ascending sequence in M is d-Cauchy
(A-reg) $(M, d, \preceq)$ is *asymptotic regular*:
each $(\preceq)$-ascending sequence in M is d-asymptotic.

As each d-Cauchy sequence is d-asymptotic, too, it follows that (C-reg) $\Longrightarrow$ (A-reg). The reverse implication also holds, in the sense

Proposition 11. *We have in (ZF-AC), for each* $M \in \exp(X)$,

(A-reg) $\Longrightarrow$ *(C-reg); whence, (A-reg)* $\Longleftrightarrow$ *(C-reg).*

Proof. Suppose that (A-reg) holds; but, for some $(\preceq)$-ascending (x_n), the d-Cauchy property fails; i.e. (for some $\varepsilon > 0$)

$$C(n) = \{(p, q) \in N \times N; n < p < q, d(x_p, x_q) > \varepsilon\} \neq \emptyset, \forall n.$$

Denote, for simplicity

$$p(n) = \min \mathrm{Dom}(C(n)),\ q(n) = \max(C(n)(p(n)),\ n \in N,$$

clearly, no choice techniques are used in this construction. Fix some rank $i(0)$. By this assumption, there exist $i(1) = p(i(0))$, $i(2) = q(i(0))$ with $i(0) < i(1) < i(2)$, $d(x_{i(1)}, x_{i(2)}) > \varepsilon$. Further, given the rank $i(2)$, there exist $i(3) = p(i(2))$, $i(4) = q(i(2))$ with $i(2) < i(3) < i(4)$, $d(x_{i(3)}, x_{i(4)}) > \varepsilon$; and so on. By induction, we get a $(\preceq)$-ascending subsequence $(y_n = x_{i(n)})$ of (x_n) with $d(y_{2n+1}, y_{2n+2}) > \varepsilon$, for all n. This contradicts (A-reg); hence, the claim. □

Having these precise, define for each $M \in \exp(X)$,

$(M, d, \preceq)$ is *regular*, when it fulfills
one of the equivalent properties (C-reg) or (A-reg).

A strongly related notion with respect to this one is

$(M, d, \preceq)$ is weakly regular: $\forall x \in M, \forall \varepsilon > 0, \exists y = y(x, \varepsilon) \in M(x, \preceq)$,
such that: $(u, v \in M,\ y \preceq u \preceq v) \Longrightarrow d(u, v) \leq \varepsilon$.

The connection between these is discussed in

Proposition 12. *We have in (ZF-AC+DC), for each* $M \in \exp(X)$

$(M, d, \preceq)$ *is regular implies* $(M, d, \preceq)$ *is weakly regular.*

Proof. Assume this would be false; that is (for some $x \in M$, $\varepsilon > 0$)

for each $y \in M(x, \preceq)$ there exist $u, v \in M$ with $y \preceq u \preceq v$, $d(u, v) > \varepsilon$.

This, by definition, yields (for the same (x, ε)):

$$\forall y \in M(x, \preceq),\ \exists\ (u, v) \in (\preceq):\ \ y \preceq u,\ d(u, v) > \varepsilon,$$

where $(\preceq) := \{(a, b) \in M \times M; a \preceq b\}$. Put $Q := \{(a, b) \in (\preceq); x \preceq a\}$; and fix $(y_0, y_1) \in Q$; for example, $y_0 = y_1 = x$. Define a relation $\mathcal{R} = \mathcal{R}(\varepsilon)$ on Q as

$$(a_1, b_1)\mathcal{R}(a_2, b_2) \text{ if and only if } b_1 \preceq a_2,\ d(a_2, b_2) > \varepsilon.$$

From the imposed condition, $Q((a,b),\mathcal{R}) \neq \emptyset$, $\forall (a,b) \in Q$. So, by (DC), it follows that, for the starting point $w_0 = (y_0, y_1)$ in Q there exists a sequence $(w_n := (y_{2n}, y_{2n+1}); n \geq 0)$ in Q, with

$$w_n \mathcal{R} w_{n+1}, \text{ for all } n; \text{ hence, by definition,}$$
$$y_{2n+1} \preceq y_{2n+2}, \quad d(y_{2n+2}, y_{2n+3}) > \varepsilon, \quad \text{for all } n.$$

As a consequence, $(y_n; n \geq 0)$ is $(\preceq)$-ascending and not d-asymptotic; in contradiction with the regularity of $(M, d, \preceq)$; hence, the claim. □

Returning to the regularity concept, it must be noted that it is a strong one; because, under a (relatively weak) rs-pseudometric condition upon d, it induces a strong property of our quasi-order over the ambient set.

Proposition 13. *Let the quasi-ordered rs-pseudometric space $(X, d, \preceq)$ and the (non-empty) subset M of X be such that $(M, d, \preceq)$ is regular. Then,*

$(\preceq)$ is antisymmetric (hence, a (partial) order) on M.

Proof. (*cf.* Hamel [39, Ch 4, Sect 4.1]). Let $u, v \in M$ be such that $u \preceq v$ and $v \preceq u$. The sequence $(y_{2n} = u,\ y_{2n+1} = v;\ n \geq 0)$ is $(\preceq)$-ascending; hence, d-asymptotic by hypothesis. This yields $d(u, v) = 0$; whence (as d is rs-pseudometric), $u = v$. □

(B) Let again $(X, d, \preceq)$ be a quasi-ordered r-pseudometric space. In the following, various convergence structures are defined, as well as their attached closure operators. Then, the relationships between these are discussed.

Define a d-convergence structure on X under the precise way. Precisely, let us say that the sequence (x_n) in X, *d-converges* to $x \in X$ (and write: $x_n \xrightarrow{d} x$) iff $d(x_n, x) \to 0$ as $n \to \infty$; that is

$$\forall \varepsilon > 0, \exists p = p(\varepsilon), \forall n\text{:} \ \ (p \leq n \Longrightarrow d(x_n, x) < \varepsilon); \text{ or, equivalently:}$$
$$\forall \varepsilon > 0, \exists p = p(\varepsilon), \forall n\text{:} \ \ (p \leq n \Longrightarrow d(x_n, x) \leq \varepsilon).$$

Then x is called a *d-limit* of (x_n); the set of all these will be denoted as $d - \lim_n(x_n)$ [or, $\lim_n(x_n)$, when d is understood]; if such elements exist, we say that (x_n) is *d-convergent*.

Given $Y \in \exp[X]$, let us say that $w \in X$ is a *$(d, \preceq)$-adherence* point of it when

$$w \in d - \lim_n(z_n), \text{ for some } (\preceq)\text{-ascending sequence } (z_n) \text{ of } Y;$$

the class of all these will be denoted as $\mathrm{cl}_{(\preceq)}(Y)$. It is not hard to see that $Y \mapsto \mathrm{cl}_{(\preceq)}(Y)$ is a semi-closure over X, in the sense

(docl-1) $\mathrm{cl}_{(\preceq)}(\emptyset) = \emptyset$, $\mathrm{cl}_{(\preceq)}(X) = X$,
(docl-2) $\mathrm{cl}_{(\preceq)}(U \cup V) = \mathrm{cl}_{(\preceq)}(U) \cup \mathrm{cl}_{(\preceq)}(V)$, $U, V \in \exp[X]$
(docl-3) $Y \subseteq \mathrm{cl}_{(\preceq)}(Y)$, $\forall Y \in \exp[X]$.

In fact, (docl-1) and (docl-3) are clear. Concerning (docl-2), let U, V be a couple of subsets of X; and let $w \in \mathrm{cl}_{(\preceq)}(U \cup V)$ be arbitrary fixed; hence,

$w \in d - \lim_n(x_n)$, for some $(\preceq)$-ascending sequence (x_n) of $U \cup V$.

By definition, there exists a subsequence $(y_n := x_{i(n)}; n \geq 0)$ of $(x_n; n \geq 0)$ with

(y_n) is an $(\preceq)$-ascending sequence in U or V.

This yields

$w \in \mathrm{cl}_{(\preceq)}(U)$ or $w \in \mathrm{cl}_{(\preceq)}(V)$, hence, $w \in \mathrm{cl}_{(\preceq)}(U) \cup \mathrm{cl}_{(\preceq)}(V)$,

proving our assertion. Unfortunately, $Y \mapsto \mathrm{cl}_{(\preceq)}(Y)$ is not involutive; i.e.

(docl-inv) $\mathrm{cl}_{(\preceq)}(\mathrm{cl}_{(\preceq)}(Y)) = \mathrm{cl}_{(\preceq)}(Y)$, for each $Y \in \exp[X]$,

is not in general true; so that, $Y \mapsto \mathrm{cl}_{(\preceq)}(Y)$ is not a closure over X as in Kuratowski [33, Ch I, Sect 4].

For each subset $Y \in \exp[X]$ let us introduce the concept

Y is $(d, \preceq)$-*closed*:
the d-limit of each $(\preceq)$-ascending sequence in Y is included in Y.

It is not hard to see that

(ocl-char) (for each $Y \in \exp[X]$): Y is $(d, \preceq)$-closed iff $Y = \mathrm{cl}_{(\preceq)}(Y)$.

In fact, the right to left inclusion is clear, by definition. For the left to right inclusion, let $w \in \mathrm{cl}_{(\preceq)}(Y)$ be arbitrary fixed. By definition,

$w \in d - \lim_n(z_n)$, for some $(\preceq)$-ascending sequence (z_n) of Y,

and this, by the imposed condition, yields $w \in Y$. By the arbitrariness of our underlying point, one gets

$Y \supseteq \mathrm{cl}_{(\preceq)}(Y)$; hence, $Y = \mathrm{cl}_{(\preceq)}(Y)$; proving our assertion.

Denote, for simplicity

$\mathcal{K}[d, \preceq]$=the class of all $(d, \preceq)$-closed subsets in X.

Some basic properties of this class are listed as follows:

(cotop-1) $\emptyset, X \in \mathcal{K}[d, \preceq]$
(cotop-2) the intersection of any subset in $\mathcal{K}[d, \preceq]$ belongs to $\mathcal{K}[d, \preceq]$
(cotop-3) the union of any finite subset of $\mathcal{K}[d, \preceq]$ is in $\mathcal{K}[d, \preceq]$.

The family $\mathcal{K}[d, \preceq]$ is therefore a *cotopology* on X, So, it may generate a closure operator, as: for each $M \in \exp[X]$,

$\mathrm{kl}_{(\preceq)}(M)$= the intersection of all $Y \in \mathcal{K}[d, \preceq]$ with $M \subseteq Y$.

The basic properties of this closure operator are concentrated in

Proposition 14. *Under these conventions, we have*
(45-1) *the selfmap* $Y \mapsto \mathrm{kl}_{(\preceq)}(Y)$ *is a closure operator, as in Kuratowski* [33, *Ch* I, *Sect* 4]
(45-2) *in addition, the following inclusion is valid:*

(cl-kl) $\mathrm{cl}_{(\preceq)}(M) \subseteq \mathrm{kl}_{(\preceq)}(M)$, *for each* $M \in \exp[X]$.

Proof. (i) We have to establish that the following are true:

(kl-1) $\mathrm{kl}_{(\preceq)}(\emptyset) = \emptyset$, $\mathrm{kl}_{(\preceq)}(X) = X$,
(kl-2) $\mathrm{kl}_{(\preceq)}(U \cup V) = \mathrm{kl}_{(\preceq)}(U) \cup \mathrm{kl}_{(\preceq)}(V)$, $U, V \in \exp[X]$
(kl-3) $Y \subseteq \mathrm{kl}_{(\preceq)}(Y)$, $\forall Y \in \exp[X]$.
(kl-4) $\mathrm{kl}_{(\preceq)}(\mathrm{kl}_{(\preceq)}(Y)) = \mathrm{kl}_{(\preceq)}(Y)$, for each $Y \in \exp[X]$.

The relations (kl-1) + (kl-3) are immediate; and (kl-4) follows in view of $\mathrm{kl}_{(\preceq)}(Y)$ being an element of $\mathcal{K}[d, \preceq]$. Concerning (kl-2), let $U, V \in \exp[X]$ be arbitrary fixed. By (kl-4),

$$U \subseteq \mathrm{kl}_{(\preceq)}(U),\ V \subseteq \mathrm{kl}_{(\preceq)}(V);\ \text{hence, } U \cup V \subseteq \mathrm{kl}_{(\preceq)}(U) \cup \mathrm{kl}_{(\preceq)}(V).$$

This, along with $\mathrm{kl}_{(\preceq)}(U) \cup \mathrm{kl}_{(\preceq)}(V)$ being an element of $\mathcal{K}[d, \preceq]$, gives

$$\mathrm{kl}_{(\preceq)}(U \cup V) \subseteq \mathrm{kl}_{(\preceq)}(U) \cup \mathrm{kl}_{(\preceq)}(V).$$

Since the reverse inclusion is also true, we are done.

(ii) Take $M \in \exp[X]$, and put $Y = \mathrm{kl}_{(\preceq)}(M)$; hence, $Y \in \mathcal{K}[d; \preceq]$ and $M \subseteq Y$. Given $w \in \mathrm{cl}_{(\preceq)}(M)$, there exists a $(\preceq)$-ascending sequence (x_n) in $M \subseteq Y$ with $x_n \xrightarrow{d} w$. This, by the properties of Y, gives $w \in Y$; wherefrom (by the arbitrariness of w), $\mathrm{cl}_{(\preceq)}(M) \subseteq Y$; proving the requested fact. □

Remark 3. The inclusion between these operators may be strict. This follows from the semi-closure operator $Y \mapsto \text{cl}_{(\preceq)}(Y)$ being not involutive, in general.

From a theoretical perspective, the closure operator $Y \mapsto \text{kl}_{(\preceq)}(Y)$ has an essential role in our $(\preceq)$-maximal results. However, under the described general setting, this operator cannot give us a consistent help in practice; because, for a fixed $Y \in \exp[X]$, the points of $\text{kl}_{(\preceq)}(Y)$ are not attainable from the points of Y. So, it is natural asking under which extra conditions upon d, this is somehow possible. As our previous developments suggest, d must be taken as an rs-pseudometric subjected to the scale of conditions

(F-tri) d is *Frink triangular*:
$\forall \varepsilon > 0$, $\exists \delta > 0$, such that: $d(x,y), d(y,z) < \delta$ imply $d(x,z) < \varepsilon$,
(B-tri) d is *Bakhtin triangular*: there exists $\beta \geq 1$, such that
$d(x,z) \leq \beta(d(x,y) + d(y,z))$, $\forall x,y,z \in X$.

We then say that $d(.,.)$ is a *Frink* (resp., *Bakhtin*) *quasi-metric* on X; and (X,d) is a *Frink* (resp., *Bakhtin*) *quasi-metric space.*

Proposition 15. *Let $(X,d,\preceq)$ be a quasi-ordered Frink (resp., Bakhtin) quasi-metric space. Then, the double inclusion is valid:*

(ocl-kl-cl) $\text{cl}_{(\preceq)}(M) \subseteq \text{kl}_{(\preceq)}(M) \subseteq \text{cl}(M)$, *for each* $M \in \exp[X]$.

Proof. The former inclusion is clear, by the above; so, it remains to establish the latter one. Let $M \in \exp[X]$ be arbitrary fixed. By a previous auxiliary fact,

$\text{cl}(M)$ is d-closed; hence, $(d,\preceq)$-closed.

This, under the definition of our closure operator, yields the needed conclusion. □

Returning to the r-pseudometric case, let us introduce a *quasi-order convergence* $(\xrightarrow{\preceq})$ over X as follows: given the sequence (x_n) in X and the point $x \in X$,

$$x_n \xrightarrow{\preceq} x \text{ if } \exists m(x) \in N, \text{ such that } n \geq m(x) \text{ implies } x_n \preceq x;$$

also referred to as: x is a $(\preceq)$*-limit* of (x_n). The class of all these will be denoted as $(\preceq) - \lim_n(x_n)$; when it is non-empty, we say that (x_n) is $(\preceq)$-*convergent.*

Remark 4. The following fact is useful in the sequel. Given the sequence (x_n) in X and the point $u \in X$, let us say that (x_n) is $(\preceq)$-*bounded* by u (or: $u \in X$ is an $(\preceq)$-*upper bound* of (x_n)), provided

$$(x_n \preceq x, \forall n); \text{ in short: } (x_n; n \geq 0) \preceq x.$$

The set of all such x will be denoted as $\text{ubd}(x_n)$ (the *upper bound* of (x_n)); if it is non-empty, we will say that (x_n) is $(\preceq)$-*bounded.*

Under the above conventions, we have, for each $(\preceq)$-ascending sequence (x_n) in X and each $x \in X$

(prop-1) $x_n \xrightarrow{\preceq} x$ iff (x_n) is $(\preceq)$-bounded by x (that is: $x \in \text{ubd}(x_n)$)
(prop-2) $(\preceq) - \lim_n(x_n)$ is identical with $\text{ubd}(x_n)$.

The proof is immediate, by the involved facts; so, we do not give details.

Finally, as a combination of these, define the *product* convergence structure $(\xrightarrow{d\preceq})$ over X, according to

$$x_n \xrightarrow{d\preceq} x \text{ when } x_n \xrightarrow{d} x \text{ and } x_n \xrightarrow{\preceq} x.$$

In this case, x is called a $(d \preceq)$-*limit* of (x_n); the set of all these will be denoted as $(d \preceq) - \lim_n(x_n)$; if such elements exist, we say that (x_n) is $(d \preceq)$-*convergent.*

Finally, let us say a few words about the completeness concepts to be considered here at the level of quasi-ordered r-pseudometric spaces $(M, d, \preceq)$, where $M \in \exp(X)$ is arbitrary fixed. The most useful for us are (with M as before)

(ful-com) $(M, d, \preceq)$ is *fully complete*:
each $(\preceq)$-ascending d-Cauchy sequence (x_n) in M is $(d \preceq)$-convergent: there exists $x \in X$ such that $x_n \xrightarrow{d} x$ and $x_n \xrightarrow{\preceq} x$ [that is: $(x_n \preceq x, \forall n)$],
(com) $(M, d, \preceq)$ is complete:
each $(\preceq)$-ascending d-Cauchy sequence (x_n) in M is d-convergent: there exists $x \in X$ such that $x_n \xrightarrow{d} x$.

To clarify the relationships between these, we need one more condition:

(c-dom) $(M, d, \preceq)$ is *conv-dominated*: the limit of each $(\preceq)$-ascending sequence in M is a $(\preceq)$-limit (hence, a $(\preceq)$-upper bound) of it.

Then, evidently,

(for each structure $(M, d, \preceq)$ as before):
(complete and conv-dominated) implies fully complete.

A sufficient condition for conv-dominated property writes (*cf.* Turinici [40])

$X(x, \preceq) := \{y \in X; x \preceq y\}$ is $(d, \preceq)$-closed, for each $x \in X$;
referred to as: $(\preceq)$ is d-selfclosed on X.

Since the verification is immediate, we omit giving further details.

(C) Under these preliminaries, we are now in position to state a lot of maximality statements with a practical meaning.

The first result in this series is a $(d, \preceq)$-maximal statement (referred to as: Kang-Park r-pseudometric maximal principle; in short: (KP-rp-mp)).

Theorem 4. *Let the quasi-ordered r-pseudometric space* $(X, d, \preceq)$ *and the subset* $M \in \exp(X)$ *be such that*

(41-i) $(M, d, \preceq)$ *is regular: each* $(\preceq)$*-ascending sequence in* M *is d-Cauchy*

holds, as well as one of the alternatives (41-ii)-(41-iii) *or* (41-iv), *where*

(41-ii) $(M, d, \preceq)$ *is complete:*
each $(\preceq)$*-ascending d-Cauchy sequence* (x_n) *in* M *is d-convergent*
(41-iii) $(M, d, \preceq)$ *is conv-dominated:*
the d-limit of each $(\preceq)$*-ascending sequence in* M *is a* $(\preceq)$*-upper bound of it*
(41-iv) $(M, d, \preceq)$ *is fully complete:*
each $(\preceq)$*-ascending d-Cauchy sequence* (x_n) *in* M *is* $(d \preceq)$*-convergent.*

Then, the following conclusion holds in (ZF-AC+DC)

$\forall u \in M$, $\exists v \in \mathrm{cl}_{(\preceq)}(M)$, *with* $u \preceq v$ *and* v *is* $(d, \preceq)$*-maximal over* M.

Proof. By regularity and an auxiliary fact above, $(M, d, \preceq)$ is weakly regular. Hence, given $u \in M$, one may construct a $(\preceq)$-ascending sequence (u_n) in M with

(41-a) $u \preceq u_0$, and $[(\forall n), (\forall y, z \in M)]$: $u_n \le y \le z \Longrightarrow d(y, z) \le 2^{-n}$.

In particular, this tells us that (u_n) is d-Cauchy (in M). Combining with the completeness hypothesis, there exists $v \in X$ such that

(41-b) $u_n \xrightarrow{d} v$; so that, $v \in \text{cl}_{(\preceq)}(M)$,

and this, by the conv-dominated property gives

(41-c) $u_n \xrightarrow{\preceq} v$; whence, $u_n \preceq v$, $\forall n$.

Alternatively, by the full completeness hypothesis, there exists $v \in X$ such that

(41-d) $u_n \xrightarrow{d\preceq} v$; whence, $u_n \xrightarrow{d} v$, $u_n \xrightarrow{\preceq} v$,

and (41-b)+(41-c) are again retainable. By (41-a)+(41-c), we get $u \preceq v$. On the other hand, let $y, z \in M$ be such that $v \preceq y \preceq z$. Then, by (41-a),

($\forall n$): $u_n \leq y \leq z$; whence, $d(y, z) \leq 2^{-n}$;

and, from this, $d(y, z) = 0$. Combining with the arbitrariness of our underlying couple, one derives that v is $(d, \preceq)$-maximal over M; as claimed. □

Formally, this result is comparable with the 1990 one in Kang and Park [41]. However, its basic lines were already set up in the papers by Turinici [42,43]. Note that, both these results extend the 1976 Brezis–Browder ordering principle [44]. Further aspects may be found in Altman [45].

The second result in this series is a $(\preceq)$-maximal statement (referred to as: Zorn–Bourbaki rs-pseudometric maximal principle (in short: (ZB-rsp-mp)).

Theorem 5. *Let the quasi-ordered rs-pseudometric space* $(X, d, \preceq)$ *and the subset* $Z \in \exp(X)$ *be such that*

(42-i) Z *is* $(d, \preceq)$*-closed (that is:* $Z = \text{cl}_{(\preceq)}(Z)$*)*
(42-ii) $(Z, d, \preceq)$ *is regular: each* $(\preceq)$*-ascending sequence in* Z *is* d*-Cauchy*

holds, as well one of the alternatives (42-iii)–(42-iv) *or* (42-v), *where*

(42-iii) $(Z, d, \preceq)$ *is complete:*
each $(\preceq)$*-ascending* d*-Cauchy sequence* (x_n) *in* Z *is* d*-convergent*
(42-iv) $(Z, d, \preceq)$ *is conv-dominated:*
the limit of each $(\preceq)$*-ascending sequence in* Z *is a* $(\preceq)$*-upper bound of it*
(42-v) $(Z, d, \preceq)$ *is fully complete:*
each $(\preceq)$*-ascending* d*-Cauchy sequence* (x_n) *in* Z *is* $(d \preceq)$*-convergent.*

Further, take the subset $M \subseteq Z$. Then, the following conclusion holds in (ZF-AC+DC)

$$\forall u \in M, \exists v \in Z, \text{ with } u \preceq v \text{ and } v \text{ is } (\preceq)\text{-maximal over } M.$$

Proof. By the admitted hypothesis (42-ii) and the alternatives (42-iii)–(42-iv) or (42-v) upon $(Z, d, \preceq)$, it is clear that the Kang–Park r-pseudometric maximal principle (KP-rp-mp) is applicable to these data. As a consequence of this, we have that, given the subset $M \subseteq Z$, the following conclusion holds in (ZF-AC+DC)

$$\forall u \in M \subseteq Z, \exists v \in \text{cl}_{(\preceq)}(Z), \text{ with } u \preceq v \text{ and } v \text{ is } (d, \preceq)\text{-maximal over } Z.$$

Taking (42-i) into account, we have (by the second part of this conclusion)

$v \in Z$ and v is $(d, \preceq)$-maximal over Z;
hence (by a previous auxiliary fact): v is $(\preceq)$-maximal over Z,

if we remember that d is rs-pseudometric. But then, via $M \subseteq Z$, it is clear that $v \in Z$ is $(\preceq)$-maximal in M; and the conclusion follows. □

The obtained Zorn–Bourbaki rs-pseudometric maximal principle (ZB-rsp-mp) has, essentially, a theoretical importance only. However, from a practical viewpoint, this result is not very useful — in this form — because the points of Z cannot be attained via points of M; so, we may ask whether this is possible in some way. It is our aim in the following to show that a positive answer is available, by taking our rs-pseudometric d and the underlying subset Z according to the following:

(B-tri) d is Frink triangular (hence, a Frink quasi-metric):
$\forall \varepsilon > 0, \exists \delta > 0$, such that $d(x, y), d(y, z) < \delta$ imply $d(x, z) < \varepsilon$
(kl-M) $Z = \text{kl}_{(\preceq)}(M)$, where $\text{kl}_{(\preceq)}(.)$ is the introduced closure operator.

The result of this upgrading is the following $(\preceq)$-maximal statement (referred to as: Zorn–Bourbaki Frink quasi-metric maximal principle (in short: (ZB-Fqm-mp)).

Theorem 6. *Let the quasi-ordered Frink quasi-metric space $(X, d, \preceq)$ and the subset $Y \in \exp(X)$ be such that*

(43-i) *Y is $(d, \preceq)$-closed (that is: $Y = \text{cl}_{(\preceq)}(Y)$)*
(43-ii) *$(Y, d, \preceq)$ is regular: each $(\preceq)$-ascending sequence in Y is d-Cauchy*

holds, as well as one of the alternatives (43-iii)–(43-iv) or (43-v), where

(43-iii) $(Y, d, \preceq)$ *is complete:*
each $(\preceq)$*-ascending* d*-Cauchy sequence* (x_n) *in* Y *is* d*-convergent (in* X*)*
(43-iv) $(Y, d, \preceq)$ *is conv-dominated:*
the limit of each $(\preceq)$*-ascending sequence in* Y *is a* $(\preceq)$*-upper bound of it*
(43-v) $(Y, d, \preceq)$ *is fully complete:*
each $(\preceq)$*-ascending* d*-Cauchy sequence* (x_n) *in* Y *is* $(d\preceq)$*-convergent.*

Further, take the subset $M \subseteq Y$ *and denote* $Z := \text{kl}_{(\preceq)}(M)$*. Then,*
(43-a) $M \subseteq Z$, $Z \subseteq Y$ *and* $Z \subseteq \text{cl}(M)$*; hence,* $M \subseteq Z \subseteq Y \cap \text{cl}(M)$
(43-b) *for each* $u \in M$*, there exists some other point* $v \in Z$*, with*

(43-b1) $u \preceq v$ *and* $v \in Y \cap \text{cl}(M)$
(43-b2) v *is* $(\preceq)$*-maximal over* Z
(43-b3) v *is* $(\preceq)$*-maximal over* M.

Proof. (i) Clearly, $M \subseteq Z$, by definition. Further,

$Z \subseteq Y$; because $M \subseteq Y$ and Y is $(d, \preceq)$-closed,

On the other hand, by an auxiliary fact above,

$Z \subseteq \text{cl}(M)$ (because $M \subseteq \text{cl}(M)$ and $\text{cl}(M)$ is $(d, \preceq)$-closed),

so, by simply combining these, the first part is proved.

(ii) Let (cond;$(Y, d, \preceq)$) stand for one of the conditions (42-ii)–(42-v). By a hereditary argument we have (under $Z \subseteq Y$) the generic inclusion

(cond;$(Y, d, \preceq)$) implies (cond;$(Z, d, \preceq)$).

And then, under the Zorn–Bourbaki rs-pseudometric maximal principle (ZB-rspm-mp) we are done. However, for practical reasons, it would be useful for us to supply an argument for this.

By the regularity condition (modulo $(Z, d, \preceq)$) and an auxiliary fact above, $(Z, d, \preceq)$ is weakly regular. Hence, given $u \in M \subseteq Z$, one may construct a $(\preceq)$-ascending sequence (u_n) in Z with

(43-c) $u \preceq u_0$, and $[(\forall n), (\forall y, z \in Z)]$: $u_n \preceq y \preceq z \Longrightarrow d(y, z) \leq 2^{-n}$.

In particular, this tells us that (u_n) is d-Cauchy (in Z). Combining with the completeness hypothesis, (modulo $(Z, d, \preceq)$), there exists $v \in X$ such that

(43-d) $u_n \xrightarrow{d} v$; so that, $v \in \text{cl}_{(\preceq)}(Z) = Z$, and this, by the conv-dominated property (modulo $(Z, d, \preceq)$), gives

(43-e) $u_n \stackrel{\preceq}{\longrightarrow} v$; whence, $u_n \preceq v$, $\forall n$.

Alternatively, by the full completeness hypothesis (modulo $(Z, d, \preceq)$), there exists $v \in X$ such that

(43-f) $u_n \stackrel{d\preceq}{\longrightarrow} v$; whence, $u_n \stackrel{d}{\longrightarrow} v$, $u_n \stackrel{\preceq}{\longrightarrow} v$,

and (43-d)+(43-e) are again retainable. By (43-c)+(43-e), we get $u \preceq v$. On the other hand, let $y, z \in Z$ be such that $v \preceq y \preceq z$. Then, by (43-c) again,

$(\forall n)$: $u_n \leq y \leq z$; whence, $d(y, z) \leq 2^{-n}$,

and, from this, $d(y, z) = 0$. Combining with the arbitrariness of the underlying couple, one derives that

$v \in Z$ is $(d, \preceq)$-maximal over Z; hence, $(\preceq)$-maximal in Z,

if we take a preceding auxiliary fact into account. But then, via $M \subseteq Z$, it is clear that $v \in Z$ is $(\preceq)$-maximal in M; and the conclusion follows. □

(C) Some basic particular cases of these correspond to the Frink triangular property imposed on our rs-pseudometric d being assured by some related conditions involving such objects.

(I) Suppose in the following that our ambient rs-pseudometric d fulfills

(B-tri) d is *Bakhtin triangular*: there exists $\beta \geq 1$, such that $d(x, z) \leq \beta(d(x, y) + d(y, z))$, $\forall x, y, z \in X$;

when (see above) d is a *Bakhtin quasi-metric* on X. As a direct consequence of Zorn–Bourbaki Frink quasi-metric maximal principle (ZB-Fqm-mp), one gets the following statement, referred to as Zorn–Bourbaki Bakhtin quasi-metric maximal principle (in short: (ZB-Bqm-mp)).

Theorem 7. *Let the quasi-ordered Bakhtin quasi-metric space* $(X, d, \preceq)$ *and the subset* $Y \in \exp(X)$ *be such that*

(44-i) Y *is* $(d, \preceq)$*-closed (that is:* $Y = \text{cl}_{(\preceq)}(Y)$*).*
(44-ii) $(Y, d, \preceq)$ *is regular: each* $(\preceq)$*-ascending sequence in* Y *is* d*-Cauchy*

holds, as well as one of the alternatives (44-iii)–(44-iv)] *or* (44-v), *where*

(44-iii) $(Y, d, \preceq)$ *is complete:*
each $(\preceq)$*-ascending d-Cauchy sequence* (x_n) *in* Y *is d-convergent (in* X*)*
(44-iv) $(Y, d, \preceq)$ *is conv-dominated:*
the limit of each $(\preceq)$*-ascending sequence in* Y *is a* $(\preceq)$*-upper bound of it*
(44-v) $(Y, d, \preceq)$ *is fully complete:*
each $(\preceq)$*-ascending d-Cauchy sequence* (x_n) *in* Y *is* $(d \preceq)$*-convergent.*

Further, take the subset $M \subseteq Y$ *and denote* $Z := \mathrm{kl}_{(\preceq)}(M)$. *Then,*
(44-a) $M \subseteq Z$, $Z \subseteq Y$ *and* $Z \subseteq \mathrm{cl}(M)$; *hence,* $M \subseteq Z \subseteq Y \cap \mathrm{cl}(M)$
(44-b) *for each* $u \in M$, *there exists some other point* $v \in Z$, *with*

(44-b1) $u \preceq v$ *and* $v \in Y \cap \mathrm{cl}(M)$
(44-b2) v *is* $(\preceq)$*-maximal over* Z
(44-b3) v *is* $(\preceq)$*-maximal over* M.

(II) A basic particular case of this corresponds to the rs-pseudometric d fulfilling

(tri) d is *triangular*: $d(x, z) \leq d(x, y) + d(y, z)$, $\forall x, y, z \in X$;

when (see above) d is a *quasi-metric* on X. As a direct consequence of Zorn–Bourbaki Bakhtin quasi-metric maximal principle (ZB-Bqm-mp), one gets the following statement, referred to as Zorn–Bourbaki quasi-metric maximal principle (in short: (ZB-qm-mp)).

Theorem 8. *Let the quasi-ordered quasi-metric space* $(X, d, \preceq)$ *and the subset* $Y \in \exp(X)$ *be such that*

(45-i) Y *is* $(d, \preceq)$*-closed (that is:* $Y = \mathrm{cl}_{(\preceq)}(Y)$*)*
(45-ii) $(Y, d, \preceq)$ *is regular*: *each* $(\preceq)$*-ascending sequence in* Y *is d-Cauchy*

holds, as well as one of the alternatives (45-iii)–(45-iv) *or* (45-v), *where*

(45-iii) $(Y, d, \preceq)$ *is complete:*
each $(\preceq)$*-ascending d-Cauchy sequence* (x_n) *in* Y *is d-convergent (in* X*)*
(45-iv) $(Y, d, \preceq)$ *is conv-dominated:*
the limit of each $(\preceq)$*-ascending sequence in* Y *is a* $(\preceq)$*-upper bound of it*

(45-v) $(Y, d, \preceq)$ *is fully complete*:
each $(\preceq)$*-ascending* d*-Cauchy sequence* (x_n) *in* Y *is* $(d \preceq)$*-convergent.*

Further, take the subset $M \subseteq Y$ *and denote* $Z := \mathrm{kl}_{(\preceq)}(M)$. *Then,*
(45-a) $M \subseteq Z$, $Z \subseteq Y$ *and* $Z \subseteq \mathrm{cl}(M)$; *hence,* $M \subseteq Z \subseteq Y \cap \mathrm{cl}(M)$
(45-b) *for each* $u \in M$, *there exists some other point* $v \in Z$, *with*

(45-b1) $u \preceq v$ *and* $v \in Y \cap \mathrm{cl}(M)$
(45-b2) v *is* $(\preceq)$*-maximal over* Z
(45-b3) v *is* $(\preceq)$*-maximal over* M.

In particular when, in addition

(sym) d is *symmetric*: $d(x, y) = d(y, x)$, $\forall x, y \in X$;
hence, $d(.,.)$ is a *metric* on X,

this last result may be viewed as a metrical variant of the Zorn–Bourbaki maximal principle [46]. An early version of it was formulated in Turinici [47]; note that it includes the one due to Dancs *et al.* [48]. In particular, when $M = X$, the underlying result is just the ordering principle in Kang and Park [41]. Moreover, by the characterization of regularity condition, this statement includes as well the related one in Turinici [40]. Note that all these are ultimately equivalent with Brezis–Browder's ordering principle [44]; we do not give details. Further aspects may be found in the papers by Turinici [49,50].

(D) A basic application of these facts is to *local* monotone variational principles. Let X be a non-empty set; and $(\leq)$ be a quasi-order on it. Then, let $d(.,.)$ be a pseudometric over X; endowed with

(r-s) d is *reflexive sufficient*: $d(x, y) = 0$ iff $x = y$;
(tri) d is *triangular*: $d(x, z) \leq d(x, y) + d(y, z)$, $\forall x, y, z \in X$;

as precise, it will be referred to as a *quasi-metric* over X. All quasi-order notions to be used refer to the *dual* $(\geq)$; but these may also be formulated in terms of our initial relation. Precisely, take the structure $(X, d, \geq)$ according to

(com) $(X, d, \geq)$ is complete:
each $(\geq)$-ascending d-Cauchy sequence (x_n) in X is d-convergent;
also referred to as: d is $(\geq)$*-complete*

(conv-dom) $(X, d, \geq)$ is conv-dominated:
the limit of each $(\geq)$-ascending sequence in X is a $(\geq)$-upper bound of it; also referred to as: $(\geq)$ is *d-selfclosed.*

In addition, let the function $\varphi : X \to R \cup \{\infty\}$ be such that

(inf-prop) φ is inf-proper ($\mathrm{Dom}(\varphi) \neq \emptyset$ and $\inf[\varphi(X)] > -\infty$)
(ge-lsc) φ is $(d, \geq)$-lsc on X:
$[\varphi \leq t] := \{x \in X; \varphi(x) \leq t\}$ is $(d, \geq)$-closed, $\forall t \in R$.

A basic particular case of this last condition upon φ is the one described as

Proposition 16. *Suppose that $(X, d, \geq)$ is conv-dominated and*

φ is $(\leq)$-increasing: $x \leq y$ implies $\varphi(x) \leq \varphi(y)$.

Then, necessarily, φ is $(d, \geq)$-lsc on X.

Proof. Let the number $t \in R$, the sequence (x_n) in $[\varphi \leq t]$ and the point $x \in X$ be such that

(x_n) is $(\geq)$-ascending and $x_n \xrightarrow{d} x$ as $n \to \infty$.

By the conv-dominated assumption,

$x_n \geq x, \forall n$; whence, $\varphi(x_n) \geq \varphi(x), \forall n$,

if we take the $(\leq)$-increasing property of φ into account. But then, $\varphi(x) \leq t$; and the conclusion follows. □

Returning to our general setting, let the (non-empty) subset G of X be taken according to

(phi-adm) G is *φ-admissible:* $G \cap \mathrm{Dom}(\varphi) \neq \emptyset$.

The following (local-type) variational statement [referred to as: quasi-order Ekeland Variational Principle on quasi-metric spaces; in short: (EVP-qo-qms)] is our basic tool for our future developments.

Theorem 9. *Under the precise general assumptions, let the (non-empty) subset G of X be φ-admissible; and fix $u \in G \cap \mathrm{Dom}(\varphi)$. There exists then some point $v \in X$, with the properties*

(46-a) *$v \in \mathrm{cl}(G)$ and $v \in \mathrm{Dom}(\varphi)$; hence, $v \in \mathrm{cl}(G) \cap \mathrm{Dom}(\varphi)$*
(46-b) *$u \geq v, d(u, v) \leq \varphi(u) - \varphi(v)$ (hence, $\varphi(u) \geq \varphi(v)$)*
(46-c) *$d(v, x) > \varphi(v) - \varphi(x)$, for each $x \in G(v, \geq) \setminus \{v\}$.*

Proof. Let $(\sqsubseteq)$ stand for the quasi-order on X

$$x \sqsubseteq y \text{ iff } x \leq y,\ \varphi(x) \leq \varphi(y).$$

As usual, we denote by $(\sqsupseteq)$ the dual of $(\sqsubseteq)$:

$$x \sqsupseteq y \text{ iff } y \sqsubseteq x; \text{ that is: } x \geq y,\ \varphi(x) \geq \varphi(y).$$

Then, let $(\preceq)$ stand for the relation (over X):

$$x \preceq y \text{ iff } x \geq y,\ d(x,y) + \varphi(y) \leq \varphi(x).$$

Clearly, $(\preceq)$ is a quasi-order on X; with, in addition,

$(\preceq)$ is coarser than $(\sqsupseteq)$: $x \preceq y$ implies $x \sqsupseteq y$.

Moreover, $(\preceq)$ is antisymmetric — hence, a (partial) order — on $\mathrm{Dom}(\varphi)$, as it can be directly seen. Denote

$$Y = X(u, \sqsupseteq) := \{x \in X; u \sqsupseteq x\}; \text{ hence, } u \in Y \subseteq \mathrm{Dom}(\varphi),$$

note that, by the above, $(\preceq)$ is an order on Y.

We claim that conditions of Zorn–Bourbaki quasi-metric maximal principle (ZB-qm-mp) are fulfilled over $(X, d, \preceq)$, with respect to the subset Y.

Step 1. Let (x_n) be a $(\preceq)$-ascending sequence in Y; that is (via $Y \subseteq \mathrm{Dom}(\varphi)$)

(asc-1) $i \leq j$ implies $x_i \preceq x_j$; hence, $x_i \geq x_j$, $d(x_i, x_j) \leq \varphi(x_i) - \varphi(x_j)$.

(1-1) The sequence $(\varphi(x_n))$ is descending and (via φ = inf-proper) bounded from below; hence a Cauchy one. This, along with (asc-1) shows that (x_n) is a $(\preceq)$-ascending d-Cauchy sequence; wherefrom $(Y, d, \preceq)$ is regular.

(1-2) Under the same premise, (x_n) is a $(\geq)$-ascending sequence in Y. This, along with $(X, d, \geq)$ being complete, gives us an element $x \in X$ with $x_n \xrightarrow{d} x$ as $n \to \infty$; which proves that $(Y, d, \preceq)$ is complete (in X).

Step 2. Let (x_n) be a $(\preceq)$-ascending sequence in Y; that is, condition (asc-1) holds, as well as

(asc-2) $(\forall n)$: $u \geq x_n$ and $\varphi(u) \geq \varphi(x_n)$.

Further, let $x \in X$ be an element of X with $x_n \xrightarrow{d} x$ as $n \to \infty$.

(2-1) By the remarks above, (x_n) is a $(\geq)$-ascending sequence in Y with $x_n \xrightarrow{d} x$ as $n \to \infty$. Further, let the index i be arbitrary fixed. Clearly,

(rela-1) $(\forall m \geq i)$: $x_i \geq x_m$, $d(x_i, x_m) \leq \varphi(x_i) - \varphi(x_m)$
(whence, $\varphi(x_i) \geq \varphi(x_m)$).

As $(X, d, \geq)$ is conv-dominated and φ is $(d, \geq)$-lsc, one derives (via (asc-1))

(rela-2) $x_i \geq x$ and $\varphi(x_i) \geq \varphi(x)$; that is: $x_i \sqsupseteq x$;

which, in the particular case of $i = 0$, yields

$u \sqsupseteq x$; that is: $x \in Y$; proving that Y is $(d, \preceq)$-closed.

(2-2) On the other hand, a combination of (rela-1) and (rela-2) yields

$(\forall m \geq i)$: $d(x_i, x_m) \leq \varphi(x_i) - \varphi(x)$,

whence, by the triangular inequality,

$(\forall m \geq i)$: $d(x_i, x) \leq d(x_i, x_m) + d(x_m, x) \leq \varphi(x_i) - \varphi(x) + d(x_m, x)$.

Passing to limit as $m \to \infty$, one derives

$(\forall i)$: $x_i \geq x$ and $d(x_i, x) \leq \varphi(x_i) - \varphi(x)$, that is: $x_i \preceq x$,

proving that $(Y, d, \preceq)$ is conv-dominated.

Summing up, all conditions in Zorn–Bourbaki quasi-metric maximal principle (ZB-qm-mp) are indeed fulfilled over $(X, d, \preceq)$, with respect to the subset $Y = X(u, \sqsupseteq)$. In this case, letting G be as above, put

$M = G \cap Y$, $Z = \mathrm{kl}_{(\preceq)}(M)$; hence, $u \in M$ (by the choice of Y).

By the first conclusion of the underlying maximal principle,

(con-1) $M \subseteq Z$, $Z \subseteq Y$ and $Z \subseteq \mathrm{cl}(M)$; hence, $M \subseteq Z \subseteq Y \cap \mathrm{cl}(M)$.

Moreover, by the second conclusion of the underlying maximal principle it follows that, for the starting $u \in M$, there exists some other point $v \in Z$, with

(con-2a) $u \preceq v$ and $v \in Y \cap \mathrm{cl}(M)$
(con-2b) v is $(\preceq)$-maximal over Z
(con-2c) v is $(\preceq)$-maximal over M.

We now show that, from these, one derives all needed conclusions.

(C-1) The first conclusion in the statement is clear, by (con-2a) (the second part), and the immediate relations

$Y \subseteq \mathrm{Dom}(\varphi)$ and $\mathrm{cl}(M) \subseteq \mathrm{cl}(G)$.

(C-2) The second conclusion in the statement is again clear, by means of (con-2a) (the first part).

(C-3) To establish the third conclusion in the statement, we proceed by contradiction. Precisely, assume that the underlying conclusion would be false:

there exists $x \in G(v, \geq) \setminus \{v\}$ with $d(v, x) \leq \varphi(v) - \varphi(x)$
(hence, $\varphi(v) \geq \varphi(x)$).

By the first half, the obtained point $x \in G$ has the properties

$v \geq x$ and $v \neq x$ (hence, $d(v, x) > 0$, as $d(.,.)$ is a quasi-metric);

and this, combined with the (consequence of) second half, yields

$v \sqsupseteq x$; whence (via $u \sqsupseteq v$), $u \sqsupseteq x$; that is: $x \in Y$; so, $x \in G \cap Y = M$.

But then, in view of $v \neq x$, the maximality of v in M would be contradicted. Hence, the working hypothesis cannot be accepted; and our third conclusion in the statement is retainable as well. The proof is thereby complete. □

Finally, as already precise, the $(d, \geq)$-lsc condition upon φ is obtainable (under the conv-dominated assumption) under

φ is $(\leq)$-increasing ($x \leq y \Longrightarrow \varphi(x) \leq \varphi(y)$).

In this case, Theorem 9 may be viewed as a (local and) quasi-metric version of the monotone variational principle in Turinici [16]. Further aspects may be found in Hyers *et al.* [51, Ch 5].

(E) A basic particular case corresponds to the choice

(triv-qo) $(\leq) = (\geq) = X \times X$ (=the trivial quasi-order on X).

Let X be a non-empty set; and $d(.,.)$ be a pseudometric over X, endowed with

(r-s) d is *reflexive sufficient*: $d(x, y) = 0$ iff $x = y$,
(tri) d is *triangular*: $d(x, z) \leq d(x, y) + d(y, z)$, $\forall x, y, z \in X$,

as precise, it will be referred to as a *quasi-metric* over X. Suppose that the general condition holds

(com) (X, d) is complete:
each d-Cauchy sequence (x_n) in X is d-convergent;
also referred to as: d is *complete*.

In addition, let the function $\varphi : X \to R \cup \{\infty\}$ be such that

(inf-prop) φ is inf-proper ($\text{Dom}(\varphi) \neq \emptyset$ and $\inf[\varphi(X)] > -\infty$)
(lsc) φ is d-lsc on X:
$[\varphi \leq t] := \{x \in X; \varphi(x) \leq t\}$ is d-closed, $\forall t \in R$.

Further, let the (non-empty) subset G of X be taken according to

(phi-adm) G is *φ-admissible*: $G \cap \text{Dom}(\varphi) \neq \emptyset$.

The following (local-type) variational statement [referred to as: Ekeland Variational Principle on quasi-metric spaces; in short: (EVP-qms)] is available.

Theorem 10. *Under the precise general assumptions, let the (non-empty) subset G of X be φ-admissible; and fix $u \in G \cap \text{Dom}(\varphi)$. There exists then some point $v \in X$, with the properties*

(47-a) $v \in \text{cl}(G)$ *and* $v \in \text{Dom}(\varphi)$; *hence,* $v \in \text{cl}(G) \cap \text{Dom}(\varphi)$
(47-b) $d(u, v) \leq \varphi(u) - \varphi(v)$ (*hence,* $\varphi(u) \geq \varphi(v)$)
(47-c) $d(v, x) > \varphi(v) - \varphi(x)$, *for each* $x \in G \setminus \{v\}$.

Proof. We show that the preceding statement is applicable here (under the trivial choice of $(\leq)$). In fact, under such a context,

(47-d) $(X, d, X \times X)$ is complete and conv-dominated
(47-e) φ is inf-proper and $(d, X \times X)$-lsc;

hence, the claim. Adding the conclusions of that statement, we are done. □

Denote for simplicity

(EVP-m)= the particular version of (EVP-qms)
under the choices d=metric and $G = X$.

This (particular) principle, due to Ekeland [13], found some basic applications in control and optimization, generalized differential calculus, critical point theory and global analysis; we refer to the quoted paper for a survey of these. So, it cannot be surprising that, soon after its formulation, many extensions of (EVP-m) were proposed. For example, the *dimensional* way of extension refers to the ambient space (R) of $\varphi(X)$ being substituted by a (topological or not) vector space. An account of the results in this area is to be found in Goepfert *et al.* [52, Ch 3]. The metrical extension of the same consists in conditions imposed upon our metric being relaxed. Some of these extensions were already stated; for the remaining ones, we refer to Hyers *et al.* [51, Ch 5]; see also Turinici [16].

By the developments above, we therefore have the following implications:

$$\text{(DC)} \Longrightarrow \text{(KP-rp-mp)} \Longrightarrow \text{(ZB-rsp-mp)} \Longrightarrow \text{(ZB-Fqm-mp)}$$
$$\text{(ZB-Fqm-mp)} \Longrightarrow \text{(ZB-Bqm-mp)} \Longrightarrow \text{(ZB-qm-mp)}$$
$$\text{(ZB-qm-mp)} \Longrightarrow \text{(EVP-qo-qms)} \Longrightarrow \text{(EVP-qms)} \Longrightarrow \text{(EVP-m)}.$$

So, we may ask whether these may be reversed. Clearly, the natural setting for solving this problem is the strongly reduced system (ZF-AC).

Let X be a non-empty set; and $(\leq)$ be a (partial) order on it. We say that $(\leq)$ has the *inf-lattice* property, provided: $\inf\{x, y\}$ exists, for all $x, y \in X$. Remember that $z \in X$ is a $(\leq)$-*maximal* element if $X(z, \leq) = \{z\}$; the class of all these points will be denoted as $\max(X, \leq)$. Call $(\leq)$, a *Zorn order* when

$\max(X, \leq)$ is non-empty and *cofinal* in X
(for each $u \in X$ there exists a $(\leq)$-maximal $v \in X$ with $u \leq v$).

Further aspects are to be described in a metric setting. Let $d(.,.)$ be a metric over X; and $\varphi : X \to R_+$ be some function. Then, the natural choice for $(\leq)$ above is

$$x \leq_{(d,\varphi)} y \text{ iff } d(x, y) \leq \varphi(x) - \varphi(y);$$

referred to as the *Brøndsted order* [53] attached to (d, φ). Denote

$$X(x, \rho) = \{u \in X; d(x, u) < \rho\},\ x \in X,\ \rho > 0$$

[the *open* sphere with center x and radius ρ]. Call (X, d), *discrete* when

for each $x \in X$ there exists $\rho = \rho(x) > 0$ such that $X(x, \rho) = \{x\}$.

Note that, under such an assumption, any function $\psi : X \to R$ is continuous over X. However, this is not extendable to the *d-Lipschitz property*

$$|\psi(x) - \psi(y)| \leq Ld(x, y),\ x, y \in X, \text{ for some } L > 0;$$

hence, all the more, to the *d-non-expansive property* ($L = 1$).

Now, the following statement is a particular case of (EVP-m):

Theorem 11. *Let the metric space* (X, d) *and the function* $\varphi : X \to R_+$ *satisfy the following:*

(48-i) (X, d) *is discrete bounded and complete*
(48-ii) $(\leq_{(d,\varphi)})$ *has the inf-lattice property*
(48-iii) φ *is d-non-expansive and* $\varphi(X)$ *is countable.*

Then, $(\leq_{(d,\varphi)})$ *is a Zorn order.*

We shall refer to it as: the discrete Lipschitz countable version of (EVP-m) (in short: (EVP-m-dLc)). Clearly, (EVP-m) $\Longrightarrow$ (EVP-m-dLc). The remarkable fact is that this last principle yields (DC); and completes the circle between all these.

Proposition 17. *We have the inclusion (EVP-m-dLc)* $\Longrightarrow$ *(DC) [in the strongly reduced system (ZF-AC)]. So (by the above),*

(48-1) *the maximal/variational principles (KP-rp-mp), (ZB-rsp-mp), (ZB-Fqm-mp), (ZB-Bqm-mp), (ZB-qm-mp), (EVP-qo-qms), (EVP-qms) and (EVP-m) are all equivalent with (DC); hence, mutually equivalent*

(48-2) *each intermediary maximal/variational statement (VP), fulfilling (DC)* $\Longrightarrow$ *(VP)* $\Longrightarrow$ *(EVP-m), is equivalent with both (DC) and (EVP-m).*

For a complete proof, see Turinici [40]. In particular, when the inf-lattice, nonexpansive and countable properties are ignored in (EVP-m-dLc), the last result above reduces to the one in Brunner [54]. Note that, in the same particular setting, a different proof of the underlying inclusion was provided in Dodu and Morillon [55]. Further aspects may be found in Schechter [29, Ch 19, Sect 19.51].

5. Main Results

Let X be a non-empty set. Take a quasi-order $(\leq)$ as well as a quasi-metric d on X. The basic conditions to be imposed upon these objects are

(bcond-1) $(X, d, \geq)$ is complete:
each $(\geq)$-ascending d-Cauchy sequence (x_n) in X is d-convergent;
also referred to as: d is $(\geq)$-*complete*
(bcond-2) $(X, d, \geq)$ is conv-dominated:
the limit of each $(\geq)$-ascending sequence in X is a $(\geq)$-upper bound of it;
also referred to as: $(\geq)$ is d-*selfclosed.*

In addition, let us accept the local condition

(bcond-3) $(X, d, \geq)$ is *full*:
$(\forall u \in X, \forall \delta > 0)$: $X[u, \delta; \geq] := X[u, \delta] \cap X(u, \geq)$
has a non-empty intersection with $X \setminus \{u\}$;
where $X[u, \delta] = \{x \in X; d(u, x) \leq \delta\}$, $X(u, \geq) = \{x \in X; u \geq x\}$.

Further, take some map $\Gamma : X \to R_+$ with the properties

(ga-1) $(\exists(\lambda, \mu), 0 < \lambda < 1 < \mu)$, such that
$d(x, y) \leq \lambda \Longrightarrow \Gamma(x) - \Gamma(y) \leq \mu$
(ga-2) $\sup[\Gamma(X)] = \infty$; hence
$[\Gamma \geq \sigma] := \{x \in X; \Gamma(x) \geq \sigma\}$ is non-empty, $\forall \sigma \geq 0$.

Note that a useful consequence of these facts is

(501) $\mathrm{cl}[\Gamma \geq \rho] \subseteq [\Gamma \geq \rho - \mu]$, $\forall \rho \geq \mu$.

In fact, let $v \in \mathrm{cl}[\Gamma \geq \rho]$ be arbitrary fixed. By definition, there exists a sequence (u_n) in $[\Gamma \geq \rho]$ with $u_n \xrightarrow{d} v$. In particular, there exists some rank $m = m(\lambda)$ with

$(\forall n \geq m)$: $d(u_n, v) < \lambda$; hence (by (ga-1)) $\Gamma(u_n) - \Gamma(v) \leq \mu$.

But then, $\Gamma(v) \geq \Gamma(u_n) - \mu \geq \rho - \mu$; and the claim follows.

Finally, pick some functional $F : X \to R \cup \{\infty\}$ with

(F-1) F is inf-proper: $\mathrm{Dom}(F) \neq \emptyset$ and $F_* := \inf[F(X)] > -\infty$
(F-2) F is $(d, \geq)$-lsc on X:
$[F \leq t] := \{x \in X; F(x) \leq t\}$ is $(d, \geq)$-closed, $\forall t \in R$.

The (extended) function (from R_+ to $R \cup \{\infty\}$)

$$m(\Gamma, F)(\sigma) := \inf F([\Gamma \geq \sigma]),\ \sigma \geq 0,$$

is well defined, via (ga-2). Moreover,

$m(\Gamma, F)(.)$ is increasing: $\sigma_1 \leq \sigma_2$ implies $m(\Gamma, F)(\sigma_1) \leq m(\Gamma, F)(\sigma_2)$;

wherefrom, the asymptotic quantity

$$\liminf_{\Gamma(u) \to \infty} F(u) := \sup_{\sigma \geq 0} m(\Gamma, F)(\sigma)\ (= \lim_{\sigma \to \infty} m(\Gamma, F)(\sigma)),$$

exists, as an element of $R \cup \{\infty\}$, in view of

(502) $F_* \leq m(\Gamma, F)(\sigma) \leq \alpha(\Gamma, F) := \liminf_{\Gamma(u) \to \infty} F(u) \leq \infty$, $\forall \sigma > 0$.

When $\alpha(\Gamma, F) = \infty$, the functional F will be referred to as Γ-*coercive*. It is our aim in the following to get sufficient conditions in order that such a property be attained. These, as a rule, require a *differential* setting. Denote, for each $u \in \mathrm{Dom}(F)$,

$$\text{(slo-1) } \nabla_{(\geq)}F(u) = \limsup\left\{\frac{F(u)-F(x)}{d(u,x)};\ d(u,x) \to 0+, u \geq x\right\},$$

that is,

$$\nabla_{(\geq)}F(u) = \inf_{\delta>0}\sup\left\{\frac{F(u)-F(x)}{d(u,x)}; x \in X[u,\delta;\geq] \setminus \{u\}\right\},$$

as well as

$$\text{(slo-2) } |\nabla_{(\geq)}|F(u) = \max\{0, \nabla_{(\geq)}F(u)\}.$$

Note that the quantity in the right member of (slo-1) is meaningful, by (bcond-3); and this tells us that (slo-1) and (slo-2) are well defined. The objects in question are a quasi-metric version of the ones introduced (in a trivial quasi-order and Banach space context) by DeGiorgi *et al.* [21]; we call these the *relative/absolute* $(d, \geq)$-*slope* of F at u. The usefulness of these differential tools for the critical point theory (over Banach spaces) was underlined by Corvellec *et al.* [56]. Here, we shall establish that the introduced slope concepts are appropriate for our quasi-order quasi-metrical setting as well.

The following asymptotic type statement is an essential step towards the answer we are looking for. Let $\mathrm{fin}(X)$ stand for the class of all finite parts of X; and (x_n) be a sequence in X. Given $Q \in \mathrm{fin}(X)$, we say that (x_n) *avoids* Q when there exists an index $m = m(Q)$, such that: $x_n \in X \setminus Q$, for all $n \geq m$.

If this holds for all $Q \in \mathrm{fin}(X)$, the obtained property will be referred to as: (x_n) *avoids* $\mathrm{fin}(X)$.

Theorem 12. *Suppose that*

(con-1) $\alpha(\Gamma, F) < \infty$ *(hence,* $-\infty < F_* \leq \alpha(\Gamma, F) < \infty$*).*

There exists then, in (ZF-AC+DC), a sequence (v_n) *in* $\mathrm{Dom}(F)$*, with*

(51-a) $\Gamma(v_n) \to \infty$ *as* $n \to \infty$ *(hence,* (v_n) *avoids* $\mathrm{fin}(X)$*)*

(51-b) $F(v_n) \to \alpha(\Gamma, F)$ *and* $|\nabla_{(\geq)}|F(v_n) \to 0$ *as* $n \to \infty$.

Proof. There are two basic steps to be passed.

Step 1. We claim that the following is true:

(51-c) for each $\eta \in]0, \lambda/2\mu[$, the set $A(\eta)$ of all $(r, v) \in R^0_+ \times \mathrm{Dom}(F)$ with

(51-c1) $r \geq 1/\eta$ and $m(\Gamma, F)(r) > \alpha(\Gamma, F) - \eta^2$,
(51-c2) $\Gamma(v) \geq r$, $|F(v) - \alpha(\Gamma, F)| < \eta^2$, $|\nabla_{(\geq)}|F(v) \leq \eta$

is not empty.

In fact, let $\eta > 0$ be taken according to

(con-2) $\eta < \dfrac{\lambda}{2\mu}$; hence, (*cf.* (ga-1)) $\dfrac{1}{\eta} > \mu > \dfrac{\lambda}{2} > \eta$.

Clearly, the family of relations (51-c1) is fulfilled, in view of

$$\lim_{r \to \infty} m(\Gamma, F)(r) = \sup_{r \geq 0} m(\Gamma, F)(r) = \alpha(\Gamma, F) > \alpha(\Gamma, F) - \eta^2.$$

The family of relations (51-c2) is fulfilled as well, in view of the following argument. Let $r > 0$ be as in (51-c1). By this, and the increasing property of $r \mapsto m(\Gamma, F)(r)$,

(503) $\alpha(\Gamma, F) - \eta^2 < m(\Gamma, F)(r) \leq m(\Gamma, F)(4r) \leq$
$\alpha(\Gamma, F) < \alpha(\Gamma, F) + \eta^2$; wherefrom (by the infimum definition)
there exists $u \in [\Gamma \geq 4r]$, such that $F(u) < \alpha(\Gamma, F) + \eta^2$;
hence, $u \in [\Gamma \geq 4r] \cap \mathrm{Dom}(F) \subseteq [\Gamma \geq 2r] \cap \mathrm{Dom}(F)$.

Taking (bcond-1)+(bcond-2) and (F-1)+(F-2) into account, it results that the quasi-order Ekeland Variational Principle on quasi-metric spaces (EVP-qo-qms) applies to the data

$(X, d, \geq)$, $\varphi := (1/\eta)F$, $G := [\Gamma \geq 2r]$.

So, for the starting point $u \in G \cap \mathrm{Dom}(F)$, there must be some $v \in X$, with

(504) $v \in \mathrm{cl}(G)$ and $v \in \mathrm{Dom}(F)$; hence, $v \in \mathrm{cl}(G) \cap \mathrm{Dom}(F)$
(505) $u \geq v$, $\eta d(u, v) \leq F(u) - F(v)$ (hence, $F(u) \geq F(v)$)
(506) $\eta d(v, x) > F(v) - F(x)$, for all $x \in G(v, \geq) \setminus \{v\}$.

We claim that v fulfills the family of relations (51-c2). In fact, (504) gives (by a previous observation and $r \geq 1/\eta > \mu$)

(507) $v \in [\Gamma \geq 2r - \mu] \subseteq [\Gamma \geq r)]$,

so that, v is an element of $\mathrm{Dom}(F)$ fulfilling the first part of (51-c2). Combining with (503)+(505),

$$\alpha(\Gamma, F) - \eta^2 < F(v) \leq F(u) < \alpha(\Gamma, F) + \eta^2,$$

which tells us that the second part of (51-c2) holds, too. This, again coupled with (505) yields (via (con-2))

$$d(u,v) \leq (1/\eta)2\eta^2 < \lambda; \text{ so [by (ga-1)]}, v \in [\Gamma \geq 4r-\mu] \subseteq [\Gamma \geq 3r] \subseteq G,$$

which improves the previous evaluations (504) and (507) of v. Finally, by (ga-1) and (con-2),

$$(\forall \delta \in]0,\lambda]) \colon X[v,\delta] \subseteq X[v,\lambda] \subseteq [\Gamma \geq 4r-2\mu] \subseteq [\Gamma \geq 2r] = G;$$
whence, $X[v,\delta;\geq] \subseteq G(v,\geq)$.

This, along with (506), gives

$$\eta \geq \sup\left\{\frac{F(v)-F(x)}{d(v,x)}; x \in X[v,\delta;\geq] \setminus \{v\}\right\}, \ \forall \delta \in]0,\lambda];$$

whence

$$\eta \geq \nabla_{(\geq)}F(v) = \inf_{\delta>0} \sup\left\{\frac{F(v)-F(x)}{d(v,x)}; x \in X[v,\delta;\geq] \setminus \{v\}\right\},$$

and the third part of (51-c2) holds as well.

Step 2. Let (η_n) be a descending to zero sequence with $]0,\lambda/2\mu[$; and put

$(A_n = A(\eta_n); n \geq 0)$; hence, A_n is non-empty, for each n.

By the Denumerable Axiom of Choice (AC(N)) (deductible in (ZF-AC+DC)), there exists a sequence (r_n) in R^0_+ and a sequence (v_n) in Dom(F), with

$(r_n, v_n) \in A(\eta_n)$, for all n; that is

(51-c1-N) $(\forall n)$: $r_n \geq 1/\eta_n$ and $m(\Gamma,F)(r_n) > \alpha(\Gamma,F) - \eta_n^2$,
(51-c2-N) $(\forall n)$: $\Gamma(v_n) \geq r_n$, $|F(v_n) - \alpha(\Gamma,F)| < \eta_n^2$, $|\nabla_{(\geq)}|F(v_n) \leq \eta_n$.

By (51-c1-N), $r_n \to \infty$ as $n \to \infty$. But, from this, conclusions (51-a)+(51-b) are clear. The proof is thereby complete. □

We are now in position to give the promised answer to our coercivity question. For the arbitrary fixed $Q \in \text{fin}(X)$, let us consider the "hybrid" condition:

(PS-Q) each avoiding Q sequence $(x_n) \subseteq \text{Dom}(F)$ for which $(F(x_n))$ converges and $|\nabla_{(\geq)}|F(x_n) \to 0$ has a subsequence (y_n) with $(\Gamma(y_n)) =$ bounded.

This will be referred to as a Palais–Smale condition (modulo $(Q; \geq)$) upon F. When $Q \in \text{fin}(X)$ is generic in such a convention, the resulting property will be referred to as: Palais–Smale condition [modulo $(\text{fin}(X); \geq)$] upon F.

Theorem 13. *Suppose that (in addition) F satisfies a Palais–Smale condition* [*modulo* $(\text{fin}(X); \geq)$]. *Then, F is Γ-coercive.*

Proof. By definition, there exists $Q \in \text{fin}(X)$ such that (PS-Q) holds. If, by absurd, F is not Γ-coercive, the relation (con-1) must be true. By Theorem 12, we have promised a sequence (v_n) in $\text{Dom}(F)$ with the properties (51-a) + (51-b); note that, as a consequence of this, (v_n) avoids Q, $(F(v_n))$ is convergent and $|\nabla_{(\geq)}|F(v_n) \to 0$. Combining with (PS-Q), one deduces that (v_n) must have a subsequence (y_n) with $(\Gamma(y_n))$ = bounded. On the other hand, $\Gamma(y_n) \to \infty$, by (51-a). The contradiction at which we arrived shows that our working assumption is not acceptable; and the conclusion follows. □

In particular, when d is a metric on X, Theorem 13 includes a statement due to Motreanu *et al.* [17]. Further aspects may be found in Turinici [12].

6. Linear Aspects

The obtained results are quasi-metric in nature; so, it would be useful having linear variants of them, expressed in a quasi-norm differential setting.

(A) Let X be a (real) vector space. We say that the function $\Delta : X \to R_+$ is a *quasi-norm* on X, when

(qn-1) Δ is reflexive-sufficient: $\Delta(x) = 0$ iff $x = 0$
(qn-2) Δ is *subadditive*: $\Delta(x + y) \leq \Delta(x) + \Delta(y)$, for all $x, y \in X$.

In addition to these, an extra condition is imposed upon this object:

(qn-3) Δ is *directional continuous*: $\Delta(\tau x) \to 0$ as $\tau \to 0+$, $\forall x \in X$.

Its usefulness will become clear a bit further.

Clearly, each norm $x \mapsto ||x||$ on X is a quasi-norm. The reciprocal of this is not in general true. An appropriate example certifying our claim is in fact the linear version of a previously discussed one, due to Alsulami *et al.* [37]. For completeness reasons, we will rephrase it, under our linear setting.

Example 3. Take $X = R$ (the real axis); and let $x \mapsto |x|$ stand for the usual modulus function; that acts as a norm over X. Further, take some $\beta \in]0,1[$; and define a mapping $\Delta : R \to R_+$ as

$$(\Delta(t) = \beta|t|, \text{ if } t < 0), \ \ (\Delta(t) = |t|, \text{ if } t \geq 0).$$

Clearly, $\Delta(.)$ is a quasi-norm over X. However, it is not a norm; because, e.g.

$$\Delta(-1) = \beta < 1 = \Delta(1).$$

Nevertheless, Δ is *directional continuous*, in view of

$$\Delta(\tau x) = \beta\tau|x| \to 0 \text{ as } \tau \to 0+, \text{ when } x < 0,$$
$$\Delta(\tau x) = \tau|x| \to 0 \text{ as } \tau \to 0+, \text{ when } x \geq 0;$$

hence the claim.

Given the quasi-norm $\Delta(.)$ over X, the mapping (in $\mathcal{F}(X \times X, R_+)$)

$$d(x,y) = \Delta(x-y), \ x,y \in X$$

defines a quasi-metric on X; which, in addition, fulfills

d is invariant to translations: $d(x,y) = d(x+a, y+a)$, $x,y,a \in X$.

Then, let K be some part of X with

K is a convex cone of X ($\xi K + \eta K \subseteq K$, for all $\xi, \eta \geq 0$),
with the non-degenerate property ($K \neq \{0\}$).

Denote by $(\leq)$ and $(\geq)$ the quasi-orders associated to K:

$$x \leq y \text{ iff } y - x \in K, \ x \geq y \text{ iff } x - y \in K.$$

We are now ready to formulate the basic conditions needed in the sequel; these are the ones appearing in the formulation of the previous local variational principle. As a rule, the quasi-order $(\geq)$ will be essentially used here.

(I) The first condition to be considered reads

(X;ge;com) $(X, d, \geq)$ is complete:
each $(\geq)$-ascending d-Cauchy sequence (x_n) in X is d-convergent.

Note that, by the invariance to translation property, this is equivalent with

(−K;ge;com) $(-K, d, \geq)$ is complete:
each $(\geq)$-ascending d-Cauchy sequence (y_n) in $(-K)$ is d-convergent;

just pass to the translated sequence $(y_n = x_n - x_0; n \geq 0)$ to establish this.

Remark 5. The dual conditions attached to these are, resp.,

(X;le;com) $(X, d, \leq)$ is complete:
each $(\leq)$-ascending d-Cauchy sequence (x_n) in X is d-convergent
(K;le;com) $(K, d, \leq)$ is complete:
each $(\leq)$-ascending d-Cauchy sequence (y_n) in K is d-convergent.

It is worth noting that these are rather distinct from the previous ones. The motivation of this consists in the sequential observation

(z_n) is d-Cauchy does not imply (in general) that $(-z_n)$ is d-Cauchy.

(II) The second condition to be considered reads

(X;conv-dom) $(X, d, \geq)$ is conv-dominated: the d-limit
of each $(\geq)$-ascending sequence in X is a $(\geq)$-upper bound of it.

Again by the invariance to translation property, this condition is equivalent with

(−K;conv-dom) $(-K, d, \geq)$ is conv-dominated: the d-limit
of each $(\geq)$-ascending sequence in $(-K)$ is a $(\geq)$-upper bound of it.

In fact, a stronger conclusion holds in our context.

Proposition 18. *The following are equivalent:*

(61-1) $(X, d, \geq)$ is conv-dominated (see above)
(61-2) $(-K)$ is $(d, \geq)$-closed:
the d-limit of each $(\geq)$-ascending sequence in $(-K)$ is in $(-K)$.

Proof. (i) Suppose that $(X, d, \geq)$ is conv-dominated. Given the $(\geq)$-ascending sequence (x_n) in $-K$ and the point $x \in X$ be such that $x_n \xrightarrow{d} x$ as $n \to \infty$, we have (by the imposed condition) that

$(0 \geq)x_n \geq x$, $\forall n$; whence, $0 \geq x$; i.e. $x \in -K$;

proving that $(-K)$ is $(d, \geq)$-closed.

(ii) Conversely, suppose that $(-K)$ is $(d, \geq)$-closed. Given the $(\geq)$-ascending sequence (x_n) in X and the point $x \in X$ with $x_n \xrightarrow{d} x$, we have that

(for each $i \geq 0$): the translated sequence $(y_i^n := x_{n+i} - x_i; n \geq 0)$ is in $(-K)$ and converges to $y_i := x - x_i$.

This, by the imposed condition, yields

(for each $i \geq 0$): $y_i \in -K$; that is: $x_i \geq x$;

proving that $(X, d, \geq)$ is conv-dominated. □

(III) Finally, let us assume that

Δ is directional continuous.

Concerning this aspect, let us consider the directional function (in $\mathcal{F}(R_+)$)

$\Delta_x(\tau) = \Delta(\tau x)$, $\tau \in R_+$, $x \in X$.

By the properties of Δ, we have

$(\forall x \in X)$: $\Delta_x(0) = 0$ and $\Delta_x(.)$ is sub-additive $(\Delta_x(\tau_1 + \tau_2) \leq \Delta_x(\tau_1) + \Delta_x(\tau_2)$, $\forall \tau_1, \tau_2 \in R_+)$.

Then, the directional continuous property means

$(\forall x \in X)$: $\Delta_x(\tau) \to 0 = \Delta_x(0)$ as $\tau \to 0+$.

In this perspective, it would be natural to ask whether the subadditive property suffices for the limit condition above. Some related aspects may be found in Matkowski and Swiatkowski [57].

Having these precise, denote for simplicity

$K_0 = K \cap X_0$, where $X_0 = X \setminus \{0\}$.

We claim that the following property holds:

(full) $(X, d, \geq)$ is *full*:
$(\forall u \in X, \forall \delta > 0)$: $X[u, \delta; \geq] := X[u, \delta] \cap X(u, \geq)$
has a non-empty intersection with $X \setminus \{u\}$;
where $X[u, \delta] = \{x \in X; d(u, x) \leq \delta\}$, $X(u, \geq) = \{x \in X; u \geq x\}$.

In fact, given $h \in K_0$, we have

(p-1) $u - \tau h \in X[u, \delta]$, $\forall \tau \in [0, \eta]$, where $\eta > 0$ is small enough
(p-2) $u - \tau h \in X(u, \geq) \cap (X \setminus \{u\})$, $\forall \tau > 0$;

so, putting these together,

(p-3) $u - \tau h \in X[u, \delta] \cap X(u, \geq) \cap (X \setminus \{u\})$, $\forall \tau \in]0, \eta]$,

proving the assertion. This tells us that, necessarily, the third basic condition of the preceding section is fulfilled.

Having these precise, take some map $\Gamma : X \to R_+$ with the properties

(ga-1-qn) $(\exists(\lambda, \mu),\ 0 < \lambda < 1 < \mu)$, such that
$\Delta(x - y) \leq \lambda \Longrightarrow \Gamma(x) - \Gamma(y) \leq \mu$
(ga-2-qn) $\sup[\Gamma(X)] = \infty$; hence,
$[\Gamma \geq \sigma] := \{x \in X; \Gamma(x) \geq \sigma\}$ is non-empty, $\forall \sigma \geq 0$.

Finally, pick some functional $F : X \to R \cup \{\infty\}$ with

(F-1-qn) F is inf-proper:
$\mathrm{Dom}(F) \neq \emptyset$ and $F_* := \inf[F(X)] > -\infty$
(F-2-qn) F is $(d, \geq)$-lsc on X:
$[F \leq t] := \{x \in X; F(x) \leq t\}$ is $(d, \geq)$-closed, $\forall t \in R$.

The problem to be solved is the one in our preceding section. Note that, by the imposed conditions, Theorems 12 and 13 hold for our data.

(B) In the light of these, the addressed question amounts to expressing this quasi-normed version of Theorem 13 in a standard differential setting. Denote

$$\Theta F(u)(h) = \limsup_{\tau \to 0+} \frac{1}{\Delta(\tau h)}[F(u) - F(u - \tau h)],\ h \in K_0.$$

This object [referred to as the *h-directional* derivative of F at u] always exists, as an element of $R \cup \{-\infty, \infty\}$. As a consequence, for each $u \in K_0$, the quantities

(n-slo-1) $\Lambda_{(-K)}F(u) = \sup\{\Theta F(u)(h); h \in K_0\}$,
(n-slo-2) $|\Lambda_{(-K)}|F(u) = \max\{0, \Lambda_{(-K)}F(u)\}$

are meaningful; after a previous convention, we call these the *quasi-normed relative/absolute* $(-K)$*-slope* of F at u.

Now, for the arbitrary fixed $Q \in \text{fin}(X)$, let us consider the "hybrid" condition:

(PS-Q-K) each avoiding Q sequence $(x_n) \subseteq \text{Dom}(F)$ with $(F(x_n)) =$ convergent and $|\Lambda_{(-K)}|F(x_n) \to 0$, has a subsequence (y_n) with $(\Gamma(y_n)) =$ bounded.

This is referred to as a Palais–Smale differential condition (modulo $(Q; K)$) upon F. When $Q \in \text{fin}(X)$ is generic in such a convention, we shall term the resulting property as: a Palais-Smale differential condition (modulo $(\text{fin}(X), K)$) upon F.

Theorem 14. *Suppose that (in addition) F satisfies a Palais-Smale differential condition (modulo $(\text{fin}(X); K)$). Then, F is Γ-coercive.*

To establish this, the following auxiliary fact will be useful:

Proposition 19. *Let $u \in \text{Dom}(F)$ be arbitrary fixed. Then,*

$$\Lambda_{(-K)}F(u) \le \nabla_{(\ge)}F(u);\ \textit{hence},\ |\Lambda_{(-K)}|F(u) \le |\nabla_{(\ge)}|F(u).$$

Proof (Proposition 19). Let $h \in K_0$ be arbitrary fixed. Given $\delta > 0$, there exists — by a previous observation – some $\eta \in]0, \delta[$, with

$$\forall \tau \in]0, \eta[: 0 < d(u, u - \tau h) = \Delta(\tau h) < \delta, \text{ and } u \ge u - \tau h.$$

This, by definition, yields

$$\begin{aligned}\Theta F(u)(h) &\le \sup\left\{\frac{F(u) - F(u - \tau h)}{\Delta(\tau h)}; 0 < \tau < \eta\right\} \\ &\le \sup\left\{\frac{F(u) - F(x)}{d(u, x)}; x \in X[u, \delta; \ge] \setminus \{u\}\right\}.\end{aligned}$$

Passing to supremum upon $h \in K_0$ gives

$$\Lambda_{(-K)}F(u) \le \sup\left\{\frac{F(u) - F(x)}{d(u, x)}; x \in X[u, \delta; \ge] \setminus \{u\}\right\}, \text{ for all } \delta > 0;$$

wherefrom, passing to infimum upon $\delta > 0$, we derive the desired fact. □

Proof (Theorem 14). By definition, there exists $Q \in \text{fin}(X)$ such that (PS-Q-K) holds. Combining with the auxiliary statement above, we have

Palais–Smale differential condition (modulo $(Q; K)$) upon F implies Palais–Smale condition (modulo $(Q; \ge)$) upon F.

In other words, conditions of Theorem 13 are fulfilled by these data; and this concludes the argument. □

In particular, when the quasi-norm $\Delta(.)$ is a norm on X, the obtained result includes the one in Motreanu and Turinici [15] proved under similar methods (involving the monotone variational principle in Turinici [16]). An extension of these facts to quasi-ordered gauge spaces is possible by following the lines in Bae *et al.* [7]; further aspects will be discussed elsewhere.

References

[1] R.S. Palais and S. Smale, A generalized Morse theory, *Bull. Amer. Math. Soc.* **70**, 165–171, (1964).

[2] L. Caklovic, S. Li, and M. Willem, A note on Palais–Smale condition and coercivity, *Diff. Int. Equations* **3**, 799–800, (1990).

[3] H. Brezis and L. Nirenberg, Remarks on finding critical points, *Commun. Pure Appl. Math.* **44**, 939–963, (1991).

[4] D. Goeleven, A note on Palais–Smale condition in the sense of Szulkin, *Diff. Int. Equations* **6**, 1041–1043, (1993).

[5] D. Motreanu and V.V. Motreanu, Coerciveness property for a class of nonsmooth functionals, *Zeitschr. Anal. Anwendungen (J. Analysis Appl.)* **19**, 1087–1093, (2000).

[6] D. Motreanu and P.D. Panagiotopoulos, *Minimax Theorems and Qualitative Properties of the Solutions of Hemivariational Inequalities* (Kluwer Acad. Publ., Dordrecht, 1999).

[7] J.-S. Bae, S.-H. Cho, and J.-J. Kim, An Ekeland type variational principle on gauge spaces with applications to fixed point theory, drop theory and coercivity, *Bull. Korean Math. Soc.* **48**, 1023–1032, (2011).

[8] H.-K. Xu, On the Palais–Smale condition for nondifferentiable functionals, *Taiwanese J. Math.* **4**, 627–634, (2000).

[9] M. Turinici, Function variational principles and coercivity, *J. Math. Anal. Appl.* **304**, 236–248, (2005).

[10] M. Turinici, Gauge variational principles and normed coercivity, *An. Şt. Univ. A. I. Cuza Iaşi (Sect I-a: Mat.)* **52**, 251–275, (2006).

[11] M. Turinici, Maximality principles: theory and practice, *Sc. Annals UASVM Iaşi* **49**, 323–360, (2006).

[12] M. Turinici, Function variational principles and coercivity over normed spaces, *Optimization* **59**, 199–222, (2010).

[13] I. Ekeland, Nonconvex minimization problems, *Bull. Amer. Math. Soc. (New Series)* **1**, 443–474, (1979).

[14] C.K. Zhong, A generalization of Ekeland's variational principle and application to the study of the relation between the weak P.S. condition and coercivity, *Nonlin. Anal.* **29**, 1421–1431, (1997).

[15] V.V. Motreanu and M. Turinici, Coercivity properties for monotone functionals, *Note Mat.* **21**, 83–91, (2002).
[16] M. Turinici, A monotone version of the variational Ekeland's principle, *An. Şt. Univ. A. I. Cuza Iaşi (Sect I-a: Mat.)* **36**, 329–352, (1990).
[17] D. Motreanu, V.V. Motreanu, and M. Turinici, Coerciveness property for conical nonsmooth functionals, *J. Optim. Th. Appl.* **145**, 148–163, (2010).
[18] M. Turinici, Functional monotone VP and metrical coercivity, *Libertas Math.* **27**, 1–16, (2007).
[19] M. Turinici, Coercivity properties for order nonsmooth functionals, *An. Şt. Univ. Ovidius Constanţa (Mat.)* **19**, 297–312, (2011).
[20] M. Turinici, Normed coercivity for monotone functionals, *Romai J.* **7**(2), 169–179, (2011).
[21] E. DeGiorgi, A. Marino, and M. Tosques, Problemi di evoluzione in spazi metrici e curve di massima pendenza, *Rend. Accad. Naz. Lincei* (*Serie 8*) **68**, 180–187, (1980).
[22] M. Turinici, Multivalued variational principles and normed coercivity, *Libertas Math.* **30**, 1–17, (2010).
[23] M. Turinici, Function variational principles and normed minimizers. In: N.J. Daras and T.M. Rassias, (Eds.), *Computational Mathematics and Variational Analysis*, pp. 513–536 (Springer Nature, Switzerland, 2020).
[24] P.J. Cohen, *Set Theory and the Continuum Hypothesis* (Benjamin, New York, 1966).
[25] P. Bernays, A system of axiomatic set theory: Part III. Infinity and enumerability analysis, *J. Symbolic Logic* **7**, 65–89, (1942).
[26] A. Tarski, Axiomatic and algebraic aspects of two theorems on sums of cardinals, *Fund. Math.* **35**, 79–104, (1948).
[27] E.S. Wolk, On the principle of dependent choices and some forms of Zorn's lemma, *Canad. Math. Bull.* **26**, 365–367, (1983).
[28] Y. Moskhovakis, *Notes on Set Theory.* (Springer, New York, 2006).
[29] E. Schechter, *Handbook of Analysis and Its Foundation* (Academic Press, New York, 1997).
[30] G.H. Moore, *Zermelo's Axiom of Choice: Its Origin, Development and Influence* (Springer, New York, 1982).
[31] S. Kasahara, On some generalizations of the Banach contraction theorem, *Publ. Res. Inst. Math. Sci. Kyoto Univ.* **12**, 427–437, (1976).
[32] M. Turinici, Function pseudometric VP and applications, *Bul. Inst. Polit. Iaşi (Sect: Mat., Mec. Teor., Fiz.)* **53**(57), 393–411, (2007).
[33] K. Kuratowski, *Topology* (Academic Press, New York, 1966).
[34] P. Hitzler, *Generalized Metrics and Topology in Logic Programming Semantics.* PhD Thesis (Natl. Univ. Ireland, Univ. College Cork, 2001).
[35] M. Turinici, Pseudometric versions of the Caristi–Kirk fixed point theorem, *Fixed Point Th.* **5**, 147–161, (2004).
[36] S. Cobzaş, Fixed points and completeness in metric and in generalized metric spaces, arxiv: 1508-05173-v5, (7 October 2019).
[37] H.H. Alsulami, E. Karapinar, F. Khojasteh, and A.-F. Roldán, A proposal to the study of contractions in quasi-metric spaces, *Discrete Dyn. Nat. Soc.* **2014**, Article ID 269286.

[38] G. Bantaş and M. Turinici, Mean value theorems via division methods, *An. Şt. Univ. A. I. Cuza Iaşi (Sect I-a: Mat.)* **40**, 135–150, (1994).

[39] A. Hamel, *Variational Principles on Metric and Uniform Spaces*. Habilitation Thesis (Martin-Luther University, Halle-Wittenberg, 2005).

[40] M. Turinici, Sequential maximality principles, In: T.M. Rassias and P.M. Pardalos (Eds.), *Mathematics Without Boundaries*, pp. 515–548 (Springer, New York, 2014).

[41] B.G. Kang and S. Park, On generalized ordering principles in nonlinear analysis, *Nonlinear Anal.* **14**, 159–165, (1990).

[42] M. Turinici, A generalization of Brezis–Browder's ordering principle, *An. Şt. Univ. A. I. Cuza Iaşi (Sect. I-a: Mat)* **28**, 11–16, (1982).

[43] M. Turinici, Pseudometric extensions of the Brezis–Browder ordering principle, *Math. Nachrichten* **130**, 91–103, (1987).

[44] H. Brezis and F.E. Browder, A general principle on ordered sets in nonlinear functional analysis, *Advances Math.* **21**, 355–364, (1976).

[45] M. Altman, A generalization of the Brezis-Browder principle on ordered sets, *Nonlin. Anal.* **6**, 157–165, (1982).

[46] N. Bourbaki, Sur le théorème de Zorn, *Archiv Math.* **2**, 434–437, (1949/1950).

[47] M. Turinici, Maximality principles and mean value theorems, *An. Acad. Brasil. Cienc.* **53**, 653–655, (1981).

[48] S. Dancs, M. Hegedus, and P. Medvegyev, A general ordering and fixed-point principle in complete metric space, *Acta Sci. Math. (Szeged)* **46**, 381–388, (1983).

[49] M. Turinici, Maximal elements in a class of order complete metric spaces, *Math. Japonica* **25**, 511–517, (1980).

[50] M. Turinici, A generalization of Altman's ordering principle, *Proc. Amer. Math. Soc.* **90**, 128–132, (1984).

[51] D.H. Hyers, G. Isac and T.M. Rassias, *Topics in Nonlinear Analysis and Applications* (World Sci. Publ., Singapore, 1997).

[52] A. Goepfert, H. Riahi, C. Tammer and C. Zălinescu, *Variational Methods in Partially Ordered Spaces. Canad. Math. Soc. Books Math*, Vol. 17 (Springer, New York, 2003).

[53] A. Brøndsted, Fixed points and partial orders, *Proc. Amer. Math. Soc.* **60**, 365–366, (1976).

[54] N. Brunner, Topologische Maximalprinzipien, *Zeitschr. Math. Logik Grundl. Math.* **33**, 135–139, (1987).

[55] J. Dodu and M. Morillon, The Hahn–Banach property and the Axiom of Choice, *Math. Logic Quarterly* **45**, 299–314, (1999).

[56] J.N. Corvellec, M. DeGiovanni, and M. Marzocchi, Deformation properties for continuous functionals and critical point theory, *Topol. Meth. Nonlin. Anal.* **1**, 151–171, (1993).

[57] J. Matkowski and T. Swiatkowski, On subadditive functions, *Proc. Amer. Math. Soc.* **119**, 187–197, (1993).

https://doi.org/10.1142/9789811261572_0029

Chapter 29

Motion Around the Equilibrium Points in the Photogravitational R3BP under the Effects of Poynting–Robertson Drag, Circumbinary Belt and Triaxial Primaries with an Oblate Infinitesimal Body: Application on Achird Binary System

Aguda Ekele Vincent* and Vassilis S. Kalantonis†,‡

**Department of Mathematics, School of Basic Sciences, Nigeria Maritime University, Okerenkoko, Delta State, Nigeria*
†Department of Electrical and Computer Engineering, University of Patras, GR-26504, Patras, Greece
‡kalantonis@upatras.gr

In the present work, we study the motion of an oblate infinitesimal mass body near the equilibrium points (EPs) of the circular restricted three-body problem (R3BP) in which the radiation pressure, Poynting–Robertson (P–R) drag effect, and triaxiality of the two primary bodies are considered in the case where both of them are enclosed by a belt of homogeneous circular cluster of material points centered at the mass center of the system. We have found numerically that five or seven EPs may lie on the plane of motion depending on the values of the parameters of the system and have examined their stability character, too. In particular, the numerical exploration is performed using the binary system Achird to compute the positions of the equilibria and the eigenvalues of the characteristic equation. It is observed that the existence, location and number of equilibria of the problem depend on the values of the parameters of the problem. We have found both numerically and analytically that under constant P–R drag effect, collinear equilibrium solutions cease to exist. The linear stability of each equilibrium point is also studied and it is found that in the case where seven equilibria exist, the new point L_{N2} is always linearly stable while the other six are always linearly unstable. In the case where five equilibria exist, all of them are always linearly unstable due to P–R drag effect.

1. Introduction

The circular restricted three-body problem (R3BP) consists of two finite bodies, known as primaries, which rotate in circular orbits around their common center of mass and a massless body which moves in the plane of motion of the primaries under their gravitational attraction and does not affect their motion. The circular R3BP has been the well-known studied problem in Celestial Mechanics. In the classical R3BP there are five equilibrium points (EPs). Three of them lie on the x-axis and are called collinear while the other two are away from the x-axis and are called triangular (non-collinear) EPs. The three collinear points are generally unstable while the triangular points are stable for the mass ratio $\mu \leqslant 0.03850...$ [1]. These EPs and the periodic orbits in the immediate neighborhood have enabled several space mission explorations [2–4] while some operations are still in progress.

During the past, several variants of this classical problem have been proposed by many scientists and astronomers in order to make it more realistic for real systems of dynamical astronomy. The dynamics of a small mass point around one star with a planet or two stars may be considered as a generalization of the classical R3BP. In this case we have the so-called photogravitational restricted problem of three bodies. Radzievskii [5] was the first in considering this problem, and because of the great importance of radiation pressure, several researchers have included radiation pressure force of either one or both primaries in the study of the R3BP [6–12]. In estimating the light radiation force, all the above studies of photogravitational R3BPs have taken into account just one of the three components of the light pressure field, which is due to the central force: the gravitation and the radiation pressure. The other two components which we are also interested in arise from the Doppler shift and the absorption and subsequent re-emission of the incident radiation. These last two components constitute the so-called Poynting–Robertson (P–R) effect [13,14]. This choice is motivated by the fact that P–R effect is considered as the most important non-gravitational effect acting on dust particles. The P–R effect clears negligible particles of the cosmic dust grain at a cosmically rapid rate in the solar system. Many authors have discussed the effects of radiation pressure and P–R drag forces on the equilibrium points and their properties such as their linear stability [13–19].

As it is well known [20], many celestial bodies have irregular shapes (they are either oblate or triaxial). For example, in our Solar System, the Earth, Jupiter, Saturn, Regulus, Neutron stars and black dwarfs are oblate while

Pluto and its moon Charon are triaxial. The lack of sphericity, triaxiality or oblateness of the celestial bodies causes substantial perturbations in the two–body orbits of the system. This inspired several researchers [21–31] to include non-sphericity of the bodies in their studies of the R3BP. Extensions to more realistic problems appeared in many previous works on the R3BP by considering the primaries oblate or triaxial planets with additional terms, such as the radiation pressure and the P–R drag [32–35]. Furthermore, the effects of small perturbations in the Coriolis and centrifugal forces together with P–R drag and oblateness have been discussed recently [36]. It was observed that under constant P–R drag effect, the well-known collinear EPs of the circular R3BP cease to exist numerically and of course analytically. It was established that the P–R effect renders unstable the EPs which are conditionally stable in the classical case.

Studies of planetary and stellar systems have revealed discs of dust particles which are regarded as young analogues of the Kuiper Belt in the Solar System [37]. These discs play important roles in the origin of planets' orbital elements. The importance of the problem in astronomy has been addressed by Jiang and Yeh [38,39] where it was shown that the presence of disc resulted in additional EPs of the system. Other works took into account the gravitational potential from the belt/disc under certain different assumptions [40–45]. Recently, Amuda *et al.* [46] explored the existence and linear stability of the EPs in the framework of circular R3BP with the postulation that the third body is oblate and the primaries emit radiation pressure together with P-R drag, taking also into account circumbinary disc. They have found that eight equilibria exist at most; six of them lie in the axis which connects the primaries while two form triangles with the line under some conditions involving the disc, mass and radiation pressure parameters. They have shown that these points are unstable due to P–R effect. However, these results contradict the studies by Ragos and Zafiropoulos [14] and Vincent and Perdiou [36] (and references therein), that the points known as "collinear" are positioned off the x-axis due to P–R effect. Moreover, the triaxiality parameters that control the shape of the primary bodies were ignored in their study.

The present work extends the research performed by Amuda *et al.* [46]. We examine the existence and stability of EPs in the R3BP when the primaries are modeled as triaxial rigid bodies enclosed by a disc as well as sources of radiation pressure with its P–R drag effect while the infinitesimal body is taken to be an oblate spheroid. With respect to that paper by Amuda *et al.* [46], the current work displays a number of interesting features which were not apparent. Additionally, the equilibrium solutions under general drag (e.g. Stokes drag, P–R drag, solar wind drag) are only poorly studied in the literature by analytical means (with some exception found in Murray [47]). More specifically, the content of this chapter is organized as follows: In Section 2, we define the mathematical model and the governing equations of motion that we use for our study. In Section 3, we determine numerically the existence and locations of the EPs and verify them graphically for various values of the parameters under consideration while their linear stability is analyzed in Section 4. Finally, Section 5 summarizes the discussion and conclusion of our study.

2. Mathematical Formulation and Equations of Motion

We consider a barycentric coordinate system $Oxyz$ rotating relative to an inertial reference system with angular velocity ω about a common z-axis. Let the two massive bodies, with masses $m_1 = 1-\mu$ and $m_2 = \mu$ $(0 < \mu \leqslant 1/2)$ where μ is the mass-ratio parameter, have fixed positions at the Ox-axis. Also, the massive bodies are considered to be sources of radiation and triaxial in nature with inclusion of P–R drag effect and a disc. In this premise, the model also accounts for oblateness of an infinitesimal third body. The equations of motion of the infinitesimal mass under the influence of its shape (oblateness) in the three-dimensional restricted three-body problem with the origin resting at the center of mass, in a barycentric rotating coordinate system under the gravitational influence of two main bodies emitting radiation pressure with the P–R drag present, have the form [14,25,44]

$$\begin{aligned}
\ddot{x} - 2\dot{y} = x &- \frac{(1-\mu)(x+\mu)q_1}{r_1^3} - \frac{\mu(x+\mu-1)q_2}{r_2^3} - \frac{3(1-\mu)(x+\mu)A_3}{2r_1^5} \\
&- \frac{3\mu(x+\mu-1)A_3}{2r_2^5} - \frac{W_1}{r_1^2}\left\{\frac{x+\mu}{r_1^2}[(x+\mu)\dot{x} + y\dot{y} + z\dot{z}] + \dot{x} - y\right\} \\
&- \frac{W_2}{r_2^2}\left\{\frac{x+\mu-1}{r_2^2}[(x+\mu-1)\dot{x} + y\dot{y} + z\dot{z}] + \dot{x} - y\right\}, \\
\ddot{y} + 2\dot{x} = y &- \frac{(1-\mu)yq_1}{r_1^3} - \frac{\mu y q_2}{r_2^3} - \frac{3(1-\mu)yA_3}{2r_1^5} - \frac{3\mu y A_3}{2r_2^5} \\
&- \frac{W_1}{r_1^2}\left[\frac{y}{r_1^2}\{(x+\mu)\dot{x} + y\dot{y} + z\dot{z}\} + \dot{y} + (x+\mu)\right]
\end{aligned} \tag{1}$$

$$-\frac{W_2}{r_2^2}\left[\frac{y}{r_2^2}\left\{(x+\mu-1)\dot{x}+y\dot{y}+z\dot{z}\right\}+\dot{y}+(x+\mu-1)\right],$$

$$\ddot{z}=-\frac{(1-\mu)zq_1}{r_1^3}-\frac{\mu zq_2}{r_2^3}-\frac{3(1-\mu)zA_3}{2r_1^5}-\frac{3\mu zA_3}{2r_2^5}$$

$$-\frac{W_1}{r_1^2}\left[\frac{z}{r_1^2}\left\{(x+\mu)\dot{x}+y\dot{y}+z\dot{z}\right\}+\dot{z}\right]$$

$$-\frac{W_2}{r_2^2}\left[\frac{z}{r_2^2}\left\{(x+\mu-1)\dot{x}+y\dot{y}+z\dot{z}\right\}+\dot{z}\right],$$

where

$$\begin{aligned} r_1^2 &= (x+\mu)^2+y^2+z^2, \qquad r_2^2=(x+\mu-1)^2+y^2+z^2,\\ W_1 &= \tfrac{(1-\mu)(1-q_1)}{c_d}, \qquad W_2=\tfrac{\mu(1-q_2)}{c_d}, \end{aligned} \tag{2}$$

with r_i, $i=1,2$, being the distances of the third body from the two primaries m_1 and m_2, resp., q_1, q_2 ($q_i \leqslant 1$, $i=1,2$) and W_1, W_2 ($W_i << 1$, $i=1,2$) stand for radiation pressure and P–R drag of the bigger and smaller primaries, resp., c_d is the non-dimensional velocity of light while the dots denote differentiation with respect to time t. Note that when radiation force is absent, there will be no P–R drag force. Also, A_3 is the oblateness coefficient of the infinitesimal body defined by the formula $A_3=(A_E^2-A_P^2)/5R^2 << 1$ where A_E and A_P are the equatorial and polar radii of the said third body, resp., and R is the distance between the primaries. The triaxiality of the first and second primaries are included with the help of the parameters $\sigma_i << 1$, $\sigma_i' << 1$ ($i=1,2$), resp., where $\sigma_1=(h^2-f^2)/5R^2$, $\sigma_2=(b^2-f^2)/5R^2$, $\sigma_1'=(h^{'2}-f^{'2})/5R^2$, $\sigma_2'=(b^{'2}-f^{'2})/5R^2$ where h,b,f represent the lengths of the semi-axis of the first primary and h',b',f' those of the second one.

More so, for a gravitational potential from the disc, the flattened potential is given as [48]

$$V(r,z)=-\frac{M_b}{\sqrt{r^2+a+(\sqrt{z^2+b^2})^2}}, \tag{3}$$

where M_b ($M_b << 1$) is the mass of the disc, r is the radial distance of the dust particle so that $r^2=x^2+y^2$, a is the flatness parameter, while b is the core parameter. Restricting ourselves to the $xy-$ plane, equation (3) becomes

$$V(r) = -\frac{M_b}{\sqrt{r^2+T^2}}, \tag{4}$$

where $T = a + b$ is the density profile of the disc chosen to the value $T = 0.01$ [49].

Denoting by n the mean motion of the third body, from Singh *et al.* [30] and Singh and Taura [49] we have the following expression:

$$n = \sqrt{1 + \frac{3}{2}(2\sigma_1 - \sigma_2) + \frac{3}{2}(2\sigma_1' - \sigma_2') + \frac{2M_b r_c}{(r_c^2+T^2)^{\frac{3}{2}}}}, \tag{5}$$

where $r_c^2 = 1 - \mu + \mu^2$ is the radial distance of the test particle. Therefore, using equations (4) and (5) and following the approach by Singh and Emmanuel [34] and Amuda *et al.* [46], the pertinent equations of model (1) in the xy-plane are finally written in the form

$$\begin{aligned}
\ddot{x} - 2n\dot{y} &= U_x - \frac{W_1}{r_1^2}\left\{\frac{x+\mu}{r_1^2}[(x+\mu)\dot{x} + y\dot{y}] + \dot{x} - ny\right\} \\
&\quad - \frac{W_2}{r_2^2}\left\{\frac{x+\mu-1}{r_2^2}[(x+\mu-1)\dot{x} + y\dot{y}] + \dot{x} - ny\right\}, \\
\ddot{y} + 2n\dot{x} &= U_y - \frac{W_1}{r_1^2}\left\{\frac{y}{r_1^2}[(x+\mu)\dot{x} + y\dot{y}] + \dot{y} + n(x+\mu)\right\} \\
&\quad - \frac{W_2}{r_2^2}\left\{\frac{y}{r_2^2}[(x+\mu-1)\dot{x} + y\dot{y}] + \dot{y} + n(x+\mu-1)\right\},
\end{aligned} \tag{6}$$

where

$$\begin{aligned}
U_x &= n^2 x - \frac{(1-\mu)(x+\mu)q_1}{r_1^3} - \frac{\mu(x+\mu-1)q_2}{r_2^3} \\
&\quad - \frac{3(1-\mu)(2\sigma_1-\sigma_2)(x+\mu)q_1}{2r_1^5} - \frac{3\mu(2\sigma_1'-\sigma_2')(x+\mu-1)q_2}{2r_2^5} \\
&\quad + \frac{15(1-\mu)(\sigma_1-\sigma_2)(x+\mu)y^2 q_1}{2r_1^7} + \frac{15\mu(\sigma_1'-\sigma_2')(x+\mu-1)y^2 q_2}{2r_2^7} \\
&\quad - \frac{3(1-\mu)(x+\mu)A_3}{2r_1^5} - \frac{3\mu(x+\mu-1)A_3}{2r_2^5} - \frac{M_b x}{(r^2+T^2)^{\frac{3}{2}}},
\end{aligned} \tag{7}$$

$$U_y = n^2 y - \frac{(1-\mu)yq_1}{r_1^3} - \frac{\mu y q_2}{r_2^3} - \frac{3(1-\mu)(\sigma_1-\sigma_2)yq_1}{r_1^5}$$
$$-\frac{3\mu(\sigma_1'-\sigma_2')yq_2}{r_2^5} - \frac{3(1-\mu)(2\sigma_1-\sigma_2)yq_1}{2r_1^5}$$
$$-\frac{3\mu(2\sigma_1'-\sigma_2')yq_2}{2r_2^5} + \frac{15(1-\mu)(\sigma_1-\sigma_2)y^3q_1}{2r_1^7}$$
$$+\frac{15\mu(\sigma_1'-\sigma_2')y^3q_2}{2r_2^7} - \frac{3(1-\mu)yA_3}{2r_1^5} - \frac{3\mu y A_3}{2r_2^5} - \frac{M_b y}{(r^2+T^2)^{\frac{3}{2}}},$$

with

$$U = \frac{n^2}{2}(x^2+y^2) + \frac{(1-\mu)q_1}{r_1} + \frac{\mu q_2}{r_2} + \frac{(1-\mu)(2\sigma_1-\sigma_2)q_1}{2r_1^3}$$
$$+\frac{\mu(2\sigma_1'-\sigma_2')q_2}{2r_2^3} - \frac{3(1-\mu)(\sigma_1-\sigma_2)q_1y^2}{2r_1^5} - \frac{3\mu(\sigma_1'-\sigma_2')q_2y^2}{2r_2^5}$$
$$+\frac{(1-\mu)A_3}{2r_1^3} + \frac{\mu A_3}{2r_2^3} + \frac{M_b}{(r^2+T^2)^{\frac{1}{2}}},$$

and

$$r_1^2 = (x+\mu)^2 + y^2, \qquad r_2^2 = (x+\mu-1)^2 + y^2. \tag{8}$$

3. Existence and Location of Equilibrium Points

It is well-known that the acceleration and the velocity of the infinitesimal body at the Lagrangian points are zero, so that $\ddot{x} = \ddot{y} = \dot{x} = \dot{y} = 0$. It thus follows from equation (6) that the equilibrium points are solutions of equations:

$$n^2x - \frac{(1-\mu)(x+\mu)q_1}{r_1^3} - \frac{\mu q_2(x+\mu-1)}{r_2^3} - \frac{3(1-\mu)(2\sigma_1-\sigma_2)(x+\mu)q_1}{2r_1^5}$$
$$-\frac{3\mu q_2(2\sigma_1'-\sigma_2')(x+\mu-1)}{2r_2^5} + \frac{15(1-\mu)(\sigma_1-\sigma_2)y^2q_1(x+\mu)}{2r_1^7}$$
$$+\frac{15\mu(\sigma_1'-\sigma_2')y^2q_2(x+\mu-1)}{2r_2^7} - \frac{3(1-\mu)(x+\mu)A_3}{2r_1^5}$$
$$-\frac{3\mu(x+\mu-1)A_3}{2r_2^5} - \frac{M_b x}{(r^2+T^2)^{\frac{3}{2}}} + \frac{W_1 n y}{r_1^2} + \frac{W_2 n y}{r_2^2} = 0, \tag{9}$$

and

$$n^2y - \frac{(1-\mu)q_1y}{r_1^3} - \frac{\mu q_2y}{r_2^3} - \frac{3(1-\mu)(2\sigma_1-\sigma_2)q_1y}{2r_1^5} - \frac{3(\sigma_1-\sigma_2)(1-\mu)q_1y}{r_1^5} - \frac{3\mu(2\sigma_1'-\sigma_2')q_2y}{2r_2^5} - \frac{3(\sigma_1'-\sigma_2')\mu q_2y}{r_2^5} + \frac{15(1-\mu)(\sigma_1-\sigma_2)q_1y^3}{2r_1^7} + \frac{15\mu(\sigma_1'-\sigma_2')q_2y^3}{2r_2^7} - \frac{W_1n(x+\mu)}{r_1^2} - \frac{W_2n(x+\mu-1)}{r_2^2} - \frac{3(1-\mu)A_3y}{2r_1^5} - \frac{3\mu yA_3}{2r_2^5} - \frac{M_by}{(r^2+T^2)^{\frac{3}{2}}} = 0. \tag{10}$$

It is also known that in the classical R3BP there are two types of equilibria or solutions depending on whether $y = 0$ or $y \neq 0$. The points for which $y = 0$, denoted by $L_{1,2,3}$, are called collinear EPs and lie on the line connecting the primaries, i.e. the x-axis of the synodic system, while the points for which $y \neq 0$, denoted by $L_{4,5}$, are called triangular EPs (non-collinear) and lie away from the x-axis of the synodic system. It is important to stress that the case where $M_d = \sigma_1 = \sigma_2 = \sigma_1' = \sigma_2' = A_3 = 0$, that is, the photogravitational version of the R3BP, has been studied in full detail [14] where, for $q_i \in (-\infty, 1]$, $i = 1, 2$, the authors discussed the existence and stability of the equilibrium points.

In Fig. 1, we present five non-collinear EPs in the case when $\mu = 0.025$, $q_1 = 0.5$, $q_2 = 0.999$, $\sigma_1 = 0.0004$, $\sigma_2 = 0.0003$, $\sigma_1' = 0.0002$, $\sigma_2' = 0.0001$, $A_3 = 0.02$, $M_b = 0.001$, $T = 0.01$ and $c_d = 299792458$. It is worth mentioning that the equilibria in the xy-plane are given by mutual intersections of both curves denoted by green dots. Besides, the intersection points of these curves are the coordinates (x_0, y_0) of the equilibria by means of an iterative process. First, we note that the y components of the equilibrium points $L_{1,2,3}$ are close to zero but not zero. It is interesting to note that the points $L_{1,2,3}$ no longer lie on the x-axis, since the condition $y = 0$ is not fulfilled for the equilibrium solutions associated to equations (9) and (10) but are positioned off the x-axis. Moreover, this can be easily seen from top right and bottom frames in Fig. 1 where we zoomed the area close to $L_{1,2}$ and L_3, resp. Using the same chosen parametric values when $\mu = 0.45$ and $M_b = 0.05$, we exemplify the case of seven equilibria as shown in Fig. 2. In this case the two additional points named L_{N1} and L_{N2} located on the negative x-axis between m_1 and the origin exist simultaneously depending on the values of μ and M_b. Similarly, enlargement of areas close to L_{N1}, L_{N2}, L_1, L_2 and L_3 of Fig. 2 are displayed in the top right, bottom left and bottom right frames, resp. Apparently, these

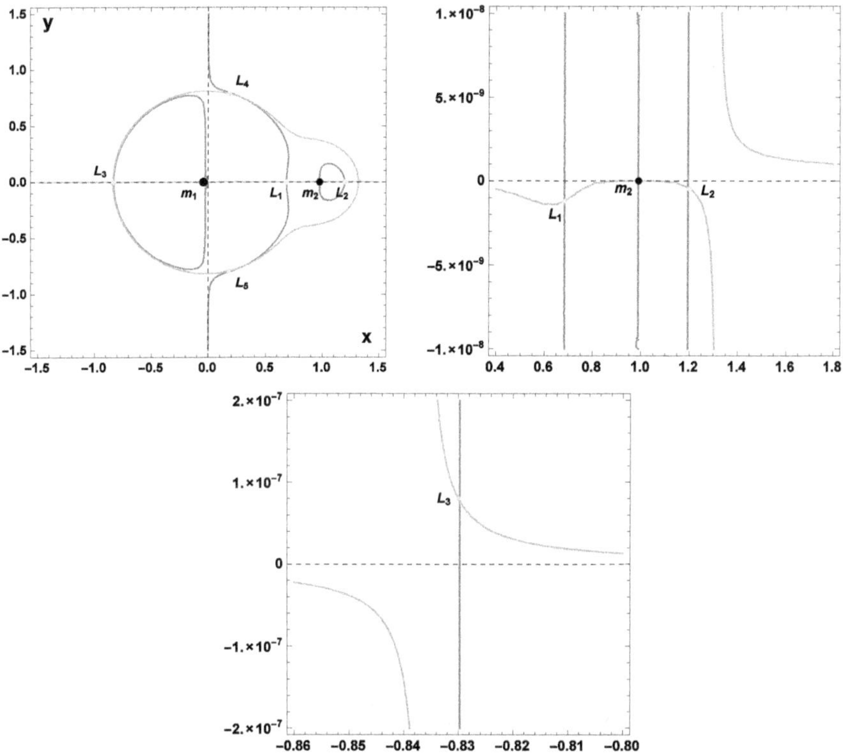

Fig. 1: (Color online) Diagrams showing the positions (green dots) for the case of five equilibrium on the xy-plane for $\mu = 0.025$, $q_1 = 0.5$, $q_2 = 0.999$, $\sigma_1 = 0.0004$, $\sigma_2 = 0.0003$, $\sigma_1' = 0.0002$, $\sigma_2' = 0.0001$, $A_3 = 0.02$, $M_b = 0.001$, $T = 0.01$ and $c_d = 299792458$. Top right and bottom frames depict zoomed images of $L_{1,2}$ and L_3, resp., with intersections of the curves. Black dots indicate the position of the bodies m_i.

numerical results suggest that the additional EPs do not in general exist even when there is a disc to interact with. For this reason, it would be interesting to carry out the investigation of the equilibrium points where more than five critical points exist.

Furthermore, for the triangular solutions, the relevant condition is satisfied, therefore, the triangular EPs are obtained from the solutions of equations (9) and (10) when $x, y \neq 0$. To evaluate analytically the occurrence of solutions, we can make use of the perturbation method. Precisely, if we take $\sigma_1 = \sigma_2 = \sigma_1' = \sigma_2' = M_b = W_1 = W_2 = A_3 = 0$ and $q_1 = q_2 = 1$ in equations (9) and (10) with $n = 1$, the solutions yield $r_1 = r_2 = 1$, which implies that the third body can be at either of these two

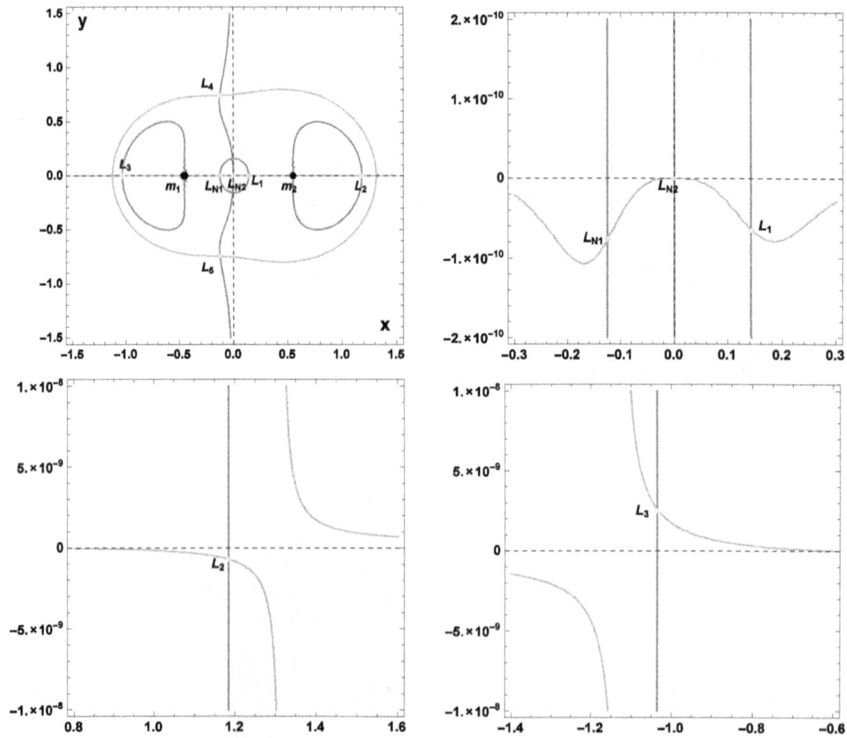

Fig. 2: (Color online) $\mu = 0.45$, $M_b = 0.05$. Similar to Fig. 1, but for the case of seven equilibria. Top right, bottom left and bottom right frames depict zoomed images of L_{N1}, L_{N2}, L_1, L_2 and L_3, resp., with intersections of the curves.

points located at the same distance from the primaries as the two primaries are from each other.

Now, with the presence of the perturbation parameters A_3, M_b, σ_i, σ_i', W_i, q_i, $i = 1, 2$, solutions of equations (9) and (10) are then sought in the form

$$r_i = 1 + \varepsilon_i, \quad \varepsilon_i << 1, \qquad i = 1, 2 \tag{11}$$

where ε_i are the artificial small parameters. Substituting the solutions of equations (11) into equation (8) and solving for x and y retaining only linear terms in ε_1 and ε_2, we obtain

$$x = \frac{1}{2} - \mu + \varepsilon_1 - \varepsilon_2, \qquad y = \pm\frac{\sqrt{3}}{2}\left[1 + \frac{2}{3}(\varepsilon_1 + \varepsilon_2)\right]. \tag{12}$$

Substituting the values of $r_{1,2}$, x and y from equations (11) and (12), resp., and n^2 from equation (5) into equations (9) and (10) appropriately, and for simplicity putting $q_i = 1 - \delta_i$ and neglecting terms of second and higher powers of σ_i, σ_i', W_i, δ_i, $i = 1, 2$, M_b, A_3 and their products, we obtain after some computations

$$\varepsilon_1 = \frac{A_3}{2} - \frac{W_1}{3\sqrt{3}(1-\mu)} - \frac{2W_2}{3\sqrt{3}(1-\mu)} - \frac{M_b(2r_c - 1)}{3(r_c^2 + T^2)^{\frac{3}{2}}} - \frac{\delta_1}{3} - \frac{11\sigma_1}{8} + \frac{11\sigma_2}{8}$$
$$- \frac{(2-3\mu)\sigma_1'}{2(1-\mu)} + \frac{(1-2\mu)\sigma_2'}{2(1-\mu)},$$
$$\varepsilon_2 = \frac{A_3}{2} + \frac{2W_1}{3\sqrt{3}\mu} + \frac{W_2}{3\sqrt{3}\mu} - \frac{M_b(2r_c - 1)}{3(r_c^2 + T^2)^{\frac{3}{2}}} - \frac{\delta_2}{3} + \frac{(1-3\mu)\sigma_1}{2\mu} - \frac{(1-2\mu)\sigma_2}{2\mu}$$
$$- \frac{11\sigma_1'}{8} + \frac{11\sigma_2'}{8}, \tag{13}$$

from which evidently $r_1 \neq r_2$. Substituting these solutions into equations (12), we may obtain the positions of the triangular equilibrium points (TEPs), $L_{4,5}$ as

$$x = \frac{1}{2} - \mu - \frac{1-q_1}{3} + \frac{1-q_2}{3} - \frac{W_1(2-\mu)}{3\mu\sqrt{3}(1-\mu)} - \frac{W_2(1+\mu)}{3\mu\sqrt{3}(1-\mu)}$$
$$- \frac{1}{2}\left(\frac{1}{\mu} - \frac{1}{4}\right)\sigma_1 + \frac{1}{2}\left(\frac{1}{\mu} + \frac{3}{4}\right)\sigma_2$$
$$+ \frac{1}{2}\left(\frac{\mu}{1-\mu} + \frac{3}{4}\right)\sigma_1' - \frac{1}{2}\left(\frac{\mu}{1-\mu} + \frac{7}{4}\right)\sigma_2',$$
$$y = \pm\frac{\sqrt{3}}{2}\left\{1 - \frac{2(1-q_1)}{9} - \frac{2(1-q_2)}{9} - \frac{4M_b(2r_c - 1)}{9(r_c^2 + T^2)^{\frac{3}{2}}} + \frac{2W_1(2-3\mu)}{9\mu\sqrt{3}(1-\mu)}\right.$$
$$+ \frac{2W_2(1-3\mu)}{9\mu\sqrt{3}(1-\mu)} + \frac{1}{3}\left[\left(\frac{-23}{4} + \frac{1}{\mu}\right)\sigma_1 + \left(\frac{19}{4} - \frac{1}{\mu}\right)\sigma_2\right.$$
$$\left.\left. + \left(\frac{\mu}{1-\mu} - \frac{19}{4}\right)\sigma_1' + \left(\frac{15}{4} - \frac{\mu}{1-\mu}\right)\sigma_2' + 2A3\right]\right\}. \tag{14}$$

Our results agree with those obtained by Amuda *et al.* [46] if we set $\sigma_1 = \sigma_2 = \sigma_1' = \sigma_2' = 0$ in equations (14) except for their y coordinates which were due to erroneous positive in the coefficient of M_b in their equation (18). In general, in the limiting case where $W_i = 0$ and $\neq 0$, $i = 1, 2$, and also by setting appropriate parameter(s) to zero, our results are in agreement with those obtained previously.

Although the restriction to linear terms in the involved parameters is made for mathematical convenience, it is also a limit of practical relevance [47]. We remark that in certain cases, due to the asymmetry of the problem, the method is not enough to determine the exact locations of the equilibria. This means that, we may have $x(L_4) \neq x(L_5)$ and $y(L_4) \neq -y(L_5)$ where the equilibria exist and that the stability close to L_4 is different from the stability close to L_5 [36,47]. We note also the resulting asymmetry of the above equilibria L_4 and L_5 in the presence of P–R effect in previous studies [13,14,19]. At this point, it is worth noting that the numerical solutions of equations (9) and (10) show that these points may not be symmetrical with respect to the primaries depending on the parameters.

Unfortunately, there is an extra difficulty to find the general analytical solutions for the system of equations (9) and (10), which give the exact locations of the points of equilibrium. However, they can be approximated by using any numerical method for solving nonlinear algebraic systems. One of the easiest ways of solving a nonlinear system of equations is by deploying the classical Newton iterative method. To date, Newton's method has been successfully applied to a wide variety of nonlinear dynamical systems for the determination of equilibrium points (see, for example, [36,50,51] and references therein). Therefore, we have selected this method for the numerical results.

We shall now discuss the positions of the non-collinear equilibria of the test body as the involved parameters vary. Firstly, in Fig. 3, we give the position diagram of Ragos and Zafiropoulos [14] for L_3 when the effects of potential from the disc, triaxiality of both primaries as well as oblateness of the infinitesimal body are neglected, which was reproduced here for checking purposes and for comparison. For this special case, the numerical solution depicted in Fig. 3 agrees with the result presented in Ref. [14] (see Figs. 1(a) and 1(b) of that paper). Note, the different scales for the horizontal and the vertical axes are due to the machine precision.

Next, we compute numerically using (9) and (10) the effects of various parameters involved on the positions of the equilibria for the binary system Achird. The physical parameters of the binary Achird system are shown in Table 1 [35,46,52]. The coordinates of the numerically determined non-collinear equilibria for the binary system Achird are shown in Table 2 for fixed values of the chosen model parameters and for various values of the mass of the disc (M_b) and the oblate infinitesimal (A_3) body. We observe that with the increase of A_3 (0 to 0.05) and M_b (0.06 to 0.08), for fixed q_1, q_2, σ_1, σ_2, σ_1', σ_2', T and μ, the coordinates of the seven non-collinear equilibria

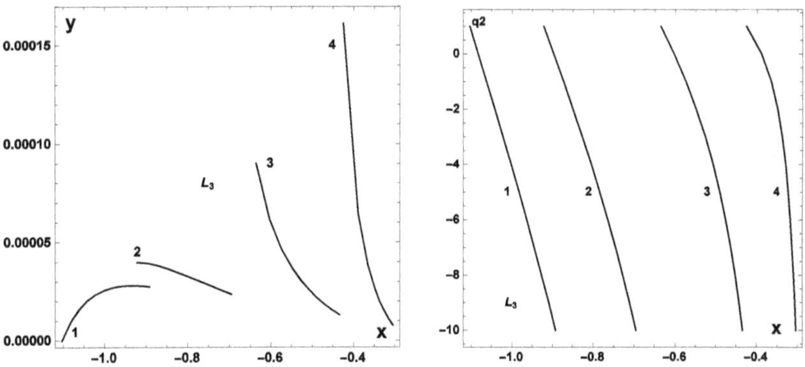

(a) (b)

Fig. 3: (a) Position of L_3 in the xy-plane and (b) Position of $x(L_3)$ as a function of q_2 in the interval $[-10, 1]$ for various values of the radiation parameter (curve 1 $\rightarrow$ $\mathrm{q}_1 = 1$, curve 2 $\rightarrow$ $\mathrm{q}_1 = 0.5$, curve 3 $\rightarrow$ $\mathrm{q}_1 = 0.1$, curve 4 $\rightarrow$ $\mathrm{q}_1 = 0.01$). Note: the values of $\mu = 0.25$ and $C_d = 48002.33$ are fixed in both cases [14]. The shape and dimension of the bifurcation lines has been lightly deformed for the sake of clarity.

Table 1: Numerical data for the binary Achird system.

	Mass ($M_{\otimes}$)		Radiation pressure				
Binary	m_1	m_2	q_1	q_2	Binary separation a	Dimensionless speed of light c_d	Mass ratio μ
Achird	0.95	0.62	0.9971	0.9997	71	67,675.52	0.3949

increase or decrease. In particular, the x-coordinates of equilibrium points L_1, L_4 (the situation is same at the symmetric point L_5) increase while at the same time the y-coordinates decrease; both the x- and y-coordinates of the point L_{N1} increase, while both x and y coordinates of L_2, L_3 and L_{N2} decrease. Figure 4 is a plot of the positions of the equilibria when the oblate infinitesimal body and mass of the disc vary for fixed values of the chosen parameters model. We observe that with the increase of A_3 and M_b, the positions L_1, L_2, L_3, L_{N2} and $L_{4(5)}$ move closer to the Ox-axis while the positions of L_{N1} move away from the Ox-axis.

The influence of triaxiality of both primaries on the positions of the non-collinear equilibria for the binary system is shown in Table 3 for fixed values of the chosen parameters. We observe that with the increase of triaxiality of the bigger primary ($\sigma_1 = 0.07$ to 0.15, $\sigma_2 = 0.04$ to 0.12) and triaxiality of the smaller one ($\sigma_1' = 0.01$ to 0.04, $\sigma_2' = 0.005$ to 0.035), for fixed q_1, q_2,

Table 2: Positions of the seven non-collinear equilibrium points for varying oblate infinitesimal bodies and varying mass disc for the binary Achird system when $\mu = 0.3949$, $q_1 = 0.9971$, $q_2 = 0.9997$, $\sigma_1 = 0.0005$, $\sigma_2 = 0.0003$, $\sigma_1' = 0.0001$, $\sigma_2' = 0.00005$, $c_d = 67,675.52$.

$M_b = 0.06$			
A_3	L_1	L_2	L_3
0.0	$0.217305, -2.88969 \times 10^{-9}$	$1.20084, -2.31203 \times 10^{-8}$	$-1.12714, 7.83542 \times 10^{-8}$
0.01	$0.211657, -2.73047 \times 10^{-9}$	$1.20967, -2.32133 \times 10^{-8}$	$-1.13414, 7.81958 \times 10^{-8}$
0.02	$0.206943, -2.59469 \times 10^{-9}$	$1.21784, -2.33004 \times 10^{-8}$	$-1.14078, 7.80509 \times 10^{-8}$
0.03	$0.202913, -2.47665 \times 10^{-9}$	$1.22546, -2.33824 \times 10^{-8}$	$-1.14711, 7.79177 \times 10^{-8}$
0.04	$0.199405, -2.37252 \times 10^{-9}$	$1.23261, -2.34602 \times 10^{-8}$	$-1.15316, 7.77946 \times 10^{-8}$
0.05	$0.196310, -2.27961 \times 10^{-9}$	$1.23936, -2.35343 \times 10^{-8}$	$-1.15895, 7.76806 \times 10^{-8}$
A_3	L_{N1}	L_{N2}	$L_{4(5)}$
0.0	$-0.097488, -1.04686 \times 10^{-9}$	$-4.696 \times 10^{-5}, -1.12618 \times 10^{-12}$	$0.104202, \pm 0.842922$
0.01	$-0.091842, -8.78628 \times 10^{-10}$	$-5.245 \times 10^{-5}, -1.12618 \times 10^{-12}$	$0.104216, \pm 0.848134$
0.02	$-0.087379, -7.58967 \times 10^{-10}$	$-5.794 \times 10^{-5}, -1.12619 \times 10^{-12}$	$0.104230, \pm 0.853206$
0.03	$-0.083693, -6.68671 \times 10^{-10}$	$-6.343 \times 10^{-5}, -1.12620 \times 10^{-12}$	$0.104242, \pm 0.858144$
0.04	$-0.080559, -5.97743 \times 10^{-10}$	$-6.892 \times 10^{-5}, -1.12621 \times 10^{-12}$	$0.104254, \pm 0.862957$
0.05	$-0.077839, -5.40371 \times 10^{-10}$	$-7.442 \times 10^{-5}, -1.12622 \times 10^{-12}$	$0.104265, \pm 0.867652$
$M_b = 0.07$			
A_3	L_1	L_2	L_3
0.0	$0.223471, -2.75593 \times 10^{-9}$	$1.19624, -2.28199 \times 10^{-8}$	$-1.12237, 7.77914 \times 10^{-8}$
0.01	$0.217525, -2.60392 \times 10^{-9}$	$1.20512, -2.29113 \times 10^{-8}$	$-1.12938, 7.76240 \times 10^{-8}$
0.02	$0.212572, -2.47468 \times 10^{-9}$	$1.21332, -2.29968 \times 10^{-8}$	$-1.13604, 7.74711 \times 10^{-8}$
0.03	$0.208342, -2.36254 \times 10^{-9}$	$1.22097, -2.30775 \times 10^{-8}$	$-1.14238, 7.73305 \times 10^{-8}$
0.04	$0.204661, -2.26374 \times 10^{-9}$	$1.22814, -2.31539 \times 10^{-8}$	$-1.14843, 7.72007 \times 10^{-8}$
0.05	$0.201413, -2.17567 \times 10^{-9}$	$1.23491, -2.32266 \times 10^{-8}$	$-1.15424, 7.70804 \times 10^{-8}$

A_3	L_{N1}	L_{N2}	$L_{4(5)}$
0.0	$-0.103018, -1.07096 \times 10^{-9}$	$-4.025 \times 10^{-5}, -9.76174 \times 10^{-13}$	$0.104207, \pm 0.839822$
0.01	$-0.097008, -8.97302 \times 10^{-10}$	$-4.495 \times 10^{-5}, -9.76178 \times 10^{-13}$	$0.104221, \pm 0.844958$
0.02	$-0.092288, -7.74734 \times 10^{-10}$	$-4.966 \times 10^{-5}, -9.76183 \times 10^{-13}$	$0.104235, \pm 0.849958$
0.03	$-0.088407, -6.82636 \times 10^{-10}$	$-5.436 \times 10^{-5}, -9.76188 \times 10^{-13}$	$0.104248, \pm 0.854828$
0.04	$-0.085116, -6.10471 \times 10^{-10}$	$-5.907 \times 10^{-5}, -9.76195 \times 10^{-13}$	$0.104260, \pm 0.859577$
0.05	$-0.082265, -5.52184 \times 10^{-10}$	$-6.378 \times 10^{-5}, -9.76201 \times 10^{-13}$	$0.104271, \pm 0.864211$

$M_b = 0.08$

A_3	L_1	L_2	L_3
0.0	$0.229075 - 2.64356 \times 10^{-9}$	$1.19178, -2.25309 \times 10^{-8}$	$-1.11774, 7.72501 \times 10^{-8}$
0.01	$0.222858 - 2.49768 \times 10^{-9}$	$1.20070, -2.26207 \times 10^{-8}$	$-1.12477, 7.70740 \times 10^{-8}$
0.02	$0.217691 - 2.37400 \times 10^{-9}$	$1.20894, -2.27048 \times 10^{-8}$	$-1.13144, 7.69130 \times 10^{-8}$
0.03	$0.213283 - 2.26686 \times 10^{-9}$	$1.21662, -2.27840 \times 10^{-8}$	$-1.13779, 7.67652 \times 10^{-8}$
0.04	$0.209450 - 2.17257 \times 10^{-9}$	$1.22382, -2.28590 \times 10^{-8}$	$-1.14386, 7.66287 \times 10^{-8}$
0.05	$0.206067 - 2.08858 \times 10^{-9}$	$1.23060, -2.29305 \times 10^{-8}$	$-1.14966, 7.65023 \times 10^{-8}$

A_3	L_{N1}	L_{N2}	$L_{4(5)}$
0.0	$-0.107965, -1.09104 \times 10^{-9}$	$-3.521 \times 10^{-5}, -8.63568 \times 10^{-13}$	$0.104212, \pm 0.836861$
0.01	$-0.101621, -9.12639 \times 10^{-10}$	$-3.933 \times 10^{-5}, -8.63570 \times 10^{-13}$	$0.104226, \pm 0.841923$
0.02	$-0.096671, -7.87628 \times 10^{-10}$	$-4.345 \times 10^{-5}, -8.63574 \times 10^{-13}$	$0.104240, \pm 0.846852$
0.03	$-0.092616, -6.94060 \times 10^{-10}$	$-4.756 \times 10^{-5}, -8.63577 \times 10^{-13}$	$0.104253, \pm 0.851656$
0.04	$-0.089185, -6.20907 \times 10^{-10}$	$-5.168 \times 10^{-5}, -8.63581 \times 10^{-13}$	$0.104265, \pm 0.856341$
0.05	$-0.086217, -5.61900 \times 10^{-10}$	$-5.580 \times 10^{-5}, -8.63586 \times 10^{-13}$	$0.104276, \pm 0.860915$

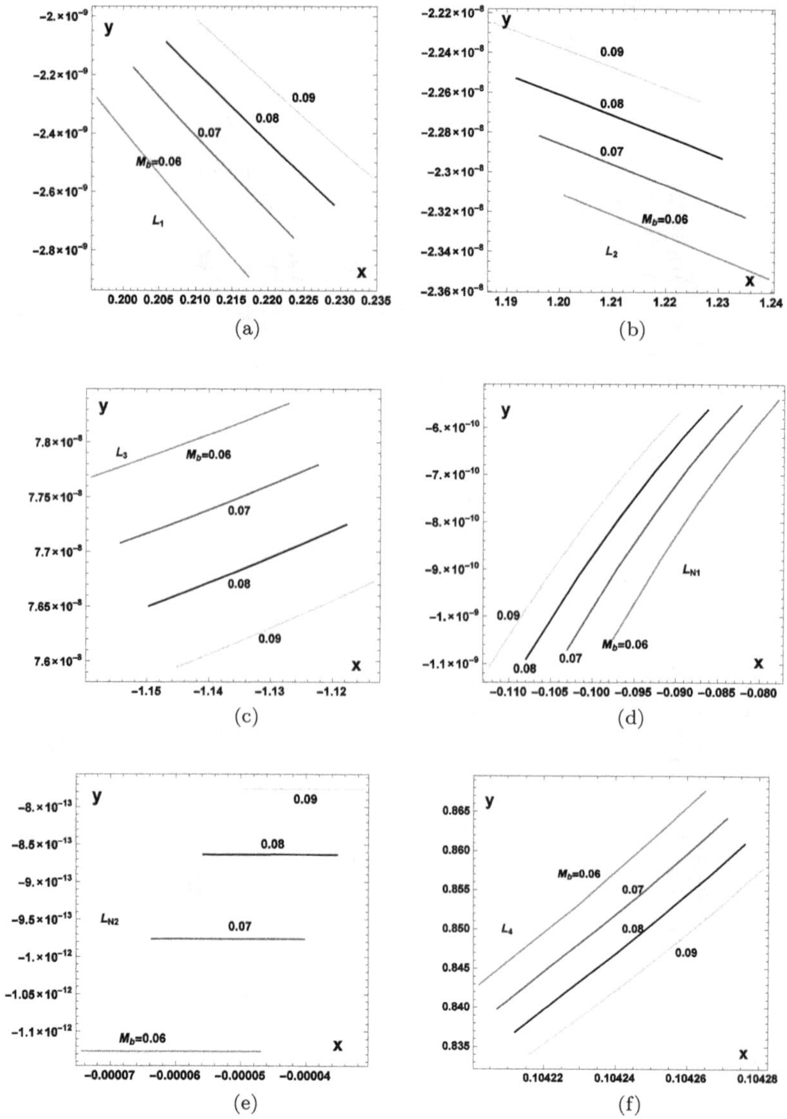

Fig. 4: Positions of the seven non-collinear equilibrium points of Achird system for fixed values of $\mu = 0.3949$, $q_1 = 0.9971$, $q_2 = 0.9997$, $\sigma_1 = 0.0005$, $\sigma_2 = 0.0003$, $\sigma_1' = 0.0001$, $\sigma_2' = 0.00005$, $c_d = 67{,}675.52$ and various values of the mass of the disc M_b and the oblateness coefficient A_3 (see Table 2). Each panel is for different equilibrium points, i.e. (a)–(f) equilibrium points L_1, L_2, L_3, L_{N1}, L_{N2} and L_4 (positive equilibrium point) correspondingly. In each panel, *Red, blue, black* and *green lines* correspond to the case where $M_b = 0.06, 0.07, 0.08$ and 0.09, resp.

M_b, A_3, T and μ, x-coordinates of L_2 increase while at the same time the y-coordinates decrease; both the x and y coordinates of L_1, L_3 and L_4 (the situation is the same at the symmetric point L_5) decrease while the x and y coordinates of L_{N2} decrease and increase, resp. In Fig. 5, we have evaluated numerically the positions of the seven non-collinear equilibrium points for different values of the parameters σ_1, σ_2, and σ_1', σ_2', for the binary system. One can observe that for increasing values of the triaxiality of both primaries L_1, L_2, L_3, $L_{4,5}$ move closer to the Ox-axis while L_{N1} and L_{N2} move away from the Ox-axis.

4. Stability of the Equilibrium Points

Knowing the exact positions (x_0, y_0) of the equilibrium points, we can easily determine their linear stability, through the nature of the four roots of the characteristic equation. In doing this, we will follow the approach by Ragos and Zafiropoulos [14] and Vincent and Perdiou [36]. We suppose ξ and η are coordinates of the equilibrium point (x_0, y_0) such that

$$\xi = x - x_0, \quad \eta = y - y_0. \tag{15}$$

Denoting the right-hand side of equation (6) by $\Omega_x = \partial\Omega/\partial x$ and $\Omega_y = \partial\Omega/\partial y$, resp., the variational form of the equations of motion is derived as

$$\begin{aligned} \ddot{\xi} - 2n\dot{\eta} &= \Omega^{(0)}_{x\dot{x}}\dot{\xi} + \Omega^{(0)}_{x\dot{y}}\dot{\eta} + \Omega^{(0)}_{xx}\xi + \Omega^{(0)}_{xy}\eta, \\ \ddot{\eta} + 2n\dot{\xi} &= \Omega^{(0)}_{y\dot{x}}\dot{\xi} + \Omega^{(0)}_{y\dot{y}}\dot{\eta} + \Omega^{(0)}_{yx}\xi + \Omega^{(0)}_{yy}\eta, \end{aligned} \tag{16}$$

where the dots are the derivatives with respect to time t and only the linear terms in ξ and η have been taken. Then, the form of the corresponding characteristic polynomial is given as

$$\lambda^4 + a\lambda^3 + b\lambda^2 + c\lambda + d = 0, \tag{17}$$

where

$$\begin{aligned} a &= -\left[\Omega^{(0)}_{y\dot{y}} + \Omega^{(0)}_{x\dot{x}}\right], \\ b &= 4n^2 + \Omega^{(0)}_{x\dot{x}}\Omega^{(0)}_{y\dot{y}} - \Omega^{(0)}_{xx} - \Omega^{(0)}_{yy} - \left[\Omega^{(0)}_{x\dot{y}}\right]^2, \\ c &= \Omega^{(0)}_{x\dot{x}}\Omega^{(0)}_{yy} + \Omega^{(0)}_{xx}\Omega^{(0)}_{y\dot{y}} + 2n\Omega^{(0)}_{xy} - 2n\Omega^{(0)}_{yx} - \Omega^{(0)}_{y\dot{x}}\Omega^{(0)}_{xy} - \Omega^{(0)}_{yx}\Omega^{(0)}_{x\dot{y}}, \\ d &= \Omega^{(0)}_{xx}\Omega^{(0)}_{yy} - \Omega^{(0)}_{yx}\Omega^{(0)}_{xy}. \end{aligned} \tag{18}$$

The second-order partial derivatives of Ω are denoted by subscripts, while the superscript (0) denotes evaluation at the coordinates (x_0, y_0) of the

Table 3: Positions of the seven non-collinear equilibrium points for varying triaxiality of the bigger and smaller primaries for the binary Achird system when $\mu = 0.3949$, $q_1 = 0.9971$, $q_2 = 0.9997$, $A_3 = 0.05$, $M_b = 0.07$, $c_d = 67{,}675.52$.

$\sigma_1' = 0.01,\ \sigma_2' = 0.005$				
σ_1	σ_2	L_1	L_2	L_3
0.07	0.04	$0.207419, -2.01267 \times 10^{-9}$	$1.21386, -1.96369 \times 10^{-8}$	$-1.16580, 5.19036 \times 10^{-8}$
0.086	0.056	$0.209228, -2.01497 \times 10^{-9}$	$1.21012, -1.93817 \times 10^{-8}$	$-1.16782, 5.16540 \times 10^{-8}$
0.102	0.072	$0.210991, -2.01644 \times 10^{-9}$	$1.20647, -1.91346 \times 10^{-8}$	$-1.16971, 5.13991 \times 10^{-8}$
0.118	0.088	$0.212710, -2.01717 \times 10^{-9}$	$1.20292, -1.88953 \times 10^{-8}$	$-1.17149, 5.11402 \times 10^{-8}$
0.134	0.104	$0.214388, -2.01723 \times 10^{-9}$	$1.19945, -1.86633 \times 10^{-8}$	$-1.17317, 5.08783 \times 10^{-8}$
0.15	0.12	$0.216026, -2.01667 \times 10^{-9}$	$1.19606, -1.84384 \times 10^{-8}$	$-1.17474, 5.06145 \times 10^{-8}$
σ_1	σ_2	L_{N1}	L_{N2}	$L_{4(5)}$
0.07	0.04	$-0.06494, -2.83191 \times 10^{-10}$	$-0.0001157, -1.04441 \times 10^{-12}$	$0.093325, \pm 0.832783$
0.086	0.056	$-0.06316, -2.64597 \times 10^{-10}$	$-0.0001243, -1.05364 \times 10^{-12}$	$0.100668, \pm 0.830209$
0.102	0.072	$-0.06153, -2.48287 \times 10^{-10}$	$-0.0001328, -1.06280 \times 10^{-12}$	$0.107735, \pm 0.827615$
0.118	0.088	$-0.06004, -2.33867 \times 10^{-10}$	$-0.0001414, -1.07188 \times 10^{-12}$	$0.114543, \pm 0.825008$
0.134	0.104	$-0.05866, -2.21027 \times 10^{-10}$	$-0.0001499, -1.08089 \times 10^{-12}$	$0.121109, \pm 0.822393$
0.15	0.12	$-0.05738, -2.09520 \times 10^{-10}$	$-0.0001584, -1.08983 \times 10^{-12}$	$0.127446, \pm 0.819776$
$\sigma_1' = 0.02,\ \sigma_2' = 0.015$				
σ_1	σ_2	L_1	L_2	L_3
0.07	0.04	$0.203405, -1.97458 \times 10^{-9}$	$1.21725, -1.95365 \times 10^{-8}$	$-1.16310, 5.15597 \times 10^{-8}$
0.086	0.056	$0.205191, -1.97763 \times 10^{-9}$	$1.21359, -1.92872 \times 10^{-8}$	$-1.16520, 5.13212 \times 10^{-8}$
0.102	0.072	$0.206933, -1.97985 \times 10^{-9}$	$1.21001, -1.90457 \times 10^{-8}$	$-1.16716, 5.10767 \times 10^{-8}$
0.118	0.088	$0.208633, -1.98130 \times 10^{-9}$	$1.20653, -1.88117 \times 10^{-8}$	$-1.16901, 5.08275 \times 10^{-8}$
0.134	0.104	$0.210292, -1.98206 \times 10^{-9}$	$1.20313, -1.85848 \times 10^{-8}$	$-1.17074, 5.05748 \times 10^{-8}$
0.15	0.12	$0.211912, -1.98218 \times 10^{-9}$	$1.19981, -1.83647 \times 10^{-8}$	$-1.17238, 5.03197 \times 10^{-8}$

σ_1	σ_2	L_{N1}	L_{N2}	$L_{4(5)}$
0.07	0.04	$-0.06499, -2.85172 \times 10^{-10}$	$-0.0001151, -1.05016 \times 10^{-12}$	$0.0890583, \pm 0.831232$
0.086	0.056	$-0.06320, -2.66414 \times 10^{-10}$	$-0.0001236, -1.05935 \times 10^{-12}$	$0.0963961, \pm 0.828793$
0.102	0.072	$-0.06157, -2.49963 \times 10^{-10}$	$-0.0001322, -1.06845 \times 10^{-12}$	$0.1034570, \pm 0.826329$
0.118	0.088	$-0.06007, -2.35418 \times 10^{-10}$	$-0.0001407, -1.07749 \times 10^{-12}$	$0.1102580, \pm 0.823847$
0.134	0.104	$-0.05869, -2.22467 \times 10^{-10}$	$-0.0001493, -1.08645 \times 10^{-12}$	$0.1168160, \pm 0.821352$
0.15	0.12	$-0.05741, -2.10863 \times 10^{-10}$	$-0.0001578, -1.09535 \times 10^{-12}$	$0.1231460, \pm 0.818850$

$\sigma_1' = 0.03, \ \sigma_2' = 0.025$

σ_1	σ_2	L_1	L_2	L_3
0.07	0.04	$0.199692, -1.93870 \times 10^{-9}$	$1.22043, -1.94351 \times 10^{-8}$	$-1.16045, 5.12223 \times 10^{-8}$
0.086	0.056	$0.201458, -1.94246 \times 10^{-9}$	$1.21684, -1.91914 \times 10^{-8}$	$-1.16262, 5.09945 \times 10^{-8}$
0.102	0.072	$0.203181, -1.94535 \times 10^{-9}$	$1.21333, -1.89553 \times 10^{-8}$	$-1.16465, 5.07600 \times 10^{-8}$
0.118	0.088	$0.204863, -1.94747 \times 10^{-9}$	$1.20992, -1.87264 \times 10^{-8}$	$-1.16656, 5.05203 \times 10^{-8}$
0.134	0.104	$0.206505, -1.94887 \times 10^{-9}$	$1.20658, -1.85044 \times 10^{-8}$	$-1.16835, 5.02766 \times 10^{-8}$
0.15	0.12	$0.208109, -1.94962 \times 10^{-9}$	$1.20332, -1.82889 \times 10^{-8}$	$-1.17004, 5.00299 \times 10^{-8}$

σ_1	σ_2	L_{N1}	L_{N2}	$L_{4(5)}$
0.07	0.04	$-0.06503, -2.87148 \times 10^{-10}$	$-0.0001145, -1.05589 \times 10^{-12}$	$0.0848913, \pm 0.829675$
0.086	0.056	$-0.06324, -2.68228 \times 10^{-10}$	$-0.0001230, -1.06502 \times 10^{-12}$	$0.0922245, \pm 0.827368$
0.102	0.072	$-0.06161, -2.51635 \times 10^{-10}$	$-0.0001316, -1.07408 \times 10^{-12}$	$0.0992797, \pm 0.825030$
0.118	0.088	$-0.06011, -2.36966 \times 10^{-10}$	$-0.0001401, -1.08307 \times 10^{-12}$	$0.1060750, \pm 0.822668$
0.134	0.104	$-0.05873, -2.23905 \times 10^{-10}$	$-0.0001486, -1.09199 \times 10^{-12}$	$0.1126270, \pm 0.820290$
0.15	0.12	$-0.05744, -2.12203 \times 10^{-10}$	$-0.0001572, -1.10084 \times 10^{-12}$	$0.1189500, \pm 0.817900$

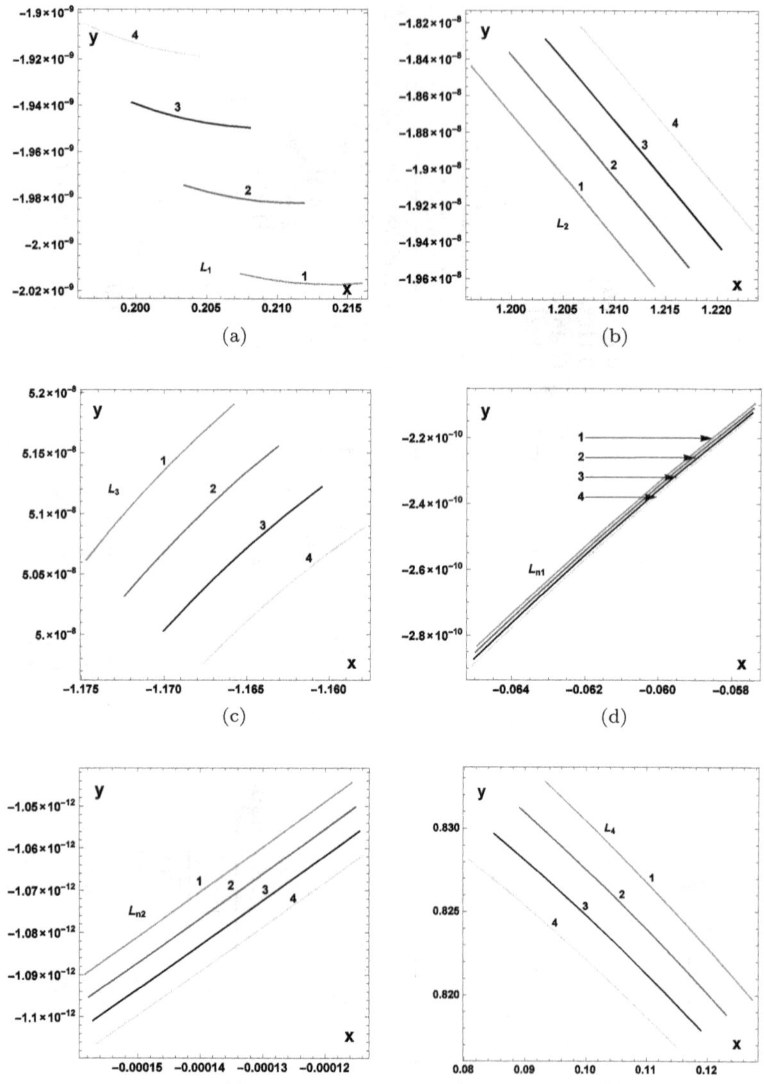

Fig. 5: Positions of the seven non-collinear equilibrium points of Achird system for fixed values of $\mu = 0.3949$, $q_1 = 0.9971$, $q_2 = 0.9997$, $A_3 = 0.05$, $M_b = 0.07$, $c_d = 67,675.52$ and various values of the triaxiality of the smaller primary σ_1', σ_2' when the triaxiality of the bigger primary σ_1, σ_2 vary (see Table 3). Each panel is for different equilibrium points, i.e. (a)–(f) equilibrium points L_1, L_2, L_3, L_{N1}, L_{N2} and L_4 (positive equilibrium point) correspondingly. In each panel, (1) is for $\sigma_1' = 0.01$, $\sigma_2' = 0.005$ (*red*), (2) is for $\sigma_1' = 0.02$, $\sigma_2' = 0.015$ (*blue*), (3) is for $\sigma_1' = 0.030$, $\sigma_2' = 0.025$ (*black*) and (4) is for $\sigma_1' = 0.04$, $\sigma_2' = 0.035$ (*green*).

equilibrium points which result from the solution of equations (9) and (10) and these can be written in simplified form as follows

$$\Omega_{xx}^{(0)} = n^2 - \frac{q_1(1-\mu)}{r_{10}^3} - \frac{q_2\mu}{r_{20}^3} + \frac{3q_1(1-\mu)(x_0+\mu)^2}{r_{10}^5} - \frac{3A_3(1-\mu)}{2r_{10}^5}$$
$$- \frac{3(2\sigma_1-\sigma_2)q_1(1-\mu)}{2r_{10}^5} + \frac{3\mu q_2(x_0+\mu-1)^2}{r_{20}^5} - \frac{3A_3\mu}{2r_{20}^5}$$
$$- \frac{3(2\sigma_3-\sigma_4)q_2\mu}{2r_{20}^5} + \frac{15(\sigma_1-\sigma_2)q_1y_0^2(1-\mu)}{2r_{10}^7} + \frac{15A_3(1-\mu)(x_0+\mu)^2}{2r_{10}^7}$$
$$+ \frac{15(2\sigma_1-\sigma_2)q_1(1-\mu)(x_0+\mu)^2}{2r_{10}^7} + \frac{15(\sigma_3-\sigma_4)q_2\mu y_0^2}{2r_{20}^7}$$
$$+ \frac{15A_3\mu(x_0+\mu-1)^2}{2r_{20}^7} + \frac{15(2\sigma_3-\sigma_4)q_2\mu(x_0+\mu-1)^2}{2r_{20}^7}$$
$$- \frac{105(\sigma_1-\sigma_2)q_1y_0^2(1-\mu)(x_0+\mu)^2}{2r_{10}^9}$$
$$- \frac{105(\sigma_3-\sigma_4)q_2\mu y_0^2(x_0+\mu-1)^2}{2r_{20}^9}$$
$$- \frac{M_b}{(T^2+r_0^2)^{\frac{3}{2}}} + \frac{3M_b x_0^2}{(T^2+r_0^2)^{\frac{5}{2}}} - \frac{2nW_2y_0(x_0+\mu-1)}{r_{20}^4}$$
$$- \frac{2nW_1y_0(x_0+\mu)}{r_{10}^4}, \tag{19}$$

$$\Omega_{yy}^{(0)} = n^2 - \frac{q_1(1-\mu)}{r_{10}^3} - \frac{q_2\mu}{r_{20}^3} + \frac{3q_1(1-\mu)y_0^2}{r_{10}^5} - \frac{3A_3(1-\mu)}{2r_{10}^5}$$
$$- \frac{3(\sigma_1-\sigma_2)q_1(1-\mu)}{r_{10}^5} - \frac{3(2\sigma_1-\sigma_2)q_1(1-\mu)}{2r_{20}^5} + \frac{3q_2\mu y_0^2}{r_{20}^5}$$
$$- \frac{3A_3\mu}{2r_{20}^5} - \frac{3(\sigma_3-\sigma_4)q_2\mu}{r_{20}^5} - \frac{3(2\sigma_3-\sigma_4)q_2\mu}{2r_{10}^5}$$
$$+ \frac{15A_3(1-\mu)y_0^2}{2r_{10}^7} + \frac{75(\sigma_1-\sigma_2)q_1y_0^2(1-\mu)}{2r_{10}^7}$$
$$+ \frac{15(2\sigma_1-\sigma_2)q_1(1-\mu)y_0^2}{2r_{10}^7} + \frac{15A_3\mu y_0^2}{2r_{20}^7} + \frac{75(\sigma_3-\sigma_4)q_2\mu y_0^2}{2r_{20}^7}$$
$$+ \frac{15(2\sigma_3-\sigma_4)q_2\mu y_0^2}{2r_{20}^7} - \frac{105(\sigma_3-\sigma_4)q_2\mu y_0^4}{2r_{20}^9}$$

$$-\frac{105(\sigma_1-\sigma_2)q_1y_0^4(1-\mu)}{2r_{10}^9}-\frac{M_b}{(T^2+r_0^2)^{\frac{3}{2}}}$$

$$+\frac{3M_by_0^2}{(T^2+r_0^2)^{\frac{5}{2}}}+\frac{2nW_2y_0(x_0+\mu-1)}{r_{20}^4}+\frac{2nW_1y_0(x_0+\mu)}{r_{10}^4}, \tag{20}$$

$$\Omega_{xy}^{(0)}=\mathcal{D}+\frac{nW_1}{r_{10}^2}+\frac{nW_2}{r_{20}^2}-\frac{2nW_1y_0^2}{r_{10}^4}-\frac{2nW_2y_0^2}{r_{20}^4}, \tag{21}$$

$$\Omega_{yx}^{(0)}=\mathcal{D}-\frac{nW_1}{r_{10}^2}-\frac{nW_2}{r_{20}^2}+\frac{2nW_1(x_0+\mu)^2}{r_{10}^4}+\frac{2nW_2(x_0+\mu-1)^2}{r_{20}^4}, \tag{22}$$

$$\Omega_{x\dot{x}}^{(0)}=-\frac{W_1}{r_{10}^2}\left[\left(1+\frac{x_0^2}{r_{10}^2}\right)+\frac{\mu}{r_{10}^2}(\mu+2x_0)\right]$$

$$-\frac{W_2}{r_{20}^2}\left[\left(1+\frac{1}{r_{20}^2}\right)-\frac{x_0}{r_{20}^2}(2-x_0)-\frac{\mu}{r_{20}^2}(2(1-x_0)-\mu)\right], \tag{23}$$

$$\Omega_{y\dot{y}}^{(0)}=-\frac{W_1}{r_{10}^2}\left(1+\frac{y_0^2}{r_{10}^2}\right)-\frac{W_2}{r_{20}^2}\left(1+\frac{y_0^2}{r_{20}^2}\right), \tag{24}$$

$$\Omega_{x\dot{y}}^{(0)}=\Omega_{y\dot{x}}^{(0)}=-\frac{W_1y_0}{r_{10}^4}(\mu+x_0)+\frac{W_2y_0}{r_{20}^4}[1-(x_0+\mu)], \tag{25}$$

where

$$\mathcal{D}=\frac{3q_1y_0(1-\mu)(x_0+\mu)}{r_{10}^5}+\frac{3q_2\mu y_0(x_0+\mu-1)}{r_{20}^5}$$

$$+\frac{15(\sigma_1-\sigma_2)q_1y_0(1-\mu)(x_0+\mu)}{r_{10}^7}+\frac{15A_3y_0(1-\mu)(x_0+\mu)}{2r_{10}^7}$$

$$+\frac{15(2\sigma_1-\sigma_2)q_1y_0(1-\mu)(x_0+\mu)}{2r_{10}^7}+\frac{15A_3\mu y_0(x_0+\mu-1)}{2r_{20}^7}$$

$$+\frac{15(\sigma_3-\sigma_4)q_2\mu y_0(x_0+\mu-1)}{r_{20}^7}+\frac{3M_bx_0y_0}{(T^2+r_0^2)^{\frac{5}{2}}}$$

$$+\frac{15(2\sigma_3-\sigma_4)q_2\mu y_0(x_0+\mu-1)}{2r_{20}^7}-\frac{105(\sigma_1-\sigma_2)q_1y_0^3(1-\mu)(x_0+\mu)}{2r_{10}^9}$$

$$-\frac{105(\sigma_3-\sigma_4)q_2\mu y_0^3(x_0+\mu-1)}{2r_{20}^9}, \tag{26}$$

and

$$r_{10}^2=(x_0+\mu)^2+y_0^2, \qquad r_{20}^2=(x_0+\mu-1)^2+y_0^2. \tag{27}$$

The determination of the stability or instability of motion around the equilibrium points can be accomplished through the computation of the characteristic roots of the variational equations (16). An equilibrium point (x_0, y_0) will be stable if equation (17), evaluated at the equilibrium, has four pure imaginary roots or four complex roots with each of them having negative real parts; otherwise, it is unstable. In Tables 4–6, we compute the characteristic roots $\lambda_{i,i+1}$, $i = 1, 3$, of each equilibrium point for different values of M_b and A_3 and for fixed values of the remaining parameters q_1, q_2, σ_1, σ_2, σ'_1, σ'_2 and μ. In this case, our numerical exploration in the computation of these roots as shown in what follows reveals that the motion of the test particle within the neighborhood of each of the equilibria L_1, L_2, L_3, L_{N1}, $L_{4(5)}$ is linearly unstable due to the presence of at least a complex root having a positive real part while point L_{N2} is linearly stable due to four complex roots with each of them having negative real parts. For the following results we conclude that P–R drag effect induced by the two radiating main bodies on a moving test particle destroys the stability of all the non-collinear equilibria except L_{N2} arising due to the mass of the disc. Similar phenomenon we observe in the photogravitational three-body problem under effects of P–R drag forces and circumbinary disc where for certain values of the radiation factors, the instability of the triangular points is greatly influenced by the radiation and P–R drag forces as they dominate the stabilizing tendency of the mass of the disc [33]. Moreover, additional computations (not shown here for economy reasons) indicate that the linear stability of the seven equilibrium points for varying triaxiality of the primaries and for fixed values of the remaining parameters are qualitatively the same as those shown in Tables 4–6.

5. Discussion and Conclusion

A generalization of the R3BP under the influence of radiation pressure, Poynting–Robertson (P–R) drag, triaxial bodies together with gravitational potential from the belt and oblateness of infinitesimal body was studied. We found that five or seven equilibrium points may lie on the plane of motion depending on the values of the parameters. It was observed that the involved parameters of the problem not only affect the number and positions of the corresponding equilibria, but they also play a significant role on their stability since it was identified that there are values of these parameters for which the points may be linearly stable. It was also observed that, in the presence of P–R drag effect, the well-known collinear equilibrium points

Table 4: The eigenvalues $\lambda_{1,2,3,4}$ of equation (17) and the corresponding positions of the seven equilibria for Achird system when $\mu = 0.3949$, $q_1 = 0.9971$, $q_2 = 0.9997$, $\sigma_1 = 0.0005$, $\sigma_2 = 0.0003$, $\sigma_1' = 0.0001$, $\sigma_2' = 0.00005$, $T = 0.01$ and $c_d = 67{,}675.52$.

$M_b = 0.06$	$A_3 = 0.05$	$A_3 = 0.1$
L_1	$-6.633180250, 6.633180088,$ $-4.630544376 \times 10^{-8} \pm 4.480131600i$ $(x, y) = (0.19631, -2.27961 \times 10^{-9})$	$-7.562097722, 7.562097554,$ $-4.671919424 \times 10^{-8} \pm 4.489805480i$ $(x, y) = (0.184865, -1.9282 \times 10^{-9})$
L_2	$-1.536291097, 1.536291087,$ $-1.631294020 \times 10^{-8} \pm 1.442776254i$ $(x, y) = (1.23936, -2.35343 \times 10^{-8})$	$-1.597392311, 1.597392302,$ $-1.521669879 \times 10^{-8} \pm 1.397174356i$ $(x, y) = (1.26858, -2.38619 \times 10^{-8})$
L_3	$-1.171435031, 1.171435027,$ $-6.537186920 \times 10^{-8} \pm 1.301137590i$ $(x, y) = (-1.15895, 7.76806 \times 10^{-8})$	$-1.222163566, 1.222163560,$ $-6.036536703 \times 10^{-8} \pm 1.258212810i$ $(x, y) = (-1.18488, 7.7214 \times 10^{-8})$
L_{N1}	$-18.40652680, 18.40652628,$ $-1.324484846 \times 10^{-7} \pm 12.62106252i$ $(x, y) = (-0.077839, -5.4037 \times 10^{-10})$	$-22.19618313, 22.19618264,$ $-1.242915193 \times 10^{-7} \pm 15.12291422i$ $(x, y) = (-0.0679961, -3.63446 \times 10^{-10})$
L_{N2}	$-1.30741103 \times 10^{-7} \pm 243.8587744i,$ $-1.25930897 \times 10^{-7} \pm 245.9428757i,$ $(x, y) = (-0.000074417, -1.1262 \times 10^{-12})$	$-1.318344807 \times 10^{-7} \pm 243.7466431i,$ $-1.248715193 \times 10^{-7} \pm 245.9889713i,$ $(x, y) = (-0.00010190, -1.12629 \times 10^{-12})$
$L_{4(5)}$	$-0.7027269737 \pm 0.9992473325i,$ $0.7027269323 \pm 0.9992472714i,$ $(x, y) = (0.104265, \pm 0.867652)$	$-0.7359246264 \pm 0.9925621301i,$ $0.7359245866 \pm 0.9925620740i,$ $(x, y) = (0.104313, \pm 0.88957)$

Table 5: Eigenvalues $\lambda_{1,2,3,4}$ of (17) and the corresponding positions of the equilibria for Achird system for the parameter values given in the caption of Table 4.

$M_b = 0.085$	$A_3 = 0.05$	$A_3 = 0.1$
L_1	$-6.999379521, 6.999379364,$ $-4.487124265 \times 10^{-8} \pm 4.720912262i$ $(x, y) = (0.20825, -2.04997 \times 10^{-9})$	$-7.957931807, 7.957931643,$ $-4.538131663 \times 10^{-8} \pm 5.158524725i$ $(x, y) = (0.195474, -1.73997 \times 10^{-9})$
L_2	$-1.591783431, 1.591783421,$ $-1.654006198 \times 10^{-8} \pm 1.486106145i$ $(x, y) = (1.2285, -2.27865 \times 10^{-8})$	$-1.655636544, 1.655636534,$ $-1.540699001 \times 10^{-8} \pm 1.438922635i$ $(x, y) = (1.25782, -2.30995 \times 10^{-8})$
L_3	$-1.209219892, 1.209219888,$ $-6.717789607 \times 10^{-8} \pm 1.337967392i$ $(x, y) = (-1.14743, 7.6221 \times 10^{-8})$	$-1.262478761, 1.262478755,$ $-6.193157669 \times 10^{-8} \pm 1.293666810i$ $(x, y) = (-1.17337, 7.56913 \times 10^{-8})$
L_{N1}	$-18.70949613, 18.70949558,$ $-1.412669993 \times 10^{-7} \pm 12.71889284i$ $(x, y) = (-0.0880479, -5.66154 \times 10^{-10})$	$-22.45526360, 22.45526308,$ $-1.314514015 \times 10^{-7} \pm 15.14737111i$ $(x, y) = (-0.0771773, -3.8379 \times 10^{-10})$
L_{N2}	$-1.301585294 \times 10^{-7} \pm 290.4474307i,$ $-1.264864706 \times 10^{-7} \pm 292.5769814i,$ $(x, y) = (-0.000052516, -8.1718 \times 10^{-13})$	$-1.310109301 \times 10^{-7} \pm 290.3495388i,$ $-1.256580699 \times 10^{-7} \pm 292.6279982i,$ $(x, y) = (-0.00007190, -8.17203 \times 10^{-13})$
$L_{4(5)}$	$-0.7180849533 \pm 1.024898060i,$ $0.7180849113 \pm 1.024897998i,$ $(x, y) = (0.104279, \pm 0.859319)$	$-0.7525729001 \pm 1.018567375i,$ $0.7525728596 \pm 1.018567318i,$ $(x, y) = (0.104328, \pm 0.880581)$

Table 6: Eigenvalues $\lambda_{1,2,3,4}$ of (17) and the corresponding positions of the equilibria for Achird system for the parameter values given in the caption of Table 4.

$M_b = 0.1$	$A_3 = 0.05$	$A_3 = 0.1$
L_1	$-7.190260126, 7.190259971,$ $-4.421408338 \times 10^{-8} \pm 4.844753819i$ $(x, y) = (0.214315, -1.94935 \times 10^{-9})$	$-8.164396029, 8.164395867,$ $-4.475319406 \times 10^{-8} \pm 5.293464007i$ $(x, y) = (0.200909, -1.65765 \times 10^{-9})$
L_2	$-1.624561075, 1.624561065,$ $-1.667156761 \times 10^{-8} \pm 1.511590444i$ $(x, y) = (1.22236, -2.23701 \times 10^{-8})$	$-1.689987361, 1.689987351,$ $-1.551724295 \times 10^{-8} \pm 1.463461056i$ $(x, y) = (1.25174, -2.26748 \times 10^{-8})$
L_3	$-1.231423969, 1.231423964,$ $-6.823485829 \times 10^{-8} \pm 1.359582283i$ $(x, y) = (-1.14092, 7.54071 \times 10^{-8})$	$-1.286125691, 1.286125685,$ $-6.284819456 \times 10^{-8} \pm 1.314456384i$ $(x, y) = (-1.16687, 7.48406 \times 10^{-8})$
L_{N1}	$-18.91823744, 18.91823686,$ $-1.459322913 \times 10^{-7} \pm 12.80850096i$ $(x, y) = (-0.0930634, -5.77081 \times 10^{-10})$	$-22.64987093, 22.64987040,$ $-1.352173164 \times 10^{-7} \pm 15.20837405i$ $(x, y) = (-0.0817043, -3.92695 \times 10^{-10})$
L_{N2}	$-1.297459987 \times 10^{-7} \pm 315.0240653i,$ $-1.268890013 \times 10^{-7} \pm 317.3701912i,$ $(x, y) = (-0.00004463, -7.05687 \times 10^{-13})$	$-1.306994115 \times 10^{-7} \pm 315.0243313i,$ $-1.259560885 \times 10^{-7} \pm 317.3289629i,$ $(x, y) = (-0.00006111, -7.05702 \times 10^{-13})$
$L_{4(5)}$	$-0.7268522804 \pm 1.039821750i,$ $0.7268522381 \pm 1.039821687i,$ $(x, y) = (0.104286, \pm 0.854726)$	$-0.7620980950 \pm 1.033713711i,$ $0.7620980542 \pm 1.033713652i,$ $(x, y) = (0.104336, \pm 0.875605)$

of the circular restricted three-body problem cease to exist, while in the absence of the P–R drag force, collinear equilibria always exist whereas the distribution of equilibria on the plane of motion differs as a result of the disc and mass parameter, as shown in Figs. 2 and 3.

The model has been applied to the binary system Achird (see Table 1 for concrete values) as seen in Figs. 4 and 5 and Tables 2 and 3. We linearized the equations of motion and computed the characteristic equation as a function of the same astrophysical parameters. For the stability of the five equilibria, the four roots of the characteristic polynomial were determined numerically and it was found that they are unstable due to the existence of at least one positive real root or a complex root with positive real part whereas, for the seven equilibria, all points of equilibrium are always linearly unstable, except equilibrium point L_{N2}, which is stable (Tables 4–6).

Similar behavior for the positions and stability of the five equilibria when the larger primary is an oblate spheroid and the smaller primary is a radiation source with its P–R drag, under the effects of small perturbations in the Coriolis and centrifugal forces has also been observed by Vincent and Perdiou [36], who found that collinear equilibria cease to exist and stable motion is not possible for all the equilibria due to the P–R drag effect, a fact that has been overseen by previous studies [13–15,17] in which the photogravitational R3BP with P–R drag effect had been investigated. Note here that, in the case where both primaries are sources of radiation with angular velocity variation, it has been found that these five equilibria may be stable [10]. For the present problem, it was observed that collinear equilibria cease to exist and stability of the equilibria L_4 and L_5 is destroyed, in the case of five EPs, primarily due to the P–R effect. However, the presence of the P–R effect cannot alter the stability state of L_{N2} in the case of seven EPs as it remains stable. Finally, it would be interesting to investigate the effect of other dissipative forces to equilibrium solutions in the restricted three-body problem.

Acknowledgments

This research was supported by the Tertiary Education Trust Fund (tetfund), Nigeria. The first author, Aguda E. Vincent, has received this research Grant.

References

[1] V. Szebehely, *Theory of Orbits: The Restricted Problem of Three Bodies* (Academic press, New York, 1967).

[2] P. Sharer and T. Harrington, Trajectory optimization for the ACE halo orbit mission, In: *AIAA paper 96-3601, AIAA/AAS Astrodynamics Conference*, San Diego, pp. 257–267, (1996).

[3] R.W. Farquhar, The flight of ISEE-3/ICE: Origins, mission history, and a legacy, *J. Astronaut. Sci.* **49**, 23–73, (2001).

[4] M. Giard and L. Montier, Investigating clusters of galaxies with Planck and Herschel, *Astrophys. Space Sci.* **290**, 159–166, (2004).

[5] R. Radzievskii, The restricted problem of three bodies taking account of light pressure, *Astron. Zh.* **27**, 250–256, (1950).

[6] J.F.L. Simmons, A.J.C. McDonald and J.C. Brown, The restricted 3–body problem with radiation pressure, *Celest. Mech.* **35**, 145–187, (1985).

[7] K.E. Papadakis, Families of periodic orbits in the photogravitational three-body problem, *Astrophys. Space Sci.* **245**, 1–13, (1996).

[8] A. Elipe and M. Lara, Periodic orbits in the restricted three body problem with radiation pressure, *Celest. Mech. Dyn. Astron.* **68**, 1–11, (1997).

[9] E.A. Perdios, O. Ragos, A.E. Perdiou and M.N. Vrahatis, Symmetric doubly asymptotic orbits in the photogravitational restricted three-body problem, *Nonlinear Anal. Theory Methods Appl.* **47**, 3443–3448, (2001).

[10] E.A. Perdios, V.S. Kalantonis, A.E. Perdiou and A.A. Nikaki, Equilibrium points and related periodic motions in the restricted three-body problem with angular velocity and radiation effects, *Adv. Astron.* **2015**, 473483, (2015).

[11] E.E. Zotos, Unveiling the influence of the radiation pressure in nature of orbits in the photogravitational restricted three-body problem, *Astrophys. Space Sci.* **360**(1), (2015).

[12] R. Mia and B.S. Kushvah, Stability and Fourier-series periodic solution in the binary stellar systems, *Few-Body Syst.* **57**, 851–867, (2016).

[13] J.A. Chernikov, The photogravitational restricted three-body problem, *Sov. Astron.* **14**, 176–181, (1970).

[14] O. Ragos and F.A. Zafiropoulos, A numerical study of the influence of the Poynting–Robertson effect on the equilibrium points of the photogravitational restricted three-body problem. I. Coplanar case, *Astron. Astrophys.* **300**, 568–578, (1995).

[15] D.W. Schuerman, The restricted three-body problem including radiation pressure, *Astrophys. J.* **238**, 337–342, (1980).

[16] J.C. Liou, H.A. Zook and A.A. Jackson, Radiation pressure, Poytning–Robertson drag, and solar wind drag in the restricted three-body problem, *Icarus* **116**, 186–201, (1995).

[17] B.S. Kushvah, The effect of radiation pressure on the equilibrium points in the generalized photogravitational restricted three body problem, *Astrophys. Space Sci.* **315**, 231–241, (2008).

[18] D.P. Rubincam, The solar Poynting–Robertson effect on particles orbiting Solar System bodies: Circular orbits, *Icarus* **226**, 1618–1623, (2013).

[19] C. Lhotka and A. Celletti, The effect of Poynting-Robertson drag on the triangular Lagrangian points, *Icarus* **250**, 249–261, (2015).

[20] J.K. Beatty, C.C. Petersen and A. Chaikin, *The New Solar System* (Cambridge University Press, 1999).

[21] P.V. SubbaRao and R.K. Sharma, A note on the stability of the triangular points of equilibrium in the restricted three-body, *Astron. Astrophys.* **43**, 381–383, (1975).

[22] A. Elipe and S. Ferrer, On the equilibrium solutions in the circular planar restricted three rigid bodies problem, *Celest. Mech.* **37**, 59–70, (1985).

[23] V.S. Kalantonis, E.A. Perdios and A.E. Perdiou, The Sitnikov family and the associated families of 3D periodic orbits in the photogravitational RTBP with oblateness, *Astrophys. Space Sci.* **315**, 323–334, (2008).

[24] S. Jain, A. Kumar, and K.B. Bhatnagar, Periodic orbits around the collinear libration points in the restricted three body problem when the smaller primary is a triaxial rigid body and bigger primary is a source of radiation pressure, *Indian J. Phys.* **83**, 171–184, (2009).

[25] E.I. Abouelmagd and S.M. El-Shaboury, Periodic orbits under combined effects of oblateness and radiation in the restricted problem of three bodies, *Astrophys. Space Sci.* **341**, 331–341, (2012).

[26] J. Singh and O. Leke, Analytic and numerical treatment of motion of dust grain particle around triangular equilibrium points with post-AGB binary star and disc, *Adv. Space Res.* **54**, 1659–1677, (2014).

[27] E.I. Abouelmagd, F. Alzahrani, J.L.G. Guirao and A. Hobiny, Periodic orbits around the collinear libration points, *J. Nonlinear Sci. Appl.* **9**, 1716–1727, (2016).

[28] S.M. Elshaboury, E.I. Abouelmagd, V.S.Kalantonis and E.A. Perdios, The planar restricted three-body problem when both primaries are triaxial rigid bodies: Equilibrium points and periodic orbits, *Astrophys. Space Sci.* **361**, 315, (2016).

[29] J. Singh, A.E. Perdiou, J.M. Gyegwe and V.S. Kalantonis, Periodic orbits around the collinear equilibrium points for binary Sirius, Procyon, Luhman 16, α-Centuari and Luyten 726–8 systems: The spatial case, *J. Phys. Commun.* **1**, 025008, (2017).

[30] J. Singh, A.E. Perdiou, J.M. Gyegwe and E.A. Perdios, Periodic solutions around the collinear equilibrium points in the perturbed restricted three-body problem with triaxial and radiating primaries for binary HD191408, Kruger 60 and HD 155876 systems, *Appl. Math. Comput.* **325**, 358–374, (2018).

[31] E.E. Zotos, Md S. Suraj, R. Aggarwal, and A. Mittal, Orbit classification in the Copenhagen problem with oblate primaries, *Astron. Nachr.* **340**, 760–770, (2019).

[32] J. Singh and A. Aminu, Instability of triangular libration points in the perturbed photogravitational R3BP with Poynting–Robertson (P–R) drag, *Astrophys. Space Sci.* **351**, 473–482, (2014).

[33] J. Singh and T.O. Amuda, Effects of Poynting–Robertson (P–R) drag, radiation, and oblateness on motion around the $L_{4,5}$ equilibrium points in the CR3BP, *J. Dyn. Syst. Geom. Theor.* **15**, 177, (2017).

[34] J. Singh and A.B. Emmanuel, Stability of triangular points in the photogravitational CR3BP with Poynting–Robertson drag and a smaller triaxial primary, *Astrophys. Space Sci.* **353**, 97–103, (2014).
[35] J. Singh and A.M. Simeon, Motion around the triangular equilibrium points in the circular restricted three-body problem under triaxial luminous primaries with Poynting-Robertson drag, *Int. Front. Sci. Lett.* **12**, 1–21, (2017).
[36] A.E. Vincent and A.E. Perdiou, Poynting–Robertson and oblateness effects on the equilibrium points of the perturbed R3BP: Application on Cen X-4 binary system. In: T.M. Rassias (Ed.), *Nonlinear Analysis, Differential Equations, and Applications*, pp. 131–147 (Springer Optimization and Its Applications, Springer, 2021).
[37] J.S. Greaves, W.S. Holland, G. Moriarty-Schieven, T. Jenness, W.R.F. Dent, B. Zuckerman, C. McCarthy, R.A. Webb, H.M. Butner, W.K. Gear and H.J. Walker, A dust ring around Eridani: Analog to the young solar system, *Astrophys. J.* **506**, 133–137, (1998).
[38] I.G. Jiang and L.C. Yeh, The drag-induced resonant capture for Kuiper Belt objects, *Mon. Not. R. Astron. Soc.* **355**, 29–32, (2004).
[39] I.G. Jiang and L.C. Yeh, On the Chermnykh-like problem: The equilibrium points, *Astrophys. Space Sci.* **305**, 341–345, (2006).
[40] B.S. Kushvah, Linear stability of equilibrium points in the generalized photogravitational Chermnykh's problem, *Astrophys. Space Sci.* **318**, 41–50, (2008).
[41] R. Kishor and B.S. Kushvah, Linear stability and resonances in the generalized photogravitational Chermnykh-like problem with a disc, *Mon. Not. R. Astron. Soc.* **436**, 1741–1749, (2013).
[42] J. Singh and J.J. Taura, Motion in the generalized restricted three-body problem, *Astrophys. Space Sci.* **343**, 95–106, (2013).
[43] I.G. Jiang and L.C. Yeh, Galaxies with super massive binary black holes: (I) A possible model for the centers of core galaxies, *Astrophys. Space Sci.* **349**, 881–893, (2014).
[44] J. Singh and O. Leke, Motion in a modified Chermnykh's restricted three-body problem with oblateness, *Astrophys. Space Sci.* **350**, 143–154, (2014).
[45] J. Singh and T.O. Amuda, Stability analysis of triangular equilibrium points in restricted three-body problem under effects of circumbinary disc, radiation and drag forces, *J. Astrophys. Astron.* **40**, 5, (2019).
[46] T.O. Amuda, J. Singh, and L. Oni, Motion around equilibrium points of an oblate body in the PR3BP with disc, *Indian J Phys.* **95**(7), 1305–1315, (2021).
[47] C.D. Murray, Dynamical effects of drag in the circular restricted three-body problem. I: Location and stability of the Lagrangian equilibrium points, *Icarus* **112**, 465–484, (1994).
[48] M. Miyamoto and R. Nagai, Three-dimensional models for the distribution of mass in galaxies, *Publ. Astron. Soc. Jpn.* **27**, 533–543, (1975).

[49] J. Singh and J.J. Taura, Stability of equilibrium points in a CR3BP with oblate bodies enclosed by a circular cluster of material points, *Astrophys. Space Sci.* **349**, 681–691, (2014).

[50] T.N. Stenborg, Collinear Lagrange point solutions in the circular restricted three-body problem with radiation pressure using Fortran. In: R.W. Argyle, P.S. Bunclark, and J.R. Lewis, (Eds.), *Astronomical Data Analysis Software and Systems XVII, Astronomical Society of the Pacific Conference Series*, Vol. 394, pp. 734–737 (2008).

[51] H.I. Alrebdi, F.L. Dubeibe, K.E. Papadakis and E.E. Zotos, Equilibrium dynamics of a circular restricted three-body problem with Kerr-like primaries, *Nonlinear Dyn.* **107**, 433–456, (2022).

[52] J. Singh and A. Umar, On stability of triangular points in the elliptic R3BP under radiating and oblate primaries, *Astrophys. Space Sci.* **341**, 349–358, (2012).

https://doi.org/10.1142/9789811261572_bmatter

Index

A

$\Diamond$−absorbing, 64
(x_n) avoids fin(X), 826
ϕ-admissible, 822
$(d, \leq)$-adherence point, 806
AGL_1(C)-representation varieties, 439–440
AGL_2(C)-representation variety, 441
acceptable, 118
Achird binary system, 839, 850–852, 854, 856–857, 862–865
adherence, 795
adjoint action, 449
adjoint boundary-value problems, 11
adjoint problem, 10–11
affine simplex, 555
Alexandroff, P., 64
algebraic and topological structures, 55
almost d-non-expansive, 801
almost metric, 773
alternative variational inequality, 202
analytic function, 280–282, 288, 295, 298
analytic methods, 398
analytic solutions, 331
annihilator, 440
ansatz, 172–174, 176, 178–179, 184
antisymmetric, 44
Apostol-type numbers and polynomials, 677, 695–696
approximate methods, 398
ascending and residual subsets, 51
ascending-valued, 103
associativity, 68
asymptotic behavior, 625
asymptotic regular, 804
automatically continuous, 485
award criterion, 190
Axiom of choice, 44, 754, 790

B

$(\leq)$-bounded, 810
backward Euler scheme, 417, 419
Bakhtin triangular, 794
balanced incomplete block design, 546
Banach fixed point theorem, 215–216
Banach space, 216, 218–219, 223, 226, 230, 245
bases of polynomials, 170
BB-variational, 761, 781
Beltrami–Klein geometry, 543–544
Bezier curve, 551
big limit and adherence relations, 53
bilinear form, 423
binary operation, 43
block matrix, 167
block symmetric linearizations, 184
border of set, 53
Bose–Einstein statistics, 537
boundary conditions, 671

boundary integral equation of the first kind, 487
boundary of set, 53
boundary value problem, 419
Bourbak, N., 64
box product, 44
Brøndsted order, 823

C

◊−compatible, 64
(d,≤)-closed, 807
Γ-coercive, 826
canonical form, 545
cardiac hemodynamics, 397
Caristi–Kirk theorem, 780, 782
Caristi's fixed point theorem, 221
Cauchy–Bunyakovsky–Schwarz inequality, 271
Cauchy *d*-lsc in the second variable, 768
Cauchy regular, 804
Cauchy–Schwarz inequality, 484
Cauchy semi-subordinated, 768
Cauchy structure, 758
chain-subordinated, 772
characteristic polynomial, 134
characters, 435
character variety, 436, 449–451
circulant double stochastic matrix, 134
circumbinary belt, 839, 841, 861
circumcevian triangle, 141
circumcircle midarc triangle, 130
closedness-preserving, 42, 90
closure (projection), 66, 795
closure and interior relations, 48
closure-interior Galois connection, 49
closure operation, 58, 62, 65, 67, 76, 106
closure operator, 796, 808
closure preserving, 42
closures, 44, 64
closure spaces, 41
c-multivalued weakly Picard operator (c-MWP), 234
coercive, 481, 487–488, 787
coerciveness, 482–483
common fixed-point, 216, 243–244, 246
compact injection, 427
companion form, 158–159, 185
companion-like structure, 167
companion matrices, 158, 164–166
compatible closure (projection) operations, 66
complement relation, 44
complete, 795, 810, 824
complete asymptotic expansion, 146, 153–154
complex coordinate, 130
complex exponential, 279
complex logarithm, 279, 300
composition-compatibility, 75, 77, 107
composition relation, 44
compositions, 65
computational procedure, 202–205
cone metric space, 216, 218–221, 235, 245
connected components, 118
connectedness, 51
conservation laws, 398
continuous, 103
continuous mapping, 422
contraction, 217
contraction map, 429
conv-Cauchy structure, 758–759
conv-dominated, 810, 824
convergence criteria, 652–653
convergence spaces, 41
convergence structure, 757
convex cone, 830
convex continuous operator, 421
convex envelope, 118
convex polyhedron, 117
convolution, 460, 660–661, 663, 666–667, 673
Cops and Robbers, 492
correct operator, 668
correct problem, 662, 668
cover (partition), 45–46
Császár, A., 45

cubic polynomial eigenvalue problem, 182
Culler, M., 435
cyclic polygon, 141
Cyclic Pursuit, 516, 525, 528
Czerwik generalized metric space, 246–249
Czerwik vector valued metric, 246

D

◊−dominating, 64
d-adherent points, 795
d-Cauchy, 794
d-closed, 796
d-compatible, 802
d-convergent, 794
d-Lipschitz property, 823
d-lsc, 751
d-non-expansive property, 823
ω-distance, 216, 221, 234, 238, 250
Davis, A. S., 45
Dedekind sums, 677, 736–739
denumerable Axiom of choice, 757, 793
dependent choice, 754–755, 791
detrended fluctuation analysis, 603
diagonal dependent choice, 756, 792
diagonalizable matrix, 442
differential operator, 661
digamma functions, 677, 722–723
directed graph, 247
directional continuous, 829
direct methods, 47
direct unary operation, 47, 102, 106
discounted stochastic game, 492
discrete, 823
discrete Lipschitz countable version of (EVP), 780, 824
distance, 116
distributed-parameter, 2
distributed-parameter object, 1
divisor equivalence, 164
domain, 43
Dontchev, J., 79
double ratios, 556–557
duality algorithm, 417
duality methods, 429
dual of a polynomial matrix, 162
dynamic geometry, 129

E

e-Cauchy, 759
e-convergent, 758
F-equivalent, 47
Edelstein's fixed point theorem, 216
effective temperature, 4
eflexive, 46
Efremovič, V. A., 53
Efremovič's proximity, 48
eigenvalue, 134, 181, 550
eigenvalue problem, 160, 171
eigenvector, 175, 177
Ekeland variational principle on almost metric spaces, 777
Ekeland variational principle on quasi-metric spaces, 822
Ekeland's variationals principle, 221, 751, 788
elator space, 45
elementary matrices, 159, 167–168
elementwise complement, 46
elementwise operation, 59, 65
elementwise right □–continuous, 103
Elez, N., 53
elliptic type, 118–119
energy balance, 418
enthalpy, 418
enthalpy function, 419
equations of motion, 842, 855, 865
equilibrium points (EPs), 839–840, 842, 845–861, 865
equivalence, 46
equivalence (partial order) relation, 44
equivalence relator, 52, 63
Euler formulas, 55
Euler's inequality, 146
evader, 491–492
E-variational, 764, 781
even lattice, 117

exact solution, 397, 660, 673
exponential and trigonometric functions, 55
extensive, 65
extremal closure operations, 59

F

□-fine, 47
family of linearizations, 165–166
fat and dense sets, 49
fat and dense subsets, 48
fat sets, 51
feedback, 1, 4
feedback control, 1–2
feedback parameter, 8
Fermi–Dirac statistics, 537
fibration, 438
Fiedler linearizations, 165–166
Fiedler pencils, 159, 170
finite dependent choice, 754, 791
finite eigenvalue, 161–162, 178
finite geometry designs, 547
finite index, 117
finite sums, 677–742
first and second Frobenius companion forms, 162–163, 171, 183
first companion form, 171–172
first order of convergence, 147
first variable continuous, 798
first variable lsc, 799
Fisher's information, 550
formal contexts, 45
formula, 485
Fourier cosine transform, 583
Fourier sine transform, 582
fractional integral operator, 565
fractional Laplace-type integral operator, 571–572
Fredholm and Volterra integro-differential equations, 659
Fredholm integro-differential equations, 659–660
free boundary problem, 417
Frink triangular, 793
Frobenius companion forms, 184
full, 825, 832
fully complete, 810
function, 42
functional equation, 590, 625
fundamental chamber, 118
fundamental domain of finite volume, 116
fundamental group, 435, 437
fundamental polyhedron, 117
fundamental solution, 486
furnace, 3

G

G-character variety, 435–436
G-normed space, 591
G-representation variety, 449
$\mathrm{GL}_2(\mathrm{C})$-representation, 448
Galois adjoint, 47
Galois connection, 46, 49, 55, 65, 102
Galois property, 63
game theory model, 189–191, 211
gap distance, 222
Gauss–Ostrogradsky, 255, 259, 261, 263
Gelfand triple, 487
generalized Laplace transform, 572
generalized lattice Weyl vector, 119
generalized metric space, 215–216, 218, 223, 225–227
generalized ordered sets, 70
generalized proximities, 64
generalized Schur form, 180
generalized sequences, 53
generalized Stieltjes transform, 572
generalized topologies, 64
generalized uniformities, 64
generalized ω-distance, 235
generating functions, 677–679, 681, 683–687, 690, 699–713
geometric invariant theory (GIT), 436
geometric realization, 118
Gerretsen inequalities, 144
global complementations, 46, 51
global optimization, 492
gosets, 45, 70

gradient, 6
gradient functional, 1
gradient projection method, 5
graph pursuit game, 494, 498
Green's formula, 485–486
Gritsenko–Nikulin, 121
Grotendieck, A., 438
Grothendieck group, 438
group of automorphisms, 116
group of symmetries, 119

H

h-directional derivative, 833
half spaces, 116
Hankel transform, 572
Hardy-Rogers contraction, 243
Hausdorff, F., 53
heating system, 3
heat transfers, 417
Hermite–Hadamard, 256, 258
Hilbert space, 429
Hölder, 279–280, 290
Hölder's inequalities, 270
holomorphic, 280
Hunsaker–Lindgren relation, 45
Hunsaker, W., 64
Hurwitz zeta function, 677–678, 707, 739–742
Hyers–Ulam stability, 590, 625, 627
Hyers–Ulam stability problem, 590
hyperbolic, 115
hyperbolic cosine, 116
hyperbolic equation, 1
hyperbolic integral quadratic form, 115
hyper relation, 44

I

□-invariant, 64, 66
idempotent, 65, 67
identity operation, 67
identity relation, 43
images, 43
increasing, 65, 103
infinite eigenstructure, 164
infinite eigenvalue, 162, 177
infinite elementary divisors, 162
infinitesimal closure, 48
infinitesimal mass, 839, 842
inf-lattice, 779
inf-lattice property, 823
inf-proper, 818
initial conditions, 670
initial value problem, 667
injective operator, 662
inner product relation, 467–468
integral conditions, 660–661
interior preserving, 42
interpolating sequence, 145
intersection-preserving structures, 58
invariant polynomials, 161
invariant to translations, 830
inverse Laplace transform, 665
inverse mapping theorem, 484
inverse operator, 662
inverse relation, 44
inversion-compatible, 47, 59, 64, 104–105
inversion formula, 468–473
inviscid Burgers, 398
involution, 66
involution operation, 47, 65, 76, 107
involutive, 65
irreducible representation, 437
irreducible stratum, 441–444
iteration process, 130
iterative methods, 641–642, 647–652

J

Jacobian determinants, 259
jump conditions, 10–11

K

Kang-Park r-pseudometric maximal principle, 811
Kasner polygons, 140
Kasner triangles, 131
KdV, 397
KdV–B, 397
Kelley, J. L., 64

Kenyon, H., 57
knots, 436
Kriging method, 551
Kronecker canonical form, 163
KST-metric, 769
KST-semimetric, 769
k^{th} root of unity, 150
Kuratowski, K., 53
Kurdics, J., 46, 51

L

(d,≥)-lsc, 818
Lagrange multipliers, 202
Lagrangian function, 203
Laplace equation, 482
Laplace transform, 660, 663–664, 667–668, 673
Laplacian, 485
latin square, 537
lattice Weyl vector, 119
Lax–Milgram theorem, 485–486
Lebesgue measure, 421
Lebesgue theorem, 422
Lefschetz motive, 438
left composition-compatible, 75–77
left continuity properties, 102
left-subordinated, 772
Levine, N., 46
Lindgre, W., 64
linear dynamical systems, 159, 180, 184
linear form, 422
linearization, 158–160, 162–165, 167, 170, 174–176, 183–185
linear pursuit, 516–517, 521, 523, 526
Lipschitz, 431
Lipschitz continuous, 429
Lipschitzian, 298
loaded boundary-value problems, 12
loaded differential equation, 5
loaded system, 5
locally non-partial, 69–70
Lorentzian Kac–Moody algebras, 114
lower continuous, 42
lower mixed continuous, 41–42
lower right □–continuous, 102, 104–107
lower semicontinuous, 102, 428, 787
lower semi-idempoten, 71

M

(d,≤)-maximal, 803
(≤)-maximal element, 823
Z-module, 115, 119
Mala, J., 46, 63
mapping, 431
mathematical model, 1
mathematical physics, 398
matrix pencil, 172, 174, 176
matrix polynomial, 177–179
maximal monotone, 429
Maxwell–Boltzman Statistics, 554
mean motion, 844
metric space, 751
midpoint-stretching polygon, 141
mildly left □–continuous, 105
mildly right □–continuous, 105–107
mildly vaguely right □–continuous, 105
MIMO, 158
minimal indices, 170
minimal structures, 64
minimal surface, 552
mirror of symmetry, 116
mixed wavelet multifractal detrended fluctuation analysis, 612–613
modeling, 159
modification, 47
modified cyclic pursuit, 516, 525–528
modified KdV, 397
modified linear pursuit, 516–524, 526–527
monotone, 423
monotone Ekeland variational principle, 774, 788
monotone maps, 429
monotonous maximal operator, 429
Moreau–Yosida approximation, 429–430

most economically advantageous tender (MEAT) criterion 190, 211
multifunction, 41–42
multipoint conditions, 660–661
multi-start, 493
multi-start multi-value iteration, 511–514
multivalued Hardy–Rogers-type operator, 240–243, 245–246
multivalued left A-contraction, 223, 228
multivalued left Perov contraction, 225
multivalued map, 429
multivalued operator with closed graph, 227, 233
multivalued right A-contraction, 225–226
multivalued weakly Picard (MWP) operator, 224–228, 234–239, 241
multivalued ω-left A-contraction, 237, 239
multi-value iteration, 506–511

N

N-player games, 491, 501
N-player graph pursuit games, 491–492, 532
Nagata, M., 450
Nakano, H., 57
Nakano, K., 57
narrow pseudometic Ekeland variational principle, 765
narrow strong pseudometric Ekeland variational principle, 765
Nash equilibrium, 189–190, 195, 202, 205, 212, 491–492, 501–506
Navier–Stokes equations, 182
necessary optimality conditions, 2
neighborhood and closure spaces, 53
neighborhoods, 43, 49
Neumann, F., 417
noise, 16
non-degenerated, 70, 118
non-Euclidean geometry, 536
non-expansive mapping, 217
nonlinear problems, 397
non-local boundary conditions, 660, 673
non-local boundary value problem, 660, 662, 667
non-paratopologically simple, 70
non-partial relator, 61, 69, 88
non-partial relator space, 54
non-void, 94
non-void relator, 60, 63
normal, 800
normalization function, 191, 193, 197–202, 205, 208, 211–212
normal vector, 559
norm of operator, 569
nowhere dense, 50
numerical methods, 659–660
numerical solutions, 398

O

objective functional, 1
oblateness, 841–843, 850, 854, 861
openness-preserving, 42, 90
open sphere, 823
operation-free product, 167–169
operator, 423
opological, 79, 84
optimal control synthesis, 2
optimal design, 541
orbit, 444, 449–451
ordered sets, 45
ordinary least squares, 538
Orr–Sommerfeld equation, 182
orthic triangle, 139
orthogonal, 117
orthogonal projection, 540
Ostrowski, 256–258, 279, 281

P

∇-progressive, 781
p-adic integral, 677–678, 696–698, 700–707, 713–719
Palais–Smale condition, 787, 829

Palais–Smale differential condition, 834
Papaz, Q., 53
parabola, 548
parametrization, 274, 279
paratopological closures, 57
paratopologically simple, 70
Parseval's formula, 459–460
partial differential equations, 2, 417
partial order, 790
Pataki connection, 58, 63, 65
Pataki, G., 46, 63, 70
payoff function, 497
pencil, 165, 173, 175
perfect information, 492, 498
Perov fixed point theorem, 215–216
Pervin preorder relator, 72
Pervin relation, 45, 89
Pervin, W. J., 46, 64
phase-change, 418
phase-change materials, 417
phase-change process, 417
phase plane, 398
phase transition, 417–418
planar waveguide, 182
plane configuration, 129
plane transformation, 129
plasma drift, 182
Poincaré function, 543
Poincaré inequality, 425, 486
poisson boundary value problem, 486
polygamma functions, 677, 731
polynomial bases, 185
polynomial eigenvalue problems, 179–180, 183
polynomial matrix, 158–160, 164, 166, 170, 172, 174, 181, 183–185
Pompeiu–Hausdorff generalized distance, 222
potential, 481
Poynting–Robertson (P–R) drag effect, 839–843, 850, 861, 865
preclosure operation, 63, 76
preclosure, semiclosure and modification operations, 66
Premakumari, S., 41
preorder, 46
preorder (tolerance) relation, 44, 63, 89
primitive lattice, 117
primitive root, 117
projection operation, 47, 63, 106
projective geometry, 543
properly simple, 47
properly upper right continuous, 103
proset (preordered set), 53
proximal, 73
proximal and topological interiors, 48
proximal closures, 57
proximally open sets, 50
proximities, 44
proximity spaces, 41
pseudometric, 45, 758, 793
pseudometric Ekeland variational principle, 762
public procurement tender, 195
public requirement tenders, 191
public tender, 189, 211
pursuer, 491–492

Q

1-quasiconvex function, 19
quadratic eigenvalue problem, 180–181
qualitative analysis, 398
quartic polynomial eigenvalue problem, 182
quasi-continuity, 103
quasi-metric, 797, 817
quasi-norm, 829
quasi-order, 789
quasiorder Ekeland variational principle on quasi-metric spaces, 818
quasi-topological, 71–73
quasi-topologically lower continuous, 78–81, 83–84, 86, 88–89, 95
quasi-topologically mixed lower continuous, 86, 89, 91, 93, 95

quasi-topologically mixed upper continuous, 86–91, 94–95
quasi-topologically upper continuous, 78–80, 82–83, 85, 88–89, 95–96
quaternion algebra, 455–456
quaternionic fractional Fourier transform, 457–461
quaternionic fractional Wavelet transform, 453–478

R

radial distance, 843–844
radiation pressure, 839–840, 842–843, 861
railtrack, 181
randomized update, 512
range, 43
Rassias's approach, 589
Rassias theorem, 589
receiver operating characteristic (ROC), 549
recovery of eigenvectors, 178
rectangle, 14
recurrent system, 132, 136
reduced Zermelo–Fraenkel system, 755, 792
reducibility/irreducibility, 441
reducible representation, 437
reducible stratum, 444–449
reductive, 436
refinement, 47, 57, 73
reflection group, 117
reflection in a hyperplane, 116
reflexive, 44, 69, 73, 117
reflexive relator, 53
reflexive-sufficient, 797, 829
regular, 161, 795, 805
relation, 41, 43, 62, 73, 78, 103, 753, 789
relative/absolute (–K)-slope, 833
relative/absolute (d,≥)-slope, 826
relative functional pseudometric Caristi–Kirk, 783
relative narrow pseudometric Ekeland variational principle, 770
relative narrow strong pseudometric Ekeland variational principle, 771
relative pseudometric Caristi–Kirk, 783
relator, 45, 54, 57, 59, 68–69, 71, 103
relator (generalized uniform) spaces, 41, 43, 78, 95, 102
remainder, 281–282, 285–286, 288–289, 292–293, 295–299
representation, 437, 441, 445, 449–451
representation variety, 437–439, 447, 450–451
residual and cofinasubsets, 50
residue of set, 53
resolvent, 429–430
resolvent technique, 647, 654–655
restricted arithmetic type, 119
restricted domains, 625
restricted three-body problem (R3BP), 839–841, 846, 861, 865
reverse matrix polynomials, 175
Riemann–Lebesgue lemma, 458–459
Riemann zeta function, 677, 707–709
Riez theorem, 422
right composition-compatible, 75–76
right □–continuous, 103
right-subordinated, 772
ROC curves, 557–558
rolling curves, 548
r-pseudometric, 794
rs-pseudometric, 803

S

F-simple, 47
second companion form, 176
second-kind, 423
second order of convergence, 147
selected dependent choice, 756, 792
selectionally right □–continuous, 103
selection function, 44
self-adjoint case, 481–485
Selfish cops, 500–501, 528
semi-analytic, 398
semi-direct product, 119
semi-discretized method, 417

semilinear inclusion system, 226
sensors, 3
separated, 798
sequence, 754, 790
set endowed with two metrics, 232
set-theoretic operation, 46
set-to-set functions, 47
set-valued functions, 44, 103
set-valued graph, 420
Shalen, P. B., 435
signature $(n, 1)$, 116
simple, 45
single-layer potential, 487
singular, 761
SISO, 158
Slater condition, 203–204
slope, 788
small interior and closure relations, 49
Smirnov's strong inclusion, 48
smooth path, 279–282, 288, 293, 295, 298, 300–303
Sobolev space, 425
solid–liquid boundary, 417
special numbers and polynomials, 677–742
spectral computational approach, 350–352
spectral Fourier analysis, 398
spectral method, 354
square of a vector, 116
stability, 840–842, 846, 850, 855–861, 865
stacks (ascending systems), 64, 70
standard eigenvalue problem, 180
starting, 761
Stefan, J., 418
Stefan problem, 418
stereographic projection, 558
strategies, 497, 501, 504
stratum, 443–447, 449
strict equivalence, 163–164
strictly monotone, 481, 483
strict monotonicity, 482–483
strict order, 789
strong convergence, 428
strong linearization, 163–164
strongly reduced system, 791
strongly reduced Zermelo–Fraenkel system, 779
strongly topological, 74
structure, 47, 50, 58, 65, 68
subadditive, 829
subdifferential, 420
subgroup, 117
subquadratic, 19
successive differences of average, 19
superquadratic, 19
super relation, 44
surface integral, 259–261
surrounding, 45
Švarc, A. S., 53
symmetric, 44, 89
symmetric invariants, 143
symmetric relator, 47
symmetrizable generalized Cartan matrix, 115
symmetry, 69
system analysis, 158
system of operatorial inclusions, 223, 232

T

(m, n)-torus knot, 436, 439–440, 449
temperature measurements, 3
Thangavelu, P., 41–42
threat strategies, 503
three-dimensional ball, 255, 274
three-dimensional space, 255, 258, 262, 265, 267–268
Tietze, H., 64
time-harmonic wave equation, 181
tolerance, 46
topological, 71–74, 81–83, 90, 93, 96, 98
topological closedness preserving, 79, 83, 95, 97
topological closure (refinement), 52, 57, 73
topological closure preserving, 98, 100–101
topological interior preserving, 97, 99

topological interior reversing, 97
topologically closedness preserving, 81
topologically equivalent, 72, 74
topologically fine, 49
topologically openness preserving, 81
topologically open sets, 50
topologically regular open subset, 70
topologically simple, 59
topologically transitive, 72
topological openness preserving, 79, 84, 95
topological openness reversing, 79
topological or proximal, 52
topological quantum field theory, 436
topological relator space, 71
topological space, 41–42
torus, 551
torus knot, 436–440, 442, 449–451
total (or non-partial) relation, 43
trace theorem, 487–488
transitive, 44
transitively sufficient, 764, 782
transitivity, 69
triangular, 794, 797
triangular pseudometric, 758
triaxiality, 839, 841, 843, 850–851, 855–857, 861
turn payoff, 498
twisted coefficient, 119
two-point boundary value problems, 660

U

Ulam–Hyers stability, 240, 244–245
ultimate stable unary operation, 59
unary operation, 43, 64, 74
uncertainty principle, 473–478
uniform closures, 57
uniformly filtered, 54
uniformly, proximally, topologically and paratopologically upper right continuous, 103
uniform spaces, 41, 45
unimodular, 161
unimodular equivalence, 161–162
union-preserving, 47, 50, 52, 63, 65–66
unique solution, 487
unique solvability, 481–484
untwisted, 119
upper bound, 810
upper continuous, 42
upper mixed continuous, 41–42
upper right □–continuous, 102–107
upper right □–F–regular, 104
upper right □–Galois connection, 104
upper right □–Pataki connection, 104
upper right □–V–normal, 104
upper semicontinuous, 102
upper semi-idempotent, 71

V

vaguely right □–continuous, 102, 105–107
value-iteration, 491–492, 504
variational inclusions, 641–655
variational inequality, 196–197, 202–206, 208, 210, 417, 420–421, 423, 429
variational inequality formulation, 190–191, 195, 212
vector field, 255, 259–261
vector space, 176–178
vector spaces of linearizations, 171
vector-valued metric, 217
virtual classes, 438
Volterra integro-differential equation, 659–661, 664, 667, 670–671, 673

W

wavelet multifractal detrended fluctuation analysis, 608–609
weak composition-compatibility, 75
weak convergence, 427–428
weak formulation, 427
weakly (strongly) quasi-topological, 71
weakly regular, 805
weakly (strongly) topological, 71

weak solution, 419
well chainedness, 51
well-posed problem, 228, 240, 244–246

X

Xavier, P., 41

Z

Zermelo–Fraenkel, 788
Zhong variational principle, 788
Zorn–Bourbaki Bakhtin quasi-metric maximal principle, 815
Zorn–Bourbaki Frink quasi-metric maximal principle, 813
Zorn–Bourbaki quasi-metric maximal principle, 816
Zorn–Bourbaki rs-pseudometric maximal principle, 812
Zorn order, 779, 823

CPSIA information can be obtained
at www.ICGtesting.com
Printed in the USA
JSHW040149050423
39455JS00001B/5

9 789811 261565